Lecture Notes in Networks and Systems

Volume 128

AF824015

The series "Lecture Notes in Networks and Systems" publishes the latest developments in Networks and Systems—quickly, informally and with high quality. Original research reported in proceedings and post-proceedings represents the core of LNNS.

Volumes published in LNNS embrace all aspects and subfields of, as well as new challenges in, Networks and Systems.

The series contains proceedings and edited volumes in systems and networks, spanning the areas of Cyber-Physical Systems, Autonomous Systems, Sensor Networks, Control Systems, Energy Systems, Automotive Systems, Biological Systems, Vehicular Networking and Connected Vehicles, Aerospace Systems, Automation, Manufacturing, Smart Grids, Nonlinear Systems, Power Systems, Robotics, Social Systems, Economic Systems and other. Of particular value to both the contributors and the readership are the short publication timeframe and the world-wide distribution and exposure which enable both a wide and rapid dissemination of research output.

The series covers the theory, applications, and perspectives on the state of the art and future developments relevant to systems and networks, decision making, control, complex processes and related areas, as embedded in the fields of interdisciplinary and applied sciences, engineering, computer science, physics, economics, social, and life sciences, as well as the paradigms and methodologies behind them.

** **Indexing: The books of this series are submitted to ISI Proceedings, SCOPUS, Google Scholar and Springerlink** **

More information about this series at http://www.springer.com/series/15179

Isak Karabegović
Editor

New Technologies, Development and Application III

Springer

Editor
Isak Karabegović
Academy of Sciences and Arts
of Bosnia and Herzegovina
Sarajevo, Bosnia and Herzegovina

ISSN 2367-3370 ISSN 2367-3389 (electronic)
Lecture Notes in Networks and Systems
ISBN 978-3-030-46816-3 ISBN 978-3-030-46817-0 (eBook)
https://doi.org/10.1007/978-3-030-46817-0

This Springer imprint is published by the registered company Springer Nature Switzerland AG
The registered company address is: Gewerbestrasse 11, 6330 Cham, Switzerland

Interdisciplinary Research of New Technologies, Their Development and Application

This book features papers focusing on the implementation of new and future technologies, which were presented at the International Conference on New Technologies, Development and Application, held at the Academy of Science and Arts of Bosnia and Herzegovina in Sarajevo from 25 to 27 June 2020. It covers a wide range of future technologies and technical disciplines, including complex systems such as Industry 4.0; patents in Industry 4.0; robotics; mechatronics systems; automation; manufacturing; cyber-physical and autonomous systems; sensors; networks; control, energy and renewable energy sources; automotive and biological systems; vehicular networking and connected vehicles; effectiveness and logistics systems, smart grids, nonlinear systems, power, social and economic systems, education and IoT. Majority of organized conferences are usually focusing on a narrow part of the issues within a certain discipline while conferences such as these are rare. There is a need to hold such conferences. The value of this conference is that various researchers, programmers, engineers and practitioners come to the same place where ideas and latest technology achievements are exchanged. Such events lead to the creation of new ideas, solutions and applications in the manufacturing processes of various technologies. New coexistence is emerging, horizons are expanding, and unexpected changes and analogies arise. Best solutions and applications in technologies are critically evaluated.

The first chapter covers mechanical design, Industry 4.0, robotics, self engineering, cyber-physical systems, mechatronic systems, automation of production processes, advanced production and metallurgy. The first article gives an analysis of the dissemination of patents in the Fourth Industrial Revolution. The second article presents the new concept of technological challenges in the form of self engineering. One of the articles gives an overview of aircraft mechanical system development and their mechatronic modelling. Another one discusses technologies which are a vital part of a smart factory that utilizes the Internet of things (IoT) to realize intelligent manufacturing concepts like predictive maintenance or an extensive machine to machine communication. The last article is about monitoring of physical properties of transmission–hydraulic fluid by simulating the operation load of the agricultural machine's hydraulic pump under laboratory conditions.

The second chapter covers computer science, information and communication technologies, Internet of things and cyber security. The first article explains optimized controller for mobile path tracking robot by particle swarm optimization. The second article is about theoretical and practical aspects of the Internet of things technology. The third article gives a brief overview of the network society. The one article provides an analysis of the cyber security of the electric power information system in Montenegro. One of the articles gives information about the green technology approach to comb-based decimators design.

The third chapter is devoted to traffic and transport systems, logistics and intelligent systems. Chapter starts with cyber security of autonomous and connected vehicles. In the second article, investigation of vehicle driving cycles in urban traffic condition has been performed. Another article explains the improvement of variable speed limit control of connected vehicles. One article suggests improving driver response efficiency in risk prevention from the traffic environment. Last article gives theoretical aspects of diagnosing car engine at the time of acceleration.

The fourth chapter is devoted to new technologies in the renewable energy, power quality and advanced electric power systems. The first article is about analysis of groundwater source heat pump operation with improvement suggestions. The second is about air staging and reburning to achieve low emissions during co-firing coal and biomass. Another article explains how to use renewable energy sources in constructions of energy-efficient residential buildings. One work is about computer studies of the tightness of the drill string connector depending on the profile of its tapered thread.

Chapter Five is devoted to new methods in the agriculture, ecology and chemical processes of a wide range of topics: carbon-based aerogels and xerogels for removing of toxic organic compounds, supercritical fluid extraction of lycopene and omega-3, then assessment of cancerogenic health risk of As, Cd, Pb and Ni from tobacco smoke or work with a topic comparative composition of Spreča river fauna (Bosnia and Herzegovina).

The sixth chapter focuses on the field of geodesy, construction, new materials and sustainable innovation and others. The first article is about the proposed new quality control plan procedure for road bridges. The second article focuses on the problem of characteristic snow load in the eastern part of Bosnia and Herzegovina. The following article analyses the influence of wind load to the billboards at Mostar valley. One of the articles describes wood as a building material as renewable source to preserve the environment and health in B&H.

The seventh chapter covers economics, E-business and entrepreneurship. The chapter starts with techno-entrepreneurship or interdisciplinary curricula challenges for 4IR in Bosnia and Herzegovina. Second article gives a mechanism to stimulate the growth of highly competitive technology business. One article is about measuring youth attitudes on materialism as a consequence of a consumer society. Last article is about new technologies in the financial industry.

The whole content of this book is intended to a wide range of technical systems and different technical disciplines in order to apply the latest solutions and achievements in technologies and to improve manufacturing processes in all disciplines where systemic thinking has a very important role in the successful understanding and building of human, natural and social systems. We hope this content will be the first in a series of publications that are intended for the development and implementation of new technologies in all industries.

Isak Karabegović

Contents

New Technologies in Civil Engineering, Architecture, Quality Control

Economics, E-Business, Entrepreneurships

List of Contributors

Adem Abdić School of Economics and Business, University of Sarajevo, Sarajevo, Bosnia and Herzegovina

Ademir Abdić School of Economics and Business, University of Sarajevo, Sarajevo, Bosnia and Herzegovina

Oday I. Abdullah College of Engineering, University of Baghdad, Baghdad, Iraq; Hamburg University of Technology, Hamburg, Germany

Ivan Abramenko Kharkiv Petro Vasylenko National Technical University of Agriculture, Kharkiv, Ukraine

Naida Ademović Faculty of Civil Engineering, University of Sarajevo, Sarajevo, Bosnia and Herzegovina

Avdul Adrović Faculty of Natural Sciences and Mathematics, University of Tuzla, Tuzla, Bosnia and Herzegovina

Ismar Alagić TRA Tešanj Development Agency, Tešanj, Bosnia and Herzegovina;
Faculty of Mechanical Engineering, University of Zenica, Zenica, Bosnia and Herzegovina;
Faculty of Engineering and Natural Sciences, International University of Sarajevo, Ilidža, Sarajevo, Bosnia and Herzegovina

Bojana Aleksić Faculty of Technology and Metallurgy, Innovation Centre, Belgrade, Serbia

Vujadin Aleksić Institute for Testing of Materials, Belgrade, Serbia

Amer Alić International Burch University, Sarajevo, Bosnia and Herzegovina

Dejan Antanasković Faculty of Management Zaječar (FMZ), Megatrend University of Belgrade, Zaječar, Serbia

Bohdan Antypenko Faculty of Electronics and Information Technologies, Sumy State University, Sumy, Ukraine

Viktoriia Antypenko Faculty of Electronics and Information Technologies, Sumy State University, Sumy, Ukraine

Maja Anđelković Faculty of Information Technology and Engineering (FITI), Novi Beograd, Serbia

Aleksandar M. Anđić Ministry of Public Administration, Directorate of Electronic Governance and Information Security, Podgorica, Montenegro

Tun Lin Aung Department of Electronic Technologies in Mechanical Engineering, Bauman Moscow State Technical University, Moscow, Russia

Alen Bajrić Faculty of Natural Sciences and Mathematics, University of Tuzla, Tuzla, Bosnia and Herzegovina

Azra Baluković KV Team, Sarajevo, Bosnia and Herzegovina

Lejla Banjanovic-Mehmedovic Faculty of Electrical Engineering, Department of Control Systems, Automation and Robotics, University of Tuzla, Tuzla, Bosnia and Herzegovina

Bakhriddin Tilovkabulovich Berdiyarov Department of Metallurgy, Tashkent State Technical University, Tashkent, Uzbekistan

Mirza Berković Faculty of Traffic and Communications, University of Sarajevo, Sarajevo, Bosnia and Herzegovina

Saša Bermanec Faculty of Traffic and Transport Sciences, Zagreb, Croatia

Muamer Bezdrob Faculty of Economics, American University in Bosnia and Herzegovina, Sarajevo, Bosnia and Herzegovina

Ernad Bešlagić Mechanical Engineering Faculty, University of Zenica, Zenica, Bosnia and Herzegovina

Dževad Bibić Faculty of Mechanical Engineering, University of Sarajevo, Sarajevo, Bosnia and Herzegovina

Arian Bisha Universiteti Politeknik i Tiranes, Tirana, Albania

Mirha Bičo Ćar School of Economics and Business, University of Sarajevo, Sarajevo, Bosnia and Herzegovina

Almir Blažević Faculty of Mechanical Engineering, University of Sarajevo, Sarajevo, Bosnia and Herzegovina

Radomir Bodiroga Faculty of Agriculture, University of East Sarajevo, East Sarajevo, Bosnia and Herzegovina

Nedić Bogdan Faculty of Engineering, University of Kragujevac, Kragujevac, Serbia

Marija Bojovic Faculty of Architecture, University of Montenegro, Podgorica, Montenegro

Savković Borislav Department for Production Engineering, Faculty of Technical Sciences, University of Novi Sad, Novi Sad, Serbia

Lubomyr Borushchak Computerized Machine Building Department, Ivano-Frankivsk National Technical University of Oil and Gas, Ivano-Frankivsk, Ukraine

Stanislav Bovchaliuk Kharkiv Petro Vasilenko National Technical University of Agriculture, Kharkiv, Ukraine

Liubov Bovnegra Odessa National Polytechnic University (ONPU), Institute of Industrial Technologies, Design and Management, Odessa, Ukraine

Amra Bratovcic Department of Physical Chemistry and Electrochemistry, Faculty of Technology, University of Tuzla, Tuzla, Bosnia and Herzegovina

Lucija Brezočnik Faculty of Electrical Engineering and Computer Science, University of Maribor, Maribor, Slovenia

Tin Brlić Faculty of Metallurgy, University of Zagreb, Sisak, Croatia

Alla E. Brom Bauman Moscow State Technical University (BMSTU), Moscow, Russia

Sam Brooks School of Mathematics, Computer Science and Engineering, City, University of London, London, UK

Y. V. Bulik Department of Cars and Transport Technologies, Lutsk National Technical University, Lutsk, Ukraine

Senad Burak Faculty of Mechanical Engineering, University of Sarajevo, Sarajevo, Bosnia and Herzegovina

Adisa Burić Institute "Kemal Kapetanović" of Zenica, University of Zenica, Zenica, Bosnia and Herzegovina

Aleksandr I. Burya Dniprovsk State Technical University (DSTU), Kamianskoe, Ukraine; Academy of Technological Sciences of Ukraine, Kiev, Ukraine

Jasmin Ćatić Faculty of Civil Engineering, University of Sarajevo, Sarajevo, Bosnia and Herzegovina

Fuad Ćatović Faculty of Civil Engineering, University "DžemalBijedić", Mostar, Bosnia and Herzegovina

Samir Čaušević Faculty of Traffic and Communication, University of Sarajevo, Sarajevo, Bosnia and Herzegovina

Giampiero Celenta MEID4 Academic Spin-Off of the University of Salerno, Fisciano, SA, Italy

Mehmed Cero Federal Ministry of the Environment and Tourism, Sarajevo, Bosnia and Herzegovina

Mirjana Čičak Mechanical Engineering Faculty in Slavonski Brod, Josip Juraj Strossmayer University of Osijek, Slavonski Brod, Croatia

Elvir Čizmić School of Economics and Business, University of Sarajevo, Sarajevo, Bosnia and Herzegovina

Alem Čolaković Faculty of Traffic and Communication, University of Sarajevo, Sarajevo, Bosnia and Herzegovina

Jovan Dašić SaTCIP Publisher Ltd., Vrnjačka Banja, Serbia

Predrag Dašić High Technical Mechanical School of Professional Studies, Trstenik, Serbia; SaTCIP Publisher Ltd., Vrnjačka Banja, Serbia

Marco Claudio De Simone Department of Industrial Engineering, University of Salerno, Fisciano, SA, Italy

Luka Dedic Faculty of Traffic and Transport Sciences, University of Zagreb, Zagreb, Croatia

Adela Delalić School of Economics and Business, University of Sarajevo, Sarajevo, Bosnia and Herzegovina

Cristiana Delprete Politecnico di Torino, Turin, Italy

B. M. Dembitskyi Department of Cars and Transport Technologies, Lutsk National Technical University, Lutsk, Ukraine

Evgeny A. Deulin Mechanical Engineering Department, MT-11, Bauman Moscow State Technical University (BMSTU), Moscow, Russia

Tome Dimovski Faculty of ICTs, St. Kliment Ohridski University–Bitola, Bitola, North Macedonia

Milena Djukanovic Faculty of Electrical Engineering, University of Montenegro, Podgorica, Montenegro

Yuriy A. Doroshenko Belgorod State Technological University named after V. G. Shukhov (BSTU), Belgorod, Russia

Himzo Đukić University of Mostar, FSRE, Mostar, Bosnia and Herzegovina

Golubović Dušan Faculty of Mechanical Engineering, University of East Sarajevo, Lukavica, Bosnia and Herzegovina

Ješić Dušan International Technology and Management Academy – MTMA, Novi Sad, Serbia

Ejub Dzaferovic Faculty of Mechanical Engineering, University of Sarajevo, Sarajevo, Bosnia and Herzegovina

Vitaliy Dzhemelinkyi Laser Systems and Physical Technologies Department, National Technical University of Ukraine "Igor Sikorsky Kyiv Polytechnic Institute", Kyiv, Ukraine

Edin Džiho Faculty of Mechanical Engineering, University "Džemal Bijedić" in Mostar, Mostar, Bosnia and Herzegovina

Šunje Edin Faculty of Mechanical Engineering, University "Džemal Bijedić" in Mostar, Mostar, Bosnia and Herzegovina

Ahmed El Sayed Faculty of Engineering and Natural Sciences, International Burch University, Ilidža, Sarajevo, Bosnia and Herzegovina

Belma Fakić Institute "Kemal Kapetanović" of Zenica, University of Zenica, Zenica, Bosnia and Herzegovina

Irina Fedosova Priazovsky State Technical University, Mariupol, Ukraine

Mirko Ficko Faculty of Mechanical Engineering, University of Maribor, Maribor, Slovenia

Iztok Fister Jr. Faculty of Electrical Engineering and Computer Science, University of Maribor, Maribor, Slovenia

Iqra Abdul Ghaffar Department of Computer Engineering, UET Lahore, Lahore, Pakistan

Nataliia N. Glagoleva Belgorod University of Cooperation, Economics and Law, Belgorod, Russia

Jasna Glišović Faculty of Engineering, University of Kragujevac, Kragujevac, Serbia

Martin Gregurić Faculty of Traffic and Transport Sciences, University of Zagreb, Zagreb, Croatia

Ivan Grujić Faculty of Engineering, Department for Motor Vehicles and Motors, University of Kragujevac, Kragujevac, Serbia

Domenico Guida Department of Industrial Engineering, University of Salerno, Fisciano, Salerno, Italy

Fuad Hadžikadunić Mechanical Engineering Faculty, University of Zenica, Zenica, Bosnia and Herzegovina

Rašid Hadžović Faculty of Civil Engineering, "Džemal Bijedić" University of Mostar, Mostar, Bosnia and Herzegovina

Edina Hajdarević Faculty of Natural Sciences and Mathematics, University of Tuzla, Tuzla, Bosnia and Herzegovina

Faisal Hayat Department of Computer Engineering, UET Lahore, Lahore, Pakistan

Nihad Hodzic Faculty of Mechanical Engineering Sarajevo, University of Sarajevo, Sarajevo, Bosnia and Herzegovina

Dženana Hodžić Committee on Education, Science, Culture and Sport, Parliament of the Federation of Bosnia and Herzegovina, Sarajevo, Bosnia and Herzegovina

Edib Horoz Institute "Kemal Kapetanović" of Zenica, University of Zenica, Zenica, Bosnia and Herzegovina

Elvis Hozdić Department of Control and Manufacturing Systems, Faculty of Mechanical Engineering, University of Ljubljana, Ljubljana, Slovenia

Ilija Hristoski Faculty of Economics, St. Kliment Ohridski University–Bitola, Prilep, North Macedonia

Ľubomír Hujo Faculty of Engineering, Department of Transport and Handling, Slovak University of Agriculture in Nitra, Nitra, Slovak Republic

Ermin Husak University of Bihać, Bihać, Bosnia and Herzegovina

Želimir Husnić The Boeing Company, Mechanical Systems and Hydraulic Design, Ridley Park, PA, USA

Jusuf Ibrulj Faculty of Mechanical Engineering, University of Sarajevo, Sarajevo, Bosnia and Herzegovina

Ekaterina I. Ikonnikova Mechanical Engineering Department, MT-11, Bauman Moscow State Technical University (BMSTU), Moscow, Russia

Safet Isić Faculty of Mechanical Engineering, University "Džemal Bijedić" in Mostar, Mostar, Bosnia and Herzegovina

Svitlana Ivanova South Ukrainian National Pedagogical University named after K. D. Ushynsky, Odesa, Ukraine

Viktor Ivanov Institute of Mechanical Engineering, Odessa National Polytechnic University, Odesa, Ukraine

Vitalii Ivanov Department of Manufacturing Engineering, Machines and Tools, Sumy State University, Sumy, Ukraine

Juraj Jablonický Faculty of Engineering, Department of Transport and Handling, Slovak University of Agriculture in Nitra, Nitra, Slovak Republic

Anđela Jakšič-Stojanović University Donja Gorica, Podgorica, Montenegro

Nudžejma Jamaković Faculty of Agriculture and Food Science, University of Sarajevo, Sarajevo, Bosnia and Herzegovina

Romana Janoušková Faculty of Engineering, Department of Transport and Handling, Slovak University of Agriculture in Nitra, Nitra, Slovak Republic

Naida Jažić-Asotić School of Economics and Business, University of Sarajevo, Sarajevo, Bosnia and Herzegovina

Zlata Jelačić Faculty of Mechanical Engineering, University of Sarajevo, Sarajevo, Bosnia and Herzegovina

Vladeta Jevremović High Technical Mechanical School of Professional Studies, Trstenik, Serbia

Jovanović Jovana Faculty of Civil Engineering and Management, University Union Nikola Tesla, Belgrade, Serbia

Gordana Jovanovic Dolecek Department of Electronics, Institute INAOE, Puebla, PUE, Mexico

Mihailo Jovanovic Faculty of Business and Law, University of Union, Belgrade, Serbia

Zlatan Jukic Faculty of Technical, HTL Rankweil, Rankweil, Austria; TU Wien, Wien, Austria

Josip Kacmarcik Faculty of Mechanical Engineering, University of Zenica, Zenica, Bosnia and Herzegovina

Kenan Kadic JP Elektroprivreda BiH d.d. - Sarajevo Power Utility, Sarajevo, Bosnia and Herzegovina

S. V. Kalinichenko Dniprovsk State Technical University (DSTU), Kamianskoe, Ukraine

Isak Karabegović Akademy of Sciences and Arts, Bosnia and Herzegovina, Sarajevo, Bosnia and Herzegovina;
University of Bihać, Bihać, Bosnia and Herzegovina

Edina Karabegović University of Bihać, Bihać, Bosnia and Herzegovina

Anes Kazagic JP Elektroprivreda BiH d.d. - Sarajevo Power Utility, Sarajevo, Bosnia and Herzegovina

Simon Klančnik Faculty of Mechanical Engineering, University of Maribor, Maribor, Slovenia

Sanela Klarić Faculty of Architecture, International Burch University, Sarajevo, Bosnia and Herzegovina

Daniela Koltovska Nechoska Faculty of Technical Sciences, St. Kliment Ohridski University–Bitola, Bitola, North Macedonia

Aleksandar Konatar Faculty of Electrical Engineering, University of Montenegro, Podgorica, Montenegro

Volodymyr Kopei Computerized Machine Building Department, Ivano-Frankivsk National Technical University of Oil and Gas, Ivano-Frankivsk, Ukraine

Andriy Koshevoy Department of Machine Building Technology, Donbass State Engineering Academy (DSEA), Kramatorsk, Ukraine

Ján Kosiba Faculty of Engineering, Department of Transport and Handling, Slovak University of Agriculture in Nitra, Nitra, Slovak Republic

Amel Kosovac Faculty of Traffic and Communications, University of Sarajevo, Sarajevo, Bosnia and Herzegovina

Olena Kovalevska Department of Machine Building Technology, Donbass State Engineering Academy (DSEA), Kramatorsk, Ukraine

Sergiy Kovalevskyy Department of Machine Building Technology, Donbass State Engineering Academy (DSEA), Kramatorsk, Ukraine

Bakir Krajinović Federal Hydrometeorological Service, Sarajevo, Bosnia and Herzegovina

Ismar Kreso "SMED Engineering BH", Mostar, Bosnia and Herzegovina

Victor Kurgan Odessa National Polytechnic University (ONPU), Institute of Industrial Technologies, Design and Management, Odessa, Ukraine

Emir Kurtović School of Economics and Business, University of Sarajevo, Sarajevo, Bosnia and Herzegovina

Natalya N. Kutina Bauman Moscow State Technical University (BMSTU), Moscow, Russia

Aitzol Lamikiz Aeronautics Advanced Manufacturing Center, University of the Basque Country, Zamudio, Spain

Samir Lemes Faculty of Mechanical Engineering, University of Zenica, Zenica, Bosnia and Herzegovina

Samir Lemeš Polytechnic Faculty, University of Zenica, Zenica, Bosnia and Herzegovina

Tone Lerher Faculty of Mechanical Engineering, University of Maribor, Maribor, Slovenia

Dmytro Lesyk Laser Systems and Physical Technologies Department, National Technical University of Ukraine "Igor Sikorsky Kyiv Polytechnic Institute", Kyiv, Ukraine

Alma Leto Džemal Bijedić University of Mostar, Mostar, Bosnia and Herzegovina

Tetiana Levytska Priazovsky State Technical University, Mariupol, Ukraine

Oleksandr Liaposhchenko Department of Manufacturing Engineering, Machines and Tools, Sumy State University, Sumy, Ukraine

Osman Lindov Faculty of Traffic and Communications, University of Sarajevo, Sarajevo, Bosnia and Herzegovina

Darko Lovrec Faculty of Mechanical Engineering, University of Maribor, Maribor, Slovenia

Tetiana Lukan Computerized Machine Building Department, Ivano-Frankivsk National Technical University of Oil and Gas, Ivano-Frankivsk, Ukraine

Tetiana Lysenko Odessa National Polytechnic University (ONPU), Institute of Industrial Technologies, Design and Management, Odessa, Ukraine

Mehmed Mahmić University of Bihać, Bihać, Bosnia and Herzegovina

Franc Majdič Laboratory for Fluid Power and Controls, Faculty of Mechanical Engineering, University of Ljubljana, Ljubljana, Slovenia

Irina O. Malykhina Belgorod State Technological University named after V. G. Shukhov (BSTU), Belgorod, Russia

Adriano G. Manca MEID4 Academic Spin-Off of the University of Salerno, Fisciano, SA, Italy

Sadko Mandzuka Faculty of Traffic and Transport Sciences, University of Zagreb, Zagreb, Croatia

Sadko Mandžuka Faculty of Traffic and Transport Sciences, University of Zagreb, Zagreb, Croatia

Camilo Andrés Manrique Escobar MEID4 Academic Spin-Off of the University of Salerno, Fisciano, SA, Italy

Anna Marchenko Faculty of Electronics and Information Technologies, Sumy State University, Sumy, Ukraine

Maria Margotta MEID4 Academic Spin-Off of the University of Salerno, Fisciano, SA, Italy

Silvia Martinez Aeronautics Advanced Manufacturing Center, University of the Basque Country, Zamudio, Spain

Yuliya L. Maslennikova Bauman Moscow State Technical University (BMSTU), Moscow, Russia

Sokhibjon Turdaliyevich Matkarimov Department of Metallurgy, Tashkent State Technical University, Tashkent, Uzbekistan

P. V. Mazyliuk Department of Cars and Transport Technologies, Lutsk National Technical University, Lutsk, Ukraine

Pavlo Mazyliuk Department of Cars and Transport Technologies, Lutsk National Technical University, Lutsk, Ukraine

Iulia Medvid Computerized Machine Building Department, Ivano-Frankivsk National Technical University of Oil and Gas, Ivano-Frankivsk, Ukraine

Anida Memić Faculty of Mechanical Engineering, University of "DžemalBijedić" in Mostar, Mostar, Bosnia and Herzegovina

Olga Miclea University of Petrosani, Petrosani, Romania

Djukanović Milena Faculty of Electrical Engineering, University of Montenegro, Podgorica, Montenegro

Danijela Miloradović Faculty of Engineering, Department for Motor Vehicles and Motors, University of Kragujevac, Kragujevac, Serbia

Ljubica Milović Faculty of Technology and Metallurgy, University of Belgrade, Belgrade, Serbia

Ludmila A. Minaeva Belgorod State Technological University named after V. G. Shukhov (BSTU), Belgorod, Russia

Tarić Mirfad Srednja škola metalskih zanimanja, Sarajevo, Bosnia and Herzegovina

Alma Mičijević Džemal Bijedić University of Mostar, Mostar, Bosnia and Herzegovina

Fatima Muhamedagić Biotechnical Faculty, University of Bihac, Bihac, Bosnia and Herzegovina

Ermin Muharemović Faculty of Traffic and Communication, University of Sarajevo, Sarajevo, Bosnia and Herzegovina

Admir Mulahusić Faculty of Civil Engineering, University of Sarajevo, Sarajevo, Bosnia and Herzegovina

Adis J. Muminovic Faculty of Mechanical Engineering, Department of Mechanical Design, University of Sarajevo, Sarajevo, Bosnia and Herzegovina

Enis Muratovic Department of Mechanical Design, Faculty of Mechanical Engineering, University of Sarajevo, Sarajevo, Bosnia and Herzegovina

Harun Music Department of Mechanical Design, Faculty of Mechanical Engineering, University of Sarajevo, Sarajevo, Bosnia and Herzegovina; PPW Engineering, Sarajevo, Bosnia and Herzegovina

Almedina Mustafić Faculty of Engineering and Natural Sciences, International Burch University, Ilidža, Sarajevo, Bosnia and Herzegovina

Dmitriy Muzylyov Department Transport Technology and Logistics, Kharkiv Petro Vasylenko National Technical University of Agriculture, Kharkiv, Ukraine

Yakiv Nemyrovskyi Department of Exploitation and Repairing Machines, Central Ukrainian National Technical University, Kropyvnytskyi, Ukraine

Viktor Nenia Faculty of Electronics and Information Technologies, Sumy State University, Sumy, Ukraine

Emir Nezirić Faculty of Mechanical Engineering, University "Džemal Bijedić" in Mostar, Mostar, Bosnia and Herzegovina

Adnan Novalic Faculty of Engineering and Natural Sciences, International Burch University, Ilidža, Sarajevo, Bosnia and Herzegovina

Adnan Novalić Faculty of Engineering and Natural Sciences, International Burch University, Ilidža, Sarajevo, Bosnia and Herzegovina

Lazar Novicevic Faculty of Electrical Engineering, University of Montenegro, Podgorica, Montenegro

Mirna Nožić Faculty of Mechanical Engineering, University "DžemalBijedić" of Mostar, Mostar, Bosnia and Herzegovina

Murco Obucina Faculty of Mechanical Engineering, University of Sarajevo, Sarajevo, Bosnia and Herzegovina

Ivana Ognjanović University of Donja Gorica, Podgorica, Montenegro

Ruslan Okopnyu Faculty of Electronics and Information Technologies, Sumy State University, Sumy, Ukraine

Piatykop Olena Department of Computer Science, Faculty of Information Technology, Priazovsky State Technical University, Mariupol, Ukraine

Salah-Eldien Omer Technical Faculty Bihać, University of Bihać, Bihać, Bosnia and Herzegovina; Zagreb, Croatia

Adnan Omerhodžić Faculty of Traffic and Communications, University of Sarajevo, Sarajevo, Bosnia and Herzegovina

Oleh Onysko Computerized Machine Building Department, Ivano-Frankivsk National Technical University of Oil and Gas, Ivano-Frankivsk, Ukraine

Iztok Palčič Faculty of Mechanical Engineering, University of Maribor, Maribor, Slovenia

Carmine M. Pappalardo Department of Industrial Engineering, University of Salerno, Fisciano, SA, Italy

Carmine Maria Pappalardo Department of Industrial Engineering, University of Salerno, Fisciano, Salerno, Italy

Kovač Pavel Department for Production Engineering, Faculty of Technical Sciences, University of Novi Sad, Novi Sad, Serbia

Nina Pavićević Faculty of Management Zaječar (FMZ), Megatrend University of Belgrade, Zaječar, Serbia

Ivan Pavlenko Department of Manufacturing Engineering, Machines and Tools, Sumy State University, Sumy, Ukraine

Olexiy Pavlenko Kharkiv National Automobile and Highway University, Kharkiv, Ukraine

Mikhailov Valeriy Pavlovich Department of Electronic Technologies in Mechanical Engineering, Bauman Moscow State Technical University, Moscow, Russia

Sead Pašić Faculty of Mechanical Engineering, University "Džemal Bijedić" in Mostar, Mostar, Bosnia and Herzegovina

Ekrem Pehlić Faculty of Health Studies, University of Bihać, Bihać, Bosnia and Herzegovina

Irena Petrinic Faculty of Chemistry and Chemical Engineering, University of Maribor, Maribor, Slovenia

Boran Pikula Faculty of Mechanical Engineering, University of Sarajevo, Sarajevo, Bosnia and Herzegovina

Dmytro Pirogov Department of Manufacturing Engineering, Machines and Tools, Sumy State University, Sumy, Ukraine

Vili Podgorelec Faculty of Electrical Engineering and Computer Science, University of Maribor, Maribor, Slovenia

Giorgi Popkhadze Faculty of Transportation and Mechanical Engineering, Georgian Technical University (GTU), Tbilisi, Georgia

Maja Popovac Faculty of Civil Engineering, University "DžemalBijedić", Mostar, Bosnia and Herzegovina

Eduard Posviatenko National Transport University, Kiev, Ukraine

Ana Prodanović Faculty of Technology and Metallurgy, Innovation Centre, Belgrade, Serbia

Ihor Prokopovich Odessa National Polytechnic University (ONPU), Institute of Industrial Technologies, Design and Management, Odessa, Ukraine

Olha Pronina Department of Computer Science, Faculty of Information Technology, Priazovsky State Technical University, Mariupol, Ukraine

Milan Radosavljević Faculty of Business Studies and Law (FPSP), Novi Beograd, Serbia

Luka Radunovic Faculty of Mechanical Engineering, University of Montenegro, Podgorica, Montenegro

Senad Rahimić Faculty of Mechanical Engineering, University of "DžemalBijedić" in Mostar, Mostar, Bosnia and Herzegovina

Zijada Rahimić School of Economics and Business, University of Sarajevo, Sarajevo, Bosnia and Herzegovina

Stoja Rešković Faculty of Metallurgy, University of Zagreb, Sisak, Croatia

Adnan Rovčanin School of Economics and Business, University of Sarajevo, Sarajevo, Bosnia and Herzegovina

Rajkumar Roy School of Mathematics, Computer Science and Engineering, City, University of London, London, UK

Sarina Šabotić Faculty of Architecture, International Burch University, Sarajevo, Bosnia and Herzegovina

Aida Šapčanin Faculty of Pharmacy, University of Sarajevo, Sarajevo, Bosnia and Herzegovina

Mirsada Salihović Faculty of Pharmacy, University of Sarajevo, Sarajevo, Bosnia and Herzegovina

Sergii Sardak Dniprovskii University of the Humanities, Dnipro, Ukraine

Isad Saric Faculty of Mechanical Engineering, Department of Mechanical Design, University of Sarajevo, Sarajevo, Bosnia and Herzegovina

Lamija Šćeta School of Economics and Business, University of Sarajevo, Sarajevo, Bosnia and Herzegovina

Munira Šestić School of Economics and Business, University of Sarajevo, Sarajevo, Bosnia and Herzegovina

Ramo Šendelj University of Donja Gorica, Podgorica, Montenegro

Neven Šerić Faculty of Economics, University of Split, Split, Croatia

Erjon Selmani Department of Mechanics, Universiteti Politeknik i Tiranes, Tirana, Albania

Vira Shendrik Information Technologies of Design, Sumy State University, Sumy, Ukraine

Sergii Shendryk Information Technologies of Design, Sumy State University, Sumy, Ukraine

Vira Shendryk Computer Science Department, Sumy State University, Sumy, Ukraine

Ihor Shepelenko Department of Exploitation and Repairing Machines, Central Ukrainian National Technical University, Kropyvnytskyi, Ukraine

Natalya Shramenko Department of Transport Technology and Logistics, Kharkiv Petro Vasylenko National Technical University of Agriculture, Kharkiv, Ukraine; Department of Operational Work Management, Ukrainian State University of Railway Transport, Kharkiv, Ukraine

Vladyslav Shramenko Faculty of Mathematics and Computer Science, V. N. Karazin Kharkiv National University, Kharkiv, Ukraine; Faculty of Technological Systems and Logistics, Kharkiv Petro Vasylenko National Technical University of Agriculture, Kharkiv, Ukraine

Ramiz Sijamhodžić Technical Faculty Bihac, University of Bihac, Bihać, Bosnia and Herzegovina

O. P. Sitovskyi Department of Cars and Transport Technologies, Lutsk National Technical University, Lutsk, Ukraine

Pero Škorput Faculty of Traffic and Transport Sciences, Zagreb, Croatia

Halid Skaka SKYMedia, Sarajevo, Bosnia and Herzegovina

Filip Skender Faculty of Metallurgy, University of Zagreb, Sisak, Croatia

Isat Skenderović Faculty of Natural Sciences and Mathematics, University of Tuzla, Tuzla, Bosnia and Herzegovina

Jasmin Smajic Make3D, Sarajevo, Bosnia and Herzegovina

Alisa Smajović Faculty of Pharmacy, University of Sarajevo, Sarajevo, Bosnia and Herzegovina

Halima Sofradžija Faculty of Political Sciences, University of Sarajevo, Sarajevo, Bosnia and Herzegovina

Irina V. Somina Belgorod State Technological University named after V.G. Shukhov (BSTU), Belgorod, Russia

Damir Špago Faculty of Mechanical Engineering, "Džemal Bijedić" University of Mostar, Mostar, Bosnia and Herzegovina

Suad Špago Faculty of Civil Engineering, University "DžemalBijedić", Mostar, Bosnia and Herzegovina

Nermina Spaho Faculty of Agriculture and Food Science, University of Sarajevo, Sarajevo, Bosnia and Herzegovina

Nadica Stojanović Faculty of Engineering, Department for Motor Vehicles and Motors, University of Kragujevac, Kragujevac, Serbia

Savo Stupar School of Economics and Business, University of Sarajevo, Sarajevo, Bosnia and Herzegovina

Aida Šukalić Džemal Bijedić University of Mostar, Mostar, Bosnia and Herzegovina

Edin Šunje Faculty of Mechanical Engineering, University "Džemal Bijedić" in Mostar, Mostar, Bosnia and Herzegovina

Suad Sućeska Sarajevo, Bosnia and Herzegovina

Igor Sydorenko Odessa National Polytechnic University (ONPU), Institute of Industrial Technologies, Design and Management, Odessa, Ukraine

Ihor Sydorenko Odessa National Polytechnic University (ONPU), Odessa, Ukraine

Osman Tibo Babić doo Jablanica, Jablanica, Bosnia and Herzegovina

Dragi Tiro Faculty of Mechanical Engineering, Džemal Bijedić University of Mostar, Mostar, Bosnia and Herzegovina

Vito Tič Faculty of Mechanical Engineering, University of Maribor, Maribor, Slovenia

Zdenko Tkáč Faculty of Engineering, Department of Transport and Handling, Slovak University of Agriculture in Nitra, Nitra, Slovak Republic

Volodymyr Tonkonogyi Odessa National Polytechnic University (ONPU), Odessa, Ukraine

Jusuf Topoljak Faculty of Civil Engineering, University of Sarajevo, Sarajevo, Bosnia and Herzegovina

Miron Torlo Faculty of Mechanical Engineering, University "Džemal Bijedić" in Mostar, Mostar, Bosnia and Herzegovina

Mirsad Trobradović Faculty of Mechanical Engineering, University of Sarajevo, Sarajevo, Bosnia and Herzegovina

Yuri Tsekhanov Voronezh State Technical University, Voronezh, Russia

Juraj Tulík Faculty of Engineering, Department of Transport and Handling, Slovak University of Agriculture in Nitra, Nitra, Slovak Republic

Nedim Tuno Faculty of Civil Engineering, University of Sarajevo, Sarajevo, Bosnia and Herzegovina

Raul Turmanidze Faculty of Transportation and Mechanical Engineering, Georgian Technical University (GTU), Tbilisi, Georgia

Sergii Tymchuk Kharkiv Petro Vasylenko National Technical University of Agriculture, Kharkiv, Ukraine

Serhii Tymchuk Kharkiv Petro Vasilenko National Technical University of Agriculture, Kharkiv, Ukraine

Camelia Lavinia Unguras University of Petrosani, Petrosani, Romania

Galyna Urum South Ukrainian National Pedagogical University named after K. D. Ushynsky, Odesa, Ukraine

Saša Vasiljević High Technical School of Professional Studies, Technical College of Applied Studies in Kragujevac, Kragujevac, Serbia

Florin Vedinas University of Petrosani, Petrosani, Romania

Fikret Veljović Faculty of Mechanical Engineering, University of Sarajevo, Sarajevo, Bosnia and Herzegovina

Michail Vereskun Priazovsky State Technical University, Mariupol, Ukraine

Grujica Vico Faculty of Agriculture, University of East Sarajevo, East Sarajevo, Bosnia and Herzegovina

Kazakov Alexandr Viktorovich Department of Electronic Technologies in Mechanical Engineering, Bauman Moscow State Technical University, Moscow, Russia

Haralambie Vochitoiu University of Petrosani, Petrosani, Romania

Samir Vojić Technical Faculty Bihac, University of Bihac, Bihać, Bosnia and Herzegovina

Hrvoje Vojvodić Faculty of Traffic and Transport Sciences, Zagreb, Croatia

Avdo Voloder Faculty of Mechanical Engineering, University of Sarajevo, Sarajevo, Bosnia and Herzegovina

Viktor Vryukalo Computerized Machine Building Department, Ivano-Frankivsk National Technical University of Oil and Gas, Ivano-Frankivsk, Ukraine

Miroslav Vujic Faculty of Traffic and Transport Sciences, University of Zagreb, Zagreb, Croatia

Miroslav Vujić Faculty of Traffic and Transport Sciences, University of Zagreb, Zagreb, Croatia

Petar Vujovic Faculty of Electrical Engineering, University of Montenegro, Podgorica, Montenegro

Sun Xiaoqin Faculty of Civil Engineering, University of Science and Technology, Changsha, China

Ye. A. Yeriomina Dniprovsk State Technical University (DSTU), Kamianskoe, Ukraine

Yousef Zaarir Faculty of Engineering and Natural Sciences, International Burch University, Ilidža, Sarajevo, Bosnia and Herzegovina

Katerina Zahumenna Kharkiv Petro Vasylenko National Technical University of Agriculture, Kharkiv, Ukraine

Nermina Zaimovic-Uzunovic Faculty of Mechanical Engineering, University of Zenica, Zenica, Bosnia and Herzegovina

Viliam Zaloga Department of Manufacturing Engineering, Machines and Tools, Sumy State University, Sumy, Ukraine

Gordana Zeba Mechanical Engineering Faculty in Slavonski Brod, Josip Juraj Strossmayer University of Osijek, Slavonski Brod, Croatia

Yiheng Zhang Odessa National Polytechnic University (ONPU), Odessa, Ukraine

Milutin Živković High Technical Mechanical School of Professional Studies, Trstenik, Serbia

New Technologies in Mechanical Engineering, Metallurgy, Mechatronics, Robotics and Embedded Systems

Dissemination of Patents of the Base Technologies of the Fourth Industrial Revolution - Industry 4.0

Isak Karabegović[1,2](✉), Edina Karabegović[2], Mehmed Mahmić[2], and Ermin Husak[2]

[1] Akademy of Sciences and Arts, Bosnia and Herzegovina, St. Bistrik 7, 71000 Sarajevo, Bosnia and Herzegovina
isak1910@hotmail.com
[2] University of Bihać, 77000 Bihać, Bosnia and Herzegovina

Abstract. The fourth industrial revolution presents a vision of advanced industrial production that is already being implemented by applying new technologies in the automation of production processes, and the exchange and processing of data. In other words, Industry 4.0 integrates the fourth-generation innovations in production processes and in all segments of society. The patents from Industry 4.0 and their implementation are responsible for Industry 4.0 application. The paper analyzes the application and implementation of Industry 4.0 patents in the world, with special attention to the European Union and its developed countries. The analysis was also conducted by industry sectors and leading companies in the application and implementation of Industry 4.0 patents. The paper provides the conclusion of the implementation of Industry 4.0 patents.

Keywords: Innovations · Patents · Industry 4.0 · Implementation · Industry 4.0 base technologies · Infrastructure · Automotive industry

1 Introduction

It is well known that the past three industrial revolutions have introduced greater automation of repetitive physical labour. The term fourth industrial revolution is derived from German and it first was first announced at the Hanover Fair in 2011 [1–3]. The following year, an industry working group introduced the concept to the German government, causing the term Industry 4.0 to expand worldwide ever since. The fourth industrial revolution presents a vision of advanced industrial production that is already being implemented by applying new technologies in the automation of production processes, and the exchange and processing of data. Industry 4.0 integrates the following fourth generation innovations in production processes: mechanical devices controlled by computer algorithms (Cyber-Physical System), interconnectivity within the production process (Internet of Things – IoT), a new way of accessing personal data and applications no longer hosted on a computer "Cloud" – which means that you can access the program, records and documents from multiple devices, at any time, from different locations (Cloud Computing), internet service (Internet of Service), large collection of data

I. Karabegović (Ed.): NT 2020, LNNS 128, pp. 3–15, 2020.
https://doi.org/10.1007/978-3-030-46817-0_1

(Big Data) and autonomous intelligent systems (Autonomous Intelligent System) in all segments of society: new education, new ecology, connectivity, individuality, new health paradigm, urbanization, mobility and globalization [4–6].

Cyber-Physical System (CPS) in the production environment includes intelligent production processes, intelligent machines, intelligent logistics systems and intelligent products that independently exchange information, independently control and initiate activities to improve the production process. The analysis and discussion of Industry 4.0 aims to increase the awareness of its multiple impacts, comprehensiveness and the complete speed of the technological revolution. Industry 4.0 will significantly increase the efficiency and flexibility of production processes and increase the value of production [7, 11, 12]. The transition to "smart" production processes that operate autonomously has already been recognized as an important challenge by industry and policymakers. It is necessary to create a framework for thinking about "Industry 4.0" which outlines key questions and highlights possible answers. In other words, it is necessary to create a platform to achieve public-private cooperation and partnerships on emerging issues related to the technological revolution.

2 Dissemination of Patents and Technologies that are the Basis of Industry 4.0 Worldwide

The dynamics of patent applications from Industry 4.0 base technologies clearly indicates that innovation in Industry 4.0 is much faster than in other technologies and fields [8–10]. The major innovation centers for Industry 4.0 base technologies since the 1990s are located in the USA, Japan, and European Union, while China has played a significant role in the recent years. In order to show a real image of patent dissemination from the Industry 4.0 base technologies, we will show the trend published by the European Patent Office (EPO), which is one of the leading suppliers of patent information, as shown in Fig. 1.

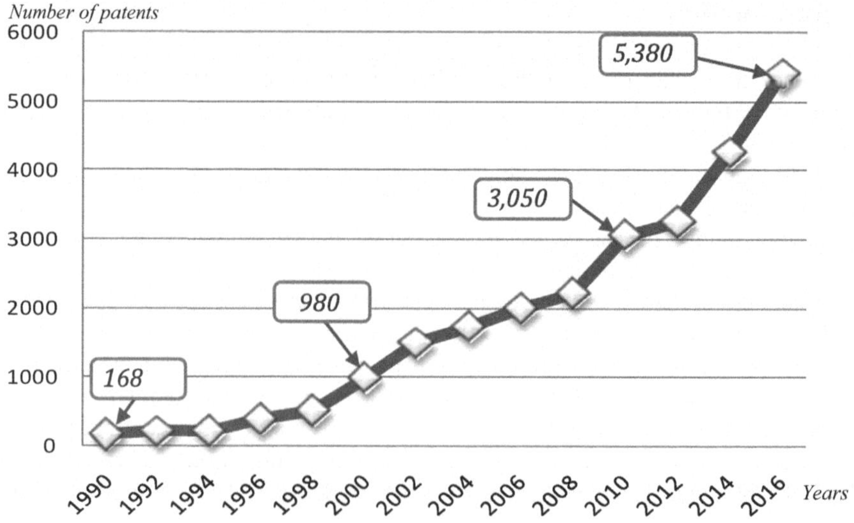

Fig. 1. The trend of patent applications of the fourth industrial revolution – Industry 4.0 [9]

The analysis of Fig. 1 shows that we have a continuous growth of patents from the fourth industrial revolution base technologies. It can be concluded that in the period from 1990 to 2016, the trend of patent registrations and applications has an exponential function, so that in 2016 the number of patents reached 5.380. We can see that the latest available patent application data signals the implementation of Industry 4.0. Massive use of Internet of Things (IoT) is currently in progress, and the estimates are that by 2025, up to 30 billion devices will collect and share data autonomously, as they will be equipped with smart sensors, processors and embedded software, and connected to the Internet (IoT). It is precisely this trend of patent growth from Industry 4.0 that shows us that the deepest changes are underway, leading us into a "smart" environment. According to [9] reports, more than 5.000 patent applications in 2016 were related to autonomous devices, and the growth rate of Industry 4.0 patent applications in the last three years was 54%. In order to get a more complete image of the number of patents of the Industry 4.0 base technologies, Fig. 2 shows the trend of patents (pending and approved) from five countries in the world: China, USA, Japan, Korea and Germany.

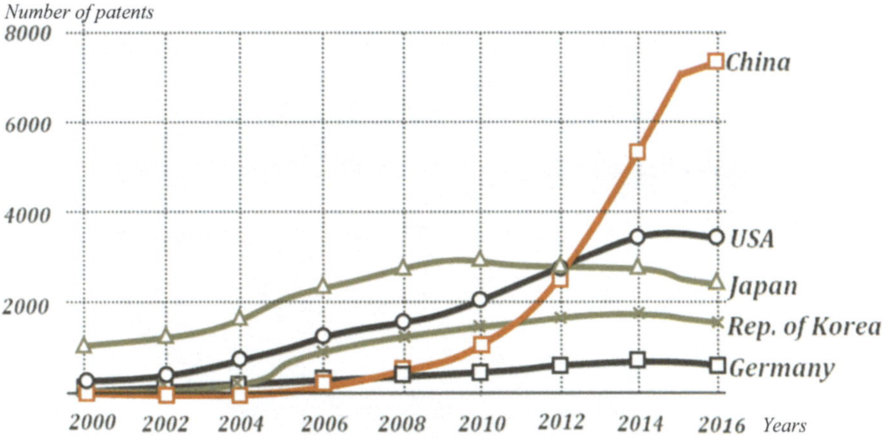

Fig. 2. The trend of patent applications in the Industry 4.0 base technologies in China, USA, Japan, Korea and Germany (pending and approved) [4, 5, 10]

The trend of patents in the Industry 4.0 base technologies (Fig. 2) shows impressive data for Industry 4.0 patent application in China. China was ranked first in patent applications in 2016, and this trend has been growing exponentially since 2010. In USA, Japan, Korea and Germany, the trend of innovation application was increasing slightly over the same period. We can conclude that the USA is ahead of Japan, which held the first place in the world until 2012. The start of Industry 4.0 innovations in Korea and China began ten years ago, while at the same time Japan and USA had a remarkable number of innovations from Industry 4.0. However, over the last ten years, the trend of innovation growth in China and Korea is faster than in developed countries, as shown in Fig. 2. For example, in Korea only two companies, Samsung and LG, apply for over 90% of patents in Industry 4.0, while almost 70% of patents in China originate from two companies, Huawei and ZTE, that focus on Industry 4.0 base technologies. The trend of innovation applications from Industry 4.0 allows only limited conclusions about patents and their implementation, as there are a huge number of low-quality patents from Industry 4.0 in China. The second reason why this trend of innovation in China should be regarded with doubt is a dysfunctional government policy that rewards innovation registration regardless of its quality. The third reason is that many applied innovations from industry in China have inferior quality in relation to those in Germany or USA. In addition, their acceptance rate is also much lower. Here we have to mention that the patents are classified into three main sectors, each of which is divided into several technological fields: core technologies that enable the transformation of any device into a smart device connected to the Internet (hardware, software, connectivity), enabling technologies used in combination with connected devices (analytics, security, artificial intelligence, positioning, power, 3D systems, user interfaces), as well as application domains where the binding potential can be utilized (personal, home, company, production processes, vehicles, infrastructure, etc.) [3–5, 9].

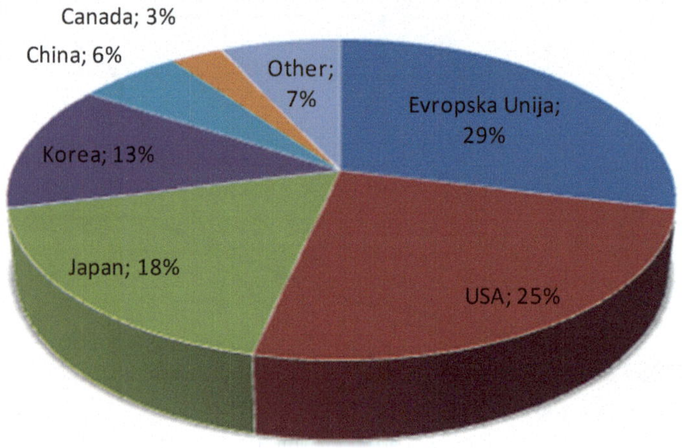

Fig. 3. The percentage of patent applications for Industry 4.0 base technologies for the period 2011–2016 [9]

The analysis of the trend of Industry 4.0 patents for the period 2011–2016 (Fig. 3) has shown that the leaders in the world are the European Union, which holds 29% of all patents in Industry 4.0, followed by the USA with 25%, and Japan with 18%. This trend is logical as these countries have started to develop the Industry 4.0 base technologies since the 1990s, which can be deduced from the trend shown in Fig. 2. These three countries still hold the major centers of innovation, as they operate in large companies from different sectors of the industry, which, due to the global marketplace, are investing in development to remain the leaders in the application of the latest innovations. The fourth place for the same period is held by Korea with 13%, followed by China with 6%, which both aim to reach the developed countries in the world, and in the recent years have a faster trend of growth of Industry 4.0 innovations than the developed countries, as we previously noted. The technologies that have been credited with the emergence and implementation of Industry 4.0 are Information Communication Technologies (ICT), and the trend of patent applications from ICTs for the period 1990–2016 is shown in Fig. 4.

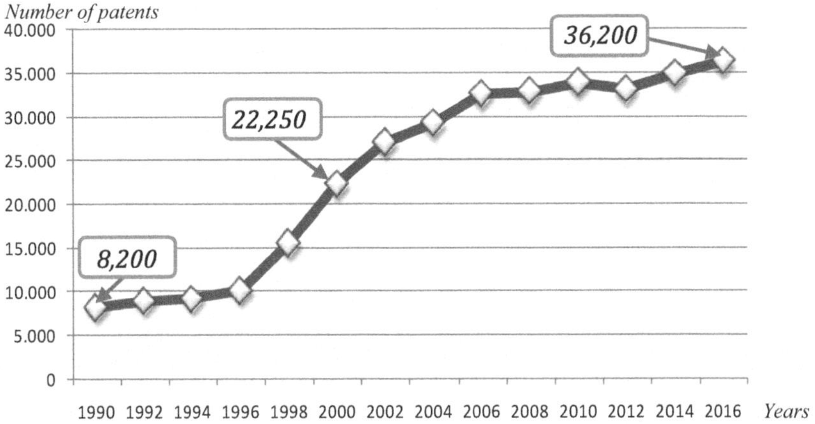

Fig. 4. The trend of patent application from information communication technologies (ICT) for the period 1990–2016 [9]

Patents in information and communication technology have led to the presence of computers and Internet access in virtually every space, from home to the factory or workplace, and later to the integration of computers into mobile communication devices. Industry 4.0 opens a new cycle of innovation, in which each device and facility will be equipped with computing capabilities and connected to network communication. Recent technical advances have played an important role in this development, enabling the integration of chips into almost all devices and their networking and use for everyday needs. Based on Fig. 4, we can conclude that the trend of patent applications in information-communication technologies is increasing, with around 8.200 patent applications in 1990 and reaching the value of 36.200 patent applications in the next twenty-six years, which means that it has increased by about 4, 5 times. This trend of ICT patent application growth is a result of the full integration of information and communication technology in the production processes of all industries, as well as applications in all segments of the environment such as home, personal, vehicles, companies, infrastructure, etc. Unlike the early years of digitalization, the growing trend in ICT patent applications has been driven by other emerging technologies such as cloud computing, robotics & automation, intelligent sensors, 3D printers and radio frequency identification - RFID, and smart intelligence, which have been implemented in most industries, enabling the development of interconnected smart devices that work autonomously [5, 10–13]. ICT has become a major driver of innovation, both in the production processes and in the environment in all segments of society. This growing trend in patent applications for information and communication technology (ICT) will continue in the years to come. For a better image of the industry 4.0 base technology innovations, patents from selected technologies and countries in 2016 are shown in Table 1.

Table 1. Patent application in five base technologies of Industry 4.0 in Germany, Korea, Japan, USA and China in 2016 (analysis of 30 top companies) [10]

	Germany	*Rep.Korea*	*Japan*	*USA*	*China*
Cloud computing	*70*	*314*	*132*	*1,559*	*1,144*
Robotics	*349*	*142*	*638*	*164*	*2,700*
Sensors	*132*	*88*	*83*	*126*	*2,586*
3D Printing	*0*	*52*	*307*	*536*	*504*
RFID	*98*	*1,039*	*1,243*	*1,055*	*384*
Total	*649*	*1,635*	*2,403*	*3,440*	*7,327*

The analysis of Table 1 provides us with the conclusion that the first place in the world by registered and implemented innovations is held by robotic technology, where China and Japan hold the first and second place. The second place by reported and implemented innovations is held by RFID technology, followed by cloud computing, smart sensors and 3D printing technology. We need to emphasize the fact, as we have already mentioned, that the innovation application in China is also valid for 2016. Figure 5 shows which companies apply for the largest number of patents.

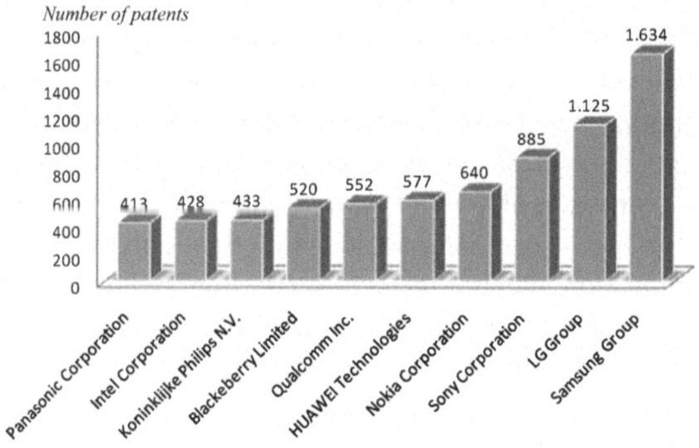

Panasonic Corp.	*Japan*
Intel Corp.	*USA*
Koninklijke Philip.	*Europa*
Blackeberry Limi.	*Canada*
Qualcomm Inc.	*USA*
HUAWEI Techn.	*China*
Nokia Corp.	*Europa*
Sony Corp.	*Japan*
LG Group	*Korea*
Samsung Group	*Korea*

Fig. 5. The trend of patent application for Industry 4.0 base technologies in top ten companies in the world by for the period 2011–2016 [9]

The ten major patent applicants for Industry 4.0 worldwide for the period 2011–2016 are the following companies: Panasonic Corporation, Intel Corporation, Koninklijkee Philips NV, Blackberry Limited, Qualcomm Inc., HUAWEI Technologies, Nokia Corporation, Sony Corporation, LG Group, and Samsung Group. The two major patent holders in Industry 4.0 in the world are companies: the LG Group with 1.125 filed patents and the Samsung Group with 1.634 filed patents for the period 2011–2016. Both companies are from Korea. Of the ten top countries, five are from Asia, three from America and two from the European Union. During this period, these ten top countries filed about 4.634 patents in Asia, about 1.500 patents in the Americas, and about 1.073 patents in the European Union. The analysis of the ten top countries has shown that the Asian companies are the leaders in the development of Industry 4.0 base technologies, and their implementation in the global market. The implementation of patents from Industry 4.0 base technologies for the period 2000–2016 in different industries and areas is shown in Fig. 6.

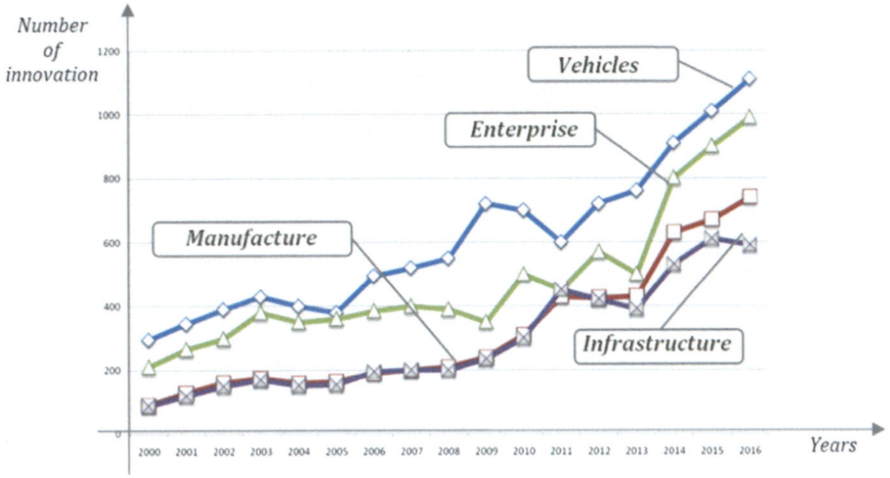

Fig. 6. The trends in patent applications for Industry 4.0 base technologies in the automotive industry, companies, production processes and logistics for the period 2000–2016

The trend of innovation application from the mentioned Industry 4.0 technologies for the period 2000–2016 is shown in Fig. 6. It gives us the opportunity to see that the automotive industry is the first in the implementation of innovations from the mentioned technologies. Likewise, it can be concluded that this trend is growing every year. Such trend is expected because there is a lot of global competition in the automotive industry, since all companies tend to invest heavily in research, development, and implementation of Industry 4.0 base technologies in order to remain competitive in the market. In addition, their goal in the coming years is to develop an autonomous vehicle, and ensure the increased and greater safety of traffic users. The second place in implementation and application of innovations from Industry 4.0 is held by companies due to global competition, but also due to the wide range of Industry 4.0 application in companies such as production process, suppliers, customers and environment of the company itself. The third place in the application of innovations in the world is held by production processes, because the goal is to achieve flexible automation, which leads companies to a smart production process, with which companies would become more competitive in the global market. The final place is reserved for logistics, where the trend of application of innovative solutions in the future will be much higher. For all four of these examples, we see that from 2000 to 2016 there is a growing trend of implementing innovative solutions from Industry 4.0, and it is expected that the upward trend will continue in the future.

3 Overview of the Dissemination of Industry 4.0 Patents in the European Union

The statement that the European Union ranked first in the patent applications from Industry 4.0 for the period 2011–2016 with 29% (Fig. 3), gives us the right to conduct a more detailed analysis of the patent application trend of Industry 4.0 base technologies, in order to get an insight into the European Union countries that invest in the development and deployment of Industry 4.0 base technologies. The other reason for the analysis is the fact that the term Industry 4.0 appeared for the first time in the European Union, more specifically at the Hanover Fair in Germany in 2011.

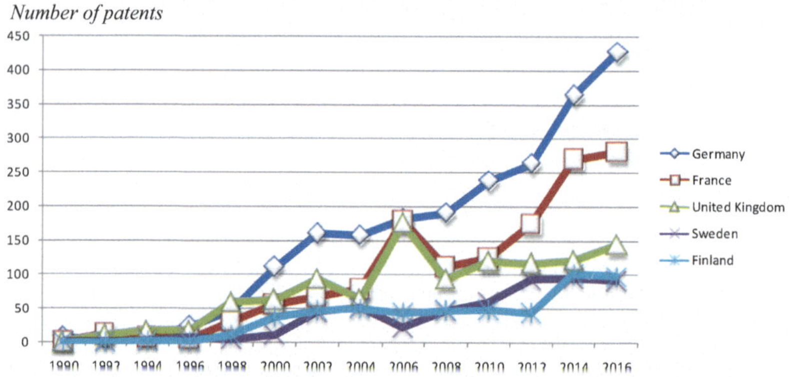

Fig. 7. The trend of patent applications from Industry 4.0 in top five countries in the European Union: Germany, France, United Kingdom, Sweden and Finland for the period 1990–2016 [9]

The analysis of the trend of patent applications from Industry 4.0 in the top four European Union countries for the period 1990–2016 is shown in Fig. 7. We can see that technologically advanced countries like Germany, France, Great Britain, Sweden and Finland are prominent in the European Union. The first place is held by Germany, where the trend of patent applications from Industry 4.0 is constantly on the rise, so that with a couple of patents in 1990 it reached the value of over 400 patent applications in 2016. The number of patent applications in Germany has abruptly increased since 2010, the reason being the arrival of Industry 4.0, which was announced in 2011 in Hannover, Germany. The second place in patent applications from Industry 4.0 is held by France, where the trend of patent applications has been on the rise since 2008, and reached over 250 patent applications in 2016. In the third place is the United Kingdom, whose patent application trend in Industry 4.0 up to 2004 was higher than French, but started to fall back as of 2008. However, the trend is slightly increasing in the following years, and in 2016 it reached around 148 patents. Sweden and Finland take the fourth and fifth places, lagging behind Germany, France and the United Kingdom. The trend of patent applications in these two countries is approximately identical, because one year one country has a higher patent application in Industry 4.0, which decreases the following year. Both countries have reached about 100 patents filed in 2016. The fact that Industry 4.0 is already present in all segments of the society, with an increasing trend of implementation, leads to the conclusion that the trend of patent applications in these top five EU countries will continue to grow.

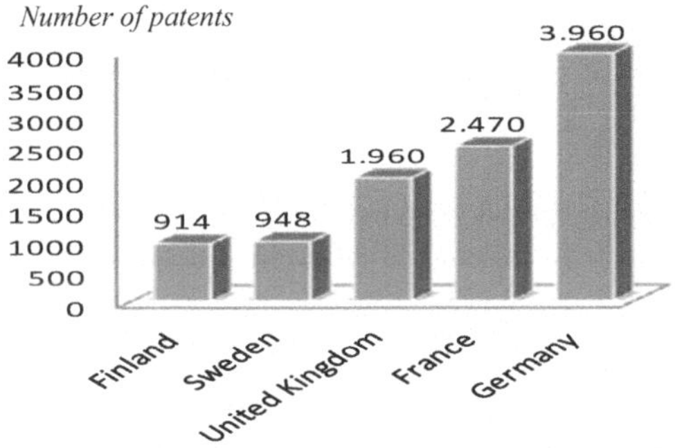

Fig. 8. The trend of Industry 4.0 patent applications in the top five countries in the European Union: Finland, Sweden, United Kingdom, France and Germany for the period 1997–2016 [5, 9, 10]

The analysis of the overall trend of patent applications from Industry 4.0 for the period 2011–2016 (Fig. 3) indicates that the European Union holds the first place in the world with 29%, which is about 14.000 patent applications from Industry 4.0. The following countries of the European Union are credited for this result of patent applications from Industry 4.0: Finland, Sweden, the United Kingdom, France and Germany. Germany has the leading position, with about 4.000 completed patent applications for this period, as shown in Fig. 8. France is in the second place with about 2.500 patent applications, followed by the United Kingdom with over 2.000 patent applications for the specified period. The fourth and fifth places are held by Finland and Sweden, each with about 1.000 patents in total. Such trend of patent applications from Industry 4.0 is expected, as the analysis of patent applications on an annual level (Fig. 7) was exactly the same [14–16]. These are all industrialized countries that are continually investing in development, and whose companies want to be global market leaders, especially the automotive industry, which has a strong competition in the world. In order to obtain a full picture which industries and segments of society in the European Union implement Industry 4.0, we have conducted a percentage analysis of patent applications in infrastructure, production vehicle production, and companies Fig. 9.

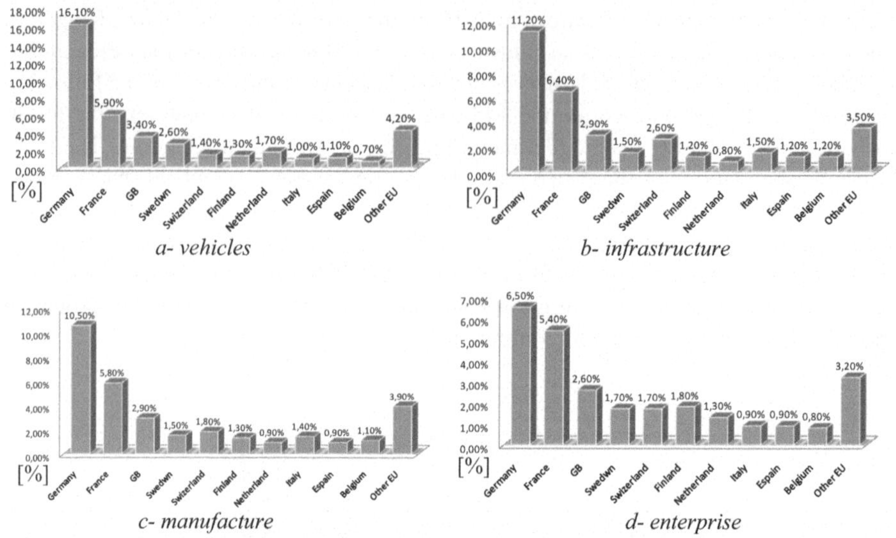

Fig. 9. Percentage of patent applications in Industry 4.0 in the ten top countries in the European Union in the areas of vehicle production, infrastructure, production and companies

The analysis of the patent application from Industry 4.0 in the ten top countries of the European Union, provided us with the conclusion that the inventions in vehicle production ranked first, which places Germany as the leader because it has the most developed automotive industry in the European Union, and holds 16.1% of all patent applications. The second place is held by France, followed by the Great Britain. In terms of infrastructure, Germany is also ranked first, with 11.2% of patent applications. The analysis of all segments has shown that Germany is the leading country of the European Union overall. With 10.5% of the application of Industry 4.0 patents, Germany is also largely responsible for the European Union's leading position in the production industry, reflecting the country's efforts to increase its application of Industry 4.0 base technologies in industrial production. Following Germany, the second place in the Industry 4.0 patent application in all segments listed in Fig. 9 is held by France. These two countries stand out in regard to the application of Industry 4.0 patents, compared to other countries of the European Union. The following countries have the most patent application from Industry 4.0: United Kingdom, Sweden Switzerland, Finland, Netherlands, Italy, Spain and Belgium. We can conclude that in the coming years, such trend in the application of Industry 4.0 patents in the European Union will continue to grow.

4 Conclusion

The analysis of the patent applications from Industry 4.0 base technologies and their application in the world and the European Union has resulted in the following conclusions:

- Patent applications are increasing every year, and Industry 4.0 is credited for integrating the following fourth-generation innovations in production processes: mechanical devices that are controlled by computer algorithms (Cyber-Physical System), interconnectivity within the production process (Internet of Things - IoT), a new way to access personal data and applications that are no longer hosted on a computer but in the cloud - meaning you can access software, records and documents from multiple devices, anytime, from different locations (Cloud computing), Internet service (Internet of Service), large amount of data (Big Data) and Autonomous Intelligent Systems.
- It can be approximately concluded that in the period between 1990 to 2016, the trend of patent applications and implementations moved by an exponential function, so that in 2016 the number of patents reached 5.380 applications from the Industry 4.0.
- The world leaders are the European Union, which holds 29% of all patents in Industry 4.0, followed by the USA with 25% and Japan with 18%. This trend is logical as these countries have started to develop the Industry 4.0 base technologies since the 1990s.
- The analysis of patent applications of the base technologies in the world has shown that the first place in the applied and implemented innovations is held by the robotic technology, followed by RFID technology, cloud computing, smart sensors and 3D printing technology.
- Two major patent holders in Industry 4.0 in the world are companies from Korea: the LG Group with 1.125 registered patents and the Samsung Group with 1.634 registered patents for the period 2011–2016.
- Of all industries, the automotive industry is the first in the application and implementation of the innovations in the mentioned technologies, followed by companies. The first reason for such order is global competition, and the second being the wide range of applications of Industry 4.0 in companies including the production process, suppliers, purchases and the environment of the company itself. The third place in the application of innovations in the world is held by the production processes, because the goal is to achieve flexible automation, which would lead companies to a smart production process, thus enabling them to become more competitive in the global market. The last place is reserved for the logistics, where the trend of application of innovative solutions in the future will become larger.
- Dissemination of Industry 4.0 patents in the European Union shows that Germany is the leader in the European Union in all segments of the application of Industry 4.0 patents, with the automotive industry coming first, followed by infrastructure. France is in the second place, followed by the developed countries of the European Union, the United Kingdom, Sweden, Switzerland, Finland, Netherlands, Italy, Spain and Belgium.
- In the following years, the trend of Industry 4.0 patent applications and their implantation of in the world and in the European Union will continue to grow.
- Despite the huge potential of Industry 4.0 in production process, there are currently numerous obstacles to implementation, some of which are the following:

- reliability issues with machine-to-machine (M2M) communication, which has not fully reached the performance and stability levels predicted by Industry 4.0 implementers,
- concern for the security of information and communication technologies,
- insufficient set of skill to implement Industry 4.0 among company engineers,
- fears that Industry 4.0, fully implemented, could cause massive layoffs in the industrial sector, leaving many low educated workers out of work.

- Regardless of all the fears that exist, in spite of the above listed, the number of patent applications and their implementation will continue to grow, resulting inhighly automated and "smart" production processes.

References

1. Schwab, K.: The Fourth Industrial Revolution. World Economic Forum, Geneva (2016). https://luminariaz.files.wordpress.com/2017/11/the-fourth-industrial-revolution-2016-21.pdf
2. Bunse, B., Kagermann, H., Wahlster, W.: Smart Manufacturing for the Future. Germany Trade & Invest, Berlin (2017)
3. Chand, S.: The Smart Factory – Risk Management Perspectives. CRO Forum, Amsterdam (2015)
4. Karabegović, I., Karabegović, E., Mahmić, M., Husak, E.: Implementation of Industry 4.0 and industrial robots in production processes. In: Karabegović, V. (ed.) New Technologies, Development and Application II 2019. Lecture Notes in Networks and Systems, vol. 76, pp. 96–102. Springer, Cham (2019)
5. Karabegović, I., Husak, E.: Industry 4.0 based on industrial and service robots with application in China. J. Mobility Veh. **44**(4), 59–71 (2018)
6. Karabegović, I.: The role of industrial and service robots in fourth industrial revolution with focus on China. J. Eng. Archit. **5**(2), 110–117 (2017). https://doi.org/10.15640/jea.v5n2a9. Published by American Research Institute for Policy Development, USA, (2018), ISSN: 2334-2986 (Print), 2334–2994 (Online). http://jea-net.com/journals/jea/Vol_6_No_1_June_2018/7.pdf
7. Schlechtendahl, J., Keinert, M., Kretschmer, F., Lechler, A., Verl, A.: Making existing production systems Industry 4.0-ready. Prod. Eng. Res. Dev. **9**, 143–148 (2015)
8. Trappey, A.J., Trappey, C.V., Govindarajan, U.H., Sun, J.J., Chuang, A.C.: A review of technology standards and patent portfolios for enabling cyber-physical systems in advanced manufacturing. Open Access J. **4**, 7336–7382 (2016). https://doi.org/10.1109/access.2016.2619360
9. Meniere, X.-Y., Rudyk, I., Valdes, J.: Patents and the Fourth Industrial Revolution. European Patent Office, Munich, Germany (2017). https://www.epo.org. https://www.epo.org/espacenet. https://www.epo.org/register
10. INDUSTRY 4.0: A review of the key technologies and trends. Cipher Report London (2016). https://www.discovercipher.com
11. Lin, K.C., Shyu, J.Z., Ding, K.: A cross-strait comparison of innovation policy under Industry 4.0 and sustainability development transition. J. Sustain. **9**, 1–17 (2017). https://doi.org/10.3390/su9050786
12. Wang, S., Wang, J., Li, D., Zhang, C.: Implementing smart factory of Industrie 4.0: an outlook. Int. J. Distrib. Sens. Netw. **4**, 1–10 (2016). https://www.researchgate.net/publication/291385881_Implementing_Smart_Factory_of_Industrie_40_An_Outlook

13. Crnjac, M., Veža, I., Banduka, N.: From concept to the introduction of Industry 4.0. Int. J. Ind. Eng. Manag. (IJIEM) **8**(1), 21–30 (2017). https://www.researchgate.net/publication/319007861_From_concept_to_the_introduction_of_industry_40
14. Aquin, M.: Data analytics beyond data processing and how it affects Industry 4.0. Insight Centre for Data Analytic, Dublin, Irska (2017)
15. Ecker, C.: Advantages and Challenges for Small manufactureres. International Collaborative Robots Workshop, Columbia, USA (2015)
16. Ermel, U., Hülsebusch, M.: Industrial IoT Risk Assessment of Smart Factories. PLUS-2016, Munchen, Germany (2016)

Self-engineering – Technological Challenges

Rajkumar Roy and Sam Brooks(✉)

School of Mathematics, Computer Science and Engineering,
City, University of London, Northampton Square, London EC1V 0HB, UK
sam.brooks@city.ac.uk

Abstract. Engineered products are becoming more complex and need longer lifetime availability; there is a need for new approaches in maintaining, repairing and overhaul (MRO). This paper presents the concept of self-engineering; the aim is to preserve the functions of a product or system and extend its lifetime and automate MRO processes. New developments in self-healing materials, self-reconfiguring electronics and robotics, which are already or could be self-engineering systems, are reviewed. Biological healing and repair mechanisms are discussed as a potential source of inspiration for new self-engineering systems. Examples of biological self-engineering are presented. Key technological challenges and research questions which need to be addressed in future self-engineering research are discussed throughout.

Keywords: Self-engineering · Through-life engineering services · Repair · Maintenance · Self-healing

1 Introduction

Everything engineered will eventually break. Maintenance Repair and Overhaul (MRO) services can delay and extend product life and fix problems when they occur. However, in some systems, MRO is difficult to implement because it is too costly, or systems are inaccessible. This paper presents the concept of a self-engineering (SE) system that aims to deliver zero-maintenance products. This approach is ambitious but can draw on inspiration from existing man made and biological mechanisms, such as self-healing, self-reconfiguration, self-adaptation and self-repair. SE can be implemented at a system, sub-system or component level, solutions from different levels have been discussed in this paper. However, the authors current work focuses mainly of a system level SE solution.

The objective of this paper is to present an overview of current SE methods and technologies relating to MRO. Research questions which need to be addressed in future work are presented throughout the paper and offer many potential areas of research.

1.1 Overview of Through-Life Engineering Services (TES)

Servitisation and Product-Service System (PSS) business models require sustained and optimum product availability to maximise income. Through-Life Engineering Services (TES) supports this requirement, enabling the development and application of PSS and

I. Karabegović (Ed.): NT 2020, LNNS 128, pp. 16–30, 2020.
https://doi.org/10.1007/978-3-030-46817-0_2

servitization for complex engineering products or systems [1]. Monitoring, diagnostics, and prognostics technologies can be used to gather data and knowledge on performance, degradation and failures and inform services such as continuous maintenance [2]. When combined with new MRO practices, methodologies, and strategies a product (or systems) functional life can be extended and failures prevented. TES is especially important for complex interdisciplinary products and services and has a key focus on minimising total life-cycle cost [3]. One growing area for TES highlighted in a recent report is to support the development of Mobility as a Service (MaaS) [4].

TES is a growing area of research with increasing publications. SE is a strategy which can fall under TES because it can support *servitisation* and PSS businesses. However, it attempts to automate the processes and remove the need for human control from the services. PSS and *servitization* businesses are a key market which could benefit from SE systems.

2 What Is Self-engineering?

2.1 Definition

A working definition of a self-engineering system is: *a system is self-engineering when it registers and responds to a loss in function or operation capability, and automatically takes action to return the functionality.*

Some key characteristics of a SE system include:

1. There must be no human/user intervention, and system response/behaviour should be automatic.
2. It must have the ability to restore or partially restore its lost function(s).
3. It must be built into the system, not added later when required.
4. The aim should be to avoid/reduce maintenance, prolong life and/or increase the system robustness.

The concept of SE systems is not completely new; the concept of a self-maintaining system in software and computer science and zero-maintenance in electronics [5] have previously been presented. Many useful inspection and repair techniques have already been developed which could be utilised in SE systems [2]. However, SE aims to encompass a wider range of technologies and initiate collaboration between different engineering disciplines.

2.2 Processes Involved

Some of the key stages involved in a SE system are:

Monitoring: the system has a sensor or procedure that enables it to register a loss of function.
Trigger: this initiates the SE response, it could be damage, degradation, or a reduction in function or performance.
Response: this is the action the system takes to restore function or repair itself.

Figure 1 outlines the stages in a biological and engineering SE process. For the engineering response the stages have all been demonstrated individually but not combined or in an automated process [6]. Both examples use a similar method to repair material with a fatigue crack. Bioinspired SE is discussed further in Sect. 4.

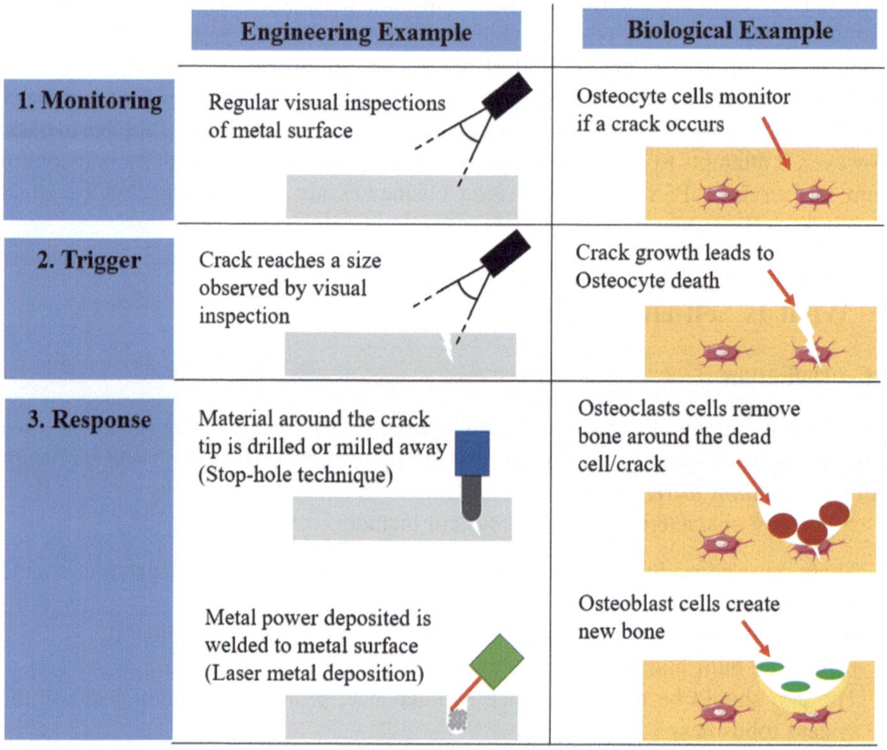

Fig. 1. Diagram showing key stages of SE process in biology and enineering. The biological example shows the process of bone repair. The engineering exmaple shows visual inspection, stop-hole crack repair and laser metal deposition [6].

2.3 Summary of Key Methods

There are several different methods of SE which are referred to in this paper; a definition of them is given here for clarity. These definitions change depending on the sector being looked at, for example, electronics systems referred to as self-healing in previous literature are actually self-reconfiguring and have been grouped as such in this paper.

- **Self-healing -** Self-healing refers to a system which, when a part or assembly is damaged, can return to close to its original state. No new parts or components are utilised the original one is 'healed'.
- **Self-repairing -** Healing requires rehabilitation of components, but a repair can include adding new materials or changing the ones already there. The repair does not leave the system in the original state; there may be damage patched using other parts.

- **Self-adapting -** A system is self-adapting if it can adjust in response to changing conditions or environments and maintain/improve its function. Self-folding and self-assembly mechanism can also fall under this definition, though they are more specialised self-adapting methods.
- **Self-reconfiguring -** A system is capable of changing its arrangement to meet new challenges, component damage or preserve its function. It is very similar to self-adapting, though the new system is a different configuration or arrangement than the original.
- **Built-in redundancy -** A system containing unused parts which the system utilises to replace parts with damage or reduced functionality. To utilise these parts, some reconfiguration of the system is often required.
- **Self-sealing -** A system can close leaks to prevent things (normally fluid) passing in or out of itself.
- **Self-organising -** A system can adapt itself without external direction to meet its needs, there is little or no centralised or hierarchical control, which may be observed in other systems.
- **Self-optimising -** The system ensures maximum utilisation of resources to meet the system requirements.
- **Self-assembly –** The system has the ability to configure from parts into an operating system autonomously.

3 Current Technology

Many different SE systems exist already, examples of ones found in research papers and patents are shown in Table 1.

3.1 Self-healing Materials

Self-healing is a large and growing area of research. Autonomic self-healing occurs without the need for additional stimulus (e.g. no external heat, light or voltage is needed). Non-autonomic systems rely on outside stimuli such as heat or light to trigger the self-healing process [7]. Another key division within self-healing materials is intrinsic and extrinsic healing property. Intrinsic self-healing materials can heal due to non-covalent chemistry or dynamic covalent chemistry [8]. Diels-Alder reactions are frequently used to make intrinsic self-healing polymers. Key examples of materials with added extrinsic healing properties include the following:

1. Microcapsules – Capsules containing a liquid healing agent are embedded within or on the surface of a material, when capsules are damaged the healing agent is released and solidifies [9].
2. Vascular - Micro-tubes filled with healing liquid agent are embedded within or on the surface of the material. Cracks or damage break the tubes, releasing the healing agent [10].

3. Shape memory materials (SMM) – Used to make a composite material which contracts when heated, pulling cracks closed and making it easier for chemical bonds to reform and heal a crack [11]. However, it requires outside stimulus and intrinsic material healing properties to heal fully.
4. Embedded bacteria – Bacteria is added within concrete material to seal cracks and prevent water ingress by creating calcium carbonate [12]. This technology is well developed, with many trials taking place and patents filed.

Self-healing metals are much more difficult due to the stronger internal bonding; however, there has been some success with metal composites or preserving metal surfaces [13]. A growing area of self-healing materials is the textile industry where extending product life is a growing issue. Coating created from proteins found in squid can heal material samples using just water [14], this is currently being sold as a commercial product.

Material fatigue, damage, corrosion or deformation, are often the cause of mechanical failures in systems. A self-healing material can help mitigate against this and keep parts operating for longer, though more work is needed on integration into commercial products. Many self-healing materials also have a limited number of operations (often only one) which limits their possible applications.

3.2 Self-reconfiguring Electronics

Electronic systems have regularly utilised self-reconfiguration and redundancy (normally together) to make fault-tolerant systems. One of the early solutions (from 1980s), was a field programmable gate array (FPGA), which contain programmable logic blocks and memory elements which can be re-configured when needed [15]. More recently, random-access memory (RAM) devices were repaired by reconfiguration. Data in faulty memory cells of a RAM can be stored at new spare addresses and the system self-reconfigures to the address change [16]. This is also referred to as built-in self-repair (BISR).

MEMS devices are relatively cheap components but form critical parts which are difficult to replace when damaged. Designs for a MEMS piezoelectric energy harvester [17], and an accelerometer MEMS device [18] have been presented with redundant modules which can reconfigure to account for the loss or damage of other modules.

3.3 Self-repairing Systems

Self-sharpening plough-shears created during the industrial revolution are an early example of a self-repairing system [19]. One side of a blade is harder, and one is softer and more vulnerable to erosion, resulting in a sharpening mechanism. Bell et al. [20] investigated if a 4-bar linkage mechanisms could be self-repaired to maintain close to the original actuation path when one joint was damaged [20].

Many companies have published patents for self-healing (self-sealing) tires [28]. Other research on making self-sealing components created a foam coating inspired by the way plants self-seal when cut [37, 38]. Internal pressure and a flexible material block holes in the surface.

Table 1. Table summarising current self-engineering solutions, the response, how it is initiated and what products it has been applied to.

SE method	Category details	Response initiated by	Response	Applied to	Ref.
Self-healing	Micro-capsules	Damage, cracks or wear	Release healing agent	Polymer and concreate	[9]
	Micro-tubes	Damage, cracks or wear	Release healing agent	Polymer composites	[10]
	SMM	Applied heat	Pulls cracks closed	Polymer composites	[14, 30]
	Embedded bacteria	Damage, cracks or wear	Fills voids in material	Concreate	[12, 23]
	Stimulus triggered	Applied stimulus	Heal chemical bonds	Polymer metals	[7, 8]*
Self-reconfiguration and Redundancy	Electrical parts	Failure of cell, module or component	Reconfigure to utilise spares modules	RAM, FPGA and MEMS	[15, 16]
Self-reconfiguration	Swarm robots	Robot removed or added	Reconfigure to keep shape	Swarm Robots	[24, 25]
Self-sharpening	Blade	Wear on blunt blade	Faster erosion on one side	Ploughshear, knifes, tools	[19, 26]
Self-repairing	4-bar linkage	damage to a linkage	Adjustment to other links	–	[27]
Self-sealing	Material	Small puncture to surface	Expansion or movement to reseal hole	Tires and pneumatic structures	[28]
Self-adapting (and self-modelling)	Robot	Loss or damage of limb	Model, evaluate and trial solutions	Starfish and 6 legged walker	[29]
Self-adapting	Robot gripper	No solid surface to grip	Flex around the object	Robotic hand gripper	[30, 31]
	Escape slide	Slide angle too shallow or steep	Inflate/deflate segment to adjust length	Aeroplane escape slide	[32]
	Flexible antenna	Antenna receiver flexes	Antenna re-tunes to receive signals	Flexible antenna	[33]
Self-adapting and Self-assembly	Photomorphogenesis	Light source	Growth/adaptation towards light	Robot swarms	[34]
Self-folding and Self-assembly	Electronics	Heating of joints	Expansion and folding of joints	Resistor, strain sensor and robot	[35, 36]

*Review paper of methods

3.4 Self-adapting Robots

Self-adapting robots aim to adapt to damage or changes in their environment. Self-adapting modular robots have been a particular area of interest, with cube and tetrahedral modules used in previous work. A starfish like robot (with 4 arms) was taught to move using built-in self-modelling and trial and error; it evaluated different actions and distance travelled in each trail to find the best way of moving. Once it learnt how to move the system was able to adapt and continue moving when a limb was removed [39]. A similar approach with a six-legged robot was tested more recently with optimised algorithms, reducing the time needed to adapt to minutes [29]. Robots which could self-adapt have many uses and could operate in harsh environments without the need for human intervention, even when damaged.

3.5 Robotic Maintenance and Inspection

Robotic inspection, MRO or servicing tasks are research subjects of interest for industry and academics, recent work in some industries is outlined below.

- Ships form a vital role in global trade and require regular inspection of hulls. Currently, ships are removed from the water for cleaning, visual inspection and repairs, taking it out of operation for weeks. The HISMAR and AURORA projects built fully automated underwater robots, but they were only able to perform inspection and basic cleaning duties unaided [40, 41].
- Airlines and manufacturers are increasingly looking to automate required inspection, servicing and maintenance. A snake-like robot developed by Nottingham University and Rolls Royce was used inspect and repair inaccessible areas of engines [42]. Another project, CompInnova, aims to create a robot to move around an aircraft's outer body, inspecting and repairing composite parts [43].
- Electricity is carried along 1000 s of miles of cables, wet cold weather increases the chance of cable degradation. LineScout is an automated power line inspection and repair robot which has to be attached to a cable but is then able to inspect and re-join lose cables autonomously [44].
- Wind turbines are growing bigger to facilitate greater efficiency and are moving to more remote locations. Drones have previously been utilised for visual inspection of wind turbines [45]. Alternative robots have been developed, which can climb the towers or blades to inspect parts closely [46]. The recently funded MIMRee and Wind TTRo projects in the EU aim to go further and include repair processes in these systems.

Most of these robotic systems employ only visual inspection techniques, repair is often left to a human operator. Combining both processes is needed to make SE systems. Repairing a system is much harder and can require specialist training, though this could change in the next few years with the growth of artificial intelligence. Another key issue is equipment size, making a suitable robot to carry inspection equipment is hard enough, adding the weight of tools and repair parts will increase the weight and complexity of the design even further.

4 Self-engineering in Biology

Biology is full of excellent examples of SE, only a few examples have been outlined in this section. Table 2 summarises the SE method used in many biological processes and indicates where this has inspired and been utilised in an engineering system.

4.1 Bioinspired and Biomimicry Design Process

There are two key approaches to create biomimicry and bioispired solutions.

Top-down (or problem-driven) - This approach begins with an engineering problem. Biological systems are searched for a suitable role-model; the search could focus on looking for a particular function, feature or process. Once an appropriate role model has been identified, this can then be investigated further and applied to solve the problem. *Bottom-up* (or solution-driven) - This approach starts with a biological solution and looks closely at behaviour, response, functions and mechanism involved. A particular solution is identified, investigated further and extrapolated. Finally, there is a search for an engineering problem to apply the solution to.

Both approaches are useful and can be utilised for different aspects of a product's design, Flectofin is an example of this [47]. It can often be difficult to replicate all biological mechanism as some will only work on a smaller scale and not the scale required by engineers [48].

Table 2. Summary table of some biological self-engineering systems and methods used. Where the system has been utilised in a product (bioinspired) it has been indicated. Green boxes indicated methods used.

		Self-engineering method used											
Organism	**Description**	**Self-healing**	**Self-repairing**	**Self-sealing**	**Self-organising**	**Redundancy**	**Self-reconfiguring**	**Self-adapting**	**Self-optimising**	**Self-sharpening**	**Self-assembly**	**Utilised in bioinspired product**	**Ref.**
Human	Skin healing process	■	■	■								Yes	[49]
	Blood vessels constriction							■	■			No	[50]
	Kidney redundancy					■		■				No	[51]
	Bone repair process	■	■									No	[52]
	Neurons (synaptic plasticity)								■			No	[53]
Sea creatures	Sea urchin teeth		■							■		No	[54]
	Molluscs shell layersstructure	■	■									Yes	[55]
	Moon jellyfish re-organise limbs				■		■					No	[56]
Plants	Seed capsules expand/open in water										■	Yes	[57]
	Photomorphogenesis										■	Yes	[34]
	Stem wound sealing with latex	■	■	■								Yes	[58], [59]
	Internal cell sealing mechanism		■	■								Yes	[37], [60]
	Pre-tensioned structure for wound sealing			■								Yes	[61]
	Regeneration from cutting				■	■		■				No	[62]

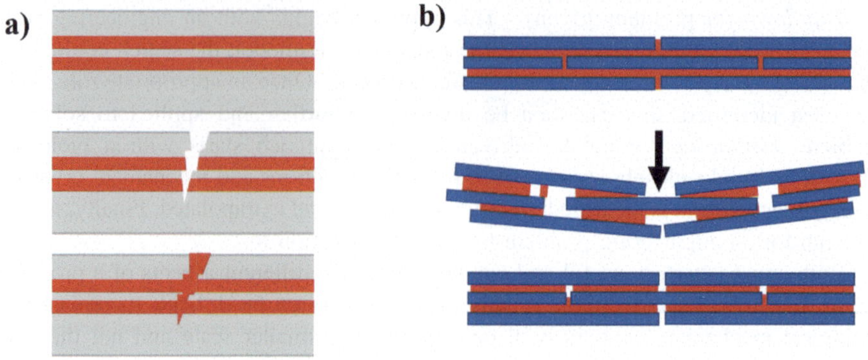

Fig. 2. Diagram of bioinspired methods of self-helaing; (a) shows vascuar micro-tubes used to self-heal a crack, with healing agent in red (see [10] for more information); (b) shows a block and mortar stucture inspired by mollusc shell with breakable bonds which can be repaired when force is removed, see [55] for more information.

4.2 Examples of Biology Inspired SE Work

Biology has inspired many SE mechanisms, some of these are listed below.

- Photomorphogenesis, the growth and movement of plants towards light, was mimicked with a robotic swarm creating a light-responsive, self-assembly and self-adapting mechanism [34].
- Vascular structures used in our bodies inspired a self-healing polymer composite with hollow glass fibres which released a healing agent when damaged [10], see Fig. 2(a) for a diagram.
- Mollusc shells, inspired a self-healing polymer and glass brick composite. Sacrificial healable bonds between layer allowed the structure to deform and return to its original shape [55], see Fig. 2(b) for a diagram.
- Delosperma cooperi plants (see Fig. 3(a)) have pre-tensioned structures which self-seal wounds in the plant, to prevent water being lost. Shape memory materials were used to replicate the effect in a composite material [61].
- Hardide coating has been applied to knives and tool blades to make them self-sharpening, it was inspired by sharks and sea urchin teeth (see Fig. 3(b)) which have a softer and harder side [54]. However, a similar mechanism was invented earlier with self-sharpening plough shears, without bio inspiration.

Biology can inspire many innovative new products and solutions, but designers should be careful as biology is not always the optimum solution. Nature does not contain wheels or gears, but they have been a vital engineering component for hundreds of years.

Fig. 3. Two sources of biological inspiration for SE systems. (A) Delosperma cooperi plants leaves which bend to self-seal when cut due to pre-tension, © 2018 Speck et al.; licensee Beilstein-Institut [61]; (B) Close up picture of self-sharpening sea urchin teeth, reproduced with permision from [64].

5 Technological Challenges and Research Questions

There are many challenges and research opportunities with SE systems. Some of the key research questions which need addressing in future work are outlined in this section. Subheadings are the key stages in a SE system outlined in Sect. 2.2.

5.1 Monitoring

- Where should the monitoring be built-in to a component or added to the system? For a new product it could be built-in but existing system need monitoring which can be added on.
- What should be monitored, and why? Only critical functions or all functions and operations?
- Should monitoring be continuous, intermittent or based on the age of a product?
- How can monitoring take place without interfering with the products function?

5.2 Trigger

- Products will continuously degrade and lose function throughout their life at what point should SE be triggered?
- Can the trigger be reset or repeated, does it occur only once?
- Should there be different levels of a trigger? For example, should it be a binary yes-no response or should it be a scale 1 to 10 based on severity?
- How do you verify if the trigger is correct?
- What other back up sensors can you use to verify the initial trigger?
- Does the trigger determine the SE method used, how is it chosen? Is the chosen SE response the best one?
- If a trigger is not verified, should the system return to the monitoring stage? Should monitoring be increased, is there a risk of a genuine trigger being missed?

5.3 Response

- How is the best response chosen? Can previous SE actions, updates and upgrades be taken into account?
- The response needs to account for factors such as available resources, time, dam-ageseverity. How are these best monitored or recorded in the system?
- When should the response be implemented, can all operations be paused if needed? Is the system vulnerable during the implementation of the response?
- What is the best was to utilised and transported resourced where needed?
- Is the system stable and functionality fully or partially restored by the SE response?
- Was the SE response appropriate, is a further response or adjustment required?

6 Evolution of Self-engineering

6.1 Evolving Automatic Control (Industry 4.0)

Many systems utilise automated monitoring to help predict when maintenance is required as part of Industry 4.0 [63]. To create successful SE systems further work is needed to automate all the stages. Automating many repair processes which are dependent on human experience and training is a significant challenge and one that needs to be addressed for SE systems to be realised.

6.2 Evolving Beyond Automatic Control

In some systems, the trigger for the SE mechanism starts the repair response without any processing; a good example is microcapsules and vascular materials discussed in Sect. 3.1. These systems have the advantage of requiring no decision making or processing and can occur reactively. Currently, most human-made SE systems require some form of control to help determine when it should be implemented. Reducing the need for control in the system could have many benefits, such as reducing the complexity and cost of running a SE system. However, it could mean that SE responses have to be limited to one method and a system would not develop or change to meet new challenges.

7 Summary

In summary, there is a need for significant and long-term research to create complete SE systems with zero-maintenance. They could greatly benefits safety-critical, inaccessible and productivity-critical systems. Many different techniques are being developed to identify material degradation and enable predictions of failure automatically. Alongside this, there has been significant developments in some SE response methods, such as self-healing materials and self-reconfigurable electronics. Considerable research is still needed outside these areas especially combining automated monitoring and automated MRO tasks. The sectors with the most well-developed SE solutions are electrical and computer systems.

Biological systems can provide a source of inspiration for SE mechanisms or methods and have already inspired developments in robotics and self-healing materials. Even some human made solutions designed without biological inspiration share similar characteristics, as shown in Fig. 1. It should also be noted that biological systems are often not limited to one SE method, while human made ones are and could potentially benefit from utilising multiple methods.

SE systems are likely to be most useful in high value industries where a *servitisation*or PSS business model is utilised. Identifying what parts or sub-systems should be made SE is another challenge which could be helped by utilising techniques used in continuous maintenance services, such as degradation mechanics, monitoring diagnostics and prognostics, and repair mechanics.

References

1. Redding, L., Roy, R.: Through-Life Engineering Services: Motivation, Theory, & Practice, 1st edn. Springer (2015)
2. Roy, R., Stark, R., Tracht, K., Takata, S., Mori, M.: Continuous maintenance and the future – foundations and technological challenges. CIRP Ann. - Manuf. Technol. **65**(2), 667–688 (2016)
3. Redding, L.E., Tjahjono, B.: State of the art in through-life engineering services. Comput. Ind. **103**, 111–131 (2018)
4. Elsy, D., Jennings, P., Roy, R.: Through-Life Engineering Services for Mobility as a Service (TES for MaaS), HVM Capatpult (2018)
5. McWilliam, R., Khan, S., Farnsworth, M., Bell, C.: Zero-maintenance of electronic systems: Perspectives, challenges, and opportunities. Microelectron. Reliab. **14**(8), June 2017
6. Marazani, T., Madyira, D.M., Akinlabi, E.T.: Repair of cracks in metals: a review. Procedia Manuf. **8**(March), 673–679 (2017)
7. Kanu, N.J., Gupta, E., Vates, U.K., Singh, G.K.: Self-healing composites: a state-of-the-art review. Compos. Part A Appl. Sci. Manuf. **121**, 474–486 (2019)
8. Wei, Z., et al.: Self-healing gels based on constitutional dynamic chemistry and their potential applications. Chem. Soc. Rev. **43**(23), 8114–8131 (2014)
9. White, S.R., et al.: Autonomic healing of polymer composites. Nature **409**(6822), 794–797 (2001)
10. Pang, J.W.C., Bond, I.P.: A hollow fibre reinforced polymer composite encompassing self-healing and enhanced damage visibility. Compos. Sci. Technol. **65**(11–12), 1791–1799 (2005)
11. Kirkby, E.L., Rule, J.D., Michaud, V.J., Sottos, N.R., White, S.R., Månson, J.A.E.: Embedded shape-memory alloy wires for improved performance of self-healing polymers. Adv. Funct. Mater. **18**(15), 2253–2260 (2008)
12. Tziviloglou, E., Wiktor, V., Jonkers, H.M., Schlangen, E.: Bacteria-based self-healing concrete to increase liquid tightness of cracks. Constr. Build. Mater. **122**, 118–125 (2016)
13. van Dijk, N., van der Zwaag, S.: Self-healing phenomena in metals. Adv. Mater. Interfaces **5** (17), 1–13 (2018)
14. Gaddes, D., et al.: Self-healing textile: enzyme encapsulated layer-by-layer structural proteins. ACS Appl. Mater. Interfaces. **8**(31), 20371–20378 (2016)

15. Frei, R., McWilliam, R., Derrick, B., Purvis, A., Tiwari, A., Di Marzo Serugendo, G.: Self-healing and self-repairing technologies. Int. J. Adv. Manuf. Technol. **69**(5–8), 1033–1061 (2013)
16. Nair, A.S., Bonifus, P.L.: An efficient built-in self-repair scheme for multiple RAMs. In: RTEICT 2017 - 2nd IEEE International Conference on Recent Trends in Electronics, Information & Communication Technology Proceedings, vol. 2018-Janua, no. Ic, pp. 2076–2080 (2018)
17. Farnsworth, M., Tiwari, A.: Modelling, simulation and analysis of a self-healing energy harvester. Procedia CIRP **38**, 271–276 (2015)
18. Xiong, X., Wu, Y.L., Ben Jone, W.: Design and analysis of self-repairable MEMS accelerometer. In:Proceedings of IEEE International Symposium on Defect and Fault Tolerance in VLSI Systems, pp. 21–29 (2005)
19. Brunt, L.: Mechanical innovation in the industrial revolution: the case of plough design. Econ. Hist. Rev. **LVI**, 23–28 (2003)
20. Bell, C., Farnsworth, M., Tiwari, A., Dorey, R.: Theoretical design of a self-rectifying 4-bar linkage mechanism. In: 2nd International Through-life Engineering Services Conference, vol. 11, pp. 385–389 (2013)
21. Yang, Y., Davydovich, D., Hornat, C.C., Liu, X., Urban, M.W.: Leaf-inspired self-healing polymers. Chem **4**(8), 1928–1936 (2018)
22. Wang, X., et al.: Improved self-healing of polyethylene/carbon black nanocomposites by their shape memory effect. J. Phys. Chem. B **117**(5), 1467–1474 (2013)
23. Ramachandran, S.K., Ramakrishnan, V., Bang, S.: Remediation of concrete using micro-organisms. ACI Mater. J. **98**(1), 3–9 (2001)
24. Rubenstein, M., Shen, W.M.: Scalable self-assembly and self-repair in a collective of robots. In: 2009 IEEE/RSJ International Conference on Intelligent Robots and Systems, pp. 1484–1489 (2009)
25. Levi, P., Meister, E., Schlachter, F.: Reconfigurable swarm robots produce self-assembling and self-repairing organisms. Rob. Auton. Syst. **62**(10), 1371–1376 (2014)
26. Zhuk, Y.N.: Hardide: advanced nano-structured CVD coating. Int. J. Microstruct. Mater. Prop. **2**(1), 90–98 (2007)
27. Bell, C., Farnsworth, M., Knowles, J., Tiwari, A.: Self-repairing design process applied to a 4-bar linkage mechanism. Proc. Inst. Mech. Eng. Part B J. Eng. Manuf. **231**(13), 2291–2301 (2017)
28. Mruk, R., Kaes, C.J.-M., Roskamp, R.F.: Self-healing rubber composition and tire (US8962730B2) (2015)
29. Cully, A., Clune, J., Tarapore, D., Mouret, J.B.: Robots that can adapt like animals. Nature **521**(7553), 503–507 (2015)
30. Sun, J., Zhang, W.: A novel coupled and self-adaptive under-actuated multi-fingered hand with gearrackslider mechanism. J. Manuf. Syst. **31**(1), 42–49 (2012)
31. Gao, D., Shi, J., Ransom, L., Janis, R.C.: Reconfigurable Gripping Devices (US20140021731A1) (2014)
32. Alberts, J.h.: Self-adapting slide (EP1306304B1) (2003)
33. Braaten, B.D., et al.: A self-adapting flexible (SELFLEX) antenna array for changing conformal surface applications. IEEE Trans. Antennas Propag. **61**(2), 655–665 (2013)
34. Divband Soorati, M., Heinrich, M.K., Ghofrani, J., Zahadat, P., Hamann, H.: Photomorphogenesis for robot self-assembly: adaptivity, collective decision-making, and self-repair. Bioinspir. Biomim. **14**(5), 056006 (2019)
35. Miyashita, S., Meeker, L., Tolley, M.T., Wood, R.J., Rus, D.: Self-folding miniature elastic electric devices. Smart Mater. Struct. **23**(9), 094005 (2014)

36. Felton, S., Tolley, M., Demaine, E., Rus, D., Wood, R.: A method for building self-folding machines. Science (80-.) **345**(6197), 644–646 (2014)
37. Rampf, M., Speck, O., Speck, T., Luchsinger, R.H.: Investigation of a fast mechanical self-repair mechanism for inflatable structures. Int. J. Eng. Sci. **63**, 61–70 (2013)
38. Rampf, M., Speck, O., Speck, T., Luchsinger, R.H.: Self-repairing membranes for inflatable structures inspired by a rapid wound sealing process of climbing plants. J. Bionic Eng. **8**(3), 242–250 (2011)
39. Bongard, J., Zykov, V., Lipson, H.: Resilient machines through continuous self-modeling. Sci. New Ser. Am. Assoc. Adv. Sci. **314**, 1118–1121 (2006)
40. Narewski, M.: Hismar - underwater hull inspection and cleaning system as a tool for ship propulsion system performance increase. J. Polish CIMAC **4**(2), 227–234 (2009)
41. Akinfiev, T.S., Armada, M.A., Fernandez, R.: Nondestructive testing of the state of a ship's hull with an underwater robot. Russ. J. Nondestruct. Test. **44**(9), 626–633 (2008)
42. Dong, X., Palmer, D., Axinte, D., Kell, J.: In-situ repair/maintenance with a continuum robotic machine tool in confined space. J. Manuf. Process. **38**(January), 313–318 (2019)
43. Papadimitriou, A., Andrikopoulos, G., Nikolakopoulos, G.: Development and control of a differential wall climbing robot based on vortex adhesion. In: 2019 18th European Control Conference, pp. 1610–1615 (2019)
44. Pouliot, N., Richard, P.-L., Montambault, S.: Linescout technology opens the way to robotic inspection and maintenance of high-voltage power lines. IEEE Power Energy Technol. Syst. J. **2**(1), 1–11 (2015)
45. Shihavuddin, A.S.M., et al.: Wind turbine surface damage detection by deep learning aided drone inspection analysis. Energies **12**(4), 1–15 (2019)
46. Sahbel, A., Abbas, A., Sattar, T.: System design and implementation of wall climbing robot for wind turbine blade inspection. In: Proceedings of 2019 International Conference on Innovative Trends in Computer Engineering, ITCE 2019, no. February, pp. 242–247 (2019)
47. Lienhard, J., et al.: Flectofin: a hingeless flapping mechanism inspired by nature. Bioinspir. Biomim. **6**(4), 045001 (2011)
48. Byrne, G., Dimitrov, D., Monostori, L., Teti, R., van Houten, F., Wertheim, R.: Biologicalisation: biological transformation in manufacturing. CIRP J. Manuf. Sci. Technol. **21**, 1–32 (2018)
49. Lorenz, H.P., Longaker, M.T.: Wounds: biology, pathology, and management. In: Essential Practice of Surgery, pp. 77–88 (2006)
50. Sessler, D.I., Moayeri, A., Støen, R., Glosten, B., Hynson, J., McGuire, J.: Thermoregulatory vasoconstriction decreases cutaneous heat loss. Anesthesiology **73**(4), 656–660 (1990)
51. Sobh, M., et al.: Long-term follow-up of the remaining kidney in living related kidney donors. Int. Urol. Nephrol. **21**(5), 547–553 (1989)
52. Taylor, D., Hazenberg, J.G., Lee, T.C.: Living with cracks: damage and repair in human bone. Nat. Mater. **6**(4), 263–268 (2007)
53. Turrigiano, G.G.: The self-tuning neuron: synaptic scaling of excitatory synapses. Cell **135**(3), 422–435 (2008)
54. Killian, C.E., et al.: Self-sharpening mechanism of the sea urchin tooth. Adv. Funct. Mater. **21**(4), 682–690 (2011)
55. D'Elia, E., Eslava, S., Miranda, M., Georgiou, T.K., Saiz, E.: Autonomous self-healing structural composites with bio-inspired design. Sci. Rep. **6**(April), 1–11 (2016)
56. Abrams, M.J., Basinger, T., Yuan, W., Guo, C.L., Goentoro, L.: Self-repairing symmetry in jellyfish through mechanically driven reorganization. Proc. Natl. Acad. Sci. U. S. A. **112**(26), E3365–E3373 (2015)
57. Guiducci, L., et al.: Honeycomb actuators inspired by the unfolding of ice plant seed capsules. PLoS ONE **11**(11), 1–21 (2016)

58. Speck, O., Speck, T.: An overview of bioinspired and biomimetic self-repairing materials. Biomimetics **4**(1), 26 (2019)
59. Bauer, G., Nellesen, A., Speck, T.: Biological lattices in fast self-repair mechanisms in plants and the development of bio-inspired self-healing polymers. WIT Trans. Ecol. Environ. **138**, 453–459 (2010)
60. Busch, S., Seidel, R., Speck, O., Speck, T.: Morphological aspects of self-repair of lesions caused by internal growth stresses in stems of Aristolochia macrophylla and Aristolochia ringens. Proc. R. Soc. B Biol. Sci. **277**(1691), 2113–2120 (2010)
61. Speck, O., Schlechtendahl, M., Borm, F., Kampowski, T., Speck, T.: Humidity-dependent wound sealing in succulent leaves of delosperma cooperi - an adaptation to seasonal drought stress. Beilstein J. Nanotechnol. **9**(1), 175–186 (2018)
62. Ikeuchi, M., Ogawa, Y., Iwase, A., Sugimoto, K.: Plant regeneration: cellular origins and molecular mechanisms. Development **143**(9), 1442–1451 (2016)
63. Hermann, M., Pentek, T., Otto, B.: Design principles for industrie 4.0 scenarios. In: Proceedings of Annual Hawaii International Conference on System Sciences, vol. 2016-March, pp. 3928–3937 (2016)
64. Wang, R.Z., Addadi, L., Weiner, S.: Design strategies of sea urchin teeth: structure, composition and micromechanical relations to function. Philos. Trans. R. Soc. B Biol. Sci. **352**(1352), 469–480 (1997)

A Reinforcement Learning Controller for the Swing-Up of the Furuta Pendulum

Domenico Guida[1], Camilo Andrés Manrique Escobar[2], and Carmine Maria Pappalardo[1(✉)]

[1] Department of Industrial Engineering, University of Salerno, Via Giovanni Paolo II, 132, 84084 Fisciano, Salerno, Italy
cpappalardo@unisa.it

[2] MEID4 Academic Spin-Off of the University of Salerno, Via Giovanni Paolo II, 132, 84084 Fisciano, SA, Italy

Abstract. In this paper, a nonlinear control strategy is developed by applying the Reinforcement Learning (RL) approach to control engineering tasks. To this end, the swing-up problem of the Furuta pendulum is solved as a benchmark example considering the presence of dry friction as a function of the instantaneous reaction forces. In the paper, a detailed description of the mechanical system is provided, including the equations of motion and the reward function used in the control algorithm based on continuous and sparse signals. The performance of the deep deterministic policy gradient algorithm in the proposed environment is also evaluated by means of numerical experiments.

Keywords: Reinforcement learning · Furuta pendulum · Dry friction · Multibody dynamics

1 Introduction

Reinforcement Learning (RL) is a type of machine learning technique dealing with the search of the optimal sequential decision-making. This computational approach is aimed at learning from interaction and is inspired by human and animal learning psychology, which has been an attractive research field since its inception. In the workflow of RL, the agent learning takes place through a tentative evaluation process guided by the reward feedback. Deep RL is the combination of RL and deep learning, with application in diverse areas such as portfolio management, self-driving cars, where it has become a turning point with respect to classic approaches [1, 2], control engineering, and robotics. In industrial applications of mechanical engineering [3–5], the investigation of the nonlinear dynamical response of mechanical systems due to external forces is of great interest [6–9], in particular for the control of dynamical systems [10–12], one of the main applications of RL, which provides an optimal control strategy that is learned rather than programmed by a human designer. In contrast with the traditional control design [13–15] that relies on a high level of system understanding, usually expressed in terms of an exact [16–18] or approximated physical model [19, 20], the data-driven approach of RL is a promising alternative and can potentially complement the traditional approach [21]. The increasing interest for

I. Karabegović (Ed.): NT 2020, LNNS 128, pp. 31–38, 2020.
https://doi.org/10.1007/978-3-030-46817-0_3

applying RL in real-world problems, robotics in particular, which can be interpreted as a RL problem where a robot seeks to optimize a cost function encoding a task by utilizing data collected through interaction, has led to the rising of several issues. For instance, the complexity of continuous state and acting spaces, the discontinuity of the control signals, the implementation of the agent in real hardware for its training, where the convergence of the value function and the learning time are a limiting factor since multiple physical iterations result infeasible, which has led to the development of data-efficient methods such as NFQ iteration and PILCO, are common examples. Other challenges that led to the development of the present approach are the reality gap when deploying the trained agent on real hardware, the interest of learning from raw data samples of sensor measurements, and the question for robustness of the learned policy. These issues are not addressed by traditional approaches like Q-learning and actor-critic methods, resulting of theoretical interest but little commercial importance. Therefore, multiple methods have been proposed to cope with these challenges, resulting in the need to have an appropriate procedure to evaluate and compare the performance of each algorithm. This led to the development of platforms such as the Arcade Learning Environment, considering mainly discretized observation-action spaces problems, and OpenAI Gym, which includes benchmark problems such as the cart-pole swing-up. Although several classic control problems have long been used to evaluate reinforcement learning algorithms [22], it results difficult to quantify progress in the domain of continuous control due to the lack of a commonly adopted benchmarks [23]. In this work, the swing-up task applied to the Furuta pendulum in presence of dry friction [24] is assumed as a benchmark problem since its moderate level of complexity results advantageous due to the potential generalization, in contrast with RL benchmarks such as the locomotion task [25], which requires highly specific algorithmic adaptations.

2 Research Methodology

The equations of motion of a general multibody system can be obtained by exploiting its topology and employing the Lagrangian approach, which yields a system of the following form:

$$\begin{cases} \mathbf{M}\ddot{\mathbf{q}} = \mathbf{Q}_v + \mathbf{Q}_e - \mathbf{C}_{\mathbf{q}}^T\boldsymbol{\lambda} \\ \mathbf{C}_{\mathbf{q}}\ddot{\mathbf{q}} = \mathbf{Q}_d \end{cases} \tag{1}$$

where $\mathbf{M}$ is the mass matrix, $\mathbf{q}$ is the vector of Lagrangian coordinates, $\mathbf{Q}_v$ is the quadratic velocity vector, $\mathbf{Q}_e$ is the generalized external forces vector, $\mathbf{C}_{\mathbf{q}}$ is the constraint Jacobian matrix, $\boldsymbol{\lambda}$ is the vector of Lagrange multipliers, and $\mathbf{Q}_d$ is the constraints quadratic velocity vector. The controller for the system is developed through the deep deterministic policy gradient (DDPG), an off-policy approach developed for continuous observation and action spaces. The DDPG algorithm employs two neural networks, Q_θ and π_ϕ, to approximate the value function and the policy as well as their correspondingtarget lagged copies, namely $Q_{\theta'}$ and $\pi_{\phi'}$. Each of these and their corresponding targets are randomly initialized with the same parameters. The algorithm works as follows. For each training time step, given an instantaneous observation $\mathbf{s}$, the following action is determined by the current policy $a = \pi_\phi(\mathbf{s}) + \boldsymbol{\eta}$, where $\boldsymbol{\eta}$ is an stochastic noise vector. Then a new observation $\mathbf{s}'$ is obtained along with a reward r, and this set is stored in an experience memory in tuples $(\mathbf{s},\ a,\ r,\ \mathbf{s}')$, from which a random subset, named the replay buffer, with dimension D, is taken each m episodes. Subsequently, the target value function for each component is computed as:

$$y_i = r_i + \gamma Q_{\theta'}\left(\mathbf{s}'_i,\ \pi_{\phi'}(\mathbf{s}'_i)\right) \tag{2}$$

where $\gamma \in (0,\ 1)$ is the discount factor. Subsequently, the mean-squared Bellman error (MSBE) is computed as follows:

$$L = \frac{1}{D}\sum_{i=1}^{D}\left(y_i - Q_\theta(\mathbf{s}_i,\ a_i)\right)^2 \tag{3}$$

This error parameter is used as a gradient to update the parameters of Q_θ. Then, for the policy update, it is assumed a continuous differentiability of the critic with respect to the actor parameters, allowing for finding the gradient using the following formula:

$$\nabla_{\pi_\phi} = \frac{1}{D}\sum_{i=1}^{D}\nabla_\phi Q_\theta\left(\mathbf{s}_i,\ \pi_\phi(\mathbf{s}_i)\right)\nabla_\phi \pi_\phi(\mathbf{s}_i) \tag{4}$$

where the parameters of Q_θ are treated as constants. This gradient is used to update the policy of the neural network. Finally, the parameters of the target networks are set by Polyak averaging as follows

$$\theta' = \tau\theta + (1-\tau)\theta', \quad \phi' = \tau\phi + (1-\tau)\phi' \tag{5}$$

This procedure is repeated until the optimal policy is found.

3 Numerical Results

The Furuta pendulum system considered as an example is shown in Fig. 1. This multibody system consists of two cylindrical bars of radius R. The horizontal pendulum has mass m_1 and length equal to $2L_1$. This bar is driven by a controlled moment. The vertical bar has mass m_2 and length $2L_2$. This bar is linked with the previous one by a revolute joint and rotates freely as a pendulum under the effect of gravity force. This second bar is subjected to dry friction in its pivot. The equations of motion of the system are obtained considering the vector of Lagrangian coordinates $\mathbf{q} = [\theta_1 \quad \theta_2]^T$ as follows:

$$\mathbf{M}\ddot{\mathbf{q}} = \mathbf{Q}_v + \mathbf{Q}_g + \mathbf{Q}_{e,nc} \tag{6}$$

where $\mathbf{M}$ is the generalized mass matrix given by:

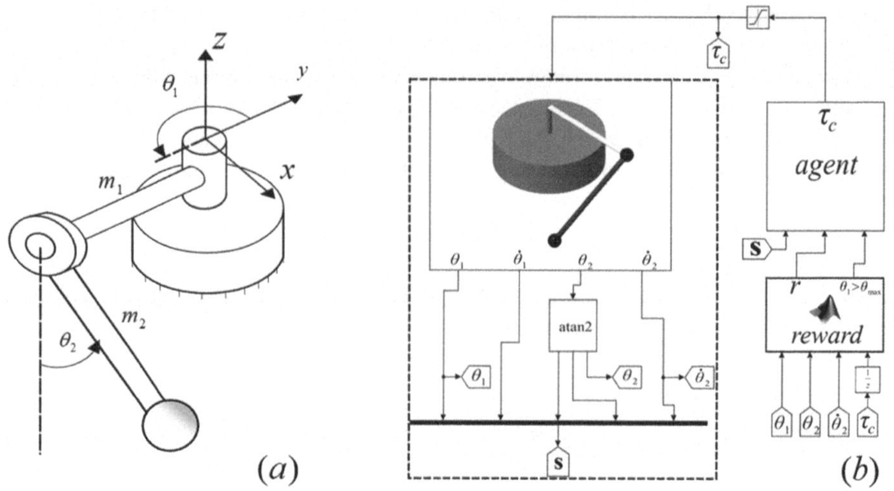

Fig. 1. (a) Furuta pendulum system, (b) Simulink block scheme of the RL controller, where the square with the dashed line represents the physical system considered.

$$\mathbf{M} = \begin{bmatrix} \frac{4}{3}L_1^2(m_1+3m_2)+\frac{4}{3}L_2^2m_2\left(1-c_{\theta_2}^2\right)+\frac{m_2}{2}R^2c_{\theta_2}^2 & 2L_1L_2m_2c_{\theta_2} \\ 2L_1L_2m_2c_{\theta_2} & \frac{4}{3}L_2^2m_2 \end{bmatrix} \tag{7}$$

where the abbreviations $s_\theta = \sin(\theta)$ and $c_\theta = \cos(\theta)$ are used. The generalized force vectors denoted with $\mathbf{Q}_v$ and $\mathbf{Q}_g$ are the system quadratic velocity vector and gravitational force vectors that are respectively defined as:

$$\mathbf{Q}_v = \begin{bmatrix} \dot{\theta}_2m_2\left(\dot{\theta}_1s_{2\theta_2}\left(\frac{1}{2}R^2-\frac{4}{3}L_2^2\right)+2L_1L_2\dot{\theta}_2s_{\theta_2}\right) \\ m_2\dot{\theta}_1^2s_{2\theta_2}\left(\frac{2}{3}L_2^2-\frac{1}{4}R^2\right) \end{bmatrix}, \quad \mathbf{Q}_g = \begin{bmatrix} 0 \\ -m_2gL_2s_{\theta_2} \end{bmatrix} \tag{8}$$

The generalized force vector $\mathbf{Q}_{e,nc} = \begin{bmatrix}\tau_c & \tau_{fr}\end{bmatrix}^T$ is the vector of external non conservative forces, where τ_c is the mechanical moment of the controlled actuator and τ_{fr} is the dry friction moment. The mechanical moment induced by the presence of dry friction is a function of the instantaneous reaction forces, that are produced in the revolute joint, and of the angular velocity of the bar. The dry friction moment is given by:

$$\tau_{fr} = -R\mu N\mathrm{sgn}\left(\dot{\theta}_2\right) \tag{9}$$

where R is the radius of the bar, μ is the coefficient of friction, N represents the magnitude of the instantaneous reaction force components perpendicular to the axis of rotation of the vertical pendulum, and the term $\mathrm{sgn}\left(\dot{\theta}_2\right)$ is the sign function of $\dot{\theta}_2$. The components of the vector of observations $\mathbf{s}$, the input of neural network π_ϕ, is set as

$\mathbf{s} = \begin{bmatrix} \theta_1 & \dot{\theta}_1 & s_{\theta_2} & c_{\theta_2} & \dot{\theta}_2 \end{bmatrix}^T$. Finally, the reward function designed for this environment combines continuous and sparse signals as follows:

$$r = c_1(\theta_1)^2 + c_{\text{lim}}(|\theta_1| > 2\pi) + c_2(\pi - |\theta_2|)^2 + c_\tau(\tau_c)^2 + c_{\dot{\theta}_2}\left(\dot{\theta}_2\right)^2 \\ + \left((\pi - |\theta_2|) < \theta_{2,\min}\right)\left(\left|\dot{\theta}_2\right| < \dot{\theta}_{2,\min}\right)\left(\frac{\dot{\theta}_{2,\min} - \left|\dot{\theta}_2\right|}{\dot{\theta}_{2,\min}}\right) c_{balance} \tag{10}$$

where the constants are set as $c_1 = -1$, $c_{\text{lim}} = -10000$, $c_2 = -5$, $c_\tau = -0.05$, $c_{\dot{\theta}_2} = -0.5$, $\theta_{2,\min} = \pi/2$, $\dot{\theta}_{2,\min} = 5(rad/s)$, and $c_{balance} = 5$. Thus, the system is modeled considering the physical constants of the multibody model and hyper parameters of the DDPG algorithm that are included in Table 1. The maximum duration of a training episode is 10 s and the actuator has been limited to a maximum torque $\tau_{\max}$. The swing-up task training curves are shown in Figs. 2(a) and (b), where the optimal policy is found in the episode 299. The training process stops when the average reward in a window of 5 episodes called $r_{average}$ meets the condition $r_{average} \leq -150$.

Figure 3(a) and (b) show the path described for the tip of the second bar during the

Table 1. Numerical value of the physical properties and hyperparameters employed.

Symbol	Meaning	Value
R	Radius of the bars	0.025 (m)
L_1, L_2	Half the length of the bars	0.5 (m)
m_1, m_2	Mass of the bars	1 (kg)
μ	Coefficient of friction	0.03
$\tau_{\max}$	Maximum torque of the controller	200 (N m)
D	Replay buffer size	128
m	Target networks update frequency	4
γ	Value function discount factor	0.99
t_s	Sampling time	0.02 (s)

swing-up process and the reward function output for that trajectory. Figure 4 represents a qualitative perspective of the controller performance by showing successive frames of optimal policy behavior. Finally, the analysis of Figs. 5(a) and (b) shows the proportionality in the magnitude of the dry friction torque and angular velocity, as well as the opposition in their directions, being in accordance with the physical intuition of the phenomenon.

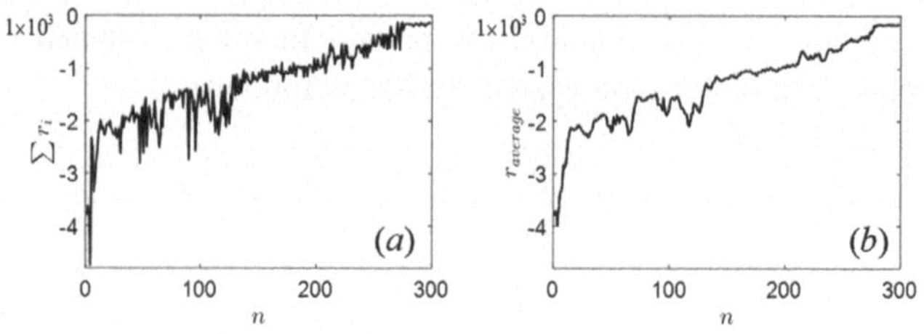

Fig. 2. (a) Cumulative reward $\sum r_i$ per episode of training n and (b) the average reward $r_{average}$ per episode of training n.

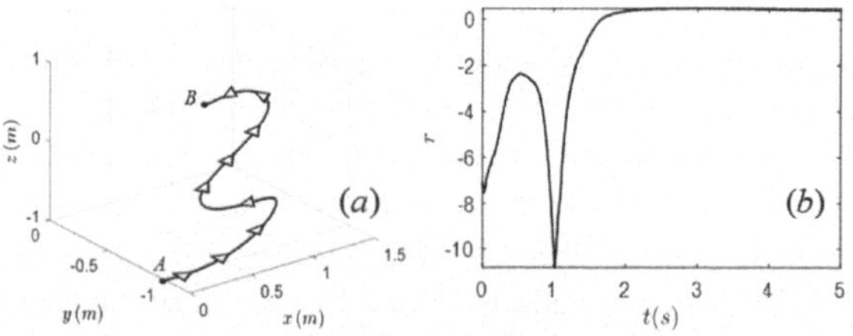

Fig. 3. (a) Pendulum tip path during the swing-up task. Point A is the initial downward position with $\theta_2 = 0$ (rad) and B, where $\theta_2 = \pi$ (rad) is the swing-up position. (b) Reward function output during the performance of the optimal policy.

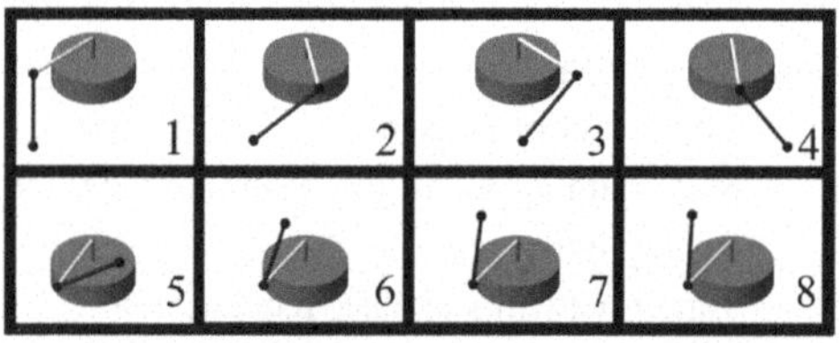

Fig. 4. Sequential frames of the system behavior during the swing-up task.

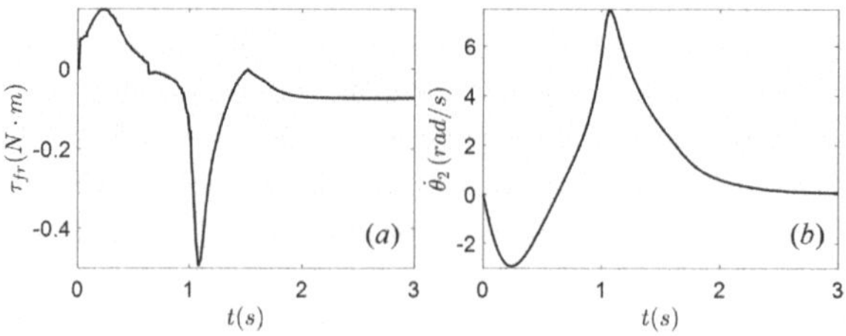

Fig. 5. (a) Dry friction torque in the revolute joint. (b) Angular velocity of the vertical bar.

4 Conclusions

In this work, the swing-up problem of the Furuta pendulum in the presence of dry friction modeled as a function of the instantaneous reaction force is proposed as a benchmark problem for the RL control approach applied to continuous observation and action spaces. The dynamic behavior of the system of interest is described by deriving the equations of motion of the open-chain mechanism applying the Lagrangian approach. On the other hand, the reward function considered in the optimization process features continuous and sparse signals. In particular, the agent learning is carried out using the DDPG algorithm. As shown by the dynamical simulations performed in this work, this algorithm proved to be general enough to deal with a system with an accurate friction model. The authors believe that the control strategy based on the data-driven approach, such as the RL discussed in this investigation, represents a promising path for the development of nonlinear controllers applicable to complex underactuated systems. Future works will focus on the implementation of the present approach in systems with a higher level of complexity.

References

1. De Simone, M.C., Guida, D.: Identification and control of a unmanned ground vehicle by using Arduino. UPB Sci. Bull. Ser. D: Mech. Eng. **80**(1), 141–154 (2018)
2. De Simone, M.C., Rivera, Z., Guida, D.: Obstacle avoidance system for unmanned ground vehicles by using ultrasonic sensors. Machines **6**(2), 18 (2018)
3. De Simone, M.C., Rivera, Z.B., Guida, D.: Finite element analysis on squeal-noise in railway applications. FME Trans. **46**(1), 93–100 (2018)
4. Quatrano, A., De Simone, M.C., Rivera, Z.B., Guida, D.: Development and implementation of a control system for a retrofitted CNC machine by using Arduino. FME Trans. **45**(4), 565–571 (2017)
5. Karabegović, I., Karabegović, E., Mahmić, M., Husak, E.: Implementation of industry 4.0 and industrial robots in the manufacturing processes. In: Lecture Notes in Networks and Systems, pp. 3–14 (2019)
6. Pappalardo, C.M., Guida, D.: Development of a new inertial-based vibration absorber for the active vibration control of flexible structures. Eng. Lett. **26**(3), 372–385 (2018)
7. Isić, S., Mehremić, S., Karabegović, I., Husak, E.: Systems for passive and active vibration damping. In: Karabegović, I. (ed.) New Technologies, Development and Application II, NT 2019, Lecture Notes in Networks and Systems, vol 76. Springer, Cham (2020)
8. Colucci, F., De Simone, M.C., Guida, D.: TLD design and development for vibration mitigation in structures. In: Lecture Notes in Networks and Systems, vol. 76, pp. 59–72 (2020)
9. Concilio, A., De Simone, M.C., Rivera, Z.B., Guida, D.: A new semi-active suspension system for racing vehicles. FME Trans. **45**(4), 578–584 (2017)
10. Pappalardo, C.M., Guida, D.: On the dynamics and control of underactuated nonholonomic mechanical systems and applications to mobile robots. Arch. Appl. Mech. **89**(4), 669–698 (2019)
11. De Simone, M.C., Guida, D.: Control design for an under-actuated UAV model. FME Trans. **46**(4), 443–452 (2018)

12. Pappalardo, C.M., Guida, D.: Use of the adjoint method for controlling the mechanical vibrations of nonlinear systems. Machines **6**(2), 19 (2018)
13. Pappalardo, C.M., Guida, D.: A time-domain system identification numerical procedure for obtaining linear dynamical models of multibody mechanical systems. Arch. Appl. Mech. **88**(8), 1325–1347 (2018)
14. Rivera, Z.B., De Simone, M.C., Guida, D.: Unmanned ground vehicle modelling in Gazebo/ROS-based environments. Machines **7**(2), 1–21 (2019)
15. Pappalardo, C.M., Guida, D.: Forward and inverse dynamics of a unicycle-like mobile robot. Machines **7**(1), 5 (2019)
16. Pappalardo, C.M., Guida, D.: On the lagrange multipliers of the intrinsic constraint equations of rigid multibody mechanical systems. Arch. Appl. Mech. **88**(3), 419–451 (2018)
17. Pappalardo, C.M., Guida, D.: A comparative study of the principal methods for the analytical formulation and the numerical solution of the equations of motion of rigid multibody systems. Arch. Appl. Mech. **88**(12), 2153–2177 (2018)
18. Pappalardo, C.M., Guida, D.: On the computational methods for solving the differential-algebraic equations of motion of multibody systems. Machines **6**(2), 20 (2018)
19. Pappalardo, C.M., Guida, D.: System identification algorithm for computing the modal parameters of linear mechanical systems. Machines **6**(2), 12 (2018)
20. Pappalardo, C.M., Guida, D.: System identification and experimental modal analysis of a frame structure. Eng. Lett. **26**(1), 56–68 (2018)
21. Shipman, W.J., Coetzee, L.C.: Reinforcement learning and deep neural networks for PI controller tuning. IFAC-PapersOnLine **52**, 111–116 (2019)
22. François-Lavet, V., Henderson, P., Islam, R., Bellemare, M.G., Pineau, J.: An introduction to deep reinforcement learning. Found. Trends® Mach. Learn. **11**, 219–354 (2018)
23. Duan, Y., Chen, X., Houthooft, R., Schulman, J., Abbeel, P.: Benchmarking deep reinforcement learning for continuous control. In: 33rd International Conference on Machine Learning, ICML 2016, vo. 3, pp. 2001–2014 (2016)
24. De Simone, M., Guida, D.: Modal coupling in presence of dry friction. Machines **6**(1), 8 (2018)
25. Schulman, J., Moritz, P., Levine, S., Jordan, M., Abbeel, P.: High-Dimensional Continuous Control Using Generalized Advantage Estimation, arXiv preprint arXiv:1506.02438 (2015)

Impact Force Localization Using Eigenfrequency Measurement in Plane Frames

Emir Nezirić(✉) and Safet Isić

Faculty of Mechanical Engineering, "DžemalBijedić" University of Mostar, Sjevernilogorb.b, 88104 Mostar, Bosnia and Herzegovina
emir.neziric@unmo.ba

Abstract. Vibration measurement is one of the most used procedures for machine and structure testings and diagnostics. Structures subjected to impact force should vibrate at eigenvibration frequencies. Which eigenfrequencies are excited should be dependent on impact force location. This paper would show the possibility of how vibration measurement on plane frames could be used to determine the location of the impact forces.

Keywords: Structural vibration · Impact force · Vibration measurement · Eigenmodes · Eigenfrequencies

1 Introduction

Impact forces are a common source of the structure damages or excitation source for structure vibrations. Localization and quantification of the impact force is an important part of structure health monitoring. Beside impact forces, there are also continuous sources of excitation in structures (i.e. engines, fans, mills etc.) [1].

There are some of the proposed methods of the impact force identification in the structures. Straszevski et al. have used triangulation of the wave propagation in the structure to determine its location [2]. Problem with this method was to apply it to the large and complex structures.

Saleem and Jo have proposed application of the augmented Kalman filter and Generic algorithm combination for impact force identification [3]. This approach has shown good theoretical concurrence between predicted and simulated data, but it still had to be confirmed with the experimental setup.

A couple of the researches have been conducted with machine learning application for impact force identification [4, 5]. This method has shown that it is required to know a lot of the pretesting data for its practical application in structure monitoring.

In this paper would be presented research on the possibility of the eigenmode application on the impact force localization in the frame structure through eigenfrequency measurements.

I. Karabegović (Ed.): NT 2020, LNNS 128, pp. 39–47, 2020.
https://doi.org/10.1007/978-3-030-46817-0_4

2 Structural Vibrations

Structures which are part of as buildings, machines, ships, platforms are constructed using different parts which are supporting its framework. Different loads on structures cause vibrations, which are usually unwanted and undesirable. Some of the loads can cause fatal vibration which leads to structural damages and sometimes structure fatality. Loads on structures could be harmonic, shock or impulse, and random. The response of the structure is mainly dependant on its natural frequencies and impact force characteristics.

2.1 Governing Equations

Vibrations of the MDOF structures usually could be described by the equation of motion in matrix form:

$$[M]\{\ddot{q}\} + [K]\{q\} = \{F\} \tag{1}$$

where $[M]$ is mass matrix, $[C]$ is damping matrix, $[K]$ is stiffness matrix, $\{F\}$ is force vector and $\{q\}$ is generalized coordinate vector. If the harmonic motion is assumed, then the characteristic equation of the system is determinant equated to zero:

$$\det\left|[M]^{-1}[K] - \lambda[I]\right| = 0 \tag{2}$$

where λ are the roots of the characteristic equation (or eigenvalues), and $[I]$ is unit matrix. The natural frequencies are determined from the eigenvalues [6].

$$\omega_i = \sqrt{\lambda_i} \tag{3}$$

Substitution of the eigenvalue λ_i in the matrix equation of motion, the corresponding mode shape called eigenvector is obtained. Number of the eigenvectors and eigenvalues is the same as the number of freedom of the structure.

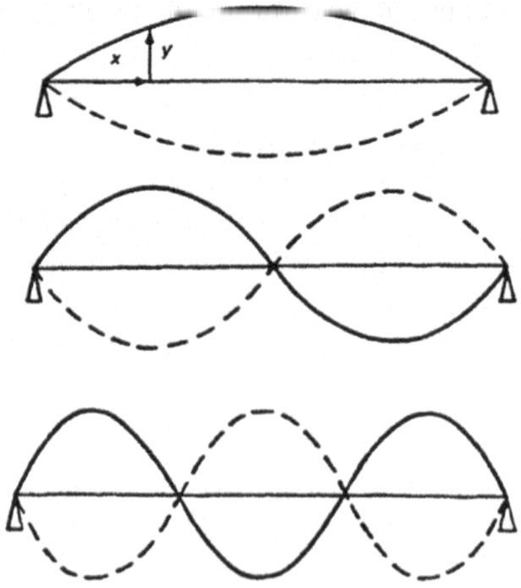

Fig. 1. Simple pinned-pinned beam lateral vibration modes

The simplest example is lateral vibrations of the simple pinned-pinned beam. If the beam is observed as 3DOF system, it could be possible to obtain its first three mode shapes and eigenvalues.

As it could be seen in Fig. 1, there are points called nodes on the beam where displacements in each mode are equal to zero. In practical meaning, there would not be measured lateral vibrations if vibrations are measured in those exact points. On the other hand, if the impact force is introduced on the same point in the lateral direction on the beam, that point should be displaced and that mode of vibrations should not be present.

3 Fem Model of the Plane Frame

Before the experimental investigation of the plane frame behaviour, preliminary research is conducted on the finite element model of the 3-member plane frame (Fig. 2).

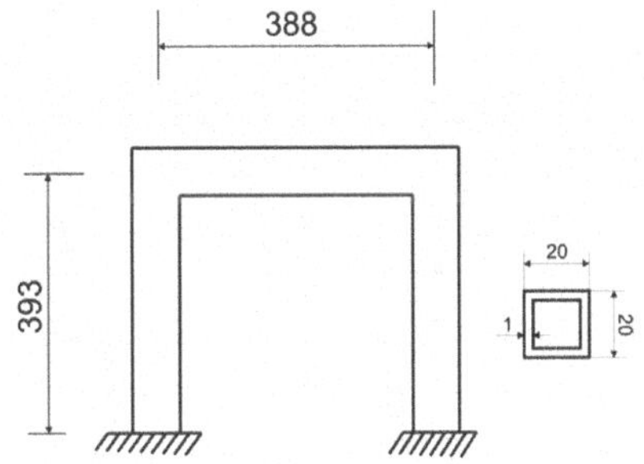

Fig. 2. 3-member plane frame

The frame is modelled in ANSYS software as line body with rectangular tube cross-section, as shown in Fig. 2. Deflection for the frame is only allowed in the frame plane, so bending in the directions perpendicular to the frame plane is prohibited. Modal analysis is conducted for the first 10 eigenfrequencies. Values of the first 10 eigenfrequencies are shown in Table 1.

Table 1. First 10 eigenfrequencies for the plane frame

Mode	Frequency [Hz]
1	128.28
2	502.11
3	800.38
4	847.03
5	1663.2
6	1991.9
7	2213.3
8	2743.1
9	3142.2
10	3342.9

First six modes are shown on the Fig. 3.

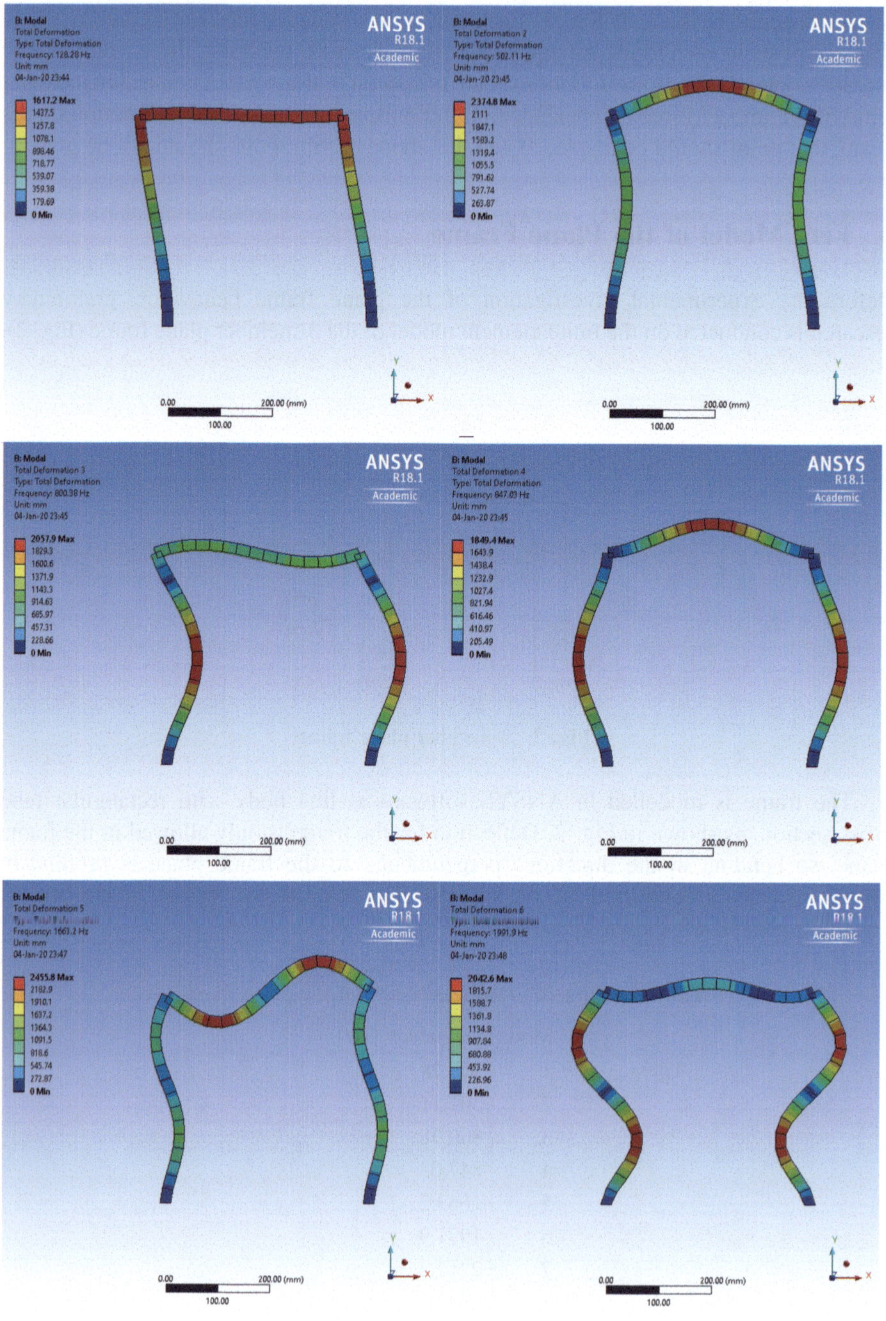

Fig. 3. First six modes of the plane frame

Line body of the frame now would be introduced in the transient structural module in ANSYS. The impact force would be modelled as time-dependent value, which is applied for only 0.001 s, and then it drops to zero. The acceleration of the different points on the frame would be discussed as the second part of the research on the FEM model. On Fig. 4 are shown locations of the impact force and the analysed acceleration points.

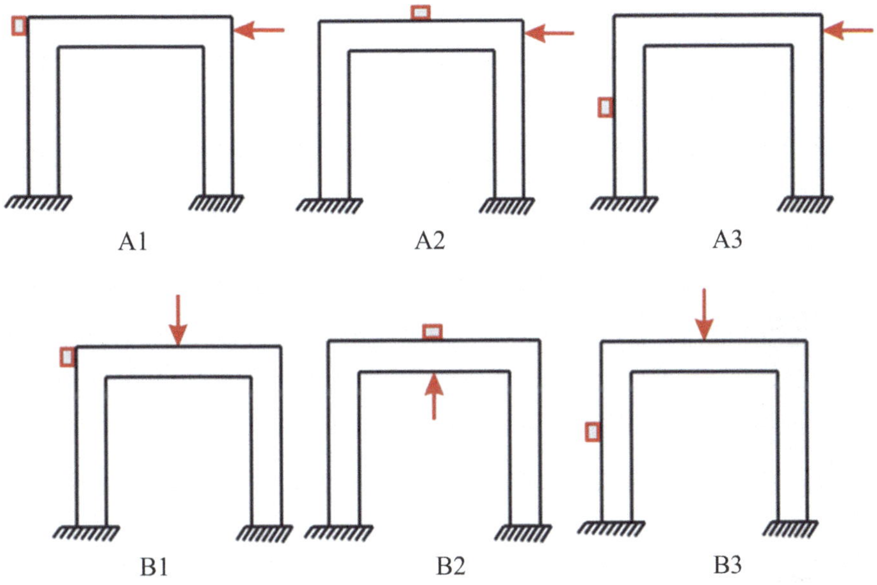

Fig. 4. Locations of the impact force and the analysed point acceleration

Accelerations of the points marked on Fig. 4 are exported to Origin software, where frequency spectrums are obtained. In Fig. 5 are shown accelerogram and frequency spectrum for the A1 combination. Peak frequencies are recorded from the frequency spectrums for all combinations and they are shown in Table 2. They are organized that they are joined to the mode with the closest eigenfrequency.

Table 2. Recorded frequencies for all combinations

Combination	Mode 1 [128 Hz]	Mode 2 [502 Hz]	Mode 3 [800 Hz]	Mode 4 [847 Hz]	Mode 5 [1663 Hz]	Mode 6 [1992 Hz]
A1	130	–	780	–	–	–
A2	–	500	–	820	–	–
A3	130	500	780	820	–	–
B1	–	500	–	820	–	–
B2	–	500	–	820	–	–
B3	–	500	–	820	–	–

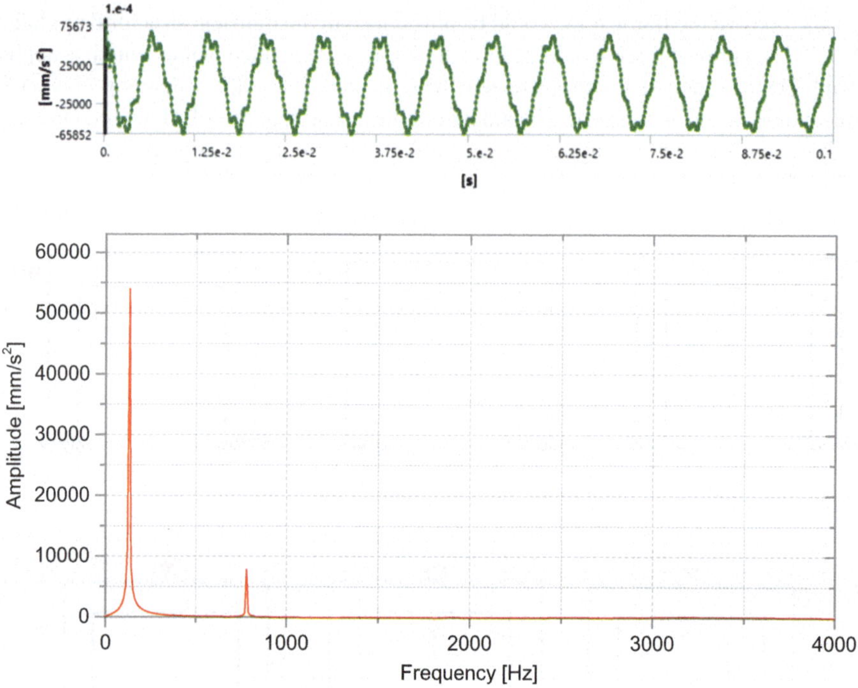

Fig. 5. Accelerogram and frequency spectrum simulated for the combination A1

Recorded frequencies are up to 3.1% smaller than frequencies obtained by modal analysis. This difference is probably result of a computational numerical error and data processing.

As it could be seen from the Table 2, the first location of the impact force initiates all first four modes, but some of the modes could not be noticed in the measurement on different locations and directions on the frame. As seen from Fig. 3, node location of observed frame motion on combination A1 does not move at all for modes 2 and 4 in the observed direction, so it could not be recorded on that point. A similar observation is for combination A2, where only modes 2 and 4 are recorded for that nodal point since modes 1 and 3 does not have a displacement in the vertical direction. For combination A3, impact force initialize all first four modes which could be recorded on that point.

For combinations B1, B2 and B3 it is shown that for this impact force location it is not possible to initiate modes 1 and 3 since these modes do not have movement of the point where impact force is introduced. In all measurements, only mode 2 and 4 are noticed.

Also, all measurements measured for point 3 (combinations A3 and B3) have captured all possible modes for that impact force location, since measurement point is not on the nodal point od the frame.

From the statements above, it gives a hint that it is possible to obtain location from the measurement of the vibrations. The first step is to know all eigenfrequencies and eigenmodes. This could be done by modelling of the analyzed system to obtain approximate values of eigenfrequencies. The second step should be to measure vibrations on different points of construction to exclude possible impact force locations. Points of measurement should be chosen to be one of the possible nodes (i.e. bearings or midspan point for shafts, supports or midspan of the beam, nodal points of the frame or truss construction) so it is possible to exclude some of the modes.

These assumptions would be tested on the experimental setting with the same geometrical data of frame and the same material.

4 Experimental Research

An experimental setup was examined in the same condition as the FEM model. The frame is made with the same dimensions and material characteristics as in preliminary research (Fig. 6).

Fig. 6. The frame used in experimental research

Impact force was applied by rubber hammer impact and by measurements on the chosen points on the frame (as shown in Fig. 4). For measurement, the microaccelerometer was used to avoid its mass impact on the total mass of the frame. Accelerations are recorded and the frequency of the motion was captured by frequency spectrum peak read (Fig. 7).

As it could be seen in Table 3, experimental research has shown some difference from the simulation in the FEM model. First, it has not captured higher eigenfrequencies than the second mode eigenfrequency. All measurement have captured the frequency of the first mode but measurement B2. The reason could be that it is the easiest mode to excite, so the smallest deviation from the impact force direction could excite other modes. The second difference is that measurement B1 did not capture second mode eigenfrequency. Reason for that could be that direction of the measurement for B1 should not have motion in the measurement direction.

Table 3. Recorded frequencies for all combinations

Combination	Mode 1 frequency [128 Hz]	Mode 1 amplitude [m/s^2]	Mode 2 frequency [502 Hz]	Mode 2 amplitude [m/s^2]
A1	118,9	13,10	–	–
A2	118,9	0,07	484,4	0,58
A3	119,4	21,08	499,0	0,60
B1	118,9	2,81	–	–
B2	–	–	485,3	0,44
B3	119,4	2,98	494,8	0,74

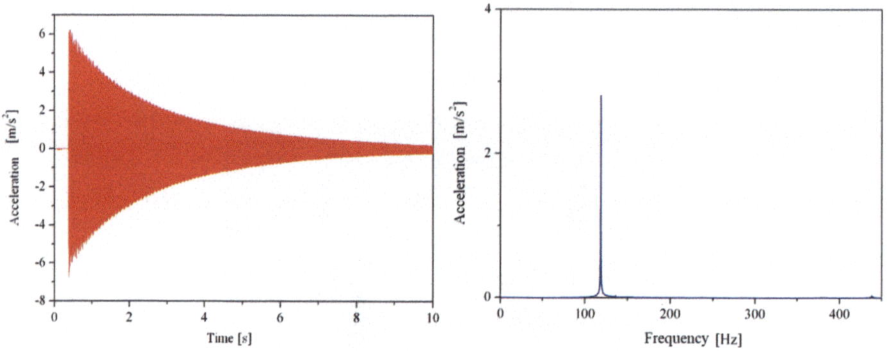

Fig. 7. Accelerogram and frequency spectrum recorded for the combination B1

The possible human mistake for this kind of testing could be avoided by introducing amplitude analysis of the peaks in the frequency spectrum, which could show the analyst which mode is dominant in the structure. For the measurement where are noticed both of the mode frequencies, it is possible to conclude which mode is dominant. For example, for combination A2 it is noticed that mode 2 is dominant, and for A3 mode 1 is dominant. The dominance of the amplitude could lead to conclude that the dominant peak is also recorded in the simulation. As a conclusion, analysis of the presence of the eigenfrequency peak in the frequency spectrum would not be enough to perform localization of the impact force. The amplitude of the peak could be possible additional data to analyse as well to help the analyst to distinct possible impact force locations.

5 Conclusions

After the short review of the current researches on this topic and conducted simulation and experimental research, the following conclusions could be made:

- Measurement of the eigenfrequencies of the eigenmodes is a simple approach for the determining structure vibrational characteristic.
- Impact force should initiate only some of the eigenmodes.
- Eigenmode shapes have characteristic points on the structures which do not have displacement during the oscillations of the structure. That points are called nodes.
- Finite element method is an appropriate tool for predetermining the eigenmodes, so it is easier to determine locations for the measurement.
- By measuring vibrations in the nodes, it is possible to determine which modes are not excited by impact forces. Multiple measuring points could help to reach the conclusions on the impact force location.
- The amplitude of the eigenfrequency peaks could be helpful to determine which mode has been more excited by the impact force location.

For the future researches, it is suggested to expand this investigation on the other structures (i.e. space frames and trusses) and to determine the complete procedure for the application of this approach for the practical use.

References

1. Nezirić, E., Džiho, E., Šunje, E.: Vibration analysis of rotating machinery as excitation of concrete structure vibrations. In: Karabegović, I. (ed.) NT 2019. Lecture Notes in Networks and Systems, vol. 76, pp. 73–81. Springer, Cham (2020). https://doi.org/10.1007/978-3-030-18072-0_8
2. Staszewski, W.J., Mahzan, S., Traynor, R.: Health monitoring of aerospace composite structures–active and passive approach. Compos. Sci. Technol. **69**(11), 1678–1685 (2009)
3. Saleem, M., Jo, H.: Heterogeneous data fusion for impact force identification in truss structures. In: Proceedings of SPIE, Sensors and Smart Structures Technologies for Civil, Mechanical, and Aerospace Systems 2018, vol. 10598, p. 105981X, 27 March 2018. https://doi.org/10.1117/12.2296763
4. Worden, K., Staszewski, W.J.: Impact location and quantification on a composite panel using neural networks and a genetic algorithm. Strain **36**(2), 61–68 (2000)
5. Sung, D.-U., Oh, J.-H., Kim, C.-G., Hong, C.-S.: Impact monitoring of smart composite laminates using neural network and wavelet analysis. J. Intell. Mater. Syst. Struct. **11**(3), 180–190 (2000). https://doi.org/10.1106/n5e7-m37y-3mar-2kfh
6. Thomson, W.T., Dahleh, M.D.: Theory of Vibration with Applications. Prentice Hall, New Jersey (2010)

Design and Testing of a Two-Stage Water-Hydraulics Pressure-Relief Valve

Franc Majdič(✉)

Laboratory for Fluid Power and Controls, Faculty of Mechanical Engineering, University of Ljubljana, Aškerčeva 6, 1000 Ljubljana, Slovenia
franc.majdic@fs.uni-lj.si

Abstract. Power-control hydraulics is an important technology used in many industries, but it can be problematic due to its impact on the environment. Using water instead of mineral hydraulic oil is one of the ways to protect our environment. However, there is a lack of components suitable for water hydraulics.

The aim of this work was to design and test a two-stage pressure-relief valve, appropriate for use in a water-hydraulics system. The valve was designed for pressures up to 40 MPa and for flow rates up to 50 L/min. In the design process we used both analytical and numerical approaches. One of the important goals was for the valve to beas simple as possible to manufacture. The prototype of the valve was made and the basic measurements were taken. Based on the first results obtained from the experiments, the valve design was modified in the direction of a simpler valve-manufacturing process with higher quality.

Keywords: Water hydraulics · Pressure-relief valve · Design · Analysis · Measurements

1 Introduction

Nowadays we cannot imagine life without technology; it is almost universal. One of the important ways that power is transferred is with hydraulics, where a fluid is used for the power transfer [1, 2]. Different kinds of hydraulic fluids are in use, the most used are hydraulic mineral oils (a more than 80% share) [5]. Most of these hydraulic fluids are hazardous to the natural environment and to humans. Unfortunately, hazardous hydraulic fluids spill into the environment frequently. There are three possible solutions, the first is to use biodegradable oils, the second is to use ecology-friendly ionic liquids and the third is to use water as a hydraulic fluid [5, 6]. If we want to change from oil to water hydraulics, a lot of different hydraulic components are needed [1, 3, 4]. An important missing component is the two-stage pressure-relief valve.

1.1 Pressure-Relief Valve

Pressure-relief valves are primarily used to protect hydraulic systems against overloading, which can lead to failure of the hydraulic system [2, 7, 8]. Almost all relief

I. Karabegović (Ed.): NT 2020, LNNS 128, pp. 48–58, 2020.
https://doi.org/10.1007/978-3-030-46817-0_5

valves are settable. For lower flows (in general up to 10 L/min) a one-stage pressure-relief valve is used; higher flows require a two-stage pressure-relief valve. A one-stage valve consists of a conical closing element, a spring, a seat and a valve body. A two-stage pressure-relief valve consists of a one-stage valve on top of its construction, while on the lower part of the valve there is a main body housing with a larger conical closing element and a spring. The advantage of a two-stage pressure-relief valve is good pressure-oscillation damping; the disadvantage is its slower response.

1.2 Water as a Hydraulic Pressure Medium

The most important properties of water as a hydraulic pressure medium are power transfer, lubrication, sealing, cooling, kinematic viscosity, compressibility, evaporation pressure, and air consumption [1, 5, 6]. ***Power transfer*** is the basic function of a hydraulic fluid. The hydraulic pump transfers the mechanical energy to the hydraulics, which transfers it with fluid through the hydraulic system to the actuators. Water has no significant differences in terms of power transfer compared to the most commonly used hydraulic mineral oil. ***Lubrication*** is an important property of a hydraulic fluid as it reduces the friction and wear in sliding contacts inside the hydraulic components. Water has very poor lubrication properties in the contacts of steel surfaces as it is in general used in oil hydraulics. ***Sealing***, which is connected to the ***kinematic viscosity*** of the fluid, has an important role in low-gap sliding contacts inside the hydraulic components. Due to the significantly lower kinematics viscosity of the water compared to mineral hydraulic oil, the gaps in the water hydraulic components should be at least half the size. *The* ***kinematic viscosity*** of water is fifteen times lower than mineral oil (ISO VG46 at 50 °C). *The* ***cooling*** properties of water are significantly better than those of mineral hydraulic oil. *The* ***compressibility modulus*** of water is $2.1 \cdot 10^5$ MPa, which is 46% lower than for hydraulic mineral oil. A very important advantage of water is its availability everywhere in nature and its friendless to the environment (no pollution).

2 Design of a Two-Stage WH Pressure-Relief Valve

A two-stage, water-hydraulics, pressure-relief valve was designed [9]. Figure 1a provides an external view of the assembled valve. A cross-section of the two-stage pressure-relief prototype valve is shown in Fig. 1b.

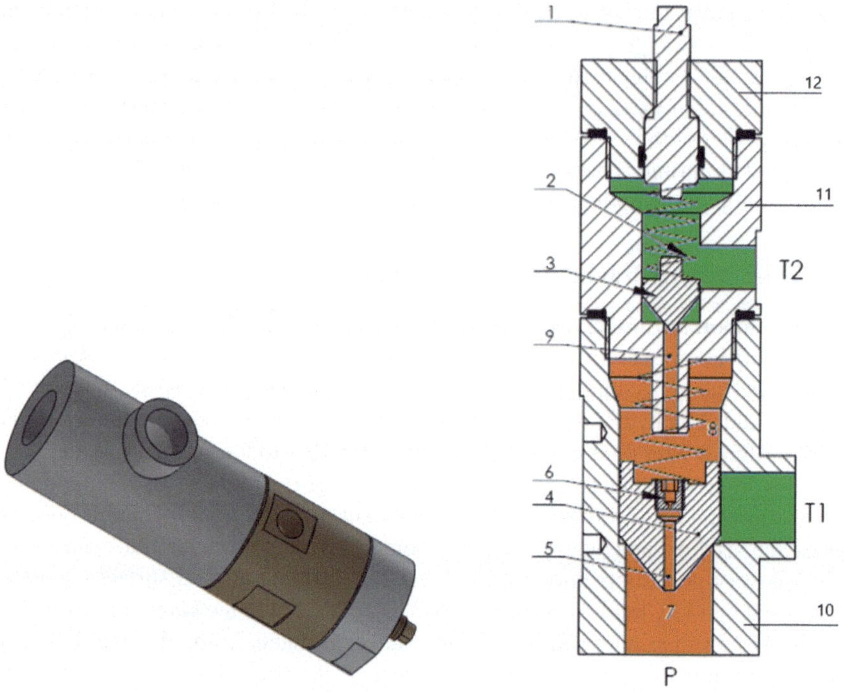

Fig. 1. New WH pressure-relief valve: 3D model (left), cross-section (Φ50 mm × 175 mm) (right)

It is composed of the main body (10), the middle body (11), the smaller upper body (12), the main conical spool (4), the nozzle with a small diameter hole (6), the main non-adjustable spring (8), the control spool (3), the adjustable spring (2) to set up the desired pressure and the adjustable screw to set up the desired opening pressure of the valve. The valve is connected to the hydraulic pressure line in parallel through the port P. As long as the actual water pressure is lower than that set with the screw (1) on the spring (2), the main (4) and the control spool (3) are closed.

The control spool (3) on the first valve stage has control of the pressure through the hole (5), through the nozzle (6) and through hole (9). When the system pressure rises above the upper spring setting, the control piston first opens a gap to the tank connector T2 (Fig. 2a). As a result, the pressure on the upper surface of the plunger piston drops, which increases and connects the pressure line to the tank (T1). As a result, the pressure in the pressure line (P) is lowered (Fig. 2b).

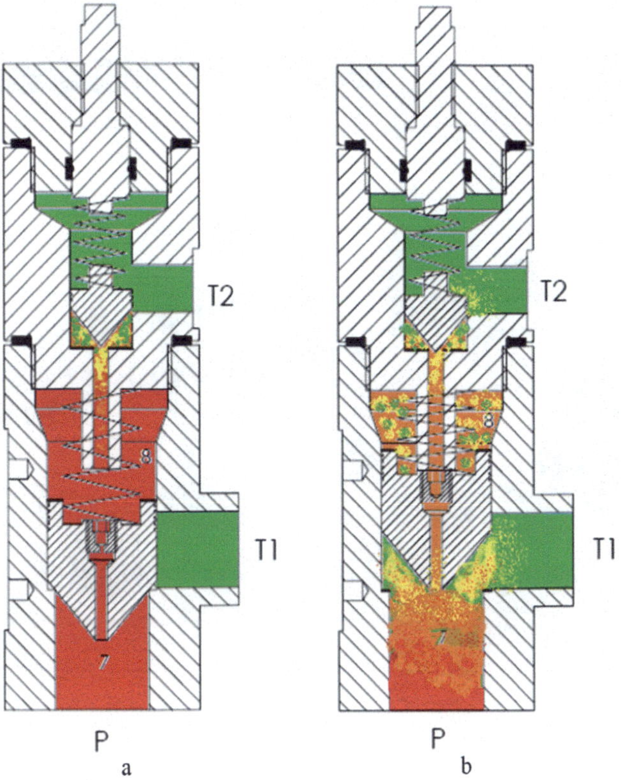

Fig. 2. Working principle of the WH pressure-relief valve: (a) open first stage, (b) open both stages

3 Experimental

3.1 Specimen Pressure-Relief Valve

Figure 3 shows parts of the prototype of the two-stage pressure-relief valve. Four different main conical spools and four different conical control spools are included. The main and the control spool were made from different materials: stainless steel, bronze, PEEK and Teflon. The connector with internal threads for the port T1 was welded to the main valve housing made of stainless steel. The middle and upper housings are also made of stainless steel, while the adjustable screw was made from bronze.

Fig. 3. Parts of the prototype of two-stage pressure-relief valve for water hydraulics

Tested spool materials:

1. The first tested configuration of the pressure valve was assembled with the main and the control spool made from polyamide (PA 66 GF30).
2. The second tested configuration of the pressure valve was assembled with the main and the control spool made from polyoxymethylene (POM).
3. The third configuration of the pressure valve was assembled with the main spool made from polyoxymethylene (POM) and the control spool made from polyamide (PA 66 GF30).

3.2 Test Rig

Figure 4 shows the hydraulic circuit of the test rig. The in-line water hydraulic piston pump (1) with an 11-kW electric motor (2) sucks water from the tank (3) and pushes it into the pressure line through the adjustable flow-control valve with pressure compensation (4) past the first pressure sensor (5) to the inlet of the tested two-stage pressure-relief valve (6). The flow of water is reduced from the main pressure line through the nozzle (6.c). The reduced flow is connected to the control spool (6.a) and the pre-tensioned spring is adjustable with a screw. When the system pressure exceeds the upper control valve's spring setting, the control spool compresses its spring and opens the flow to the second exit port (T2). Consequently, the pressure in the hydraulic control line drops and the main spool (6.b) opens. The main water flows to the first exit port (T1). The pressure in the control-hydraulics line is measured with the second pressure sensor (7). The pressure in the common return line is measured with the third sensor (8). The water flow rate through the tested pressure valve is measured with the flow sensor (9). At the end of the hydraulic circuit the water is pushed through the 3-μm return filter (10) with a Beta ratio of 1000 (for 6-μm particles) and the integrated by-pass valve back to the tank (3). The 2/2 directional control valve (11) is used for an unloaded start-up of the electric motor (2). The system pressure should not exceed 150 bar, and this is maintained with the pressure-relief valve (12).

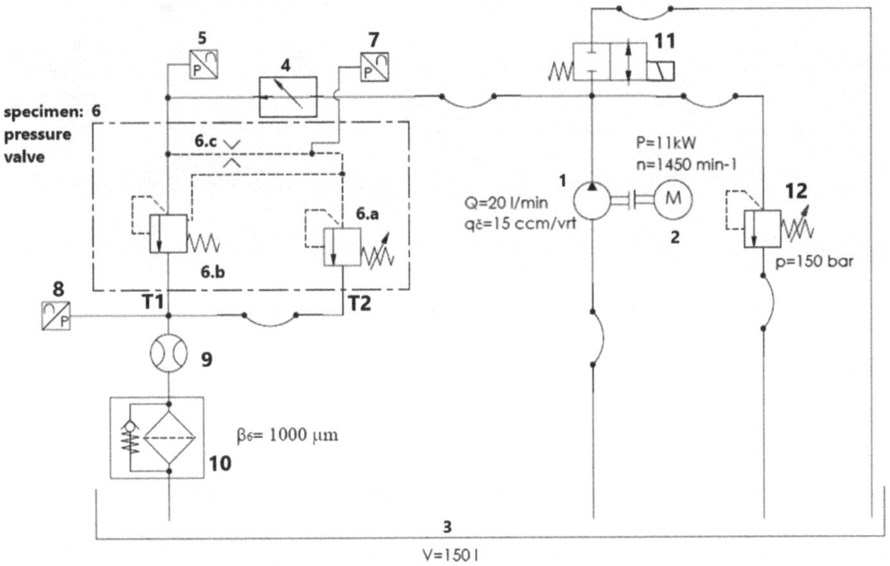

Fig. 4. Hydraulic test rig for WH pressure-relief valve

Figure 5 shows the main part of the water hydraulic test rig, the hydraulic in-line piston pump with an electric motor (1, 2), the inlet pressure line with flow-control valve (4), the inlet pressure sensor (5), the specimen two-stage pressure-relief valve (6), the pressure sensor for the hydraulic control line (7), the outlet pressure sensor (8), the flow sensor (9) on the return line and the 2/2 solenoid directional unloading valve (12).

Distilled water with a volume of 150 L was used for the hydraulic tests of the pressure valve. The temperature of the water was maintained at 40 °C with a cooler during the tests.

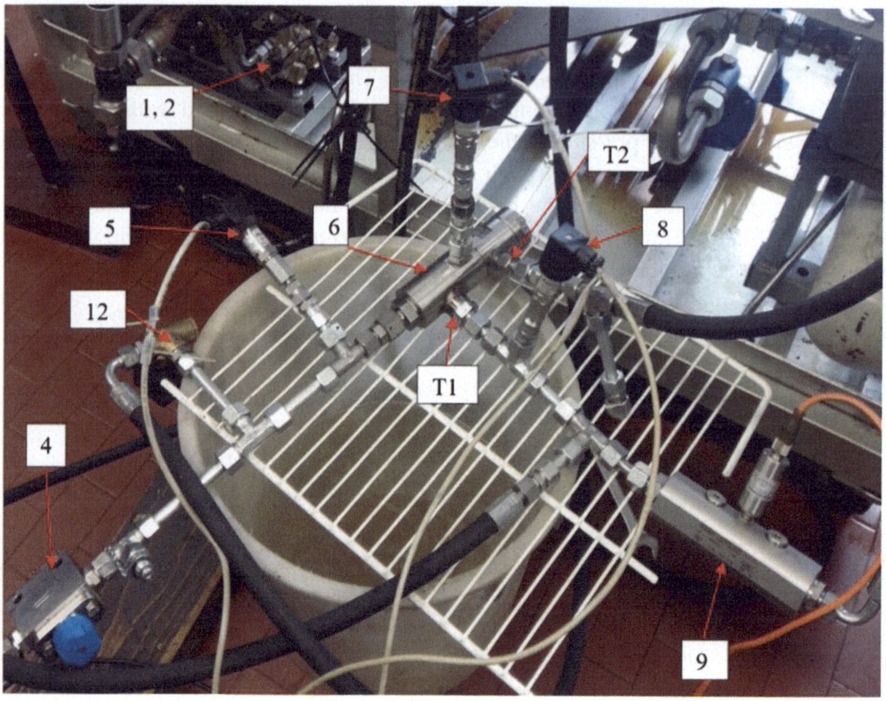

Fig. 5. WH pressure relief valve on test rig in laboratory

4 Results

The results of the first test with the control and the main spool made of PA 66 GF30 polymer are shown in Fig. 6. It is obvious that the pressures between the inlet and the controlling chamber were nearly 20 bar different at the start of the measurement. After increasing the setting of the spring, both pressures increased in parallel and the difference between these pressures decreased. After the start of deceasing the pressure setting of the control spool, the difference in the pressure starts to increase. At the end to the first measurement the difference between the inlet pressure and the pressure in the middle chamber in the valve was 20 bar. The highest achieved pressure was 76 bar at a leakage flow rate in the range of 10 L/min. The pressure pulsation of the inlet pressure is connected to the three-piston pump. A higher pressure of water was not possible due to the too high leakage of the investigated pressure-relief valve.

Figure 7 shows the results of the measurement of the pressure-relief valve containing both spools of the POM polymer. The behaviour of the second configuration of the investigated pressure valve was like the first one. Also, during the second measurement an unacceptable leakage occurred. The maximum pressure was only 64 bar at a leakage flow rate in the range of 6 L/min.

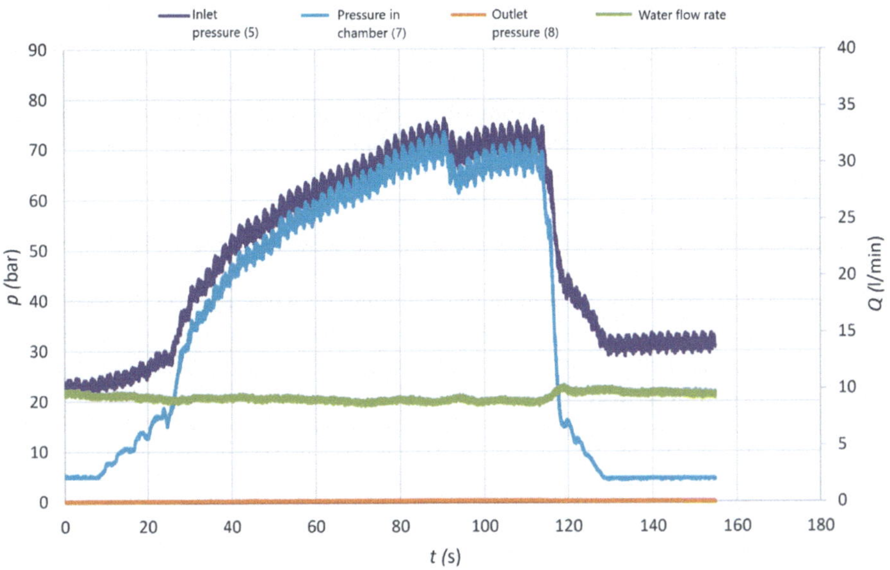

Fig. 6. Results with closing conical elements made of PA66GF30

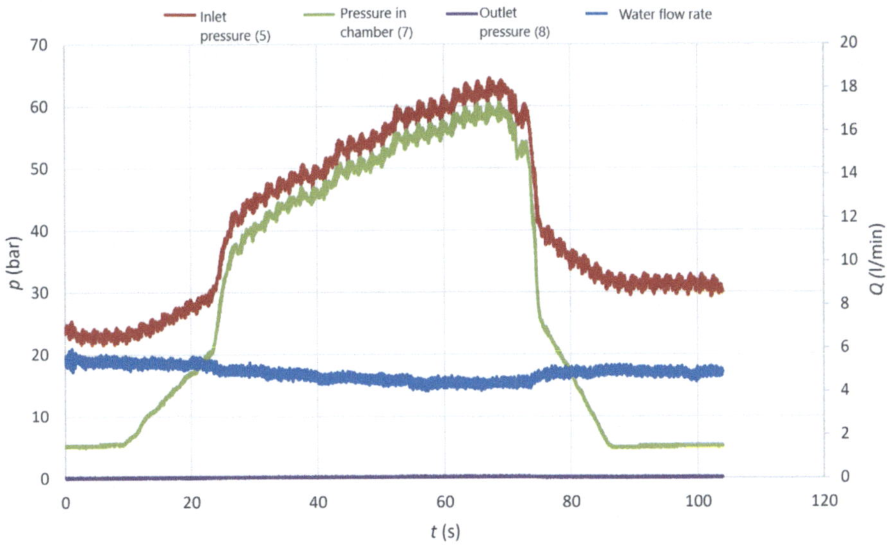

Fig. 7. Results with closing the conical element made of POM

Figure 8 shows the third results for the relief valve containing the main spool of PA66GF30 polymer and the control spool made of POM polymer. The combination of PA and POM materials and the smaller gaps between the spools and the holes/seats gave the best results. In this combination of two different materials for the main and the control spool, small gaps between the spools and the holes played the most important role in the connection to the lowest measured leakage (0.8 L/min) for these measurements. In the third measurement (Fig. 8) the pressure valve started to work properly. At the start there was just a small leakage. After increasing the system pressure to the setting pressure of the tested valve, it opened. The first step increase of the pressure to 30 bar opened the main spool and the flow was open to the tank. After the next increase of the pressure setting of the valve, the pressure in the middle chamber (7) increased in parallel with the system pressure. The third measurement was made at a pressure of 47 bar.

Figure 9 shows low leakage of the valve during the measurement.

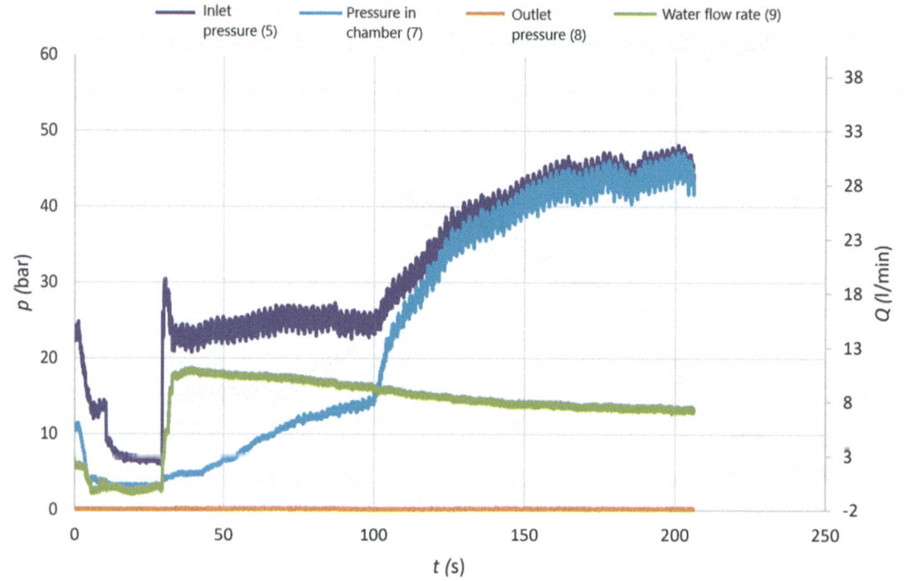

Fig. 8. Results with control spool made of POM and the main spool made of PA66GF30

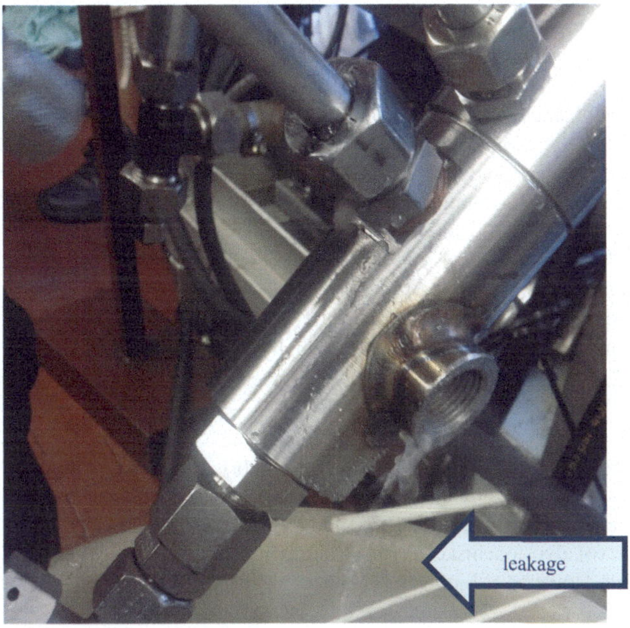

Fig. 9. Measuring of leakage for third configuration of the pressure valve (POM/PA)

5 Conclusion

Using water in hydraulics is one of the possible solutions to prevent environmental pollution and preserve our sources of drinking water. However, water as a hydraulic pressure medium offers many development challenges. This paper presents the research and development of a two-stage pressure-relief water valve as follows:

1. A few different concepts, designs and a 3d model of the valve were completed.
2. A prototype of the valve with different spool (the control and the main) materials was completed.
3. A water-hydraulics test rig was prepared.
4. Experiments with three different material combinations for the control and the main spool were made.
5. The most important parameters for the correct functionality of the tested valve are the dynamic gaps between the two spools and the housing/holes.

Suggestions for further work: geometric and fabrication optimizations and further measurements of the valve are required. The presented work shows that water can be used as a hydraulic pressure medium, but a lot of effort, money and time is needed to develop each component.

References

1. Majdič, F.: Voda kot kapljevina v pogonsko-krmilni hidravliki: doktorska disertacija. Univerza v Ljubljani, Fakulteta za strojništvo (2010)
2. Fiendeisen, D.: Ölyhydraulik: Handbuchfür die hydrostatische Leistungsübertragung in der Fluidtechnik. Springer, Heidelberg (2006). Technische Universität Berlin, Germany
3. Suzuki, K., Urata, E.: Development of a direct pressure-sensing water hydraulic relief valve. Int. J. Fluid Power **9**, 5–13 (2008)
4. Suzuki, K., Urata, E.: Development of a direct pressure-sensing pressure-reducing valve for water hydraulics. Department of Mechanical Engineering, Yokohama, Japan (2008)
5. Trostmann, E.: Water Hydraulics Control Technology. Marcel Dekker Inc., New York (1996). Technical University of Denmark, Lyngby
6. Trostmann, E., Frolund, B., Olesen, B.H., Hilbrecht, B.: Tap Water as a HydraulicsPressureMedium. Marcel Dekker, New York (2001)
7. Dasgupta, K., Karmakar, R.: Modelling and dynamics of single-stage pressure relief valve with directional damping. Simul. Model. Pract. Theory **10**, 51–67 (2002)
8. Yinshui, L., Xiaojun, R.: Simulation and analysis of a seawater hydraulic relief valve in deep-sea environment. Ocean Eng. **125**, 182–190 (2016)
9. Alif, D.: Razvoj dvostopenjskega vodno-hidravličnega tlačnega omejilnega ventila: *magistrska naloga*. Univerza v Ljubljani, Fakulteta za strojništvo (2017)

Integration of CAD, MBD, and FEA Programs for the Topology Optimization of Aircraft Components

Adriano G. Manca[1] and Carmine M. Pappalardo[2(✉)]

[1] MEID4 Academic Spin-Off of the University of Salerno, Via Giovanni Paolo II, 132, 84084 Fisciano, SA, Italy
[2] Department of Industrial Engineering, University of Salerno, Via Giovanni Paolo II, 132, 84084 Fisciano, SA, Italy
cpappalardo@unisa.it

Abstract. This research paper proposes a general integration approach between a CAD (Computer-Aided Design) software, an MBD (Multi-Body Dynamics) software, and an FEA (Finite Element Analysis) software. The proposed method is used in the design of a cargo hatch. For this purpose, the programs used are SOLIDWORKS, MSC ADAMS, and ANSYS. To achieve the topological optimization, the proposed integration technique is applied to a flexible component of a cargo hatch opening system. To find the loading conditions, the mechanical parts of the opening mechanism are considered as rigid bodies, while the support element of the opening system is assumed to be deformable. Finally, the topological optimization of the component under study is carried out obtaining a reduction of its mass with a change in its structural strength, but under a certain limit. The results found in this study are considered significant because applying the proposed method to various parts of the aircraft can lead to a redesigned opening system with a reduction in the total weight of the aircraft.

Keywords: Topology optimization · CAD · Multibody · FEA · Aircraft components

1 Introduction

In the world of modern engineering, the number of programs for the computer aided design and analysis is enormous. These are used as decision support systems in several workplace environments. Mechanical engineering is also pervaded by simulation software and the lack of communication existing between the different programs is a subject that is the object of current research since it is proved that this issue leads to a substantial waste of money and time. In particular, the reference research framework of this paper is multibody dynamics [1–10] and nonlinear control [11–13]. An important problem that is studied by some branches of engineering is the dynamic response of mechanical systems described by nonlinear equations [14]. For example, in the design of a new structure, the support of 3D design software helps to study the behavior of mechanical systems [15]. In the field of finite element analysis, FEM models are used to study mechanical systems [16]. Furthermore, with the help of design software together

I. Karabegović (Ed.): NT 2020, LNNS 128, pp. 59–65, 2020.
https://doi.org/10.1007/978-3-030-46817-0_6

with the multibody analysis software [17–20] or finite element analysis software [21], the study of various mechanical systems can be realized too. This paper proposes a general topology optimization method based on the integration of computer aided design programs and rigid-flexible multibody dynamics programs. When talking about industry 4.0 [22], one of the technologies that comes into play is Additive Manufacturing, which is a process that allows complex parts to be manufactured. The problem is that, in general, the results of topology optimization provides parts that are difficult to manufacture [23]. Therefore, many authors devote some studies to solve these issues [24]. The topology optimization can be applied in the aeronautical sector [25]. It is for these reasons that the object of this investigation is to develop a methodology suitable for optimizing aircraft mechanical components.

2 Research Methodology

The conceptual sequence on which this research work is based can be summarized in the flowchart represented in Fig. 1.

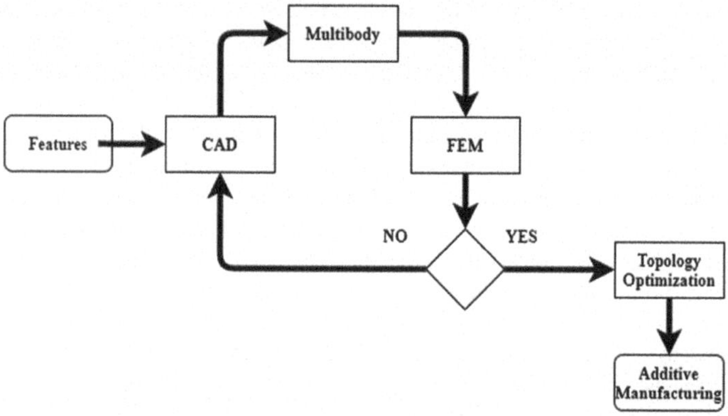

Fig. 1. Conceptual flowchart from the desired features to the additive manufacturing process.

To start this investigation, it is necessary defined some analytical terms:

- $\mathbf{M}$: Rigid multibody system mass matrix;
- $\mathbf{q}$: Generalized coordinate vector of the rigid multibody system;
- $\mathbf{C}$: Constraint vector;
- $\mathbf{C_q}$: Jacobian matrix of the constraint equations;
- $\mathbf{Q}_e$: External generalized force vector;
- $\mathbf{Q}_v$: Inertia quadratic velocity vector;
- $\mathbf{Q}_d$: Constraint quadratic velocity vector;

These terms conform to a rigid multibody system, the index-one differential-algebraic equations of motion, which are defined as follows:

$$\begin{cases} \mathbf{M}\ddot{\mathbf{q}} = \mathbf{Q}_v + \mathbf{Q}_e - \mathbf{C}_\mathbf{q}^T \boldsymbol{\lambda} \\ \mathbf{C}_\mathbf{q}\ddot{\mathbf{q}} = \mathbf{Q}_d \end{cases} \tag{1}$$

On the other hand, the definition of the following mathematical terms is necessary in the structural analysis:

- $\mathbf{M}_f$: Flexible multibody system mass matrix;
- $\mathbf{D}_f$: System damping matrix;
- $\mathbf{K}_f$: System stiffness matrix;
- $\mathbf{q}_f$: Vector of nodal coordinates of the flexible multibody system;
- $\mathbf{Q}_f$: External generalized force vector applied to the material nodes;

In the case of a flexible multibody system, by employing a finite element approach, the equations of motion are given by:

$$\mathbf{M}_f\ddot{\mathbf{q}}_f + \mathbf{D}_f\dot{\mathbf{q}}_f + \mathbf{K}_f\mathbf{q}_f = \mathbf{Q}_f \tag{2}$$

The proposed integration method involves the following sequential phases (Fig. 2):

Fig. 2. Proposed integration method from CAD to FEA.

These phases should be iterated until the model passes the structural analysis according to the established operating conditions.

3 Numerical Results

A cargo aircraft, which represents the case study considered in this work, is a fixed-wing aircraft designed for the transport of goods, rather than passengers. Generally, this system is endowed with large doors for loading and unloading operations. The object of verification of this investigation is the upper support connected to the opening/closing mechanism of the cargo hatch. The analysis is carried out by importing the CAD model shown in Fig. 3a into the ANSYS environment. The material selected for the analysis is structural steel with a yield strength equal to 250 MPa. The mesh, shown in Fig. 3c, is made using 18951 tetrahedral elements with edges of dimension equal to 4,5 mm. The system is constrained with two fixed joints applied on the faces A and B of Fig. 3c. Two external forces are applied at points C and D, each equal to half of the maximum constraint reaction that occurs in that hinge joint during motion. The net reaction force is obtained through a dynamic analysis in MSC ADAMS View Environment, using the model shown in Fig. 3b. The support is then imported into

ANSYS, as shown in Fig. 3c, and the constraint reaction force, whose components are $F_y = -1347.5(N)$ and $F_z = 1230.5(N)$, is assumed as the leading loading condition of the structural analysis.

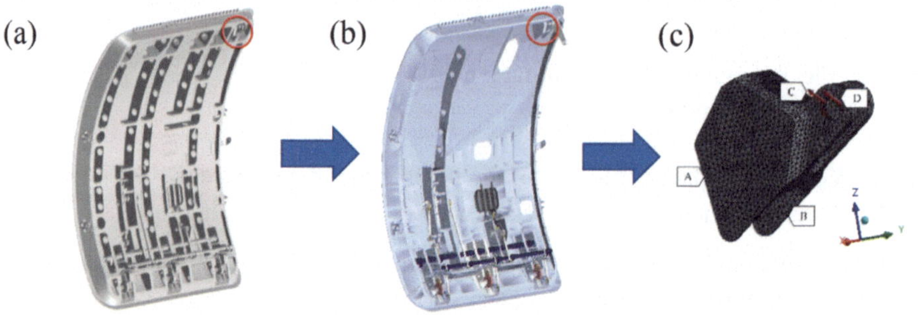

Fig. 3. Cargo hatch: (a) CAD environment, (b) MBD environment, (c) FEM environment. The red circle indicates the position of the component in the hatch.

Subsequently, a topological optimization of the component of interest is realized using the ANSYS tool done for this purpose. The optimization procedure followed in the paper is based on the following steps:

- Static structural analysis;
- Topology optimization execution;
- Redesign of the component;
- New static structural analysis for verification purposes.

The results of the second structural analysis are confronted with the yield strength of the material to understand if they are satisfactory or the analysis should be redone. To perform the topology optimization, 50% of the mass is imposed to be retained. The original component weights 1.4175 kg and the result obtained, shown in Fig. 4, is a new component with a mass equal to 1.0135 kg. Therefore, 71.407% of the original mass actually is retained in the redesigned mechanical part.

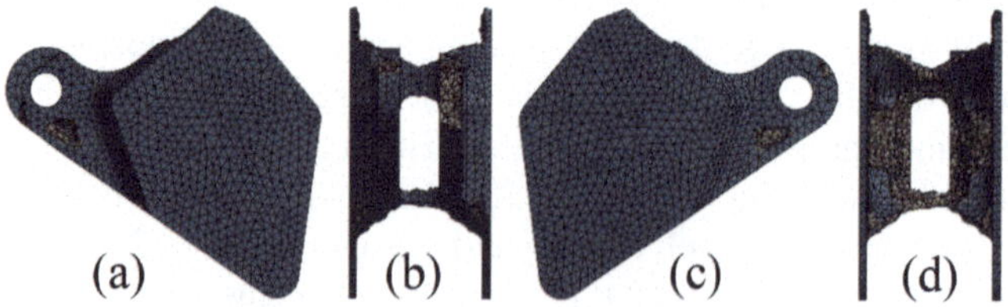

Fig. 4. Topology optimization results on the support: (a) right side view, (b) front view, (c) left side view, (d) back view.

The result of the topological optimization is a new component that is difficult to manufacture due to the poor quality of its surface resulting from the mesh made of tetrahedral elements. Therefore, as shown in Fig. 5, the original mechanical part was redesigned in ANSYS SpaceClaim environment, following the topology optimization results as a guide. By doing so, the final component has a mass of 1.0775 kg.

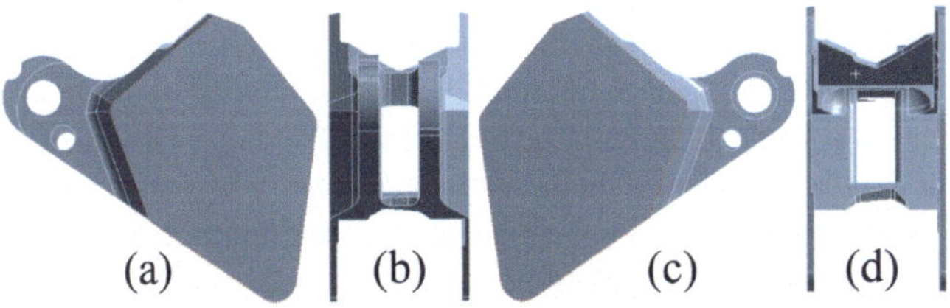

Fig. 5. Redesigned support: (a) right side view, (b) front view, (c) left side view, (d) back view.

The last task to perform is a new structural analysis to evaluate the new distribution of stress. The comparison of the results obtained before and after the topological optimization can be seen in Figs. 6a and b, respectively.

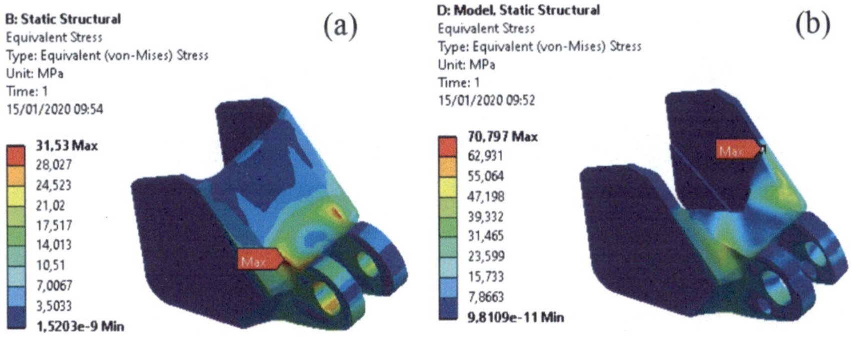

Fig. 6. Distribution of equivalent (von-Mises) stress on the original support (a) and redesigned support (b).

In numerical terms, the reduction of the component weight is 23.98%, with an increase of approximately 100% of the maximum von Mises stress, which still leads to a safety factor greater of two and, therefore, is considered acceptable.

4 Conclusions

The central goal of this paper is to develop a general method to perform a topological optimization of aircraft mechanical parts. To explain the proposed procedure, the hatch of a cargo aircraft is taken as an illustrative example. The integrated frameworks employed in this work are based on the CAD, MBD, and FEM environments. To achieve the objective of the paper, the programs used are SOLIDWORKS, MSC ADAMS, and

ANSYS, respectively. In the conversion of the models from the CAD environment to the multibody environment, only some components are considered as rigid bodies to simplify the dynamical simulations. At the end of this phase, the constraint reactions are obtained as outputs of the multibody analysis and are subsequently used as inputs of the structural analysis. To this end, in the ANSYS Workbench environment, the component under study is imported and loaded with the forces obtained to perform static structural analysis. Starting from the stress distribution obtained in the mechanical component of interest, a shape optimization is carried out with the help of the Topology Optimization tool. After redesigning the original part based on the optimized form, a new static structural analysis is carried out to determine if the new configuration of the component can resist the stress to which it is subjected. The numerical results found in this study show a considerable reduction of the mass together with an increase in the maximum stress value. However, considering a safety factor greater than two, the maximum stress value is far from the yield strength of the structural steel used to manufacture the component, thereby demonstrating the feasibility of the proposed redesign.

Acknowledgments. This research work was supported by the project DOOR 4.0.

References

1. De Simone, M.C., Guida, D.: Identification and control of a unmanned ground vehicle by using Arduino. UPB Sci. Bull. Ser. D Mech. Eng. **80**(1), 141–154 (2018)
2. De Simone, M.C., Rivera, Z., Guida, D.: Obstacle avoidance system for unmanned ground vehicles by using ultrasonic sensors. Machines **6**(2), 18 (2018)
3. Pappalardo, C.M., Guida, D.: Forward and inverse dynamics of a unicycle-like mobile robot. Machines **7**(1), 5 (2019)
4. Pappalardo, C.M., Guida, D.: A time-domain system identification numerical procedure for obtaining linear dynamical models of multibody mechanical systems. Arch. Appl. Mech. **88** (8), 1325–1347 (2018)
5. Pappalardo, C.M., Guida, D.: On the Lagrange multipliers of the intrinsic constraint equations of rigid multibody mechanical systems. Arch. Appl. Mech. **88**(3), 419–451 (2018)
6. Pappalardo, C.M., Guida, D.: On the dynamics and control of underactuated nonholonomic mechanical systems and applications to mobile robots. Arch. Appl. Mech. **89**(4), 669–698 (2019)
7. Pappalardo, C.M., Guida, D.: A comparative study of the principal methods for the analytical formulation and the numerical solution of the equations of motion of rigid multibody systems. Arch. Appl. Mech. **88**(12), 2153–2177 (2018)
8. Pappalardo, C.M., Guida, D.: On the computational methods for solving the differential-algebraic equations of motion of multibody systems. Machines **6**(2), 20 (2018)
9. Pappalardo, C.M., Guida, D.: System identification algorithm for computing the modal parameters of linear mechanical systems. Machines **6**(2), 12 (2018)
10. Pappalardo, C.M., Guida, D.: System identification and experimental modal analysis of a frame structure. Eng. Lett. **26**(1), 56–68 (2018)
11. Quatrano, A., De Simone, M.C., Rivera, Z.B., Guida, D.: Development and implementation of a control system for a retrofitted CNC machine by using Arduino. FME Trans. **45**(4), 565–571 (2017)

12. Pappalardo, C.M., Guida, D.: Development of a new inertial-based vibration absorber for the active vibration control of flexible structures. Eng. Lett. **26**(3), 372–385 (2018)
13. Pappalardo, C.M., Guida, D.: Use of the adjoint method for controlling the mechanical vibrations of nonlinear systems. Machines **6**(2), 19 (2018)
14. Colucci, F., De Simone, M.C., Guida, D.: TLD design and development for vibration mitigation in structures. In: Karabegović, I. (ed.) New Technologies, Development and Application II, NT 2019. Lecture Notes in Networks and Systems, vol. 76, pp. 59–72. Springer, Cham (2020)
15. De Simone, M., Guida, D.: Modal coupling in presence of dry friction. Machines **6**(1), 8 (2018)
16. Nezirić, E., Isić, S., Karabegović, I., Voloder, A.: FEM model of misaligned rotational system with rotating looseness. In: Karabegović, I. (ed.) New Technologies, Development and Application NT 2018. Lecture Notes in Networks and Systems, vol. 42. Springer, Cham (2019)
17. Concilio, A., De Simone, M.C., Rivera, Z.B., Guida, D.: A new semi-active suspension system for racing vehicles. FME Trans. **45**(4), 578–584 (2017)
18. De Simone, M.C., Guida, D.: Control design for an under-actuated UAV model. FME Trans. **46**(4), 443–452 (2018)
19. Guida, R., De Simone, M.C., Dašić, P., Guida, D.: Modeling techniques for kinematic analysis of a six-axis robotic arm. IOP Conf. Ser. Mater. Sci. Eng. **568**(1), 012115 (2019)
20. Rivera, Z.B., De Simone, M.C., Guida, D.: Unmanned ground vehicle modelling in Gazebo/ROS-based environments. Machines **7**(2), 1–21 (2019)
21. De Simone, M.C., Rivera, Z.B., Guida, D.: Finite element analysis on squeal-noise in railway applications. FME Trans. **46**(1), 93–100 (2018)
22. Karabegović, I., Karabegović, E., Mahmić, M., Husak, E.: Implementation of industry 4.0 and industrial robots in the manufacturing processes. In: Karabegović, I. (ed.) New Technologies, Development and Application II, NT 2019. Lecture Notes in Networks and Systems, pp. 3–14. Springer, Cham (2019)
23. Jankovics, D., Gohari, H., Tayefeh, M., Barari, A.: Developing topology optimization with additive manufacturing constraints in ANSYS®. IFAC-PapersOnLine **51**(11), 1359–1364 (2018)
24. Plocher, J., Panesar, A.: Review on design and structural optimisation in additive manufacturing: towards next-generation lightweight structures. Mater. Des. **183**, 1–20 (2019)
25. Wang, J., Li, Y., Hu, G., Yang, M.: Lightweight research in engineering: a review. Appl. Sci. (Switzerland) **9**(24), 1–24 (2019)

Technological Features of Locating Charts in Fixture Design

Vitalii Ivanov(✉), Ivan Pavlenko, Viliam Zaloga, Oleksandr Liaposhchenko, and Dmytro Pirogov

Department of Manufacturing Engineering, Machines and Tools, Sumy State University, 2, Rymskogo-Korsakova Street, Sumy 40007, Ukraine
ivanov@tmvi.sumdu.edu.ua

Abstract. According to modern technological requirements, the problem of increasing the efficiency of drilling-milling-boring machine tools is the first priority in multiproduct manufacturing. In this regard, the use of fixtures is an obvious stage in ensuring the quality and accuracy of engineering products. In this paper, the main objectives are related to the choice of locating charts in fixture design. Based on the developed algorithm, a reasonable choice of locating elements for parts is implemented using the 3-2-1 locating chart. Additionally, the decision-making sequence is given in the process of choosing the functional elements of the fixture using the database. The related algorithm considers the design and technological features of the workpiece. The practical significance of the research is in the application of CAFD for the proper choice of fixture locating elements. As a result, the proposed approach allows increasing the efficiency of designing technological equipment and reducing the complexity of technological preparation of the manufacturing process.

Keywords: CAFD · Flexibility · Automation · Algorithm · Accuracy

1 Introduction

Modern machine-building production is characterized by multifaceted products and instability of production volumes. The wide variety of parts requires machining to achieve the accuracy of size, shape and relative positioning, as well as the quality of the surfaces according to the technological requirements. All types of parts require accurate locating and reliable clamping during machining to provide the designer-specified parameters, which will ensure the efficiency and reliability of the final product. In this regard, considerable attention is paid to fixtures [1]. Given the advanced technological capabilities of modern machine tools, their high cost and the need for a large number of adjustments for machining parts of another size, rapid development is the intensification of production [2]. This provides increasing the flexibility of fixtures, reducing the preparatory and final time, and the possibility of multitool and multiaxis machining of parts in a single setup [3, 4]. All the mentioned reduces the preparatory time and makes an effect on decreasing the production costs [5, 6].

I. Karabegović (Ed.): NT 2020, LNNS 128, pp. 66–74, 2020.
https://doi.org/10.1007/978-3-030-46817-0_7

2 Literature Review

Fixture design is a comprehensive and time-consuming process. It requires a relatively large amount of information to be considered at the same time [7]. The locating chart for prismatic parts on the three planes according to the principle "3–2–1" provides maximum rigidity and a minimum number of functional elements. The proper positioning of the workpiece in a fixture is an important step to ensure machining accuracy. In this regard, the locating elements are located as far apart as possible under the condition that the center of mass is situated between them to provide a stable workpiece position. The clamping elements ensure the spatial position of the workpiece in the machining process. It is important to identify the optimal position of the locating and clamping elements [8].

Particularly, Menassa and DeVries [9] used the minimization of the workpiece deflections at specific points as the main criterion for determining the position of the locating elements. Meyer and Liou [10] proposed an approach based on linear programming for synthesizing fixture layouts considering dynamic conditions of the machining process. In addition, a solution to minimize the clamping force and the reaction force of the functional elements was proposed. Li and Melkote [11] proved that the locating error is minimized due to the localization of elastic deformations. Roy and Liao [12] developed a heuristic method for positioning the locating and clamping elements of fixtures. Tao et al. [13] presented a methodology of geometric substantiation for the optimal positions of the clamping elements for spatially complicated workpieces.

Liao and Hu [14] presented a system for analyzing the fixture layout based on the developed dynamic model of the system "fixture – workpiece" depending on the nonstationary loads. Li and Melkote [15] proposed a combined method of the optimal synthesis of a fixture considering the workpiece dynamics during machining. Amaral et al. [16] used finite element modeling to solve the optimization problem for testing the reliability of fixtures. Tan et al. [17] verified the most advantageous layouts of fixtures in terms of clamping forces by means of numerical simulation and optimization techniques. Mihaylov and Nikolcheva [18] developed a method of positioning the locating elements for prismatic and cylindrical workpieces for the case of the locating principle "3–2–1". According to this method, the positions of the functional elements are evaluated considering the mass center of the workpiece. This approach allows for increasing the rigidity of the system "fixture – workpiece".

Wan et al. [19] proposed an effective method of designing clamping elements during machining. Particularly, the number of restrictions of degrees of freedom was substantiated analytically depending on the workpiece machining modes. Additionally, the algorithms of the automatic generation of the locating chart and realizing the static analysis under the action of cutting and clamping forces during machining were developed. Retfalvi [20] developed an approach to accelerate the fixture design process. This approach includes the choice of functional elements, their locating and consequent assembly. In this research, the input data is a CAD model and features of its machining. Moreover, the system independently selects the locating surfaces.

Hunter Alarcon et al. [21] developed and implemented the program software for fixture design. The peculiarity of their software is in independent generating the accompanying design and technological documentation and providing the interaction between the direct contractors during the fixture design process. Nelaturi et al. [22] developed an algorithm of designing fixture configurations considering the stability of the workpiece, contact points between the workpiece and its functional elements, and the permanency of clamping forces. The system selects the proper fixture layout using the available elements satisfying the design requirements.

Peng et al. [23] created a system of virtual assembling of modular fixtures. The methodology of positioning functional elements was created based on the analysis of the degrees of freedom of interacting elements. Wu et al. [24] developed an algorithm for determining the location of functional elements on the supporting plate for locating the complex-shape parts. The algorithm is based on the theory of linkage mechanism theory, which allows optimal positioning both for the workpiece and functional elements. Additionally, this algorithm allows analyzing the locating modes under the production conditions. Vichare et al. [25] described a technique to identify the interaction between functional elements of modular fixtures using a single production resource model. This model allows creating an information environment with logical relationships between the functional elements for the consequent identification of the proper assembly of the fixture. In the research [26], the parameter identification approach for the evaluation of forces was developed for the mechanical system "fixture – workpiece".

Due to the above-mentioned analysis, the aim of the presented research is to develop an algorithm for automated selection of the locating elements of fixtures to ensure the reliability of the designing procedure and increasing the machining accuracy on drilling-milling-boring machine tools.

3 Research Methodology

In order to reduce the time spent on the fixture design procedure, an algorithm for automated selection of fixture locating elements is developed. This algorithm is used to locate the prismatic parts on the drilling-milling-boring machine tools according to the design and technological features of the workpieces. It allows selecting the locating elements (e.g. supporting plates or supports) for the workpiece according to the three-plane locating chart (3-2-1 locating chart). The algorithm is implemented in three main steps (Fig. 1) for primary, secondary, and tertiary datums. The selection is realized by performing logical operations based on the developed rules.

At the next step, the information about the roughness of the base surface of the workpiece, as well as the type of the locating element are specified (Fig. 2).

4 Results

A surface with the largest area is selected as a primary datum. In the case of supporting plates, the geometric conditions are checked considering the length and width of the workpiece. Additionally, the use of two plates is considered as a sufficient condition for the implementation of the primary datum. The selection of the supporting plates from the library of locating elements is made considering the length l_p and width bp of the

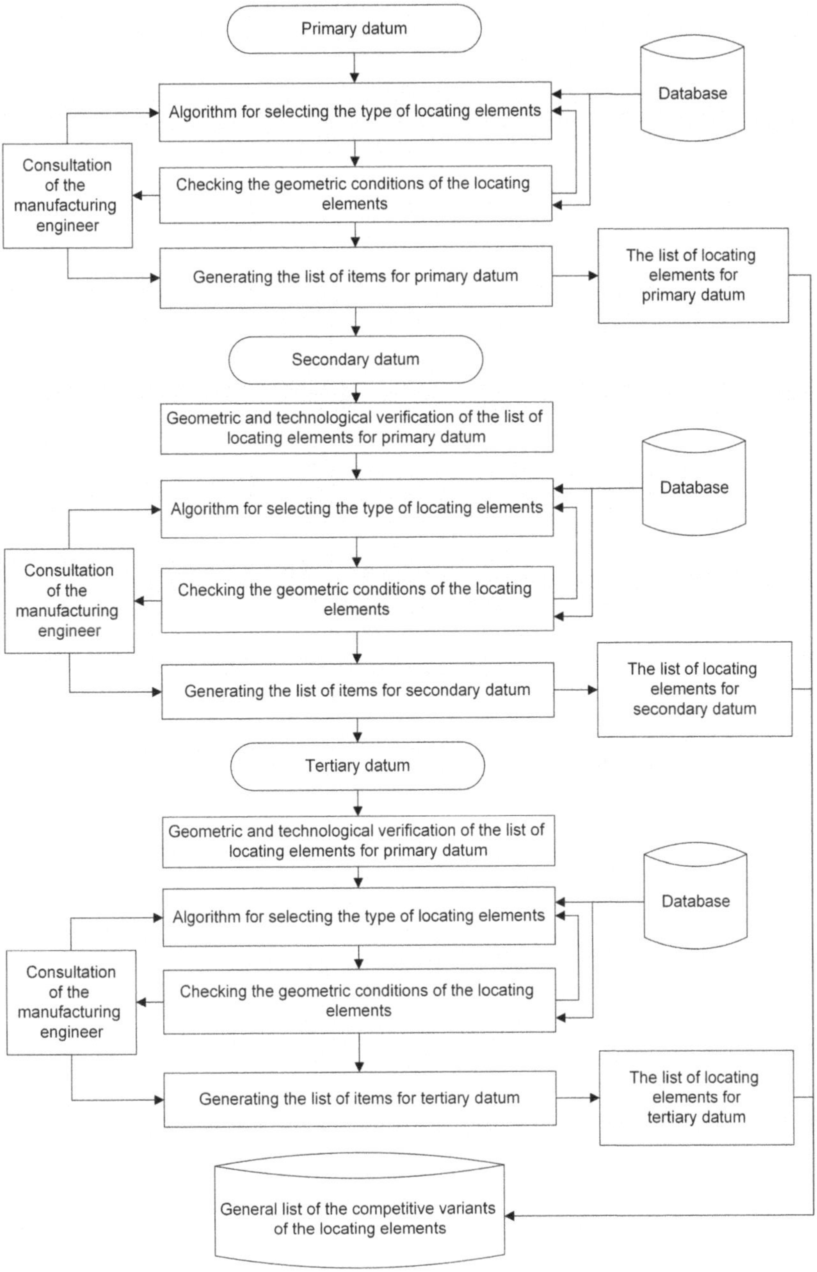

Fig. 1. Algorithm for selecting the locating elements of fixtures according to the locating chart "3–2–1"

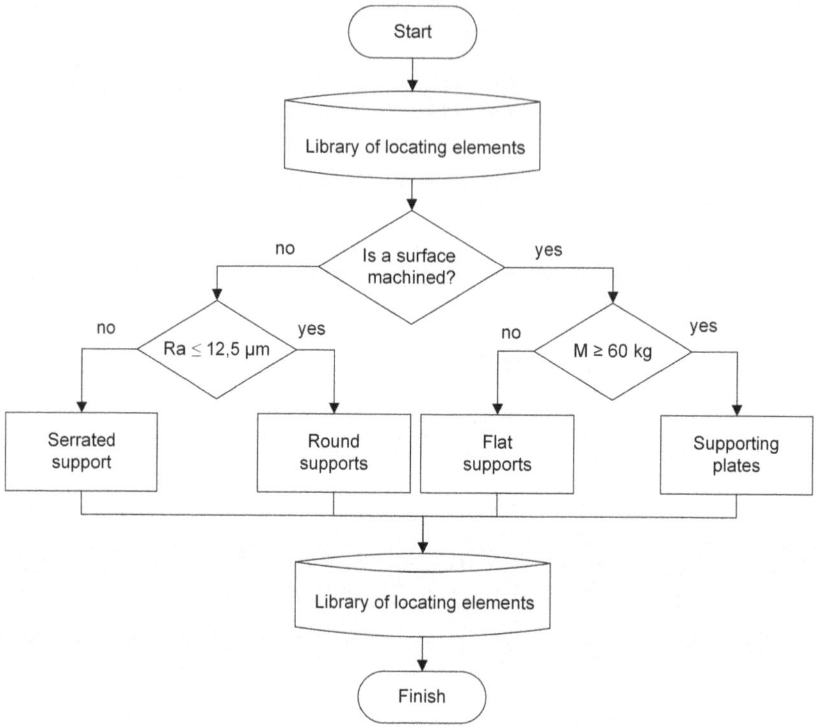

Fig. 2. Algorithm for selecting the type of locating elements

plate, as well as the length l_{wp} and width b_{wp} of the locating base according to the following equations:

$$l_p \leq (1 - \eta)b_{wp}; b_p \leq (1 - \eta)\frac{l_{wp}}{2}, \tag{1}$$

where η – safety factor, which is usually equal to 0.1.

When using supports, geometric conditions are checked for the length l_{wp} and width b_{wp} of the primary datum. The diameter of the support D_s is taken at least twice the width of the workpiece and four times the length of the workpiece:

$$D_s \leq (1 - \eta)\min\left\{\frac{b_{wp}}{2}; \frac{l_{wp}}{4}\right\}. \tag{2}$$

If no supporting plate or support meets the geometric conditions of the primary datum, the consultation of the manufacturing engineer is mandatory according to the developed scheme (Fig. 1). The result of the choice of locating elements is a list of supporting plates or supports satisfying all the geometric and technological conditions of the primary datum.

Typically, the secondary datum is a surface with the largest length. To improve the algorithm's performance over the choice of locating elements, to implement the consequent bases, a check of the established locating elements is carried out according to technological and geometric conditions. The verification of the list items that do not satisfy this stage in terms of surface roughness is applied. After that, a geometric check is performed along the length of the support plate l_p depending on the length of the workpiece l_{wp}. Additionally, the width of the plate b_p depends on the height of the secondary datum surface. To ensure the reliability of the locating chart, one support plate is sufficient, and its dimensions are calculated considering the safety factor:

$$l_p \leq (1 - \eta) b_{wp}; b_p \leq (1 - \eta) h_{wp}, \tag{3}$$

Unlike the plates, for providing the secondary datum, it is sufficient two supports, the diameter of which at least twice smaller than the length of the workpiece:

$$D_s \leq (1 - \eta) min\left\{ h_{wp}; \frac{l_{wp}}{2} \right\}. \tag{4}$$

As a result of the list checking of the secondary datum according to geometric and technological features, a list of elements is generated. These elements can be additionally included in the list of the primary datum. If none of the items in the list of the primary datum satisfies the geometric and/or technological conditions of the secondary datum, the algorithm will be executed, but only for the case of a new secondary datum.

Analogously with the selection of elements for the primary datum, if it is not impossible to select the locating elements according to their geometrical features, the manufacturing engineer independently adjusts the input data or enters the required set of elements. As a result of the second stage of the algorithm, a list of locating elements for the implementation of the secondary datum is generated.

The third stage is characterized by the fact that a single supporting plate or support is sufficient to realize the tertiary datum. At this stage, the list of locating elements of the primary datum is checked according to the conditions of the tertiary datum (similar to the stage of verification the list of the primary datum for the locating conditions). Verification is carried out using the following equations:

$$l_p \leq (1 - \eta) b_{wp}; b_p \leq (1 - \eta) b_{wp}, \tag{5}$$

$$D_s \leq (1 - \eta) min\{ h_{wp}; l_{wp} \}. \tag{6}$$

Analogously with the first and second stages, in case of non-compliance with technological and geometric conditions, the manufacturing engineer adjusts and/or form a list. The result of the third stage of the algorithm is the formation of a list of the locating elements for the tertiary datum.

The result of the whole algorithm is a list of competitive variants of the locating elements, which are suitable for the implementation of technological datums of the prismatic parts including lists of elements for the primary, secondary, and tertiary datums.

To systematize the data when selecting the locating elements, the basic condition is to create a database containing a complete list of functional elements, as well as their technological and geometric features. One of the main technological characteristics is the ability to implement a particular datum with a specific element. Additionally, the possibility of using functional elements for the machined or untreated surfaces of the workpiece, surface with a certain roughness and a certain form is specified. Each functional element has a unique number.

To create a list of elements needed to implement the locating chart, the corresponding algorithm is developed. It allows selecting functional elements from the general list by the following technological features of the workpiece: surface treatment, roughness, weight, and geometric dimensions.

This can significantly shorten the list of elements and allocate them in terms of the realization of a particular datum. The specific requirements for the functional elements are formed for selecting the necessary set of functional elements, which fully meet the technological requirements. The developed algorithm is integrated into the CAFD system [27, 28].

5 Conclusions

As a result of the implementation of the proposed approach, an algorithm of selection of locating elements of fixtures has been developed based on the design and technological features of the workpiece. On the example of the locating chart in three planes for the prismatic parts, it was found that the choice of the locating elements is influenced by technological (treatment mode, surface roughness, etc.) and structural (dimensions and the shape of the workpiece) parameters. The selection of functional elements is realized in an automated mode. However, in the case of disputes, the engineer can adjust the computer-selected options and assign the combinations of functional elements based on his(her) own experience.

References

1. Ivanov, V., Zajac, J.: Flexible fixtures for CNC machining centers in multiproduct manufacturing. EAI Endorsed Trans. Ind. Netw. Intell. Syst. **4**(12), e4 (2018)
2. Ivanov, V., Dehtiarov, I., Pavlenko, I., Kosov, I., Kosov, M.: Technology for complex parts machining in multiproduct manufacturing. Manage. Prod. Eng. Rev. **10**(2), 25–36 (2019). https://doi.org/10.24425/mper.2019.129566
3. Bakker, O.J., Papastathis, T.N., Ratchev, S.M., Popov, A.A.: Recent research on flexible fixtures for manufacturing processes. Recent Pat. Mech. Eng. **6**(2), 107–121 (2013)
4. Tohidi, H., AlGeddawy, T.: Planning of modular fixtures in a robotic assembly system. Proc. CIRP **41**, 252–257 (2016). https://doi.org/10.1016/j.procir.2015.12.090
5. Li, H., Chen, W., Shi, S.: Design and application of flexible fixture. Proc. CIRP **56**, 528–532 (2016). https://doi.org/10.1016/j.procir.2016.10.104
6. Gothwal, S., Raj, T.: Different aspects in design and development of flexible fixtures: review and future directions. Int. J. Serv. Oper. Manage. **26**(3), 386–410 (2017)

7. Ivanov, V.: Process-oriented approach to fixture design. In: Ivanov, V., et al. (eds.) Advances in Design, Simulation and Manufacturing, DSMIE-2018 50. Lecture Notes in Mechanical Engineering, pp. 42–50. Springer, Cham (2019). https://doi.org/10.1007/978-3-319-93587-4_5
8. Ivanov, V., Pavlenko, I.: Fundamental approach for analysis of dynamic characteristics of fixtures. EAI Endorsed Trans. Ind. Netw. Intell. Syst. **4**(13), e1 (2018). https://doi.org/10.4108/eai.20-3-2018.154366
9. Menassa, R.J., Devries, W.R.: Optimization methods applied to selecting support positions in fixture design. J. Eng. Ind. Trans. ASME **113**(4), 412–418 (1991)
10. Meyer, R.T., Liou, F.W.: Fixture analysis under dynamic machining. Int. J. Prod. Res. **35**(5), 1471–1489 (1997)
11. Li, B., Melkote, S.N.: Improved workpiece location accuracy through fixture layout optimization. Int. J. Mach. Tools Manuf **39**(6), 871–883 (1999). https://doi.org/10.1016/S0890-6955(98)00072-8
12. Roy, U., Liou, J.: Geometric reasoning for re-allocation of supporting and clamping positions in the automated fixture design system. IEEE Trans. **31**, 313–322 (1999)
13. Tao, Z.J., Kumar, A.S., Nee, A.Y.C.: A computational geometry approach to optimum clamping synthesis of machining fixtures. Int. J. Prod. Res. **37**(15), 3495–3517 (1999). https://doi.org/10.1080/002075499190158
14. Liao, Y.G., Hu, J.: Flexible multibody dynamics based fixture-workpiece analysis model for fixturing stability. Int. Machine Tools Manuf. **40**(3), 343–362 (2000)
15. Li, B., Melkote, S.N.: Optimal fixture design accounting for the effect of workpiece dynamics. Int. J. Adv. Manuf. Technol. **18**(10), 701–707 (2001). https://doi.org/10.1007/pl00003951
16. Amaral, N., Rencis, J.J., Rong, Y.M.: Development of a finite element analysis tool for fixture design integrity verification and optimisation. Int. J. Adv. Manuf. Technol. **25**(5–6), 409–419 (2005). https://doi.org/10.1007/s00170-003-1796-6
17. Tan, E.Y.T., Kumar, A.S., Fuh, J.Y.H., Nee, A.Y.C.: Modeling, analysis, and verification of optimal fixturing design. IEEE Trans. Autom. Sci. Eng. **1**(2), 121–132 (2004). https://doi.org/10.1109/tase.2004.835601
18. Mihaylov, O., Nikolcheva, G.: An integrated RBR fixture design system. In: CBU International Conference, Praha, vol. 5, (2017)
19. Wan, N., Wang, Z., Mo, R.: An intelligent fixture design method based on smart modular fixture unit. Int. J. Adv. Manuf. Technol. **69**(9–12), 2629–2649 (2013). https://doi.org/10.1007/s00170-013-5134-3
20. Attila, R.: Fixture design system with automatic generation and modification of complementary elements for modular fixtures. Acta Polytechnica Hungarica **12**(7), 163–182 (2015)
21. Hunter Alarcon, R., RıosChueco, J., Perez Garcıa, J.M., VizanIdoipe, A.: Fixture knowledge model development and implementation based on a functional design approach. Robot. Comput. Integr. Manuf. **10**, 56–66 (2010)
22. Nelaturi, S., Rangarajan, A., Fritz, C., Kurtoglu, T.: Automated fixture configuration for rapid manufacturing planning. Comput. Aided Des. **46**, 160–169 (2014). https://doi.org/10.1016/j.cad.2013.08.028
23. Peng, G., Wang, G., Liu, W., Yu, H.: A desktop virtual reality-based interactive modular fixture configuration design system. Comput. Aided Des. **42**(5), 432–444 (2010). https://doi.org/10.1016/j.cad.2009.02.003
24. Wu, Y., Gao, S., Chen, Z.: Automated modular fixture planning based on linkage mechanism theory. Robot. Comput. Integr. Manuf. **24**(1), 38–49 (2008). https://doi.org/10.1016/j.rcim.2006.06.003

25. Vichare, P., Nassehi, A., Newman, S.T.: Unified representation of fixtures: clamping, locating and supporting elements in CNC manufacture. Int. J. Prod. Res. **49**(16), 5017–5032 (2011). https://doi.org/10.1080/00207543.2010.518992
26. Pavlenko, I., Trojanowska, J., Ivanov, V., Liaposhchenko, O.: Parameter identification of hydro-mechanical processes using artificial intelligence systems. Int. J. Mechatron. Appl. Mech. **2019**(5), 19–26 (2019)
27. Ivanov, V., Vashchenko, S., Rong, Y.K.: Information support of the computer-aided fixture design system. In: 12th International Conference on ICT In: Ermolayev, V. et al. (ed.), Education, Research and Industrial Applications. Integration, Harmonization and Knowledge Transfer, (CEUR-WS.org), vol. 1614, pp. 73–86 (2016) http://ceur-ws.org/Vol-1614/paper_37.pdf
28. Ivanov, V., Pavlenko, I., Vashchenko, S., Zajac, J.: Information system for computer-aided fixture design. In: Knapcikova, L., Balog, M. (eds.) Industry 4.0: trends in management of intelligent manufacturing systems, pp. 121–132. Springer, Cham (2019). https://doi.org/10.1007/978-3-030-14011-3_11

Analysis of Accuracy and Adequacy of Dynamic Models of Objects

Sergiy Kovalevskyy[1(✉)], Olena Kovalevska[1], Andriy Koshevoy[1], and Vladeta Jevremović[2]

[1] Department of Machine Building Technology, Donbass State Engineering Academy (DSEA), Academic str., 72, Kramatorsk 84313, Ukraine
kovalevskii@i.ua

[2] High Technical Mechanical School of Professional Studies, 37240 Trstenik, Serbia

Abstract. The article formulates the main problems of ensuring the accuracy and adequacy of dynamic models of objects. It is shown that such problems are characteristic of conditions that are distinguished by the need to maintain the current state of control systems. A technique for assessing dynamic models of objects based on the parametrization of general laws to achieve accuracy and adequacy of mathematical models is presented. Recommendations on the formation of training and test sets formed from constantly renewed sets of precedents for dynamic modeling are given.

Keywords: Identification · Modeling · Modeling accuracy · Adequacy of mathematical models · Neural networks (NN)

1 Introduction

The problem of dynamic identification of objects is the development of a mathematical modeling tool that would have the capabilities that provide the required accuracy characteristics *d(t)* and adequacy *A(t)* dynamic model. This issue is devoted to many papers [1–7]. However, the issue of relevance of a multitude of precedents needs further discussion and additional research in field intelligent and robotic machine tools [8, 9]. Mathematical models of differential dynamical objects is given in books [10–13] and papers [14].

2 Formulation of the Problem

Simulation accuracy is usually inversely related to simulation time (Fig. 1):

$$d(t) = D \cdot \left(1 - e^{\frac{-t}{\alpha}}\right) \tag{1}$$

where t – the time it takes to create a model once; α – measure of the intensity of achieving the accuracy of the mathematical model; D – the accuracy coefficient of the dynamic model.

I. Karabegović (Ed.): NT 2020, LNNS 128, pp. 75–80, 2020.
https://doi.org/10.1007/978-3-030-46817-0_8

However, the adequacy of the model, in general, decreases with time (Fig. 1):

$$A(t) = A_1 \cdot \left(1 - e^{\frac{-t}{\beta_1}}\right) - A_2 \cdot \left(1 - e^{\frac{-t}{\beta_2}}\right) \tag{2}$$

where β_1 and β_2 indicators of the intensity of the adequacy of the model; A_1 and A_2 – dynamic model adequacy ratios.

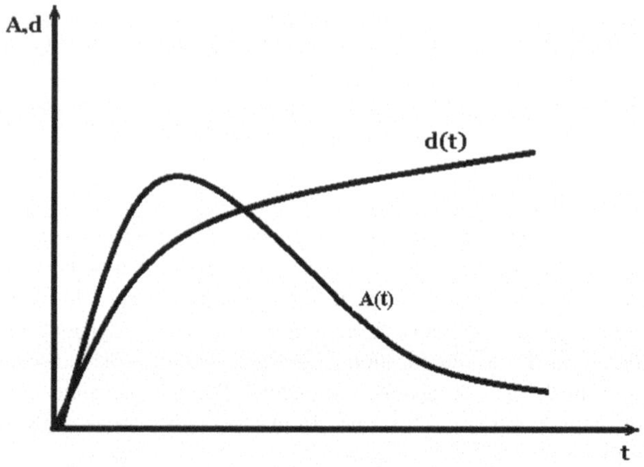

Fig. 1. The overall nature of the change over time indicators *d(t)* and *A(t)*

3 Methods and Results of Research

It can be assumed that, at best, the accuracy and adequacy of the model should tend to the maximum value. However, quantification can be judged only if these estimates have reasonable numerical values with the following initial data (Fig. 2):

$$S(t) = k_1 \cdot A(t) + k_2 \cdot d(t) \to max, \tag{3}$$

under restrictions:

$D = 1$; $A_1 = 1$; $A_2 = 1$; $\alpha = 2$; $\beta_1 = 2$ and $\beta_2 = 1$.where: *S(t)* – objective function of optimizing the reliability of a dynamic model; k_1 and k_2 – parameters of the objective function.

Analytic expression for maximum value *S(t):*

$$\frac{dS(t)}{dt} = 0, \tag{4}$$

From the expression (4) it follows that for dynamic models it is necessary to provide a condition for updating models, in which:

$$S(t) = k_1 \cdot A_1 \cdot \left(1 - e^{\frac{-t}{\beta_1}}\right) - k_2 \cdot A_2 \cdot \left(1 - e^{\frac{-t}{\beta_2}}\right) + k_2 \cdot D \cdot \left(1 - e^{\frac{-t}{\alpha}}\right) \tag{5}$$

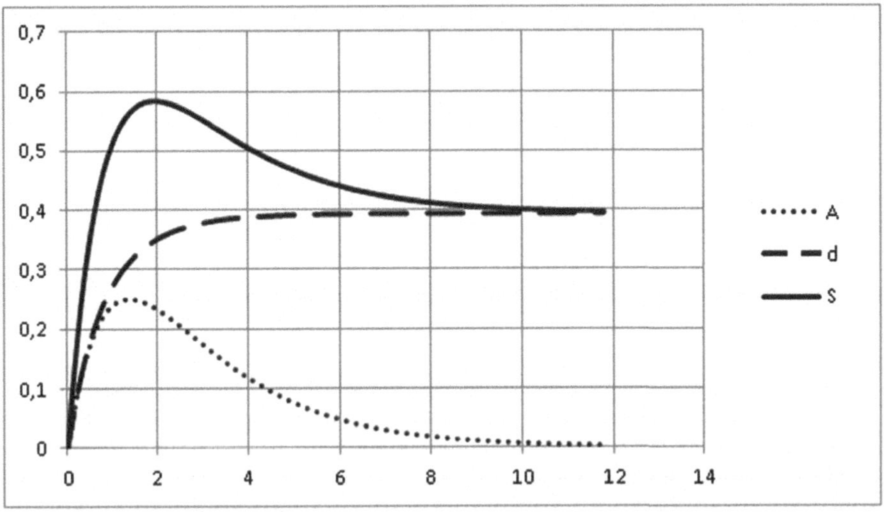

Fig. 2. Formation of a complex indicator *S(t)* with raw data

Numerical solutions found allowed to create an approximate base of precedents for the neural network model (Table 1).

Table 1. Maximum values *S(t)* for neural network modeling

D	A_1	A_2	α	β_1	β_2	k_1	k_2	S(t)max
1	2	2	2	2	1.5	1	1	2.9
2	2	2	2	2	1.5	1	1	3.7
3	2	2	2	2	1.5	1	1	4.8
1	1	2	2	2	1.5	1	1	6.8
2	1.5	2	2	2	1.5	1	1	5.8
1	1.5	2	2	2	1.5	1	1	4.8
3	1.5	2	2	2	1.5	1	1	5.8
1	1.5	2	1	2	1.5	1	1	4.8
2	1.5	2	1	2	1.5	1	1	4.8
0.5	1.5	2	1	2	1.5	1	1	3.8
0.5	1.5	1.5	1	2	1.5	1	1	2.5
0.5	1.5	1.5	1	2	1	1	1	1.7
1	1.5	1.5	1	2	1	1	1	1.9
2	1.5	1.5	1	2	1	1	1	2.2
3	1.5	1.5	1	2	1	1	1	2.5

(continued)

Table 1. (*continued*)

D	A_1	A_2	α	β_1	β_2	k_1	k_2	S(t)max
3	1.5	1.5	1	3	1	1	1	2.5
3	1.5	1.5	1.5	3	1	1	1	2.5
3	1.5	1.5	2	3	1	1	1	2.5
2	1.5	1.5	2	3	1	1	1	2.3
2	1.5	1.5	2	2	1	1	1	2.2
2	1.5	1.5	2	3	1	1	1	2.3
1	1.5	1.5	2	3	1	1	1	2
3	1.5	1.5	2	3	1	1	1	2.5
3	1.5	1.5	2	3	2	1	1	5.8
3	1.5	1.5	0.5	3	2	1	1	4.8
3	1	1.5	0.5	3	2	1	1	6.8

For the created neural network model, the verbal description made in the environment NeuroPro 0,25 [10, 11] has the following form:

Database fields (initial symptoms): D; A_1; A_2; α; β_1; β_2.

Database fields (end syndromes): S_{max}.

Preprocessing of the input fields for the database to feed the network:

$$\begin{aligned} \mathrm{D} &= (D - 1.75)/1.25 \\ \mathrm{A}_1 &= (A_1 - 1.5)/0.5 \\ \mathrm{A}_2 &= (A_2 - 1.75)/0.25 \\ \alpha &= (\alpha - 1.25)/0.75 \\ \beta_2 &= (\beta_2 - 1.5)/0.5 \end{aligned} \tag{6}$$

Functional converters:

Sigmoid1(A)=A/(0,1+|A|)
Syndromes 1-ro level:
Syndrome1_1=Sigmoid1(0,009517628*D-0,05518812*A1 +0,005194922*b2)
Syndrome1_2=Sigmoid1(0,612629*D-0,517355*A1+0,6975336*A2-0,1852509*a-0,1394655)
Syndrome1_3=Sigmoid1(-D+A2-0,8100481)
Syndrome1_4=Sigmoid1(0,06461254* a+0,145597*b2-0,08538858)

Final syndrome:

$$\begin{aligned} S_{max} &= \text{Syndrome1\_1} + 0,4465326 * \text{Syndrome1\_2} + 0,07499535 * \text{Syndrome1\_3} \\ &\quad + \text{Syndrome1\_4} + 0,3638409 \end{aligned} \tag{7}$$

Postprocessing end syndromes:

$$S_{max} = ((S_{max} * 5,10000014305115) + 8,50000023841858)/2) \quad (8)$$

Significance of inputs reflects the contribution of each variable to the total quality indicator of a dynamic model:

$$\begin{aligned} D &= 0.6566159 \\ A_1 &= 0.2404453 \\ A_2 &= 1 \\ \alpha &= 0.3158262 \\ \beta_2 &= 0.4155805. \end{aligned}$$

Thus, expressions (7) and (8) allow you to assess the methods of mathematical modeling, ranking them relative to the conditions of maximum efficiency of the method. Decision (4) regarding the Eq. (5) in the general case, it will expand the range of evaluation methods for creating dynamic models of objects in the process of their implementation.

4 Conclusions

The presented approach allows to provide a comparative analysis of dynamic models in general. For example, for a variety of precedents Table 1 indicator A_2 is the most significant in comparison with the indicator D. This means that when creating dynamic models, the adequacy of models should be preferred over their accuracy. The adequacy of the model should be assessed by test samples that represent 1/3 part of the total amount of use cases available for modeling. At the same time, many precedents should be considered as constantly replenished - each newly added precedent requires to exclude one of the precedents from the initial set of precedents. At the same time, it is necessary to choose the excluded case from the set of initial ones. As a selection criterion, you should take the one that will be the most distant from the new precedent. Such a selection should be implemented in software, which will certainly lead to a decrease in the dynamics of updating the mathematical model. This concerns the original set of precedents.

References

1. Duvenaud, D., Lloyd, J.R., Grosse, R., Tenenbaum, J.B., Ghahramani, Z.: Structure discovery in nonparametric regression through compositional kernel search. In: Proceedings of the 30th International Conference on Machine Learning, pp. 1166–1174 (2013)
2. Giannakis, G., Pichler, M., Kontes, G., Schranzhofer, H., Rovas, D.: Simulation speedup techniques for computationally demanding tasks. In: Proceedings of the 13th Conference Building Simulation of International Building Performance Simulation Association, pp. 3761–3768 (2013)

3. Gilan, S.S., Goyal, N., Dilkina, B.: Active learning in multi-objective evolutionary algorithms for sustainable building design. In: Proceedings of the Genetic and Evolutionary Computation Conference, GECCO 2016, pp. 589–596 (2016)
4. Maimon, O., Rokach, L. (eds.): Data Mining and Knowledge Discovery Handbook. Springer, Boston (2005)
5. Mezić, I.: Spectral properties of dynamical systems, model reduction and decompositions. Nonlinear Dyn. **41**(1–3), 309–325 (2005)
6. Symonds, P., Taylor, J., Chalabi, Z., Mavrogianni, A., Davies, M., Hamilton, I., Vardoulakis, S., Heaviside, C., Macintyre, H.: Development of an England-wide indoor overheating and air pollution model using artificial neural networks. J. Build. Perform. Simul. **9**(6), 1–14 (2016)
7. Sokolov, V.I., Tavanyuk, T.Ya.: Opredelenie peredatochnyh funkcij elektrogidravlicheskogo sledyashego privoda oborudovaniya dlya obrabotki davleniem. Visnik Shidnoukrayinskogo Nacionalnogo Universitetu im. Volodimira Dalya **155**, 130–135(2011)
8. Kovalevska, O.S., Kovalevskyy, S.V.: Application of acoustic analysis in control systems of robotic machine tools. Naukovij Zhurnal «Radioelektronika, Informatika, Upravlinnya» «Radio Electron. Comput. Sci. Control» **45**(2), 51–59 (2018)
9. Kovalevskyy, S., Kovalevska, O., Dašić, P., Ješić, D., Kovač, P.: Engineering consulting technology in production engineering intelligent mobile machines. Int. J. Ind. Eng. Manag. (IJIEM) **8**(4), 203–208 (2017)
10. Dašić, P.: Reliability of technical systems: Selected scientific-professional papers. SaTCIP Publisher Ltd., Vrnjačka Banja, Serbia (2019). 308 pp. (in Serbian)
11. Dašić, P.: Response surface methodology: Selected scientific-professional papers. SaTCIP Publisher Ltd., Vrnjačka Banja, Serbia (2019). 306 pp. (in Serbian)
12. Kovalevskyy, S.V., Kovalevskaya, E.S., Tulupov, V.I.: Razvitie metodov akusticheskoj diagnostiki v mashinostroenii. Kramatorsk, Ukraine, Donbass State Engineering Academy (DSEA) (2014). 91 pp. (in Russian)
13. Kovalevskyy, S.V., Kovalevska, O.S., Korzhov, Ye.O., Koshevoj, A.O.: Diagnostika tehnologichnih sistem i virobiv mashinobuduvannya (z vikoristannyam nejromerezhevogo pidhodu). Donbass State Engineering Academy (DSEA), Kramatorsk, Ukraine (2016). 186 pp.
14. Sokolova, Ya.V., Tvanyuk, T.Ya., Sokolov, V.I.: Nelinejnaya matematicheskaya model elektrogidravlicheskogo sledyashego privoda s drosselnym regulirovaniem. Visnik Shidnoukrayinskogo Nacionalnogo Universitetu im. Volodimira Dalya **152**(10), 168–175 (2010)

Conditions on Full Rotation of the Drive Member of the Four-Joint Mechanism

Avdo Voloder(✉), Fikret Veljović, and Senad Burak

Faculty of Mechanical Engineering, University of Sarajevo, Sarajevo, Bosnia and Herzegovina
voloder@mef.unsa.ba

Abstract. The conditions that must be satisfied for the driving member of the four-joint mechanism to be able to make full rotation and to prevent the movement of this mechanism are analyzed. The analysis is based on observation of the end positions of this mechanism. Mathematical expressions have been obtained that describe these conditions and these expressions are generalizable to any geometry of a four-joint mechanism. These results are also described graphically in the form of diagrams. Several numerical examples of combinations of the lengths of the individual members of the mechanism during its motion are also presented.

Keywords: Four-joint mechanism · Mechanism member movement restriction · Simulations

1 Introduction

One of the important problems of plane mechanism dynamics is the problem of solving the inverse dynamics [1, 2]. Sometimes, some kinematic problems also need to be solved. One of them is the possibility of blocking the operation of the mechanism due to the unfavorable ratio of the lengths of its members. When it comes to the four-joint mechanism, this problem is somewhat addressed in [3, 4].

One of the common problems with the movement of a mechanism is the fact that, for some combinations of the length of its members, the mechanism may be blocked. In case the driving member of the mechanism is a rotating member, then that member cannot fully rotate. If the driving forces exceed a certain limit, some members of the mechanism break, which has considerable consequences.

The aim of the analysis in this paper is to obtain general relations for the interdependence of the lengths of individual members of the four-joint mechanism in order to avoid blocking its motion. In this way, one of the problems of synthesis of this mechanism could be effectively addressed.

2 Analysis of the Problem

We will observe the two extreme positions of the four-link mechanism, as shown in Fig. 1. We will introduce the following tags:

I. Karabegović (Ed.): NT 2020, LNNS 128, pp. 81–88, 2020.
https://doi.org/10.1007/978-3-030-46817-0_9

$$\overline{AB} = a, \overline{BC} = b, \overline{CD} = c, \overline{DA} = d. \tag{1}$$

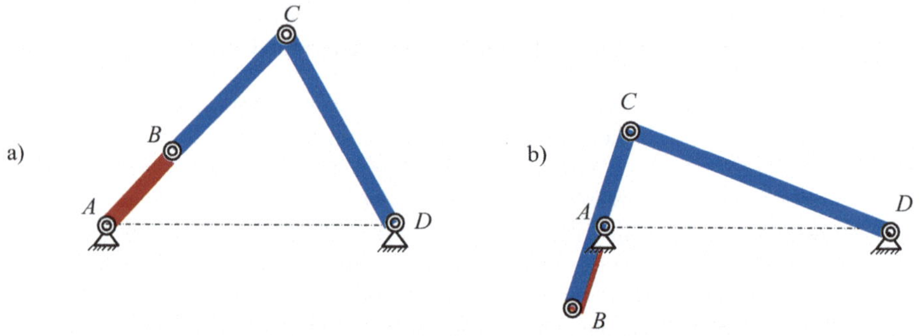

Fig. 1. Two extreme positions of the four-joint mechanism (a and b)

According to the position of the mechanism members in Fig. 1a, the following inequalities must be satisfied

$$a + b \leq c + d, \tag{2}$$

$$c \leq a + b + d, \tag{3}$$

$$d \leq a + b + c. \tag{4}$$

Taking into account the positions of the mechanism members in Fig. 1b, the above relations take the form

$$b - a \leq c + d, \tag{5}$$

$$c \leq d + b - a, \tag{6}$$

$$d \leq b - a + c, \tag{7}$$

resulting in

$$a + c \leq b + d, \tag{8}$$

$$a + d \leq b + c. \tag{9}$$

From (2), (8) and (9) we obtain

$$a \leq c + d - b, \tag{10}$$

$$a \leq b + d - c, \tag{11}$$

$$a \leq b + c - d. \tag{12}$$

Solving system (10), (11) and (12) we get the condition that the length of the working member is not greater than any other member of the four-joint mechanism

$$a \leq b,\ a \leq c,\ a \leq d. \tag{13}$$

The unique solution to inequalities (10)–(12) is

$$0 \leq \mathrm{a} \leq \min(\mathrm{b,c,d}) + \mathrm{med}(\mathrm{b,c,d}) - \max(\mathrm{b,c,d}), \tag{14}$$

from which it follows

$$\min(\mathrm{b,c,d}) + \mathrm{med}(\mathrm{b,c,d}) \geq \max(\mathrm{b,c,d}). \tag{15}$$

If condition (15) is not satisfied, then a solution for length **a** does not exist.

According to the expression (14), diagrams of the dependence of the lengths of the working member of the mechanism in function of the lengths of the other members (shaded area) can be obtained, at which the working member of the mechanism will be able to rotate (Figs. 2, 3 and 4).

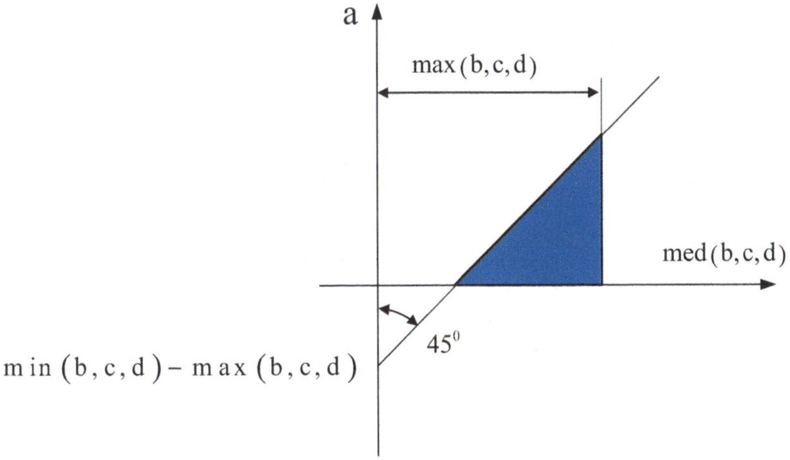

Fig. 2. Dependence of length of the member 1 on size med(b, c, d), in case of full rotation of the working member of the mechanism (shaded area)

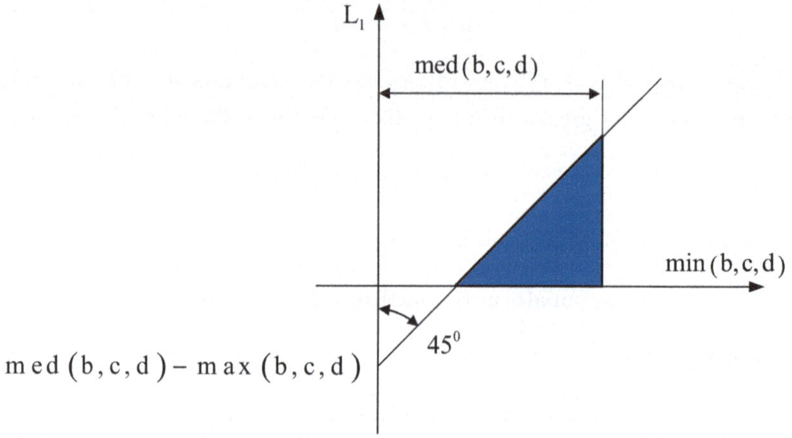

Fig. 3. Dependence of length of the member 1 on size min(b, c, d), in case of full rotation of the working member of the mechanism (shaded area)

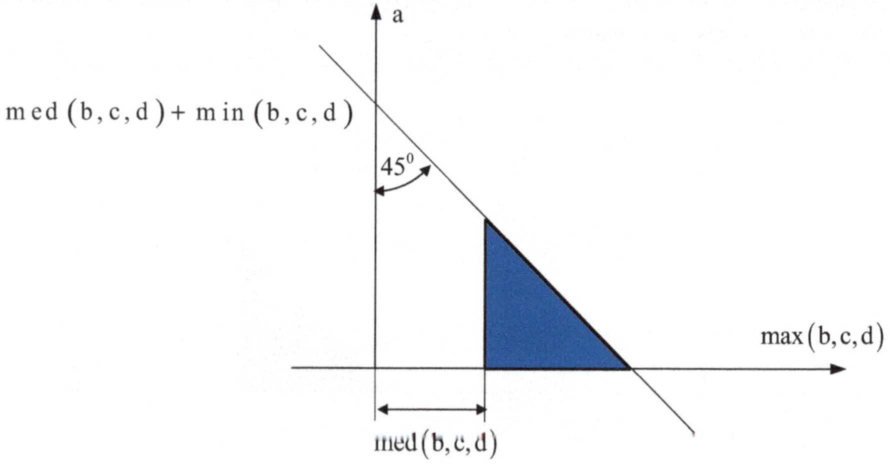

Fig. 4. Dependence of length of the member 1 on size max(b, c, d), in case of full rotation of the working member of the mechanism (shaded area)

For variable L_2, system (10)–(12) gives

$$b \leq c + d - a, \tag{16}$$

$$b \geq a + c - d, \tag{17}$$

$$b \geq a + d - c. \tag{18}$$

The unique solution of system (16)–(18) is:

$$\max(a,c,d) + \min(a,c,d) - \mathrm{med}(a,c,d) \le b \le \max(a,c,d) + \mathrm{med}(a,c,d) - \min(a,c,d). \tag{19}$$

Analogously, for variables **c** and **d** it follows:

$$\max(a,b,d) + \min(a,b,d) - \mathrm{med}(a,b,d) \le c \le \max(a,b,d) + \mathrm{med}(a,b,d) - \min(a,b,d), \tag{20}$$

$$\max(a,b,c) + \min(a,b,c) - \mathrm{med}(a,b,c) \le d \le \max(a,b,c) + \mathrm{med}(a,b,c) - \min(a,b,c). \tag{21}$$

According to expression (19), dependency diagrams of the member 2 are obtained as a function of the lengths of the other members (shaded area) at which the working member of the mechanism will be able to move (Figs. 5, 6 and 7).

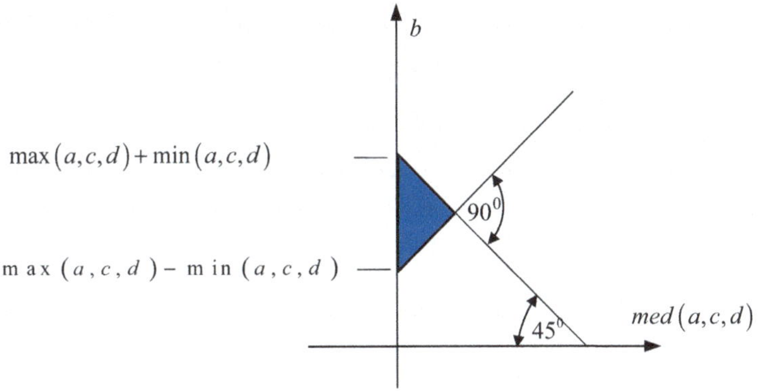

Fig. 5. Dependence of length of the member 2 on size med(a, c, d), in case of full rotation of the working member of the mechanism (shaded area)

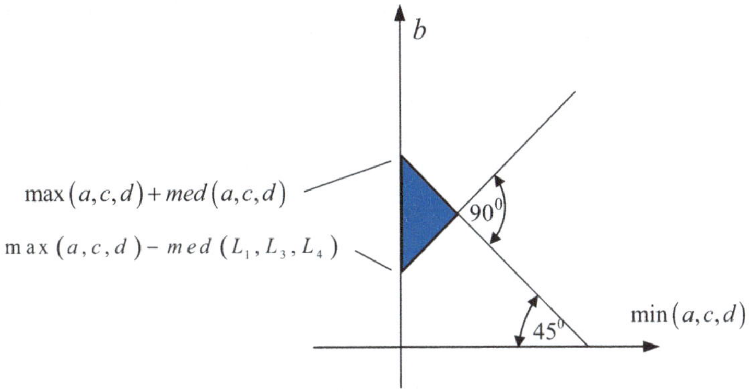

Fig. 6. Dependence of length of the member 2 on size min(a, c, d), in case of full rotation of the working member of the mechanism (shaded area)

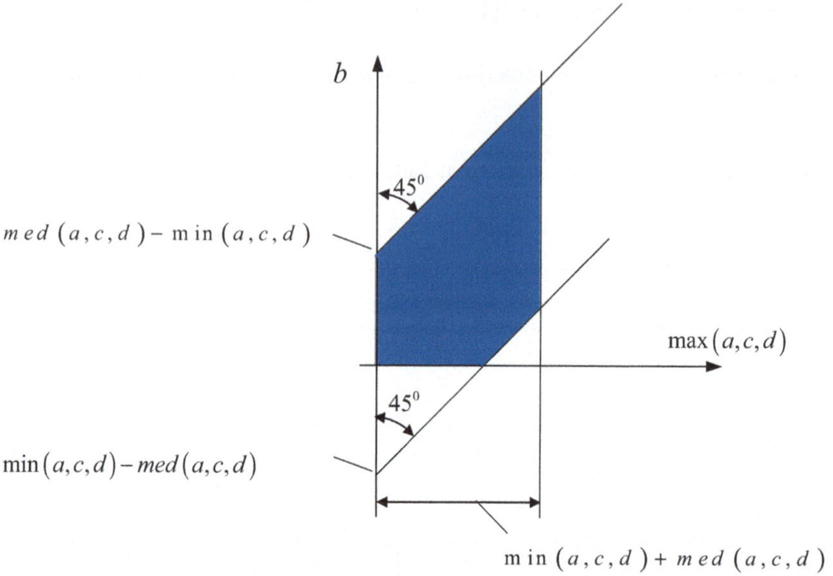

Fig. 7. Dependence of length of the member 2 on size max(a, c, d), in case of full rotation of the working member of the mechanism (shaded area)

By analogy, according to expression (20), with the condition of full rotation of the driving member 1, similar diagrams can be obtained for the length of the member 3, depending on $med(a, b, d)$, $min(a, b, d)$ i $max(a, b, d)$.

Similarly, according to expression (21) and with the condition of full rotation of member 1, corresponding diagrams can be obtained for the length of member 4, depending on $med(a, b, c)$, $min(a, b, c)$ and $max(a, b, c)$.

It can be seen from above analysis that the areas of possible movement of the mechanism are limited.

As an example of the application of the obtained expressions or diagrams, some combinations of lengths of any three members of the mechanism were analyzed, with intervals of length values for the remaining member being obtained. The results of the analysis are presented in Tables 1, 2, 3 and 4. In cases where no solution exists, the (*) mark was used.

Table 1. Length range of member 1 depending on the length of the other three members for some cases at full rotation of the driving member of the mechanism

b	c	d	**a**
3	2	5	**0**
2	6	6	**0–2**
2	4	3	*
5	2	4	**0–1**
5	5	6	**0–4**
7	4	6	**0–3**

Table 2. Length range of member 2 depending on the length of the other three members for some cases at full rotation of the driving member of the mechanism

a	c	d	**b**
2	3	6	**5–7**
3	11	5	**9–13**
4	5	6	**5–7**
5	11	16	**10–22**
3	12	13	**4–22**
10	1	3	**8–12**

Table 3. Length range of member 3 depending on the length of the other three members for some cases at full rotation of the driving member of the mechanism

a	b	d	**c**
2	4	4	**2–6**
12	3	5	**10–14**
3	15	2	**14–16**
4	10	9	**5–15**
6	11	7	**10–12**
6	7	12	**11–13**

Table 4. Length range of member 4 depending on the length of the other three members for some cases at full rotation of the driving member of the mechanism

a	b	c	**d**
2	6	4	**4–8**
2	4	11	**9–13**
3	3	7	**7–7**
4	7	15	**12–18**
15	12	10	**12–17**
8	12	19	**19–23**

3 Conclusion

General terms have been derived to describe the conditions under which the working member of a four-jointed flat mechanism can be fully rotated. Solving the problem, among other things, gives the condition that the length of the working member is not greater than any other member of the four-joint flat mechanism. According to the calculated relations, appropriate diagrams were obtained. The working areas of rotation of the driving member of the mechanism in these diagrams are limited by the corresponding directions. The expressions obtained are general, so that their use can determine the possibility or limitation of the full mobility of the driving member of the mechanism for any length of the members of the mechanism. In order to illustrate the application of the obtained expressions, some numerical examples of combinations of the lengths of the individual members of the mechanism in its possible movement are presented.

References

1. Chaudhary, K., Chaudhary, H.: Optimal dynamic design of planar mechanisms using teaching–learning-based optimization algorithm. In: Proceedings of the Institution of Mechanical Engineers, vol. 230, no. 19, 3442–3456 (2015)
2. Gramblička, S., Kohár, R., Stopka, M.: Dynamic analysis of mechanical conveyor drive system. Elseiver Procedia Eng. **192**(2017), 259–264 (2017)
3. Grashof, F.: Theoretische Maschinenlehre, vol. 2. Voss, Hamburg (1983)
4. Norton, R.L.: Design of Machinery. McGraw-Hill Inc. Company, New York (1992). ISBN 0-07-909702-2
5. Loerch, R.J., Erdman, A.G., Sandor, G.N., Mihda, A.: Synthesis of four bar linkages with specified ground. In: Proceedings - OSU Applied Mechanisms Conference, Oklahoma State University (1975)

Achieving Crowning Contact of Spur Bevel Gears Through Deliberately Introduced Mounting Errors

Viktor Ivanov[1(✉)], Galyna Urum[2], and Svitlana Ivanova[2]

[1] Institute of Mechanical Engineering, Odessa National Polytechnic University, Odesa, Ukraine
ivv@opu.ua

[2] South Ukrainian National Pedagogical University named after K. D. Ushynsky, Odesa, Ukraine

Abstract. Spur bevel gears are characterized by a load concentration at the ends of the teeth, where gear teeth are at their weakest. To reduce load concentration at the ends of the teeth, the modification is used by longitudinal crowning of a flank profile. The crowning contact in a spur bevel gear can be achieved through deliberately introduced mounting errors. With a deviation of the shaft angle along with the deviation mounting distance, it is possible to achieve initial contact in the middle of gear tooth surfaces. The problem of the theory of gearing is solved - dependencies are obtained to determine the initial contactpoint of the teeth. For this, the intersection line of an octoid surface of the tooth and a plane of gearing was found, in the presence of mounting errors. The trajectory of the contact point on the surface of the teeth during mating was determined. The maximum achievable contact crowning values are given depending on the degree of transmission accuracy and the total number of teeth. Crowning is calculated for the range of modules 6.3… 10. A fixture design for controlling axial displacement during gear assembly has been developed.

Keywords: Crowning profile · Spur bevel gears · Deliberately mounting deviation

1 Introduction

Bevel gears, compared with cylindrical ones, have a more complex design, manufacturing and mounting technology. Because of this, it is difficult to achieve the theoretically correct contact of the teeth. This is especially true for spur bevel gears. In this train, any manufacturing and mounting errors lead to the edge contact of the teeth. In spur gears, a crowning contact is used to avoid edge contact. Crowning contact creates certain technological problems. There are a number of manufacturing methods for crowning contact in spur gears. In spur bevel gears, the manufacture of a crowning contact is more expensive and it has not been widely used. Advances in the gear manufacturing technology have led to decrease in the amount of gears, manufactured through cutting [1]. Modern machinery utilizes gears manufactured through casting,

I. Karabegović (Ed.): NT 2020, LNNS 128, pp. 89–97, 2020.
https://doi.org/10.1007/978-3-030-46817-0_10

stamping, forging, plastic molding, sintering and 3D printing [2]. The use of multitasking machine with CNC enables the manufacturing of a random tooth profile [3]. The tooth surfaces are modeled using a CAD; it is finished on the basis of a CAM and transmitted to a CNC machine [4].

Studies devoted to calculating the shape of a crowning profile take into account specific factors that lead to an error in the position of the gears. Basically crowning profile is investigated, which eliminates the influence of a pinion shaft deformation and the shafts misalignment due to the mounting errors or bearing deformations [5]. The main factors which introduce the largest error in the deviation from the nominal position of the teeth are the machining and mounting errors.

Combined influence of shaft misalignments and crowning profile on the load distribution and tooth bending stresses of helical gear pairs are presented in article [6]. This study demonstrate the direct link between the crowning profile and gear misalignments as well as the effectiveness of the crowning profile in preventing edge loading conditions due to misalignment. These errors are probabilistic in nature. Therefore, the optimal crowning profile will be different for each of the gears of the same type. Instead of crowning contact often use the term barreling. The profile is convex with the initial point of contact in a middle of the tooth face. This profile prevents the edge contact of the teeth with any combination of errors. The shape of a barreling tooth in the form of a circle arc, cycloid and, logarithmic curve is discussed [7].

All machining and mounting errors, deformation of the cantilever shaft, as well as deformation of the bearings lead to the initial contact at the heel end of the bevel gear tooth. Therefore under light load, the contact pattern should be located near toe of the teeth and under operating load; the contact pattern should be shifted at the central part of the tooth width.

The purpose of the study is to create a cheap method of localized point contact that does not require separate manufacturing operations or to make changes to the manufacturing process.

2 Determination of the Initial Contact Point of the Teeth and the Value of the Crowning Magnitude

The presence of mounting distance errors f_{AMri} leads to edge contact of teeth. With negative mounting distance errors, the edge contact is at the heel end, and with a positive mounting distance errors, the edge contact is at the toe end. A shaft angle error also leads to edge contact - at the end of the heel if the angle is greater than 90° and on the toe if the angle is less than 90°. In the presence of a mounting errors combination, the teeth rotate relative to each other around a point lying on the contact line. Linear contact is broken and converted to localized point contact. Parameters of localized point contact are searched.

In Fig. 1 the octoid surface of the gear tooth is shown, the pitch cone of which has a mounting distance error. The origin of the Cartesian coordinate system *xyz* coincides with the apexes of this cone. In the section of the cone by plane $z = const$, the octoid formation scheme is shown (Fig. 1). The in volute roll angle φ defines the point P. The line is drawn perpendicular to the radius from point P to the intersection with the spherical involute 1 at point C. The OC and OP lines form an angle ψ. A beam from point C at an angle of 70° to the line segment CP is drawn. A point B, which belongs to octoid 2, is located at the base of the perpendicular, which lowered from a point P on a line CB. OPB points specify the position of the plane normal to the tooth surface $\sum$, along the generatrixOB. Similarly, a plane is defined containing the generatrix OK.

The origin of the coordinate system $x'y'z'$ is belonged to plane $\prod$. This origin is corresponded to the nominal position of the gear pitch cone. Plane $\prod$ is normal to the nominal position surface of the octoid. O'K is the generatrix of the octoid surface. $a'b'$ is the line of intersection of a plane $\prod$ with a surface $\sum$. K is the intersection point of the lines $a'b'$ and O'K, in which the initial contact of the tooth surfaces occurs.

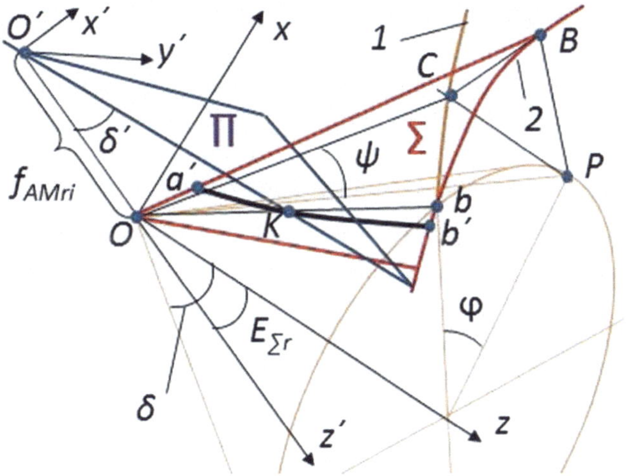

Fig. 1. Scheme to determine the value of the crowning and the initial contact point

The equations of the octoid surface in parametric form were used to determine the coordinates of the point K [8]

$$\begin{cases} x = \dfrac{sin\varphi\, sin\psi\, cos\psi ctg^2\alpha + cos\varphi\, sin\psi\, ctg\alpha cos\delta + Acos\varphi\, sin\delta}{-\, sin\delta\, ctg\alpha sin\psi + A\, cos\delta} z \\ y = \dfrac{-\, cos\varphi\, sin\psi\, cos\psi\, ctg^2\alpha + sin\varphi\, sin\psi\, ctg\alpha cos\delta + A\, cos\varphi\, sin\delta}{-\, sin\delta\, ctg\alpha sin\psi + A\, cos\delta} z \end{cases} \tag{1}$$

And equation of plane $\prod$:

$$\begin{aligned} (sin\varphi' - cos\varphi' cos\psi' ctg\alpha cos\delta' - (cos\varphi' - sin\varphi' cos\psi' ctg\alpha cos\delta') \\ + \cos\varphi' ctg\alpha \sin\delta'(z - f_{AMri}) = 0, \end{aligned} \tag{2}$$

where $A = 1 + ctg^2\alpha cos^2\psi$, φ' and ψ' - parameters of the intersection line of the surface $\sum$ and the plane $\prod$, $\delta' = \delta + E_{\sum r}$.

At point K, the conditions are satisfied $\psi = \psi'$ and $\varphi = \varphi'$.

So that

$$z = \frac{f_{AMri} sin\delta'}{sin\delta' - \frac{sin\psi\, ctg\alpha(1 - cos\delta\, cos\delta') - Asin\delta\, cos\delta'}{-sin\delta\, ctg\alpha sin\psi + Acos\delta}} \tag{3}$$

The z coordinates for the point K, which lying on the pitch cone, can be found with the formula:

$$z = \frac{f_{AMri} sin\delta'}{sin\delta' - tg\delta cos\delta'} \tag{4}$$

which is obtained from formula (3) by substituting $\psi = 0$.

In the case, the tooth profiles are created by a spherical involute, this formula is valid for any phase of meshing. With octoidal teeth, the z coordinate of the point at which the conditions $\psi = \psi'$ and $\varphi = \varphi'$ are satisfied varies depending on the phase. The trajectories of the initial contact point on the tooth surface of pinion 5 and wheel 1 were found for train with a gear ratio u = 1.76 (Fig. 2). The distance error for the pinion f_{AMr1} is taken positive, and the angle is taken $\delta_1' = \delta_1 + E_{\sum r}$; for the wheel, the distance error f_{AMr2} is taken negative, and the angle is taken $\delta_2' = \delta_2 - E_{\sum r}$. For any values of f_{AMri} and $E_{\sum r}$ are associated by relation (3), for $z = const$ the trajectories 1 and 5 will be the same. Along with the indicated combination of errors, another is possible: for pinion f_{AMr1} is negative and $\delta_1' = \delta_1 - E_{\sum r}$; for wheel f_{AMr2} is positive and $\delta_2' = \delta_2 + E_{\sum r}$.

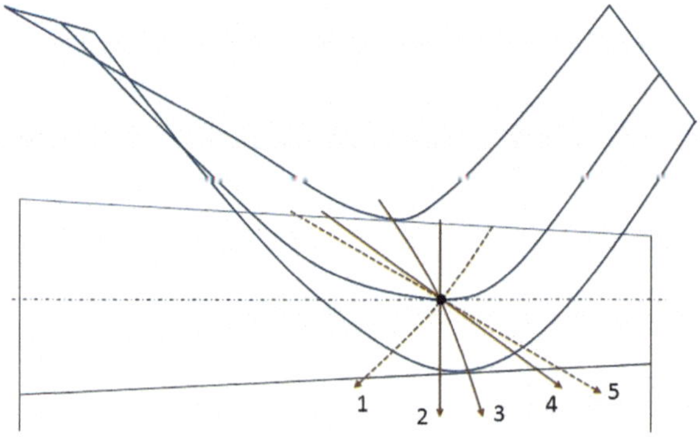

Fig. 2. The trajectories of the initial contact point of the teeth and the gap in meshing.

In order to match the parameters ψ' and φ', for which a gap is found, the coordinate z, is necessary, substituting ψ and φ in the system of Eqs. (1), to find $x(z)$ and $y(z)$. By substituting $x(z)$ and $y(z)$ for Eqs. (2), we find the $z = f(\psi', \varphi')$. Given ψ' and φ', which determine the position of the point by the height of the profile, we find z. When the point K is on the pitch cone, the described transformations can be performed analytically.

$$z = \frac{f_{AMri} ctg\alpha sin\delta'}{-\frac{x}{z} ctg\alpha cos\delta' - \frac{y}{z} + ctg\alpha sin\delta'} \tag{5}$$

The gap between any two points b and b' belonging to the octoid surface is the distance between point b and the projection of point b' on the normal to the tooth surface at point b. Point b corresponds to the coordinates of x and y, at point $b' - x'$, and y'. Then the distance between them is:

$$r = \sqrt{(x' - x)^2 + (y' - y)^2} \tag{6}$$

And gap in meshing Δ is:

$$\Delta = r \sin(\varphi' - \varphi) \tag{7}$$

In the case when the point lies on the pitch cone in the formula (7), it is necessary to set $\varphi = 0$.

To determine the position of the instantaneous axis of the tooth contact area, it is necessary to find a cross section in which the rise rate of the gap is smallest. For a certain section $z = const$, the gaps between the pinion tooth surface and the plane containing the OK generatrix and tangent to the tooth profiles, as well as between the wheel tooth surface and this plane, were found numerically. By one-dimensional minimization of the function of the total gap $\Delta'' = \Delta''(\delta'')$, we found the value of the angle δ'' defining the position of the plane in which the gap rise rate is the smallest (Fig. 2). An analysis of the calculated values of the angle δ' showed that in most cases, with sufficient accuracy, it can be calculated by the following formula

$$\delta'' = \frac{\delta\prime - \delta}{u + 1} \tag{8}$$

As can be seen from Fig. 2, in a case of the teeth contact along a generatrix belonging to the pitch cone, the position of the initial contact point K is known, since lines 1 and 5 intersect at this point. For other phases of mesh, we find the initial contact point by solving the system of equations

$$\begin{cases} \Delta_1\left(z_1, \psi_1', \varphi_1'\right) + \Delta_2\left(\frac{z_1}{u}, \psi_1', \varphi_2'\right) = 0 \\ \frac{\partial \Delta_1}{\partial Z_1} + \frac{\partial \Delta_2}{\partial Z_1} = 0 \end{cases} \tag{9}$$

Trajectory 2,3,4 the initial contact point are calculated for gear ratio values $u = 1$; 1,75; 2,5 (Fig. 2). Gaps in various phases of mesh are found as well. If there is an offset error of shafts f_{ar}, the z coordinate of the point K is determined from the system (9); the right of the first equation must be set equal to $-f_{ar}\cos\alpha$. To determine the gap, taking into account machining errors, it is necessary to present the function of the gaps array in the coordinate's $\Delta_u = \Delta_u(\varphi_1', z_1)$ and add up with the gaps caused by the mounting errors. Finally, the initial contact point was found from the system

$$\begin{cases} \Delta_1(z_1, \psi_1', \varphi_1') + \Delta_2\left(\frac{z_1}{u}, \psi_2', \varphi_2'\right) + \Delta_u(\varphi_1, z_1) + f_{ar}\cos\alpha = 0 \\ \frac{\partial\Delta_1}{\partial Z_1} + \frac{\partial\Delta_2}{\partial Z_1} + \frac{\partial\Delta_u}{\partial Z_1} = 0 \end{cases} \quad (10)$$

3 Bevel Gear Assembly Method

The localization that occurs with a certain combination of mounting errors can be used to prevent the appearance of an edge contact resulting from deformation of the gearbox elements. The gap at the tooth end, which is achieved by a combination of mounting errors, is limited by the size of the root clearance [9]. The maximum gap that can be achieved is calculated for most common gear parameters (Table 1).

Table 1. Maximum gap at the end of the teeth for the range of modules 6.3… 10

Gear accuracy grades	Maximum gap (mcm)			
	<40	40…44	45…49	50…55
6	4,5	5,5	6,5	7,0
7	6,0	7.0	8,0	9,0
8	7.5	9,0	11,5	13,0

Given the design of the gearbox, you can find the optimal ratio of mounting errors, which compensates for the deformation of the transmission elements. As the crowning gaps along tooth face width are relatively small thus it is assumed that the theoretical point contact under load spreads over a surface along the whole or part of the "potential" contact line made up of the points of the meshing teeth surfaces in which the geometrical separations of these surfaces along the tooth face width are minimal [10]. In spur bevel gears with localized contact, the longitudinal line of the teeth spot coincides with the contact line in normal spur bevel gears. Thus, the load distribution is calculated for linear contact, taking into account the gaps due to crowning.

For train with parameters: $z_1 = 17$, $z_2 = 30$, $m_e = 10$, the fifth accuracy grade, simple supported shaft of pinion, at the outer end of pinion set double row tapered roller bearing, at the inner end of pinion set cylindrical roller bearing, optimal combination of mounting errors is: $f_{AMr2} = +755$ mcm, $f_{ar} = +9$ mcm, $E_{\sum r} = +0{,}52°$. For the same transmission with the same accuracy grade, cantilever shaft of pinion (L/a = 3,0 where a - distance from the middle pinion facewidth to the middle of the bearing located at the outer end, L - distance between the middle of the bearings), at the outer end of pinion set double row tapered roller bearing, another bearing is cylindrical roller, optimal combination of mounting errors is: f_{AMr2} = +827 mcm, f_{ar} = −17 mcm, $E_{\sum_r}$ = +0,59°. The maximum value of the load distribution over the tooth facewidth is calculated. For an aligned contact, the maximum load value is 235.4 N/mm, for misaligned contact (crowning contact) maximum value of load is reduced to 180.8 N/mm.

It is difficult to manufacture gearbox housing with the changed shaft angle $E_{\sum_r}$ and offset error f_{ar}. It is advisable to set the mounting distance of the pinion or wheel, as is usually done, but take into account the existing errors in the manufacture of the housing - f_{ar}, $E_{\sum_r}$. In this case, the task is not to achieve the correct mounting distance of the wheels. The mounting distance is specially calculated and must be set in accordance with the calculation in order to achieve a crowning contact. To accomplish this task, a fixture design for the assembly of bevel gear has been developed, which allows one gear to be displaced relative to another and to achieve a given value of mounting distance. $f_{AMr} = f_{AMr1} sin\delta_1 + f_{AMr2} cos\delta_1$.

The fixture includes a shell 1, the inner conical surface of which 1a is mated to the pinion back cone 2 and the wheel (Fig. 3). On the side of the shell there are protrusions 1b, to which, with the help of screws 3, a conical sleeve 4 is fastened, mated with the pinion face cone.

Figure 4 shows the gear wheel shaft 1, on which the clamp 2 is mounted, consisting of halves 2a and 2b, connected with bolts 3 and nuts 4. On one of the clamp halves (2a) made scale 2c; the axis (2d) of the scale outer cylindrical surface is eccentric to the axis of the shaft.

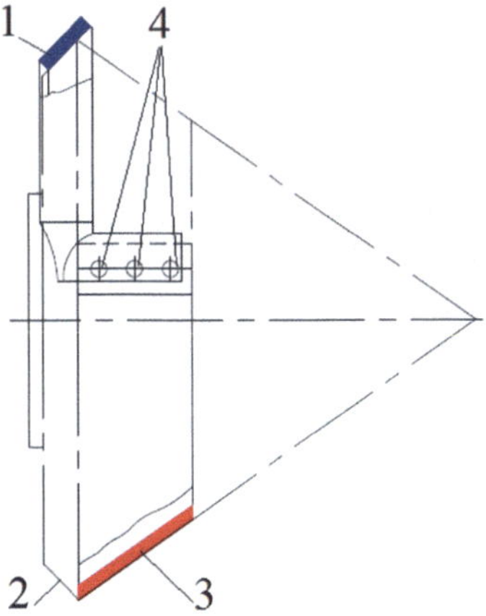

Fig. 3. Fixture design

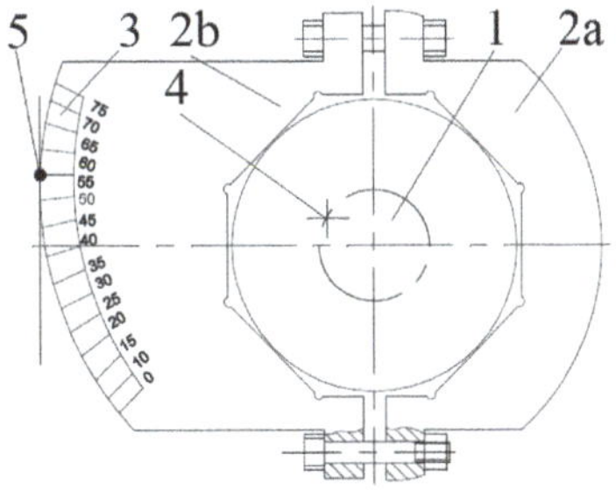

Fig. 4. Clamp design

The clamp contacts the toe end of pinion. To assemble a gearbox around the back cone surface, a bevel sleeve 4 is mounted. A shell 1 is attached to it. Then the bevel wheel is axially displaced so that the wheel back cone coincidence with the inner surface, the shell 1. This coincidence is a condition for the theoretically correct positioning of bevel gears, in which the mounting distance error of the pinion and wheels are equal to zero.

Both halves of the clamp are attached to the wheel shaft so that they are opposite the pinion inner end, and the wheel shaft is rotated together with the clamp until the scale contacts the pinion inner surface.

Then the pinion is shifted in the axial direction and the distance to the shaft axis is measured again. The difference in the scale readings must correspond to a deliberately introduced mounting error.f_{AMr2} = +755 mcm, f_{ar} = +9 mcm, $E_{\sum r}$ = +0,52°.

4 Conclusion

It is established that with a certain combination of mounting errors, localized contact in a spur bevel train is possible. The size of the gap, which is achieved by a combination of mounting errors commensurate with the gaps provided by the crowning of a flank profile. The load distribution is calculated for localization that achieved by the proposed manner. The maximum value of the distributed load is reduced by 23%. In some cases, the localization provided during assembly can be a substitute for the operation of manufacturing tooth profile crowning. Localization through deliberately mounting errors requires a change in assembly technology. When assembling, axial displacement is established, in contrast to usual technology, which ensures the nominal position of the gears or the correct contact spot in an unloaded gear train. In each gearbox, an distance errors is introduced deliberately, which corresponds to the existing shaft angle error. A fixture design is developed that implements the proposed assembly method.

References

1. Fuentes-Aznar, A., Gonzalez-Perez, I.: Mathematical definition and computerized modeling of spherical involute and octoidal bevel gears generated by crown gear. Mech. Mach. Theory **106**, 94–114 (2016)
2. Fuentes, A., Iserte, J.L., Gonzalez-Perez, I., Sanchez-Marin, F.T.: Computerized design of advanced straight and skew bevel gears produced by precision forging. Comput. Methods Appl. Mech. Eng. **200**(29–32), 2363–2377 (2011)
3. Ivanov, V., Karaivanov, D., Ivanova, S., Volkova, M.: Gear mesh geometry effect on performance improvement for external gear pumps. In: MATEC Web of Conferences, vol. 287, p. 01007. EDP Sciences (2019)
4. Tsuji, I., Kawasaki, K., Gunbara, H., Houjoh, H., Matsumura, S.: Tooth contact analysis and manufacture on multitasking machine of large-sized straight bevel gears with equi-depth teeth. J. Mech. Des. **135**(3), 034504 (2013)
5. Sainte-Marie, N.: A transmission-error-based gear dynamic model: Applications to single- and multi-mesh transmissions. Doctoral dissertation (2016)
6. Hotait, M., Kahraman, A.: Experiments on root stresses of helical gears with lead crown and misalignments. J. Mech. Des. **130**(7), 074502 (2008)
7. Wang, W.S., Fong, Z.H.: A dual face-hobbing method for the cycloidal crowning of spur gears. Mech. Mach. Theory **43**(11), 1416–1430 (2008)

8. Litvin, F.L., Fuentes, A.: Gear Geometry and Applied Theory. Cambridge University Press, Cambridge (2004)
9. Ivanov, V., Urum, G., Ivanova, S., Naleva, G.: Analysis of matrix and graph models of transmissions for optimization their design. East. Eur. J. Enterp. Technol. **4**(1), 11–17 (2017)
10. Simon, V.: Computer simulation of tooth contact analysis of mismatched spiral bevel gears. Mech. Mach. Theory **42**(3), 365–381 (2007)

Integrated Development and Design of Gears Reduction Drive

Isad Saric[1], Jasmin Smajic[2], and Adis J. Muminovic[1(✉)]

[1] Faculty of Mechanical Engineering, Department of Mechanical Design, University of Sarajevo, Vilsonovo setaliste 9, 71000 Sarajevo, Bosnia and Herzegovina
{saric,adis.muminovic}@mef.unsa.ba
[2] Make3D, Sarajevo, Bosnia and Herzegovina
smajicjasmin036@gmail.com

Abstract. This paper presents the methodology for integrated product development and design. Gears reduction drive was chosen as example. For design process, advance 3D methods and technologies were used, technologies such as: 3D geometric parametric modeling of virtual prototype, simulation of working principles, manufacturing of real prototype using additive manufacturing (3D printing) and usage of augmented reality. In conclusion part of the paper, using results from this research, advantages of integrated product development and design are highlighted, with special focus on influence of this methodology on students education at Faculty of Mechanical Engineering in the field of mechanical power transmissions.

Keywords: Gears reduction drive · Simulation · Additive manufacturing (3D printing) · Augmented reality

1 Introduction

Holland company Ultimaker, one of the biggest companies for development and manufacturing of 3D printers, in correlation with European investment Bank had a competition "Ultimaker Education Challenge" for giveaway of 50 Ultimaker 3D printers. Education institutions from all over the Europe could apply. Goal of the competition was to find innovative ways to improve education curriculums using 3D printers and to find teachers, professors and students who are passionate about gaining and transferring knowledge and improving education. Around 250 education institutions from all over the Europe apply on this competitions. Student project form Faculty of Mechanical Engineering in Sarajevo was one of the best 50 projects which won the 3D printer Ultimaker 3. Ultimaker 3 is one of the best 3D printers in the semiprofessional 3D printer range. The announcement of the winner was held at Bett Show in London. Bett Show is the biggest event in the world, related to the education technologies. Students from Faculty of Mechanical Engineering in Sarajevo apply with the project "Improvement of teaching 'How machines works' using Ultimaker 3D printer". Goal of the project was to develop and manufacture real size gear reduction drive using Ultimaker 3D printer. On this way they acquired the prerequisites for implementation the research carried out in this paper.

I. Karabegović (Ed.): NT 2020, LNNS 128, pp. 98–103, 2020.
https://doi.org/10.1007/978-3-030-46817-0_11

2 Methodology for Development and Design of Gears Reduction Drive

For development and design of gears reduction drive, methodology showed at Fig. 1, was used. Special attention, in this research, was on phase IV and V, because in that two phases real size gears reduction drive, in a form of prototype, was manufactured using additive manufacturing (3D printing). Real size prototype is tested to find out was he fulfilled crucial requirements.

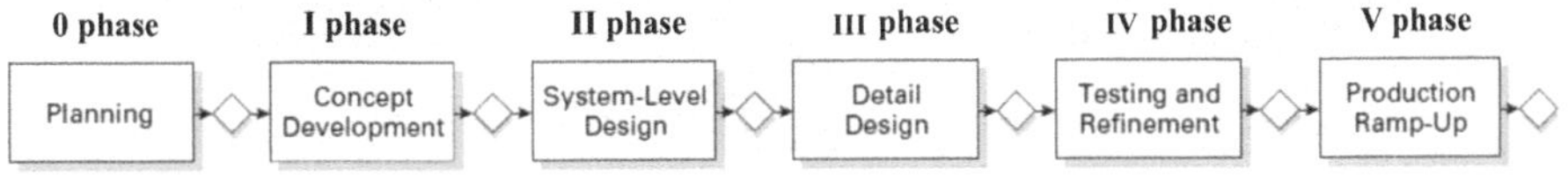

Fig. 1. Methodology of product development and design, showed in phases [1, 2]

2.1 Calculations of Gears Reduction Drive

Before 3D geometrical parametric modeling phase it is necessary to carry out calculations of all parts of gears reduction drive (gears, shafts, bearings, etc.) [3]. For calculations of gears and shafts integrated intelligent systems IICADkmpsandIICADv [4, 5] was used. These systems are developed at Department of Mechanical Design from University of Sarajevo - Faculty of Mechanical Engineering. Most important geometrical characteristics of gears are shown in Table 1.

Table 1. Geometrical characteristics of gears

Geometrical characteristics	Straight bevel gears		Spur gears		Worm gears	
	1	2	1	2	Worm	Worm gear
Module m (mm)	4		4		3	
Number of teeth's z (mm)	40		25	50	1	34
Diameters D (mm)	160		100	200	27	102
Width b (mm)	30		30		130	30

2.2 3D Geometrical Parametric Modeling of Gears Reduction Drive

3D geometrical parametric modeling of individual parts of gears reduction drive and their assembly was done using SolidWorks and CATIA software systems (Fig. 2) [6–9]. 3D models of bearings are chosen as standard parts from online database. After detail force, stress calculation and analysis of shafts, standard bearings are chosen: ball bearing 63008-2RS1, cylindrical roller bearing NJ 208 ECPH and tapered roller bearing 32008 X/Q.

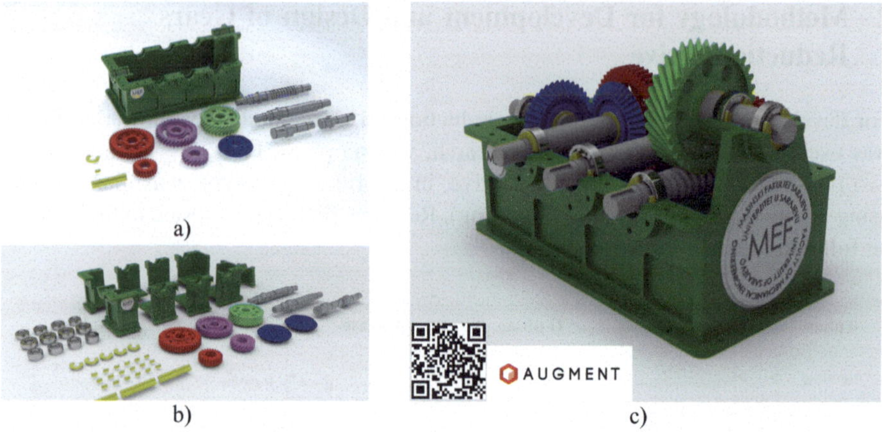

Fig. 2. 3D geometrical models of gears reduction drive: (a) individual parts, (b) redesigned parts for additive manufacturing, (c) assembly which can be shown in 3D environment using augmented reality apps.

2.3 Simulation of Working Principles of Gear Reduction Drive

Simulation of working principles of gear reduction drive is carried out using Digital Mockup (DMU Kinematics) module of CATIA software system.

3 3D Printing and Augmented Reality

Parts of gears reduction drive, which was developed in SolidWorks and CATIA systems are manufactured using Ultimaker 3 additive manufacturing machine and assembled in real size working prototype (Fig. 3). All parts of gear reduction drive (gears, pins, Seger rings, bearings and housing for electric motor) are manufactured using PLA (polylactic acid) material. Support material PVA (polyvinyl alcohol) is used where it was necessary.

Ultimaker 3 additive manufacturing machine have size limitation for manufactured parts in amount of 215 × 215 × 200 mm and gears reduction drive housing have length 640 mm and height of 300 mm. Because of that housing shafts were divided in more parts and then manufactured. After manufacturing, these parts were combined together using additionally manufactured inside pins for connections using strong glue [10–12].

3D models of individual parts and whole assembly of gears reduction drive are integrated in Android and iOS AR (Augmented Reality) app AUGMENT (Fig. 3b and Fig. 2c). Using this app user can show 3D model on his mobile or tablet in augmented reality. Mobile or tablet must have internet connection.

Figure 4a shows one gear of gears reduction drive and Fig. 4b shows whole assembly in augmented reality. Uploads of these models can be carried out only using

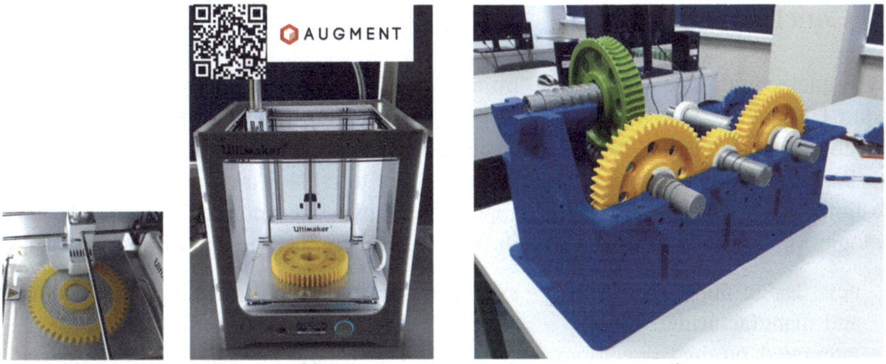

a) b) c)

Fig. 3. Additive manufacturing of real size fully functional prototype of gears reduction drive (Laboratory for Computer Aided Design - LabCAD, Faculty of Mechanical Engineering Sarajevo) [2]

AUGMENT–3D Augmented Reality application, which can be downloaded and installed from Google Play Store (Android) or App Store (iOS). After installation of application, it is necessary to scan QR code shown on Fig. 3b and Fig. 2c using mobile phone or tablet computer [2, 13].

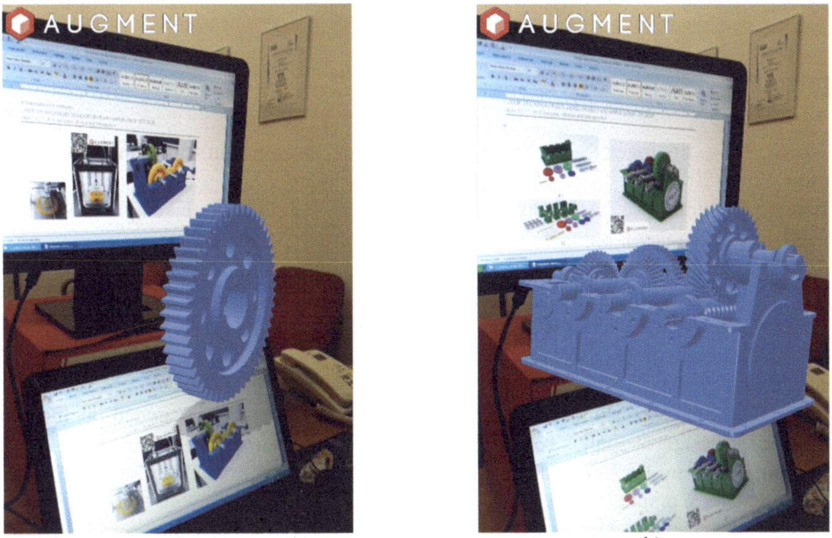

a) b)

Fig. 4. 3D model of gear (a) and whole assembly (b) shown on user mobile phone in augmented reality application

4 Conclusions

By implementation of the project "Improvement of teaching 'How machines works' using Ultimaker 3D printer", described in introduction of this paper, conditions for research about integrated development and design of gears reduction drive are fulfilled.

Paper want to demonstrate why is it necessary to replace traditional (conventional) design approach with new design approaches which incorporates modern 3D technologies and methodologies. Through this research following conclusions are made:

- new design approaches enables systematic design which integrate product design and manufacturing,
- integrated product development and design of a new product (in this case gears reduction drive) directly influences to the shortening of the duration time needed for development and design of a new product.
- parts manufactured this way (using polymers) are lightweight and it is easy to move them around, which is very important in the area of education. Real metal parts are heavy and because of that professors cannot carry them to the classrooms.
- using this types of manufactured prototypes it is easy to explain working principles of gears and gears reduction drives to the students. Education of students is significantly improved in comparison to the standard education practice which include drawings, pictures and PowerPoint presentations.

Additive manufacturing enables easy manufacturing of various real size machines and parts. These machines and parts manufactured on this way can be used to improve education processes at faculties and engineering schools.

References

1. Ulrich, K.T., Eppinger, S.D.: Product Design and Development, 5th edn. McGraw-Hill, New York (2012)
2. Saric, I., Muminovic, A.J.: Product Development and Design. Faculty of Mechanical Engineering Sarajevo, Sarajevo (2018)
3. Karayel, D., Ozkan, S.S., Vatansever, F.: Integrated knowledge-based system for machine design. Adv. Mech. Eng. **5**, 1–10 (2013)
4. Saric, I., Muminovic, A., Colic, M., Rahimic, S.: Development of integrated intelligent computer-aided design system for mechanical power-transmitting mechanism design. Adv. Mech. Eng. **9**(7), 1–16 (2017)
5. Saric, I., Pervan, N., Muminovic, A., Colic, M.: Development of integrated intelligent cad system for design of shafts. Tehnički Vjesnik **25**(Supplement 1), 99–104 (2018)
6. Bodein, Y., Rose, B., Caillaud, E.: Explicit reference modeling methodology in parametric CAD system. Comput. Ind. **65**(1), 136–147 (2014)
7. Saric, I., Repcic, N., Muminovic, A.: Mechanical transmissions parameter modelling. In Gokcek, M. (ed.) Mechanical Engineering, pp. 3–16 (2012). InTech
8. Saric, I., Repcic, N.: Parameter Modelling of Gears. TECH TECHNOL EDUC MANAG-TTEM - Technics Technologies Education Management-Ttem **7**(3), 968–974 (2012)

9. Saric, I., Muminovic, A.J.: Development of integrated intelligent CAD system for synthesis and stress-deformation analysis of pressure vessels. Int. J. Eng. Technol. **7**(1), 147–151 (2018)
10. Camburn, B., Dunlap, B., Gurjar, T., et al.: A systematic method for design prototyping. J. Mech. Des. **137**(8), 081102 (2015)
11. Gibson, I., Rosen, D.W., Stucker, B.: Additive Manufacturing Technologies: Rapid Prototyping to Direct Digital Manufacturing. Springer, Boston (2010)
12. Muminovic, A.J., Muminovic, A., Mesic, E., Saric, I., Pervan, N.: Spur gear tooth topology optimization: finding optimal shell thickness for spur gear tooth produced using additive manufacturing. TEM J. **8**(3), 788–794 (2019)
13. Frizziero, L., Liverani, A., Caligiana, G., Donnici, G., Chinaglia, L.: Design for disassembly (DfD) and augmented reality (AR): case study applied to a gearbox. Machines **7**(2), 1–26 (2019)

Development of Linear Servo Hydraulic Drive for Material Testing

Vito Tič(✉) and Darko Lovrec(✉)

Faculty of Mechanical Engineering, University of Maribor,
Smetanova 17, 2000 Maribor, Slovenia
{vito.tic,darko.lovrec}@um.si

Abstract. Material testing devices are used in laboratory and industrial environments for testing and research in fracture mechanics and are typically operated by servo-hydraulic or servo-pneumatic actuators. They contain an increasing number of electronic and microprocessor-controlled components, with which they achieve appropriate dynamics and the ability to store and process signals.

The paper presents the development, implementation and operation of a flexible device for testing the dynamic strength of materials based on linear electro-hydraulic servo drive with closed loop force control. To control the components, a Beckhoff multi-core controller is used, which simultaneously runs a control program with real time force control, as well as a human-machine interface in the Windows environment. The presented device is capable of achieving forces up to 40 kN at test speeds up to 20 Hz.

Keywords: Servo · Hydraulic · Linear drive · Material testing

1 Introduction

The development of new materials requires an understanding of their mechanical properties, one of the most important being their fatigue behaviour. The main purpose of such tests is to ensure the safety of the components and devices when used [1–3]. Bearing in mind that theoretical equations do not always correspond to the material's real behaviour, determining the durability of materials by experimental procedures is all the more important [4]. In order to investigate the main causes of material fraction (failure of engineering component) subjected to uniaxial and/or multi-axial loading, mechanical fatigue tests must be carried out in different regimes and in a wide variety of materials [5].

Several specialized devices have been identified in the literature that authors have designed and built for specific materials testing applications [6–10]. In the field of fatigue testing machines, however, the market is saturated with expensive dedicated test devices that can operate in very different frequency regimes [10]. Recent research is focused on the development of devices that achieve material fatigue through resonant systems or actuators [11, 12]. The design and implementation of a self-designed device for testing the dynamic strength of materials based on linear electro-hydraulic servo and closed-loop force control will be presented below.

I. Karabegović (Ed.): NT 2020, LNNS 128, pp. 104–113, 2020.
https://doi.org/10.1007/978-3-030-46817-0_12

Electro-hydraulic servo systems are servomechanisms that allow movement of heavy loads with high inertia, with high response speed and with very precise control. This is made possible by electro-hydraulic servo valves, which have been developed for a very large increase in power. Thus, with a weak electrical signal, high hydraulic power can be controlled very quickly and precisely.

Microprocessor systems must be used to operate servo valves. The Programmable Logic Controllers (PLCs) are well-suited for various automation tasks in the industry, so they are widely used. The biggest advantages of PLCs are modular construction, programmability, easy maintenance, resistance to environmental influences, etc.

2 Device Design

The device for testing the dynamic strength of materials is controlled by a servo valve in closed loop force control. Each control system consists of four basic elements: a closed loop control element, a measuring element, an actuating element and a control object. In our case, the control element is Beckhoff's industrial PLC, the measuring element is a force sensor, the actuator is a servo valve and the control object is a hydraulic cylinder.

2.1 Device Construction

The construction of the device shown in Fig. 1 is made of high quality materials and adapted to pre-fabricated hydraulic cylinders and a support table [13]. The design is also adapted to perform highly dynamic tests with closed loop force control, with both amplitude and frequency adjustable.

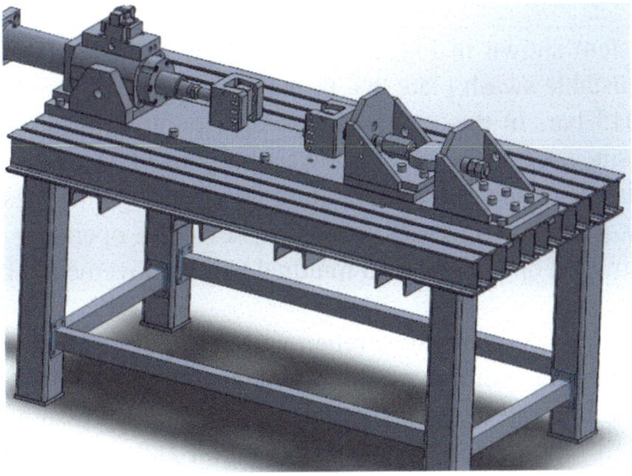

Fig. 1. Design of device construction in SolidWorks software environment

The design of the device itself allows for compressive, tensile and compressive-tensile static and dynamic tests, whereas it is necessary to properly design and construct the clamping device for mounting the test piece. Due to further needs of research in the field of fatigue material tests, we initially limited ourselves to tensile tests, which will be performed later on aluminum alloys.

On one side of the device is a hydraulic cylinder having a dedicated wedge clamp mounted on the piston rod, while the other wedge clamp is mounted rigidly on a welded bracket via a force measuring cell. However, the distance between the fixtures can be adjusted by adjusting the welded supports by screws. Particular attention was given to the construction of wedge-shaped self-tightening clamps containing movable jaws with ribbed surface for better grip. The fixtures are made of 115CrV3 (DIN) steel, with the fixture being tempered to 850 ± 20 MPa and the jaws have a hardness of 52 ± 2 HRC. As mentioned earlier, the force sensor is mounted on the opposite side of the hydraulic cylinder (Fig. 2), while the displacement sensor is in the cylinder extension itself, as shown in Fig. 4.

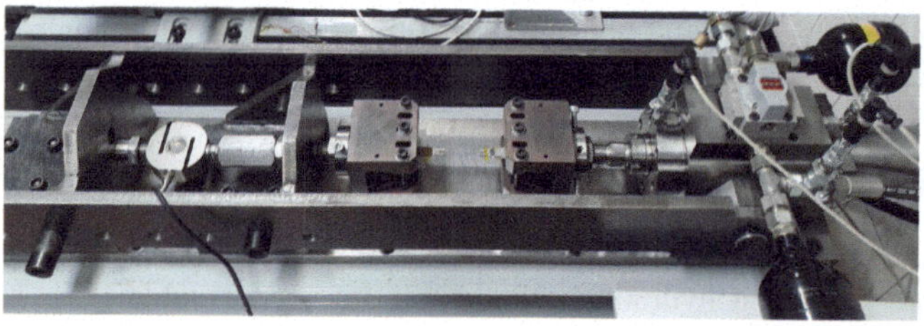

Fig. 2. Designed device for testing the dynamic strength of materials

3 Hydraulic System

The hydraulic system shown in Fig. 3 is powered by a Parker PAV 32 axial piston pump with an adjustable swash plate that provides up to 45 L/min of flow at operating pressures up to 315 bar. In this case, the use of a pump with an adjustable flow is almost essential, since it is necessary to lower the pump flow for each test to such a degree that a minimal amount of hydraulic fluid flows through the pressure relief valve. This is especially important in cases when the device is operating in high-pressure mode (high dynamics of operation) with minimal piston movements (the test pieces are usually very rigid).

In addition to the safety pressure relief valve, the electrically controlled Bosch Rexroth DBEM pressure relief valve is used to adjust the working pressure, which can be used to programmatically adjust the desired operating pressure or to operate in bypass mode. Due to the high dynamics of operation and the associated high fluctuations in pressure, hydraulic accumulators were installed on both, the pressure and the return line, in the immediate vicinity of the servo valves. The hydraulic accumulators were carefully calculated and selected and then properly charged with nitrogen to the optimum pressure [14].

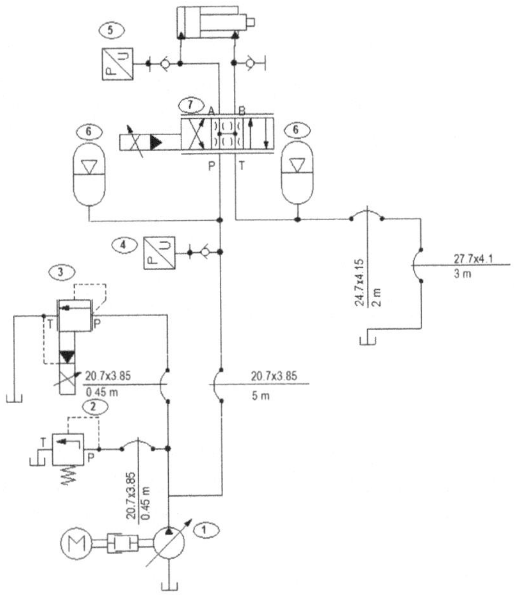

Fig. 3. Schematic diagram of the hydraulic system of the device

A two-stage hydraulic servo valve MOOG G-761, consisting of an electric torque motor and two stages of hydraulic amplification, is used to control the hydraulic cylinder. The main actuator used is a double acting cylinder with double-sided piston, shown in Fig. 4. Compared to the single-sided cylinder the speed of movement and the force in both directions are the same, due to the same piston surfaces in both chambers.

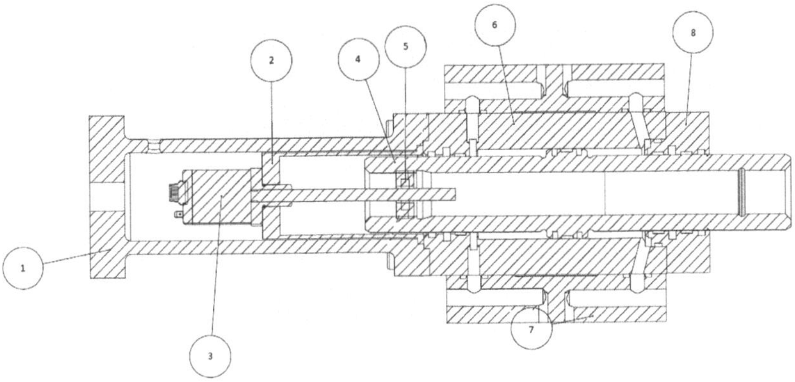

1	Holder	5	Magnetic ring
2	Sensor protection	6	Cylinder
3	Piston position sensor	7	Valve plate
4	Piston	8	Cylinder head

Fig. 4. Hydraulic cylinder with integrated displacement and speed sensor

Also, such an implementation has certain other advantages, such as: better guided piston rod, sensors can be mounted on either side,… In this case, too, the displacement sensor was mounted in a special housing behind the cylinder, thus protecting it against external influences.

When installing the hydraulic cylinder, it is necessary to ensure a sufficiently high rigidity of the system, so that the cylinder is secured from the bottom and from both sides. The cylinder also has connecting lines on the underside for an additional valve, which could double the inlet and outlet flows of hydraulic fluid (consequently the travel speed), but in current configuration we use only one servo valve, so these lines were properly sealed.

4 Control System

The control system is based on the Beckhoff CX5140 quad-core industrial PLC, which runs TwinCAT 3 (a real-time PLC part) and Windows 7 Embedded Standard simultaneously. The key advantage of this kind of computer controller over conventional PLCs is that a single controller can perform very fast operations, such as PID control (which in our case is implemented in about 20 µs), as well as executes user interface program simultaneously. Despite the controller's extraordinary processing capacity, the control cycle is limited by the speed of the input/output modules, which in our case had an average analog/digital and digital/analog conversion time of about 0.4 ms. Therefore, we set the main PLC program clock to a period of 0.1 ms (Fig. 5).

Fig. 5. Beckhoff CX5140 controller with I/O modules

A special feature of reading data from sensors is the connection of a force measuring cell, which is connected directly to the EL3356-0010 input module. In this way, we greatly improved the quality and frequency of the captured force signal. The module also allows very easy calibration of the force cell with the control of the pressure in the cylinder chamber and calculation of the force applied to the cell. Namely: over time and with use, the properties of the force measuring cell and its signal can change slightly (signal drift).

4.1 Control Program (PLC)

The PLC control program created in the TwinCAT 3 programming environment was reasonably divided into several subroutines. If we focus on the main part of the subroutine for the implementation of closed loop force control, this is done in the following order [15]:

The first function (*Parameter input*) defines the initial settings of the variable values. Here we set the values of the variables that we want to set only when the program starts.

In the second function (*Init*), we initialize the parameters for the sine wave generator and the PID controller and count the cycles. Cycle counting is performed by observing the generated sine wave. It is performed at each positive pass through the mean of the sine wave (virtual zero).

The following function (*Calculation*) performs value calculations. The frequency of the sine period is calculated from the entered frequency on the control panel, the PID controller links are switched on and off, and the force is calculated.

The following two functions contain signal generator and PID controller blocks, which are standard blocks of the TwinCAT 3 Controller ToolBox. The inputs and outputs of the blocks are linked to variables (Fig. 6), whose parameters are set on the control panel.

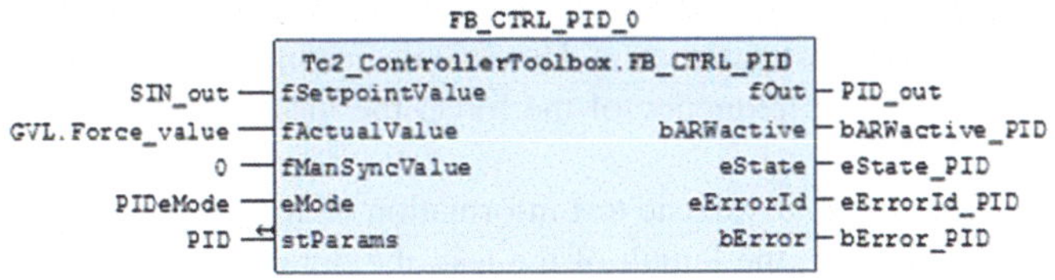

Fig. 6. Control program – PID block

The following *Output function* scales the value from the output of the PID controller to a value suitable for controlling the servos. Before passing the value to the output, the condition that the sine wave generator is active must be fulfilled. This prevents the controller from operating the servo valves when the sine generator does not operate.

The following function (*Variables update*) converts the actual and desired force values into a format suitable for display on chart. It also calculates the minimum and maximum values of position, force and pressure in the current cycle. When a new cycle occurs, the values of the variables are reset, and the extreme values are transferred to the control panel (HMI) where they are displayed and recorded as needed.

The last function (*Fail control*) checks if the piston position value is less than 5 mm. This is only possible if the test specimen has broken and the piston has moved to the extreme inner position. This is when the test conclusion procedure is initiated.

When the controller executes the last function, it returns to the *Init* function and resumes executing the program. We return to the *Init* function because this allows us to change the parameters of the signal generator and the PID controller even during the test, which is desirable under certain conditions.

4.2 User Interface (HMI)

Control panel or the user interface is designed in the Visual Studio C# environment, with which we create a classic Windows application, which gives us a lot of freedom in programming and upgrading the system. An additional advantage of using the C# environment to create an HMI interface is the ability to use *ScopeView*, which allows us to draw charts of our variables and then save the recordings. In this case, communication between the real-time part (PLC program) and the user interface is via the TwinCAT ADS protocol, which allows the variables to be read and written in the PLC part, thus influencing the implementation of the real-time control program.

The designed user interface is shown in Fig. 7 and is implemented in conjunction with a real-time PLC program on a Windows Beckhoff industrial controller. Although the interface looks a bit complex, it is quite easy to handle. On the left side it offers pump on/off and setting of the working pressure, which can also be changed during the test. Next to this is the *Regulator Parameters* group box with PID controller settings, with the recommended values already entered. It is also possible to switch the integral and/or differential part of the controller on or off. The *Test Parameters* group box allows us to define basic test parameters. Here we determine the length of the test (with the number of cycles), the frequency of the force, the mean value of the force and its amplitude.

In the *Test Data* group box, basic test information is displayed during the test, such as the start time of the test, the length of the test, the duration of the test and the time until the test ends.

The other part of the user interface is represented by elements that allow you to see the test as it runs. So we have a graph (*ScopeView*) that draws the desired and actual force value at the rate of the basic PLC cycle of 0.1 ms. In this way, we have the possibility of automated periodic storage of material fatigue data, which is essential for understanding fracture mechanics research. At the top right is another *Test Data* group box that shows the lowest and highest values of position, force and pressure in each cycle.

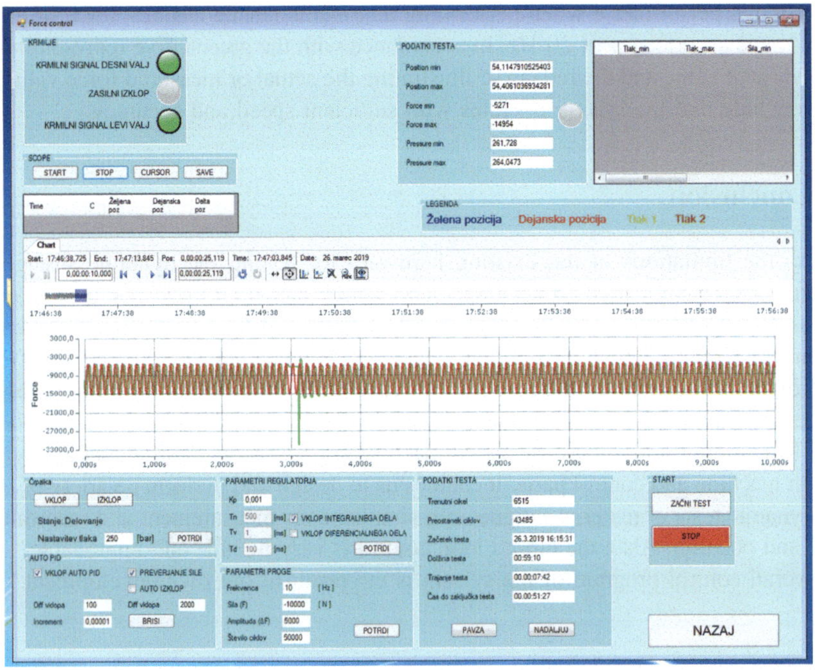

Fig. 7. The user interface that runs on Windows

5 Device Operation – Results

After successfully starting the device for the first time, we began real material fatigue testing. During the tests, the working pressure as well as the frequency and force of fatigue were gradually increased. In doing so, we manually adjusted the parameters of the PID controller to achieve the desired system response. Figure 8 shows a snapshot of

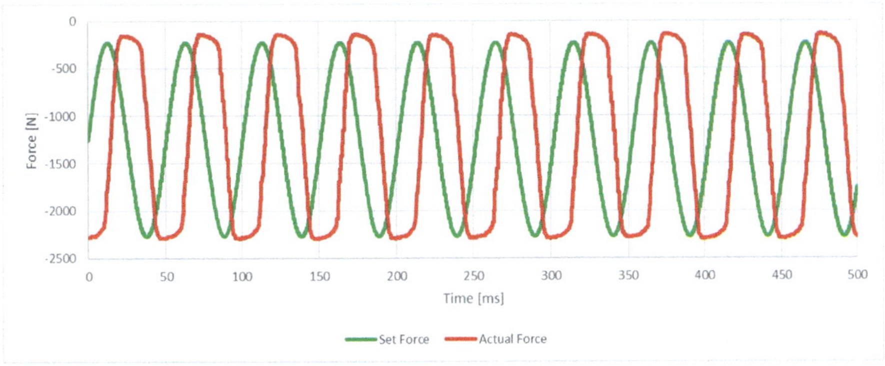

Fig. 8. Snapshot of device operation at 20 Hz frequency

the system's performance when performing a dynamic tensile test at a force of 250 to 2250 N and a frequency of 20 Hz. In accordance with the green curve representing the desired force value and the red curve illustrating the actual or measured force value, we can conclude that the system operates with sufficient speed and accuracy.

6 Conclusion

Due to the limitations of the existing hardware for testing the dynamic strength of materials, we have started to develop, manufacture and implement our own material fatigue testing machine, which is based on an advanced high-performance hydraulic system. The self-designed system has many advantages over the existing INSTRON 1255 dynamic material fatigue machine, because it allows us to test materials at higher frequencies, up to 20 Hz, which means that individual tests are performed in a shorter time. More importantly, a self-designed machine is much more flexible and upgradable in both machine and control parts. It enables us to design and perform a variety of static and dynamic tests of material fatigue with simultaneous measurement and recording of force and position. Also, the closed loop control is very flexible and upgradable, since the overall control program of the system is the result of our own development.

References

1. Pao, P.S., Gill, S.J., Feng, C.R.: On fatigue crack initiation from corrosion pits in 7075-T7351 aluminum alloy. Scripta Mater. **43**(5), 391–396 (2000)
2. Weißgraeber, P., Felger, J., Geipel, D., Becker, W.: Cracks at elliptical holes: stress intensity factor and finite fracture mechanics solution. Eur. J. Mech. A/Solids **55**, 192–198 (2016)
3. DuQuesnay, D.L., Underhill, P.R., Britt, H.J.: Fatigue crack growth from corrosion damage in 7075-T6511 aluminium alloy under aircraft loading. Int. J. Fatigue **25**(5), 371–377 (2003)
4. Cui, W.: A state-of-the-art review on fatigue life prediction methods for metal structures. J. Mar. Sci. Technol. **7**(1), 43–56 (2002)
5. Doquet, V., De Greef, V.: Dwell-fatigue of a titanium alloy at room temperature under uniaxial or biaxial tension. Int. J. Fatigue **38**(2012), 118–129 (2011)
6. Marcelo, F., Bustos, P.: Design and construction of a torsional fatigue testing machine operated by inertial load. Dyna **172**, 46–55 (2012)
7. Petersen, D., Link, R., Fletcher, D., Beynon, J.: Development of a machine for closely controlled rolling contact Fatigue and wear testing. J. Test. Eval. **28**(4), 267–275 (2000)
8. Feng, M., Li, M.: Development of a computerized electrodynamic resonant fatigue test machine and its applications to automotive components. In: SAE Technical Paper Series (2010)
9. Bathias, C.: Piezoelectric fatigue testing machines and devices. Int. J. Fatigue **28**(11), 1438–1445 (2006)
10. Ambriz, J.L.A., Almaraz, G.M.D., Julio, C., Juarez, V., Gomez, E.C., Zuñiga, I.F.: Design and construction of a torsion fatigue machine : torsion fatigue tests on two industrial aluminum alloys, vol. 79, (2017). ISSN 1454-2358
11. Zuñiga, G.M., Ishvari, T., Dominguez Almaraz, F., Guzman Tapia, M., Avila Ambriz, J.L.: Controlled Pre-Corrosion Attack and Ultrasonic Fatigue Endurance of Titanium Alloy Ti – 6Al – 4V. Lat. Am. J. Solids Struct. **14**(3), 512–527 (2017)

12. Ouarabi, M., Mora, R.P., Bathias, C.: Very high cycle fatigue strength and crack growth of thin steel sheets, vol. 36, pp. 112–118 (2016)
13. Štefane, M.: Implementacija in prvi zagon elektrohidravlične linearne servoosi. Univerza v Mariboru, Fakulteta za strojništvo (2017)
14. Jurgec, T.: Vpliv hidravličnega cevovoda na delovanje linearne hidravlične servoosi. Univerza v Mariboru, Fakulteta za strojništvo (2017)
15. Jerebic, L.: Regulacija sile na linearni hidravlični servoosi. Univerza v Mariboru, Fakulteta za strojništvo (2018)

Structural Analysis of Direct Passive Pressure Reducing Valves Using Modified Kinematic Graphs

Yiheng Zhang[1(✉)], Ihor Sydorenko[1(✉)], Volodymyr Tonkonogyi[1(✉)], Liubov Bovnegra[1(✉)], and Predrag Dašić[2(✉)]

[1] Odessa National Polytechnic University (ONPU), Shevchenko Ave 1, Odessa 65044, Ukraine
iheng_1101@163.com, igs.ods@gmail.com, {vmt47,dlv5}@ukr.net

[2] High Technical Mechanical School of Professional Studies, Str. Radoja Krstica, 19, 37240 Trstenik, Serbia
dasicp58@gmail.com

Abstract. Structural analysis is the primary stage in the design of any mechanical system, which completely determines the effectiveness of its application in various fields of technology. One of the most effective tools for solving this problem is graph theory, which provides a convenient device for modeling structural properties of systems and relationships between objects of diverse nature. In this paper has conducted research related to testing the theory of modified kinematic graphs for the analysis of existing designs of direct pressure reducing valves with manual control. The methodology of the transition from the actual design of the pressure reducing valve to its model in the form of a modified kinematic graph is shown. Based on the equilibrium equation for the device in question, which determines its operability, the adequacy of the obtained model was verified. It is shown that the specificity of the object under consideration requires adjustment of its model, as a result of which it is rational to switch from the adjacency matrix to its extended version, which is the assembly matrix. The parameters that determine the structure of the simulated device are highlighted. One of the criteria for choosing the optimal solution for the synthesis of new direct pressure reducing valves using the theory presented is indicated.

Keywords: Pressure reducing valve · Mechanical structure · Structural analysis · Kinematic graph · Adjacency matrix

1 Introduction

Structural analysis is the primary stage in the design of any mechanical system, which completely determines the effectiveness of its application in various fields of technology [1].

Modern methods of structural analysis of technical solutions and systems are based on a morphological approach, the main features of which can be considered:

I. Karabegović (Ed.): NT 2020, LNNS 128, pp. 114–121, 2020.
https://doi.org/10.1007/978-3-030-46817-0_13

- genetic approach to the classification, description, evolution of development and prediction of various kinds of mechanical products as complex technical systems of natural - anthropogenic origin;
- multilevel morphological approach to the construction and synthesis of complex technical systems;
- aggregate - modular principle of the layout of technical systems;
- the use of advanced information technology and intelligent computer systems.

By analyzing the structures of controlled technical devices to identify elements of the same purpose and, as a consequence, the same properties, we can distinguish elements whose functional purpose is determined only by the specifics of the design. In this case, an additional study of these elements is required from the point of view of their functional expediency and possible substitution or association with existing structural elements. One of the effective tools for solving this problem is graph theory, which provides a convenient device for modeling structural properties of systems and relationships between objects of diverse nature. Graphs are widely used as structural models of physical systems that admit an idealized representation in the form of circuits with lumped components [2].

2 Theoretical Research

The purpose of the study is approbation the theory of modified kinematic graphs to analyze existing designs of direct-acting pressure reducing valves with manual control, determination of parameters determining their structural differences, selection and assessment of the criteria for choosing the optimal solution, which will allow you to choose the optimal solution from the many possible.

2.1 Literature Review

One of the effective tools for solving a wide range of scientific problems is graph theory, which provides a convenient device for modeling the structural properties of systems of almost any kind and modeling structural relations between objects of various nature [3, 4]. The existing experience of using graph models in various scientific fields indicates that the graph is a very flexible logical system [5, 6].

Graphs are widely used as structural models of physical systems that admit an idealized representation in the form of circuits with lumped components [7, 8]. The use of graphs as the basis of the apparatus for the analysis and synthesis of mechanical devices is a fairly common scientific approach [9–11].

The use of graphs allows one to fairly accurately describe almost any technical system, the functional interaction of its constituent elements, as well as the processes occurring in it [12, 13]. Some scientific papers indicate that the specifics of problems solved with the help of graphs can also determine their own specifics [14].

2.2 Research Methodology

The studies are based on the theory and the corresponding methodology for the analysis and synthesis of elastic systems based on modified kinematic graphs, which is a development of the theory of synthesis of mechanisms based on kinematic graphs [15, 16].

The results of studies related to the approbation of the theory presented above made it possible to formulate the basic principles of structural analysis and synthesis of such elastic systems [16].

According to one of the principles of the theory used, an idealized representation of the structure of an ordinary planar passive device, with one parameter that is subject to controlled change, is a modified kinematic graph (not bipolar).

The modification of the graph is that an additional "marked" edge is introduced into the kinematic graph that corresponds to a certain parameter (elasticity, dispersion, etc.), the Change of which is directly related to the kinematics, and sometimes determines this.

In addition, from the point of view of kinematics, this parameter is set by a virtual analogue of the kinematic pair of the 4th class.

The degree of mobility of the obtained model, in the form of a graph, based on a flat task, should characterize the zero mobility of the device in question by the following equation:

$$W = 3(p - 1) - 2q_5 - p_4 - p_M < 0 \tag{1}$$

where p is the total number of graph vertices corresponding to the number of hard parts of the device; q_i is the number of edges of the graph equal to the number of kinematic pairs of the i-th class; q_M is the number of "marked" edges of the graph, equal to the number of non-kinematic bonds (elastic, dissipative, etc.) between the hard parts of the device.

According to another principle announced in the theory used, the nature of the kinematic control of the properties of the device when it is modeled using a modified kinematic graph can be judged using an indicator in the form of a cyclomatic number:

$$\sigma = q - p + 1 \geq 2 \tag{2}$$

where q is the number of edges of the graph, regardless of their type; p is the number of vertices of the graph, regardless of their type.

In addition, the theory used indicates that more detailed information about the mechanism to be analyzed or synthesized is provided by the assembly matrix, which is a more detailed analogue of the adjacency matrix.

Considering that any of the existing passive pressure-reducing valves with adjustable pressure can be represented as a flat passive device with one parameter subject to controlled change, in this paper we approbation the presented theory, in the form of a solution to a new problem - analysis structures of hydraulic equipment.

3 Results and Discussion

Let us analyze the basic device in the form of a direct passive pressure reducing valve RD-120 (Fig. 1a).

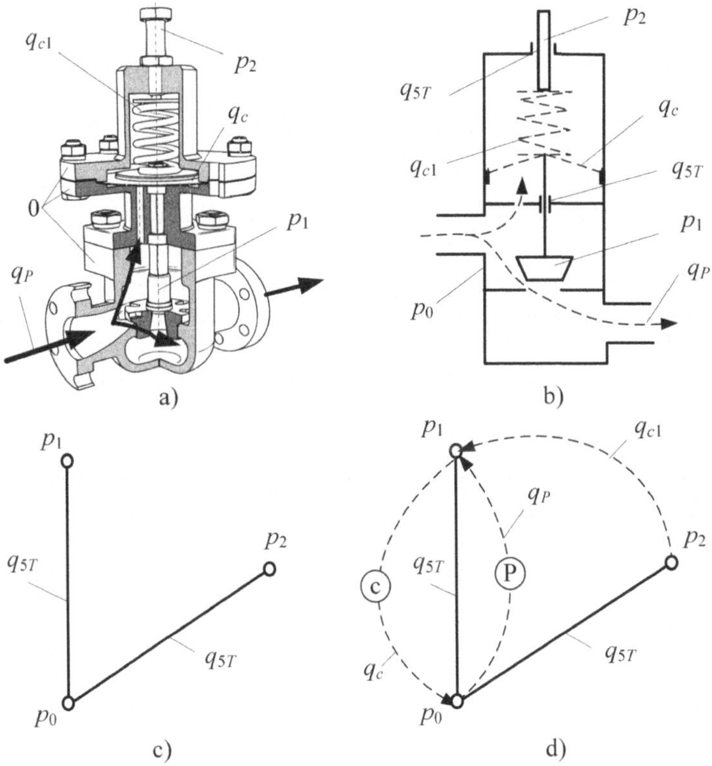

Fig. 1. Simulation of a direct pressure reducing valve: simulated device (a); kinematic scheme (b); model in the form of a kinematic graph (c); model in the form of a modified kinematic graph (d)

At the initial stage, it is advisable to switch from the actual design of the device to its kinematic scheme (Fig. 1b). Such a transition makes it quite simple to obtain a model of the basic device in the form of a kinematic graph (Fig. 1c).

The analysis was carried out under the condition that: the mechanism under consideration is flat; pi is the vertex of the graph from their total number p, which determines the ith element of the device with the corresponding numbering, q5 is the edge of the graph, which determines the functional contact interaction between the two elements of the device, and corresponds to the kinematic pair of the 5th class. For a more detailed structuring of the model, the possible types of this edge of the graph corresponding to a rotational or translational kinematic pair are accepted (q_{5R} and q_{5T}, respectively). With this in mind, the analogue of the adjacency matrix for the obtained kinematic graph is called the assembly matrix, has the form:

$$M = \begin{array}{c|c|c|c} & p_0 & p_1 & p_2 \\ \hline p_0 & 0 & q_{5T} & q_{5T} \\ \hline p_1 & q_{5T} & 0 & 0 \\ \hline p_2 & q_{5T} & 0 & 0 \end{array} \qquad (3)$$

The characteristic model of the basic device in the form of a kinematic graph (see Fig. 1b) in terms of mobility:

$$W = 3(p - 1) - 2q_{5(R,T)} - q_4 = 3(3 - 1) - 2 \cdot 2 - 0 = 2. \qquad (4)$$

This result indicates that the model under consideration in terms of structure defines a mechanism with excessive mobility. The element is an adjusting screw, which corresponds to the top of the p_2 graph in the model, acts as excess mobility.

According to the definition of excess mobility, this element does not affect the kinematics of the mechanism, but is introduced into its structure for constructive reasons [1, 9].

In this case, the adjusting screw serves to change a certain kinematic parameter (in the form of displacement), which determines the analyzer rigidity (additional bending of the membrane) by compressing the spring in contact with it. The nature of the kinematic control of the device, when it is modeled by a kinematic graph, can be judged by the indicator in the form of a cyclomatic number. Cyclomatic base unit number:

$$\sigma = q - p + 1 = 2 - 3 + 1 = 0 \qquad (5)$$

This result indicates the absence of any cycles in the kinematic graph, which is typical for mechanical devices with the complete absence of a kinematic control system.

From the analysis above, it follows that the kinematic graph does not allow us to track the relationship between the kinematics of the device and the change in its other than mechanical characteristics, since it is simply not intended for this. However, the advantage of the graph lies in the fact that it allows you to combine physically related parameters of various nature, for example, "displacement - force" for elastic systems. It is this circumstance that underlies the theory of modification of the kinematic graph under consideration.

Modification of the kinematic graph defining the basic device was performed by introducing additional "marked" edges into its composition: q_P — defining the controlled parameter in the form of pressure; q_c - determining the rigidity of the analyzer - membrane; q_{c1} - determining the stiffness of the adjustment element - springs (see Fig. 1d). Then the characteristics of the basic model of the device in the form of a modified kinematic graph in terms of a cyclomatic number:

$$\sigma = q - p + 1 = 5 - 3 + 1 = 3 \qquad (6)$$

which determines the presence of three cycles in the resulting graph.

Consider the cycles in the obtained modified kinematic graph, which must correspond to certain functional interactions of the elements of the simulated device, as well as have adequacy to the equilibrium equation that determines the operability of the device in question. We write the equation of equilibrium in the following form:

$$PS - cx = c_1(x_{1T} + x_1) \quad (7)$$

where P is the pressure acting on the membrane; S is the membrane area; c is the rigidity of the membrane; c_1 - spring stiffness; x is the deformation of the membrane under the action of the pressure force of the medium; x_{1p} - preliminary compression of the spring when installing in the device; x_1 - adjusting compression of the spring when adjusting the device.

Considering that the order of the vertices for the edges qP and qc is different, the existing pair of cycles with a common edge q_{5T} (p_0 - q_{5T} - p_1 - q_c and p_0 - q_{5T} - p_1 - q_P), which the digraph defines [2]. This digraph shows that the pressure change (rib q_P) causing translational movement (rib q_{5T}) of the moving rod (vertex p_1) on one side will be compensated by the elastic resistance of the membrane (vertex p_c) on the rod (vertex p_1), the value of which is determined by the same translational displacement (rib q_{5T}). Therefore, the presented part of the modified kinematic graph describes the main functional interactions between the elements of the basic device, which are major and adequate to the left side of Eq. (7). The cycle p_0 - q_{5T} - p_2 - q_{c1} - p_1 - q_{5T} in this case shows the functional interaction between the main (vertices p_0, p_1) and additional (vertex p_2) elements of the basic device, which determines the change in its properties and is adequate to the right side of Eq. (7).

This result suggests that the modified kinematic graph of a device related to hydraulic elements with variable characteristics should have at least three cycles, two of which determine the main functional interaction of the device elements, and the remaining one determines their change.

It should be noted that the calculation of the degree of mobility of the model in the form of a modified kinematic graph with initially accepted assumptions:

$$W = 3(p - 1) - 2q_{5(RT)} - q_4 - q_M = 3(3 - 1) - 2 \cdot 2 - 0 - 3 = -1 \quad (8)$$

is not entirely true and needs to be adjusted. To do this, we use the assembly matrix of this model:

$M =$ (9)

	p_0	p_1	p_2
p_0	0	q_{5T}, q_P, q_c	q_{5T}
p_1	q_{5T}, q_P, q_c	0	q_c
p_2	q_{5T}	q_c	0

It is easy to see that in the adjacency between the poles (vertices) p_0 and p_1 there are edges q_{5T} - defining a pair of the 5th class and q_c, q_P - defining virtual pairs of the 4th class. Obviously, in such a neighborhood, the number of mobilities can only be determined by a kinematic pair of the 5th class. Based on this, the calculation of the degree of mobility should be performed as follows:

$$W = 3(p - 1) - 2q_{5(RT)} - q_4 - q_M = 3(3 - 1) - 2 \cdot 2 - 0 - 1 = 1 \quad (10)$$

The result obtained by expression (10) shows that the system will be self-governing. In this system, a change in the controlled parameter that causes a certain movement leads to the kinematic interaction of all elements of the device, without exception.

4 Conclusion

Based on the results of the studies established the following:

(1) Verification of the accepted theory and the corresponding methodology for the analysis and synthesis of elastic systems based on modified kinematic graphs in relation to the analysis of hydraulic equipment elements should be considered successful, since the results adequately reflect the device considered in the work, the functional interaction between its constituent elements, as well as those occurring in it processes.

(2) The direct passive pressure reducing valve RD-120, considered in the section, is self-controlled, since it is established that a change in the controlled parameter that causes a certain movement leads to the kinematic interaction of all elements of the device, without exception.

(3) It has been established that the modified kinematic graph of a device related to hydraulic equipment elements with variable characteristics must have at least three cycles, two of which determine the main functional interaction of the device elements, and the remaining one determines their change.

(4) The calculation of the degree of mobility of the model in the form of a modified kinematic graph, with the initially accepted assumptions, needs to be adjusted based on the assembly matrix of this model.

The results obtained make it possible to further use the technique presented in the work for the synthesis of new self-regulating direct-acting pressure reducing valves. Given that the solution to the synthesis problem usually has many options, then to find the optimal solution, you can use the efficiency indicator in the form of a graph of energy.

References

1. Kraynev, A.F.: Mechanics (The Art of Building) Machines. Fundamental Dictionary Engineering, Moscow (2014). (in Russian)
2. Yan, H.-S., Hung, C.-C.: Identifying and counting the number of mechanisms from kinematic chains subject to design constraints. ASME J. Mech. Des. **128**(5), 1177–1182 (2006)
3. Soni, A.H.: Structural analysis of two general constraint kinematic chains and their practical application. J. Eng. Ind. **93**(1), 231–238 (1971)

4. Bendsoe, M.P., Sigmund, O.: Topology Optimization: Theory, Methods, and Applications. Springer, Berlin (2013)
5. Dao, T.-S., McPhee, J.: Dynamic modeling of electrochemical systems using linear graph theory. J. Power Sources **196**(23), 10442–10454 (2011)
6. Chiu, T.: An improved algorithm for the construction of generalized kinematic chains. Mech. Mach. Theory **78**, 229–247 (2014)
7. Yang, F., Deng, Z., Tao, J., Li, L.: A new method for isomorphism identification in topological graphs using incident matrices. Mech. Mach. Theory **49**, 298–307 (2012)
8. Ding, H., Zi, B., Huang, P., Kecskeméthy, A.: The whole family of kinematic structures for planar 2- and 3-DOF fractionated kinematic chains. Mech. Mach. Theory **70**, 74–90 (2013)
9. Schlick, T.: Adventures with RNA graphs. Methods **143**, 16–33 (2018)
10. Periyasamy, B.: Solution of linear differential equations in chemical kinetics through flow graph theory approach. J. Taiwan Inst. Chem. Eng. **57**, 9–17 (2015)
11. Simoni, R., Carboni, A.P., Martins, D.: Enumeration of kinematic chains and mechanisms. Proc. Inst. Mech. Eng. Part C J. Mech. Eng. Sci. **223**(4), 1017–1024 (2009)
12. Tian, C., Fang, Y., Ge Jeffrey, Q.: Structural synthesis of parallel manipulators with coupling sub-chains. Mech. Mach. Theory **118**, 84–99 (2017)
13. Mirafzal, S.M.: More odd graph theory from another point of view. Discrete Math. **341**(1), 217–220 (2018)
14. Martins, D., Carboni, A.P.: Variety and connectivity in kinematic chains. Mech. Mach. Theory **43**(10), 1236–1252 (2008)
15. Sydorenko, I.: Passive Vibration Isolation Device with Elements of Active Systems. Palmarium Academic Publishing, Saarbrücken (2011)
16. Sydorenko, I.: Analysis of the structure of the relaxation shock absorbers and modified the traditional kinematic graphs. ASB Mech. Mach. Tools Mater. **1**(26), 88–91 (2014)

The Study of the Elastic Characteristics of the Coupling with Nonlinear Feedback When Starting the Motor

Victor Kurgan(✉), Igor Sydorenko, Ihor Prokopovich, Liubov Bovnegra, and Tetiana Lysenko

Odessa National Polytechnic University (ONPU), Institute of Industrial Technologies, Design and Management, Shevchenko Avenue 1, Odessa 65044, Ukraine
kurgan@opu.ua, igs.ods@gmail.com, dlv5@ukr.net, tvl12odessa@gmail.com

Abstract. The most difficult moment in the work with an asynchronous motor is the launch. And the more powerful drive is the more difficult launch. This is due to certain features of the asynchronous motors: a limited starting torque and starting throws of the current of the stator motor chain. The mathematical modeling of oscillating process of actuation of the actuator with an asynchronous motor, which includes an elastic coupling with nonlinear mechanical feedback, is carried out. The influence of the type of elastic characteristics of the coupling on the magnitude of the amplitude and frequency of the oscillation process and its time was studied. A single-mass rotational system model was used for the studies. According to the Runge-Kutta method, the oscillation processes of starting the transmission of a machine unit with an induction motor were investigated. To determine the coefficient of vibration isolation, a system with an elastic coupling having a linear elastic characteristic was calculated. A study was also conducted in the case where the coupling determines the elastic characteristics of the Duffing type "soft" and "hard" type. The results of the calculations show that it is advisable to use a nonlinear coupling with a combined characteristic. On the basis of this, a synthesis of the target elastic characteristic and the study of the oscillatory process in the application of the proposed elastic coupling.

Keywords: Elastic coupling · Mechanical feedback · Oscillation process · Rotational mass · Starting torque

1 Introduction

In modern machine-building, elastic couplings with metal elastic elements have become widespread. This is facilitated by the ability of these devices not only to transmit torque, but also to prevent negative oscillations in the technical system. This is achieved by introducing into the design an elastic coupling of mechanical feedback, which provides a wide range of elastic characteristics, including also nonlinear. Studies in this direction have shown that the nonlinearity of the elastic characteristics of one of

I. Karabegović (Ed.): NT 2020, LNNS 128, pp. 122–130, 2020.
https://doi.org/10.1007/978-3-030-46817-0_14

the components of the machine aggregate can significantly change the nature of the oscillating processes, which occur.

2 Scientific Research

Studies using mathematical model proved that elastic couplings with a nonlinear elastic characteristic show the most positive results [1]. However, already existing elastic couplings do not fully meet the stated requirements due to their narrow working range [2–4]. Up to now created potential designs of elastic couplings that implement a nonlinear elastic characteristic are not widely used due to the small number of their actual mechanical constructions.

2.1 Review of Modern Information Sources on the Subject of the Paper

At this stage, most of drives use asynchronous motors. The features of their operation, specifically the start-up of the engine, cause the considerable oscillatory load on the drive, this is due to the large and short-term starting torque. Because of this, there is a significant number of works devoted to oscillating processes in technical systems [5–9]. A mathematical modeling of the start of an asynchronous electric motor was carried out by using software packages [10, 11]. Developed the promising designs of nonlinear elastic couplings, which reduce the load on the drive and prevent negative oscillations [12–14]. The following studies show the feasibility of using elastic couplings with nonlinear mechanical feedback.

2.2 Objectives and Problems of Research

Mathematical modeling of oscillatory process of transmission starting of a machine assembly with an asynchronous electric motor, which consist of an elastic coupling with nonlinear mechanical feedback and studying the effect of elastic characteristics on the magnitude of the amplitude, frequency of the oscillatory process and its time.

2.3 Main Material Presentation

The chosen aim of the research is based on the fact that the results of the researches carried out in the field of nonlinear oscillation mechanics indicate that the nonlinearity of the elastic characteristics of one of the components of the machine assembly can significantly change the nature of oscillatory processes.

In the given research area it is believed that the starting torque M_{srart} of the asynchronous motor shaft is a torque that advances on the shaft of an asynchronous electric motor under the following conditions: the speed of rotation is equal to 0, the current has a constant value, the electric motor windings are connected to rated supply frequency and voltage, the winding connection corresponds to the rated operating mode of the electric motor.

In mathematical modeling of the oscillatory processes of the machine assembly, the starting torque $M_s(t)$ is modeled by a function of time characterized by two time intervals: the build-up time t_1 to the maximum value and the time of decrease to the rated value t_2. In order to calculate the maximum starting torque the following expression is used.

$$M_{\max} = M_r \cdot k_{tr} \tag{1}$$

where M_r – rating moment on the electric motor shaft; k_{tr} – starting torque ratio. The value of this parameter varies within 1, 5…6 for different types of engines and loads.

Duration of the starting torque is determined experimentally, depending on the type of engine and the type of its load. Usually the value of this parameter varies within 0, 5…1, 6 s.

In order to achieve this goal a two-mass rotatory mechanical system (J_1 – main rotating mass, subject to protection against the negative demonstration of the starting torque), which includes the proposed passive elastic coupling with a nonlinear mechanical linkage (J_2 – the second rotating mass, which is the mass of the coupling) should be subject to mathematical modeling. In this case the system of differential equations has the following form:

$$\begin{cases} J_1\ddot{\varphi}_1 + b_1\ddot{\varphi}_1 - b_2(\dot{\varphi}_2 - \dot{\varphi}_1) + c_2(\varphi_2 - \varphi_1) = 0 \\ J_2\ddot{\varphi}_2 + b_2(\dot{\varphi}_2 - \dot{\varphi}_1) + c_2(\varphi_2 - \varphi_1) = M_s(t) \end{cases} \tag{2}$$

However, the rotating mass of the J_2 coupling in several cases is less than the rotating mass of J_1 transmission objects ($J_2 >> J_1$) and the stiffness of the shaft sections, which determines the torsion angle φ_1, is several times greater than the stiffness of the elastic coupling, which determines the torsion angle φ_2 ($\varphi_2 >> \varphi_1$).

Taking this into account it is advisable to carry out mathematical studies of the process of transmission starting of a machine assembly with an asynchronous electric motor, which includes the proposed elastic coupling, using a mathematical model of a single-mass rotatory system. In this case, the model treats the rotating mass J_1 as an object to be protected from the negative demonstration of the starting torque, and the elastic coupling is considered as an elastic linkage between it and the engine. Then the corresponding differential equation will have the following form:

$$J\ddot{\varphi} + M_{el}(\varphi) + M(\dot{\varphi}) = M_s(t) \tag{3}$$

where J – moment of inertia of the rotating mass; $M_{el}(\varphi)$ – elastic characteristic, which depends on the stiffness of elastic elements applied in the coupling; $M(\dot{\varphi})$ – moment of dissipation, which determines the irreversible energy dissipation; $\dot{\varphi}$ and $\ddot{\varphi}$ – corresponding derivatives of the angular displacement in time t.

Initial conditions are as follows:

$$\varphi(0) = 0,\ \dot{\varphi}(0) = 0,\ M_s(0) = 0 \tag{4}$$

On the basis of the Eq. (3) mathematical modeling of the oscillatory processes of transmission starting of a machine assembly with an asynchronous electric motor AIR112MV6 with the following characteristics was carried out: $P = 4\ kW$,

n = *1000 rpm*; rating moment M_{rat} = *34.5 N·m*; starting torque ratio k_{tr} = *1.8*; time of the starting torque t_s = *0.8 s*.

In calculations the starting torque is presented in the form of two non-linear sections associated with time and has a maximum $M_{smax}(0.026)$ = *61 N·m* (Fig. 1a)

$$\begin{array}{ll} t_{s1} = 0\ldots 0,26\text{s}, & M_{mot} = M_{s1}(t) = 16299t^2 + 6312,\ 1t + 3,4857; \\ t_{s2} = 0,26\ldots 0,8\text{s}, & M_{mot} = M_{s2}(t) = 16299t^2 - 4512,\ 1t - 2,5734; \\ t_s > 0,8\text{s} & M_{mot} = M_r = 34,5\text{N}\cdot\text{m} \end{array} \quad (5)$$

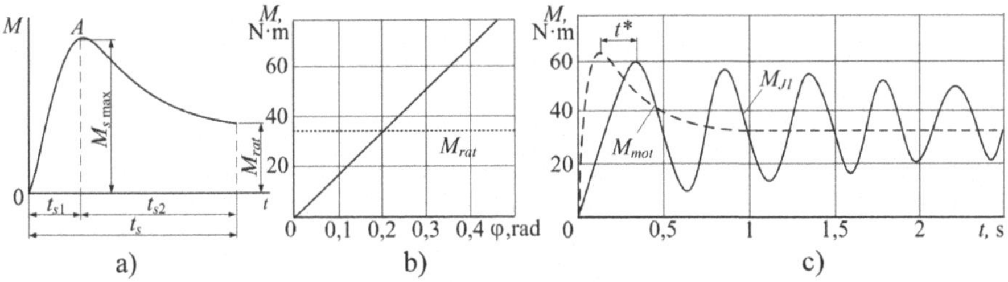

Fig. 1. Modelling of oscillatory processes of transmission starting of a machine assembly: starting torque (a); linear elastic characteristic (b); oscillatory process of transmission starting of a machine assembly with a coupling that possesses linear elastic characteristic (c)

For the possibility of conducting a comparative analysis in order to determine the appropriate efficiency ratios the calculation of the accepted conditions of the system, which containsan elastic coupling with a linear elastic characteristic, was carried out (Fig. 1b). Using the Maple 18 mathematical package, where the corresponding function implements the Runge-Kutta method, the solution of Eq. (3) was carried out in numerical form taking into account the initial conditions (4) and external load (5), which made it possible to state the following. Emerging at the process of starting a transmission of a machine assembly with an asynchronous electric motor, the oscillatory process is fading and low frequency with a constant frequency T = *2 Hz* (Fig. 1c). Oscillatory processes with the frequency of the first frequency octave, that is T = 2, 4, 8, 16, 31.5 and 63 Hz, refer to the low-frequency oscillatory process. The response of the system to external disturbance in the form of M_{J1} = *59.3 N·m* occurs with the delay after the appearance of the maximum external load equal to t^* = *0.18 s*, which is due to the presence of an elastic linkage. Oscillation decay time under condition of $M_{J1} = M_r$ equals t = *6.7 s*.

The coefficient that determines the efficiency of using an elastic coupling with a linear elastic characteristic is the coefficient of vibration isolation:

$$k_R = \frac{M_0}{A_0} \quad (6)$$

where M_0 – amplitude of the moment behind the coupling; A_0 – amplitude of the moment of disturbance.

In this case the coefficient of vibration isolation is:

$$k_R = \frac{M_0}{A_0} = \frac{59,3}{61} = 0,97 \quad (7)$$

Numerical solution of the Eq. (3), taking into account the general parameters of the system, the initial conditions (4) and the external load (5), is carried out in cases where the elastic characteristic of the coupling is nonlinear.

In the first case the coupling determined an elastic characteristic of a "soft" Duffing type. The value of the elastic torque at a certain nominal torsion angle of half-couplings $\varphi = 0,\ 2$ rad was equal to the value of the elastic torque of the previously considered linear characteristic $M = 34.5\ N{\cdot}m$ (Fig. 2a).

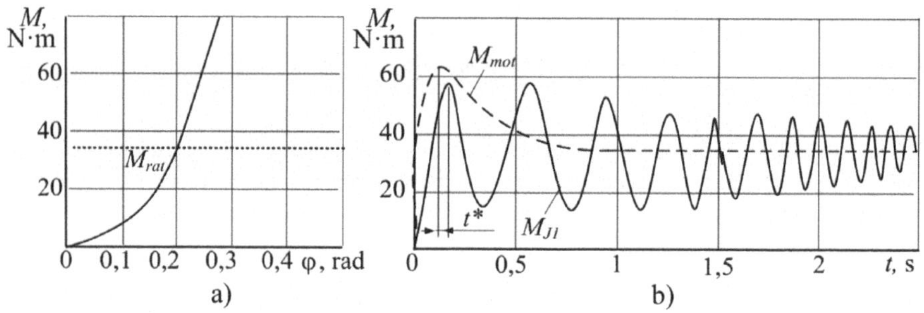

Fig. 2. Oscillatory process of asynchronous motor starting: elastic characteristic of a "soft" Duffing type coupling (a); oscillatory process (b)

Emerging at the start of the transmission of the machine assembly with an asynchronous electric motor, the oscillatory process is fading and low frequency with frequency *T*, which increases over time (Fig. 2b).

The response of the system to external disturbance in the form of $M_{J1} = 57.23\ N{\cdot}m$ occurs with the delay after the appearance of the maximum external load equal to $t^* = 0.38\ s$, which is defined by the value of the elastic torque, that is less than the similar one in the linear system, and lays in the range of the torsion angle of the half-couplings equal to $\varphi = 0,2\ldots0,\ 6$ rad. Oscillation decay time, which is determined by $M_{J1} = M_r$, equals $t = 3.8\ s$. The coefficient of vibration isolation k_R in this case is

$$k_R = \frac{M_0}{A_0} = \frac{57,34}{61} = 0,94 \quad (8)$$

In the second case the coupling determined an elastic characteristic of a "hard" Duffing type. The value of the elastic torque at a certain nominal torsion angle of half-couplings $\varphi = 0,2$ rad was equal to the value of the elastic torque of the previously considered linear characteristic $M = 34.5\ N{\cdot}m$ (Fig. 3a).

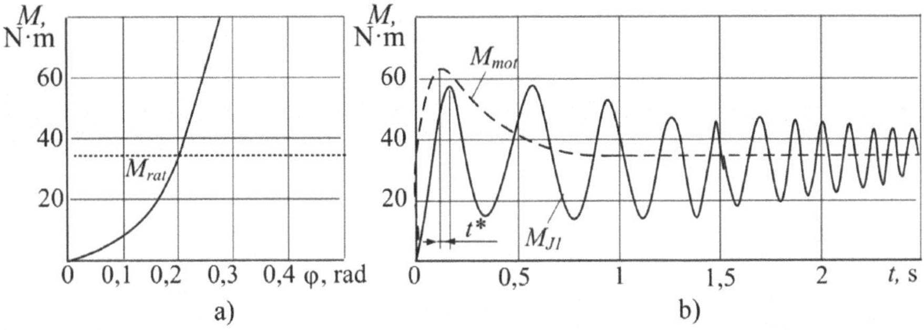

Fig. 3. Oscillatory process of asynchronous motor starting: elastic characteristic of a "hard" Duffing type coupling (a); oscillatory process (b)

Emerging at the start of the transmission of a machine assembly with an asynchronous electric motor, the oscillatory process is fading and low frequency with a frequency T decreasing over time (Fig. 3b). The response of the system to external disturbance in the form of M_{J1} = *59.16 N·m* occurs with the delay after the appearance of the maximum external load equal to t^* = *0.16 s*, which is defined by the value of the elastic torque, that is higher than the similar one in the linear system, and lays in the range of the torsion angle of the half-couplings equal to $\varphi = 0,2\ldots0,6$ rad. The decay time of the oscillatory process is determined by $M_{J1} = M_r$ and equals t = *3.4 s*. The coefficient of vibration isolation k_R in this case is:

$$k_R = \frac{M_0}{A_0} = \frac{59,13}{61} = 0,96. \tag{9}$$

Results of mechanical studies conducted to optimize the oscillatory process during the starting of transmissions with an asynchronous motor show that the use of nonlinear couplings with elastic characteristics of a "hard" Duffing type can reduce the time of the oscillatory process, however it determines the transmission load close to the starting torque. Application of nonlinear couplings with elastic characteristics of a "soft" Duffing type allows slight reduction of the transmission load, at the same time it lengthens the time of the oscillatory process. Taking this into account it is proposed to use nonlinear couplings with a combined characteristic in order to solve such a problem. The basis for such a proposal is the results of research done by Arkhangelskiy [14]. It has been established that optimization of the oscillatory process occurring at the start of transmission with an asynchronous motor can be obtained by applying a nonlinear elastic coupling in the transmission, which implements a combined characteristic with two sections, determined by the value of the rating rotary moment. The first section ($M = 0.....M_r$) must correspond the elastic characteristic of the "soft" Duffing type and the second section ($M = M_r.....1.3\ M_s$) must relate with the elastic characteristic of the "hard" Duffing type. The researcher has proposed a specialized design of an elastic coupling that implements a similar characteristic, but because of structural constraints its elastic characteristic corresponds to the target characteristic with a compliance coefficient equal to k_c = 0.89 and is fragmentarily linear (line 1, Fig. 4a).

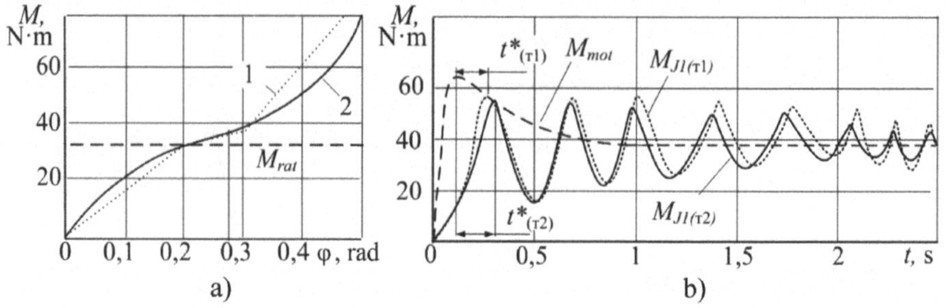

Fig. 4. Oscillatory process of asynchronous motor starting: combined elastic characteristics (a); oscillatory process (b)

From this perspective the calculations of the oscillatory process during the start of the transmission with the asynchronous motor, while applying the proposed coupling both with mentioned above elastic characteristic (combined, type 1) and with the synthesized target characteristic with the coefficient of compliance $k_c = 0.99$ (hereafter combined, type 2) have been carried out [12]. The synthesized elastic characteristic consists of the corresponding nonlinear sections that share borders at a certain value of the elastic torque and determine the rating rotary moment of half-couplings $\varphi = 0,2$ rad. (curve 2, Fig. 4a). Emerging at the start of the transmission of a machine assembly with an asynchronous motor in two calculation cases the oscillatory process is fading and low frequency with the frequency *T*, which varies over time (Fig. 4b). The response of the system to external disturbance in the form of $M_{J1\,(t1)} = 56.26\ N{\cdot}m$ occurs with the delay after the appearance of the maximum external load in the first case equal to $t^*_{(t1)} = 0.18$ s and in the second $M_{J1(t2)} = 56.26\ N{\cdot}m$ with the delay equal to $t^*_{(t2)} = 0.21$ s. Thus it is established that the average value of the elastic torque lays in the range of the torsion angle of the half-coupling $\varphi = 0,2\ldots0,6$ rad, which is higher than in the linear system and less than in a system with a "hard" Duffing type characteristic. The decay time of the oscillatory process is determined by $M_{J1} = M_r$ and in the first case $t_{(t1)} = 3.18$ s, while in the second case $t_{(t2)} = 0.28$ s, being the smallest indicators in the performed calculations. This is due to the fact that at high amplitudes of oscillations elastic characteristics cause an increase in their frequency. This, in turn, indicates the presence of high velocities and the greater effect of dissipative forces than in the previously considered variants. In this case the coefficient of vibration isolation k_R for the first calculation is as follows:

$$k_R = \frac{M_0}{A_0} = \frac{56,79}{61} = 0,931, \tag{10}$$

$$k_R = \frac{M_0}{A_0} = \frac{56,13}{61} = 0,92. \tag{11}$$

The results of the conducted analytical studies are presented in the Table 1.

Table 1. Coefficients of vibration isolation k_R and oscillation decay time at the start of the transmission with asynchronous motor with an elastic coupling

Type of elastic characteristic of coupling	Coefficient of vibration isolation k_R	Oscillation decay time $t\ (s)$
Linear	0,98	6,7
"Soft" Duffing type	0,94	8,3
"Hard" Duffing type	0,96	3,4
Combined, type 1	0,93	3,2
Combined, type 2	0,92	2,8

3 Conclusion

Implementation of elastic characteristics of the "soft" Duffing type of the coupling in comparison with the case of implementation of a linear elastic characteristic of the coupling enables reduction of negative demonstrations of oscillations by 3…4%, however it leads to an increase in duration of oscillatory process 1, 5…2 times. Implementation of elastic characteristics of the "hard" Duffing type of the coupling in comparison with the case of implementation of a linear elastic characteristic of the coupling allows to reduce the negative demonstrations of oscillations by 2…3% and leads to a decrease in duration of oscillatory process 1, 5…2 times.

Results of research of prof. G.V. Arkhangelskiy concerning optimization of oscillatory process in case of start of transmission of a machine assembly with an asynchronous motor using elastic coupling with a combined nonlinear elastic characteristic have been confirmed. Mathematical modeling of the system starting with the proposed coupling structure, which implements the target characteristic in the form of a fragmentarily linear characteristic with a compliance coefficient $k_c = 0.89$, resulted in a decrease of negative demonstration of oscillations by 5…6% and reduction of the time of oscillatory process 1, 5…2, 5 times, comparing with the case of realization of a linear elastic characteristic by a coupling.

Mathematical modeling of the system starting using an elastic coupling with a mechanical feedback that implements a nonlinear target combined characteristic with a compliance coefficient $k_c = 0.98$ caused a decrease of negative demonstration of oscillations by 7…10%, and reduction of the time of oscillatory process 2, 8…3 times, compared with the case of realization of a linear elastic characteristic by a coupling.

References

1. Sydorenko, I., Kurgan, V.: Synthesis of target elastic characteristics on the basis of elastic move with nonlinear mechanical refractory connection. Bull. Khmelnytsky Natl Univ. **5**(2), 26–31 (2017)
2. Szolc, T., Konowrocki, R., Pisarski, D., Pochanke, A.: Influence of various control strategies on transient torsional vibrations of rotor-machines driven by asynchronous motors. Mech. Mach. Sci. **63**, 205–220 (2019)

3. Wang, L., Jia, Z., Zhang, L., Liu, H.: Analysis of the soft starting of adjustable speed asynchronous magnetic coupling used in belt conveyor. Commun. Comput. Inf. Sci. **923**, 382–393 (2018)
4. Zhang, Q., Lu, Q.: Analysis on rigid-elastic coupling characteristics of planar 3-RRR flexible parallel mechanisms. Lect. Notes Comput. Sci. **10463**, 394–404 (2017)
5. Piotr Serkie, S.: Comparison of the control methods of electrical drives with an elastic coupling allowing to limit the torsional torque amplitude. Eksploatacja i Niezawodność **19** (2), 203–210 (2017)
6. Olshanskiy, V., Burlaka, V., Slipchenko, M.: Free oscillations of an oscillator with nonlinear positional friction. Ukrainian J. Mech. Eng. Mater. Sci. **4**(2), 50–57 (2018)
7. Hmida, A., Hammami, A., Chaari, F., Khabou, M.T., Haddar, M.: Modal analysis of spur gearbox with an elastic coupling. Appl. Cond. Monit. **5**, 153–163 (2017)
8. Chen, X., Hu, J., Peng, Z., Yuan, C.: Bifurcation and chaos analysis of torsional vibration in a PMSM-based driven system considering electromechanically coupled effect. Nonlinear Dyn. **88**(1), 277–292 (2017)
9. Andrukhiv, A., Sokil, B., Sokil, M.: Asymptotic method in investigation of complex nonlinear oscillations of elastic bodies. Ukrainian J. Mech. Eng. Mater. Sci. **4**(2), 58–67 (2018)
10. Andrukhiv, A., Sokil, B., Sokil, M.: Resonant phenomena of elastic bodies that perform bending and torsion vibrations. Ukrainian J. Mech. Eng. Mater. Sci. **4**(1), 65–73 (2018)
11. Wang, C., Sun, Q., Cao, G., Zeng, J.: Analysis on asynchronous start permanent magnet synchronous motor cogging torque optimization based on equivalent magnetic motive force. Lect. Notes Electr. Eng. **423**, 977–987 (2018)
12. Sydorenko, I., Kurgan, V.: Synthesis of nonlinear elastic couplings on the basis of modified kinematic graphs. Proc. Odessa Polytech. Univ. **3**(53), 5–11 (2017)
13. Nitzan, S.H., Zega, V., Li, M., Ahn, C.H., Corigliano, A., Kenny, T.W., Horsley, D.A.: Self-induced parametric amplification arising from nonlinear elastic coupling in a micromechanical resonating disk gyroscope. Sci. Rep. **5**, 9036 (2015)
14. Arkhangelsk, G.: Efficiency of using an elastic coupling with an extended section of quasi-zero stiffness. Mach. Parts **51**, 17–22 (2015)

Finite Element Analysis (FEA) of Automotive Parts Design as Important Issue of Assembly Technology Designing of Passenger Vehicle

Ismar Alagić[1,2,3](✉)

[1] TRA Tešanj Development Agency,
Trg Alije Izetbegovića 1, 74260 Tešanj, Bosnia and Herzegovina
ismar.alagic@gmail.com
[2] Faculty of Mechanical Engineering, University of Zenica,
Fakultetska 1, 72000 Zenica, Bosnia and Herzegovina
[3] Faculty of Engineering and Natural Sciences, International University
of Sarajevo, Hrasnička cesta 15, 71210 Ilidža, Sarajevo, Bosnia and Herzegovina

Abstract. The purpose of this article is to present a procedure to exploitation characteristics optimization of three different automotive parts (wheel hub, ball joint and tie rod) from production programme of Prevent FAD company based on the finite element formulation made by Algor software. Advanced solid modelling in Algor FEM software is ideally suited for testing functionality of automobile parts.

This article shows that Finite Element Method (FEM) helps designer asses the effects of flexible components on full system performance, improve the accuracy of simulations and thus bring it closer to the system-level design. The conducted research has begun with creation of 3D-CAD solid approximate model in the form of a multi-body system, after that solid mesh was generated where all meshed elements assumed to be perfectly rigid, and in final stage of testing finite element analysis was performed using Algor software package.

In order to determinate the tensile test of three different subassemblies from production programme of company Prevent FAD, designed for assembling steering and breaking system of passenger vehicle, finite element calculation has been carried out using Algor 3D CAD software. Sufficiently accurate stress distribution, maximum warp angle and displacement (Z component) of above mentioned parts have been obtained through the whole tensile process.

The simulation process has been carried out based on reaction forces and compared with experimentally obtained results. Material characteristics, load and constraint information, surface boundary conditions, nonlinear material models and load curve information are also covered by Mechanical Event Simulation (MES) Module of Algor software. The accuracy of the simulation results, stress distribution and displacement after unloading are discussed through the comparison with the experimental results.

This paper's goal is to show that application of FEA is successful in solving assembly technology issue of three different parts form production programme of company Prevent FAD. These results have the great influence on working function of above mentioned three parts and their application, especially in order to design of assembly technology of steering and breaking system of passenger vehicle.

I. Karabegović (Ed.): NT 2020, LNNS 128, pp. 131–146, 2020.
https://doi.org/10.1007/978-3-030-46817-0_15

Keywords: Wheel hub · Ball joint · Tie rod · Tensile process · Finite Element Analyse · Assembly technology

1 Introduction

Company Prevent FAD is renowned OEM supplier of worldwide known automakers. This company produces wide range of different parts of steering and breaking system for passenger vehicle. The main goal of investigation carried out in this paper, was to determine the effects of three different parts (wheel hub, ball joint and tie rod) design flexibility on appropriate static load.

Wheel hub as part of brake system is primarily used in automobile industry in wide variety of exploitation characteristics during actual driving conditions. The purpose of this article is to present a procedure to exploitation characteristics optimization of wheel hub based on the finite element formulation made by Algor software. The proposed methodology allows that this optimization simulate testing of loading conditions in exploitation environment of wheel hubs.

These method allows in the short time and with high accuracy to obtain optimal working characteristics of the wheel hub. These characteristics have the great influence on working function and its application. This work is the introduction to the finite element analysis optimization of exploitation characteristics of wheel hub during actual driving conditions. One of the matters regarding the research in this area are numeric stress analysis and strength evaluation of wheel hubs. The research conducted covered the following matters: stress von mises, intermediate stress, nodal displacement and max warp angle. The geometry of research set as well as finite element analysis were made by means of the Algor system.

The main goal of investigation carried out in case of ball joint, was to determine the effects of the ball joint design flexibility on its static load value of 25 KN. Computed static loads of the ball joint were used as FEM boundary conditions. Static load was used for ball joint stress and displacement validation.

This article shows that coupling CAD 3D-FEA-MES design approach helps designer asses the effects of flexible components on full system performance, improve the accuracy of simulations and such bring him closer to the system-level design. Integrating system of three components, first CAD 3D design work, second FEA model and last one MES (Mechanical Event Simulation) as level of motion simulation, improve the reliability of load prediction by using FEA model with complete time histories of component deformation. Providing this data to designer can give added confidence in subsequent analyses of stress, fatigue and other desired component variables. This method allows in the short time and with high accuracy to obtain optimal structure characteristics of the ball joint. One of the matters regarding the research in this area is its stiffness with allowed value of stress and displacement. These characteristics have the great influence on working function of ball joint and its application. These characteristics have the great influence on working function of ball joint and its application, especially in order to design of assembly technology of steering system of passenger vehicle. In automobiles a tie rod is part of the steering mechanism. The tie rods connect the centre link to the steering knuckle on cars with conventional suspension systems and recirculation ball steering gears. The tie rod transmits force from the steering centre link or the rack gear to the steering knuckle, causing the wheels to turn. The real problem with the tie rod assembly is its strength and durability. The tie rod assembly consists of four following separate parts: tie bar, housing, bushing and bellow. Several methods have been proposed for estimating the actual tensile forces in tie rods. It is imperative that tie rod operate reliably during exploitation and under severe heavy working conditions, since cars are entrusted with the care of precious human life which would be endangered if an accident occurs. Tie rod as part of steering system is primarily used in automobile industry in wide variety of exploitation conditions during actual driving conditions [1].

This article shows that Finite Element Method (FEM) helps designer asses the effects of flexible components on full system performance, improve the accuracy of simulations and thus bring it closer to the system-level design. The conducted research has begun with creation of 3D-CAD solid approximate model in the form of a multi-body system, after solid mesh was generated where all meshed elements assumed to be perfectly rigid, and in final stage of testing finite element analysis was performed using Algor software package.

2 Braking and Steering System of Passenger Vehicle

The braking system is the most important system in car. If brakes fail, the result can be disastrous. Brakes are actually energy conversion devices, which convert the kinetic energy (momentum) of vehicle into thermal energy (heat).

The brake system is composed of the following basic components: The "master cylinder" which is located under the hood, and is directly connected to the brake pedal, converts foot's mechanical pressure into hydraulic pressure. Steel "brake lines" and flexible "brake hoses" connect the master cylinder to the "slave cylinders" located at each wheel. Brake fluid, specially designed to work in extreme conditions, fills the system. "Shoes" and "pads" are pushed by the slave cylinders to contact the "drums" and "wheel hub" thus causing drag, which (hopefully) slows the car (Fig. 2).

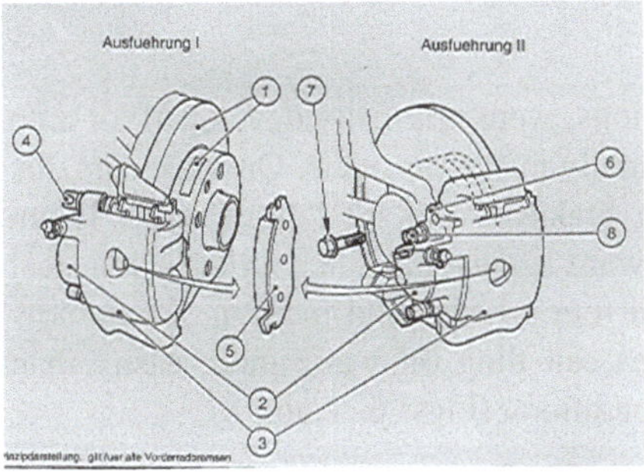

Fig. 1. Brake system of vehicle.

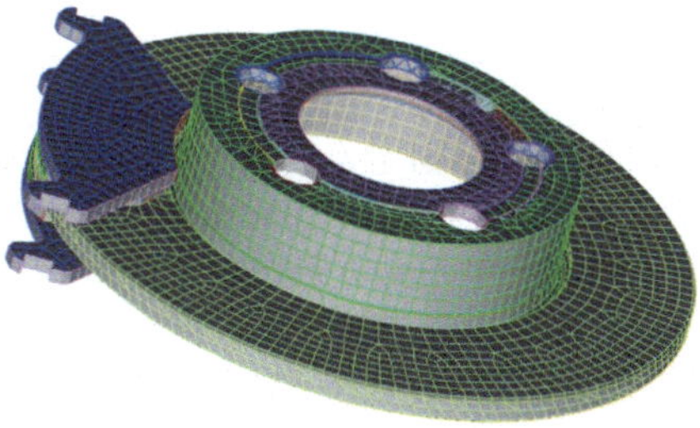

Fig. 2. Mesh of Brake disc carried out by Algor software.

In recent years, brakes have changed greatly in design. Disc brakes, used for years for front wheel applications, are fast replacing drum brakes on the rear wheels of modern cars. This is generally due to their simpler design, lighter weight and better braking performance. The greatest advantage of disc brakes is that they provide significantly better resistance to "brake fade" compared to drum type braking systems. Brake fade is a temporary condition caused by high temperatures generated by repeated hard braking. It occurs when the pads or shoes "glaze" due to the great pressure and heat of hard use. Once they cool, the condition subsides. Disc brakes allow greater air ventilation (cooling) compared to drum brakes. Drum brakes are not internally ventilated because if they were, water could accumulate in them. Disc brakes can rapidly fling off any water that they are exposed to, and so they can be well ventilated. Disc brakes use a clamping action to produce friction between the "wheel hub" and the "pads" mounted in the "caliper" attached to the suspension members. Inside the calipers, pistons press against the pads due to pressure generated in the master cylinder. The pads then rub against the rotor, slowing the vehicle. Disc brakes work using much the same basic principle as the brakes on a bicycle; as the caliper pinches the wheel with pads on both sides, it slows the bicycle. Disc brakes offer higher performance braking, simpler design, lighter weight, and better resistance to water interference than drum brakes (Fig. 3).

Fig. 3. 3D model of wheel hub.

Fig. 4. Mesh of wheel hub FEA model.

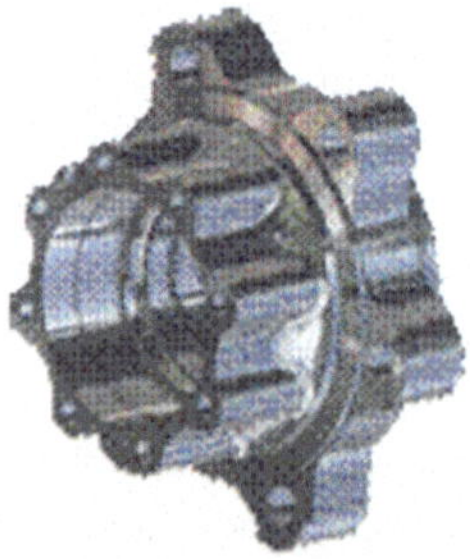

Fig. 5. 3D model of analyzed wheel hub.

Disc brakes, like many automotive innovations, were originally developed for auto racing, but are now standard equipment on virtually every car made. On most cars, the front brakes are of the disc type, and the rear brakes are of the "drum" type. Drum brakes use two semi-circular shoes to press outward against the inner surfaces of a steel drum. Older cars often had drum brakes on all four wheels, and many new cars now have 4-wheel disc brakes. Because disc brakes can fling off water more easily than drum brakes, they work much better in wet conditions (Figs. 6, 7 and 8).

Fig. 6. Steering system of vehicle.

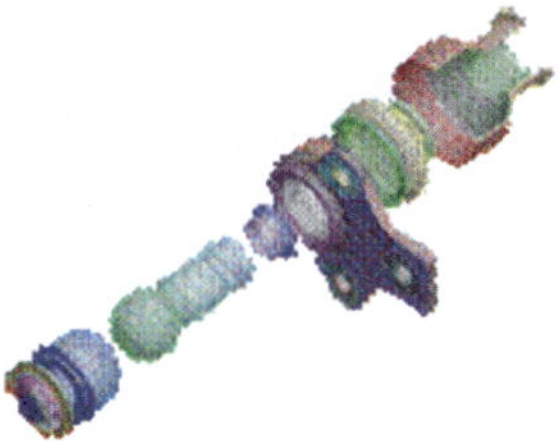

Fig. 7. Components of ball joint assembly.

Fig. 8. Subassembly of housing and ball journal.

Disc brakes also allow better airflow cooling, which also increases their effectiveness. Some high performance disc brakes have drilled or slotted holes through the face of the wheel hub, which helps to prevent the pads from "glazing" (becoming hardened due to heat). The manual steering system incorporates [11]: steering wheel and column; manual gearbox and pitman arm or a rack and pinion assembly; linkages; steering knuckles and ball joints; and the wheel spindle assemblies. The steering linkage is made of interconnected parts which move every time the steering wheel is turned. The rotating movement of the steering column activates mechanisms inside the steering box [3] (Figs. 9, 10 and 11).

Fig. 9. Ball joint - version a.

Fig. 10. Ball joint - version b.

Fig. 11. Statical model of ball joint.

Tie rod ends, which join the key parts, pass on the steering wheel's motion no matter what the angle of the linkage or the vibration from the road. In a pitman arm steering setup, the movement inside the steering box causes the Pitman shaft and arm to rotate, applying leverage to the relay rod, which passes the movement to the tie rods. Definition of ball joint is as follows: A flexible coupling in a vehicle's suspension that connects the control arm to the steering knuckle. A ball joint is so named because of its ball-and-socket construction (Figs. 12, 13 and 14).

Fig. 12. Tie rod-tie rod end assembly

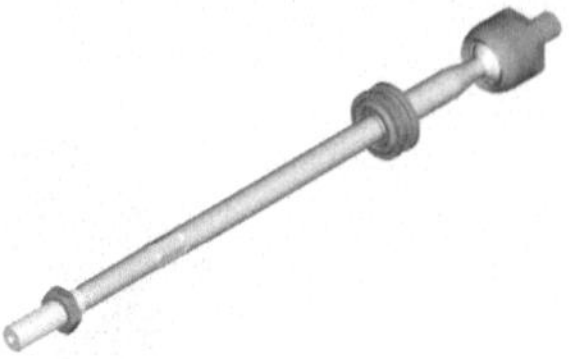

Fig. 13. Tie rod assembly

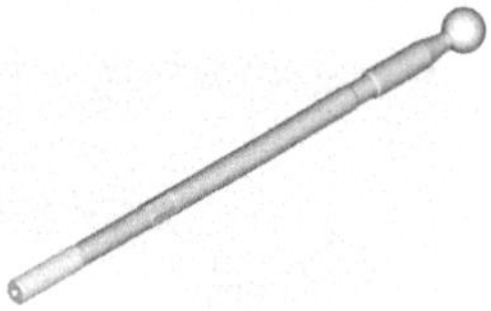

Fig. 14. Tie rod.

3 Description of the Finite Element Model

Brake systems and wheel hub assemblies are critical safety-relevant components of the vehicle chassis exposed to severe loads. Their reliability can only be safeguarded by performing tests in the laboratory under conditions which resemble actual loading conditions as closely as possible.

The wheel and brake assembly testing rig enables the simulation of all loads acting on the wheel-brake assembly which have a bearing on its service strength. All relevant parameters and service loads acting on the overall system are derived from loads occurring during actual driving conditions. These are:

- driving speed equivalent to vehicle speeds up to approx. 160 km/h, depending on the size of drum;
- side loads resulting from the centrifugal forces acting on the wheel/brake assembly during cornering maneuvers;
- wheel contact force resulting from vehicle weight and impacts resulting from the road surface;
- braking pressure;
- drive torque.

The wheel hub model was made according of the Mitsibushi wheel hub prototype "Mitsibushi Caterpilar No. 91B3301400-001". Due to the high level of wheel hub geometry configuration complication some simplicities were introduced allowing the analysis's time shortening.

The goal of numerical analysis carried out by Algor software were to determine mode shape of wheel hub in exploitation stage as well as to evaluate value of exploitation characteristic effects such as: stress, displacement, max warp angle etc. Three-dimensional models of wheel hub consisted of brick type elements. The models were build from scratch with Algor MES with Nonlinear Material Models. Total number of finite elements was 31882 elements for all analysed models. The distribution, direction and value of the static force presumed to the analysis were similar to the value and direction, which were observed at the maximal exploitation load. The total resultant force value equaled 40000 N. The material properties of the analyzed wheel hub was defined as follows:

Material	Mass density	Modulus of elasticity	Poisson's ratio	Shear modulus of elasticity
Grey cast iron	6914 kg/m^3	0,6895·10^5 MPa	0,22	0,2829·10^5 MPa

In a typical stress analysis using FEA Algor, there is a basic set of steps that the analysis usually follows:

(1) Create a mesh (a grid of nodes and elements) that represents the model;
(2) Define a unit system;
(3) Define the model's global properties;
(4) Define the element properties;
(5) Apply the loads and the boundary conditions which prevent the model from moving;
(6) Assemble the element stiffness matrices;
(7) Solve the system of linear algebraic equations;
(8) Calculate the results;
(9) Review the results;
(10) Presentation of results.

These steps are usually broken up into three stages:

(a) Pre-processing: Steps 1–5;
(b) Processing: Step 6–8 (These steps are automatically performed by Algor);
(c) Post-processing: Steps 9 and 10.

Finite element analysis is a computerized method for predicting how a real world object will react to forces, heat, vibration, etc., in terms of whether it will break, wear out, or work the way it was designed [8]. The finite element method works by breaking a real object down into a large number (1000 s to 100 000 s) of elements, such as little cubes. The behaviour of each little element, which is regular in shape, is readily predicted by set mathematical equations [9] (Table 1).

Table 1. Technical characteristics of ball joint assembly components.

Part ID	Part name	Material name	Element type	Number of elements
1	Housing	Steel (ASTM-A242)	Brick	20540
2	Cover	Steel (ASTM-A36)	Brick	2824
3	Ball socket	Poliacetal, TL VW 557	Brick	2936
4	Bushing	Plastics Nylon, Type 6/6	Brick	7717
5	Gasket I	Spring wire, DIN 17 223	Brick	390
6	Gasket II	Spring wire, DIN 17 223	Brick	648
7	Steering knuckle cap	Hard PE	Brick	27170
8	Ball journal	Steel (ASTM-A242)	Brick	17547

The distribution, direction and value of the static force presumed to the analysis were similar to the value and direction, which were observed at the maximal exploitation load. The total resultant force value equalled 25000 N (Tables 2 and 3).

Table 2. Loads, surface force.

ID	Description	Part ID	Surface ID	Magnitude	V_x	V_y	V_z	Follows moving surface	Load curve
1	Pull (tensile) force	8	7	250000	0	0	1	No	1

Table 3. Constraints, surface boundary conditions.

ID	Part ID	Surface ID	T_X	T_Y	T_Z	R_X	R_Y	R_Z
1	1	22	Yes	Yes	Yes	Yes	Yes	Yes
2	1	37	Yes	Yes	Yes	Yes	Yes	Yes
3	1	52	Yes	Yes	Yes	Yes	Yes	Yes
4	1	32	Yes	Yes	Yes	Yes	Yes	Yes
5	1	42	Yes	Yes	Yes	Yes	Yes	Yes
6	1	62	Yes	Yes	Yes	Yes	Yes	Yes

The brick model was used in research. Total number of finite elements was 79772 elements for all analysed models.

Table 4. Experimental results obtained by testing on control device MR 96, tensile force 25000 N.

Diameter d_1 of ball journal before test (mm)	17,4	17,4	17,45
Diameter d_1 of ball journal after test (mm)	17,42	17,41	17,46
Diameter d_2 of ball journal before test (mm)	19,3	19,3	19,3
Diameter d_2 of ball journal after test (mm)	19,31	19,32	19,32

The experimental tensile test of ball journal from housing of ball joint (pull force, magnitude 25 kN) was performed by control device MR 96 based on diagram presented in Fig. 15. The results of testing were presented in Table 4 (Fig. 16).

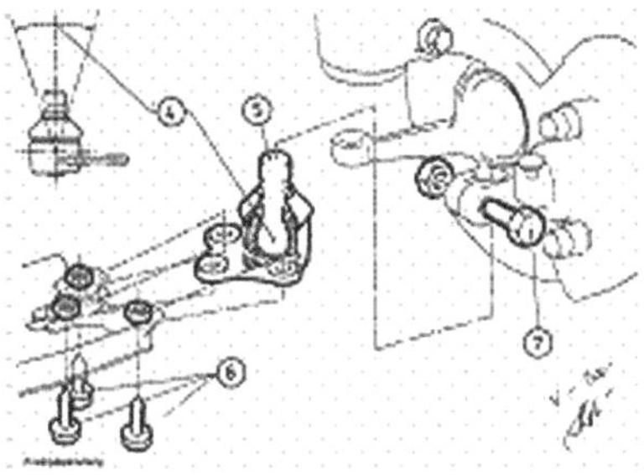

Fig. 15. Position of ball joint in steering system of vehicle.

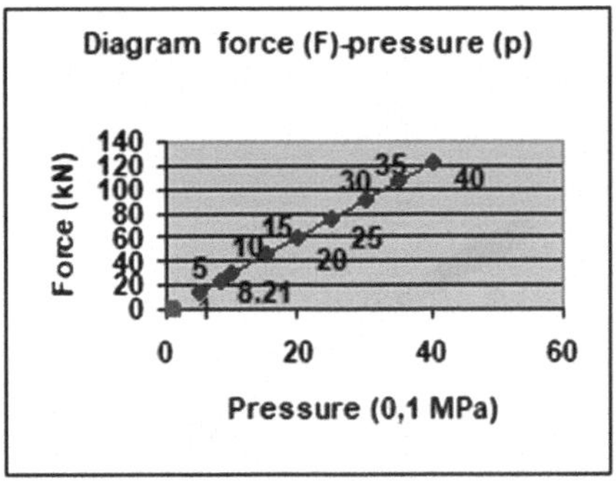

Fig. 16. Result of testing carried out by device MR 96.

In order to conduct the tensile test of tie bar from housing of tie rod assembly (tensile force, F = 30000 N) designed for assembling wheel transmission system of passenger vehicles, finite element calculation has been carried out using Algor software. The aim of this research is simulation of tensile test (tensile force, magnitude 30 000 N) using finite element method. The accuracy of finite element model depends on the assumptions made and the correlation between the computer models and testing application. The constructing model depends on material properties as well as testing conditions and testing equipment.

Integrated system of three components, firstly CAD three dimensional design work is shown in Fig. 1 (performed by Autodesk Mechanical Desktop), Finite element (FE) model (software Algor Version 16), and on Mechanical Event Simulation (MES) as a level of motion simulation, improve the reliability of load prediction by using FE-model [2]. This model allows obtaining optimal structure characteristics of the tie rod in the short time and with high accuracy.
Design documentation requires the following allowed technical characteristics:

(a) Maximum allowed displacement 0,5 mm;
(b) Maximum allowed warp angle 38 + 6°;
(c) Ultimate stress (material of tie bar) 1100–1300 MPa (Figs. 17 and 18).

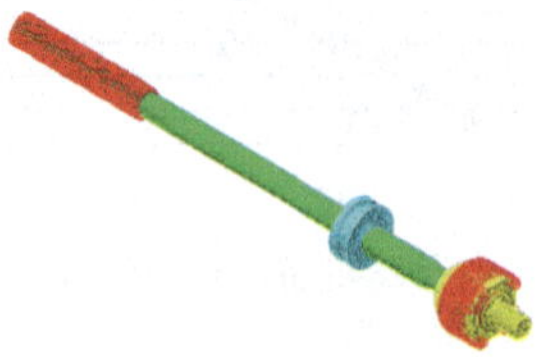

Fig. 17. Mesh of tie rod assembly.

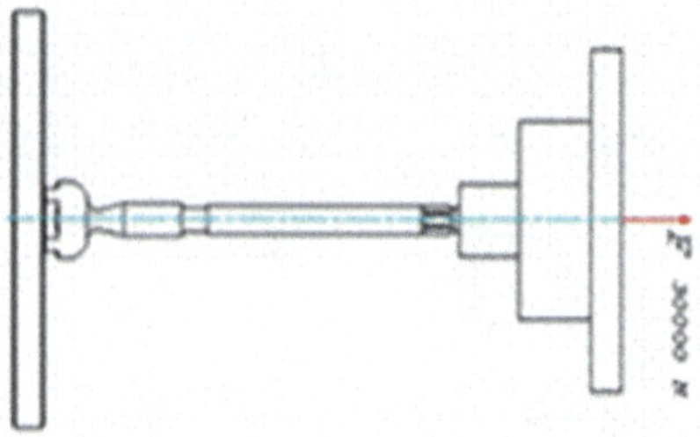

Fig. 18. Experiment performed by device MR 96.

Numerical calculation is performed with aid of the FEM (software Algor) (Table 5).

Table 5. Material properties.

Material	Mass density, kg/m^3	Modulus of elasticity, MPa	Poisson's ratio	Shear modulus of elasticity, MPa
Steel, ASTM-A242	7854,8	199950	0,29	77221
Plastics Nylon, Type 6/6	1143,5	2757,9	0,35	1021,4

Three dimensional models of tie rod consisted of tetrahedral solid elements. The models were built with Algor R Version 16, Static Stress with Linear Material Models. Total number of finite elements was 85 875 elements for all analyzed models. Tensile loads of tie rod were used as FEM boundary conditions (Table 6).

Table 6. Technical characteristic of model elements.

Part ID	Part name	Material name	Element type	No. of elements
1	Tie bar	Steel (ASTM-A242)	Tetrahedral	43002
2	Bushing	Plastics Nylon, Type 6/6	Tetrahedral	10043
3	Housing	Steel (ASTM-A242)	Tetrahedral	20407
4	Bellow	Plastics Nylon, Type 6/6	Tetrahedral	12423

Tensile load was used for tie rod stress, strain and displacement validation. All necessary input information regarding FEM analysis is shown below (Table 7).

Table 7. Surface force.

ID	Description	Part ID	Surface ID	Magnitude	V_x	V_y	V_z
1	Tensile force	3	29	30000	1	0	0

4 FE-Simulation Research and Experimental Results

Control device MR 96 is the horizontal testing machine suitable for tensile test of long materials such as steering tie rods. The tie rod is loaded in tension. The tensile test on three tie rod sample, of 21 mm inner diameter and 26 mm outer diameter and 329 mm of length, have provided the following results presented in Table 4. During the tensile test, all results were recorded. The tensile force is calculated using: $F_t = A \cdot E \cdot \varepsilon_t$, where: A is the cross section, E is the Young modulus of the tie rod and ε_t is the tension strain. To prevent any damage, each of three tie rods (three samples) will be tested at 120% of its maximum exploitation loads (tensile load, magnitude 24 kN). The tie rods have been pulled at 30 000 N. No damages have been noticed.

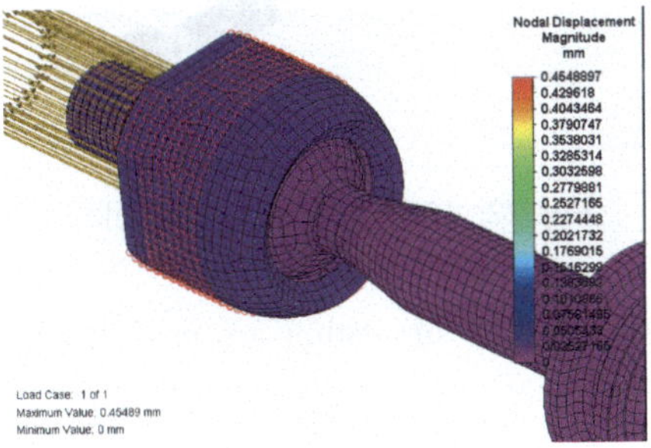

Fig. 19. FEM model - displacement (magnitude).

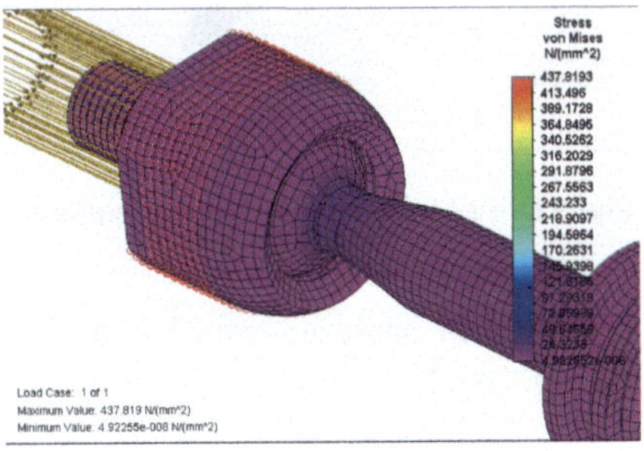

Fig. 20. FEM model - stress distribution.

The distribution, direction and value of the tensile force presumed in the analysis were similar to value and the direction, which were recorded at the maximal exploitation load. As a matter of fact, the total produced tensile force F = 30000 N, was approximately 20% above the value of maximum exploitation load applied during actual driving conditions. Therefore testing procedures provides reliably assumption of this problem.

Table 8. Experimental results performed by control device MR 96, tensile force F = 30 000 N.

Sample	I (mm)	II (mm)	III (mm)
Inner diameter of tie bar before tensile test	ϕ21,40	ϕ21,40	ϕ21,45
Inner diameter of tie bar after tensile test/elongation	ϕ21,42/0,02	ϕ21,45/0,05	ϕ21,50/0,05
Outer diameter of tie bar before tensile test	ϕ26,30	ϕ26,30	ϕ26,30
Outer diameter of tie bar after tensile test/elongation	ϕ26,40/0,1	ϕ26,32/0,02	ϕ26,35/0,05

A 3D-CAD study has been performed by FEM with Algor. The results of FEM simulation are provided in this document (Figs. 19, 20, 21 and 22). The safety margin of the tie rod is defined as material strength/maximum stress. The experimental tensile test of tie bar from housing of tie rod assembly is performed by control device MR 96 (see Table 8). The deformation achieved as result of finite element analysis is similar to the results of the tensile test performed by control device MR 96. The maximum appeared displacement (0,455 < allowed value 0, 5 mm, see Fig. 4) and maximum stress value is lower than ultimate stress allowed by documentation request (437,8 < 1100 MPa, see Fig. 5). Also, analysis of strain distribution (see Fig. 21) and warp angle (41,4 < allowed value 44°, see Fig. 22) through the whole tensile test showed to us lower angle than allowed value required by design documentation.

Fig. 21. FEM model - strain distribution.

Fig. 22. FEM model - warp angle.

In case of analyzed wheel hub, the results of FEM simulation are provided in Figs. 23, 24, 25 and 26.

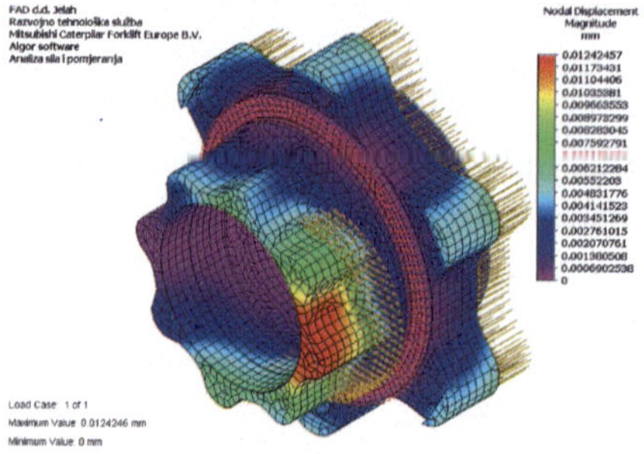

Fig. 23. Nodal displacement validation.

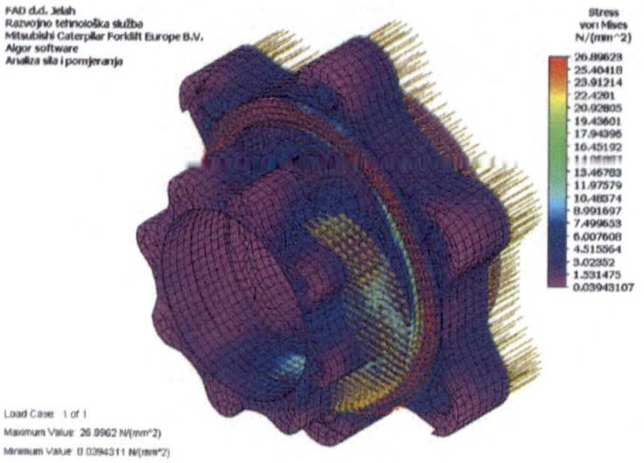

Fig. 24. Stress validation of FEA model.

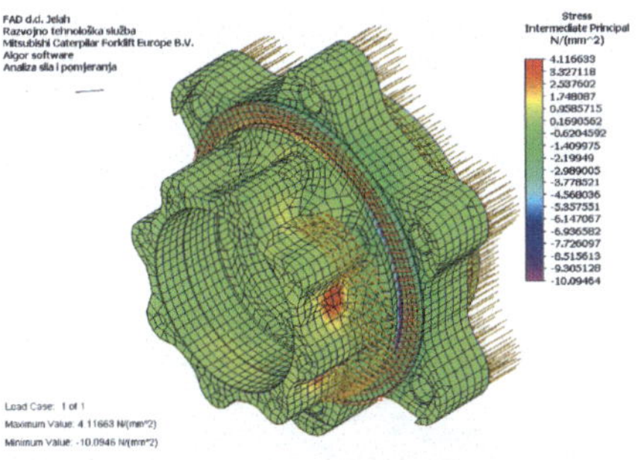

Fig. 25. Stress intermediate principal validation of FEA model.

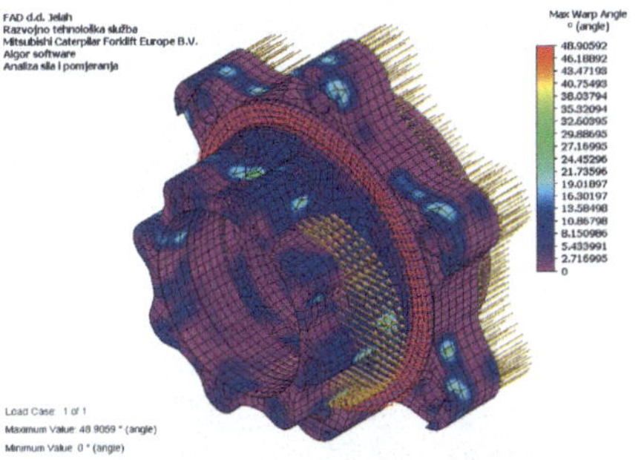

Fig. 26. Max warp angle validation of FEA model.

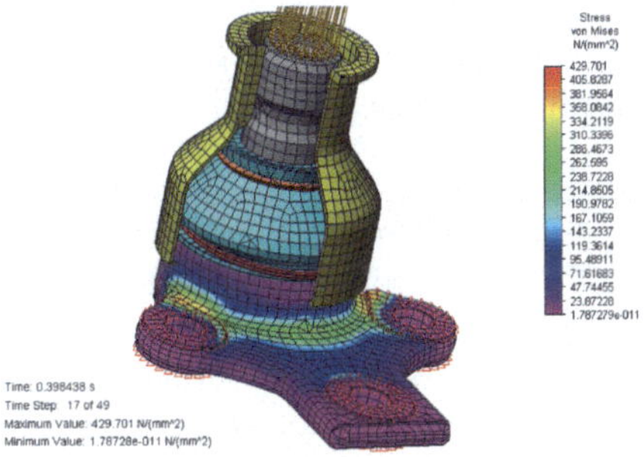

Fig. 27. Stress validation of FEA model of ball joint.

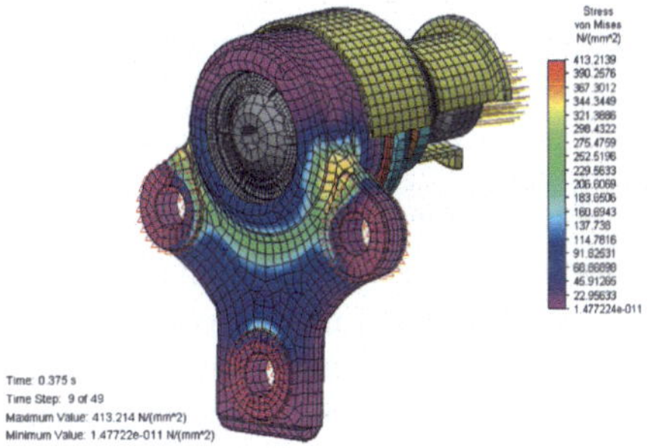

Fig. 28. Stress validation of FEA model.

In case of ball joint, the results of stress distribution and allowed displacement are presented in Figs. 27, 28, 29, 30, 31 and 32.

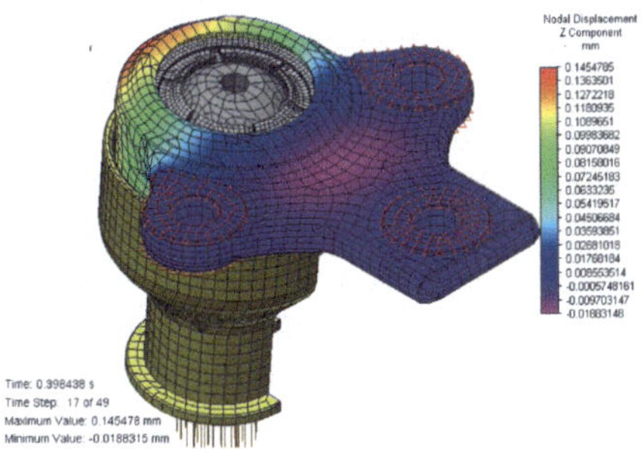

Fig. 29. Nodal displacement (Z component) validation of FEA model of ball joint.

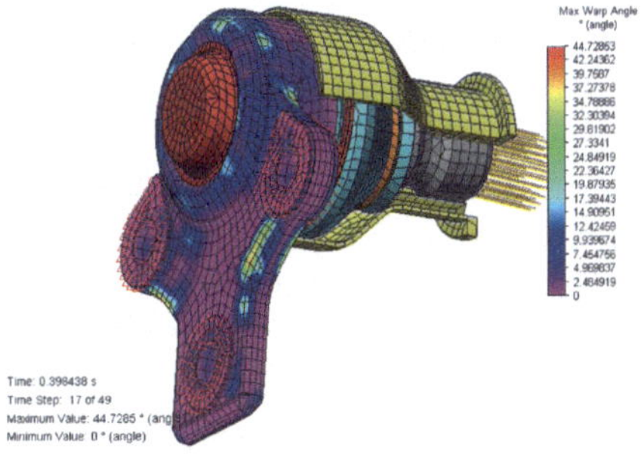

Fig. 30. Max warp angle validation of FEA model.

The displacement achieved as a result of finite element analysis doesn't much differ from results of the laboratory-test performed by control device MR 96. The maximum displacement appeared into Z direction 0,145 [mm].

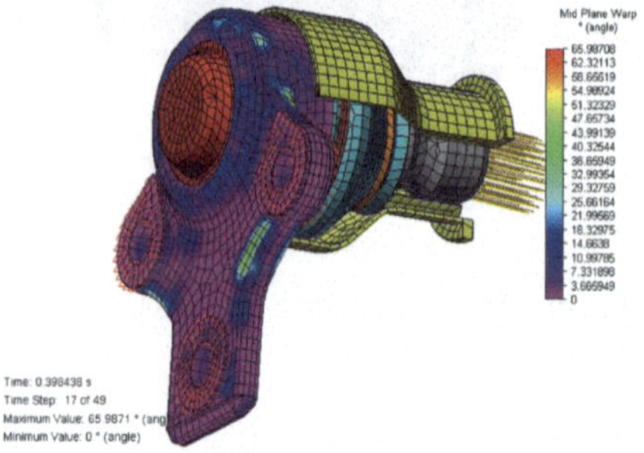

Fig. 31. Plane warp validation of FEA model.

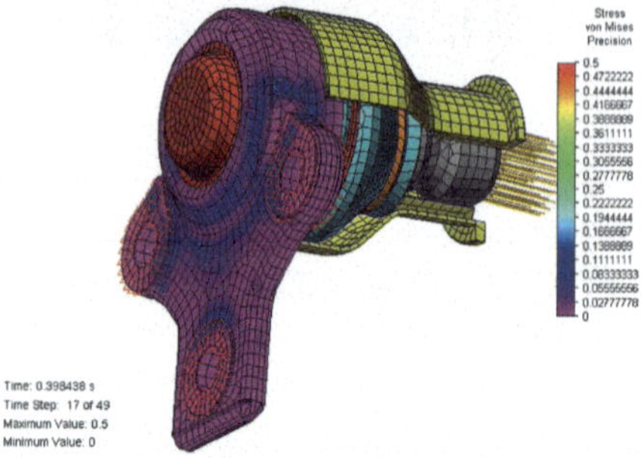

Fig. 32. Stress von misses precision validation of FEA model.

On the basis of conducted simulations were possible to affirm that the magnitude of deformations depends on model geometry. The largest concentration of stresses appeared in places near of cover of ball joint. Through change of model geometry it was possible to influence on expansion of stresses and displacement distribution.

To analysis of maximum warp angle of ball joint using FEA solver was very low compared to its allowed value 58 ± 6°.

5 Conclusion

It can be seen from the obtained results that analysis with FEM and the software Algor gave good predictions of stress distribution during actual exploitations conditions of wheel hub. The simulation process has been carried out using MES module of Algor software. The numerical predictions of automobile parts behaviour can give us a lot of information and help with defining recent load increase in hub wheels with insurance of their high exploitation characteristics in the increasingly complex conditions of durable exploitation. Application of Algor software decisively supports numeric stress analysis and strength evaluation of wheel hubs since it effectively allows the simulation of the complex exploitation conditions under actual service loads by having installed the data and methods used and approved by experimental stress analysis. It therefore is a strong tool to improve the reliability of numeric strength evaluation. Modeling of exploitation characteristics of wheel hub and other automobile parts can be optimized by using the FEA. The finite element analysis method is becoming a common tool in automobile parts development.

Based on presented experimental research and results analysis, following can be concluded:

- Models of ball joint were designed in three-dimensional software Mechanical Desktop. Designed parts of model were transformed into FEA software Algor, where solid mesh was generated on the basis of surface mesh.
- With FEM it has become possible to predict both the magnitude and the distribution of the stresses and displacement in ball joint assembly due to the tensile test of ball journal from housing of ball joint part No. 181 16 001 0 (pull force, magnitude 25 kN).
- The quality of the mesh proved to be essential factor in performing successfully FEM analysis. The number, size and shape of the elements are of importance for the solution accuracy. The number of nodes influences especially regarding the simulation of tensile force.
- The results of laboratory test performed by control device MR 96 were similar to the results obtained as result of finite element analysis. There are the effects of fact that laboratory conditions of testing were known during the FEA modelling and design of ball joint, especially designing of assembly technology for steering system of vehicle. The results of tests were allowed application of ball joint as proper part of suspension and steering system for passenger vehicle [10].
- The improvement of design and test of ball joint as well as its application can be achieved by using FEA combined with actual testing procedures as useful information for designing od assembly technology of the whole steering system of passenger vehicle [12].
- The conducted research has begun with creation of 3D-CAD solid approximate model in the form of a multi-body system, after that solid mesh was generated where all meshed elements assumed to be perfectly rigid, and in final stage of testing finite element analysis was performed using Algor software package.
- From the presented results we can conclude that the distribution of deformation and stress do not exceed the upper limit value and that there are neither damages nor surface defects after performed tensile test.
- The results of tensile test performed by control device MR 96 were closer to the results of FEM simulation.
- Using FEM made possible to predict the whole tensile test of tie rod assembly.
- The correctness and accuracy of computed results is still dependent on the selection related to various modelling parameters. Some of the most important aspects, such as boundary conditions or correct mesh and type of elements are performing a decisive role in achieving of correct results.
- The mentioned conclusion is only valid for above defined working conditions and incorrigible estimated value of tensile force (F = 30000 N).

References

1. Alagić, I.: Industrial Engineering & Maintenance: Lean Production - Six Sigma with application of tools and methods in specific working conditions, CIP - Katalogizacija u publikaciji, Nacionalna i univerzitetska biblioteka, 629.331:[658.5:005.6(075.8), COBISS. BH-ID 24232454, Štamparija-S, Tešanj (2017). ISBN 978-9958-074-09-7
2. Alagić, I., Božičković, S.: Preduzetnička infrastruktura, Saobraćajni fakultet u Doboju, 410 str., ilustr.; 25 cm, Biografije autora: str. 331-332. - Prilozi: str. 333-342. - Napomene i bibliografske reference uz tekst. - Bibliografija: str. 369-394, COBISS.RS-ID 8161048, Doboj (2019). ISBN 978-99955-36-74-9
3. Alagić, I.: Lokalna uprava i lokalni ekonomski razvoj suočeni sa izazovima koncepata pametni gradovi i industrija 5.0, Planjax komerc, Tešanj (2018). ISBN 978-9958-34-342-1
4. Alagic, I.: Using the finite element method (FEM) in order to optimize of clamping force for jaw-chucks during machining process of the brake disc. In: 9th International Research/Expert Conference Trends in the Development of Machinery and Associated Technology, TMT 2005, Antalya, Turkey, 26–30 September 2005 (2005)
5. Alagić, I.: FE-simulation research of static stiffness of braking disc No. 1J0 615 601 under the influence of the chucking process. In: Međunarodni skup Informatika u proizvodnom i poslovnom okruženju IPOM, Zbornik radova, Viša tehnička škola, Doboj, Bosna i Hercegovina (2004)
6. Alagić, I.: Finite element modelling of wheel hub as a part of brake system. In: 8th International Research/Expert Conference Trends in the Development of Machinery and Associated Technology, TMT 2004, Proceedings, Neum, B&H (2004)
7. Alagic, I.: Tensile process test of tie rod using FEM simulation. In: 10th International Research/Expert Conference Trends in the Development of Machinery and Associated Technology, TMT 2006, University of Zenica, Faculty of Mechanical Engineering in Zenica, UPC Universitat Politecnica de Catalunya, Barcelona, Spain, BAHÇEŞEHIR ÜNIVERSITESI Istanbul, Turkey, Barcelona-Lloret de Mar, Spain, 11–15 September 2006 (2006)
8. Bathe, K.J.: Finite Element Procedures in Engineering Analysis. Prentice Hall, Englewood Cliffs (1996). ISBN 0-13-349679-X
9. Crisfield, M.A.: Non-linear Finite Element Analysis of Solids and Structures, vol. 2. Wiley, New York (1997). ISBN 0-471-95649-X
10. Hajdić, N., Alagić, I., Petković, D.: Primjena GT koncepta u proizvodnji automobilskih dijelova/Application of GT concept in car's parts production. In: VI Studentska konferencija "TECHNO-EDUCA 2012", Tema: Preduzetničkim idejama ka ljepšoj budućnosti, Univerzitet u Zenici, Centar za preduzetništvo I inovacije CIP EIC, BSC Zeničko-dobojskog kantona, Zenica, B&H, 07 i 08, Novembar 2012 (2012)
11. Piljug, E., Alagić, I.: Disassembly and assembly technology of tie rod. In: VI Studentska konferencija "TECHNO-EDUCA 2012", Tema: Preduzetničkim idejama ka ljepšoj budućnosti, Univerzitet u Zenici, Centar za preduzetništvo I inovacije CIP EIC, BSC Zeničko-dobojskog kantona, Zenica, B&H, 07 i 08, Novembar 2012 (2012)
12. Alagić, I.: Opšte odrednice lanca dobavljača - model sistema upravljanja kvalitetom u upravljanju lancem dobavljača automobilske industrije. In: The 2nd Conference ODRŽAVANJE 2012/MAINTENANCE, Zenica, B&H, June 13–16 2012 (2012)

Development of Aircraft Mechanical Systems and Mechatronics Modeling

Želimir Husnić(✉)

The Boeing Company, Mechanical Systems and Hydraulic Design, Rt. 291 & Stewart Avenue, Ridley Park, PA 19078, USA
zhusnic@yahoo.com, zelimir.husnic@boeing.com

Abstract. Preliminary design of aircraft mechanical systems is complex process. Aircraft fly-by-wire controls associated with mechanical systems models need to be draw together in an appropriate sequence in order to perform various design studies with high accuracy and efficiency. The paper discuss Mechatronic approach related to this process. Mechatronic approach in design, modeling and simulation offer improvement in overall quality, while decrease cost and time required for development.

Keywords: Aircraft · Mechanical systems · Hydraulics · Development · Modelling · Simulation · Control system · Mechatronic · Validation

1 Introduction

The physical properties of the product are primary goal for engineering design. Engineering design is the process of designing a system, component, or process to meet desired needs.

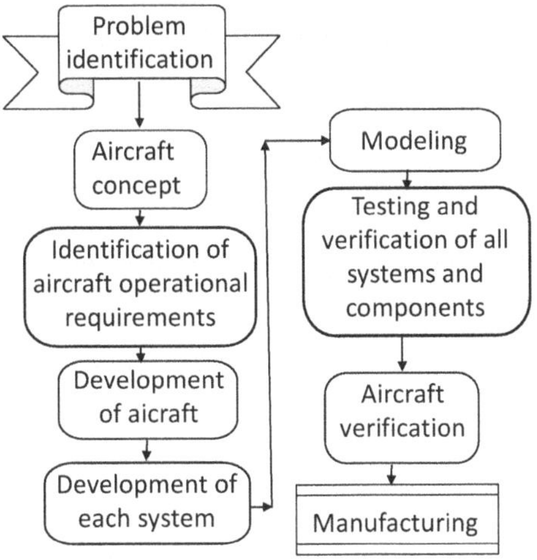

Fig. 1. Aircraft concept design and development process

Design Process include steps as follows: recognize the need, problem definition, gathering of data, concept, design and analysis, development, testing and manufacturing.

Selection of the aircraft configuration is fundamental to the aircraft conceptual design process including identification of aircrafts functional requirements and interfaces.

The aircraft flight control surfaces allowing regulation and control of the aircraft's flight attitude and permit stable flight. The aircraft mechanical systems enable the flight control surfaces operating and regulation.

The aircraft concept design and development process is shown in Fig. 1.

I. Karabegović (Ed.): NT 2020, LNNS 128, pp. 147–153, 2020.
https://doi.org/10.1007/978-3-030-46817-0_16

The aircrafts mechanical system development and integration should consider flight control surfaces and other functions as follows: flaperons, elevators, rudder, engine air particle separator, rotor and swashplate, rotor brake control, rotor phasing, engine start motor, nacelle conversion control, wing rotation, power steering, landing gear, cargo winch, cargo door, cargo handling.

The system development specify the system architecture to the item level and the system function and safety requirements.

Aircraft Function Development include following: structural aspects of the aircraft, avionics system and vehicle systems.

The aircraft vehicle systems that enable the aircraft to fly safely are Fuel, Propulsion, Flight Controls, Hydraulics, and Mechanical Controls.

Vehicle systems management provide an integrated communication for interfacing with system components, built in test, control functions, power demands, and communicating with the pilot's cockpit.

2 Aircraft Hydraulics System

Hydraulic system create power from engine driven pumps or electric motor driven pumps and distribute power to all connected components. The hydraulic supply must satisfy requirement for pressure and flow under all demand conditions. Hydraulic fluid must be clean and system shall be monitored to detect and isolate leaks.

The hydraulic systems shall be configured such that failure of any systems which cause loss of fluid or pressure will not result in complete loss of flight control. Ground rules regarding reliability and redundancy do change dependent of type of aircraft and overall requirements [1, 2].

Usually minimum two independent hydraulic systems are required for aircrafts.

If the aircraft's hydraulic system supplies all the primary and secondary flight controls and the aircraft cannot be controlled without hydraulic power, three hydraulic systems should be considered. Even this would not be required it nevertheless represents wise design practice.

Each hydraulic system shall be separated from all other systems and fluid shall not be transferred between systems.

In general, fixed wing aircraft shall maintain level one flying qualities of with one fluid system failure (including the power source) and level three flying qualities including carrier landing, with two fluid system failures. Rotary wing aircraft shall meet flying qualities with two fluid system failures for return to the intended landing area.

The most significant interface issue is the level of redundancy of the incoming command signal and the actuator valve configuration in terms of type of valve and the redundancy levels throughout the actuator.

2.1 System Development

Regardless of the hydraulics system configuration engineers with different skills and backgrounds must work together and communicate with each other to insure that the system is fundamentally successful.

The System development process include two major milestone for preliminary and critical design review [3].

Preliminary design review include items as follows: design synthesis, trade-studies, functional requirements, layout, environmental and weight analysis, risk mitigation, specifications for components or subsystems, stress analysis, development test, failure modes.

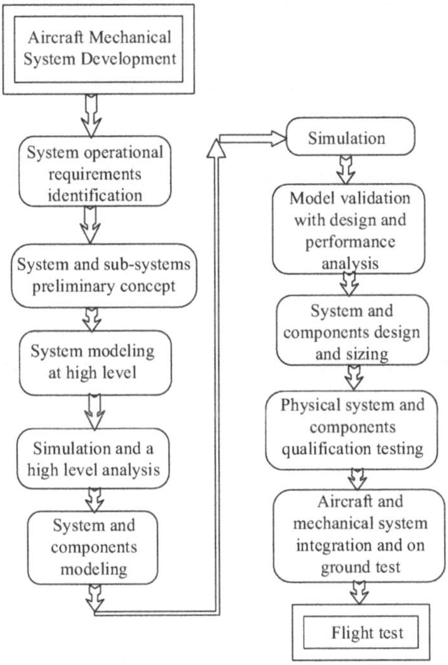

Fig. 2. Aircraft Hydraulics System Modeling

Critical design review include items as follows: design analysis, design studies results, model validation, performance analysis, engineering drawings, mechanical and electrical design, detailed design, stress analysis, environmental and thermal aspects, electrical and mechanical interfaces, reliability and maintainability, produceability, interface control drawings, development test results, manufacturing readiness.

A system Performance requirements review will contain system safety, redundancy, interface, digital, analog architecture, servo-electronics, failure detection, degraded/fail safe modes, hydro mechanical and electromechanical actuators, servo valves characteristics, stability, control valve characteristics, sensor, sizing of system components, electric characteristics, hydraulics characteristics, vibration, fatigue, reparability.

Mechatronic design modeling and simulation shown in Fig. 2 propose great improvement in overall development methodology.

2.2 Hydraulics System Characteristics

The basic determination regarding the hydraulic system, redundancy and system arrangement shall be made during the preliminary design phase.

In order to control the hydraulics system interfaces must be designed to meet a widespread range of sensors and actuators. Example of the aircraft hydraulic system shown in Fig. 3 include relay, switch, rotational speed, linear position, rotary position, actuator position, temperature, pressure, flow, level, proximity, cleanliness and fluid contamination sensor.

During work on system concept is required involvement of engineers with excellent knowledge in aircraft hydraulics systems.

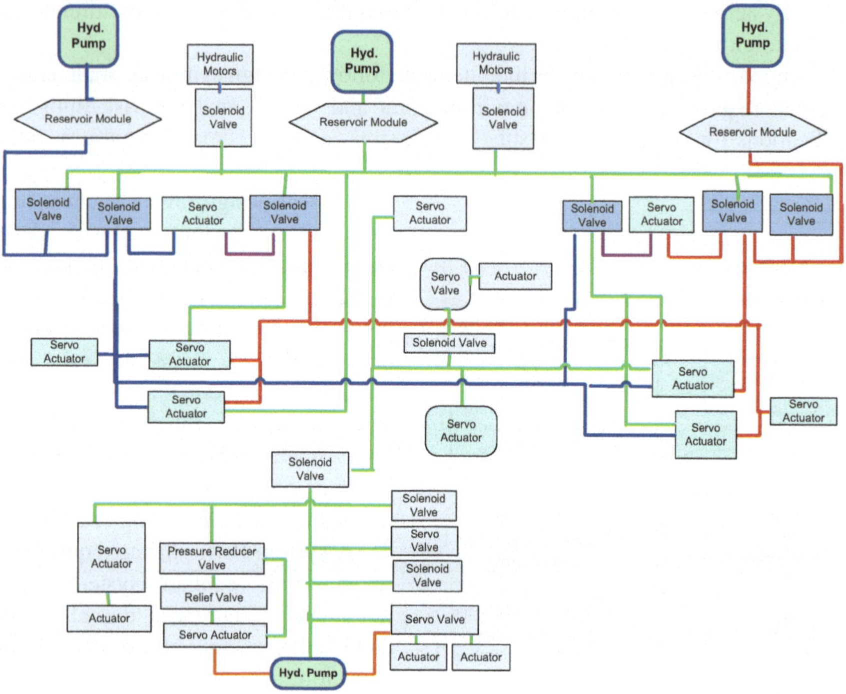

Fig. 3. Aircraft Hydraulics System

The System requirements, type of fluid, number of hydraulic systems, system pressure, temperatures, flow, main pump, backup pump, hydraulic motors, hand valves, solenoid valves, servo valves, actuators, servo actuators, accumulators, reservoirs, relief valves, filtration, tubing, fittings, schematic and system architecture, power system, heat exchangers, hand pump, seals, material, etc. [4].

The redundancy approaches shall use the minimum levels required to satisfy the requirements of aerospace industry standards. It could take the form of duplicate or alternate components, channels, or subsystems; each capable of performing the given function. The flight safety and mission reliability requirements of aerospace industry standards to be met. Redundant subsystems shall employ isolation of the redundant elements to preclude failure of one portion of the system from affecting any other part of the system.

The fault features that are demonstrated by analysis and simulation to provide a safe reaction to, and recovery from, single failures or failure combinations.

2.3 Modeling and Simulation

The goal of modeling is to create mathematical models that accurately simulate the real physical world.

All true physical systems are complex. Creating a mathematical model means that we have to simplify the system to the point that it can be analyzed.

After outlining possible solutions to the problem, the design team reviews the list, considers the pros and cons of each possible proposal, and then comes to an agreement on the best solution to continue the project.

Design constraints are taken into account when determining the best solution.

The team working on the project modeling and deciding the solution determines what must be done to turn the concept into a workable solution.

Real Time Aircraft Hydraulics System Models should be generated for each subsystem and for the entire aircraft hydraulics system. Model need to be based on requirements and physics.

Primary uses of models are for hydraulics system design and development, controls system design and development, system evaluation and modifications, accident investigation and failure troubleshooting, training and familiarization with aircraft system.

Mechatronic Design Approach shown in Fig. 4 provides numerical and graphical analysis, simulation with design tools that stimulate innovation, integration, communication and collaboration within design teams (feedback between teams). This reduces the risk of non-compliance with functional requirements and allows for early and constant verification throughout the entire work process.

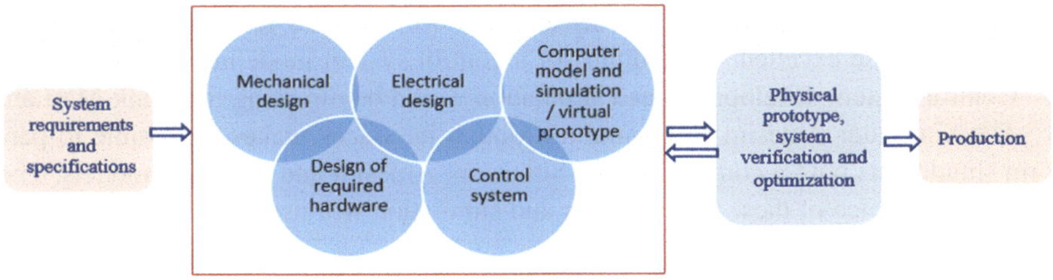

Fig. 4. Mechatronic design and development approach

This is all done before making a physical prototype, verifying and optimizing the system.

The modeling requirements should include analysis requirements, computer-aided engineering, simulation, model reliability, assumptions, model design, validation, and deliverables for the integral interfacing systems (mechanical, electric, electronic, thermal, and hydraulic).

Mechatronic represent combination of electrical and non-electrical inputs, outputs, or actuators, sensors, or electrical components.

The modeling and simulation tools may support synchronized electrical circuit simulation, mechanical dynamics simulation, heat and fluid mass transport simulation.

The simulation could include modeling and simulation to varying degrees of reliability. Also, co-simulation using multiple modeling environments may be required.

Validation of requirements is the process of ensuring that the specified requirements are sufficiently correct and complete. Validation Methods include traceability, analysis, modeling, test, similarity, and engineering review. Special tests, simulations, or demonstrations may be used to validate requirements.

The validation for the simulation model assume that the model corresponds to a physical system or component. The test procedures simulation will verify that the model performances were implemented as required and properly collect quantitative data for purposes of component and system description. The validation will define the accuracy of the model relative to the empirical data attained from the physical system.

2.4 Examples

Design engineers need to use different tools in order to achieve optimum design. Dependent of system different tools could be used, for instance for the aircraft hydraulics system shown in Fig. 3 could be used numbers of software for either system or component level modeling. In this work we will mention Amesim and Matlab-Simulink as examples of software that could be used for hydraulic system modeling.

The Simcenter Amesim is good toll for modeling and a power analysis of the hydraulic system [5, 6].

The purpose of Amesim model is to perform a high level power analysis, in order to assess the flow balance between the hydraulic power generators and consumers and the pressure available. The integration of components results from designing the mechatronic system and implanting the sensors, actuators and other components.

The complex hydraulic system as shown in Fig. 3 with servo valves and/or servo actuators require the controller and sensors that will provide feedback in real time. Simulink provide excellent capabilities for a control system modeling and simulation.

Control System development and simulation would be run using Simulink-MatLab.

The Simcenter Amesim and Simulink Matlab interfaces makes it possible to perform simulations with a combination of Simcenter Amesim and Simulink models. The user has the choice of these two methods and simulation will use solvers from the both software packages.

Hydraulic System Amesim Model would be transferred to Matlab-Simulink to run simulation. Simulation can be run where the hydraulic system is disturbed from initial or nominal condition by from external forces, flow, pressure demand or control command inputs. As result of simulation a controller could be designed for the hydraulic system.

Components developments, modeling and simulation need to be completed prior final the hydraulics system simulation and validation. There are number of software that could be used for this phase in process, for instance it could be used Amesim, Simulink, Easy5, Mathematica, CFD Software ANSYS and FloMaster, etc… [7].

It's crucial that engineers have exceptional understanding of the system and components. Also, team work with modeling and simulation software experts is essential.

Different types of simulation methods could be used, but validation will confirm that the specified model requirements are acceptable and complete.

3 Conclusion

The intensive use of computers and software in design, theoretical and experimental modelling, digital simulation, hardware-in-the-loop simulation, control systems design, computational fluid dynamics (CFD) analysis, verification, and relationship with the physical system design represent a challenge of mechatronic systems development. During system and components development engineers with different skills and backgrounds must work together and focal personnel, responsible for integration and coordination, are engineers with mechatronic background or education.

Hydraulic System Computer Models offer great benefits during development phase, for system modifications, troubleshooting, predictive maintenance, model based system engineering and applicable data base, training and simulation of fault modes.

Note. The Boeing Company is not associated with this paper

References

1. SAE AS5440 "Hydraulic Systems, Aircraft, Design and Installation Requirements For", Aerospace Standards, SAE, Warrendale, PA
2. SAE ARP4754 "Guidelines for Development of Civil Aircraft and Systems", Aerospace Standards, SAE, Warrendale, PA
3. Moir, I., Seabridge, A.: Design and Development of Aircraft Systems. Wiley, Hoboken (2013)
4. Husnić, Z., Dedić, R.: Uljna Hidraulika, Univerzitet, "Džemal Bijedić" Mašinski fakultet Mostar (2016)
5. Simulink - MATLAB, MathWorks (mathworks.com)
6. Simcenter Amesim, Siemens PLM Software (plm.automation.siemens.com)
7. FloMASTER, Mentor, a Siemens Business (mentor.com)

Chair Production and Robots Usage

Salah-Eldien Omer[1,2(✉)]

[1] Technical Faculty Bihać, University of Bihać, St.Irfana Ljubujankića bb, 77 000 Bihać, Bosnia and Herzegovina
[2] Zagreb, Croatia
sagzagreb@hi.htnet.hr

Abstract. The chair production is one of the complicated wood processing. The design of chairs from the old times is developed many times according to the agronomic recommendations to the chair construction generally a especially for the usage in certain places. The chair construction is also developed from certain shapes to modern ones which could guarantee the good quality and conformability of the product. In the production phases from the mechanical processing to the finishing process many production phases are very complicated and needs very well operating machines with high precision and good quality. Robots were introduced to the chair production before many years but not in most phases. When they started to be used in this process they brought a lot of high productivity and quality in final products.

Keywords: Chair construction · High quality production of chair elements and less production cost in many chair designs

1 Introduction

Production of chairs as a technology and process we consider it as the queen of processing in wood industry or furniture production. Especially the production of chairs from solid hard wood which a much requested product all over the world. We produced big quantities of chairs from different massive wood and exported all over the world with low prices according to the quality and price of wood used in their production. A lot of wood destroyed to produce many chairs with the high quality which was requested from all the buyers. The maximization of wood usage is requiring high quality processing machines plus very qualified workers.

Businesses of all sizes are quickly realizing that robotic automation is a highly cost-effective and efficient way to improve profits. There are a lot of ways to save money in a manufacturing line, but here we will explain how industrial robots in chair production address the biggest ways to save you money. In chair processing, no matter what application you are trying to automate, industrial robots decrease waste, provide a quick return on investment (ROI), improve space-efficiency, increase product throughput, and enhance the safety environment. Industrial robots perform applications with precision and consistency, saving valuable materials and reducing waste. The precision that robots bring creates fewer mistakes on the production line, providing

I. Karabegović (Ed.): NT 2020, LNNS 128, pp. 154–163, 2020.
https://doi.org/10.1007/978-3-030-46817-0_17

a higher quality product and loyal customer base, while also reducing the costs associated with waste removal.

Many chair producers are hesitant to the idea of purchasing a robot for their chair production, because they think it will cost too much. It is an investment, but one that pays off! Industrial robots offer a very fast ROI with their efficiency and reduced need for breaks, vacation, or sick leave. Industrial robots work continuously and help reduce overhead on your production line. Also, there are alternatives to cut the costs associated with buying a robot. Robots offers used refurbished robots and customized systems to help alleviate initial costs, while not sacrificing robot quality or efficiency. Bringing flexibility and space savings to the floor, industrial robots can fit into confined spaces and can be installed on shelves, pedestals, walls, ceilings, or on rails. Collaborative robot environments can also help reduce floor space by reducing the need for extra fencing. Though there are some limitations to perform some applications by these robots, it is part of the future of automation. Robots help bring greater profits by reducing part cycle time. There is no denying that an industrial robot can perform tasks at higher speeds, lift heavier products, and perform more precise applications and movements. It is important to identify which steps in your process are the most beneficial to automate. Redundant and hazardous work takes a lot of time, effort and energy and integrating a robot can easily remove bottlenecks in your manufacturing line. Industrial robots take over dangerous and dirty work. They reduce worker exposure to fumes, sparks, arc glare, or dust particles. They decrease the amount of injuries and health problems to employees by handling those heavy-to-lift products or tools and remove the worker from unsafe applications and environments.

Since the chair production is a special process, with many wood processing phases from dimensioning the input wood elements to their final shaping, Robots shows a very high efficiency in all these phases.

Also Chair design considers intended usage, ergonomics (how comfortable it is for the occupant) as well as non-ergonomic functional requirements such as size, stacking ability, folding ability, weight, durability, stain resistance, and artistic design. Intended usage determines the desired seating position. “Task chairs”, or any chair intended for people to work at a desk or table, including dining chairs, can only recline very slightly; otherwise the occupant is too far away from the desk or table. Dental chairs are necessarily reclined. Easy chairs for watching television or movies are somewhere in between depending on the height of the screen, children chairs are static. All these facts need a very precise work and production to assure the quality of these products to a very demanded market [1].

2 Chair Construction and Processing

Ergonomic design of all types of chairs, which is a very important factor in chairs constructions, distributes the weight of the occupant to various parts of the body. A seat that is higher results in dangling feet and increased pressure on the underside of the knees. It may also result in no weight on the feet which means more weight elsewhere. A lower seat may shift too much weight to the "seat bones". All other important dimensions in the chair constructions offer certain elements which will assure the needed ergonomic design of the chair.

A reclining seat and back will shift weight to the occupant's back. This may be more comfortable for some in reducing weight on the seat area but may be problematic for others who have bad backs. In general, if the occupant is supposed to sit for a long time, weight needs to be taken off the seat area and thus "easy" chairs intended for long periods of sitting are generally at least slightly reclined. However, reclining may not be suitable for chairs intended for work or eating at table.

The back of the chair will support some of the weight of the occupant, reducing the weight on other parts of the body. In general, backrests come in three heights: Lower back backrests support only the lumbar region. Shoulder height backrests support the entire back and shoulders. Headrests support the head as well and are important in vehicles for preventing "whiplash" neck injuries in rear-end collisions where the head is jerked back suddenly. Reclining chairs typically have at least shoulder-height backrests to shift weight to the shoulders instead of just the lower back.

Some chairs have footrests. A stool or other simple chair may have a simple straight or curved bar near the bottom for the sitter to place his or her feet on.

Some chairs have two curved bands of wood (also known as rockers) attached to the bottom of the legs. They are called rocking chairs.

A kneeling chair adds an additional body part, the knees, to support the weight of the body. A sit-stand chair distributes most of the weight of the occupant to the feet. Many chairs are padded or have cushions. Padding can be on the seat of the chair only, on the seat and back, or also on any arm rests or footrest the chair may have. Padding will not shift the weight to different parts of the body (unless the chair is so soft that the shape is altered). However, padding does distribute the height adjustment and more comfort to the user. Weight is increasing the area of contact between the chair and the body. A hard wood chair feels hard because the contact point between the occupant and the chair is small. The same body weight over a smaller area means greater pressure on that area. Spreading the area reduces the pressure at any given point. In lieu of padding, flexible materials, such as wicker, may be used instead with similar effects of distributing the weight. Since most of the body weight is supported in the back of the seat, padding there should be firmer than the front of the seat which only has the weight of the legs to support. Chairs that have padding that is the same density front and back will feel soft in the back area and hard to the underside of the knees [3].

There may be cases where padding is not desirable. For example, in chairs that are intended primarily for outdoor use. Where padding is not desirable, contouring may be used instead. A contoured seat pan attempts to distribute weight without padding. By matching the shape of the occupant's buttocks, weight is distributed, and maximum pressure is reduced.

Actual chair dimensions are determined by measurements of the human body or anthropometric measurements. The two most relevant anthropometric measurements for chair design is the popliteal height and buttock popliteal length.

For someone seated, the popliteal height is the distance from the underside of the foot to the underside of the thigh at the knees. It is sometimes called the "stool height". The term "sitting height" is reserved for the height to the top of the head when seated. The popliteal height, after adjusting for heels, clothing and other issues, is used to determine the height of the chair seat. Mass-produced chairs are typically 430 m high. For someone seated, the buttock popliteal length is the horizontal distance from the back most part of the buttocks to the back of the lower leg. This anthropometric measurement is used to determine the seat depth. Mass-produced chairs are typically 375–425 mm deep (Fig. 1).

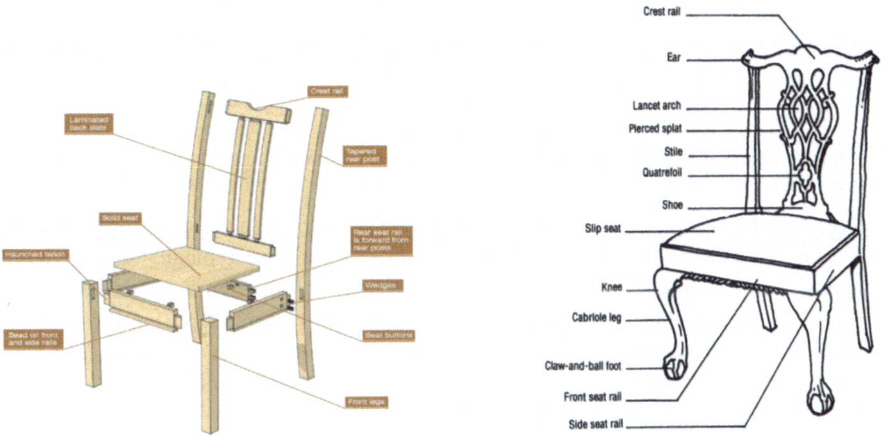

Fig. 1. Two basic different construction of chairs.

Additional anthropometric measurements may be relevant to designing a chair. Hip breadth is used for chair width and armrest width. Elbow rest height is used to determine the height of the armrests. The buttock-knee length is used to determine "leg room" between rows of chairs. "Seat pitch" is the distance between rows of seats. In some airplanes and stadiums, the leg room (the seat pitch less the thickness of the seat at thigh level) is so small that it is sometimes insufficient for the average person.

For adjustable chairs, such as an office chair, the aforementioned principles are applied in adjusting the chair to the individual occupant. Caster wheels are attached to the feet of chairs to give more mobility. Gas springs are attached to the body of the chair in order to give height adjustment and more comfort to the user. In certain cases of designing special product as chair, the designer follows the elements needed to full fill the purpose or the attended place of usage specially when they mend for children usage [3] (Fig. 2).

Fig. 2. Special construction and design of chair different modern chairs.

3 Robots in Chair Production

Robots have been proven to deliver a host of benefits in a wide variety of applications. End users introducing robots to their production process have typically seen a significant transformation their productivity and efficiency, with higher levels of output, product quality and flexibility amongst the many improvements reported [2].

Ten reasons which were generally verified by using robots in wood industry processing where also show the same effect in chair processing industry specially the production of chairs from solid hard wood:

- They reduce operating costs (energy saving about 8% on every 1 OC reduction in heating levels, and 20% by turning off unnecessary lighting. Also, can reduce or eliminate the costs associated with manual workers,
- Improve product quality and consistency, the machine will not suffer from tiredness distraction and similar, and you can count on high quality finish of products,
- Improve quality of work for employees; workers will no longer have to work in dusty hot or hazardous environments. They will learn also programming skills,
- Increase production output rates. They can be left running over night and during weekends with little supervision. They will not time out for breaks, sickness and similar,
- Increase product manufacturing flexibility, after programming the machine, you can easily switch from one to another phases of production,
- Reduce material waste and increase yield, by precise processing the input material,
- Comply with safety rules and improve workplace health and safety,
- Reduce labor turnover and difficulty of recruiting workers,
- Reduce capital costs (inventory, work in progress). Reduce the cost of consumables used and reduce wastage, moving products faster in production,
- Save space in high value manufacturing areas. They can be placed on shelf systems, on walls or even in ceilings.

Many years ago, chair manufacture was primarily manual work. But production techniques have changed dramatically, and today is probably the most technologically advanced manufacturer of furniture and chairs specially [4].

Entering the factory, the first thing that strikes you is what doesn't strike you now when using sophisticated technology. The smell of the glue, the din of the machines, the dust from the wood: all that is history.

Today the factories are quiet and clean, even though many chairs are made in the modern chair factory per year.

Today, each chair is sanded by two robots (instead of many workers) that reach every part of its surface, while the lacquering process uses a high-speed disk. So, the quality of polishing and lacquering is even higher than before, and the working environment has improved markedly because the machines have been placed inside cabins with efficient dust extraction and noise insulation.

The first and effective usage of Robots shows in the first phases of mechanical work preparing the dry elements in certain shapes and curves (Figs. 3 and 4).

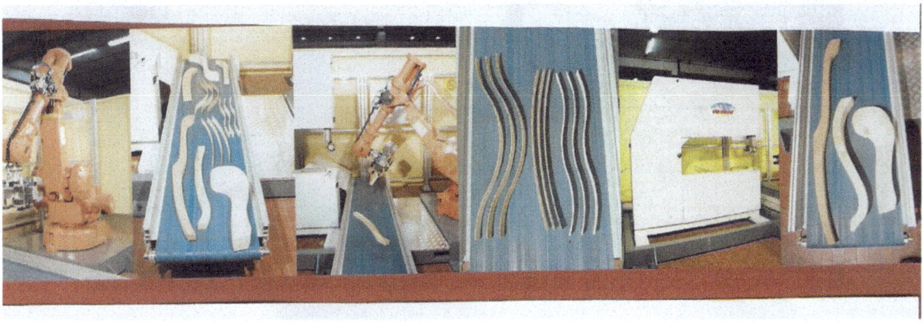

Fig. 3. Preparation of different curved chairs elements by Robot.

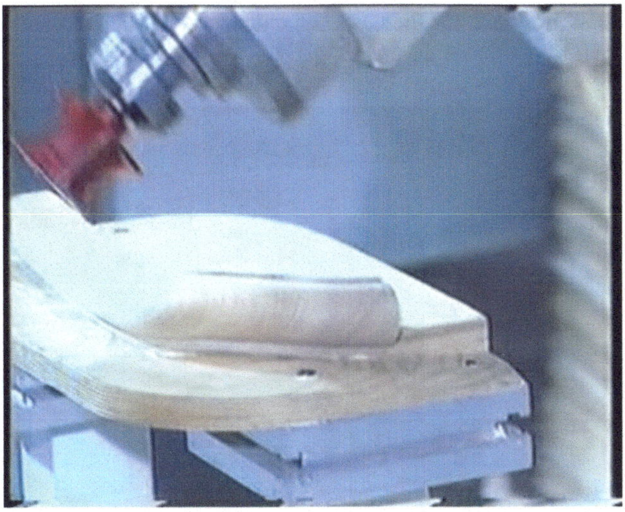

Fig. 4. Sanding process of chair elements by Robot unit.

The second phase of processing chair elements is sanding and preparation to assembling phase. This phase is mostly done by CNC machines or if there is a Robot center containing sanding units. The Robot centers are very expensive, and they are not justified in medium or small production capacities.

The assembling phase is now changed from manual work to Robot usage. The Robots are programmed to work in special working area or sometimes in the track of production line with the help of qualified workers (Fig. 5).

Fig. 5. Small Robert units for assembling chairs from solid wood.

The more sophisticated Roberts contain units for gluing the joints before assembling, with special programs for the designed chairs and the defined area for the glues, and then continue to assemble the chairs parts as planned (Fig. 6).

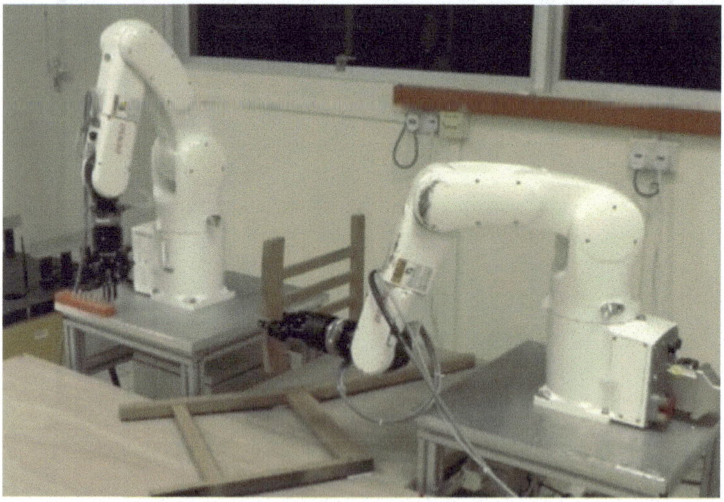

Fig. 6. Assembling Robots units with pumps for spraying glues, before assembling the parts.

The chairs after assembling and control the drive to lacquering units individually or more than one chair as the capacity of the lacquering Robot units design (Fig. 7).

Fig. 7. Three different types of lacquering units designed according to the type of chair needed to lacquer.

The last phase using Robots in the chair production is the packing, depending on the requested type of packing (Fig. 8).

Fig. 8. Universal packing Robots, program as needed.

4 Conclusions

Concerning the above mentioned material which is taken from the planned development program for two factories which produced chairs from solid hard wood, we can conclude the following:

- The process of producing solid wood chairs is one of the advanced technology of wood processing into the upgraded products and in it there is a lot of phases which could be advanced by using Robots,
- The design of many requested chairs in the market is varied from product to product depending on the place of usage and needed characteristics of chair,
- There is a lot of special construction of chairs from solid wood or laminated plywood are better to produce with sophisticated technologies which bring accuracies in the designed shape and assure the final quality,
- Some working placeless in the technical processing of solid wood in the chair production could organize to in high productivity phases by using robots or CNC centers with special tools meant for that process,
- In the phases of sanding cure elements of chairs in massive production, the Robots are welcome for their designed and programs specially for that fine work,
- The assembling process in chair production is a very delicate one and it needs well programmed pressure and time based on the glue the producer use, which is easy to get by programmed Robots,
- The lacquering process for the chairs is a very sensitive one and it needs equal lacquering implementation bots with adjustable pressure where Robots show very high precise and satisfied quality.

References

1. The international eCAADe 2015 conference the Robotic Woodcraft team, TU Vienna
2. Landscheidta, S: Opportunities for robotic automation in wood product industries: the supplier and system integrators′ perspective. In: 27th International Conference on Flexible Automation and Intelligent Manufacturing, FAIM2017, Modena, Italy, June 2017
3. Omer, S.-E., Majdak, I.: Inventive Product construction way to International Market, Sarajevo, RIM 2017
4. Omer, S.-E.: Justiified Usage of Robots in Wood industry, Saint Peterson University (2017)

Programming Robot KUKA KR 16-2 for a Palletizing Application

Samir Vojić(✉) and Ramiz Sijamhodžić

Technical Faculty Bihac, University of Bihac, Dr. IrfanaLjubijankićabb, 77000 Bihać, Bosnia and Herzegovina
samir.vojic@unbi.ba

Abstract. This paper presents the robot programming process in the palletization process, as well as an analysis of the results obtained after palletization. The KUKA robot along with an example of programming of this type of robot in the palletization is described. Robot programming is designed to take arbitrarily ordered work pieces onto and stack them on a pallet.

Keywords: Robot · KUKA robot · Palletization · Programming

1 Introduction

From the earliest occurrences of manufacturing plants, industrial plants and factories, there was a need to transport raw materials, semi-finished products and products. Initially, there was a major problem with transportation during the loading of goods and commodities. This process was time-consuming because it was done manually. With the advent of pallets, the process began to accelerate. However, again, the problem occurred when loading heavy parts. The palletization process required a great deal of human force and was reluctantly accepted by the workers. With the advent of palletizing robots, this problem came to an end. Not only did the robots perform the process in less time, they also made huge financial savings. Their advantage is reflected in the fact that they do not get tired and rarely make mistakes. For most robots, the principle of operation in the palletization process is similar: the workpiece activates the sensor that drives the robot, the robot picks it up from the conveyor belt and transfers it to the pallet, where it releases it to the appropriate position. When palletizing, the robots also have a counting sensor built-in. After a certain number of transferred workpieces to the pallet, the robot receives a signal that the palletization process starts from scratch, that is, to a new pallet.

2 Development and Application of Palletization

Palletization is a set of organizationally connected technological processes for the purpose of automation, manipulation and transportation from the place of production to the point of consumption or further transport. Palletization has been known for more than 90 years, but it began to be applied only about 30 years ago. The pallet is actually

I. Karabegović (Ed.): NT 2020, LNNS 128, pp. 164–171, 2020.
https://doi.org/10.1007/978-3-030-46817-0_18

a full-length wooden plateau that serves as a loading area for larger loads and more products. It handles and transports a large number of goods at once. Massive application of pallets began during the Second World War [5].

The palletization rate and the trend of its growth are highest in industrialized countries. With the introduction of this system, the first phase of larger, manipulative transport units was created, which resulted in faster movement of goods, a significant reduction in damage to goods and a significant reduction in production costs. All the benefits of the pallet were only noticed when the pallet started circulating from the manufacturer to the consumer and vice versa.

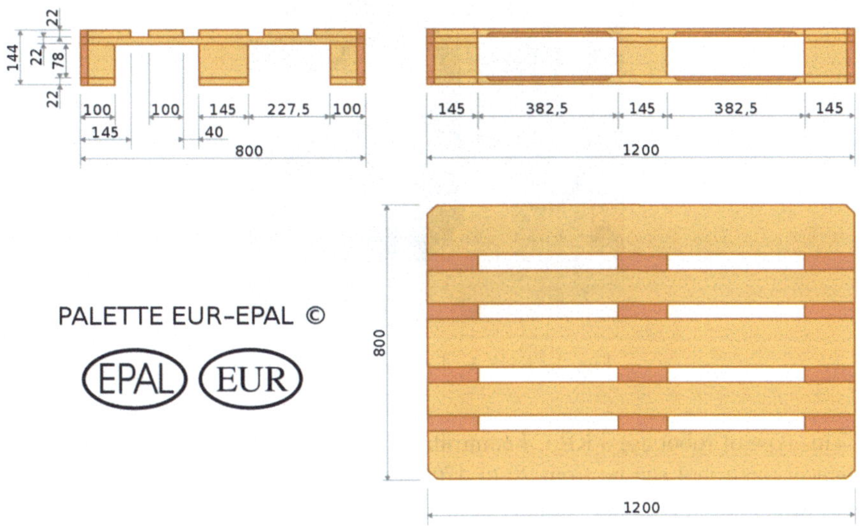

Fig. 1. Standard euro pallet [4]

European standard pallets are mostly used-euro pallets. Euro pallets are wooden flat pallets measuring 1200 mm × 800 mm × 144 mm. The weight of euro pallets is 20–24 kg. Figure 1 shows a standard Euro pallet with dimensions [5].

3 Robot KUKA KR16-2

KUKA is a German manufacturer of industrial robots. In 1973, KUKA designed and produced its first robot called FAMULUS, which had six electromechanical axes. Just as most robot manufacturers today have their own programming language for robot programming, KUKA also has its own KRL programming language. Robot programming can be both direct and indirect [2].

In order for the robot system, equipped with sensors and actuators, to achieve the desired movement with the ultimate goal of completing the task, a control action is required [1].

KUKA Robot is an industrial robot manufactured to operate at high speed and high accuracy during operation. KUKA robot KR 16-2 is a type of robot that is attached to the floor, with 6 degrees of freedom of movement, is shown in Fig. 2.

Fig. 2. KUKA of robot KR 16-2 [3]

This type of robot use a KR C4 controller, the supply voltage is 180 to 230 V. The weight of this robot can be from 20 to 4700 kg, and the precision in the cycle repeat from 0.015 mm to 0.2 mm.

4 Programming Robot in the Palletization Process

Robot programming can be direct and indirect. In this example, direct programming, or programming using a smartPAD pendant, was used for on-site learning. This is a drawback of this type of programming because the robot cannot work productively

Fig. 3. Workpiece measuring 100 × 100 × 90 mm

while programming. The task is to simulate the transfer of workpieces to a pallet using a vacuum clamp. The starting positions of the workpieces are defined so that the robot can run smoothly. Figures 3 and 4 show the workpiece and pallet used in this process.

Fig. 4. Pallet of dimensions 200 × 300 mm

The operator programmed a robot with smartPAD, which was out of productive use during that time. This mode of programming is also called PlayBack programming. The robot, using a vacuum force, takes one workpiece one at a time, in a predetermined order and transfers it to the pallet.

Figure 5 shows the vacuum gripper, manufactured by SCHUNK, which was used in this process.

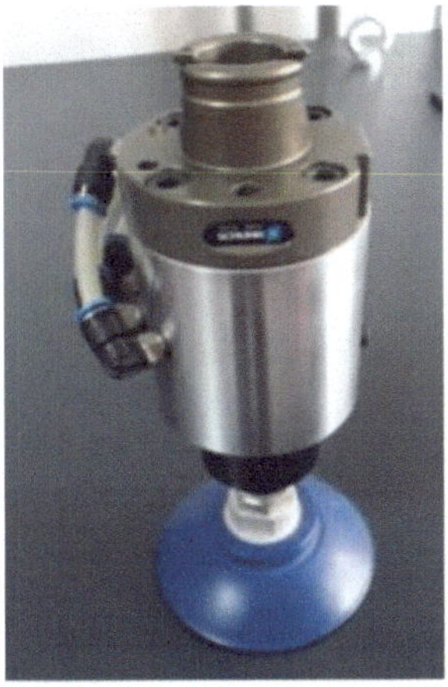

Fig. 5. Vacuum gripper

The vacuum gripper exerts a grip force on the vacuum principle. The carrying force F is equal to the difference in pressure p, between the pressure in the suction bowl and the surrounding one, and the size of the suction bowl suction surface A:

$$F = \Delta p \cdot A \quad (1)$$

4.1 KUKA Robot Programming

Before programming the path, we select the "Program" folder on the learning pendant, and then take a new program, which we have named "vak". Figure 6 shows a navigational file manager, consisting of:

- header (1),
- directory structure (2),
- file list (3) and
- status bar (4).

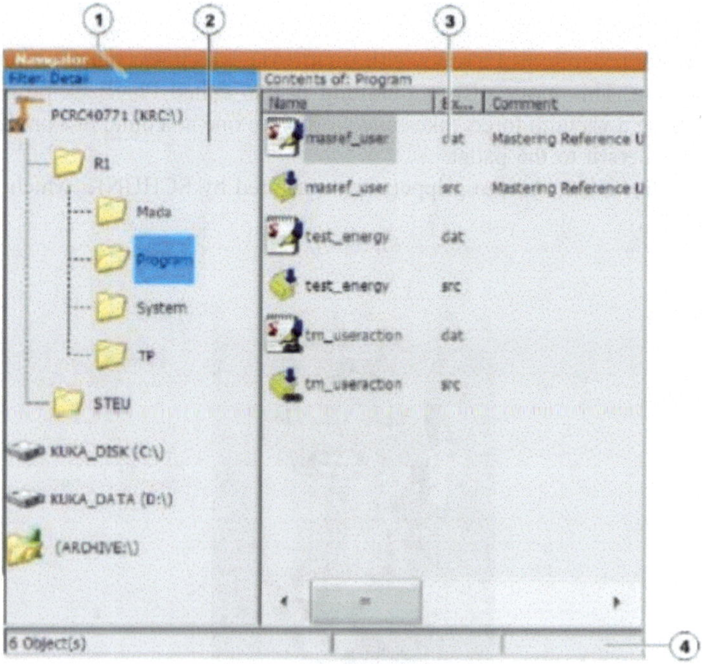

Fig. 6. Navigator file manager

Using the KRL programming language, we do robot programming. We store each robot receiver position in the controller. The robot is programmed to move at different speeds between different positions. Thus, when gripping and releasing a workpiece, the receiver moves at a slower speed so that the robot can grasp the workpiece precisely, while when carrying the piece and returning to new pieces, the robot moves at a higher speed. Part of the code is given in Fig. 7.

```
DEF vak( ) INI
CHECK vak State=OFF at START Delay=0 ms SET vak
State=VACUUM GDAT3
SET vak State=OFF CONT at START Delay=0 ms loop
PTP HOME Vel=100 % DEFAULT
PTP p14 Vel=100 % PDAT8 Tool[1]:Spitze Base[0] LIN
p15 Vel=2 m/s CPDAT8 Tool[1]:Spitze Base[0] SET GRP 3
State=OPN GDAT13
LIN p16 Vel=2 m/s CPDAT9 Tool[1]:Spitze Base[0] PTP
p17 Vel=100 % PDAT9 Tool[1]:Spitze Base[0] LIN p18
Vel=2 m/s CPDAT10 Tool[1]:Spitze Base[0] SET GRP 3
State=CLO GDAT14
LIN p19 Vel=2 m/s CPDAT11 Tool[1]:Spitze Base[0] PTP
p20 Vel=100 % PDAT10 Tool[1]:Spitze Base[0] LIN p21
Vel=2 m/s CPDAT12 Tool[1]:Spitze Base[0] SET GRP 3
State=OPN GDAT15
LIN p22 Vel=2 m/s CPDAT13 Tool[1]:Spitze Base[0] PTP
p23 Vel=100 % PDAT11 Tool[1]:Spitze Base[0] LIN p24
Vel=2 m/s CPDAT14 Tool[1]:Spitze Base[0] SET GRP 3
State=CLO GDAT16
LIN p37 Vel=2 m/s CPDAT23 Tool[1]:Spitze Base[0] PTP
p38 Vel=100 % PDAT16 Tool[1]:Spitze Base[0] LIN p39
Vel=2 m/s CPDAT24 Tool[1]:Spitze Base[0] SET GRP 3
State=OPN GDAT21
LIN p40 Vel=2 m/s CPDAT25 Tool[1]:Spitze Base[0] PTP
p41 Vel=100 % PDAT17 Tool[1]:Spitze Base[0] LIN p42
Vel=2 m/s CPDAT26 Tool[1]:Spitze Base[0] SET GRP 3
State=CLO GDAT22
LIN p46 Vel=2 m/s CPDAT29 Tool[1]:Spitze Base[0] PTP
p47 Vel=100 % PDAT19 Tool[1]:Spitze Base[0] LIN p48
Vel=2 m/s CPDAT30 Tool[1]:Spitze Base[0] SET GRP 3
State=CLO GDAT24
LIN p49 Vel=2 m/s CPDAT31 Tool[1]:Spitze Base[0]
endloop
PTP HOME    Vel= 100 % DEFAULT END
```

Fig. 7. Part of the programming code

4.2 Analysis of the Palletization Process

This paper shows how the KUKA robot performs the palletization process, that is, how randomly ordered workpieces are transferred and stacked on a pallet, Fig. 8.

Fig. 8. Starting and ending state of work pieces

At the initial position, the workpieces are arbitrarily arranged with sufficient spacing to allow the robot to run smoothly. Each work piece has its own position and its ordinal number. The robot picks them up one by one in a programmed order and places them at their destination, or pallet.

After the palletizing process, we can conclude that this type of robot performs such tasks without major problems, which shows its justified presence in the palletization process. Today, robots are also manufactured that have integrated various sensors and cameras in the hand of the robot so that if any deviation from the set standards occurs, the robot controller itself makes necessary corrections.

5 Conclusion

This 21st century will be dominated by robots, which will be used for almost all the work that man does. It is predicted that in the first half of the century they will be employed in programmed jobs, and later they will be able to engage in those intellectual and creative jobs. In the process of palletizing, for a company to do all the tasks on time, a large number of workforce is required, which certainly creates high costs for the employer. The robots have proved to be a full hit in terms of humanity and cost-effectiveness.

References

1. Karabegović, I., Doleček, V.: Robotika, Bihać (2002)
2. KUKA System Software 8, Training Documentation, KUKA Roboter GmbH (2013)
3. https://www.robots.com/robots/kuka-kr-16-2-f. Accessed 08 Dec 2019
4. https://www.epal-pallets.org/eu-en/load-carriers/epal-euro-palletEuropaleta. Accessed 08 Dec 2019
5. https://www.robotics.org/content-detail.cfm/Industrial-Robotics-Industry-Insights/Palletizing-and-De-PalletizingApplications/content_id/3411. Accessed 08 Dec 2019
6. Karabegović, I., Vojić, S., Doleček, V.: Programiranje industrijskih robota u realnom i virtualnom okruženju, Zbornik radova RIM 2003, Bihać (2003
7. Karabegović, I., Vojić, S., Doleček, V.: Simulation of the process of sorting and storage of finished products using intelligent systems. In: Proceedings of the International Conference MECHANIKA 2005, Lithuania (2005)
8. Karabegović, I., Vojić, S., Doleček, V.: Programiranje industrijskih robota u procesu sortiranja i skladištenja gotovih proizvoda, Zbornik radova RIM 2005, Bihać (2005)
9. Karabegović, I., Vojić, S., Doleček, V.: Robot gripper path programming while movement with exact positioning. In: Proceedings of the International Conference MECHANIKA 2006, Lithuania (2006)
10. Luan, N., Zhang, H., Tong, S.: Optimum motion control of palletizing robots based on iterative learning. Ind. Robot Int. J. **39**(2), 162–168 (2012). https://doi.org/10.1108/01439911211201627
11. Pedro, T., Lima, J., Costa, P.: Double A path planning for industrial manipulators. In: Robot 2015: Second Iberian Robotics Conference: Advances in Robotics, Lisbon, vol. 2, pp. 119–130 (2016)

Reinforcement Learning Based Human-Prosthetic Robot Interaction Control in Movement Therapy

Zlata Jelačić(✉)

Faculty of Mechanical Engineering, University of Sarajevo,
Vilsonovošetalište 9, 71000 Sarajevo, Bosnia and Herzegovina
jelacic@mef.unsa.ba

Abstract. A human–robot interactive control is proposed to control a lower limb prosthetic robot for amputee patients in the movement therapy training. The developed rehabilitation prosthetic robot is driven by the hydraulic system and has two rotational degrees of freedom. An adaptive admittance model is adopted considering its suitability for human–robot interaction.

The combined effects of flexibility of linear hydraulic actuators and compliance provided by the controller should contribute to the training comfort, safety, and therapeutic outcome in the gait rehabilitation.

In this paper, we present the human-robot interaction admittance model where individual admittance parameters suitable for patients are obtained by reinforcement learning.

Keywords: Movement therapy · Prosthetic robot · Human–robot interactive control · Admittance model · Reinforcement learning

1 Introduction

When focussing on motor activities, it is widely accepted that the repetition of sessions of movement therapy plays a key role in the modification of motor outcomes [1]. In this context, robotic devices, due to their ability to improve patients' compliance to treatments, are increasingly being exploited to strengthen and validate training programs directed to improve motor outcomes.

The role of robotic devices in improving motor outcomes during rehabilitation is quite obvious, however the rationale behind clinical efficacy in rehabilitation programs which include robotic devices, when compared to the conventional ones [2], has not been developed in a theoretical framework.

Certainly, device-based rehabilitation is more effective due to its "more engaging" nature. In this framework, it may be helpful to take into consideration the theory behind machine learning and artificial intelligent systems, for which the outputs of motor re-training represent a valid replica: it is indeed demonstrated that repetitive task-oriented practice in an un-supervised scenario (which mimics a learning-by-doing paradigm) in motor re-training represents a viable and valid method for rehabilitation, and that another means to convey rehabilitation is usually obtained by either a real teacher

I. Karabegović (Ed.): NT 2020, LNNS 128, pp. 172–181, 2020.
https://doi.org/10.1007/978-3-030-46817-0_19

(e.g. the physical therapist), or a virtual teacher (such as in virtual reality-based rehabilitation), thus triggering an instance of the learning-by-imitation scheme. If there habilitation exercises are administered in a *game* fashion, which means asking the player to choose a solution in terms of movement execution out of a number of discrete possible alternatives, we can hypothesize that this is an example of a reinforcement-learning scheme [3].

2 Reinforcement Learning

The internal model hypothesis, i.e. the presence of an internal representation of the dynamics and kinematics of movement at the neural level, is acknowledged as one of the leading theories of motor control and learning in neuroscience [4], and many authors proposed different mathematical proxies of this empirical paradigm (see e.g. [5]). In the framework of rehabilitation, it is assumed that, in the brain (the controller), this internal representation is retrained by changing its functional structure over time through interaction with modifications either in the environment or in the affordances of the body. It is acknowledged that most of the computational models that can simulate this ability towards change are based on the presence of rules (needed to adjust the parameters of the controller), where the interaction between the controller and the controlled plant (the body) allows some kind of modification. Among them, reinforcement learning is one of those machine-learning approaches where an entity decides a strategy that will maximize some reward function, as a result of the actions [6]. The controller, which is called the agent in machine learning definitions, is not actually told which actions need to be taken, but learns to adapt the policy, based on the obtained reward.

Algorithms able to analytically provide this kind of training are now at hand [7], and the mechanism itself has been fully studied in behavioural analyses of games (see e.g. [8]), where simple reinforcement learning schemes have been demonstrated as efficient in respecting both the law of effect, for which choices that carry positive effects in the short term tend to be repeated in the future, and the power law of practice, for which the performance of a natural system when facing unseen conditions tend to increase with a decreasing rate over time [9].

In the framework of motor control, this approach has been hypothesized as the only one that grants consistent improvement in general learning-by-doing schemes. At the same time, the implementation of reinforcement learning is generally based on the presence of a discrete number of different choices (a number of primitives), among which the agent is able to choose. The availability of (non-strictly) determined games that involve decision making and execution of a discrete and finite number of motor tasks is fostering new scenarios for enhancing re-training based on reinforcement. It is indeed acknowledged that the prefrontal motor cortex serves as the command generator that sends input to the descending pathways. The way this system chooses the pattern of commands based on the desired action can be represented as the function of a planner, whereas a controller at a lower hierarchical level transforms these neural commands into actual movers. If no mechanism intended to modify the parameters of the planner is present, the controller will only be able to adapt to modifications based on the nature of the planner, in a short-term paradigm based on the difference between the intended action and the executed one. If, instead, some measure of long-term reward is used to let the planner modify its behaviour, the latter will adapt its parameters based on how much the reward differs from the reward it was expecting. Once this difference is negligible, the planner will be trained to drive the lower hierarchy blocks in such a way that the actions are the ones needed to correctly perform the task and obtain the reward which was expected.

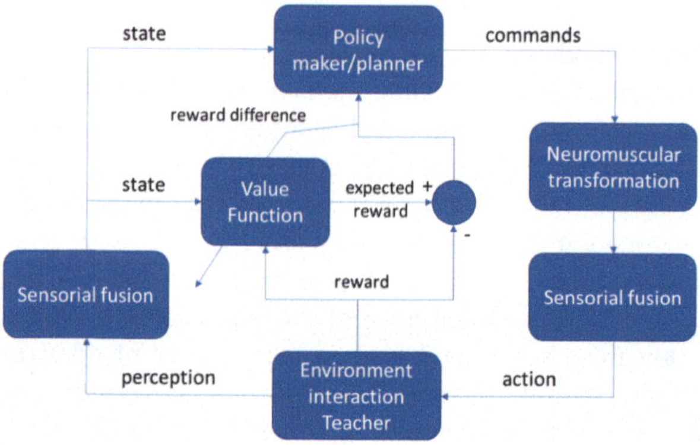

Fig. 1. Schema of the reinforcement learning structure for motor training

Through the interaction with the teacher or with the environment, humans sense some variables and estimate the state; this is input to the Value Function block, which is able to estimate the expected reward based on the state. The error-signal between the expected reward and the obtained reward is used both to adapt the planner parameters to generate the required commands, and to train the Value Function estimator.

This mechanism is the basis of training in games, where the Value Function inner block in Fig. 1 replicates the mechanism of "acting to maximize the reward". If no games are present, the training is performed based on the difference between the intended action and the obtained one, with no long-term goal. As a result, if this were the case of rehabilitation programs for humans, the absence of reward mechanisms can lead to a "less adaptive" behaviour where no changes in the parameters of the planner appear due to long-term rewards. On the contrary, a system able to excite the reward mechanism (such as the case of games in a training session) will facilitate modifications of the planner, improve retention of the actions that grant a reward, and possibly facilitate generalization ability. It is here speculated that, also in real life activities, the same mechanism appears, where the human brain is able to extract some sort of reward function based on the performance of the executed task, in terms of accomplishment of the desired objective (see e.g. [10]).

A new adaptation law is proposed in this study. Reinforcement learning is introduced into the parameters adjusting process. Based on the learning algorithm, an admittance model including patient's personal characteristics is established for the interactive control in active rehabilitation training. Our SmartLeg group developed a hydraulic wearable lower limb prosthetic robot. Hydraulic actuators provide a certain degree of passive compliance, while the proposed admittance control algorithm based on reinforcement learning provides active compliance by adapting robot behaviour to patients. The combined effects of flexibility of actuators and compliance of control can improve the training comfort, safety, and therapeutic outcome in the gait rehabilitation.

3 Interactive Active Compliance Control Strategy

3.1 Determination of Human-Robot Interactive Model

Seen from the instantaneous power flow between two or more physical systems, physical systems come in only two types: admittances, which accept effort (e.g. force) inputs and yield flow (e.g. motion) outputs; and impedances, which accept flow (e.g. motion) inputs and yield effort (e.g. force) outputs. For two physical systems in contact with each other, the emphasis of their fusion characteristics is mutual compensation: if the environment is impedance, the robot should show admittance characteristics and viceversa. The patient is regarded as the environment to robot in the rehabilitation task. When the patient has active walking consciousness, the affected lower limb has certain movement ability. The uncoordinated movements between the robot and the patient exert the contact force, so the environment has impedance characteristics. Further, the design of the robot controller should eliminate the contact force, takes the contact force as input and reacts as the motion adjustment. Therefore, the interactive model between the two physical systems (e.g. the patient and the robot) is of admittance characteristics, as shown in Fig. 2.

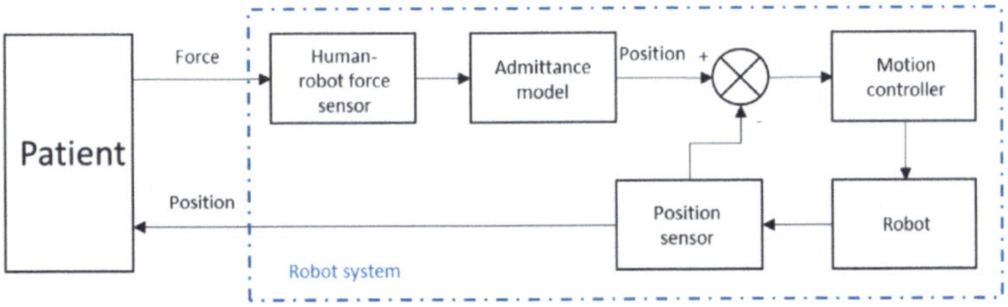

Fig. 2. Human–robot interaction in gait rehabilitation training

It can be seen that there is no force closed-loop in interactive control. The compliance in interaction is embodied in the dynamic relationship of the force and the position, that is, the characteristics of the admittance model.

The output of admittance model is the adjustment of position denoted as $\Delta\theta(t)$, and the desired joint trajectory $\theta_d(t)$ for the robot is calculated in Eq. (1), where $\theta_r(t)$ represents the reference trajectory

$$\Delta\theta(t) + \theta_r(t) = \theta_d(t) \tag{1}$$

The admittance model is the dynamic relationship between the human–robot interaction force F_{int} and the adjustment of position $\Delta\theta(t)$. It is given as follows

$$F_{int} = M\Delta\ddot{\theta}(t) + B\Delta\dot{\theta}(t) + K\Delta\theta(t) \tag{2}$$

where M, B, and K are mass, damping, and stiffness coefficient, respectively.

The admittance model transfer function (3) is obtained through the Laplace transformation from Eq. (2)

$$\frac{\Delta\theta(s)}{F_{int}(s)} = \frac{1}{M(s)s^2 + B(s)s + K} \tag{3}$$

It can be seen from formula (3) that M, K, and B in the admittance model can be used as adjustable parameters. To determine the appropriate adjustable admittance parameters, the influence of each admittance parameter on the control characteristics is studied by simulation, as shown in Fig. 3.

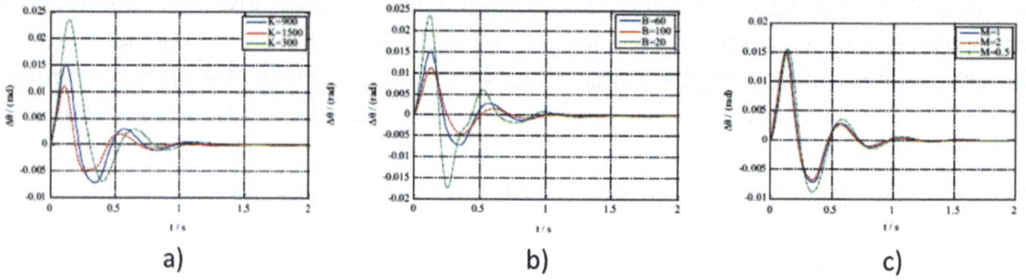

Fig. 3. Effects of admittance parameters. (a) Effect of stiffness coefficient K, (b) effect of damping coefficient B, and (c) effect of inertia coefficient M

From the simulation results, we can see that stiffness coefficient K reflects the adaptability of the adjustment of position. The greater the stiffness, the smaller the adjustment of position is. Damping coefficient B reflects the adaptive speed of the adjustment of position. The greater the damping, the longer the rising time is. And with high B value, the same amount of position adjustment needs more energy. The adjustable range of mass coefficient M is limited, because it causes the rapid change of the damping ratio and natural frequency of the system at the same time, which influences the stability of the system. It has little influence on the changeable amount of the position in the limited adjustable range. Therefore, it is not selected as an adjustable parameter of the admittance model. Stiffness coefficient K and damping coefficient B are the adjustable parameters of the admittance model to meet the requirement of the patients during different rehabilitation period.

3.2 Admittance Parameter Adjustment Algorithm Based on Reinforcement Learning

In active rehabilitation training, the adaptive parameters (K and B) algorithm is adopted according to the degree of participation of patients. It allows the robot to adapt itself to the patient in the range of stiffness (K_{up}, K_{dn}) and the range of damping (B_{up}, B_{dn}) that are set to ensure the stability of the system. In addition, human walking is a coordinated high-level planning process, and patients have individual walking habits before the impairment of lower limb function. Therefore, reinforcement learning algorithm is applied to the personalized learning of admittance parameters. The purpose of learning algorithm is to find optimal strategies for parameter adjustment suitable for different patients using the collected interactive data of themselves. Using this method, the adjusting process is according to the human control performance and walking habits and adapts to the different disability levels of patients.

Sigmoid function is used in the design of adaptive law. It is continuous, mono-incremental, and bounded in the definition domain $(-\infty, +\infty)$, and the range is from 0 to 1, as shown in Fig. 4(a). It is defined as

$$S(x) = \frac{1}{1 + e^{-x}} \tag{4}$$

Based on Sigmoid function, the used function is

$$S'(x) = \frac{h}{1 + e^{c|x|}} \tag{5}$$

where the parameter h is used to adjust the peak value, and the parameter c is used to adjust the concentration of the function, as presented in Fig. 4(b).

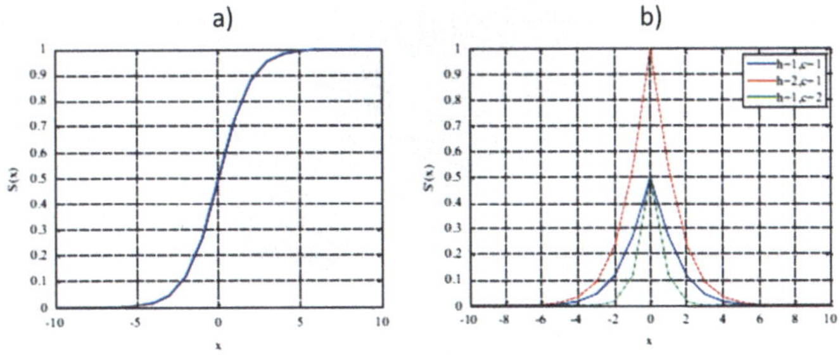

Fig. 4. Functions' graphic representation: (a) sigmoid function $S(x)$ and (b) $S_0(x)$

An adaptive law is proposed to tune the admittance parameters based on Eq. (5), and the adjusting values of ΔB and ΔK are defined in Eq. (6), where B_{up} and B_{dn} are the upper and lower limits of damping, respectively; K_{lup} andK_{ldn} are the upper and lower limits of stiffness, respectively; F_d is the desired force set according to the evaluation onpatient's condition; h_1, c_1, h_2, and c_2 are the parameters of Sigmoid deformation function defined in Eq. (5).

$$\Delta B = \begin{cases} B_{up}, & B < B_{up} \\ h_1 \frac{1}{1+e^{c_1 |F_{int} - F_d|}}, & B_{up} \leq B \leq B_{dn} \\ B_{ldn}, & B > B_{dn} \end{cases}$$
$$\Delta K = \begin{cases} K_{up}, & K < K_{up} \\ h_2 \frac{1}{1+e^{c_2 |F_{int} - F_d|}}, & K_{up} \leq K \leq K_{dn} \\ K_{dn}, & K > K_{dn} \end{cases} \quad (6)$$

4 Experimental Results

The interactive control experiments are conducted on the prototype of gait rehabilitation prosthetic robot (Fig. 5) under two modes of passive rehabilitation training and active rehabilitation training.

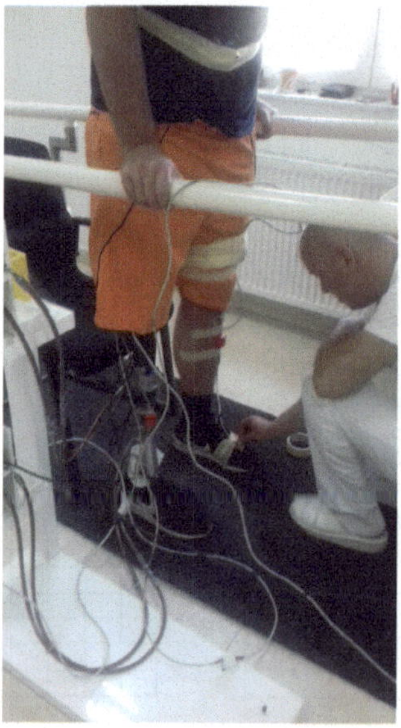

Fig. 5. Gait rehabilitation prosthetic robot

4.1 Experimental System and Protocol

At present, it is in the prototype stage. A healthy person (male, 26 years, 1.85 m height) participated in this study. Before the training, he wore the exoskeleton robot prototype, which was adjusted in accordance with the length of his lower limbs and tied the hip and knee joints together with a strap. The inner ring of the human–robot contact force detection device we developed closely surrounded the tester's thigh. Along with the walking, the human–robot contact force was measured in real time.

4.2 Active Rehabilitation Training Experiment

In the active rehabilitation training experiment, but the lower limbs walked on the treadmill. When the treadmill speed is set to 0.8 km/h, the gait cycle of human walking is 4 s. The reference trajectories of the hip and knee cylinders and the control parameters of the position controller are consistent with those set in the passive rehabilitation training experiment. The measured data from the beginning to the four gait cycles are selected for analysis purpose, presented in Fig. 6.

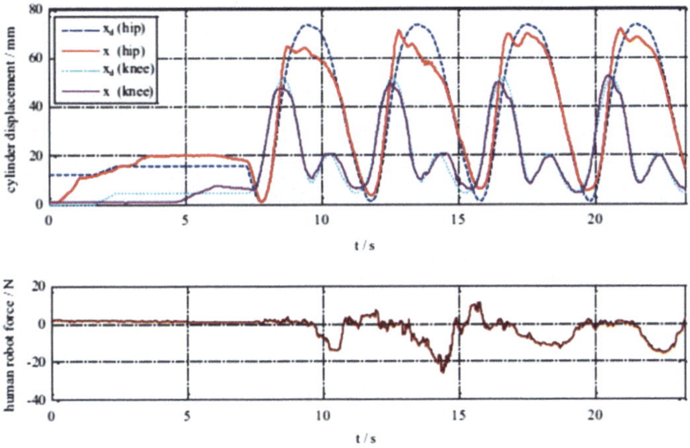

Fig. 6. Experimental curves of active rehabilitation training

The tester's initiative leads to the increase of human–robot contact force in active gait rehabilitation training. In the preparation stage when the prosthetic robot drives the lower limb from the upright state to the leg-lifting state, the human–robot contact force is only (0.61, 2.15) N, because the joint motion angle varies slowly as the passive training. With the beginning of the normal gait training, the human–robot contact force increases, and the range of human–machine contact force is (−26.21, 11.06) N. The purpose of active rehabilitation training is to encourage patients to participate actively with the robot providing part of the help and adapting to human movements to make the human body feel comfortable. At present, the human–robot contact force detection device is only installed on the thigh; so, just the hip motion control is verified.

5 Discussion and Conclusion

Whether impedance control or admittance control is employed in the process of human–robot interaction, the self-adaptation of model parameters is the focus of current research. The self-adaptation is mainly realized by discrete selection and continuous adjustment. The velocity threshold and the corresponding damping parameter empirical values were chosen according to the experience. The principle of the variable impedance control strategy was simple, which reflected the adaptability adjustment of the model parameters to a certain extent. However, the discrete impedance parameter reduced the controllability and fluency of operation. Most of the adaptive impedance or admittance interactive control strategies mentioned above adopt the method of constructing functions to adjust the parameters without considering the individualized characteristics of patients or manipulators in the interactive process. Manual modification of parameters in functions requires experience and a certain amount of time to explore. It is also inconvenient for computer control. To overcome the abovementioned limitations, the interactive control based on reinforcement learning is proposed in this study.

An adaptive admittance model is used in the interactive control strategy to stimulate the training initiative of patients. The parameter adaptive law is designed using the deformation formula of sigmoid function to meet the needs of a dynamic and time-varying interaction process. In addition, a personalized parameter learning method based on reinforcement learning is proposed. The proposed techniques in this study contribute to personal adaption and active compliance in robot-assisted rehabilitation training. To the best of the authors' knowledge, the interactive control based on reinforcement learning with the hydraulically driven lower limb rehabilitation training prosthetic robot has not been reported in literature.

The next step is to realize the active resistance rehabilitation training. Instead of setting the expected human–robot contact force to be zero, the desired force is given according to the patient's need. Through the admittance model, the stable contact force is always needed to be overcome to provide resistance for the patient's rehabilitation training. At the same time, clinical experiments will be carried out as soon as possible to further optimize the design of the system.

References

1. Kwakkel, G., van Peppen, R., Wagenaar, R.C., WoodDauphinee, S., Richards, C., Ashburn, A., Miller, K., Lincoln, N., Partridge, C., Wellwood, I., Langhorne, P.: Effects of augmented exercise therapy time after stroke: a meta-analysis. Stroke **35**, 2529–2539 (2004)
2. Broeren, J., Claesson, L., Goude, D., Rydmark, M., Sunnerhagen, K.S.: Virtual rehabilitation in an activity centre for community-dwelling persons with stroke: the possibilities of 3-dimensional computer games. Cerebrovasc. Dis. **26**, 289–296 (2008)
3. Sutton, R.S., Barto, A.G.: Reinforcement Learning: An Introduction. MITPress, Cambridge (1998)
4. Jordan, M.I., Rumelhart, D.E.: Forward models: supervised learning with a distal teacher. Cogn. Sci. **16**, 307–354 (1992)
5. Tin, C., Poon, C.S.: Internal models in sensorimotor integration: perspectives from adaptive control theory. J. Neural Eng. **2**, S147–S163 (2005)
6. Minsky, M.L.: Steps toward artificial intelligence. Proc. IRE **49**, 8–30 (1961)
7. Williams, R.J.: Simple statistical gradient-following algorithms for connectionist reinforcement learning. Mach. Learn. **8**, 229–256 (1992)

8. Zaghloul, K.A., Blanco, J.A., Weidemann, C.T., McGill, K., Jaggi, J.L., Baltuch, G.H., Kahana, M.J.: Human substantia nigra neuron sencode unexpected financial rewards. Science **323**, 1496–1499 (2009)
9. Erev, I., Roth, A.E.: Predicting how people play games: reinforcement learning in experimental games with unique, mixed strategy equilibria. Am. Econ. Rev. **88**, 848–881 (1998)
10. Cohen, M.X.: Neurocomputational mechanisms of reinforcement-guided learning in humans: a review. Cogn. Affect. Behav. Neurosci. **8**, 113–125 (2008)

Modeling, Analysis and Simulation of Work for the Punching and Cutting Operations on Inner Plate of the Front Car Door

Isad Saric[1], Enis Muratovic[1(✉)], and Harun Music[1,2]

[1] Department of Mechanical Design, Faculty of Mechanical Engineering, University of Sarajevo, Vilsonovo setaliste 9, 71000 Sarajevo, Bosnia and Herzegovina
muratovic@mef.unsa.ba

[2] PPW Engineering, Sarajevo, Bosnia and Herzegovina

Abstract. In this paper we displayed the problematic of modeling, analysis and simulating work process of specific tool used for punching and cutting operations on inner plate of the front car door. Basic principles of treating material with deformations are displayed, to easily understand operations that are performed during the work of the tool. Principles of geometric modeling, application of Boolean operations, parametrized individual standard parts and databases are very important factors for tool modeling. Beside modeling it is very important to calculate the specific parts based on standards which are used in car industry. After geometrical modeling of tool, simulation and work analysis have been performed to control the movement speed of the tool during production. Simulation also enables faster design process for the tool, increases work safety etc.

Keywords: Tool · Modeling · Analysis · Simulation · CATIA

1 Introduction

Modeling of solid objects in three dimensions (3D) is used to define geometry and volume that enables realistic representation of an object on computer with necessary accuracy for precise construction. In addition, it also possesses great potential for creating data bases that can be used to entirely define objects.

Most of mechanical elements are variant and reproducible. It isn't rational to model these parts interactively so it is much more efficient to use parametric modeling that allows object to be automatically modeled and saved in a form of an program. Application of parametric 3D modeling also allows control of created geometry via: parameters integrated in formulas, parametric laws, tables etc. [1–3] Parametric modeling is founded on concepts of knowledge, creating and using parametric parts and assemblies. Creation of 3D parametric models is most frequently combined with manufacturing characteristics (***F****eatures* ***B****ased* ***D****esign* - FBD) and Boolean approach (***C****onstructive* ***S****olid* ***G****eometry* - CSG). Solid modeling is necessary for parts that need to be subjected to status analysis of inner points, calculations etc. Models attained from

I. Karabegović (Ed.): NT 2020, LNNS 128, pp. 182–189, 2020.
https://doi.org/10.1007/978-3-030-46817-0_20

this procedure contain informations such as inner points coordinates. Mechanical elements are formed by adding, departing or subtracting basic geometrical bodies. Shape forming of these mechanical elements is achieved with Boolean operations over set of an inner points inside basic geometrical bodies.

Most recently, construction methods that are used are relied on simulation (***Sim-ulation Based Design*** – SBD), where complete development cycle is simulated before rapid prototyping [4–7]. SBD plays major role in assembling/disassembling instructions or assembling/disassembling capability in specific environment.

2 Processing by Punching and Cutting

Work process of the specific tool that will be displayed in this research is based on punching and cutting operations. Material separation is performed on closed contour for punching process, and on opened contour for cutting process. For the first process punch core represents scrap, but for the second process it represents cut part. Due the very complicated manufacturing, these tools are specifically used in serial processing [8, 9]. Goal of this type of processing is to manufacture the workpiece with least scrap and work operations so it can be immediately used or built in assembly. This can be accomplished with material reaching yield stress, which means it has to be loaded beyond point of elasticity [10, 11]. Necessary force is acquired with form of two tools on press. The tool that is mounted on press has the task of shaping the workpiece. Force is transferred through tool on workpiece. As a representative that takes special place in manufacturing process, inner plate of front car door was considered, which enables subsequential inserting and mounting of other parts.

2.1 Construction Parts of the Tool

Every tool that is constructed is consisted of upper and bottom part of the tool [12]. These parts are completely detachable one from the other. Their appearance is shown on Fig. 1. Bottom part of the tool is mounted on press table. Connection is achieved with fitting and bolt tightening process. Main guidance of the upper part of the tool is accomplished with plates and cylindrical slideways. Main guidance needs to be acquired before punching and cutting operations. Beside main guidance, there is also sheet metal holder guidance, that is performed by relating to upper part of the tool. Sheet metal holder, as movable component of the tool, has to secure enough amount of pressure on sheet metal before the punchers and/or cutters performed their function. Most important parts of sheet metal holder are pressure plugs, that enable force distribution from press to sheet metal holder. Pressure plugs represent standard parts and they are ordered from manufacturers catalogue, based on their length and number of bolts needed to provide clamp with sheet metal holder.

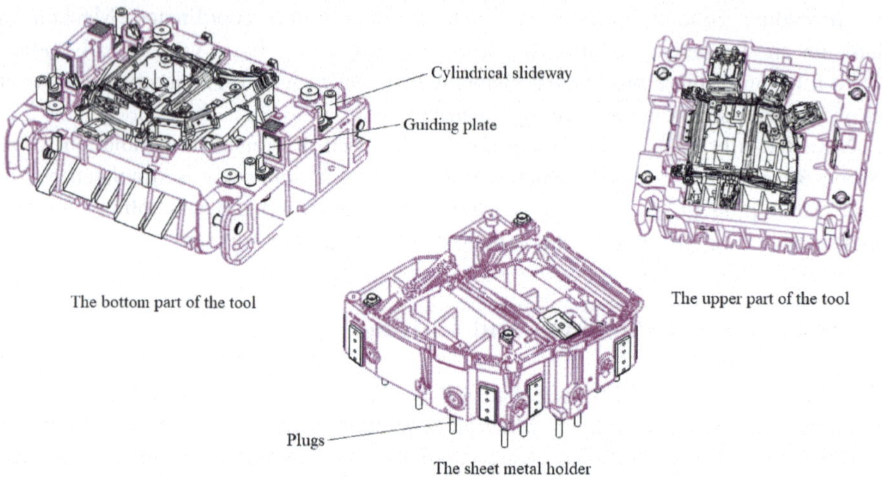

Fig. 1. Punching and cutting tool

Sheet metal scrap segments must be removed from the tool. This process is serviced with special sheet metal parts that expelled scrap segments. The minimum angle that needs to be acquired for scrap to be expelled must be greater than friction angle.

2.2 Calculation of the Punching and Cutting Force

In calculations that are used in technical practice, the most significant element for determining punching and cutting force is shear stress of material (τ_m). Shear stress within these operations does not have constant value, but it is variably dependent on the relative depth of punching and cutting ($\varepsilon = z/s$), type of material and temperature deformation. For nominal and actual shear stress same conclusions are admitted, same as for streching. Nominal stress in the moment of material separation is equal to:

$$\sigma_m = \frac{F_m}{A_0}, A_0 = \frac{d_0^2 \pi}{4} \tag{1}$$

$$\tau_m = \frac{F_m}{A_0}, A_0 = d\pi s \tag{2}$$

The actual stress in the same phase of the strain equals to:

$$\sigma_{mst} = \frac{F_m}{A_m} = \frac{F_m}{A_0(1 - \psi_m)} \tag{3}$$

$$\tau_{mst} = \frac{F_m}{A_m} = \frac{F_m}{d\pi(s - z_{ot})} = \frac{F_m}{A_0(1 - \varepsilon_{ot})} \tag{4}$$

whereas:

F_m – force in the moment of material separation,
s – thickness of material,
$\psi_m = \frac{A_0 - A_m}{A_0}$ – transversal contraction of test piece in the starting point of deformity localization, and
$\varepsilon_{ot} = \frac{z_{ot}}{s}$ – relative depth of the punch.

Actual stress is greater than nominal:

$$\sigma_{mst} > \sigma_m, \tau_{mst} > \tau_m, A_m < A_0 \tag{5}$$

The punching and cutting force for tools with parallel cutting edges is determined as it follows:

$$F = Ls\tau_m \tag{6}$$

whereas:

L – total length of all punched and cut contours,
Such calculated force, due to non-uniform thickness of material and bluntness of cutting edges of the tool (unavoidable during exploitation), is increased for 30%, therefore the actual cutting force on basis of which the press is set is:

$$F_M = 1{,}3F = 1{,}3Ls\tau_m \tag{7}$$

Figure 2 shows measured lengths of contours for punching and cutting process, acquired with the application of the Measure tool from the software system CATIA. Lengths of these contours are than used to calculate punching and cutting force according to expression (7).

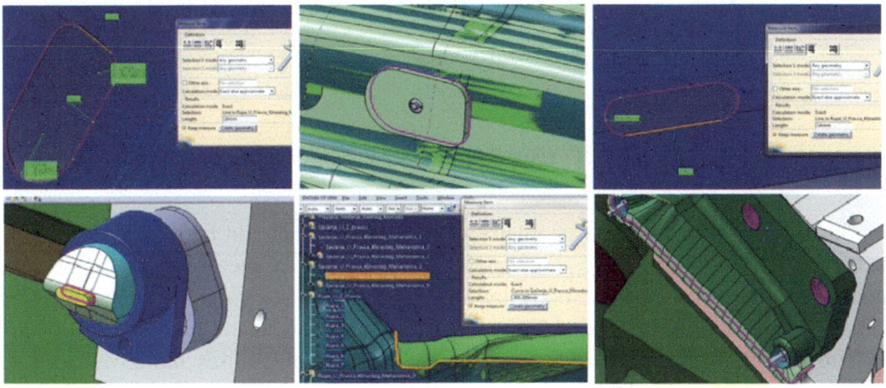

Fig. 2. Meassuring of different contours

Total contour length in the direction of z axis is equal to:

$$L = 72{,}22 + 20{,}41 + 119{,}32 + 50{,}24 + 155{,}71 + 176{,}93 + 347{,}98 + 176{,}92 + 344{,}44 + 305{,}3 = 1769{,}47\,\text{mm} \tag{8}$$

Based on material characteristics ($\tau_m = 300\,\text{MPa}, s = 1{,}2\,\text{mm}$) total punching and cutting force is calculated:

$$F_M = 1{,}3 \cdot 1{,}2 \cdot 1769{,}47 \cdot 300 = 828111{,}96\,\text{N} \approx 828\,\text{kN} \tag{9}$$

Force of holding sheet metal amounts 10% of total punching and cutting force:

$$F_D = 0{,}1 \cdot 828 \approx 83\,\text{kN} \tag{10}$$

3 3D Geometric Modeling and Setting of Parameters of the Tool

Set of all input data that enables tool modeling represent so called method. Surface of the workpiece is the most important input information, because it is used for modeling of all other parts that are in contact with workpiece. 3D geometric modeling of specific tool has been performed in software system CATIA. Course of 3D modeling of the tool for punching and cutting operations on inner plate of front car door using Boolean operations is exposed on Fig. 3.

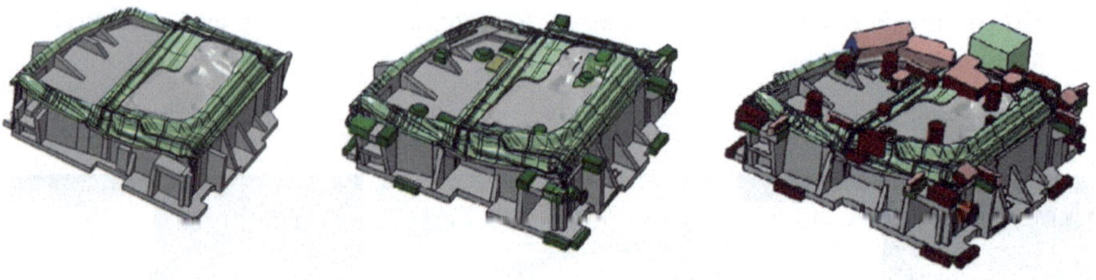

(a) (b) (c)

Fig. 3. Course of 3D geometric modeling using Boolean operations

Basic body contains basic geometry of tool model, that can be subsequently corrected by adding, milling etc. Figure 3a shows the appearance of basic body after modeling. Correction of basic body by adding (Fig. 3b) is existent of standard parts additives, which achieve needed fittings. Figure 3c shows all milling operations that need to be done on concerned workpiece. This procedure utilises the advantages of Boolean operations. Model of the tool is very complex, but procedures of machine treatments that need to be managed are very clearly specified.

After the 3D geometric modeling of the tool, setting up of its parameters took place. Figure 4 shows parametrization of plate slideway [13, 14].

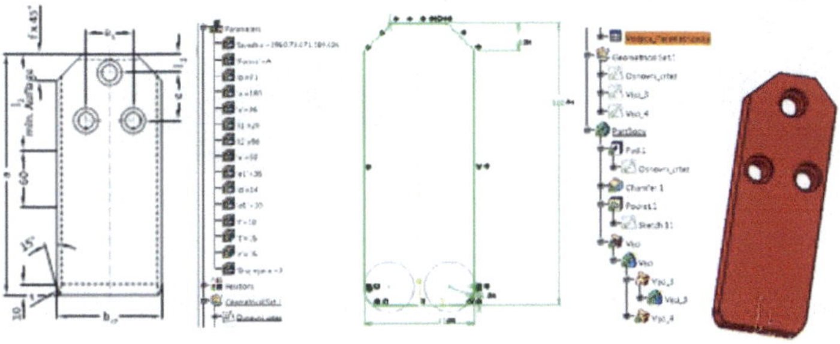

Fig. 4. Main quantities and appearance of bearing board

After linking parameters with specific dimensions, a table is generated and later processed when quantities pursuant to the corresponding standard are entered.

4 Analysis and Simulation of Operation of the Tool

Work simulation of the specific tool has been performed in DMU Fitting module of the CATIA software system. Major role of Fitting simulation is enabling reliable simulation of assembly/disassembly process. This can be very useful while creating videos for instructing employees in charge of assembly/disassembly process. Fitting simulation does not require usage of Constrains, accept for defining positions for individual parts of the tool.

Movement simulation of the upper part of the tool requires two informations: value of velocity of press movement (0,45 m/s) and range of motion of press (1,35 m). After activating Track tool from DMU Simulation palette, movement speed of press is entered. Compass is than brought on upper surface, so that W axis is perpendicular on surface, which is followed with entering value of range of motion. When the value of range of motion is entered and when upper part of the tool has moved, option Record saves his position. Thereby, the simulation of upper part of tool is easily completed (Fig. 5a). After the position of upper part of the tool is recorded, line is shown as movement trajectory.

The movement of the sheet metal holder is simulated in the same way (Fig. 5b). Only difference between these two motions is reflected in that the sheet metal holder's motion is shorter than the motion of the upper part of the tool for the value of the holder's motion (1350-50 mm). Again, line is shown as movement trajectory.

Role of the sheet metal holder is to hold the sheet metal before punching operation, and afterwards cutting operation. Thus, sheet metal is held until upper part of the tool is moved for 50 mm (holder's range of motion), after which the holder starts its own movement.

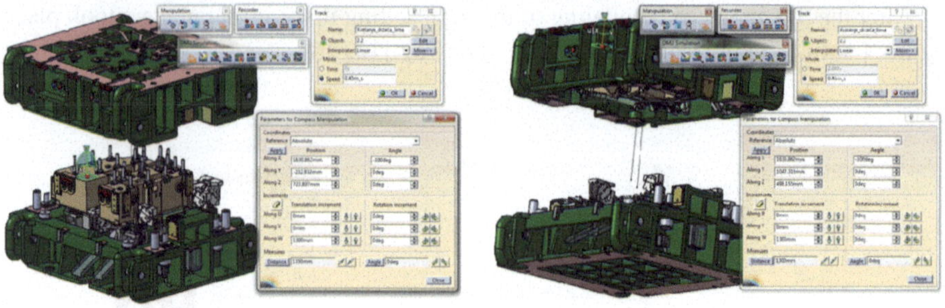

(a) (b)

Fig. 5. Simulation of tool motion

5 Conclusion

This paper demonstrates advantages of modern tool modeling approach. This approach enables faster and efficient modeling. For example, these tools used to be modeled in a classic manner without computer usage, so the modeling process lasted over six months. This time is now reduced to (5–10) weeks, depending on tool complexity. In order of accomplishing modeling process in shortest time possible, it is necessary to have parameterized data bases. Realisation of one workpiece requires (3–5) tools, where all of these told have same parts differently implanted.

Considering that casted workpieces are complex it is necessary to use many operations in modeling process, which results in long and vast program tree. By separating program's tree on units and using Boolean operations for manufacturing processes, transparency of model is increased.

Simulation usage has great impact on understanding work processes of the tool. It is often used for instructing new engineers. Without simulation it would be impossible to notice if one of the parts will get in a collision with other parts. Simulation of tool movement in real time enables specifying shortest times needed for accomplishing particular operations on specific tool.

References

1. Saric, I., Muminovic, A., Colic, M., Rahimic, S.: Development of integrated intelligent computer-aided design system for mechanical power-transmitting mechanism design. Adv. Mech. Eng. **9**(7), 1–16 (2017)
2. Saric, I., Pervan, N., Muminovic, A., Colic, M.: Development of integrated intelligent CAD system for design of shafts. Tehnički vjesnik **25**(Suppl. 1), 99–104 (2018)
3. Saric, I., Muminovic, A.J.: Development of integrated intelligent CAD system for synthesis and stress-deformation analysis of pressure vessels. Int. J. Eng. Technol. **7**(1), 147–151 (2018)
4. Tekkaya, A.E.: State-of-the-art of simulation of sheet metal forming. J. Mater. Process. Technol. **103**(1), 14–22 (2000)

5. Makinouchi, A.: Sheet metal forming simulation in industry. J. Mater. Process. Technol. **60** (1–4), 19–26 (1996)
6. Tisza, M., Lukács, Zs., Gál, G.: Integrated process simulation and die-design in sheet metal forming. Int. J. Mater. Form. **1**(Suppl. 1), 185–188 (2008)
7. Tisza, M.: Numerical modeling and simulation in sheet metal forming academic and industrial perspectives. Mater. Sci. Forum **473–474**, 407–414 (2005)
8. Birkert, A., Haage, S., Straub, M.: Umformtechnische Herstellung komplexer Karosserieteile. Springer, Berlin (2013)
9. Wagner, S.: Werkzeuge der Blechumformung. Universität Stuttgart, Stuttgart (2015)
10. Suchy, I.: Handbook of Die Design, 2nd edn. McGraw-Hill, New York (2006)
11. Nee, J.G.: Fundamentals of Tool Design, 6th edn. Society of Manufacturing Engineers (SME), Michigan (2010)
12. Saric, I., Pervan, N., Colic, M., Muratovic, E.: Recent developments and design of tools for punching and cutting. Int. J. Mech. Eng. Technol. **10**(5), 373–382 (2019)
13. Alberti, N., Cannizzaro, L., Micari, F.: Knowledge-based systems and F.E. simulations in metal-forming processes design an integrated approach. CIRP Ann. **40**(1), 295–298 (1991)
14. Sitaraman, S.K., Kinzel, G.L., Altan, T.: A knowledge-based system for process-sequence design in axisymmetric sheet-metal forming. J. Mater. Process. Technol. **25**(3), 247–271 (1991)

Intelligent CAD Systems for Generation G Code

Senad Rahimić(✉) and Anida Memić

Faculty of Mechanical Engineering, University of "DžemalBijedić" in Mostar, Mostar, Bosnia and Herzegovina
senad.rahimic@unmo.ba

Abstract. Computer Aided Design (CAD) systems are now indispensable in many industries, such as mechanical, aeronautical, electrical, architectural, and automobile industries. In this paper show link between a CAD application and an excel file that generates G code. Particularly in mechanical engineering where geometrical information is dominant, CAD has become not optional but necessary. The excel file will use an object-oriented programming method to define the stages of making a finished piece. This paper will show the connection of 3D CAD models with the defined stages of technological process development in order to obtain G code.

Keywords: Intelligent CAD systems · G code · Object-oriented programming

1 Introduction

Intelligent machines are basically defined as hierarchical structures in the order of intelligence and inverse order of precision [1]. Intelligent systems are systems, which operating autonomously or semi-autonomously in uncertain environments with minimum supervision and interaction with a human operator [2]. This intelligence called machine intelligence is embedded in the machine's hierarchical structure mapping its abilities into the space of tasks it is asked to perform with minimum interaction and supervision with a human operator. Intelligent control has been developed to implement the functions of hierarchical systems and machines and may be considered as a fusion between the mathematical and linguistic methods and algorithms applied to systems and processes. Contemporary designers need to be trained to work with new tools that can meet the high requirements of design agency, where the application of computers in the process of developing new products is necessary at all stages [3]. For many activities in the design process of products, computer support within the conventional CAD systems is not sufficiently developed. For such activities, it is possible to develop appropriate support systems. The technology for tool-path programming in CAD/CAM systems is today still based on data exchange model from 1950's (Fig. 1). For introducing the automatic and intelligent way of tool path generation new data exchange protocol is needed. The worldwide IMS research initiative is going on to develop a new data model entitled STEP-NC (ISO 14646 standard). Paper provides a future view how this standard could be used in intelligent CAD/CAM systems. Intelligent, artificial neural network based system for autonomous planning of turning

I. Karabegović (Ed.): NT 2020, LNNS 128, pp. 190–193, 2020.
https://doi.org/10.1007/978-3-030-46817-0_21

operation is proposed in [15]. This system optimizes cutting conditions taking into account cutting tools, material of the workpiece and machine tool characteristics.

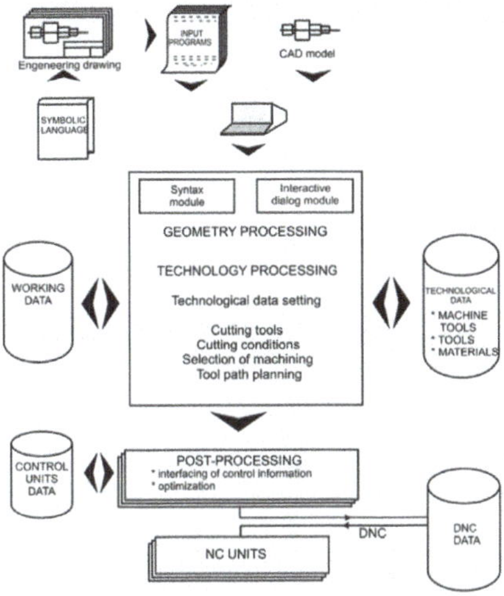

Fig. 1. Conventional CAD/CAM system.

2 Intelligent Model of G Code

Basic idea of the system was developed in past research work and is shown on Fig. 2. The first step is geometrical feature recognition and classification. It is described in more detail in references. Recognition and optimisation system consists of two main parts, and works in two stages. The process starts with processing of the CAD part model in order to analyse the shape and all characteristics of geometrical features.

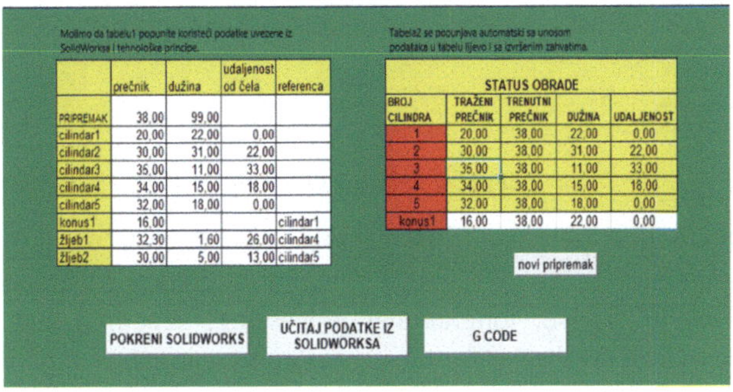

Fig. 2. Application on CAD/CAM system.

In Fig. 3 is shown part of Visual C code to connect to the CAD application SolidWorks wherein the rotating element is modeled and defined on all the objects that are associated with the external application.

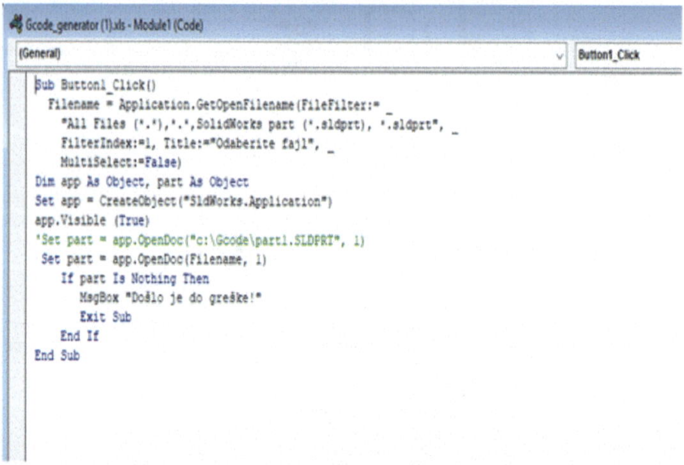

Fig. 3. Visual C programming.

The application generates for each stage of technological proce rotary pieces G code and the output is formed tekstlnu file that can be used for the CNC machine (Fig. 4).

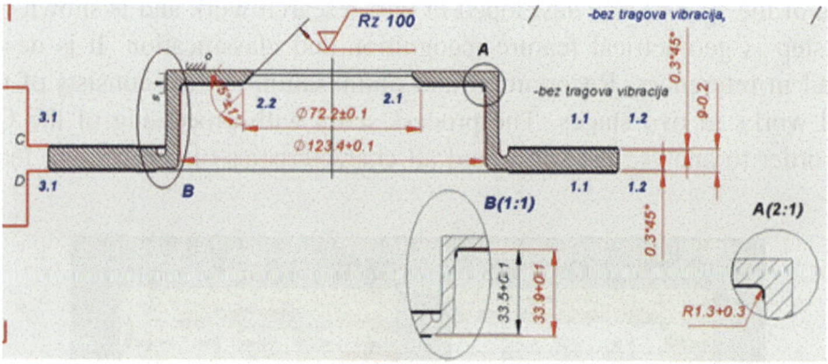

Fig. 4. 2D rotation part

Figure 5 shows the application that generates the G code for a given rotational piece. The application consists of separate parts for defining the technological process of processing the transverse, longitudinal processing, penciling and grooves. The knowledge base built special rules for those operations using the supports rule methods of artificial intelligence. For each phase of the entries on automatically Geometric information and interactive technology to bring information. The application generates for each stage of technological proce rotary pieces G code and the output is formed tekstlnu file that can be used for the CNC machine.

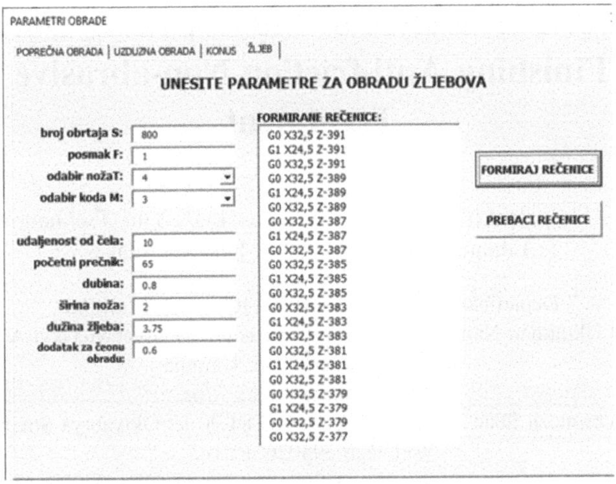

Fig. 5. Application on CAD/CAM system

3 Conclusion

In this paper showed link between a CAD application and an excel file that generates G code. Particularly in mechanical engineering where geometrical information is dominant, CAD has become not optional but necessary. The excel file will use an object-oriented programming method to define the stages of making a finished piece. This paper will show the connection of 3D CAD models with the defined stages of technological process development in order to obtain G code. The knowledge base built special rules for those operations using the supports rule methods of artificial intelligence.

References

1. Balic, J.M., Valavanis, K.P., Tsour veloudis, N., Ioannidis, S.: Intelligent manufacturing systems: programming and control, Maribor, Faculty of Mechanical Engineering, Laboratory for Intelligent Manufacturing, Chania, Technical University of Crete, Department of Production Engineering and Management, Maribor, Chania (2003)
2. Valavanis, K.P., Saridis, G.N.: Intelligent Robotic Systems: Theory, Design and Applications. Kluwer Academic Publishers, Berlin (1992)
3. Jiang, B., Lau, H., Chan, F.T.S., Jiang, H.: An automatic process planning system for the quick generation of manufacturing process s plans directly from CAD drawings. J. Mater. Process. Technol. **87**, 97–106 (1999)

Power Parameters of Micro-cutting During Finishing Anti-friction Non-abrasive Treatment

Ihor Shepelenko[1(✉)], Yakiv Nemyrovskyi[1], Yuri Tsekhanov[2], Eduard Posviatenko[3], and Sergii Sardak[4]

[1] Department of Exploitation and Repairing Machines, Central Ukrainian National Technical University, 7 Universytetskyi Avenue, Kropyvnytskyi 25006, Ukraine
kntucpfzk@gmail.com
[2] Voronezh State Technical University, 84 20 let Oktyabrya Street, Voronezh 394026, Russia
[3] National Transport University, 1 Suvorova Street, Kiev 02000, Ukraine
[4] Dniprovskii University of the Humanities, 35 Ermolova Street, Dnipro 49000, Ukraine

Abstract. Based on the method of the theory of similarity and dimensions using a model experiment, we studied the effect of the force parameters of the finishing anti-friction non-abrasive treatment (FANT) on the micro-cutting process depending on the conditions of contact interaction, which allows a reasonable approach to the formation of an anti-friction coating by the friction-mechanical method. From the standpoint of process mechanics, the role of power parameters on the patterns of micro-cutting is considered. Methods and devices have been developed that allow the simulation of the FANT process, with a phased study of micro-cutting and adhesive sticking of antifriction material on the treated surface. A relationship is established between the force exerted on the surface being machined, the geometry of the microprotrusion, as well as the cutting depth, the application of which allows the use of the methods of the theory of similarity and dimensions to obtain a dimensionless value of force. An analytical relationship has been established to determine the strength for various materials and cutting conditions when applying FANT coatings. Ways to increase the efficiency of the micro-cutting process by providing optimal values of the cutting front angle are indicated. It has been experimentally proved that with an increase in the force of pressing the tool against the work surface to the set value, the continuity of the antifriction coating increases.

Keywords: Finishing anti-friction non-abrasive treatment · Micro-cutting · Power · Front angle · Contact interaction · Coating

1 Introduction

One of the most important and urgent problems of modern engineering is the development and widespread use of new, scientifically sound, environmental, technically and economically feasible processing technologies in order to improve the quality of

I. Karabegović (Ed.): NT 2020, LNNS 128, pp. 194–201, 2020.
https://doi.org/10.1007/978-3-030-46817-0_22

work surfaces by achieving optimal performance properties. The improvement of existing and the creation of new technological methods is a creative task and should be aimed at improving the quality of the working surfaces of parts, the main indicators of which include the physical, mechanical and geometric characteristics of the surface layer, formed at the finishing operations of the technological process.

Increasing demands on the operational properties of work surfaces of parts stimulates the development of surface modification methods and coating. In the global engineering even a priority area has developed - the application of multifunctional coatings and in the field of science - the direction of thin-film materials science [1].

An important requirement for such coatings is minimal friction and high wear resistance, which will ensure stable operation of the contacting friction pairs in the conditions of boundary lubrication and without lubrication. These properties are possessed by antifriction coatings obtained by finishing antifriction non-abrasive treatment (FANT) due to the frictional interaction of a copper-containing tool with the surface of the workpiece. Rubbing the friction surface with a tool made of copper and its alloys in the presence of a process fluid allows to provide the transfer of material and the formation of an antifriction coating with a thickness up to 5 μm on the surface of the part, as well as hardening the surface of the base material to a depth of 70–80 μm [2].

The use of FANT helps to reduce the running-in time and the exclusion of scuffing of friction surfaces, increase the bearing capacity of parts and joints, protect the friction surface from hydrogen wear, reduce the friction temperature and increase the operation period of the assembly [3].

The formation of an antifriction coating during FANT depends on a number of factors, including the conditions of contacting the tool with the surface to be treated, and the shape and size of microroughnesses determine the quality of the resulting coating and its continuity [4]. The study of the features of contacting surfaces, as well as the basic parameters of the FANT process, their laws will improve the quality of the antifriction coating, and hence the operational properties of the part. In this regard, studies of the influence of the basic parameters of the FANT process on the formation of an antifriction coating, depending on the conditions of contact interaction of a copper-containing tool with the surface being treated, are very relevant.

2 Literature Review

The FANT method, known due to the works [2, 3, 5, etc.], is widely used in various branches of mechanical engineering, including the processing of parts (cylinder liners, crankshafts, etc.), which limit the life of internal combustion engines.

However, despite certain successes in the practical application of FANT, there are significant gaps in the study of the patterns of coating formation; the role and features of plastic deformation and physico-mechanical processes in the coating, in the transition and surface layers of the base metal are not fully disclosed.

The lack of consensus explaining the mechanism of coating formation during FANT gives rise to various hypotheses in interpreting the process of coating formation by the friction-mechanical method.

According to [6], the following conditions contribute to the application of FANT coatings: micro-cutting, plastic contact, as well as the optimal parameters for coating application.

The authors of [2] believe that the formation of an antifriction coating by the friction-mechanical method has characteristic features similar to those processes that occur during friction welding. At the same time, the need for micro-cutting as an integral part of the process is emphasized.

According to [7], the issue of the formation of a FANT coating should be considered in three stages. At the first stage, a surface-active medium is applied to the surface of the part, which, having good surface wettability, helps soften and dissolve oxide films on the surface of the part and tool. At the second stage, a solid body (workpiece) is in contact with a soft counterbody (tool). In this case, tool wear occurs due to micro-cutting by the surface roughness of the workpiece. In the third stage, when the hollows of the treated surface are filled, an increase in the thickness of the coating layer occurs under the influence of adhesive interaction.

It should be noted that, despite the difference in approaches that explain the formation of an antifriction coating by the friction-mechanical method, the authors of [1–7] note the importance and necessity of micro-cutting as the basis for the formation of the entire coating.

In our opinion, an important reserve for improving the quality of the anti-friction coating applied by FANT is to increase the efficiency of this process precisely at the stage of micro-cutting. In [4], from the standpoint of process mechanics using a model experiment, the FANT process was described at the micro-cutting stage, which made it possible to establish ways to increase the efficiency of applying an antifriction coating. The authors [4] proved the feasibility of creating a regular microrelief with positive values of the cutting front angle on the treated surface to create favorable conditions for micro cutting. In this regard, the study of the influence of the main, and above all, the power characteristics of the FANT and their influence on the process of micro-cutting is of scientific and practical interest. This is also necessary to justify the conditions for the formation of an antifriction coating, to select the coating technology and to improve its quality by optimizing processing conditions.

The aim of the work is to study the power parameters of the FANT when modeling micro-cutting based on the conditions of contact interaction of the tool with the treated surface.

3 Research Methodology

Experimental studies of the contact interaction of surfaces were carried out on special samples of gray cast iron SCh20 and brass L63. To study the process of micro cutting, the method of the theory of similarity and dimensions was used [8]. Cast iron cutters were made, the geometry of the cutting part of which simulated a separate microroughness of the surface of the workpiece. The front angle of the microroughness models varied within $\gamma = +5° \div -15°$. Micro-cutting of a plate made of brass L63 was modeled using the developed device [4], which was mounted on a model 676P milling machine (Fig. 1a).

Friction mechanical coating was also carried out by the model 676P milling machine. In this case, the test sample of cast iron SCh20 in the form of a disk was rigidly fixed on the working table of the milling machine by the polished supporting end (Fig. 1b).

Fig. 1. The working area of the machine when modeling micro-cutting (a) and coating (b) FANT

The coating was applied to a surface treated with a process fluid (glycerin) due to the longitudinal movement of the machine table manually. At the same time, moving under load, an anti-friction coating is formed on the treated surface. To apply the coating under other conditions (changing the load on the antifriction block, coating material, etc.), it is enough to turn the test sample relative to the edge by a certain angle. This technique provides reusable use of the test sample while maintaining the information obtained during previous experiments.

4 Results

The experimental determination of the dependence of the power parameter - the force $\boldsymbol{P_e}$, which largely determines the conditions for the passing of micro-cutting, at different angles γ, showed the following. At positive angles γ, it is possible to achieve large depths of micro-cutting with significantly less effort $\boldsymbol{P_e}$ (Fig. 2).

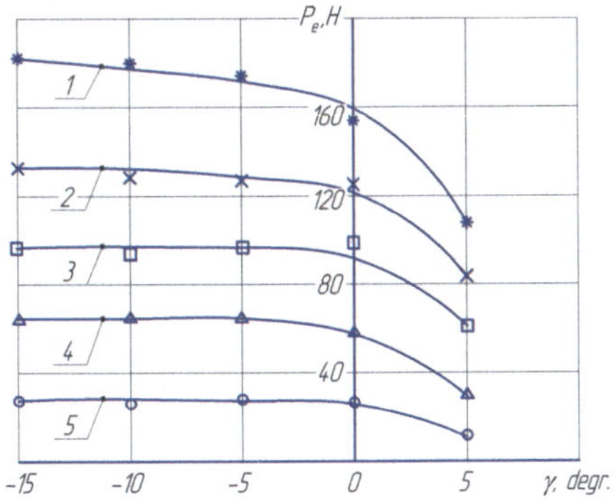

Fig. 2. The dependence of the effort $\boldsymbol{P_e}$ on the angle γ when modeling cutting with a cast iron cutter SCh20 of a specimen made of brass L63 (cutting depth $\boldsymbol{t_c}$, mm: 1–0, 6; 2–0, 4; 3–0, 3; 4–0, 2; 5–0, 1)

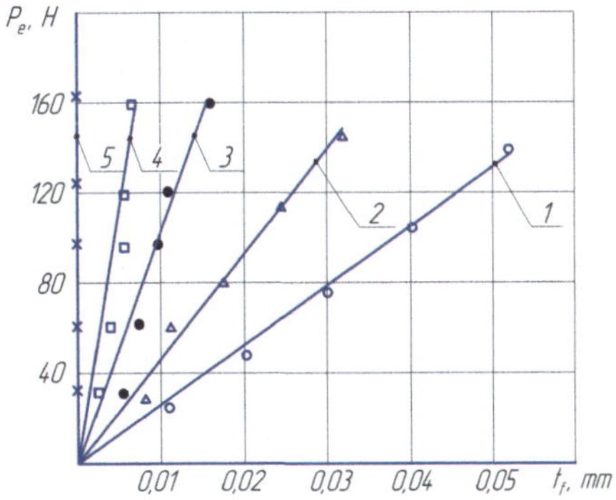

Fig. 3. The dependence of the effort $\boldsymbol{P_e}$ on the actual depth of cut $\boldsymbol{t_f}$ when modeling micro-cutting with a cutter from cast iron SCh20 of a sample of brass L63 at a cutter angle γ: 1–5°; 2–0°; 3–−5°; 4–−10°; 5–−15°

As follows from Fig. 3, the dependence $\boldsymbol{P_e}$ on angle γ has a qualitatively similar character at different cutting thicknesses. The minimum value $\boldsymbol{P_e}$, regardless of the nominal cutting thickness, is observed at angle $\gamma = +5°$. With a decrease in angle γ to 0°, the value $\boldsymbol{P_e}$ increases and a further decrease in angle γ has little effect on the growth of force $\boldsymbol{P_e}$. It should be noted that for large thicknesses of the cut ($\boldsymbol{t_c}$ = 0.4 − 0.6 mm), the intensity of change of $\boldsymbol{P_e}$ at γ from 0° to −15° increases slightly.

The dependences $\boldsymbol{P_e}$ on the actual cut thickness $\boldsymbol{t_f}$ are linear (Fig. 3).

With decreasing angle γ with constant effort, the value of $\boldsymbol{t_f}$ decreases, which is explained by the deterioration of the cutting conditions, the wear of the cutting edge increases sharply with the formation of the blunting radius $\boldsymbol{r}$ of the tip of the cutter, and this happens most intensively at negative angles γ.

Based on Fig. 3 the relationship of the force $\boldsymbol{P_e}$ with the cutting thickness $\boldsymbol{t_f}$ can be expressed as the following dependence

$$P_e = C_f \cdot t_f, \tag{1}$$

where $\boldsymbol{C_f}$ - coefficient taking into account the influence of the angle, approximated by the following dependence:

$$C_f = 86,9\gamma^2 - 883\gamma + 4896. \tag{2}$$

To obtain dimensionless $\overline{P_n}$, methods of the theory of similarity and dimensions were used [8]. The normal component of the force $\boldsymbol{P_n}$, which determines the technological force of pressing the bar against the workpiece, depends on: the type of material to be rubbed, micro-cutting depth $\boldsymbol{t}$, chip width $\boldsymbol{B}$, microprotrusion geometry (primarily the cutting front angle γ and blunting radius $\boldsymbol{r}$), friction conditions, which are determined by the friction coefficient f.

The mechanical properties of the hardened material are well defined by such generalizing parameters as hardness $\boldsymbol{HV}$. In [9], it was used to analyze the power parameters during deforming broaching, and in [10] when cutting.

We write the functional dependence

$$P_n = P_n(HV, t, , r, \gamma, \alpha). \tag{3}$$

In accordance with the dimensional theory [8], we select the main (determining) parameters: $\boldsymbol{t}$ (m), $\boldsymbol{HV}$ (N/m^2). The remaining parameters are expressed through these two main ones. We express linear dimensions in units of $\boldsymbol{t}$, and the angles γ and the friction coefficient f are already dimensionless. Then, in accordance with [8]:

$$P_n = HV^a \cdot t^b \cdot \overline{P_n}\left[1, 1, \frac{B}{t}, \frac{r}{t}, \gamma, \alpha, f\right]. \tag{4}$$

Here $\overline{P_n}$ is a dimensionless (relative) force, which must be determined from a model experiment.

In order for the dimensions of the left and right sides of (4) to be the same, we choose the exponents: $\boldsymbol{a}$ = 1; $\boldsymbol{b}$ = 2:

$$P_n = HV \cdot t^2 \cdot \overline{P_n}. \tag{5}$$

From the principles of superposition, it is obvious that $\overline{P_n}$ should depend linearly on the cutting width $\frac{B}{t}$. Then (5) takes the form

$$P_n = HV \cdot t \cdot B \cdot \overline{P_n}\left[\gamma, \alpha, \frac{r}{t}, f\right]. \tag{6}$$

In the model experiment, the angle α was constant, as well as the friction coefficient $\boldsymbol{f}$. The blunting radius $\boldsymbol{r}$ was determined experimentally and, as shown below, it depends only on the cutting front angle γ. Width $\boldsymbol{B}$ = 3 mm was constant in all cases.

Then the dimensionless effort

$$\overline{P_n}\left[\frac{r}{t}, \gamma\right] = \frac{P_n}{HV \cdot t \cdot B}. \tag{7}$$

From (7) it follows that the dimensionless force for each angle γ is a constant value. The hardness of the sample from brass L63 $\boldsymbol{HV}$ = 690 MPa. The dependence of $\boldsymbol{P_n}$ on γ, determined by (7), is presented in Fig. 4.

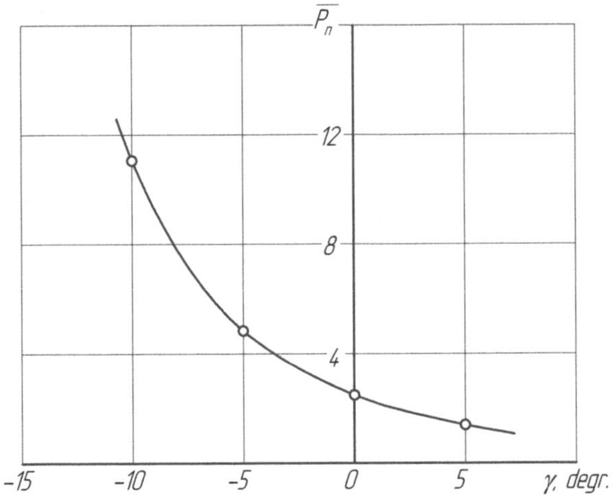

Fig. 4. The dependence of the dimensionless force $\overline{P_n}$ on angle γ when modeling cutting of a sample of brass L63 with a SCh20 cast iron cutter

This dependence, taking into account (2), can be approximated as

$$P_n = 0,042\gamma^2 - 0,433\gamma + 2,4 \quad (8)$$

It allows one to determine the force $\boldsymbol{P_e}$ for any material and cutting conditions using dependence (7).

The final stage in the study of the power parameters of the FANT process was the application of a brass antifriction coating on the surface of the SCh20 cast iron samples. For this purpose, based on the simulation results, a microrelief with microprotrusions was preliminarily applied to the surface to be treated with an angle of γ = 5° and a pitch of $\boldsymbol{S}$ = 0.175 mm.

The force of pressing the antifriction tool to the surface to be treated was calculated according to dependence (8), taking into account the contact area, the number of microroughnesses on it and the magnitude of the contact pressure. Figure 5 shows a fragment of the surface of cast iron SCh20 sample with a brass coating obtained at various contact pressures.

a) b) c)

Fig. 5. Sample surface (×20) made of cast iron SCh20, ***HV*** 1.7 MPa; tool - brass L63 with angle γ = 5°, pitch of the microrelief of the base $\boldsymbol{S}$ = 0.175 mm and contact pressures: (a) $\boldsymbol{q}$ = 4,6 MPa; (b) $\boldsymbol{q}$ = 9,2 MPa; (c) $\boldsymbol{q}$ = 13,6 MPa

According to metallographic studies, in all cases a non-continuous antifriction skin is formed on the surface of the sample, while the area occupied by the antifriction material depends on the force of pressing the antifriction bar to the surface to be treated. With an increase in the pressing force, the concentration of the antifriction product on the surface increases markedly.

5 Conclusion

Studies of the power parameters of FANT during the simulation of micro-cutting allowed us to formulate the following conclusions:

1. A relationship is established between the angle γ, the thickness of the micro-cutting t and the force $\boldsymbol{P_e}$ acting on a single microroughness. With positive angles γ, it is possible to achieve large depths of micro-cutting, and hence the effectiveness of this

process with much less effort $\boldsymbol{P_e}$. The dependence of $\boldsymbol{P_e}$ on angle γ has a qualitatively similar character at different thicknesses of micro-cutting $\boldsymbol{t}$.

2. The proportionality of the relationship between the force $\boldsymbol{P_e}$ and the thickness of the micro-cutting t_f has been established, which allows using the methods of the theory of similarity and dimensions to obtain the dimensionless value of force $\overline{P_n}$.
3. An analytical dependence has been obtained to determine the dimensionless force $\overline{P_n}$, which makes it possible to determine the value of the $\boldsymbol{P_e}$ force for any material and micro-cutting conditions.
4. The increase in the continuity of the antifriction coating is achieved by increasing the contact loads to 13.6 MPa, as well as with a pre-prepared surface for the coating by turning with microprotrusions at an angle of $\gamma = 5°$ and pitch $\boldsymbol{S} = 0.175$ mm.

References

1. Solovyih, E.K.: Tendentsii razvitiya tehnologiy poverhnostnogo uprochneniya v mashinostroenii, 92 (2012). [in Russian]
2. Balabanov, V.I., Bolgov, V.J., Ishhenko, S.A.: Friction application of nanoscale antifriction coatings on parts. Nanotechnol. Ecol. Prod. **1**(3), 104–107 (2010). [in Russian]
3. Garkunov, D.N.: Finishing antifriction non-abrasive treatment (FABO) of friction surfaces of parts. RVM (Repair Restor. Modernization) **3**, 36–41 (2009). [in Russian]
4. Shepelenko, I., Nemyrovskyi, Y., Tsekhanov, Y., Posviatenko, E.: E. Modeling of contact interaction of micro roughness at FANP. In: I-th International Scientific and Technical Conference «Prospects for the Development of Mechanical Engineering and Transport-2019», pp. 218–219 (2019)
5. Polcer, G.: Osnovy frikcionnogo naneseniya pokrytiya v usloviyah selektiv-noj peredachi. RVM (Repair Restor. Modernization) **10**, 23–28 (2010). [in Russian]
6. Pogonyshev, V.A., Panov, M.V.: Theoretical and experimental basis for increasing the wear resistance of machine parts. In: Mechanics and Physics of Processes on the Surface and in Contact of Solids, Parts of Technological and Energy Equipment, pp. 78–84 (2011). [in Russian]
7. Shepelenko, I.V., Posviatenko, E.K., Cherkun, V.V.: The mechanism of formation of antifriction coatings by employing friction-mechanical method. Probl. Tribol. **1**, 35–39 (2019). https://doi.org/10.31891/2079-1372-2019-91-1-35-39
8. Sedov, L.I.: Metodypodobija i razmernosti v mehanike, 423 (1987). [in Russian]
9. Tsehanov, J.A., Shejkin, S.E.: Mehanikaformoobrazovanijazagotovokprideformirujuschemprotjagivanii, 203 (2001). [in Russian]
10. Tsehanov, J.A.: Modelirovanie' energosilovyhparametrovrezanijametodamiteoriipodobija i razmernostej, 30–33 (2015). [in Russian]

Overview of Mobile Applications for CNC Programming

Dragi Tiro[1(✉)] and Anida Memić[2]

[1] Faculty of Mechanical Engineering, Džemal Bijedić University of Mostar, Sjevernilogor bb., 88 000 Mostar, Bosnia and Herzegovina
dragi.tiro@unmo.ba

[2] Džemal Bijedić University of Mostar, Mostar, Bosnia and Herzegovina

Abstract. This paper describes free application software designed for smartphones running Android. In recent years, more such applications have emerged and are classified here into three groups: simulation software, learning software and CNC engraving software. The paper describes some typical applications for these three types of software.

Keywords: CNC programming · Application software · Smartphone

1 Introduction

In the last fifteen years, there has been a rapid development of CNC machines [2–4] and use of smartphones, which are real computers because they have a processor, memory, input and output hardware devices, as well as software. Of course, engineers are increasingly using these devices in practice. Application software (mobile applications) has been increasingly used in recent years with the development of smartphones. A large number of mobile applications for CNC programming [1] have also emerged. This paper will analyse free apps for the Android operating system, which is the most popular mobile OS today.

2 Mobile Applications for CNC

Mobile applications for CNC programming can be found on known software download platforms, such as Play Store, Google Play (Fig. 1), etc.

There are many free apps available here. We can generally divide them into 3 groups: software with simulation, CNC engraving software, and educational software.

2.1 Software with Simulation

This group of mobile applications allows the user to enter the CNC code using a virtual keyboard and to monitor the processing in very realistic 3D simulations. You can open pre-made CNC programs stored on a smartphone as a file or many sample programs that come with the application at installation. A characteristic representative of this group of applications is the CNC Simulator Free product of VirtLabs Software from Russia.

I. Karabegović (Ed.): NT 2020, LNNS 128, pp. 202–207, 2020.
https://doi.org/10.1007/978-3-030-46817-0_23

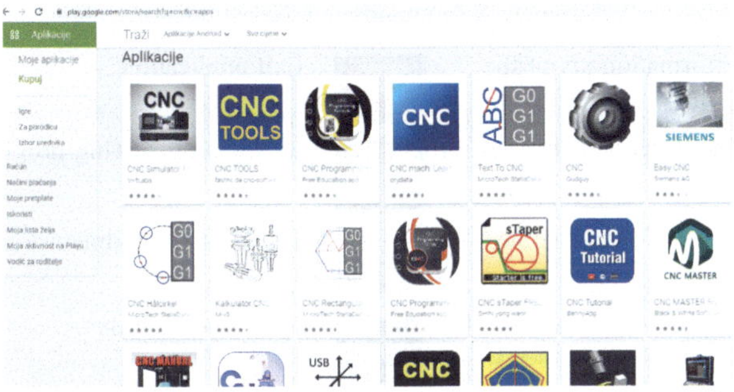

Fig. 1. Application software on Google Play

2.1.1 CNC Simulator Free

The basis of the three-dimensional simulation model is a lathe with a classical arrangement of units, equipped with CNC system, clamping head, coolant supply system and others [5]. The material is machined on two axes in a horizontal plane.

The functionality of this application consists of preparing the texts of control programs in the standard G code format, checking the drivers for syntax and technological errors, playing on the computer screen (or other computer device) and metal cutting tools to simulate the metal turning process, three-dimensional visualization of the design process [5]. Opening the application gives the window shown in Fig. 2.

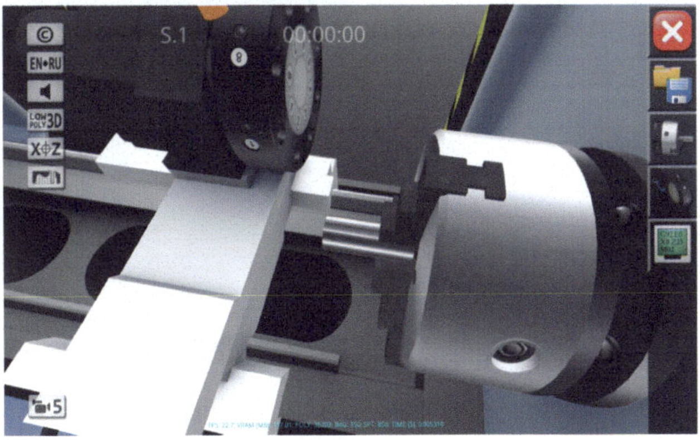

Fig. 2. User interface of the program

The interface consists of commands on the left and right side as well as commands on the bottom in the left corner (Fig. 2). Some of the commands are explained in Table 1. We can start and stop the entered processing program at any time by following each processing step. If an error is detected, we may correct it. This application can also be used for educational purposes, e.g. students who first encounter CNC lathe and CNC programming.

Table 1. Commands in CNC simulator

	Brief information about the manufacturer		Exit application
	Language selection (English or Russian)		import a processing program
	The sound of the machine on/off		Tool and stock dimensions
	Turns the 3D machine on/off		opens the toolbox
	XZ coordinate system on/off		Opens the program entry window and 3D simulation
	Tool trajectory (path) on/off		Views from five different angles for better viewing

2.2 CNC Engraving Software

It is often necessary to engrave a mark on the machine parts (Fig. 3). This group of application software enables the user to obtain a CNC engraving program quickly and easily. All you need to do is type a mark, and the app generates a CNC code. The characteristic representatives of this application group are: Text To CNC and CNC Tools.

Fig. 3. The counter with engraved mark.

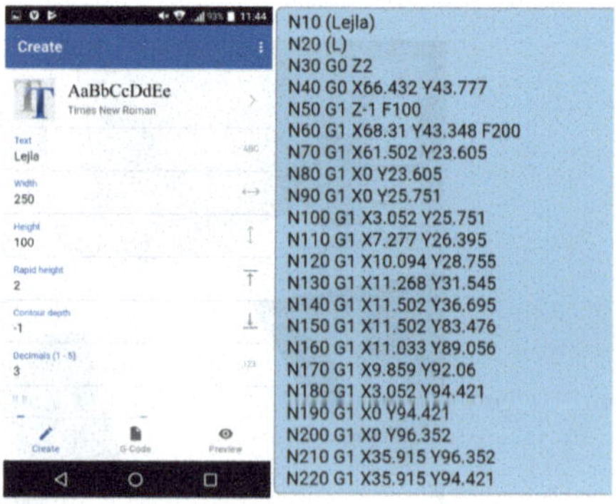

Fig. 4. Application's interface and part of the generated CNC code

2.2.1 Text to CNC

The application is used to make CNC code for engraving letters into the material. The application is easy to use and you only need to write the mark, to enter the letter type, height, width, etc., then simply click on the button G-Code to list the CNC code. Internet access is required to use the application. Figure 4 shows the interface of the application and the part of the generated code. At the top of the window is the option for selecting the type of letter used for printing. There are many types of letters to choose from.

For the example shown in Fig. 4, the steps were as follows:

1. Select the type of letter
2. Click on Create (enter text that is later engraved), define height, width, depth, speed etc.
3. Click on the G-Code (the program will appear, with characteristics for the desired text)

For an example of the word "Lejla" made according to the parameters defined in Fig. 4 left, a generated code is obtained - Fig. 4 right. Also, the tool trajectory is obtained.

2.2.2 CNC Tools

This application generates CNC code for engraving of characters and letters based on selected parameters. But in addition to engraving, CNC Tools is also an application that gives us a code for milling, drilling, turning and threading (Fig. 5).

Clicking on the operations offered in the application adjusts certain parameters, such as cutting speed, diameter, pitch, etc. depending on what type of processing it is. The generated program can be saved to a mobile phone, or shared via social networks, bluetooth or cable to a PC computer.

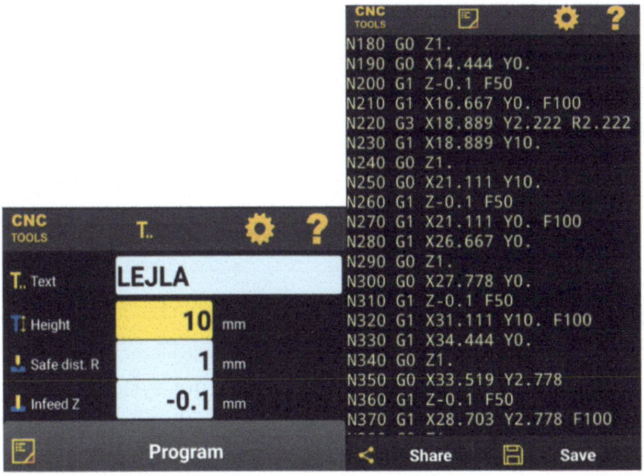

a) b)

Fig. 5. (a) Engraving parameters; (b) the code generated

2.3 Educational Software

These applications are intended for beginners and intermediate users. They teach students how to make CNC program. These applications are useful for those who are beginning to learn CNC programming. Instructions are included in the application's installation, such as CNC basics, CNC programming basics, CNC management. These apps give users the ability to learn from anywhere, and they are optimized for most devices from Android OS. One of the characteristic applications in this group is the CNC Program *Example.*

2.3.1 CNC Program Example

CNC Program Example is an application that helps users learn about writing and executing CNC code through a wide variety of workpiece processing examples. The application takes up 7.95 MB. When the application starts, a window with 80 different options opens. The first options are for beginners in CNC programming, and most options are examples of processing different details on workpieces.

As can be seen from Fig. 6, this application provides solved examples of machining different parts so that engineers, when they want to make a CNC program for machining parts on a workpiece, can find and solve a similar problem. The application is very good for practical use.

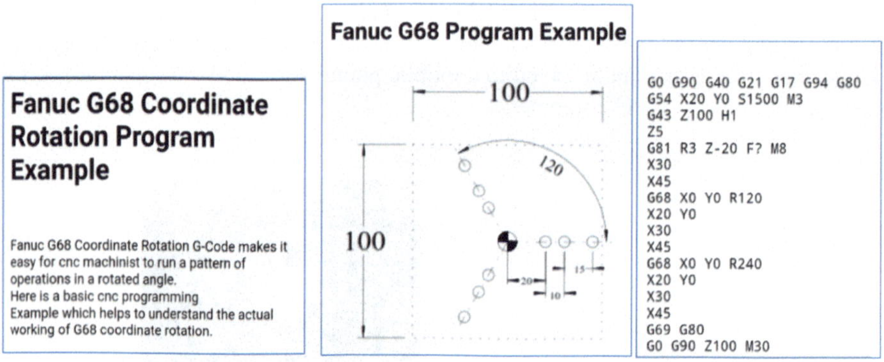

Fig. 6. Example of using the main function G68 Coordinate Rotation Program

3 Conclusion

There are many free mobile apps for Android operating systems. With them we can learn CNC programming, to program, simulate a machining process, or generate a CNC engraving code. They have the advantage of being able to help practicing engineers as well as students. In addition to education, some of these applications are also useful for practical applications in the industry.

Entering the code is difficult on the virtual keyboard, and the screen of the mobile phone is limited in size, which is a disadvantage of these applications. Therefore, it is better to use these applications on larger screen devices, for example on tablets or smart TVs. Android OS can be installed on PCs (desktop or laptop), so these applications can be used that way. The disadvantage is that these software have limited capacities and options, so for example. CAD/CAM programming is not possible. The testing also concludes that in most applications we do not have access to all the tools (they are locked and need to be purchased).

It is expected that these applications will be used more and that their new versions will have even more functionality, due to the high popularity of smartphones and tablets running Android.

References

1. Bošnjaković, M., Stoić, A.: Programiranje CNC strojeva. Veleučilište, SlavonskiBrod (2014). ISBN 978-953-7847-08-1
2. Yao, Y., Liu, M., Du, J., Zhou, L.: Design of a machine tool control system for function reconfiguration and reuse in network environment. J. Robot. Comput.-Integr. Manuf. **56**, 117–126 (2019)
3. Jia, X.W., Zhang, L.L.: Application of Bluetooth in wireless communication of CNC manufacturing. Appl. Mech. Mater. **422**, 232–236 (2013)
4. Güler, O., Yücedağ, I.: Developing an CNC lathe augmented reality application for industrial maintenance training. In: 2nd International Symposium on Multidisciplinary Studies and Innovative Technologies (ISMSIT), Ankara, Turkey (2018)
5. https://virtlabs.tech/cnc-simulator/

Modelling of Cutting Forces in Hard Steel Turning

Kovač Pavel[1(✉)], Tarić Mirfad[2], Nedić Bogdan[3], Savković Borislav[1], Golubović Dušan[4], and Ješić Dušan[5]

[1] Department for Production Engineering, Faculty of Technical Sciences, University of Novi Sad, Trg Dositeja Obradovića 6, 21000 Novi Sad, Serbia
pkovac@uns.ac.rs
[2] Srednja škola metalskih zanimanja, Sarajevo, Bosnia and Herzegovina
[3] Faculty of Engineering, University of Kragujevac, Kragujevac, Serbia
[4] Faculty of Mechanical Engineering, University of East Sarajevo, Lukavica, Bosnia and Herzegovina
[5] International Technology and Management Academy – MTMA, Novi Sad, Serbia

Abstract. The aim of this study is to investigate the effect of machining parameters on cutting resistance during hard steel turning. This paper describes a novel approach in cutting resistance modeling using a design experiment. The output of the processing should take into account the cutting resistance they measured when testing the range of speeds, displacements, and depth of bursting. Mathematical models have been obtained that have been shown to have high confidence that they can approximate the measurement results. The connection between cutting forces in terms of changing the parameters of the cutting mode, as well as the undesirable occurrence and wear of the tool, which is certainly present in the process of removing hard-working material, is given.

Keywords: Hard steel · Turning · Design experiment · Cutting forces

1 Introduction

Machining of hardened steels by turning offers advantages such as: high flexibility, complete machining in one clamp and the ability to optimize machining processes with energy savings, shorter production time, higher productivity, reduced number of machines required, improved machined surface integrity, elimination of cooling and lubrication fluids means (in many cases), reducing the number of finishing operations, more favorable forms of chip, and eliminating any thermal deformation of the work-piece due to heat treatment and cutting forces.

Machine tools must have a stiffness capable of absorbing a large dynamic penetration force, which can be twice as high as that of typical machining operations. From a metallurgical point of view, materials with a small hardness deviation (less than 2 HRC units) by cutting depth allow for the best machining process.

The most commonly used hardened steels in machining (high strength and hardness steels) depend to a large extent on the chemical composition which, by thermal

I. Karabegović (Ed.): NT 2020, LNNS 128, pp. 208–220, 2020.
https://doi.org/10.1007/978-3-030-46817-0_24

treatment, provides the required exploitation properties of these materials, such as high temperature resistance, high strength and hardness, etc.

Structure of materials with very hard carbides and nitrides ensure high content of alloying elements (Cr, Ni, Co, Mo, W, Ti) intermetallic phases of Ni_3Al and Ni_3 type are also present, which results in intensive abrasive wear of the tool, which is a consequence of low stability and/or reduction of cutting ability when machining.

The problem of efficient machining of hard-to-process materials is the topic of much research today, given the constant introduction of new materials as well as the development of new processing procedures, machine tools and materials for cutting tools. The development of new materials for hard metal and cubic boronitride cutting tools has improved the efficiency and cost-effective processing of structural and tool steels in heat-treated condition (hardness of 30–67 HRC and high strength). In this way it is possible to perform machining on the lathe, with high accuracy and low roughness of the machined surface, as a substitute for more cost effective machining process-grinding. In doing so, significant techno-economic effects are achieved, among which the most important are [1]:

- Reduction of processing time and cost (relative to standard finishing),
- Machining by turning multiple surfaces in one clamp is feasible, which is rarely possible during grinding, avoiding the effects that occur during grinding (structural changes due to overheating of the surface layer, residual stresses and cracks) and improving the performance characteristics of parts.

1.1 Cutting Force When Machining Hardened Materials

During the cutting process, cutting forces (resistances) occur that can be broken down into three orthogonal directions, Fig. 1:

- $F_v = F_1$ main cutting resistance (tangential, in the direction of speed of main movement),
- $F_p = F_2$ penetration resistance (radial),
- $F_s = F_3$ auxiliary movement resistance (axial).

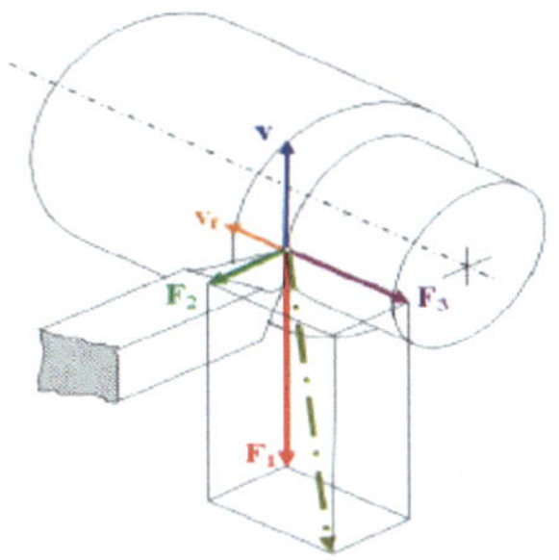

Fig. 1. Decomposition of cutting resistance in turning to orthogonal components

As the cutting speed increases, the components of the cutting resistance decrease due to the increase in temperature and the softening of the workpiece material, while with the increase of the steps the components of the cutting resistance increase. With the increase of the hardness of the workpiece material, the component of the cutting resistance remains approximately constant while the component increases significantly, and with a hardness of about 58 is reached.

High strength and hardness steel cutting machining is characterized by the following characteristics [1]:

- The cutting resistance is about 1.5–2 times higher than when machining steel in a heat-treated state, and the penetration resistance F2 is greater than the main cutting resistance F1. For this reason, machining of hardened steels can only be effected on machines with increased rigidity (especially in the direction of penetration resistance).
- The machining should be carried out at an increased cutting speed, which results in a high cutting temperature, sufficient to ensure a significant reduction in the strength and hardness of the workpiece material in the shear zone (the formation of a hot shavings is a proof of the high cutting temperature). Due to the very high thermal and mechanical loading of the tool, the machining of hardened steels can only be efficiently performed using very high-quality tool materials.

Figure 2 gives a comparison of the value of the cutting resistance components obtained by machining on a 16MnCrS5 hardened steel for cementing (hardness C4381) with a hardness of 62 HRC tools with K10 hard metal cutting inserts, cubic boronitride PKB and mixed Al2O3 + TiC ceramics, namely without cooling [2].

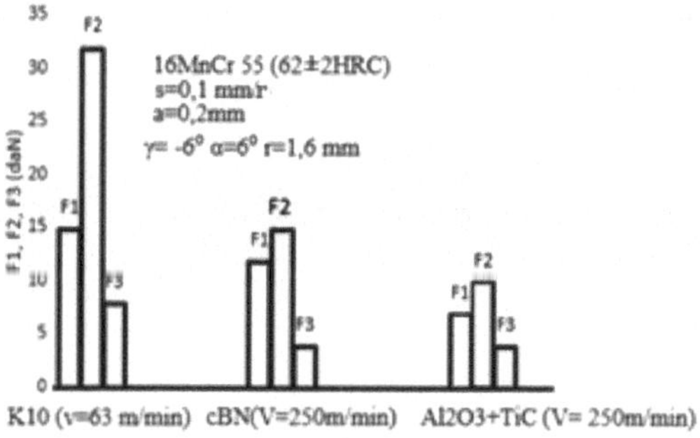

Fig. 2. Comparison of the components of the cutting resistance, and obtained by machining on a lathe of steel for cementation in hardened state with tools with tiles of hard metal, cubic boronitride and mixed ceramics [2]

2 Modeling the Cutting Process

New advancements in technology for machine tools, together with advanced cutting tools, all with the aim of improved production in terms of productivity, product quality and cost reduction, require certain performance models to be applied in the design process of the cutting process system. In order to achieve a significant degree of quality in parts/products, it is important to choose: (1) optimal cutting conditions; (2) refrigerants/type of lubricant; (3) cutting tools, etc. The basic goal of modelling is to obtain information about the behaviour of a system not by studying the system itself, but by studying the behaviour of the system by creating a model, and then, by experimenting on such a model, one becomes aware of the behaviour of the existing system.

The development of advanced models makes it possible to meet accelerated demands in terms of productivity, product quality and reduction of production costs [3]. Process modelling is a support for performing activities that make up one process more efficiently. In creating a process model, it is first necessary to define the goal or objectives of the process [4]. A model is an approximate representation of a system or process that serves to understand and modify or manage a system. Models provide: a description of complex phenomena that enables them to better understand and communicate those who solve problems [5].

The mathematical model describes the behaviour of the system using appropriate mathematical procedures and operations, based on certain data. For a mathematical model to serve a purpose, it must be conceptually defined, easy to use, able to be continually upgraded or developed in accordance with new knowledge and requirements. The basic purpose of applying mathematical models is to describe the behaviour of the real system as a whole or its individual segments under certain conditions [6].

3 Experiment

Workpiece material standard EN 10027-1 (EN 90MnV8) or DIN V 17006-100 (DIN 90MnCrV8) diameter 36 mm, hardness of 55 HRC was used for the experiment. Tile cutting tool is a cubic boron nitride with negative geometry. Machine tool on which cutting was made is universal lathe - Prvomajska DK480, which has the following characteristics: power 10 kW, maximum turning diameter is 320 mm, maximum machining length is 2000 mm.

Based on the literature sources and the availability of tools, two types of tool materials were selected, namely: two types of tiles; hard metal P 30 (HM) code CCMW 12 04 04 manufactured by ORION from Germany [7].

The markings of the cutting tips according to DIN 4983 more closely define the geometry, as follows: the shape of the plate C → rhomb; the rake angle N → = 0°, C → = 7; tolerance class M; Type of tile → with opening A, W and G; length of cutting blade → 12.7 mm (12); cutting edge thickness → 4.76 mm (04); radius of tool tip → 0.4 mm (04). All tiles have a rake angle ($\gamma = -0°$).

The processing regime included the following elements:

- cutting speed v, (m/min),
- feed f, (mm/rev),
- cutting depth, a (mm)

Figure 3 shows the measurement of the cutting force in the turning process. The cutting force was measured in the turning process using a Kistler 9265 A1 dynamometer, which was installed in place of the existing tool carrier. The tool with the plate is mounted in a dynamometer. The signal from the dynamometer was amplified with a Kistler type CA 5001 amplifier. The signal was then introduced to the computer by the A/D converter where the data was stored and processed. The scheme of cutting resistance measurement is shown in Fig. 3. Three orthogonal resistance components were measured: F_1, F_2, and F_3.

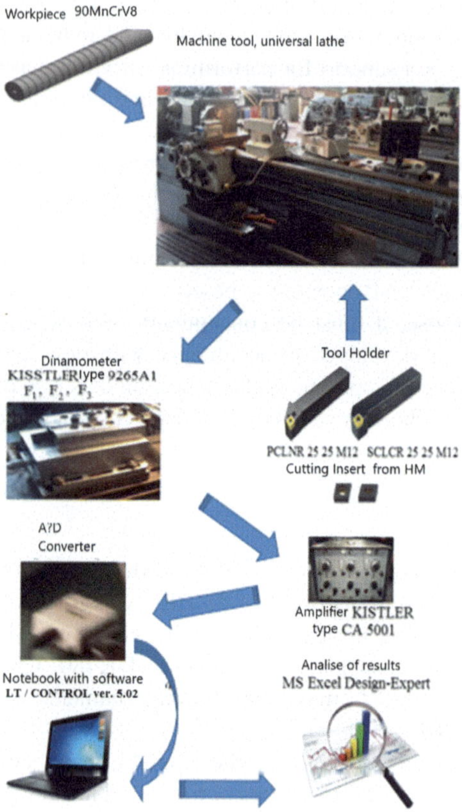

Fig. 3. Measurement of cutting force in the process of turning

Prior to the experimental embodiment, the preparation of the workpiece was done. In order to get out and credible measurement results due to the workpiece is gripped in the chuck at one end and the other end rests on the cusp. It is necessary to download a specific material layer in order to avoid a throw-oval results were credible. At a length of 500 mm rods was to be divided on 24 fields in 20 mm length where the processing was performed without cooling and lubrication. Each field is planned for one experimental point. In this way, a treatment in a single clamping, and thus have provided the same conditions for all the experimental points. Table 1 shows the levels of the factors with which the experiment was performed.

Table 1. Levels of factors

Level (code)	v (m/s)	s (mm/r)	a (mm)
Upper (+sqrt (2))	180	0,25	0,7
Upper medium (+1)	160	0,2	0,5
Medium (0)	120	0,1	0,22
Lower medium (−1)	90	0,05	0,1
Lower (−sqrt (2))	80	0,045	0,07

3.1 Measurement Results

Table 2 shows the measurement results according to the experiment plan for the HM tool material and tool geometry.

Table 2. Input parameters of the experiment v, s, a, and the output parameter F_1, F_2, F_3

No.	v (m/s)	s (mm/r)	a (mm)	F_1 [N]	F_2 [N]	F_3 [N]
1	90	0,05	0,1	55	141	46
2	160	0,05	0,1	54	138	41
3	90	0,2	0,1	116	300	60
4	160	0,2	0,1	110	401	80
5	90	0,05	0,5	170	350	285
6	160	0,05	0,5	155	360	195
7	90	0,2	0,5	280	710	440
8	160	0,2	0,5	240	680	245
9	120	0,1	0,22	120	280	130
10	120	0,1	0,22	121	300	133
11	120	0,1	0,22	130	390	186
12	120	0,1	0,22	132	400	190
13	80	0,1	0,22	116	255	102
14	180	0,1	0,22	130	650	225
15	120	0,045	0,22	92	398	177
16	120	0,25	0,22	194	710	190
17	120	0,1	0,07	57	152	46
18	120	0,1	0,7	250	770	340
19	80	0,1	0,22	120	300	111
20	180	0,1	0,22	131	600	260
21	120	0,045	0,22	98	398	180
22	120	0,25	0,22	200	650	200
23	120	0,1	0,07	53	141	42
24	120	0,1	0,7	265	784	360

4 Modeling the Output Performance of the Machining Process

In this part of the paper, an analysis of the measurement results was performed and mathematical models of output quantities were formed using a factorial experiment plan. An experiment plan for successful experimentation would include: identifying problems, selecting input parameters and belonging (level) functions, selecting an experiment plan, performing the experiment, then statistically processing the experimental data obtained, and finally, conclusions and future research.

4.1 View of the Data Obtained from the Experimental Three-Factor Plan

Data processing according to the three-factor plan of the second-order experiment was performed for the main cutting resistance, penetration resistance, auxiliary movement resistance shown in Table 2.

4.1.1 Three-Factor Model for Main Cutting Resistance for HM Cutting Insert with $(\gamma = 0°, \alpha = 7°, \lambda = -6°)$

The model is in the form: $F_1 = Cv^{p1}s^{p2}a^{p3}$

Table 3 provides data on measurement and calculation values for the machining conditions with the tool from HM and Table 4 gives regression coefficients, results of significance estimation and model adequacy assessment.

Table 3. Measured and calculated values for the main cutting resistance F_1

No.	Measured values	Modelled values		
	Main cutting force F_1 [N]	Model without interactions		
		F_{1rac} [N]	F_{1min} [N]	F_{1max} [N]
1.	55	57.39	52.22	63.07
2.	54	57.15	52.00	62.81
3.	116	100.19	91.16	110.11
4.	110	99.78	90.79	109.66
5.	170	160.17	145.74	176.02
6.	155	159.51	145.14	175.30
7.	280	279.62	254.43	307.31
8.	240	278.48	253.39	306.05
9.	120	125.11	120.97	132.10
10.	121	125.11	120.97	132.10
11.	130	125.11	120.97	132.10
12.	132	125.11	120.97	132.10
13.	116	125.47	112.85	142.44
14.	130	124.75	112.19	141.62
15.	92	90.76	75.88	95.78
16.	194	180.82	166.86	210.62
17.	57	60.28	54.45	68.74
18.	250	261.74	232.50	293.48
19.	120	125.47	112.85	142.44
20.	131	124.75	112.19	141.62
21.	98	90.76	75.88	95.78
22.	200	180.82	166.86	210.62
23.	53	60.28	54.45	68.74
24.	265	261.74	232.50	293.48

Table 4. Regression coefficients, estimation of significance and model adequacy for main cutting resistance

Model	Coefficients			
	Regression		Significance	
	Designation	Value	Designation	Value
Model without interactions	C	857.9038	F_t	6.61
	p_0	6.75449	F_{r0}	259473.19
	p_1	−0.00713	F_{r1}	0.03 (*)
	p_2	0.40196	F_{r2}	573.33
	p_3	0.63773	F_{r3}	1945.15
	Adequacy		F_a	4.3077

Note: Table values for significance: >=6.61; For adequacy: <=4.47

Figure 4 is a comparison diagram of the experiment - model when machining by turning with a HM CCMW cutting insert and tool carrier SCLCR 25 25 M12 for the main cutting resistance.

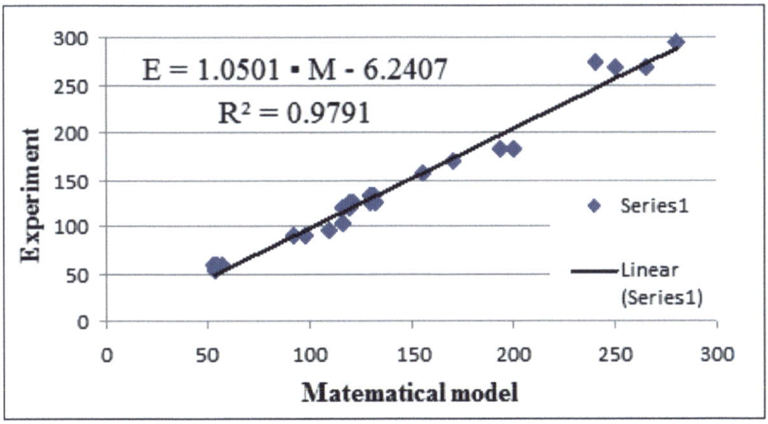

Fig. 4. Comparison diagram of an experimental model for machining by cutting a HM cutting insert for the main cutting resistance F_1

Since and the cutting regime element (v) are unaffected, and the cutting regime elements (s and a) are influential, and the model is adequate.

On the basis of the experimental measured and calculated values given in Table 3. Figure 5 shows a comparison model of an experimental model for machining by cutting with a cutting insert of HM CNMA 120404, for the main cutting resistance.

In the same way, results were obtained for the other two components of the cutting resistance F_2 and F_3 where the experimental conditions were not changed.

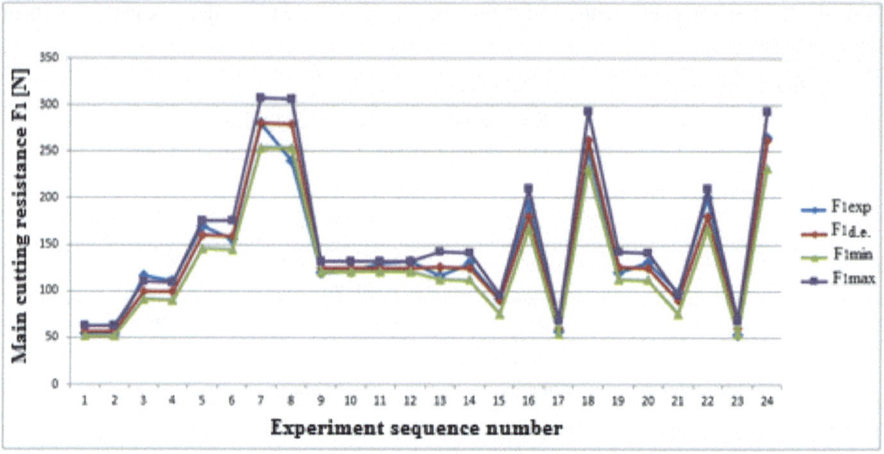

Fig. 5. Graphical representation of the obtained values for the main cutting resistance

Figures 6 and 7 give a comparison diagram of an experiment for penetration resistance and F_3. In Eqs. 1, 2 and 3, certain mathematical models were obtained based on the regression analysis performed:

$$F_1 = 857.9038 \cdot v^{-0.00713} \cdot s^{0,40196} \cdot a^{0.63773} \tag{1}$$

$$F_2 = 174,3623 \cdot v^{0.55557} \cdot s^{0.42175} \cdot a^{0.61967} \tag{2}$$

$$F_3 = 191.1774 \cdot v^{0.33243} \cdot s^{0.16645} \cdot a^{0.96013} \tag{3}$$

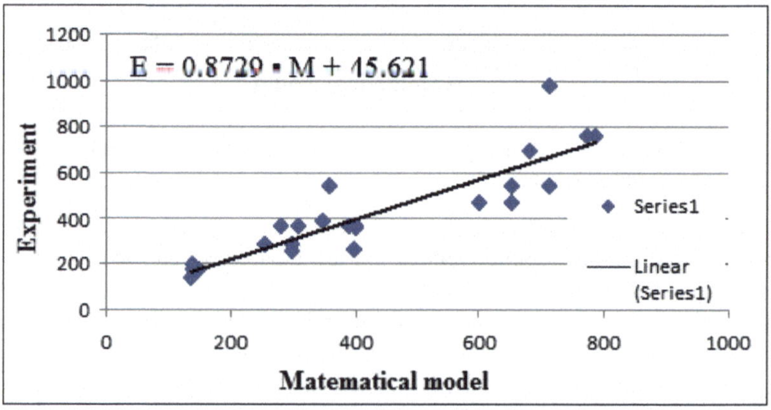

Fig. 6. Comparison diagram of an experimental model when machining a HM cutting insert for F_2 penetration resistance

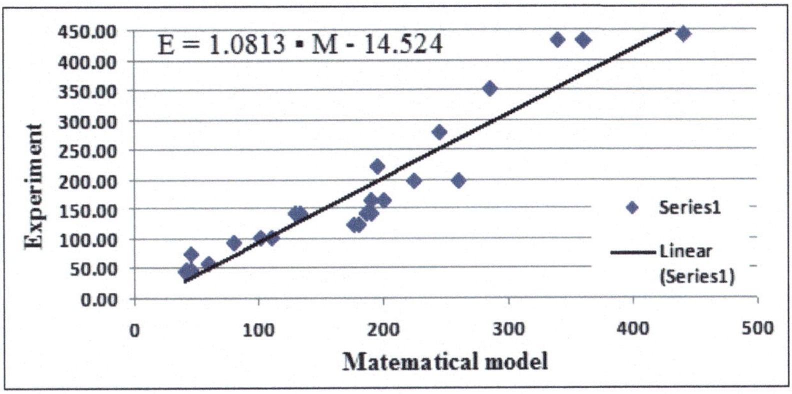

Fig. 7. Comparison diagram of an experimental model when machining a HM cutting insert for resistance to auxiliary movement F_3

5 Discussion

Cutting resistance during machining by hardened steel is different from that of conventional machining. For hardened steel when machining, the highest penetration resistance is F_2, while for conventional turning, the highest cutting resistance is F_1. As the cutting length increases, there is an increase in the VB bandwidth and thus a significant increase in the penetration force value. During the machining process of hardened steel, a complex interaction between the tool and the workpiece occurs on the contact surfaces. As a consequence, high cutting resistances and high cutting temperatures result in wear and sometimes tool breakage. Most often, these phenomena lead to damage and changes in the mechanical characteristics of the treated surface.

Figure 8 shows the change in cutting resistance as a function of cutting speed, Fig. 9 offset, and Fig. 10 on cutting depth.

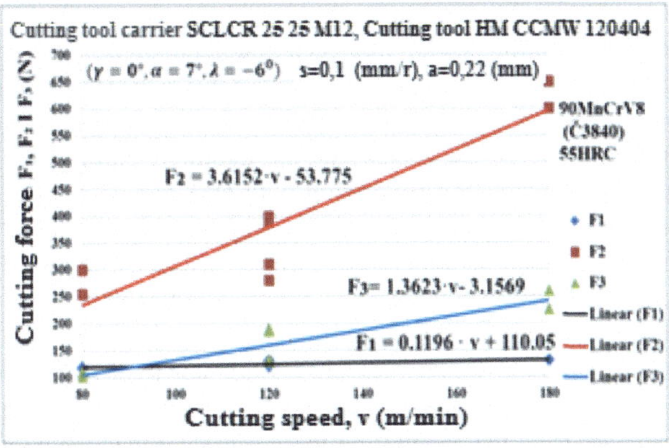

Fig. 8. Dependence of the cutting resistance components (F_1, F_2, F_3) on the cutting speed (v)

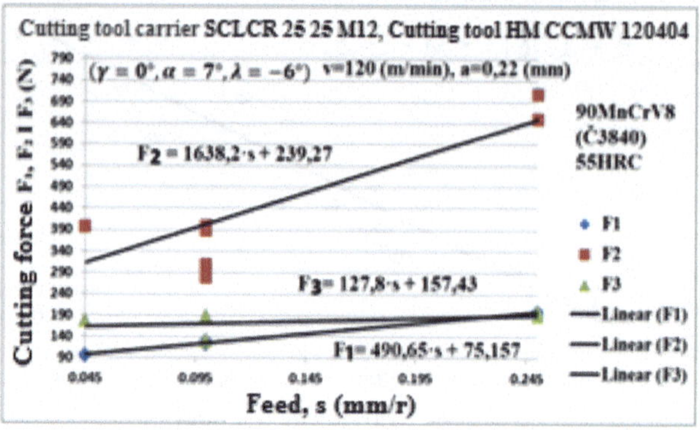

Fig. 9. Dependence of the cutting resistance components (F_1, F_2, F_3) on the feed (f)

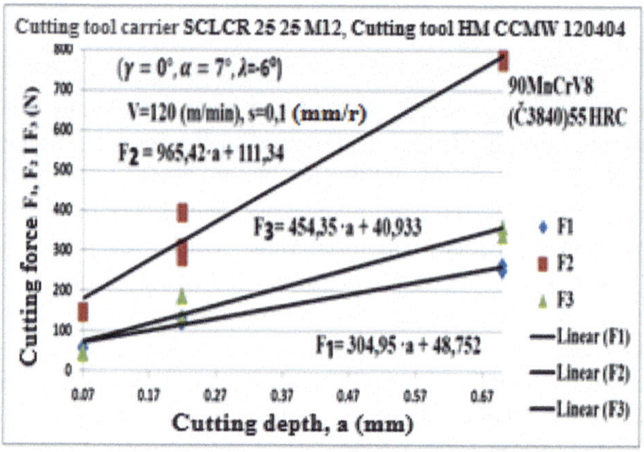

Fig. 10. Dependence of the cutting resistance components (F_1, F_2, F_3) on the cutting depth (a)

With the increase in cutting speed, there is a gradual increase in the value of cutting resistance in cutting tools from HM. When it comes to HM cutting discs with increasing cutting speed, the quality of the machined surface roughness. The same is the case with the increase of the steps and the depth of cutting the quality of the machined surface when machining by cutting with HM cutting insert.

At a cutting depth of 0.2 mm, the radial component (penetration resistance F_2) becomes more dominant than the other two cutting resistance components. This can be explained by the voltage acting solely on the rounded part of the cutting tip of the tool. The increase in cutting resistance is proportional to the increase in pitch. In all step-change experiments, the penetration resistance is greater than the main cutting resistance. The increase in cutting resistance is proportional to the increase in displacement. In all experiments with feed, the penetration force is greater than the main cutting force.

It is known from the literature that with increasing the hardness of the workpiece material, the cutting resistance component increases more slowly compared to the component to establish a relation at a hardness of about 58 HRC that is greater than and as confirmed by the studies depending on the cutting mode.

For HM cutting insert, the maximum force is F_2, while the smaller forces are F_1 and F_3. The size of the cutting force components depends on the increase in tool wear, so that the surface roughness slightly deteriorates (increases) with the beginning of tool wear. Wear of the cutting tool affects the cutting forces and vice versa. Research on tool wear shows that cutting forces, surface roughness and cutting temperature tend to increase with tool wear when machining by scraping hardened steel. Figure 11 shows a diagram of the dependence of the cutting resistance F_1, F_2 and F_3 and the width of the wear band (VB) when machining on a 90MnCrV8 toughened steel blade with HM cutting discs.

From the diagram in Fig. 11, it can be seen that there is a correlation between the components of the cutting resistance and the bandwidth that can be represented in the form of a linear correlation with a high value of correlation coefficients.

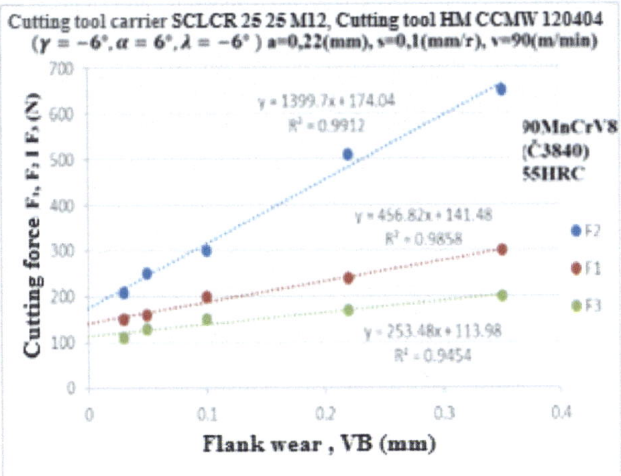

Fig. 11. Diagram of the dependence of the cutting resistance F_1, F_2 and F_3 and the width of flank wear band (VB) during machining on the toughened steel lathe by cutting inserts from HM

6 Conclusion

All mathematical models satisfy the model adequacy condition. Regarding the resistance component, the greatest influence is shown by the cutting depth, then the step, and finally the cutting speed. The analysis of the force model shows that the greatest impact is the step and then the depth of cut. According to the significance estimation, the cutting speed did not show any significant influence on the resistance model. The resistance model according to the significance assessment is influenced by all three factors in the model. Based on the resistance model, it can be seen that the depth of cut, then the displacement and the depth of cut, have the greatest influence, while the cutting speed has the least significant effect on the model.

The issue of tool wear in the process of machining hard steel is very complex. Thus, an explanation of the condition of the treated surface can be related to the phenomena of tool wear. Based on the results obtained, conditions can be determined.

References

1. Jovičić, M.: Obrada rezanjem teškoobradljivih metala i legura, Beograd (1992)
2. Tönshoff, H.K., Chryssolouris, G.: Einsatz kubischen Bornitrids (CBN) beim Drehen gehärteter Stähle. Werkstatt und Betrieb **114**(1), S.45–S.49 (1981)
3. Van Luttervelt, S.A., Childs, T., Jawahir, I.S., Klocke, F., Venuvinod, P.K.: The state-of-the-art of modeling in machining processes. CIRP Ann. Manuf. Technol. **47**(2), 587–626 (1998)
4. Scheer, A.W., Thomas, O., Adam, O.: Process modeling using event-driven process chains. In: Dumas, M., Van der Aalst, W.M., der Hofstede, A.H. (eds.) Process-Aware Information Systems: Bridging People and Software Through Process Technology, pp. 119–145. Wiley, Hoboken (2005)
5. Ogunnaike, B.A., Ray, W.H.: Process Dynamics, Modeling, and Control. Oxford University Press, Oxford (1994)
6. Invester, R.W., Kenedy, M., Davies, M., Stevenson, R., Thiele, J., Furness, R., Athavale, S.: Assessment of machining models: progress report. Mach. Sci. Technol. **4**(3), 511–538 (2000)
7. Cataloge HAHN+KOLB: Werkzeuge Maschinen Betriebseinrichtungen ATORN ORION (2015)

Hardness Prediction Model for Drawing with Wall Thickness Reduction

Himzo Đukić[1(✉)] and Mirna Nožić[2]

[1] University of Mostar, FSRE, Ul. Matice hrvatske bb, 88 000 Mostar, Bosnia and Herzegovina
himzo.djukic@fsre.sum.ba
[2] Faculty of Mechanical Engineering, University "DžemalBijedić" of Mostar, Mostar, Bosnia and Herzegovina

Abstract. Deep drawing with a reduction in the thickness of the wall with a change in the cross section of the workpiece, starting from its bottom to the top, leads to hardening of the material and increase its hardness. As the thickness of the wall decreases, a constant increase in hardness ensures a stable process without interruption of workpieces. By measuring the hardness at the characteristic points at the outer diameter, experimental values were obtained, on the basis of which a model for hardness prediction was derived. The model makes it possible to determine the hardness in all sections to the top of the workpiece, if its geometry is known.

Keywords: Model · Hardness · Deep drawing with reduction of wall thickness · Cross section contraction · Section geometry

1 Introduction

Deep draw with wall thickness reduction is the most commonly used for ammunition production. Workpieces are made in multiple drawing operations, with changes in their plastic properties, changes in hardness and grain size [5, 7]. In order to ensure the production of quality workpieces without interruption, it is necessary to ensure that the hardening of the material is accompanied by a steady increase in hardness. For every finished product, the lawfulness of surface hardness growth must be established, going from its bottom to the top. Due to the specific exploitation of the product, its output hardness is prescribed. Its change can be influenced by several different parameters such as: heat treatment mode, degree of deformation, deformation rate, etc.

2 Influence of Reducing Surface Size Decrease on Surface Strength Growth

All types of ammunition have a variable thickness of the sheath wall going from the bottom to the top of the workpiece. Due to the decrease in the thickness of the wall, the degree of deformation in height increases. The difference in cross-sectional area of the envelope at the bottom crossing in the envelope relative to the surface at the top of

I. Karabegović (Ed.): NT 2020, LNNS 128, pp. 221–226, 2020.
https://doi.org/10.1007/978-3-030-46817-0_25

the workpiece may be over 70% [1, 6]. As the degree of deformation increases from the bottom to the top, it is normal to expect that as the deformation increases, the surface hardness also increases. In order not to break the workpiece in the force transfer zone, it is necessary to ensure a uniform growth of surface hardness going from the bottom to the top of the workpiece. If the surface hardness does not increase evenly or there are points with a decrease in surface hardness, there is a possibility that the workpiece may break. The uniform growth of surface hardness is significantly influenced by the proper geometry of the extractor. Sudden changes in the diameter of the puller lead to a sudden increase in the degree of deformation, which causes the workpiece to be interrupted in the drawing process. When designing new products, special attention is paid to the last extraction operation and defining the surface hardness growth, going from the bottom to the top of the workpiece. In order to find a model for predicting surface hardness growth, extensive surface hardness measurements were performed on a workpiece with a diameter of 20 mm and a height of 110 mm, made of CuZn28 brass, Fig. 1.

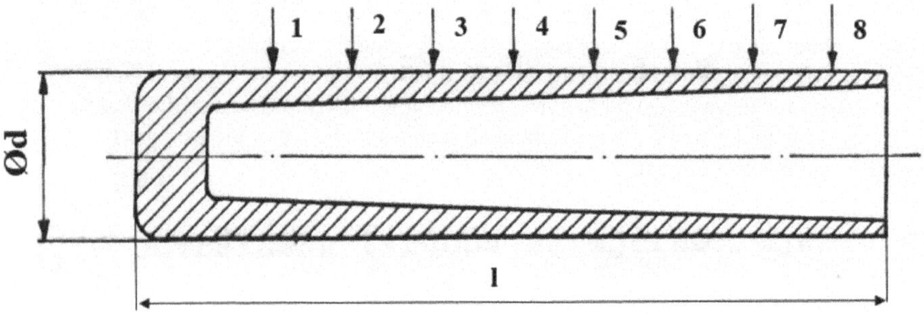

Fig. 1. Work item with measuring points

The surface hardness was measured at eight points. The results of measuring the cross-sectional area A [mm^2] at eight points and the calculated cross-sectional curvature values by height ψ [%] are given in Table 1. The same table shows the mean values of surface hardness measured at 60 work items (HV3) and the differences between hardness at measuring points 2 to 8 with hardness at measuring location 1 (ΔHV3).

Table 1. Results of measurement of hardness and cross-sectional

Label	Measuring location							
	1	2	3	4	5	6	7	8
A [mm^2]	81,4	61,7	50,3	33,3	27,3	24,6	24,3	24,0
$\Psi = (A_1 - A_i)/A_1$	0	18	24	38	59	63	67	70
HV_3	188	199	207	213	217	223	227	227
ΔHV_3	0	11	19	25	29	35	39	39

3 Hardness Prediction Model

By analyzing the hardness growth results given in Table 1, it can be assumed that the surface hardness, going from bottom to top, increases in function [2]:

$$HV_i = HV_1 + C \cdot \psi^n \quad (1)$$

where are they:

HV1 - surface hardness at measuring point 1,

HVi - surface hardness at measuring points 2 to 8,

C - constant,

$\Psi = (A1 - Ai)/A1$ [%] - contraction of the cross-section in height with respect to the cross-sectional area at the measuring point of crossing the sheath into the bottom of the workpiece,

n- exponent.

In order to facilitate finding unknown quantities C and n, Eq. (1) is transformed into the form:

$$HV_i - HV_1 = C \cdot \psi^n \quad (2)$$

Logarithmic expression (2) yields:

$$\log \Delta HV = \log C + n \cdot \log \psi \quad (3)$$

This represents the equation of direction in a double logarithmic system.

Using the measured values of the hardness and the calculated values of the contractions of the cross-sections given in Table 1, we form a system of equations whose solution yields unknown quantities for C and n in the form:

$$\begin{aligned} C &= 1,49 \\ n &= 0,76 \end{aligned} \quad (4)$$

The assumed growth function of surface hardness in height (1) takes the form of:

$$HV_M = 188 + 1,49 \cdot \psi^{0,76} \quad (5)$$

The calculated values of surface hardness according to model (5) are given in Table 2. By comparing the experimental hardness and the calculated hardness by model, the maximum difference was found at four points from.

The experimental values of the surface hardness given in Table 1 were processed by the mathematical package in Excel. The experimental points and the mathematical form of the third degree curve, which most likely describes the same, are given in Fig. 2.

Table 2. Comparison of experimental hardness values with the model

Label	Comparison of experimental results with the model							
	Measuring location							
	1	2	3	4	5	6	7	8
HV_3	188	199	207	213	217	223	227	227
HV_M	188	201	205	212	221	223	224	226
$\Delta = (HV_3 - HV_M)/HV_3$	–	−1	+1	–	−1	–	+1	–

The mathematical form of a curve describing the experimental values with $R^2 = 0,9781$ is $y = 0,0001x^3 - 0,0172x^2 + 1,0689x + 187,46$. Substituting y with HV, x with ψ gives the expression:

$$HV_i = 0,0001 \cdot \psi^3 - 0,0172 \cdot \psi^2 + 1,0689 \cdot \psi + 187,46 \quad (6)$$

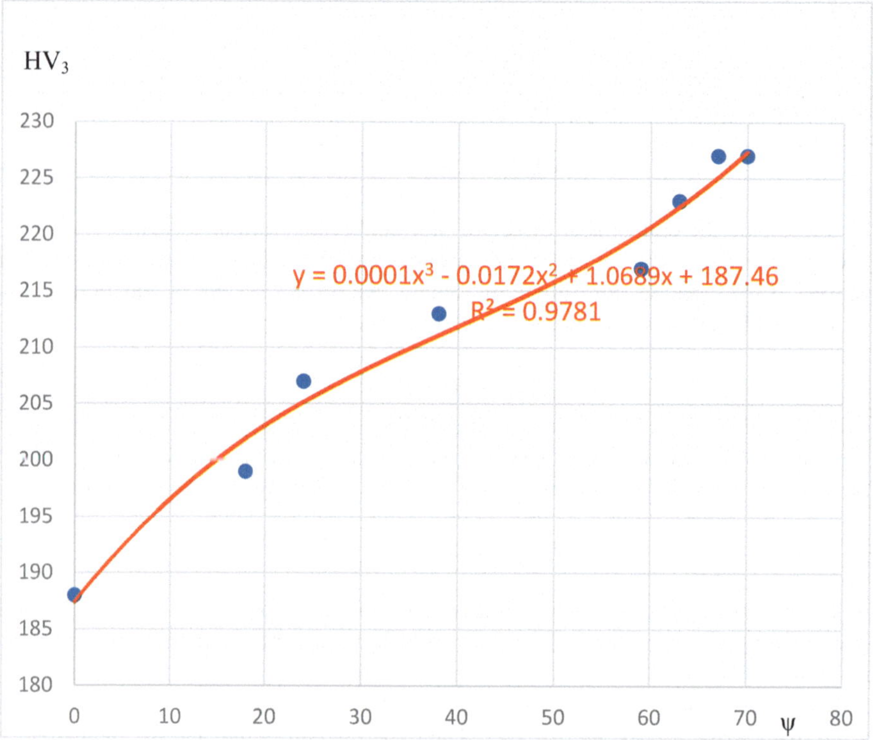

Fig. 2. The diagram of the dependence of hardness on the cross-section contraction

Comparison of hardness values calculated by function (6) and experimental values is given in Table 3. The maximum difference is 6.6%, and unacceptable values are found in 6 out of 8 points.

Based on the results given in Table 3, it can be concluded that no mathematical model is most likely to describe experimental results that are acceptable for practice [4].

Table 3. Comparison of excel results with experimental results

Label	Comparison of excel results with experimental results							
	Measuring location							
	1	2	3	4	5	6	7	8
HV_3	188	199	207	213	217	223	227	227
HV_{EXC}	188	202	205	209	211	211	212	212
$\Delta = (HV_3 - HV_{EXC})/HV_3$	-	+1,5	−1	−1,9	−2,8	−5,4	−6,6	−6,6

4 Conclusion

The proposed model of surface hardness prediction by workpiece height when drawing with reduction of wall thickness gives very good results, so that based on the cross-sectional geometry and the measured hardness at the bottom of the sheath, all surface hardness in all cross sections going from the bottom can be calculated with great accuracy to the top. The application of the model is important when designing new products to ensure proper growth of surface hardness by the height of the workpiece.

References

1. Đukić, H., Nožić, M.: Limit values of maximal logarithmic strain in multi-stage cold forming operations. J. Technol. Plast. **40**(1) (2015)
2. Nožić, M., Đukić, H.: The impact of the deformation redistribution on the special narrowing force. In: 5th International Conference, New Technologies, Development and Application, NT-2019, Sarajevo, Bosnia and Herzegovina (2019)
3. Wang, G., Gu, Z., Xu, H., Yu, G.: Numerical analysis and experimental investigation of surface defects in die hemming process. Int. J. Mater. Form. **13**(1), 91–102 (2020)
4. Nasab, A.S., Mashayekhi, M.: Application of an efficient anisotropic damage model to prediction of the failure of metal forming process. Int. J. Damage Mech **28**(10), 1556–1579 (2019)
5. Lee, C., Hong, S.: Curvature area prediction for the deep drawing-ironing process of a cylindrical cup using finite element method and regression analysis. J. Mech. Sci. Technol. **32**(12), 5913–5918 (2018)
6. Parida, A.K., Soren, S., Jha, R.N., Sadhukhan, S.: 12 mm thick circular blanks of Al-killed AISI 1020 steel-applied for cylindrical cup manufacturing by multistage deep drawing with simultaneous ironing. Mater. Res. **19**(5), 965–976 (2016)

7. Đukić, H., Nožić, M.: Obrada deformisanjem, drugo izdanje, Univerzitet “Džemal Bijedić”, Mašinski fakultet Mostar (2018)
8. Nožić, M., Đukić, H.: Projektovanje tehnologija obrade deformisanjem, Univerzitet “Džemal Bijedić” Mašinski fakultet Mostar (2016)
9. Adamović, D., Stefanović, M., Aleksandrović, S., Gulipija, Z., Živković, M., Đačić, S., Živić, F.: Analysis of the surface state of tools influence on the ironing process. J. Technol. Plast. **36**(2), 111–120 (2011)
10. Adamović, D., Mandić, V., Živković, M., Gulipija, Z., Stefanović, M., Topalović, M., Aleksandrović, S.: Numerical modeling of ironing process. J. Technol. Plast. **38**(2), 109–124 (2013)

The Use of Digital Factory Technologies in Slovenian Manufacturing Companies

Iztok Palčič(✉), Simon Klančnik, Tone Lerher, and Mirko Ficko

Faculty of Mechanical Engineering, University of Maribor,
Smetanova ulica 17, Maribor, Slovenia
iztok.palcic@um.si

Abstract. This paper deals with the use of advanced manufacturing technologies in Slovenian manufacturing companies in the light of Industry 4.0. The main objective of the paper is to determine the adoption of specific advanced manufacturing technologies, such as digital factory technologies, 3D manufacturing technologies and use of specific types of robots. At the same time, we present a possible Industry 4.0 readiness index and assess Industry 4.0 readiness of Slovenian manufacturing companies. Results are based on a sample of 118 Slovenian manufacturing companies, whose data were obtained through the 2018/19 European Manufacturing Survey edition. The results are presented with the use of descriptive statistics. Results show that the use of specific advanced manufacturing technologies in Slovenian manufacturing companies is quite diverse and that Industry 4.0 readiness increases.

Keywords: Manufacturing company · Advanced manufacturing technology · Industry 4.0 · Industry 4.0 readiness index · European Manufacturing Survey · Slovenia

1 Introduction

Manufacturing is continuously evolving from concept development to methods and tools available for the production of goods for use or sale. Traditionally, manufacturing refers to an industrial production process through which raw materials are transformed into finished products to be sold in the market. However, these days manufacturing is considered to be an integrated concept at all levels from machines to production systems to an entire business level operation [1].

When we speak about manufacturing, we must think about manufacturing technologies, nowadays especially about advanced manufacturing technologies (AMT) and information-communication technologies (ICT). These technologies are a vital part of smart factory that utilize the internet of things (IoT) to realize intelligent manufacturing concepts like predictive maintenance or extensive machine to machine communication [2]. In smart factory products, resources and processes are characterized by cyber-physical systems (CPS) [3]. The digital factory is a model of a planned or real factory used for design, planning and operations. In the smart factory, the digital factory developed during engineering should be integrated with the smart factory with its real time data and inferred statistics and information [4].

I. Karabegović (Ed.): NT 2020, LNNS 128, pp. 227–237, 2020.
https://doi.org/10.1007/978-3-030-46817-0_26

There are many representatives of AMT and ICT (that are focus of digital factory) and it is a fact that the dispersion of these technologies heavily depends on industry and company size. It is also true that technologies are very context related and are not suitable for all manufacturing companies. On the other hand, we have analyzed several technologies from the "digital factory" area and "automation/robots" area that have a potential and the need to find their way in all manufacturing environments. Therefore, our research presents the diffusion of 16 selected AMT that are characteristic for the Industry 4.0 era in the Slovenian manufacturing industry. Slovenia has a well-developed manufacturing industry, especially in the Rubber and Plastics, Metal Works, Mechanical Engineering, Electrical Engineering and other metal sectors. Slovenian manufacturing companies are producers of final products in different sectors, as well as world-wide suppliers, especially in the Automotive industry. Therefore, it is of vital importance to monitor the development of these industries in terms of AMT and ICT use.

Besides analyzing diffusion of AMT and ICT this paper also presents a possible Industry 4.0 readiness index and assess Industry 4.0 readiness of Slovenian manufacturing companies. Some models have been proposed in the past to determine the degree of maturity of a company that measures the progress made by the company in the implementation of Industry 4.0. However, no model has yet been proposed that determines the degree of readiness of a company, which is an earlier step in the maturation process [5]. Therefore, we will present an Industry 4.0 readiness index, developed by Fraunhofer Institute for Systems and Innovation Research – ISI [6], who is a coordinator of the largest European Manufacturing Survey (EMS). The model was applied to Slovenian manufacturing industry.

This paper is organized as follows: first, we will introduce the concept of Industry 4.0 with focus on the proposed Industry 4.0 readiness index. The methodological section explains the characteristics of the EMS. After that, we present the use of selected technologies in Slovenian manufacturing companies and the results, obtained with the use of Industry 4.0 readiness model. Finally, a concluding discussion is provided for the findings, where some managerial implications, research limitation and directions for future research are given.

2 Industry 4.0 Readiness Index

According to [7] companies have serious problems to grasp the overall idea of Industry 4.0. One of the problems is that they experience problems in determining their state-of-development with regard to the Industry 4.0 vision and therefore fail to identify concrete fields of action. To overcome growing uncertainty and dissatisfaction in manufacturing companies regarding the idea of Industry 4.0, new methods and tools are needed to provide guidance and support to align business strategies and operations [7]. Therefore, in the last few years different maturity and readiness models are being developed [8–12]. There is a difference between readiness and maturity: readiness assessment takes place before engaging in the maturing process (staring point), whereas maturity assessment aims for capturing the as-it-is state whilst the maturing process.

Unfortunately, these models generally treat maturity and readiness as synonyms [5], which is especially problem for real readiness models.

The proposed Industry 4.0 readiness index was developed by Fraunhofer ISI [6] to analyze data collected within our research: EMS. The logic of the Fraunhofer Industry 4.0 readiness index is presented in Fig. 1 and it is based on the selected Industry 4.0 enabling technologies (AMT). Since the different technologies are highly process and operation-dependent and come from different technology fields, a simple counting of the technologies used is not sufficient for an Industry 4.0 readiness index. Therefore, these AMT are divided into three technology fields: Digital management systems, Wireless human-machine communication and Cyber-physical system (CPS)-related processes. While the first two technology fields cover IT-related processes (Industry 4.0 basic technologies) and still have a clear distance from Industry 4.0, technology field CPS already contains the first approaches to networked/digital production and can therefore be classified as Industry 4.0 closer than the other two technology fields [6].

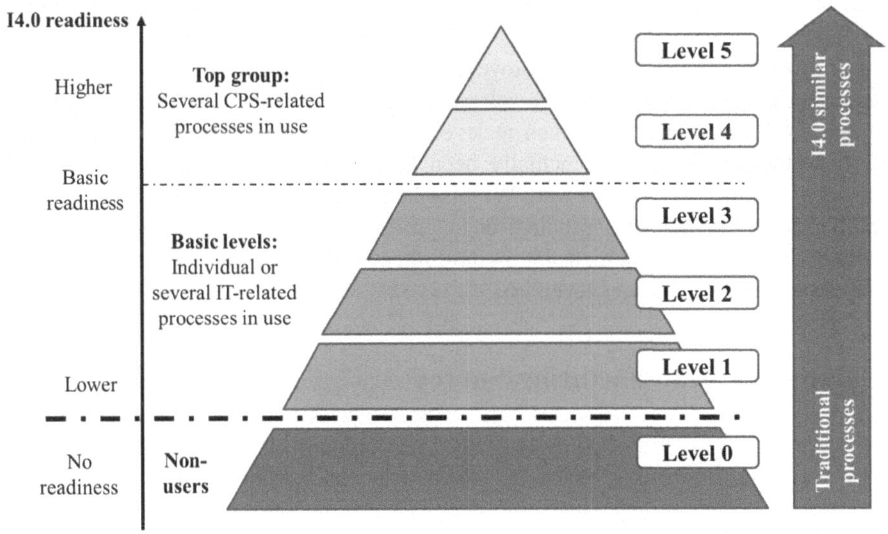

Fig. 1. Industry 4.0 readiness index [6]

Using this grouping, the companies can be classified as Industry 4.0-close companies, which, on the one hand, use and combine several technology fields in production and, on the other hand, use several of the CPS-related processes in their production. Accordingly, Industry 4.0 readiness index results with the following main groups and levels:

Non-users who are not (yet) ready for Industry 4.0:

- Level 0: Companies that do not use any of the Industry 4.0 enabling technologies and tend to still rely on traditional production processes.

Basic levels, as the basis on the way to Industry 4.0, with little readiness:

- Level 1 (beginners): Companies that use IT-related processes in one of the three technology fields.
- Level 2 (advanced beginners): Companies that use IT-related processes in two of the three technology fields.
- Level 3 (advanced users): Companies that are active in all three technology fields and use both IT-related processes and CPS-related processes.

Top group, as a pioneer on the way to Industry 4.0, with a slightly higher readiness:

- Level 4: Companies that are active in all technology fields and use at least two technologies of CPS-related processes.
- Level 5: Companies that are active in all technology fields and use at least three technologies of the CPS-related processes.

With each level, the Industry 4.0 readiness status increases or the distance to networked production decreases. While there is no readiness for Industry 4.0 in level 0, companies in levels 1 to 5 have a basic readiness. Companies that already use IT-related processes (levels 1 and 2), however, have a greater Industry 4.0 distance than companies in levels 3 to 5 that are already implementing the first elements of networked production. However, even at levels 4 and 5 it cannot be assumed that the threshold to Industry 4.0 has actually been breached. Rather, only the distance to networked production has narrowed. With the help of this Industry 4.0 readiness index, the change from traditional production to production close to Industry 4.0 can be mapped. Companies with a higher level have already made the transition more strongly than companies in the lower levels [6].

3 European Manufacturing Survey

The research data was collected using the EMS, coordinated by the Fraunhofer Institute for Systems and Innovation Research – ISI, the largest European survey of manufacturing activities. The survey's questions deal with manufacturing strategies, application of innovative organizational and technological concepts in production, cooperation issues, production offshoring and backshoring, servitisation, and questions of personnel deployment and qualification. In addition, data on performance indicators such as productivity, flexibility, quality and returns is collected. The survey takes place every three years. In most countries, EMS is organized as a paper-based survey at company level (the core questionnaire has six pages). The persons contacted to fill in the questionnaires are the production manager or the CEO of the manufacturing companies. The responding companies present a cross-section of the main manufacturing industries. Included are producers of rubber and plastics, metal works, mechanical engineering, electrical engineering, textile and others. The survey is conducted among manufacturing companies (NACE Revision 2 codes from 13 to 32) having at least 20 employees. The main objectives of the EMS project are to find out more about the use of production and information technologies, new organizational approaches in manufacturing and the implementation of best management practices.

Our research is based on EMS data from a Slovenian subsample from the year 2018/19 round. We sent 847 questionnaires and received 127 responses – a 15% response rate. For the purpose of our presented study, we have excluded companies from the textile industry (NACE 13, 14 and 15) and included 118 responses from other industries into the analysis. The response rate is still a little above 15%. In our 2018/19 subsample, manufacturing companies from NACE divisions 22, 25 and 28 are represented most widely. with around 25% of companies from NACE 25, around 20% from NACE 28, and around 16% from the NACE 22 division. We classified manufacturing companies in three size classed based on the number of employees. The largest share of respondents is from medium sized companies (around 43%), followed by small companies (32%) and large companies (25%).

4 Results and Discussion

4.1 Adoption of AMT in Slovenian Manufacturing Companies

In our last EMS 2018/19 research round, we have analyzed the use of 16 AMT and ICT. Technologies were divided into four groups:

- Digital factory (9 technologies);
- Automation and robotics (2 technologies);
- Additive Manufacturing Technologies (2 technologies);
- Energy efficiency technologies (3 technologies).

For each technology, we have asked for the following information:

- Use of technology (yes/no);
- Use planned in the upcoming period 2019–2021;
- Year in which this technology was used for the first time in your factory;
- Extent of actual utilization compared to the most reasonable potential utilization in the factory: Extent of utilized potential "low" for an initial attempt to utilize, "medium" for partly utilized and "high" for extensive utilization;
- Upgrade of the already implemented technology in the last three years – Follow-on investment since 2015 (yes/no).

In this paper, we are focusing on Industry 4.0 specific AMT from the digital/smart factory area, industrial robots and additive technologies (we are skipping energy efficient technologies). Analysis shows that software for production planning and scheduling (e.g. the ERP system) is the most frequently used technology, but other "Digital factory" technologies are catching up (Table 2). This is especially evident for digital technologies to provide drawings, work schedules or work instructions directly on the shop floor, for product and process data exchange with suppliers and customers (Electronic Data Interchange – EDI). There is also a rise in implementation of real-time production control systems, and introduction of mobile/wireless devices for programming and controlling facilities and machinery.

When we looked at the use of both types of discussed robots together, we found that at least one type of robots is present in 64% of companies, making industrial robots the most widely used technology from our list. If we combine both types of additive manufacturing technologies, we can observe that at least one of them is used in around one third of the companies. We added one "future" AMT, namely artificial intelligence (AI) and we can see the first installations of AI, as this technology is present in 5% of the observed population of manufacturing companies (Table 1).

Table 1. AMT adoption in Slovenian manufacturing companies

AMT	Share [%]
Digital factory	
Mobile/wireless devices for programming and controlling facilities and machinery (e.g. tablets)	32,2%
Digital solutions to provide drawings, work schedules or work instructions directly on the shop floor	54,2%
Software for production planning and scheduling (e.g. the ERP system)	62,7%
Digital exchange of product/process data with suppliers/customers (Electronic Data Interchange EDI)	51,7%
Near real-time production control system (e.g. systems of centralised operating and machine data acquisition, MES)	39,8%
Systems for automation and management of internal logistics (e.g. Warehouse management systems, RFID)	20,3%
Product-Lifecycle-Management-Systems (PLM) or Product/Process Data Management (PDM)	19,5%
Virtual reality, or simulation for product design, or product development (e.g. FEM, digital prototyping, computer models)	38,1%
Artificial Intelligence (e.g. Deep Learning, Machine Learning or Neural Networks)	5,1%
Automation and robotics	
Industrial robots for manufacturing processes (e.g. welding, painting, cutting)	50,0%
Industrial robots for handling processes (e.g. depositing, assembling, sorting, packing processes, AGV)	35,6%
Additive manufacturing technologies	
3D printing technologies for prototyping (prototypes, demonstration models, 0 series)	32,2%
3D printing technologies for manufacturing of products, components and forms, tools, etc.	23,7%

We asked manufacturing companies about their first implementation of AMT, planned use of specific AMT in the period 2019–2021, and upgrade of the existing AMT in the period from 2016 to 2018 (Table 2). We can observe that both types of industrial robots are among the AMTs that have been used in manufacturing companies for the longest time. We can observe that the majority of AMT is upgraded continuously, on average, from 30 to 55% of implementation cases. Especially, AMTs from the "Digital factory" group are very frequently subject to follow-up investments/upgrades. The share of companies planning to invest in AMT in the next 3 years has risen dramatically, compared to our previous research rounds. The column "Planned use" presents the share of companies that do not possess a specific AMT, but are planning to introduce it in the period from 2019 to 2021. A third of companies, currently without any ERP system, digital solutions to provide info directly on the shop floor, near real-time production control systems and mobile/wireless devices in production, are planning to implement these technologies in the following years. 20% of companies without industrial robots for manufacturing processes or handling processes, plan to introduce them by the end of 2021. Around 10 to 15% of companies want to acquire different additive technologies, and 10% of companies will introduce AI solutions, which would triple the existing share of companies that use AI.

Table 2. Characteristics of AMT in Slovenian manufacturing companies

AMT	Use [%]	Planned use [%]	Year of first use	Upgrade [%]
Digital factory				
Mobile/wireless devices	32,2	30,0	2011	55,3
Digital solutions to provide info directly on the shop floor	54,2	33,3	2010	37,5
ERP system	62,7	34,1	2008	43,2
Digital Exchange of data with suppliers	51,7	22,8	2011	29,5
Near real-time production control system	39,8	28,2	2011	51,1
Systems for automation and management of internal logistics	20,3	23,4	2014	33,3
PLM, PDM	19,5	15,8	2012	30,4
Virtual reality or simulation for product design	38,1	15,1	2010	40,0
Artificial intelligence	5,1	10,7	2014	16,7
Automation and robotics				
Industrial robots for manufacturing processes	50,0	20,3	2008	42,4
Industrial robots for handling processes	35,6	19,7	2010	38,1
Additive manufacturing technologies				
3D printing technologies for prototyping	32,2	12,5	2006	36,8
3D printing technologies for manufacturing	23,7	14,4	2009	35,7

Next, we looked at the extent of potential use or actual utilization (low, medium, high) of the 10 most frequently used AMTs (Fig. 2). This assessment is partly subjective, but still provides some interesting information.

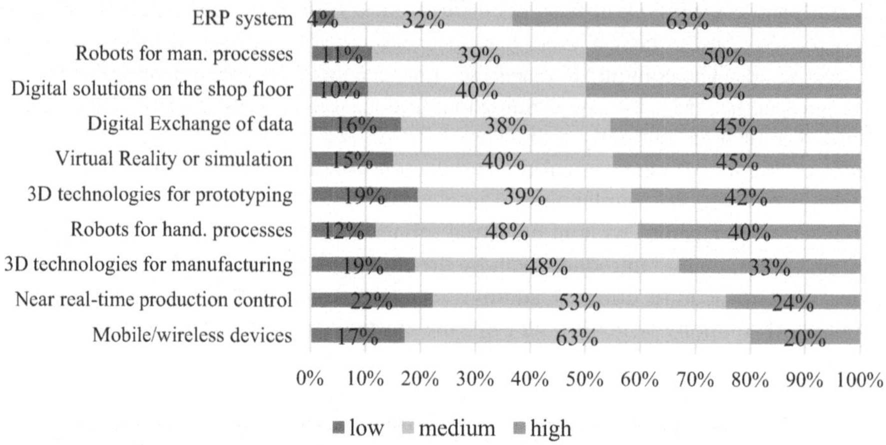

Fig. 2. Extent of use potential for the 10 mostly widely used AMTs

Figure 2 presents that the majority of AMT manufacturing companies admits, that they are not exploiting them to full/high potential. This indicates that there is still a lot of room for improvement in terms of exploiting the use potential for all of the AMTs discussed. ERP is the only AMT where more than half of manufacturing companies believe that they use this technology to high potential. For the majority of other analyzed AMTs, it can be seen that they are fully exploited in less than half of cases. We can argue that some AMTs, such as ERP, industrial robots, digital exchange of data, have been present in manufacturing companies for many years, and this longer time period offers better opportunities to use them in a way to maximize their potential.

We calculated the average values for extent of use, where "low use" received value 1, "medium use" received value 2, and "full/high" use received value 3. The highest extent of use is for ERP solutions with the value of 2,6 (out of a maximal value 3). The lowest is, as expected, for extremely new technologies, such as AI (1,4), proving that these technologies are still at the beginning of their use in manufacturing companies.

When comparing our findings to previous research rounds, we can see clearly that the share of companies that are able to exploit the potential of specific AMT fully is slightly higher for the majority of technologies from both research rounds. This points to the fact that companies have less and less time to learn how to use new technologies, and have to exploit their potential very quickly.

4.2 Industry 4.0 Readiness Index in Slovenian Manufacturing Companies

Figure 3 presents all three technology fields in Industry 4.0 readiness index and AMT from our research:

- Digital management systems: The first technology field is formed by software systems for production planning and scheduling (ERP) and the product lifecycle management systems. These are classified as the basic technologies of IT and digitization and are thus assigned to IT-related processes.

- Wireless human-machine communication: In the second technology field, the digital visualization is summarized with the mobile devices. This field is also assigned to the I4.0 basic technologies and thus to IT-related processes.
- Cyber-physical system (CPS)-related processes: The third technology field takes into account the near real-time production control system, the automation of logistics and the digital data exchange with suppliers and customers. These technologies already had production elements in cyber-physical systems and are therefore considered to be among the advanced I4.0 technologies.

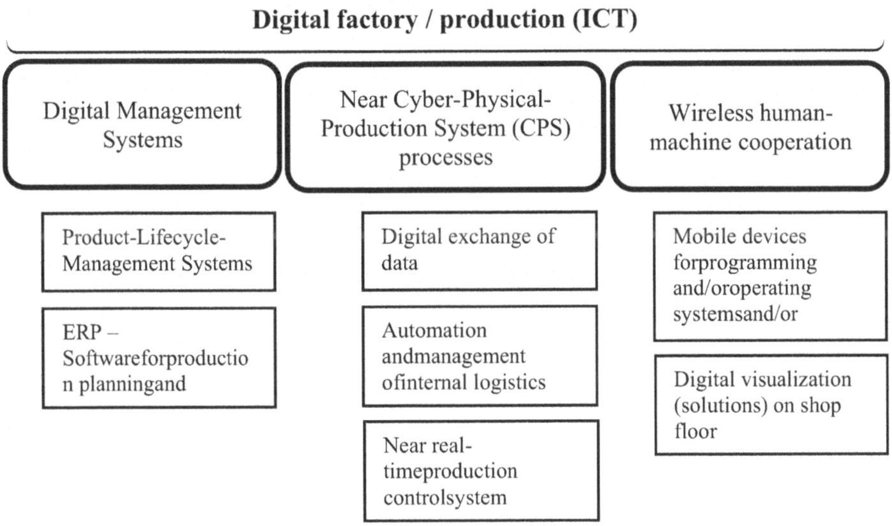

Fig. 3. Industry 4.0 readiness index with selected AMT [6]

Figure 4 depicts the distribution of all six described Industry 4.0 readiness levels for Slovenian manufacturing companies. A fairly high proportion, around 17% of all companies have so far not implemented any AMT in production. Around 57% of all companies already have IT-related processes in their production and form the basic levels. However, this group covers a big share of companies (nearly two thirds). This basic users group includes the group of beginners who only use technologies from one field (almost 20%; level 1), the advanced beginners who operate in two technology fields (almost 24%; level 2), but also the already advanced companies that are combining technologies from all three technology fields (almost 14%; level 3).

In the two highest levels 4 and 5, this top group consist of a total of 26,3% of all companies. About every fourth company is consequently active in all three technology fields, and not only uses IT-related processes, but also several CPS-related processes simultaneously. Levels 4 and 5 have very similar share of companies.

A look at Slovenian manufacturing sector shows that there is still a certain share of companies that heavily relies on traditional production processes (non-users). The main group of Slovenian manufacturing companies has slowly started to use IT-related processes, but there is a big difference between beginners and advanced users. The former are clearly closer to the non-users in terms of the type of production processes, the advanced companies are slowly preparing to enter the top group. The top group is not only active in each of the three technology fields, but also uses several CPS-related processes. There is a certain readiness to digitize their production, where the level 5 companies (13,6%) in particular seem to be preparing for Industry 4.0 related production or are already trying to implement it.

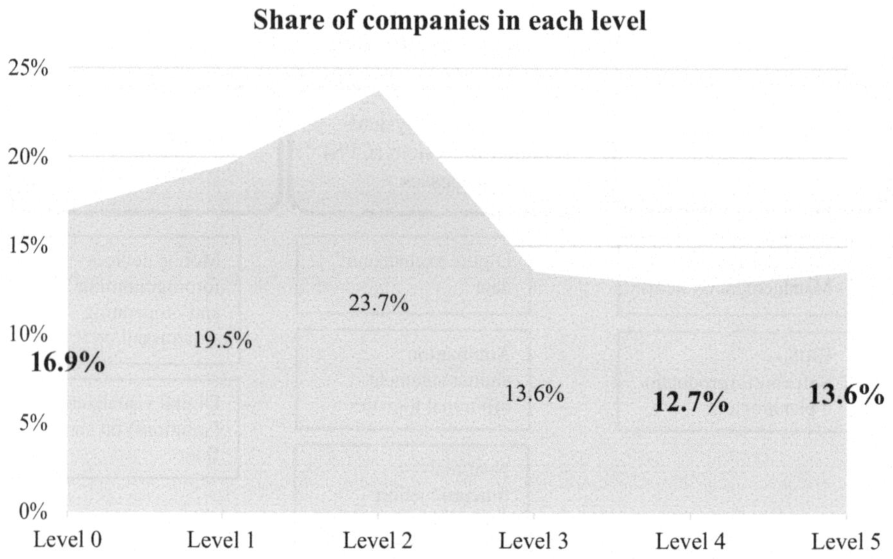

Fig. 4. Industry 4.0 readiness index in Slovenian manufacturing companies

5 Conclusions

Our results show that the use of AMT in Slovenian manufacturing companies is very diverse and context related. Comparison with several western European countries/ partners from EMS show that Slovenia is not lagging behind countries, such as Germany, Austria and Switzerland. Slovenian manufacturing industry is highly integrated in global production networks. Slovenian suppliers are part of global value networks and as such, they have to use AMT to cooperate with global companies.

As is the case with all research, some issues must be taken into account when considering the reliability, significance, and general use of the obtained results. First, the data from Slovenia contains 118 companies in the EMS 2018 round. Although the sample is not small, further research should go towards the direction of a larger sample of more countries. The paper present latest results of our research with the use of simple descriptive statistics. Our findings will be tested in the future with the use of advanced regression models.

We will also make more in depth analysis regarding the use of AMT where we will consider company size, technological intensity of the industry they belong to, their status as the final producer for consumers or business customers (OEM) or supplier (system supplier or supplier of parts or components). We will also look into the relationship between introduced AMT and specific characteristics, such as product complexity, production type and ability to introduce new products.

This research also has practical and managerial implications. Our results on the adoption of AMT indicate the current state of technology implementation in Slovenian manufacturing companies, which can serve managers as a roadmap for future investments. Similar with Industry 4.0 readiness index managers can on one hand get familiar with one of the possible models to measure company readiness for Industry 4.0 and on the other hand compare their company to Slovenian average Industry 4.0 readiness.

References

1. Esmaeilian, B., Behdadb, S., Wang, B.: The evolution and future of manufacturing: a review. J. Manuf. Syst. **39**, 79–100 (2016)
2. Egger, J., Masood, T.: Augmented reality in support of intelligent manufacturing – a systematic literature review. Comput. Ind. Eng. **140**, 106195 (2020)
3. Hermann, M., Pentek, T., Otto, B.: Design principles for Industry 4.0 scenarios: a literature review. Working Paper (2015). https://www.researchgate.net/publication/307864150_Design_Principles_for_Industrie_40_Scenarios_A_Literature_Review
4. Shariatzadeh, N., Lundholm, T., Lindberg, L., Sivard, G.: Integration of digital factory with smart factory based on Internet of Things. In: 26th CIRP Design Conference. Procedia CIRP, vol. 50 (2016)
5. Pacchini, A.P.T., Lucato, W.C., Facchini, F., Mummolo, G.: The degree of readiness for the implementation of Industry 4.0. Comput. Ind. **113**, 1–8 (2019)
6. Lerch, C., Jaeger, A., Meyer, N.: I4.0-Readiness – Baden-Württemberg auf dem Weg zur Industrie 4.0? Fraunhofer ISI Working Paper (2016)
7. Schumacher, A., Erol, S., Sihn, W.: A maturity model for assessing Industry 4.0 readiness and maturity of manufacturing enterprises. Procedia CIRP **52**, 161–166 (2016)
8. Basll, J., Doucek, P.: A metamodel for evaluating enterprise readiness in the context Industry 4.0. Information **10**(89), 1–13 (2019)
9. Lichtblau, K., Stich, V., Bertenrath, R., Blum, M., Bleider, M., Millack, A., Schmitt, K., Schmitz, E., Schröter, M.: IMPULS - Industrie 4.0-Readiness. Impuls-Stiftung des VDMA, Aachen-Köln, Germany (2015)
10. Sarvari, P.A., Ustundag, A., Cevikcan, E., Kaya, I., Cebi, S.: Technology roadmap for Industry 4.0. In: Industry 4.0: Managing the Digital Transformation. Springer Series in Advanced Manufacturing, pp. 95–103. Springer, Cham (2018)
11. Stefan, L., Wienbruch, T.W., Kreimeier, L.D., Bernd, D.K.: Concept for an evolutionary maturity based Industrie 4.0 migration approach. Procedia CIRP **72**, 404–409 (2018)
12. Ganzarain, J., Errasti, N.: Three stage maturity model in SME's towards Industry 4.0. J. Ind. Eng. Manag. **9**(5), 1119–1128 (2016)

Autonomous Work Systems in the Cyber-Physical Production Systems Concept

Elvis Hozdić(✉)

Department of Control and Manufacturing Systems, Faculty of Mechanical Engineering, University of Ljubljana, Aškerčeva 6, 1000 Ljubljana, Slovenia
ehozdic@yahoo.com

Abstract. Under the influence of globalization which brings many changes in all spheres of life and in the sphere of industrial production, manufacturing enterprises are forced to adapt their manufacturing structure these challenges in order to promptly and effectively respond to the complex demands of today's market, which is global and less national. Over time, it's developed different concepts of manufacturing systems. They were intended to respond to market demands in a time in which existed. Today, thanks to the developed information-communication technologies (ICT), manufacturing enterprises tend to structure their systems in a spirit of cyber-physical systems (CPS).

This work presents a new concept of the autonomous work system (AWS) in a spirit of the CPS. The development AWS in a spirit of the CPS represents an evolution extension of the organizational structure of production of complex and adaptive distributed production systems. The introducing additional communications between the elements of the AWS in a spirit of the CPS enables a better production performance with respect to the selected performance measures.

The possibility of implementing the concept of AWS in a spirit of the CPS is presented through industrial example in the production area of high pressure die casting (HPDC).

Keywords: Autonomous work system · Cyber-physical production systems · High pressure die casting

1 Introduction

Under the influence of globalization which brings many changes in all spheres of life and in the sphere of industrial production, manufacturing enterprises are forced to adapt their manufacturing structure these challenges in order to promptly and effectively respond to the complex demands of today's market, which is global and less national. Over time, it's developed different concepts of manufacturing systems. They were intended to respond to market demands in a time in which existed. Today, thanks to the developed information-communication technologies (ICT), manufacturing enterprises tend to structure their systems in a spirit of cyber-physical systems (CPS).

The term "*cyber-physical systems*" coined in 2006 in a high-level working group [1] composed of selected experts from the USA and European Union, advocates the co-existence of cyber and physical elements with a common goal. Embedded systems have

I. Karabegović (Ed.): NT 2020, LNNS 128, pp. 238–249, 2020.
https://doi.org/10.1007/978-3-030-46817-0_27

been developed over the past decades, however CPS explicitly pose a focus on the integration of computation with physical processes [2].

Using the appropriate sensor and actuator technologies, these systems are able to receive direct physical signals and convert them into digital data and, based on such information, enable management of systems in real time. They can share collected data in the form of information and make them available through digital networks, making in that way the Internet of Things [3].

The Internet of Things (IoT) is a combination of technological push and a human pull for more and ever–increasing connectivity with anything in the immediate and wider environment. This push–pull combination makes the development very strong, unstoppable, fast and extremely disruptive. It is foreseen that within the next decade the internet will exist as a seamless fabric of classic computer networks and networked objects. Content and services will be all around us, always available, paving the way to new applications, enabling new ways of working, interacting, entertaining and living. This innovation is enabled through embedding of digital electronics into everyday physical objects, making them *"smart"* and letting them seamlessly integrate within the global resulting cyber–physical infrastructure [4].

In the last two or three decades, several influencing attempts to reshaping industrial production have been published, which can be recognized as cornerstones of Industry 4.0. With the effort of adapting manufacturing and its systems to the time in which they exist, new manufacturing approaches integrating new manufacturing systems with new business models and architectures have emerged [5]. A review of the listed approaches and concepts of manufacturing systems are presented in the paper [6].

Recently, a new approach is emerging in terms of a cyber-physical production system (CPPS) [6], which upgrades in a way the mentioned attempts. It originates from the concept of cyber–physical systems (CPS) [1], and links physical and virtual components of a production system into a coherent whole.

The aim of this paper is presentation of autonomous work systems (AWS)in a spirit of the CPS, definition of its building blocks and characterization of its relations to production, service, and social networks, as well as smart environments.

The paper first discusses the cyber-physical production systems issues. Next, it introduces the concept of the AWS in a spirit of the CPS. Further on, the conceptual model and structure of AWS in a spirit of the CPS are presented. Finally, the industrial examples and potential applications are examined in the Sect. 4.

2 Cyber-Physical Production Systems

Systematically combination of CPS and advanced manufacturing technology enables the structuring of the CPPS [6–9].

In the paper [10], the conceptual model of cyber-physical production systems (CPPSs) is presented. The model consists of three elements: (1) the human *Subject* as a social element, (2) the cyber system (CS) as a cyber element, and (3) the physical work system (PWS) as a physical element, Fig. 1. Each of the elements exists in its own space: the *Subject* in the social space, the CS in the cyberspace and the PWS in the physical space. The connectivity between these elements and thus between the different spaces is enabled via the corresponding communication interfaces, which also bridge the different spaces. Each element has the corresponding relations with its specific environment: the *Subject* with the business and social environment, the CS with the cyber environment and the PWS with the physical environment. The PWS structure is based on the elementary work system (EWS) structure [11]. Its core represents the process, which is implemented on a process implementation device (e.g., machine-tool) and controlled by a logic controller. The latter one is connected on a network via a communication interface, which serves for downloading program code (e.g., NC code) on the controller and for communicating commands in the downward direction and statuses as feedback. An important part of PWS is a monitoring system. It collects data from sensors, which measure characteristic input and output parameters related to process, PWS resources, workpieces, operations and physical environment [12]. The important distinction between the EWS and the PWS is that in the latter the *Subject* is moved from the physical space to the social space of the CPPS.

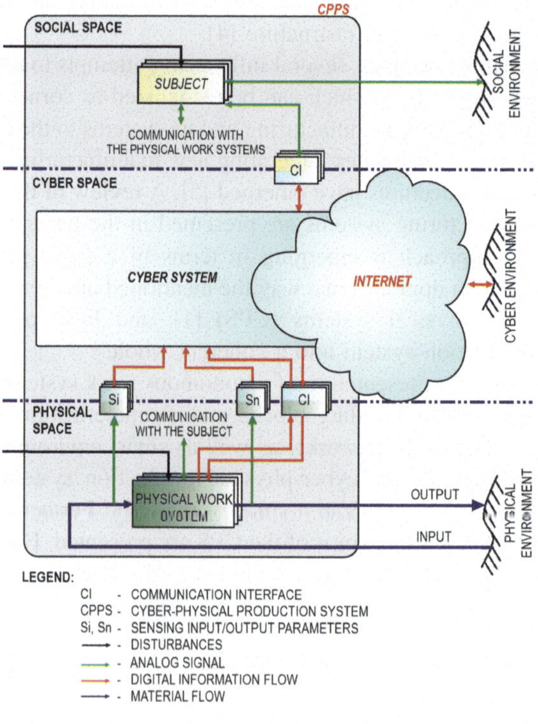

Fig. 1. Generical model of the CPPS [10]

The CS of the CPPS [10, 13] is structured into three hierarchical levels: (1) operational level (EWS cyber system), (2) coordination & collaboration level (AWS cyber system), and (3) business level (CAMS cyber system).

The infrastructure of the CS enables vertical connectivity between the cyber manufacturing structures and the horizontal connectivity of the cyber manufacturing structures in the network structures [10].

In a CPPS, the horizontal integration is accomplished through value networks, and the vertical integration is achieved through networked manufacturing systems [8]. The integration can be realized by the use of different technologies, such as Multi-Agent Systems (MASs), Service-Oriented Architecture (SOA), Internet of Things (IoT), cloud computing, augmented reality, Big Data, Machine-to-Machine (M2M) communication, and similar [8].

In the next, the autonomous work system in the cyber-physical production systems concept is presented.

3 Autonomous Work Systems

Autonomous work system (AWS) is a system with rounded technological functionality and corresponding management functionality [14]. The technological functionality is built upon its core element EWS, with, according to [11], consists of a process, process implementation device (PID), and a human *Subject*. The management functionality of AWS covers management and control of internal operations as well as external coordination on the network [14]. The functional competence and the management functionality are the key enablers of autonomy of AWS. Autonomy AWS is characterized with its specific competences in the sense of knowledge, experience and resources, and with the responsibility to make decisions in their own work, without external guidance and instructions from the outside world.

The conceptual model of the AWS in a spirit of the CPS based on the CPPS concept presented in [10]. The conceptual model of the AWS in a spirit of the CPS is depicted in Fig. 2.

The EWS cyber system enables (1) connecting the physical and social spaces in the EWS, (2) digitalization of the functions (e.g. monitoring and control) and the cybernation and work processes in the EWS, (3) development and implementation of new digitalized functions in the EWS (e.g., self-organization, self-adaption, self-diagnostic, self-learning, etc.), (4) vertical connection of the EWS in the integrated work structure, and (5) horizontal connection of the EWS in a network within the Internet of Things. The EWS cyber system structure is shown in [10].

The AWS cyber system enables (1) vertical connection with subordinated EWS cyber systems, as well as with superior cyber systems, (2) implementation of the digitalized autonomous function in the AWS (management of resources, scheduling, quality control, performance measurements, prognostics, self-learning, etc.), and (3) horizontal connection of the AWS and adaptive distributed manufacturing systems (ADMS) within the Internet of Services. The AWS cyber system is connected with the superior system through the coordination function, while it is connected to the subordinate systems (the EWS cyber systems) through the monitoring and control functions [10]. The AWS cyber system structure is shown in Fig. 3.

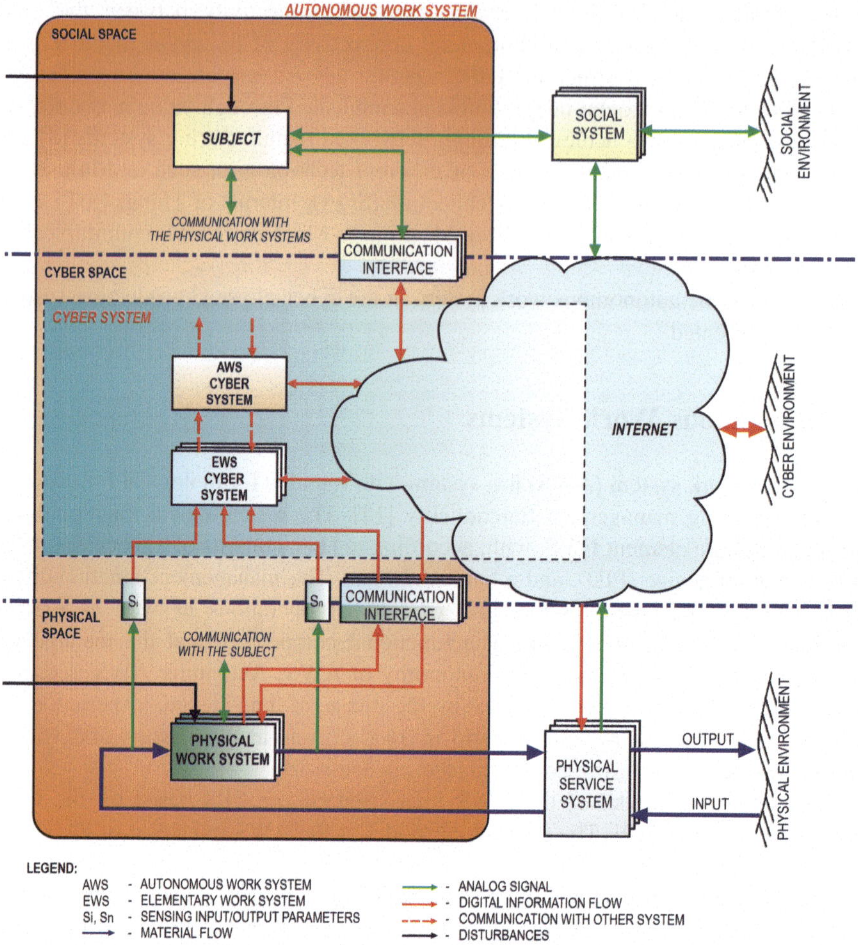

Fig. 2. Model of AWS in the CPPS concept [15]

The basic part of the cyber system of AWS is management & control of AWS. The management and control enable communication cyber system of AWS with cyber systems of EWSs and communication with other systems in cyber system. Also, it enables communication with the *Subject* in the collaborative layer of cyber system. The data connection cyber system of AWS and cyber systems of EWSs is based on connectivity individual D&K base of EWSs with the data & knowledge management in the cyber system of AWS. The outputs from the management & control (actual data from cyber system of EWSs, the data from D&K base of AWS, information from *Subject* or other systems) represent input for the functionality of AWS.

The functionality of AWS are based on the corresponding management functionality that is described in [14] (such as performance evaluation, diagnostics, prognostics, planning and scheduling, management of resources, etc.).

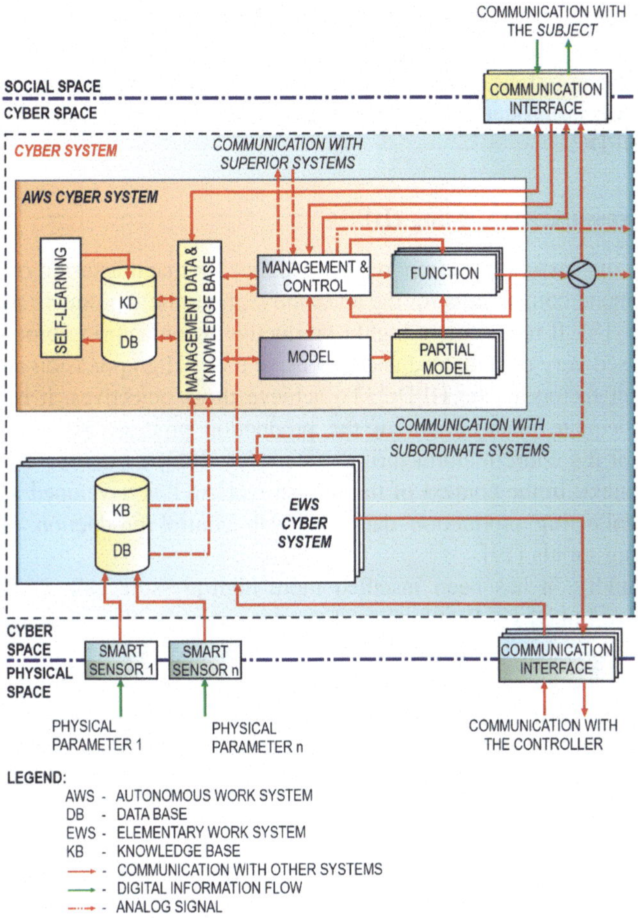

Fig. 3. Structure of the AWS cyber system [10]

The support for realization functions presented models respectively partial models for corresponding function from knowledge base of AWS. Outputs of a function are information that needed for decision-making *Subject*, management and control EWSs, and participation AWS in the service networks.

The functions in cyber system of AWS enable collaborative participation *Subject* and cyber elements (such as agents and multi-agent's technology, expert systems, genetic algorithms, neural networks, web service, etc.) in realized certain tasks.

In the next, the possibility of introducing the developed concept of AWS in a spirit of the CPS in the field of industrial production system is presented.

4 Industrial Example and Potential Application

To illustrate the possibility of introducing the concept of AWS in the field of contemporary industrial production in Sect. 4 shows the production area of high pressure die casting (HPDC).

4.1 High Pressure Die Casting (HPDC)

High pressure die casting is an important process in the manufacturing of high volume and low cost metal components for the automotive, household products and electronics industries [16–18]. It represents a highly-productive manufacturing process for which execution is used very complicated and expensive tools. High precision and quality are two key, target factors in the HPDC. To achieve these objectives, it is necessary to timely and adequate management of the production process, as well as advanced management of the entire manufacturing system for HPDC.

In this context, in the context of the project CoCAST is developed and integrated module for collecting production data and web control production cell for high-pressure casting molds [19].

In the foundry, it has been installed more high-pressure cells (AWS for a die casting) that are linked via Profibus and Ethernet networks. The presence of Profibus enabled communication in real time, while the presence of a network Ethernet connection enabled foundries to higher business levels. The implementation of the project CoCAST, functionality of the program logical controller (PLC) cells was extend for process parameters, machine data and production data. Information structure of the cell for a die casting is shown at Fig. 4.

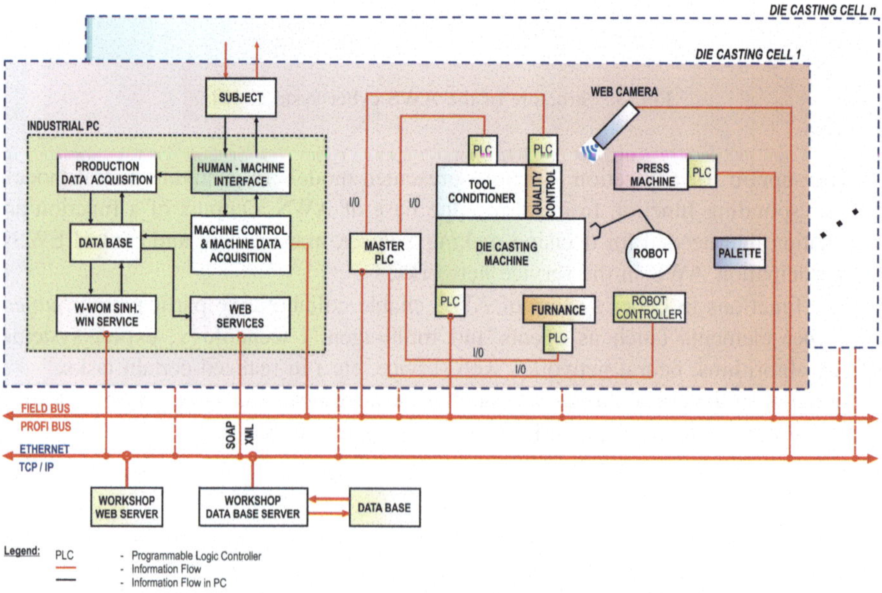

Fig. 4. The cells for a die casting, adopted according to [20]

For the analysis of the current situation in cell for a die casting in this study points to the indicative data from CoCAST base data. The project CoCAST is developed and integrated module for collecting production data and web control production cell for high-pressure casting molds [19].

Analysis of the data collected in the base CoCAST gives good information and knowledge on the state cells for a die casting. For the analysis of cell for a die casting significant data recorded in the form of an alarm (different delays of the production process). Alarms are described from a different point of view, there is a longer hour period, their frequency and intensity of the stochastic. For the analysis of alarms used data on alarm conditions, which are collected in the period from 07 January to 30 June for cell for a die casting.

Data analysis of alarms shown in Table 1, is indicative of the number of alarms that have arisen due to interference on the stove (a total of 762 alarms for a period of six months) and alarm caused by the improper performance of a function by a robot (a total of 702 alarms). Such indication suggests the need for the introduction of advanced means of communication between individual elementary work system (EWS) within AWS for a die casting to manage and control software were at a higher level.

Table 1. The structure and number of alarms associated with the operation of machines and their auxiliary functions

Code	Alarm description	Jan	Feb	Mar	Apr	May	Jun	Amount
51	Tool_51	57	27	48	15	51	30	228
136	Tool_136	27	6	0	54	63	21	171
137	Furnance_137	63	84	33	57	72	18	327
141	Robot_141	39	27	6	18	15	9	114
142	Robot_142	69	213	24	18	132	132	588
144	Robot_144	21	42	30	30	30	12	165
202	Furnance_202	186	174	54	82	176	90	762
220	Tool_cindit_220	12	21	33	33	6	30	135
	Amount	**474**	**594**	**228**	**307**	**545**	**342**	**2490**

In Fig. 4 shows that the communication between the PLC individual EWS takes place on the principle of I/0 exchange signals. In this communication, the EWS is not getting feedback from the other EWS, which was previously delivered signal to perform a function, and whether or not this feature is forwarded properly.

With the aim of improving processes, reducing the number of outages caused by inadequate means of communication between participants in the process, better responsiveness and greater and more productive EWSs and AWS for a die casting there is a need for more advanced means of communication. The use of advanced communication methods, based on the CPS and the CPPS concept (based on M2M communication), possible to perform the reduction of unnecessary stopping the production process, which consequently leads to increased productivity, as individual EWSs and overall AWS for a die casting.

In the extension, the possibility of introducing the CPPS concept in the field of HPDC is described.

4.2 The Model of AWS for a Die Casting Based on the CPPS Concept

On the principles for connecting elements of the physical, cyber, and social system, defined in [10, 13, 21], below Fig. 5 illustrated the structure of AWS for a die casting in a spirit of the CPS.

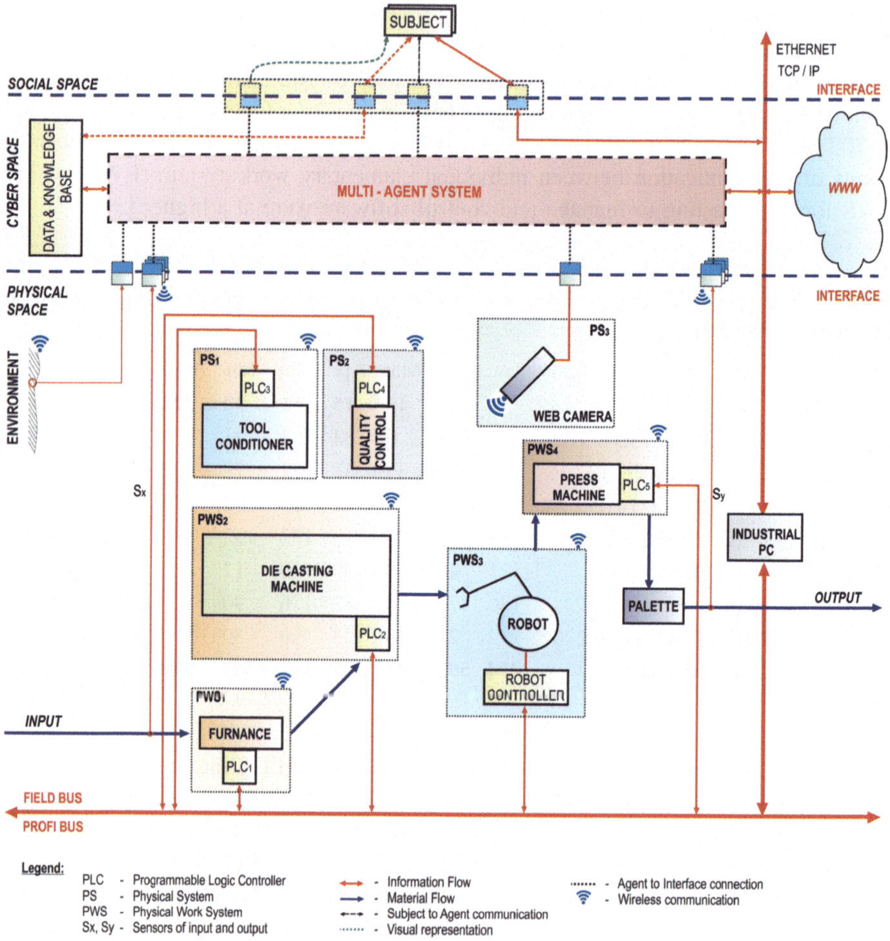

Fig. 5. The model of AWS for a die casting based on the CPPS concept

Physical elements of AWS for a die casting AWS_{HPDC} are: PWS of the Furnace (PWS_1), PWS of a die casting machine (PWS_2), PWS of the robot (PWS_3), PWS of the Press (PWS_4), PS of a tool conditioner (PS_1), PS of a quality control (PS_2), and PS of a web camera (PS_3). PWS and PS are in physical space connected with material flow as well as Field Bus and Profibus communication channels.

To connect PWS in the CS, the multi-agent system (MAS) is established with the corresponding multi-agent platform (MAP) and agent structures and applied modern communication technology based on TCP/IP protocols and 802.15.4 standard for wireless local area network. Through industrial PC a two-way communication of each PLC with MAS. The concept of CPPS enables real-time communication between the PLC defined IEC 61131-3 and IEC 61499 standards [22], as well as inter-agent communication defined FIPA standards [23].

Modern ICT enabled the collection and transmission of data from the physical world in the cyber world. The exploitation of such information in the cyber system, through the implementation of MAS, provides access to and distribution of all relevant information on the manufacturing system in real time.

The role of the MAS in the cyber system of the AWS for a die casting based on the CPPS concept is management and control through the dynamic interaction of the agents in real time. The ontological definition of an agent in the CPPS concept is defined in [24].

Communication based on a modification of the signal in the cell for a die casting represents a solid and very inflexible communication. Applying machine-to-machine (M2M) communication creates the assumption for better, faster and safer participation between PWS in the AWS for a die casting based on the CPPS concept. In such communication, unnecessary delays and timely responds to alarm conditions that occur in AWS for a die casting based on the CPPS concept during the execution process are avoided.

5 Conclusion

The cyber-physical production systems open new possibilities and potentials in design, development and production of high-tech products.

This paper introduces the new concept of AWS in a spirit of the CPS. Based on the developed AWS concept enabled the integration of physical, cyber, and social elements of the functional environment. The main advantage of a new model AWS is that the elements of the cyber system of the CPPS enable the management and control of manufacturing systems in real time.

The presented approach is illustrated on industrial example based on real data collected from production. The demonstration shows the feasibility and the potential of the proposed approach.

The implementation of the AWS concept based on the CPPS concept in real industrial environment is a challenge for further research in the domain of the CPPS, thus contributing to research that defines the consultation document - *"Factory of the Future"* of the European Commission.

References

1. Gill, H.: NSF perspective and status on cyber-physical systems. In: National Workshop on Cyber-Physical Systems, Austin, TX (2006)
2. Lee, E.A., Seshia, S.A.: Introduction to Embedded Systems: A Cyber-Physical Systems Approach. Berkeley University of California, California (2011)
3. Vogel-Heuser, B., Bayrak, G., Frank, U.: Agenda CPS - scenario Smart Factory (Agenda CPS - Szenario smart factory). In: Increased Availability and Transparent Production, Kassel (2011)
4. Miorandi, D., Sicari, S., De Pellegrini, F., Chlamtac, I.: Internet of Things: vision, applications and research challenges. Ad Hoc Netw. **10**(7), 1497–1516 (2012)
5. Jovane, F., Koren, Y., Boer, C.R.: Present and future of flexible automation: towards new paradigms. CIRP Ann. Manuf. Technol. **52**(1), 543–560 (2003)
6. Monostori, L., et al.: Cyber-physical systems in manufacturing. CIRP Ann. Manuf. Technol. **65**(2), 621–641 (2016)
7. Uhlemann, T.H.-J., Lehmann, C., Steinhilper, R.: The digital twin: realizing the cyber-physical production system for Industry 4.0. Procedia CIRP **61**, 335–340 (2017)
8. Liu, Q., et al.: An application of horizontal and vertical integration in cyber-physical production systems. In: 2015 International Conference on Cyber-Enabled Distributed Computing and Knowledge Discovery, pp. 110–113 (2015)
9. Jiang, Z., Jin, Y., E, M., Li, Q.: Distributed dynamic scheduling for cyber-physical production systems based on a multi-agent system. IEEE Access **6**, 1855–1869 (2018)
10. Hozdić, E., Kozjek, D., Butala, P.: A cyber-physical approach to the management and control of manufacturing systems. StrojniškiVestn. J. Mech. Eng. **65**(1) (2020)
11. Peklenik, J.: Complexity in manufacturing systems. CIRP J. Manuf. Syst. **24**, 17–25 (1995)
12. Hozdić, E.: Integrating cyber and physical environment for adaptive process control in work systems. In: Karabegović, I., Kovačević, A., Banjanović, M.L., Dašić, P. (eds.) Handbook of Research on Integrating Industry 4.0 in Business and Manufacturing. Springer (2020, in press)
13. Hozdić, E.: Socio-cyber-physical systems alternative for traditional manufacturing structures. In: Karabegović, I. (ed.) New Technologies, Development and Application II. NT 2019. Lecture Notes in Networks and Systems, vol. 76. Springer, Cham (2020)
14. Butala, P., Sluga, A.: Autonomous work systems in manufacturing networks. CIRP Ann. Manuf. Technol. **55**(1), 521–524 (2006)
15. Hozdić, E.: Model of cyber-physical manufacturing systems. Doctoral dissertation, University of Ljubljana (2020)
16. Žapčević, S., Butala, P.: Adaptive process control based on a self-learning mechanism in autonomous manufacturing systems. Int. J. Adv. Manuf. Technol. **66**(9–12), 1725–1743 (2012)
17. Chiang, K.-T., Liu, N.-M., Chou, C.-C.: Machining parameters optimization on the die casting process of magnesium alloy using the grey-based fuzzy algorithm. Int. J. Adv. Manuf. Technol. **38**(3–4), 229–237 (2008)
18. Cleary, P.W., Savage, G., Ha, J., Prakash, M.: Flow analysis and validation of numerical modelling for a thin walled high pressure die casting using SPH. Comput. Part. Mech. **1**(3), 229–243 (2014)
19. Sluga, A., Butala, P., Peklenik, J.: A conceptual framework for collaborative design and operations of manufacturing work systems. CIRP Ann. Manuf. Technol. **54**(1), 437–440 (2005)

20. Žapčević, S.: Model of a self-learning manufacturing work system. Doctoral dissertation, University of Ljubljana (2013)
21. Hozdić, E., Butala, P.: Concept of socio-cyber-physical work systems for Industry 4.0. Tech. Gaz. **27**(2) (2020)
22. Marik, V., McFarlane, D.: Industrial adoption of agent-based technologies. IEEE Intell. Syst. **20**(1), 27–35 (2005)
23. Bellifemine, F.L., Caire, G., Greenwood, D.: Developing Multi-Agent Systems with JADE. Wiley, Hoboken (2007)
24. Hozdić, E.: An ontology based approach for production planning and control of cyber-physical production systems. In: Karabegović, I., Kovačević, A., Banjanović, M.L., Dašić, P. (eds.) Handbook of Research on Integrating Industry 4.0 in Business and Manufacturing. Springer (2020, in press)

New Technology of Printed Circuit Board Microdrilling

Mikhailov Valeriy Pavlovich, Tun Lin Aung(✉), and Kazakov Alexandr Viktorovich

Department of Electronic Technologies in Mechanical Engineering, Bauman Moscow State Technical University, MT-11, 2nd Baumanskaya st. 5, Moscow 105005, Russia
tunlinaung52salay@gmail.com

Abstract. The paper describes the new technology principle of high precision drive for blind holes microdrilling into circuit boards. The experimental results of high precision drive based on active magnetorheological elastomer damper usage being included into closed control system.

Keywords: Printed circuit board · Microdrilling · High-precision drive · Active damper · Magnetorheological elastomer · Closed control system · Positioning error · Time of the transition process

1 Introduction

In the technology of multilayer printed circuit board manufacturing the minimal demanded diameter of connecting holes is about 50 μ. Nowadays the main drilling instrument manufacturers: (Union Tool (Japan), IND-Sphinx (India), HAM Micro-prazision (Germany), HPtec (Switzerland) show the drilling instruments for the demanded parameters forming [1].

Nowadays the instruments for glass-fibre plastic drilling with the mentioned parameters may be used only for FR-4 plastic and for holes wit diameter about 150 μ, and the spindle rotor drive is based on air bearing, that rotates with rotor speed 300–325 thousand rev/min The main manufacturers are: Posalux PS300 и PS325 (Switzerland) [2], and the drill average durability is about 1000 holes in glass-fibre plastic.

From the another side the bland holes forming is a difficult procedure because the bottom form of the hole leads to high intensity of drill wear. The hole bottom form is the emergency parameter of the technology, and the drill average durability becomes about 500–700 holes [2].

2 Experimental Equipment

The new technology of blind holes micro drilling problem consists in small depth of hole also as in the necessity to drill cupper foil that covers the circuit boards. We must remember, that drill diameter is 50 μm, depth if blind hole is ±5 μm, demanded error

I. Karabegović (Ed.): NT 2020, LNNS 128, pp. 250–255, 2020.
https://doi.org/10.1007/978-3-030-46817-0_28

of positioning in the direction of Z axis is less ±0.5 μm. The view of blind holes at optimal depth of the hole in the body of circuit board is shown on Fig. 1.

Fig. 1. The view of optimal depth blind holes in multilayer circuit boards.

To ensure high depth precision of the bland holes it is necessary to use the positioning mechanisms with high precision, and this precision must be on nano–range scale [1–11]. It is necessary know, that initial instrument transference system use positioning for rude processing, and precision instrument transference system use positioning for precise final processing.

In our case the additional high precision positioning table use chasing-milling system CAM 1020 Active Pro company VHF (Germany) with positioning error about 10 μ (Fig. 2). The positioning error is on axis X and Y is about ±50 μm, and on axis Z ±10 μm.

The error of initial positioning on axis X and Y is about ±50 μm and the speed of transference is 180 mm/s [3].

So the experiments show, that magnetorheological elastomer is a perspective material for active damping system design and new productive technology manufacturing [12–14].

Fig. 2. Chasing-milling system of CAM 1020 Active Pro Company (Germany) being supplied with positioning table

The positioning table consists of two plates being join with four elastic devices with mass correctors also as with four drive based on magnetorheological elastomer usage. The overview of positioning table based on magnetorheological elastomer damper usage is shown on Fig. 3 without upper plate of the table.

Fig. 3. The overview of positioning table based on magnetorheological elastomer damper usage (without upper table plate)

The drive (Fig. 4) includes magnetorheological elastomer membrane 1 with mobile rigid center 2, housing 3, electromagnetic coil 4, core 5, and base 6. Together with the rigid center, the core forms an air gap. The drive operates as follows. The supply of the control current to the coil in the electromagnetic system creates an enclosed magnetic field. A radial magnetic field with induction is formed in the membrane; its maximum is close to the rigid center. Under the action of the magnetic induction, membrane 1, with a rigid center, moves axially within the air gap. The magnetorheological elastomer consists of a silicone rubber matrix, within which spherical carbonyl-iron particles (1–10 μm) are distributed [12]. Under the action of the magnetic field, the magnetorheological elastomer may be reversibly deformed and its viscoelastic properties may change. That permits improvement in the damping in comparison with ordinary viscoelastic systems. In addition, the drive in the platform may be used as active damper for the vibration insulated object.

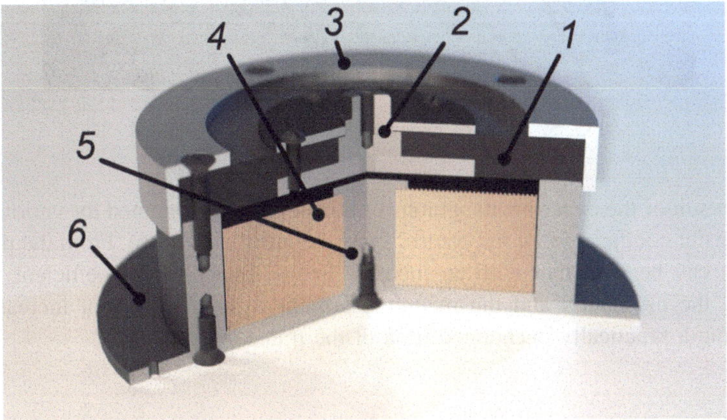

Fig. 4. The view of positioning drive based on magnetorheological elastomer usage.

The results of experimental studies of the parameters of transients during the operation of a high-precision drive based on the MR elastomer with a closed control system, which is also an active damper (Fig. 5) are presented. The amplitude-frequency characteristics of the damper and vibration-isolating platform with an open-loop control system in semi-active mode are shown based on four dampers and four elastic suspension nodes.

3 Experimental Results

The purpose of experiments was to evaluate the quality parameters of transients of a closed control system of the high-precision drive, depending on the transmission coefficient of the regulator. To control the high-precision drive, a personal computer with a Lab View program was used, a National Instruments USB-6009 ADC/DAC unit, an amplifier unit consisting of 4 amplifying modules based on operational amplifiers, a DL6220/ECL2 capacitive displacement sensor with a measurement error of 0.02 μ.

Fig. 5. High-precision drive with sensor installed

As a result of the experiments, graphs of transients were obtained for various values of the transfer coefficient k of the control system controller (Fig. 6). From the presented graphs, it can be seen that with an increase in the transmission coefficient k of the regulator, the oscillation and the magnitude of the system overshoot increase while maintaining a practically unchanged time of the transition process.

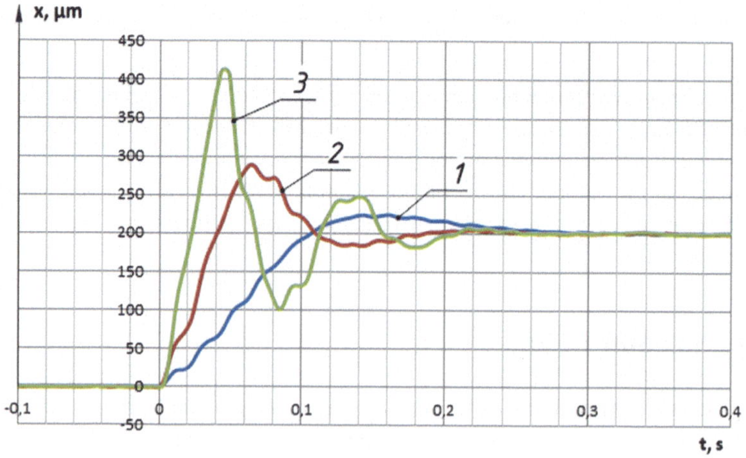

Fig. 6. Transient graphs when moving the high-precision drive with a closed control system of 200 µ for different values of the transfer coefficient k of the controller: 1 – 0,002; 2 – 0,005; 3 – 0,01

4 Conclusion

It was shown, that for blind holes drilling into multilayer circuit boards with high precision at nano-scale range it was necessary to use the additional drive based on magnetorheological elastomer.

It was shown also, that magnetorheological elastomer usage is perspective for active damper, positioning and vibration insulators devices in the technology of precise cutting (drilling).

It was shown also, that elastomer materials under action of magnetic field variety it's rheological properties, that may be used for damper properties and parameters of positioning table improving, for example precision of drilling process.

So we may use the magnetorheological drive in the role of active damper.

With an increase in the transmission coefficient of the control system regulator, the oscillation and the magnitude of the system overshoot increase while maintaining a practically unchanged time of the transition process.

References

1. Hidehito, W., Hideo, T., Masami, M.: Microdrilling for printed circuit boards (PCBs)—influence of radial run-out of microdrills on hole quality. Precis. Eng. **32**, 329–335 (2008)
2. https://www.posalux.com/micro-drilling-and-routing/
3. https://www.vhf.de/cnc-fraesmaschinen/maschinenbaureihen/active-pro.html
4. Wigglesworth, W., Jordan, S.: Semicond. Int. **32**(10), 4–26 (2009)
5. Active Vibration Isolation. Accurion. http://www.accurion.com
6. Würzburg, B.H., Grossenlupnitz, R.R.: US Patent No. 20080318045 A1. Appl. No. 11/574397, 25 August 2005, Date of Patent 27 August 2004
7. Gruzevich, Yu.K., Soldatenkov, V.A., Achildiev, V.M., Levkovich, A.D., Bedro, A.N., Komarova, M.N., Voronin, I.V.: J. Opt. Technol. **85**(5), 308–313 (2018). https://doi.org/10.1364/JOT.85.000308
8. Ovchinnikov, I., Brancevich, P.: Procedia Engineering **176**, 610–617 (2017). https://doi.org/10.1016/j.proeng.2017.02.304
9. Krestnikovskiy, K.V., Panovko, G.Ya., Shokhin, A.E.: Vibroengineering Procedia **8**, 208–212 (2016)
10. Panovko, G., Shokhin, A., Eremeykin, S.: Vibroengineering Procedia **8**, 174–178 (2016). (in Russian)
11. Chernikov, S.A.: J. Mach. Manuf. Reliab. **44**, 439 (2015). https://doi.org/10.3103/S1052618815050052
12. Mikhailov, V.P., Bazinenkov, A.M.: J. Magn. Magn. Mater. **431**, 266–268 (2017). https://doi.org/10.1016/j.jmmm.2016.10.007
13. Mikhailov, V.P., Bazinenkov, A.M., Dolinin, P.A., Stepanov, G.V.: Instrum. Exp. Tech. **61**(3), 427–432 (2018)
14. Mikhailov, V.P., Bazinenkov, A.M., Dolinin, P.A., Stepanov, G.V.: Russ. Eng. Res. **38**(6), 434–437 (2018). https://doi.org/10.3103/S1068798X18060138

Determination of Relaxation and Creep Modulus of Polymer Materials Obtained by 3D Printing

Jusuf Ibrulj(✉), Ejub Dzaferovic, and Murco Obucina

Faculty of Mechanical Engineering, University of Sarajevo, Vilsonovošetalište 9, 71000 Sarajevo, Bosnia and Herzegovina
j_ibrulj@yahoo.com

Abstract. Typically, Prony series viscoelastic models are used to obtain the so-called. “Master” curves in analytical form from experimental results. However, the process of approximating experimental results (“fitting”) with the Prony model requires prior knowledge of the model’s time parameters, in order to fit into the appropriate module. In this paper, parameter optimization was used in such a way that all the parameters in the Prony series for (E_i, τ_i determined by a simple least squares method. The Prony model was chosen so that the number of members in the Prony series is initially smaller and increases in proportion. An example was chosen for this paper to demonstrate parameter optimization for Prony’s relaxation module model and the data was used from an available article, while an experiment using a constant ramp test was performed to determine the creep module. The previous experiment was done for real ABS plastic material (Acrylonitrile butadiene styrene).

Keywords: Viscoelasticity · Relaxation · ABS plastic · 3D printing

1 Introduction

Viscoelastic behavior is a function of time and temperature. Generally, this dependence requires the definition of a set of principal curves at different temperatures that fully characterize the material under all possible operating conditions. These major curves can be obtained experimentally [1, 2] or by viscoelastic interconversions [3, 4]. In any case, the experimental data must be fitted to a viscoelastic model, e.g. Prony series [5], to perform further calculations with viscoelastic material. With this condition, there are several major methods, such as procedure X [6, 7], collocation method [8], multi-data method [9], or the Emri and Tschoegl iterative algorithm [10, 11] used to derive the coefficients viscoelastic model.

The basis of the method is to establish a set of discrete times and to approximate them by different criteria to the remaining coefficients of the model. Experimental data are usually obtained by logarithmic sampling and, also, the viscoelastic function of the material spans several time logarithmic decades. Different methods select discrete times (τi or τp in some specific way) so that they are eventually equally distributed on a logarithmic time axis (homogeneous distribution). Although some of these methods have

I. Karabegović (Ed.): NT 2020, LNNS 128, pp. 256–266, 2020.
https://doi.org/10.1007/978-3-030-46817-0_29

some disadvantages in the approximation process, such as negative spectral lines, it can generally be argued that a good approximation can be achieved by reducing the discrete time space or increasing the number of members in the corresponding Prony series.

In this paper, a method of approximating viscoelastic experimental functions, such as the relaxation modulus and the creep modulus, is used, including the possibility of inhomogeneous distribution of discrete times. The model was validated by experimental creep curves on ABS (Acrylonitrile butadiene styrene) plastic samples printed on a suitable 3D printer, and the results compared with those obtained numerically.

3D printing is a way of quickly prototyping which devices based on patented 3D technology "print", i.e. directly from 3D CAD programs, they create physical objects in space. Three-dimensional printing, especially color printing, gives development engineers and designers a clear insight into the design process, the ability to highlight various parameters, the ability to easily and early detect possible errors and correct them quickly and efficiently. In order to meet market demands, more modern processing processes such as additive processes (Fused Deposition Modeling - FDM, Laminated Object Manufacturing - LOM, Stereolithography - SLA, Digital Light Processing, PolyJet, Selective Laser Sintering - SLS, Selective Laser Melting are being developed and applied (SLM), 3D Printing - 3DP, Electron Beam Melting (EBM) etc. [12]. The main difference between the different types of 3D devices available is the way in which layers are formed and parts of the structure we want to get are created. Some of them soften the materials to form layers, others layer liquid duromers that harden in different ways, while others cut thin layers of material and then bond them together.

Using these capabilities of 3D printers significantly shortens the build time and increases the quality level of models and prototypes. Areas of application of 3D-printers are: architecture, design, informatics, education, mechanical engineering, aeronautics, metallurgy, modeling, medicine, bio-modeling, geo-info systems, molecular chemistry, etc.

3D printers are three-axis (x, y and z) mini-systems with numerical control (NC). The system software first converts the 3D CAD design into cross sections, that is, thin layers of 0.089–0.203 mm thick, which is selected depending on the accuracy to be achieved. The object (model) is then made in such a way that special powder is applied in the workspace in programmed thin layers and secured with a powder-binding binder ("prints") to standard printer components, starting with the lowest layer. Such "printing" can be monochromatic or in color.

By selecting the type of powder and filling the model with different components (fasteners), users can create items of different properties depending on the technical requirements that the model needs to meet (strength, elasticity, temperature durability, etc.).

The greater the complexity of the element, the greater the advantage of 3D printing over the conventional production system. In addition, 3D printing uses raw material efficiently, resulting in very low or zero waste [13]. Reducing the amount of waste material is particularly important from an environmental point of view. During the layer-by-layer process, the material undergoes physical and thermal changes concerning its mechanical properties. Using mechanical analysis and simple relaxation and creep tests, it is possible to determine how the dynamic properties of the material change depending on the process parameters and how such parameters affect the mechanical properties [14].

2 Theoretical Background

Viscoelastic materials can be easily understood as materials whose properties lie somewhere between an elastic solid and Newtonian fluid. Although its features are generally more complex, these settings make it easier understanding of viscoelastic mathematical models for these types of materials. The resilient solid is represented by a spring while Newton's fluid is represented by a slider, in other words viscoelastic materials can be represented by combinations of springs and sliders. The simplest the models are Maxwell and Kelvin Model, where the former is used (with elements in a row) to represent relaxing viscoelastic functions, E(t), while the other (with elements in parallel) is used to represent viscoelastic creep functions, D(t), [15–17].

Dobijeni rezultati s pravim materijalima i ovim jednostavnijim modelima ne mogu biti zadovoljavajući, tako da se generalizovane verzije prethodno pomenutih modela, moraju koristiti. Kod Prony-jeve serije svaki član serije identificira se sa jednim pojedinačnim Maxwell-ovim elementom ili u slučaju funkcije puzanja sa pojedinačnim Kelvin-ovim elementom.

Prony serija za generalizovani Maxwell-ov model je [18]:

$$E(t) = E_\infty + \sum_{m=1}^{M} E_m \cdot \exp\left(-\frac{t}{\tau_m}\right) \tag{1}$$

Where are: E_∞, E_m, τ_m so called. "Rubber" modulus of relaxation, Prony series coefficients and relaxation times, respectively. While the corresponding series for the Kelvin model [18]:

$$D(t) = D_o + \sum_{m=1}^{M} D_m[1 - \exp(-t/\rho_m)] \tag{2}$$

Where are: D_o, D_m, ρ_m coefficients of Prony series respectively for the case of determining the creep function.

2.1 Nonlinear Regression

Several methodologies are used to solve the problem of nonlinear approximation of experimental data, the most important being the Levenberg-Marquardt method, the Region Trust method, and other self-developed algorithms. The algorithm always starts with the current assumptions of the solution of the optimization problem, and a model is built near this current (assumed) point. The algorithm adjusts from iteration to iteration and allows the model to cover the original problem. Many non-linear optimization routines are already available in the MATLAB Optimization Toolbox. In this paper, the MATLAB function lsqcurvefit, is intended to solve the problem of nonlinear approximation (fit) using a least-squares method with an interface designed specifically for data approximation problems.

2.2 Examples of Application

2.2.1 Determination of Relaxation Modulus

The data from the available paper [19] were taken to determine the relaxation modulus and the algorithm described above was used and results are presented graphically in the following figures (Figs. 1, 2, 3 and 4).

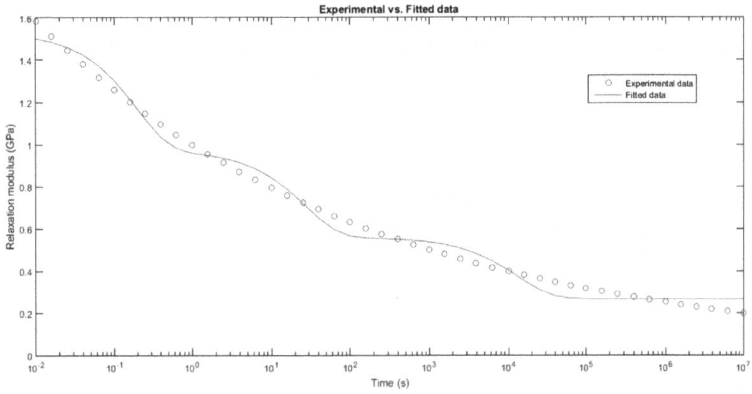

Fig. 1. Results of the approximation of the relaxation modulus with 7 parameters (3 members in the Prony series and one free member).

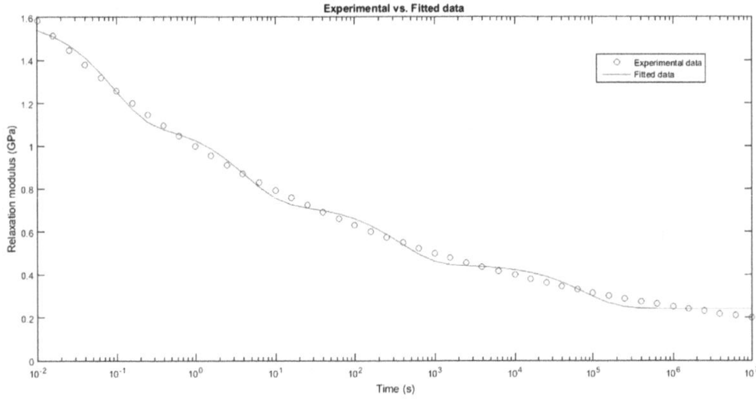

Fig. 2. Results of the approximation of the relaxation module with 9 parameters (4 members in the Prony series and one free member).

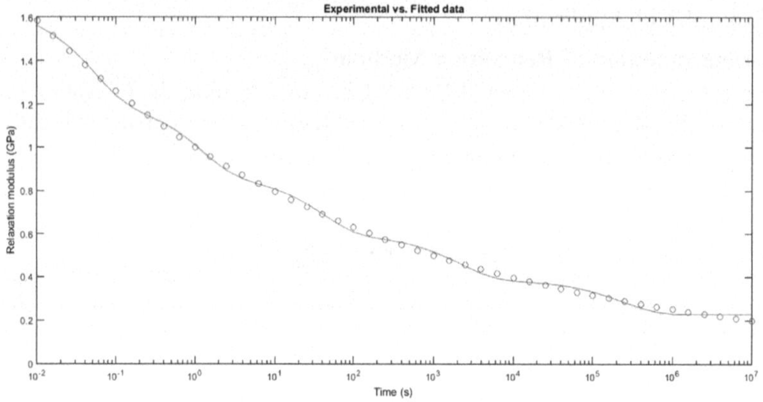

Fig. 3. Results of the approximation of the relaxation modulus with 11 parameters (5 members in the Prony series and one free member).

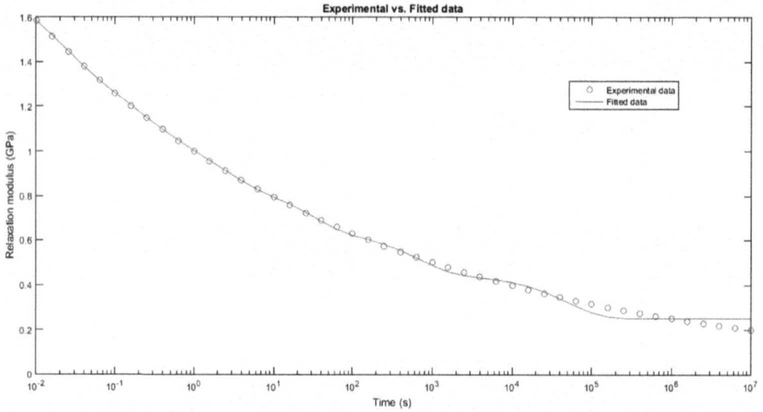

Fig. 4. Results of the approximation of the relaxation module with 19 parameters (9 members in the Prony series and one free member).

Values for individual members in the Prony series are given in tables (Table 1, Figs. 5 and 6).

Table 1. Members in Prony's series.

Optimizirani koeficijenti Prony-jeve serije								
	3 člana		4 člana		5 članova		9 članova	
i	Ei	ti	Ei	ti	Ei	ti	Ei	ti
0	2.7E−01		2.3E−01		2.3E−01		2.5E−01	
1	2.9E−01	1.3E+04	4.3E−01	4.9E−02	4.3E−01	4.9E−02	9.6E−02	2.5E+00
2	4.1E−01	2.7E+01	2.1E−01	2.2E+03	2.1E−01	2.2E+03	1.3E−01	2.2E+00
3	5.6E−01	1.9E−01	3.4E−01	1.2E+00	3.4E−01	1.2E+00	2.1E−01	2.9E+01
4			2.7E−01	4.0E+01	2.7E−01	4.0E+01	4.7E−03	5.5E−02
5					1.6E−01	2.4E+05	1.7E−01	9.3E−02
6							1.8E−01	3.2E−01
7							3.1E−01	1.7E−02
8							2.0E−01	5.0E+04
9							2.0E−01	6.6E+02

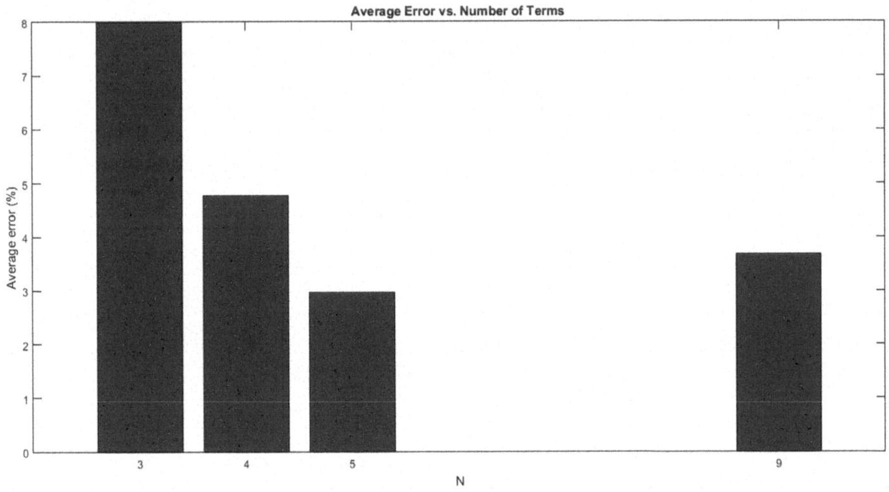

Fig. 5. Average deviation of the optimized values in the Prony series with respect to the experimental results.

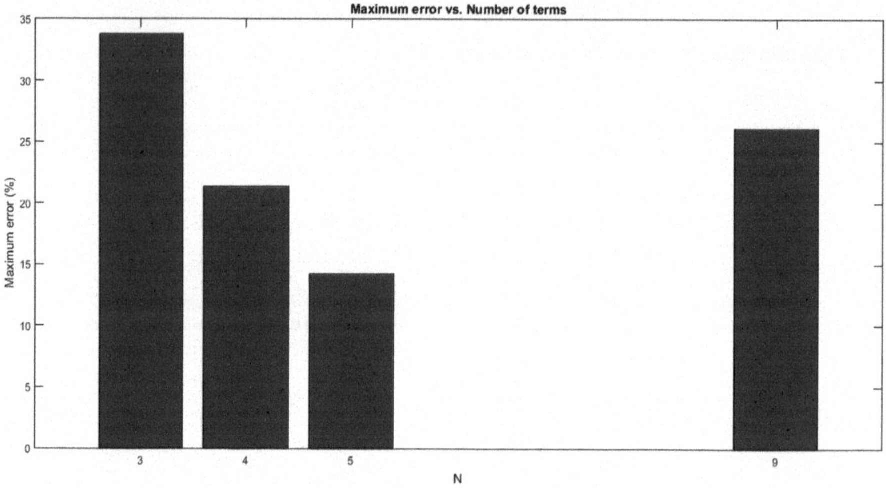

Fig. 6. Maximum deviation of the optimized values in the Prony series with respect to the experimental results.

2.2.2 Determination of the Creep Compliance

To demonstrate the usefulness and applicability of the proposed procedure, we use real experimental data measured at constant load increase for the creep test (Figs. 7, 8, 9, 10, 11 and Table 2).

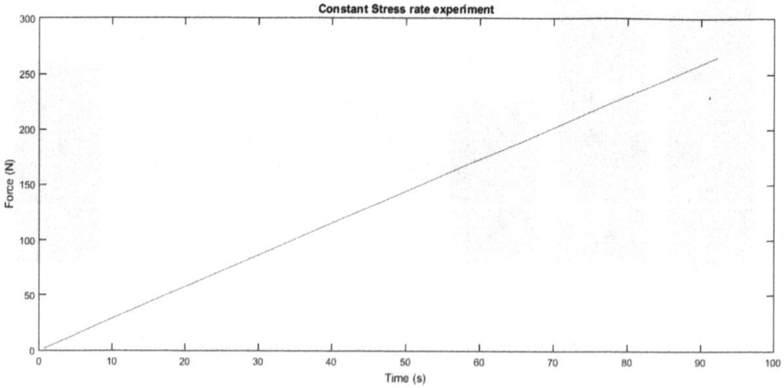

Fig. 7. Creep test with linear load function.

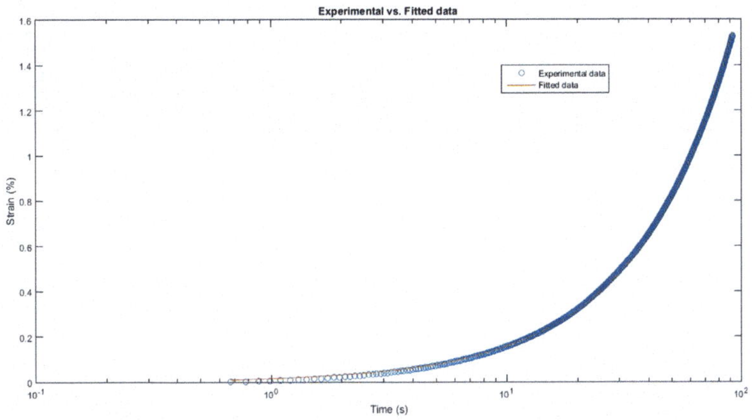

Fig. 8. Results of the approximation of the creep function with 3 parameters (1 member in series and one free member).

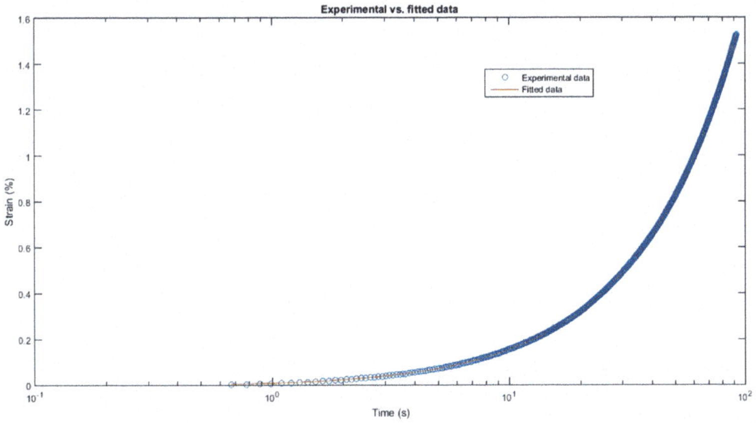

Fig. 9. Results of the approximation of the creep function with 5 parameters (2 members in series and one free member).

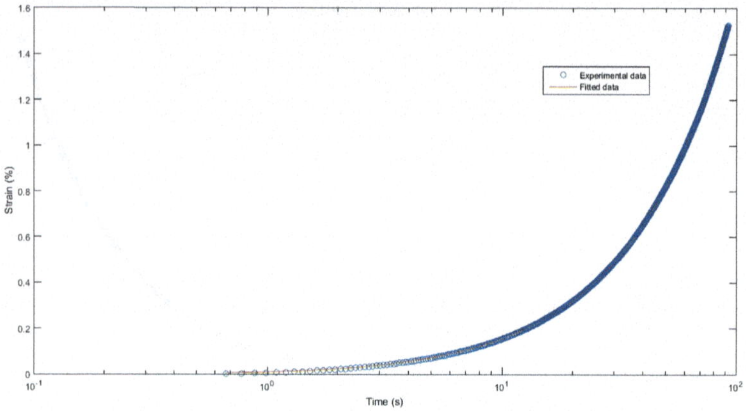

Fig. 10. Results of the approximation of the creep function with 7 parameters (3 members in series and one free member).

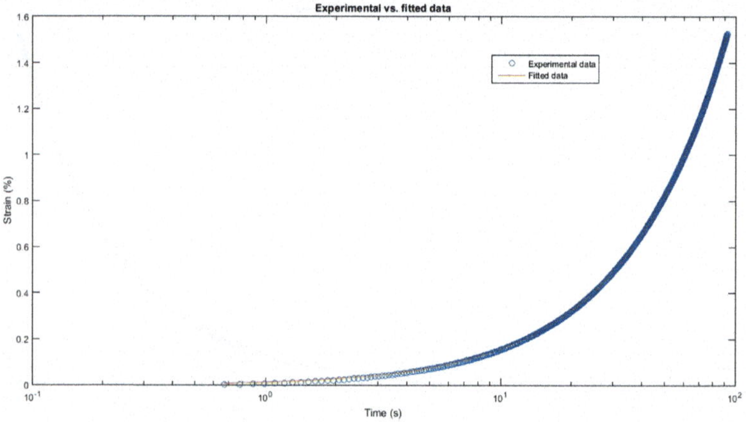

Fig. 11. Results of the approximation of the creep function with 9 parameters (4 members in series and one free member).

Table 2. Coefficients in Prony's series.

Optimizirani koeficijenti funkcije puzanja								
	1 član		2 člana		3 člana		4 člana	
i	D_i	r_i	D_i	r_i	D_i	r_i	D_i	r_i
0	2.2E−07		1.2E−08		5.1E−08		1.3E−07	
1	7.3E−04	2.1E+06	1.5E−07	6.3E−01	1.1E−07	9.9E−01	7.6E−09	2.9E+00
2			6.7E−08	9.1E−01	1.5E−08	2.7E−01	3.2E−09	5.3E−01
3					5.8E−08	3.7E−01	7.9E−08	5.5E−01
4							1.5E−08	6.9E−01

As can be seen in the graphs, there is good agreement between experimental and fitted results, especially when determining creep compliance. The number of members in the Prony series also affects the accuracy of the approximation, i.e. the value of the average and maximum error of the approximation. The approximation error decreases with the increase in the number of Prony series members as expected, while in the case of "too many" members may increase again.

3 Conclusion

This article describes a method for determining the relaxation and creep modules. The individual formulation of linear viscoelastic constitutive law and Prony's model (series) are used to describe the behavior of the material. Examples of determining these functions in mathematical form in this study have shown that this method can be successfully used to obtain analytical expressions of the relaxation and creep modules. The results of the fit show a good match with the experimentally obtained results. Unlike the prevailing practice of determining the relaxation and creep modulus from simple impulse-based tests, the so-called a test of a proportional (linear) increase in stress, defined in analytical form as a linear dependence of stress over time. Such tests are quite demanding for performing the most in terms of load control in order to keep it linear over time and given that the polymeric material relaxes under load at all times. Given that we had at our disposal a very reliable rheometer device that allows load control, the problem was much easier on that side.

References

1. Fernández, P., Lamela-Rey, M.J., Fernández-Canteli, A., García-Barruetabeña, J., Cortés, F.: VI International Materials Symposium, Guimarães, Portugal, p. 149 (2011)
2. García-Barruetabeña, J., Cortés, F., Abete, J.M., Pelayo, F., Lamela-Rey, M.J., Fernández-Canteli, A.: Mater. Des. **32**(5), 2783 (2011)
3. Fernández, P., Rodríguez, D., Lamela, M.J., Fernández-Canteli, A.: Mech. Time-Depend. Mater., 169 (2011)
4. Emri, I., Von Bernstorff, B.S., Cvelbar, R., Nikonov, A.: J. Non-Newton Fluid Mech. **129**(2), 75 (2005)
5. Chen, T.: Determining a Prony series for a viscoelastic material from time varying strain data. NASA/TM-2000-210123. ARL-TR-2206 (2000)
6. Tobolsky, A.V., Murakami, K.: J. Polymer Sci. **40**, 443 (1959)
7. Tobolsky, A.V.: Properties and Structures of Polymers, p. 186. Wiley, New York (1960)
8. Schapery, R.A.: Proceedings of the 4th US National Congress of Applied Mechanics, vol. 2, p. 1075 (1962)
9. Cost, T.L., Becker, E.B.: Int. J. Numer. Methods Eng. **2**, 207 (1970)
10. Emri, I., Tschoegl, N.W.: Rheol. Acta **32**, 311 (1993)
11. Tshoegl, N.W., Emri, I.: Rheol. Acta **32**, 322 (1993)
12. Berman, B.: 3-D printing: the new industrial revolution. Bus. Horiz. **55**, 155–162 (2012)
13. Wimmer, R., Steyrer, B., Woes, J., Koddenberg, T., Mundigler, N.: 3D printing and wood. Pro Ligno **11**, 144–149 (2015)

14. Arivazhagan, A., Masood, S.H.: Dynamic mechanical properties of ABS material processed by fused deposition modelling. Int. J. Eng. Res. Appl. (IJERA) **2**, 2009–2014 (2012)
15. Tshoegl, N.W.: The Phenomenological Theory of Linear Viscoelastic Behaviour: An Introduction. Springer, Heidelberg (1989)
16. Ferry, J.D.: Viscoelastic Properties of Polymers, 3rd edn. Wiley, New York (1980)
17. Lakes, R.S.: Viscoelastic Solids. CRC Press, Boca Ratón (1998)
18. Mun, S., Chehab, G.R., Kim, Y.R.: Determination of time-domain viscoelastic functions using optimized interconversion techniques. Road Mater. Pavement Des. **8**(2), 351–365 (2007)
19. Chambers, R.S.: RELFIT: A Program for Determining Prony Series Fits to Measured Relaxation Data. Sandia National Laboratories Albuquerque, New Mexico (1997)

Additive Manufacturing of the Superalloy Turbine Blades by Selective Laser Melting: Surface Quality, Microstructure and Porosity

Dmytro Lesyk[1(✉)], Silvia Martinez[2], Vitaliy Dzhemelinkyi[1], and Aitzol Lamikiz[2]

[1] Laser Systems and Physical Technologies Department, National Technical University of Ukraine "Igor Sikorsky Kyiv Polytechnic Institute", Kyiv 03056, Ukraine
lesyk_d@ukr.net

[2] Aeronautics Advanced Manufacturing Center, University of the Basque Country, Zamudio, Spain

Abstract. The laser additive manufacturing technology is studied that applies lasers to fabricate the complexly shaped parts by successive selective melting of the metal powder in a layer by layer process. The turbine blade test parts were manufactured by the selective laser melting (SLM) process using an Inconel 718 powder. A multidirectional scanning strategy was used during SLM. The paper is focused on the study of the surface features and material characteristics of the SLM-printed parts. The surface texture supplemented by roughness profile parameters is analyzed. The microstructure, material compositions, material phases, porosity, and hardness in the near-surface layers are also addressed. Microstructural studies were performed by the X-ray diffraction analysis, optical microscopy, and scanning electron microscopy. Results indicated that partially melted powder particles, balling, and signs of the laser tracks are formed on the side surfaces of SLM-built turbine blades. The columnar dendritic microstructure mainly consists of a base γ-phase matrix and $\gamma^{||}$-phase precipitates. The SLM-built turbine blade parts show the accumulation of porosity close to the surface.

Keywords: Metal additive manufacturing · Selective laser melting · Inconel 718 powder · Turbine blade · Surface texture · Microstructure · Porosity · Hardness

1 Introduction

The metal additive manufacturing (or three dimensional (3D) printing) technology is applied for producing 3D objects by melting fine metal powders or wire layer by layer from a computer-aided design (CAD) model of the part. The additive manufacturing techniques allow producing the complex 3D part geometries with high precision, which is difficult or impossible to realize using conventional methods such as machining, casting or forging metals. The microstructure and properties formed by additive manufacturing technology differ from the conventionally manufactured parts [1–3].

I. Karabegović (Ed.): NT 2020, LNNS 128, pp. 267–275, 2020.
https://doi.org/10.1007/978-3-030-46817-0_30

There are a number of different technologies (powder bed fusion, directed energy deposition, and binder jetting) used in metal additive manufacturing [2, 3]. Most of the metal additive manufacturing technologies melt powder or wire by a concentrated heat source such as a laser or electron beam. The quality of the 3D-printed parts is defined by surface roughness/waviness, and dimensional tolerance, microstructure, porosity, hardness, and residual stress.

The most common additive manufacturing techniques used with metals are powder bed fusion (PBF) and directed energy deposition (DED) methods via the ability to fabricate near fully dense metal components with mechanical properties that are comparable to the conventionally manufactured parts [1, 2]. The DED methods are a more complex printing process commonly used to repair or add additional material to existing components. The DED methods are usually applied for producing large-scale parts while the PBF methods can be used for the manufacturing of small-sized parts. The PBF methods are metal 3D printing production technology designed to melt the metal powders together layer by layer using a laser beam or electron beam.

The parts fabricated by the selective laser melting (SLM) or electron beam melting (EBM) process are extremely dense and strong. As a consequence, the SLM and EBM processes may be applied in aerospace and medical applications [4–6]. In the aviation and aerospace industries, it is known that the nickel and titanium alloys are widely used, which meet the stringent requirements of aircraft engine manufacturers for metallographic and mechanical properties [7]. It should also be noted that the SLM process works well with titanium and superalloys, manufacturing components difficult to machine [6–8]. As compared to the casting approach, the SLM process can reduce the time to manufacture turbine blades [9]. Thus, the production of lightweight and complexly shaped metal parts for the aerospace sector is considered to be feasible only using the SLM process.

The effects of the treatment strategy and the SLM parameters on the microstructure and mechanical properties of titanium and superalloys were widely reported, and the optimized SLM parameters can usually lead to better mechanical properties [7, 10–12]. However, the application of laser 3D printing techniques for manufacturing aircraft components still requires a large range of research and tests to confirm the stability of the characteristics with the subsequent procedure of mandatory certification of production technology and materials for compliance with the stringent requirements to aircraft components.

The aim of this work is to study the surface texture, microstructure/phase state, hardness, and residual porosity of Inconel 718 turbine blade parts fabricated by the SLM process.

2 Experimental Procedures

2.1 Powder Material

The turbine blade test parts with base dimensions of 85.4 mm long, 32.5 mm wide, and 60 mm height were fabricated using a nickel-based pre-alloyed Inconel (IN) 718 powder (the average diameter the powder was 32.4 μm). The chemical composition of the SLM-built specimens was the following (wt.%): Ni 51.34, Cr 19.08, Nb 5.33, Mo 3.28, Ti 0.98, Al 0.46, Mn 0.27, Si 0.18, C 0.08, and Fe balance.

2.2 Selective Laser Melting

The turbine blade specimens were printed by the SLM process using an industrial Renishaw AM400 machine (the system build volume is 248 mm × 248 mm × 285 mm), which was equipped with a maximum power 400 W ytterbium fiber laser (1070 nm wavelength) and a scanning optics with a maximum scanning speed of 7000 mm/s (Fig. 1) [8]. The SLM-built turbine blade parts were manufactured in continuous laser mode using a laser power of 200 W, and a laser scanning speed of 700 mm/s (Table 1).

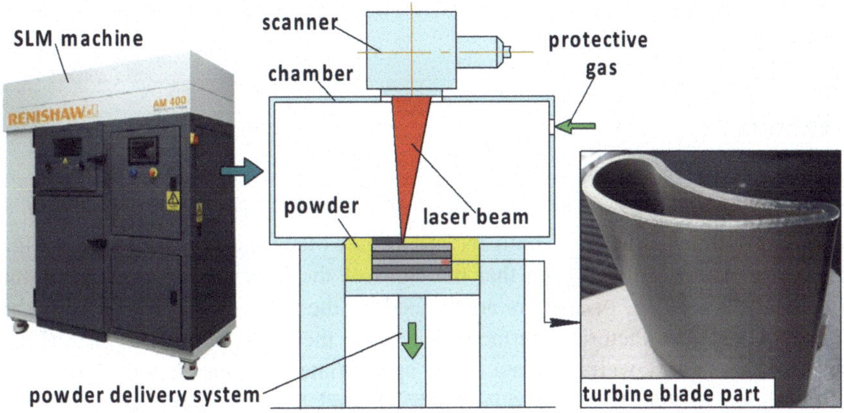

Fig. 1. General view of Renishaw AM400 machine and scheme for the SLM process of the IN 718 turbine blade part

Table 1. Parameters of the SLM process

Power, W	Scanning speed, mm/s	Laser spot size, μm	Stripes spacing, mm	Hatch angle, °	Layer thickness, μm
200	700	70	5	67	60

The laser beam (70 μm) was scanned in a single direction within 5 mm wide stripes. A multidirectional scanning strategy was used when the biaxial laser scanning patterns were rotated by 67° for each consecutive layer. The parameters of the SLM process applied for producing the turbine blade parts are given in Table 1.

2.3 Specimen Surface Characterization

The surface texture was studied using a 3D Taylor Hobson Form Talysurf 120 surface profilometer.

The cross-section microstructure and porosity were observed by a Leica MEF4A optical microscope (OM) and an FEI XL30 ESEM scanning electron microscope (SEM). The X-ray diffraction (XRD) analysis of the specimens was performed using a Rigaku Ultima IV diffractometer in a CuK_{α}-radiation.

The microhardness in the specimen cross-sections was measured using a Leica VMHT tester with a Vickers indenter loaded by 0.025 kgf. Additionally, the surface hardness was registered using a DIA TESTOR 2Rc hardness tester at a load of 10 kgf.

3 Results

3.1 Surface Texture

The surface texture of the SLM-built IN 718 turbine blade specimen is given in Fig. 2. The experimental results showed that the signs of the laser passes, balling, partially melted powder particles or spatters are formed on the side surfaces of the SLM-built specimen. The manufacturing surface defects of the SLM-built specimen (Fig. 2) observed in this study well correlate to the appropriate literature data [5, 13].

In addition to that, the surface is rough (*Ra* and *Sa* parameters are ~5 μm, *Wa* parameter is ~1.88 μm) having the irregularities of relatively high height both of the external and internal surfaces.

3.2 Microstructural Characteristics

The microstructure in the build direction is characterized by the overlapped semi-circular track patterns and mainly consists of elongated colonies of cells across multiple layers (Fig. 3a–d), which grow epitaxially from the substrate into the direction determined by the heat flow direction.

The fine columnar-shaped randomly oriented grains and cuboidal interdendritic precipitates are formed by a layer-by-layer building of the bulk due to the rapid cooling (10^4–10^6 K/s) of the melt pool [2]. The melt pool width was varied in the range of 60 to 120 μm. As seen in Fig. 3e, f, several individual laser passes are presented where the laser passes are rotated by 67° for the melting of the following layer of powder. The scan direction microstructure contains the transversely cut dendritic cells (Fig. 3g). Thus, the fine columnar-shaped randomly oriented dendritic (dark-colored) and cuboidal interdendritic precipitates (appeared as bright islands) are formed (Fig. 3h). The average surface macrohardness of the SLM-built specimens was 32.5 HRC_{10} (Fig. 2).

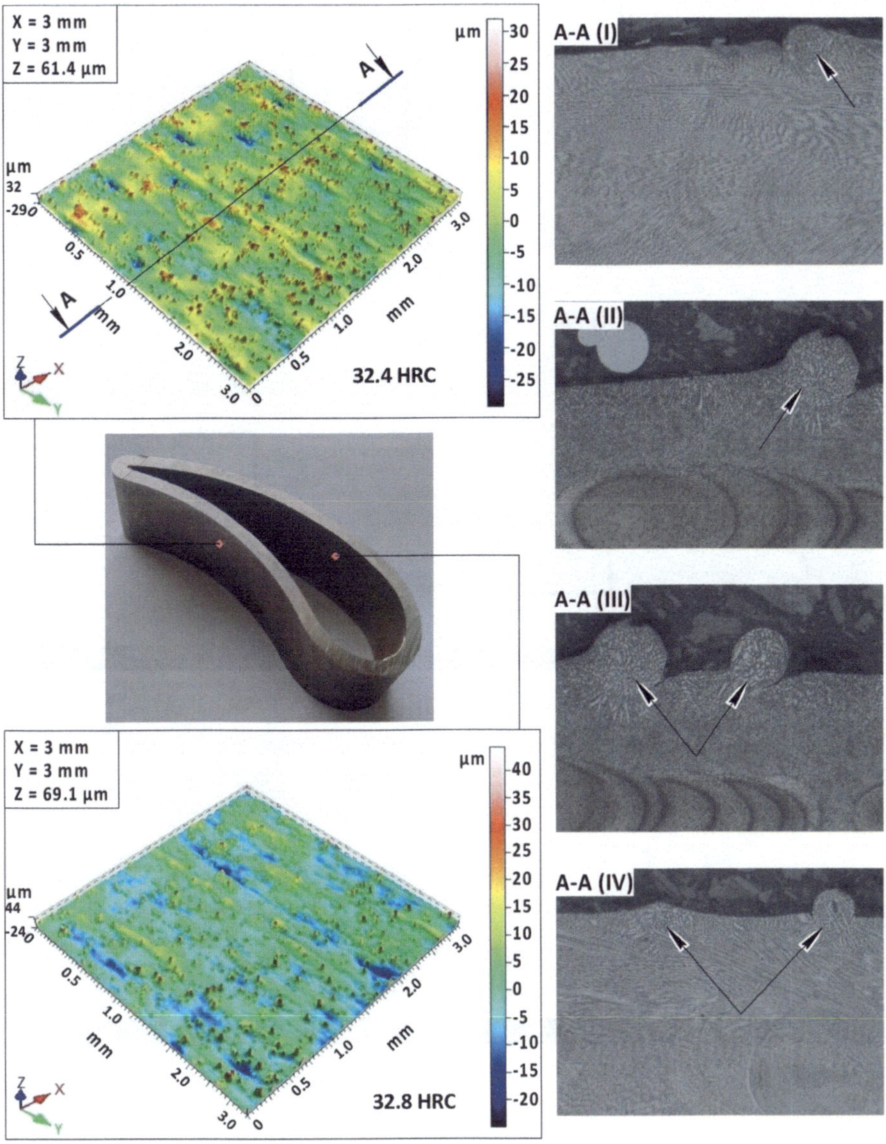

Fig. 2. Texture and macrohardness of the side surfaces of SLM-built IN 718 turbine blade specimen (the arrows in Fig. show the manufacturing surface defects in the cross-section A-A (I-IV))

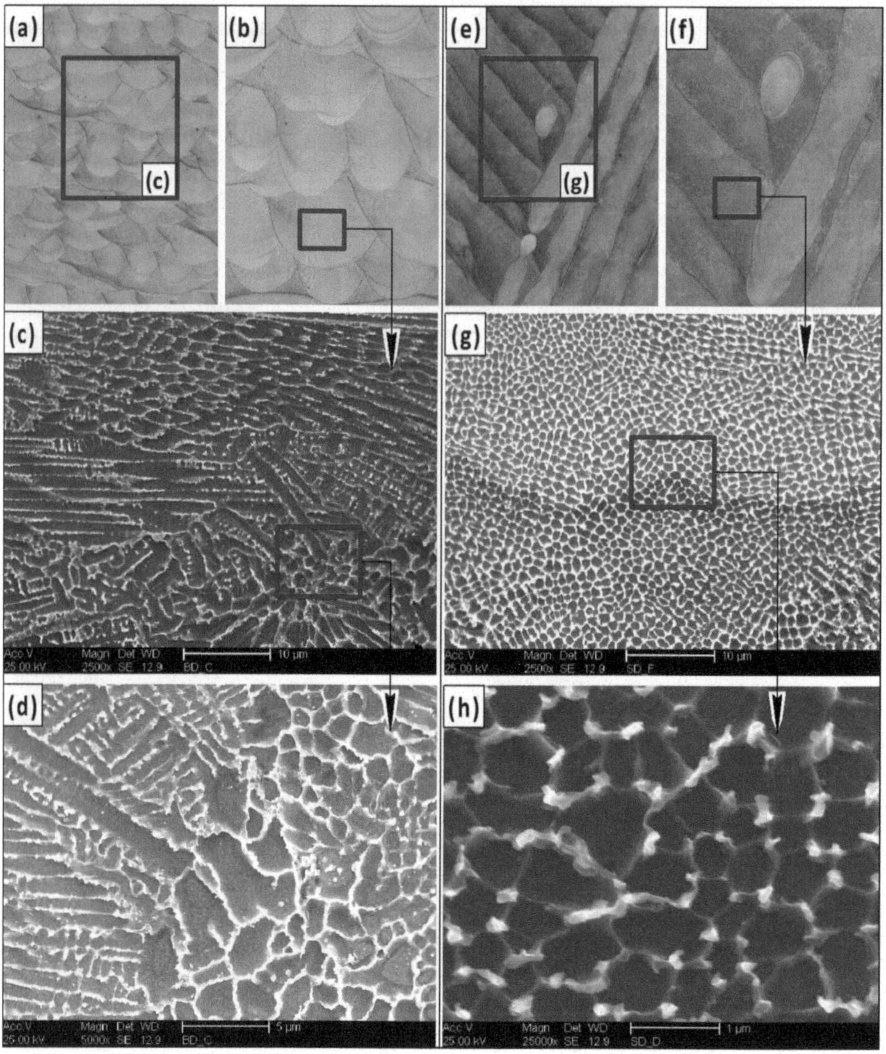

Fig. 3. OM and SEM imagines of microstructure in the build direction (a–b) and scan direction (c–h) of the SLM-built IN 718 specimen

The XRD analysis confirmed that the microstructure of the SLM-built IN 718 specimen mainly consists of a base γ-phase (Ni-Cr-Fe) matrix with a face-centered cubic crystal structure and disk-shaped $\gamma^{||}$-phase precipitates with a body-centered tetragonal crystal structure (Fig. 4a). The spheroidal $\gamma^{|}$-phase, needle-like δ-phase, island-like Laves phase (Fig. 3h), and carbides are usually encountered in this nickel-based alloy [10, 15].

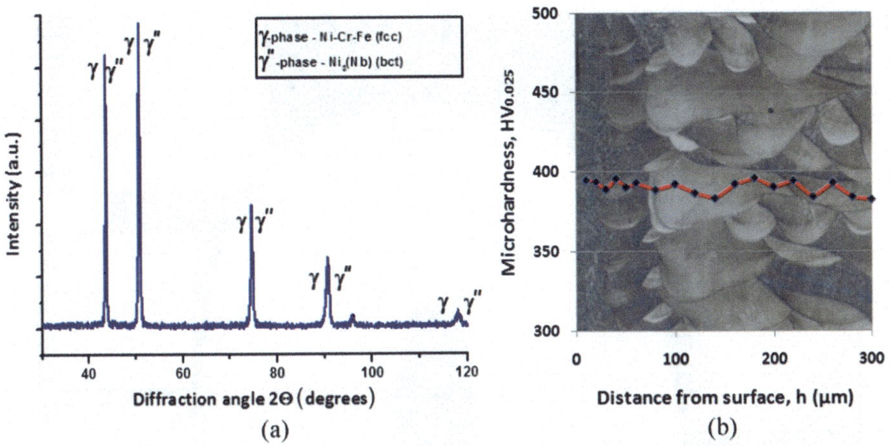

Fig. 4. Microhardness distribution of the near-surface layer in the build direction of the SLM-built IN 718 turbine blade specimen

A near-surface microhardness depth profile measured in the polished cross-section of the SLM-built specimen is shown in Fig. 4b. Analyzing the HV depth profiles one can conclude that surface microhardness ($\sim$390 $HV_{0.025}$) in this study well correlate to the appropriate literature data [15, 16].

3.3 Porosity

Along with the surface texture parameters and microstructure, the residual porosity is also of special importance for the wear/corrosion and fatigue resistance of a material. Considering the near-surface layers, a small number of large-sized spherical pores (up to 50 μm in size) were found in the SLM-built turbine blade specimen, which can be explained to the inclusion of gases in the deposited layers during the melting and solidification processes (Fig. 5). In our experimental conditions, except the spherical pores (gas pores), several small-sized aspherical pores (lack-of-fusion defects) were revealed as well. It should be noted that the lack-of-fusion pores can be formed in the build direction via insufficient penetration of the melt pool into the previous layer [14]. The average surface porosity percentage in the build direction of the SLM-built specimen was 0.697%.

Thus, based on the above-mentioned results, it should be noted that the SLM-built metal parts suffer from the presence of manufacturing surface defects (Fig. 2) and residual subsurface porosity (Fig. 5), and tensile residual stresses in the near-surface layers as well [8]. As a result, post-processing of SLM-built turbine blade parts is required to reduce/eliminate both the superficial and volume defects. Currently, the post-processing technologies (thermal, mechanical, and chemical treatments) are developed to improve the fatigue life and wear resistance of metal components fabricated by the SLM process [17].

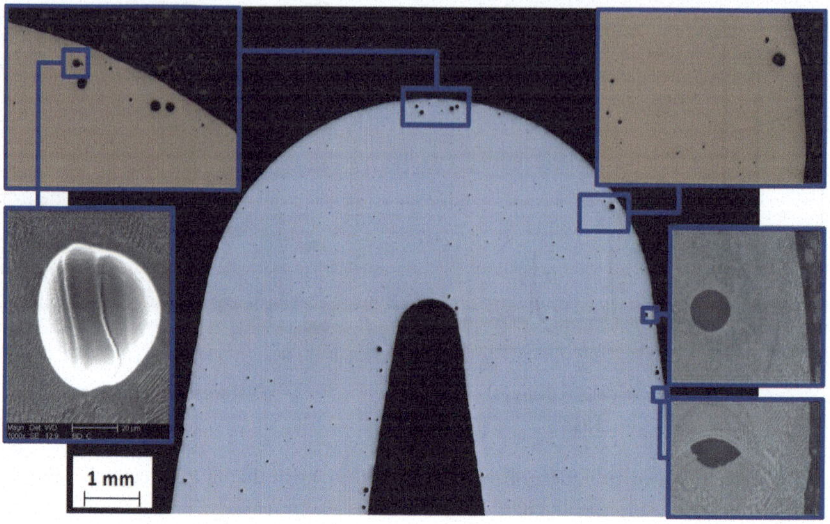

Fig. 5. Porosity in the build direction of the SLM-built IN 718 turbine blade specimen

4 Conclusion

The surface features and material characteristics of the IN 718 turbine blade parts fabricated by the SLM process are studied. The following conclusions can be drawn:

1. The SLM-built turbine blade test parts were manufactured with the laser power of 200 W, the laser scanning speed of 700 mm/s, and the laser beam diameter of 70 μm using the multidirectional scanning strategy.
2. The surface roughness is rough (*Ra* and *Sa* parameters are ~5 μm) due to the presence of balling, partially melted powder particles or spatters, and signs of the laser passes.
3. The fine dendritic microstructure consisted of a mixture of columnar (elongated) and equiaxed cells is formed in the SLM-built IN 718 part.
4. The uniform microhardness distribution in the near-surface layer with an average magnitude of 390 $HV_{0.025}$ was revealed. The SLM-built turbine blade parts show the accumulation of porosity close to the surface (0.697%).

References

1. Kumar, S., et al.: A comparison of additive manufacturing technologies. Int. J. Innov. Res. Sci. Eng. Technol. **3**, 147–152 (2016)
2. Sames, W.J., et al.: The metallurgy and processing science of metal additive manufacturing. Int. Mater. Rev. **61**, 315–360 (2016)
3. Zhang, Y., et al.: Additive manufacturing of metallic materials: a review. J. Mater. Eng. Perform. **27**, 1–13 (2018)

4. Yakout, M., et al.: On the characterization of stainless steel 316L parts produced by selective laser melting. Int. J. Adv. Manuf. Technol. **95**, 1953–1974 (2018)
5. Koutiri, I.: Influence of SLM process parameters on the surface finish, porosity rate and fatigue behavior of as-built Inconel 625 parts. J. Mater. Process. Technol. **255**, 536–546 (2018)
6. Caiazzo, F., et al.: Laser powder-bed fusion of Inconel 718 to manufacture turbine blades. Int. J. Adv. Manuf. Technol. **93**, 4023–4031 (2017)
7. Akca, E., et al.: A review on superalloys and IN718 nickel-based superalloy. Period. Eng. Nat. Sci. **3**, 15–27 (2015)
8. Lesyk, D.A., et al.: Post-processing of the Inconel 718 alloy parts fabricated by selective laser melting: Effects of mechanical surface treatments on surface topography, porosity, hardness and residual stress. Surf. Coat. Technol. **381**, 125136 (2020)
9. Ealy, B., et al.: Characterization of LAM-fabricated porous superalloys for turbine components. J. Eng. Gas Turbines Power **6**, 58080 (2016)
10. Mostafa, A., et al.: Structure, texture and phases in 3D printed IN718 alloy subjected to homogenization and HIP treatments. Metals **7**, 196–219 (2017)
11. Ali, H., et al.: Effect of scanning strategies on residual stress and mechanical properties of selective laser melted Ti6Al4V. Mater. Sci. Eng., A **712**, 175–187 (2018)
12. Liu, S., et al.: Additive manufacturing of Ti6Al4V alloy: a review. Mater. Des. **164**, 107552 (2019)
13. Nasab, M.H., et al.: On morphological surface features of the parts printed by selective laser melting (SLM). Adv. Manuf. J. **24**, 373–377 (2018)
14. Sabzi, H.E., et al.: Defect prevention in selective laser melting components: compositional and process effects. Mater. **12**, 3791 (2019)
15. Idell, Y., et al.: Unexpected δ-phase formation in additive-manufactured Ni-based superalloy. J. Miner. Met. Mater. Soc. **68**, 950–959 (2016)
16. Wang, X., et al.: Review on powder-bed laser additive manufacturing of Inconel 718 parts. Proc. ImechE Part B. **23**, 1890–1903 (2017)
17. Lesyk, D.A., et al.: Surface finishing of complexly shaped parts fabricated by selective laser melting. In: Tonkonogyi, V. et al. (eds.) Advanced Manufacturing Processes. InterPartner-2019. LNME, pp. 186–195. Springer, Cham (2020)

Importance of Additive Manufacturing Technology for Startup Launching: A Case Study

Milena Djukanovic[1(✉)], Luka Radunovic[2], Petar Vujovic[1], and Aleksandar Konatar[1]

[1] Faculty of Electrical Engineering, University of Montenegro, Podgorica, Montenegro
milenadj@ucg.ac.me
[2] Faculty of Mechanical Engineering, University of Montenegro, Podgorica, Montenegro

Abstract. Additive manufacturing or 3D printing technology enables building objects using a computer-aided design (CAD) model while successively adding material layer by layer. Recently the number of world's successfully launched startups based on additive 3D printing process is growing. In this paper we present startup launching in Montenegro based on 3D printing while creating sustainable zero-waste fashion accessories supported by a mobile application that will enable customers to create and order custom-made products.

Keywords: Additive manufacturing · 3D printing · 3D application · Startup

1 Introduction

The additive manufacturing process more often called 3D printing is one of the most revolutionary technologies of this century. The 3D printing process builds three-dimensional objects using a computer-aided design (CAD) model while successively adding material layer by layer and for that reason it is also called additive manufacturing. Recently 3D printing performs crucial roles in many applications, with the most important being manufacturing, medicine, architecture, art and design [1–4]. The main disadvantage of 3D printing, for a long time, was high entry costs not allowing a profitable implementation to mass-manufacturers. However, the market for 3D printing has shown some of the quickest growth within the manufacturing industry.

The initial idea of employing 3D printing was to create a faster alternative of prototyping. A big advantage of using 3D printing is its versatility and flexibility, which makes it perfect for small-scale manufacturing, for example of spare parts. Investing in 3D printing technology can help iterate and improve prototype design. Also, it allows companies to produce products on a case-by-case basis.

This paper is structured to begin with a review on 3D software, 3D printers and materials used in additive manufacturing processes. Following that, the design and development of a 3D printed product which can be used as a fashion gadget designed by the corresponding author and further developed with other authors will be presented.

I. Karabegović (Ed.): NT 2020, LNNS 128, pp. 276–284, 2020.
https://doi.org/10.1007/978-3-030-46817-0_31

In that sense, the development of 3D printed models will be shown, as well as final 3D printed fashion products, and a mobile application that allows the custom-made purchase of these products. This product will show the importance of knowing and using additive manufacturing technology in order to develop a startup based on 3D printed products.

2 Softwares, Printers and Materials Used in Additive Manufacturing Process

The first part of utilizing 3D printing technology is designing a 3D model in a computer environment. Nowadays, there are two kinds of 3D software - ones are designed for creating precise 3D models used in engineering and the others are used for creating models where precision is not that important and is sacrificed in order to increase ease and freedom of creating objects which give models that are sculpted like pieces of clay (Fig. 1, Fig. 2) [5].

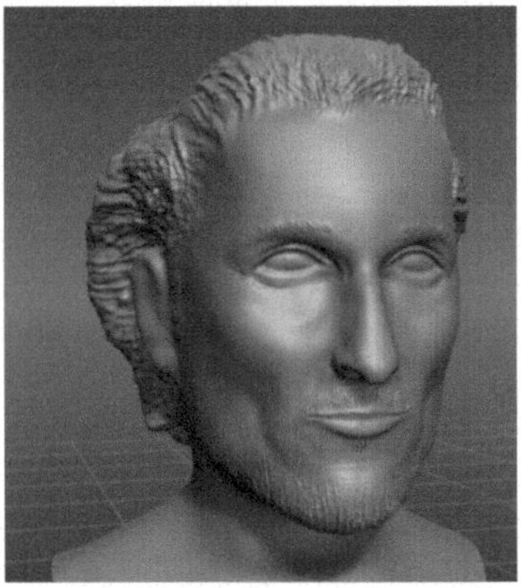

Fig. 1. Example of 3D sculpting

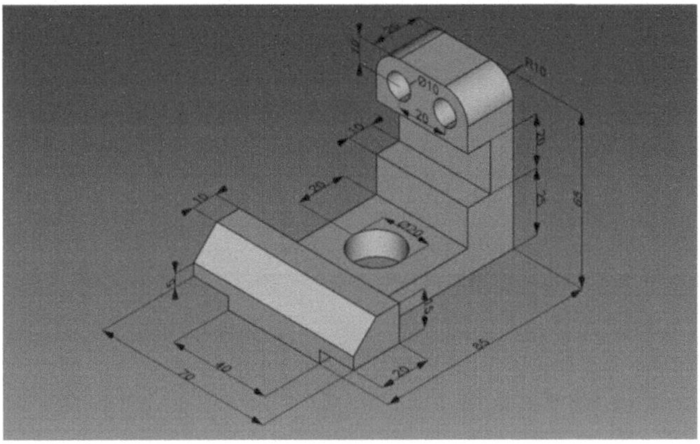

Fig. 2. Example of precise 3D modeling

To put it more specifically, modeling is working with polygonal shapes and can manipulate objects in order to create very precise and specific modifications to it, but this comes at a cost of a much slower work. On the other hand sculpting of a 3D model allows us to create more organic looking models, and is often utilized in the 3D animation business in order to create different types of characters. Tools present in the 3D sculpting environment allows us to create these shapes and characters much faster and more organic looking, and with more curves than we would be able to do in a 3D modeling environment. But this comes at a price of precision and is often more taxing on computers [6].

When it comes to the most suitable process that can produce real models out of the software 3D model, leaving almost zero waste of material, additive manufacturing found its purpose. The most common use of this technique on the market is the 3D printer, as it can be easy to use and quite fast for simple applications. Although there are many types of 3D printers on the market, for various types of models that can be printed, only nine types of printing technology currently exist [7]. Most common is Fused Deposition Modeling (FDM)- thermoplastic material is heated and deposited in-plane coordinates layer by layer, Stereolithography (STA)- liquid resin is hardened by laser layer by layer and Selective Laser Sintering (SLS)- bonding grains of various materials by melting it with laser layer by layer (Fig. 3) [8].

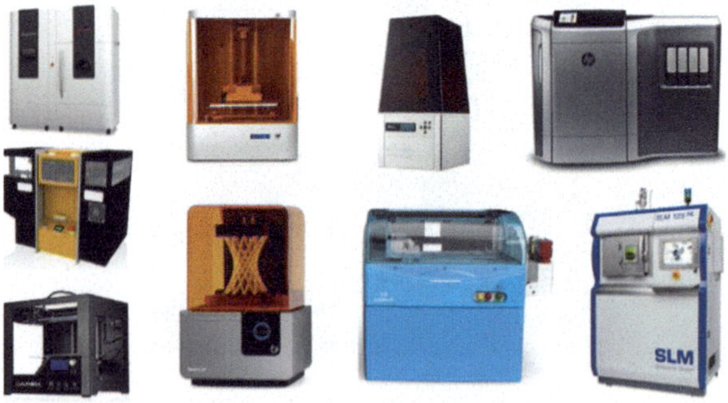

Fig. 3. Types of 3D printers

When it comes to materials used for printing, there are more and more options each day, but the most popular ones are PLA and ABS.PLA (Polylactic Acid) is a biodegradable material which is produced from things like corn starch or sugarcane (Fig. 4). One of the main strengths of it is that it is simple to print with. Outside of 3D printing, it has uses in fields such as medical implementation and packaging of the food [9].

Fig. 4. PLA filament

On the other hand, there is ABS (Acrylonitrile butadiene styrene) which is oil-based material (Fig. 5). This material is often used in protective headgear and toys. The advantage that ABS has over PLA is that objects made from it have higher durability, strength and are more flexible. To offset this, the print process is a bit more complicated and there are fumes throughout the process [10].

Fig. 5. ABS filament

3 Modeling and Development of a 3D Printed Product

The most common file formats that are used in all 3D printing software are STL (stereolithography) and VRML (virtual reality modelling language). STL format, as a monochrome format with no texture information is quite useful for most affordable and commonly found printers, and especially those that can only print one type of coloured plastic at a time as it is light for processing and rendering in 3D software. This type of format is used mostly in FDM 3D printing, where g-code generating procedure is based on slicing 3D cad model in a large number of 2D cuts parallel to the printing surface, and because of this procedure (2D printing in one plane, then lifting the heating-extrusion head for one value of layer height) whole FDM procedure is also called 2.5D printing [11]. For more complex types of 3D printing (inkjet, sl., sls. etc.) other 3D model input formats are preferred, because those types of printing use laser technology and resin/powder, allowing prints to have multi-coloured layers with higher precision than in FDM type [12].

All those input formats are converted to G-code using cam software suite. Although G-code is used as a universal programming language for controlling CNC machines and 3D printers, many 3D printer manufactures are making their own "slicing" software for generating G-code [13]. The reason for this is not only the imposition of their software on the market, but rather the printer safety. Specifically, when creating G-code itself, the working volume of the printer, the temperature of the bed and the hot end, machine zero must be taken into consideration. The machine itself doesn't know its limits, so all of that must be regulated using slicing software, which would not allow its user to generate G-code that would harm the machine in the process itself. Most popular cam software for "slicing" 3D models is UltimakerCura, made to be universal tool for wide variety of 3D printer shapes, sizes, materials, g-code flavour which has big impact in machine movement (micro-controller, as the brain of the whole printer has built-in G-code interpreter, which must be in sync with G-code flavour set in Cura software) [14].

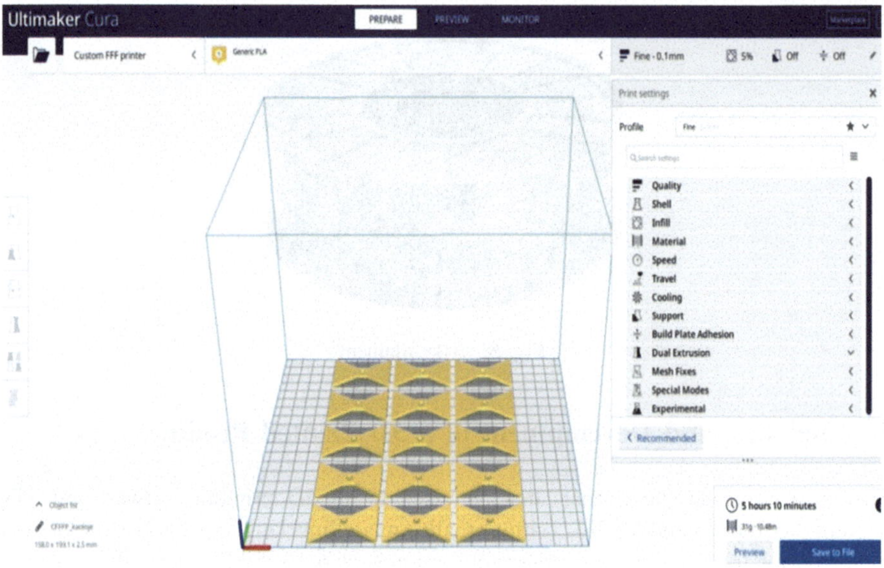

Fig. 6. Ultimaker Cura's G-code visual preview of the 3D bowtie model

Ultimaker Cura is open source software with extensions that allow the user to program desired and special functions in order to publish them as an experimental function (Fig. 6). The community of 3D printing rises rapidly, and so do the experimental functions as well at the same speed, pushing the limits of standard 3D printing. This creates the opportunity for all users to have more stable, more accurate and even better-looking 3D printed products (Fig. 7).

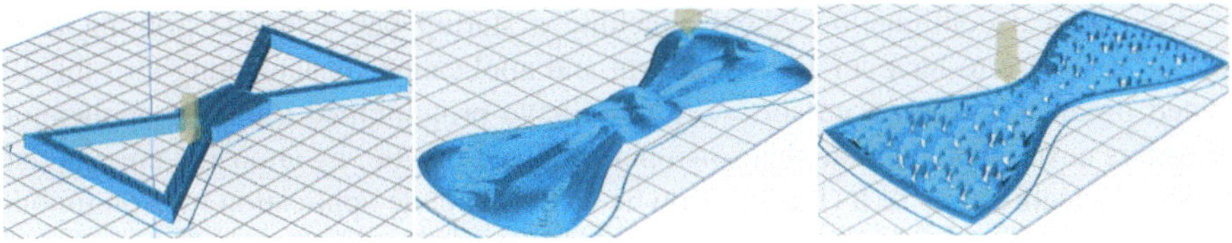

Fig. 7. Bowties models sliced in Cura software

4 Startup Launching Using Additive Manufacturing Technology

In order to bring more innovative solutions to the world's market, 3D printing technology can be a whole product or part of the product for launching a new startup [15–17]. There are already successful stories for a startups based on 3D printing promising to change the world: in the area of digitizing prosthetics and orthotics (artificial supports for the limbs or spine) using 3D printing - Mercuris [18], printing plastic objects, as well as self-replicating -RepRap [19] or providing customized, high-precision 3D-printed silicone components for use in industries from fashion to robotics, medicine, and art – Spectroplast [20].

Additive manufacturing technology has helped us create our own startup product by designing various 3D printed bowtie tendrils followed by adding different types of zero-waste core and belt textiles. Result of this innovative process are 3D printed fashion accessories ready to be put in the market, as shown in Fig. 8.

Fig. 8. Bowties gallery

As a logical continuation when launching a startup based on 3D printing technology, it is wise to have a mobile platform and application for launching also the 3D printed products/fashion accessories. Authors have programmed a mobile application that offers customers to create their own personalized bowtie, by choosing colours and shapes of specific parts of it (Fig. 9). Bowtie's client mobile application runs on mobile devices and requires a connection with the Internet. The service requires users to provide their name, surname, and address. Currently, this application supports only devices with Android OS and it is compatible with older devices as it was built in Android studio application development program.

The code is based on giving special IDs to every possible element. Since there are three main elements of 3D printed bowtie accessory, there are also three main categories A, B and C. Every element of those respected categories has the specific number next to the letter which represents its category, for example A1 is the bowtie with red core. After the user selects the specific parts for his/her bowtie, there will be created a new unique ID (now for the complete bowtie). For example, if a user chooses a red core with a blue body (in and out) and yellow tendrils, the ID that will be sent to the designer is "A1B2C3". With this ID designer will know which variation of 3D printed bowtie the user has chosen.

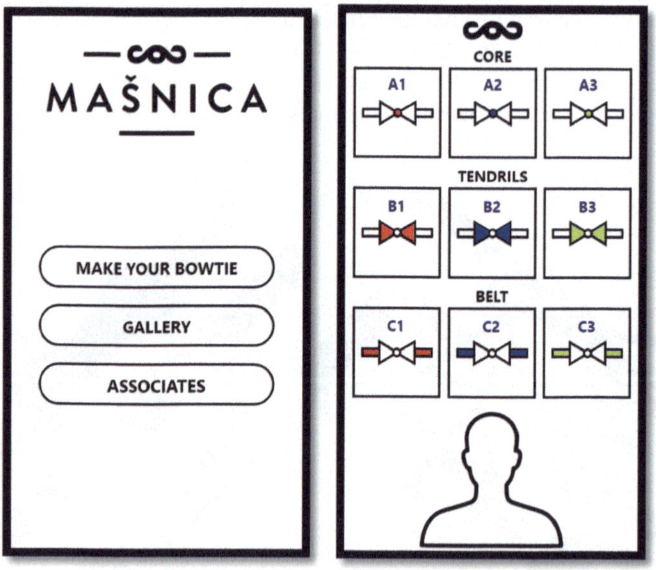

Fig. 9. Android application of 3D printed bowties

5 Conclusion

The popularity of 3D printing applications is increasing day by day, while creating more space in startup portfolios. In this paper we have presented startup launching in Montenegro based on 3D printing while creating sustainable zero-waste fashion accessories supported by a mobile application that will enable customers to create and order custom-made products. Added-value of this startup is using 3D printing technologies for making sustainable, recyclable and zero-waste fashion accessory, with the possibility of online custom-made creation and order.

In the future we will work on the improvement of the mobile app by adding user design capabilities allowing the user to sketch his own design of desired 3D printed bow tie post it on the ordering platform and receive that 3D product at his mailing address.

Acknowledgment. This paper is prepared within two projects focused on the integration of 3D Printing Technology in startups and education - "Montenegrin Startup School 2019" and "Montenegrin Engineering School 2020", supported by the Ministry of Science in Montenegro.

References

1. Gibson, I., Rosen, D.W., Stucker, B.: Additive Manufacturing Technologies. Springer, Heidelberg (2010)
2. Redwood, B., Schoffer, F., Garret, B.: The 3D Printing Handbook: Technologies, Design and Applications. 3D Hubs B.V., Amsterdam (2017)
3. Kumar, L.J., Pandey, P.M., Wimpenny, D.I.: 3D Printing and Additive Manufacturing Technologies. Springer Nature Singapore Pte Ltd., Singapore (2019)
4. Djukanovic, M., Grujicic, R., Radunovic, L., Boskovic, V.: Conceptual solution of the robotic arm/plotter. In: New technologies, Development and Application. LNNS, vol. 42, pp 170–179. Springer (2019)
5. Lansdown, H.: Digital Model Making: Laser Cutting, 3D Printing and Reverse Engineering. Crowood Publishing, Ramsbury (2019). ISBN: 1785005855
6. Mongeon, B.: 3D Technology in Fine Art and Craft: Exploration of 3D Printing, Scanning, Sculpting and Milling. Routledge Publishing, Abingdon (2015). ISBN: 9781315730455
7. Evans, B.: Practical 3D Printers: The Science and Art of 3D Printers. Springer Science +Business Media, New York (2012)
8. Nandi, C., Caspi, A., Grossman, D., Tatlock, Z.: Programming Language Tools and Techniques for 3D Printing. In: 2nd Summit on Advances in Programming Languages (SNAPL 2017). SchlossDagstuhl-Leibniz-ZentrumfuerInformatik (2017)
9. Bergström, J.S., Hayman, D.: An overview of mechanical properties and material modeling of Polylactide (PLA) for medical applications. Ann. Biomed. Eng. **44**, 330–340 (2016)
10. Vishwakarma, S.K.: Characterization of ABS material: a review. Quest J. J. Res. Mech. Eng. **3**(5), 13–16 (2016)
11. Nidagundi, V.B., Keshavamurthy, R., Prakash, C.P.S.: Studies on parametric optimization for fused deposition modelling process. Mater. Today Proc. **2**(4–5), 1691–1699 (2015)
12. Paul, R., Anand, S.: A new Steiner patch-based file format for additive manufacturing processes. Comput. Aided Des. **63**, 86–100 (2015)

13. Djukanovic, M., Grujicic, R., Radunovic, L., Boskovic, V.: Programming of the robotic arm/plotter system. In: Advanced Technologies, Systems, and Applications III. LNNS, vol. 60, pp 342–354. Springer (2019)
14. Cura Software. https://ultimaker.com/en/products/cura-software
15. Thiel, P., Masters, B.: Zero to One: Notes on Startups, or How to Build the Future. Crown Business, New York (2014)
16. de Jong, J.P.J., De Bruijn, E.: Innovation lessons from 3D printing. MIT Sloan Manage. Rev. **54**, 43 (2013)
17. Lee, Y., Koo, S.H.: Introduction to special collection on 3D printing and wearable technology in fashion. Fash. Text. **5**, 34 (2018). Springer Nature
18. https://www.eu-startups.com/tag/mercuris/
19. https://www.eu-startups.com/tag/reprap/
20. https://spectroplast.com/

Reverse Engineering Using 3D Scanning and FEM Analysis

Josip Kacmarcik(✉), Nermina Zaimovic-Uzunovic, and Samir Lemes

Faculty of Mechanical Engineering, University of Zenica,
Fakultetska 1, 72000 Zenica, Bosnia and Herzegovina
kjosip@mf.unze

Abstract. A reverse engineering process of a part including 3D scanning, CAD modeling and FEM analysis is presented in the paper. A cast bearing support has been selected as an example part for the investigation. The final goal of the investigation was to determine maximal allowable radial bearing load of the part. The parametric solid model necessary for the FEM was designed in CAD software using polygonal surface model of a part obtained by 3D scanning and subsequent data processing in 3D scanning software. A handheld scanner working on structured light principle was used for the scanning. The direction and intensity of the maximal allowable force was determined via FEM simulations in which different loading configurations were varied.

Keywords: 3D scanning · CAD · FEM · Reverse engineering

1 Introduction

Often in engineering practice a need emerges for some form of reverse engineering, i.e. a need to create technical data for a part or an assembly for which technical documentation is incomplete or completely missing. Reversible engineering is almost inevitable in engineering practice, and for a successful process knowledge from different engineering fields need to be combined and implemented [1–3].

In this paper, different engineering fields were combined in a reverse engineering of a part without any previous technical documentation, where the goal was two determine the maximum loading capacity of the part. Finite element method (FEM) analysis was used to obtain the loading capacity. A parametric solid model necessary for FEM was modeled using polygonal surface model (STL file) [3], obtained by 3D scanning, as a base.

2 3D Scanning

3D scanning of a part was done with Artec Eva handheld 3D scanner (Fig. 1), working on structured light principle. Artec Eva scanner besides geometry also captures a surface texture [4]. Artec Studio 14 Professional [5], provided with the Artec Eva scanner, was used for scanning data processing. The cast bearing support showed in Fig. 2 was chosen as an example for the investigation in the paper.

I. Karabegović (Ed.): NT 2020, LNNS 128, pp. 285–291, 2020.
https://doi.org/10.1007/978-3-030-46817-0_32

First step was scanning of the part from different angles in order to cover the all outside surface of it. In a scanning process it is necessary to keep a measured object in a focus of a scanner, what is obtained with the correct holding position and direction. For Artec Eva scanner, the focus is from 520 mm up to 880 mm distance to a part, Fig. 3. As a results from the scanning, four different scans (points clouds) were obtained (Fig. 4), by scanning of the part positioned in different orientations.

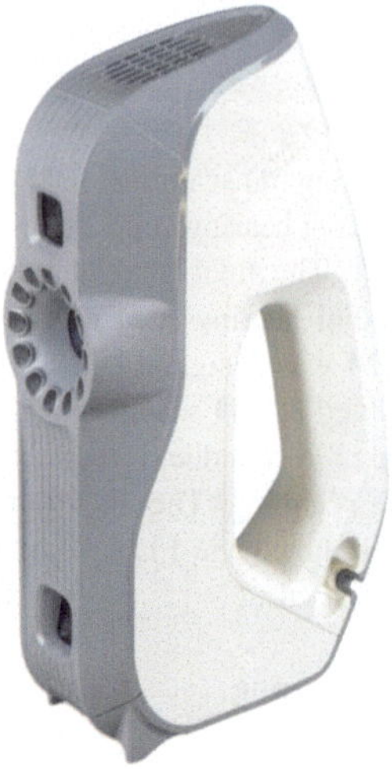

Fig. 1. Artec EVA 3D scanner.

Fig. 2. The scanned part.

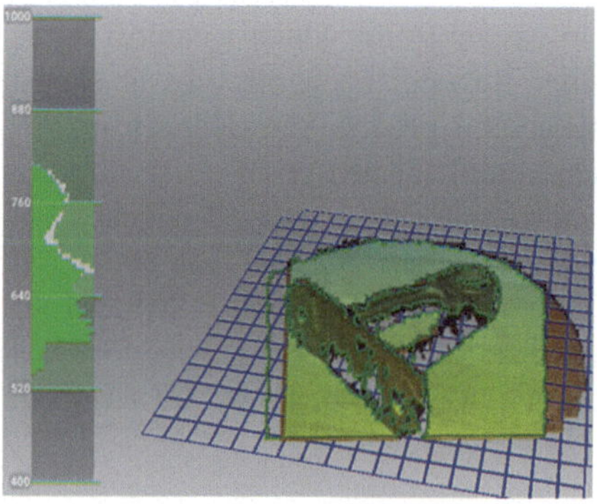

Fig. 3. 3D scanning of a part.

Fig. 4. Scans of the part.

Every scan on Fig. 4 is trimmed, where base on which part was laid and visible measurement noise (major outliers) registered together with the part surfaces, were removed. For the base removal an automatic algorithm available in the software was used and major and visible outliers were removed manually. As it can be seen on Fig. 4, different scans were misaligned, so the next step was alignment of the scans.

The scans are aligned by defining point pars between different scans (Fig. 5), and that points were used to bring scans close together and align surfaces. For that purpose, white marks were made on the part before the measurement (see Fig. 2). After alignment, on the obtained aligned scans, different data processing algorithms available among software editing tools are used, namely *global registration*, *outlier removal* and *sharp fusion* [5]. The *global registration* is done after scans alignment in order to bring all registered points into a single coordinate system. The *global registration* algorithm find pair matches between the points registered in scanning. The *outlier removal* algorithm removes small surfaces nor connected to the main surfaces. In the fusion process a polygonal surface model is created, by joining all the points from scans. The *sharp fusion* algorithm was chosen among available in the software as it is recommended for industrial objects. In the *sharp fusion* the option *fill holes* is available, which fill gaps in the mesh being reconstructed. The *fill holes* option was necessary here because the internal surfaces of the two small holes in the part could not be completely scanned, and gaps in them were closed using *fill holes* algorithm. The final results of the scanning process, the polygonal surface model of the part is shown in Fig. 6. The polygonal model was exported in stl. file format for further work.

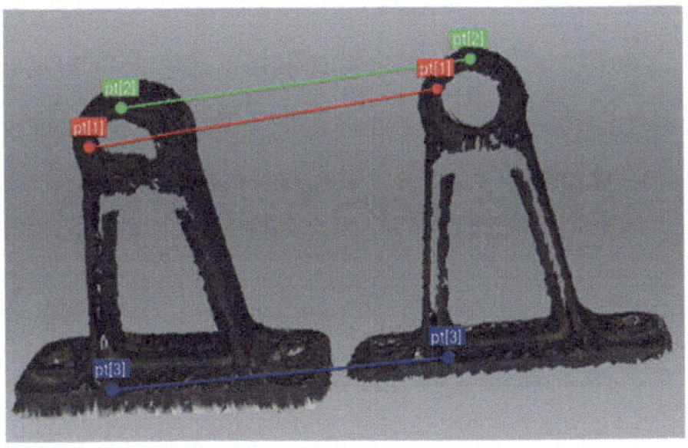

Fig. 5. Scans alignment using points.

Fig. 6. Polygonal 3D model with coordinate system orientation.

3 Cad Modeling

Polygonal surface model of the part (Fig. 6) obtained by scanning was used for creation of parametric solid model in commercial CAD software Solidworks [6]. After the import of stl. file in Solidworks, accessory geometries (points, lines and planes) that are aligned to part features were created. The accessory geometries were used for definition of two planes and a point (origin), which were finally used to define the coordinate systems which axis are coincident with the idealized symmetry planes of the part, Fig. 7. The creation of the new coordinate system was necessary because the exiting one was not aligned with the part features (see Fig. 6).

Afterwards, a common modeling tools like extrude, cut or fillet are used in Solidworks to create 3D model of the part. The sketches for modeling are created using polygonal model in a background as a base. The dimension for sketches and features definition were adopted so that good coincidence with polygonal model was achieved. Also, idealized symmetry of the model is respected (about *xz* an *yz* plane, Fig. 7). In Fig. 8 and 9 two of the sketches used for model creation are shown. During modelling some of the geometrical features (casting drafts and some fillets of the original part) are skipped as it is concluded that their influence on the results of the subsequent numerical analysis would be minor, and additionally they would make the modelling and numerical solving more complex.

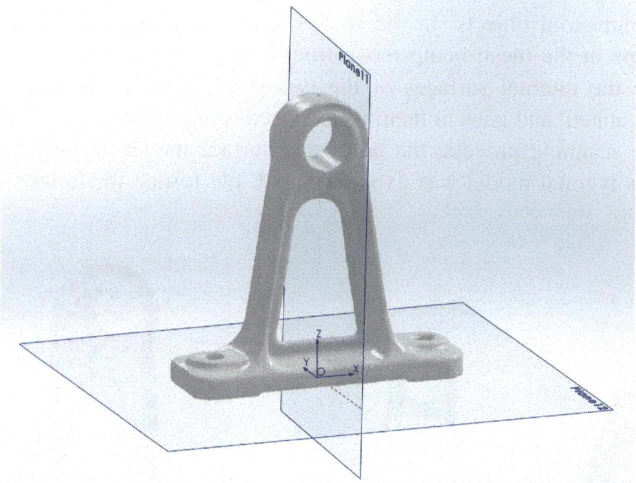

Fig. 7. Coordinate system aligned with the part.

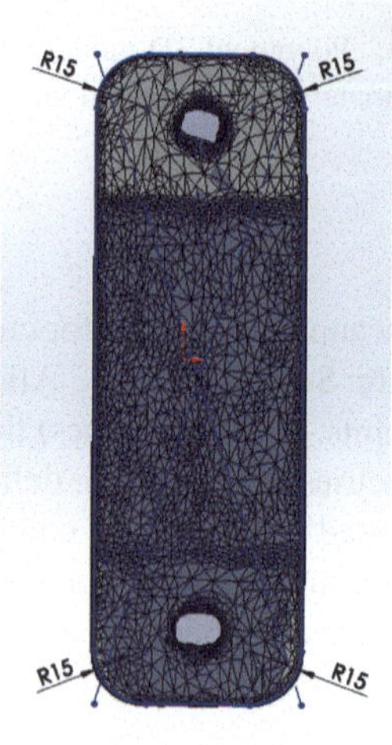

Fig. 8. Sketch for the part base.

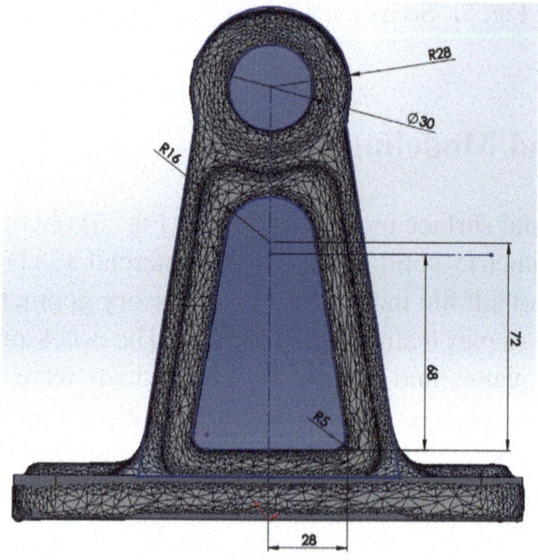

Fig. 9. Sketch for the top segment of the part.

Figure 10 shows finalized parametric 3D model, and Fig. 11 its comparison with initial polygonal model. Idealizations and approximations inevitable in a reverse engineering process lead to the deviations between the solid and initial polygonal file, visible in the Fig. 11.

Fig. 10. Final shape of parametric solid model.

Fig. 11. Comparison of parametric (blue) and polygonal 3D model (black mesh).

4 Fem Analysis

The goal of FEM analysis was to determine the maximal loading capacity for the radial bearing force. Solidworks Simulation module [6] was used for the analysis which was conducted in two stages. In the first stage of the numerical investigation the least favorable direction of the radial force was determined and in the second one the maximal allowable load. In both stages, supports of the part were numerically defined on two bottom flat and cylindrical surfaces, Fig. 12, thus simulating the bolted connection with the surface. The material of the part was defined as cast carbon steel from the Solidworks material database (with 200 GPa modulus of elasticity and 248 MPa yield strength).

In the first stage, six FEM simulations were conducted in which a direction of radial force was varied. The load had arbitrary intensity of 1000 N, acting on a half of cylindrical bearing surface. Directions of forces were rotated for angle of 45° in subsequent simulations (see lines in Fig. 12). In order to simplify the analysis, the load was defined with uniform distribution, although it does not agree with a real contact condition in a bearing. It is believed that the load distribution would not affect a final outcome in this stage. Registered maximum von Misses stresses for different angles are presented in Fig. 13. The horizontal direction of a force resulted in the highest value of the maximal von Misses stress in the part, hence the horizontal direction of a radial force is the least favorable loading condition.

In the second stage of the investigation with FEM, the maximum allowable horizontal radial force is determined. Here, the load is defined more realistically with a sinusoidal distribution, as a bearing load acting on a half of bearing surface. The maximum allowable load was determined with Solidworks design study module in which load intensity is varied in multiple simulations, where a factor of safety in value of 1,5 for von Misses stresses is used as a constraint for an optimal design. The load intensity is varied in the range from 4000 N to 7000 N, in steps of 100 N. The optimal design study results are shown in Table 1, with the value of the force and related values of maximal von Misses stress and factor of safety. The value of 5,1 kN is determined as a maximal allowable radial force for the part.

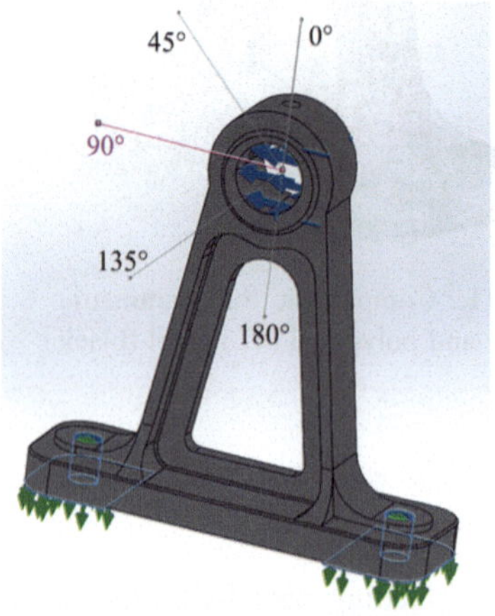

Fig. 12. Boundary conditions for FEM analysis

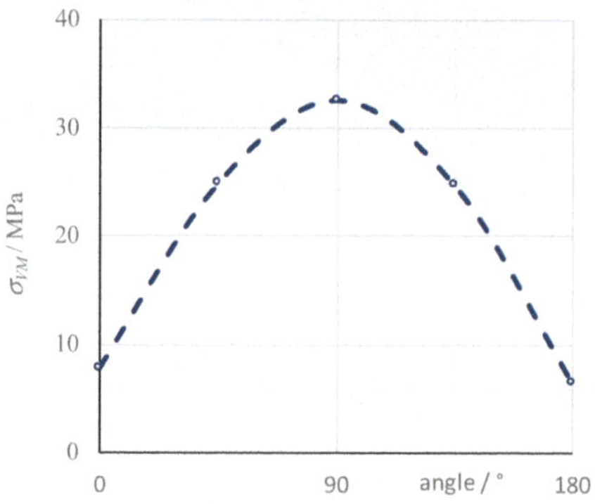

Fig. 13. Maximal Von Misses stress in the part with angle of force (1000 N)

Table 1. Design study results for the optimal factor of safety.

Force/N	Maximal von Misses stress/MPa	Factor of safety
5100	164,84	1,506

5 Conclusion

Reverse engineering can be a very important part of an engineer practice and it often needs implementation of knowledge from different engineering fields and modern equipment and software. In the paper an example of a reverse engineering process involving use of a modern measuring device, optical 3D scanner, and software for scan data processing, CAD and FEM, is presented. A cast bearing support is reverse engineered, and as a result from the process a parametric solid model and the maximal allowable load for the part are obtained.

References

1. Messler, R.W.: Reverse Engineering: Mechanisms, Structures, Systems & Materials. McGraw-Hill Education, New York (2013)
2. Alexandru, T.C. (ed.): Reverse Engineering: Recent Advances and Applications. BoD–Books on Demand, Norderstedt (2012)
3. Vinesh, R., Kiran, J. (eds.): Reverse Engineering: An Industrial Perspective. Springer Science & Business Media (2017)
4. ARTEC 3D. Artec Eva: Fast 3D scanner for professionals. https://www.artec3d.com/portable-3d-scanners/artec-eva. Accessed 16 Dec 2019
5. ARTEC3D. User Guide ARTEC STUDIO 11, Luxembourg: Artec Industry
6. SolidWorks 2018, Dassault Systèmes. www.solidworks.com

Microstructure Assessment of Steel for Elevated Temperatures by the Replica

Belma Fakić(✉), Adisa Burić, and Edib Horoz

Institute "Kemal Kapetanović" of Zenica, University of Zenica,
Ulica: Travnička cesta br. 7, 72 000 Zenica, Bosnia and Herzegovina
belma.fakic@ikk.unze.ba

Abstract. Materials used in thermal power plants are exposed to elevated temperatures in their daily work. Changes in the state of the microstructure of a material are monitored by non-destructive testing methods, that is, by taking a microstructure impression using a replica method. The elevated temperature leads to changes in the material that occur in the appearance of pores along the grain boundaries. Changes in the material and the degree of degradation of material will be presented in this paper.

Keywords: Steel · Microstructure · Replica · Elevated temperature · Creep · Micropore

1 Introduction

During long-term exploitation at elevated temperatures, the microstructure of the material is degraded and, consequently, its influence on the mechanical properties of the material. Estimates of material depletion based on microstructure imprints are increasingly represented in observing material behavior during long-term exploitation.

2 Technique of Surface Examination by Metallographic Replica

Microscopic examination of the metal surface using transparent films or plastic materials for the detection of mechanical and metallurgical inhomogeneities on the metal surface is described by the standard non - destructive imprint method.

Macroscopic preparation of the surface to be examined is performed by degreasing and cleaning of impurities. Microscopic preparation for surface condition detection is performed by grinding, multiple grinding operations, each of which is finer than the previous one. During these steps, no more than 0.2 mm thickness should be removed. After sanding the surface is polished. The final stage of surface preparation is erosion with the appropriate reagent. The plastic or transparencies are glued to the wrinkled surface to transfer the microstructure imprint [1]. After some time, the film is carefully removed and placed between the glasses and observed at magnification 500 times on an optical light microscope.

I. Karabegović (Ed.): NT 2020, LNNS 128, pp. 292–298, 2020.
https://doi.org/10.1007/978-3-030-46817-0_33

3 Microstructural Damage by Creep

In order to obtain a understanding of the creep damage and rupture process, analysis on the changes of microstructure and the cavity nucleation, growth and coalescence process under different stress levels was carried out and a there is guidelines for rating the microstructural composition and creep rupture damage. It is well known that the final rupture life of such material operated at elevated temperature has been influenced by both of the micro-structural changes and development of cavities, however, large amount of experiments indicate that the dominant accumulated damage is caused by cavity development [2].

Under low stress, the distribution of the cavity was observed perpendicular (90°) to the external applied stress; this has been confirmed in, which claimed that with a low applied uniaxial stress (50–80 MPa) specimens failure is resulted from the coalescence of discrete cavities on prior austenite grain boundaries which were oriented approximately to the tensile stress axis. Moreover, Kawashima stated that the cavity has also been seen at 45° to the tension stress [2, 4]. The creep cavities have been observed to nucleate at carbides, second phase particles, a ridge, a triple point and inclusions along the grain boundaries [2, 5]. For instance, the sulphide particles which are the primary nucleation sites in the overheated condition tend to nucleate cavities early in creep life [2, 3, 5]. The reports indicated that the cavity preferred to nucleate at stress concentration point on grain boundaries under low stress level [2].

Typical initial creep cavity shape under different stress levels are given in Table 1. The observation from literatures experiments found that at lower stresses, the shape of the cavity is spherical [2, 3]; for the higher stress regime, the cavity shape normally presents as circular during most of the secondary creep stage [2, 3] due to the influence of local stress concentration.

Table 1. Typical initial creep cavity shape under different stress levels

Stress level	Low stress (0,2–0,4 Yield stress)	High stress (> 0,5 Yield stress)
Typical cavity shape	Spherical voids	Circular voids

Typical damage of microstructure [2, 3] from has been associated with the typical curve, reflecting the dominant creep damage processes (cavitation) with typical creep curve for low Cr-Mo alloy steel at lower stress (0.2–0.4 Yield stress). Initiation behaviour about the development of cavity has included five different processes: cavity nucleation, growth, and coalesce, micro-cracks growth and macro-cracks growth processes; these five process has been summarized associated with time to rupture in Table 2.

As Table 2 shows the majority of life span is involved with the cavity nucleation from about $0.2Tf$ to $0.7Tf$ (stage 1), which takes about more than 50% of the total lifetime; nevertheless, the final rupture process has only taken account about less than 30% of life to rupture (stage 3) [2, 3].

Table 2. Initiation time for the development of cavitation process [2, 3]

Initiation behaviour at time to rupture: T/*Tf*	0,0–0,2	0,2–0,4	0,4–0,6	0,6–0,8	0,8–1,0
Cavity nucleation	–	≥ 0,25/0,26	0,4–0,6	–	–
Cavity growth	–	–	0,38–0,56	–	–
Cavity coalesce	–	–	≥ 0,4	–	–
Micro- crack growth	–	–	–	≥ 0,78/0,85	–
Macro- crack growth	–	–	–	≥ 0,7	0,96–1

Based on the experimental data of the stress versus time to the rupture, the mechanical relationship could be approximately assumed [2] as:

$$T_f \propto \frac{1}{\sigma} \tag{1}$$

where T_f is rupture time, and σ is external applied stress.

Figure 1 shows the relationship between plastic creep deformation over time and damage when metallic materials are used in the creep range.

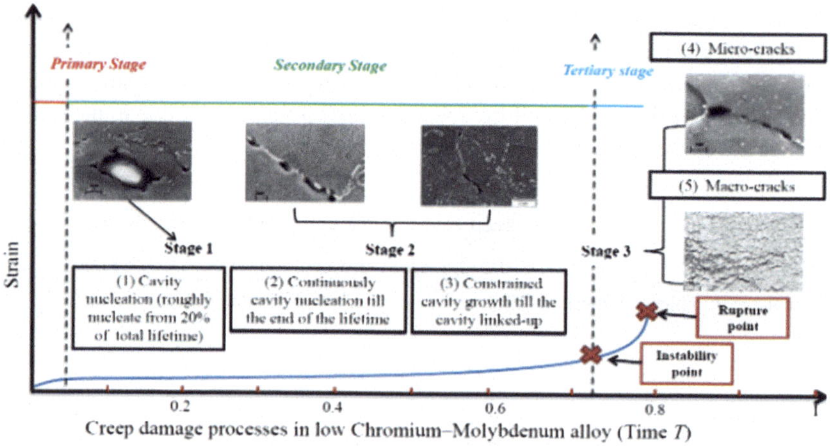

Fig. 1. Relationship between plastic creep deformation over time and damage when metallic materials are used in the creep range [2]

Primary creep stage is the rate of creep reduces, and is regarded as being due to an increase in the dislocation density. Secondary creep stage is characterized by the equilibrium between the solidification and softening processes, i.e. the rate of creep is roughly constant. In addition to the changes in the precipitation structure, other thermally activated processes such as the decomposition of pearlite, coagulation and precipitation of carbides, etc. can also occur in the microstructure.

These processes are material, time and temperature dependent. All processes that take place in the microstructure up to this point are considered reversible, i.e. they can be undone through appropriate heat treatment measures. Irreversible creep damage manifests itself in the form of pores, the extent of which depends on the material and stressing (stress, temperature and time). In combination with the metallurgical processes (subgrain growth, grain coarsening, greater particle distance), the rate of creep increases significantly. Another consequence is the occurrence of linear porosity and microcracks, primarily on grain boundaries, that grow under the effect of stressing.

This range is known as the tertiary creep stage. The initial stages of creep damage (visible by light optical means) in the form of creep pores depends on the material (microstructure), temperature and stressing condition (stress and multiaxiality) [6]. Note that the multiaxiality of the stressing reduces the material's deformation capability and so promotes the formation of pores.

4 Material Data

Chemical composition of steel 14MoV63 and some mechanical properties in accordance to standards are given in Table 3. Adding of alloying elements Cr, Mo and V and with adequate heat treatment this alloy is designed to have work period about 1000 000 h [7].

Table 3. Chemical composition of steel 14MoV63 and mechanical properties

Standard BAS EN 10216-2 14MoV63 WN. 1.7715	Chemical composition, mas %						Mechanical properties	
	C	Mn	Si	Cr	Mo	V	Hardness HB	$R_{p0,2}$, at 450 °C, MPa
	0,10–0,15	0,40–0,70	0,15–0,35	0,30–0,60	0,50–0,70	0,22–0,28	141–208[a]	203

[a]ČSN 415128

Microstructure of steel mainly consisting of ferrite with a content of tempered bainite and pearlite 10 to 15%.

Microstructure can be seen from TTT diagram given at Fig. 2 [7].

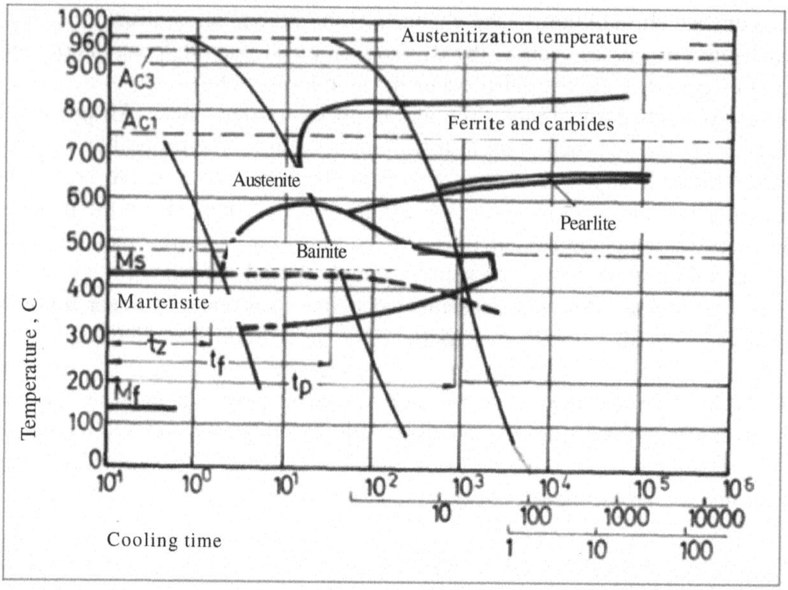

Fig. 2. TTT diagram of steel 14MoV63

Alloying elements, whether dissolved in a ferrite matrix or forming coherent or incoherent carbides, affect the course of material degradation in the creep process, via interaction in dislocation processes. Primarily formed coherent carbides under the long-term influence of temperature change to another incoherent form, and the process of forming new carbides from elements dissolved in solid ferrite solution begins. In the further process, the coagulation and coarsening of the carbides occur, which in the processes of recovery and recrystallization leads to degradation of the structure and the decline of mechanical properties [2, 3].

Molybdenum, either in solid solution or through coherently precipitated carbides, has a particularly beneficial effect on reducing the creep rate and degradation process. The described processes take place in the stationary creep region and are observed when examined on an optical microscope and much clearer. Degradation processes in the stationary region dominantly affect the creep flow in the tertiary phase in which microcracks and fractures occur [7–9].

Microphotographs of damaged microstructure after work hours 200 000 h and approximately 275 000 h are given at Fig. 3.

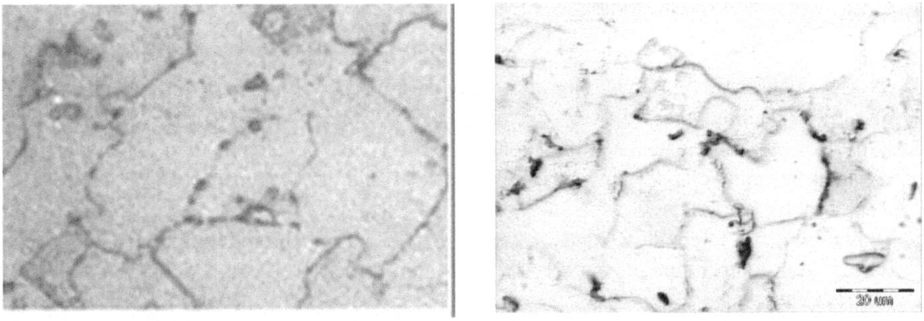

Assessment class: 2a [7] Assessment class: 3a [10]

Fig. 3. Microstructure after work hours 200 000 h and approximately 275 000 h, respectively

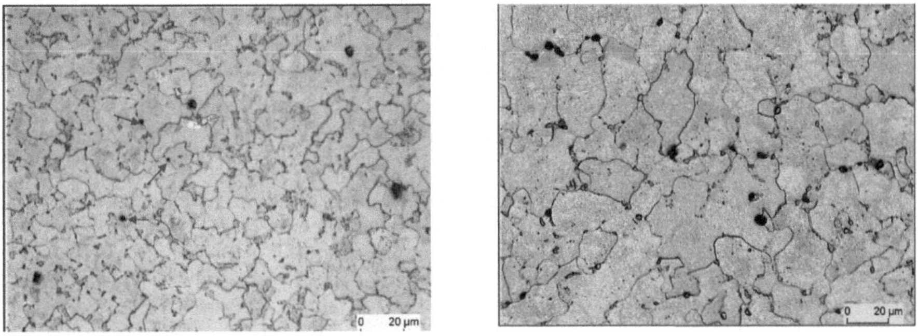

Assessment class: 2a Assessment class: 3a

Fig. 4. Microstructure rating charts VGB S-517 [6]

5 Conclusion

The test results of the material that was in operation at an elevated temperature of 200 000 h are consistent with the described degradation of the microstructure in scattered creep pores due to prolonged exposure to high temperature up to 545 °C.

After 275 000 h at an elevated temperature up to 545 °C, testing by replica shows more degradation of microstructure in numerous creep pores with orientation what is confirmed using rating charts (Fig. 4).

This part of power plant is in high risk of exploitation, and have to be check in short period, because test result shows that material is in second creep stage which is close to initiate microcracks.

References

1. STANDARD BAS ISO 3057: Non-destructive testing — Metallographic replica techniques of surface examination (1998)
2. https://www.researchgate.net/publication/271587463_The_development_of_creep_damage_constitutive_equations_for_low_Cr-Mo_alloy_steel_and_its_weldment_at_low_stress. Accessed 10 Jan 2020
3. Dobrzański, J.: Internal damage processes in low alloy chromium–molybdenum steels during high-temperature creep service. J. Mater. Process. Technol. **157–158**, 297–303 (2004)
4. Kawashima, F., et al.: Micro-macro combined simulation of the damage progress in low-alloy steel welds subject to type IV creep failure. JSME Int. J. **47**, 410–418 (2004)
5. Needham, N.G.: Cavitation and fracture in creep resisting steels: final report. Commission of the European Communities (1983)
6. STANDARD VGB-S-517: Guidelines for rating the microstructural composition and creep rupture damage of creep-resistant steel for high pressure pipe-lines and boiler components and their weld connections, Essen, Germany (2014)
7. Kratina, E., Fakić, B., Husika, A.: The process of the materials microstructure degradation and the influence on properties in the power plant components in long time exposure to elevated temperature exploitation. In: 7th Scientific/Research Symposium with International Participation, Metallic and Nonmetallic Materials, pp. 237– 242 (2008)
8. Pavlović, P.: Materijal čelik. SKTH, Zagreb (1990)
9. Promisel, N.E., et al.: Metals Handbook, Selection and Properties, vol. 1. ASM, Ohio (1978)
10. Test report of Metallographic laboratory 129/19 dated 01 July 2019

Application the Method Digital Image Correlation for Measurement of Small Deformations at the Beginning of Plastic Flow of Materials

Stoja Rešković(✉), Tin Brlić, and Filip Skender

Faculty of Metallurgy, University of Zagreb, Aleja narodnih heroja 3, 44000 Sisak, Croatia
reskovic@simet.hr

Abstract. This paper presents the results of the investigation of very small deformations at the very beginning of the plastic flow of the material. The studies were performed on low carbon microalloyed steel with 0.035% niobium. The tests using the digital image correlation (DIC) method were performed. The measurements by static tensile test using a digital camera at variable strain rates were carried out. The measurement results by MatchID software were analyzed. Studies have shown that at the beginning of plastic flow deformations do not occur evenly throughout the deformation zone. The stretching rates at the beginning of the plastic flow are affecting the amount and deformation distribution in the deformation zone.

Keywords: Digital image correlation (DIC) · Deformation zone · Small deformation · Microalloyed steel

1 Introduction

When as examining the behavior of steels during cold deformation, that inhomogeneous deformations occur at the transition from the elastic to the plastic region in some steels it is observed. The amounts of these deformations are very small, they are not evenly distributed in the deformation zone and can't be fully tested by classical methods [1, 2].

Today, with the development of digital technology, new test methods have been developed that give great opportunities to test the behavior of steel during deformation. The digital image correlation method (DIC) is a relatively new optical-numerical method that allows the deformation changes to be determined over a period of deformation time from randomly applied markers [3, 4]. Using the appropriate software, it is possible to determine their amount at each point of the deformation zone.

The digital image correlation method has proven to be an ideal method for recognizing the mechanical behavior of metals in deformation [5]. With this method, very useful qualitative and quantitative information on the distribution and amounts of deformation in the deformation zone is possible to obtain [6, 7].

I. Karabegović (Ed.): NT 2020, LNNS 128, pp. 299–304, 2020.
https://doi.org/10.1007/978-3-030-46817-0_34

The DIC method is based on the calculation of marker displacement on the surface of the test specimen. For proper use of this method, the test specimens must be properly prepared. Sample preparation consists of arbitrarily applying markers to the surface of the sample. Markers are the most common coating of appropriate paint. It is necessary a good contrast between the marker and the substrate to get. Usually on the black surface white markers are randomly applied. [6].

A digital camera is a basic tool for the digital image correlation method [7]. It must be perpendicular to the test specimen and be able to record the specimen throughout the duration of the test.

The analysis of the measurement results is performed in such a way that the measurements in the area of a certain degree of deformation are compared with the image without deformation. The analysis of the measurement results was performed within the appropriate software in which there are several different correlation functions. Qualitative and quantitative analysis of deformation is possible.

The aim of this paper is to investigate the possibility of determining very small deformations and their distribution in the deformation zone using the DIC method.

2 Experimental

The tests were performed on samples of low carbon microalloyed steel with addition of 0.035% niobium. The chemical composition of the sample was tested on an ARL 8660 quantometer, Table 1.

Table 1. Chemical composition of the examined micro-alloyed steel (wt%)

	C	Mn	Si	P	S	Al	Nb	Ni
(wt%)	0,09	0,75	0,05	0,018	0,014	0,02	0,035	0,0081

Samples for testing using the digital image correlation method were first grinded and polished. In this way, the oxide layer was removed from the sample surface, Fig. 1a. A black matte finish was first applied to the sanded specimen. After drying, white markers were applied, Fig. 1b.

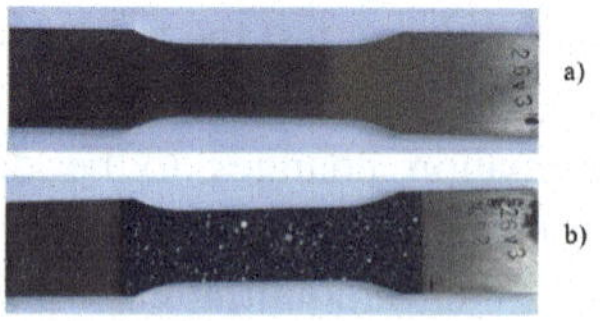

Fig. 1. Sample for DIC testing: (a) after sanding and polishing, (b) after coating and marker application

A static tensile test was performed on the tensile testing machine Zwic 50 KN. The tests were carried out at three stretching rates: 5, 20 and 50 m/min. For the digital image correlation test, a Panasonic HDC SD9 with 2.1 Mpixel with CCD sensor was used that was positioned perpendicular to the test specimen, Fig. 2. Before the very start of the test, the digital camera was calibrated. The analysis of obtained results with digital image correlation was carried out with the MatchID software package.

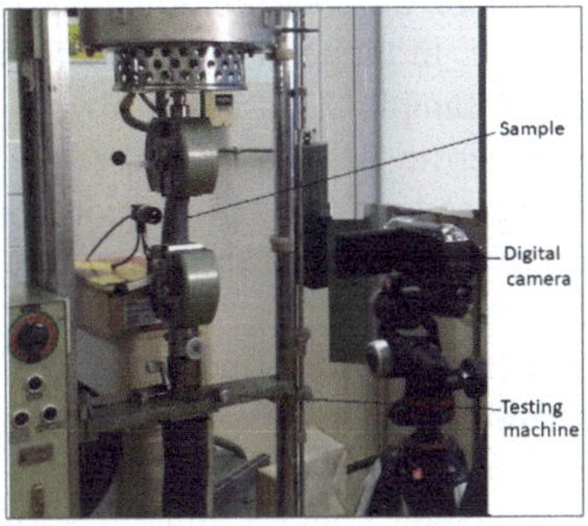

Fig. 2. Placement of equipment during testing

3 Results and Discussion

The method of digital image correlation carried out a qualitative and quantitative analysis of the change in the amount of deformation at the beginning of the plastic flow of microalloyed steel with 0.035% niobium at stretching rates of 5, 20 and 50 m/min.

The qualitative analysis, Fig. 3, was performed at the point Rp, and after 5% and 10% of the total deformation, respectively.

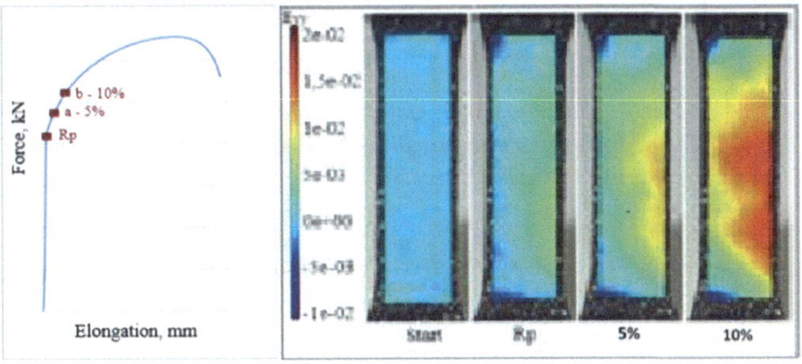

a b

Fig. 3. Deformation maps at the beginning of the plastic flow at a stretching speed of 20 m/min

At point Rp, plastic deformation begins. It is not uniform throughout the deformation zone. The plastic deformation first begins in crystalline grains whose planes are sliding at an angle of 45° with respect to the stretching direction of the specimen. In other grains, elastic deformation takes place. The effect of the tensile force changes the orientation in the crystal grain and the deformation continues again in the grains whose sliding planes are at an angle or closest to the angle of 45° relative to the direction of the tensile force.

The deformation maps shown in Fig. 3b clearly show that plastic flow is initiated in Rp. Elastic deformation takes place at the same time as plastic deformation. At 5% of the total deformation, the progress of the plastic flow of the material in the deformation zone is clearly visible. At 10% of the total deformation, plastic deformation occurs throughout the deformation zone. In the tests, plastic deformation started on one side and spread to the other side of sample. This is not always the case. For low carbon steels having a homogeneous ferrite pearlite structure, the deformation starts in the middle and spreads towards the ends of the deformation zone [4, 7].

Comparing the deformation changes with the scale, Fig. 3b, it can be clearly seen that the deformation amounts are very small.

Quantitative line DIC analysis was performed to determine their exact amount. It shows the change in deformations along the length of the specimen.

Quantitative line analysis was performed at the point Rp, and at 5% and 10% of the total deformation at the at stretching rates: 5 mm/min, 20 mm/min and 50 mm/min which corresponding to the strain rates of 0.0018 s^{-1}, 0.007 s^{-1} and 0.0185 s^{-1}. The results of the analysis are shown in Figs. 4, 5 and 6. The distribution of deformations along the line in $x_1 - x_2$ of the deformation zone at the time of the onset of plastic flow, point Rp, is shown in Fig. 4.

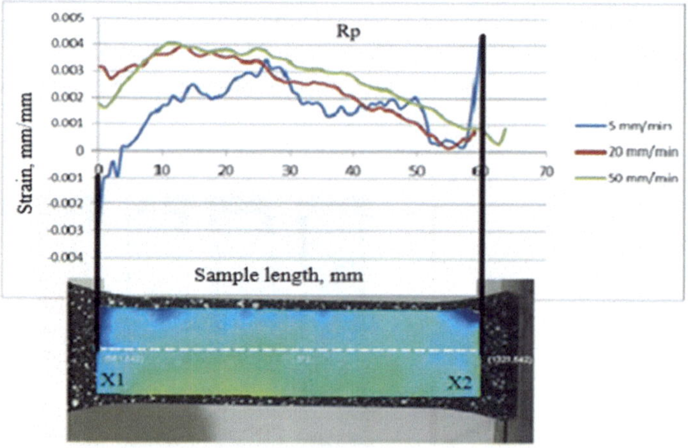

Fig. 4. Strain distribution at the beginning of plastic material flow (Rp)

The deformations are inhomogeneous and distributed over the entire length of the specimen. The amounts of deformation increase with the increase in the stretching rates. At the lowest stretching rate (5 m/min), the deformation is smallest at the ends of the deformation zone and amount to only 0.0001 mm/mm. The maximum amount of deformation at this velocity was measured at the center of the sample and it is

0.0034 mm/mm. With increasing deformation rate, deformation occurs faster in the deformation zone. At a stretching rates of 20 m/min, the deformation is higher, is 0.0039 mm/mm, and is more homogeneous along the length of the deformation zone. Deformation occurs more at one end of the sample. At a stretching rates of 50 m/min, the maximum deformation increases again to 0.0041 mm/mm.

The contribution of the further force action to the amounts of total deformations is clearly visible in Figs. 5 and 6. At 5% of total deformation, Fig. 5, at all velocities, the areas in which plastic deformation takes place are more homogeneous with respect to the beginning of the plastic flow. The contribution of the increase in the deformation is clearly visible. At stretching speed 5 mm/min, the maximum amount of deformation rises to 0.0078 mm/mm. At a sample stretching rates of 20 mm/min, the maximum deformation amount is 0.0087 mm/mm.

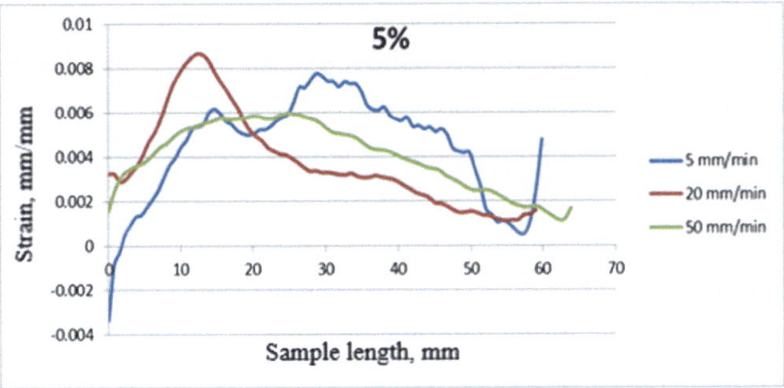

Fig. 5. Strain distribution at 5% of the total deformation

At a stretching rates of 50 mm/min it also increases due to an increase in the degree of deformation, but this increase is smaller and amounts to 0.006 mm/mm. The deformation is more homogeneous and spreads uniformly through the deformation zone.

As can be seen in the deformation maps, Fig. 3, at 10% of the total deformation, deformation occurs throughout the deformation zone. Figure 6 clearly shows that the deformations are significantly more homogeneous with respect to the onset of plastic flow of the material. At a stretching rates of 50 m/min, the deformation along of the deformation zone is homogeneous and smallest, being 0.0092 mm/mm.

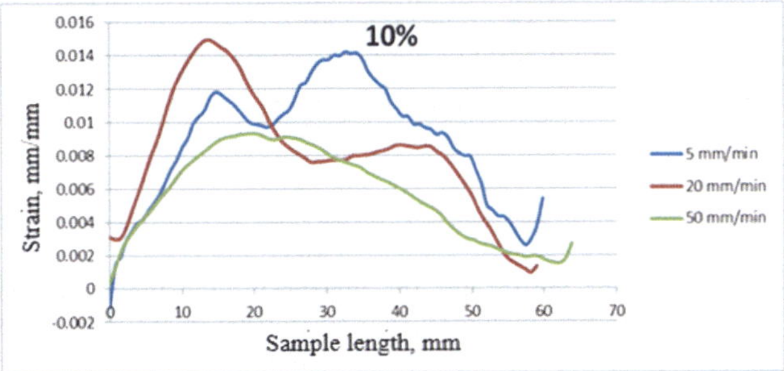

Fig. 6. Strain distribution at 10% of the total deformation

The Figures 5, 6 and 7 clearly show that when reaching the beginning of the plastic flow of the material, the deformation is greatest at the highest velocity of extension. It is not homogeneously distributed in the deformation zone. When the deformation begins to occur throughout the deformation zone, Fig. 6, at 10% of total deformation, the influence of stretching rate changes. After there, the deformation increases at the lowest stretching rates. At the highest stretching rate, the deformation is smaller and more homogeneous in the deformation zone. At lower stretching rates, the deformation is larger and more in homogeneous.

4 Conclusion

Studies have shown that it is possible to measure a very small deformation of o, ooo1 mm/mm by the DIC method. It is possible to accurately detect the amounts of deformation at each point of the deformation zone.

The performed research on Nb microalloyed steel showed that at the very beginning of the plastic flow of the material, the deformation does not occur homogeneously in the deformation zone. At 10% of the total degree of deformation, deformation occurs throughout the deformation zone. The strain rates has a significant effect on the onset of deformation and plastic flow of the material at the beginning of the plastic flow of the material.

In further studies, it is necessary to investigate the influence of microstructure on the plastic flow of materials in the deformation zone.

Acknowledgements. This work has been fully supported by the Croatian Science Foundation under the project number IP-2016-06-1270. Principal investigator: prof. S. Rešković.

References

1. Skobir, D.A.: High strength low alloy HSLA steels. Mater. Technol. **45**(4), 295–301 (2011)
2. Moon, U.J., Kima, S., Jang, J., Lee, J., Lee, C.: Orowan strengthening effect on the nanoindentation hardness of the ferrite matrix in microalloyed steels. Mater. Sci. Eng., A **487** (2), 552–557 (2008)
3. Lu, H., Cary, P.D.: Deformation measurements by digital image correlation: implementation of a second-order displacement gradient. Exp. Mech. **40**(4), 393–400 (2000)
4. Rešković, S., Jandrlić, I.: Influence of niobium on the beginning of the plastic flow of material during cold deformation. Sci. World J. **2013**, 5 (2013). Article ID 723725
5. Eskandari, M., Zarei-Hanzaki, A., Yadegari, M., Soltani, N., Asghari, A.: In situ identification of elastic-plastic strain distribution in a microalloyed transformation induced plasticity steel using digital image correlation. Opt. Lasers Eng. **54**, 79–87 (2014)
6. Palanivelu, S., De Pauw, S., Van Paepegem, W., Degrieck, J., Van Ackeren, J., Kakogiannis, D., Wastiels, J., Van Hemelrijck, D., Vantomme, J.: Validation of digital image correlation technique for impact loading applications. In: DYMAT, pp. 373–379 (2009)
7. Rešković, S., Jandrlić, I., Brlić, T.: The influence of niobium content and initial microstructure of steel on the occurrence of Lüders band at the start of the plastic flow during cold deformation. Mater. Sci. Eng. IOP Conf. Ser. **461**, 1–6 (2018)

Development of the Wasteless Technology of Processing of Steel-Smelting Slags

Sokhibjon Turdaliyevich Matkarimov(✉)
and Bakhriddin Tilovkabulovich Berdiyarov

Department of Metallurgy, Tashkent State Technical University,
#2 University street, Almazar district, 100098 Tashkent, Uzbekistan
sohibtm@gmail.com

Abstract. In article questions of development low-waste technologies of processing of steel-smelting slag are considered, gland allowing by extraction and its connections from steel-smelting slag to receive additional raw materials for production became, and the remains to use in building industry. Studying of gravitational methods of enrichment of steel-smelting slag and heat treatment the ore-fuel of pellets is the basis for work.

Keywords: Slag · Metallurgical dust · Rolling scale · Tails of dressing-works · Iron · Magnetite · Fusion mixture · Melting

1 Introduction

The main problem of modern steel-smelting production in the world is complex processing of slag, utilization of valuable components and increase in complexity of use of raw materials. Over time the given problem will increase since every possible exhaustion of the rich and easily opened ore fields, increase in requirements to environmental protection, growth of needs for ferrous metals is observed. Production of ferrous metals is followed by formation of slag and other types of technogenic waste. Their quantity, in connection with rapid growth of smelting of cast iron and steel, continuously increases and now is hundreds of millions tons. It is easy to present what huge areas can be busy with slag dumps. And it when the deficiency of the fertile land plots occupied by the industrial enterprises sharply increases and becomes aggravated situation with protection of natural resources. But it is only one sides of the problem, the second consists in potential opportunities of use of slag in the national economy [1].

Complex processing of steel-smelting slag as the main way of engineering-ecological development of society, were offered in the mid-sixties the past century, generally by scientists from the CIS countries. On the basis of their application it was supposed not only it is most useful to use the consumed raw material resources, but also to try to process completely formed waste.

When choosing effective ways of processing of slag seek for possibly fuller extraction of metal, increase in degree of purity of metal-products and their wider use. A metal-product, comprising 60–85% of iron, use in steel-smelting production,

I. Karabegović (Ed.): NT 2020, LNNS 128, pp. 305–313, 2020.
https://doi.org/10.1007/978-3-030-46817-0_35

50–60% of iron—in domain furnace charge and 30–50% of iron - in an agglomerative charge. The mineral component of slag is also used in metallurgical production as a iron-flux and also processed in breakstone, applied in concrete, a road and railway construction.

From the analysis fortunes of processing of slag abroad can be made conclusion that processing of steel-smelting slag quickly increases and they are used as a reverse product in metallurgy, road construction and agriculture.

For effective application of gravitational methods for division of components of steel-smelting slag, it is necessary to have preliminary data on the chemical composition, physical and physic-mechanical properties of the processed slag. Also, at gravitational enrichment such indicators as density, fractional and particle size distribution of the minerals which are a part of slag [2] are of great importance. Taking into account it we will consider in more detail influence of these parameters on indicators of gravitational enrichment.

Density of minerals – the relation of mass of grain in m monolith to its volume W:

$$\delta = \frac{m}{W'} \tag{1}$$

where δ – density of mineral, kg/m^3;

m – mass of grain, kg;

W′ – the volume of grain, m^3.

Fineness of separate grain is designated by its linear size.

For simplification of calculations exclude influence of a form of particles and then use the concept “equivalent diameter”, i.e. diameter of spherical de grain which volume is equal to the volume of grain of irregular shape:

$$W = \frac{\pi d_{\ni}^{3}}{6} \tag{2}$$

$$d_{\ni}^{3} = \sqrt[3]{\frac{6W}{\pi}} = \sqrt[3]{\frac{6m}{\pi\delta}} = 1,24\sqrt[3]{\frac{m}{\delta}} \tag{3}$$

The form of grains depends by nature minerals. The form of grains is characterized by the form coefficient representing the relation of a surface of an equal sphere (equivalent diameter of grain) to the surface of grain of irregular shape (the size of grain of irregular shape):

$$\varphi = \frac{S_s}{S} = \frac{4{,}87W^{2/3}}{S} \tag{4}$$

where φ – form coefficient;

S_s is a sphere surface;

S – surface of grain of irregular shape.

Values of coefficient of a form of particles φ are given below:
spherical.................................. 1.0
rounded, rounded.................... 0.8–0.9
angular..................................... 0.6–0.7
lamellar.................................... 0.6–0.7

2 Fractional Analysis of Steel-Smelting Slag

Division of components of steel-smelting slag into fractions of various density for definition of characteristics of their concentrating was the purpose of carrying out the fractional analysis.

Subject to the fractional analysis various classes of a size. As dividing environments are used: water solution of mineral salts, organic liquids, suspensions.

The analysis of minerals banding by fineness up to 1 mm in static conditions of division with application the middle of capacities with environments of various density, and fineness less than 1 mm - in dynamic conditions of division with use of centrifuges.

For stratification of tests of ores, and in our case - steel-smelting slag - it is possible to use heavy liquids and suspensions with a density of 2400, 2700, 3000, 3300, 3500, 3700, 4000, 4200 kg/m^3.

No more than 10 kg consistently immerse test in portions in the prepared liquids or suspensions by means of a reticular tank, removing every time and separating a mesh scoop easy fractions, floating on a surface. The number of fractions form n 1 where n— amount of liquids or suspensions. The received fractions wash with water, dry up, weigh, subject to cutting then direct to a chemical analyze.

On the basis of data on weight and quality of fractions tables are formed, and then curve enrichments are under construction.

The fractional analysis of three tests of steel-smelting slag of JSC "Uzbeksteel" (Uzbekistan) was in vitro made. The chemical composition of tests is given in Table 1.

Table 1. Chemical composition of steel-smelting slag.

# Tests	Connections, %									
	Fe	FeO	Fe_2O_3	CaO	SiO_2	Al_2O_3	MnO	MgO	P_2O_5	Other
Test of #1	3,7	2,0	15,2	25,6	21,6	5,8	5,3	13,8	1,7	5,3
Test of #2	4,1	2,3	14,7	24,3	23,0	5,2	5,4	13,9	1,2	5,9
Test of #3	3,6	1,7	14,5	25,7	22,5	5,7	5,5	13,4	1,3	6,1
Average structure	3,8	2,0	14,8	25,2	22,4	5,6	5,4	13,7	1,4	5,8

Results of the fractional analysis showed that density of the components which are a part of slag fluctuates from 2.5 to 5.2 g/cm^3. Depending on density of components they can be divided into the following fractions: heavy, average and easy fractions.

To heavy fraction components of slag with a density more than 4.0 g/cm^3, belong to average fraction from 2.7 to 4.0 g/cm^3 and to easy - with a density less than 2.7 g/cm^3.

The fractional analysis of slag of JSC "Uzbeksteel" (Uzbekistan) showed the following distribution of components on fractions:

- heavy fraction MnO - 4.7–5.0 of g/cm^3, FeO Fe_2O_3 - 5.0–5.2 g/cm^3;
- average fraction Al_2O_3 - 3.9–4.0 g/cm^3; MgO - 3.2–3.7 g/cm^3; CaO - 3.0–3.5 g/cm^3.
- easy fraction SiO_2 - 2.5–2.7 g/cm^3.

The received results of the fractional analysis are reduced in Fig. 1.

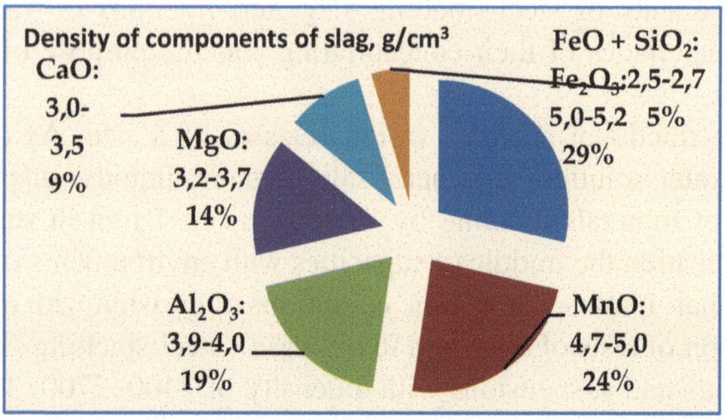

Fig. 1. Fractional analysis of steel-smelting slag of JSC "Uzbeksteel" (Uzbekistan).

3 Determination of Particle Size Distribution of Steel-Smelting Slag

Particle size distribution of steel-smelting slag of a define in the way of a dry sieving. Applied set to determination of particle size distribution sit, the device for stirring sit, doing 140–150 courses a minute.

For determination of particle size distribution mm were used a sieve with a size of cells: 3.0; 2.5; 2.0; 1.5; 1.2 mm. Previous weighed hinge plate of steel-smelting slag was located on the top sieve of set. Duration of stirring was 5–10 min. the Device stopped, from it was taken out a sieve, and weighing of the rest on each sieve was made. The hinge plate of slag weighing 5.0 kg was exposed to the graduation analysis.

Results of the graduation analysis are given in Table 2.

Table 2. Particle size distribution of initial steel-smelting slag

Class, mm	The weight, kg	The exit on classes, %	The total exit on "+", %
+3,0	0,859	17,18	17,18
−3,0 + 2,5	0,280	5,6	22,78
−2,5 + 2,0	0,320	6,4	29,18
−2,0 + 1,5	0,470	9,4	38,58
−1,5 + 1,2	0,243	4,86	43,44
−1,2	2,828	56,56	56,56
	5,0	100	100

For the maximum division of grains when using gravitational methods of enrichment, one of necessary conditions is, division of particles from the dead containing rock, for achievement of this condition researches on crushing of steel-smelting slag with determination of particle size distribution of the crushed slag were conducted.

4 Definition of the Key Process Parameters of Allocation of Valuable Components Using Differences in Their Physic-Mechanical Properties

Results of the carried-out analyses allowed to choose the enrichment methods based on the difference of density of minerals. Researches on enrichment of slag by methods of hydraulic classification, a jigging, the screw separator and enrichments on a concentration table were conducted. These methods found broad practical application and give rather objective information about technical and economic indicators of process [4].

Laboratory researches of hydraulic classification were conducted on the one-spiral qualifier with a length of trough of 1.6 m, 0.28 m wide and with a diameter of spiral of 0.28 m.

For carrying out experiments the pulp with T was prepared: = 1:10 till 1:20 which was given to a trough of the spiral qualifier. Studied efficiency of division of particles depending on duration of stay of material in the device and sitovy composition of initial material. Results of pilot studies are presented in Fig. 2.

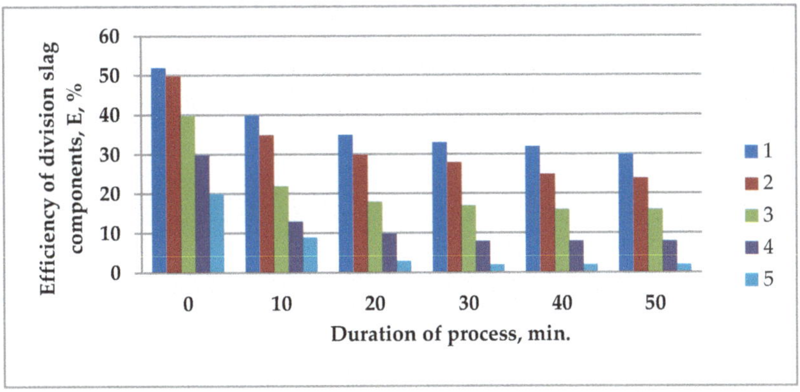

Fig. 2. Dependence of efficiency of division of components of the processed steel-smelting slag on process duration: (1) dmax = 4 mm; (2) dmax = 3 mm; (3) dmax = 2 mm; (4) dmax = 1 mm; (5) dmax $\leq$ 0.5 mm

Apparently from data in Fig. 2 division of grains on the hydraulic spiral qualifier proceeds in 15–20 min, at the same time the efficiency of division is not high, and the initial size of grains has no great influence. It means that this method is ineffective for the solution of tasks of release of iron and its oxidic connections in an independent product. However the method of hydraulic classification can be used at preparatory processes, for example, for washing and desludging of initial material.

Researches on an enrichment of the processed steel-smelting slag by method of a jigging were conducted in the laboratory two-chamber diaphragm jigger MOD-2 vehicle. Average values of results of 5–6 pilot studies are presented in Table 3. In experiments tails of the first jigging loaded again therefore received two concentrates.

Table 3. Qualitative quantitative indices of division of valuable components of steel-smelting slag at a jigging

Products	Exit		Contents, %			Extraction, %		
	kg	%	$FeO+Fe_2O_3$	Fe	MnO	$FeO+Fe_2O_3$	Fe	MnO
Initial slag	10,0	100	17,6	3,8	5,4	100	100	100
Concentrate 1	4,9	49	15,2	2,8	4,7	42,0	36,8	43,1
Concentrate 2	2,1	21	31,5	6,2	9,4	37,5	34,2	36,5
Tails	3,0	30	12,08	3,7	3,7	20,5	29,0	20,4

These Table 3 show that extraction of valuable components in a concentrate low though an exit of a concentrate made 49–50%. The maintenance of useful components rather big (up to 29%) that will lead to irrevocable losses of metal. Probably, physic-mechanical properties of slag-forming components do not differ at so big size that it is effective to divide them by a jigging method. Use of this technique does not give effective technical and economic indicators and it cannot be recommended for industrial introduction [5].

One of possible methods of division of components of the processed steel-smelting slag is use of the screw separator. Results of the conducted researches are presented in Table 4.

Table 4. Qualitative quantitative indices of division of valuable components of steel-smelting slag on the screw separator

Products	Exit		Contents, %			Extraction, %		
	kg	%	$FeO+Fe_2O_3$	Fe	MnO	$FeO+Fe_2O_3$	Fe	MnO
Initial slag	10,0	100	17,6	3,8	5,4	100	100	100
Concentrate 1	4,2	42	17,85	4,5	5,5	42,6	50,0	42,1
Concentrate 2	3,5	35	14,75	4,2	4,3	29,6	39,5	27,8
Tails	2,3	23	21,3	1,8	6,7	27,8	10,5	29,6

Analysis of data of Table 4 shows that extraction of valuable components when using the screw separator is slightly higher, than at the previous series of experiments. At the same time total extraction made, %: FeO Fe_2O_3 – 72.2; Fe – 89.5; MnO – 70.4. However losses with tails are considerable size that significantly reduces technical and economic indicators of process and can hardly be an effective method of processing of the fulfilled steel-smelting slag [6].

One more method of processing of steel-smelting slag is enrichment on a concentration table.

Pilot studies were conducted on a laboratory single-tier concentration table of LKS – 1Ya. The table is intended for material enrichment by fineness – 3 mm, productivity of a table of 15–20 kg/h; the size of the course of a sound board is regulated in repartitions of 8–16 mm; number of the courses in a minute 275–325; inclination of a sound board from 0 to 100; consumption of water of 0.5 m^3/h.

After establishment of technical indicators of a table began performance of a pilot study.

At first water in the quantity sufficient for a covering was pumped by a thin layer of all surface of a table. Steel-smelting slag moved in a loading box of a table in the form of the pulp received after crushing with the relation: T = 2:1. The quantity firm in a pulp was 1 kg.

Watching formation of a fan on a table, regulated an inclination and an amount of water in a middle and trailer part of a table. The inclination was regulated so that the border of coarse-grained fraction of tails got to the first tail receiver.

Having passed all material, reduced an inclination of a table a little and a brush washed away the material which accumulated on a sound board, without stopping a table. Having washed away all material, the table was stopped, each of the received fractions was dehydrated, dried.

In the Central laboratory JSC "Uzbeksteel" (Uzbekistan) the analysis of the chemical composition of products was made and the maintenance of Femet, FeO, Fe_2O_3 is defined and also, carried out microscopic and UV analysis. Which results are given in Table 5, Fig. 3.

Qualitative quantitative indices of enrichment of the crushed steel-smelting slag on a concentration table are given in Table 5.

Table 5. Average values qualitatively - quantitative indices of division of valuable components of steel-smelting slag on a concentration table

Products	Exit, %		Contents, %		Extraction, %	
	kg	%	$FeO+Fe_2O_3$	Fe	$FeO+Fe_2O_3$	Fe
Initial slag	5,0	100	17,6	3,8	100	100
Concentrate of I	1,46	29,3	37,9	10,2	63,0	78,2
Concentrate of II	1,2	24,0	16,8	1,73	22,8	11,2
Tails	2,34	46,7	5,3	0,9	14,2	10,6

The comparative analysis of indicators of extraction of FeO, Fe_2O_3 was made for the choice of an optimum way of gravitational enrichment of steel-smelting slag, Fe results of which are given in Table 6.

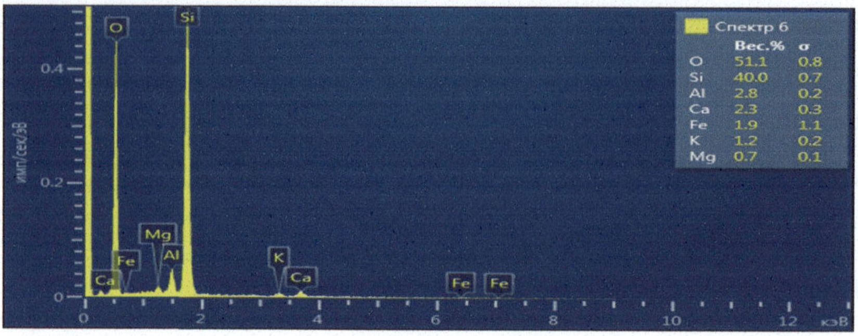

Fig. 3. The electronic and microscopic picture of a concentrate after enrichment on a concentration table

Table 6. The comparative analysis of extraction of useful components from steel-smelting slag by method of gravitational enrichment

Enrichment methods	Extraction valuable components	Extraction, %		
		Concentrate-1	Concentrate-2	Tails
Jigging	$FeO+Fe_2O_3$	42,3	20,0	37,7
	Fe	28,6	12,8	58,6
Screw	$FeO+Fe_2O_3$	38,6	18,6	42,8
	Fe	32,7	21,4	45,9
Concentration table	$FeO+Fe_2O_3$	63,0	22,8	14,2
	Fe	78,2	11,2	10,6

Comparing the received results to results of a jigging it is possible to tell that enrichment on a concentration table is more preferable than other methods of enrichment.

5 Conclusion

By results of the conducted researches it is visible that at enrichment of steel-smelting slag on a concentration table extraction of valuable components makes, %: a concentrate 1 – FeO - Fe2O3 - 63.0; Fe – 78.2; concentrate 2 – FeO - Fe2O3 - 22.8; Fe – 11.2. Content of iron, its oxides and MnO is in the limits quite available to their subsequent use in furnace charge of metallurgical furnaces. They can be used as oxidizers when smelting became, or the ore-fuel of pellets, their metallization and loadings in ars-smelting furnaces can be applied to preparation.

References

1. Matkarimov, S.T., Berdiyarov, B.T., Yusupkhodjayev, A.A.: Technological parameters of the process of producing metallized iron concentrates from poor raw material. Int. J. Innov. Technol. Explor. Eng. **8**(11) (2019)
2. Matkarimov, S.T., Nosirkhudjayev, S.Q.U., Ochildiyev, Q.T., Nuraliyev, O.U.U., Karimdjonov, B.R.: Technological processes of receiving metals in the conditions of moderate temperatures. Int. J. Innov. Technol. Explor. Eng. **8**(12), 1826–1828 (2019)
3. Yusupkhodjayev, A.A., Valiyev, X.R., Ochildiyev, Q.T., Matkarimov, S.T., Nuraliyev, O.U., Jumankulov, A.A.: Development of technology of receiving quality raw materials for smelting of steel from local iron ore fields. Int. J. Adv. Sci. Eng. Technol. **6**(2) (2019)
4. Use of slag of ferrous metallurgy/Under the editorship of V.I. Dovgopo-la. - the 2nd prod., additional - M.: Metallurgy, 168 p. (1978)
5. Matkarimov, S.T.: Extraction iron and its compounds from steel-smelting slags SC 'Uzbeksteel' by using gravitation methods. Eur. Sci. Rev. **11–12**, 67–70 (2018)
6. Matkarimov, S.T., Samadov, A.U., Xoliqulov, D.B.: Extraction iron and its compounds from slags by using gravitation methods. Eur. Sci. Rev. **9–10**, 231–234 (2018)

HSLA Steel - Simulation of Fatigue

Vujadin Aleksić[1(✉)], Bojana Aleksić[2], Ana Prodanović[2], and Ljubica Milović[3]

[1] Institute for Testing of Materials, Bulevar vojvode Mišića 43, 11000 Belgrade, Serbia
vujadin.aleksic@institutims.rs
[2] Faculty of Technology and Metallurgy, Innovation Centre, Belgrade, Serbia
[3] Faculty of Technology and Metallurgy, University of Belgrade, Belgrade, Serbia

Abstract. In present paper, based on the results of the experimental study of the behaviour of the samples made of a HSLA steel, in the form of round smooth specimens (RSS) under LCF conditions (with controlled and completely reversible strain $\Delta\varepsilon/2$ = const, $R_\varepsilon = \varepsilon_{min}/\varepsilon_{max} = -1$, according to ISO 12106: 2003 (E)) and square specimens (SqS) under conditions of HCF conditions (with controlled one-way variable load on tension with σ_{min} = const, according to ISO 12110-1:2013), a computational stress analysis was performed using the FE method and the behaviour of this steel in the fatigue behaviour simulation (LCF and HCF) in SolidWorks programme was shown.

On the basis of the performed analysis of the results of the stress-strain state and the determination of the life using the life isolines for a particular load cycle involving the entire RSS ligament, i.e. SqS for a particular load in a wide range of LCF and HCF loads and comparison with the results of experimental tests, a figure i.e. graphical presentation was given and conclusions were drawn that justify the efforts for numerical assessment of the life cycle of the loaded machine parts.

Keywords: HSLA · LCF · HCF · FEM · Simulation · Ligament

1 Introduction

A large number of structures' damages caused by fatigue lead to catastrophic fractures. It is estimated that such damages represent 50 to 90% of all damages in operation life [1, 2]. For this reason, for more than 160 years extensive studies have been devoted to the study of fractures caused by fatigue load and the conditions of crack initiation and propagation [3, 4].

Despite the fact that more than 10 publications related to fatigue of materials are published every day in the world, the economic impact of fracture of machine parts and structures in the world is significant and accounts for approximately 4% of gross national product (GNP) [2]. Rapid progress in the area of material fatigue can be tracked by various databases that may contain millions of records [2, 5, 6].

I. Karabegović (Ed.): NT 2020, LNNS 128, pp. 314–321, 2020.
https://doi.org/10.1007/978-3-030-46817-0_36

2 ON Nionikral 70 Steel

HSLA steel Nionikral 70 (NN-70) that is discussed in present paper, is a Yugoslav version of the US Steel HY-100 and, thanks to its exceptional mechanical properties, it is designed for manufacture of ship structures, submarines and pressure vessels by welding, requiring toughness which depends on the accumulated strain that in the extreme case may turn into plastic strain [7, 8], which is extremely important. Due to exposure to complex loading with constant cycles during operation, this equipment is susceptible to damage caused by fatigue. Therefore, in order to predict the lifetime of the components of the equipment made of steel NN-70, it is essential to understand the behaviour of the material and the mechanism of damage in fatigue conditions. Tables 1 and 2 show the chemical composition and mechanical properties of the NN-70 steel, respectively.

Table 1. Chemical composition of NN-70 steel, mass % [9–15]

C	Si	Mn	P	S	Cr	Ni	Mo
0.106	0.209	0.220	0.005	0.0172	1.2575	2.361	0.305

Table 2. Mechanical properties of NN-70 steel, 20 °C, [9–15]

Ultimate tensile strength, R_m, MPa	855	
Yield stress, $R_{p\ 0.2}$, MPa	813	
Modulus of elasticity E, GPa	Static	Dynamic
	211.5	221.4

3 On HCF and LCF

In cases of interest in the field of engineering structures exposed to variable voltages, two types of fatigue are distinguished, Fig. 1, high-cycle fatigue (HCF) with a large number of variations to fracture that is lower than the boundary state $R_{p\ 0.2}$ and low-cycle fatigue (LCF) with a small number of cycles to fracture, but within the range of plastic stress.

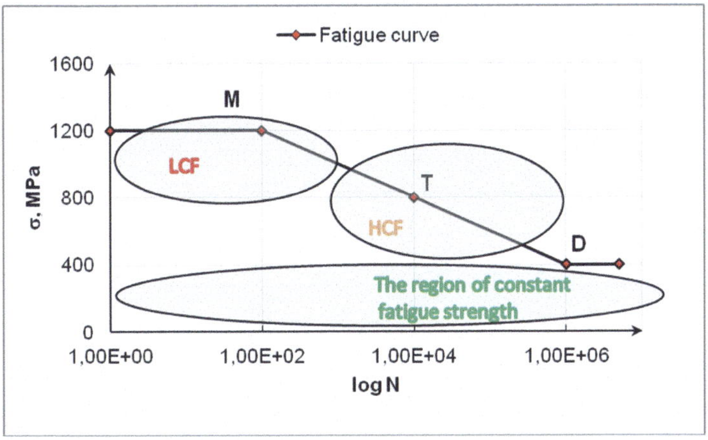

Fig. 1. Regions of low- and high-cycle fatigue of NN-70 steel

On the diagram, Fig. 1, in accordance with the data on the yield point (T), tensile strength (M) and constant endurance limit (D), the regions of LCF and HCF are designated in single logarithmic coordinates, log N vs. σ, at variable tensile stress of NN-70 steel [9].

LCF is the low-frequency fatigue of material, in which the micro-cracks and fractures occurs in the case of repeated plastic strain with the number of cycles to fracture $N = 5 \times 10^4$. LCF is often referred to as statistical durability-stability under repeated static loads. Characteristics of the fatigue process at LCF differ from the characteristics of the fatigue process at HCF for the same load levels. Therefore the assessment of the suitability of the material for longer operation must include two types of testing: at high-frequency fatigue with high frequency (high value of frequency) and at low-cycle fatigue with lower values of frequency.

Fatigue fracture occurs under the influence of cyclically variable stresses, resulting in a progressive crack or cracks initiation, until the remaining cross section (ligament) loses the ability to transfer the load. As a rule, fatigue begins from a stress concentrators such as discontinuity or notch are, which cause a stress state triaxiality in the material. Experience has shown that the time of crack initiation is relatively short, and therefore the structure life is usually determined by the time of the cracks growth, or more precisely, according to the time of crack growth up to the critical crack length.

3.1 Testing of LCF and HCF of NN-70

LCF tests were performed on 10 smooth round specimens, Fig. 2a, made of square samples 11 × 11 × 95 mm from the plate of made of NN-70 steel, with a semi-amplitude of controlled strain, $\Delta\varepsilon/2 = 0.40 - 0.80$.

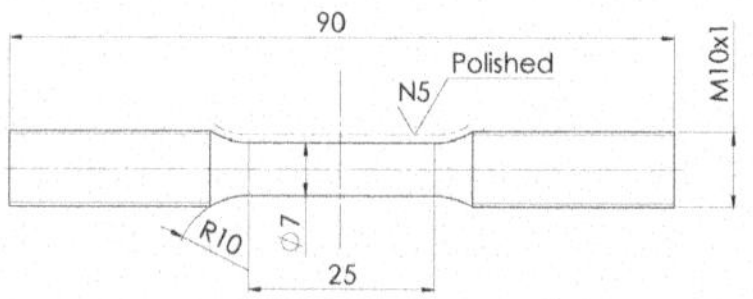

a) Round smooth specimen (RSS) for LCF

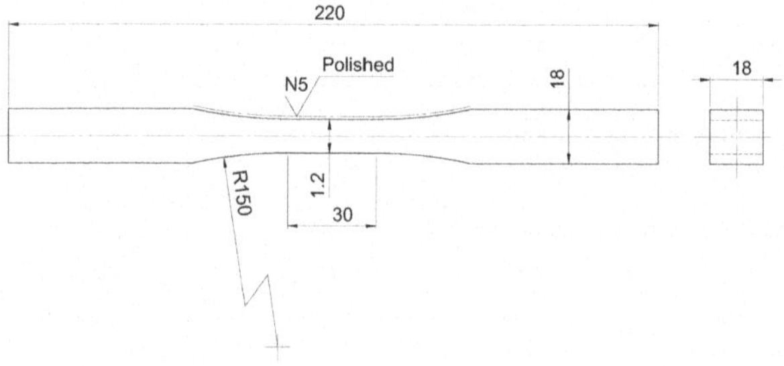

b) Square specimen (SqS) for HCF

Fig. 2. Specimen for testing of LCF - HCF [7, 9, 10]

For testing of HCF, 20 specimens were made with dimensions shown in Fig. 2b. Samples were cut out of 18 mm-thick plate. The specimens were polished on the critical cross-section of longer sides of the squares, and on the shorter sides, the oxide layer originating from rolling and heat treatment of the plate was kept.

The HCF testing was performed at the Laboratory for Machine Materials of the Faculty of Mechanical Engineering in Belgrade on the pulsator ZD 20 Pu of 200 kN, according to ISO 12110-1:2013 [16] standard, and LCF testing according to the standard ISO 12106: 2003 (E) [17] was carried out at the Military Technical Institute in Žarkovo, on the universal MTS system of 500 kN [9–15].

One of the results of individual testing of LCF and HCF of NN-70 steel are the *S-N* curves. The diagrams are connected in a single *S-N* LCF and HCF curve of NN-70 steel, Fig. 3, which was used as one of the input data for computational analysis of static stress using the Finite Element Method (FEM) and fatigue-behaviour simulation (LCF and HCF) of this steel in the SolidWorks programme.

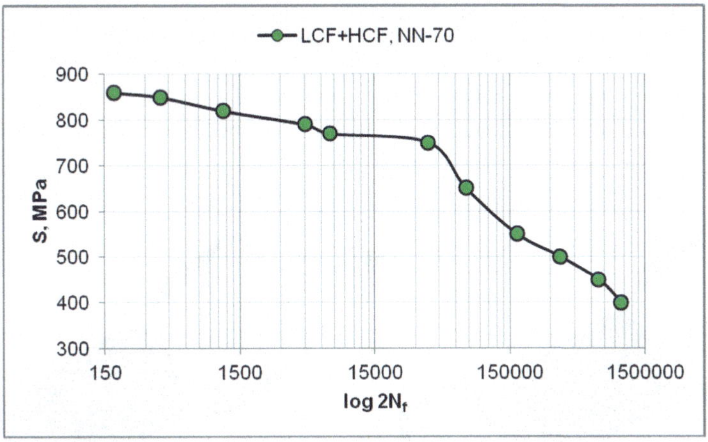

Fig. 3. S-N diagrams of LCF and HCF of steel NN-70, obtained by tests with two types of specimens

3.2 Preparation for Static and Fatigue FEM Calculation in SolidWorks

For the static and fatigue FEM calculation of the specimen models, RSS and SqS, in SolidWorks, the material data obtained by static and LCF-HCF tests, are given in Fig. 4.

Illustration of the results of FEM fatigue testing simulation is shown in Figs. 5 and 6.

In addition the analysis of distribution of von Mises stresses, in static FEM calculation of the SolidWorks programme enables the analysis of the results of calculation of normal stresses, Table 3, as well as the analysis of strain and elongation.

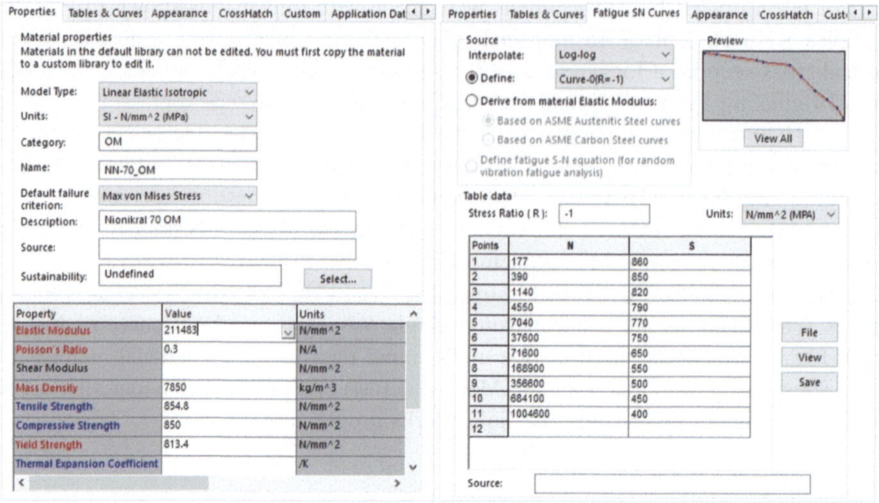

Fig. 4. Input data for FEM calculation of fatigue behaviour of NN-70.

In addition to minimum and maximum number of iso-cross-sections, fatigue calculation gives us a percentage of damage of a specific section of the specimen as well.

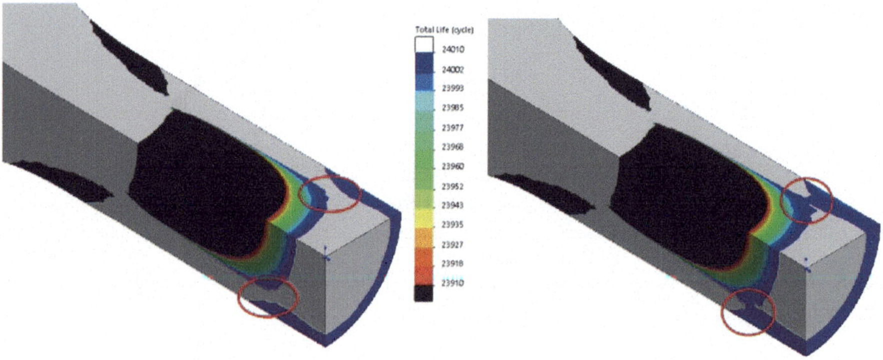

2 cycles before fracture and cycle of fracture

Fig. 5. Determination of fracture cycle of RSS for load of 29 kN, see Table 3.

The methodology for determination of the number of cycles (N_{fSqS} for square and N_{fRSS} for round smooth specimen) at which the fracture occurs for a particular load, is shown in Figs. 5 and 6, and the results of the applied methodology are shown in Table 3. Iso-section of a life cycle that involves the entire ligament of the specimen, which is between minimum and maximum number of the lifetime cycles, is the cycle in which fracture of the specimen occurs.

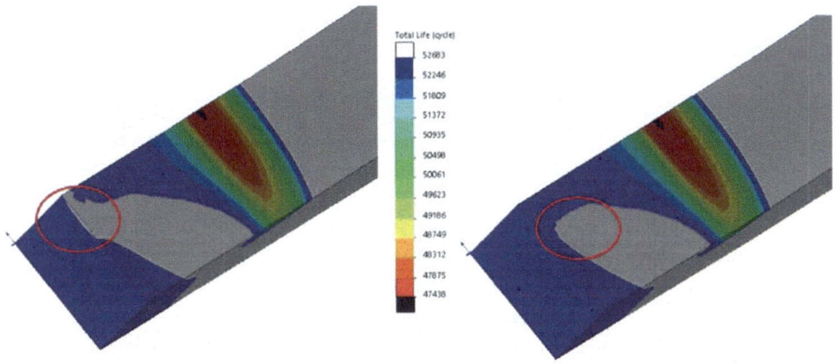

2 cycles before fracture and cycle of fracture

Fig. 6. Determination of fracture cycle of SqS for load of 140 kN, see Table 3

Table 3. The results of static calculation and fatigue simulation of FEM in SolidWorks

Testing				FEM			
Fatigue type	Spec., cross-section, mm^2	N_f	S_{Nf}, MPa	$*N_{fSqS}$	$**N_{fRSS}$	Specimen load	
						RSS**, F, S kN, MPa	SqS*, F, S kN, MPa
LCF	RSS**, 38.5	177	860	177	177		
				177			173.858
					181	33.857	
		390	850	331	329		
				461			170.843
					698	32.832	
		1,140	820	1,063	1,056		
				1,124			165.818
					2,017	31.806	
		4,550	790	4,180	4,173		
				5,014			158.784
					5,482	30.780	
		7,040	770	6,778	6,774		
				15,385			153.759
					24,010	**29.754**	**←Fig. 5**
HCF	SqS*, 201.6	37,600	750	32,732	32,693		
			Fig. 6→	**52,683**		**Fig. 6→**	**140.694**
					59,606	26.676	
		71,600	650	70,989	71,008		
				127,038			117.580
					137,207	22.572	
		168,900	550	167,489	167,228		
				205,029			108.536
					259,963	20.520	
		356,600	500	352,438	352,383		
				509,932			95.471
					533,636	18.468	
		684,100	450	678,365	677,312		
				842,502			85.420
					881,618	16.416	
		1,004,600	400	1,000,001	1,000,001		

Based on the results given in Table 3, the *S-N* diagrams are made that are compared with one another in Fig. 7.

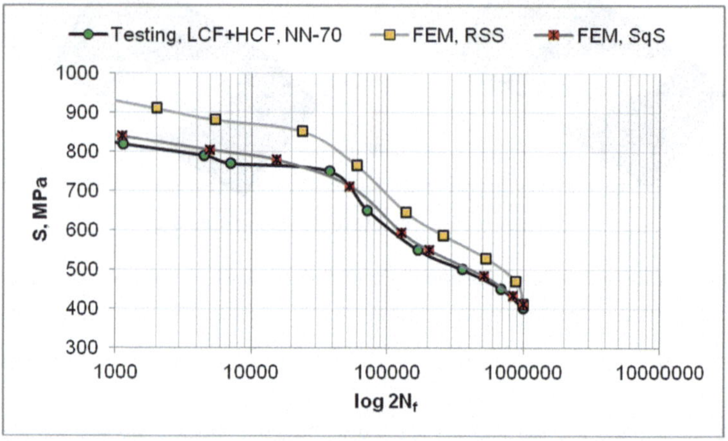

Fig. 7. *S-N* fatigue curves obtained by testing and FEM simulation, for two types of specimens, RSS and SqS

The results obtained show better agreement of the experimental results with the results of the simulation and calculation of the fatigue behaviour of NN-70 steel for SqS than for RSS. An explanation could be in different types of specimen loads: SqS was loaded with unidirectional cyclic load, while RSS was loaded by fully reversed cyclic load.

4 Conclusion

The paper presents the methodologies for processing of the results of the fatigue testing and the results of fatigue simulation and the FEM calculation for two types of test specimens, the RSS for the examination of LCF and SqS for the examination of HCF in order to obtain comparative results of fatigue testing.

The methodology of merging the LCF test results and HCF results in the result of the *S-N* curve of NN-70 steel fatigue curve provided us with input data for the FEM calculation of fatigue behaviour of the NN-70 steel.

The methodology for determination of the number of cycles leading to separation of the specimen parts, i.e. to the specimen fracture, applied in this paper makes it possible to determine the calculation of FEM fracture cycle on the other elements made of NN-70 steel exposed to the effects of fatigue load as well.

Acknowledgements. This work is a contribution to the Ministry of Education and Science of the Republic of Serbia funded Project TR 35011.

References

1. Fuchs, H.O., Stephens, R.I.: Metal Fatigue in Engineering. A Wiley - Interscience Publication, John Wiley and Sons, New York (1980)
2. Tóth, L.: Reliability assessment of cracked structural elements under cyclic loading. In: Carpinteri, A. (ed.) Handbook of Fatigue Crack Propagation in Metallic Structures, vol. 2, pp. 1643–1683. Elsevier (1994)
3. Smith, R.A.: The versailles railway accident of 1842 and the first research into metal fatlgue. In: Kitagawa, H., Tanaka, T. (eds.) Proceedings of 4th International Conference on Fatigue and Fatigue Thresholds, Honolulu, 15–20 July 1990, vol. 4, pp. 2033–2041. Materials and Component Engineering Publications Ltd., Birmingham (1990)
4. Messadi, G.: Great Inventions through History. Chambers (1991). 237 p.
5. Richard, H.A., Sander, M.: Fatigue Crack Growth. Solid Mechanics and Its Applications, vol. 227. Springer, Cham (2016)
6. Berto, F., et al.: Review of the influence of non-singular higher order terms on the stress field of thin welded lap joints and small inclined cracks in plates. Frattura Integr. Strutt. **41**, 260–268 (2017)
7. Zrilić, M., et al.: Static and impact crack properties of a high-strength steel welded joint. Int. J. Press. Vessels Pip. **84**, 139–150 (2007)
8. Das, A., Chowdhury, T., Tarafder, S.: Ductile fracture micro-mechanisms of high strength low alloy steels. Mater. Des. **54**, 1002–1009 (2014)
9. Aleksić, V.: Low cycle fatigue of high strength low alloy steels. D.Sc. thesis, University of Belgrade (2019). (in Serbian)
10. Aleksić, V., et al.: Behaviour of Nionikral-70 in low-cycle fatigue. Struct. Integrity Life **17** (1), 61–73 (2017)
11. Milović, L., et al.: Structural integrity assessment of welded pressure vessel produced of HSLA steel. J. Iron Steel Res. Int. **18**(1–2), 888–892 (2011)
12. Milović, L., et al.: Assessment of the behaviour of fatigue loaded HSLA welded steel joint by applying fracture mechanics parameters. Integritet i vek konstrukcija **12**(3), 175–181 (2012)
13. Milović, L., et al.: Assessment of pressure vessel load capacity in the presence of cracks. Integritet i vek konstrukcija **13**(1), 9–16 (2013)
14. Bulatović, S., et al.: Identification of low cycle fatigue parameters of high strength low-alloy (HSLA) steel at room temperature. Metalurgija **53**(4), 466–468 (2014)
15. Bulatović, S., et al.: Impact of choice of stabilized hysteresis loop on the end result of investigation of high-strength low-alloy (HSLA) steel on low cycle fatigue. Metalurgija **53** (4), 477–480 (2014)
16. ISO 12110-1:2013: Metallic materials - Fatigue testing - Variable amplitude fatigue testing - Part 1: General principles, test method and reporting requirements
17. ISO 12106:2003(E): Metallic materials-fatigue testing-axial-strain-controlled method

Optimization of Polychlorotrifluoroethylene Processing Technology by the Response Surface Methodology

S. V. Kalinichenko[1], Ye. A. Yeriomina[1], Aleksandr I. Burya[1,2], and Predrag Dašić[3,4(✉)]

[1] Dniprovsk State Technical University (DSTU), Street Dniprostrobudivs'ka 2, 51918 Kamianskoe, Ukraine

[2] Academy of Technological Sciences of Ukraine, Street Glushkova 42, 80000 Kiev, Ukraine

[3] High Technical Mechanical School of Professional Studies, 37240 Trstenik, Serbia
dasicp58@gmail.com

[4] SaTCIP Publisher Ltd., 36210 Vrnjačka Banja, Serbia

Abstract. The influence of the processing conditions on the impact elasticity of polychlorotrifluoroethylene was considered in this article by the response surface methodology (RSM) with design of experiment (DoE). Impact elasticity is chosen as a parameter of optimization, since it is the most sensitive indicator to the changes in parameters of materials formation. It is found that hardening behavior of polychlorotrifluoroethylene is the most important factor that affects the parameter of optimization, because it significantly changes the characteristics of finished products since it reduces the degree of crystallinity of polymer. It is showed that it is possible to control different properties of composites based on polychlorotrifluoroethylene by varying processing conditions and knowing the patterns of the impact of degree of crystallinity on the technical properties of polymer. Extreme manufacturing conditions of polychlorotrifluoroethylene and the mathematical model that adequately describes the dependence of impact elasticity on the technology of processing of the material are found.

Keywords: Polychlorotrifluoroethylene (PCTFE) · Mathematical modeling · Response Surface Methodology (RSM) · Design of Experiment (DoE)

1 Introduction

Fluoropolymers occupy an important place in the polymer world because of their unique properties. For example, their chemical resistance to many substances contributed to the introduction of these materials in the nuclear and chemical industries, and their high electrical insulting properties contributed to the introduction in the electrical engineering and electronics. Climatic properties and the lack of aging of fluoropolymers made it possible to use them in construction. Except of the above industries, fluoropolymers are used in aviation and space technologies, energetics and mining branches.

I. Karabegović (Ed.): NT 2020, LNNS 128, pp. 322–330, 2020.
https://doi.org/10.1007/978-3-030-46817-0_37

We can judge about the profitability of the use of fluoropolymer products by numerous examples, however, they have been closed materials over the years, that's why their properties are almost unexplored [1]. For example, it is known from the literature [1–4] that the temperature history of the manufacture of fluoropolymer products significantly affects the thickness, physical and mechanical, tribological and thermodynamic properties. However, the system studies are necessary to find the patterns of influence of technological parameters of formation on the characteristics of fluoropolymers. That was the purpose of the work, because knowing the optimal processing regime and the patterns of influence we can control certain properties of the finished product.

2 Materials and Methods

2.1 Materials

A bright representative of fluoropolymers – the properties of products from them depends significantly on the temperature history – is fluoroplastic-3 (polychlorotrifluoroethylene) (PCTFE or PTFCE), the chemical structure of which is shown in Fig. 1.

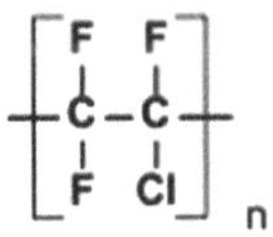

Fig. 1. The chemical structure of polychlorotrifluoroethylene (PCTFE or PTFCE)

Fluoroplastic-3 is a thin fluffy white powder which consists of polymer particles of regular shape in the form of balls with an average size of about 200 μm. It's a crystalline polymer which, due to its asymmetric structure, has a slight tendency to crystallize: 85–90% at slow cooling and 35–40% at hardening.

According to Kaufman [4], the polychlorotrifluoroethylene unit cell has a hexagonal structure with such dimensions: $a = 6.5Å$; $c = 34\ Å$. The chains are packed screw-like, there're 14 monomer units in the period. Melting temperature of the fluoroplastic-3 crystals is 481–483 K. The maximum crystallization rate is observed at 463 K, and the minimum one at 423 K.

Polychlorotrifluoroethylene was formed into the products by compression molding method. The powdered polymer was pelleted at room temperature and pressure 40 MPa. The prepregs were loaded into a mold heated to 423 K, and then the temperature was raised to 510–526 K. After this they were held for 10 min without pressure, then the sample was loaded (40 MPa). To fix the shape of the product, it was cooled under pressure to a temperature of 480–496 K after which it was pushed out of the mold into water or grease for hardening or cooled in air.

The optimization of polychlorotrifluoroethylene processing technology was carried out by the method of mathematical experiment planning. Impact strength (a_n) was chosen as a parameter of the optimization, because it is the most sensitive indicator of changes in the formation of the materials [5, 6]. The determination of impact number was made on the samples with a length of 50 mm and a cross-section of 6 × 4 mm by the method of Charpy (GOST (State Standard) 4647-80) that means a dynamic test of one-time hit for bending using pendulum of the model MK 0,5-1.

2.2 Methods

The investigated process was considered as a functional: $y = f(x_1, x_2, x_3)$, where varying independent factors were the formation temperature (X_1), mold removal temperature (X_2) and cooling medium (X_3). To simplify the calculations, the dosage values of the studied factors were converted into conventional units and set in such way that when converting to a conditional scale they would correspond to −1; 0; +1 according to the formula:

$$x_i = \frac{X_i - X_{i0}}{h} \tag{1}$$

where: x_i is a coded value of the factor, X_i and X_{i0} are upper and main levels of variation of the factors respectively, h is factor variation step.

A reasonable choice of parameters that affect the studied process and the intervals of their variation is the pledge of successful work. Also it should be taken into account that if the variation step of the variable is too small the values of some coefficients of the equation may turn out to be insignificant, that is, to coincide with the measurement error. That's why the variation step was 10 units. The results of the dosage calculations are given in Table 1.

Table 1. Initial data for planning the experiment

Parameter	Symbol		The step of variation	Levels of variation				
				+1,68	+1	0	−1	−1,68
Formation temperature	X_1	$T_{f.}$, K	5	526	523	518	513	510
Extract temperature	X_2	$T_{e.}$, K	5	496	493	488	483	480
Cooling medium	X_3	–	–	Water	Water	Oil	Air	Air

Experiment and mathematical modeling was ralized out on the basis of design of experiment (DoE) and response surface methodology (RSM) which are described in the papers [7–16].

3 Results and Discussion

At the first stage of the study, a full factorial experiment of type 2^3 (Table 2) was performed. It allows to get separate estimates for the coefficients of the regression equation of such form:

$$y = b_0 + b_1x_1 + b_2x_2 + b_3x_3 + b_{12}x_1x_2 + b_{13}x_1x_3 + b_{23}x_2x_3 + b_{123}x_1x_2x_3 \tag{2}$$

where: y is the calculated value of the optimization parameter, b_i and b_{ij} are the coefficients of the regression equation.

The coefficients of the regression equation are found using these formulas:

$$b_i = \frac{1}{N}\sum_{i=1}^{N} \bar{y}_i \tag{3}$$

Table 2. Type 2^3 experiment planning matrix

No. experiment	x_0	x_1	x_2	x_3	x_1x_2	x_1x_3	x_2x_3	$x_1x_2x_3$	$\bar{y}$
1	1	1	1	1	1	1	1	1	54.10
2	1	−1	1	1	−1	−1	1	−1	86.20
3	1	1	−1	1	−1	1	−1	−1	87.30
4	1	−1	−1	1	1	−1	−1	1	101.30
5	1	1	1	−1	1	−1	−1	−1	43.60
6	1	−1	1	−1	−1	1	−1	1	31.35
7	1	1	−1	−1	−1	−1	1	1	14.10
8	1	−1	−1	−1	1	1	1	−1	13.50

As a result of calculations, the following values of the coefficients were obtained: $b_0 = 53.93$, $b_1 = -4.16$, $b_2 = -0.12$, $b_3 = 28.29$, $b_{12} = -0.81$, $b_{13} = -7.37$, $b_{23} = -11.96$, $b_{123} = -3.72$.

The Eq. (2) took this form after substituting the values of the coefficients:

$$\begin{aligned} y = 53.93 - 4.16 \cdot x_1 - 0.12x_2 + 28.29 \cdot x_3 - 0.81 \cdot x_1 \cdot x_2 - 7.37 \cdot x_1 \cdot x_3 \\ -11.96 \cdot x_2 \cdot x_3 - 3.72 \cdot x_1 \cdot x_2 \cdot x_3 \end{aligned} \tag{4}$$

To check the adequacy of the Eq. (4) and determine the coefficient dispersion, it is necessary to determine the error mean square of the experiment S_y^2. Let's find it by the results of six experiments which are set in the center of the plan (Table 3, №. 1-6).

Table 3. The results of the experiments in the center of the plan and in the "star" points

The content of the plan	No. experiment	x_0	x_1	x_2	x_3	x_1^2	x_2^2	x_3^2	$\bar{y}$
Experiments in "star" points	1	1	1.68	0	0	2,83	0	0	43.5
	2	1	−1.68	0	0	2,83	0	0	59.7
	3	1	0	1.68	0	0	2.83	0	39.8
	4	1	0	−1.68	0	0	2.83	0	39.1
	5	1	0	0	1.68	0	0	2.83	88.5
	6	1	0	0	−1.68	0	0	2.83	6.46

(continued)

Table 3. (*continued*)

The content of the plan	No. experiment	x_0	x_1	x_2	x_3	x_1^2	x_2^2	x_3^2	$\bar{y}$
Experiments in the center of the plan	7	1	0	0	0	0	0	0	15.5
	8	1	0	0	0	0	0	0	11.3
	9	1	0	0	0	0	0	0	10.4
	10	1	0	0	0	0	0	0	10.4
	11	1	0	0	0	0	0	0	14.2
	12	1	0	0	0	0	0	0	15.5

There is the simple average of the optimization parameter $\bar{y}$ in the center of the plan:

$$\bar{y}_0 = \frac{1}{N_0}\sum_{i=1}^{N_0} y_i = 12.88 \tag{5}$$

where: y_i is the value of the optimization parameter in the u-th experiment, $\bar{y}_0$ is the simple average of the optimization parameter in N_O experiments, N_O is the quantity of the experiments with zero coordinates in the center of the plan, u is a parallel experiment number in the center of the plan.

The error mean square S_y^2 of the experiment is calculated by this formula:

$$S_y^2 = \frac{S_E}{N_0 - 1} = \frac{30.3}{6 - 1} = 6.06. \tag{6}$$

where: S_E is the sum of squares in the center of the plan which is calculated by the formula:

$$S_E = \sum_{i=1}^{N_0} (y_i - \bar{y}_0) = 30.3. \tag{7}$$

The number of degrees of freedom is connected with the error mean square of the experiment: $df_O = N_0 - 1 = 5$.

The obtained difference is many times greater than the measurement error of the experiment S_y:

$$S_y = \sqrt{S_y^2} = \sqrt{6.06} = 2.46. \tag{8}$$

There's the difference between the value of the optimization parameter $\bar{y}$ in the center of the plan and the value of the free term b_O:

$$|\bar{y}_0 - b_0| = 41.05. \tag{9}$$

It appears from this that the coefficients for the quadratic terms of the equation are significantly different from zero and the studied dependence cannot be approximated with sufficient accuracy by the Eq. (4). Since the description of the response surface by a first-order polynomial turned out to be insufficient, we moved on to planning a second-order one. The measurements were carried out at three levels of each of the parameters. Under these conditions, the full factorial experiment contains too large number of experiments which is equivalent to 3^3. In order to reduce the number of experiments, let's use so-called composite or sequential plan developed by Box and Wilson [9]: rotatable central composite design of the experiment of the second order of type 3^2 [7, 9].

For three factors it consists of an implemented plan of a full factorial experiment of type 2^3 (Table 2) and additionally contains research in the "star points" (Table 3, №. 1-6) and in the center of the plan (№. 7-12).

According to the accepted plan of the experiment, in general $N = N_c + 2{\cdot}k + N_0 = 20$ experiments were conducted, where N_c – is the number of experiments at the core of the plan, k is the number of factors. Each experiment was repeated twice ($m = 2$) in a randomized order to exclude systematic errors. The averaged experimental data of the experiments are presented in Tables 2 and 3.

A mathematical description of this process was proposed to be sought in the form of the regression equation which determined the dependence of the studied process on the formation and mold removal temperatures and cooling medium and was presented in the form of a second-order polynomial:

$$y_i = b_0 + \sum_{i=1}^{n} b_i \cdot x_i + \sum_{i<j}^{n} b_{i,j} \cdot x_i \cdot x_j + \sum_{i=1}^{n} b_{ii}^2 \cdot x_i^2. \tag{10}$$

On the basis of the factorial experiment, the regression coefficients were recalculated according to the formulas which are described in the papers [10, 14–17].

After calculating all the coefficients the regression equation takes the form:

$$\begin{aligned} y = 12.41 &- 4.43x_1 + 0.017x_2 + 26.67x_3 - 0.8x_1 \cdot x_2 - 7.37x_1 \cdot x_3 - 12x_2 \cdot x_3 \\ &-3.7x_1 \cdot x_2 \cdot x_3 + 14.9x_1 \cdot x_1 + 10.6x_2 \cdot x_2 + 13.44x_3 \cdot x_3. \end{aligned} \tag{11}$$

The dispersions of the coefficients of the equations were found using these formulas:

$$\begin{aligned} S_{b_0}^2 &= \tfrac{2 \cdot A \cdot \lambda^2 \cdot (k+2)}{N} \cdot S_y^2 = 1.0049, \\ S_{bi}^2 &= \tfrac{c^2}{N} \cdot S_y^2 = 0.4431, \\ S_{bij}^2 &= \tfrac{c^2}{\lambda \cdot N} \cdot S_y^2 = 0.7567, \\ S_{b_{ii}}^2 &= \tfrac{A \cdot c^2 \cdot [(k+1) \cdot \lambda - (k-1)]}{N} \cdot S_y^2 = 0.4171. \end{aligned} \tag{12}$$

The check of the static significance of the coefficients in the regression equation b_0, b_1, b_2, b_3, b_{12}, b_{13}, b_{23}, b_{123}, b_{11}, b_{22}, b_{33} was carried out on the basis of the calculation of confidence intervals taking into account error mean square that characterized the error in determining the coefficients of the equation. A confidence interval was calculated by Student's t-test (t_{cr} = 2.015) [18] which is chosen according to the degrees of freedom (df = 5) and significance level (α = 5%). For the rotatable central composite design of the experiment confidence intervals are calculated by the formulas:

$$\begin{aligned}\varDelta b_0 &= \pm t_{cr} \cdot S^2_{b_0} = 2.58,\\ \varDelta b_i &= \pm t_{cr} \cdot S^2_{b_i} = 1.71,\\ \varDelta b_{ij} &= \pm t_{cr} \cdot S^2_{b_{ij}} = 2.24,\\ \varDelta b_{ii} &= \pm t_{cr} \cdot S^2_{b_i i} = 1.66.\end{aligned} \tag{13}$$

It is considered that the regression coefficient is significant if the condition $\Delta b_i < |b_i|$ is met.

The coefficients of the Eq. (11), except of b_2 and b_{12}, are statistically significant, that's why the equation which describes the studied process will have this form:

$$\begin{aligned}y = 12.4 - 4.43x_1 + 26.67x_3 - 7.37x_1 \cdot x_3 - 12x_2 \cdot x_3 - 3.72x_1 \cdot x_2 \cdot x_3\\ + 14.9x_1 \cdot x_1 + 10.6x_2 \cdot x_2 + 13.44x_3 \cdot x_3.\end{aligned} \tag{14}$$

To determine the adequacy of the mathematical description (11) after calculating the regression coefficients, the degree of accordance of the obtained model to the theoretical form of the connection between the studied and initial parameters was checked. For this purpose, the variance of the adequacy was determined by the formula:

$$S^2_{ad} = \frac{S_R - S_E}{N - B - (N_0 - 1)} = 41.158, \tag{15}$$

where B is the number of significant coefficients of the equation, S_R is the residual sum of squares that is calculated by the formula:

$$S_R = \sum_{i=1}^{N} (y_i - \bar{y}_i) = 194.9. \tag{16}$$

The number of degrees of freedom is connected with it: df_{ad} = N, Let's find the estimated value of the Fisher's ratio test (F_{cal}) that is the ratio of variance of adequacy S^2_{ad} to error mean square S^2_z and is calculated by the formula:

$$F_{cal} = \frac{S^2_{ad}}{S^2_y} = 6.7989. \tag{17}$$

As a significance level $\alpha = 2.5$ and the degrees of freedom for the numerator is 6 and for the denominator is 5, the table value $F_t = 6.9777$ [18]. The values $F_{cal} < F_{t.}$, that's why the model (14) describes the studied process adequately.

Analyzing the obtained mathematical model in the studied range of variation of factors, we can see clearly that that the cooling medium most affects the impact strength of polychlorotrifluoroethylene. It should be noted that the optimization parameter reaches the maximum values ($a_n = 98.05$) under the following processing modes:: processing temperature is 513 K, the temperature of removal of material from the mold is 483 K, cooling medium is water. Minimum values ($a_n = 5.0$) are reached when the processing temperature is 515,5 K, temperature of removal of material from the mold is 485,5 K, cooling medium is air.

4 Conclusions

The effect of processing regimes on impact strength of polychlorotrifluoroethylene is studied. The factors that most affect optimization parameter are determined by the method of mathematical experiment planning. Extreme regimes of manufacturing polychlorotrifluoroethylene and the mathematical model that adequately describes the dependence of the impact strength on the technology of processing the material are found.

References

1. Loginov, B.A.: Wonderful world of fluoropolymers. 2nd ed, Moscow (2009). (in Russian)
2. Pugachev, A.K., Pirog, O.A., Melnikova, K.P., Yu Syityiy, V.: Composite materials based on fluoroplastics. Himiya, Lviv (1980). (in Russian)
3. Pashin, Y.A., Malkevich, S.G., Dunaevskaya, T.S.: Fluoroplastic. Himiya, Lviv (1987). (in Russian)
4. Goryainov, A.V., Bozhkov, G.K., Tihontsov, M.S.: Fluoroplastic in mechanical engineering, Moscow (2009. (in Russian)
5. Burya, A.I., Ye Yeriomina, A., Qian, X.Y., Feng, X.M.: The influence of carbonyl nickel content on thermophysical properties of metal polymer materials based on aromatic polyamide phenylon. Adv. Mater. Res. **1004–1005**, 509–512 (2014)
6. Burya, A.I., Ye Yeriomina, A.: The effect of various metallic filling materials on the wear resistance of aromatic polyamide-based composite materials. J. Frict. Wear **37**(2), 151–154 (2016)
7. Box, G.E.P., Wilson, K.B.: On the experimental attainment of optimum conditions. J. Roy. Stat. Soc. Ser. B (Methodological) **13**(1), 1–45 (1951)
8. Dašić, P.: Comparative analysis of different regression models of the surface roughness in finishing turning of hardened steel with mixed ceramic cutting tools. J. Res. Dev. Mech. Ind. **5**(2), 101–180 (2013)
9. Dašić, P.: Research of processed surface roughness for turning hardened steel by means of ceramic cutting tools. In: Proceedings of the 2nd World Tribology Congress (WTC-2001), Vienna, Austria, 3–7 September 2001, Österreichische Tribologische Gesellschaft – The Austrian Tribology Society (ÖTG), Vienna (2001). Article no. M-51-29-055

10. Dašić, P.: Response Surface Methodology: Selected Scientific-Professional Papers, p. 306. SaTCIP Publisher Ltd., Vrnjačka Banja (2019). (in Serbian)
11. Dašić, P.: The choice of regression equation in fields metalworking. In: Proceedings of the 3rd International Conference Research and Development in Mechanical Industry (RaDMI-2003), Herceg Novi, Serbia and Montenegro, 19–23 September 2003, pp. 147-158 (2003)
12. Flaig, J.J.: A new classification of variables in design of experiments. Qual. Technol. Quant. Manage. **3**(1), 103–110 (2006)
13. Milenković, D., Dašić, P., Nedeff, V., Moşnegutu, E.: Analysis of regression models of maximal adsorption on high-porosity adsorbents with and without ultrasound application. Rev. Chim. **60**(7), 706–710 (2009)
14. Montgomery, D.C.: Design and Analysis of Experiments, 8th edn. Wiley, Hoboken (2012)
15. Myers, R.H., Montgomery, D.C., Vining, G.G., Borror, C.M., Kowalski, S.M.: Response surface methodology: a retrospective and literature review. Qual. Technol. Quant. Manage. **36**(1), 53–77 (2004)
16. Spiridonov, A.A., Vasilev, N.G.: Planning an experiment, Sverdlovsk (1975). (in Russian)
17. Burya A.I., Yeriomina Y.A., Volokh V.I., Dašić, P.: Study of the effect of transducer thickness and direction on the coercive force magnitude. In: Karabegović, I. (ed.) New Technologies, Development and Application II. NT 2019. LNNS, vol. 76, pp. 229–237. Springer, Cham (2019)
18. Bolshev, L.N., Smirnov, N.V.: Tables of mathematical statistics. Nauka, Moscow (1983). (in Russian)

Procedure for Determining the Wind Tunnel Blockage Correction Factor

Ernad Bešlagić[1], Samir Lemeš[2](✉), and Fuad Hadžikadunić[1]

[1] Mechanical Engineering Faculty, University of Zenica, Zenica, Bosnia and Herzegovina
[2] Polytechnic Faculty, University of Zenica, Fakultetska 1, 72000 Zenica, Bosnia and Herzegovina
slemes@unze.ba

Abstract. Before launching a new wind turbine, it is necessary to determine its output performance. The performance can be achieved either by testing the complete wind energy converter in actual operating conditions in the natural environment or by testing the wind turbine in a wind tunnel. In addition to the many advantages offered by wind tunnel testing, there are also some disadvantages. One of them is the appearance of a blockage effect that occurs when wind turbines are tested in a closed test section of a wind tunnel.

The paper briefly explains the existing methods for determining the correction factor which corrects the measured wind turbine output in a wind tunnel in order to eliminate the blockage effect. This paper is part of a project aimed at determining the blockage correction factor for helicoidal Darrieus wind turbines. This stage of the project consists of an experimental design for testing which will be carried out in an open wind tunnel with a closed test section, constructed from scratch at the university laboratory.

Keywords: Wind tunnel · Vertical axis wind turbine · Blockage effect

1 Introduction

Mitigation of climate changes, caused primarily by the overuse of fossil fuels, including the exploration of the renewable energy sources, has become one of the greatest challenges of the century. Various designs of wind turbines became one of the prevailing sustainable energy sources. The wind turbines convert the wind kinetic energy into rotational energy of the generator which produces electrical energy. Therefore, the major challenge the industry is facing is to design and manufacture the wind turbine sets as efficient and cost-effective as possible.

A wind turbine rotor is the first element in a system by which the wind energy is converted into the electricity. The rotor's ability to convert the maximum amount of wind energy passing through the rotor into mechanical energy is a direct result of its aerodynamic properties, which largely determine the overall efficiency of the entire wind turbine.

Wind turbine efficiency can be increased either by improving aerodynamic efficiency or by increasing its dimensions. The most common wind turbine design

I. Karabegović (Ed.): NT 2020, LNNS 128, pp. 331–339, 2020.
https://doi.org/10.1007/978-3-030-46817-0_38

nowadays has three blades rotating around the horizontal axis. The aerodynamic efficiency of this design has almost reached its maximum. Therefore, a further increase in the efficiency of the horizontal wind turbine can be obtained only by an increase in its dimensions, which makes the processes of production, transportation, installation and maintenance of such wind energy converter harder.

As the number of large wind turbines and large wind farms increases around the world, the recent years brought more attention to small wind turbines that can provide independent power supply for telecommunication, monitoring or transporting equipment. Wind turbines with a vertical axis of rotation are more common in these applications. This type of wind turbine has certain advantages (wind direction independence, easier installation and maintenance) which, in most cases, are not enough to compensate the basic disadvantage to horizontal wind turbines. The lower utilisation factor of wind energy compared to horizontal wind turbines is the main reason why this type of wind turbine is currently less represented.

Increasing the wind energy utilisation factor, among other things, requires improving the aerodynamic characteristics of the wind turbine. The aerodynamic design of the rotor requires knowledge much more beyond the basic physical laws of energy conversion. The designer needs to find a relation between the actual rotor geometry, e.g. the shape, size, number of blades, or the blade air profile shape in one hand, and its aerodynamic properties in the other. The designer can determine this relation by using analytical methods, numerical simulations and experimental researches. Although the results of computer simulations are becoming more accurate and reliable, many believe that Computational Fluid Dynamics (CFD) will never be able to completely eliminate the need for experimental verification during the development of these products. If possible, the most successful approach to developing the aerodynamic design of a new rotor or redesigning an existing rotor includes the optimum combination of experimental, analytical and computational results.

When a new wind turbine design is developed, its characteristics must be checked before releasing it to the market. There are two possible options. The first one involves installing a complete wind turbine set in its operating environment and directly measuring the wind turbine output characteristics. The second method involves experimental testing in wind tunnels, measuring the desired rotor information from models or prototypes using special measuring equipment. The wind tunnel is a basic tool of experimental aerodynamics that enables the study of the wind flow around the test object, determining the forces acting on it and its interaction with the wind flow.

Testing the wind turbine rotor performance in a wind tunnel has an advantage over the field measurements due to much shorter time and less costs. In addition, the test parameters in a wind tunnels can be precisely controlled, allowing experiments to be performed at different wind speeds and with low turbulence intensities [1].

Testing of wind turbine characteristics in wind tunnels also has some disadvantages. The first disadvantage is the inability to test the real prototypes due to the large dimensions of the wind turbine rotor. Instead of real scale prototype testing, the reduced models are used. In order to get the valid model tests, a geometric, kinematic and dynamic similarity between the flow around the model and the flow around a real wind turbine rotor is required. The model has to be an exact replica of the actual rotor, and the dimensionless parameters has to be the same for both the real-time atmospheric conditions flow and the flow in a wind tunnel. The Reynolds number is a dimensionless parameter relevant to incompressible subsonic tests in wind tunnels [2].

Another disadvantage is the limited size of the wind tunnel test section. The test section walls cause the flow conditions in the tunnel to differ from those present in a free flow. Therefore, it is necessary to make certain corrections to the measured output characteristics when research is performed in wind tunnels. One of these corrections takes into account the influence of the obstacle on the flow velocity causing the wake appearance in the air flow.

2 Blockage Effects

When a rotor is tested in a wind tunnel, the test section walls are usually closer than any obstacles present in actual operating conditions in the natural environment. This causes the blockage effect, having two causes.

The first causative factor is the reduction of air passage space, caused by the insertion of a wind turbine model into the test section. This part of the blockage effect is called the solid blockage. The test rotor is a physical obstacle to flow and there is a temporary increase in the flow velocity near the rotor, where there is a decreased space for air flow. Moving away from the test rotor, the solid blockage decreases, and the flow velocity returns to its initial value. The solid blockage is directly related to the ratio between the wind turbine model projection surface, and the cross-sectional area of the wind tunnel test section perpendicular to the air flow direction (blockage ratio B_R).

The test section walls affect the smooth development of backward vortices that extend several diameters behind the wind turbine, causing the wake blockage effect. The size of the wake is directly related to the resistance which counteracts the wind turbine to the flow of air, i.e., to the energy taken from the air stream. The higher the wake behind the wind turbine, the greater the wake effect of the blockage. As the wake blockage increases the wind velocity in the wind tunnels with closed test section, it has to be considered.

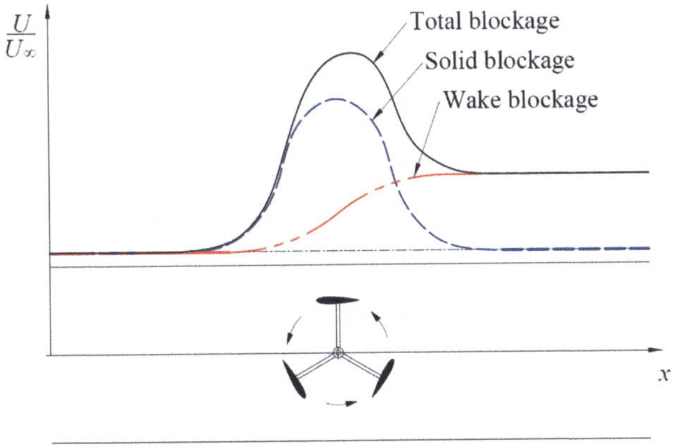

Fig. 1. Blockage effects in a wind tunnel test

Figure 1 shows the individual and cumulative impact of the two causes of the blockage effect on the airflow velocity along the test section of the air tunnel [1]. It can be concluded from the figure that the solid blockage effect influences the flow velocity more in the zone around the wind turbine itself, while the wake blockage influences the flow velocity more in the zone downstream the wind turbine.

3 Methods for the Blockage Effect Corrections

Previous studies have shown that the overall blockage effect depends most on the blockage ratio and the power produced by the rotor (which depends on the air flow velocity, the velocity of the rotor, the number of blades, the inclination angle of the blades, etc.).

In all cases where the blockage ratio B_R exceeds 10%, the results obtained must be corrected by the blockage correction factor. In this case, the correction factor varies depending on the shape and characteristics of the test rotor, and it is necessary to carry out tests in the wind tunnel for each specific shape of the wind turbine rotor, in order to determine the blockage correction factor. The blockage correction factor should be applied to the overall velocity increase due to the vortex and solid blockage effect.

3.1 Basic Parameters for the Wind Turbine Characteristics Description

Blockage ratio (B_R) is defined as:

$$B_R = \frac{A_s}{A_t} \tag{1}$$

where A_s is the projection surface of the wind turbine model, while A_t is the cross-sectional area of the air tunnel test section perpendicular to the air flow direction.

The type speed ratio (*TSR*) is the ratio between the peripheral velocity of the blade tip caused by the wind turbine rotational motion and the wind velocity:

$$\lambda = \frac{\omega R}{U_\infty} = \frac{\pi n R}{30 U_\infty} \tag{2}$$

where ω is the wind turbine angular velocity, R is the wind turbine radius and U_∞ is the free-wind velocity. This ratio is very important because it is directly related to the amount of energy that the wind turbine generates at some wind velocity. For each wind turbine, there is an optimal *TSR* at which the maximum portion of the kinetic wind energy is converted into mechanical energy and at which the maximum power factor is achieved:

$$C_P = \frac{T\omega}{\frac{1}{2}\rho A_s U_\infty^3} \tag{3}$$

where T is the torque and ρ is the air density.

The drag coefficient is defined by the expression:

$$C_D = \frac{F_D}{\frac{1}{2}\rho A_s U_\infty^2} \tag{4}$$

where F_D is the drag force.

3.2 Existing Methods for Determining the Blockage Correction Factor

There are several different methods for correcting the blockage effect. *Pope and Harper* [3] have suggested that the effect of the blockade is corrected by multiplying the wind velocity by a correction factor equal to 1/4 of the blockage ratio B_R. *Bahaj and others* [4] recommend a method based on an actuator disc model. This method is used to correct the output of the wind turbine power, and the *TSR* using a blockage correction factor equal to the ratio between the wind velocities in the blank test section, and the test section with the wind turbine rotor. *Maskell* [5] measured the drag coefficient and the base pressures at four different square planes perpendicular to the flow direction and developed a new semi-empirical factor that serves to correct the blockage effect. *Alexander* [6] and *Alexander and Holownia* [7] have successfully applied the method proposed by *Maskell* [5] to the Savonius turbine. Using the method recommended by *Maskell* [5], *Jeong et al.* [1] proposed an expression for the blockage correction coefficient for a Darrieus flat-blade turbine, the so-called Giromill wind turbine. This term is applicable where $B_R < 30\%$.

These studies have shown that the correction factor varies depending on the shape and characteristics of the sample tested. This supports the assertion that it is necessary to carry out wind tunnel tests for each specific sample shape, in order to determine the correction factor due to the blockage [1].

4 Determining the Blockage Effect Correction Factor

4.1 Tests with Different B_R Values

Wind turbine output performance can be determined by directly measuring the electricity on the generator. This method is suitable in the final development stage when all wind turbine components have been developed and assembled. To determine the mechanical efficiency of the wind turbine itself, it is more appropriate to measure the angular velocity and torque on the rotor shaft and to determine the power coefficient from the obtained results and expressions (3).

In order to determine the correction factor of the blockage effect during wind tunnel testing, the wind turbine output performances should be determined for different values of the surface ratio B_R. For at least three B_R values, tests are performed at different wind velocities and different *TSR* velocity ratios. In this case, one of the tests must be performed with $B_R < 10\%$. This way, the reference values of the wind turbine output parameters are determined without influence of the blockage effect.

It would be ideal to test the same wind turbine prototype in wind tunnels with different cross section sizes of the test section and thus eliminate the influence of any parameter other than B_R on the output performance (Fig. 2a). In this case:

$$B_{R_i} = \frac{A_s}{A_{t_i}}, \quad i = 1\ldots n \tag{5}$$

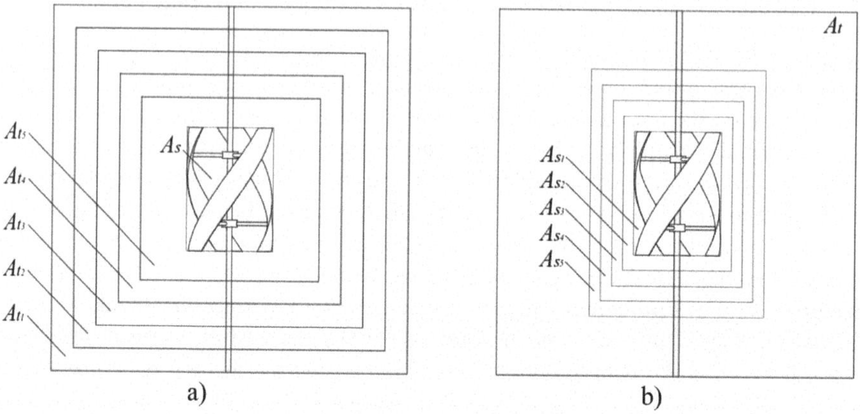

Fig. 2. Change of BR by resizing: (a) test sections; (b) wind turbines

When only one wind tunnel is available, different values of B_R can only be achieved by testing different sizes of wind turbine (Fig. 2b). The B_R values are determined according to the following expression:

$$B_{R_i} = \frac{A_{s_i}}{A_t}, \quad i = 1\ldots n \tag{6}$$

In such cases, it is crucial to achieve geometric, kinematic and dynamic similarity of the flows for all different models and flows around the real wind turbine rotor. This means that the model must be an exact replica of the actual rotor and the particular dimensionless parameters must be the same for both real atmospheric and wind tunnel conditions.

4.2 Determination of Dimensionless Parameters for Aerodynamic Similarity

Knowing that torque and resistance force must be determined during testing, it is necessary to apply the Buckingham π procedure on these two parameters in order to determine the dimensionless parameters that must be satisfied in a wind turbine prototypes caling [8].

For this purpose, torque and resistance force are expressed as functions of independent parameters:

$$T = f(U_\infty, \rho, \mu, \omega, c, D, H) \tag{7}$$

$$F_D = f(U_\infty, \rho, \mu, \omega, c, D, H) \tag{8}$$

In the expressions (7) and (8), c is the chord length of the wind turbine blade aerodynamic profile, D is the diameter and H is the height of the wind turbine (Fig. 3).

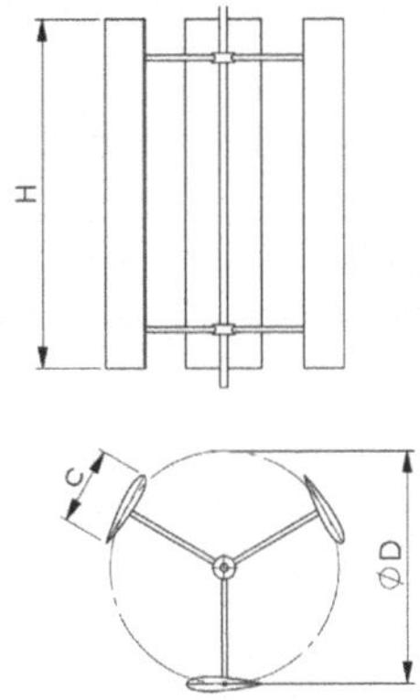

Fig. 3. Basic geometric parameters of a wind turbine

The Buckingham π procedure brings that, in order to achieve geometric similarity to the wind turbine prototype, the following two parameters must correspond:

$$\frac{c}{D} \text{ and } \frac{H}{D} \tag{9}$$

In order to achieve the dynamic similarity between wind turbine models, the following two parameters must correspond:

$$\frac{U_\infty}{\omega D} \text{ and } \frac{\rho \omega D^2}{\mu} \tag{10}$$

Since it can be assumed that the density ρ and the dynamic viscosity μ of the air in all tests will be the same, these two parameters can be excluded from the above expression.

The wind tunnel used for this experiment has a 0.64 m^2 square test section. The testing will be performed on five wind turbine models whose geometrical characteristics are given in Table 1. In determining the wind turbine dimensions, the attention was paid to satisfy the geometric similarity conditions defined by the expression (9).

Table 1. Geometric characteristics of five wind turbine models

	I	II	III	IV	V
Diameter, D [mm]	180	225	270	315	360
Height, H [mm]	270	337,5	405	472,5	540
Chord, c [mm]	60	75	90	105	120
Ratio B_R [%]	7,59	11,87	17,09	23,25	30,38

The reference wind turbine model has surface ratio 7.59%. The tests will be performed at wind velocities in the range of 5 to 8 m/s and *TSR* velocity ratios in the range from 0.2 to 2.2. The *TSR* determines the wind turbine angular velocity, which has to be strictly controlled to satisfy the dynamic conditions of flow similarity for all five models.

Based on these given ranges and expressions (10), the values shown in Table 2 have been created and can be considered as part of the design of experiment.

Table 2. Values of test parameters for air tunnel tests

B_{R1} 7,59								B_{R2} 11,87	B_{R3} 17,09	B_{R4} 23,25	B_{R5} 30,38
U_{∞_1} [m/s]	TSR	0,2	0,6	1,0	1,4	1,8	2,2	U_{∞_2} [m/s]	U_{∞_3} [m/s]	U_{∞_4} [m/s]	U_{∞_5} [m/s]
5,00	n $[\mathrm{min}^{-1}]$	106,10	318,31	530,52	742,72	954,93	1167,14	6,25	7,50	8,75	10,00
6,00		127,32	381,97	636,62	891,27	1145,92	1400,56	7,50	9,00	10,50	12,00
7,00		148,54	445,63	742,72	1039,81	1336,90	1633,99	8,75	10,50	12,25	14,00
8,00		169,77	509,30	848,83	1188,36	1527,89	1867,42	10,00	12,00	14,00	16,00

Table 2 contains the values of the wind velocities that must be achieved in order to satisfy the criteria of dynamic similarity of the flow, compared to the reference wind turbine model. For all models, the same rotation numbers should be achieved in order to achieve the same *TSR* values as for the reference model.

5 Conclusion

This paper reveals the first phase of testing of helicoid Darrieus wind turbines, used to determine the correction factor of the blockage effect that occurs in closed test sections of a wind tunnel. To ensure the geometric and dynamic similarity, the testing parameters for the five wind turbine sizes were determined for the desired range of wind velocities and *TSRs*. The experiment will be carried out in one wind tunnel, therefore special attention must be paid to satisfy the conditions of aerodynamic similarity.

The values obtained reveal that, at *TSR* values greater than 1.4, the turbine angular velocity exceeds the value of 1200 rpm. This information should be kept in mind when designing the rotor. The wind turbine blades will be produced with 3D printing technology. The previous tests have shown that there is a high possibility of bursting the 3D printed rotor bladesat higher rotational velocities.

Previous practice in determining the correction factor of the blockage effect involved testing one rotor in multiple wind tunnels with different test section sizes in order to obtain different B_R ratios. This paper uses an inverse concept which tests the wind turbines of different size in a single wind tunnel. The numerical and experimental validation of this concept will follow in the next steps of the project.

References

1. Jeong, H., Lee, S., Kwon, S.-D.: Blockage corrections for wind tunnel tests conducted on a Darrieus wind turbine. J. Wind Eng. Ind. Aerodyn. **179**, 229–239 (2018)
2. Tourn, S., et al.: Characterization of a new open jet wind tunnel to optimize and test vertical axis wind turbines. J. Renew. Sustain. Energy **9**(3), 033302 (2017)
3. Pope, A., Harper, J.J.: Low-Speed Wind Tunnel Testing. Wiley, New York (1966)
4. Bahaj, A.S., et al.: Power and thrust measurements of marine current turbines under various hydrodynamic flow conditions in a cavitation tunnel and a towing tank. Renew. Energy **32**(3), 407–426 (2007)
5. Maskell, E.C.: A theory on the blockage effects on bluff bodies and stalled wings in a closed wind tunnel. Aeronautical Research Council, R&M. 3400 (1963)
6. Alexander, A.J.: Wind tunnel corrections for Savonius rotors. In: 2nd International Symposium on Wind Energy Systems, Amsterdam, Netherlands (1978)
7. Alexander, A.J.: Wind tunnel tests on a Savonius rotor. J. Wind Eng. Ind. Aerodyn. **3**(4), 343–351 (1978)
8. Wright, C.R., Hansen, M.C.: Optimization of helical turbine in low head applications. Master thesis, The University of Utah, USA (2011)

Effect of Rear Spoiler Shape on Car Aerodynamics and Stability

Nadica Stojanović[1(✉)], Danijela Miloradović[1], Oday I. Abdullah[2,3], Ivan Grujić[1], and Saša Vasiljević[4]

[1] Faculty of Engineering, Department for Motor Vehicles and Motors, University of Kragujevac, SestreJanjić 6, 34000 Kragujevac, Serbia
nadica.stojanovic@kg.ac.rs
[2] College of Engineering, University of Baghdad, Baghdad, Iraq
[3] Hamburg University of Technology, Hamburg, Germany
[4] Technical College of Applied Studies in Kragujevac, Kragujevac, Serbia

Abstract. Computational dynamics is widely used when designing cars in order to obtain the optimum design and reduce the time and costs. Car design plays the main key in decision on purchasing the car. Car manufacturers have a very difficult task which is build a car that will be attracted potential buyers with its design, which does not deteriorate the car's aerodynamics or has the stability at high level of speeds. In order to enhance the car's stability, it's necessary to add the rear spoilers to the car body. This paper presents deep study for effects of fitting different types of rear spoilers on the car's aerodynamics and stability. Determination of corresponding aerodynamic drag and lift coefficients was performed using ANSYS software 14.5/Fluid Flow CFX module. It was found that the aerodynamic drag coefficient increased slightly with (3.4%) when fixed the rear spoilers, but a considerable reduction of the aerodynamic lift coefficient value (around 32%).

Keywords: Rear spoiler · Car aerodynamics · Drag coefficient · Lift coefficient · Car stability

1 Introduction

Car aerodynamics is one of the most important factors in the design and development of the new cars according to the new criteria of design to reduce the fuel consumption and improve the performance. In addition, when car travels at higher speeds, its aerodynamic properties have negative effects on car's performance, handling and stability [1]. The aerodynamic lift force should also be taken into account in order not to deteriorate the car's stability at high speeds. This force presents the vertical component of air resistance force, which tends to decrease the pressure between the wheel and the road, thus reduce the handling performance of the steered axle and the traction force [2]. The lift force will be increasedat high level of speed, which leads to lifting of the front or the rear axle, depending on the longitudinal distribution of the lift force.

The range of aerodynamic drag coefficients for the new cars is between 0.22 and 0.275 [3]. The rear spoilers are increasingly used in order to improve the car

I. Karabegović (Ed.): NT 2020, LNNS 128, pp. 340–347, 2020.
https://doi.org/10.1007/978-3-030-46817-0_39

aerodynamics. However, the use of rear spoilers may lead to increase the aerodynamic drag coefficient, which further leads to increase the consumption of fuel. While, the values of drag coefficient may be reduce when applied the rear spoiler, where the drag coefficient is function of the shape of the rear spoiler [4]. The position of the rear spoiler, height and angle relative to horizontal will be specify if the value of aerodynamic drag will be increased or decreased [5, 6]. The aerodynamic lift coefficient decreases in any case, which is good because the car's stability is not endangered at high speeds [7]. Also, allows moving at higher speeds in curves [8]. The using of advanced software such ANSYS will be provided the accurate results within short time [9], so it's possible to perform various numerical tests without major investments. Where, these reasons were the main ones that leaded to study this problem numerically.

The goal of the research paper is to determine the effects of using the rear spoilers on values of aerodynamic drag and lift coefficients. It was build three dimensional models of cars using SOLIDWORKS and then exported these models to ANSYS software 14.5. The final step is analysing numerically these models to obtain the aerodynamic characteristics using ANSYS/Fluid Flow CFX module.

1.1 3D Car Models

Figure 1 shows the 3D car model without the rear spoiler was created in full scale, using SOLIDWORKS software package.

Fig. 1. 3D car model without a spoiler

In order to determine the effects of rear spoiler on aerodynamic drag and lift coefficients, it was proposed two different pedestal spoiler types which were created and mounted on the same car as shown in Fig. 2.

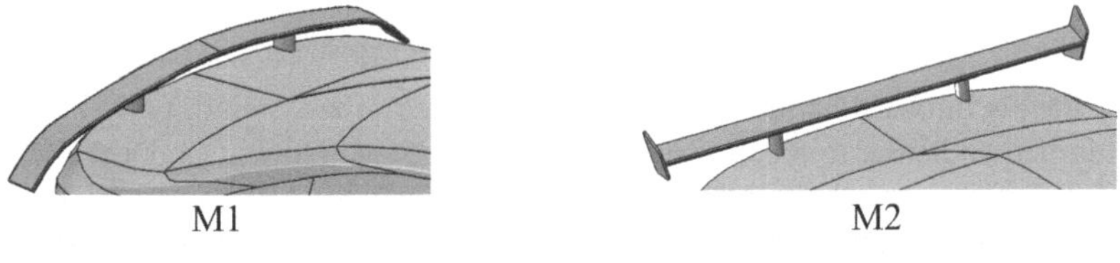

Fig. 2. The proposed models for pedestal spoiler (M1 and M2)

It will be determined the aerodynamic characteristics (drag and lift coefficients) for the two different cases, in the first case the car is without additional spoiler and the second case where fix two different types of spoilers on the same car.

2 Modeling Using Finite Elements

This section presents the details about the load and boundary conditions, also the information about the finite element models. Values of aerodynamic drag and lift coefficients were determined using ANSYS software 14.5/Fluid Flow CFX module. Table 1 shows the type of element (computational mesh type/tetrahedral) and numbers of elements and nodes that used to build all numerical models. It can be seen that the fine mesh was used near the car surface in order to obtain accurate results for loads, streamlines and turbulence that occur during car travel.

Table 1. Number of elements and nodes of the computational mesh

Case	*Computational mesh*	*Number of elements*	*Number of nodes*
Car without a spoiler	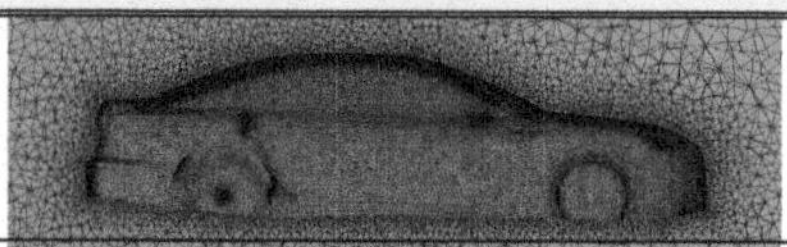	2988527	555375
Car with spoiler (M1)	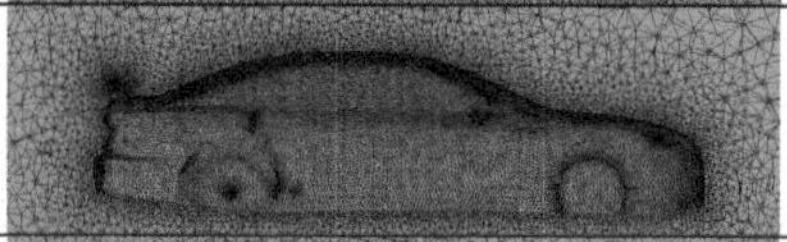	3317332	613768
Car with spoiler (M2)	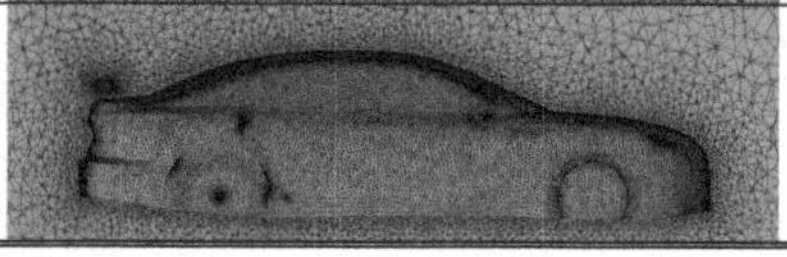	3323161	614457

Boundary conditions that used in the numerical simulation:

- Air temperature, 25 °C,
- Environmental pressure, 101325 Pa,
- Pressure difference between the observed area and outside that area, 0 Pa,
- Air density, 1.225 kg/m^3,
- Car speed, 100 km/h.

The car travels through calm air, without the presence of crosswind. Only the case of air acting perpendicular to the car surface was considered. The k-ε model was applied, because it can simulate different flows in very good manner [10]. It can be found the values of k and ε based on the following equations [11]:

$$\frac{\partial(\rho k)}{\partial t} + \frac{\partial}{\partial x_j}(\rho U_j k) = \frac{\partial}{\partial x_j}\left[\left(\mu + \frac{\mu_t}{\sigma_k}\right)\frac{\partial k}{\partial x_j}\right] + P_k + \rho\varepsilon + P_{kb}, \tag{1}$$

$$\frac{\partial(\rho k)}{\partial t} + \frac{\partial}{\partial x_j}(\rho U_j k) = \frac{\partial}{\partial x_j}\left[\left(\mu + \frac{\mu_t}{\sigma_\varepsilon}\right)\frac{\partial \varepsilon}{\partial x_j}\right] + \frac{\varepsilon}{k}(C_{\varepsilon 1}P_k - C_{\varepsilon 2}\rho\varepsilon + C_{\varepsilon 1}P_{\varepsilon b}), \tag{2}$$

where: $C_{\varepsilon 1} = 1.44$, $C_{\varepsilon 2} = 1.92$ and $\sigma_\varepsilon = 1.3$ (the constants adopted from [9]). P_{kb} and $P_{\varepsilon b}$ represent the effect of the lift forces and P_k is the turbulence due to viscous forces. The other variables are: density ρ, velocity vector U_j, molecular (dynamic) viscosity μ and turbulence viscosity μ_t.

3 Results and Discussions

During car travel, different values of airflow velocity are obtained, depending on the observed cross planes. Figure 3 shows the cross planes in which airflow velocity was analysed for the defined boundary conditions.

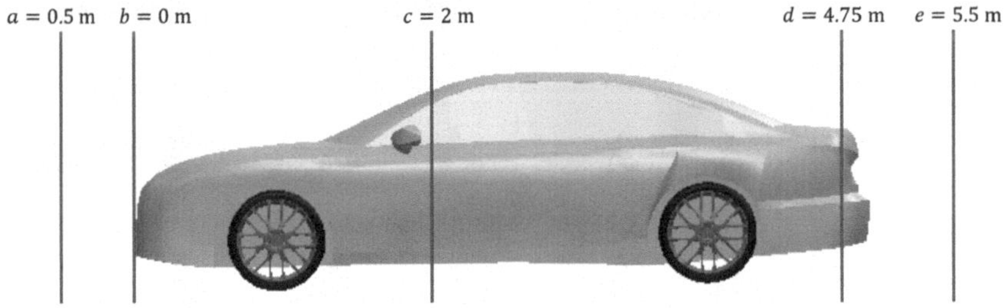

Fig. 3. Spatial positions of the observed cross planes

Table 2 present the obtained airflow velocity contours. It is obvious that the airflow velocity around the car is higher for the car equipped with rear spoiler than the car without it. The highest airflow velocity is obtained in case when the car equipped with spoiler (M2).

Regardless of the fact that the car travelled through the calm air, without the cross wind, there was an airflow in plane *a* as shown in Table 2. This existing because the car "pushes" the air and it can be seen clearly in Fig. 4, where the highest air pressure values occurred at the frontal car area due to the car travel, where similar results were obtained in earlier research by Stojanovic et al. [12]. Maximum pressure levels at the car surface were the same for all the observed cases, because the car speed was the same and other boundary conditions were also the same for all the observed cases. At the plane *b*, where the front plane of the car is, airflow speed was 0 due to friction between the car surface and the airflow.

The greatest turbulence occurred in the zone of the rearview mirrors and behind the car as shown in Fig. 5. During the travel of the car, the difference of pressure in front of the car and behind it occurred, as well as the difference of the airflow velocities around the car (Table 2, planes *c* and *e*).

Table 2. Airflow velocity contours in the selected planes

	Car without spoiler	Car with M1 spoiler	Car with M2 spoiler
Airflow velocity, m/s	0 4.75 9.50 14.26 19.01 23.76 28.51 33.26 38.01 42.77 47.52	0 4.79 9.59 14.40 19.20 23.99 28.79 33.59 38.39 43.19 47.99	0 4.82 9.65 14.47 19.30 24.12 28.94 33.77 38.59 43.41 48.24
Longitudinalplane			
Plane a			
Plane b	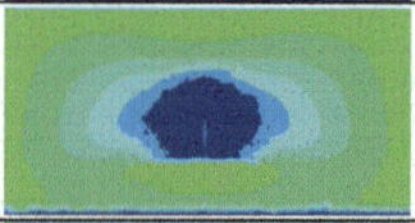	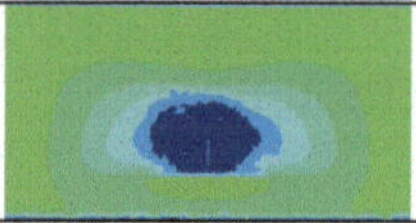	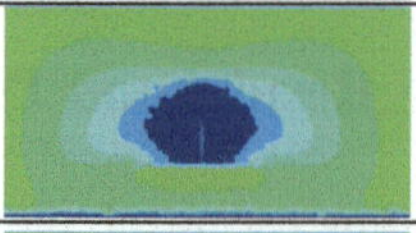
Plane c		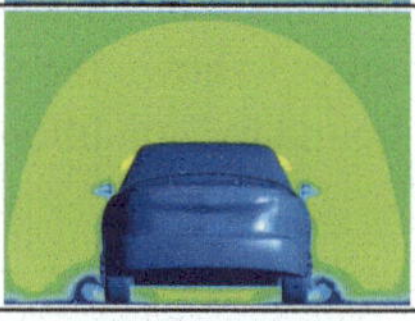	
Plane d			
Plane e			

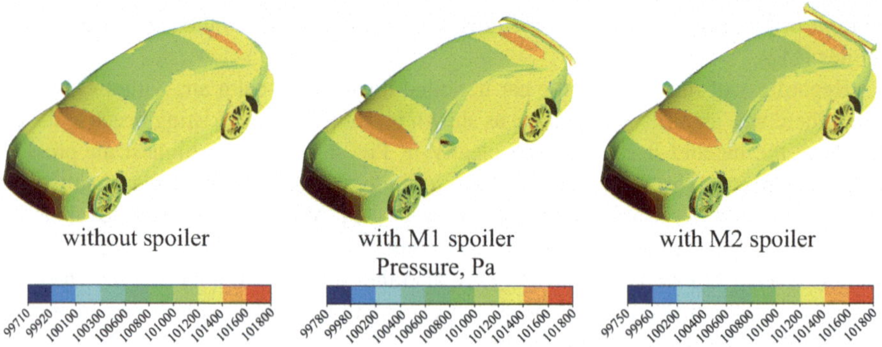

Fig. 4. Airflow pressure on the car surface

Values of aerodynamic drag and lift coefficients for the car without and with

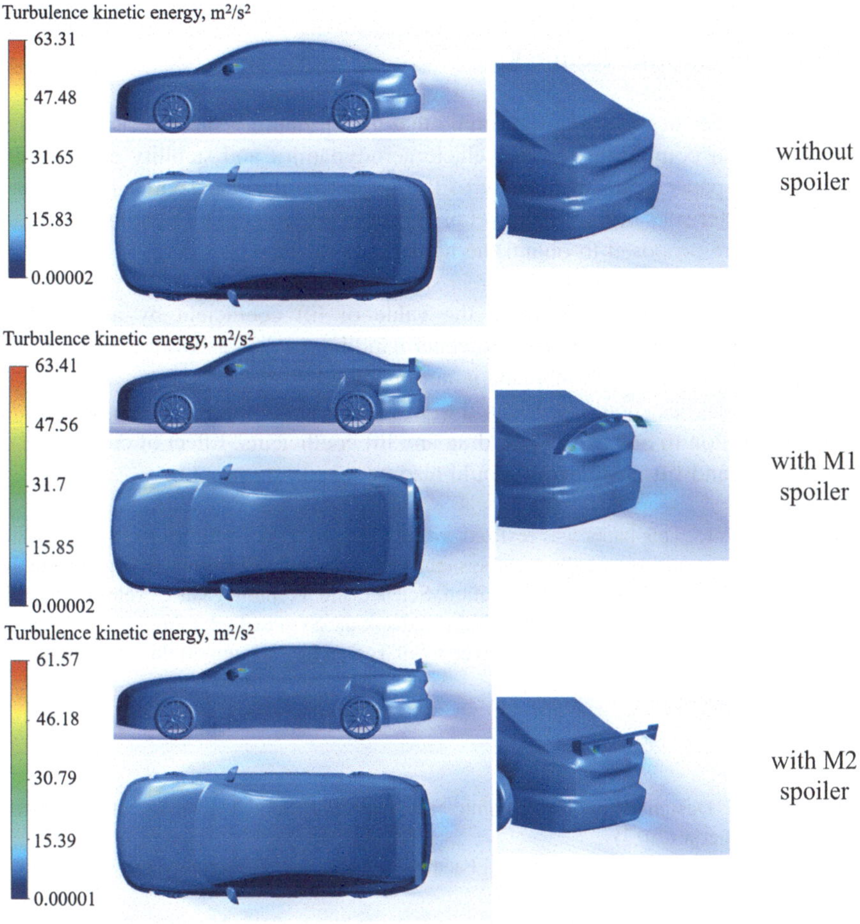

Fig. 5. Turbulence kinetic energy

spoilers (M1 and M2) list in Table 3. The lowest value of the drag coefficient was obtained for the car without the rear spoiler, where similar results were obtained by Hamut et al. [13]. While the lowest value of the lift coefficient was obtained for the car equipped with M1 spoiler.

Table 3. The values of aerodynamic drag and lift coefficients

	Drag coefficient, -	*Lift coefficient, -*
Car without spoiler	0.291	0.071
Car with M1 spoiler	0.292	0.044
Car with M2 spoiler	0.301	0.048

Considering the obtained values of drag and lift coefficients, the car equipped with M1 rear spoiler exhibits the best aerodynamic behaviour.

4 Conclusions and Remarks

Modern tendencies in vehicle design imply that the optimum vehicle design should be obtained without compromising the vehicle aerodynamics and stability even at high speeds. In this paper, a detailed study of the impact of the rear spoiler on the aerodynamic characteristics during the travel of the car is presented. Different models of the rear spoiler were proposed to obtain the numerical results.

It was found that the effects of the rear spoilers leaded to increase the value of drag coefficient by 3.4% and to decrease the value of lift coefficient by almost 32%. Reduction of the lift coefficient leads to better handling of the car, even at high speeds, where the safety of traffic participants is not degraded.

Further researches should be focused to find the optimal design of rear spoiler and its optimal position in order to reduce drag and lift coefficients. Effect of cross wind on values of drag and lift coefficients should also be taken into account.

Acknowledgement. This paper was realized within the framework of the project "The research of vehicle safety as part of a cybernetic system: Driver-Vehicle-Environment", grant number TR35041, funded by the Ministry of Education, Science and Technological Development of the Republic of Serbia. Also, the authors would like to thank the System Technology and Mechanical De-sign Methodology Group/Hamburg University of Technology to support this re-search paper.

References

1. James, A.E.: Design of an aerodynamic rear spoiler, Federal University of Agriculture, Makurdi, Nigeria (2013)
2. Janković, A.: Automotive dynamics, Faculty of Mechanical Engineering, University of Kragujevac, Kragujevac, Serbia (2008)
3. Redfern, J.: The most aerodynamic new cars you can buy in 2018, Motoring Research (2018). https://www.motoringresearch.com/car-news/most-aerodynamic-new-cars-2018/. Accessed 10 Nov 2019
4. Hu, X.-X., Wong, E.T.T.: A numerical study on rear-spoiler of passenger vehicle. Int. J. Mech. Mechatron. Eng. **5**(2), 636–641 (2011)
5. Das, R.C., Riyad, M.: CFD analysis of passenger vehicle at various angle of rear end spoiler. Proc. Eng. **194**, 160–165 (2017)
6. Tomar, A.S., Prajapati, A., Sharma, A., Shrivastava, S.: CFD analysis on the aerodynamic effects of spoiler at different angle on car body. Int. J. Innov. Technol. Explor. Eng. (IJITEE) **8**(7), 2845–2848 (2019)
7. Ipilakyaa, T.D., Tuleun, L.T., Kekung, M.O.: Computational fluid dynamics modelling of an aerodynamic rear spoiler on cars. Niger. J. Technol. (NIJOTECH) **37**(4), 975–980 (2018)
8. Bhatt, P.: Dynamic rear car spoiler for high speed turning stability. Int. J. Mech. Prod. Eng. **3** (12), 48–49 (2015)
9. Yuan, Z., Wang, Y.: Effect of underbody structure on aerodynamic drag and optimization. J. Meas. Eng. **5**(3), 194–204 (2017)

10. Shih, T.H., Liou, W.W., Shabbir, A., Yang, Z., Zhu, J.: A new k-e eddy viscosity model for high Reynolds number turbulent flows. Comput. Fluids **24**(3), 227–238 (1995)
11. ANSYS Inc: ANSYS CFX-Solver Theory Guide (2011). http://read.pudn.com/downloads500/ebook/2077964/cfx_thry.pdf. Accessed 18 Nov 2019
12. Stojanovic, N., Abdullah, O.I., Schlattmann, J., Grujic, I., Miloradovic, D.: 3D aerodynamic simulation of the heavy duty commercial vehicle. In: Dumitru, I., Covaciu, D., Racila, L., Rosca, A. (eds.) The 30th SIAR International Congress of Automotive and Transport Engineering, SMAT 2019, pp. 68–81. Springer, Cham (2020). https://doi.org/10.1007/978-3-030-32564-0_9
13. Hamut, H.S., El-Emam, R.S., Aydin, M., Dincer, I.: Effects of rear spoilers on ground vehicle aerodynamic drag. Int. J. Numer. Meth. Heat Fluid Flow **24**(3), 627–642 (2014)

Ionic Liquids as Wide Operating Temperature Range Lubricant

Darko Lovrec(✉) and Vito Tič

Faculty of Mechanical Engineering, University of Maribor, Smetanova 17, 2000 Maribor, Slovenia
darko.lovrec@um.si

Abstract. Over the past few years, the term 'ionic liquid as super lubricant' has been appeared into lubricant expert conversations with growing importance. They have been designed as green, environment friendly technical fluid for use under harsh operational condition and to cope with the environment conditions. In addition to exceptional lubricating and other important physical and chemical properties, they also have excellent temperature stability. For this reason they are very suitable for use as a liquid lubricant, e.g. as a hydraulic fluid, with a wide operating temperature range.

In the forefront of discussion are the latest findings linked to the excellent viscosity-temperature behaviour of tested ionic liquids – on the viscosity and Viscosity Index. Their excellent viscosity-temperature properties are confirmed on the basis of standard test procedure. In comparison to conventional mineral based oils or synthetic lubricants, the results regarding viscosity dependence on temperature are much better.

Keywords: Lubricants · Ionic liquids · Viscosity · Viscosity index · Temperature range

1 Introduction

Lubricants used within different technical systems, machines and devices, are often discussed as a most important machine part. They are an extremely important component of a variety of gear drives, bearing systems, as well as hydraulic systems, used widely in stationary and mobile machinery, operating under different temperature conditions. When selecting the proper lubricant for a specific application, the tribological system must be identified to its fullest extent. This includes the type of motion, speeds, loads, and the temperatures operating environment. The latest is of the utmost importance when it comes to hydraulic fluids.

Lubricants used as a hydraulic fluid need to be suited for gear, piston and vane hydraulic pumps used within stationary production machines, and for marine, woodlands, mining and other mobile hydraulic systems. Moreover, they are supposed to ensure minimal fluid friction at low start-up temperatures and maintain optimum viscosity at high operating temperatures; in some cases even at extreme ambient temperatures.

I. Karabegović (Ed.): NT 2020, LNNS 128, pp. 348–359, 2020.
https://doi.org/10.1007/978-3-030-46817-0_40

Requirements for today's hydraulic fluids arise from trends in the development of hydraulic components and systems, as well as from operating conditions, taking into account operating trends:

- Higher operating pressures: Typical operating pressures for hydraulic equipment range between 350 bar and 450 bar,
- Smaller and lighter components: Reduced fluid volume, higher circulation number and less residence time for cooling,
- Higher fluid operating temperature: e.g. 80 °C, common for mobile hydraulics with more than 100 °C peak temperature, or low ambient temperatures,
- Wider operating temperature range,
- Increasing energy efficiency of the plant, including structural components' design, implementation of energy-saving control concepts, and use of energy-efficient hydraulic fluids, with excellent lubricating properties over the wide operating temperature range.

Most of these requirements relate to the hydraulic fluid properties, especially those related to fluid viscosity. The fluid must have the appropriate viscosity on the one hand, and this should be, as little as possible, dependent on the temperature, on the other. The influence of the viscosity on the operation and performance of the hydraulic system is shown in Fig. 1.

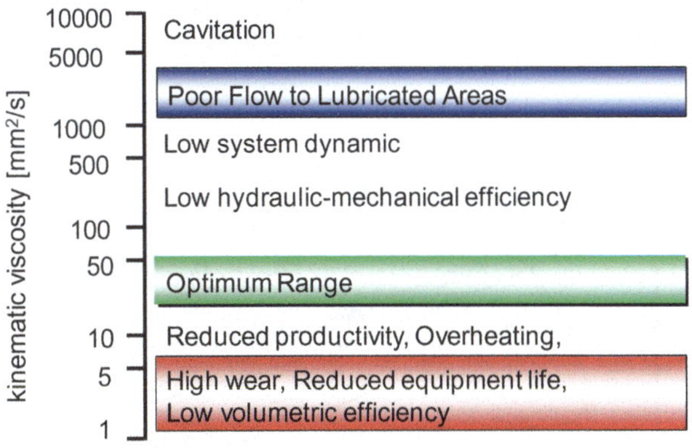

Fig. 1. Importance of fluid viscosity within hydraulic system

Losses inside hydraulic components (and accordingly efficiencies), can be divided into two main groups:

- Hydraulic-mechanical losses – energy loss due to fluid friction, as a resistance of the motion of the component parts in the fluid itself, non-stationary flow condition…, which is reflected in lower available force on the actuator, and
- Volumetric losses – energy loss as the result of internal leakage within all component parts with gaps: Pumps, valves, hydromotors… which is reflected in lower actuator speed.

In the case of too high fluid viscosity at low temperature, cavitation, insufficient lubrication and low system dynamic can occur, leading not only to reduced hydraulic-mechanical efficiency (see also Fig. 2), but also to component damage (e.g. at insufficient filling of the pump). Cavitation can also occur when the liquid in a pump turns to a vapour at low pressure.

On the other hand, too low fluid viscosity at elevated temperatures leads to excessive leakage and, consequently, poor volumetric efficiency (see also Fig. 2), which leads to insufficient lubrication. Consequently, due to the interruption of the lubricant film, this leads to increased friction, and, thus, wear, which is again reflected in increased leakage.

In order to maximise energy efficiencies in hydraulic systems, hydraulic-mechanical and volumetric losses must be balanced, so that the sum of these losses is minimised. Since the hydraulic-mechanical losses in sealing gaps are proportional to the fluid viscosity, and volumetric losses inversely proportional to viscosity, it is clear that an optimal viscosity must be selected.

Figure 2 shows, as an example, the effect of viscosity on the hydraulic pump efficiency. Apart from the above-mentioned risks in the operation of the device, both too low and too high viscosity are reflected in the efficiency, both volumetric, hydraulic-mechanical, and, consequently, the overall efficiency. The resulting requirements must be fulfilled by the lubricant, in general: The ability of a lubricant to maintain optimum viscosity under a wide operating temperature range can be achieved with a shear stable fluid, with temperature independent viscosity, and with a low friction coefficient at start- and operating temperatures.

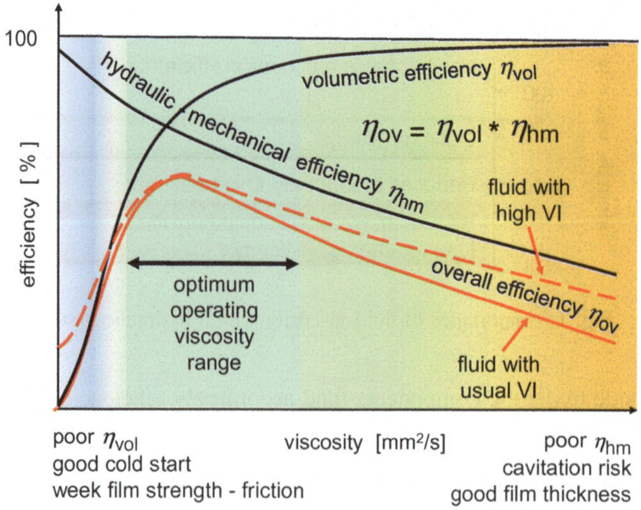

Fig. 2. Effects of viscosity on efficiencies

Fluid(kinematic) viscosity, defined by viscosity grade VG (with a deviation of the standard value of ±10%), is not the only parameter affecting hydraulic system energy efficiency and the operating temperature range. Also, the Viscosity Index (VI) of the fluid is a very important parameter. In the case of using high speed pumps, e.g. in mobile hydraulics, or in the case of speed controlled constant pumps in stationary hydraulic systems, as well as in the case of long hydraulic tubes, a lower viscosity will increase the efficiency. Using a low viscosity fluid (VG 32 or VG 22), it is also very important to take into account the Viscosity Index. This is especially true for mobile machines operating at low temperatures.

In mobile applications where the fluid temperature variations can be very high, the VI has a high impact on energy efficiency, especially on hydraulic pump efficiency [1]. Research has shown that the efficiency of the pump is increased considerably if a hydraulic fluid with a higher Viscosity Index (for example VI > 160) is used, instead of a conventional mineral hydraulic oil with a typical viscosity index about VI = 100. In the case of the use of fluid with VI = 200, with respect to the baseline operating temperature of the fluid $T = 50$ °C (measured in the tank), in the area of lower operating temperatures, e.g. 20 °C, the total efficiency increases by approximately 13% in the area of higher operating temperatures, e.g. 90 °C by 17%. The percentage value of efficiency improvement depends on the type of pump used, the pump speed range and the system pressure range. However, the efficiency is, in any case, higher than in the case of the use of mineral oil. The effect of using a fluid with a higher VI is shown, in principle, in Fig. 2 – a dotted line.

Also, the lower viscosity is preferred in the case of fluid flowing through the hydraulic pipes and hoses. At lower viscous fluid, the pressure drop at the same pipeline dimensions (dimension, length) and operating conditions (velocity, operating pressure, temperature) is lower. Consequently, the efficiency of the pipeline is higher, due to the lower pressure drop [2]. The same conclusions, with regard to the favourable effect of fluid with lower viscosity on efficiency, also apply to hydraulic pumps, hydraulic motors and hydraulic transmissions. This is particularly pronounced in the case of higher drive speeds of the components [3].

Lubricant manufacturers try to approach these requirements with special additives that otherwise improve certain properties (e.g. lower temperature dependence of viscosity) but, on the other hand, other properties may deteriorate (e.g. the tendency to foaming, low shear stability…). In addition, such a fluid is not necessarily environmentally friendly. A completely new option that solves most of these problems successfully represent Ionic liquids (ILs), used as novel, high efficiency lubricants.

2 Ionic Liquids as Lubricants

Ionic liquids represent a new class of liquid high-tech materials with extraordinary properties, among them nanostructural organisation. A number of research findings have opened up many new areas of application for the use of these liquids. Based on their unconventional properties, ionic liquids allow fundamentally new approaches to technical challenges. They have the potential to open doors to radical innovations [4].

Ionic liquids are defined as molten salts with a melting temperature of 100 °C or lower in the liquid state, a large number of them being liquid even at room temperature, and then being called “room temperature molten salts” [5, 6]. Table 1 shows some of their main properties, being of strategic importance for their application as a lubricant or hydraulic fluid.

Table 1. Some main properties of ILs for use as a lubricant or hydraulic fluid

Property	Characteristic
Low friction coefficient	Low wear, low energy consumption, high efficiency
High viscosity index	Low dependency from temperature
Adjustable viscosity	Use of low viscosity, energy efficient fluid
Excellent thermo-oxidative stability	Low temperature influence, durable
Medium to low corrosion	No analogy with inorganic salts
Very low vapour pressure	No cavitation
Very low compressibility	More rigid transfer of forces and signals, low elasticity
Other important properties	Non-flammable, no boiling point, no biofouling, adjustable density, melting point down to −60 °C, higher heat capacity…

Ionic liquids, typically, consist of an organic cation and an inorganic or organic anion. The combination of cation and anion, as well as the design of the ion's side chains, determines the fundamental properties of ionic liquids. Figure 3 shows the large number of types of cations and anions, resulting in countless possible combinations with a broad scope of different properties, including attractive tribological properties. This fundamentally opens up the option for tailor-made lubricants and lubricant additives.

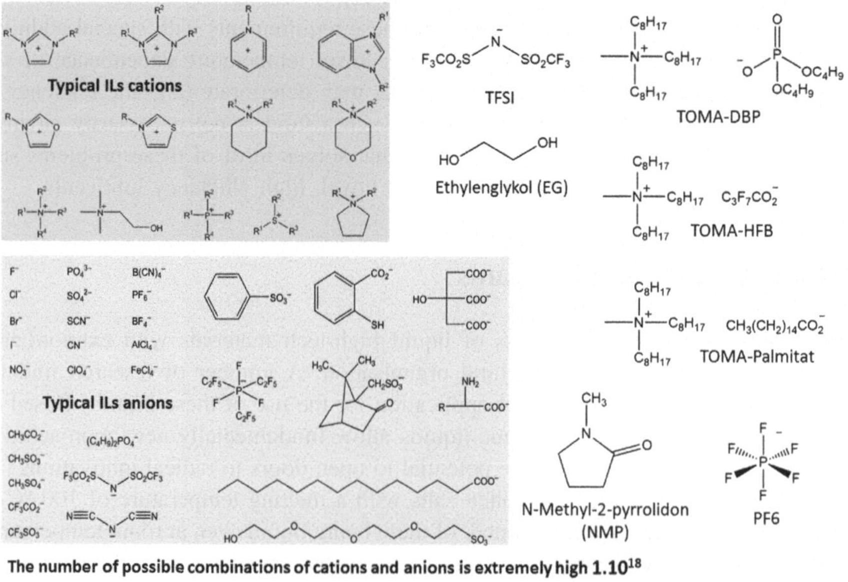

Fig. 3. Enormous combination of anions and cations and discussed IL-structures

Since the 1990s, thousands of review papers and dozens of books have been published describing the extraordinary physical and chemical properties of Ionic Liquids, so ionic liquids have found application in the fields of Lubricants [6], as Heat Transfer and Storage Fluids [7], at Heating, Ventilation, Air Conditioning [8, 9], as Sealing Fluids [10], Cutting and Drilling Fluids [11], as Pressure Transmission Fluids (Hydraulics) [12] and generally as operating fluids [13], in process machines, summarised as Ionic Engineering Liquids.

This research was focused to those Ionic Liquids that have the characteristics of a perfect lubricant, with remarkable anti-wear capabilities, adjustable viscosity and above all the excellent viscosity temperature behaviour.

3 Kinematic Viscosity and Viscosity Index Tests

Kinematic viscosity is the most important physical property of lubricating and hydraulic fluids – is a criterion for resistance to flowing of the liquid under active pressure. Its value is given by the ratio between the active shear stress and shear velocity gradient. Viscosity is an essential property of lubricating oil, as it affects its capacity to form a lubricating film, or to reduce friction and wear. It depends on the temperature, pressure and shear velocity.

Kinematic viscosity, as a criterion for the resistance of the fluid to flowing under the influence of gravity, was measured by capillary viscometers, and is given in mm^2/s or cSt. The measurement is simple, fast and precise, and performed at two standard temperatures of measuring, 40 °C and 100 °C. The two kinematic viscosity data at both temperatures are, at the same time, the basis for determining the Viscosity Index. For the kinematic viscosity measuring, a Cannon-Fenske viscometer with the necessary peripheral equipment (tempered bath, stopwatch) was used, and the measurement performed according the ASTM D445 Standard [14].

The Viscosity Index VI, can be calculated according to Eq. (1):

$$VI = \frac{(L - U)}{(L - H)} \cdot 100 \tag{1}$$

Where U is the fluid's kinematic viscosity in [mm^2/s] at 40 °C, and L and H are values based on the fluid's kinematic viscosity in [mm^2/s] at 100 °C. Values L and H are the values of viscosity at 40 °C for fluids of VI = 0 and VI = 100 respectively, having the same viscosity at 100 °C as the fluid whose VI we are trying to determine (L and H values can be found in the ASTM D2270 Standard and are generally valid for oils) [15]. When the data are inserted into Eq. (1), it gives the Viscosity Index of the observed fluid.

The Viscosity Index values of the more refined mineral based oils on the market are about 100, while the multi-grade and synthetic oils have a higher Viscosity Index, i.e. around 150. The importance of higher VI-Value in connection to the hydraulic oil viscosity grade is depicted clearly in Fig. 4.

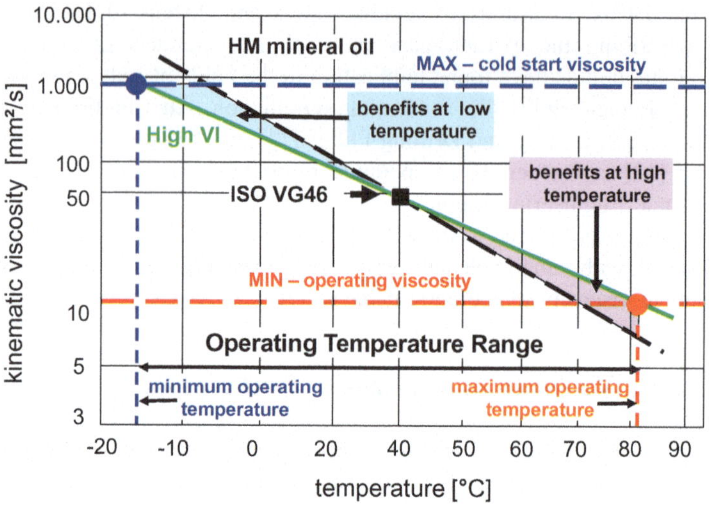

Fig. 4. Benefits of using high VI hydraulic oil

A favourable effect of expanding the operating temperature range can be achieved with a lubricant of the same viscosity class VG, but with a higher Viscosity Index. The effect is even bigger in the case of using the lubricant of the lower viscosity class and, at the same time, with a higher VI. The ionic liquids' Viscosity Indexes are, in general, higher than those of mineral based oils, wherein the viscosity grade can be adjusted.

4 Test Results

In accordance with the presented requirements for high-tech lubricants suitable for different applications, the lubricant must have a low viscosity degree VG (measured at 40 °C), and as little dependence on temperature as possible (a high Viscosity Index value). In terms of good lubricating properties, it should have e.g. a high point of welding (high welding-load value) with low wear (small wear-diameter), and for sliding properties, the friction coefficient should be as low as possible, both at low temperatures and at higher operating temperatures. Beside these requirements it should provide good corrosion protection and be compatible with the materials used within hydraulic system [16–18].

In Table 2, a part of the measurement results relating to kinematic viscosity and Viscosity Index for tested fluids – mineral based oil and ILs, are given. The values are given only for those produced and tested ILs that have proven other important properties for use as a lubricant: Non-corrosive, non-foaming, compatible with materials… and with good lubricating properties [17, 18].

Table 2. Comparisons of viscosity-related properties of tested ILs and mineral oils

Property [unit] Method	Viscosity 40 °C [mm^2/s] ASTM D 445	Viscosity index [–] ASTM D 2270
Tested fluid		
Mineral oil VG46	47,07	104
Mineral oil VG22	22,10	107
IL1 EMIM-EtSO4	42,44	168
IL2 18PI163 (TOMA-DBP + 40% NMP)	47,36	155
IL3 17PI045	46,59	155
IL4 B2002b	45,23	140
IL5 EMIM-TFSI	71,89	132
IL6 TOMA-DBP + 10% TOMA palm.	60,29	133
IL7 B2002a	19,97	220

The first analysed ILs, often used and reported in the literature, was 1-Ethyl-3-methylimidazolium ethylsulphate (EMIM-$EtSO_4$). Based on the results given in Table 2, it is evident that IL1 EMIM-$EtSO_4$ has, at the same temperature, lower viscosity than mineral oil, but higher VI, which means that its Viscosity-Temperature (VT) behaviour is less dependent on temperature. The values of these two parameters already point in the direction towards the energy-saving lubricant with wide operating temperature range. With their values, IL7 B2002a stands out, having, at 40 °C, two grades lower viscosity compared to mineral oil VG22, and with a much higher Viscosity Index. The measured viscosity at 100 °C in the case of IL B2002a amounts to 5,281 mm^2/s, and, in the case of IL B2002b, to 7,729 mm^2/s. Such values for the lowest permissible kinematic viscosity as applied to mineral oils, to the lower upper temperature limit in the amount of 80 to 90 °C, which means that, in the case of ILs, the lubricant film strength at these temperatures is much higher than in the case of mineral oil.

The viscosity class ISO VG46 (with a deviation of the VG-value of ±10%) represents the commonly used viscosity gradation. In principle, for hydraulic systems, gradations from VG22 to VG68 are used, depending on the purpose of use of the fluid (e.g. mineral oil) – the type of machine and the operational temperature range. In principle, higher viscous fluid is used for higher loads, and lower viscosity fluids in cases of lower ambient temperatures (cold start of the device) and in terms of using the fluid as an energy-efficient fluid. The impact of Viscosity Index values to the width of the operating temperature range is shown in Figs. 5 and 6.

Figure 5 shows the width of the operating temperature range of use of standard grade VG46 mineral hydraulic oil and three examples of ionic liquids of the same viscosity grade.

The limit temperature values are given for three different types of hydraulic pumps – for a conventional vane pump, axial piston pump and external gear pump. The maximum and minimum viscosity values refer to the type and version of the specific pump used – the values are listed in the pump manufacturer's data sheet. The highest viscosity

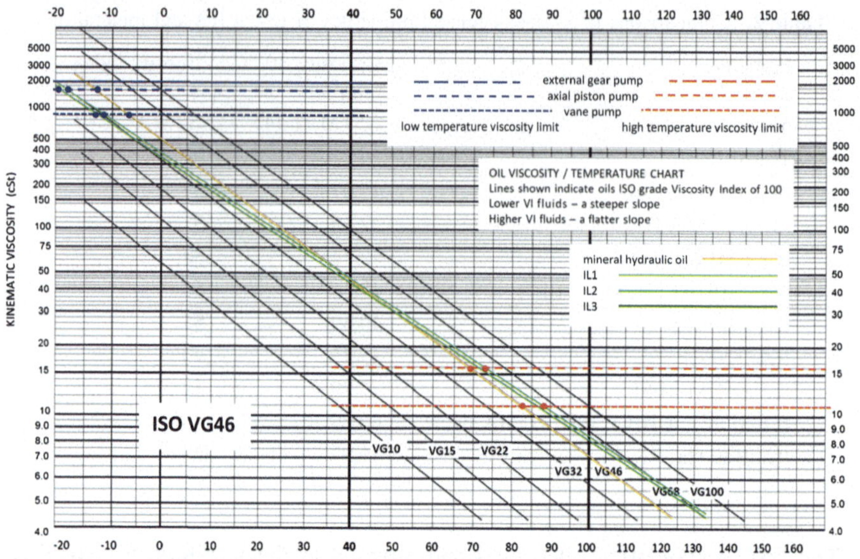

Fig. 5. Operating temperature range of VG46 lubricants, with different viscosity index

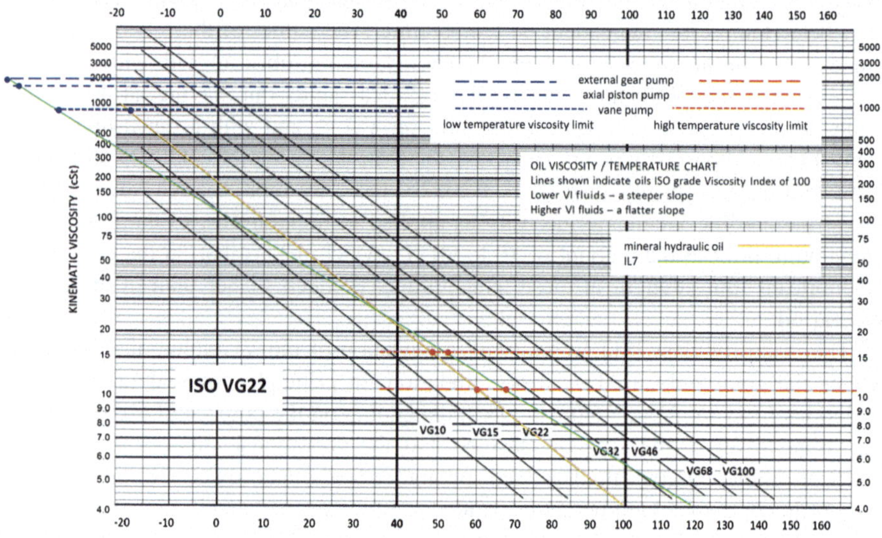

Fig. 6. Operating temperature range of VG22 lubricants, with different viscosity index

value determines the starting viscosity at the lowest starting temperature values (the so-called cold-start viscosity). The lowest viscosity value is the value that still ensures the sufficient lubrication and strength of the lubricating film in the component itself at highest operating temperature. In the present case, the limit values used are given in the datasheets of the same manufacturer of all three pump types and are generally applicable to the same pump types.

Compared to mineral hydraulic oil, ionic liquids have a much higher viscosity index (see values given in Table 2) and, accordingly, the operating temperature range is much wider – in fact, they cover three viscosity classes of mineral oils. Although this effect can be achieved by using so-called multigrade oils or oils with increased VI-value, due to various additives to improve Viscosity-Temperature behavior, we can have the side effects mentioned in Sect. 1.

A similar comparison is shown in Fig. 6, but for the case of a lower viscous fluid – class VG22. The trend in the development of hydraulic fluids is towards the use of energy-efficient hydraulic fluids, characterized by the use of fluids with lower viscosity, high viscosity indexes, and thus covering a wide range of operating temperatures, with better lubricating properties, greater environmental friendliness, smaller temperature influence on the operating device behaviour and higher overall efficiency of whole hydraulic system. Ionic liquids fully meet these requirements.

5 Conclusion

Many machines use a common lubricant across numerous frictional zones, and have varying loads, speeds and temperatures. Outdoor mobile equipment typically works under these challenging conditions. For such machines and operating conditions, the use of a high Viscosity Index lubricant is definitely a huge advantage.

Synthetic lubricants with high VI, can help operators in many industries protect their equipment from a wide range of operating environments and conditions. Using high VI lubricants, which are characteristic for ionic liquids, can deliver superior equipment protection across a wider operating temperature range than in the case of conventional, mineral-based lubricants that have a lower VI. In this way, they are helping to improve equipment reliability and performance to advance plant productivity.

If a lubricant's viscosity is too low, the lubricant will not be able to keep moving parts separated. This may lead to problems such as increased friction and wear, as well as increased heat and oxidation – and, consequently, to increased energy consumption. If viscosity is too high, the lubricant won't flow adequately. This can lead to increased drag and friction and, in turn, higher operating temperatures and energy consumption as well. Moreover, in hydraulic circuits, it is often found that the majority of friction losses are in the pipes. The pressure drop in the pipe is directly proportional to the viscosity of the lubricant. Reducing the lubricant viscosity will reduce the pressure drop across the pipes, and will require lower pressures to be delivered by the pump – which will result in energy savings.

In state-of-the-art tribological systems, including widely used hydraulic systems, the lubrication properties are becoming increasingly important due to increasing operating pressures and mechanical loads, as they guarantee long useful lifetimes of the components used. Tested ionic liquids proved to have excellent viscosity-temperature behaviour. This makes them very suitable for use as lubricants at both high and low temperatures. In the latter case, they can be classified as low-temperature liquid lubricant. Compared to the use of mineral oils, energy consumption of entire hydraulic system is also reduced. At low temperatures, there is no need for (too) long pre-heating phase of the hydraulic fluid, to the appropriate starting viscosity, or no preheating is necessary at all. At high operating temperatures, due to the still adequate viscosity, less intensive cooling is necessary. Based on the results obtained, we can summarize, that ionic liquids are excellent candidate for energy-saving, environmentally-friendly lubricants suitable for use in a wide range of operating temperatures.

Acknowledgment. This research was supported by company proionic GmbH from Grambach/Austria, providing all the samples of ionic liquids and being very cooperative partner within IL-selection process, for sharing their wisdom with us. We are also grateful to the company OLMA d.o.o. from Ljubljana /Slovenia, which has allowed us to use their equipment and facilities, as well the personal to carry out very extensive experimental work. We are thankful to all colleagues in both companies who provided their expertise and skills that greatly assisted this research.

References

1. Dustman, C.K.: Performance Benefits of High Viscosity Index Hydraulic Fluid. Lubricants-Russia, Moscow (2006)
2. Javalagi, S., Singireddy, S.: Hydraulic fluid properties and its influence on system performance. Master's thesis, LIU-IEI-TEK-A–12/01284—SE, Linkoeping University, Sweden (2012)
3. Rydberg, K.E.: Hydraulic fluid properties and their impact on energy efficiency. In: 13th Scandinavian International Conference on Fluid Power Proceedings, Linkoeping, Sweden (2013). ISBN 978-91-7519-572-8, Article no. 44, 447–453
4. Kalb, R.: Ionic liquids – a new generation of high-tech engineering liquids. In: International Fluid Power Conference 2015, Maribor, University of Maribor, Slovenia, pp. 49–77 (2015)
5. Wasserscheid, P., Welton, T.: Ionic Liquids in Synthesis, 2nd edn. Wiley-VCH Verlag GmbH & Co. KGaA, Weinheim (2008)
6. Uerdingen, M., Anastas, P.: Handbook of Green Chemistry, pp. 203–219. Wiley-VCH Verlag GmbH & Co. KGaA, Weinheim (2010)
7. Beck, H., Schmidt, C., Ahrenberg, H., Schick, C., Kragl, U., Kessler, O.: Ionic liquids as new quenching media for aluminium alloys and steels. HTM J. Heat Treatm. Mat. **70**, 73–80 (2015). https://doi.org/10.3139/105.110256
8. Zheng, D., Dong, L., Huang, W., Wu, X., Nie, N.: A review of imidazolium ionic liquids research and development towards working pair of absorption cycle. Renew. Sustain. Energy Rev. **37**, 47–68 (2014). https://doi.org/10.1016/j.rser.2014.04.046
9. Dinnage, P., Kalb, R.: Liquid sorbent, method of using a liquid sorbent, and device for sorbing a gas. Patent US20110247494 (2011)
10. Hilgers, C., Uerdingen, M., Wagner, M., Wasserscheid, P., Schlucker, E.: Patent WO2006087333 (2006)
11. Kalb, R., Hofstaetter, H.: Method of treating a borehole and drilling fluid. Patent WO2010106115 (2010)
12. Kambič, M., Kalb, R., Tašner, T., Lovrec, D.: High bulk modulus of ionic liquid and effects on performance of hydraulic system. Sci. World J., 1–10 (2014). https://doi.org/10.1155/2014/504762. Article no. 504762

13. Schluecker, E., Wasserscheid, P.: Ionic liquids in mechanical engineering. Chem. Ing. Tec. **83**(9), 1476–1484 (2011)
14. ASTM D445 – 19, Standard Test Method for Kinematic Viscosity of Transparent and Opaque Liquids (and Calculation of Dynamic Viscosity)
15. ASTM D2270 – 10(2016) Standard Practice for Calculating Viscosity Index from Kinematic Viscosity at 40 °C and 100 °C
16. Lovrec, D., Tič, V.: Excellent lubricating properties of ionic liquid - myth or truth. In: Serbiatrib '19: Proceedings on Engineering Sciences, 16th International Conference on Tribology, vol. 1, no. 1, pp. 555–562 (2019). ISSN 2620-2832
17. Kambič, M., Kalb, R., Tič, V., Lovrec, D.: Compatibility of ionic liquids with hydraulic system components. Adv. Prod. Eng. Manag. **13**(4), 492–503 (2018). https://doi.org/10.14743/apem2018.4.306. ISSN 1854-6250
18. Kambič, M., Kalb, R., Lovrec, D.: Lubrication properties of ionic liquids suitable for use within hydraulic systems. In: Proceedings of International Conference Fluid Power 2015, Congress Centre Habakuk, Maribor, Slovenia, Faculty of Mechanical Engineering, Maribor, pp. 79–93 (2015)

Simulation of Vehicle's Inertia Using a Flywheel Mass to Test Disc Brake System

Ivan Grujić[1], Saša Vasiljević[2](✉), Jasna Glišović[1], and Nadica Stojanović[1]

[1] Faculty of Engineering, University of Kragujevac, Sestre Janjić 6, 34000 Kragujevac, Serbia
[2] High Technical School of Professional Studies, Technical College of Applied Studies in Kragujevac, Kosovska 8, 34000 Kragujevac, Serbia
vasiljevic.sasa036@gmail.com

Abstract. The characteristic of all bodies having a certain mass is that during its movement, in addition to all forces acting, the force of inertia also acts. In the case of the translational motion of the body, such as the movement of a vehicle along a straight line, the force of inertia acts. However, in the case of body rotation, the term moment of inertia is introduced. In order to simulate the inertia of a quarter of a moving vehicle moving, but in a laboratory test, this paper presents a methodology for simulating the inertia of a vehicle using a flywheel mass, i.e. a moment of inertia. In this particular case, an analysis of the required rotational speed of the flywheel was performed in order to simulate a specific speed of vehicle and to test the working characteristics of the disc brakes. Knowing the speed of the translational moving of a quarter of the vehicle model and knowing the characteristics of the flywheel mass, and applying the basics of kinetic energy of these two types of motion, a solution was reached about the required rotation speed of the flywheel mass in order to simulate the movement of the vehicle. The conclusion about the rotation speed of the flywheel mass in order to simulate the movement of the vehicle was reached by deriving the equations from kinetic energy.

Keywords: Simulation · Moment of inertia · Vehicle · Braking

1 Introduction

The simulation of certain phenomena occurring in nature is usually performed in laboratory conditions, with the aim of analysing these phenomena and recognizing certain regularities of their occurrence. When simulating some phenomena, attention must be paid to ensure that all conditions and forces acting at that moment are fulfilled. The simulations in the laboratory have many advantages over the real-world experiments, allowing repeatability of the experiment under identical conditions, increased safety and independence from environmental influences that are often a problem in real-world experiments.

In the case of the translational motion of the body with a certain mass, it is influenced by the inertia force. Similar, during the body rotation, the body is affected by

I. Karabegović (Ed.): NT 2020, LNNS 128, pp. 360–367, 2020.
https://doi.org/10.1007/978-3-030-46817-0_41

inertia, but the term moment of inertia is introduced. Between the inertia and the moment of inertia, it is possible to set up a conjunction based on the kinetic energy of the body which it has during rotational motion and during translational motion. In this way, it is possible to determine the speed of rotation of a given mass in order to equalize the torque with the inertia that the vehicle would have when traveling on the road.

In order to test the disc brakes of vehicles under laboratory conditions, in this paper, an analysis of the velocity of rotation of the velocity of the flywheel mass was performed in order to simulate the inertia of a quarter of the vehicle using the flywheel mass. The inertia of a quarter of the vehicle being simulated is equated with the moment of inertia of the flywheel mass, which was set up on an experimental installation to test the disc brakes of a motor vehicle. Of course, the simulation of different vehicle speeds is done by varying the flywheel's number of revolts per minute, and therefore the goal is to determine the rotational speed at which identical kinetic energy is obtained as in the case of the kinetic energy that the vehicle has during the translation movement, which needs to be overcome by braking.

2 Literature Review

The body inertia can be defined as opposition to changes in body movement [1]. On the other hand, due to the rotational motion of the body, it also has its inertia, which depends on the shape of the body, but also on the mass of the body [2]. In rotational body movement, inertia depends on the mass of the body, the radius of rotation, the shape of the body and the position of the axis of rotation [3]. Of course, when the body moves or rotates, the body has certain energy, and that energy is called kinetic energy. In the case of the body moving in a straight line, the kinetic energy depends on the speed of movement of the vehicle (v) and its mass (m), and it can be expressed via Eq. 1 [4]. The kinetic energy of a rotating body depends on the angular velocity (ω) with which the body rotates and the moment of inertia of the body (J), in this case the kinetic energy can be expressed by Eq. 2 [5].

$$E_{Ktranslational} = \frac{1}{2} \cdot m \cdot v^{2^2} \tag{1}$$

$$E_{Krotatate} = \frac{1}{2} \cdot J \cdot \omega^2 \tag{2}$$

In various tests, at least when it comes to vehicles, often in laboratory tests, a flywheel which has the same kinetic energy as the vehicle during translational body movement is used. In this way, a certain speed of the vehicle can be simulated, so very often examples of low-mass devices can be found in the studies. Thus, according to the source [6], an example of a disc brake test installation is shown. On this initial dynamometer, there is also a flywheel mass to simulate vehicle brakes. The flywheel mass can often be derived from multiple masses that can be switched on or off from time to time, thus producing different values of kinetic energy at the same rpm, thus simulating different vehicle speeds or different vehicle masses at the same speed. Based on the source [7], one can also see an example of an inertial dynamometer with a flywheel mass in which the purpose is to test the friction coefficient of the brake pads. Furthermore, an example of an initial dynamometer for testing brake wear or particle formation (PM) whose source is the braking system is shown in sources [8] and [9]. According to these sources, a flywheel is also used which, by its rotation and kinetic energy, simulates the movement of the body along the path. In this way, i.e. simulating the kinetic energy of the vehicle, realistic braking conditions of the vehicle can be achieved. Real conditions are achieved because when braking, the vehicle in real conditions, it achieves the same value of thermal energy, bearing in mind that energy never disappears, but only changes the energy form. Kinetic energy is transformed into thermal energy generated by braking. The role of the braking system is to convert kinetic energy into thermal energy [10–12]. With such installations, it is possible to perform different tests related to the braking system.

3 Installation for Disc Brake Analysis and Flywheel Mass Position

The measuring system consists of five units interconnected for the purpose of testing the brakes of a motor vehicle. One of the elements of a measuring installation is a flywheel mass, which plays a very important role, bearing in mind that the installation should aim at testing the brakes of a motor vehicle. A schematic representation of the measurement installation is shown in Fig. 1, where all the most important elements are visible.

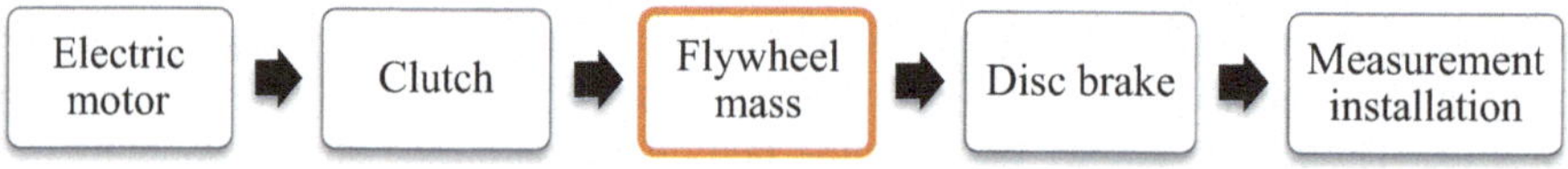

Fig. 1. Inertial dynamometer installation for brake system analysis

As noted, this paper analyses the required operating conditions that must be met by the specific flywheel mass used in this test installation in order to simulate the specific mass of the vehicle or the inertia that the vehicle would have at a given vehicle speed. The flywheel mass for which the study is performed in this case is shown in Fig. 2.

(a)

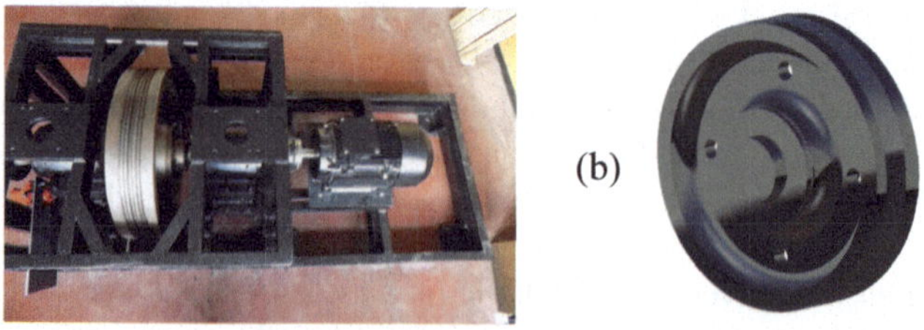

(b)

Fig. 2. Position of flywheel mass in the installation (a) and 3D model of flywheel (b)

Figure 2a shows the flywheel mass on the measurement installation, while Fig. 2b shows the flywheel mass modelled in the Autodesk Inventor software package.

Considering that it is necessary to know the characteristics of the flywheel mass, such as the mass and the moment of inertia of the flywheel mass according to its shape, the calculation of parameters according to the model of the flywheel mass was performed using software calculation that enables Autodesk Inventor. Based on the data obtained in the software package, Table 1 shows the most important parameters and characteristics of the flywheel shown in Fig. 2.

Table 1. Characteristics of flywheel mass

Dimensions	Value
Mass	372.03 kg
Surface area	2.101 m^2
Volume	0.052 m^3
The moment of inertia of the flywheel mass with respect to a specific rotation axis	24.622 kgm^2

According to the data shown in Table 1, a calculation was made in the further part of the paper, and a methodology and installation were presented based on which the inertia of a quarter of the vehicle can be simulated by the kinetic energy of a flywheel mass at different speeds.

4 Methodology

In order to achieve the same kinetic energy that the body would have during translational motion and during the effect of initial force on the body, while simulating this motion, the rotational motion of the flywheel mass can be determined by placing in the relationship the kinetic energy due to translational motion and the kinetic energy of the body due to the rotational motion of the body (Eqs. 1 and 2) followed by Eq. 3 [13, 14].

$$\frac{1}{2} \cdot m_s \cdot v^2 = \frac{1}{2} \cdot J \cdot \omega^2 \tag{3}$$

where

m_s - body mass to be simulated [kg],
v - the translational speed of the body [m/s],
J - moment of inertia of a flywheel mass [kgm^2],
ω - angular speed of rotation [rad/s].

Due to the rotation of the body, its angular velocity is ω, so it is possible to determine from Eq. 3 the required angular velocity at which the body should rotate to simulate the inertia of the body that it would have during translational motion. The calculation procedure is shown in Eq. 4.

$$\omega = \sqrt{\frac{m_s \cdot v^2}{J}} \tag{4}$$

Knowing the Eq. 5 and the method of calculating the angular velocity based on the number of revolutions of the body (n), it is possible to include the Eq. 5 in the Eq. 4.

$$\omega = \frac{n \cdot \pi}{30} \tag{5}$$

where

n - revolutions per minute,
π - Archimedes constant.

$$\frac{n \cdot \pi}{30} = \sqrt{\frac{m_s \cdot v^2}{J}} \tag{6}$$

As it is necessary for the research to know the required number of revolutions of a flywheel mass in order to simulate the movement of the vehicle at a certain speed (v) from the Eq. 6, it is possible to derive the required equation for the number of revolutions, and thus follows the Eq. 7. Based on this equation, it is possible to determine the number of revolutions of the flywheel mass in order to simulate the identical kinetic energy of a given mass to be simulated (m_s), and thus in a brake system study the system would have the identical required force to stop the flywheel mass as in the case of the translational motion of a vehicle down the road.

$$n = \frac{30 \cdot v\sqrt{\frac{m_s}{J}}}{\pi} \tag{7}$$

For the purpose of determining the required number of revolutions of the flywheel mass of the characteristics shown in Table 1, and for the purpose of simulating the movement of the vehicle along the road or quarter of the vehicle, which in this case represents a mass of 450 kg, Eq. 7. On the basis of this equation, with all known members, speeds were simulated movements of a quarter of the vehicle from 10 to 250 km/h, at intervals of 20 km/h. Based on all the data and methodology presented during the test, the flywheel mass would have the same kinetic energy as the vehicle moving along the path at the speeds mentioned.

5 Results and Discussion

Applying the methodology presented in the previous part of the paper, conclusions were reached about the magnitude of the rotational speed of a flywheel mass in order to simulate the speed of vehicle movement along the road. The results of the analysis of the required RPM of a flywheel mass in dependence on the simulated speed are shown in Fig. 3, i.e. the values of the RPM versus the desired simulated speed are shown in Table 2. It is important to note that in this case only a quarter of the vehicle mass, which in this case is 450 kg, is simulated and that the obtained values of the required rpm would be higher in the case of a larger mass of the vehicle, i.e. mass to be simulated. In this case, i.e.to test the braking system using the installation shown in Fig. 2a and simulate the speed of a quarter of the vehicle's mass, which in this case is 450 kg, translational speeds of 10 to 230 km/h require the flywheel mass to turn at speeds in range of 113.46 to 2836.44 rpm.

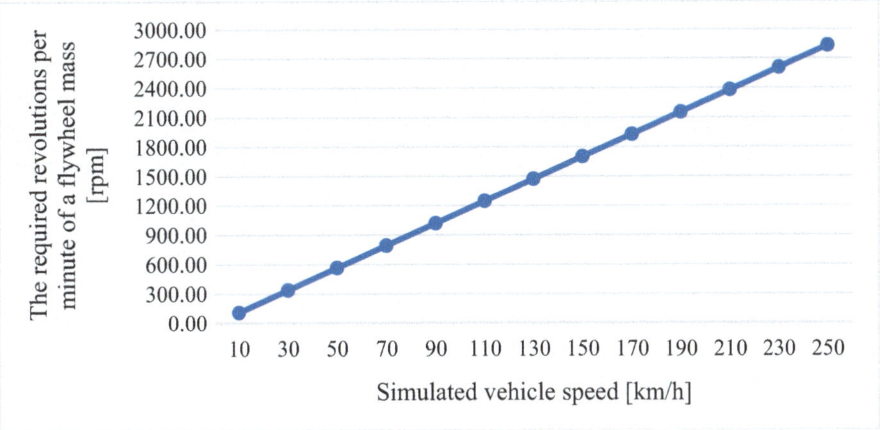

Fig. 3. Graphical representation of the required number of revolutions of flywheel mass in order to simulate the speed of the vehicle due to movement along the road

Table 2. Required RPM to simulate vehicle speed due to movement

Simulated vehicle speed [km/h]	Required RPM to simulate vehicle speed [rpm]
10	113.46
30	340.37
50	567.29
70	794.20
90	1021.12
110	1248.03
130	1474.95
150	1701.86
170	1928.78
190	2155.69
210	2382.61
230	2609.53
250	2836.44

6 Conclusion

Body movements can be characterized by different physical variables and physical laws. Due to the translational movement of the body and also the rotational movement of the body, the body has certain energy. In the case of simulation of the translational motion of the body under laboratory conditions, the kinetic energy possessed by the body can be simulated by flywheel mass. In this case, the characteristics of the flywheel mass and the desired simulation of vehicle movement should be taken into account. If the characteristics of the flywheel are known, it is possible to determine the required RPM of the flywheel based on the kinetic energy to simulate the speed of the vehicle. This paper determines the number of rotations of the flywheel mass back mass in order to simulate the movement of the vehicle, i.e. a quarter of the vehicle at speeds from 10 to 250 km/h. The analysis found that in order for the flywheel to have identical kinetic energy as a quarter of the vehicle moving at the stated speeds, it is necessary for RPM of the flywheel mass to be in range of 113.46 to 2836.44 rpm. Of course, if the desired mass to be simulated is changed, the same procedure is used to determine the required number of revolution of a flywheel as shown in the paper.

References

1. National Aeronautics and Space Administration. https://www.nasa.gov/pdf/563410main_FTNL_Instructor_Manual.pdf. Accessed 10 Dec 2019
2. Bauer, W., Westfall, G.D.: University Physics with Modern Physics. McGraw Hill, New York (2011)
3. Muvdi, B.B., Al-Khafaji, A.W., McNabb, J.W.: Dynamics for Engineers. Springer, New York (1977)
4. Boston University, Kinetic and Potential Energy/Conservation of Energy. http://www.bu.edu/gk12/kai/Lesson%204/E_Back.pdf. Accessed 11 Dec 2019
5. Östergård, R.: Flywheel energy storage a conceptual study (2011). https://uu.diva-portal.org/smash/get/diva2:476114/FULLTEXT01. Accessed 14 Dec 2019
6. Ćirović, V., Aleksendrić, D.: Dynamic modelling of disc brake contact phenomena. FME Trans. **39**(4), 177–183 (2011)
7. Glisovic, J., Radonjic, R., Babic, M., Miloradovic, D., Catic, D.: Design of vehicle road testing method for determination of brake pad friction characteristics. J. Balk. Tribol. Assoc. **17**(4), 513–525 (2011)
8. Kukutschová, J., Moravec, P., Tomášek, V., Matějka, V., Smolík, J., Schwarz, J., Filip, P.: On airborne nano/micro-sized wear particles released from low-metallic automotive brakes. Environ. Pollut. **159**(4), 998–1006 (2011)
9. Iijima, A., Sat, K., Yan, K., Kato, M., Kozawa, K., Furuta, N.: Emission factor for antimony in brake abrasion dusts as one of the major atmospheric antimony sources. Environ. Sci. Technol. **42**(8), 2937–2942 (2008)
10. Laws of Energy. https://www.eia.gov/energyexplained/what-is-energy/laws-of-energy.php. Accessed 14 Dec 2019
11. Thermal Energy. https://www.khanacademy.org/science/physics/work-and-energy/work-and-energy-tutorial/a/what-is-thermal-energy. Accessed 15 Dec 2019

12. Talati, F., Jalalifar, S.: Analysis of heat conduction in a disk brake system. Heat Mass Transf. **45**(8), 1047–1059 (2009)
13. Energy and Power in Waves. http://users.physics.harvard.edu/~schwartz/15cFiles/Lecture10-Power.pdf. Accessed 15 Dec 2019
14. Rotational Motion. http://scipp.ucsc.edu/outreach/09RotationalMotion.pdf. Accessed 16 Dec 2019

Engine Speed and Load Effect on the Sealing Capacity of a Piston Ring-Pack

Erjon Selmani[1(✉)], Cristiana Delprete[2], and Arian Bisha[3]

[1] Department of Mechanics, Universiteti Politeknik i Tiranes, Mother Teresa square 1, 1001 Tirana, Albania
erjon.s@hotmail.it
[2] Politecnico di Torino, Turin, Italy
[3] Universiteti Politeknik i Tiranes, Tirana, Albania

Abstract. The combustion chamber is ought to be perfectly sealed, however, part of the air and fuel mixture can escape from it. Among the several losses there is the gas flow from the inter-ring crevices, which is always present. This leakage is known as blow-by, and affects efficiency, correct lubrication and emissions. The amount of leakage is dependent on many factors, and among the most important are the engine speed and load, which are able to affect the system through the forces applied on it. The aim of this paper was to understand in a more detailed way how the engine speed and load could affect the sealing efficiency of a ring-pack. For this purpose, a complete range of speeds and loads were used in the simulations. The equations of the ring motions and gas dynamics has been implemented and solved in ©Ricardo RINGPAK solver. The results showed that inertia and inter-ring gas pressures drives the sealing behavior of the rings. The blow-by trend showed to decrease with the speed and increase with the load, exception made for the idle condition where the values were different to the other cases, especially at higher speeds. Among the two parameters, the engine speed resulted to affect more significantly the blow-by trend.

Keywords: Blow-by · Internal combustion engines · Speed · Load · Ring dynamics

Nomenclature

m [kg]	mass of gas in the ith region
p [Pa]	pressure in the regions
Q_{in}, Q_{out}[kg/s]	mass flow entering and exiting a crevice
R [J/kgK]	ideal gas constant
T [K]	temperature
mr [kg]	mass of the ring
$F_{\_friction}$ [N]	oil friction force
$F_{\_i}$ [N]	inertia force
$F_{gas\_f/b}$ [N]	gas pressure in radial direction on the ring section
$F_{\_P}$ [N]	gas pressure force in axial direction
$F_{\_oil}$ [N]	hydrodynamic oil pressure in radial direction

I. Karabegović (Ed.): NT 2020, LNNS 128, pp. 368–381, 2020.
https://doi.org/10.1007/978-3-030-46817-0_42

F_s[N]	oil squeeze force in the grove
$F\_{Tension}$ [N]	radial force due to ring installation in the cylinder liner
c.a.	crank angle

1 Introduction

The piston and piston-rod mechanism, are powered by the gas pressure obtained from the combustion inside the chamber. In order to allow the piston to move and slide over the cylinder liner, a pack of rings is mounted into the piston crown. The piston rings main duty is to seal off the combustion chamber, however, their tightness is not perfect and part of the intake gas mixture is lost toward the crankcase. This phenomenon is commonly known as blow-by gas, and was recognized to brink negative effects on performances, lubrication and emissions [1–5]. Initial studies [6, 7] have analysed the ring motions in the axial direction, further studies [8, 9] have also included the motions in the other directions. When the ring moves inside its groove, the ability to seal the combustion chamber is expected to decrease. This motion is called radial collapse [10], and was seen to occur when the ring lifts-off from the contact with the cylinder liner. In the above mentioned references, the collapse tended to increase with speed and load. In references [11–14], the ring was accounted also with the third degree of freedom, the twist around the axis of the ring cross section. In all cases, the ring motions and gas dynamics are related to a wide set of parameters affecting them. Some are static, such as design, materials and surfaces. Some others are dynamic, such as lubrication, temperature variation and engine speed and load. In the previous studies, Curtis [18] states that the engine speed is a parameter which may affect the ring dynamics and gas flow. Kim et al. in [19] states that the blow-by tends to decrease with speed. Tamminen in [20] made some experiments and saw that the blow-by and the inter-ring pressures increased with the engine load. Zottin et al. in [21] saw that the blow-by tended to increase with the engine load, but its peak value was obtained at the medium load. Arnault in [5] referring to a diesel engine, saw that the blow-by increased with speed and load, and the maximum value was seen at full load and medium speed condition. Furthermore, according to the results of Irimescu et al. [22], the blow-by decreased with the engine speed and increased with the compression ratio. Rabuté et al. in [13] saw that the blow by had peak values at high speed and zero load. This study was focused in understanding the behavior of the ring-pack and the ongoing of the blow-by for the complete range of speed and load. In literature, similar analyses have mainly considered one of the two parameters, or has had few concern to investigate the combined effect between them. In addition, in some of the references, the results seems to oppose the others. However, in any of the references wasn't indicated how the blow-by varies within the entire engine map, or which of the two parameters may be more important. The present work features a crossed analysis of working conditions of the engine, ranging from idle to maximum load and from minimum to maximum speed. The mathematical model of the problem was formulated in terms of ring and gas dynamics and solved in ©Ricardo RINGPAK. The analysis was applied to a turbo diesel engine and the performance of the ring-pack was evaluated in terms of inter-ring pressures, ring axial and radial dynamics and blow-by gases. The results are then discussed and compared with the reference literature. The work ends with the conclusions stemming from the discussions.

2 Approach for Ring and Gas Dynamics

Piston rings are curved beams with an end gap, Fig. 1. This allows the mounting operation and ensures a radial force toward the cylinder liner. This force is an important element of the sealing capability of a ring, while the gap is the main escape route for the gas. Typically there are three rings, two compression rings and one oil ring, Fig. 2.

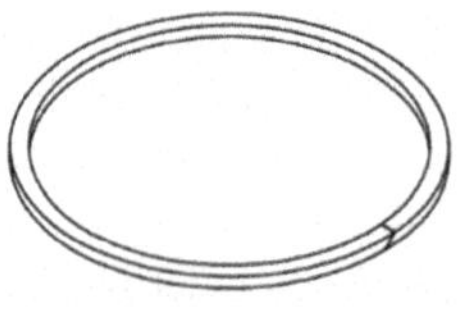

Fig. 1. Piston ring

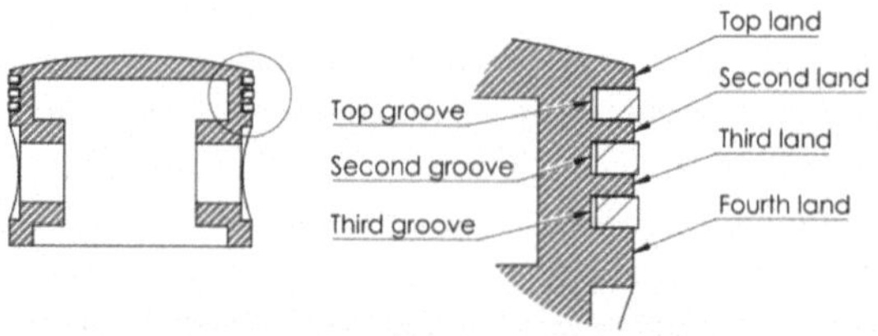

Fig. 2. Piston and ring-pack

To compute the pressure distribution in each zone, the ideal gas Eq. (1) is coupled with the continuity Eq. (2) [9].

$$\left(\frac{dp}{dt}v + \frac{dv}{dt}p\right) = \left(\frac{dm}{dt}\right)RT + mR\left(\frac{dT}{dt}\right) \quad (1)$$

$$\frac{dm}{dt} = Q_{in} - Q_{out} \quad (2)$$

The mass flow through the gap and through ring-groove clearance can be modeled using the orifice-flow model for a laminar and compressible flow [9]. The ring has three degree of freedom, and hence, the Newton's second law of equilibrium can be applied in axial and radial direction and in the twist motion of the ring [8, 9]. According to the schemes of Figs. 3 and 4, the following equations can be written:

$$m_r\frac{d^2x}{dt^2} = F_{\_P} - F_{\_i} - F_{\_s} - F_{friction} \quad (3)$$

$$m_r\frac{d^2y}{dt^2} = F_{\_gas\_back} + F_{\_tension} - F_{\_gas\_front} - F_{\_oil} \quad (4)$$

$$I\frac{d^2\alpha}{dt^2} = M_{gas} + M_{oil\_groove} + M_{oil\_cylinder} - M_{twist} \quad (5)$$

The forces and moments, their calculations are described in great details in references [8, 9].

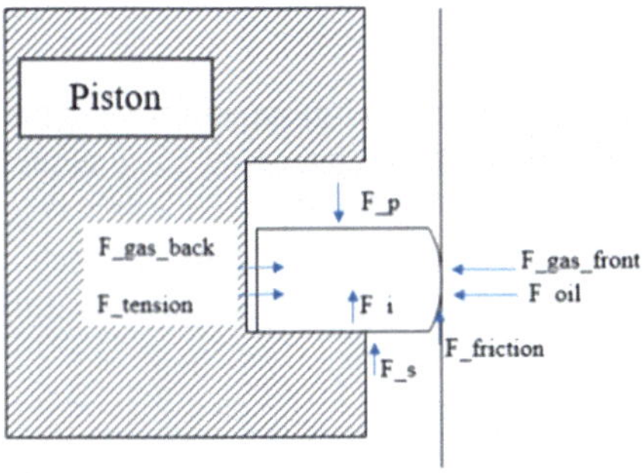

Fig. 3. Forces applied on the ring section

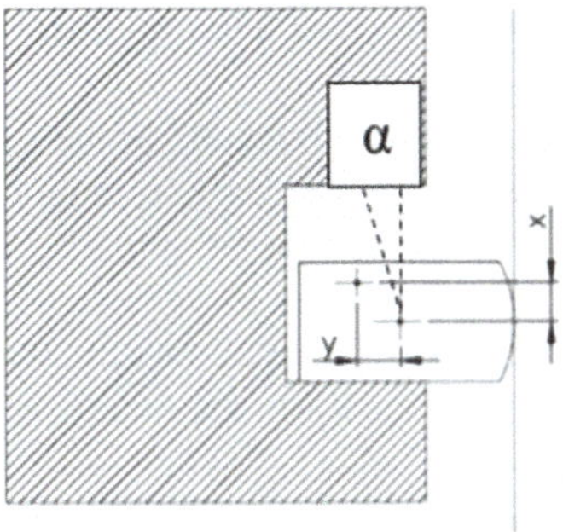

Fig. 4. Axial and radial ring locations from the groove center, twist angle

3 Simulation

The number of parameters affecting the sealing efficiency of a ring-pack is high. In previous works [15–17], authors investigated the effect of the ring gaps and elastic forces, the effect of the distortion orders and the effect of the inter-ring volumes on the sealing efficiency of the system. In the present work has been proposed to study the effect of the speed and load of the engine on the behavior of the ring-pack, and on the total amount of gas lost in the crankcase as blow-by. The range of speeds and loads are reported in Table 1, and covers the entire map of values for the engine under exam. The speed varies with steps of 500 rpm, while the load with steps of 20% increment nearly. In literature the effect of those parameters has usually been considered separately, or for a limited range of values. In the present work, the analysis for the entire engine map will allow to understand which of the two parameters is more significant for the blow-by control.

Simulations are performed considering a direct injection turbo diesel engine with 0.120 m bore and 0.26 m stroke. The ring pack is composed of three rings; a barrel face top ring, a taper face second ring and a mono-piece and spring-loaded oil ring. The minimum oil film thickness is assumed to be 5.3 μm. The input data are implemented in ©Ricardo and solved in the RINGPAK solver.

Table 1. Speed and load values used in the simulations

Load	Idle	30%	50%	70%	100%
Speed	750				
	1000	1000	1000	1000	1000
	1500	1500	1500	1500	1500
	2000	2000	2000	2000	2000
	2500	2500	2500	2500	2500
	3000	3000	3000	3000	3000
	3500	3500	3500	3500	3500
	4000	4000	4000	4000	4000

4 Results and Discussion

In this section, the results of the simulations are presented and grouped for each case. Results will include the ring axial and radial dynamics and the inter-ring gas pressures. The cumulated blow-by for each case will be given at the end of the section in a single graph, with the aim to understand the differences between cases. In the explanations only the speeds 1000, 2000, 3000 and 4000 rpm will be discussed, due to the fact that there is a linearity in the behavior of the system.

4.1 Idle

Figure 5 shows the results of the simulations for the first two idle load conditions, where each row correspond to the results of a case. In the first row we can see the behavior of the system for 750 rpm and no load. All the rings experience axial motion and no one of the rings did show axial flutter. The top ring motion is driven by inertia in the first phase, and by the gas pressure of the second land in the phase after the 100 c.a. The second ring is driven by the pressure of the third land in the first phase (up to 0° c.a.) and by inertia from 300 c.a. and on. The third ring is driven only by inertia during the entire cycle. The radial motion of the second shows a normal contact between the ring faces and the liner, where no radial collapse is encountered for this case. In the third figure are given the inter-ring pressures and their maximum values. As we can see, both the second and the third land pressures overcome the top land pressure during the cycle, in addition, the pressure increase in these lands is smooth. This behavior can be explained with the flow pattern of the gas which can flow mainly through the ring gap. In the second row are given the results for 1000 rpm, the trend is similar to the upper case except for the reduced motion of the top ring in the range −100:100 c.a. which corresponds to the power phase of the cycle. Due to this reason, the second land peak pressure is lower than the previous case.

The first row of Fig. 6 depicts the 2000 rpm case. The top ring axial motion is slightly different from the previous cases, because it is stable at the groove floor for a long time and lifts up only between 300 and 500 c.a. This lift, which occurs at the same time with the second ring, is initially promoted by the gas pressure and then from the inertia force. This relationship also applies to the first lift of the second ring, while the third ring motion is strictly related to inertia.

The second land pressure is related to the second ring motion, because it marks an instant reduction at the time when the second ring moves from the top to the lower groove flank. The radial position of all the rings is stable, except for the second ring which seems to suffer the increase of speed.

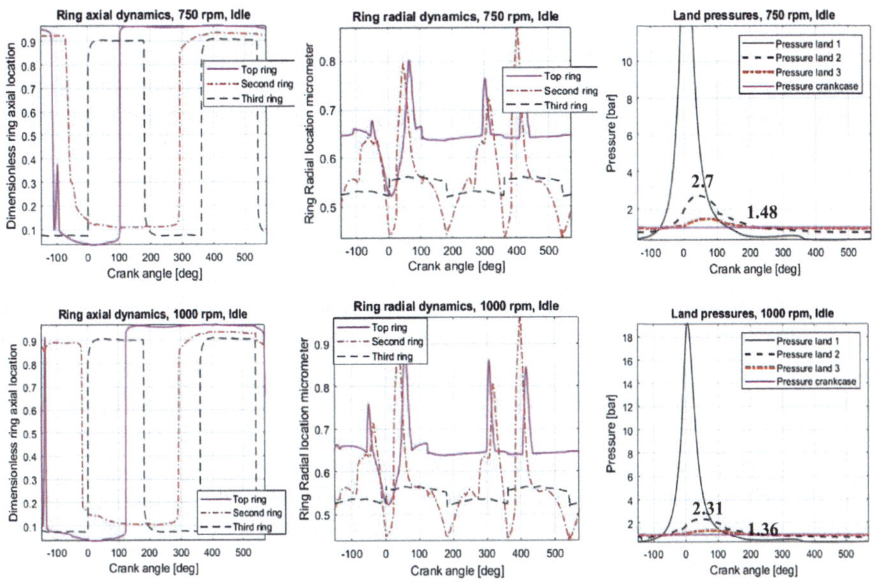

Fig. 5. Results for the first two idle conditions

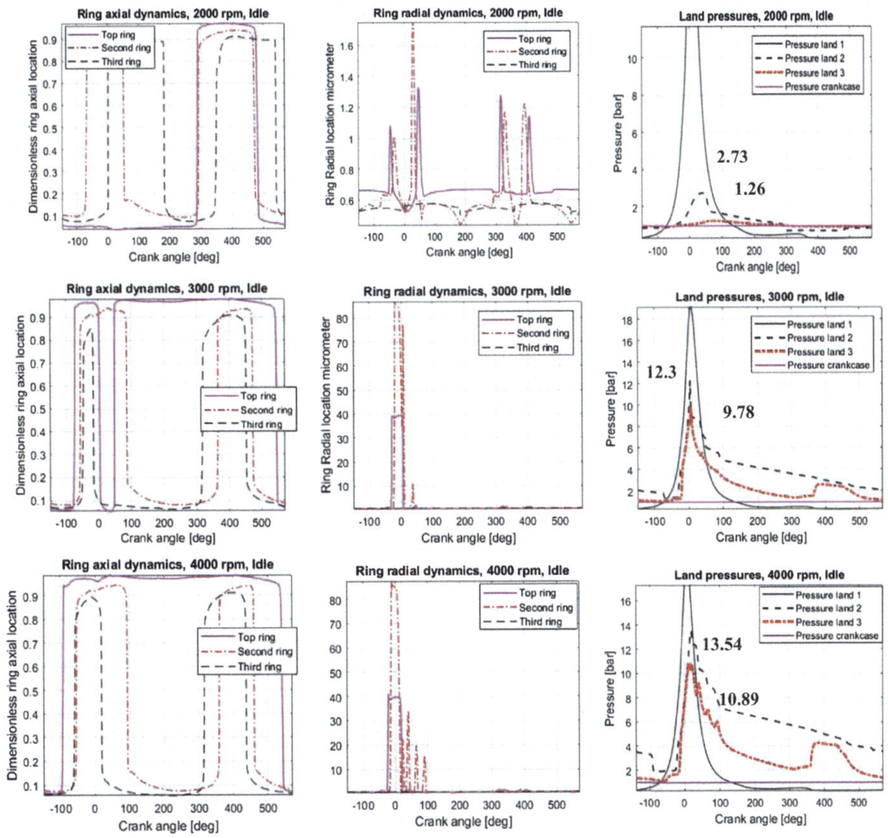

Fig. 6. Results for the remaining idle conditions

In comparison to the previous cases, at 3000 rpm the top ring is almost always lifted up in the groove, except for a short fall after the peak combustion pressure. The second and the third ring axial motions seems to replicate the inertia force, but with different periods, related to the high pressures in their lands. Both the compression rings undergo to collapse in the radial direction, and both of the times it occurs when the rings are lifted up. The gas pressure curves show a sharp rise when the rings are collapsed.

At 4000 rpm the behavior is similar, exception made for the top ring which is always lifted in the groove ceiling. The reason is related to an increased inertia force, which can overcome the gas pressure force even after the combustion peak.

Both the cases shows similar trend between them, where the top ring is almost always lifted up in its groove, while the other rings show some motions during the power phase and during the exhaust-intake phase (300–450 c.a.). The gas pressures in the second and third lands are abnormally high, and are mainly related to the radial collapse of the top and second rings, which occur in all the cases of this figure.

There are also two cases worthy of note in this figure:

In the first row, the top ring motion from top-to-bottom of the groove reduces the amount of radial collapse for this ring, this makes the pressure in the second and third land to be lower than the other graph.

In the second row, the top ring is stable at the upper flank of its groove but the second ring suffers multiple radial collapses, even after the peak firing pressure. As a consequence, the continued radial collapse of the second ring, combined with the axial position of the top ring, proves to be ineffective in the reduction of the gas flow.

4.2 30% Load

Figure 7 show the results of the simulations for the 30% load condition. According to the first row, when the load is 30% and the engine is running at 100 rpm, the top ring lifts to the upper groove flank between 150 and 250 c.a. when the piston is between the expansion and the exhaust stroke. This motion is caused by the gas pressure of the second land, which is high and acts underneath the top ring. After this point the pressure under the ring falls and inertia overcomes it, forcing the ring to change position in the groove. The last stroke of the cycle is the intake stroke, the gas is close to the ambient pressure and the inertia forces the ring to lift up. Due to these motions, the top ring leaves a large amount of gas to flow downwards and this makes the second land pressure to increase and reach a high value.

Due to this pressure, a high force is generated above the second ring and its motion is prevented until the last stroke of the cycle, when the pressure has fallen and inertia can overcome against it. The third ring has an axial flutter during the first stroke (compression), this is a result of the effectiveness of the second ring to prevent the gas flow due to its stable position. No one of the rings shows problems with the radial position.

The second row gives the results of the 2000 rpm case. The engine speed has doubled but the inertia has increased with the square of the speed. Due to this, the ring positions are now more similar to the inertia graph. The top ring lifts only at the last part of the cycle while the second ring shows a short lift before the peak firing pressure.

The lift is caused by the inertia, but while the pressure increases, the gas flowing from above forces the ring to sit down. The third ring lifts only when the inertia force changes direction. Due to the late lift of the top ring, the gas pressure on the second land is low in comparison to the previous case, which indicates that the top ring axial position is crucial on this matter.

In the remaining rows of Fig. 7 are given the 3000 and 4000 speed cases, which are both similar in their trends. The axial motion of the rings is entirely driven by the inertia force acting on them, exception made for the first ring, which is subjected by the gas pressure force until 300 c.a. The main changes with respect to the previous cases are encountered in the radial direction where the second ring suffers a radial collapse, which occurs several times few crank angles after the peak combustion pressures. As can be seen from the graphs, the second land pressure is lower than the previous cases but the second ring encounters problems with the radial collapse. This behavior is related to the axial position of the second ring, which cannot move from the top to the bottom of the groove due to the very high inertia force, and so, the gas pressure can only push the ring inwards into the groove and flow into the third land.

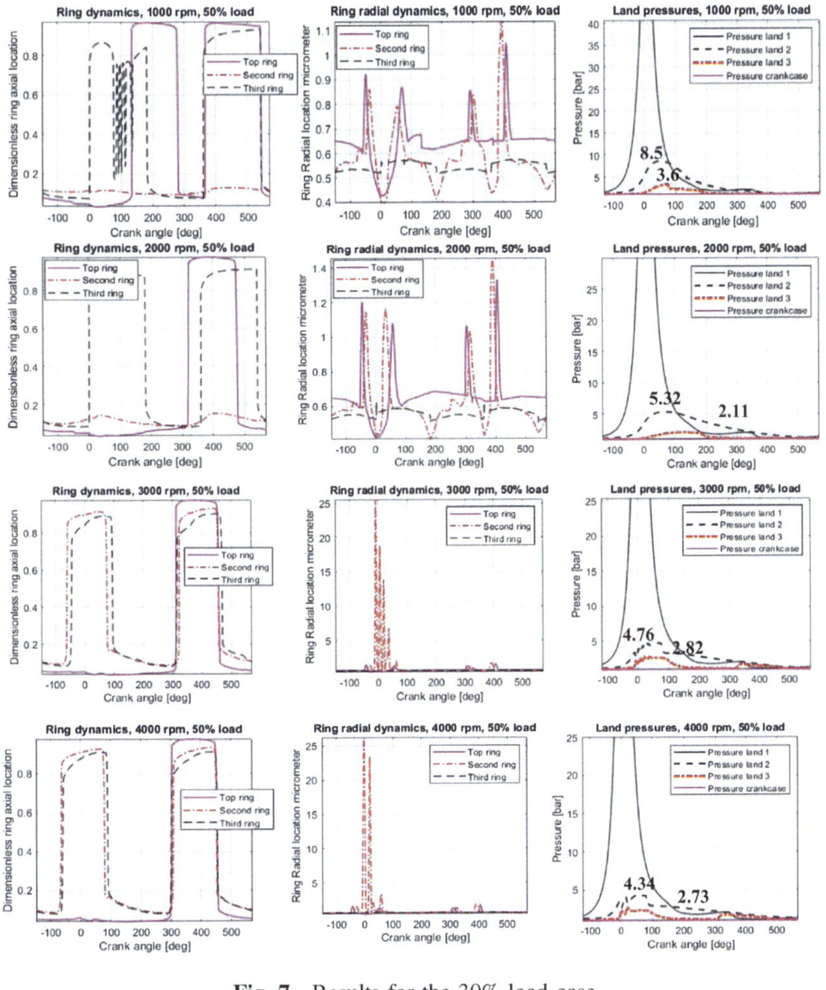

Fig. 7. Results for the 30% load case

4.3 50% Load

In Fig. 8 are given the results for the 50% load case. The results are similar to the results of Fig. 7. The ring axial and radial positions are almost the same and the main difference lays on the higher inter-ring pressures in the third graph, but this fact is normal because of the higher peak pressure into the combustion chamber in comparison to the previous case. Similarly to the previous case, the inter-ring pressures are higher at the lower speeds and tends to reduce when the speed increases. Furthermore, when the second ring collapses, the pressure curves for the second and third land shows a stepped rise.

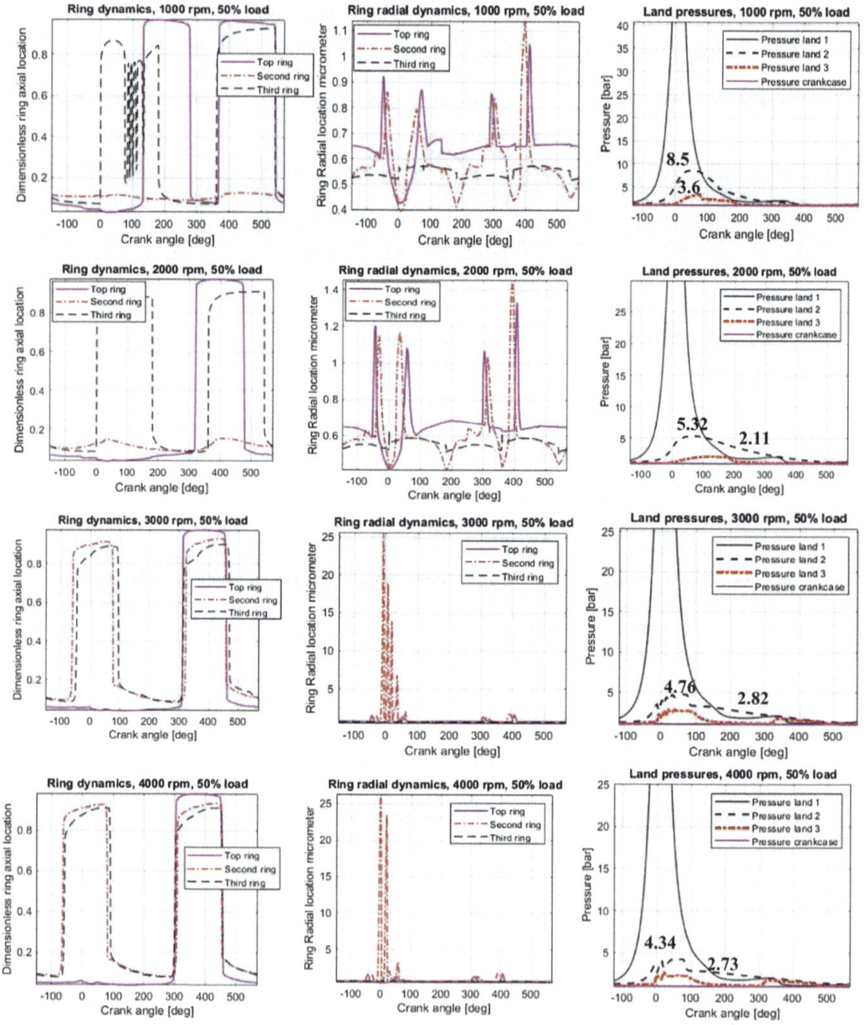

Fig. 8. Results for the 50% load case

4.4 70% and 100% Load

In Fig. 9 given only the results for the 100% load case since the graphs of the 70% case are very similar to those depicted here with small differences. The behavior of the system for each level of speed and for both levels of load are similar to the 50% load case, however, there are few differences between the two.

The main difference can be found in the peak pressure of the second land, for the 70% load and 1000 rpm, which is higher than the respective pressure of the 100% load. This difference is related to the time at which the top ring lifts from the bottom of the groove, which in the 70% case occurs earlier. In addition, at the speed of 2000 rpm, the top ring lift occurs earlier for the 70% load case, and this makes the second land pressure to be lower than the 100% load case.

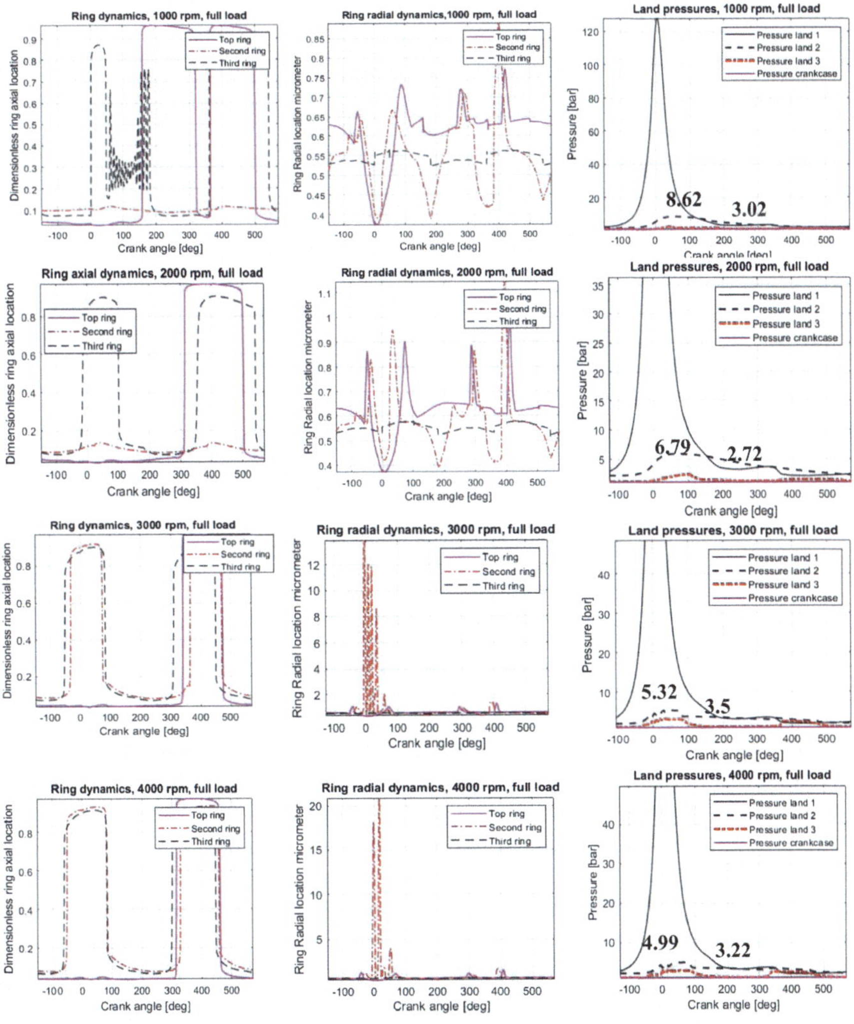

Fig. 9. Results for the full load case

5 Comparative Analysis

In Fig. 10 are given all the results obtained from the simulations in terms of cumulated blow-by. The cumulated blow-by is calculated making the integral of the mass flow of gas through the third land-groove, which ends up into the crankcase.

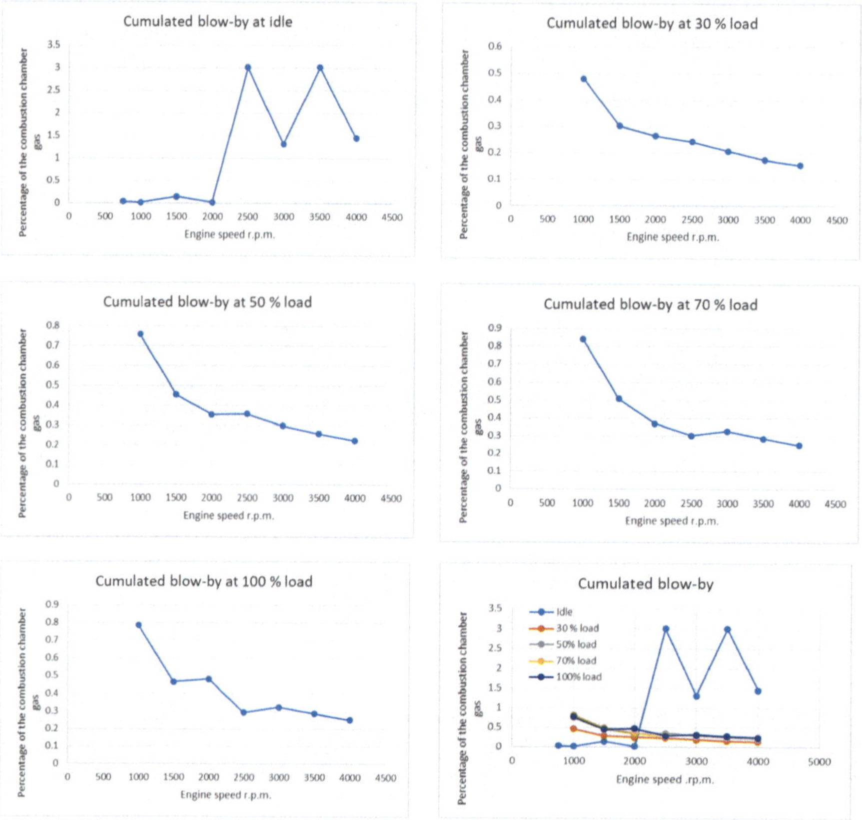

Fig. 10. Cumulated blow-by

According to the these graphs, except for the idle case, the highest values of blow-by is obtained at the lowest speeds. In addition, the obtained trend shows that the value decreases as long as speed increases. At idle, the values are close to zero until 2000 rpm and then sharply increase. According to discussions, at these speeds, the second ring radial collapse reaches a value which is double in comparison to all the other cases, furthermore, the top ring is continuously located at the upper groove flank. In addition to this, at speeds higher than 2000 rpm, also the top ring suffers the radial collapse. Another trend, which can be seen from figures, is related to the smoothness of the curves and the values between the lowest and the highest speed points. While load increases, the curve smoothness decreases and the extreme point values increases.

The last figure shows all the curves plotted on the same graph. From this figure is possible to understand in a concise and clear way, how the blow-by could be related to both speed and load. Exception made for the idle case, the blow-by shows to clearly decrease as the speed increases, and to slightly increase as the load increases. These results are in good correlation with the ones of Kim [19] and Irimescu [22], in partial correlation with the results of Arnault [5] but in opposition with the ones of Rabuté [13]. However, in reference [13] the engine was spark ignited, and thus the comparison must be made for that type of engine.

Furthermore, from the figure is possible to see that the variation of the blow-by gasses with speed and load is more relevant at the lower speeds. As long as the speed increases, the points are more close to each other and the difference in value reduces. After 2500 rpm, the 70% and 100% load cases have almost the same values.

If we neglect the idle case, which is not a useful condition for the normal drive, the highest blow-by for the engine under exam is obtained at 1000 rpm.

6 Conclusions

The present paper features the tightness analysis of a piston ring-pack for automotive application. The system behavior is analyzed when the full engine speed and load levels are varied. The obtained results are in very good correlation with the literature. However, the following important outcome emerged from the discussions:

- The results proved that the system is highly dependent on the engine speed and load.
- Apart from the idle condition, the blow-by resulted to decrease with speed and increase with load.
- The highest blow-by values were seen at high speed and idle condition, however, this point is not a usual operating point in everyday use of engines.
- The variation of blow-by between the lowest and the highest speed is higher than the variation between idle and full load.
- Between the two parameters, the engine speed showed to have a more relevant impact on the blow-by compared to the load.
- The inter-ring pressures are responsible for the inter-ring dynamics, in particular at low speed, when the inertia force acting on the ring is of low magnitude. As a result, the highest values of gas escape arise at low speed and high load.
- The inertia force is proportional to the ring mass and the ring acceleration. According to results, the blow-by values decrease as the inertia increase. Hence, heavier rings may produce larger benefits starting from the lower speeds.
- The paper provided a complete map of the blow-by against the engine speed and load for a diesel engine.
- The positive correlation with some of the reference literature may be an indication of the goodness of the model, and the analysis with the employed software. However, experimental tests could allow to fully validate the model and understand the physical behavior of the rings.

Conflict of Interests. The authors declare that there is no conflict of interest with third parties or institutions. And the outcome of the present work is the result of personal research. No funds have been granted for this works.

References

1. Zhao, J.X., Chia-fon, F.L.: Modeling of blow-by in a small-bore high-speed direct-injection optically accessible diesel engine. No. 2006-01-0649. SAE Technical Paper (2006)
2. Froelund, K., Ertan, Y.: Impact of engine oil consumption on particulate emissions. In: ICAT International Conference on Automotive Technology, Istanbul, Turkey (2004)
3. Green, R.M., Cloutman, L.D.: Planar LIF observations of unburned fuel escaping the upper ring-land crevice in an SI engine, SAE paper 970823 (1997)
4. Alkidas, A.C.: Combustion-chamber crevices: the major source of engine-out hydrocarbon emissions under fully warmed conditions. Prog. Energy Combust. Sci. **25**(3), 253–273 (1999)
5. Arnault, N., Samuel, B.: Engine lube-oil consumption stakes and benefits from significant blow-by oil mist reduction. No. 2012-01-1617. SAE Technical Paper (2012)
6. Furuhama, S., Tosio, T.A.D.A.: On the flow of gas through the piston-rings: 2nd report, the character of gas leakage. Bull. JSME **4**(16), 691–698 (1961)
7. Namazian, M., Heywood, J.B.: Flow in the piston-cylinder-ring crevices of a spark-ignition engine: effect on hydrocarbon emissions, efficiency and power. No. 820088. SAE Technical Paper (1982)
8. Keribar, R., Dursunkaya, Z., Flemming, M.F.: An integrated model of ring pack performance. ASME Trans. J. Eng. Gas Turbines Power **113**, 382–389 (1991)
9. Wannatong, K., Chanchaona, S., Sanitjai, S.: Simulation algorithm for piston ring dynamics. Simul. Model. Pract. Theory **16**(1), 127–146 (2008)
10. Iijima, N., et al.: An experimental study on phenomena of piston ring collapse. No. 2002-01-0483. SAE Technical Paper (2002)
11. Tian, T.: Dynamic behaviours of piston rings and their practical impact. Part 1: ring flutter and ring collapse and their effects on gas flow and oil transport. Proc. Inst. Mech. Eng. Part J J. Eng. Tribol. **216**(4), 209–228 (2002)
12. Tian, T.: Dynamic behaviours of piston rings and their practical impact. Part 2: oil transport, friction and wear of ring/liner interface and the effects of piston and ring dynamics. Proc. Inst. Mech. Eng. Part J J. Eng. Tribol. **216**(4), 229–248 (2002)
13. Rabute, R., Tian, T: Challenges involved in piston top ring designs for modern SI engines. Trans.-Am. Soc. Mech. Eng. J. Eng. Gas Turbines Power **123**(2), 448–459 (2001)
14. Cheng, C., Schock, H., Richardson, D.: The dynamics of second ring flutter and collapse in modern diesel engines. J. Eng. Gas Turbines Power **137**(11), 111504 (2015)
15. Delprete, C., Selmani, E., Bisha, A.: Gas escape to crankcase: impact of system parameters on sealing behavior of a piston cylinder ring pack. Int. J. Energy Environ. Eng. **10**(2), 207–220 (2019)
16. Selmani, E., Delprete, C., Bisha, A.: Simulation of the cylinder bore distortion and effect on the sealing capacity of the ringpack. SN Appl. Sci. **1**(4), 314 (2019)
17. Erjon, S.: The geometrical parameters of a piston-ringpack assembly and their impact on the blow by gasses. In: Fifth International Conference on "Social and Natural Sciences – Global Challenge 2019" (ICSNS V-2019), Vienna 13 April 2019 (2019). ISBN: 978-9928-259-15-8
18. Curtis, J.M.: Piston ring dynamics and its influence on the power cylinder performance. No. 810935. SAE Technical Paper (1981)

19. Kim, C., Bae, C., Choi, S.: Gas flows through the inter-ring crevice and their influence on UHC emissions. No. 1999-01-1533. SAE Technical Paper (1999)
20. Tamminen, J., Sandström, C.-E., Nurmi, H.: Influence of the piston inter-ring pressure on the ring pack behaviour in a medium speed diesel engine. No. 2005-01-3847. SAE Technical Paper (2005)
21. Zottin, W., Clemente, M., Ribeiro, C.: Application of a piston ring dynamics simulation code to solve blow-by fluctuation of a heavy-duty diesel engine in transient load operation. No. 942392. SAE Technical Paper (1994)
22. Irimescu, A., et al.: Compression ratio and blow-by rates estimation based on motored pressure trace analysis for an optical spark ignition engine. Appl. Therm. Eng. **61**(2), 101–109 (2013)

Application of RFID Technology for Better Efficiency of Resource Planning

Gordana Zeba(✉) and Mirjana Čičak

Mechanical Engineering Faculty in Slavonski Brod, Josip Juraj Strossmayer University of Osijek, Trg Ivane Brlić-Mažuranić 2, 35000 Slavonski Brod, Croatia
gzeba@sfsb.hr

Abstract. Industry 4.0 has a significant impact on global manufacturing and the transformation of business processes. Manufacturers have to adapt fast to the requirements of Industry 4.0, due to increasing competition in the global market and the necessity for mass customized products. The main goal of Industry 4.0 is a smart factory and to achieve this goal is a particularly important integration (horizontal, vertical, end-to-end). RFID technology (as wireless automatic identification technology that facilitates tracking and traces of resources) is one of the technologies that support the Internet of Things which is crucial for Industry 4.0 and integration. Efficient resource planning is one of the challenges of Industry 4.0, and integration of the Manufacturing Execution System (MES) and Enterprise Resource Planning (ERP) within the enterprise, as well as the integration with the business partners (suppliers and customers), is indispensable. Modern ERP systems facilitate efficient management of business processes and resources (such as people, machines, materials), concerning all data within the enterprise and throughout the value chain. The application of RFID technology is of great importance for the acquisition of real-time information needed to make management decisions. The goal of this paper is to present trends in the research and advantages of the application of Radio-frequency Identification (RFID) technology for better efficiency of resource planning.

Keywords: RFID · ERP · Industry 4.0 · Integration

1 Introduction

The rapid progress of manufacturing and information and communication technologies and their introduction into production has led to the fourth industrial revolution. New technologies such as Big Data Analytics, Cloud Computing, Artificial Intelligence, Wireless Sensor Networks, Internet of Things and Internet of Services form the basis of the Industry 4.0 paradigm. The Internet of Things is one of the key technologies that support Industry 4.0. Industry 4.0 is a technology framework that emphasized the importance of facilitating vertical, horizontal and end-to-end integration for Smart Factory.

The manufacturing industry has undergone three significant transformations throughout history, and one is in progress, which are referred to as industrial

I. Karabegović (Ed.): NT 2020, LNNS 128, pp. 382–387, 2020.
https://doi.org/10.1007/978-3-030-46817-0_43

(technological) revolutions. At present, manufacturing is undergoing a digital transformation within the fourth industrial revolution, called Industry 4.0, using Cyber-Physical Systems, Artificial Intelligence and other innovative technologies (Fig. 1).

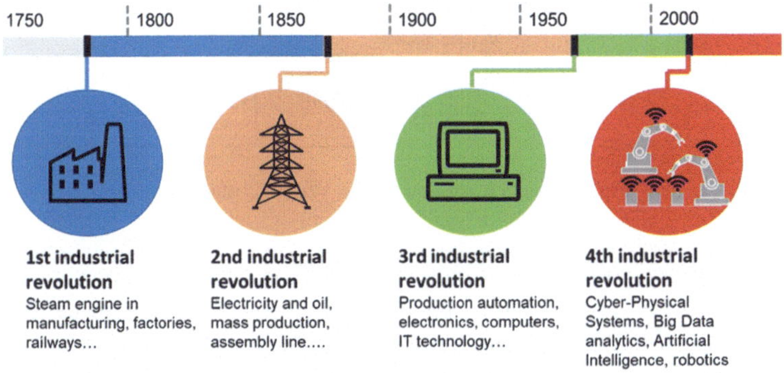

Fig. 1. Industrial revolutions throughout history

The new technologies of Industry 4.0 facilitate digital transformation and integration of the traditional industry. Manufacturing has several challenges within the Industry 4.0 framework, one of which is resource planning [1]. Efficient resource planning is crucial because it ensures the optimal use of resources in the enterprise. An important problem with resource planning and decision making is the lack of necessary information about resources. For the decision-making process, it is necessary to provide accurate and timely information on the production status and availability of all resources. It is necessary to provide real-time information that is collected in the production by Radio Frequency Identification (RFID) technology and sensors network, and through Manufacturing Execution System (MES) delivered to Enterprise Resource Planning (ERP) systems as the highest level of management. The authors of the paper [2] believe that the application of RFID technology provides a good foundation for the challenges of manufacturing under Industry 4.0.

2 Trends in Literature

We searched the platform of citation databases Web of Science Core Collection (WoSCC) by keyword "resource planning", within timespan: 1955 - December 2019, which yielded 4960 results (only conference papers and journal articles did search). After refining the results with the keywords "Internet of Things" and "RFID", we obtained 28 and 3 results, respectively (Fig. 2). That indicates the need for further research in the field of using RFID technology in the management of resources.

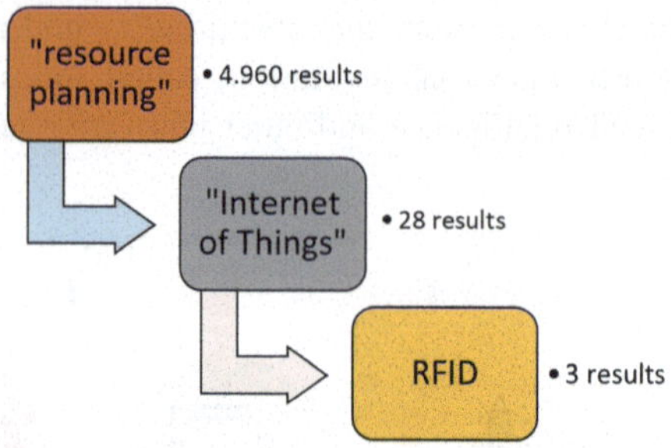

Fig. 2. Search results on the WoSCC platform

We also analyzed publications within the WoSCC databases by keywords "Industry 4.0", "manufacturing" and "resource planning" to determine trends in the researches, by the frequency of keywords being repeated. We used Atlas.ti software to process the publication from WoSCC platform and the result is shown in the word cloud with the most prominent terms appearing (Fig. 3).

Fig. 3. Word Cloud with the prominent words for topic "Industry 4.0", "manufacturing" and "resource planning"

The most common keywords are manufacturing, systems, control, industry, information, enterprise, planning, management, smart, data, resource. Furthermore, there are keywords: Internet, things, RFID, real, time, integrated. The font size is proportional to the frequency of occurrence of a word. These words highlight trends in the researches on resource planning within the Industry 4.0 framework.

3 Enterprise Resource Planning

Enterprise Resource Planning (ERP) systems improve efficiency and effectiveness by planning the use of resources [3]. ERP system is a software that automates and integrates business processes and provides a basis for making decisions based on real-time information. The development of the ERP systems was carried out of the first software packages for inventory control '60s via MRP and MRP II system '70s and '80s, to the intelligent ERP today. Intelligent ERP systems are a new generation of ERP systems that can automate and optimize business processes and manage resources using real-time information. New technologies that are significant for modern ERP systems are Cloud Computing, Big Data Analytics, Internet of Things, In-Memory Computing, Software as a Service, etc. These technologies are transforming traditional ERP systems into modern intelligent ERP systems that are an integral part of Industry 4.0 and form the backbone of businesses through innovative real-time resource management. Application of SaaS (Software as a Service) technology combined with the Internet of Things technology facilitates the integration of physical resources and services in the cloud [4].

4 RFID Technology

Radio Frequency Identification (RFID) technology is part of a group of automated data collection technologies that facilitates resource planning. It is used to uniquely identify and track and trace objects with attached RFID transponders in real-time. All resources in an enterprise with attached RFID transponders are smart objects thanks to carrying information about themselves and have the ability to communicate with their environment (other objects, control systems, machines, workers, etc.) wirelessly via radio waves. Besides unique serial numbers (EPC Electronic Product Code), RFID transponders may also contain other information about the object, relevant to operations management. The RFID system includes three primary components: a transponder (with a chip and antenna), an interrogator (reader) with an antenna for communication between the reader and the transponder and server.

The main problems associated with the use of RFID systems is the high cost, and the possibility of electronic interference depending on the environment, which may affect the reliability of the system. The consequence of the application of RFID technology with an ERP system has improved visibility on different levels of the manufacturing system (with the aim of vertical integration) and the exchange of information with business partners along the value chain (horizontal integration). The use of RFID technology for production management has become necessary [5].

5 Enterprise Integration

Manufacturing must be flexible to the requirements of customers for individualized products and mass-customized manufacturing. Manufacturers must respond quickly to changing customer requirements, changes in inventory levels, or changes occurring in production due to unforeseen circumstances (such as changes in production order, machine failures, errors, quality fluctuations). For such a rapid response, the manufacturer must have real-time data whose collection is facilitated by the use of smart sensor technologies, including RFID technology. RFID technology, Wireless Sensor Network and Internet of Things technology ensure the integration of all levels of management within the enterprise and between enterprises in the value chain. Smart products and resources are connected, and materials and parts needed can be located at any time [6].

Horizontal integration connects enterprise information systems of different enterprises, thus transferring information along the entire value chain (from supplier to customer). This way of integration facilitates resource planning in enterprises along the value chain (as they exchange information on stock status, planned deliveries). Vertical integration is the integration of levels of an industrial enterprise. There are four levels according to the standard ISA-95 that focuses on levels 3 and 4, respectively on collaboration between MES and ERP systems (levels 0, 1 and 2 are the levels of process control).

By integrating all processes and using sensors, RFID transponders, actuators, the programmable logic controller (PLC) at each stage of production, resource planning in real-time is facilitating. ERP systems must support real-time decision making and integration, which is crucial in the context of Industry 4.0. Real-time integration using RFID technology and Big Data analytics lead to the optimization of resources in the manufacturing processes and better performance [7]. It is necessary to ensure the vertical integration of enterprises from the lowest level of production processes to which RFID transponders are applied, through the MES system, to the highest level of enterprise management, i.e. ERP system (Fig. 4).

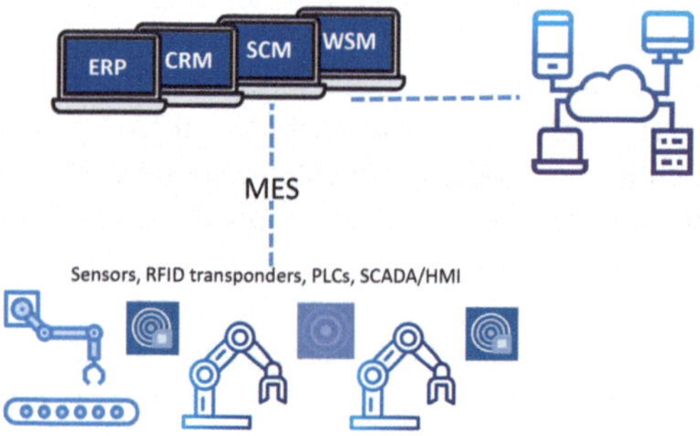

Fig. 4. Integration of an enterprise

Horizontal integration of enterprise with other enterprises in the value chain through Supply Chain Management (SCM) and Customer Relationship Management (CRM) applications (data sharing between customers and suppliers to make the data available in real-time) is also very important. Data in the modern ERP system can be accessed via mobile devices from different locations and at any time for faster response and decision making.

6 Conclusion

In a smart factory, which is the goal of Industry 4.0, manufacturing resources and products are smart and interact with each other, thanks to the usage of RFID and other innovative technologies. Timely and accurate data collected with RFID technology facilitate production monitoring and management, thus reducing losses. Modern ERP systems can efficiently manage resources thanks to RFID technology that automatically collects accurate data on the state of production and the status of resources, in real-time. Enterprises must invest in new innovative technologies in the context of Industry 4.0 to achieve competitive advantages.

References

1. Karabegović, I., Karabegović, E., Mahmić, M., Husak, E.: Implementation of Industry 4.0 and industrial robots in the manufacturing processes. In: Karabegović, I. (ed.) New Technologies, Development and Application II. NT 2019. Lecture Notes in Networks and Systems, vol. 76, pp. 3–14. Springer, Cham (2019)
2. Mladineo, M., Veza, I., Gjeldum, N., et al.: Integration and testing of the RFID-enabled smart factory concept within the learning factory. Procedia Manuf. **31**, 384–389 (2019)
3. Beheshti, H.M., Beheshti, C.M.: Improving productivity and firm performance with enterprise resource planning. Enterp. Inf. Syst. **4**(4), 445–472 (2010)
4. Chen, S.-L., Chen, Y.-Y., Hsu, C.: A new approach to integrate internet-of-things and software-as-a-service model for logistic systems: a case study. Sensors **14**(4), 6144–6164 (2014)
5. Kamigaki, T.: Object-oriented RFID with IoT: a design concept of information systems in manufacturing. Electronics **6**(1), 14 (2017)
6. Tjahjono, B., Esplugues, C., Ares, E., Pelaez, G.: What does Industry 4.0 mean to supply chain? Procedia Manuf. **13**, 1175–1182 (2017)
7. Lu, Y.: Industry 4.0: a survey on technologies, applications and open research issues. J. Ind. Inf. Integr. **6**, 1–10 (2017)

Retrofitting Techniques for Agricultural Machines

Giampiero Celenta[1] and Marco Claudio De Simone[2(✉)]

[1] MEID4 Academic Spin-Off of the University of Salerno, Via Giovanni Paolo II, 132, 84084 Fisciano, SA, Italy
[2] Department of Industrial Engineering, University of Salerno, Via Giovanni Paolo II, 84084 Fisciano, SA, Italy
mdesimone@unisa.it

Abstract. Intelligent machines and systems are critical elements for future integrated infrastructures based on human-machine interaction and information sharing. To achieve this, it is necessary to design and build a new generation of machines and systems capable of interacting as nodes of the same network. However, this revolution has the limitation of condemning all the machines already on the market to obsolescence. In this work, a retrofitting activity is reported on a tracked vehicle to make the vehicle capable of carrying out operations independently. The goal is to create a machine for agricultural applications, capable of assisting harvesting operations in mountainous areas in the presence of steep terrain, preserving the operator from any risk.

Keywords: Retrofitting · Machine design · Automation · Control systems · Robotics

1 Introduction

The interest in unmanned guided vehicles (UGVs) or systems (USs), more in general, is growing at a constant rate. Many universities and companies are investing a lot of effort in R&D of UVSs for many sectors, and new applications are presented nearly every day. Among all, the authors are particularly interested in the use of autonomous systems for security and monitoring activities and applications in agriculture. In literature, it is possible to find many examples of robotic applications for agriculture for medium-large machinery for applications in large plots of land. In fact, for some time now, machines capable of working in the open field independently and systems for the automation of stables have been available on the market [1]. These machines, with a non-trivial cost, allow increasing the level of automation of farms allowing them to be competitive on the market.

However, very little has been done concerning small machines for steep terrain or mountainous areas where economies of scale cannot be exploited, and the risk for the operator is very high. For this second type of machine, the market offers few solutions with a not very reduced cost, which, therefore, cannot make space in the economic contexts where the labor cost is decidedly low. Moreover, the development of autonomous versions often involves a total redesign of the machine, leaving out systems

I. Karabegović (Ed.): NT 2020, LNNS 128, pp. 388–396, 2020.
https://doi.org/10.1007/978-3-030-46817-0_44

already on the market, forcing those who want to use such technology to rebuy the equipment. In summary, this work describes the retrofitting activity of an agricultural machine for steep terrain, capable of carrying out autonomous operations at low cost.

This paper is organized as follows. In Sect. 2, a short literature survey is reported. In Sect. 3, the research methodology and the machine used for the experimental activity are described in detail, followed by the results (Sect. 4) and finally by the conclusions.

2 Literature Review

The intense competitiveness introduced by globalization, the accelerated population growth, and the continuous shortage of labor in the agricultural sector, are the primary motivations for the growing interest in agricultural robotics and precision farming. Nowadays, agricultural vehicles play a fundamental role in the farming sector, and a lot of research activities related to autonomous navigation has increasingly taken place in recent years [2]. The technical developments achieved in the last two decades and the constant reduction in the cost of electronic devices makes it possible to develop autonomous tractors and implements for arable farming. Despite safety aspects, such machines open new possibilities in the design of new tools and cropping systems. Furthermore, as the use of artificial intelligence becomes more common in agriculture, the industry is going through enormous changes. Taranis represents an example of an agriculture intelligence platform selected to be part of John Deere's startup collaboration. It uses data science, computer vision, and deep learning techniques for enabling farmers to make informed decisions [3]. Such technologies can play a critically important role in improving agricultural productivity while at the same time reducing or maintaining land area used for agricultural purposes. For example, drones have the potential to scout crops and to reduce the hard work involved in seed planting and fertilization, or automated irrigation systems can enhance precision and reduction of water use and manual labor. Moreover, low productivity has impacts that extend beyond the agricultural sector. Low productivity prevents households and countries from accumulating assets, diversifying their economies, and making use of new technologies to move into higher value-added areas. Furthermore, a consequence of not adopting the latest technologies, makes countries rely only on expanding the area of land under cultivation to increase agricultural output, exacerbating deforestation, and environmental sustainability [4]. Moreover, the possible use of small machines for areas that are difficult to access could make remote areas, challenging to cultivate, attractive again, thus guaranteeing control of the territory and a lowering of the hydrogeological risk. Obviously, the characteristics that the machine must have, will have to be substantially different compared to open field working machines. Moreover, the possible use of small machines for areas that are difficult to access could make remote areas, challenging to cultivate, attractive again, thus guaranteeing control of the territory and a lowering of the hydrogeological risk. Obviously, the characteristics that the machine must have will have to be substantially different compared to open field working machines. For the design of off-road machines, the study of the interaction between wheel and ground is of primary importance. The subject that studies the characterization of the interaction between the vehicle and the ground is defined as Terramechanics. The types of stresses that are present when a vehicle is in motion are mainly of two kinds. One is due to the weight of the vehicle that acts perpendicularly to the ground, and the other is due to the torque applied to the sprocket that provides the motion of the vehicle [5]. The normal response to the ground, modeled with the pressure-sinkage relationship proposed by Bekker is reported in (1):

$$p = \left(\frac{k_c}{b} + k_\phi\right) z^n \tag{1}$$

with p the pressure due to the weight of the vehicle, b is the width of the track, z is the sinkage and n, k_c and k_ϕ are pressure sinkage parameters [6]. The shear stress, instead, allows evaluating the shearing action between the running gear and the torque provided by the sprocket. Once evaluated such parameters, it is possible to assess the motion resistance of the vehicle, function of the terrain parameters reported in (2).

$$R_v = \frac{1}{(n+1) b^{1/n} (k_c/b + k_\phi)^{1/n}} \left(\frac{W}{l}\right)^{(n+1)/n} \tag{2}$$

where W and l are respectively, the weight and the distance traveled by the vehicle. Another essential aspect of developing an autonomous vehicle is the propulsion system. An electric motor allows easy control of the actuators but introduces non-trivial problems such as the need to install batteries that will reduce the payload and the distances that can be carried out in total autonomy. Once the functional requirements have been established, through the use of 3D design software and the use of rigid and flexible multibody simulators, it is possible to view and analyze the dynamic behavior of the vehicle by validating the design choices [7, 8]. This paper, however, presents a hybrid solution designed to recover machines already on the market by installing devices thought specifically to retrofit machines and systems for low-cost agricultural applications [9]. In this way, it will be possible to reach a vast fleet of machines still in good condition and take advantage of the high autonomy allowed by internal combustion engines to carry out missions autonomously.

3 Research Methodology

The vehicle used for this experimental activity is a Minitransporter Tekna TT350 equipped with two rubber tracks driven by a 2 HP 4 Strokes petrol engine, with a maximum load capacity of 350 kg equipped with a SUBARU 4-stroke engine that develops a maximum power of 4.5 Hp at 4000 rpm. The torque curve of this engine, typical of agricultural machines, allows the vehicle to be operated in harsh conditions such as the presence of mud or obstacles. Furthermore, thanks to the low position of the center of gravity, the machine can tackle average slopes of 20%. The manual control of the vehicles is ensured by two operating levers and a driving bar. The gear bar engages the main clutch, which has the purpose of transferring the motion from the crankshaft to the gearbox unit as well as to the differential mechanism necessary for the movement of the tracks. The operating levers, on the other hand, stop the left or right axle of the differential lock respectively, allowing the vehicle to rotate. The 4-speed gearbox consists of 3 different reduction gears for moving forward and one reverse gear for moving backward. In order to act on different track linkages, we installed a four-module hydraulic system (see Fig. 1a). Each module consists of a hydraulic cylinder controlled by proportional solenoid valves. The module, part of the hydraulic block, has the function of pushing or pulling the command to which it is hooked thanks with the aid of a double-acting piston. In order to perform its function, the manifold of the hydraulic block needs to be connected to a pressurized oil circuit with an attached hydraulic pump. The cartridge acts as a shutter in the delivery valve to the double-acting piston. The actuation of the cartridge, ensured by the solenoid, allows the shutter to open thanks to the magnetic field generated. The displacement of the piston is proportional to the current value supplied to the solenoid.

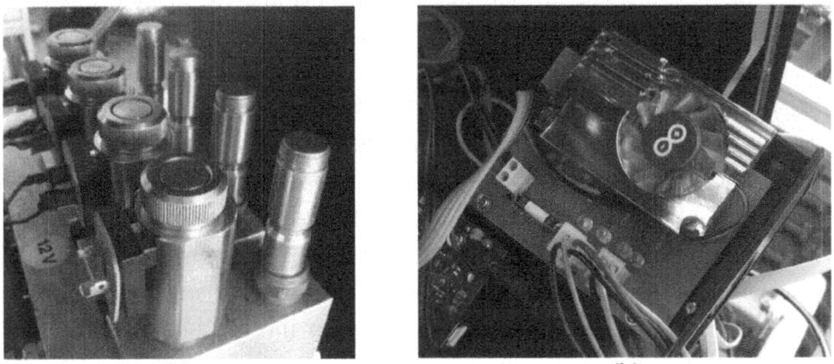

(a) (b)

Fig. 1. The UTV: (a) the retrofitting control device; (b) The C.S.V. board; (c) The electro-hydraulic actuator installed

The current value I_{max} equal to 1.8A is the one necessary to move the piston to the end of the stroke. For lower current values, the piston will stop in intermediate positions proportionally to the current value. When the solenoid current fails, the piston is called to the rest position by the spring inside the module. The C.S.V. (control solenoid valve) reported in Fig. 1b is a power circuit whose purpose is the interface between the Arduino prototyping board and a group of solenoids for the control of hydraulic actuators. The need to create this interface is forced by the maximum current limit of the prototyping board. Arduino supports maximum currents of 40 mA per I/O port with voltages not exceeding 5 V, while the group of solenoids to be managed works with 12 V voltages and maximum absorptions of 1.8 A.

The discrepancy between the voltage and current values has made the simple circuitry necessary. In order to implement the mechanisms necessary to control the vehicle and process the data from the sensors, it was necessary to use an Arduino prototyping board with an ATmega2560 microcontroller. Arduino is an open-source prototyping platform electronics based on flexible hardware and software. It can

Table 1. Tracked vehicle properties

Technical details		Value
Vehicle dimension	Height	1090 mm
	Length	1610 mm
Track dimension	Width	180 mm
	Length	580 mm
Transmission	Mechanical	3+1 gears
Weight		190 kg

perceive the surrounding environment through different inputs and can interact with it, thanks to as many outputs. The microcontroller on the board is programmed using the Arduino programming language (based on Wiring) and the Arduino development environment (based on Processing). The Arduino Mega 2560 board is an ATmega2560 based microcontroller. It has 54 input/output pins (of which 14 can be used as PWM outputs), 16 analog inputs, 4 UART (hardware serial ports), a USB connection, a power jack, an ICSP header, and a reset button. Various sensors have been installed on the machine for autonomous navigation. SRF05 ultrasonic sensors are used for identifying any obstacles along the way. Such sensors are used primarily for vehicle safety but can also be used to identify objects or for chasing activities. An inertial platform, a GPS sensor for vehicle geolocation and a datalink for data transmission have also been installed on the machine. In Fig. 2a shows the Tekna TT350 mini transporter before the retrofitting process. The hydraulic system used for maneuvering the bucket influenced the choice to install the group of solenoid valves for the implementation of the levers.

(a) (b)

Fig. 2. The UTV: (a) the custom vehicle; (b) the retrofitted vehicle

Figure 2b, instead, shows the retrofitted vehicle in which the bucket has been removed. From the photo, it is possible to notice the protective plate of the solenoid valve unit and the ultrasonic sensors used for navigation. The retrofitted machine has been designed for applications in vineyards (see Fig. 3a), where the plants generally have a regular shape but could also be used for different contexts such as mountainous or hilly ones.

4 Results

The idea was to design a machine capable of navigating autonomously within a vineyard assisting the operator by following him, as shown in Fig. 3b, during his duties in the vineyard or navigating autonomously through the use of waypoints [10, 11]. For safety reasons, the machine must always be kept at a distance of 1 m from the operator and must be able to distinguish him from the support poles of the vines and the plants themselves. Recognition is always possible thanks to the use of ultrasonic sensors and techniques based on neural networks [12].

(a) (b)

Fig. 3. Experimental activities: (a) navigation scheme for the vineyard; (b) prey tracking activity

The machine is also designed for precision agriculture applications [13]. Specifically, the system must be able to move independently within the vineyard to analyze the vegetative state of the plants, the possible need for fertilizers, or protective treatments against diseases. For this reason, through a ground station created ad hoc, it is possible to send the GPS coordinates of the field to be investigated to the machine [14]. Through the positioning sensor on the machine, the vehicle will be able to move within the vines, developing a real-time map of the environment that surrounds it, indicating the critical issues encountered [15, 16].

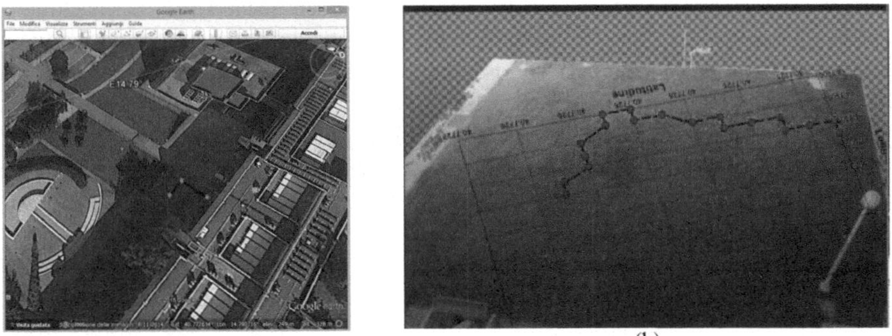

(a) (b)

Fig. 4. Autonomous navigation: (a) the google hearth waypoint setting; (b) Gps readings during autonomous navigation

Figure 4a shows the graphic interface created for the control station. From this window it is possible to select the GPS coordinates for the definition of the trajectory to be covered. The experimental campaigns were conducted in the campus of the University of Salerno. Figure 4b instead shows the readings of the gps sensor positioned on the machine during autonomous navigation [17]. The position detected by the sensor is obviously characterized by an average error of about 50 cm. This error is compensated by differential navigation techniques while driving [18]. For controlled navigation in narrow environments such as those found in a vineyard, autonomous navigation is obtained by using sensor fusion techniques by merging the measurements given by all the sensors present on the machine [19, 20].

5 Conclusions

The efforts of academia and industry to find sustainable solutions from an economic and environmental point of view involve more and more sectorsdelete. The several benefits that can be achieved by using intelligent machines and systems in agriculture can be summarized in the optimization of the use of land and water resources, the reduction of the use of pesticides and fertilizers, and a reduction in costsdelete. Another non-trivial aspect is the strong reduction in the risk of accidents in extreme scenarios such as mountainous contexts or on steep slopes. Producing an unmanned version of a machine often requires a total redesign of the system by the manufacturer, starting from the propulsion that is generally preferred electric. However, this transformation condemns all the models currently used in agriculture. For this reason, it was decided to invest in the retrofitting of existing machines by using low-cost components and open-source softwaredelete.

The retrofitting device is designed ad-hoc on the characteristics of the machine. In this case, the vehicle is a tracked mini transport powered by an internal combustion engine and endowed with a hydraulic circuit for maneuvering the bucket. For this reason, electro-hydraulic actuators have been installed for managing the leverages that govern the machine using PWM signals. The controller used for this activity is an Arduino Mega2560, an open-source controller. The activity required by the vehicle is the continuous monitoring to be carried out in a vineyard through autonomous inspections. For this reason, ultrasonic sensors have been installed on the machine to identify obstacles and objects, inertial platforms, and a GPS sensor for autonomous navigation. For this specific application, the GA-003 NEO-M8 based GNSS receiver was installed on the vehicle. The experimental tests proved the ability of the retrofitted system to move in dynamic contexts in complete safety. In addition, the choice of installed sensors allows the machine to work with the operator in complete safety. The next step will be to test the vehicle in extreme conditions such as forest scenarios or in case of steep slopes.

References

1. Gradetsky, V., Knyazkov, M.: Unmanned vehicle systems in unstructured environments: challenge and current status. In: 21st International Symposium on Measurement and Control in Robotics, ISMCR 2018, pp. 110–131 (2019)
2. Fernandez, B., Herrera, P.J., Cerrada, J.A.: A simplified optimal path following controller for an agricultural skid-steering robot. IEEE Access **7**, 95932–95940 (2019). Art. no. 8763950
3. Bandoim, L.: How self-driving tractors and AI are changing agriculture. Forbes, 27 April 2019
4. Sanders, A.K., Falcão, T., Haider, A., Jambeck, J., LaPointe, C., Vickers, C., Ziebarth, N.: World Economic and Social Survey 2018: Frontier Technologies for Sustainable Development (2018)
5. Wong, J.W.: Terramechanics and Off-Road Vehicle Engineering: Terrain Behaviour, Off-Road Vehicle Performance and Design. Butterworth-Heinemann, Oxford (2009)
6. De Simone, M.C., Guida, D.: On the development of a low-cost device for retrofitting tracked vehicles for autonomous navigation. In: AIMETA 2017 - Proceedings of the 23rd Conference of the Italian Association of Theoretical and Applied Mechanics, vol. 4, pp. 71–82 (2017)
7. Formato, A., Ianniello, D., Villecco, F., Lenza, T.L.L., Guida, D.: Design optimization of the plough working surface by computerized mathematical model. Emirates J. Food Agric. **29** (1), 36–44 (2017)
8. Pappalardo, C.M., Guida, D.: On the use of the Udwadia-Kalaba equations for the nonlinear control of a generalized Van Der Pol-Duffing oscillator. In: International Conference "New Technologies, Development and Applications", pp. 82–95. Springer, Cham (2019)
9. Pappalardo, C.M., Lombardi, N., Dašić, P.V., Guida, D.: Design and development of a virtual model of an electric vehicle of category L7. In: IOP Conference Series: Materials Science and Engineering, vol. 568, no. 1, p. 012114. IOP Publishing (2019)
10. De Simone, M.C., Rivera, Z.B., Guida, D.: Obstacle avoidance system for unmanned ground vehicles by using ultrasonic sensors. Machines **6**(2), art. no. 18 (2018)
11. Formato, A., Ianniello, D., Romano, R., Pellegrino, A., Villecco, F.: Design and development of a new press for grape marc. Machines **7**(3), art. no. 51 (2019)
12. Formato, A., Guida, D., Ianniello, D., Villecco, F., Lenza, T., Pellegrino, A.: Design of delivery valve for hydraulic pumps. Machines **6**(4), 44 (2018)
13. Pappalardo, C.M., Guida, D.: On the dynamics and control of underactuated nonholonomic mechanical systems and applications to mobile robots. Arch. Appl. Mech. **89**(4), 669–698 (2019)
14. Pappalardo, C.M., Guida, D.: Development of a new inertial-based vibration absorber for the active vibration control of flexible structures. Eng. Lett. **26**(3), 372–385 (2018)
15. De Simone, M.C., Guida, D.: Identification and control of a unmanned ground vehicle by using Arduino. UPB Sci. Bull. Ser. D **80**, 141–154 (2018)
16. Jegathesh Amalraj, J., Banumathi, S., Jereena John, J.: A study on smart irrigation systems for agriculture using IoT. Int. J. Sci. Technol. Res. **8**(12), 1935–1938 (2019)
17. Asok Kumar, A., Bindu, G.R., Cherian, E., Parvathy, M.L.: Energy saving and economic analysis of switched reluctance motor in agricultural applications. Technol. Econ. Smart Grids Sustain. Energy **5**(1), 2 (2020)
18. Domínguez-Niño, J.M., Oliver-Manera, J., Girona, J., Casadesús, J.: Differential irrigation scheduling by an automated algorithm of water balance tuned by capacitance-type soil moisture sensors. Agric. Water Manag. **228**, 105880 (2020)

19. Mistry, I., Tanwar, S., Tyagi, S., Kumar, N.: Blockchain for 5G-enabled IoT for industrial automation: a systematic review, solutions, and challenges. Mech. Syst. Sig. Process. **135**, 106382 (2020)
20. Molotkova, N.V., Makeeva, M.N., Blium, M.A., Gerasimov, B.I., Gerasimova, E.B.: Improving Russian agribusiness competitiveness within the digital transformation framework. Lecture Notes in Networks and Systems, vol. 84, pp. 342–350 (2020)

Improving the Automated Design Efficiency: A Case Study

Alla E. Brom(✉), Natalya N. Kutina, and Yuliya L. Maslennikova(✉)

Bauman Moscow State Technical University (BMSTU),
2-nd Baumanskaya No. 5, 105005 Moscow, Russia
abrom@yandex.ru, maslennikova.yuliya@yandex.ru

Abstract. Automation and digitalization has many advantages for engineering, but the authors raise the problem that automation of design processes is not always effective. The purpose of the article is to investigate the cycle generated by design automation and to determine the management impacts that will lead to increasing of automation efficiency. The authors constructed a cognitive model of automotive design processes on the example of PJSC "Avtodiesel", calculated changes of the system components in the dynamics and identified the factor that needs management influence.

Keywords: Automation and digitalization · Automotive design · Cognitive model

1 Introduction

It is the era of automation and digitalization of all processes, starting with the origin of an idea, design and creation of a product, its operation and disposal. Digital engineering integrates with digital manufacturing, forming a common digital design and production cycle, which forms the core of Industry 4.0 in turn [1–3]. Digital and automated engineering is based on CAD/CAM/CAE/PDM systems. Active implementation of these technologies leads to the fact that today the design is a 3-D modeling, and the production and testing of the product layout can be done by prototyping [4, 5].

The positive effect of automation and digitalization of engineering is undeniable [6, 7]. However, introducing automated and digital systems, performance indicators do not always grow. Cost and duration of the design cycle are not reduced enough, productivity and product quality are not increased. There are errors at the stages of design study of the product, manufacturing layout and design of tooling. There are design, production and operational defects, which leads to problems with uninterrupted and long service life of products. Consequently, there is a question, what reasons computer-aided design does not give the maximum possible results, what management decisions need to be taken to improve its efficiency (Fig. 1).

Thus, there is a need to study the cycle which generates by automation of design processes. It is important to determine which factors make up the system, how they interact with each other, how we can improve the efficiency of design automation.

I. Karabegović (Ed.): NT 2020, LNNS 128, pp. 397–402, 2020.
https://doi.org/10.1007/978-3-030-46817-0_45

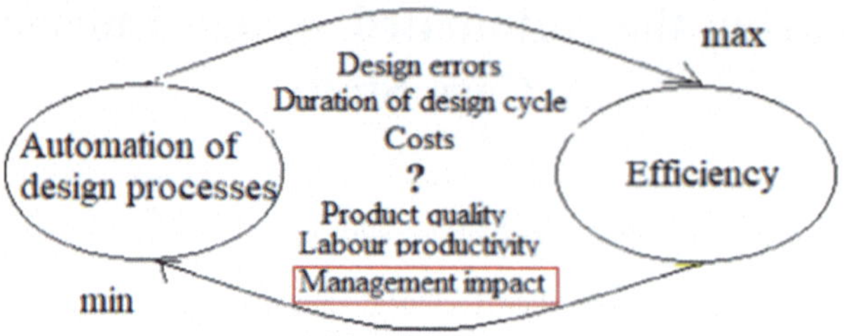

Fig. 1. Reduced efficiency in automated design processes

2 Methods

For the analysis and modeling of the situation, the authors apply the method of cognitive modeling, first used by the American scientist Axelrod in 1976. The Method is based on an expert view of a system, including the construction of the sign-oriented graph and the calculation of factors changes in the dynamics by using management actions. A sign-oriented graph is a cognitive map consisting of factors and cause-and-effect relationships between them.

$$G = \langle V, E \rangle . \tag{1}$$

Where: *V* is a vertex of the graph, the system factors; E is the set of arcs describing the relationship between vertices. The cognitive model is represented by a vector functional graph:

$$F = \langle G, X, F \rangle \Phi. \tag{2}$$

where: *X* is the set of parameters of the vertex *V; F = F(X,E)* is a functional transformation of arcs, which puts in conformity to each arc, a weight coefficient w_{ij} denotes the strength of the interaction.

If it is not possible to represent factors quantitatively, linguistic values of estimation of force of communication are chosen. They can be set using the digits corresponding to the interval [0;1]. The scale describing the power of influence is presented in Table 1.

Table 1. Scale of interpretation of the influence power

Weight	Interpretation of the influence power
0	No influence
0,1	Virtually absent
0,3	Weak
0,5	Average
0,7	Significant
0,9	Strong
1	Maximum possible
0,2; 0,4; 0,6; 0,8	Intermediate levels

Interaction of factors is based on perturbation (impulse) and its transmission. The increment of the factor value $p_i\,(0) = x_j\,(1) - x_i\,(0)$ is the initial pulse. The components of the vector $P(t + 1)$ are computed:

$$p_i(t+1) = \sum w_j \cdot p_j(t) \tag{3}$$

In this method, the initial values of all factors can be considered zero, then the predictive vector of factor values is determined by the accumulated increment of factor values for n steps of the model:

$$x_j(n) = \sum_{j=0}^{n} p_j(t) \tag{4}$$

The cognitive model gives an idea of what factors and with what force, in what direction it is necessary to influence in order to get the desired change in the target factors while minimizing costs.

3 Results

We carry out cognitive modeling of automation of design processes system on the example of the Russian enterprise PJSC "Avtodiesel" design department. The company's products are multi-purpose diesel engines, clutches, gearboxes, spare parts, as well as stationary units based on them.

The first step is to identify the main elements or factors that describe the system of design processes automation. (Table 2).

Table 2. Main factors of design process automation system

№	Name of the factors
1.	The breadth of automation design processes
2.	Production cost
3.	Design costs
4.	Product quality
5.	Duration of the design cycle
6.	Rate of response to changes in demand
7.	Level of designers' qualification
8.	Availability of experimental stands and mock - ups for testing
9.	Level of the company competitiveness
10.	Sustainability of industry boundaries
11.	Integration of design, technological and commercial cycles
12.	Costs of design errors
13.	Ability to create complex and highly complex systems
14.	Ability to create unique products
15.	Emergence of standard solutions
16.	Transparency of control
17.	Labour productivity in design processes
18.	Company profitability

Next, we construct a cognitive model of the system, which presents the relationship of these factors and the strength of their influence on each other (Fig. 2).

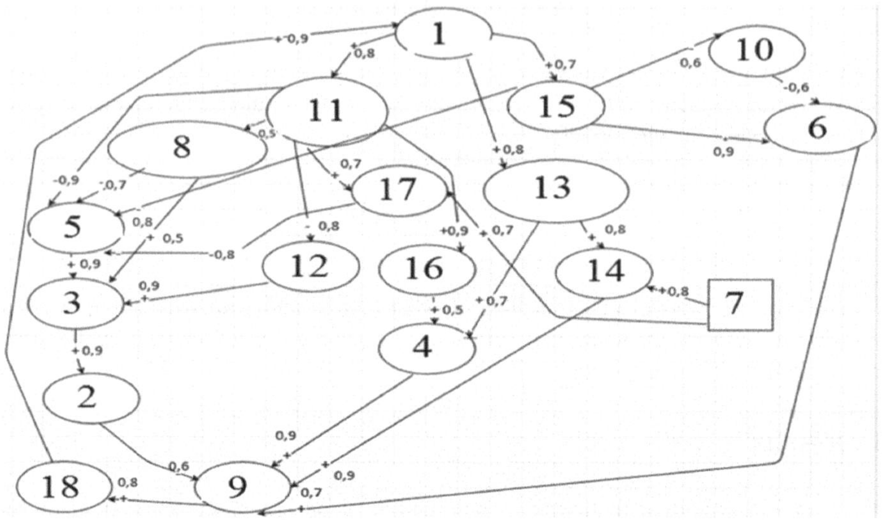

Fig. 2. Cognitive model of the system of design processes automation

The cognitive model shows that all factors, including design costs, cost of design errors, product quality, labor productivity, duration of the design cycle are interrelated and affect each other within the automation cycle, however, the "Level of designers' qualification" factor has no influence within the system. This suggests that it is a component which needs management influence from the outside.

Let us calculate the performance of the model in the dynamics by increasing the "Level of designers' qualification" by 10% (Fig. 3).

The graph clearly shows that in addition to the standard solutions appearance, reducing the necessity of experimental models and simulators, the emergence of opportunities to create unique products, with raising of designers professional skills by 10% at 15 step, profits grows by 67.7%, the design cost is reduced by 14.5%, the number of design errors and costs associated with them fall by 9.6%, productivity grows by 13.7%, the quality of the product is increasing by 20% the duration of the design cycle reduces by 15%.

It means that for the effective operation of automated design systems the management of PJSC "Avtodiesel" needs to develop a strategy for personnel advanced training.

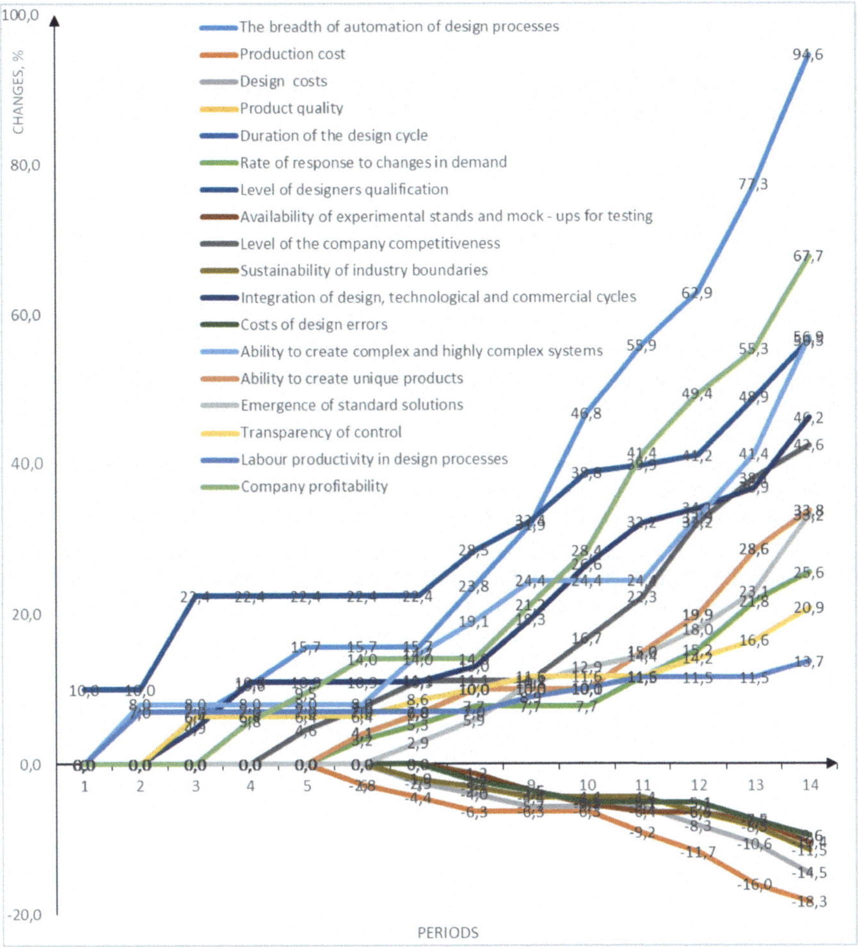

Fig. 3. Factors of design process automation system in dynamics

4 Conclusion

Cognitive modeling allows top management of enterprises to assess the situation and analyze the interaction of existing factors, to detect negative links and structures to be replaced by new objects, processes and relationships that eliminate the negative impact and create a pronounced positive effect.

The use of cognitive technologies allows to develop and justify the development strategy of the enterprise, to determine the possible options for the development of the situation, taking into account the consequences of making important decisions and compare them in a short time.

References

1. Stock, T., Seliger, G.: Opportunities of sustainable manufacturing in industry 4.0. Procedia CIRP **40**, 536–541 (2016)
2. Vdovin, D., Levenkov, Y., Chichekin, V.: Light frame design for quad bike using topology optimization. In: IOP Conference Series: Materials Science and Engineering, vol. 589, no. 1, p. 012026 (2019)
3. Schluse, M., Priggemeyer, M., Atorf, L., Rossmann, J.: Experimentable digital twins – streamlining simulation-based systems engineering for industry 4.0. IEEE Trans. Ind. Inform. **14**(4), 1722–1731 (2018)
4. Manzhirov, A.V., Murashkin, E.V., Parshin, D.A.: Modeling of additive manufacturing and surface growth processes. In: AIP Conference Proceedings, vol. 2116, p. 38001 (2019)
5. Bozhko, A.: Math modeling of sequential coherent and linear assembly plans in CAD systems. In: Proceedings of the Global Smart Industry Conference (GloSIC-2018), p. 8570090 (2018)
6. Leonov, V., Grishko, D., Efendiyev, M.: Software package design for partial automatization of the design process of re-entry interplanetary modules. In: Proceedings of the International Astronautical Congress (IAC-2018) (2018)
7. Uss, A.Yu., Chernyshe, A.B., Atamasov, N.V.: Development of the calculation method and designing of a vortex jet device for gas flow regulation purposes. In: AIP Conference Proceedings, vol. 2141, p. 030028 (2019)

Monitoring of Physical Properties of Transmission-Hydraulic Fluid by Simulating the Operation Load of Agricultural Machine'S Hydraulic Pump Under Laboratory Conditions

Ľubomír Hujo, Juraj Jablonický, Romana Janoušková(✉), Ján Kosiba, Zdenko Tkáč, and Juraj Tulík

Faculty of Engineering, Department of Transport and Handling, Slovak University of Agriculture in Nitra, Tr. A. Hlinku 2, 949 76 Nitra, Slovak Republic
xjanouskova@uniag.sk

Abstract. This work describes the results of changes in physical properties of transmission-hydraulic fluid during simulation of operating load of FHD 17 under laboratory conditions, while monitoring the effect of quality changes of the used fluid on the flow efficiency of the hydraulic pump. Aforementioned hydraulic pump is used in hydraulic system of agricultural machine. Laboratory testing equipment allows us to simulate the operational load of the working circuit of the agriculturemachine's hydraulic system with the possibility of testing the operating fluids andthe main elements of the hydraulic system. Hydraulic pump flow rate changes were monitored at precisely determined intervals and the influence of the physical properties of the fluid on the flow efficiency was rewieved and consequently evaluated by mathematical – statistical analysis.

Keywords: Laboratory testing · Hydraulic pump · Hydraulic system · Ecological fluid · Hydraulic pump flow

1 Introduction

Hydraulic fluids are power carriers in hydraulic system of agricultural machinery and manufacturers of these fluids must take into account specific requirements for improving their quality while reducing the environmental burden while using them in the mobile machinery [1, 2]. Production of the fluid of required quality with acceptable price is becoming increasingly difficult thus creates the opportunity to use fluids with accordance to strict performance specification. Simulateously, the mean time of the oil filling change is monitored with the aim of extending the oil filling change interval [3, 4]. The change of the physical – chemical properties of transmission – hydraulic fluid is an important indicator of the technical condition of the hydraulic pump and fundamentally influences the operation of the individual elements of the hydraulic circuit. The evaluation of the oil filling condition is performed either in pre–scheduled intervals or continuously [5–7]. Accelerated fluid testing is used to optimize testing time. After the testing, physical properties and the effect of the working fluid on

I. Karabegović (Ed.): NT 2020, LNNS 128, pp. 403–410, 2020.
https://doi.org/10.1007/978-3-030-46817-0_46

technical state of hydraulic pump are analysed focusing on the changes of the flow characteristics. The basic assumption of proper function and effective care about hydraulic fluids is suitably chosen methodology of liquid testing while monitoring the level of contamination [6, 8–10]. Time needed for realization of operational measurements shortens in experimental setting of pressure measurements with numerical simulation in engines under laboratory conditions [11, 12].

2 Materials and Methods

Laboratory equipment allowing multiple types of hydraulic fluid tests was developed to simulate the running load of hydraulic circuit components and simulation of pressure surges [13, 14]. The measurement equipment contained in hydraulic circuit consists of Hydac ETS 4144 – A000 temperature transducer, Hydac EVS 3100–1 flow sensor and Hydac HDA 3744 – A- 400–000 pressure transducer connected to HMG 3010 recording unit. This allows to measure and record three variables simultaneously. The recording unit allows the flow rate measurement from 6 to 60 $dm^3.min^{-1}$, from 15 to 300 $dm^3.min^{-1}$ and from 40 to 600 $dm^3.min^{-1}$ by pressure up to 40 MPa and temperature up to 100 °C. To evaluate the properties of the used hydraulic fluid (its specification can be seen in Table 1), integrated lubricant and oil analyzer was used to measure the particle counting, filtration, elementary spectometry with ICP spectometer and viscometer. The equipment tests the condition of abrasion metals, contamination and lubricants condition by measuring of the small sample of the oil without the use of solvents. Technical parameters of lubricants and oils analyzer can be seen in Table 2.

Table 1. Specification of tested hydraulic fluid Mol Farm NH Ultra

Characteristics	Unit	Values
Bulk density at 15 °C	$kg.m^{-3}$	875
Kinematic viscosity at 40 °C	$mm^2.s^{-1}$	64.2
Kinematic viscosity at 100 °C	$mm^2.s^{-1}$	10.9
Viscosity index	–	162
Pour point	°C	−36
Flash point in open crucible	°C	210

Table 2. Technical parameters of the lubricants and oils analyser

Parameter	Specification
Application	Mineral and synthetic lubricants including transmission and engine oils, hydraulic mixtures, turbines and biodiesel
Outputs	Si, Al, Cr, Ti, Fe, Ni, Pb, Cu, Sn, Mo, Ag, Zn, V, TAN and TBN ($mg. KOH.g^{-1}$), oxidation (abs/mm^2), nitridation (abs/mm^2), sulphation (abs/mm^2), water (number of particles/million), glycol (% weight), carbon black (% weight), viscosity at 40 °C, 100 °C, particle counting > 4 μm, solid abrasion > 4 μm per filter

(*continued*)

Table 2. (*continued*)

Parameter	Specification
Methodology	ASTM E 242 (IR), ASTM D7279 (visk.), ISO 21018 – 3 (particle counting)
Accuracy	≤ ±15% from measure value
Sample volume	10 ml
Operating temperature	From 0 °C to + 40 °C
Relative humidity	From 10% to 80%
Homologation	RoHS CE, FCC, CSA, IEC 61010–1:2001/EN 61010–1:2001, CE, cTÜVus, ICE 61010–1:2002

Differential Scanning Calorimeter (DSC) was used for the thermal analysis measurement of the samples. This device measures enthalpy changes in the samples due to physical and chemical properties changes as function of temperature or time. In order to objectively assess the accuracy of measured values of the flow and pressure, it was needed to determine the minimal number of measurements with utilization of mathematical – statistical analysis. Input data to determine the number of measurements were obtained from the flow and pressure measurement at 500, 750, 1000, 1250, 1500, 1750, 2000, 2250, 2500, 2750 rpm. The minimal number of measurements was determined at sampling time 0 h and susequently after every 125 h of operation up to 500 h of operation. Methodological procedure of tested hydraulic fluid sampling was governed according to STN 65 6207 (Hydraulic oils and fluids. Sampling for determination of the content of mechanical impurities).

3 Results and Discussion

Changes in density of tested transmission – hydraulic fluid at 40 °C in depencence on the operation hours during the operational load simulation were monitored. Furthermore, the change of kinematic viscosity and the flow of the hydraulic pump in dependence on operation hours with thermal analysis was monitored as well. Based on the measured data and the average values of the flow efficiency in relation to operation hours (Fig. 1), it can be concluded that tested hydraulic fluid and its physical properties do not negatively affect the flow efficiency of hydraulic pump QHD 17. Maximal decrease value of the flow efficiency reached 0.34% after 375 operation hours. After 500 operation hours the decrease value of the flow efficiency compared to the reference value at 0 operation hours was 0.03%.

In accordance with ISO 15380: 2011 norm, it is possible to confirm that tested fluid does not affect the flow characteristics of QHD 17 hydraulic pump due to the absence of the flow efficiency reduction higher than 20%. Simultaneously, at individual intervals of operation hours, the hydraulic fluid samples were taken to assess the physical properties of hydraulic fluid (Table 3) and to determine the concentration of important chemical elementrs (Table 4).

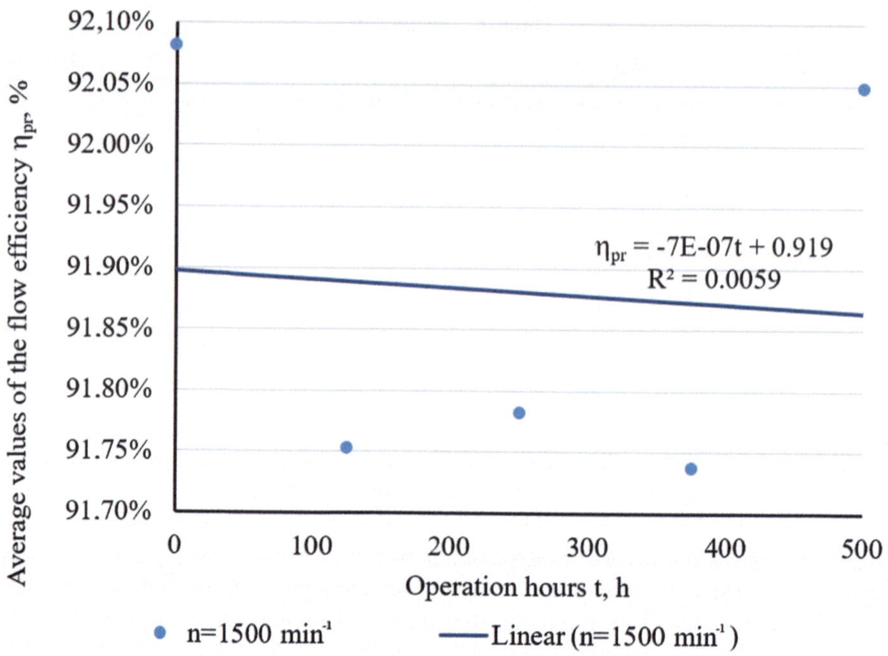

Fig. 1. Dependence of the average flow efficiency values of QHD 17 hydraulic pump at 1500 rpm

Table 3. Physical properties of the hydraulic fluid –MOL Farm NH Ultra

Method	Parameters	Unit	Sampling interval				
			0 h	125 h	250 h	375 h	500 h
Appearence	Colour	–	Blue				
	Transparency	–	Transparent				
	Structure	–	Homogenic				
	Consistency	–	Liquid				
ASTM D7042	Bulk density - 40 °C	$kg.m^{-3}$	860.7	860.65	860.66	860.6	860.62
	Bulk density - 100 °C	$kg.m^{-3}$	827	826.2	826	826	826
	Dynamic viscosity – 40 °C	mPa.s	49	47.1	46.5	46.56	46.05
	Dynamic viscosity – 100 °C	mPa.s	8.42	7.99	7.87	7.85	7.78
	Kinematic viscosity - 40 ° C	$mm^2.s^{-1}$	56.9	54.8	54	54.1	53.5
	Kinematic viscosity - 100 ° C	$mm^2.s^{-1}$	10.2	9.71	9.57	9.54	9.46
	Viscosity index	–	170	163.9	163	161.8	161.6
Acid number ASTM D664 A		$mg.KOHg^{-1}$	2.91	3.29	3.36	3.59	3.67
Water content DIN 51777		%	0.07	0.05	0.04	0.05	0.04

ICP (**I**nductively **C**oupled **P**lasma) spectometry aimed to determine the concentration of important chemical elements in tested hydraulic fluid. During operation, there is an abrasive wear in hydraulic circuit, so the formation of abrasive particles cannot be avoided. A small amount of abrasive particles may also appear in the new transmission – hydraulic fluid as part of additives. If the particle size increases significantly, this may indicate a malfunction of any part of hydraulic circuit.

Table 4. ICP spectrometry results of tested transmission – hydraulic fluid

ICP spectrometry: PN 002, PN012, PN014								
	Chemical element	Symbol	Unit	Sampling interval				
				0 h	125 h	250 h	375 h	500 h
Abrasive metals	Aluminum	Al	$mg.kg^{-1}$	<10.0	<10.0	<10.0	<10.0	<10.0
	Molybdenum*	Mb	$mg.kg^{-1}$	<10.0	<10.0	<10.0	<10.0	<10.0
	Tin	Sn	$mg.kg^{-1}$	<10.0	<10.0	<10.0	<10.0	<10.0
	Titan	Ti	$mg.kg^{-1}$	<10.0	<10.0	<10.0	<10.0	<10.0
	Nickel	Ni	$mg.kg^{-1}$	<10.0	<10.0	<10.0	<10.0	<10.0
	Chrome	Cr	$mg.kg^{-1}$	<10.0	<10.0	<10.0	<10.0	<10.0
	Copper	Cu	$mg.kg^{-1}$	<10.0	<10.0	<10.0	<10.0	<10.0
	Iron	Fe	$mg.kg^{-1}$	<10.0	<10.0	<10.0	<10.0	<10.0
	Plumbum	Pb	$mg.kg^{-1}$	<10.0	<10.0	<10.0	<10.0	<10.0
	Zinc*	Zn	$mg.kg^{-1}$	1450	1440	1430	1470	1430
Contaminants	Silicium*	Si	$mg.kg^{-1}$	<10.0	<10.0	<10.0	<10.0	<10.0
	Sodium	Na	$mg.kg^{-1}$	<50.0	<50.0	<50.0	<50.0	<50.0
	Potassium	N	$mg.kg^{-1}$	<10.0	<10.0	<10.0	<10.0	<10.0
Additives	Barium	Ba	$mg.kg^{-1}$	<10.0	<10.0	<10.0	<10.0	<10.0
	Magnesium	Mg	$mg.kg^{-1}$	15	12	13	13	18
	Phosphorus	F	$mg.kg^{-1}$	1290	1300	1300	1310	1290
	Calcium	Ca	$mg.kg^{-1}$	3650	3670	3650	3710	3640
Other important elements	Manganese	Mn	$mg.kg^{-1}$	<10.0	<10.0	<10.0	<10.0	<10.0
	Sulphur	S	$mg.kg^{-1}$	4650	4740	4750	4780	4680

* can be used as additivum

To evaluate the transmission – hydraulic fluid from the point of view of abrasive metals quantity, it is necessary to know the limit values. Standard D6595 – 00 sets the limit values of abrasion metals in transmission – hydraulic fluid. The results of the ICP spectrometry (Table 4) carried out by accredited laboratory Klüber, tested transmission - hydraulic fluid did not show higher values of abrasion metals after 500 h of operation compared to the new oil sample. Based on measurement and hydraulic fluid analysis, tested transmission – hydraulic fluid does not aggressively affect the components of the hydraulic circuit which simulated the running load. The only element that exceeds the limit value is zinc where it is necessary to follow the reference sample analyzes at 0 h of operation. Under the results of the ICP spectrometry of tested transmission - hydraulic fluid from the point of view of detected contaminants, it is necessary to know the limit values whereby in tested transmission – hydraulic fluid had not been observed higher values of contaminants after 500 h of operation. The results of the ICP spectrometry shown in Table 4 bring the information about activation of individual additives. Activation of additives can occure due to higher temperature or pressure by additives activation of calcium (corrosion protection), phosphorus (wear protection) and in lesser degree magnesium. An additive element, barium, using as corrosion protection had not been activated. Author [15] reports on studies suggesting that approximately 70% of operating failures of hydraulic systems are causing the presence of contaminants. Prior to hydraulic system failure detection due to increased level of contaminants, hydraulic efficiency is reduced up to 20%. Test of responsiveness to contaminants makes the wear process fast to obtain the results in much shorter period than most tests used to evaluate the hydraulic fluid last. In this case, it could be consider question about utilization of constant or cyclic loading during testing.

4 Conclusions

Fluids designed for common transmission and hydraulic systems of tractros are exposed to specific conditions characteristic for agricultural production. Fluid contamination causes a number of side effects such as special wear, steel surfaces corrosion, oil oxidation and change of its physical and chemical properties. Contamination is particularly harmful to ecological fluids as it speed up the degradation processes [16]. Several authors [7, 17] deals with the evaluation of hydraulic fluid during operational test with susequent analysis of contaminants. As stated [4], machines working in agricultre are characterized by demanding operation, working in dusty and humid environment what negatively affects the fluid fillings contamination. Therefore, it is necessary to examine the factors influence of ecological fluid application on reliable operation of tractors. The use of ecological fluid requires a high degree of the oil filling purity and immeasurable concentration of water [16]. For that reason, accurately machinery industry of individual hydraulic circuit components is very important. To monitor the accuracy of CNC machines is necessary, using the new methods and trends in development and planning of products with multiriterial diagnostics of CNC machines [18–20]. Laboratory tests were aimed at assessing of the properties of tested transmission – hydraulic fluid MOL Farm NH Ultra and its impact on technical and operational characteristics of QHD 17 hydraulic pump. Laboratory testing equipment allows simulating variable testing conditions to simulate the real condition of the hydraulic circuit operation of agricultural wheeled tractor. Tested transmission – hydraulic fluid was evaluated on the basis of its physical properties specified from the samples taken at precisely determined intervals. ICP spectometry recorded values of abrasion metals, contaminants and additive elements. The results of physical properties of tested transmission – hydraulic fluid did not negatively affect the flow characteristics of QHD 17 hydraulic pump at cyclic loading which simulated the real operating conditions of the hydraulic circuit of agricultural tractor.

Acknowledgments. This work was supported by project VEGA 1/0155/18 "Applied research of the use of ecological energy carriers in agricultural, forestry and transport technology."

This work was supported by project KEGA 028SPU-4/2019 "Practical utilization of design and testing knowledge of transmission systems of hydraulic mechanisms of mobile agricultural and forestry machinery."

References

1. Janoško, I., Černecký, J., Brodnianska, Z., Hujo, Ľ.: Environmental Technologies and Technology. 1. press. SUA in Nitra, p. 306 (2016). (in Slovak)
2. Simikić, M., Dedović, N., Savin, L., Tomić, M., Ponjičan, O.: Power delivery efficiency of a wheeled tractor at obligue drawbar force. Soil Tillage Res. **141**, 32–43 (2014)
3. Hujo, Ľ., Čorňák, Š., Tkáč, Z., Jánošová, M.: Laboratory research of transmission – Hydraulic fluid. In: 7 th International Conference on Trends in Agricultural Engineering (2019)
4. Tkáč, Z., Čorňák, Š., Cviklovič, V., Kosiba, J., Glos, J., Jablonický, J., Bernát, R.: Research of biodegradable fluid impacts on operation of tractor hydraulic system. Acta Technologica Agric. **20**(2), 42–45 (2017)
5. Hujo, Ľ., Tkáč, Z., Tulík, J., Kosiba, J., Uhrinová, D., Jánošová, M.: Monitoring of operation loading of three-point linkage during ploughing. Research in agricultural engineering **62**(1), 24–29 (2016)
6. Kučera, M., Aleš, Z., Pexa, M.: Detection and characterization of wear particles of universal tractor oil using of particles size analyser. Agron. Res. **14**(4), 1351–1360 (2016)
7. Kučera, M., Majdan, R., Abrahám, R., Kučera, M., Haas, P.: Analysis of the effect of loading process on tribological system properties. Acta Universitatis Agriculturae et Silviculturae Mendelianae Brunensis **64**(3), 825–833 (2016)
8. Helebrant, F., Ziegler, J., Marasosová, D.: Technical diagnostics and reliability I. Tribodiagnostics. VŠB, Technická univerzita, Ostrava, p. 155 (2001). (in Czech)
9. Janoško, I., Polonec, T., Lindák, S.: Performance parameters monitoring of the hydraulic system with bio-oil. Res. Agric. Eng. **60**, 37–43 (2014)
10. Majdan, R., Jánošová, M., Halenár, M.: Increasing the purity of hydraulic fluids in tractors. Spravodaj ATD SR **14**(2), 15–21 (2017)
11. Puškár, M., Brestovič, T., Jasminská, N.: Numerical simulation and experimental analysis of acoustic wave influences on brake mean effective pressure in thrust-ejector inlet pipe of combustion engine. Int. J. Veh. Des. (IJVD) **67**(1), 1 (2015)
12. Zastempowski, M.: Test stands with energy recovery system for machines and hydraulic transmissions. J. Res. Appl. Agric. Eng. **58**(2), 188–191 (2013)
13. Lindák, S., Majdan, R., Janoškoa, I., Pap, M., Szabó, M.: Hydraulic device for simulation of pressure shocks. Acta Technologica Agriculturae **2**(17), 44–48 (2014)
14. Hujo, Ľ., Jablonický, J., Tkáč, Z.: Design of innovative laboratory simulation equipment for testing hydrostatic transducers and hydraulic fluids. SUA in Nitra, p. 140 (2017). (in Slovak)
15. Tessmann, K.R.: Qualification of Qualification of Hydraulic Fluid through Pump Testing. Technology Transfer Publication. - Stilwater, Oklahoma, An FES/BarDyne (1998)
16. Majdan, R., Kurekova, M., Nosian, J., Paulenka, R.: Znečistenie univerzálnych olejov v traktoroch a návrh druhého stupňa filtrácie. In: Spravodaj ATD SR, vol. 2, roč. 15, no. 2, pp. 22–28 (2018)
17. Halenár, M., Nosian, J., Kuchar, P., Tulík, J., Furstenzeller, A.: Evaluation of hydraulic fluid during of the operating test. Acta Facultatis Technicae **23**(2), 73–80 (2018). ISSN 1336-4472

18. Košinár, M., Kuric, I.: Monitoring possibilities of CNC machin tools accuracy. In: 1st International Conference on Quality and Innovation in Engineering and Management (QIEM) (2011)
19. Kuric, I.: New methods and trends in product development and planning. In: 1st International Conference on Quality and Innovation in Engineering and Management (QIEM), pp. 453–456 (2011)
20. Kuric, I., Zajačko, I., Cisár, I.: Analytical intelligence tools for multicriterial diagnostics of CNC machines. Adv. Sci. Technol. Res. J. **10**(32), 59–64 (2016)

Computer Science, Information and Communication Technologies, Control Systems

PSO Optimized Fuzzy Controller for Mobile Robot Path Tracking

Lejla Banjanovic-Mehmedovic[1(✉)] and Azra Baluković[2]

[1] Faculty of Electrical Engineering, Department of Control Systems, Automation and Robotics, University of Tuzla, 75000 Tuzla, Bosnia and Herzegovina
lejla.mehmedovic@untz.ba
[2] KV Team, Sarajevo, Bosnia and Herzegovina

Abstract. An essential aspect of Industry 4.0 is autonomous production methods, powered by Industry of Things (IoT) based on a connected mesh of objects, devices and computer machines, which can communicate with each other. Many things can be transported across the factory via autonomous mobile robots, avoiding obstacles and coordinating with other robots. Autonomous robots are fully integrated with the manufacturing system in Industry 4.0 to a greater extent than ever before.

This paper addresses the design and implementation of PSO optimized fuzzy control for mobile robot path tracking. Firstly, fuzzy control is applied to track the desired trajectory. Secondly, this path tracking has been improved with the PSO optimized fuzzy controller. The proposed PSO fuzzy controller was investigated based on several conducted MATLAB simulation scenarios for a mobile robot. The simulation results show better performances with the proposed PSO optimized fuzzy controller when it is compared with the fuzzy controller without optimization.

Keywords: Fuzzy control · Industry 4.0 · Mobile robot navigation · Particle swarm optimization (PSO) · Path tracking

1 Introduction

In many static and dynamic environments in the scope of smart factories, mobile robots are increasingly being employed. Industry 4.0 goes beyond the typical idea of machine-to-machine communication, allowing mobile robots to transport of components and even completed products. Following a defined trajectory, the mobile robots navigate itself to avoid obstacles and move to a target point, with possibility to collaborate with the production systems like automation cells.

The different motion control techniques can be distinguished: (i) model-based methods and (ii) artificial methods based on fuzzy logic, neural networks, genetic algorithms or hybrid combination of these approaches [1].

The fuzzy logic controller is the most attractive as the form of reactive and behavior navigation. The rule-based fuzzy logic provides reasoning and decision making with uncertain and imprecise information. One of the main advantages of a fuzzy navigation strategy is the ability to extract heuristic rules from human experience.

I. Karabegović (Ed.): NT 2020, LNNS 128, pp. 413–421, 2020.
https://doi.org/10.1007/978-3-030-46817-0_47

The application of multi-input/multi-output (MIMO) fuzzy path-following and navigation control for the mobile robot is presented in the paper [1]. The strategy presents a desired performance to accurately follow the different designed trajectories for the case of working in indoor or conned environments as well as avoiding obstacles and going to a target.

By using Fuzzy Logic Controllers, it is possible to achieve very good results in path tracking. However, it is possible to get better results with optimized Fuzzy Logic Controllers, using the optimal topology of the membership functions. In the paper [2], a fuzzy logic controller has been developed for the path tracking of a rescue robot. In order to tune the membership functions and the rule weights, the Genetic Algorithm method was used to achieve minimum deviation from a desired path.

Our purpose is to control a mobile robot for certain transport operations in a factory. In our paper, evaluation of autonomous system fuzzy control ability using the selected metaheuristic strategy - Particle Swarm Optimization Technique (PSO) - is presented in order to find the best parameters of the membership function and to achieve better control. In Sect. 1 the introduction is presented, the basic elements of a mobile robot kinematic model is provided in Sect. 2. The basic fuzzy logic controller is presented in Sect. 3 and the theory of the PSO is mentioned in Sect. 4. In Sect. 5, the description of the proposed optimized fuzzy controller with simulation results are shown; and finally the conclusions are presented in Sect. 6.

2 Mobile Robot Kinematic Model

The new era of manufacturing bears the inter-connectivity of machines, processes and products. The mobile robot with the ability to self-navigate is becoming more common in industry environment, Fig. 1.

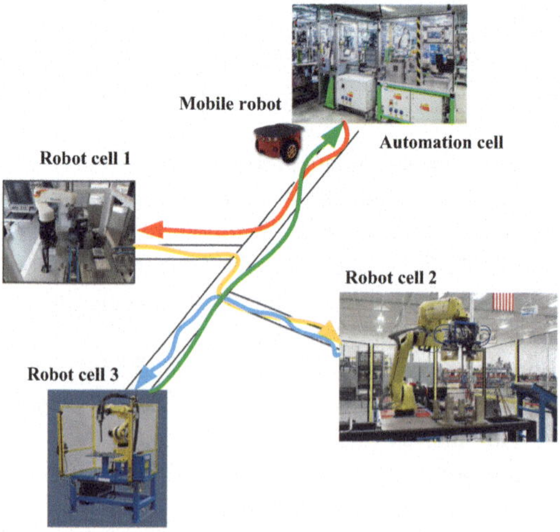

Fig. 1. Mobile robot navigation in industry environment

In order to analyze the system and develop controllers, a differential drive is considered for the kinematic model of mobile robot. We used a discrete form for the robot model, for simulation in a Matlab environment. The discrete form of the kinematic model is given by the following equations [3, 4]:

$$x_{k+1} = x_k + T\frac{V_{Rk} + V_{Lk}}{2}\cos\theta_k \tag{1}$$

$$y_{k+1} = y_k + T\frac{V_{Rk} + V_{Lk}}{2}\sin\theta_k \tag{2}$$

$$\theta_{k+1} = \theta_k + T\frac{V_{Rk} - V_{Lk}}{D} \tag{3}$$

where (x_{k+1}, y_{k+1}) is the robot position in next step, calculated using the current robot position (x_k, y_k) and velocities for both robot wheels (V_{Rk}, V_{Lk}); T-time sampling.

3 Fuzzy Logic Controller for Mobile Robot Path Tracking

Fuzzy Logic is a form of logical reasoning that can be incorporated into automation systems to operate with vague or ambiguous concepts. Quantitative conclusions can be drawn from a set of observations and qualitative rules (based on knowledge) [2]. Robot as a system has two control variables (velocities of left and right wheels), its navigation is described with three coordinates (x, y and φ). In this application, the fuzzy control function is defined as a function of the positional error (d) and the orientation error(φ). The two output variables of the fuzzy controller (the velocities of left and right motors) are determined based on the deviation between the desired and actual position and orientation, Fig. 2.

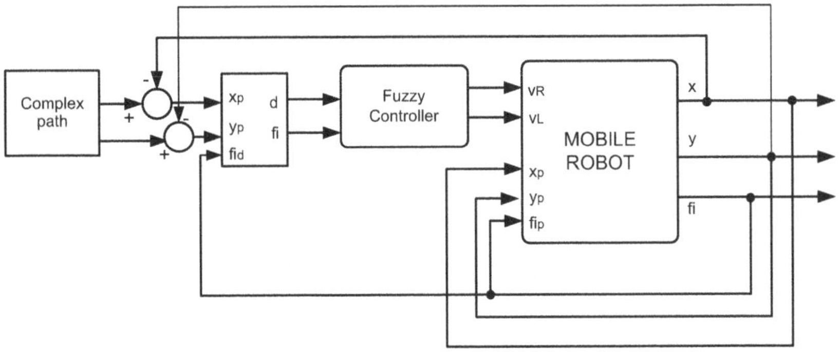

Fig. 2. Fuzzy controller for the mobile robot path tracking

The path is given with the desired coordinates from the start to the target point in the path. Every 0.5 s, the distance between the desired coordinates on the trajectory and the actual coordinates of the robot is calculated at a given time, as well as the orientation error according to the equations:

$$d = \sqrt{(x_d - x)^2 + (y_d - y)^2} \quad (4)$$

$$\varphi = tan^{-1}\frac{y_d - y}{x_d - x} \quad (5)$$

where (x_d, y_d) - the desired path coordinates and (x, y) - the actual coordinates of the mobile robot.

The first input in the fuzzy controller is the distance (D), presented by 4 fuzzy membership functions Z (zero), S (small), M (medium) and B (big) in Fig. 3. The second input in the fuzzy controller is an orientation error (OE) and it is represented by 7 triangular membership functions Z (zero), NS (negative small), NM (negative medium), NB (negative big), PS (positive small), PM (positive medium) and PB (positive big), shown in Fig. 4.

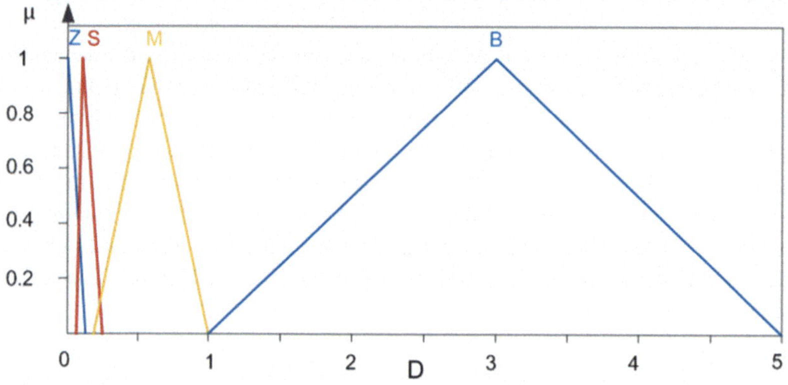

Fig. 3. Membership functions for the input Distance of the fuzzy controller

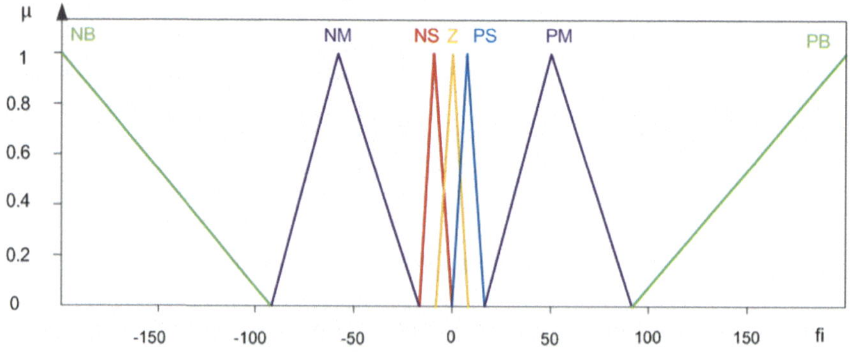

Fig. 4. Membership functions for the input Orientation error of the fuzzy controller

Fuzzy outputs are the velocities of the right and left motors. Their membership functions are given as constant values: ZE = 0, NZP = 0.5, NS = −0.7, NM = −1.2, NB = −5, VNB = −7, PS = 0.7, PM = 1.2, PB = 5 and VPB = 7.

Since the first fuzzy input have 4 membership functions and the second 7 membership functions, we can create a base of 4 × 7 = 28 fuzzy rules. The rules of the fuzzy inference system are shown in Table 1.

Table 1. Fuzzy logic rules for velocity of the right anf left motor

OE	NS	NM	NB	Z	PS	PM	PB
D	(V_R, V_L)	(V_R, V_L)	(V_R, V_L)	(V_R, V_L)	(V_R, V_L)	(V_R, V_L)	(V_R, V_L)
Z	(NS,PS)	(NM,PM)	(NB,PB)	(ZE,ZE)	(PS,NS)	(PM,NM)	(PB,NB)
S	(NS,PS)	(NM,PM)	(NB,PB)	(NZP,NZP)	(PS,NS)	(PM,NP)	(PB,NB)
M	(NS,PS)	(NB,PM)	(NB,PM)	(PS,PS)	(PS,NS)	(PM,NB)	(PM,NB)
B	(NS,PS)	(NM,PM)	(NB,PB)	(PB,PB)	(PS,NS)	(PM,NM)	(PB,NB)

The path tracking result with the fuzzy controller is presented in Fig. 5.

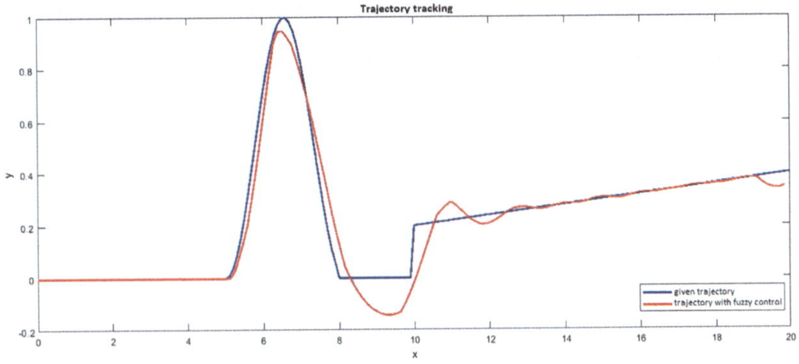

Fig. 5. Path tracking with the fuzzy controller

We can see there are some deviations from the given trajectory. Therefore, we introduce the optimized fuzzy controller using Particle Swarm Optimization (PSO).

4 Particle Swarm Optimization (PSO)

There are many optimization methods used to find optimal solutions. The best ones used for path planning are Evolutionary Methods and Meta-Heuristic Methods [5]. Many evolutionary methods have been shown to be ineffective, when search space is too large in path planning strategy. The meta-heuristic methods can overcome these drawbacks with the fascinating results in this broad field of research. The meta-heuristic approach includes different artificial intelligence techniques such as Genetic Algorithm (GA), Ant Colony Optimization (ACO), Particle Swarm Optimization (PSO), Bacterial Foraging Optimization (BFO), Bee Colony Optimization (BCO), Firefly Algorithm Optimization (FAO) and Grey Wolf Optimization (GFO).

Particle swarm optimization (PSO) is a population based stochastic optimization technique, which shares many similarities with evolutionary computation techniques such as Genetic Algorithms (GA). In the past several years, PSO has been successfully applied in many research and application areas because PSO gets better results in a faster way with only few parameters to adjust [6].

In PSO, the potential solutions, called particles, fly through the problem space by following the current optimum particles. Particles fly through the search space with velocities, which are dynamically adjusted according to their historical behaviors [7]. The system is initialized with a population of random solutions and searches for optimum by updating generations, N. The i-th particle is represented by its position as a point $X_i = (x_{i,1}, x_{i,2}, \ldots, x_{i,d}) \in S$ in an S-dimensional space. Throughout the process, each particle monitors three values: its current position X_i; the best position it reached in previous cycles $pbest_i = (pbest_{i,1}, pbest_{i,2}, pbest_{i,3}, \ldots, pbest_{i,d}) \in S$ and its flying velocity $V_i = (v_{i,1}, v_{i,2}, v_{i,3}, \ldots, v_{i,d})$. Each particle updates its position and velocity to catch up with the best particle g as:

$$X_i(t+1) = X_i(t) + V_i(t+1) \tag{6}$$

$$V_i(t+1) = V_i(t) + c_1 r_{i,1}(t)(pbest_i(t) - X_i(t)) + c_2 r_{i,2}(t)(gbest(t) - X_i(t)) \tag{7}$$

In iteration t, the best position of the whole swarm is denoted as gbest(t), the best track value of each particle is $pbest_i(t)$; c_1 and c_2 are two positive constants (acceleration coefficients), which denote the cognitive and social parameters, respectively; $r_{i,1}$ and $r_{i,2}$ are two random functions in the range [0,1].

PSO pseudocode:

```
For each particle
   Initialize the particle
Do
   For each particle
      Calculate the fitness value
      If the fitness value is better than the best fitness value (pBest) in the history,
         set the current value as the new pBest
   Choose the particle with the best fitness value of all the particles as (gBest)
   For each particle
      Calculate particle velocity according to the equation (1)
      Update particle position according to the equation (2)
While maximum iterations or minimum error criteria is not attained
```

5 Design of PSO Optimized Fuzzy Controller for Path Tracking

The Fuzzy Controller achieves good results for simple path tracking. For complex path tracking, the fuzzy controller was optimized with the PSO algorithm. One of the main factors in the design of efficient fuzzy logic controllers is the selection of the membership functions parameters. One method of removing the uncertainty associated with the tuning of membership functions parameters is the use of optimization algorithms.

In the path tracking application, the fitness function evaluates the robot's path, taking into account the distance and orientation error from the desired path [5]. The optimum membership functions differ from one path to another. The desired path is chosen in the optimization process, to assure the robustness of the optimal fuzzy controller.

In our design approach of the optimized fuzzy controller, random values of the membership functions parameters (a_i, i = 1,.. 5) for distance, (b_i, i = 1,.. 4) for orientation error and (c_i, i = 1,.. 5) for motor velocities are generated and fuzzy controller is created. After completing the simulation in the Matlab, the obtained data is further used to calculate the fitness function.

Next function is used as a fitness function to evaluate the performance of the system simulation:

$$f = \sum k_1 \cdot d_i + c \cdot k_2 \cdot alpha_i \tag{8}$$

where d_i is the difference between the desired and the actual position:

$$d_i = \sqrt{(x_{di} - x_i)^2 + (y_{di} - y_i)^2} \tag{9}$$

$alpha_i$ is difference between the desired and actual orientation (orientation error):

$$alpha_i = tan^{-1} \frac{y_{di} - y_i}{x_{di} - x_i} \tag{10}$$

k_1, k_2 are weight coefficients and $k_1 + k_2 \leq 1$; c is a factor multiplying the second part of the sum, in order not to be much higher than the first, since the distance values are very small. The extensive simulations were made in the Matlab/Simulink with different membership functions of the fuzzy controller. Using PSO as the optimization algorithm, the best path tracking is for the next PSO parameters: particle number = 50, iteration number = 50, $k_1 = 0.65, k_2 = 0.35$. The optimal values of membership functions for Distance are shown in Fig. 6.

The PSO changed the membership functions for Orientation error and they are shown in Fig. 7.

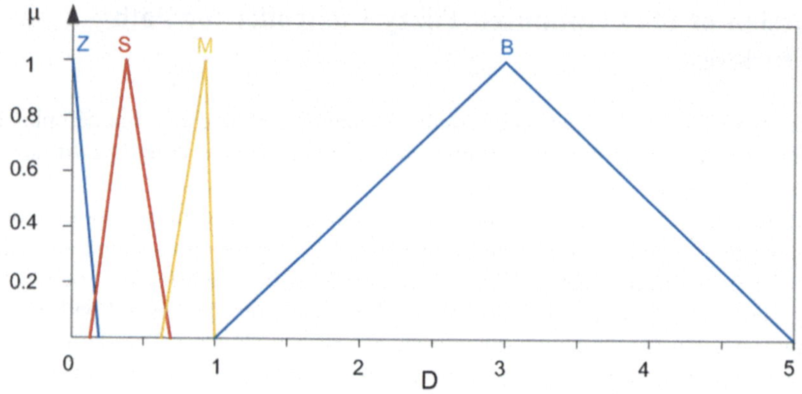

Fig. 6. Membership functions for Distance of the optimized PSO fuzzy controller

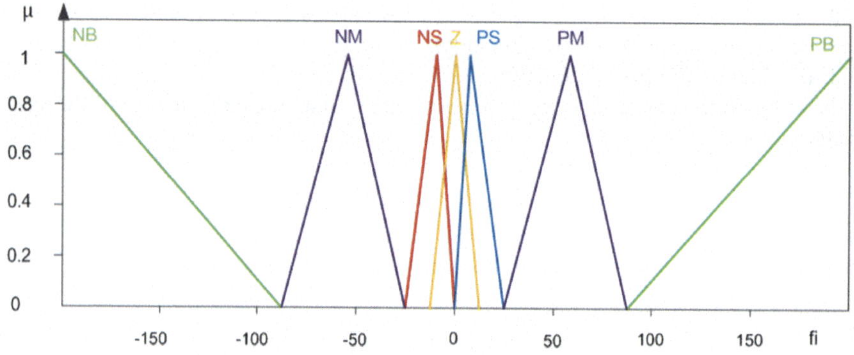

Fig. 7. Membership functions for Orientation error of the optimized PSO fuzzy controller

The simulation result is presented using a desired path including several break points to show the controller performance. Figure 8 shows the response of initial and PSO optimized fuzzy controller [4]. As it can be seen, the deviation from the desired path has been reduced.

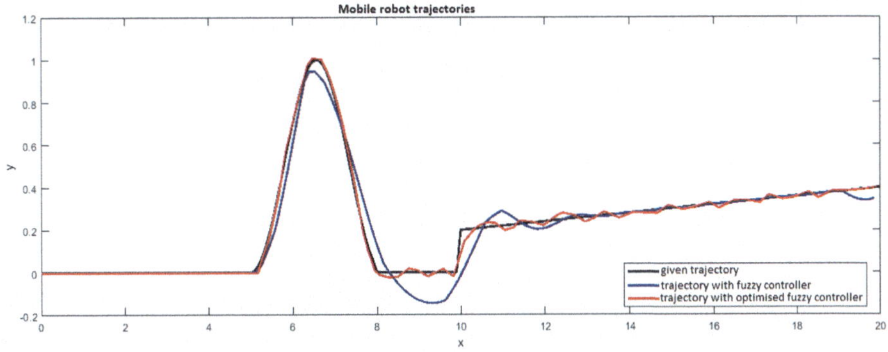

Fig. 8. Comparison of path tracking using the fuzzy controller and PSO optimized fuzzy controller

6 Conclusion

By using optimized Fuzzy Logic Controllers, it is possible to achieve very good results in path tracking. In our research, PSO was used to design the optimal topology of the membership functions. Other approaches such as the GA or ACO optimization techniques could also be used to achieve this goal. However, it is possible to get good and faster results with PSO by adjusting only a few parameters. The experimental results show better performances with the proposed PSO optimized fuzzy controller, when it is compared with the fuzzy controller without optimization.

Our future work will focus on designing fuzzy controllers with multi objective optimization to coordinate the collaborative behavior of the mobile robots in a complex factory environment for advanced Industry 4.0 applications.

References

1. Mac, T.T., Copot, C., De Keyser, R., Tran, T.D., Vu, T.: MIMO fuzzy control for autonomous mobile robot. J. Autom. Control Eng. **4**(1), 65–70 (2016)
2. Pishkenari, H.N., Mahboobi, S.H., Alasty, A.: Trajectory tracking of a mobile robot using fuzzy logic tuned by genetic algorithm, IJE Trans. A Basics, **19**(1), 1–10 (2006)
3. Kumar Malu, S., Majumdar, J.: Kinematics, localization and control of differential drive mobile robot. Global J. Res. Eng. Robot. Nano-Tech. **14**(1), 1–9 (2014). Version 1.0
4. Balukovic, A.: Bachelor Degree of Electrotechnic Engineering. University of Tuzla, Bosnia and Herzegovina (2019)
5. Zafar, M.N., Mohanta, J.C.: Methodology for path planning and optimization of mobile robots: a review. In: International Conference on Robotics and Smart Manufacturing (RoSMa2018) (2018)
6. Valdez, F., Melin, P., Castillo, O.: Particle swarm optimization for designing an optimal fuzzy logic controller of a DC motor. In: Proceedings of the 2012 Annual Meeting of the North American Fuzzy Information Processing Society, NAFIPS 2012, USA (2012)
7. Kubota, N., Adji Sulistijono, I.: A comparison of particle swarm optimization and genetic algorithm for human head tracking. In: Joint 3rd International Conference on Soft Computing and Intelligent Systems and 7th International Symposium on advanced Intelligent Systems, pp. 2204–2209 (2006)

Theoretical and Practical Aspects of Internet of Things (IoT) Technology

Savo Stupar[1(✉)], Mirha Bičo Ćar[1], Emir Kurtović[1], and Grujica Vico[2]

[1] School of Economics and Business, University of Sarajevo, Trg oslobođenja 1, 71000 Sarajevo, Bosnia and Herzegovina
savo.stupar@efsa.unsa.ba
[2] Faculty of Agriculture, University of East Sarajevo, Vuka Karadžića 30, East Sarajevo, Bosnia and Herzegovina

Abstract. One of the greatest inventions of human civilization of all time is the global (worldwide) computer network - the Internet. Like most other inventions, the Internet has its good and bad sides. When it comes to the good sides of the Internet, it first and foremost facilitates all forms of communication: Person to Person (P2P); human with the computer and the computer with the computer. With the development of the Internet, new types of communication are emerging: human communication with "smart things", which, together with human communication with a computer, represents Person to Maschine communication (P2M); communication between "smart things", which together with computer-to-computer communication and computer communication with "smart things," make up a new category of communication Maschine to Maschine (M2M). Ordinary objects (things or devices) used to need to be manually activated or programmed. This is no longer the case today, as most of these objects come automated, with pre-built "smart" components that can make their own decisions (without human presence). Internet of Things (IoT) or "Smart Things" are machines, robots, sensors, controls, etc. The aim of this paper is to explain in an understandable way the concept of IoT, its architecture, characteristics, technologies that enable the implementation of the IoT concept in practice, and the areas and security aspects of its application.

Keywords: Internet of Things · RFID · NFC · Smart Things · Wi-Fi

1 Introduction

"Everything touches everything", Jorge L. Borges
"If you want something new, you have to stop doing something old", Peter F. Drucker

The concept of connected devices first emerged in the early 1990s. At that time, researchers at the Auto-ID Center at MIT began thinking of the idea of building a system that would allow devices to be connected in the physical world through sensors and wireless signals. In 1999, Kevin Ashton, one of the founders of the Auto-ID Center at MIT, coined the term "Internet of Things." Already in 1997, working at Proctor & Gamble as an assistant brand manager, he considered the possibility of using RFID (radio frequency identification) tags to better manage the company's supply chain in a

I. Karabegović (Ed.): NT 2020, LNNS 128, pp. 422–431, 2020.
https://doi.org/10.1007/978-3-030-46817-0_48

way to count and track merchandise unnecessarily for human intervention. A little later, the IoT concept evolves and expands to connect various objects, devices, sensors and "everyday" things to the Internet [1].

Nowdays belief is that nothing happens in isolation, and that most events and phenomena are either causally or synchronistically linked to many other parts of the most complex possible system - the universe. This results in the realization that the world is small and that everything in it is connected to everything else. Although at first glance it does not seem to make any sense to draw an analogy e.g. The Internet as a global computer network with a network of neurons in the human brain, nevertheless the science of networks suggests the opposite. There are some universal laws when it comes to the operation of networks. In almost all scientific disciplines, scientists have come to the conclusion that complexity has a rigorous architecture, and in order to understand it, it is necessary to know as much as possible about networks. The notion of networking for the sake of communicating for the purpose of easier and more correct functioning, gradually and recently is beginning to refer to connecting everything with everything.

IoE connects people, data, things and processes across networks of one billion or even a trillion connections. These links create large amounts of data, and we have never had access to some of that information. When this data is analyzed and used intelligently, the possibilities seem endless [2]. It can generally be said that people, processes, data, devices, things interconnect … This makes networking more relevant and valuable than ever before, turning information into action that creates new opportunities, richer experiences and great economic chances for companies, individuals and countries. This is exactly what we can call the Internet of Everything, abbreviated IoE [3]. IoE is thus an attempt to network everything with everyone, and according to A. Banafa, the pillars of the Internet of Everything are [3]: People - Connecting people in more relevant, valuable ways; Data - Turn data into intelligence to make better decisions; Processes - Deliver the right information to the right person (or device) at the right time, and Things - physical devices and objects connected to the Internet and to each other for intelligent decision-making, often called the Internet of Things (IoT).

IoE requires a certain level of technology, and that technology and its development encounters some limitations and challenges that complicate the process of total networking. There are several shortcomings in the technology itself that limit the use of the Internet of Things [4]:

- Existing sensors are generally expensive and energy consuming, limiting the widespread use of wireless applications
- Most systems and networks are not scalable or upgradable which increases the problem of data exchange
- There is no wireless technology that optimizes all aspects: cost, range, power, architecture, various requirements and applications [4].

Similar to other key ITs of Industry 4.0 (Big Data, Cloud Computing, Blockchain) the Internet of Things (IoT) too first emerged as a theoretical concept, namely the concept of connecting more devices to the Internet, and when the rapid development of ICT technologies, first of all the Internet, created the conditions for its practical realization, it then becomes technology, one of the most significant "pillars" of Industry 4.0. The ability to connect devices wirelessly opens up new possibilities for both the interaction between different systems and their control, monitoring of functioning and provision of advanced services. By this, the term device refers to everything that has the ability to connect to a computer network: from cell phones or handsets, to household appliances, rooms, furniture, clothing, accessories, etc.

The Internet of Things allows the integration of a huge number of devices that have certain sensors built in, which to a greater or lesser degree have the ability to communicate independently, both with each other and with various applications. The key idea behind the IoT concept is to connect everything that can be connected via the Internet, and to gather information and remotely control the devices. For Industry 4.0, the IoT concept is particularly important because it enables the Internet to be brought to the lowest levels of production. The IoT concept has the potential to control the information that is generated and processed even from the smallest sensors and devices, so that information can be accessed from anywhere. Figure 1 shows the number of devices connected to the Internet over the years, with a sharp increase in the last five years and even more in the future. The Internet is being transformed from a global network of interconnected computers to the Internet of interconnected things.

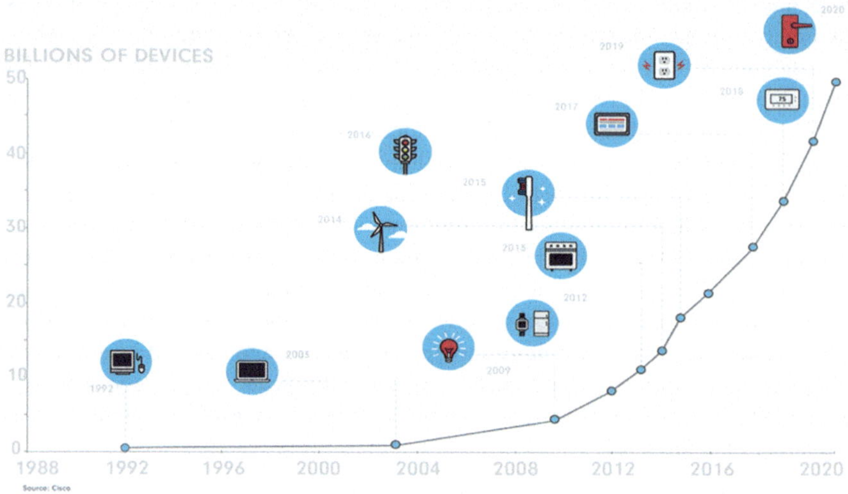

Fig. 1. Number of devices connected to the Internet. Source: Authors according to [5]

2 Characteristics of IoT Concepts

IoT is a concept characterized by several features that will be outlined below [6–8]:

- Everything communicates: smart things have the ability to communicate wirelessly between themselves, and the ability to form an ad-hoc network of interconnected objects;
- Everything is identified: smart things are identified by a digital name;

- Everything interacts: smart things can interact with each other in the local environment by reading data from sensors (passive surgery) and activating their existing capabilities (performing surgery). These two dimensions (sensor and activator) comprise two basic operations that represent the interface and coupling between the digital and the physical realms [7];
- Interconnectivity: By applying the IoT concept, everything can be connected to the global information and communication infrastructure [8].
- Connectivity Services: IoT is capable of providing connectivity services, regardless of limitations, that is, taking into account the limitations like privacy protection and providing semantic consistency between physical and virtual (intangible) things that are interconnected. The possibility of providing such services is conditioned by the necessity to change both the physical elements of things and those of information [8].
- Heterogeneity: Devices that connect with the IoT concept are very heterogeneous because they are based on different hardware platforms and different types of computer networks. They have the ability to communicate with other devices or service platforms through various types of computer networks [8].
- Dynamic changes: The state of the device can change, so smart devices can be in a state of "sleep" at one time and then in a state of "wake-up", or a state of "on" and/or "off". The speed and location of those devices that have the ability to move can be changed in the same way [8].
- Huge scales (ranges). The number of devices that need to be managed to communicate with each other will be significantly higher than the devices currently connected to the Internet. The ratio of the number of connections (communications) triggered by the devices and the number of communications initiated by people will significantly shift in favor of the devices that initiate communication. Even more critical will be managing the vast amount of data generated and interpreting it for concrete application. This applies to the semantics (interpretation of meaning) of the data as well as the effectiveness of the data processing [8].

3 IoT Architecture

IoT connects a huge number of devices, and as a consequence a huge amount of data is generated, which needs to be managed appropriately. The greater the amount of data that is generated in order to connect everyone with everyone, the harder it is to manage that data in the right way. In order to provide this, it is essential that the IoT has an appropriate architecture. In general, the basic IoT architecture model consists of 3 levels, namely: (i) perception, (ii) network, and (iii) application layer [9].

- A layer of perception, such as facial skin and five sensory loT organs, identify objects by gathering information. Its main task is to identify the object.
- The network layer is like a neural network and a loT brain. Its main function is the transfer and processing of information. The network layer includes the integration of communication and Internet networks, as well as the networks of the control center, information center and intelligent processing center, etc. The network layer will transmit and process information obtained from the perception layer.

- The application layer is a combination of social and industrial requirements in the pursuit of extensive intellectualization. The application layer is a deep integration of loT and industrial technology whose combination should realize an intellectualized ("smart") industry, similar to the social division of labor, as the ultimate form of humane society [9]. Enhanced IoT architecture models contain multiple layers [10].
- The first layer is the Perception Layer, which consists of physical objects and sensors. This layer has the function of identifying and collecting information from the sensor. The perception layer includes two-dimensional labels and bar code readers, RFID (Radio Frequency Identification) tags and camera readers/recorders, GPS devices, sensors, terminals and sensor networks. Depending on the type of sensor, this information may relate to recording the condition and tracking changes in location, weight, wind speed, vibration, temperature, direction of movement, humidity, acceleration and similar. All the information collected that way is forwarded to the next layer. Sensor devices are actually "smart" devices that the user controls using an internet application. They usually consist of one or more sensors for different purposes. In order for these devices to be remotely operated, they must have a unique address that will allow them to communicate seamlessly with authorized devices using the appropriate protocol. Most often, this unique address consists of 8 bytes (representing one of 4 parts of an IP address) in a particular network.
- The second layer is the Network Layer, or communication networks, which are responsible for connecting to other "smart" things, sensors on IoT devices, network devices and servers. This layer safely transfers information from sensor devices to processing devices. Data transmission can be done via direct wired connection (cables) or without intermediaries (wireless), and the technologies used for this are 3G, UMTS, Wifi, Bluetooth, InfraRed, ZigBee, NFC depending on the type of sensor. Thus, the network layer transmits information from the perception layer to the middleware layer.
- The third layer is the middleware layer, which may also be called the process layer. Its task is to store, analyze and process large amounts of data retrieved from the network layer. Devices implement different types of services through IoT. Each device connects and communicates with only those devices that implement the same type of service. This layer is responsible for managing the services and providing different types of services to the lower classes. All messages and commands sent from one device to another through applications through communication networks are forwarded to the database by the intermediary layer. Intermediary layer performs information processing and makes automatic decision based on results.
- The fourth layer is the Application Layer. This layer enables global application management based on object information, which is processed in the Middleware layer. The application layer manages various applications where the Internet of Things can be used, some of which are smart homes, smart cities, smart healthcare, smart farms, intelligent transportation, etc.
- The fifth layer is the Business Layer, This layer is responsible for managing the entire IoT system, including the applications and services they provide. It creates business models, charts, flowcharts, etc., based on information received from the

application layer. The true success of IoT technology depends on the choice of good business models. After analyzing the results, this layer will help define future actions and business strategies. Another name for this layer is client applications, which are actually applications that allow the user to manage devices that are connected to a single network. The application does not communicate directly, but through communication networks to which devices that have sensors are also connected, through which it achieves the desired communication. Thanks to client applications, the user has the ability to keep track of all device-related information and to keep up to date with events that may interest him and that may influence his decisions. User applications essentially serve to engage users with devices. Some of the known platforms for these applications are Web, iOS, Android etc.

4 Method of Functioning IoT

The way IoT works is based on communication between people, devices and things. The simplest way to explain how an IoT works is that people use a device to manage things online. A prerequisite for the successful functioning of IoT is the installation of sensor devices that will continuously record information about some events and pass them on for further processing, to take appropriate action to control them. For this to be possible, it is necessary to have an appropriate device (activator or controller) with which the user will connect via the user application to the communication network in which the devices to be controlled are located. Once connected to the communications network, the sensor-based IoT application management service reads and stores certain information in its database, which it will further manage. After that, through the control device, based on the information received from the sensor devices, communication with the devices we want to manage occurs, during which time the IoT application management service continues to record activities, process the data and enable the application to run successfully. Once the communication is established, there are certain actions (at the request or need of the user) and the exchange of information important for both the user and the devices, which are again in the function of the user.

For example, if we take into account the automatic switching on of heating, that is, controlling and managing the temperature in the house during the winter days, then in order to accomplish such a task, a device that regulates the temperatures is required, that is, the heating in the house and which has sensors built in to record the temperature values. To be able to remotely control the temperature, the user must own a control device (usually a smartphone) and install a client application, which is related to regulating the home temperature. With the help of the application he has installed, the user must also use some kind of communication technology to connect to the communication network. Subsequently, the IoT application management service enables communication between the smartphone and the temperature-regulating device. Upon communication between the two devices, continuous interaction between them continues. The user invokes the application on their mobile phone and connects to the thermostat (sensor and activator) in the home. The thermostat returns him the information that the current temperature in the house e.g. 16 °C, so the user orders the thermostat to reach a temperature of 24 °C at 6:00 pm. The thermostat, in the function of the actuator, that is, the control device, records this information with the help of the service and prepares to perform this action.

5 Technologies Allowing the Implementation of the IoT

The realization of IoT depends on the dynamics of technological breakthroughs in a number of different scientific disciplines. Connecting large objects and devices to large databases, different types of networks, or the Internet required a simple, efficient, and inexpensive system for identifying those things. One of the crucial technological advances that has these potentials and has made it possible to realize the concept of IoT is the ability to identify things using RFID technology. Only with the help of this or similar identification system it is possible to collect and process data about things. RFID is a system that represents wireless communication technology and is used to identify and track objects without being wired. It is also used to exchange data using short-range radio signals [11]. NFC or Near Field Communication technology is similar to RFID technology and operates on the same principle for identification and two-way communication. This is technology that works wirelessly over very short distances, allowing data to be transferred between devices by touching or approaching only to few inches [12]. Another technology that has made it possible to put the IoT concept into practice is sensor and detector technology. One of their essential characteristics is the ability to capture, store and manage information about the physical status of objects or devices. Another technology essential to the realization of the concept of IoT is the incorporation of intelligence elements ("minds") into the devices themselves, which are then easier to manage. In this way, embedded intelligence in devices also increases the potential of the network by transmitting the processed information to the end nodes of the network. The next technology from the list of technologies without which the creation and development of the IoT concept would not have been possible is nanotechnology that enables the connection of physically smaller and smaller devices. Wi-Fi is a technology that uses radio waves to transmit data, providing wireless Internet and creating wireless local area networks (WLAN). Bluetooth technology enables communication between devices and their wireless connection via BT hotspots to the voice transfer network or high-speed Internet. The maximum distances at which devices can communicate are 10 m - when the transmit power is 1 mW, and when the transmit power is increased, in perfect conditions reaching up to 100 m. GPS (Global Positioning System) is a global positioning system consisting of a network of 24 satellites in a geostationary orbit, a control and monitoring system on Earth, and user terminals. Satellites continuously transmit information via radio signals, thereby allowing the user to determine their position on Earth through the receiver. The GPS system allows you to determine the position of the receiver in three dimensions: latitude, longitude and altitude at which the receiver is located, and the accuracy is ± several meters. In the end, the list of technologies, technological advancements and potentials for realizing the IoT concept would look like this: Radio Frequency Identification (RFID and NFC), Sensor and detector technology, The incorporation of intelligent elements into things, Nanotechnology, Wi-Fi, Bluetooth, and GPS.

Due to the fact that the implementation of the IoT concept is still in its infancy, it is more than clear that this is not a comprehensive and definitive list of technologies that enable the practical implementation of the IoT concept. The synergistic application of all these technological advances will increase the potential of IoT to connect various objects and devices around the world in an intuitive and intelligent way.

6 Areas and Reasons for the Application of IoT

IoT applications are virtually limitless and cover almost all areas of human activity, starting with better management and optimization across supply chains, through more efficient control and savings in transportation and logistics, sensors and smart elements embedded in car, aircraft, trains and ships (to facilitate the management of these means of transport), up to telecommunication applications, remote medical systems, health care, animal care, children care, the elderly and infirm care, smart buildings and cities, energy systems, and assistance in agriculture, intelligence services different types of disasters, environmental monitoring, education and more. For example, in a typical mobile healthcare scenario, a nurse, physician, or patient may deploy an appropriate network of small medical wireless sensor devices (such as thermometers, heart rate, blood pressure, blood glucose, blood, oxygen saturation, and other medical devices to monitor the patient's condition) to the patient's body. These devices can communicate with each other, but also with devices such as smartphones, tablets, smart TVs, etc. via wireless short-range network technologies such as Bluetooth, ZigBee, or Wi-Fi. In such a control, for example, based on the data collected from the sensor, the insulin pump will automatically inject a certain amount of insulin into the body, and the collected data can be simultaneously transmitted via the public telecommunications network to the telemedicine system. Based on this information, your doctor may change your insulin therapy or call your doctor for an appointment. Utility workers can monitor the sanitary sewer network in real time to control the quality of the water and thus determine whether the elements within the network are functioning properly or not. In this way, the system can respond to unexpected events in a timely manner, avoiding possible damage to cities that can cause natural disasters such as floods [13].

Due to the possibility of continuous collection and processing of microclimatic and agronomic data (ambient temperature, humidity, PH value of the earth, light intensity, CO_2 concentration in the soil) for control in order to optimize and timely implement appropriate agrotechnical measures with consequent cost reduction, IoT technology has significant potential for application in agricultural production. The implementation of remote control enables agricultural, scientific and technological personnel to obtain various parameters of the production environment through a mobile phone or computer, and to improve the timeliness and accuracy of data on the growth or growth retardation of agricultural products. Using the wireless network allows them to control irrigation in a timely manner, as well as monitor the conditions for crop growth. This application of IoT technology provides the scientific basis for accurate greenhouse control, yield increase, product quality improvement, tuning of growth cycles and ultimately improving economic benefits [14]. And at the end of this chapter, we will outline another interesting and useful way of applying IoT technology, which is related to the livestock sector. To prevent animal theft, RFID tags are affixed to them, which allow animals to be monitored. The location or exact positioning of the animal can be displayed on the map in the control center via data sent wirelessly. In rural areas where there is common grazing, animals tend to get lost. Chips with radio frequency identifiers (RFID) are placed on the stock, RFID readers are placed at various monitoring points to send information to agricultural services [15].

7 Conclusion

The IoT concept enables people to interact with devices and devices with devices, integrating them into a web-based application management network. IoT technology, along with other technologies which are key to the development of Industry 4.0 (Cloud Computing, Big Data, Blockchain) is a revolutionary concept, so it is no wonder that the potential of these technologies, and especially the potential of IoT technology, compares to the cumulative potential of all other technologies. Internet of things technology can be compared to an under-explored mine from which treasures have just begun to be extracted from which the devices of the future are made. This technology provides us with incredible capabilities that we could only dream of until a few years ago. However, whether IoT provides benefits only, or as we have concluded in the paper, there are certain threats. Most IoT connoisseurs, including CISCO's Internet of Things group, predict that about 50 billion (smart) devices will be connected to the network this year (2020), and that number will grow exponentially. One of the biggest drawbacks of this technology right now is its lack of security, so this needs to be taken seriously in the future. In the future, all devices that have the ability to connect will be connected, and therefore available anytime, anywhere.

References

1. Greengard, S.: The Internet of Things. MIT Press, Cambridge (2015)
2. Evans, D.: The Internet of Things: how the next evolution of the internet is changing everything. CISCO White Pap. **1**, 14 (2011)
3. Banafa, A.: The Internet of Everything (IoE) (2016). https://www.bbvaopenmind.com/en/the-internet-of-everything-ioe/. Accessed 5 Jan 2020
4. Yeo, K.S., Chian, M.C., Wee Ng, T.C., Tuan, D.A.: Internet of Things: trends, challenges and applications. In: 2014 International Symposium on Integrated Circuits (ISIC), Singapore, pp. 568–571 (2014)
5. Afshar, V.: Cisco: Enterprises Are Leading The Internet of Things Innovation. Cisco (2017). https://www.huffpost.com/entry/cisco-enterprises-are-leading-the-internet-of-things_b_59a41fcee4b0a62d0987b0c6. Accessed 08 Nov 2019
6. Miorandi, D., et al.: Internet of Things: vision, applications and research challenges. Ad Hoc Netw. **10**, 1497–1516 (2012)
7. Akyildiz, I.F., Kasimoglu, I.H.: Wireless sensor and actor networks: research challenges. Ad Hoc Netw. J. **2**, 351–367 (2004)
8. ITU-T, Internet of Things Global Standards Initiative. http://www.itu.int/en/ITU-T/gsi/iot/Pages/default.aspx. Accessed 10 Nov 2019

9. Wu, M., Lu, T., Ling, F., Sun, J., Du, H.: Research on the architecture of Internet of Things. In: 3rd International Conference on Advanced Computer Theory and Engineering (ICACTE), August 2010
10. Khan, R., Khan, S.U., Zaheer, R., Khan, S.: Future internet: the Internet of Things architecture, possible applications and key challenges. In: 10th International Conference on Frontiers of Information Technology, Islamabad (2012)
11. Lin, J., Yu, W., Zhang, N., Yang, X., Zhang, H., Zhao, W.: A survey on Internet of Things: architecture, enabling technologies, security and privacy, and applications. IEEE IoT J. **4**(5), 1125–1142 (2017)
12. Vishwajeet, B.H.: A survey on the smart homes using Internet of Things (IoT). Int. J. Adv. Res. Comput. Sci. Manag. Stud. **2**, 243–246 (2014). http://ijarcsms.com/docs/paper/volume2/issue12/V2I12-0005.pdf. Accessed 08 Nov 2019
13. Kovačević, T., Čagalj, M., Perković, T.: Internet of Things - mogućnosti i izazovi. OPEN info Trend (2018). http://www.infotrend.hr/clanak/2017/6/internet-of-things—mogucnosti-i-izazovi,90,1314.html. Accessed 12 Dec 2019
14. Duan, Y., Zhao, C., Tian, Z.: Application of the Internet of Things technology in agriculture. Appl. Mech. Mat. **687–691**, 2395–2398 (2014)
15. Dlodlo, N., Kalezhi, J.: The Internet of Things in agriculture for sustainable rural development. In: International Conference on Emerging Trends in Networks and Computer Communications (ETNCC). IEEE (2015)

Technosociality and the Rise of the Network Society

Halima Sofradžija(✉)

Faculty of Political Sciences, University of Sarajevo,
Skenderija 72, 71000 Sarajevo, Bosnia and Herzegovina
mima8@hotmail.com

Abstract. The rapid transformation of the world influenced by innovative technologies is evidently creating an entirely new culture, technoculture, within which new values, new forms of interactivity and social substance (contents of the society) are recognized and established, which eventually changes our reality irreversibly at all levels of human experience. Application of new technologies deeply transforms the social space. The rise of the network society and contemporary global processes invite us to contemplate on and better understand the complexity of the digital age. After all, today we talk about a historical transformation of the whole society, which has become firmly embraced by technology and innovative technologies, which, of course, consequently dramatically changes social practices and shapes an entirely new quality of life in the postmodern reality.

Keywords: Technoculture · Techno-sociality · Networked society · Innovative technologies

1 Introduction

The modern world, whose capillaries are imbued with technology, and which deeply changes under the influence of further innovations in the field of technology, creates a new architecture of society. Precisely because it is so much present in the whole area of human activity, the revolution and deep transformations of information technology, as well as scientific and technological progress in general, according to Manuel Castells, should be a starting point in any serious analysis of the complexity of the new culture, economy, politics and emerging society (see subtext T*echnology, society and historical change* in *The Rise of the Network Society*, 2000:40) [1]. The spirit of the information age reveals a deep connection between science and technology because, after all, the entire "construction of the modern world is scientific and technical" [2], with continuing tireless efforts to improve the information and telecommunications technology that clearly transforms the entire society and relations inside it, inside the society.

I. Karabegović (Ed.): NT 2020, LNNS 128, pp. 432–436, 2020.
https://doi.org/10.1007/978-3-030-46817-0_49

2 Technoculture, Technosociality and Irreversible Digitization of the Society

In the era of new technologies and of the further process of technicizing the world, F. Balle's statement that the new technologies, especially digital communication technologies, have opened the world to humans, but also captured them, becomes quite clear to us, because it has simply become recognizable that technology has Janus' character, that it is marked by a specific duality in many ways. There is no doubt that we are immersed in a hypermedia environment, because after all, recent literature clearly indicates that "technology itself is a medium in a literal sense, it mediates perceptions, a way of remembering, shaping of identities" (Čačinović, Puhovski, 2001) [3]. The world we live in carried substantial features of mediation. In agreement with Piter Sloterdijk, we can say that we live in a "mediation saturated society", within which the increasingly complex forms of networked mediation continue to develop, tirelessly and continuously [4]. Life on the Network in the networked world has long since become the reality of modern man, which Manuel Castells certainly discusses in his now widely cited work *The Rise of the Network Society* (according to available data of Social Science Citation Index, Castells is mentioned as one of the most cited scientists nowadays). The networking is manifold, visible at all levels, in all substances of social life; for example, if we talk about a "network enterprise", it is the form of organization that is best suited to the global information economy (Giddens) [5], simply because in today's contemporary context, it is easier for both large corporations or small private companies to survive on the market as a part of a Network. Therefore, we can talk about network corporations, network organizations of various profiles, and about network individuals. The morphology of the Network and Janus' nature of life in the Network become achievable and possible as such only **through technology**. Many contemporary theoreticians believe that this preoccupation with, and exposure to technology inevitably influences the transformation of contemporary culture, including the human perception of reality. Manuel Castells refers to the newly emerged culture, which in its broadest sense, has come to a qualitatively whole new level of determination by technology, as the culture of real virtuality. Therefore, technoculture is a culture determined by technology, digital technologies are integrated in everyday life. One of the essential characteristics of technoculture, in addition to networking, is fluidity, therefore, fluidity is recognizable and visible from virtual digital currency - bitcoin [6] already, then from the running capital that moves from one end of the world to the other, the spreading of power as Michael Foucault talks about it (power is fluid), to the Zygmunt Baumann's analysis of understanding of the elements of *fluid life, liquid life* [7] that takes place in an environment of constant uncertainty. *Dromology* [8] is another feature of it, a new science of acceleration, many authors warn about the immediate link between the accelerated technological transformation of the world and the global uncertainty (Virillio, Baudrillard, Bauman, Beck, Giddens) [9]. In the analysis of modern times, Paul Virillio takes acceleration as an important determinant, in his attempt to explain and point out the logic and influence of speed on social movements and consequences that follow. Today, we talk about *Technoscience*,

Technoculture, Technosphere, Technocapitalism (Suarez-Villa) [10], all of these demonstrate the presence and print of technology.

Technology changed the understanding of *space* and *time* categories. Is there a need to emphasize and remind that the Internet, video conferences, video surveillance, e-commerce, nanotechnology, etc., create a *new social environment* and socio-cultural context. The omnipresence of new technologies in everyday culture powerfully changes and marks human lives. Immersed and dragged into their technologized created spaces, seduced by the siren's call and the dazzle of technology, millions of people move into parallel virtual worlds every day [11]. In this immersion in digitalized spaces, they exercise their sociability through technology; therefore, we talk about techno-sociality. Techno-sociality reveals, identifies and realizes a completely new experience of sociability - the realization of sociability and remote connectivity through technology, where technology serves as the medium. Like we said, an essential feature of contemporary culture is mediation; it is the techno-sociality that features mediated communication and interaction through the use of technical devices. Techno-sociality is our everyday reality. Meeting and encounter in virtual spaces, without the authentic encounter and physical presence in the same space, but at certain distant points, speaks of orientation to a form of alienation, or as Sherry Turkle describes this situation: Alone Together (full title of the book is *Alone Together: Why We Expect More from Technology and Less from Each Other*) [12]. Technoculture is also a "*screen culture*", where our lives are marked by the orientation to screens, *hyper-reality,* an electronic magic in the digitalized world. This electronic magic is indeed our daily experience, where one looks at the screen longer than at any human face. Being glued to the screen in all its equivocation also implies the realization that we live in a culture of images, a culture that is increasingly moving away from the written word. It is a symbolic presentation of the path from the Gutenberg Galaxy to the McLuhan Galaxy, or more clearly, from the culture of written word to the culture of images. Technoculture is also a culture of spectacle that Guy Debord wrote about (*Society of the Spectacle*). Recognizing the exposure to media activity and the existence of hypermedia experience, McLuhan indicated that the media had become a "prison without walls".

There is no doubt that the omnipresence of new technologies is noticeable in all aspects of social life. Technology reshaped the world and changed the image of the society. In the new social rationality, the process of technizing the world and the opportunities and consequences this situation brings must be the subject of reflection. We might also wonder if the advancement of technology is happening to an unprepared humankind, and whether ethical principles follow such advancement (Hans Jonas) [13]. The postmodern age is fascinated by innovation, patents, the technosphere, constant change, the surrender to risks, and accordingly, responsible action should be more intensely present than it is. Globalization as the most pronounced process of our time is seen by many as the stage of capital; as Jurgen Habermas points out, capitalism has made technical innovation a principle (*Technology and Science as Ideology*) [14]. Like Rudiger Safranski, will we ask the same question that this author asks in the very title of his paper: - *How much globalization can we bear* [15]. Technology is not an extra-social phenomenon, it is a product of the society, a product of the human mind. In this dialectic of technology and society, one can no longer escape own realization. Humans have certainly expanded the playing field with technology, as *homo ludens*, because technology itself has its seductive power, hence such an irresistible commitment to the possibilities offered by technology in its "triumphant expedition".

3 Conclusion

Application of new technologies deeply transforms the social space. Shifts in technology, communication, business, and science create a world where social practices are noticeably changing. Manuel Castells refers to the newly emerged culture, which in its broadest sense, has come to a qualitatively whole new level of determination by technology, as the culture of real virtuality. Technoculture is a culture determined by technology, within which techno-sociality materializes as a new form of sociability. Technology reshaped the world and changed the image of the society. Modern society is characterized by mediation, where technology is recognized as an indispensable and omnipresent medium in a network society. Irreversible digitization of the society rearranges the reality in all fields.

References

1. Castells, M.: Uspon umreženog društva. Golden marketing, Zagreb (2000)
2. Sofradžija, H.: Hiperpolitika i savremeno društvo – proces tehniziranja svijeta, Fakultet političkih nauka, Sarajevo (2015). (See sub text *Poznanstvljenje države*, pp. 70–89)
3. Čačinović Puhovski, N.: Doba slika u teoriji mediologije, Naklada Jesenski i Turk, Zagreb, p. 12 (2001)
4. Sofradžija, H.: Posredovana slika svijeta – mediji, umjetnost i tehnologija u umreženom društvu, Zbornik radova medunarodnog znanstvenog simpozija *Filozofija medija (Umjetnost i mediji),* Medijsko sveučilište i Centar za filozofiju medija i mediološka istraživanja, Zagreb, pp. 159–167 (2013)
5. Giddens, A.: Odbjegli svijet – kako globalizacija oblikuje naše živote, Naklada Jesenski i Turk, Zagreb (2005)
6. Stupar, S., Bičo Car, M., Šahić, E.: Challenges of applying blockchain technology. In: Karabegović, I. (eds.) New Technologies, Development and Application (IINT 2019). Lecture Notes in Networks and Systems, vol. 76. Springer, Cham (2019)
7. Bauman, Z.: Fluidni život. Mediteran Publishing, Novi Sad (2009)
8. Virillio, P.: Speed and Politics: An Essay on Dromolog. Semiotext, Los Angeles (2007)
9. Beck, U.: Moć protiv moći u doba globalizacije – Nova svjetskopolitička ekonomija, Školska knjiga, Zagreb (2004). (see Baudrillard, J.: Simulacija i zbilja, Naklada Jesenski i Turk, Zagreb (2001))
10. Suarez-Villa: Technocapitalism – A Critical Perspective on Technological Innovation and Corporatism. Temple University Press, Philadelphia (2009)
11. Lash, S.: Technological forms of life. Theory Cult. Soc. **18**, 105–120 (2001). https://journals.sagepub.com/
12. Turkle, S.: Alone Together: Why We Expect More from Technology and Less from Each Other. Basic Book, New York (2011). (see https://quote.ucsd.edu/childhood/files/2013/05/turkle-tethered.pdf)

13. Jonas, H.: Princip odgovornost – Pokušaj jedne etike za tehnološku civilizaciju (1991). (Logos, V., Masleša, S., Delić, Z., et al.: Socijalni kapital i socioekonomski razvoj Bosne i Hercegovine, In Scan Tuzla (2018))
14. Habermas, J.: Tehnika i znanost kao ideologija, Suvremena misao, Zagreb, p. 72 (1986)
15. Safranski, R.: Koliko globalizacije čovjek može podnijeti, Gekon, Zagreb (2008). http://www.matica.hr/media/knjige/proslost-i-buducnost-20-stoljeca-804/

Cyber Security Analysis of the Electric Power Information System in Montenegro

Aleksandar M. Anđić[1], Ramo Šendelj[2(✉)], and Ivana Ognjanović[3]

[1] Ministry of Public Administration, Directorate of Electronic Governance and Information Security, Podgorica, Montenegro
[2] University of Donja Gorica, Oktoih 1, 81000 Podgorica, Montenegro
ramo.sendelj@udg.edu.me
[3] University of Donja Gorica, Podgorica, Montenegro

Abstract. In most countries, the electricity sector is designated as critical infrastructure. Therefore, nowadays, the field of cyber security is a crucial component in the functioning of all factors in the electricity sector of a country. A similar situation is observed in Montenegro, where the energy sector is designated as critical and highly ranked on the priority list for ensuring high-quality cyber security. Analyzing existing models at the global level, and taking into account the standards and recommendations provided by the relevant institutions, the authors of this paper have made an effort to illustrate the current state of cyber security of the electricity system in Montenegro, using the Feature model. Based on the established model, an analysis of the cybersecurity situation of this sector was done and some conclusions were drawn.

Keywords: Power system · Cyber security · Critical infrastructure · Feature model · ES-C2M2

1 Introduction

Considering the necessity of using electricity in the modern world, there are increasing attempts at unauthorized access to and manipulation of data, which aims at disrupting the proper functioning of the system or interrupting operation. Consequently, the inability to supply electricity has great consequences on the economy and population of a country.

Initially, the critical IT infrastructure in Montenegro was defined on the basis of the Methodology for determining critical information infrastructure. During 2016, the Law on Amendments to the Law on Information Security was adopted, which formally defined the critical IT infrastructure as: "*information systems of bodies whose interruption of work or destruction would endanger the life, health, safety of citizens and the functioning of the state, and on whose functioning the pursuit of activities of public interest depends on*" [1]. At the end of 2019, a draft Law on Critical Infrastructure Designation and Protection was prepared, which would properly regulate this area [2].

Having strategic documents is of crucial importance for further developing the ability to defend against modern cyber threats. In addition, it is desirable to have cyber security models, which will be of great help in overviewing the current state of affairs

I. Karabegović (Ed.): NT 2020, LNNS 128, pp. 437–445, 2020.
https://doi.org/10.1007/978-3-030-46817-0_50

and allow for an adequate assessment of the impact of new investments in certain areas on the overall cyber security. Pursuing this goal, by analyzing models for the protection of electricity systems at the global level, comparing the practices and recommendations of relevant institutions, we are able to define key domains and analyze the current state of cyber security of the electricity system of Montenegro. Using the Feature model, we have formed a model that illustrates the organization of cyber security of electricity system in Montenegro.

2 Models of Cyber Protection of Electric Power Systems at the Global Level

Nowadays, there are a number of mature models for ensuring cyber security of information systems. However, if we focus on electricity-related information systems, it is evident that this is not the case in this area. In most cases, when it comes to models for cyber protection of electric power systems, they also imply certain standards, good practices and frameworks. Further in this paper, we will briefly outline clearly defined models, created by leading institutions in this field globally and some models created by individuals and private companies.

2.1 Electricity Subsector Cybersecurity Capability Maturity Model (ES-C2M2)

ES-C2M2 is one of three models developed by C2M2 program [3]. It has been developed by the US Department of Energy in collaboration with the US Department of Homeland Security in a way that suits all organizations in the electricity sector, regardless of size, type or structure. It is a scalable framework that can be equally used at the complete organization level or for individual parts. It is organized into 10 groups (domains), covering the areas of: risk management, asset management, change and configuration, identity and access, threats and vulnerabilities, workforce, supply chain and external dependencies, cyber security program, information sharing and communication, as well as response to incidents and business continuity. The model defines four levels of Maturity Indicator Level (MIL), labelled MIL0 to MIL3. In order to reach a certain level, all recommended practices must be implemented in it. For example, if an organization has MIL2 in the threat domain, it will mean that all practices are measured at least with the MIL2 level [3].

2.2 ICS-SCADA Cyber Security Maturity Assessment Model

This model is a three-dimensional model, which was established to determine the cyber security maturity of critical sectors in the European Union. Developed by European Union Agency for Cybersecurity (ENISA), it contains a survey of 82 questions, divided into three dimensions and a total of nine sub-dimensions, related to the process life-cycle (creation - implementation - monitoring - change), where it is feasible [4]. The answers for all questions were scaled from 1 to 5, with clearly defined evaluation criteria.

When assessing the overall situation, based on the results of the survey, the country can be classified into one of five levels, taking into account some non-standard activities related to the cyber security of the ICS-SCADA system in certain countries. The levels of assessment are: *basic* (no activities are being implemented), *developing* (activities are under development or ad-hoc), *established* (regular performance of activities at the basic level), *advanced* (implementation of activities, with a strong understanding of the specifics of ICS-SCADA requirements), and *leading* (activities are conducted at a level that exceeds current needs (are able to anticipate future needs)).

2.3 3 × 3 Matrix Model by Eric Knapp

The model was created with an aim to simplify different systems operating in critical infrastructures, so that common countermeasures to cyber-attacks can be mapped and system vulnerabilities properly identified. It is brought down to representing three unique environments in a 3 × 3 matrix [6] (Fig. 1).

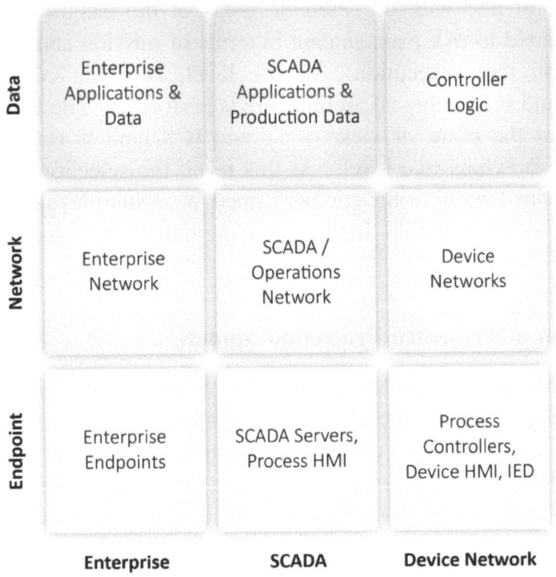

Fig. 1. A view on the environment in a 3 × 3 matrix [6]

This model is especially intended for mapping hosts, network and data security in business, SCADA, and industrial environments, defining nine unique security challenges, that can be applied to modern security controls that are widely used globally. Although each field of the model matrix represents a unique security challenge, the fact that an attack can "move" from one field to another should not be neglected [5].

2.4 SCADA Security Maturity Model

Red Tiger Security has developed its four-phase model, which in the first phase conceives site assessment, defining cyber security standards for SCADA systems, as well as a common network architecture model [7]. The second phase focuses on training administrators and users, raising awareness and financing cybersecurity related projects. Risk mitigation is the task of the third phase of the model, and is about developing preventive and detective cyber defense solutions. Finally, the fourth phase encompasses business and maintenance. More specifically, this phase refers to conducting periodic tests, reconfiguring the existing system, developing new technologies, recertifying, assessing and auditing [7].

2.5 Electricity Subsector Cybersecurity Risk Management Model

The US Department of Energy (DoE) has defined its Risk Management Model, which fosters a detailed state analysis of the organization in this section [8]. It is a three-level model, where the first level refers to the strategic goal of the organization in terms of risk management, in line with the strategic goals of the electric power sector. The second level is related to risk management in terms of mission and business processes and it focuses on their execution. At this level, two-way communication with employees in IT and ICS technical architecture is performed. The third level relates to systems risk, from the point of view of IT and ICS, and is based on information obtained from the previous two levels. At this level, the selection, development, and cyber security control monitoring are performed at system-level, which are subsequently assigned to IT and ICS based on a previously established cyber security architecture [8].

2.6 Comparison of Presented Protection Models

The presented cyber security models of electric power systems can be of great help to companies, by using self-evaluation forms, to get an impression of their maturity level in terms of cyber security. Based on a very comprehensive survey of world practices, standards, protection models and recommendations from relevant institutions, we have identified five key areas (domains). Through these domains, we have defined a number of goals that it is desirable for companies to achieve, in order to provide the preconditions for an adequate organization of cyber security.

Actually, the comparison of previously presented models will be made in a tabular form through the prism of the defined domains and goals from our research, as well as the presented recommendations for more adequate protection of the electric power information system in Montenegro.

According to Table 1, we conclude that only one of the models presented, to a certain extent, treats all five defined domains.

Considering the domain representation in the models, we can see that in most cases the risk areas and incident management are most treated, while, in the fewest cases, the focus is on the area of cooperation with other entities. However, this area is indispensable in institutional models.

Table 1. Comparative representation of defined domains in the models described

Model	Domain[a]				
	D	RM	IM	EER	COE
Electricity Subsector Cybersecurity Capability Maturity Model (ES-C2M2)	●	●	●	●	●
ICS-SCADA Cyber Security Maturity Assessment Model (ENISA)	●		●	●	●
3x3 Matrix(Eric Knapp)		●	●		
SCADA Security Maturity Model (Red Tiger)	●	●	●	●	
Electricity Subsector Cybersecurity Risk Management Model (US DoE)		●			

[a]D – Documentation; RM – Risk Management; IM – Incident Management, EER – Expertise, education and reporting; COE – Cooperation with other entities

3 Feature Model of Organizing the Security System of Electric Power System in Montenegro

As stated earlier, based on the state analysis, we have formed a model that illustrates the cyber security system of the electricity system in Montenegro. The model was created using the Feature Model (FM).

First, let us remind ourselves what the FM model is in theory.

Feature models (FMs) are models that describe variability among single-domain applications, with specification of functions and their interdependencies [9]. Relationships among characteristics can be: *mandatory features* - when child properties always go with parents, *optional features* - child properties that optionally appear with parents, *alternative feature groups* - when only one feature can be included in an allocation and *or feature groups* - when all features are optional but at least one must be selected. Functional interdependencies are determined by integrity constraints, of which the most commonly used is include - when one feature includes the other, and exclude - when one feature excludes others [9, 10].

Our model (Fig. 2) clearly emphasizes the independence of ICS (Core Business) and IT systems. Also, the project management process includes an aspect of cyber security, which is a mandatory feature of the ICS system. Cyber security is determined by predefined domains, which further include a large number of features.

The domain of documentation is determined by mandatory characteristics, such as rules, procedures and policies, as well as others that are of optimal character. Such are strategies, laws, models and frameworks, guidelines, analyses, etc.

Risk management includes network monitoring, remote access, backup and the existence of a Disaster Recovery (DR) site. When it comes to incident management, it emphasizes the necessity for an organizational unit with ICS-CIRC, cyber defense capability and 24/7 reporting service. In addition, the use of a variety of tools and proper incident recording are anticipated. In the domain of expertise, education and reporting, particular emphasis is placed on business continuity, training and expertise. Periodic cyber security reporting is anticipated. Finally, in the domain of cooperation, the mandatory cooperation of local ICS-CIRT teams is emphasized, while cooperation with National CIRT, international institutions and information sharing is encouraged.

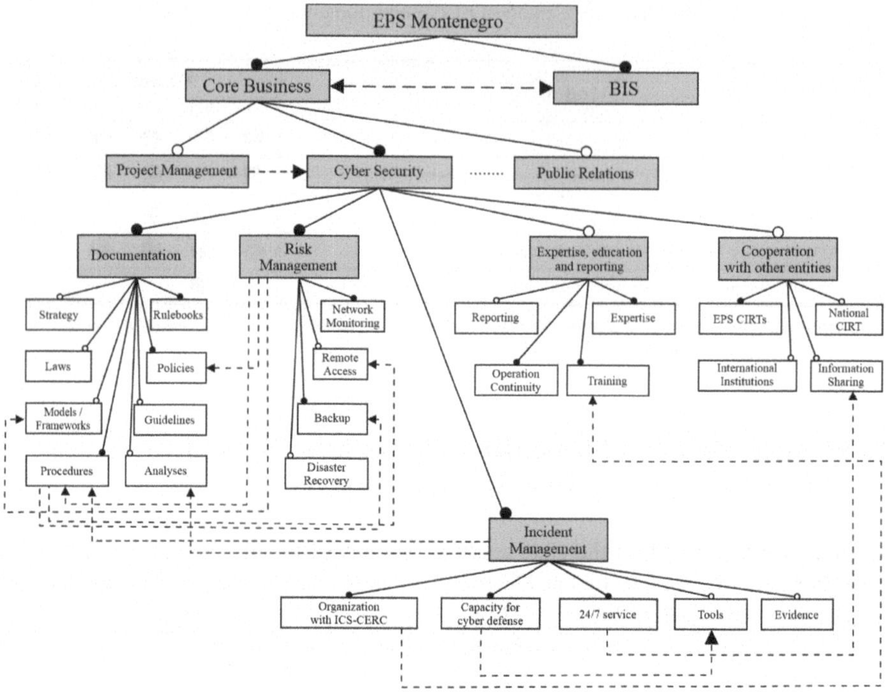

Fig. 2. Feature model of cyber security organization in Montenegro

However, there is interdependence in the defined domains. Thus, risk management requires procedures, models, frameworks, documentation policies, while procedures include backup and remote access. Also, incident management requires adherence to procedures, as well as conducting business impact analyses. Employees of an organizational unit with an ICS-CERC mandate require training, while cyber defense capabilities include a variety of tools. Finally, the 24/7 incident reporting service is tied to the information sharing feature, which is an optional item within the domain of collaboration with other entities.

4 Analysis of the Cyber Security Conditions in the Electric Power System in Montenegro

During 2019, for the purpose of this research, a questionnaire was sent to the companies that are part of the electricity system in Montenegro, containing a total of 45 questions and 31 sub-questions, divided into five chapters. The chapters referred to documentation, issues related to risk and incident management, expertise, education and reporting, and issues related to cooperation with other entities. The questionnaire was prepared in accordance with a previously conducted analysis and survey of the best practices of EU Member States and other countries of the world, as well as the recommendations of relevant institutions in the field of cyber security organization. The results obtained by providing answers to the questions asked have enabled us to define the key features in our model, which we presented graphically above. The state analysis was carried out in accordance with established practices at the global level, as well as on the expert experience of the authors in this paper.

Namely, due to the protection of information in the surveyed companies, it is almost impossible to assess the level of realization precisely. An additional dilemma is the organization's policy on prioritizing investment in a particular cyber security domain. If we take that the *unrealized measure* is graded 0 points, *partially realized* 0.5, and the *fully realized measure* 1 point, we can draw certain conclusions.

Table 2. Evaluation of implementation, based on the answers to the questions from the Questionnaire

Domain	Characteristic	Implementation coefficient
D	*Strategy*	0
D	*Rulebooks*	0.5
D	*Laws*	0
D	*Policies*	0.5
D	*Models/Frameworks*	0.5
D	*Guidelines*	0.5
D	*Procedures*	0.5
D	*Analyses*	0
RM	*Network monitoring*	1
RM	*Remoteaccess*	1
RM	*Backup*	1
RM	*DisasterRecovery*	0.5
IM	*Org. with ICS-CERC*	0
IM	*Capab. for cyberdefense*	0.5
IM	*24/7 service*	0.5
IM	*Tools*	0.5
IM	*Evidence*	0
EER	*Reporting*	0.5
EER	*Operationcontinuity*	0.5
EER	*Trainings*	0.5
EER	*Expertise*	0
COE	*EPS CIRTs*	0
COE	*National CIRT*	0
COE	*International institutions*	0.5
COE	*Information sharing*	0

In our case, the implementation by domain is equal to **D** = 31.25% (2.5/8), **RM** = 87.5% (3.5/4), **IM** = 30% (1.5/5), **EER** = 37.5% (1.5/4), **COE** = 12.5% (0.5/4).

If all domains and all features are equally important, the grading is **9.5**/25, representing **38%** of implementation. For example, the company prioritize the following scheme: **D** = 15, **RM** = 30, **IM** = 30, **EER** = 15, **COE** = 10 (**SUM** = 100). If we take into account the previously presented percentage of implementation by domains, we reach the level of realization of **46.82%**.

The values in Table 2 clearly indicate the need for additional efforts to achieve the set goals. Thus, in the domain of documentation, it is necessary to work on developing new and updating existing procedures, policies and rules. As a prerequisite for strategic management of the cyber security segment, it is necessary to have a strategy with an action plan to execute the activities. In addition, it is necessary to encourage changes to existing ones and propose new legal solutions and periodically analyze the impact of incidents on overall security.

In the domains of risk and incident management, it is advisable to carry out activities to establish a Disaster Recovery site by all actors, for the sake of scalability of key information systems in the event of a disaster. On the other hand, it is necessary to form an organizational unit that will hold the ICS-CERC mandate, therefore, to strengthen cyber defense capabilities, using a variety of tools and accurately keep records of incidents. It is very important to have a service for reporting and exchanging of information with the actors of the electricity system, which operates on a 24/7 principle.

Continuous training of employees is an obligation of every employer, because only in this way a quality professional staff that will be able to maintain business continuity can be produced. In this domain, it is necessary to pay close attention to quality incident reporting.

Finally, cooperation between local CIRT teams is an imperative, which is essential for all actors in this sector. Collaboration with the National CIRT team and international institutions in this area is extremely important, as it is the exchange of information on "lessons learned" from various incidents.

5 Conclusion

From the above facts about the state of cyber security in the electricity system of Montenegro, we can conclude that the existing security systems are not sufficiently adequate to defend against modern cyber threats. Additional efforts need to be made to achieve the goals that have been partially achieved, as well as those that have not been achieved to date. The directions of our further research are reflected in combining the presented model with the *Business Processing Model* (*BPM*). At the same time, we will consider the application of known algorithms *AHP, TCP nets, CS-AHP* and *COP-network*, which will allow us to see how much their use will advance the previously presented analysis. Therefore, the power sector subjects will be given the opportunity to self-assess their maturity level and see how much investment in a particular domain will improve the overall cybersecurity of the company.

References

1. Ministry of Public Administration: Cyber Security Strategy of Montenegro 2018–2021. Government of Montenegro, Podgorica (2018)
2. Government of Montenegro: Proposal for a Law on Determination and Protection of Critical Infrastructure, Podgorica (2019)
3. US Department of Energy, US Department of Homeland Security: Electricity Subsector Cybersecurity Capability Maturity Model v1.1 (2014)
4. European Union Agency for Cybersecurity (ENISA): Analysis of ICS-SCADA Cyber Security Maturity Levels in Critical Sectors, 11 December 2015. https://www.enisa.europa.eu/publications/maturity-levels/at_download/fullReport
5. Knapp, E., Samani, R.: Applied Cyber Security and the Smart Grid - Implementing Security Controls into the Modern Power Infrastructure. Syngress - Elsevier (2013)
6. Knapp, E.: A New Cyber Security Model for SCADA, 1 May 2012. https://www.securityweek.com/new-cyber-security-model-scada. Accessed 25 Dec 2019

7. Red Tiger Security: SCADA Security Maturity Model. Red Tiger Security. http://redtigersecurity.com/services/scadaics-security-consulting/scada-security-maturity-model/. Accessed 25 Dec 2019
8. US Department of Energy: Electricity Subsector Cybersecurity Risk Management Process (2012)
9. Šendelj, R., Ognjanović, I.: Multi-criteria decision making for optimal configuration of business process model families. Inf. Technol. Control **47**(3), 532–563 (2018)
10. Šendelj, R., Ognjanović, I.: Framework for optimal selection using meta-heuristic approach and AHP algorithm. In: Applications and Theory for Analytic Hierarchy Process - Decision making for strategic decisions, Rijeka, Croatia, pp. 193–217. InTech (2016)

Nature-Inspired Cryptoanalysis Methods for Breaking Vigenère Cipher

Lucija Brezočnik(✉), Iztok Fister Jr., and Vili Podgorelec(✉)

Faculty of Electrical Engineering and Computer Science, University of Maribor, Koroška cesta 46, 2000 Maribor, Slovenia
{lucija.brezocnik, iztok.fisterl, vili.podgorelec}@um.si

Abstract. The protection of sensitive data against unauthorized access remains a primary concern of modern life. Over time, many different approaches have been introduced to tackle this problem, from substitution ciphers in classic cryptography to post-quantum cryptography as a representative of modern cryptography. In this paper, we focus on a polyalphabetic substitution cipher, precisely the Vigenere cipher. For a cryptoanalysis of the latter, we utilized five nature-inspired algorithms, i.e., Differential Evolution, Firefly Algorithm, Particle Swarm Optimization, Artificial Bee Colony Algorithm, and Cuckoo Search, were utilized. Furthermore, different key lengths were analysed to investigate the search behaviour of the selected algorithms. The results of the experiment show that the applicability of the nature- inspired algorithms for cryptoanalysis is very promising. Out of the tested algorithms, the Differential Evolution outperformed other algorithms.

Keywords: Cryptoanalysis · Nature-inspired algorithms · Swarm intelligence · Vigenère cipher

1 Introduction

The technique of securing information against third party adversaries is called cryptology. It encompasses cryptography (the art of creating encryption schemes) and cryptoanalysis (the art of analysing and breaking encryption schemes). In this paper, we will focus exclusively on classic cryptography.

There are two main types of classical ciphers, i.e. transposition ciphers and substitution ciphers [10]. The latter group is further divided into monoalphabetic (e.g. Caesar cipher, Affine cipher, and Atbash cipher), polyalphabetic (e.g. Vigenère cipher, Gronsfeld cipher, and Beaufort cipher), and polygraphic (e.g. Hill cipher) ciphers.

In this paper, we investigate the use of different nature-inspired algorithms in a cryptoanalysis of the Vigenère cipher. This field is relatively new and therefore has not been researched extensively. Although, some methods show promise. In most cases, researchers utilize a Genetic Algorithm (GA) [7–9] for automated cryptoanalysis. Out of the swarm intelligence algorithms, a Particle swarm optimization (PSO) [12] is used in the majority of the times, followed by the Cuckoo Search algorithm (CS) [3] and Bees algorithm (BA) [1].

I. Karabegović (Ed.): NT 2020, LNNS 128, pp. 446–453, 2020.
https://doi.org/10.1007/978-3-030-46817-0_51

The primary missions of this paper are:

- to present a short overview of cryptoanalysis methods which include nature-inspired algorithms,
- to analyse a universal cryptoanalysis method that can be combined with any nature-inspired algorithm, and
- to investigate the behaviour of different nature-inspired algorithms for breaking the Vigenère cipher.

The structure of this paper is as follows: Sect. 2 outlines the Vigenère cipher, while Sect. 3 acquaints the reader with the fundamentals of nature-inspired algorithms and briefly presents used ones. Section 4 describes proposed nature-inspired cryptoanalysis methods, which were evaluated with the experiment presented in Sect. 5. The obtained results are gathered and analysed in Sect. 6, while Sect. 7 concludes the paper and provides future paths.

2 The Vigenere Cipher

The Vigenère cipher was named after Blaise de Vigenère and proposed in 1586 [10]. It is a polyalphabetic cryptosystem, which was famous for centuries as being hard to break. For example, let's take the English alphabet as a base. The English alphabet consists of 26 letters or characters, thus, there exists $26! \approx 4.033 \times 10^{26} \approx 2^{88}$ of possible keys. It is evident, that an exhaustive key search is precluded.

Algebraically, we can present the Vigenère cipher using modular arithmetic. Let's define plaintext, key, and ciphertext as $P = (P_1, P_2, \ldots, P_m)$, $K = (K_1, K_2, \ldots, K_n)$, and $C = (C_1, C_2, \ldots, C_m)$, where m and n denote the length of the plaintext/ciphertext and key, respectively.

Formally, we can now define the encryption E using the key K as:

$$C_i = E_K(P_i) = (P_i + K_i) \quad mod\ 26 \tag{1}$$

Decryption D using key K can be represented as:

$$D_K(C_i) = (E_i + K_i) \quad mod\ 26, \tag{2}$$

where $D_K(C_i)$ denotes the offset of the i-th character of the \$P\$.

The encryption of the words *NEW TECHNOLOGIES* with the key *KEY* is presented in Table 1.

Table 1. Example of an encryption with the Vigenère cipher.

Plain text	N	E	W	T	E	C	H	N	O	L	O	G	I	E	S
Key	K	E	Y	K	E	Y	K	E	Y	K	E	Y	K	E	Y
Ciphertext	X	I	U	D	I	A	R	R	M	V	S	E	S	I	Q

3 Nature-Inspired Algorithms

Nature-inspired algorithms, as the name implies, draw inspiration from nature [4]. Nowadays they are incorporated in solving many real-world optimization problems and are showing promising results. Because of that, we picked five popular algorithms from that group and utilized them in our proposed method. In the following Sects. 3.1–3.5, the selected algorithms are briefly presented. An in-depth explanation of nature-inspired algorithms can be seen in [4].

3.1 Differential Evolution

DE is a type of evolutionary algorithm and was proposed by Storn and Price in 1997 [11]. It uses mechanisms inspired by the theory of evolution, i.e. survival of the fittest. The fittest individuals of the population produce more offspring, therefore, their successors are carrying more advantageous genetic material and are more likely to survive. The pseudocode of the DE is outlined in Algorithm 1.

Algorithm 1: Pseudocode of the DE

```
init_population();
whiletermination_condition_not_metdo
random_choice_of_three_vectors();
calculate_mutation();
calculate_crossover();
calculate_selection();
update_population();
end while
```

3.2 Particle Swarm Optimization

PSO was proposed by Kennedy and Eberhart in 1995 [6] when they were inspired by the swarming behaviour of birds and fishes. In the PSO, particles are moving in the search space with given velocity and direction. The latter two change based on the personal best result of each particle and the global best of the swarm. The pseudocode of the PSO is outlined in Algorithm 2.

Algorithm 2: Pseudocode of the PSO

```
init_population();
whiletermination_condition_not_metdo
evaluate_new_solution();
save_the_local_best_solution();
save_the_global_best_solution();
update_velocity();
update_position();
end while
```

3.3 Firefly Algorithm

Yang invented FA in 2008 [14]. The algorithm mimics the behaviour of the fireflies, but since direct transformation is not possible, the following adjustments were introduced: Fireflies are all unisex, attractiveness is proportional to the light intensity and, the light intensity is determined by a fitness function. The pseudocode of the FA algorithm is presented in Algorithm 3.

Algorithm 3: Pseudocode of the FA

```
init_population();
whiletermination_condition_not_metdo
evaluate_fireflies();
order_firelies();
replace_firelies();
find_the_best_firefly();
move_fireflies();
end while
```

3.4 Artificial Bee Colony Algorithm

ABC was defined by Karaboga in 2005 [5] and was inspired by the intelligent Foraging behaviour of honey bee swarm. The ABC algorithm consists of employed bees, onlookers and scouts, where each of them has a different task. Employed bees fly to their food source until the food source is abandoned. Onlooker bees are just watching employed bees and randomly choosing food sources, while scout bees try to find new food sources. The pseudocode of the FA algorithm is presented in Algorithm 4.

Algorithm 4: Pseudocode of the ABC

```
init_population();
whiletermination_condition_not_metdo
send_employed_bees();
send_onlooker_bees();
send_scouts();
end while
```

3.5 Cuckoo Search

Parasitism of some cuckoo species was inspiration for the CS algorithm proposed by Yang and Deb in 2009 [15]. The goal of the algorithm is to transfer the best nests, i.e. nests with the highest quality of eggs, into the next generation. Cuckoos' behaviour of laying eggs into randomly chosen nests and the host bird's decision of throwing the egg away or abandoning the nest is directly drawn from biology. The pseudocode of the FA algorithm is presented in Algorithm 5.

Algorithm 5: Pseudocode of the CS

```
init_host_nests_locations();
whiletermination_condition_not_metdo
generate_new_solution();
evaluate_new_solution();
randomly_choose_a_nest();
optionally_replace_a_nest();
optionally_build_a_new_nest();
keep_the_best_nests();
end while
```

4 Nature-Inspired Cryptoanalysis Methods

To perform a cryptoanalysis of the Vigenère cipher, we first have to identify the period of the cipher, i.e. the length of the key. The repetition of the key is one of the biggest weaknesses of this cipher and can be exploited by a Friedman attack that is based on the Index of Coincidence (I_c). The calculation of the I_c is presented in Eq. 3.

$$I_c = \frac{\sum_{i=1}^{j} f_i(f_i - 1)}{n(n-1)}, \tag{3}$$

where n denotes the length of the plaintext and f_i is the frequency count of the i-th letter. Frequency is calculated for all j letters of the alphabet. For an easier calculation, we can use an approximation of the key length [2, 3] presented in the following equation:

$$key\_length = \frac{0.027n}{I_c(n-1) - 0.038n + 0.065}, \tag{4}$$

When we set on the key length, we have to find the actual key in the search space of j^{key_length} (for example, using the English alphabet, that is 26^{key_length}). For this difficult task, we utilize the nature-inspired algorithms presented in Sect. 3.

The fitness function is calculated the same for all algorithms using Eq. 5.

$$f(K) = \sum_{i=1}^{j} |FE_i - FD_i|, \tag{5}$$

where FE_i and FD_i denotes the frequency of the English language and frequency of the deciphered cipher text, respectively.

5 Experiment

The method was implemented in the Python programming language. All nature-inspired algorithms used in this method were taken from the external library NiaPy [13], which is a micro-framework for building nature-inspired algorithms. The experiment was run on a computer that has a Windows operating system and an Intel Core i7 processor with 16 GB of RAM.

The plain text length was 2,582 characters, which does not include spaces, numerals and special characters. To encode the plain text, we used four different sized keys, i.e. TEXT, CIPHER, VIGENERE, and CRYPTOGRAPHY, and the English alphabet. After the parameter tuning, the following parameter settings were used based on their best performance:

- DE ($NP = 20$, $F = 0.5$, $CR = 0.9$),
- PSO ($NP = 200$, $c_1 = 1.3$, $c_2 = 1.3$, $w = 0.7$),
- FA ($NP = 30$, $\alpha = 0.5$, $\beta_{min} = 0.2$, $\gamma = 1.0$),
- ABC ($NP = 40$, $Limit = 2$),
- CS ($NP = 20$, $pa = 0.9$, $\alpha = 0.25$).

Each configuration of the algorithms was run independently five times. The termination condition was set to 50,000 fitness evaluations (*nFES*).

6 Results

This section outlines the results of cryptoanalysis methods for breaking the Vigenère cipher.

Table 2 presents the maximum number of correctly recovered key characters (MAX_KC), the minimum number of correctly recovered key characters (MIN_KC), and the average number of correctly recovered key characters (AVG_KC) per algorithm for all key sizes. It is apparent that only the DE algorithm managed to recover all of the key characters. PSO performed the second-best (average of 3.8, 5.4, 6, and 7.8 of correctly recovered key characters), closely followed by ABC (average of 4, 5.2, 6.4, and 7 of correctly recovered key characters). The FA algorithm recovered all keys only for the key size four, but its average numbers do not differ drastically from the ABC algorithm. The worst results were obtained by the CS algorithm, which had the smallest average number of recovered key characters for all key sizes.

We also measured the time that it took an algorithm to break the cipher (Table 3). The DE algorithm not only performed the best in recovering the highest number of key characters, but also took the least amount of time doing that. Like in Table 2, PSO performed second best, followed by ABC. The CS algorithm was relatively fast but did not manage to recover even a size 4 key.

Since each algorithm was run independently five times, we wanted to see how many times an algorithm managed to recover the whole key in those five times (Fig. 1). The DE algorithm recovered the right key in every run for the key sizes four and six. At key size 8 was mistaken only once and at the key size 12, it succeeded to recover the right key one time. DE is the only algorithm that managed to do that.

Table 2. Number of correctly recovered key characters per algorithm.

Key size	MAX_KC					MIN_KC					AVG_KC				
	DE	PSO	FA	ABC	CS	DE	PSO	FA	ABC	CS	DE	PSO	FA	ABC	CS
4	4	4	4	4	3	4	3	3	4	2	4	3.8	3.8	4	2.6
6	6	6	5	6	3	6	5	5	4	3	6	5.4	5	5.2	3
8	8	8	6	8	5	6	5	4	5	2	7.6	6	5	6.4	3.8
12	12	9	8	8	5	11	6	5	6	3	11.2	7.8	6.6	7	4

Table 3. Time analysis (in seconds) of obtaining the maximum number of correctly recovered key characters. In cases when all key characters were recovered correctly, they are marked as bold.

Key size	DE	PSO	FA	ABC	CS
4	**35.7**	**178.8**	**349.7**	**180.2**	181.7
6	**38.3**	**184.9**	232.8	**187.3**	185.4
8	**40.2**	**214.9**	426.4	**246.6**	205.3
12	**52.8**	216.4	312.1	200.3	229.5

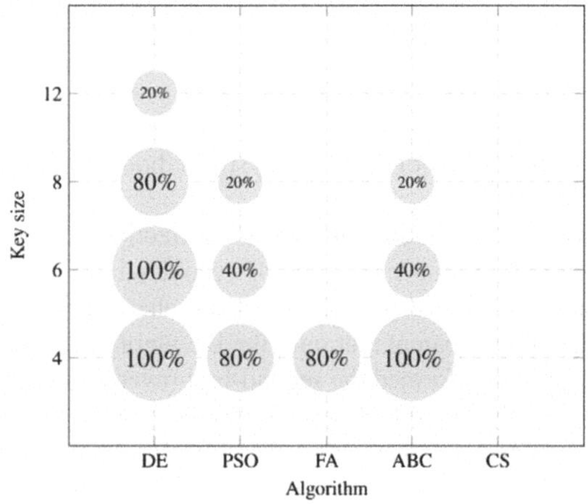

Fig. 1. Bubble plot for the percentage of all correctly obtained key characters in five independent runs regarding two variables: Algorithm (x-axis) and key size (y-axis).

7 Conclusion

A method that offers a solution for breaking the Vigenère cipher was analysed. It was tested with four different key sizes on the same plaintext. The results are revealing, in that the DE algorithm outperformed other algorithms considering the number of correctly recovered key characters and time.

The modular design of the method offers further utilization of other nature-inspired algorithms. Therefore, an in-depth comparison of the algorithms for a cryptoanalysis of the Vigenère cipher could be carried out. Furthermore, since all of the nature-inspired algorithms are sensible on their parameter settings, it would be a good way to perform parameter tuning. In the future, we are planning to introduce nature-inspired algorithms to modern cryptography.

Acknowledgements. The authors acknowledge the financial support from the Slovenian Research Agency (Research Core Funding No. P2-0057).

References

1. Ali, I.K.: Cryptanalysis of simple substitution ciphers using bees algorithm. Baghdad Coll. Econ. Sci. Univ. **36**, 373–382 (2013)
2. Beker, H., Piper, F.: Cipher Systems: The Protection of Communications. Northwood Books, Chambersburg (1982)
3. Bhateja, A.K., Bhateja, A., Chaudhury, S., Saxena, P.: Cryptanalysis of Vigenere cipher using cuckoo search. Appl. Soft Comput. **26**, 315–324 (2015)
4. Brezočnik, L., Fister, I., Podgorelec, V.: Swarm intelligence algorithms for feature selection: a review. Appl. Sci. **8**(9), 1521 (2018)
5. Karaboga, D.: An idea based on honey bee swarm for numerical optimization. Technical report, Erciyes University, Engineering Faculty, Computer Engineering Department (2005)
6. Kennedy, J., Eberhart, R.: Particle swarm optimization. In: Proceedings of International Conference on Neural Networks (ICNN 1995), vol. 4, pp. 1942–1948. IEEE, Perth (1995)
7. Mudgal, P.K., Purohit, R., Sharma, R., Jangir, M.K.: Application of genetic algorithm in cryptanalysis of mono-alphabetic substitution cipher. In: 2017 International Conference on Computing, Communication and Automation (ICCCA), pp. 400–405. IEEE (2017)
8. Omran, S., Al-Khalid, A., Al-Saady, D.: Using genetic algorithm to break a monoalphabetic substitution cipher. In: 2010 IEEE Conference on Open Systems (ICOS 2010), pp. 63–67. IEEE (2010)
9. Omran, S., Al-Khalid, A., Al-Saady, D.: A cryptanalytic attack on Vigenère cipher using genetic algorithm. In: 2011 IEEE Conference on Open Systems, pp. 59–64. IEEE (2011)
10. Stinson, D.R.: Cryptography: Theory and Practice. Chapman and Hall/CRC, London (2005)
11. Storn, R., Price, K.: Differential evolution–a simple and efficient heuristic for global optimization over continuous spaces. J. Global Optim. **11**(4), 341–359 (1997)
12. Uddin, M.F., Youssef, A.M.: Cryptanalysis of simple substitution ciphers using particle swarm optimization. In: 2006 IEEE International Conference on Evolutionary Computation, pp. 677–680. IEEE (2006)
13. Vrbančič, G., Brezočnik, L., Mlakar, U., Fister, D., Fister Jr., I.: NiaPy: Python microframework for building nature-inspired algorithms. J. Open Source Softw. **3**, 613 (2018)
14. Yang, X.-S.: Firefly Algorithm. In: Nature-Inspired Metaheuristic Algorithms, p. 128. LuniverPress (2008)
15. Yang, X.-S., Deb, S.: Cuckoo Search via Lévy flights. In: 2009 World Congress on Nature & Biologically Inspired Computing (NaBIC), pp. 210–214. IEEE (2009)

Using a Chaos Generator to Achieve Cryptographic Strength Parameters Close to Absolutely Stable Ciphers

Irina Fedosova[1], Tetiana Levytska[1], Vira Shendrik[2(✉)], Michail Vereskun[1], and Sergii Shendryk[2]

[1] Priazovsky State Technical University, Mariupol, Ukraine
[2] Information Technologies of Design, Sumy State University, Sumy, Ukraine
v.shendryk@sumdu.edu.ua

Abstract. It is possible now to implement algorithms that can be roughly comparable in reliability with absolutely stable cryptosystems. The Chua scheme is a dynamic chaos model has a complex behavior with general ease of implementation and is capable of operating over a wide range of values. So, it was chosen as a modeling object to use it to information security. This chaos generator gives a very high degree of unpredictability, close to truly random sequences. The result is practically not predictable regardless of the availability of information about the algorithm and parameters. The generated sequences of numbers, taking into account their unlimited length and uniqueness of combinations of values, are ideal from the point of view of their use as keys for data encryption by the cryptosystem. The scientific novelty of this study is the developed method of applying the mathematical model of the chaos generator "Chua scheme" as the main component of the hybrid cryptosystem. The chaos generator is used as a source of public and private keys of the asymmetric encryption algorithm and the key of the symmetric algorithm directly used for data encryption.

Keywords: Chaos generator · Cryptosystem · Chua scheme · Mathematical model · Cryptography

1 Introduction

At the present time, when common and easily accessible technical devices are used to transfer and store any data, the protecting information issue of violation, its confidentiality, integrity, and accessibility is one of the most significant questions. Transmitted data can be affected by the transmission medium or external (relative to the information system) medium, as well as various actions of intruders are aimed to intercepting, damaging information, etc.

Modern cryptography is characterized by opened encryption algorithms which are using computational tools. These crypto algorithms assume the presence of a certain length key and a relatively simple transformations set, the so-called cryptographic primitives [1, 2]. However, due to the nature of these cryptosystems, there are a large number of methods to decrypt the information encoded by them. One of the main

I. Karabegović (Ed.): NT 2020, LNNS 128, pp. 454–461, 2020.
https://doi.org/10.1007/978-3-030-46817-0_52

problems of traditional cryptosystems is that, eventually, the sequence of operations of encryption of the information flow begins to repeat (the key length is limited), and this leads that the sequence can be disclosed by a third person, and the flow is decrypted [3].

The using of models of chaos generators, which have complex, unpredictable and highly dependent on the initial parameters of behavior, as a component of the software cryptosystem can significantly increase the cryptographic stability of the cipher, relative to existing algorithms today.

The developed method of applying the mathematical model of the chaos generator "Chua scheme" as the main component of the hybrid cryptosystem, where the chaos generator is used as a source of public and private keys of the asymmetric encryption algorithm and the key of the symmetric algorithm directly used for data encryption is the scientific novelty of this study.

One of the main problems of common computationally stable systems is the length of the key associated with the method of obtaining this key. The number of operations and the length of the key are usually rigidly fixed by the crypto algorithm. The key of symmetric cryptosystems can be determined by analyzing the messages are transmitted, and the keys of public-key cryptosystems and hybrid cryptosystems can be determined by certain parameters relatively easy.

The Chua scheme is a dynamic chaos model with complex behavior with General ease of implementation and it is capable to operate over a wide range of values, so it was chosen as a modeling object to use it for information security. This chaos generator gives a very high unpredictability degree, closed to truly random sequences, where the result is almost unpredictable regardless of the information availability about the algorithm and parameters. The generated sequences of numbers, taking into account their unlimited length and uniqueness of values combinations, are ideal from the point of view of their use as keys for data encryption by the cryptosystem.

2 The Data Protection By Using the "Chua's Circuit" Chaos Generator

2.1 Overview of Existing Solutions and Programs

Information searching of such software solutions in general is limited to several foreign projects, including scientific articles [4], and patents [5]. In Ukraine, software developments and practical implementations of using chaos generators in the area of digital information protection were not found, but there were theoretical studies on the basis, which modeling was carried out and methods of synchronization of generators were put forward.

The considered projects of cryptosystems on chaos generators are used for synchronization of generators the various values transferred on secret or open channels. Secret channels using in practice implies additional encryption of these values, which complicates the implementation of the cryptosystem. You can work around this problem by reducing the number of values passed and setting the remaining values as initial constants. In this case, the chaos generator is used as an irreversible function by analogy with public key cryptosystems, but in this case it is possible to obtain synchronization data by intercepting the public key, and, accordingly, to perform decryption, substitution or distortion of the transmitted data.

In the case of traditional public key cryptosystems, the above problem is solved with certificates and a specific public keys set, but this approach will not allow the full potential of the cryptosystem to be realized on the chaos generator, where the number of public keys or sets of the initial state of the generator should not be limited. Limiting the number of possible initial values can affect the uniqueness of the generated random number sequence significantly.

2.2 A Cryptosystem Based on a Mathematical Model of Chaotic Oscillations Generated on the Basis of Differential Equations

The cryptosystem project uses a software implementation of a mathematical model of the Chua scheme, a type of chaos generator.

A Chua scheme is a simple electronic scheme demonstrating a number of bifurcation phenomena and attractors. The circuit consists of two capacitors, an inductor, a linear resistor, and a nonlinear resistor (commonly called a Chua diode) [6, 7].

The complexity of the structure of chaotic signals and their irregularity, as well as the ability to create one and the same chaotic generator completely different processes with a very slight change in the initial conditions, provides an opportunity to create chaos generators models for their application in cryptography, which is devoted to this work.

Due to the fact that the Chua scheme is one of the most simple chaos generators (its mathematical model is described by only three differential equations, but at the same time it has a complex behavior, peculiar to chaos generators), it was chosen for the information encryption system as the most effective in terms of implementation.

The Chua scheme is described by the following system of equations [7]:

$$\begin{cases} C_1 \frac{dv_{C_1}}{dt} = G(v_{C_2} - v_{C_1}) - g(v_{C_1}) \\ C_2 \frac{dv_{C_2}}{dt} = G(v_{C_1} - v_{C_2}) - i_L \\ L \frac{di_L}{dt} = -v_{C_2} \end{cases} \tag{1}$$

where $g(v_{C_1})$ is a piecewise linear function defined as:

$$g(v_{C_1}) = G_b v_{C_1} + \frac{1}{2}(G_a - G_b)(|v_{C_1} + E| - |v_{C_1} - E|) \tag{2}$$

The graph also shows the load line, three equilibrium points d, 0, and –d are formed out of intersection with it.

The system will take the following form, after replacing the coefficients in the system of Eq. 1 with dimensionless ones [8]:

$$\begin{cases} \frac{dx}{dt} = a(y - x - h(x)) \\ \frac{dy}{dt} = x - y + z \\ \frac{dz}{dt} = -\beta y \end{cases}, \quad (3)$$

where *h(x)* is defined as:

$$h(x) = m_1 x + \frac{1}{2}(m_0 - m_1)(|x + 1| - |x - 1|) \quad (4)$$

Within the framework of this work, a modified mathematical model, where the coefficient c was added, was used. It allows more flexible control of changes in the oscillation modes of the system. The modified system of equations has the following form:

$$\begin{cases} \frac{dx}{dt} = a(y - x - h(x)) \\ \frac{dy}{dt} = c(x - y + z) \\ \frac{dz}{dt} = -\beta y \end{cases} \quad (5)$$

The function *h(x)* in system 5 corresponds to its definition from formula 4.

The numerical solution of these equations shows that at certain ratios between the components of the chain, the change in the values of variables in time becomes chaotic. However, chaotic modes of oscillations are found in a rather narrow range of parameters.

We considered the attractor called "double curl" shown in Fig. 1.

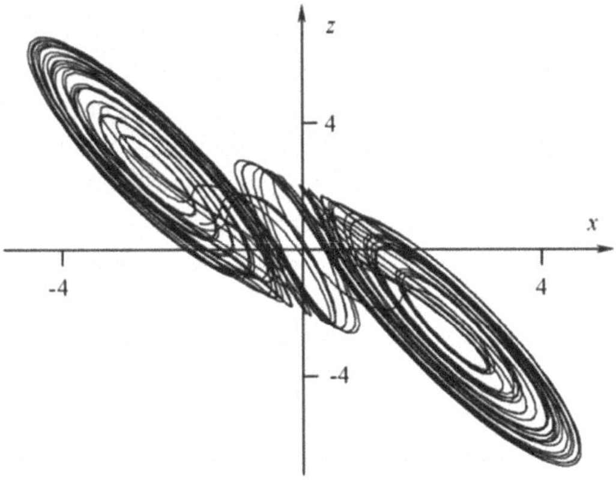

Fig. 1. Attractor of the "double curl"

The trajectory of such an attractor is non-periodic, and the mode of operation is unstable. As a result, even small deviations of the parameters cause significant changes. The nonperiodicity in time of any of the coordinates of the system, the continuous spectrum and the time-decreasing autocorrelation function are the result of this behavior [9, 10]. It causes the chaotic dynamics of strange attractors: indicates that the prediction of the trajectory, that hit the attractor, is difficult, since a small inaccuracy in the initial data after some time can lead to a big discrepancy between the prognosis and the real trajectory [11].

In this work, we apply a chaotic masking algorithm using time-varying output values of the chaos generator x. Figure 2 shows one of the possible time dependences of the change in the value of x for a model with dimensionless coefficients. To achieve a degree of unpredictability close to truly random sequences, a slice of the parameter values is performed.

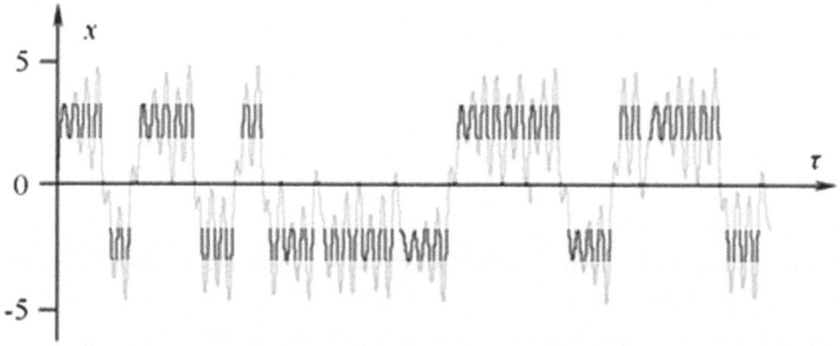

Fig. 2. Slice of values of parameter x

The time dependence of the change x is one of the parameters that can be used in the cryptosystem to encrypt the transmitted information, for example, by imposing a chaotic signal on the information, gamming.

As it can be seen on the Fig. 2, all points of the graph where the function's result is much rarer (gray areas of the graph) have been discarded, and the remaining x values (black areas) have an equal statistical frequency of occurrence in the generated key. Consulting non-periodicity and unpredictability, the chaos generator in this case acts as a generator of truly random numbers on two intervals of the value of x.

The key generation initiation parameters required by the cryptosystem to synchronize the chaos generators are randomly selected to ensure that the key is unique for each new data transfer session.

To conduct experimental studies of a specific implementation of the mathematical model of the chaos generator "Chua scheme" and cryptosystem based on this model, an application was developed in the programming environment RAD Studio (C++ programming language). The Runge-Kutta method is used to solve the system of differential equations of the chaos generator model. Figure 3 shows a fragment of the program code for an application that uses a cryptosystem based on the chaos generator Chua scheme developed as part of the thesis. The code fragment shown in Fig. 3 refers to the block for solving the system of differential equations of the chaos generator model and a cut of the obtained values, methods for setting and obtaining current parameters and generator values, a fragment of the text data encryption method.

The developed application is a pair of independent processes. The interface of these processes, EncryptorTestC and EncryptorTestS, is almost identical and clearly demonstrates the results of the cryptosystem. The EncryptorTestC leads process window is shown in Fig. 4. The EncryptorTestC process is a leading application process and contains a master chaos generator, which the slave chaos generator in the EncryptorTestS process is synchronized with.

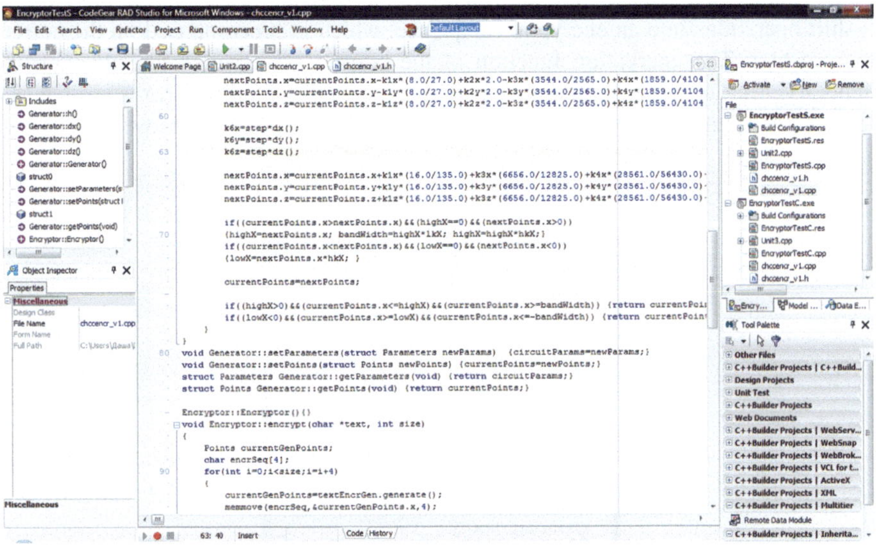

Fig. 3. A fragment of the cryptosystem program code

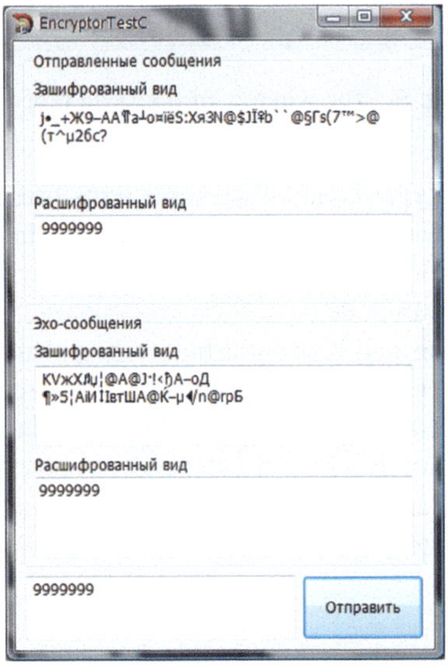

Fig. 4. The window of the lead process

The mechanism of processing the data entered into the application works as follows. The data that you want to encrypt and transfer is stored in a temporary array, based on the size of which determines how long the key is required to encrypt the text. Next, the generated key corresponding to the size of the block of length is superimposed on the blocks of the processed text using the exclusive "OR" (XOR) operation. The key is generated on the basis of t the chaos generator results "Chua scheme", changes in the values of the variable x in time. The generated numbers are converted by bit-shift operations into an encryption sequence, which is superimposed on the current block of text. The encryption function of the cryptosystem works directly with the temporary array of text transmitted to it. The encrypted text in bit form is written to a file that opens the EncryptorTestS process for processing, the window of which is shown in Fig. 5.

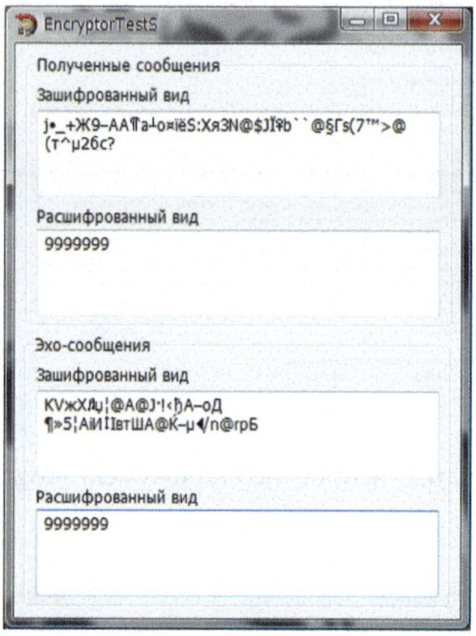

Fig. 5. The window of the slave process

After receiving the encrypted data, the reverse transformation of the encrypted text is performed in a similar way to encryption, because of the peculiarities of the XOR operation, the same function can mix and remove the key out of the encrypted data without changing its program code.

Applications use writing and reading a binary file, which was done to be able to analyze the results of the cryptosystem to transfer encrypted data among themselves. The synchronization keys transfer is carried out in a similar way.

The developed application is a pair of independent processes. The interface of these processes, EncryptorTestC and EncryptorTestS, is almost identical and it demonstrates the results of the cryptosystem clearly.

3 Conclusion

Traditional cryptosystems and projects of cryptosystems on chaos generators are considered in this work. The mathematical model of the chaos generator "Chua scheme", its behavior and modes of operation was considered on the basis of the literature. The requirements for random number generators used in cryptosystems to create encryption sequences and methods for achieving parameters that meet these requirements were also considered.

As a result, a mathematical model was developed for a specific software solution for the problem of data protection.

As part of this work, an application using the developed cryptosystem based on the chaos generator "Chua scheme" has been successfully implemented. The developed application is a convenient model for conducting experimental studies of a specific implementation of the cryptosystem. In the future, it is planned to conduct a number of static tests and test the stability of the implemented mathematical model of the chaos generator "Chua scheme" and the most developed cryptosystem.

References

1. Bennett, C.H., Bessette, F., Brassard, G., Salvail, L., Smolin, J.: Experimental quantum cryptography. J. Cryptol. **5**(1), 3–28 (1992)
2. Mao, V.: Modern Cryptography: Theory and Practice, p. 768. Hewlett-Packard, Palo Alto (2005)
3. Barichev, S.G., Goncharov, V.V., Serov, R.E.: Fundamentals of modern cryptography, 175 p. Hotline-Telecom (2002)
4. Leros, A.P., Andreatos, A.S.: The article is a steganography telecom system using a chua circuit chaotic noise generator for data cryptography. Chaotic Model. Simul. (CMSIM) **1**, 199–208 (2013)
5. Crandall, R.E., Mitchell, D.P., Krueger, S., Tribble, G.: Pat. 7587047 United States. Chaos generator for accumulation of stream entropy. Announced 22.06.09; Publ. 08.09.09. - 3 s
6. Fradkov, A.L., Evans, R.J.: Control of chaos: methods and applications in engineering. Ann. Rev. Control **29**, 33–56 (2005)
7. Chua, L.O.: Chua's circuit: ten years later. IEICE Trans. Fundam. **E77-A**(11), 1811–1822 (1994)
8. Gámez-Guzmán, L., Cruz-Hernández, C., López-Gutiérrez, R.M., García-Guerrero, E.E.: Synchronization of Chua's circuits with multi-scroll attractors: Application to communication. Commun. Nonlinear Sci. Numer. Simulat. **14**, 2765–2775 (2009)
9. Fortuna, L., Frasca, M., Xibilia, M.G.: Chua's Circuit Implementations – Yesterday, Today and Tomorrow. World Scientific Publishing Co. Pvt. Ltd, Danvers (2009). ISBN-13 978-981-283924-4
10. Hilborn, R.C.: Chaos and Nonlinear Dynamics. An Introduction for Scientists and Engineers, 2nd edn. Oxford University Press, New York (2000). ISBN 0198507232
11. Šalamon, M., Dogša, T.: The model of chaoticness detector. Informacije MIDEM – J. Microelectron. Electron. Compon. Mater. **39**(2), 93–99 (2009). ISSN 0352-9045

The Architecture of Fuzzy Logic Automat of Parallel Action for the Intelligent Smart Grid Networks

Stanislav Bovchaliuk[1], Serhii Tymchuk[1], Sergii Shendryk[2], and Vira Shendryk[2](✉)

[1] Kharkiv Petro Vasilenko National Technical University of Agriculture, Kharkiv, Ukraine
[2] Sumy State University, Sumy, Ukraine
v.shendryk@cs.sumdu.edu.ua

Abstract. Some technology introduction problems of Smart Grid in the electric networks of Ukraine are considered in the article. It is offered to apply the automats of parallel action in control system by a power supply. It is offered to make alteration in the structure of parallel action classical automat for the decision of problem vagueness detains. The general strategy is set for the construction of parallel action automats with fuzzy logic.

Keywords: Smart Grid · PLD-controller · Parallel action automat · Technology of parallel logical management · Fuzzy logic

1 Introduction

The modern state of electric networks in Ukraine needs modernization. In addition, the permanent increasing on the electric system takes place. It is conditioned by the consumers amount increasing and specific consumption of energy by consumers. An additional call for power engineering specialists is swift development of the so-called "green" energy - wind power-stations, sunny stations sources, etc. [1–3]. Energy sources characteristics are:

- mainly small generating power;
- its instability (day's, weather, etc.);
- distribution of electric networks.

These features do not allow integrating such sources to the existent electric networks. They require introduction of considerable changes to organization and maintenance of Ukrainian power economy.

The power trend progress is clearly defined in energy well – developed countries. Its basic theses are nonbreaking and co-ordination actions for providing three constituents:

- energy supply (trouble proof electric energy supply of corresponding quality);
- energy availability (energy safety and availability of price on electric power);
- to energy acceptability (minimum influence on an environment) [4].

I. Karabegović (Ed.): NT 2020, LNNS 128, pp. 462–468, 2020.
https://doi.org/10.1007/978-3-030-46817-0_53

Thus, Ukraine has a unique chance to modernize power economy to the level technically and the economically developed countries. Also, there is a necessity to move in direction of construction the intellectual new generation power system. It will allow bringing the power services grant to conformity with the standards of progressive world countries.

2 Question State Analyses

Intellectualization conception of electro energy has an aim to build fully integrated, self-regulated and self-reproduced system. Such system has a network topology and includes all generating sources, bus and distributive networks, all types of electric energy consumers. This system should follow by a single network of automated devices in real time [1, 4]. This conception got the universally recognized and generally accepted term - Smart Grid. So, Europe determination is European Technology Platform Smart Grids: (intellectual networks) are electric networks that satisfy the future requirements, in relation to energy efficiency and economy of grid functioning. It takes place due to the coordinated functioning and by means of modern bilateral communications between the elements of electric networks, electric stations, heat-sink sources and consumers [5, 6].

World leaders in development and introduction of Smart Grid conception [4] is:

- the USA - The Energy Independence and Security Act of 2007;
- countries of EU - Smart Grid Technology Roadmap Report Development;
- South Korea - Smart Country Strategy Development.

In these countries, it is set the pilot schemes of different intellectual content grade and different readiness rates. Operating data accumulate gradually, complication and volume of innovative power hubs increase.

Movement of Ukraine in direction to the electric systems intellectualization on the base of Smart Grid conception was announced by the head of HEK “Ukrenergo” Vsevolod Kovalchuk. It is planned the pilot scheme realization of Smart Grid - Vehicle - to - Grid (V2G) technology cooperative work of electric vehicle and electric system. Besides it is planned a research of possibility to use next groups of technologies:

- Demand Response;
- Virtual Power Plant.

In the authors’ opinion, one of perspective realization ways of Smart Grid elements is information technology application of parallel logical management on the base of safe PLD-controller of parallel action. This technology was developed for the management of continuous cyclic action objects. It was practically realized for the management of critical application (responsible, reliable, faultless) objects. However, processes carry, mainly, indefinite character in electric networks. In the consequence of its direct application of the indicated technology in energy cannot give considerable advantages in comparing the classic methods and technologies of successive action. However, if connect the elements of fuzzy logic with information technology of parallel logical management, it can give an effective instrument for the Smart Grid construction.

3 Definition the Goal

Form conception of logical management function realization in an unclear form on the base of parallel action automat. Form the way to build elements of intellectual power circuit within the framework of Smart Grid conception on the base of fuzzy logic automat of parallel action (FLAPA).

4 Basic Materials

There are at least two reasons for mathematical vehicle application of fuzzy logic:

- presence of vagueness input quantity;
- natural parallelism of logical induction procedure.

We shall consider these reasons.

Presence of Vagueness Input Quantity. Problems of making decisions in energy find in the conditions of incompleteness and vagueness of external information. It results in the receipt of inexact results. In accordance to [7] a vagueness takes place, if an universal set consists more than of one point. For example, a vagueness arises up in the process of parameters measuring that determine technological processes in energy. Such vagueness's can have different nature:

- inaccuracy (or error) of measuring devices,
- presence of different values forming of the plural measuring;
- second-rate measuring, for example through dispersed after a parameter or after territory, etc.

In such cases, the classic systems of logical management use determination (for example by the method of statistical average procession). Thus, the part of data is lost.

Natural Parallelism of Logical Induction Procedure. Such natural parallelism allows to integrate the elements of fuzzy logic realization in the structure of classic parallel action automats. It will allow extending their functionality and ways of application.

The first FLAPA prototype is a device for machine management a machine for casting under pressure [8]. The simplified structure of this device is shown on Fig. 1. The blocks on the indicated structure of discriminators D1 and D2 execute phasing of analog signal inputs. They lead signal inputs (Input 1, Input 2) for three levels - more (>), norm (n), less (<). Logical processing unit (LPU) sets the algorithm of technological function process. Regulators R1 - R3 form the equipment management signals (Output 1 - Output 3). They execute function of dephazification.

The actually shown structure presents the hard logic of parallel action automat. An automat did not envisage programming and was realized on an element base that time.

We shall consider the input data possible variants for the synthesis of FLAPA structure. In general view, they can be three types: (1) the analogue; (2) digital binary; (3) digital phazificated.

The analogical data types can be realized on the automat outputs.

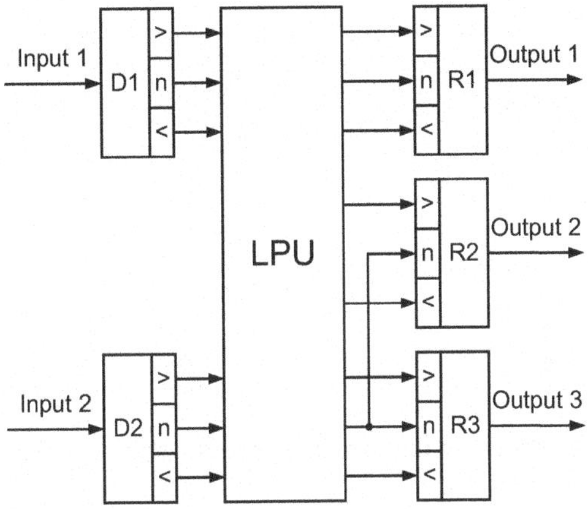

Fig. 1. Simplified structure of FLAPA prototype.

In a general view, FLAPA can be presented as follows (Fig. 2), where: $a_1 \ldots a_k$ - κ digital binary inputs; $g_1 \ldots g_p$ – p analog inputs; $c_1 \ldots c_m$ – m digital binary outputs; $f_{x1} \ldots f_{xn}$ are data phazificated outputs; $f_{y1} \ldots f_{yl}$ are data phazificated outputs. Thus, data phazificated inputs and outputs actually present an ordinary binary code.

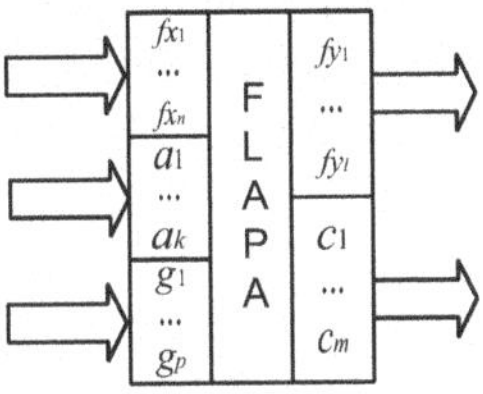

Fig. 2. The generalized FLAPA structure

If consider ordinary digital and digital phazificated data inputs, then it is possible to reach the next conclusion: binary data is possible to present as unclear, but with a value μ, only 0 and 1. For the unclear data form μ can possess any value from 0 to 1.

The example of such data membership functions is shown on Fig. 3. Digital data after procedure of phazification present the ordinary binary code of certain bit. A bit depends on necessary exactness of phazification. Thus, the synthesis of FLAPA structure both types of data input can be presented as a set of binary numbers $k + zn$, where z is a bit of data phazification.

Analog data on the automat input transform in digital form before obligatory. That is analog data as a result of analog-digital transformation will be presented by a parallel digital binary code. Thus, data inputs for FLAPA of all mentioned types actually present the parallel binary code of certain bit. The same output data of all types can be presented by an analogical code.

At the FLAPA synthesis important questions show up: what amount of input quantity in an unclear form can treat by an automat?; how many output commands of management can be formed?

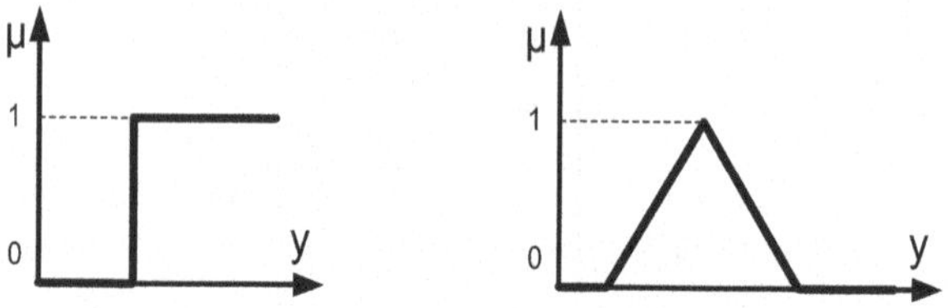

Fig. 3. General view of function μ for data in a binary and unclear form

Theoretical an amount of logical managing parallel action automat inputs-outputs (LAPA) is unlimited, thus an amount of FLAPA inputs is also unlimited. It is possible to assert from experience of industrial prototypes creation of parallel action inspectors, that the amount of inputs-outputs, limited to complication of technical realization and procedure of managing cyclogram realization.

At plenty of input quantities, it is suggested to apply FLAPA cascading. It will allow breaking up intricate problems to more simple. Realization of such approach is shown on Fig. 4. Therefore, every FLAPA works only with two unclear sizes. The result of two sizes treatment, in an unclear form, is passed to next FLAPA, where co-operates with the third unclear size. A process recurs farther. Analog and binary data are not examined in this instance, as can treat on any of the stages.

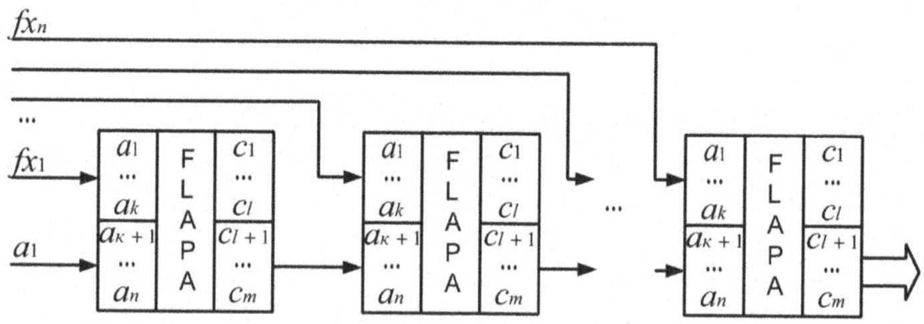

Fig. 4. FLAPA cascading

It is possible to do the next conclusion: classic structure of LAPA (Fig. 5), can be based for the synthesis of FLAPA structure.

LAPA consists of follow elements: display unit – DU; schemes of comparing unit – CU; logic operation selection block – LOSB; logic control unit – LCU; address counter – AC; output register – OR; and also blocks of states memory, commands, transitions and forbidden combinations – SMU, CMU, JMU, UCMU.

Proceeding from the tasks that belong to FLAPA, it is suggested to enter following changes to its structure:

- whereas the questions of management safety are not examined and not grounded, it is possible to eliminate the blocks of UCMU and SR. They provided non-extradition of the forbidden combinations of initial commands managements that is why they are unnecessary.
- to eliminate DU, as information delivery about current status of input data is not actual.
- other blocks remain in the FLAPA structure, though change its functional value.

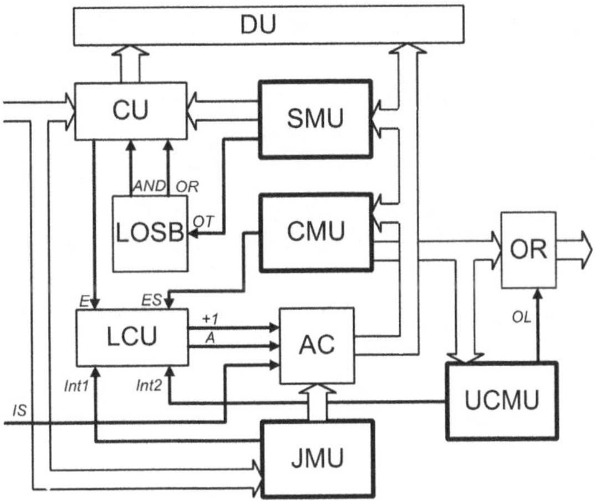

Fig. 5. LAPA structure

Proposal generalized FLAPA structure is shown on Fig. 6.

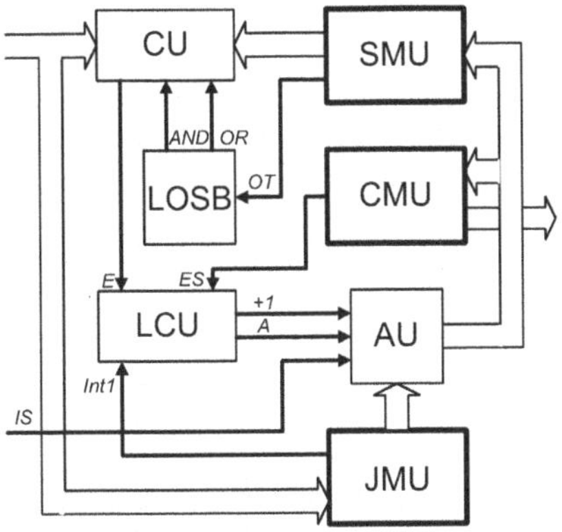

Fig. 6. FLAPA structure

FLAPA work differs from LAPA, by following features:

- JMU loses the basic functionality from address of subroutine to computation. It, both with CMU, intended for storage of forming table unclear inferencing;
- SMU contains only the fragments of successive implementation of commands, if there is a necessity in them. If such determined sequence is absent, then SMU is eliminated from a structure;
- AU executes the same functions, that AC in LAPA. Now its basic function is forming of unclear inferencing with JMU, but not successive working off the subprogram lines with SMU.

5 Conclusion

The research showed that fuzzy logic was an effective instrument for a power equipment management. It allows realizing the automats of parallel action for a management such equipment for one time of discrete automat time. It is also well proven that the base LAPA structure can be taken for basis at the synthesis of the generalized FLAPA structure. It allows passing to the practical synthesis of automats structures, its HDL-models and physical realization. Thus, elements of information technology of parallel logical management can be taken for basis for the construction of intellectual kernel modern clever Smart Grid energetics.

References

1. Shulyma, O., Shendryk, V., Baranova, I., Marchenko, A.: The features of the smart microgrid as the object of information modeling. In: Communications in Computer and Information Science, vol. 465, pp. 12–23 (2014)
2. Shendryk, V., Shulyma, O., Parfenenko, Y.: The topicality and the peculiarities of the renewable energy sources integration into the Ukrainian power grids and the heating system. In: Renewable and Alternative Energy: Concepts, Methodologies, Tools, and Applications, pp. 1317–1349. IGI Global Publisher (2016)
3. Shulyma, O., Shendryk, V., Parfenenko, Y., Shendryk, S.: The model for decision support on design of the hybrid renewable energy system. In: Proceedings of the 2017 IEEE 9th International Conference on Intelligent Data Acquisition and Advanced Computing Systems: Technology and Applications, IDAACS 2017, pp. 47–50 (2017)
4. Stognii, B., Kyrylenko, O., Prahovnyk, O., Denysiuk, S.: The evolution of intelligent electrical networks and their prospects in Ukraine. Tech. Electrodyn. **5**, 52–67 (2012)
5. Smart Grid – European Technology Platform for Electricity Networks of the Future. European Commission (2005). http://www.smartgrids.eu/
6. European Technology Platform- Smart Grids. April 2011: Strategic Deployment document for European Commission (2014). http://www.smartgrids.eu/
7. Kandel, A., Byatt, W.: Fuzzy sets, fuzzy algebra, and fuzzy statistics. IEEE Proc. **66**, 1619–1639 (1978)
8. Kostritsa ,V., Furman, I.: A s. 1026946 USSR, MKI3 B 22 D 17/32. Ustroistvo upravlenyja mashinoy dlia litija pod davlenijem. (USSR). – №3425477/22-02; zaiavl. 20.01.82; opubl. 07.07.83, Bul. № 25

Green Technology Approach to Comb-Based Decimators Design

Gordana Jovanovic Dolecek[1(✉)] and Isak Karabegovic[2]

[1] Department of Electronics, Institute INAOE, E. Erro 1, 72740 Puebla, PUE, Mexico
gordana@ieee.org

[2] University of Bihac, Bihac, Bosnia and Herzegovina

Abstract. This paper presents a simple method to design comb-based decimation filter. The goal is to get a low power design, while improving the comb alias rejection. The proposed structure has two stages. In the first stage is a comb filter in a nonrecursive form, and presented with its polyphase components. In the second stage is a comb filter cascaded with the modified combs. The cascade is also presented with the polyphase components. In that way the filters in the first stage are moved to the rate which is lesser than the high input rate, and which is determined by the decimation factor of the first stage. Similarly, the filters at second stage are moved to the rate which is M times time lesser than the high input rate, where M is a decimation factor. As a consequence, the proposed structure is a low power structure. The proposed method is compared with some methods from literature.

Keywords: Decimation · Aliasing · Decimation filter · Low power · Two-stage

1 Introduction

Green communications "extends to all branches of communications with the goal of minimizing resource use whenever possible and of selecting energy-efficient communications and technologies", [1, 2].

In this paper we consider comb-based decimation filter design with the goal to get a power efficient structure and an improved alias rejecting.

Decimation finds applications in communications, in oversampled Sigma-Delta A/D (Analog/Digital) converters, sampling rate conversion, among others, [3].

Decimation introduces aliasing which must be eliminated by filter, called decimation filter. The most simple decimation filter is a comb filter which does not require multipliers. This filter naturally provides aliasing rejection in the bands around the comb zeros, called folding bands. However, the aliasing rejection of comb filter is very limited, and must be improved.

The most effective way is obtained by introducing additional zeros into comb folding bands, and thus increase widths and attenuations in those folding bands. This method was first proposed by Presti [4]. However, the proposed RS (Rotated Sinc) requires two multipliers. Additionally, the coefficient quantization may lead to instability because the

I. Karabegović (Ed.): NT 2020, LNNS 128, pp. 469–477, 2020.
https://doi.org/10.1007/978-3-030-46817-0_54

ideal pole-zero cancellation can be lost. Taking into account that the comb filter is a multiplierless filter, the methods proposed to improve it should be also multiplierless, like [5–8].

We consider here the case, where the decimation factor can be presented as a product of two integers. The low power design is achieved by a multiplierless design and applying the polyphase decomposition in both stages.

The rest of the paper is organized in the following way. Next section describes a two-stage structure, and how to introduce the additional zeros into comb folding bands using simple multiplierless filters in the second stage. Section 3 presents the proposed method, and is illustrated with three examples. The efficient structure is also elaborated. Last section shows some comparisons with methods from literature.

2 Improving Two-Stage Comb Structure

2.1 Two-Stage Comb Structure

The overall decimation factor is presented as a product of the decimation factors for first and second stages, denoted as M_1 and M_2, respectively:

$$M = M_1 M_2. \tag{1}$$

The transfer functions of comb filters of first and second stages, at high input rate, are given as,

$$H_1(z) = \frac{1}{M_1}\frac{1-z^{-M_1}}{1-z^{-1}};\, H_2(z^{M_1}) = \frac{1}{M_2}\frac{1-z^{-M}}{1-z^{-M_1}}. \tag{2}$$

Figure 1 presents magnitude responses of combs (2) and the overall comb for $M = 12$, $M_1 = 3$ and $M_2 = 4$. Note that the comb at second stage has the same zeros as the overall comb, except in fourth folding band. ($M_2 = 4$). Similarly, the comb in the first stage has only zero in the comb fourth folding band. This is also confirmed in Fig. 2 where are presented pole-zero plots. In general case, the zeros of the comb, at the second stage, contribute to comb zeros in all folding bands except in the bands that are multiples of M_2. Similarly, the zeros of the first comb stage, contribute to comb zeros only in the comb folding bands, which are multiples of M_2.

Using this result, we show in the next section, how the modified comb in the second stage, introduces zeros into comb folding bands.

2.2 Placing the Zeros into Comb Folding Bands

Consider next modified comb filters placed in the second stage:

$$G_1(z) = 1 + a_1 \sum_{k=1}^{M_2-2} z^{-k} + z^{-(M_2-1)}. \tag{3}$$

$$G_2(z) = 1 + a_2 \sum_{k=1}^{M_2-2} z^{-k} + z^{-(M_2-1)}. \quad (4)$$

$$G_3(z) = 1 + a_3 \sum_{k=1}^{M_2-3} z^{-k} + z^{-(M_2-2)}. \quad (5)$$

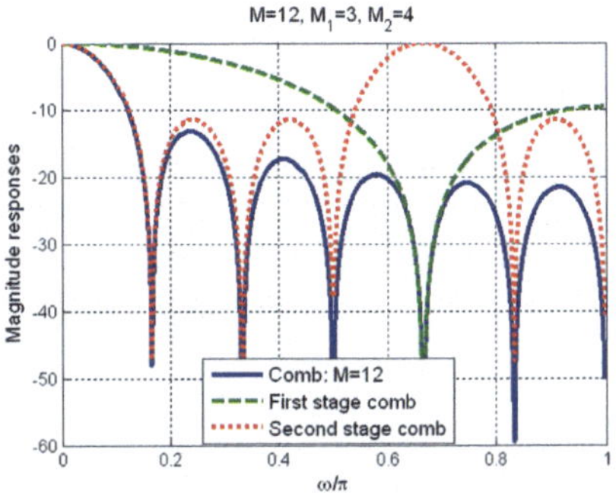

Fig. 1. Magnitude responses.

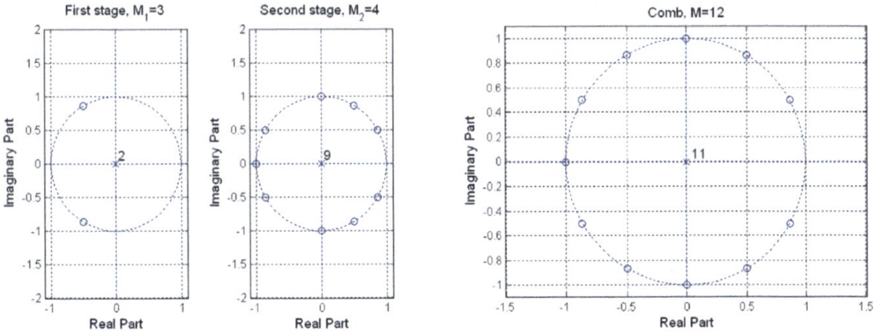

a. First and second stages. b. Overall comb.

Fig. 2. Pole-zero plots.

The modified combs (3)–(5) are derived from the certain class of the symmetrical polynomials, explored in [9]. Using MATLAB simulations, we found the coefficients a_i, $i = 1, .., 3$, in (3)–(5), in a such way that zeros of the modified combs fall into comb folding bands:

$$a_1 = {}^1\!/_2;\, a_2 = {}^3\!/_2;\, a_3 = {}^7\!/_2. \quad (6)$$

Taking (6), the corresponding transfer functions, (3)–(5), at high input rate, are:

$$G_1(z^3) = 1 + \frac{1}{2}(z^{-1\times 3} + z^{-2\times 3}) + z^{-3\times 3}. \tag{7}$$

$$G_2(z^3) = 1 + \frac{3}{2}(z^{-1\times 3} + z^{-2\times 3}) + z^{-3\times 3}. \tag{8}$$

$$G_3(z^3) = 1 + \frac{7}{4}z^{-1\times 3} + z^{-2\times 3}. \tag{9}$$

Nest example illustrates this idea.

Example 1. We consider comb from Fig. 1b with $M_1 = 3$ and $M_2 = 4$. Figure 2 contrasts the pole-zero plots for overall comb filter and the modified comb filters, (3)–(5), placed at the second comb stage. Note that the modified combs introduce additional zeros into all folding bands, except into fourth comb folding band (Fig. 3).

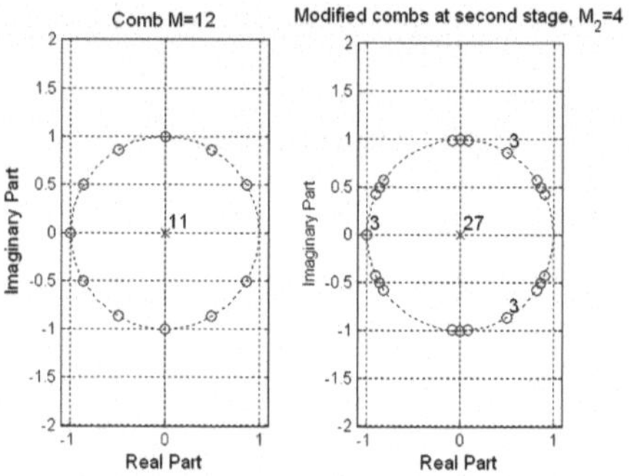

Fig. 3. Effect of modified comb zeros on comb folding bands.

This is also confirmed in Fig. 4 where are shown the magnitude responses of the overall comb and the modified combs. Note that all folding bands are improved except the fourth folding band.

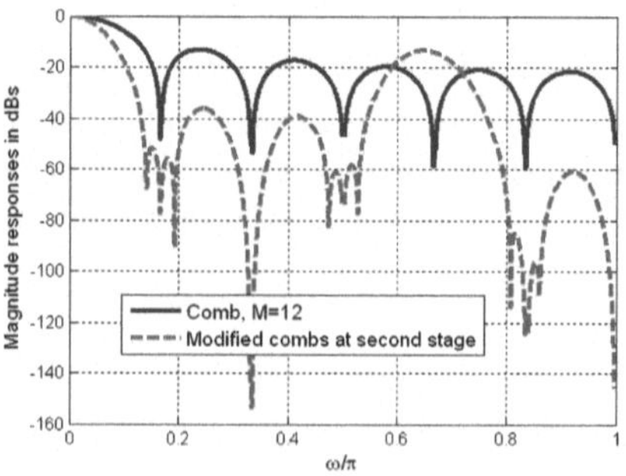

Fig. 4. Magnitude responses of comb and modified comb filters.

3 Proposed Filter

3.1 Transfer Function

From (7)–(9) we have:

$$G(z^{M_1}) = \frac{1}{N_{22}} \prod_{i=1}^{3} G_i(z^{M_1}), \tag{10}$$

where $G_i(z)$, i = 1, 2 and 3, are defined in (3)–(6),and N_{22} is the normalization factor.

Taking (2) and (10), the transfer function of the proposed filter, at high input rate, is:

$$H_p(z) = H_1^{K_1}(z) H_2^{K_2}(z^{M_1}) G(z^{M_1}), \tag{11}$$

where K_1 and K_2 are the numbers of the cascaded filters at first and second stages, respectively.

The method is illustrated in the following example.

Example 2: We consider M_1 = 3 and M_2 = 4 and K_1 = 4, and K_2 = 2 - The magnitude responses of the proposed filter and comb filter are contrasted in Fig. 5.

Note that the aliasing rejection is improved in all comb folding bands, except in fourth comb folding band.

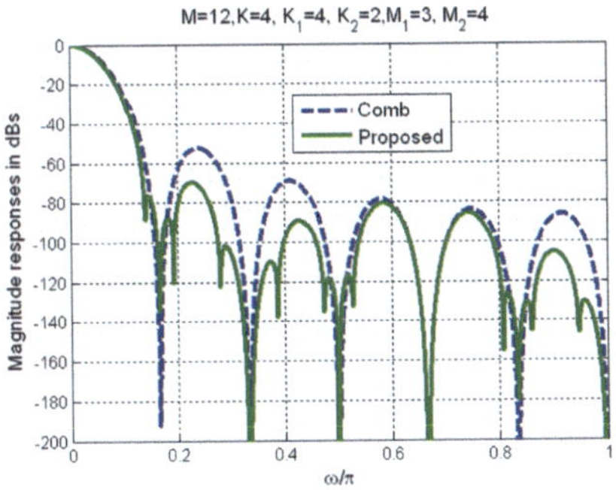

Fig. 5. Magnitude responses in Example 2.

3.2 Choice of Parameters

The parameters M_1 and M_2 are chosen depending of the design goal.

If the principal goal is to improve comb alias rejection, then chose M_2, as high as possible. In that way the aliasing rejection will be improved, as discussed in Sect. 2.

If the principal goal is to decrease the power consumption, than chose M_1 as high as possible, because the modified comb filters work at the rate which is M_1 times lesser than the high input rate.

The compromise solution is to chose the decimation factor M_1 slightly lesser in comparison with M_2.

Considering that the modified combs are in the second stage, the choice of the parameters K_1 and K_2, is as follows:

$$K_1 = K; K_2 = K - 2, \quad (12)$$

where K is the parameter of the overall comb filter.

Example 3: In this example we consider the same comb, as in Example 2, taking $M_1 = 2$ and $M_2 = 6$. The other parameters are equal as in Example 2.

The magnitude responses of the proposed filter and comb with the parameter $K = 5$, are contrasted in Fig. 6. Note that the aliasing rejection is improved in all comb folding bands, except in the last one.

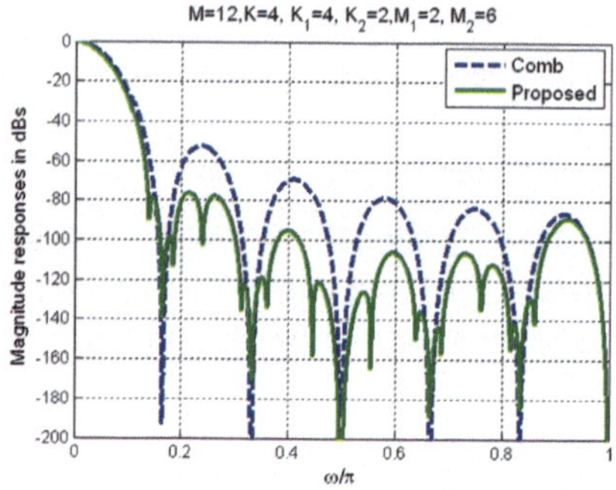

Fig. 6. Magnitude responses in Example 3.

Example 4: In this example two cases are elaborated: (a) $M = 20$, $M_1 = 4$, and $M_2 = 5$. (b) $M = 20$, $M_1 = 2$, and $M_2 = 10$. The magnitude responses are shown in Fig. 7a and b.

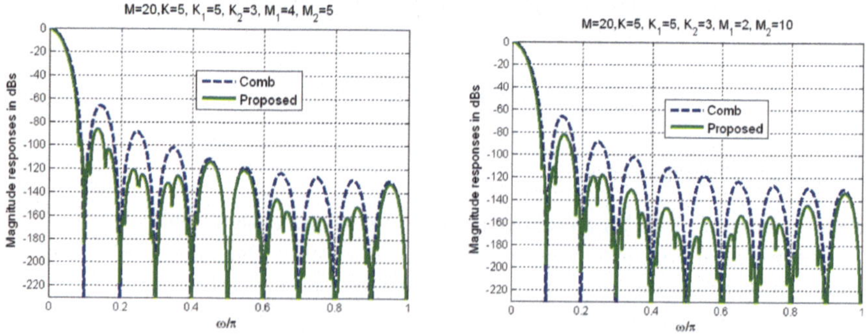

a. M_1=4, M_2=5.b. M_1=2, M_2=10.

Fig. 7. Magnitude responses in Example 4.

3.3 Structure

The two-stage structure is presented in Fig. 8a. In the first stage is the cascade of K combs of length M_1 which can be presented using the polyphase decomposition [3],

$$H_1^{K_1} = \frac{1}{N_1} \sum_{k=0}^{M_1-1} z^{-k} F_k(z^{M_1}), \tag{13}$$

where $F_k(z^{M_1})$, $k = 0, \ldots, M_1-1$, are the polyphase components and N_1 is the normalization factor, $N_1 = (M_1)^{K_1}$.

The polyphase components can be moved to lower rate, as shown in Fig. 8b. In the second stage is the cascade of the filters $H_2(z)$ and the modified combs $G(z)$, which can be also presented using the polyphase components $P_k(z)$:

$$H_{22}(z) = \frac{1}{N_2} \sum_{k=0}^{M_2-1} z^{-k} P_k(z^{M_2}), \tag{14}$$

where N_2 is the normalization factor and N_{22} is the normalization factor of $G(z)$, $N_2 = (M_2)^{K_2} N_{22}$. The expanded polyphase components in (14) can be moved to a lower rate, which is M time lesser than the high input rate. The polyphase decomposition is the same as that shown in Fig. 8b, just replacing $F_k(z)$ with $P_k(z)$, M_1 with M_2, and the normalization factor $1/N_1$ with $1/N_2$.

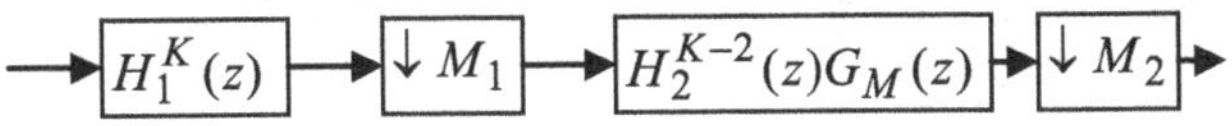

a. Two-stage structure.

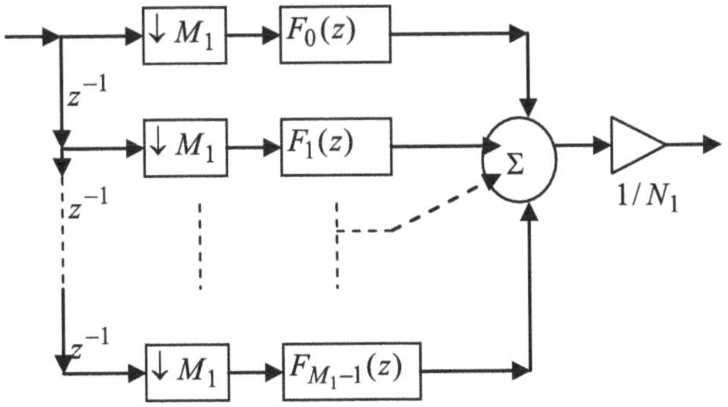

b. Polyphase decomposition in the first stage.

Fig. 8. Structure.

4 Comparisons with Methods from Literature

4.1 Comparison with Method in [4]

For the sake of comparison, we consider the parameters in [4]: $M = 16$ and $K = 4$, and the rotation parameter $\alpha = 0.1841$. In the proposed method $M_1 = 2$, $M_2 = 8$, $K_1 = 4$, $K_2 = 2$.

The magnitude responses are shown in Fig. 9. Note that the proposed method has better alias rejection while it does not require multipliers.

4.2 Comparison with Method in [8]

In [8] is introduced the one-stage structure with two modified combs. In this comparison, the parameters are: $M = 15$ and $K = 3$. In the proposed method $M_1 = 3$, and $M_2 = 5$. The magnitude responses are shown in Fig. 10. The proposed method has slightly better aliasing rejection in all folding bands except in the fifth folding band.

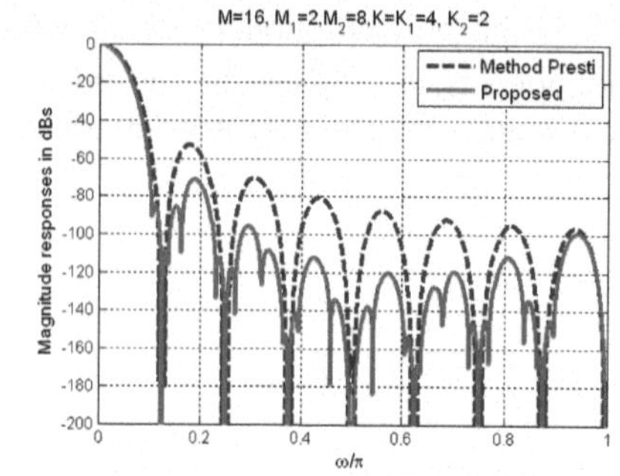

Fig. 9. Comparison with method in [4].

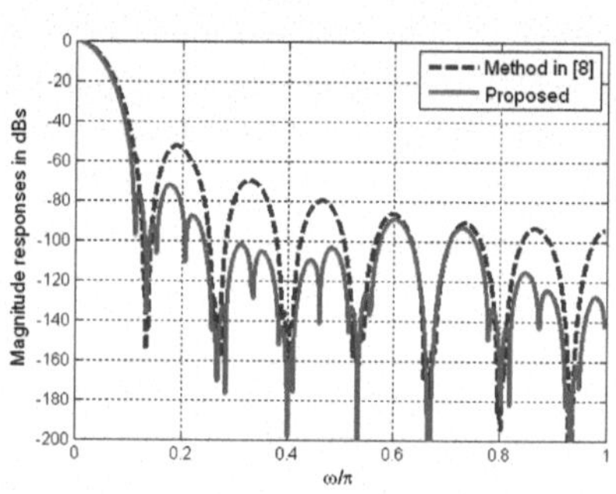

Fig. 10. Comparison with method in [8].

5 Conclusion

This paper presents a novel multiplierless two-stage comb-based decimation structure. The modified combs are introduced at the second stage. The coefficients of modified combs are chosen in the form of power of two. Consequently, the multiplications are realized by shifts and adds. Generally, there is a trade-off between decrease of power, and the improvement of alias rejection. In order to decrease power consumption, the polyphase decomposition is performed in both stages.

References

1. Kumar, A., Singh, K., Bhattacharya, D.: Green communication and wireless networking. In: Proceedings of 2013 International Conference on Green Computing, Communication and Conservation of Energy (ICGCE), Chennai, India (2013)
2. Jinseng, W., Rangan, S., Zhang, H. (eds.): Green Communications Theoretical Fundamentals, Algorithms, and Applications. CRC Pres/Taylor and Francis Group, Boca Raton (2016)
3. Dolecek, G.J. (ed.): Advances in Multirate Systems. Springer, Cham (2018)
4. Lo Presti, L.: Efficient modified sinc-filters for sigma-delta AD converters. IEEE Trans. Circ. Syst. II **47**(11), 1204–1213 (2000)
5. Coleman, J.O.: Chebyshev stopbands for CIC decimation filters and CIC-implemented array tapers in 1D and 2D. IEEE Trans. Circ. Syst. I **59**(12), 2956–2968 (2012)
6. Stošić, B.P., Pavlović, V.D.: Design of new selective CIC filter functions with passband-droop compensation. Electron. Lett. **52**(2), 115–117 (2016)
7. Jovanovic Dolecek, G., Garcia Baez, R., Laddomada, M.: Design of efficient multiplierless modified cosine-based comb decimation filters: analysis and implementation. IEEE Trans. Circ. Syst. I **36**(5), 2031–2049 (2017)
8. Jovanovic Dolecek, G.: Improving comb alias rejection by exploiting certain class of palindromic polynomials. In: Proceedings of the 23 European Conference on Circuit Theory and Design (ECCTD 2017), Catania (2017)
9. Botta, V.: Some results on zeros of palindromic and pertubed polynomials of even degree. Proc Ser. Brazilian Soc. Appl. Comp. Math. **3**(1), 1–4 (2017)

Determination of the Sampling Interval of Time Series of Measurements for Automation Systems

Sergii Tymchuk[1], Ivan Abramenko[1], Katerina Zahumenna[1], Sergii Shendryk[2], and Vira Shendryk[2,3](✉)

[1] Kharkiv Petro Vasylenko National Technical University of Agriculture, Kharkiv, Ukraine
[2] Sumy State University, Sumy, Ukraine
[3] Computer Science Department, Sumy State University, Sumy, Ukraine
v.shendryk@cs.sumdu.edu.ua

Abstract. The technique for the sampling interval of time series determining of parameter measurements for automation systems are improved. Mathematical apparatus of Discrete Fourier Transform is used to describe the frequency characteristics of signals in measuring channels of automation systems and Kotelnikov theorem. It is performed computer simulation of the proposed method of decision-making based on comparison of harmonic amplitudes of the maximum frequency of spectrum in different components of the measuring signal.

Keywords: Measuring channel · Sampling interval · Digital filtration · Kotelnikov theorem · Discrete Fourier Transform · Modeling

1 Introduction

Measurement of technological parameters is implemented by measuring channels, which are structurally or functionally separate part of information system. Information system performs a sequence of operations from perception of measured value to obtaining of result of its measurements, which is displayed by a number or a corresponding code [1–3].

Components of the measuring part make a certain error $\varepsilon(t)$ in a signal $y(t)$.

Most signals in automation systems are analog in nature. It is impossible to enter such signal into a computer and process it, because at any time interval it has an infinite set of values. Therefore, in digital processing systems, a signal is represented in discrete digital form.

The digital signal is derived from the analog signal by time sampling, level quantization, and encoding.

The sampling step in time Δt can be determined by Kotelnikov theorem (in English literature - Nyquist-Shannon theorem), which states that the analog signal $z(t)$ having a limited frequency spectrum can be uniquely and lossless restored by its samples taken with the sampling frequency f_d that must be at least double the maximum frequency of the spectrum $f_{\max}$ [4–7].

I. Karabegović (Ed.): NT 2020, LNNS 128, pp. 478–483, 2020.
https://doi.org/10.1007/978-3-030-46817-0_55

However, the sampling theorem is formulated for the case when the signal began infinitely long ago and never ends, and also has no break points. Real signals do not meet these conditions because they are finite in time and usually have discontinuities in time implementation. Accordingly, the width of their spectrum is infinite. In this case, complete signal recovery is not possible.

Therefore, in each case it is necessary to look for a compromise solution, based on the nature of the signal, the required accuracy of its recovery, the characteristics of the smoothing filter used and other factors [6, 7]. All this leads to the fact that in real automation systems the sampling rate is chosen empirically k several times higher than the maximum frequency of the spectrum - $f_{\partial} = k \cdot 2f_{\max}$.

Values k in the range 2–5 [8–10] are usually used, but there are currently no clear criteria for determining the value k for a particular time series in the literature recourses.

To improve the methodology for determining of the sampling interval of the time series of measurements of the parameter for automation systems, it is necessary to solve the problem of investigating of ways of the Discrete Fourier Transform use for signals of measuring information, taking into account Kotelnikov theorem.

2 Research Methodology

To determine the sampling intervals of the measuring signal, we use its frequency representation, which can be obtained using the methods of harmonic analysis.

Practical applications deal with finite samples of values. Therefore, for the purposes of spectral analysis of the discrete signal, the Discrete Fourier Transform (DFT) is used.

To ensure the functioning of the automation system in real time, the calculation of the DFT must be performed in a time interval, which corresponds to the formation of a single data packet. It is assumed that while the DFT of the current data packet is being calculated, the microprocessor accumulates data for the next packet.

When determining the sampling step in time Δt, it is also necessary to take into account the filtering operation, which is the most complex of the operations of primary signal processing of measuring information, and its efficiency is determined by the value Δt.

Filtering is most often performed programmatically using various filtering algorithms. Exponential smoothing algorithms are the most widely used (about 90% of filtering problems are solved with the help of these algorithms) [5, 6].

The adequacy of the smoothed time series can most simply be determined by calculating the average relative approximation error [2].

The proposed method is based on a comparative analysis of the results of using DFT for two signals - smoothed digital signal and filtered noise. To determine the sampling step in time Δt for a particular measured parameter it is proposed to determine the optimal value of the sampling step by comparing the harmonic amplitudes of the maximum frequency of the spectrum $f_{\max}$ in the named components of the measuring signal.

Given the standard accuracy of automatic control systems, the following criterion for the definition Δt can be formulated - the part of the harmonic amplitude of the maximum frequency of the spectrum in the filtered noise should be no more than 5% of the amplitude of this harmonic in the smoothed signal, that is $100 A_{\max,sum}/A_{\max,zgl} \leq 5\%$.

3 Experiments

To test the proposed criterion and compare its effectiveness with known approaches, a simulation model of the measuring channel and ADC was developed by means of the Matlab program, which includes all stages of primary processing of measuring signals.

The time series of the parameter measurements was formed from 3 harmonics by the formula

$$y(t) = y_0 + a_1 \sin(2\pi f_1 t) + a_2 \sin(2\pi f_2 t) + a_3 \sin(2\pi f_3 t)$$

where a_1, a_2, a_3, and f_1, f_2, f_3 were randomly determined from lists of permissible values of 0.8–1.9 rad/s and 0.05–0.5 Hz.

The law of distribution of probabilities of error of the measuring channel $\varepsilon(t)$ was accepted normal. Noise filtering was performed by the method of simple exponential smoothing.

The fragment of the generated time series at the ADC input and the filtered useful signal obtained at Δt = 0.02 s and α = 0.3 are shown in Fig. 1. The dependences of the part of the signal harmonics amplitudes in noise on the value of the sampling step Δt for different harmonics are shown in Fig. 2.

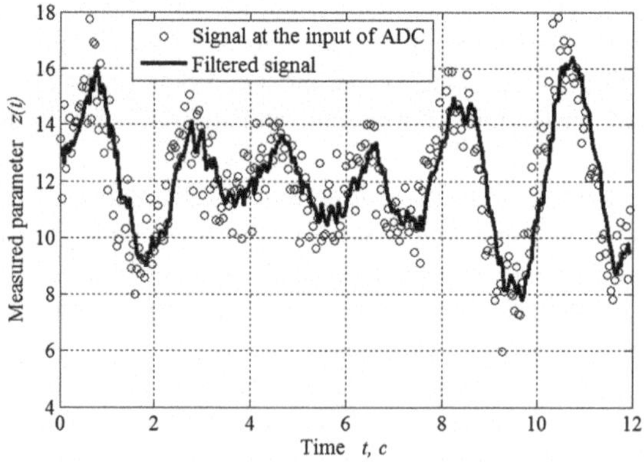

Fig. 1. Temporal series of the measured parameter $z(t)$ and filtered useful signal

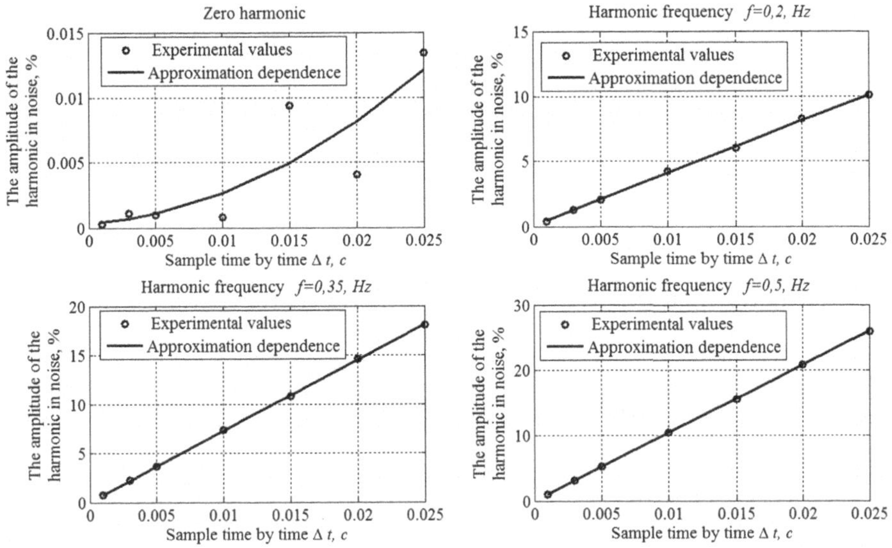

Fig. 2. The dependence of the amplitude of the harmonic of the useful signal in the noise on the sampling step

Processing of the calculations made it possible to obtain a formula for determining the value of the step Δt by known values $f_{\max}$

$$\Delta t = 0{,}0045 + 0{,}138e^{-17{,}14f_{\max}} \tag{1}$$

and the formula for determining the margin factor k for Kotelnikov theorem in the form of a record $f_{\partial} = k \cdot 2f_{\max}$.

$$k = 208{,}3\left(1 + 0{,}33e^{-125f_{\max}} - 1{,}33e^{-31{,}25f_{\max}}\right), \tag{2}$$

which correspond to the given error of 5% (at other values of the given error it is possible to receive similar dependences).

To assess the effectiveness of the filtering operation, the dependence of the average relative error of approximation $\bar{\Delta}_{ap} = f(\Delta t)$ was also determined, where Δt was calculated by the formula (1) (Fig. 3).

For comparison of accuracy of restoration of the sampled signal it was established that at the maximum value Δt = 0.1 s, which is defined with Kotelnikov theorem, harmonics by frequencies 0.1, 0.2, 0.3, 0.4 and 0.5 Hz are allocated with errors of 20.6, 41, 63, 83 and 100% respectively. With a minimum value Δt = 0.04 s, these harmonics are allocated with errors of 8.3, 16.7, 25, 33.4 and 41.8%, respectively, which is unacceptable.

The analysis of the obtained results shows that the optimal value Δt at $f_{\max}$ = 0.05–0.5 Hzvaries depending on $f_{\max}$ within 0.005–0.063 s, and the number of samples for the period of maximum frequency - within 300–406.

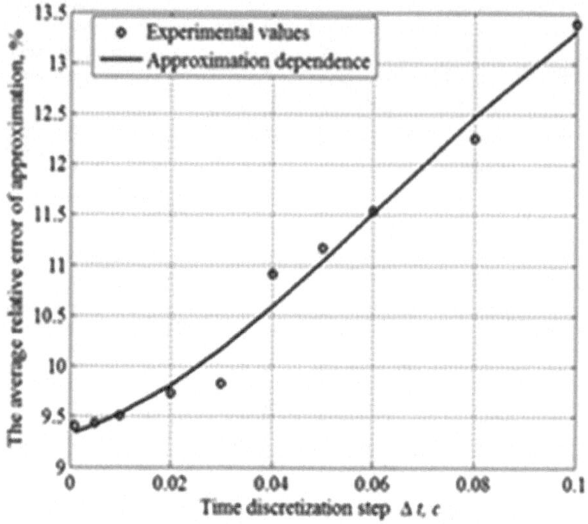

Fig. 3. Dependence of the mean relative error of filtering on the sampling step Δt

4 Conclusion

The results of simulation using Matlab software allowed to draw a conclusion about the effectiveness of the technique of determining of the sampling interval of the time series of parameter measurements for automation systems, which is based on the use of Discrete Fourier Transform for smoothed digital signal and filtered noise.

The graphs shown in Fig. 2 and dependences (1)–(2) show that the value of the polling period of analog variable sensors differently affects the amplitude distribution of the useful signal in the noise for different harmonics. This, in turn, makes it possible to search for rational relationships between the harmonious composition of a particular signal and the optimal value of the polling period depending on the required accuracy of the signal recovery.

The results obtained can be used for realizations of measurements of any real parameter and differ significantly from the generalized recommendations of known approaches.

Practical significance of the results is determined by studies of the effectiveness of the influence of the value of the polling period of analog variables on the distribution of amplitudes of the useful signal in noise for different harmonics. The obtained analytical dependences of the sampling interval of the time series of measurements on the maximum frequency of the signal spectrum and the margin factor in the Kotelnikov formula with an error of 5% can be used in development of specific automation systems.

The proposed technique can be used for selection of parameters of smoothing digital filters, as well as for comparative analysis of their efficiency.

References

1. Elperin, I.V., Pupena, O.M., Sidletskyi, V.M., Shved, S.M.: Avtomatyzatsiiavyrobnychykh-protsesiv, Lira-K, Kyiv, Ukraine (2015)
2. Tregub, V.G.: Osnovy kompiuterno-intehrovanoho keruvannia. NUKhT, Kyiv (2005)
3. ShannonClaude, E.: Communication in the presence of noise. Proc. Inst. Radio Eng. **37**(1), 10–21 (1949). https://doi.org/10.1109/jrproc.1949.232969. Reprint as classic paper in: Proc. IEEE **86**(2) (1998). Archived 2010-02-08 at the Wayback Machine. Jump up to: a b c d
4. Moshe, M., Yonina, E.C.: Blind multiband signal reconstruction: compressed sensing for analog signals. IEEE Trans. Signal Process. **57**(3), 993–1009 (2009). Cite SeerX 10.1.1.154.4255
5. Kipnis, A., Goldsmith, A.J., Yonina, E.C., Weissman, T.: Distortion rate function of sub-Nyquist sampled Gaussian sources. IEEE Trans. Inf. Theory **62**, 401–429 (2016). https://doi.org/10.1109/tit.2015.2485271. arXiv:1405.5329
6. Kipnis, A., Yonina, E.C., Goldsmith, A.: Analog-to-digital compression a new paradigm for converting signals to bits. IEEE Signal Process. Mag. **35**(3), 16–39 (2018). https://doi.org/10.1109/msp.2017.2774249. arXiv:1801.06718. Bibcode: 2018ISPM…35…16 K
7. Nyquist, H.: Certain topics in telegraph transmission theory. Trans. AIEE. **47**(2), 617–644 (1928). https://doi.org/10.1109/t-aiee.1928.5055024. Bibcode: 1928TAIEE.47.617 N. Reprint as classic paper in: Proc. IEEE **90**(2) (2002). Archived 2013-09-26 at the Wayback Machine
8. Kotelnikov, V.A.: On the carrying capacity of the ether and wire in telecommunications. In: Material for the First All-Union Conference on Questions of Communication, Izd. Red. Upr. Svyazi RKKA (1933, in Russian). (English translation, PDF)
9. Shannon, C.E.: A mathematical theory of communication. Bell Syst. Tech. J. **27**(3), 379–423 (1948). https://doi.org/10.1002/j.1538-7305.1948.tb01338.x. hdl:11858/00-001M-0000-002C-4317-B
10. Nemirovsky, J., Shimron, E.: Utilizing bochners theorem for constrained evaluation of missing fourier data (2015). arxiv:1506.03300. [physics.med-ph]

Fuzzy Model of Quality Control of the Finished Software Product

Olha Pronina(✉) and Piatykop Olena

Department of Computer Science, Faculty of Information Technology, Priazovsky State Technical University, Universitetskaya street 7, Mariupol 87500, Ukraine
pronina.lelka@gmail.com

Abstract. When developing a software product, the testing phase is the most important. Testing takes the most significant time of the entire development, so the automation of the testing process is an urgent task. Within the scope of the application, one can imagine the testing phase as a set of input linguistic variables, namely functional suitability, accuracy, interoperability, compliance, security, where the degree of confidence in the quality of the software product depends on the assessment of each variable. In this paper, you consider a model for checking the quality of a software product after its development. A fuzzy model is developed that was used to model the iterative life cycle of a software product. This approach is universal and can be applied to models of another category, for example, cascading due to a softer repetition of the testing stage. This knowledge and model can be applied in the field of automation of the testing process. This approach will significantly increase the speed of passage of this stage, reduce the influence of the human factor, minimize the possible error and optimize the development process.

Keywords: Fuzzy model · Linguistic variable · Software product quality · Software product testing

1 Introduction

In today's world the process of writing software is constantly improving products. Number of developed software products is constantly growing; his writing is relatively less time testing. Because testing is one of the major and important steps to develop transmission in productive operation.

The testing process makes it possible to find out the correspondence between the given task and its realized form, as well as reduce the number of defects and improve the quality of the software.

Testing is divided into a large number of species from different directions, for example, for program purposes, system analysis, formality and more. And, in the context of software development, the full range of types is used, the main, by prevalence, is purpose testing, which includes functional testing (suitability, accuracy, interaction, etc.) and non-functional (convenience, failure and recovery, configuration, etc.). Unlike

I. Karabegović (Ed.): NT 2020, LNNS 128, pp. 484–490, 2020.
https://doi.org/10.1007/978-3-030-46817-0_56

functional testing, where product compliance checks are performed, non-functional examines the product in terms of the running environment and environment.

Checking the correspondence between a working prototype and the expected behavior of a program by a finite set of tests, collected and grouped in a specific way, is called testing. Testing is one of the main techniques of quality control, which includes activities in Test Management, Test Design and Test Execution and Test Analysis.

2 Analysis of Research and Publications

Software Quality can be thought of as a set of software features that relate to its ability to meet established and foreseeable needs.

The set of functional testing properties determines the quality of the software product and its readiness for the operational phase. According to the international standard [1], a set of properties is divided into:

(1) Suitability – the ability of software to fulfill its functional goals;
(2) Accuracy – the closeness of the true result to the expected result;
(3) Interoperability – the ability of a software product to perform a set of functions defined in its description and meet the specified needs of users;
(4) Security – the stability of the software, the prediction of possible errors, defects and bugs, failures and error situations;
(5) Functionality compliance – the ability of software to comply with standards, agreements or regulations in laws and regulations regarding functionality.

Software development testing is the process of assessing that all parts of the program behave as expected and receive feedback on how the user interacts with the application variant. Tests perform software checks to verify that the result is in accordance with the specifications, taking into account various inputs. Each of these checks is called a test case.

The test scenarios check the requirements and characteristics of a particular functionality (functionality). They may provide certain details or steps that can be reproduced [2].

Developing the right and effective tests is a difficult task. The testing principles presented below have been developed over the last 40 years and are a general guide to testing as a whole [3].

Testing may find bugs in the software, but not prove them missing. However, it is important to find test variants that will detect as many errors as possible. This will reduce the likelihood of errors occurring in the software.

According to research by Michael Bolton, researcher and founder of Develop Sense [4], the main problem of testing lies in the analysis of the end result of testing, in their processing to make decisions about the readiness of the product, and the sequence of actions for obtaining them is not important.

The author examines the Data-Driven Testing methodology used in test automation. The peculiarity is that test scripts are executed and verified on the basis of data stored in a central data warehouse or database. The logic that will be executed in the script also depends on the data [5].

The testing principles presented by the author are classic and underpin most systems [6]. The author notes in his work the importance of the testing phase, the fragmentation of the test results directly related to the quality of the software.

Jerry Weinberg, in his book [7], notes that what we get as a result is first and foremost information and cannot be left out of potentially meaningful observations.

In [8] the authors considered the process of software testing by repetition on the principle of an iterative cycle. At each step, repetitions are performed to find new bugs that are detected and corrected, but they have not been found before. In addition, the execution time is predicted as well as the cost of the testing process, given its cyclical recurrence.

The Value of Test Automation by the author discusses why test automation is needed and what its short-term and long-term benefits are [9].

Suppose the result can be distorted by a human factor and testing of one product may be presented with various errors: inattention of the tester, his workload in the project, routine execution of the same operations, and more. Reducing the impact of the human factor will provide more reliable data on the testing phase, and systemizing indicators and constructing a final estimate based on these data will reduce the error and accelerate decision making.

The testing phase in an iterative lifecycle model can be represented as a conformity assessment according to the standard [1]. And the final evaluation, based on the evaluation of the criteria, will summarize the results obtained and presented as a degree of confidence in the quality of the software. You can evaluate many factors using fuzzy logic. Parameter estimation will be the degree of confidence in the quality of the software.

The description of the quality of a software product includes five key parameters; if you evaluate using only one parameter, it will not be a qualitative assessment. Evaluation of all parameters takes a lot of time. Therefore, the construction of a model for assessing the quality of the tested software product is an urgent scientific and practical task.

2.1 Software Product Quality Assessment Model

Software testing is one of the most important steps in development, as it captures the final quality of the code, which closely characterizes the compliance with a number of quality criteria (functionality, efficiency, reliability, mobility, practicality, ease of maintenance) between the projected in the statement of the task and the real system.

To describe the quality model of a software product, a lot of tests {T} are used, presented in the form (1).

$$T = \{(p_1, \Omega_1^j), (p_2, \Omega_2^j), (p_3, \Omega_3^j), (p_4, \Omega_4^j), (p_5, \Omega_5^j)\}, \quad (1)$$

where $A = \{p_k\}$ – many signs, $k \in [1; 5]$;
k – attribute index;
$\Omega_i^j = \{\omega_i^j | \mu(\omega_i^j)\}$ – set of attribute values p_k, representing the names of fuzzy variables;
index values $j \in [1; 5]$ denote the set of attribute values.

Set A includes the following features $\{p_i\}_{i=1}^{5}$: p_1 – suitability, p_2 – accuracy, p_3 – interoperability, p_4 – security, p_5 – functionality compliance. Signs are input variables for the model being developed; for their description, it was decided to use linguistic variables.

Since when describing the input parameters of the trip, they are more informative for users [10, 11].

The analysis made it possible to identify both the names of the terms and their number for each linguistic variable. Since all terms can be measured in a certain quantitative scale, direct methods were used to construct membership functions.

During the analysis of the types of membership functions of terms [10] empirically, the following were selected: sigmoid, for extreme terms and a generalized bell for the rest [10].

Thus, features are described, respectively, by linguistic variables:

- β_1 is determined by the tuple $\langle \beta_1, T(\beta_1), X \rangle$, where β_1 = «Suitability», $T(\beta_1)$ = {SB, ST, SS, SN, SF}, X = [0, 100]. The parameters of the terms are presented in Table 1;
- β_2 is determined by the tuple $\langle \beta_2, T(\beta_2), X \rangle$, where β_2 = «Accuracy», $T(\beta_2)$ = {AI, AP, AR}, X = [0, 10]. The parameters of the terms are presented in Table 2;
- β_3 is determined by the tuple $\langle \beta_3, T(\beta_3), X \rangle$, where β_3 = «Interoperability», $T(\beta_3)$ = {IZ, IL, IN, IF}, X = [0, 100]. The parameters of the terms are presented in Table 3;
- β_4 is determined by the tuple $\langle \beta_4, T(\beta_4), X \rangle$, where β_4 = «Security», $T(\beta_4)$ = {FU, FL, FN, FS, FF}, X = [0, 10]. The parameters of the terms are presented in Table 4;
- β_5 is determined by the tuple $\langle \beta_5, T(\beta_5), X \rangle$, where β_5 = «Functionality compliance», $T(\beta_5)$ = {RZ, RP, RC, RF}, X = [0, 100]. The parameters of the terms are presented in Table 5.

Output linguistic variable ω_1, defined by a tuple, where ω_1 = «degree of confidence in the quality of the software product», $T(\omega_1)$ = {SL, SM, SH}, X = [0;1]. The terms SL – «low confidence», SM –«medium confidence», SH – «high confidence».

Membership functions, for terms SL, SH - sigmoid, for SM - generalized bell.

Table 1. Parameters of the terms of a linguistic variable «Suitability»

Term name	Term value	Function name	Parameters			Range of universum $X = [X_1, X_2]$	
			a	b	c	X_1	X_2
$\Omega_1^1 = SB$	Badly	$\mu_{SB}(x; a, c)$	−0,2	–	24	0	30
$\Omega_1^2 = ST$	Tolerable	$\mu_{ST}(x; a, b, c)$	7	1,5	40,3	30	50
$\Omega_1^3 = SS$	Satisfactorily	$\mu_{SS}(x, a, b, c)$	7,5	1,7	59,3	50	70
$\Omega_1^4 = SN$	Normally	$\mu_{SN}(x, a, b, c)$	8,6	2,9	79	70	90
$\Omega_1^5 = SF$	Fine	$\mu_{SF}(x; a, c)$	0,9	–	90,8	90	100

Table 2. Parameters of the terms of a linguistic variable «Accuracy»

Term name	Term value	Function name	Parameters			Range of universum $X = [X_1, X_2]$	
			a	b	c	X_1	X_2
$\Omega_2^1 = AI$	Irrelevant	$\mu_{AI}(x; a, c)$	−2	–	2,3	0	3
$\Omega_2^2 = AP$	Partially	$\mu_{AP}(x; a, b, c)$	3,5	15	4,3	3	6
$\Omega_2^3 = AR$	Relevant	$\mu_{AR}(x; a, c)$	3	–	6,3	6	10

Table 3. Parameters of the terms of a linguistic variable «Interoperability»

Term name	Term value	Function name	Parameters			Range of universum $X = [X_1, X_2]$	
			a	b	c	X_1	X_2
$\Omega_3^1 = IZ$	Zero	$\mu_{IZ}(x; a, c)$	−0,4	–	31,5	0	35
$\Omega_3^2 = IL$	Little	$\mu_{IL}(x; a, b, c)$	8	1,1	47,5	35	65
$\Omega_3^3 = IN$	Normal	$\mu_{IN}(x; a, b, c)$	3,5	0,2	77,5	65	90
$\Omega_3^4 = IF$	Full	$\mu_{IF}(x; a, c)$	0,3	–	90	90	100

Table 4. Parameters of the terms of a linguistic variable «Suitability»

Term name	Term value	Function name	Parameters			Range of universum $X = [X_1, X_2]$	
			a	b	c	X_1	X_2
$\Omega_4^1 = FU$	Unacceptable	$\mu_{FU}(x; a, c)$	−2,4	–	2,6	0	3
$\Omega_4^2 = FL$	Little	$\mu_{FL}(x; a, b, c)$	0,8	1,6	4	3	5
$\Omega_4^3 = FN$	Normal	$\mu_{FN}(x, a, b, c)$	0,8	1,6	6	5	7
$\Omega_4^4 = FS$	Satisfactory	$\mu_{FS}(x, a, b, c)$	0,8	1,6	8	7	9
$\Omega_4^5 = FF$	Full	$\mu_{FF}(x; a, c)$	3	–	9,3	9	10

2.2 Experiment on Evaluating the Quality of a Software Product

A system has been developed to evaluate the quality of the software being developed. The system is based on a mathematical model of the quality of the software product. Using the developed system, an experiment was conducted. Testing experts were recruited and a collection of ten web resources was selected. Links to web resources have different themes and systems complexity. The assessment is based on four random web resources and one expert. The evaluation criteria and unification parameters are calculated in the system according to universal metrics for evaluating web resource testing. The final assessment is the degree of confidence in the software development.

This work evaluates the accuracy of testing according to the specified automatic parameters. To evaluate the task, the experts take the following steps:

- analysis and construction of a general view of the test software;
- execution and recording of test results using third-party services;
- generalization and calculation of test results for each model parameter;
- the decision of the final assessment that describes the quality of development on all the data received.

For the system, the process is similar to a real expert, provided that the manual input data is already available and will be specified in one of the execution steps. Otherwise, the automated data metric will give an incomplete assessment of the quality of the product and may lead to a decrease in the final confidence rating.

To distribute the expert's work evenly, each expert was provided with four random web resources for testing. Given the possible discrepancy between manual tests and the indication of the automated manager, it was decided to consider the number of confirmations for manual tests as reliable data from an expert.

The test results, namely the degree of confidence in the quality of the software product, made by experts and the system are shown in Fig. 1. Groups of experiments are given. Five experts participated in each group. Data from experts was averaged for clarity.

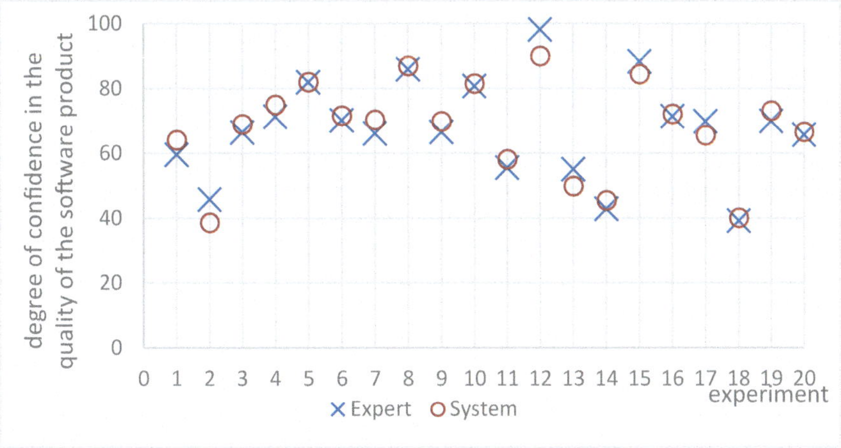

Fig. 1. Comparison of test results by an expert and a system

By analyzing the data obtained, it can be replaced that the assessments of experts and systems lie in the same ranges of terms. Moreover, the value from experts in the general experiment was significantly different.

3 Conclusion

The developed model fully describes the quality of the software product during testing with all the main parameters. This fuzzy model was used to model the iterative lifecycle of a software product. This approach is universal and can be applied to models of another category, for example, cascading due to the softer repetition of the testing phase. It is investigated that the use of the decision method of quality assessment of software developed on the basis of the fuzzy logic device doubled the reliability of the tests and the speed of deciding the product readiness on the basis of quality assessment.

The developed model can be applied in the field of automation of the testing process. This approach will at times increase the speed of passage of this stage, reduce the impact of the human factor, minimize the possible error and optimize the development process.

References

1. Software engineering: Product quality. Internal Metrics (ISO 9126: 2012, IDT): DSTU ISO 9126: 2012 [Effective from 28-11-2012], 46 (National Standard of Ukraine). State Standard of Ukraine, Kyiv (2012)
2. Black, R.: Key Testing Processes, Knowledge, Kiev, 566 p. (2014)
3. Nguyen, H.Q.: Happy About Global Software Test Automation: A Discussion of Software Testing for Executives, p. 164. Happy About, Cupertino (2016)
4. Heusser, M.: How to Reduce the Cost of Software Testing, p. 340. CRC Press, Boca Raton (2011)
5. Tarlinder, A.: Developer Testing: Building Quality in to Software, p. 342. Addison-Wesley Professional, Boston (2016)
6. Scherbakov, O.V.: Evaluation of software testing efficiency based on the analysis of the number and criticality of defects found. Inf. Process. Syst. **3**(93), 88–92 (2011)
7. Weinberg, G.: Perfect software and other illusions about testing, United States of America, p. 200 (2011)
8. Dubovoy, V.M., Pylypenko, I.V.: Modeling of software testing process as a branching cyclical technological process. Autom. Technol. Bus. Process. **7**(4), 55–64 (2015)
9. Axelrod, A.: Complete Guide to Test Automation: Techniques, Practices, and Patterns for Building and Maintaining Effective Software Projects, p. 560. Apress, Berkeley (2018)
10. Kargin, A.A.: Introduction to Intelligent Machines. Book 1. Intelligent Regulators, p. 526. Nord-Press, Donetsk (2010)
11. Leonenkov, A.V.: Fuzzy modeling in MATLAB and fuzzyTECH, St. Petersburg, Russia, p. 736 (2005)

Parallel Processing of HTTP Requests in E-Commerce: A Modeling Framework

Ilija Hristoski[1(✉)], Daniela Koltovska Nechoska[2], and Tome Dimovski[3]

[1] Faculty of Economics, St. Kliment Ohridski University–Bitola, Prilepski Braniteli St 133, 7500 Prilep, North Macedonia
ilija.hristoski@uklo.edu.mk

[2] Faculty of Technical Sciences, St. Kliment Ohridski University–Bitola, Bitola, North Macedonia

[3] Faculty of ICTs, St. Kliment Ohridski University–Bitola, Bitola, North Macedonia

Abstract. Since e-Commerce has become a prevalent form of doing business, the workload posed to e-Commerce websites, especially the global ones, increases continually on a daily basis. In order to maintain high QoS levels and keep pace with ever-increasing service demands, e-Commerce systems have to exhibit high performances. Within the paper, we propose a modeling framework for performance evaluation of generic e-Commerce systems based on the utilization of Client/Server Interaction Diagrams (CSIDs) and the class of Non-Markovian Stochastic Petri Nets (NMSPNs). CSIDs are used for capturing the interaction between the client and particular servers for specific HTTP requests, corresponding to specific e-Commerce functions that e-Customers invoke during online shopping sessions. As an example, we represent the CSID for the SEARCH function using NMSPNs, thus providing a performance model suitable for estimating the speedup gains of e-Commerce systems that utilize multicore CPUs and parallel processing of HTTP requests within their web servers.

Keywords: Parallel processing · e-Commerce · Non-Markovian Stochastic Petri Nets (NMSPNs) · Client/Server Interaction Diagrams (CSIDs) · Modeling and simulation

1 Introduction

Ten years ago, Intel and Microsoft have laid the groundwork for their continued business health by funding research into desktop and mobile parallel computing for both consumers and businesses under the motto 'The future of computing is parallel'. Induced by the fact that more and more cores were put into a single processor, the emergence of the new paradigm shift meant moving the computing industry from a single- to a many-core world. This new paradigm, known as 'parallel processing' or 'parallel computing', is based on the general idea of splitting up a large problem into smaller ones and solving them concurrently. It has been around for many years and becomes *de facto* a basis of the so-called high-performance computing. Since processor

I. Karabegović (Ed.): NT 2020, LNNS 128, pp. 491–498, 2020.
https://doi.org/10.1007/978-3-030-46817-0_57

architectures are nowadays available in multi-core configurations, the key question is how to use those cores in an efficient and sustainable way. Due to the fact that a great portion of applications today really does not benefit from multiple cores, the pursuit of new applications, new approaches, new techniques and new ideas that will take advantage of parallel processing, especially in distributed and online environments, has become mainstream these days.

On the other hand, the last decade has witnessed a continued growth of e-Business and e-Commerce activities worldwide, which has radically increased the demand for computing and communications resources, especially e-Commerce websites' resources and networks, to the limit of their capacity. As a result, such an enormous workload increase has led, in many cases and too much often, to system crashes, degraded performances and deteriorated dependability (reliability, availability, security, safety…). All of these have led not only to frustrated e-Customers and their disloyalty but also to the endangered reputation of e-Commerce companies and substantial financial losses. The importance of e-Commerce websites' speed and its impact on sales and revenues has been stressed out by many articles and studies. For instance, "a slow loading and unresponsive website is a major deterrent, as the average online user will only tolerate load time of up to 2 s" [1]. Other key reasons why improving website speed (i.e. responsiveness) brings great competitive advantage include satisfied e-Shoppers (as they are the most prominent means to retain old and attract new ones), increased sales and conversions (as reducing load times by 3 s increases revenue by 7%–12%), and higher SEO ranking (as website speed has become a significant ranking factor) [1].

Having minded the previous, quantifying and analyzing the impact of e-Commerce workload on the underlying website infrastructure has become a topic of utmost importance lately. As a result, three modeling approaches have been introduced to quantify and analyze the interaction of e-Customers and e-Commerce website infrastructure regarding the demand placed on the e-Commerce website infrastructure by e-Customers' requests. These include the Customer Behavior Model Graph (CBMG), the Customer Visit Model (CVM), and the Client/Server Interaction Diagram (CSID) [2]. In this paper, CSIDs are going to be used as basic building blocks for modeling purposes. The basic idea is to represent the CSIDs that correspond to various e-Commerce functions as a set of models built by using the class of Non-Markovian Stochastic Petri Nets (NMSPNs), convenient for carrying out performance analyses of generic e-Commerce systems running web servers with multi-core CPUs.

2 Client/Server Interaction Diagrams in E-Commerce

E-Customers initiate the execution of specific e-Commerce functions (e.g. LOGIN, REGISTER, SEARCH, BROWSE, ADD_TO_CART, REMOVE_FROM_CART, CHECKOUT, PAY, etc.) during their interaction with the e-Commerce website's servers, including, among others, the web server (WS), the application server (AS), and the database server (DB), usually organized in a two- or three-tier architecture. The interaction that starts with an HTTP request coming from the client (C), and ends with an HTML document sent back to the client (C), is known as a client/server (C/S) interaction [2]. Each identified e-Commerce function supported by a virtual store's website is implemented through a corresponding set of C/S interactions.

Client/Server Interaction Diagrams (CSIDs) provide a superb graphical notation for describing the anatomy of C/S interactions between the client and the servers during the execution of specific e-Commerce functions [2]. For instance, the representation of the possible C/S interactions during the execution of the SEARCH e-Commerce function, which is represented by the CSID depicted in Fig. 1, consists of nodes (squares that represent the visits to clients, and circles that represent visits to specific servers), interconnected by directed arcs $i \rightarrow j$, which indicate that a message is being sent from node i to node j. Each node is identified and labeled by a unique number. All C/S interactions within a given CSID start with the client node $i = 1$ and end with another client node. Arcs in the CSID are labeled by pairs $[p_{ij}, m_{ij}]$, where p_{ij} are the probabilities that the message is sent from node i to node j, and m_{ij} are the message sizes, in bytes, sent from node i to node j. In Fig. 1, three possible C/S interactions (i.e. paths from the start node 1 to each of the end nodes 3, 6, and 10) are depicted.

In this paper, the CSID for the SEARCH e-Commerce function, along with the corresponding probabilities p_{ij} between each pair of nodes i and j, will be used as a basis for building a model for performance analysis of the execution of e-Customers' HTTP requests with a generic e-Commerce system, given that clients' HTTP requests are processed in a multi-core CPU within the web server.

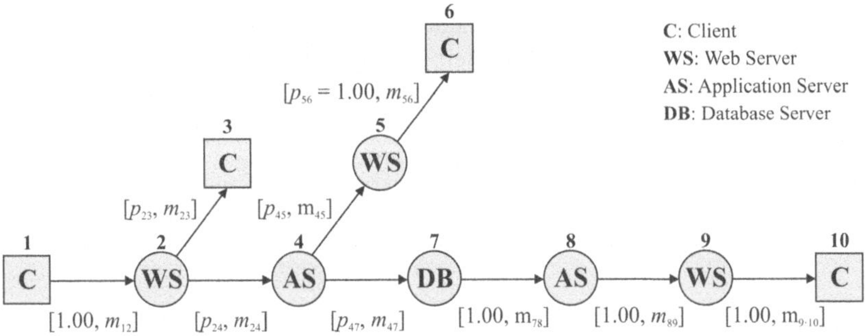

Fig. 1. Example of a C/S interaction diagram for the SEARCH e-Commerce function (Source: Adapted from Menascé & Almeida [2], p. 75)

The aim is to provide a comprehensive framework for modeling and performance evaluation of specific types of HTTP requests when they are processed in parallel. The performance evaluation based on the utilization of such a model can provide valuable insights into the performance gains due to the parallel processing within e-Commerce web servers.

3 Non-Markovian Stochastic Petri Nets

Non-Markovian Stochastic Petri Nets (NMSPNs) are a specific class of timed stochastic Petri nets where some of the timed transitions have non-exponentially distributed firing times. Similar to the classes of Stochastic Petri Nets (SPNs) and Generalized Stochastic Petri Nets (GSPNs), which are known under the common name of Markovian Stochastic Petri Nets, NMSPNs are widely recognized as a graphical tool for modeling and performance analysis of distributed systems, which utilize the graphical notation introduced by ordinary, non-timed Petri Nets (PNs): places, transitions, arcs, inhibitor arcs, and marking-dependent multiplicity of arcs.

Besides the usage of immediate and exponential transitions, NMSPNs may include either timed transitions with deterministic (i.e. fixed) firing delays (drawn as black-filled rectangles), in which case they become Deterministic and Stochastic Petri Nets (DSPNs), or general transitions (drawn as gray-filled rectangles) with firing delays that can be described by its probability mass function (pmf), which belongs to the class of expolynomial functions [3]. Such a distribution function can be piecewise-defined by exponential polynomials and can contain jumps, making it possible to mix discrete and continuous components. The class of expolynomial distributions contains many known distributions as special cases (e.g. uniform, triangular, truncated exponential, finite discrete, etc.) and allows the approximation of practically any distribution (e.g. by using splines), which is particularly well suited for numerical analysis of the dynamic behavior of wide range of parallel and distributed systems [4–6]. The analysis of an NMSPN model can be two-fold: (1) Qualitative: performed by studying the structural characteristics of the underlying Petri Net; (2) Quantitative: performed by computing the steady-state (stationary) and/or the transient (time-dependent) probability distributions of the associated stochastic model (i.e. underlying stochastic process), equivalent to a NMSPN model. Several dedicated software packages, like WebSPN [7] and TimeNET [4], offer integrated environments for modeling and performance analysis of NMSPNs.

4 The Modeling Framework

The proposed modeling framework is based on the assumption that HTTP requests, originating from the e-Customer and representing specific e-Commerce functions, are atomic, i.e. each of them can be processed entirely by a single core within a multi-core CPU. Therefore, the basic building block we use is the simplified representation of a multi-core CPU, modeled with Non-Markovian Stochastic Petri Nets (NMSPNs). This approach is similar to that one of Marsan *et al.* [8] who applied the class of Generalized Stochastic Petri Nets (GSPNs) for modeling and performance evaluation of multiprocessor systems. The NMSPN model of a web server, powered by a dual-core CPU, is depicted in Fig. 2, whilst the simplified NMSPN model of the CSID corresponding to the execution of the SEARCH function is given in Fig. 4. What follows is a brief description of the validation process *vis-à-vis* the proposed model.

In Fig. 2, HTTP requests for a specific e-Commerce function (e.g. SEARCH) arrive at the web server in time intervals that follow the Poisson distribution, with an arrival rate of λ. Each firing of the exponential transition *T_HTTP_search_arrival* puts a single token (i.e. an HTTP request) into the place *P_WS*.

These requests can be either rejected (the immediate transition *T_WS_rejected* fires if the place *P_WS_buffer_size* is empty) or accepted and stored in a buffer with a finite capacity of *K* (place *P_WS_buffer*). Each firing of the immediate transition *T_WS_buffer* takes a single token from both the places *P_WS* and *P_WS_buffer_size* and puts a single token in the place *P_WS_buffer*. Next, for each HTTP request, the web server looks up in the cache to check whether it already contains the corresponding HTML code. The immediate transition *T_WS_not_in_cache* fires with a probability of p_{cache}, which resembles the situation when the HTML code is not contained within the cache, whilst the immediate transition *T_WS_in_cache* fires with the complementary probability of $(1-p_{\mathrm{cache}})$, in the opposite case. If the HTML code is being fetched from the cache (a token in the place *P_WS_retrieve_from_cache*), the general transition *T_WS_retrieve_from_cache* fires after a random time, which, in this particular case, represents a continuous random variable *X* that follows the Uniform distribution $\mathrm{U}(a_2, b_2)$. This is made under the assumption that all values $x \in X$ within the continuous interval $[a_2, b_2]$ are equally possible. A continuous random variable *X* has a Uniform distribution on the interval $[a, b]$, denoted $\mathrm{U}(a, b)$, if for all $x \in X$, $x \in [a, b]$, the probability of *x* is equally likely (Fig. 3), and equal to $1/(b-a)$.

If the HTML code is not being fetched from the cache (a token in the place *P_WS_wait_in_queue*), then the HTTP request has to be processed either by CPU core #1 (a token in the place *P_WS_processing_1*) or by CPU core #2 (a token in the place *P_WS_processing_2*), depending on which one is going to be available first (a token in the place *P_WS_ready_to_process_1* or in the place *P_WS_ready_to_process_2*).

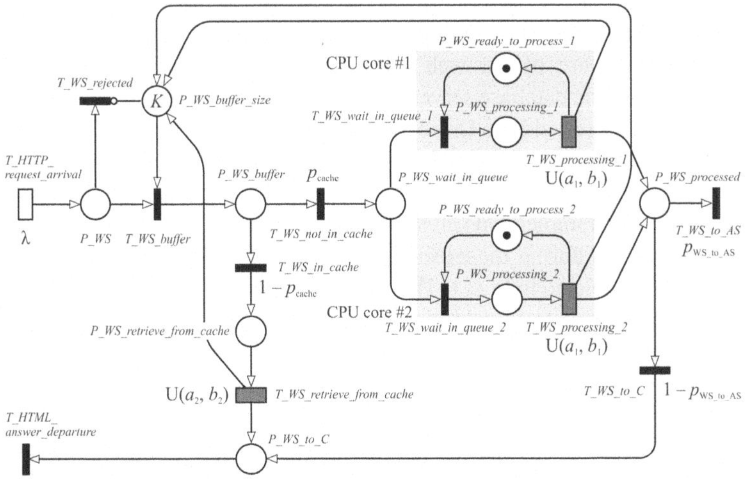

Fig. 2. NMSPN representation of a web server powered by a dual-core CPU (Source: Authors' own representation)

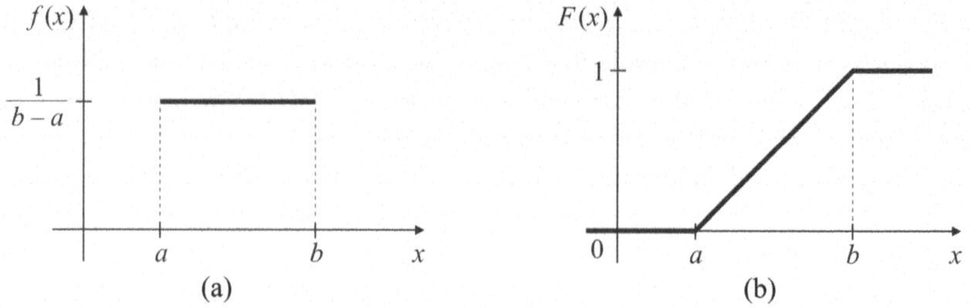

Fig. 3. Graphical representation of the (a) probability density function (pdf) and (b) cumulative distribution function (CDF) of the Uniform distribution U(a, b)

General transitions *T_WS_processing_1* and *T_WS_processing_2* fire after random times that comprise continuous random variable X that follows the Uniform distribution U(a_1, b_1). Again, this is made under the assumption that all values $x \in X$ from the continuous interval [a_1, b_1] are equally likely. Each firing of either of these general transitions puts back a single token into the place *P_WS_buffer_size*, meaning that the processing of a single HTTP request has finished, and into the place *P_WS_processed*, as well.

This enables two immediate transitions, *T_WS_to_AS* (with a firing probability of $p_{\text{WS\_to\_AS}}$) and *T_WS_to_C* (with a firing probability of $(1-p_{\text{WS\_to\_AS}})$). If the transition *T_WS_to_AS* fires, then the web server (WS) has sent a message to the application server (AS) for further processing. If the transition *T_WS_to_C* fires, the web server has completed the answer (an HTML document) to an HTTP SEARCH request, and sends it back to the client (C), denoted by a token in the place *P_WS_to_C* and firing of the immediate transition *T_HTML_answer_departure*.

The web server powered by a dual-core CPU can be considered as an *M*/*G*/2/*K* queuing system, since it is characterized by a Markovian arrival process (i.e. exponentially distributed inter-arrival times of HTTP requests), general service time distribution (i.e. U(a, b)), two servers (i.e. two CPU cores), and a finite capacity of $K > 0$.

If the inner structure of the web server (WS) is being replicated for both the application (AS) and the database server (DB), along with specific adjustments *vis-à-vis* the number of CPU cores and the Uniform distribution parameters' values, it is possible to interconnect them into a unique NMSPN structure that corresponds to a CSID suitable for performance analysis of a specific e-Commerce function. Figure 4 shows such an NMSPN model, which implements all three possible C/S interactions between the client and any of the servers during the execution of the SEARCH e-Commerce function, is based on the CSID depicted in Fig. 1. The interconnection has been done by a number of auxiliary places, immediate transitions, and both ordinary and inhibitor arcs for the purposes of implementing blocking and synchronization among various servers. For simplicity reasons, the inner structure of the three servers is omitted.

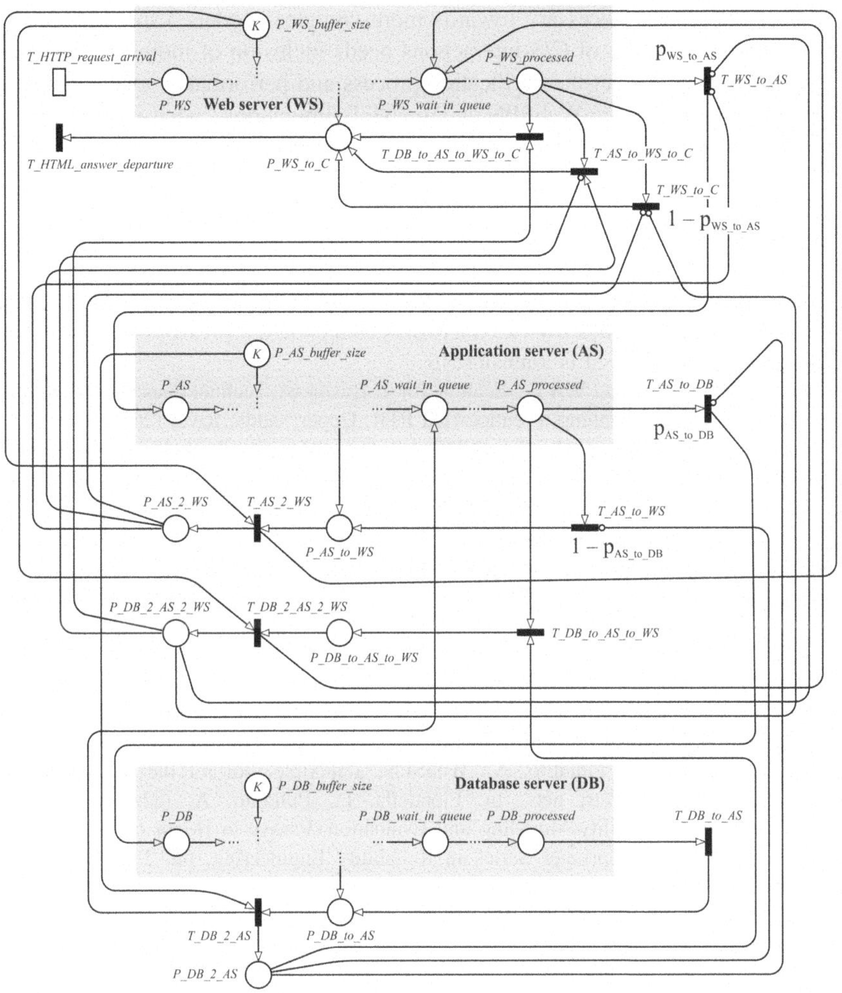

Fig. 4. NMSPN model of the CSID corresponding to the SEARCH e-Commerce function (Source: Authors' own representation)

5 Conclusion

The proposed modeling approach, based on the utilization of CSIDs and the application of the class of NMSPNs, provides a powerful and flexible framework for performance evaluation of the execution of various e-Commerce functions on multiple hardware architectures. As such, it can be successfully used for evaluation of the performance improvements gained by the parallel processing paradigm, through the usage of multi-core CPUs within web servers. However, despite the clarity and simplicity in specifying the inner structure of particular servers, which is achieved by the usage of NMSPNs, an obvious limitation is that their utilization in representing the logic of particular CSIDs leads necessary towards more complex models, especially in cases when the implementation of C/S interactions needs inclusion of more servers. Future work includes carrying out the verification process and performance evaluation of the modeling framework, using WebSPN or TimeNET simulations, 'what-if' analyses, and testing various operating scenarios *vis-à-vis* the input parameters.

References

1. CDNetworks: 3 Reasons Why Improving Ecommerce Website Speed is a Must, CDNetworks.com (2014) https://www.cdnetworks.com/en/news/3-reasons-why-improving-ecommerce-website-speed-is-a-must/4246
2. Menascé, D.A., Almeida, V.A.F.: Scaling for E-Business: Technologies, Models, Performance, and Capacity Planning. Prentice-Hall PTR, Upper Saddle River (2000)
3. Zimmermann, A., Knoke, M.: TimeNET 4.0 User Manual. A Software Tool for the Performability Evaluation with Stochastic and Colored Petri Nets. Technische Universität Berlin, Faculty of EE & CS, Berlin (2007)
4. German, R., Kelling, C., Zimmermann, A., Hommel, G.: TimeNET: a toolkit for evaluating non-Markovian stochastic Petri nets. Perform. Eval. **24**(1995), 69–87 (1995)
5. Fricks, R.M., Puliafito, A., Telek, M., Trivedi, K.S.: Applications of non-Markovian stochastic Petri nets. Perform. Eval. Rev. **26**(2), 15–27 (1998)
6. Horváth, A., Puliafito, A., Scarpa, M., Telek, M.: Analysis and evaluation of non-Markovian stochastic Petri nets. In: Haverkort, B.R., Bohnenkamp, H.C., Smith, C.U. (eds.) Computer Performance Evaluation. Modelling Techniques and Tools, TOOLS 2000. LNCS, vol. 1786, pp. 171–187. Springer, Heidelberg (2000)
7. Longo, F., Scarpa, M., Puliafito, A.: WebSPN: a flexible tool for the analysis of non-markovian stochastic Petri nets. In: Fiondella, L., Puliafito, A. (eds.) Principles of Performance and Reliability Modeling and Evaluation. Essays in Honor of Kishor Trivedi on his 70th Birthday. Springer Series in Reliability Engineering, pp. 255–285. Springer International Publishing, Switzerland (2016)
8. Marsan, M.A., Conte, G., Balbo, G.: A class of generalized stochastic Petri nets for the performance evaluation of multiprocessor systems. ACM Trans. Comput. Syst. **2**(2), 93–122 (1984)

A Dynamic Call Admission Control Scheme and Performance Modeling for 4G LTE Networks

Faisal Hayat[1], Zlatan Jukic[2,3](✉), and Iqra Abdul Ghaffar[1]

[1] Department of Computer Engineering, UET Lahore, Lahore, Pakistan
fsl.hayat@gmail.com, khaniqra356@gmail.com
[2] Faculty of Technical, HTL Rankweil, Rankweil, Austria
zlatan.jukic@htl-rankweil.at
[3] TU Wien, Wien, Austria

Abstract. There is an exponential increase in the number of high bandwidth and quality demanding users that has forced cellular service providers to offer better quality of service (QoS). In long term evolution (LTE) also called 4G networks, the number of service classes have been increased and call admission control (CAC) strategies are implemented to fulfill the QoS requirements. CAC schemes based on bandwidth reservation, bandwidth degradation and preemption have been proposed with the aim of better utilization of network resources. However, in heterogeneous network environment, high priority traffic classes sometimes demand complete protection and very efficient and dynamic call admission control (CAC) strategies are still required to guarantee quality of service while achieving better utilization of network resources. A dynamic CAC is proposed in this work with the aim to provide complete protection to high priority traffic. A combined usage threshold and total preemption mechanism have been introduced guarantying complete protection and improved quality of service for high priority traffic class. Markov chains have been used to evaluate the proposed CAC scheme along with necessary simulations for validation. Results show that the proposed CAC scheme significantly improves call dropping probability of high priority traffic, without much penalizing the low priority traffic. Numerical results imitate the experimental results with insignificant differences.

Keywords: Call admission control (CAC) · Call blocking probability (CBP) · Call dropping probability (CDP) · Handoff call (H/C) · New call (N/C) · Real-time (RT) · Non-real time (NRT) · Quality of service (QOS) · Long term evolution (LTE)

1 Introduction

The rapid advancement in technology helped 3Gevolve to 4G also known as LTE (Long Term Evolution) under 3rd Generation Partnership Project (3GPP) [1].

LTE is enhancement of Universal Mobile Telecommunication System (UMTS) which is replacement of Global System for Mobile Communication (GSM). As LTE is modified version, so it also retains the features of UMTS. LTE provides pure packet

I. Karabegović (Ed.): NT 2020, LNNS 128, pp. 499–514, 2020.
https://doi.org/10.1007/978-3-030-46817-0_58

switched network and it adopts circuit switched fallback mechanism to facilitate the voice call. Circuit Switch Fallback mechanism refers that 4G switches back to 3G to complete the voice call. The architecture of LTE network has three key components User Equipment (UE), Evolved Universal Terrestrial Radio Access Network (E-UTRAN) and Evolved Packet Core (EPC) as illustrated in Fig. 1.

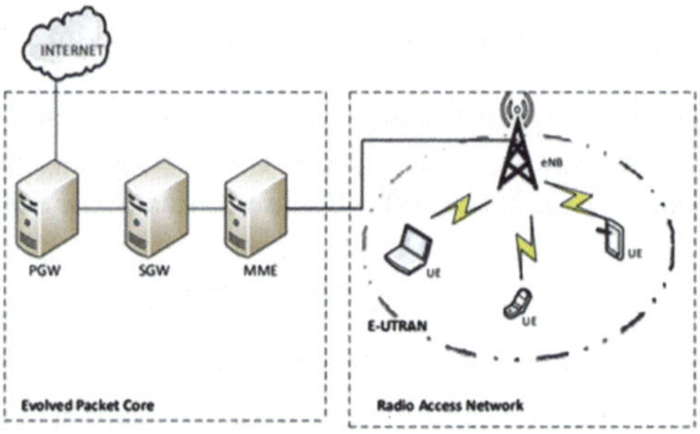

Fig. 1. Architecture of LTE networks

UE is basically a mobile device and E-UTRAN consists of base stations that are named as evolved node Bs (eNBs). eNB basically handles communication between EPC and UE. These base stations also control low-level operations like control messages to and from UEs. eNBs also maintain link with each other which is helpful in case of handover. Third component of LTE architecture is EPC which connects radio access network to packet data network. EPC consists of a home subscriber server (HSS), a mobility management equipment (MME) and a serving gateway (S-GW).

LTE has three types of channels: Logical, Transport and Physical. Logical channels map on transport channels and transport channels map on physical channels which carry user data. LTE uses Orthogonal Frequency Division Multiple Access (OFDMA) for downlink (Base station to UE) and Single Carrier Frequency Division Multiple Access (SC-FDMA) for uplink (UE to Base station) transmission [2]. It uses flexible bandwidth of 1.4 MHz to 20 MHz. 1.4 MHz bandwidth has 6, 10 MHz has 50 and 20 MHz has 100 physical resource blocks (PRBs).

PRBs are basically number of sub-carriers allocated to users. One PRB is further divided into 7 resource blocks (RB). One RB consist of 12 subcarriers and 1 symbol. It means that one RB has 12 * 1 = 12 resource elements. The bandwidth of one RE is 15 kHz so, the total bandwidth of one RB is 15 * 12 = 180 kHz. LTE assigns a 180 kHz resource block to a user for data transmission. As one PRB consist of 7 RBs, so one PRB can accommodate seven users.

Assignment of a resource block while obtaining required QoS (overall measurement of system performance) is a major challenge in 4G LTE networks. For this purpose, call admission control (CAC) schemes are used to efficiently assign a radio resource to a new user while maintaining QoS of ongoing calls [3]. QoS of calls depends upon many factors e.g. bandwidth allocated to a call and call type etc. but one of the major factor is service class of call. There are three main service classes:

unsolicited grant service (UGS), real-time (RT) service, non-real-time (NRT) service. UGS service class provides top level services e.g. voice over IP. Any kind of interruption like channel unavailability is not bearable for UGS service class. RT service has less priority as compared to UGS. Examples of RT service class are stream audio, video, voice and handoff calls. Last service class is NRT. Example of NRT is file transfer protocol (FTP) and new calls etc.

From three major service classes of calls, we used two (RT and NRT) in this work. We used two service classes because UGS and RT are considered as one service class and named as real-time (RT) service class. Second service class is NRT. QoS of NRT and RT service classes depend upon call blocking probability (CBP) and dropping (CDP) probability. That's why our primary focus is to reduce call blocking (CBP) and dropping (CDP) probability. These two parameters are mainly considered for performance evaluation of CAC techniques. CBP is denying of new calls in current cell due to insufficient resources and CDP is dropping of ongoing call during handoff due to user's mobility.

A good CAC technique balances call blocking (CBP) and call dropping probability (CDP) but call dropping probability is handled carefully because dropping of ongoing call is more frustrating for a user compared to new user. That's why most CAC techniques prioritized handoff calls by reserving resources to reduce call dropping probability and remaining resources are used by new calls. Different CAC schemes have been proposed in this context. These schemes are named as bandwidth reservation, borrowing and preemption based CAC techniques. These techniques handled two traffic classes: high priority and low priority. Channels are reserved for high priority class and low priority class can borrow channels from high priority. Later on, when high priority call arrives, it pre-empts low priority calls. These pre-empted calls are placed in a queue and they resume their service on availability of channels. In these CAC schemes, high priority calls can pre-empt resources from low priority up-to a certain threshold. As a result, high priority call suffers if arrival rate is very high.

In this work, a dynamic CAC is proposed with the aim to provide complete protection to high priority traffic class (RT). A combined usage threshold and total preemption mechanism have been introduced guarantying complete protection and improved quality of service for high priority traffic class. This paper is organized as follow: Sect. 2 contains summary of related work. In Sect. 3, markov chain model and its performance analysis is given. Section 4 presents real time simulation of proposed CAC technique. Section 5 describes results and discussion while in Sect. 6 conclusion of paper is given.

2 Related Work – Historical Development of CAC Schemes

There exists a considerable body of literature on Call Admission Control (CAC) design. This section presents a review of recent literature that focus on reservation, degradation and preemption based CAC techniques. Fang and Zhang employed new call bounding CAC scheme [4]. Focus of the design was to reduce the handoff blocking probability by reserving bandwidth resources. They differentiated traffic types into handoff and new calls. In this proposal, a fixed number of channel resources are reserved for handoff calls and rest can be used by new calls. Due to fixed reservation, handoff calls are protected over new calls even under heavy traffic load.

Another CAC scheme under fixed resource regime was proposed by authors in [5]. It was named as cut-off threshold for packet communication. This scheme classified traffic types into high and low priority classes. Queue is also introduced for low priority traffic class. High priority class can use all channels and low traffic class can use channels up to combined usage threshold. Simulation results proved that high priority traffic calls are more protected in cut-off threshold scheme because channels from common pool can be used on demand.

Charabi et al. [6] proposed a comprehensive CAC scheme with retrial of new calls. This scheme aimed to reduce the blocking probability of new calls by introducing a waiting queue. In this scheme, a handoff call can use reserved as well as common pool resources but a new call can only use common pool resources. In order to decrease the blocking, on finding all channels busy a new call waits in the queue and can retry for a channel while waiting in the queue. Results showed an improvement in blocking of new calls.

An efficient reservation based CAC scheme for protection of handoff calls as well as new calls was presented in [7]. It provided priority to handoff calls without totally neglecting new calls. In this scheme, queues are introduced for both handoff calls and new calls in case of unavailability of channels in system. This scheme resulted in reduction of handoff dropping as well as new call blocking probability due to the queues.

The authors in [8] employed user categorization and delay aware reservation based CAC scheme. In this proposal, RT and NRT traffic types are used. This scheme introduced fix resources and waiting queue for each traffic type. These queues are served according to adaptive priority (AP). The AP was calculated on the basis of delay tolerance and user categorization. Simulation results showed efficient system utilization and QOS provisioning for handoff and new calls.

The work in [9] introduced an adaptive joint CAC scheme, which reserves resources for each radio access technology (RAT). In case of unavailability of resources, if a new call arrives then scheme adopts degradation in that particular RAT. Adaptive joint CAC scheme guarantees QOS for all accepted calls, reduce CBP and CDP and efficiently utilize bandwidth in each RAT.

In [10], a degradation based CAC scheme for multimedia applications was developed by Satyanarayana, Satyanarayana and Raju. This scheme degraded bandwidth of ongoing high priority traffic class in case of bandwidth for low priority class is insufficient. After call completion of low priority traffic class, bandwidth of high priority class is upgraded again. This degradation and upgradation of bandwidth is according to user requests and traffic types. This scheme reduced blocking probability of new calls and also provides efficient utilization of bandwidth.

Another CAC degradation scheme was presented by authors in [11], which assigned resources to all traffic classes dynamically. This markov model based CAC scheme degraded bandwidth of low priority traffic class to minimum level in case of overloaded cell. This scheme provided protection to high priority class from degradation and also ensured efficient resource utilization. Authors in [12], proposed call admission control technique in HAP for enhanced QoS in wireless network.

This technique categorized traffic classes into high priority and low priority class. Upon arrival of high priority call in overloaded cell, bandwidth of low priority call is degraded to average level. In this way, high priority class is accommodated and low priority is also not much affected by degradation.

A Dynamic threshold based CAC scheme for downlink LTE networks was introduced by Mamman, Hanapi, Abdullah and Muhammed [2]. This scheme reserved bandwidth for high priority class and remaining is used by low priority class. For high priority class, resources are reserved on the basis of dynamic threshold and threshold is calculated by traffic intensity and bandwidth required for handoff calls.

The authors in [13] also employed bandwidth degradation CAC scheme using call duration. This technique generated traffic with its simulation time, average bandwidth requirement and random average call duration time. Upon deficiency of bandwidth resources, call duration time and call type is determined. If call is non-voice and call duration time is less than average call duration time then degrade bandwidth of ongoing calls otherwise not. A simulation result of this scheme shows negligible blocking probability and also improves bandwidth utilization and channel holding time.

The authors in [14] came up with idea of channel borrowing for two-tier LTE networks. This scheme divides bandwidth into two parts, one for LTE macro-cell and other for picocell. In this proposal, if call arrives in picocell and bandwidth not enough, at that point place it into waiting queue. Similarly, upon arrival of call in macro-cell, due to inaccessibility of transmission capacity, it can utilize bandwidth of picocell or place it into waiting queue. This scheme borrowed bandwidth from picocell but no preemption in case of picocell call arrival. In this way, this technique proved that call blocking of macrocell reduced but picocell may increase.

The technique introduced by Belghith in [15], was flexible CAC with preemption. This scheme sets threshold for NRT calls and they are accepted within that threshold otherwise block them.

Halabian et al. designed a reservation and preemption based call admission control scheme in LTE networks [16]. This scheme handles three type of traffic classes Handoff, NRT and BE. Channels reserved for handoff calls but BE calls can borrow from pool of handoff channels.

Another reservation and borrowing based CAC scheme was designed for heterogeneous networks [17]. In this proposal, cells are categorized into femtocell and macrocell. In macrocell, resources are fixed for handoff calls but in femtocell, both calls can equally share all resources. Upon arrival of NRT call in femtocell and finding that all channels are busy then call is transferred into macrocell. NRT call can use reserved resources of handoff call in macrocell. In the event that handoff call moved into femtocell, at that point if channel is not accessible then block it. But if handoff call moved into macrocell and all channels are busy then NRT calls are preempted and allocate those channels to handoff calls. Simulation results of this scheme showed that call dropping is reduced but call blocking may increase due to preemption.

Another comprehensive bandwidth reservation, borrowing and preemption based CAC scheme was presented by authors in [18] to reduce call blocking and dropping probability. Bandwidth/Channels are reserved for RT calls. In this proposal, NRT calls can borrow channels from reserved channels of RT calls. Later on, when RT call arrives and all channels are busy then NRT call is preempted and put into priority queue. In this way, high priority class is protected from low priority class and CDP, CBP both reduced. As one macrocell have multiple femtocells in its coverage area so in macrocell, a CAC technique which immediately assigns resources to handoff calls or large number of resources for handoff calls is necessary.

3 Proposed Work

Call Admission Control (CAC) in mobile/cellular network, deals with automatic classification of a call into one of three possible classes: serviced, blocked and dropped by the nearest base station of the user. When a call is initiated in one cell, it will request its home cell for a channel. If home cell cannot provide any free channel, the call may shift to the neighbouring cell. If the neighbours too cannot find a suitable channel then the call is dropped. Again, if the caller, while a call is in progress, moves out of a cell to any of its neighbour, a channel of the neighbouring cell will be allocated to the call. This is known as a handoff. The calls for which no channels can be assigned are dropped [19].

In this section, we introduce our proposed CAC scheme. Figure 2 show that our CAC scheme categorizes incoming calls into two traffic classes real-time (RT) and non-real time (NRT). RT is high priority traffic class (e-g: handoff calls and emergency service calls) and NRT is low priority traffic class (e-g: new calls). In proposed CAC scheme, first condition for NRT users is combined usage threshold. If total used channels are less than threshold, our system accepts NRT calls otherwise next condition is checked. Next condition is about space in queue. If queue is not full then our system places NRT call in to queue otherwise NRT call is rejected.

RT user is directly accepted in system if total used channels are less than combined threshold. If combined usage threshold is reached then next condition is about existence of NRT users in system. If NRT users exist then our CAC scheme will take resource from one of the NRT users and allocate it to RT user. The pre-empted NRT user is placed into the queue, if queue is not full otherwise our system blocked NRT user. From a queue, NRT user resumes its service if number of busy channels is less than combined threshold. In case of no NRT user exists in system then next condition for RT user is either all channels in system are busy or not. If all channels are busy then our system reject RT user otherwise RT user is accepted by our system. The previous descripted model can be better explained by queuing diagram of model.

A. Queuing Diagram of Proposed CAC Scheme

As we have already discussed, proposed CAC technique handles two types of traffic classes RT and NRT. In Fig. 2, A_1 is high priority traffic class which is RT and A_2 is low priority traffic class which is NRT. There are n servers and threshold is m. Space in queue is s. A_2 traffic class is accepted in system if total traffic $A_1 + A_2$ is less than threshold m. After this threshold, A_2 is placed into the queue. A_1 is always accepted as long as channels or A_2 traffic class is available in system. A_1 pre-empts channel of A_2 upon its arrival if combined threshold is reached. The pre-empted NRT calls are placed into the queue.

A_2 is placed into queue either A_2 is pre-empted or total traffic is more than threshold. Figure 2 is queuing diagram which explains proposed CAC technique.

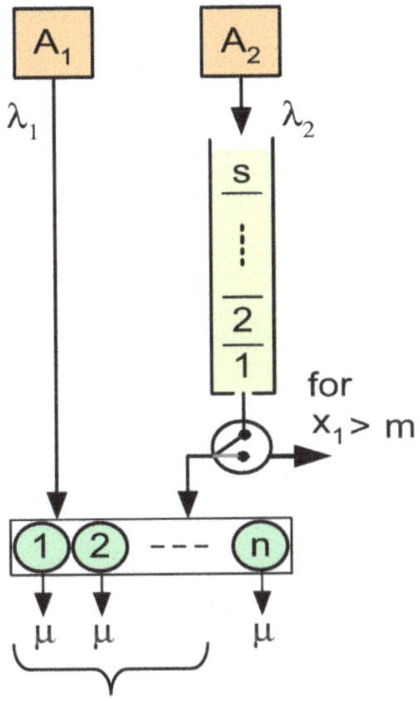

Fig. 2. Queueing diagram of proposed CAC technique

This queuing diagram can be modelled by using markov chain state diagram.

B. Markov Chain of Proposed CAC Technique

In this section, queuing diagram of proposed CAC technique is modelled with the help of markov chain. In proposed model, markov chain is used for calculations of performance measures e-g: call blocking, call dropping and call waiting probabilities. These performance measures are calculated by using state probabilities of markov chain diagram. Figure 3 illustrated Flowchart of proposed CAC scheme and Fig. 4 illustrated markov chain (state transition diagram) of proposed CAC scheme. In Fig. 4, combined usage threshold is m. System bandwidth is channelized and number of channels in system are n.

In markov chain of proposed CAC scheme, Fig. 4, n denotes number of physical network resources in a cell which should be allocated to arriving calls. In markov model of proposed CAC scheme, state is represented as (x_1, x_2, x_3) where x_1 represents high priority traffic class (RT), x_2 represents low priority traffic class and, x_3 represents queue of NRT. A fixed number of traffic classes can be defined by specifying the parameters for each class, or an infinite number of classes can be obtained by allowing any choice of parameters within some range.

From Fig. 4, it is clear that, x_1 can take all n channels and, x_2 can take channels up to combined threshold m. After m threshold, NRT calls are placed into queue. In markov chain model, there must be arrival or service of certain traffic class to jump from one state to another. Arrival of traffic classes is poisson distributed with parameters λ_{RT} and λ_{NRT}.

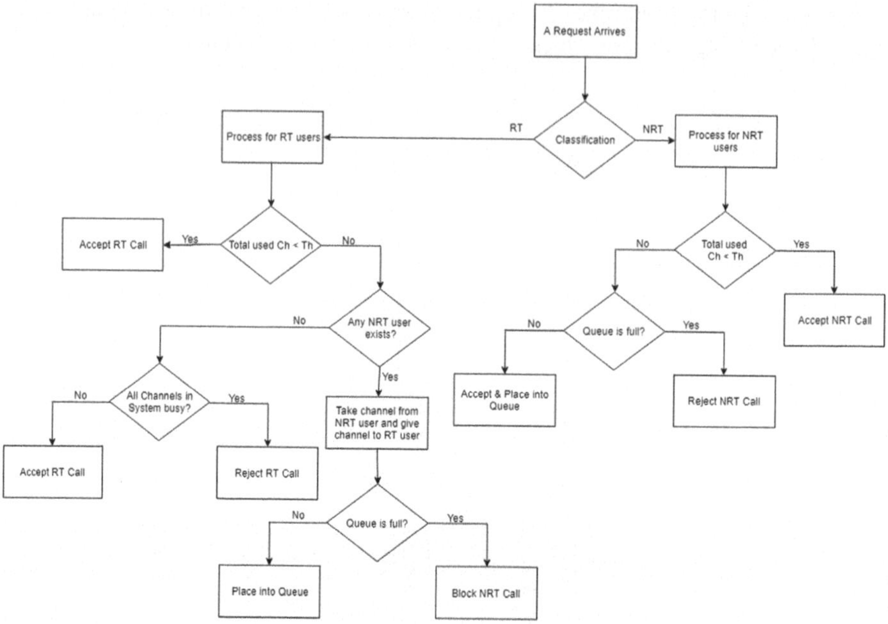

Fig. 3. Flowchart of proposed CAC scheme

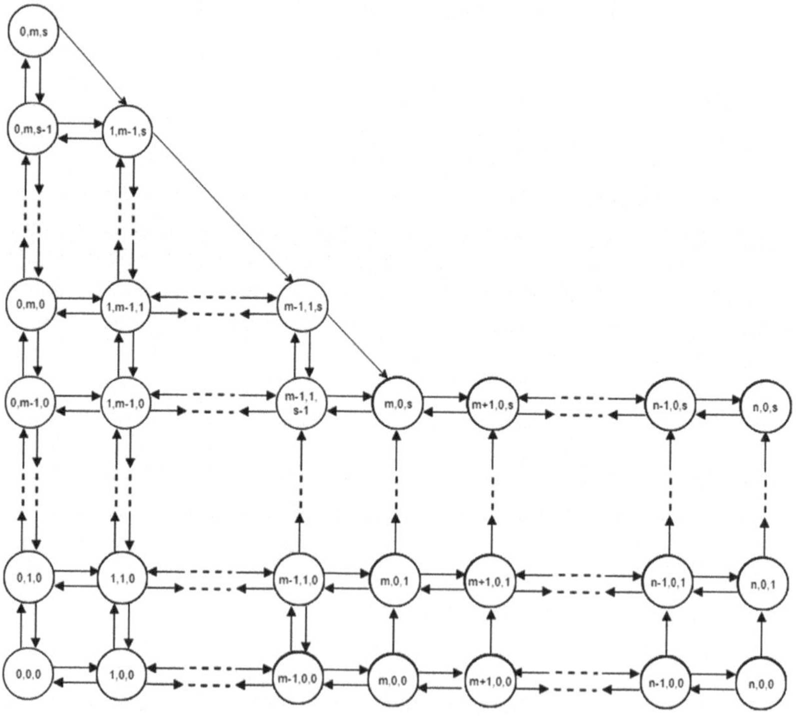

Fig. 4. Markov model of proposed CAC technique

Service of traffic classes is exponentially distributed with parameters μ_{RT} and μ_{NRT}. Transition rates from Fig. 4 can be calculated as:

$$\begin{aligned}
&q(x_1, x_2, x_3 : x_1 + 1, x_2, x_3) = \lambda_{RT} \\
&q(x_1, x_2, x_3 : x_1 + 1, x_2 - 1, x_3 + 1) = \lambda_{RT} \\
&q(x_1, x_2, x_3 : x_1 + 1, x_2 - 1, x_3) = \lambda_{RT} \\
&q(x_1, x_2, x_3 : x_1, x_2 + 1, x_3) = \lambda_{NRT} \\
&q(x_1, x_2, x_3 : x_1, x_2, x_3 + 1) = \lambda_{NRT} \\
&q(x_1, x_2, x_3 : x_1 - 1, x_2, x_3) = x_1\mu_{RT} \\
&q(x_1, x_2, x_3 : x_1 - 1, x_2 + 1, x_3 - 1) = x_1\mu_{RT} \\
&q(x_1, x_2, x_3 : x_1, x_2 - 1, x_3) = x_2\mu_{RT} \\
&q(x_1, x_2, x_3 : x_1, x_2, x_3 - 1) = x_2\mu_{RT}
\end{aligned} \tag{1}$$

A. Performance Measures of Proposed CAC Technique

In proposed CAC scheme, three performance measure characteristics are considered. These performance measures are named as call blocking probability (CBP), call dropping probability (CDP) and call waiting probability (CWP). Blocking state of RT call is known as call dropping probability. Blocking of NRT call is known as call blocking probability and waiting state of NRT call is known as call waiting probability. From Fig. 4, we have the following properties for the dropping of RT calls, blocking of NRT calls and waiting of NRT calls.

- an RT arrival will be dropped if $x_1 = n$
- an NRT arrival will be blocked if $x_1 + x_2 \geq m$ and $x_3 = s$
- an NRT arrival will wait in queue if $x_1 + x_2 \geq m$ and $x_3 < s$

Let $p(x_1, x_2, x_3)$ denote the steady-state probability of being at state (x_1, x_2, x_3). From Fig. 4, CDP is derived as:

$$\text{CDP} = \sum_{x_3=0}^{s} P(n, 0, x_3) \tag{2}$$

Equation 2 is sum of states where all n channels are busy due to real-time traffic class. From Fig. 4, CBP is derived as:

$$\text{CBP} = \sum_{x_1=n}^{0} \sum_{x_2=0}^{m} \sum_{x_3=0}^{s} P(x_1, x_2, x_3) \tag{3}$$

Equation 3 is sum of states where NRT call is blocked. From Fig. 4, call waiting probability (CWP) is derived as:

$$\mathrm{CWP} = \sum_{x_1=0}^{\mathrm{m}} \sum_{x_2=0}^{\mathrm{n}} \sum_{x_3=0}^{s-1} P(x_1, x_2, x_3) \tag{4}$$

Equation 4 is sum of states where customer is waiting in queue but queue is not full.

It is important to note that above literature focused on bandwidth reservation with static threshold [4–8], whereas [2, 9–13] introduced bandwidth reservation and degradation with static or dynamic threshold. Similarly [14–18] introduced bandwidth reservation, borrowing and preemption based CAC techniques to reduce call blocking, and dropping probability, also improve systemutilization. Each type of technique has its own benefits and shortcomings. Reservation based CAC scheme reserves resources for each traffic class and other traffic class can't use them so, efficient resource utilization is not possible in this type of techniques.

So, our scheme is applicable in heterogeneous network which has large handoffs. Heterogeneous network is comprised of macrocell and femtocell. One macrocell consist of many femtocells which have low coverage area and macrocell have large coverage area. Femtocell is useful for large.

4 Real-Time Simulation

Our CAC scheme can also be modelled using real-time simulation. As our CAC scheme considers two types of traffic classes (RT and NRT). RT can use system channels directly and NRT can use system channels through queue if system channels are busy up to threshold. The real-time simulation results of proposed CAC scheme are obtained with the help of Dev C++. The performance of the proposed scheme is measured in terms of CBP, CDP and CWP. Simulation scenario consists of event-list and simulation time. Event list consists of insert, erase, get next event and print event-list as member functions. Simulation is started by inserting first event. After that, the call request is classified into two classes based on their QoS requirement and call type.

Based on QoS requirements the calls can be categorized as RT and NRT. RT is high priority traffic class and the best example is live streaming. On the other hand, NRT is low priority traffic class and the best examples are youtube and email.

The arrival rate for both RT and NRT is in the form of a Poisson distribution with the service is exponentially distributed. The total simulation time is 100000 s. Table 1 summarized the simulation parameters. In Simulation, high priority traffic class user is inserted by preempting NRT call if threshold is reached. NRT user is inserted if total busy channels are less than combined usage threshold. After each arrival and service of event, performance measure characteristics are updated.

Table 1. Simulation parameters

Parameters	Value
Total channels	30
Threshold	15
λ_{NRT}	15
λ_{RT}	1–20
μ_{RT}	x1 μ_{RT}
μ_{NRT}	x2 μ_{NRT}
Simulation period	100000 s

5 Results and Discussion

In this section, experimental outcomes are observed in terms of probability of blocking, probability of dropping and probability of waiting. These performance measures are used to compare proposed model with New Call Bounding, Cut-off Threshold Delay Loss System and Reservation, Borrowing, Preemption based CAC scheme. We also validate our system by comparing the CDP and CBP values obtained from markov chain state diagram and the CDP and CBP values obtained from real-time simulation of the proposed model. For evaluation of proposed model, we consider different scenarios like high traffic load of RT, high traffic load of NRT etc.

A. CDP and CBP in case of High Traffic Load of Real-Time (RT)

As we have already discussed, simulation and the analytical study both are performed in our model. To compare the analytical results with the simulation results for the proposed model and then with the conventional schemes (New Call Bounding, Cut-off Threshold Delay Loss System and Reservation, Borrowing, Preemption), a system with C = 30 and T = 15 is considered. In comparison, the CDP and CBP values are calculated for the system using Eq. 2 and 3. The CBP and CDP values are calculated for $\lambda_{RT} = i$ where i = 1, 2, … 20 and $\lambda_{NRT} = 15$. We present the CDP and CBP result of these mentioned parameters in Figs. 5 and 6 respectively. Parameter setting for the analysis of this part is summarized in Table 2.

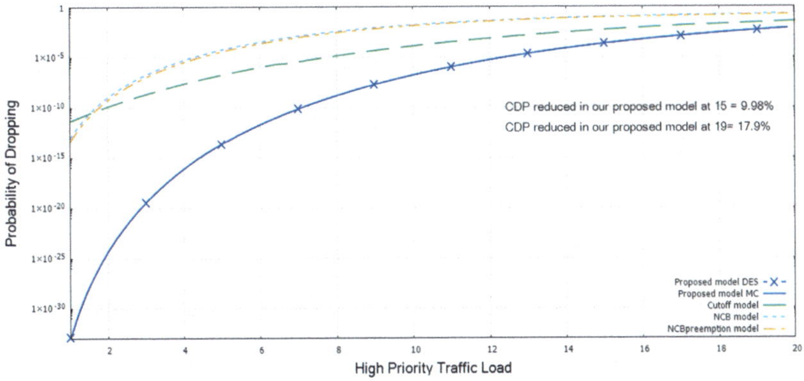

Fig. 5. CDP in case of high traffic load of RT

Table 2. Simulation parameters in case of high traffic load of RT

Sr. no	Total channels	Threshold	Traffic classes	Arrival rate	Buffer size
1	30	15	1	1–20	0
2	30	15	2	15	5

From Fig. 5, it is revealed that the results of proposed model are considerably decreases in terms of CDP compared with the conventional schemes.

This decrease in CDP is shown in Fig. 5 when traffic load of RT is 15 and 19. At traffic load 15, reduced CDP in proposed model is 9.98% but at traffic load 19, it becomes 17.9%. Figure 5 also shows that the simulation results and markov model results are in total agreement.

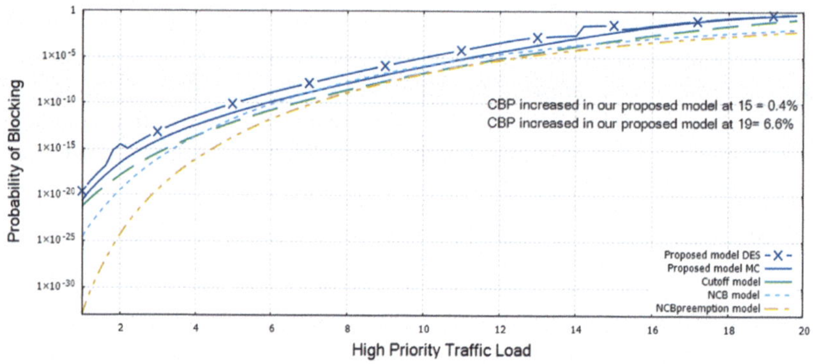

Fig. 6. CBP in case of high traffic load of RT

Figure 6 shows that the proposed model results in terms of CBP are slightly higher as compared to conventional schemes. This increase of CBP is shown in Fig. 6 when traffic load of RT is 15 and 19. However, the amount of increase for CBP in the proposed model is very small compared to the gain obtained in terms of the CDP decrease. At traffic load 15, increased CBP in proposed model is 0.4% but CDP reduced is 9.98%. Similarly at traffic load 19, increased CBP in proposed model is 6.6% but CDP reduced is 17.9%. Figure 6 also shows that the simulation results and markov model results performs closely.

B. CDP and CBP in case of High Traffic Load of Non-Real-Time (NRT)

In case of high traffic load of NRT, analytical results and simulation results of the proposed model are compared. Our proposed model is also compared with the conventional schemes (New Call Bounding, Cut-off Threshold Delay Loss System and Reservation, Borrowing, Preemption). For comparison, a system with C = 30 and T = 15 is considered. In comparison, the CDP and CBP values are calculated for the system using Eq. 2 and 3. The CBP and CDP values are calculated for $\lambda_{RT} = 15$ and

λ_{NRT} = i where i = 1, 2, … 20. We present the CDP and CBP result of these mentioned parameters in Fig. 7 and Fig. 8 respectively. Parameter setting for the analysis of this part is summarized in Table 3.

Figure 7 shows that the results of proposed model are considerably decreases in terms of CDP compared with the conventional schemes. This decrease in CDP is shown in Fig. 7 when traffic load of NRT is 10 and 15. At traffic load 10, reduced CDP in proposed model is 4.8% but at traffic load 15, it becomes 9.98%. Figure 7 also shows that the simulation results and markov model results are in total agreement.

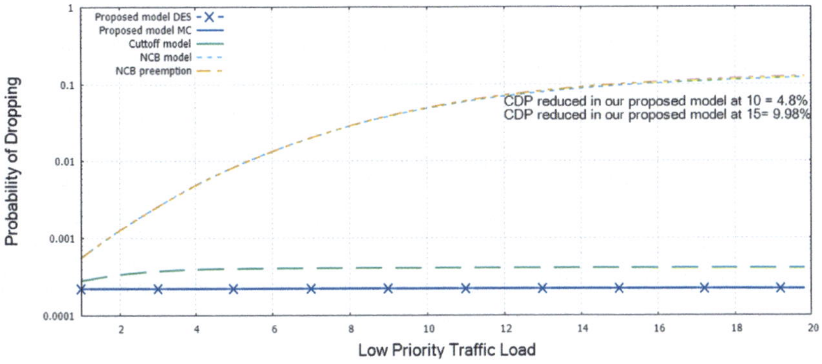

Fig. 7. CDP in case of high traffic load of NRT

From Fig. 8, it is cleared that the proposed model results in terms of CBP are slightly higher as compared to conventional schemes. This increase of CBP is shown in Fig. 8 when traffic load of NRT is 10 and 15. At traffic load 10, increased CBP in proposed model is 4% but CDP reduced is 4.48%. Similarly at traffic load 15, increased CBP in proposed model is 8% but CDP reduced is 9.98%. Figure 8 also shows that the simulation results and markov model results performs closely.

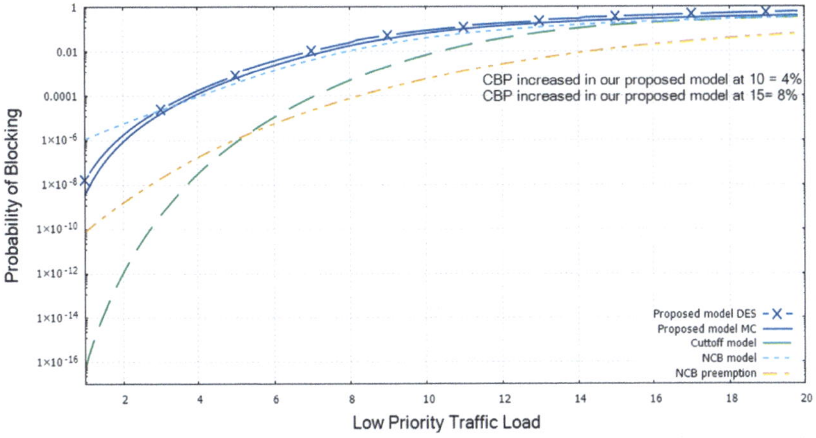

Fig. 8. CBP in case of high traffic load of NRT

C. CWP in case of High Traffic Load of Real-Time (RT)

In case of high traffic load of RT, CWP of proposed model is compared with Cut-off Threshold Delay Loss System. For comparison, parameters are summarized in Table 2 and CWP is calculated from Eq. 4. Figure 9 presents the CWP result of parameters summarized in Table 2.

Figure 9 shows that the result of proposed model is increased in terms of CWP and it is illustrated at traffic load of 15 and 19. At traffic load 15, CWP is increased 9% from Cut-off Threshold Delay Loss System but at traffic load 19, it becomes 2%. With more traffic load, waiting probability decrease because call blocking increase with more traffic load.

D. CWP in case of High Traffic Load of Non-Real-Time (NRT)

In case of high traffic load of NRT, CWP of proposed model is compared with Cut-off Threshold Delay Loss System. For comparison, parameters are summarized in Table 3 and CWP is calculated from Eq. 4. Figure 10 presents the CWP result of parameters summarized in Table 3. Figure 10 shows that the result of proposed model is increased in terms of CWP and it is illustrated at traffic load of 15 and 19. At traffic load 15, CWP is increased 9% from Cut-off Threshold Delay Loss System but at traffic load 19, it becomes 10%.

Table 3. Simulation parameters in case of high traffic load of RT

Sr. no	Total channels	Threshold	Traffic classes	Arrival rate	Buffer size
1	30	15	1	15	0
2	30	15	2	1–20	5

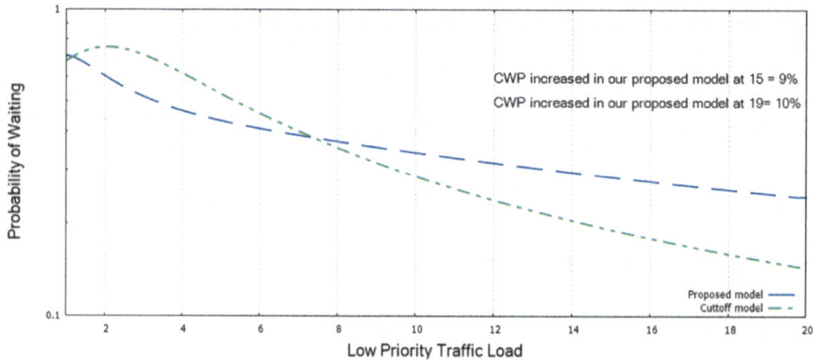

Fig. 9. CWP in case of high traffic load of RT

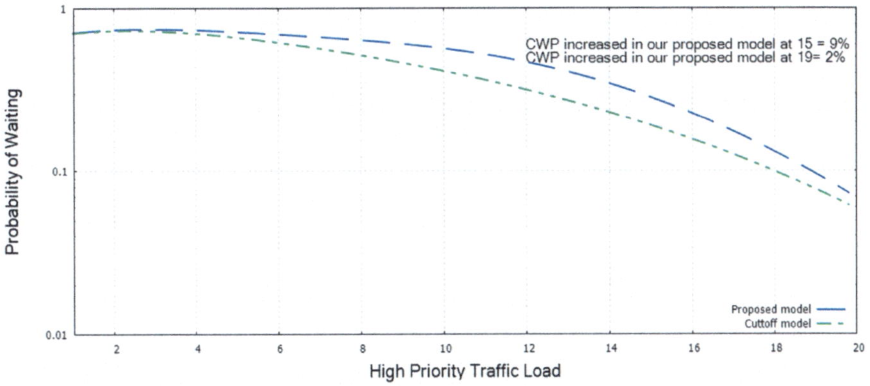

Fig. 10. CWP in case of high traffic load of NRT

6 Conclusion

In this paper, we have proposed "A Dynamic Call Admission Control scheme and Performance Modelling for 4G LTE Networks" to protect real-time calls in heterogeneous network. This CAC scheme introduced combined usage threshold and total preemption of NRT calls for protection of RT calls. A markov chain model was introduced with CBP and CDP to evaluate the effectiveness of proposed CAC scheme. Extensive simulation experiments were conducted to validate the markov model results of proposed CAC scheme. Simulation and markov model results are in total agreement with negligible differences. Results also shows the CDP of proposed scheme is considerably reduced as compared to new-call bounding, cut-off threshold and reservation, borrowing, preemption based CAC technique but CBP and CWP slightly increases. This further indicates that proposed scheme provides effective quality of service (QOS) to RT calls in LTE networks.

References

1. Lawal, M.A., Saidu, I., Abdullahi, Y.S.: Call admission control schemes in LTE networks: a survey. IJCSN Int. J. Comput. Sci. Netw. **6**, 523–532 (2017)
2. Mamman, M., Hanapi, Z.M., Abdullah, A., Muhammed, A.: An adaptive call admission control with bandwidth reservation for downlink LTE networks. IEEE Access **5**, 10986–10994 (2017)
3. Al-jawhar, Y.A.: Survey: admission control algorithms for vehicular (Wi-Fi and Wi-Max) network technologies. Res. J. Appl. Sci. **6**, 67–75 (2018)
4. Fang, Y., Zhang, Y.: Call admission control schemes and performance analysis in wireless mobile networks. IEEE Trans. Veh. Technol. **51**, 371–382 (2002)
5. Sarwar, S., Van As, H.R., Aziz, K.: On a cut-off threshold delay-loss system with n types of traffic for overload protection. IEEE Access **52**, 21–27 (2007)
6. Charabi, L., Gharbi, N., Ben-Othman, J., Mokdad, L.: Call admission control in small cell networks with retrials and guard channels. IEEE Access **2**, 70–75 (2016)

7. Khdhir, R., Mnif, K., Belghith, A., Kamoun, L.: An efficient call admission control scheme for LTE and LTE-a networks. IEEE Access **5**, 981–987 (2016)
8. Alqahtani, S.A.: Delay aware and users categorizing-based call admission control for multi-services LTE-a networks. Arab. J. Sci. Eng. **41**, 3631–3644 (2016)
9. Mohammed, A.O.G., Alroufai, M.A., Khalil, M.H., Ahmed, T.A.: Performance evaluation of call admission control strategies for LTE. Research Submitted in Partial fulfillment for the Requirements of the Degree of B.Sc. (Honors) in Electronics Engineering, pp. 1–45 (2016)
10. Satyanarayana, S.K., Satyanarayana, Ch., Raju, V.S.G.N.: Call admission control algorithm for wireless multimedia networks. Int. J. Eng. Res. **3**, 30–35 (2014)
11. Ben Ali, K., Obaidat, M.S., Zarai, F., Kamoun, L.: Markov model-based adaptive CAC scheme for 3GPP LTE femtocell networks. In: IEEE International Conference on Communication (ICC), vol. 5, pp. 6924–6928 (2015)
12. Alsamhi, H., Rajput, N.S.: Implementation of call admission control technique in HAP for enhanced QoS in wireless network deployment. Telecommun. Syst. **6**, 141–151 (2016)
13. Hassan, A.S.A., Abdalla, A.G.E., Mustafa, A.B.A.N., Khider, I.: Call duration-based bandwidth adaptation for zero blocking in wireless networks. In: International Conference on Computer, Control, Electrical, and Electronics Engineering (ICCCEEE), pp. 1–6 (2018)
14. Al Qahtani, S.A., Ahmed, N.M.: Call admission control for two-tier LTE macro/pico cellular networks. In: Proceedings of the World Congress on Engineering and Computer Science, vol. 2, pp. 20–25 (2015)
15. Belghith, A., Turki, N., Cousin, B., Obaidat, M.S.: Flexible call admission control with preemption in LTE networks. In: IEEE ICC-Communications Software, Services and Multimedia Applications Symposium, vol. 3, pp. 101–108 (2015)
16. Halabian, H., Rengaraju, P., Lung, C.H., Lambadaris, I.: A reservation-based call admission control scheme and system modelling in 4G vehicular networks. EURASIP J. Wirel. Commun. Netw. **6**, 125 (2015)
17. Omitola, O., Srivastava, V.M.: A channel borrowing CAC scheme in two-tier LTE/LTE-advance networks. In: International Conference on Advanced Computing and Communication Systems ICACCS-2017, pp. 1–5 (2017)
18. Chavan, T.A., Saras, P.: A Novel Call Admission Control Algorithm for next generation wireless mobile communication. Int. J. Rough Sets Data Anal. (IJRSDA) **4**, 981–993 (2018)
19. Ghosh, S., Konar, A.: Call Admission Control in Mobile Cellular Networks. Springer, Studies in Computational Intelligence, preface (2013)

Statistical Analysis and Modeling of Global Innovation Index (GII) of Serbia

Predrag Dašić[1,2(✉)], Jovan Dašić[2], Dejan Antanasković[3], and Nina Pavićević[3]

[1] High Technical Mechanical School of Professional Studies, Street Radoja Krstića 19, 37240 Trstenik, Serbia
dasicp58@gmail.com

[2] SaTCIP Publisher Ltd., Vrnjačka Banja, Serbia

[3] Faculty of Management Zaječar (FMZ), Megatrend University of Belgrade, 19000 Zaječar, Serbia

Abstract. The global innovation index (GII) is an indicator which annual ranking of countries by their capacity and success in innovation and innovative activities and published annually from 2007. According to the GII index, for the observed period 2009–2019, Serbia was in quartile Q2, except in 2009–2010 and 2016 when it was in Q3 and Q4. In the paper is given trend analysis and approximation of Serbia's global innovation index (GII) for period 2011–2019, Data about GII index is adequately approximated with 5th-degree polynomial regression model (PRM5) with R = 0.92085, R^2 = 0.84797 and $AdjR^2$ = 0.59459.

Keywords: Global innovation index (GII) · Statistical analysis · Polynomial regression model (PRM)

1 Introduction

The Global Innovation Index (GII) is an indicator which annual ranking of countries by their capacity and success in innovation and innovative activities [1–9]. The GII was calculated based on a large number of indicators, grouped into two sub-indices: Innovation Input Sub-Index (IISI) and Innovation Output Sub-Index (IOSI). The final value of GII index is gained by the simple average of obtained sub-index scores.

The GII index has been launched in 2007 by INSEAD (in French: Institut Européen d'Administration des Affaires - European Institute of Business Administration) and World Business (British magazine), with calculated values of GII index in the interval between 1 and 7. Three editions were published with that methodology, and that: 2007, 2008–2009 and 2009–2010. Editions from 2011 to 2019 are published by Cornell University, INSEAD and the World Intellectual Property Organization (WIPO), with calculated values of GII index in the interval between 0 and 100.

The GII 2019 edition [9] ranks 129 countries based on 80 indicators, with values of GII index which are located in intervals from 67.24 (Switzerland) to 14.49 (Yemen). Switzerland is the world's most-innovative country according to the GII 2011 to GII 2019 editions.

I. Karabegović (Ed.): NT 2020, LNNS 128, pp. 515–521, 2020.
https://doi.org/10.1007/978-3-030-46817-0_59

GII editions are available on the Web site: https://www.globalinnovationindex.org/ and Web pages the WIPO: https://www.wipo.int/global_innovation_index/en/ and the INSEAD: https://www.insead.edu/.

In papers [10–22] values for the GII index for different years and countries have been analyzed.

2 Data and Methods

Data on values of global innovation index (GII) for Serbia has been retrieved from GII editions for period 2009–2019 [1–9], with a certain calculations from the part of the authors. For the trend analysis, we used the following parameters: annual growth rate (AGR) and cumulative growth index (CGI), described in the papers [23–27].

For the calculation of descriptive statistics parameters and approximation of data about GII index for Serbia we used software package CoPRMTV [28], which is described in the papers [23–26].

3 Results and Discussions

On Table 1 are given data about global innovation index (GII) for Serbia for the period 2009–2019 [1–9], with calculation for position [%], Q (Quartile), AG (Annual Growth), CI (Chain Index), AGR (Annual Growth Rate) and CGI (Cumulative Growth Index) and CAGR (Compound Annual Growth Rate) by the authors. The Serbia are not included in the GII rankings for 2007.

Table 1. Data about global innovation index (GII) for Serbia for the period 2009–2019

Year	Score	Rank	Position [%]	Q	AG	CI	AGR	CGI
2009	2.57	91/130	70.00	Q3				
2010	2.68	101/132	76.52	Q4				
2011	36.31	55/125	44.00	Q2	–	–	–	100.00
2012	40.00	46/141	32.62	Q2	3.69	110.16	10.16	110.16
2013	37.87	54/142	38.03	Q2	−2.13	94.68	−5.33	104.30
2014	35.89	67/143	46.85	Q2	−1.98	94.77	−5.23	98.84
2015	36.47	63/141	44.68	Q2	0.58	101.62	1.62	100.44
2016	33.75	65/128	50.78	Q3	−2.72	92.54	−7.46	92.95
2017	35.34	62/127	48.82	Q2	1.59	104.71	4.71	97.33
2018	35.46	55/126	43.65	Q2	0.12	100.34	0.34	97.66
2019	35.71	57/129	44.19	Q2	0.25	100.71	0.71	98.35

Trend analysis for global innovation index (GII) with annual growth rate (AGR) for the period 2011–2019 is shown on Fig. 1. Highest growth of annual growth rate (AGR) was by 10.16 in 2012 year, while the highest fall was by −7.46 in 2016.

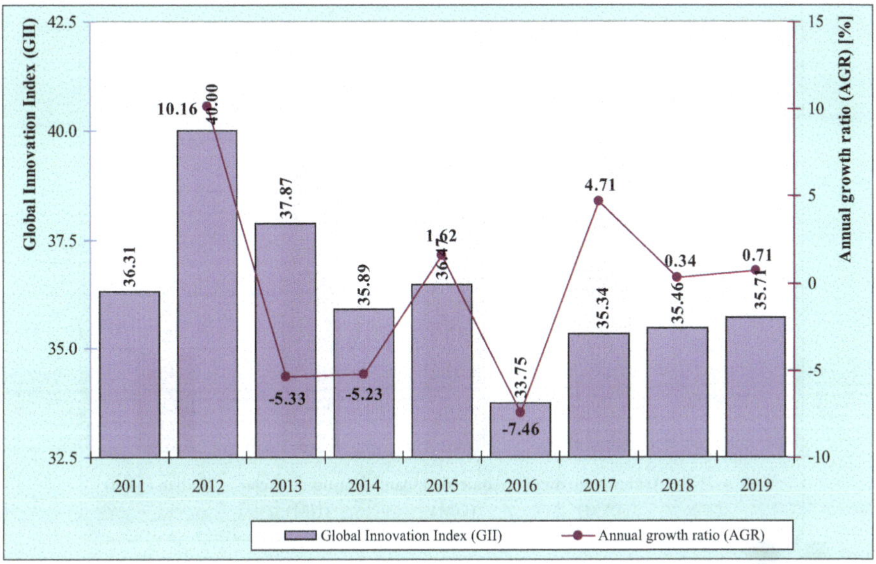

Fig. 1. Trend analysis of global innovation index (GII) with annual growth rate (AGR) for Serbia for the period 2011–2019

According to the GII index, for the observed period 2009–2019, Serbia was in quartile Q2, except in 2009–2010 and 2016 when it was in Q3 and Q4.

The data about global innovation index (GII) for Serbia for the period 2011–2019 changed in intervals from 33.75–40. The lowest value of GII index in Serbia was in 2016 (GII = 33.75) and the highest in 2012 (GII = 40), and values for the GII index varied around arithmentic mean (AM = 36.3111). Geometric mean is GM = 36.2740, harmonic mean is HM = 36.2378 and median is Med = 35.8900 (Fig. 2).

Standard deviation is SD = 1.7617, variance is Var = 3.1036, coefficient of variation is CoV = 4.8517, skewness is Skew = 1.0396 and kurtosis is Kurt = 1.9825.

Due to a large increased in the value of the GII index in 2012 (AGR = 10.16 [%]) and large decrease in the value of the GII index in 2016 (AGR = −7.46 [%]), the GII index is small decreased by −1.65 [%] (CGI = 98.35 [%] and CAGR = −0.2081 [%]) in 2019 compared to 2011.

The data about global innovation index (GII) for Serbia for observed period 2011–2019 can be approximated using 5th-degree polynomial regression model (PRM5) which has the form (Fig. 3):

$$GII = 22.37667 + 22.44629 \times y - 10.21745 \times y^2 + 1.97290 \times y^3 - 0.1742 \times y^4 + 0.00582 \times y^5 \quad (1)$$

where:

y - year and

GII - Global innovation index.

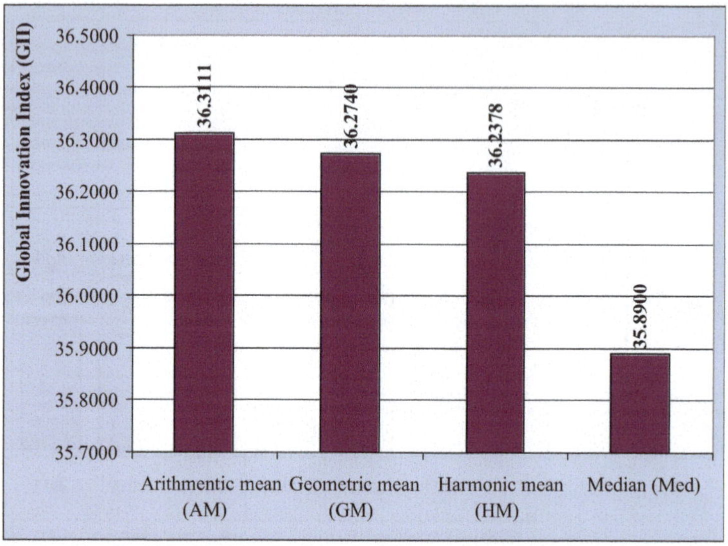

Fig. 2. Graphical representation of indicators of central tendency of GII index of Serbia for the period 2011–2019

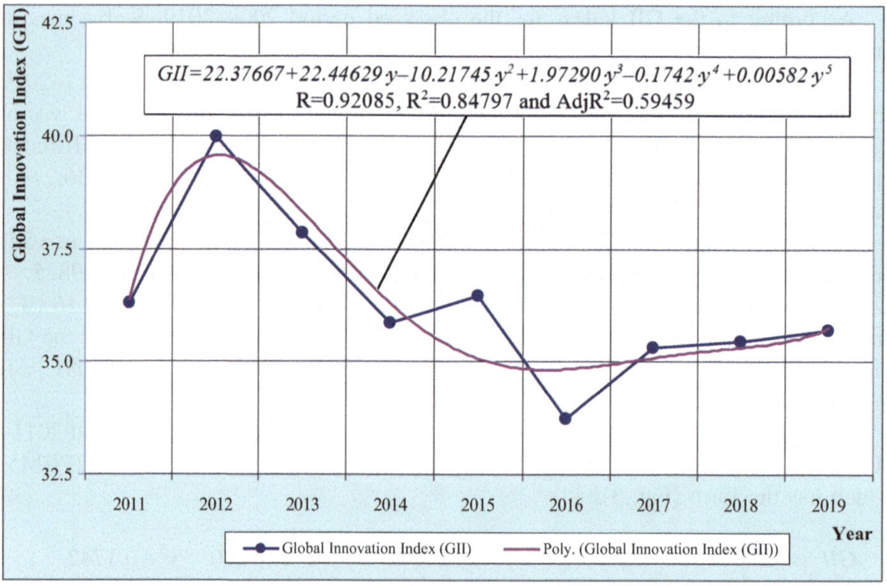

Fig. 3. Graphical representation of data of GII index and prediction value of PRM5

For the adopted 5th-degree polynomial regression model (PRM5): coefficient of correlation is R = 0.92085, coefficient of determination is $R^2 = 0.84797$ and adjustment coefficient of determination is $AdjR^2 = 0.59459$.

4 Conclusion

The data about GII index for Serbia for the period 2011–2019 changed in intervals from 33.75–40. Indicators of central tendency of GII index of Serbia for the period 2011–2019 are: AM = 36.3111, GM = 36.2740, HM = 36.2378 and Med = 35.8900.

Due to a large increased in the value of the GII index in 2012 (AGR = 10.16 [%]) and large decrease in the value of the GII index in 2016 (AGR = −7.46 [%]), the GII index is small decreased by −1.65 [%] (CGI = 98.35 [%] and CAGR = −0.2081 [%]) in 2019 compared to 2011.

Data about GII index for Serbia for the period 2011–2019 was adequately approximated with 5th-degree Polynomial Regression Model (PRM5) based on the specified values of the R = 0.92085, $R^2 = 0.84797$ and $AdjR^2 = 0.59459$.

References

1. Dutta, S. (ed.) The Global Innovation Index 2011: Accelerating growth and development. World Intellectual Property Organization (WIPO), Geneva, Switzerland and Institut Européen d'Administration des Affaires (INSEAD), p. 358. Fontainebleau, France (2011)
2. Dutta, S. (ed.) The Global Innovation Index 2012: Stronger innovation linkages for global growth. World Intellectual Property Organization (WIPO), Geneva, Switzerland and Institut Européen d'Administration des Affaires (INSEAD), p. 440. Fontainebleau, France (2012)
3. Dutta, S., Lanvin, B. (eds.) The Global Innovation Index 2013: The local dynamics of innovation. World Intellectual Property Organization (WIPO), Geneva, Switzerland and Institut Européen d'Administration des Affaires (INSEAD), p. 392. Fontainebleau, France (2013)
4. Dutta, S., Lanvin, B., Wunsch-Vincent, S. (eds.) The Global Innovation Index 2014: The human factor in innovation. World Intellectual Property Organization (WIPO), Geneva, Switzerland and Institut Européen d'Administration des Affaires (INSEAD), p. 400. Fontainebleau, France (2014)
5. Dutta, S., Lanvin, B., Wunsch-Vincent, S. (eds.) The Global Innovation Index 2015: Effective innovation policies for development. World Intellectual Property Organization (WIPO), Geneva, Switzerland and Institut Européen d'Administration des Affaires (INSEAD), p. 418. Fontainebleau, France (2015)
6. Dutta, S., Lanvin, B., Wunsch-Vincent, S. (eds.) The Global Innovation Index 2016: Winning with global innovation. World Intellectual Property Organization (WIPO), p. 422. Geneva (2016)
7. Dutta, S., Lanvin, B., Wunsch-Vincent, S. (eds.) The Global Innovation Index 2017: Innovation feeding the world. World Intellectual Property Organization (WIPO), p. 432. Geneva (2017)
8. Dutta, S., Lanvin, B., Wunsch-Vincent, S. (eds.) Global Innovation Index 2018: Energizing the world with innovation. World Intellectual Property Organization (WIPO), p. 384. Geneva (2018)

9. Dutta, S., Lanvin, B., Wunsch-Vincent, S. (eds.) The Global Innovation Index 2019: Creating healthy lives - The future of medical innovation. World Intellectual Property Organization (WIPO), p. 399. Geneva (2019)
10. Al-Sudairi, M., Bakry, S.H.: Knowledge issues in the global innovation index: Assessment of the state of Saudi Arabia versus countries with distinct development. Innovation: Management, Policy and Practice, Vol. 16, Issue 2 (2014), pp. 176–183. ISSN 1447-9338. https://doi.org/10.1080/14479338.2014.11081980
11. Crespo, N.F., Crespo, C.F.: Global innovation index: moving beyond the absolute value of ranking with a fuzzy-set analysis. J. Bus. Res. **69**(11), 5265–5271 (2016)
12. Hamidi, S., Berrado, A.: Segmentation of innovation determinants: case of the global innovation index. In: Proceedings of the 12th International Conference on Intelligent Systems: Theories and Applications (SITA 2018), Rabat, Morocco; 24–25 October 2018. Association for Computing Machinery (ACM), New York, USA, Article no. 48: pp. 1–8 (2018)
13. Iqbal, B.A., Rahman, M.N.: Chapter 13: The economics and econometrics of global innovation index. In: Handbook of Research on Strategic Innovation Management for Improved Competitive Advantage. IGI Global, Hershey, Pennsylvania, USA, pp. 236–246 (2018)
14. Jankowska, B., Matysek-Jędrych, A., Mroczek-Dabrowska, K.: Efficiency of national innovation systems - Poland and Bulgaria in the context of the global innovation index. Comparative Econ. Res. **20**(3), 77–94 (2017)
15. Morgado, E.M.: Brazil in the global innovation index 2012: Deepening the understanding of the results. Sistemas, Cibernética e Informática - Systemics, Cybernetics and Informatics (SC&I), 10(2), 33–37 (2013). (In Portuguese)
16. Oveshnikova, L., Lebedinskaya, O., Timofeev, A., Mikheykina, L., Sibirskaya, E., Lula, P.: Studying the sector of the Russian high-tech innovations on the basis of the global innovation index INSEAD. Adv. Intell. Syst. Comput. (AISC) **726**, 87–96 (2019)
17. Pençe, I., Kalkan, A., Çeşmeli, M.Ş.: Estimation of the country ranking scores on the global innovation index 2016 using the artificial neural network method. Int. J. Innov. Technol. Manage. **16**(4), 19400078 (2019)
18. Reis, D.A., De Moura, F.R., De Aragão Gomes, I.M.: The linkage between input and output in the global innovation index. In: Proceedings of the 13th European Conference on Innovation and Entrepreneurship (ECIE-2018); Aveiro, Portugal; 20–21 September 2018. Academic Conferences and Publishing International Ltd. (ACPI), Reading, United Kingdom, pp. 635–642 (2018)
19. Sohn, S.Y., Kim, D.H., Jeon, S.Y.: Re-evaluation of global innovation index based on a structural equation model. Technol. Anal. Strateg. Manage. **28**(4), 492–505 (2016)
20. Stavbunik, Y., Pelucha, M.: The innovation system of Kazakhstan in the context of the global innovation index. Innov. J. Public Sector Innov. J. (TIJ) **24**(1), 1–24 (2019)
21. Vlasova, V., Kuznetsova, T., Roud, V.: Drivers and limitations of Russia's development based on the evidence provided by the global innovation index. Voprosy Ekonomiki **22**(8), 24–41 (2017)
22. Davidson, N., Mariev, O., Pushkarev, A.: The impact of externalities on the innovation activity of Russian firms. Foresight and STI Governance **12**(3), 62–72 (2018)
23. Dašić, P.: Application of polynomial regression models for approximation of time series. J. Econ. Manage. Based on New Technol. **1**(2), 81–160 (2012)
24. Dašić, P.: Statistical analysis and modeling of global competitiveness index (GCI) of Serbia. J. Res. Dev. Mech. Ind. **10**(4) (2018)

25. Stanivuk, T., Dašić, P., Aščić, A.: Approximation of global competitiveness index (GCI) for Croatia using polynomial regression model. In: Proceedings of the 7th International Conference "Economics and Management-Based on New Technologies" (EMoNT-2017); Vrnjačka Banja, Serbia; 15–18 June 2017. pp. 22–31. SaTCIP Publisher Ltd., Vrnjačka Banja (2017)
26. Tošović, R., Dašić, P., Ristović, I.: Sustainable use of metallic mineral resources of Serbia from an environmental perspective. Environ. Eng. Manage. J. **15**(9), 2075–2084 (2016)
27. Turmanidze, R., Dašić, P., Popkhadze, G.: Analysis of some techno-economical indicators of Georgia. J. Res. Dev. Mech. Ind. **11**(4) (2019)
28. Dašić, P.: Software package CoPRMTV: Choice of Polinomial Regression Model Between Two Variables. ver. 2.0. SaTCIP Publisher Ltd., Vrnjačka Banja (2010)

AI Enhanced Services in Person-Centred Care in Neurology

Ivana Ognjanović(✉) and Ramo Šendelj

University of Donja Gorica, Oktoih 1, 81000 Podgorica, Montenegro
ivana.ognjanovic@udg.edu.me

Abstract. The Person-centred care (PCC) is an innovative approach that sees patients as equal partners in the planning, development and delivery of care, and active participants in the management of their health and wellbeing. Preliminary research over PCC applications showed advances in concordance between care provider and patient on treatment plans, improved health outcomes and increased patient satisfaction. On the other side, AI has promising potentials to support and enhance PCC services and improve their functioning. This paper presented results of preliminary research aimed on identification of AI enhanced services applied in PCC settings within neurology department, with special focus on stroke patients and their rehabilitation process.

Keywords: Artificial intelligence · Person-centred care · Neurology

1 Introduction

The person-centred care (PCC) as innovative approach in health care has been shown to advance concordance between care provider and patient on treatment plans, improve health outcomes and increase patient satisfaction [1]. The core concept behind PCC is putting the '*person before the disease- see the person with the disease not just the disease*' thus ensuring patients and professionals acting as partners in the care process. Yet, despite these and other documented benefits, there are a variety of significant challenges to putting PCC into clinical practice, naming just a few: (i) *patient narratives*- person's views about his/her life situation and condition, his/her feelings, beliefs and preferences, all linked with biological markers or images, as well as with cultural, educational and behavioural settings; (ii) *shared decision making*- creation of extensive illness management between professionals and patients (often including relatives); (iii) *consistent evaluation framework*- evaluate the outcome and assess the benefits for all (patients, professionals, health care system).

The innovative PCC approach is directly linked to existing healthcare system (it's efficiency, cost and quality of care indicators, data integration, etc.) and concrete health issues. Since the recent issues of modern medicine is to prevent death with focused on increased frequency and severity of impairment attributable to neurological disorders [2], special attention by both physicians and researchers is paid to reduce neurological disorders and support patients in rehabilitation. On the other side, artificial intelligence and machine learning have proven capabilities to explore massive amount of medical data (coming from different sources) and deliver automatic medical diagnostics. It is

I. Karabegović (Ed.): NT 2020, LNNS 128, pp. 522–529, 2020.
https://doi.org/10.1007/978-3-030-46817-0_60

not about technology itself; it is about utilizing technology for treating the patients as humans and adopting their benefits in PCC settings. That is a reason why we are focused on applying PCC in neurosurgery and developing AI enhanced support for increasing the health care in neurosurgery with simultaneous reduction of costs (which is in neurosurgery of high demand due to the nature of interventions).

The paper is organised as follows: Sect. 2 presents key challenges and issues in PCC development, while Sect. 3 gives potentials of using AI to address identified issues. Finally, Sect. 4 introduces innovative methodological approach for development of AI enhanced services in neurology settings, while Sect. 5 concludes paper with directions for further development.

2 Key Challenges in PCC Development

'Person-centred care' (PCC) is an approach that sees patients as equal partners in the planning, development and delivery of care, and active participants in the management of their health and wellbeing. First evidences about the concept in theory could be found in late 2009, after the swine influenza pandemic drama which raised questions about required reforms of already overburdened health care systems [1]. As an alternative, a PCC is advocated as a key concept for effective illness management with proven unique impacts of reduction of costs with simultaneous increase of health care quality and patients' satisfaction. The journey of scientific community towards creation of PCC concept can be described with the following key points: (i) WE-CAREproject (www.we-do-care.eu) gathered together several European research institutes, businesses and non-profit organisations to create a strategic plan and a roadmap for how research and innovation within healthcare can help reduce healthcare costs; (ii) ongoing COST 15222 project '*European Network for cost containment and improved quality of health care*' (https://costcares.eu) is focused to support scientific and technology development necessary for large-scale implementation of PCC ensuring support to EU countries to reduce their costs which is recognised as a major societal problem (EU wide 1.400 billion euros/year).

The PCC paradigm has many advantages in compare with traditional Patient-centred care. Person-centred care is a way of thinking and doing things that sees the people using health and social services as equal partners in planning, developing and monitoring care to make sure it meets their needs. This means putting people and their families at the centre of decisions and seeing them as experts, working alongside professionals to get the best outcome. Person-centred care is not just about giving people whatever they want or providing information. It is about considering people's desires, values, family situations, social circumstances and lifestyles; seeing the person as an individual, and working together to develop appropriate solutions.

From the implementation perspective, few countries like Sweden, USA, UK [3] have started with national focuses and actions aimed on implementation of PCC into health practice. Implementation frameworks are focused on establishing routines that initiate, integrate and safeguard PCC in daily practice through implementation of frameworks [1, 3] for: (i) developing partnership with patients, (ii) making shared decision making with patients and professionals, and (iii) consistent evaluation frameworks. The Gothenburg University Centre for Person Centred Care (GPCC) which has been involved in individualized patient care for almost a decade, with noticeably improved outcomes for participating patients, conducted research studies [4] on PCC which showed: (i) When chronic heart failure patients receive person-centred care during their hospital stay, they feel less uncertain about their condition and treatment. In addition, the length of hospital stays was reduced by a third; (ii) The person-centred care after acute coronary heart syndrome (infarction, instable angina) significantly improves patients' self-efficacy compared to the control group. In this randomised, controlled study the patients receiving person-centred care also had almost three times greater chance of recovery than those who did not. In addition, findings from this study demonstrated that person-centred care improved the cohesiveness of transitional care across the "care chain" for the individual patient; i.e. the transition from one care provider to another and/or to home; (iii) Still another example of what can be achieved with a person-centred approach is shown in a study of older patients with a hip fracture. This study showed:(1) significantly reduced pain compared to the control group; (2) significant reduction of pressure wounds and other medicinal complications compared to traditional (or standard) care; (3) a 50% reduction of the length of hospital stay compared to traditional care; (4) a 40% reduction in care costs compared to traditional care.

3 Potentials of Using AI in PCC

However, to the best of our knowledge, there is an absence of using intelligent processing and reasoning of massive amount of medical and health data and (semi-) automatic support in PCC, even having proven potentials to be applied for medical diagnosis and treatments [5], automatic decision-making support [6], as well as collection and sharing of data about person's health indicators (i.e. Internet-of-Things concept: sensors and other devices [7]). Furthermore, use of AI methods and techniques still rise several challenges mainly characterized as: (i) development of data-driven approach and integration of data with different meaning and roles, from different sources, and (ii) ethical and security issues due to compliance with recent standards and regulations at EU and global level

U.S. Department of Health and Human Services (HHS), with support from the Robert Wood Johnson Foundation, did research [8] to consider how AI will shape the future of public health, community health, and health care delivery. They focused on technical capabilities, limitations, and applications that can be realized within the next ten years. Their recommendation can be shortly presented as:

1. *AI Applications in Clinical Practice-* Support work to prepare AI results for the rigorous approval procedures needed for acceptance for clinical practice. Create testing and validation approaches for AI algorithms to evaluate performance of the algorithms under conditions that differ from the training set.
2. *Confluence of AI and Smart Devices for Monitoring Health and Disease*: support the development of AI applications that can enhance the performance of new mobile monitoring devices and apps; develop data infrastructure to capture and

integrate data generated from smart devices to support AI applications; track developments in foreign health care systems, looking for useful technologies and also technology failures.

3. *Create Comprehensive Training Databases of Health Data for AI Tool Development*: support the development of and access to research databases of labelled and un-labelled health data for the development of AI applications in health; support the assessment of AI algorithms trained with data labelled at levels that significantly exceed standard assessment, for instance the use of outputs from the next stage of diagnostics; support research to characterize the trade-offs between data quality, information content (complexity and diversity) and sample size, with the goal of enabling quantitative prediction of the quantity and quality of data needed to support a given AI application.

4 Innovative Approach in PCC Neurology Lab

On the bases of conducted comprehensive review of the literature, the following potentials of AI use in PCC domain are identified:

(i) **Domain-specific knowledge representation** with evidences from different resources, different representations and roles. The *semantically-enhanced families of services-oriented architectures* [9] approach which is originally routed in modern software engineering with proven potentials in different areas (including learning analytics, cyber security over clouds, etc.) is expected to provide a framework for: (i) semantically enhanced representation of domain knowledge in both neurology and patients as persons in general, (ii) representation of requirements and preferences defined either by patients (in the form patient's attitudes, expectations, and preferences) or by professionals (as proven knowledge in neurosurgery and/or hidden knowledge about clinical routines), (iii) decision making approach implemented in the form of well-known algorithms (e.g. AHP [10] TCP nets [11], cp-nets [12] etc.) adopted based on created semantic structure for knowledge representation.

(ii) AI use for **exploring data (data centric)** to characterize problems and propose solutions- Despite the increasing availability of medical data from different sources, most of the applications of AI in the medical field are still mono-modality (utilizing one type of data at the time). Thus, our approach is routed in exploring the potential of AI applications over heterogeneous data about patients and capital knowledge in neurosurgery—with the associated benefits in PCC of increased accuracy for diagnostics, therapy planning and increased patient's well-being. To this end, data coming from different external devices will be analysed (e.g. devices for monitoring physical activities, etc.)

In parallel to the development of AI, digitalization of healthcare has experienced a tremendous development. However, access and exploitation of both personal and medical data is considered a difficult endeavour, due to its sensitivity, particularly for institutions not directly linked to the healthcare sector. Therefore, our approach is also directed to facilitate collection, storage and exploitation of clinical data with emphasis on compliance with the new EU GDPR. Furthermore, **ethical issues** are of key importance due to PCC focus on patients and persons, thus providing data about each individual (directly or indirectly by using different devices for monitoring health indicators, habits, etc.). **Cybersecurity** is a key issue in the collection, storage and sharing of medical data both on the cloud and on individual computers. Machine learning provides new solutions to address the challenge of cybersecurity of medical data collections.

Summary overview of our innovative approach is given in the following Table 1 with illustrated examples in one of neurological disease- stroke and stroke rehabilitation process. The Table 1 is created by making a mapping with one of standard questionnaires used for patients who had a stroke (given in Table 2).

Table 1. Summary overview of AI techniques applied in PCC applied in Neurology department for stroke rehabilitation

Methods/techniques	Potentials to be applied in PCC	Examples in neurology (linked to Table 2)
Feature modelling [13]	Features: key elements/subjects/occurrences Concerns and qualifier tags: characteristics and potential values	Features: lifestyle; stroke; treatment; rehabilitation Concerns and qualifier tags: vascular risk (question 8), Place of residence before and after (questions 2, 13), treatment (questions 4-9); rehabilitation (questions 11-13)
Semantically enhanced architecture [9]	Process/workflow	Hospitalization (question 1); treatment (questions 4-9); modified ranking score before and after stroke (questions 3, 14); rehabilitation and monitoring the effects (questions 11-13)
Algorithms for intelligent reasoning	Categorization of patients and identification of hidden knowledge in medical records and other available data	All questions merged with medical records of the patients
Recommendation systems [6]	Rehabilitation process tailored to each specific patient aimed on maximization of treatment	Use of mobile apps with integrated communication with sensors/image analysis etc. aimed on measuring improvements and effects of each specific (question 11): • Working therapy • Physical therapy • Speaking exercises • Swallowing exercises

Table 2. Questions from standard questionnaires for stroke patients

Question	Answers	Question	Answers
1. Department of hospitalization	• Intensive care • Cardiovascular • Neurology • Neurosurgery • Rehabilitation • Other	8. Which of the following vascular risk factors do patient have?	• Atrial fibrillation • Smoking • Diabetes mellitus • Hypercholesterolemia • Hypertension
2. Patients' place of residence before stroke:	• Own household, independent from others • Own household, dependent from others • Stationary institution	9. Do patient receive any of the following medications during the hospital stay:	• Anticoagulant drugs • Antiplatelet drugs • Thrombolysis • Other
3. Modified ranking score before stroke:	• Without symptoms • Without significant disability despite symptoms • Minor disability • Moderate disability, but able to walk independency • Moderate disability, not able to walk independency • Major disability • Unknown	10. Do patient receive any of the following medications after hospitalization:	• Anticoagulant drugs • Antiplatelet drugs • Antidiabetic drugs • Lipid-lowering drugs • Antihypertensive drugs • Other
4. What neurological symptoms do patient have, immediately after hospitalization?	• Impaired consciousness • Weaknesses/paresis • Aphasia (speech disorder)	11. Which additional procedures have been processed during hospitalization?	• Working therapy • Physical therapy • Speaking exercises • Swallowing exercises
5. Stroke type:	• Ischemic • ICH • SAH • Unspecified	12. Which of the following complications have been occurred during hospitalization?	• Deep vein thrombosis • Other CV complications • Pneumonia
6. Which of listed diagnostic methods have been used?	• CT • MR • UZ carotid arteries • Lumbar puncture (NHS) • Angiography • EKG • Autopsy • Other	13. If patient is alive, write his next destination:	• Home • Other hospital/stationary institution • Nursing home • Other
7. How many hours have passed from stroke to the first CT scan:	(1) up to 24 h (2) 24 h – 7 days (3) 8 – 14 days (4) more than 14 days (5) CT scan note done (6) Other	14. If patient is alive, modified ranking score on the date of discharge of hospital was:	• Without symptoms • Without significant disability despite symptoms • Minor disability • Moderate disability, but able to walk independency • Moderate disability, not able to walk independency • Major disability • Unknown

Presented innovated mapping gives a framework for development and implementation of PCC services in neurology department, which practical use is expected to provide additional analysis and contributions to understanding benefits and implications of PCC concept [4] and technological support.

5 Conclusion

This paper presented results of preliminary research aimed on identification of AI enhanced services applied in PCC settings within neurology department, with special focus on stroke patients and their rehabilitation process. However, it clearly identified which specific elements related to stroke rehabilitation can be integrated within AI enhanced services, all aimed on developing services tailored to each specific person and aimed on maximization of improvements of his/her health conditions. The future work will be on piloting of aforementioned services, their testing and evaluation with special focus on monitoring patients' behaviours and their health indicators.

Acknowledgement. Presented research is conducted within National Scientific Research project "AI Enhanced Person-Centred Care in Neurosurgery (AI4NeuroPCC)", funded by Ministry of Science Montenegro (2019–2021).

References

1. Ekman, I., Swedberg, K., Taft, C., Lindseth, A., Norberg, A., Brink, E., Carlsson, J., Dahlin-Ivanoff, S., Johansson, I., Kjellgren, K., Lidén, E., Öhlén, J., Olsson, L., Rosén, H., Rydmark, M., Sunnerhagen, K.: Person-centred care- ready for prime time. Eur. J. Cardiovasc. Nurs. **10**(4), 248–251 (2011)
2. WHO: Public Health Principles and Neurological Disorders, WHO (2006)
3. Lloyd, H.M., Pearson, M., Sheaff, R., Asthana, S., Wheat, H., Sugavanam, T.P., Britten, N., Valderas, J., Bainbridge, M., Witts, L., Westlake, D., Horrell, J., Byng, R.: Collaborative action for person-centred coordinated care (P3C): an approach to support the development of a comprehensive system-wide solution to fragmented care. Health Res. Pol. Syst. **15**, 98 (2017)
4. Ekman, I.: The GPCC. Eur. Heart J. **39**(7), 495–496 (2018)
5. Dilsizian, S.E., Siegel, E.L.: Artificial intelligence in medicine and cardiac imaging: harnessing big data and advanced computing to provide personalized medical diagnosis and treatment. Curr. Cardiol. Rep. **16**, 441 (2015)
6. Ognjanović, I., Gašević, D., Bagheri, E.: A stratified framework for handling conditional preferences: an extension of the analytic hierarchy process. Exp. syst. Appl. **40**(4), 1094–1115 (2013)
7. Farahani, B., Firouzi, F., Chang, V., Badaroglu, M., Constant, N., Mankodiya, K.: Towards fog-driven IoT eHealth: promises and challenges of IoT in medicine and healthcare. Fut. Gener. Comput. Syst. **78**(2), 659–676 (2018)
8. Artificial Intelligence for Health and Health Care, Office of the National Coordinator for Health Information Technology (ONC) (2017)
9. Šendelj, R., Ognjanović, I.: Multi-criteria decision making for optimal configuration of business process model families. Inf. Technol. Control **47**(3), 532–563 (2018)

10. Saaty, T.L.: The Analytic Hierarchy Process. McGraw-Hill, New York (1980)
11. Domshlak, C., Brafman, R.I.: Introducing variable importance tradeoffs into CP-nets. In: The Eighteenth Conference on Uncertainty in AI (2002)
12. Chen,S., Buffett, S., Fleming, M.W.: Reasoning with conditional preferences across attributes. In: The 20th Conference of the Canadian Society for Computational Studies of Intelligence on Advances in Artifical Intelligence, Montreal, Canada (2007)
13. Nešić, D., Krüger, J., Stănciulescu, Ș., Berger, T.: Principles of feature modeling. In: ESEC/FSE 2019: Proceedings of the 2019 27th ACM Joint Meeting on European Software Engineering Conference and Symposium on the Foundations of Software Engineering (2019)

Montenegrin Dictionary Based Brute Force Attack

Milena Djukanovic[1](✉), Lazar Novicevic[1], and Mihailo Jovanovic[2]

[1] Faculty of Electrical Engineering, University of Montenegro, Dzordza Vasingtona bb, 81 000 Podgorica, Montenegro
milenadj@ucg.ac.me

[2] Faculty of Business and Law, University of Union, Belgrade, Serbia

Abstract. Cyber-attacks and cyber-crimes are growing quickly and users are becoming unsafe on the internet. Cyber threats will put users increasingly at risk. For that reason statistical analysis have been done in order to determine how strong passwords are in the area of Montenegro, and to determine if there is a way to crack them by using brute force attack. Brute force attack or attack by "force" is based on cracking the password with all possible combinations, including cracking Wi-Fi routers, social media account, e-mails, hashes, zip or pdf files protected by password, internet protocols (telnet, ftp, http, https, smb.) etc. We will explain brute force attack for cracking passwords based on statistical analysis data in Montenegro. The first reason we will be doing this is to improve security, password strength, and the conscience of the people in Montenegro. The second reason is to help the people in law enforcement who deal with digital forensics, to find digital evidence in cyber incidents in the shortest amount of time.

Keywords: Brute force attack · Cyber security · Password cracking · Password security · Dictionary attack

1 Introduction

Brute force is a technique used in penetration testing where "force" is used to crack some credentials. This happens when we try to use all possible password combinations to get access to a website, Wi-Fi router, account in social media or similar. Most people use weak passwords that can be brute forced and plaintext can be revealed. One of the most common initial attacks against website or server security is a brute force attack: guessing the username and password. This attack has a high success rate because a lot of people use easy-to-remember passwords. To crack a six character password using an average CPU today takes roughly 20 s [1, 2].

Hybrid brute force attack also known as dictionary attack uses a list of dictionary words trying to crack password. These are one of the most common types of brute force attack. Also interesting to mention is credential stuffing where a username and password pairing is known by the attacker, and this information can be used in order to gain access to multiple websites and network resources. Last, but not least, reverse brute force attack is a type of attack where hacker uses a list of common passwords or

I. Karabegović (Ed.): NT 2020, LNNS 128, pp. 530–536, 2020.
https://doi.org/10.1007/978-3-030-46817-0_61

passwords collected from leakage data against many possible usernames in a network. Table 1 shows us how much time is needed to crack a password [3].

Table 1. Amount of time for cracking password

Length of passwords	Just numbers	Combinations of lower and upper case alphabets	Combinations of numbers, lower and upper case alphabets	Combinations of numbers, lower and upper vase alphabets, symbols
3	Immediately	Immediately	Immediately	Immediately
4	Immediately	Immediately	Immediately	Immediately
5	Immediately	Immediately	5 s	6 s
6	Immediately	10 s	5 min	17 min
7	Immediately	9 min	3 h	19 h
8	Immediately	5 h	12 days	70 days
9	9 s	6 days	173 days	13 years
10	55 s	180 days	2 year	944 years
11	8 min	19 years	102 years	75k years

1.1 Tools for Brute Forcing

Implementation of brute force attack can be done either online either offline with appropriate tools and we can apply it on: wireless protocols, zip and rar files, hash files, internet protocols, Linux and Windows login passwords, social media accounts etc. (Table 2). A lot of these tools are already installed on some versions of Linux operating system [4].

One of the most successful tools for cracking wireless 802.11 is Aircrack-ng. It is used for cracking passwords and takes two parameters, one is.pcap format which contains a handshake packet and the second one is a wordlist with passwords [5]. The best option for cracking zip or rar files, which are also protected with passwords, would be John the Ripper open source tool which can be used both on Linux and Windows

Table 2. Tools for brute force attack

Brute force	
Tools	Applications
Aircrack-ng	Wireless 802.11
John the Ripper	Zip and rar files, hash
Hashcat	Hash
Rainbow table	Hash
Hydra	Internet protocols
Script	Social media

operating system [6]. Another useful tool for cracking login credentials is Hydra, especially against many protocols like ssh, telnet, ftp, http, https, databases etc. [7].

Most known hashes are MD5, Base64, SHA-1, SHA-512, Tiger, MySQL5, Salted MD5 and it should be impossible to produce same hash value. Hashcat is a free open-source program for cracking a hash with different type of attack, such as dictionary, mask, hybrid or combination of mentioned ones [8]. Also, program Rainbow table has generally the same purpose and is available for Windows and Linux OS, but with time memory trade-off algorithm [9]. Additionally, we have a brute force attack using a script usually written in python programing language and it can be used to crack a social media account (Facebook, Instagram, Twitter etc.).

2 Statistical Analysis of Data Regarding Passwords

Montenegro is not yet a target of huge cyber-attacks and for that reason it is not clear do people care enough about cyber security and their personal data. Therefore, our aim is to find out do they pay deserved attention for their cyber safety and do they use strong passwords. In order to determinate how safe residents of Montenegro are in the digital world and how strong and effective are their passwords, we have carried out a digital survey among 200 citizens of Montenegro, aged between 20 and 50 years old [10]. We have assumed that these persons actively and daily use digital data and passwords through students' life, work and hobbies. Analysis of survey's results can tell us more about computer awareness and safety of society in Montenegro when creating passwords for e-mails, internet address and applications, as well as their social networks' profiles. Hopefully, this research will draw attention of examinees and many more people in order to take measures to improve the situation of their general security level.

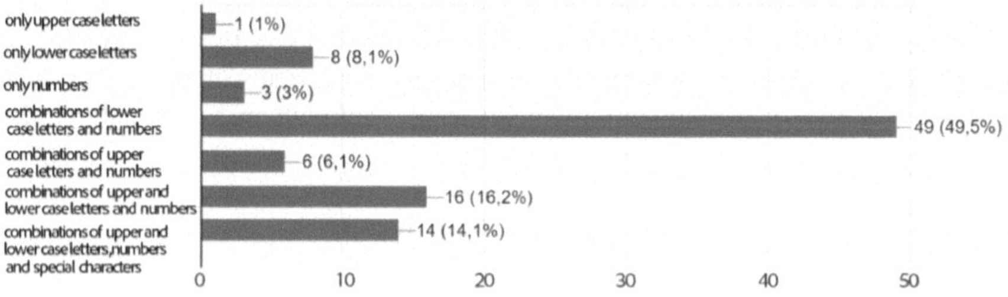

Fig. 1. Content of the used password among examinees

According with results of this analysis, we can conclude that most of them use a combination of lower-case alphabet letters and numbers for their passwords (Fig. 1). Also, average number of characters for their passwords is 8 and 71,6% of examinees are using first name, second name, date of birth or nickname for their passwords which for the brute force attack we claim as keywords, because they are often predictable and they are easily reachable (Fig. 2).

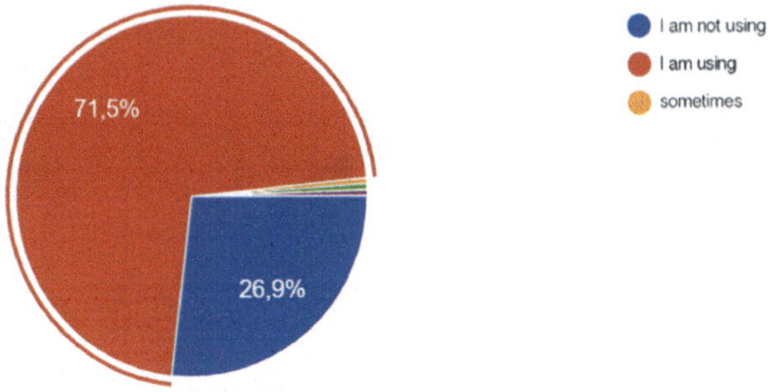

Fig. 2. Percentage of examinees using keywords for their passwords

Examinees from Montenegro also use the name of a hometown or an address as part of their password - 13% of them. Additionally, when asked, do they use a child name, their birthday or nick name, 40% of examinees said yes. They also use the name of a favorite club/sports team, and favorite player (11%). The most commonly used words which people use for their password is their first name, birthday and/or child name. A little over 37% of them save password on web browser and almost 60% use the same password for different accounts. From all the information gathered from this survey we can conclude that people in Montenegro use strong password, but we will be able to crack it with brute force attack for educational purpose or for digital investigation only.

3 Creation of a Montenegrin Dictionary Based Brute Force Attack

The idea for exploiting collected statistical data is to program a special dictionary for a targeted person and apply brute force attack in order to achieve used password. To receive all essential information from the target, we will use passive and active reconnaissance to find first name, last name, date of birth, does he/she have a child/children and similar information. Passive and active reconnaissance makes a very important part of the attack because collected information represent the basis of this attack [11]. A huge amount of this information, such as first and second name, date of birth, hometown, address etc. we are able to search for through social media networks and unfortunately, this sensitive data is available to everyone.

In order to make a „perfect" dictionary, information such as these can help us crack passwords in every segment (Fig. 3). To make a good dictionary the best way is to create python code which will enable us to build dictionary based on results of the survey. If we use data from the survey and understand what are the keywords for a certain target, by using a python code we are able to generate a password list only for that targeted person [12]. The last move is to use a tool or script and dictionary for cracking passwords [13].

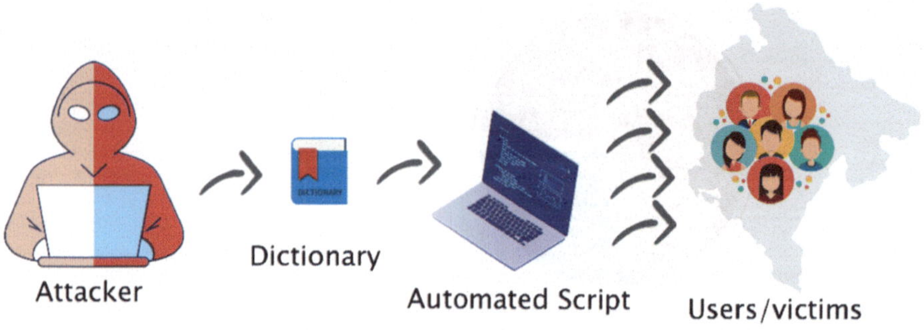

Fig. 3. Montenegrin dictionary-based brute force attack

In general, different dictionary-based attacks are one of the most common attacks using brute force which attempt to defeat an authentication mechanism by systematically entering each word from dictionary as password [14]. These attacks have a high success rate because users tend to select easy-to-remember credentials or reuse credentials, but also because system administrators frequently do not control multiple access attempts. For example, if the length of the password is 6 characters and it is made up of only lower-case alphabets then the number of possible passwords is 26^6 which are 308,915,776 words with all possible combinations [15]. Also, on Internet can be found a huge list of most used English passwords from data leakage which attackers can use for brute force attack [16].

3.1 Special Dictionary List

Difference between using a standard dictionary attack and the one we are proposing is that we use credentials from the survey in order to make a "perfect" list with possible password combinations for a certain target in Montenegro. For example, based on a survey we can notice that most of examined people use combinations of lower case alphabet letters and numbers, as well as combinations of their first name and date of their birthday as passwords. Our special dictionary has to be created by using this list of words. Python programing language is the best option to make an application for mixing words with lower case alphabet letters and numbers as the first condition, and for mixing first name and date of birth of victim as a second condition. Next conditions will be all possible combinations of input words from a certain target. Also it is very important to be aware of the sensitive target information in order to put them into dictionary. Further on, we can notice that examinees aged from 35 to 50 years put their children's name as their passwords, which leaves us with the need of very successful passive and active target recognitions in order to be able to collect enough data for dictionary.

This type of dictionary can be applied on every type of brute force attack. Huge advantage with this type of attack is that it will reduce time of cracking in very high percentage. For example, for a standard dictionary there would be approximately 15000000 possible combinations with most common words in the world, but proposed dictionary will be based in Montenegrin language and will have only possible combinations of password for certain people based on survey which will reduce time of cracking. Based on executed survey, each person who participated in it will have one special dictionary that can be used for the attack

4 Conclusion

A widespread authentication method is the usage of passwords. Despite many warnings and efforts, users still keep using weak passwords providing an entry point to attackers. In this paper we have conducted data statistical analysis in order to determine how strong passwords are in the area of Montenegro among digitally educated people. We have concluded that people in Montenegro need to pay more attention to cyber security, as well as the formulation of their passwords in order not to become a target of cyber and brute force attacks.

Acknowledgment. This paper is prepared within two projects - COST CA17124 "Digital forensics: evidence analysis via intelligent systems and practices" and "Piloting a blockchain-based capital market system in Montenegro", supported by Ministry of Science of Montenegro and Ministry of science and technology of Peoples Republic of China.

References

1. Bošnjak, L., Sreš, J. Brumen, B.: Brute-force and dictionary attack on hashed real-world passwords. In: 41st International Convention on Information and Communication Technology, Electronics and Microelectronics (MIPRO). IEEE (2018)
2. Juels, A., Rivest, R.L.: Honeywords: making password-cracking detectable. In: Proceedings of the 2013 ACM SIGSAC Conference on Computer & Communications Security, pp. 145–160. ACM, November 2013
3. Ma, W., Campbell, J., Tran, D., Kleeman, D.: Password entropy and password quality. In: 2010 Fourth International Conference on Network and System Security, pp. 583–587. IEEE, September 2010
4. Beggs, R.W.: Mastering Kali Linux for Advanced Penetration Testing. Packt Publishing Ltd., Birmingham (2014)
5. d'Otreppe, T.: Aircrack-ng (2019)
6. Solar Designer: John the ripper (2003). http://www.openwall.com/john/
7. Najera-Gutierrez, G., Ansari, J.A.: Web Penetration Testing with Kali Linux: Explore the Methods and Tools of Ethical Hacking with Kali Linux. Packt Publishing Ltd., Birmingham (2018)
8. Chester, J.A.: Analysis of Password Cracking Methods & Applications (2015)
9. Ying, H.M., Thing, V.L.: A novel rainbow table sorting method. In: International Conference on Technical and Legal Aspects of the e-Society (CYBERLAWS), February 2011
10. Wright, K.B.: Researching internet-based populations: advantages and disadvantages of online survey research, online questionnaire authoring software packages, and web survey services. J. Comput.-Mediated Commun. **10**(3), 1034 (2005)
11. Sanghvi, H.P., Dahiya, M.S.: Cyber reconnaissance: an alarm before cyber attack. Int. J. Comput. Appl. **63**(6), 36–38 (2013)

12. Zelle, J.M.: Python Programming: an Introduction to Computer Science. Franklin, Beedle & Associates Inc, Portland (2004)
13. Velu, V.K.: Mastering Kali Linux for Advanced Penetration Testing. Packt Publishing Ltd., Birmingham (2017)
14. Van Tilborg, H.C., Jajodia, S. (eds.): Encyclopedia of Cryptography and Security. Springer, Heidelberg (2014)
15. Chanda, K.: Password security: an analysis of password strengths and vulnerabilities. Int. J. Comput. Netw. Inf. Secur. **8**(7), 23 (2016)
16. Jourdan, P., Stavrou, E.: Towards designing advanced password cracking toolkits: optimizing the password cracking process (2019)

Applications eDaljinar and mDaljinar

Suad Sućeska(✉)

Brčanska 7, 71000 Sarajevo, Bosnia and Herzegovina
suad_s@bih.net.ba

Abstract. Applications eDaljinar and mDaljinar are made on the basis of the document 'DALJINAR with routes and minimum travel times for bus stations and stops in BiH'. Purpose of these applications is to enable work with distance and duration data between stations of passenger transport routes. These data include routes, stations, roads, entities, distances from previous station, distances from start station, drive duration from previous station, drive duration from start station, roads intersections. Web application eDaljinar is made in PHP, and mobile application mDaljinar is made for mobile operating system Android minimal version 21.

Keywords: Travel distances · Web application · Mobile application · Mobile user · Android

1 Introduction

The reference document for determination the distance and duration of drive between any two stations in inter-entity and international passenger transport is the Daljinar [1].

The Daljinar contains officially determined distances and duration of drive between stations of international and inter-entity routes in BiH. Sluggishness of process of updating data, data verification and obsolescence of the current procedure in comparison to modern electronic procedures is the main reason for the development of these applications. The applications, also, use data from afore cited document (Daljinar). Web application eDaljinar should enable to view, input new ones, and change entered data of distance and duration of drive between stations. Viewing data does not require login. To update existing and input new data, users have to be logged in, what means they had to be registered in application before.

Mobile application mDaljinar is made to enable mobile users to obtain up-to-date data regarding distances and duration of drive between stations of routes of BiH passenger transport. It is intended for all interested in the afore given data, and especially for passengers, companies and institutions dealing with passenger transport. It provides current data on routes and distances between stations. Mobility of the application enables to control application data in the field. Any differences between the data in the application and those in the field may be submitted to the application administrators for updating.

To use the application, a smartphone must have the Android OS with minimum version 21.

I. Karabegović (Ed.): NT 2020, LNNS 128, pp. 537–543, 2020.
https://doi.org/10.1007/978-3-030-46817-0_62

2 About Application eDaljinar

Web application eDaljinar is aimed for working with distances and durations of drive between passenger transport stations. These data include: routes, stations, roads, entities, distances from previous station, distances from starting station, drive duration from the previous station, drive duration from starting station, intersections of roads. User registration is optional. Anyone who accesses the application can view certain pages with limited functionality. The administrators enter new data for: routes, stations, roads, entities, distances and drive durations. The application is started on following address: http://www.aplikacije.suads.com.ba/eDaljinar/.

2.1 Description of Work of the Application eDaljinar

Launching of Web application eDaljinar gives starting page shown in Fig. 1. All pages of the application contain the top and bottom menus. The top menu contains buttons: 'Homepage', 'Kontakt' (Contact) and 'Uputstvo' (Manual), and if the user is logged in and button 'Novi korisnik' (New user). The click on the button 'Homepage' gives starting page 'Relacije' (Routes). The button 'Kontakt', which gives the page with the same name, is used to contact the application administrator. To send a message, you must enter: first name, last name, email address and text of the message. The logged in user only enters the message text, while the other fields are entered automatically. The message is sent by click on the button 'Send'.

The click on the button 'Uputstvo' gives you manual for application usage.

After login the top menu also contains a button 'Novi korisnik', which is necessary to click to get a page for creating a new user, where you have to enter the fields: first name, last name, password and email address, and then click on the button 'Create'. The user data will be used just for work with the application eDaljinar, what is compliant with GDPR. Users who do not want to give requested data to use the application do not need to register.

The bottom menu contains buttons: 'Relacije' (Routes), 'Stanice' (Stations) and 'Login'. The click on the button 'Relacije' gives you a page with table of routes. The click on the button 'Stanice' gives a page with tabulated list of stations.

To login to the application, click on the button 'Login'. Enter your email address and password on the page with the same name, then click on the button 'Login'. When user is logged in, this menu gets the button 'Korisnik update' and the button 'Login' is changed into 'Log out'.

The click on the button 'Korisnik update' gives a page with table containing the data of currently logged user, which can be changed using this page.

To log off the application, click on the button 'Log out'.

The starting page of the application is page 'Relacije' with list of routes. This page gives list of routes containing following data: Route ID, Route Index, Route (Start and End Route Station), the button 'Udaljenosti' (Distances). New routes are entered by application administrators. The table with the list of routes of logged user gets new column with button 'Izmjene'. The page is made with the paging. Selection of page is made using page number series above the table of routes by clicking on some page number or using the arrows.

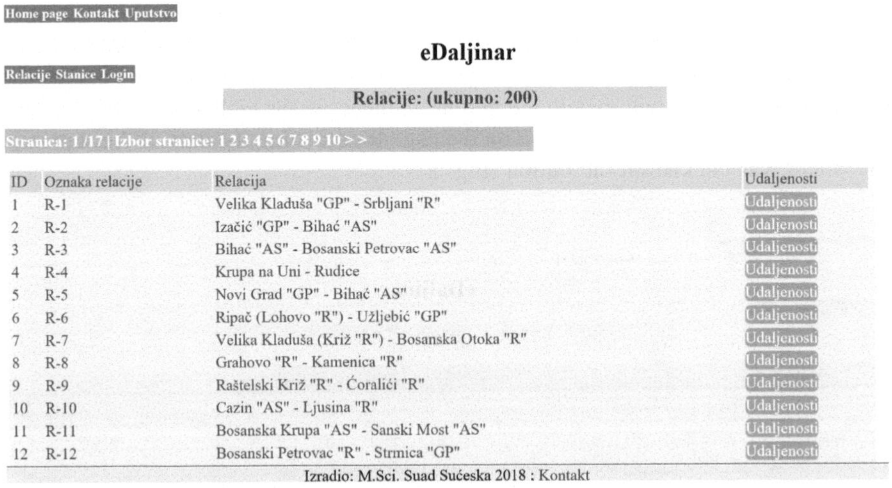

Home page Kontakt Uputstvo

Relacije Stanice Login

eDaljinar

Relacije: (ukupno: 200)

Stranica: 1 /17 | Izbor stranice: 1 2 3 4 5 6 7 8 9 10 > >

ID	Oznaka relacije	Relacija	Udaljenosti
1	R-1	Velika Kladuša "GP" - Srbljani "R"	Udaljenosti
2	R-2	Izačić "GP" - Bihać "AS"	Udaljenosti
3	R-3	Bihać "AS" - Bosanski Petrovac "AS"	Udaljenosti
4	R-4	Krupa na Uni - Rudice	Udaljenosti
5	R-5	Novi Grad "GP" - Bihać "AS"	Udaljenosti
6	R-6	Ripač (Lohovo "R") - Užljebić "GP"	Udaljenosti
7	R-7	Velika Kladuša (Križ "R") - Bosanska Otoka "R"	Udaljenosti
8	R-8	Grahovo "R" - Kamenica "R"	Udaljenosti
9	R-9	Raštelski Križ "R" - Ćoralići "R"	Udaljenosti
10	R-10	Cazin "AS" - Ljusina "R"	Udaljenosti
11	R-11	Bosanska Krupa "AS" - Sanski Most "AS"	Udaljenosti
12	R-12	Bosanski Petrovac "R" - Strmica "GP"	Udaljenosti

Izradio: M.Sci. Suad Sućeska 2018 : Kontakt

Fig. 1. Homepage of the application eDaljinar

Click on the button 'Udaljenosti' gives page 'Udaljenosti' which contains data on distances and drive duration between current and the previous station, as well as the staring station of the route.

Click on the button 'Izmjene' on the page 'Relacije', gives to the logged user a page which allows update of route data in the table row in which button was clicked.

The page 'Udaljenosti' provides tabularly arranged data on distances of all stations of the selected route. It contains the following fields: route index, station, distance from previous station (one distance), distance from starting station (distance total), drive duration from previous station (one drive duration), drive duration from starting station (drive duration total), duration of stay at the station, road name, intersections of roads, entity, international, inter-entity.

The logged user also gets column with the button 'Izmjene' in the table with list of routes.

Click on the button 'Izmjene' gives to the logged user the page to update the distance data in the row where the button 'Izmjene' was clicked (Fig. 2).

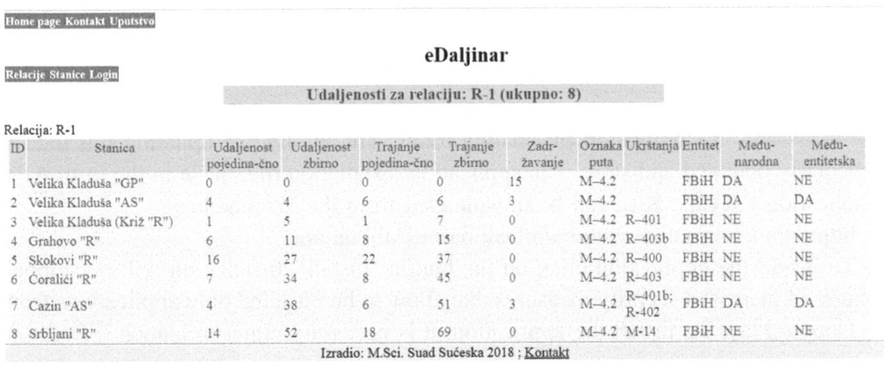

Home page Kontakt Uputstvo

Relacije Stanice Login

eDaljinar

Udaljenosti za relaciju: R-1 (ukupno: 8)

Relacija: R-1

ID	Stanica	Udaljenost pojedina-čno	Udaljenost zbirno	Trajanje pojedina-čno	Trajanje zbirno	Zadr-žavanje	Oznaka puta	Ukrštanja	Entitet	Međunarodna	Među-entitetska
1	Velika Kladuša "GP"	0	0	0	0	15	M–4.2		FBiH	DA	NE
2	Velika Kladuša "AS"	4	4	6	6	3	M–4.2		FBiH	DA	DA
3	Velika Kladuša (Križ "R")	1	5	1	7	0	M–4.2	R–401	FBiH	NE	NE
4	Grahovo "R"	6	11	8	15	0	M–4.2	R–403b	FBiH	NE	NE
5	Skokovi "R"	16	27	22	37	0	M–4.2	R–400	FBiH	NE	NE
6	Ćoralići "R"	7	34	8	45	0	M–4.2	R–403	FBiH	NE	NE
7	Cazin "AS"	4	38	6	51	3	M–4.2	R–401b; R-402	FBiH	DA	DA
8	Srbljani "R"	14	52	18	69	0	M–4.2	M-14	FBiH	NE	NE

Izradio: M.Sci. Suad Sućeska 2018 ; Kontakt

Fig. 2. Page with data for distances of route in the application eDaljinar

Click on the button 'Stanice', in bottom menu, opens the page 'Stanice' (Stations), which gives the following data in tabular form: ID, station, route, entity, international (YES/NO), inter-entity (YES/NO), road name. This page is also made with paging. The logged user also gets column with the button 'Izmjene' in the table with list of stations. Click on the button 'Izmjene' gives you the page for updating the data of the station in row where was clicked on the button (Fig. 3).

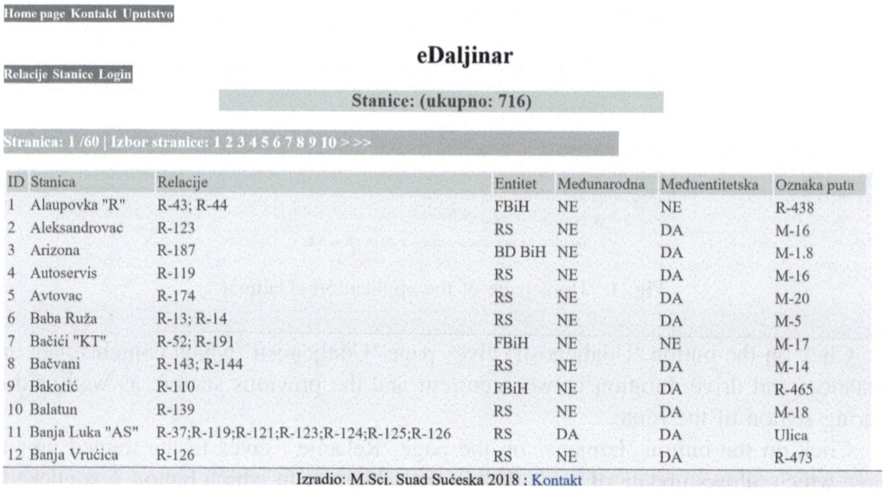
Home page Kontakt Uputstvo

Relacije Stanice Login

eDaljinar

Stanice: (ukupno: 716)

Stranica: 1 /60 | Izbor stranice: 1 2 3 4 5 6 7 8 9 10 > >>

ID	Stanica	Relacije	Entitet	Međunarodna	Međuentitetska	Oznaka puta
1	Alaupovka "R"	R-43; R-44	FBiH	NE	NE	R-438
2	Aleksandrovac	R-123	RS	NE	DA	M-16
3	Arizona	R-187	BD BiH	NE	DA	M-1.8
4	Autoservis	R-119	RS	NE	DA	M-16
5	Avtovac	R-174	RS	NE	DA	M-20
6	Baba Ruža	R-13; R-14	RS	NE	DA	M-5
7	Bačići "KT"	R-52; R-191	FBiH	NE	NE	M-17
8	Bačvani	R-143; R-144	RS	NE	DA	M-14
9	Bakotić	R-110	FBiH	NE	DA	R-465
10	Balatun	R-139	RS	NE	DA	M-18
11	Banja Luka "AS"	R-37;R-119;R-121;R-123;R-124;R-125;R-126	RS	DA	DA	Ulica
12	Banja Vrućica	R-126	RS	NE	DA	R-473

Izradio: M.Sci. Suad Sućeska 2018 : Kontakt

Fig. 3. The page with data of stations in the application eDaljinar

The click on the button 'Korisnik update' gives to the logged user the page where he can update his data. This page contains the following fields: first name, last name, email, password, role, active, input date. After data update, click on the button 'Ažuriraj' to update the data in the database.

To log out of the application click on the button 'Logout' in the bottom menu.

3 About Mobile Application mDaljinar

Aim of mobile application mDaljinar is to make available data on passenger traffic routes, distances between their stations and drive durations, using a smartphone. Changes of this data are very common today in B&H because of the intensive works on new roads and modernization of existing roads. The method of updating this data is particularly outdated and slow. The application installation file, made in the form APK (Application Package Kit), can be downloaded from the Web address:

http://aplikacije.suads.com.ba/mDaljinar/mDaljinar.apk

To install the application click on the button 'Install'. Installation will be stopped because of increased security measures that allow to be installed only applications from the Google Play. To install the application, it is necessary either to launch it from the folder 'Download' on smartphone or temporarily change the security settings of the smartphone [5].

Justification of security measures should be emphasized, because Google regularly checks applications for malware content.

The application gives the following advantages for obtaining the data: mobile application, data is always available (24/7), data is accessible from any place, even while driving, application usage and getting data is quick and easy. The application can be used by anyone interested.

3.1 Description of Work of Mobile Application mDaljinar

Mobile application mDaljinar is written in Android (JAVA). It consists of one activity:

MainActivity, which contains two dialogues: Uputstvo (Manual) and Info (Fig. 4).

Fig. 4. Main activity of mobile application mDaljinar: MainActivity – Routes [2]

The application requires turned on any one of the Internet connections (Wi-Fi,…). The initial activity of the application gives a list of bus routes. Tap on the button 'Udaljenost' (Distance), in the last column of the table, gives tabulated data on distances between stations, drive duration between stations and duration of staying at stations of the route where was clicked on the button. For each station two values of distance and drive duration are given: distance (time) to the first (starting), and to the previous station of the route. Values for distances are given in kilometers (km) and for drive duration in minutes (') (Fig. 5).

ID	Stanica	Daljina pojedi-načno	Daljina zbirno	Trajanje pojedi-načno	Trajanje zbirno	Zadr-žavanje
1	Velika Kladuša "GP"	0	0	0	0	15
2	Velika Kladuša "AS"	4	4	6	6	3
3	Velika Kladuša (Križ "R")	1	5	1	7	0
4	Grahovo "R"	6	11	8	15	0
5	Skokovi "R"	16	27	22	37	0
6	Ćoralići "R"	7	34	8	45	0
7	Cazin "AS"	4	38	6	51	3
8	Srbljani "R"	14	52	18	69	0

Fig. 5. Main activity of mobile application mDaljinar: MainActivity – Distances [3]

Tap on the button 'Nazad na Relacije' to return to the activity 'Routes'. The action bar of the application contains a Menu with options: Uputstvo, Info, Close (Manual, Info, Close). Tap on the option Uputstvo to get manual. Tap on the option Info to get general information about the application. To close the application tap on the option Close (Fig. 6) [4].

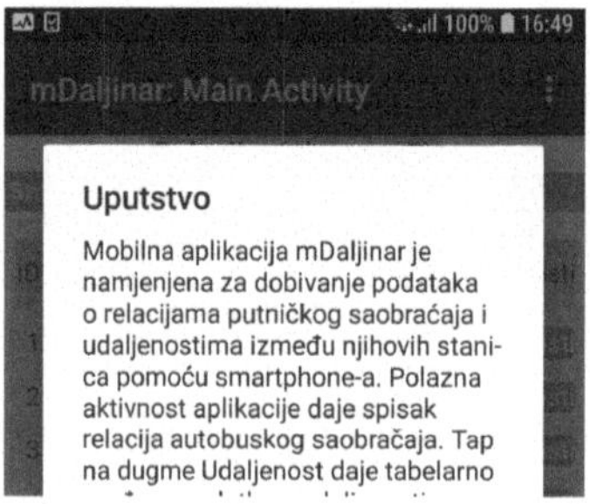

Fig. 6. Manual of mobile application mDaljinar

4 Conclusion

Web application eDaljinar enables faster review and data update on distances and minimum drive durations for bus stops in B&H for international and inter-entity traffic. These data include: routes, stations, roads, entities, distances from the previous station, distances from the starting station, drive duration from the previous station, drive duration from the starting station, intersections of roads. Official document with this data is 'DALJINAR sa relacijama i minimalnim vremenima vožnje za autobusne stanice i stajališta u BiH' (DALJINAR with distances and minimum drive duration for bus stops and stops in BiH) [6]. The existing procedure for update this data is very slow, especially considering the number and speed of changes on B&H roads. A mobile version of this application (mDaljinar) is also made to obtain the above data. It enables to get this data while traveling.

On this way, the above data can be verified and, in the case of significant differences, can be requested their modification.

References

1. DALJINAR sa relacijama i minimalnim vremenima vožnje za autobusne stanice i stajališta u BiH – Verzija 2.0, Ministarstvo komunikacija i transporta, Decembar (2017)
2. McWherter, J., Gowell, S.: Professional Mobile Application Development, John Wiley Inc. Indianapolis, Indiana (Kindle Edition. Wrox) (2015)
3. Introduction to Android Application Development, LINK group 1998 (2017)
4. Advanced Android Application Development, LINK group 1998 (2017)
5. https://pcchip.hr/helpdesk/kako-na-android-uredaju-instalirati-aplikacije-bez-google-play-trgovine/
6. http://mkt.gov.ba/dokumenti/transport/cestovni_transport/registar_redova_voznje/default.aspx?id=6311&langTag=bs-BA

Intelligent Transport Systems, Logistics, Traffic Control

Cybersecurity of Autonomous and Connected Vehicles

Pero Škorput(✉), Sadko Mandžuka, Saša Bermanec, and Hrvoje Vojvodić

Faculty of Traffic and Transport Sciences, Vukelićeva 4, 10000 Zagreb, Croatia
pero.skorput@fpz.hr

Abstract. This paper describes the cybersecurity problem area of intelligent transport systems (ITS), and especially regarding autonomous and connected vehicles. The main vulnerabilities of the ITS system are analyzed in the paper. Real cases and examples of attacks and hacking of autonomous vehicle systems are described. Also, the paper describes the goals and priorities of the transport system for protection against cyber-attacks.

Keywords: Intelligent Transport Systems · Autonomous and connected vehicles · Cybersecurity · Transport system security

1 Introduction

Intelligent Transport Systems (ITS) are fundamental to increase safety and mobility problems. They can make transport safer, more efficient and more sustainable by applying various information and communication technologies to all modes of transport [1]. The concept of Cooperative Intelligent Transport Systems (C-ITS) in terms of communication includes gradual increment of commercial solutions and systems in automotive industry on the market such as safe distance control systems, speed maintaining systems, autonomous management of vehicles within the traffic lanes, avoidance of overtaking in unsafe places, safe passage through the intersection, etc. These systems, with the benefits and their advantages that they provide to users of the transport system, could further increase the superstructure of the existing system functionality in which individual vehicles could constantly communicate with each other or communicate with the transport infrastructure [2]. ITS technologies are increasingly vulnerable to cyber-attacks, due to the rapid development and integration with IT technologies.

More and more often cyber threats and attacks are directly pointed towards the ITS traffic communications, as vehicle-to-vehicle (V2V), vehicle-infrastructure (V2I), infrastructure-infrastructure (I2I). Another traffic field that cyber threats are pointed is ITS infrastructure (automatic toll collection, meteorological stations, traffic cameras, traffic signs, speed sensors, dynamic roadblocks, autonomous car navigation aids, etc.). Cyber-attacks to ITS system can cause traffic accidents, traffic jams and lead to a domino effect that can cause financial losses of all participants in the transport system and the wider traffic network.

I. Karabegović (Ed.): NT 2020, LNNS 128, pp. 547–553, 2020.
https://doi.org/10.1007/978-3-030-46817-0_63

2 Examples of Cyber Attacks on Its Infrastructure and Autonomous Vehicles

Some of the reported attacks in recent years occurred in Texas in 2016 when a Dynamic Message Signs (DMS) was hacked, while a Ransomware attack on a transport agency was conducted in San Francisco where attacker messages were posted on travel and pretravel information systems. In just a few months after this attack on the transport agency, the attackers infected more than 70% of police surveillance cameras in Washington DC with Ransomware, and the same malware also infected Russian Railways computers, train stations and screens for displaying passenger information of German railways.

Also, in 2017 a group of university researchers has managed to hack into autonomous vehicles by putting stickers on road signs to fool the system vision of autonomous vehicles such as shown in Fig. 1 and vehicle traffic sign "Stop" interpreted as a sign to limit the speed of 70 km/h. The consequences of such a simple attack can be devastating, especially because there is a lot of vehicles with semi-autonomous capabilities that are expected to have a significant integration in autonomous traffic.

Fig. 1. Stickers used to visually manipulate road signs with the purpose of defrauding autonomous vehicles [3]

Attacker's profile on ITS infrastructure does not differ significantly from the attackers who carried out cyber-attacks aimed at computer systems of organizations, state, and critical infrastructure. In this regard, threats are developing parallel with the development of the ITS ecosystem. From a cybersecurity perspective, one of the obvious reasons for hacking an ITS system is to break into the ITS ecosystem. If an attacker can successfully use the ITS system to gain access to the corporate network, with minimal effort it can penetrate deep into the network system and cause significant damage [3].

3 Assessment of the Level of Attack Attachment Towards Intelligent Transport System

Attacks on the ITS infrastructure can be divided into three categories: physical, wireless and network attacks. Physical attacks refer to the ITS infrastructure that is physically exposed to traffic systems (roadways, curbs, etc.). Wireless attacks refer to the V2V, V2I and I2I wireless communications that will form the backbone of future ITS technologies, which will certainly pose a significant security challenge. The attacks on the network are related to the traditional information and communication systems where most of the cyber-attacks begin.

There is an overlap between all attack categories, as shown in Fig. 2. This indicates different categories that are interconnected and can successfully compromise ITS systems. It is also possible that a single attack involves all three categories simultaneously. For example, attackers can send incorrect commands to a variable traffic sign (VMS) via a wireless connection, physically connecting to a VMS device, and/or over the network using a compromised management control system.

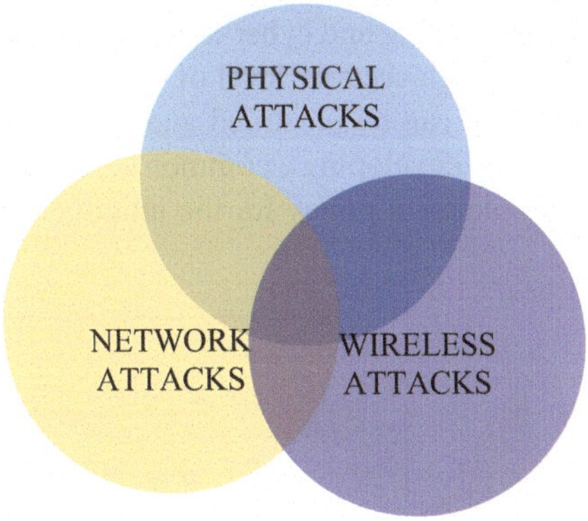

Fig. 2. Categories of attacks and their overlap [3]

One of the leading companies for the security of the cloud, "Trend Micro" 2017 conducted a study of possible vectors cyber-attacks directed at the ITS applications and systems. In doing so, realistic attack scenarios were devised and an assessment of the threat level of individual threats was made. For this purpose, ITS components were divided into six categories (vehicles, road reporting systems, traffic flow controls, payment systems and applications, application and systems management, and communication systems and applications) and by scale of impact on the ITS ecosystem and environment.

Of the total number of threats analyzed, 53.85% were rated as high-risk, 40% as medium-risk, and 6.15% of threats were rated as low-risk. About high-risk threats, represent 71.4% of network attacks, 31.4% of wireless attacks, and 25.7% physical attacks, and noted the overlap between the network, wireless and physical attacks. It was revealed that a network cyber-attacks represent the most serious threat of ITS systems, then wireless attacks and the slightest threat represent physical attacks. From six ITS categories covered by the analysis, the highest number of high-risk threats are present in two ITS categories: traffic management systems and payment systems [3].

4 The Concept of Improving Cybersecurity of Intelligent Transport Systems

The main prerequisite for implementing ITS in the transport system is the security of the ITS system itself. Therefore, it is important to protect ITS elements from attack, whether intentional or unintentional, with the aim of enabling uninterrupted service. Therefore, it is necessary to use the latest insights to identify the possible types of attacks and the risks they carry for the purpose of implementing the necessary security measures. The "Report on a European Strategy for Cooperative Intelligent Transport Systems (2017/2067 (INI))" underlines the importance of implementing high cybersecurity standards in the prevention of hacking and cyber-attacks in all member states, especially with a goal of maintaining the high-security level of C-ITS communications. It is noted that any weakness and risk that can be identified and imagined at this stage of development should be eliminated by developing a common security strategy, including strict security standards, and a certification policy for the introduction of C-ITS.

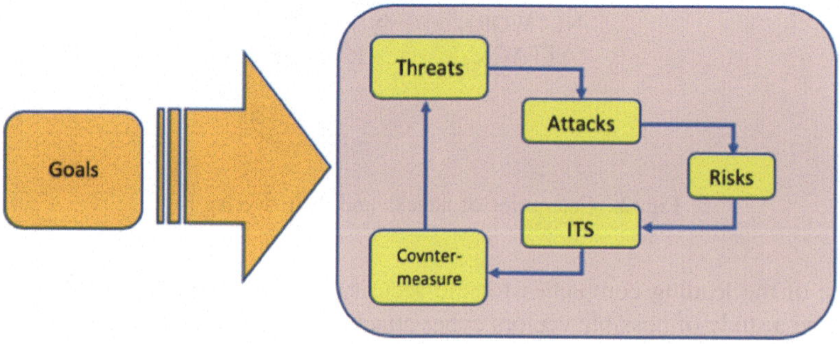

Fig. 3. Concept of enhancement ITS cybersecurity

Figure 3 is shown the relationship between all security objectives, identified security threats, and security services, apropos countermeasures selected with a view to supporting the security objectives and the protection of the ITS system of identified security threats. Taking into account the existing risks listed in this chapter, the basic principles of defense from threats are predicted, threaten and taken with appropriate countermeasures such as the quick identification of the actual violations of safety and quick response to the same, preventing further violations of safety and loss of sensitive data, prevention of attacks by securing all approaches to the subject and possible exploitation, application of lessons learned for further strengthening of defense and preventing recurrence of incidents [3]. The concept of enhanced security can be further broken down into the following steps.

Identifying security risks include analyzes of the security risk directed towards to transport infrastructure. First, the critical transport infrastructure is identified, and then various security scenarios are defined, whose task is to test the resilience and expose the vulnerabilities of the system. With the implementation of the scenario, the identification of risk is provided. Thereafter, security scenarios are ranked by probability, from the most likely to the least likely event, and by risk, from the riskiest to the least risky.

The second step is choosing the security services in order to reduce risks presenting a typical security mechanism, protective measures or countermeasures that provide different aspects of security. Guided by the security objectives and threats identified in the given scenarios, and which are expected to be negatively affected by the system or its communication. Services may include transit security and the management of evacuation and controlled return to service after an accident, etc.

The third step is the development of operational concepts in which the roles and responsibilities of all participants in the traffic of the security services are identified. Participants' requests for information and communication of data sharing tools are also identified to identify threats to traffic safety. The provision of communication is of great importance because potential attackers after the attacks can disrupt activities response to the crisis.

The development of an implementation plan is a fourth step in which the implementation timetable and the budget for key services are determined according to national road safety priorities. Also, funding sources are identified and the progress of the plan is monitored. If significant issues or deviations from the implementation timetable occur, the plan should be reviewed to avoid deviations from security initiatives.

Evaluation is the last step that needs to be implemented to ensure that measures taken in the field of traffic safety satisfy the level of security and protection. The reason is that security risks, participants, and technology may change over time [4].

The objectives related to the protection and cybersecurity of the transport sector stem from documents and legal acts covering the area of national security. For example, protection against terrorism and disasters is a common goal of all sectors.

In Table 1 the goals and priorities of the transport system are presented and their task is to support a secure transport system, allowing the free movement of people and goods without fear of possible harm and loss of all kinds of freedoms.

Table 1. Transport system security objectives and priorities [5]

Traffic infrastructure	
Goals	Priorities
Risk management focused on the material, human and cyber elements of critical transport infrastructure	Improve risk management capabilities and capabilities in the event of a terrorist attack To achieve and maintain the required level of security of cyber systems critical for mission-critical traffic activities
Utilizing the ability to respond, recover and coordinate the transport system sector in supporting the resilience of the entire society	Improve the preparedness of critical transport infrastructure to ensure the necessary level of resilience in the event of accidents Support a national certification program for emergency response personnel and the repair of critical infrastructure affected by an accident Extend cooperation to improve the resilience of society and interdependent sectors
Implementation of the process for effective cooperation with the aim of sharing key information between state departments, the judiciary and public authorities, and between public and private actors	Improve information sharing procedures between government departments, the judiciary and public authorities, and between public and private actors to maintain ongoing situational awareness Enhance and expand cooperation that will involve interdependent sectors and authorities at the state, county, city and municipal levels To improve the protection and safety of traffic reporting on perceived problems, analyzing and distributing them through national mechanisms for reporting
Improving the preparedness and resilience of the international transport system in order to respond to all hazards in order to protect the interests	Extend the risk-based security approach, including risk-sharing, to securing and managing the flow of people and goods. Support the resilience of the international supply chain

5 Conclusion

Intelligent transport systems are based on computer and information-communication systems and have to meet numerous cybersecurity requirements. Identifying threats and vulnerabilities of ITS systems and taking appropriate security measures is often an overlooked aspect. From a cybersecurity perspective, one of the obvious reasons for hacking the ITS system is to break into the ITS system that can cause traffic accidents, create traffic jams that affect basic services, freight and passenger movements, and passenger cars, and lead to a domino effect that can cause financial losses of all participants in the transport system and the wider transport network. Cyber-attacks on transport infrastructure are becoming more common as more and more C-ITS systems connect online. Therefore, cybersecurity should be considered mandatory and one of the pillars of ITS architectures and frameworks. Future research should pay particular attention to the vulnerability of the traffic management system [6].

Acknowledgment. The research presented in this paper is supported by the University of Zagreb Program Funds Support for scientific and artistic research (2019) through the project: "Research of the influence of autonomous vehicles on urban traffic flow characteristics".

References

1. Vujić, M., Škorput, P., Ćelić, J.: Wireless communication in cooperative urban traffic management. Pomorstvo **29**(2), 150–155 (2015)
2. Škorput, P., Vojvodić, H., Mandžuka, S.: Cybersecurity in cooperative intelligent transport systems. In: 2017 International Symposium ELMAR, Zadar, pp. 35–38 (2017)
3. Huq, N., Vosseler, R., Swimmer, M.: Cyberattacks Against Intelligent Transportation Systems Assessing Future Threats to ITS, Trend Micro, TrendLabs (2017)
4. Bragdon, C.: Transportation Security. Elsevier Inc., Oxford (2008)
5. United States Department of Transportation: Transportation Systems Sector-Specific Plan – 2015, Homeland Security (2015)
6. Mandžuka, S., Ivanjko, E., Vujić, M., Škorput, P., Gregurić, M.: The use of cooperative ITS in urban traffic management. In: Intelligent Transport Systems: Technologies and Applications. John Wiley (2015)

Investigation of Vehicle Driving Cycles in Urban Traffic Condition

Mirsad Trobradović(✉), Boran Pikula, Almir Blažević, and Dževad Bibić

Faculty of Mechanical Engineering, University of Sarajevo, Vilsonovo šetalište 9, 71000 Sarajevo, Bosnia and Herzegovina
trobradovic@mef.unsa.ba

Abstract. Traffic jams and resulting air pollution have become severe problem in world urban areas long time ago. Fast growing number of vehicle in recent years, and inadequate development of traffic infrastructure, made these problems even worse in cities of Bosnia and Herzegovina. To start solving these problems, it will be useful to know real traffic conditions in particular urban area. One of the important information is vehicle driving cycles in real traffic conditions which can enable to determinate typical driving pattern for particular area. Measurements were conducted in main city roads in Sarajevo during different periods of working and non-working day. Results and analysis of driving cycles are presented in the paper.

Keywords: Vehicle · Driving cycle · Urban traffic conditions

1 Introduction

Ever increasing number of road vehicles has become severe problem in world urban areas long time ago. Among other things, it produces noise and air pollution which significantly reduces quality of life of urban population.

Fast growing number of road vehicle in recent years, and inadequate development of traffic infrastructure made these problems even worse in cities of Bosnia and Herzegovina. Almost all cities in Bosnia and Herzegovina are overloaded with traffic, where traffic jams during rush hours are especially problematic.

To start solving these problems, it will be useful to know real traffic conditions in particular urban area. One of the important information is road vehicle driving cycles in real traffic conditions which can enable to determinate typical driving pattern for particular area. By definition, driving cycle is set of data points given in speed vs. time, speed vs. distance or similar diagram, in specific area or part of road [1–3]. Up to date, various driving cycles are developed, primarily for the purpose of traffic analysis or the measurement of road vehicle emissions. Review of existing test driving cycles for road vehicle emissions measurement is presented in [4]. Comprehensive review of existing methodologies and practices for developing driving cycles is given in [5].

Generally speaking, driving cycles in real traffic conditions are highly dependent on local condition such as road geometry and topology, number of vehicle (traffic volumes), speed limits, weather conditions etc. and can be very different for various areas.

I. Karabegović (Ed.): NT 2020, LNNS 128, pp. 554–559, 2020.
https://doi.org/10.1007/978-3-030-46817-0_64

For each urban area, with its own road infrastructure, there are specific traffic conditions and specific problems that require a detailed analysis of the driving cycle.

Considering aforementioned and with the purpose of determining appropriate driving cycles, measurements were conducted in main city road in Sarajevo, in real traffic condition. Measurements were conducted during different periods of working and non-working day. Measurements were made during the winter months. Results and analysis of driving cycles are presented as follows.

2 Measurements and Determination of Driving Cycles

As mention before, in order to establish a driving pattern for particular area, measurements of vehicle driving cycles in real traffic conditions were made. Measurements were conducted in main city road in Sarajevo, on route Baščaršija-Ilidža-Baščaršija (Fig. 1). This road is a major city communication road, which covers most of the city traffic. The total length of the route is approximately 20 km, with 13 points of intersection with roads with significant traffic frequency. Legal speed limit is 60 km/h.

Fig. 1. Main city road in Sarajevo where measurement was conducted (image from Google Earth)

The measurements were taken for different time periods during the working day, during the traffic peak in the morning and afternoon, and at midday. For comparison, measurements were taken during the non-working day (Sunday) at midday. Vehicle speed, vehicle acceleration (deceleration) and distance traveled were measured.

The GPS Data Logger Racelogic DriftBox was used for the measurement. It is a GPS device that allows measuring and recording the parameters of vehicle movement in the longitudinal and lateral directions, such as speed, acceleration, position of the vehicle. The device records the results during testing, and through the associated software enables subsequent processing of the results. Main specifications of measuring device are given in Table 1.

Table 1. Main specifications of GPS Data Logger Racelogic DriftBox

Output data rate	10 Hz
Operating temperature	−20 °C to +50 °C
Velocity measurement	
Resolution	0.01 km/h
Accuracy	0.1 km/h
Distance measurement	
Resolution	0.01 m
Accuracy	0.05%

Multiple measurements were conducted for every considered periods of day, during several days. One particular driving cycle (speed vs. time chart) for every period of day is presented on following figures. The results shown below represents normal traffic conditions, and do not include traffic jams caused by car accidents, car failures, etc.

Figure 2 shows driving cycle (speed vs. time) during morning hours (7:30 to 8:30), on typical working day. Results are presented for two routes separately (Ilidža-Baščaršija and Baščaršija-Ilidža).

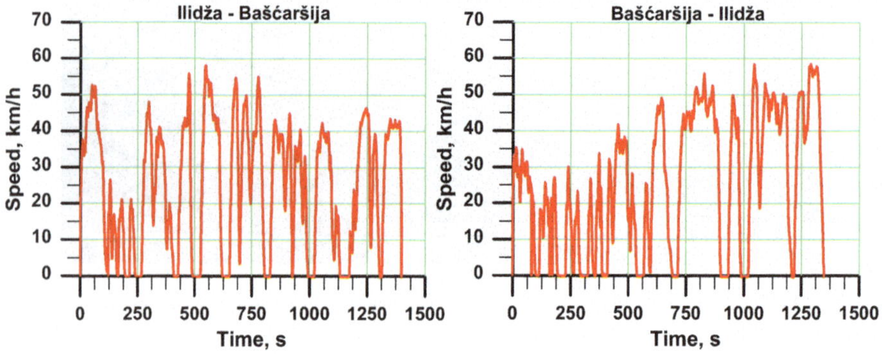

Fig. 2. Vehicle driving cycle – working day – morning

Figure 3 shows driving cycle at midday (12:00 to 13:00), on typical working day. Figure 4 shows driving cycle during afternoon hours (16:00 to 17:00), on typical working day. For comparison, measurements were also taken during the non-working day (Sunday) at midday. Driving cycle is shown on Fig. 5.

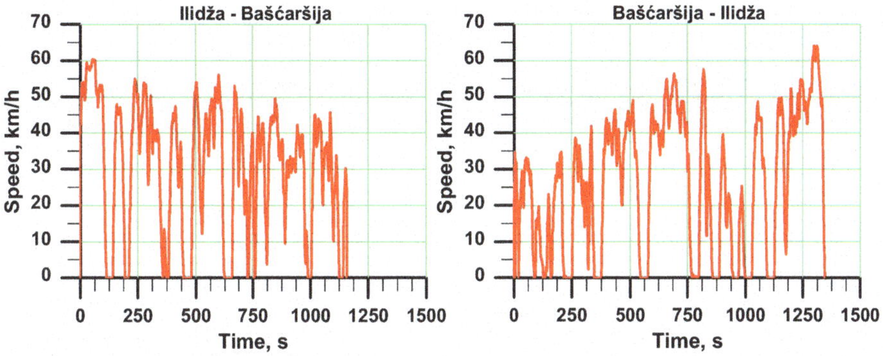

Fig. 3. Vehicle driving cycle – working day – midday

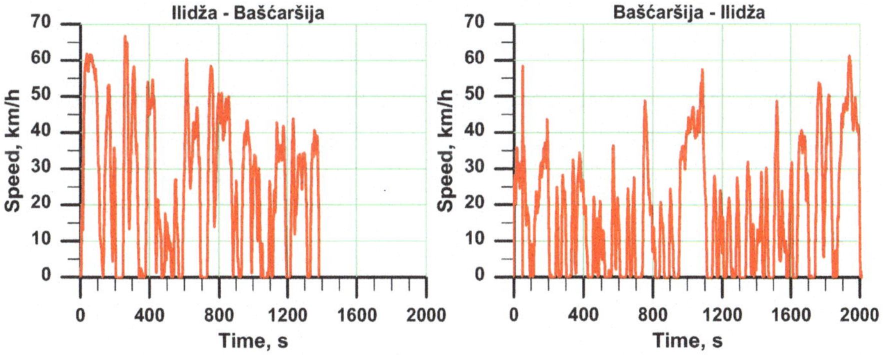

Fig. 4. Vehicle driving cycle – working day – afternoon

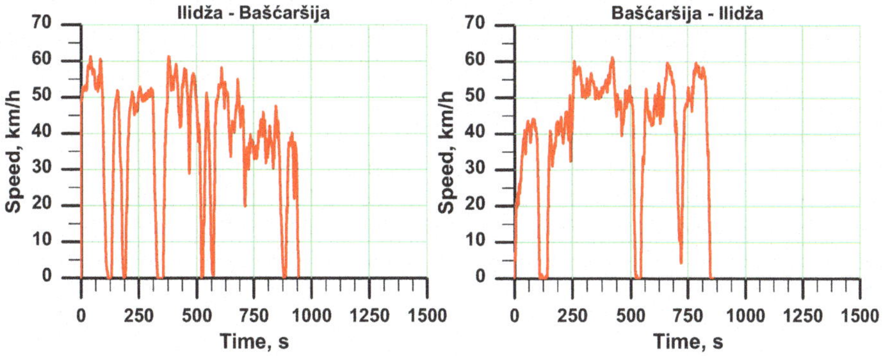

Fig. 5. Vehicle driving cycle – non-working day (Sunday) – midday

Analyzing the results obtained, it is possible to determine the average speed of the vehicle in different periods of the day. The results are shown in Fig. 6 and in Table 2.

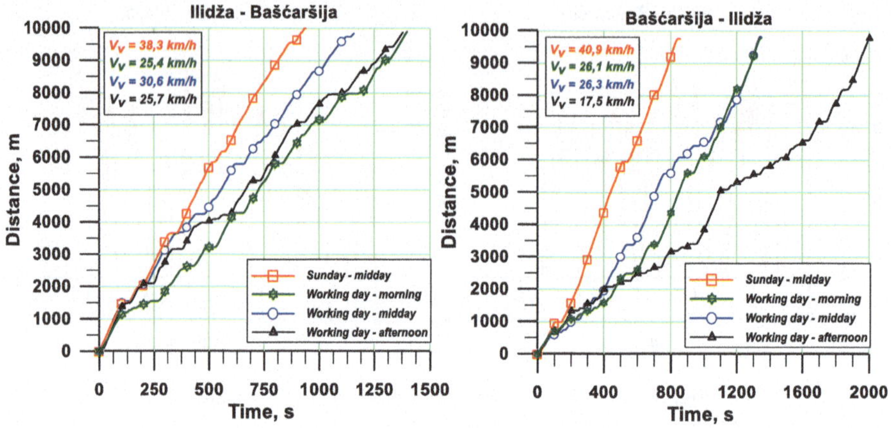

Fig. 6. Average vehicle speed for different period of day

Table 2. Average vehicle speed for different period of day on analyzed routes

Period of day/route	Ilidža-Baščaršija		Baščaršija-Ilidža	
	Average speed, km/h	Driving time, s	Average speed, km/h	Driving time, s
Working day				
Morning	25.4	1399	26.1	1349
Midday	30.6	1158	26.3	1345
Afternoon	25.7	1380	17.5	2011
Non-working day				
Midday	38.3	942	40.9	860

The analysis of the results shows a significant difference in driving cycles at different periods of the working day. This is primarily related to the number of vehicle stoppings, and locations with an increased number of stops.

Morning and afternoon driving cycles (Fig. 2 and 4) have increased number of zones of slow speed due to more extensive lateral traffic (traffic from intersecting lateral roads), which is not significantly expressed in the midday cycle. Such slow vehicle speed causes unfavorable exhaust emissions from vehicles with gasoline and diesel engines, particulate during winter month. In addition, this "stop and go" driving speed profile leads to excessive use of braking systems with brake linings and tires wear and increased particulate emission. A particularly significant deviation occurred on the afternoon cycle on the route Baščaršija-Ilidža (Fig. 4, right), where total driving time is much longer and there is increased number of vehicle stops.

On the other hand, on recorded non-working day driving cycles (Fig. 5) it can be observed that the number of stoppings is less than the number of road intersection points, which means that the vehicle can use the so-called "green wave" on certain sections of route, which is almost impossible on working days.

3 Conclusion

Investigation of driving cycles in real traffic conditions for one urban area in Bosnia and Herzegovina are presented in this paper. Measurements were conducted at the main city road in Sarajevo, for different periods of the day. One representative driving cycle for every period of day was shown.

The following conclusions can be drawn:

Available standard vehicle driving cycles are not suitable for analyzing of traffic in considered urban area (Sarajevo).

There is difference in driving profile between driving cycles at different periods of the working day, but values of average speed are similar. The exception is afternoon driving cycle on route Baščaršija-Ilidža with significantly lower average speed and longer driving time.

There is significant difference in driving profile and average speed between driving cycles for working day and for non-working day. The latter has much favorable characteristics, with almost 50% higher average speed and more than 65% shorter driving time.

Investigation of driving cycles contributes to a more realistic view of the traffic condition for each urban area and can lead to better understanding of real traffic problem in particular area. It also contributes to development of more accurate test driving cycles for road vehicle in urban areas.

It should be noted that presented study is of a preliminary nature and serve to obtain only a rough picture of the representative driving cycle and to establish basic driving patterns. To obtain a complete picture on the basis of which a representative driving cycle could be created and established a reliable driving pattern, more measurements are necessary.

References

1. Tong, H., Hung, W.: A framework for developing driving cycles with on-road driving data. Transp. Rev. **30**(5), 589–615 (2010)
2. Galgamuwa, U., Perera, L., Bandara, S.: A representative driving cycle for the southern expressway compared to existing driving cycles. Transp. Dev. Econ. **2**, 22 (2016)
3. Galgamuwa, U., Perera, L., Bandara, S.: Developing a general methodology for driving cycle construction: comparison of various established driving cycles in the world to propose a general approach. J. Transp. Technol. **5**(4), 191–203 (2015)
4. Barlow, T.J., Latham, S., McCrae, I., Boulter, P.: A reference book of driving cycles for use in the measurement of road vehicle emissions. TRL Limited (2009)
5. Zhang, X., Zhao, D.J., Shen, J.M.: A synthesis of methodologies and practice for developing driving cycles. Energy Procedia **16**(Part C), 1868–1873 (2012)

Improvement of Variable Speed Limit Control Effectiveness in Context of Connected Vehicles

Martin Gregurić(✉), Sadko Mandžuka, and Miroslav Vujić

Faculty of Traffic and Transport Sciences, University of Zagreb, Vukelićeva 4, 10000 Zagreb, Croatia
martin.greguric@fpz.hr

Abstract. The problem with Variable Speed Limit Control (VSLC), as one of powerful ITS services, is that a certain percentage of drivers do not obey posted speed limits. Mentioned behavior induces a reduction of the VSLC service effectiveness. In order to deal with this problem, the concept of Intelligent Speed Adaptation (ISA) is introduced. ISA can directly reduce the vehicle's speed according to the information obtained from the traffic control system installed on road infrastructure. Computed optimal speed limits are transmitted to the vehicles with ISA communication capabilities. Today, the connected vehicles concept is the core approach for implementation ISA system. The focus of this paper is to provide an analysis regarding the impact of the ISA on the effectiveness of VSLC in the context of connected driving.

Keywords: Variable speed limit control · Connected vehicles · Intelligent speed adaptation

1 Introduction

Urban motorways originally planned as the urban bypasses are in the last decade become the place of congestions due to increased traffic demand and lack of space for constructional build-up. Those congestions impact overall traffic throughput in the adjacent urban networks. Variable Speed Limit Control (VSLC), as the most prominent Intelligent Transportation Systems (ITS) traffic control method on motorways, enables direct control over the motorway mainstream flow. VSLC preforms homogenization of individual vehicle speeds and reduction of the mean speed of vehicles. The mean speed of vehicles is reduced under the values that can cause the occurrence of critical traffic density and consequently traffic congestion [1, 2].

Additionally, as a side-effect, VSLC improves the overall motorway traffic safety [2]. The latest versions of traffic responsive VSLC are based on methodologies such as linear control, multi-layer control, shock-wave theory and model predictive control [3]. The most prominent problem with the VSLC is a high percentage of drivers who do not obey posted speed limits. This problem can be overridden by the introduction of the connected vehicle concept. Connected vehicles by using communication equipment share information with the other vehicles with the same abilities as well as traffic infrastructure outside the vehicle. The Intelligent Speed Adaptation (ISA) is based on the concept of data exchange between vehicles and traffic infrastructure dedicated for

I. Karabegović (Ed.): NT 2020, LNNS 128, pp. 560–563, 2020.
https://doi.org/10.1007/978-3-030-46817-0_65

computation of speed limits. This system can automatically reduce the vehicle's speed according to the information obtained from the traffic management system dedicated for VSLC. An even a smaller number of connected vehicles with ISA capability can increase the effectiveness of VSLC. Those vehicles must fully obey posted speed limits so the other vehicles behind them cannot drive faster. The general idea of VSLC is shown in Sect. 2. In Sect. 3, a quick overview of the connected vehicle concept is given and in Sect. 4 ISA is set in the context of connected driving. Paper is concluded with the concluding remarks in Sect. 5.

2 Variable Speed Limit Control

The effects of VSLC on traffic flow are described in [2]. The authors report two main impacts of VSLC on traffic flow: reduction of the mean speed near-critical density and homogenization of vehicle speeds. Homogenization of vehicle speeds reduces the variation in vehicle speeds on the motorway segment controlled by VSLC. Reduction of speed differences among vehicles and the mean speed differences between lanes provides suppression of shock waves [3]. This is done by slowing down traffic before it reaches a state when queuing starts, effectively reducing the intensity of traffic jams [2, 3]. The impact of the speed limit on a fundamental diagram of the relationship between density and flow can be seen in Fig. 1.

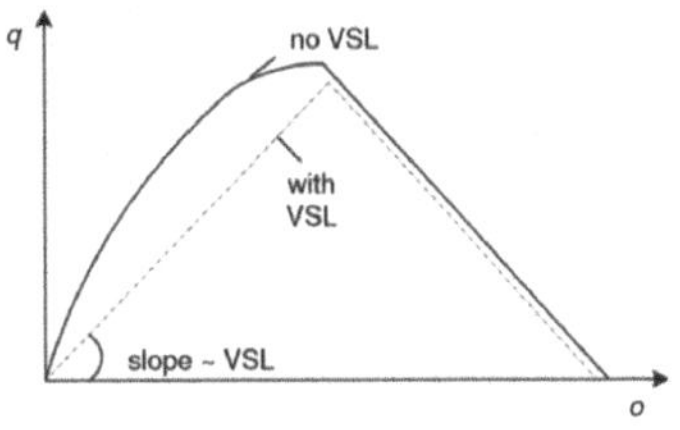

Fig. 1. Impact of the speed limit on fundamental diagram

3 Connected Vehicles

Three different parts in connected vehicle ecosystem (Fig. 2) can be distinguished: (i) urban motorway (or any other type of road traffic network); (ii) motorway management system (with communication and processing unit); and (iii) road vehicle (with OBU). The connected vehicle uses OBU which gives them the ability to receive data from other vehicles and infrastructure, transmit its real-time data (e.g., its current position, acceleration, speed, etc.) and pass computed data directly to the power train of the vehicle as well as inform the driver about its conducted action through the display device.

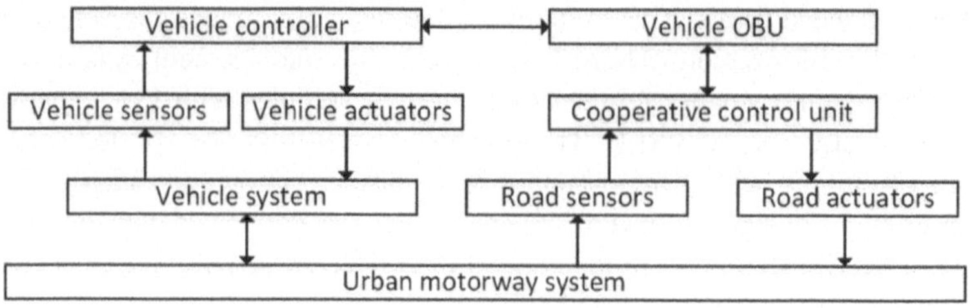

Fig. 2. Architecture for connected vehicle ecosystem [5]

Exchange of information will be based on Dedicated Short-Range Communications (DSRC) radio or/and 5GLTE wireless technology. EU has defined ITS-G5 technology to be operated on the 5.9 GHz frequency band and fulfills the requirements set out by ITS Directive 2010/40/E and its "delegated act" of requirements for Cooperative-ITS which involves connected vehicles. The motorway management system and the connected vehicles are both acting upon the urban motorway with the same goal, to use it optimally. However, this goal can only be achieved if the vehicles obey the imposed control information such as speed limit [5].

4 Intelligent Speed Adaptation

An ISA system can be considered as one type of augmented VSLC, which is based on the concept of connected vehicles, rather than the traffic signalization in the form of Variable Messaging Signs (VMS). This system uses cooperative OBU, which monitors the location and speed of the vehicle, compares it to a defined speed limit and takes corrective action such as advising the driver and/or managing the maximal speed of the vehicle [4, 5]. The road vehicle development roadmap [6] foresight that in 2022 on-board systems similar to ISA will be operational in vehicles. In paper [7], positive results are derived from the UK ISA field trials, as well as the Danish ISA field test described in [8]. The field as mentioned above test has included lower insurance rates for young drivers if they use such "incentive" ISA in their vehicles. It is noticed that the application of the "incentive" ISA system has led to a reduction of 4% on 50 km roads and of 5% on 80 km roads in the proportion of distance traveled by more than 5 km/h above the speed limit. On 110 km roads, the reduction was 9% and on 130 km roads, it was 3% [8]. In UK trial most overridden speeds by ISA are found in case of lower speed limits 40 km/h (13.3%) and at highest speed limits of 130 km/h at motorways (16.3%). On all other posted speed limits, ISA has overridden about 10% of vehicle speeds. In both experiments, ISA is found to have a distinctive effect in transforming the speed distribution from a conventional bell shape to an asymmetric distribution biased towards the high-speed end. ISA not only diminished excessive speeding but also led to a reduction in speed variation, prompting a positive implication to accident reduction [7].

5 Conclusion

ISA system based on connected vehicle data exchange can reduce vehicle speed and solve the problem of a higher percentage of vehicles that do not obey the posted speed limit. Several simulation-based analysis and real-world applications research have shown that ISA can increase VSLC effectiveness. Urban environments are especially interesting for ISA applications since this is the place where there has been a significant increase in traffic congestion in recent times, with very adverse effects in terms of pollution due to vehicle exhaust gases. Existing forecasts indicate that the use of connected (and autonomous) vehicles will help reduce these congestions in urban areas.

Acknowledgment. The research presented in this paper is supported by the University of Zagreb Program Funds Support for scientific and artistic research (2019) through the project: "Research of the influence of autonomous vehicles on urban traffic flow characteristics".

References

1. Papageorgiou, M., Diakaki, C., Dinopoulou, V., Kotsialos, A., Wang, Y.: Review of road traffic control strategies. Proc. IEEE **91**(12), 2043–2067 (2003)
2. Papageorgiou, M., Kosmatopoulos, E., Papamichail, I.: Effects of variable speed limits on motorway traffic flow. Transportation Research Record: Journal of the Transportation Research Board, No. 2047, Washington, D.C., pp. 37–48 (2008)
3. Grabec, I., Švegl, F.: Stabilization of traffic by speed limit variation. In: Proceedings of the 22nd ISEP 2014, Ljubljana, Slovenia, 24–25 March 2014, p. R7 (2014)
4. Gregurić, M., Mandžuka, S.: The use of cooperative approach in intelligent speed adaptation. In: Proceedings of TELFOR 2018, Beograd, Srbija, pp. 579–583 (2018)
5. Liu, R., Tate, J.E.: Network effects of intelligent speed adaptation systems. Transportation **31** (3), 297–325 (2004)
6. Phaal, R.: Foresight vehicle technology roadmap-technology and research directions for future road vehicles. Society of Motor Manufacturers and Traders Ltd., London (2002)
7. Agerholm, N., Waagepetersen, R., Tradisauskas, N., Harms, L., Lahrmann, H.: Preliminary results from the Danish intelligent speed adaptation project pay as you speed. IET ITS **2**(2), 143–153 (2008)
8. Lai, F., Carsten, O.: What benefit does Intelligent Speed Adaptation deliver: a close examination of its effect on vehicle speeds. Accid. Anal. Prev. **48**, 4–9 (2010)

Autonomous Vehicles in Urban Traffic

Miroslav Vujic(✉), Sadko Mandzuka, and Luka Dedic

Faculty of Traffic and Transport Sciences, University of Zagreb, Vukeliceva 4, 10000 Zagreb, Croatia
miroslav.vujic@fpz.hr

Abstract. This paper describes the implication of autonomous vehicles on urban transport system. Levels of vehicles autonomy and basic technologies as well as some impacts of autonomous vehicles on urban transport is presented. Also, there are some consequences of autonomous and connected vehicles on road infrastructure. Some implication of deployment of autonomous and connected vehicles for transport system of Republic of Croatia are highlighted.

Keywords: Autonomous vehicles · Connected vehicles · Public transport · Road infrastructure

1 Introduction

An autonomous vehicle is a vehicle operated without a driver that is, independently moving without the presence of a human being, using sensors, cameras, telematics systems and artificial intelligence [1]. Vehicle autonomy can be divided into six levels, of which the zero level is without autonomy, where a person is fully in control of the vehicle and the fifth level is complete autonomy without the interaction of the vehicle and the human [2].

According to many studies, with the introduction of autonomous vehicles on the market, most of these challenges facing the transport system should be solved. According to statistics [3], over 90% of road accidents are caused by human error. With the introduction of more autonomous vehicles, the number of accidents would be drastically reduced. In addition to reducing road accidents, the use of autonomous vehicles would also have the effect of reducing the adverse effects of transport on the environment.

Also, there are wide possible improvements in the reduction of environmental impact and energy consumption when comparing a connected and automated vehicle (CAV) with an internal combustion engine vehicle (ICEV), [4]. Also, the results of a study [5] conducted in Finland indicate the possibility of developing a lot of new mobility business models. In the study [5] is presented that autonomous vehicles could increase the percentage of shared rides, which would ultimately reduce the number of vehicles. A study related to the effects of autonomous vehicles on urban networks [6] shows that with the 100% use of autonomous vehicles, intersection capacity would increase by up to 40%.

I. Karabegović (Ed.): NT 2020, LNNS 128, pp. 564–569, 2020.
https://doi.org/10.1007/978-3-030-46817-0_66

In Sect. 2 levels of vehicles autonomy and basic technologies is described. Impact of autonomous vehicles on urban transport is given in Sect. 3. Some consequences of autonomous vehicles on road infrastructure is described in Sect. 4. Conclusion is given in Sect. 5.

2 Overview of Autonomous Vehicle Technology

2.1 Levels of Autonomy

According to the Association of Automotive Engineers – SAE, there are six levels of autonomy (0–5) as shown in Fig. 1. The level 0 (without autonomy) is the lowest and on it everything the driver does. The driver has complete control of the vehicle and no systems help the driver while driving. For the level 1 (driver assistance), the vehicles have some systems that assist the driver when driving. These can be systems that assist with braking, acceleration, steering, transmission and ignition, while for ignition some systems, such as cruise control, which maintain the vehicle speed at the one specified by the driver, require the driver's intervention and do not engage themselves during driving. The level 2 (partial autonomy) implies that two or more automated systems together make it easier for the driver to operate the vehicle. At the level 3 (conditioned autonomy), the vehicle simultaneously manages, brakes, accelerates and monitors events in the vehicle's moving environment.

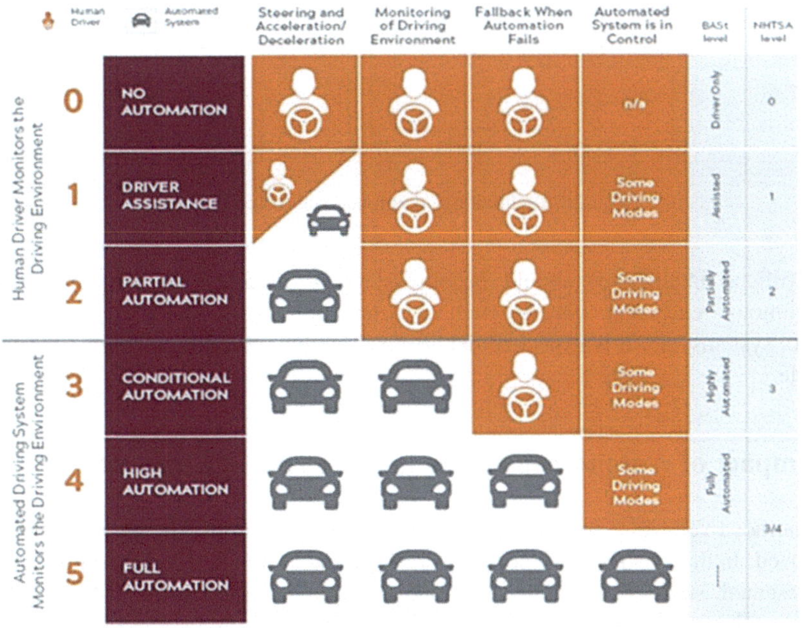

Fig. 1. Levels of autonomy [1]

At the level 4 (high autonomy), do not need a human (driver), i.e. the vehicle can drive completely independently, but under certain conditions, for example a 'smart' highway. The level 5 (complete autonomy) implies that the vehicle can drive independently in all conditions and in them the person is only a passenger.

2.2 Autonomous Vehicle Technology

Autonomous vehicles use sensors, cameras, radar and artificial intelligence to move independently without a driver (Fig. 2). Thus, they use various sensor technology for navigation, computer vision that attempts to perceive human vision, that is, the human eye, with telematics systems that monitor and analyse the environment in which the vehicle is moving, and artificial intelligence, which ultimately processes all data and controls the vehicle.

Computer vision has found an important use in autonomous driving, so in autonomous vehicles it is used to identify, analyse movements, reconstruct events and images collected from their environment.

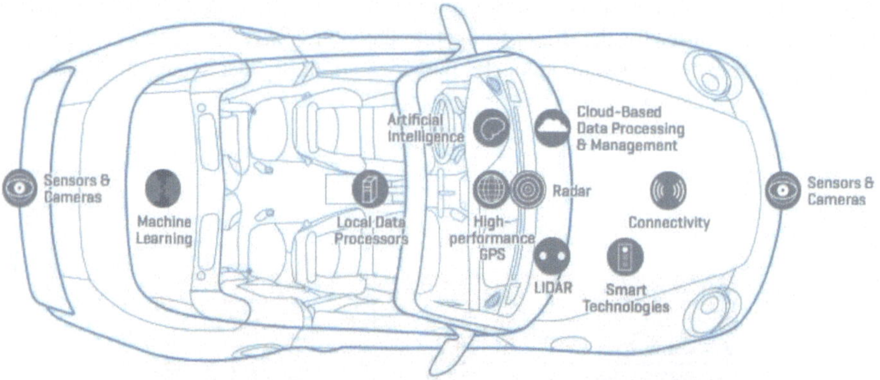

Fig. 2. Technology in autonomous vehicles [1]

Artificial intelligence should become standard in new vehicles, especially in autonomous vehicles in systems such as computer vision, radar units, driver status control systems, etc. It is expected that artificial intelligence will be able to control the vehicle.

3 Impact of Autonomous Vehicles on Urban Transport

Autonomous vehicles are basically "ideal" drivers, so the traffic parameters should be improved. In this chapter basic features of autonomous vehicles in urban transport will be presented, such as first stage of autonomy – public transport vehicles in reserved lanes in urban traffic network.

3.1 Basic Features

A study [4] conducted by Michigan, USA, shows possible improvements in greenhouse gas (GHG) reduction and energy consumption when comparing a connected and automated vehicle (CAV) and an internal combustion engine vehicle (ICEV). The study showed that the initial energy consumption would be 3–20% higher when using autonomous vehicles. However, considering the possibility of eco-driving, plateaus and connectivity, V2I results at the network level indicate a reduction of greenhouse gas emissions by up to 9%. Due to the communication of autonomous vehicles with the connected/cooperative vehicles concept (V2V and V2I), traffic flows on urban roads would become harmonized, which would ultimately result in a large reduction of unwanted harmful by-products of conventional vehicle traffic.

3.2 Autonomous Urban Public Transport

In today's cities, where economic growth is increasing and the population is growing, or cities are increasingly populated, good public transport is also needed. Public transport users in such cities want accurate, fast, efficient and accessible public transportation. With the development of autonomous vehicle technology, it has become possible for public transport to develop autonomous vehicles, that is, for public transport service providers to take advantage of the advancement of technology for autonomous vehicles and for receiving them on public transport vehicles such as buses, trams, trains, etc.

Many countries around the world, such as the US, Japan, Switzerland, Hungary, Italy, Denmark, etc., use autonomous public transport in the form of Metro, intercity trains or shorter passenger service lines between aircraft and air terminals. However, this type of JGP works in a closed system and the same technology cannot be applied to passenger transport systems such as trams or buses. Autonomous vehicle technology, that is, vehicles that can make their own decisions quickly and for all road users, is more applicable to such systems than that used for autonomous trains running in closed and semi-closed systems. The greatest savings for autonomous trams would be on drivers, i.e. because there would be no drivers because their share in the operating costs of the trams goes up to 50% [7]. An example of the movement of an autonomous tram by road can be seen in Fig. 3.

4 Road Infrastructure and Autonomous Vehicles

Autonomous vehicles are in the need of special road markings, signs and signals which must be upgraded from the existing situation. In the first phase, autonomous vehicles should have separate lanes (or dedicated lanes). Also, autonomous vehicles have the potential for the improvement of the capacity of the traffic network. Because autonomous vehicles strive for ideal driver conditions, the overall traffic network performances can be improved. According to conducted simulations to investigate the influence of cooperative adaptive cruise control on highway capacity. Their results indicated that cooperative adaptive cruise control can increase a highway capacity, of

2,868 vehicles per hour per lane to 10,720 vehicles per hour per lane, when 100% are communicating vehicles and the average speed of vehicles is fixed at 100 km/h—which is a capacity improvement of about 3.7 times.

Although many researchers have obtained simulation results confirming the capacity improvements made possible by autonomous vehicle technologies, their results are inconsistent because they are based on varying rates of implementation.

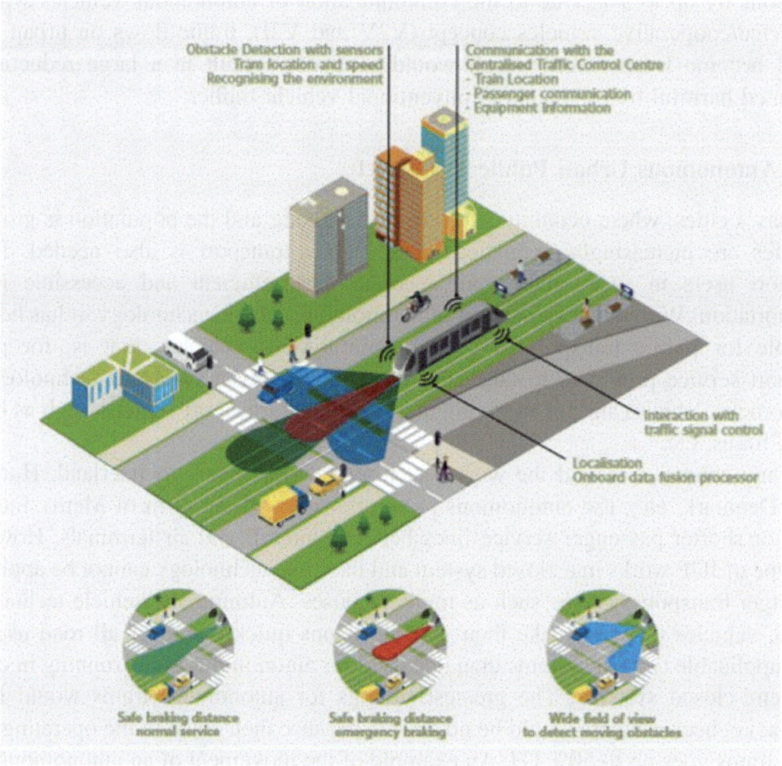

Fig. 3. An example of an autonomous tram [8]

5 Conclusion

All the studies conducted so far point to significant implications for the inclusion of autonomous vehicles in urban transport. This is primarily related to the efficacy and safety of city traffic, as well as reducing environmental impact. However, some legal and insurance challenges should not be ignored. Autonomous vehicles in the Republic of Croatia have not yet been put under the legal framework, that is, the use of autonomous vehicles has not yet been regulated, and with the progress in the development of autonomous vehicles and the emergence of vehicles of higher levels of autonomy, there has been a need to regulate their use in road transport. In conclusion, for better insight into what people think autonomous vehicles will affect public transport, respondents should be more informed about autonomous vehicles, their technology and capabilities.

Acknowledgment. The research presented in this paper is supported by the University of Zagreb Program Funds Support for scientific and artistic research (2019) through the project: "Research of the influence of autonomous vehicles on urban traffic flow characteristics".

References

1. Cheng, H.: Autonomous Intelligent Vehicles - Theory, Algorithms, and Implementation. Springer, New York (2011)
2. SAE: Levels of Driving Automation, SAE International Standard J3016. Society of Automotive Engineers (2014)
3. Green, M., Senders, J.: Human error in road accidents. Visual Expert (2004)
4. Leduc, G.: Road traffic data: collection methods and applications. Working Papers on Energy, Transport and Climate Change, pp. 1–55 (2008)
5. Matthaeia, R., et al.: Autonomous Driving: Technical, Legal and Social Aspects. Springer, Heidelberg (2015)
6. http://www.ncsl.org/research/transportation/autonomous-vehicles-self-driving-vehicles-enacted-legislation.aspx. Accessed 27 Dec 2019
7. Watzenig, D., Horn, M. (eds.): Automated Driving: Safer and More Efficient Future Driving. Springer, Basel (2016)
8. Trentesaux, D., et al.: The autonomous train. In: SYSOSE, pp. 514–520 (2018)

New Technologies for Improving Driver Response Efficiency in Risk Prevention from Traffic Environment

Osman Lindov(✉) and Adnan Omerhodžić(✉)

Faculty of Traffic and Communications, University of Sarajevo, Sarajevo, Bosnia and Herzegovina
olindov@gmail.com, adnan.omerhodzic@gmail.com

Abstract. Risk in traffic or traffic environment is constant, always present and can never be completely eliminated. In urban areas, the highest percentage of risky traffic situations is related to pedestrians and their specificities in traffic participation. Pedestrians as vulnerable road users participate in different functions, i.e. behaviors and modes of movement. Due to the flexibility and the ability to change speeds and movements relatively easily, pedestrians often cause incident and risky traffic situations. The timely detection of pedestrians by motor vehicle drivers is one of the key parameters that directly affects the available response capabilities of motor vehicle drivers to take safety actions in order to avoid pedestrian conflict. This paper presents the basics of the concept and capabilities of new technologies for monitoring and exploring drivers' views in order to generalize conclusions to improve the effectiveness of drivers' response in the prevention of traffic risks. Custom hardware and software components will be used to monitor the driver's view for research and analysis purposes.

Keywords: Tracking views · Risk in traffic · Drivers response · Traffic safety

1 Introduction

The study of eye movement monitoring has a relatively long history, which is evident from the numerous scientific and professional literature in this field, as well as a large number of developed systems and devices. The original devices and instruments that emerged at the end of the 19th century provided objective results but were uncomfortable for examiners. At the beginning of the 20th century, systems and methods were developed based on photography, and by the end of the 20th and the beginning of the 21st century, with the development of computer technologies, significant advances were made in the improvement of devices with the ability to use and analyze real-time data. Eye movement tracking systems are nowadays used in various fields: market research, marketing, web space usage research, immobile view control, psychology and vision research, medical research, entertainment and video games [1]. The use of these systems and devices for the purpose of exploring and enhancing human behavior capabilities provides multiple opportunities for examining driver behavior, that is, the ability and skills of spotting and responding to risks from the traffic environment.

I. Karabegović (Ed.): NT 2020, LNNS 128, pp. 570–580, 2020.
https://doi.org/10.1007/978-3-030-46817-0_67

Among the most important parameters for the analysis of the driver's eye view is the fixation of the view on the elements and areas of interest in the driver's field of view. The results of the previous research provide useful information and conclusions drawn about the differences in the abilities and skills of different age groups of drivers when responding to traffic risks [2]. Of particular importance for improving driving safety are activities aimed to young drivers who are novice in the process of training or early use of vehicles [3]. Research shows the significant contribution of view-tracking technologies and the formation of inferences about driver behavior in real-world traffic conditions. The synthesis of the results of relevant studies focused on monitoring the driver's view allows the formation of generalized and validated conclusions that can be used to improve the driver's ability and response skills during driving [4].

The interaction between human perception and autonomous driving has not been sufficiently explored. That is why analyzing human perception during the driving, while the driver is making driving decisions can provide great benefits for studying autonomous driving. Initial experiments were conducted to investigate the possible connection between the driving behavior patterns and the driver's sight fixation, where devices and systems for tracking eye movement were used [5]. Eye movement monitoring systems and devices were used in a study of traffic safety analysis in school zones, analyzing the impact of the environment on driver behavior [6]. In this study, simulators were used to create different environmental scenarios, such as different pedestrian behaviors, parked vehicles, obstacles on and off the road.

A significant contribution to the study of driver behavior and skills to respond to potential risks from the traffic environment is provided by machine learning techniques. According to a survey of over 80 research projects and papers published in the last ten years and eight effective machine learning models, high levels of performance and machine learning models have been proven to assess driver behavior [7]. However, the application of machine learning models in the analysis of driver behavior is still limited and more effort is required to obtain well-formed and generalized results.

2 Abilities of Timely Reaction in Traffic

Human behavior in traffic is influenced by structure and dynamics of personality, experience and skill, age, gender, socio-demographic characteristics, influence of intoxicants, fatigue, illness and etc. The abilities of a person as a participant in traffic and as a factor of safety in traffic are: the ability of the sensory organs, mental abilities and psychomotor abilities. Human differences in ability are expressed in the perception of danger, the speed of response and the ability to overcome the perceived danger. For the safe traffic operation, it is necessary to harmonize the psychophysical capabilities of the person, the technical characteristics of the vehicle and the characteristics of the road and the immediate environment.

From the traffic environment come predictable and unpredictable elements (incentives) that affects drivers reaction. Predictable elements that affects the way how driver responds can be classified into the following groups: road condition, vehicle condition, weather conditions and the ambient of environment. The unpredictable elements that affects the way how driver responds cannot be classified and are the most dangerous for traffic safety. Man with his psychophysical abilities which are individual reacts to danger through the following abilities: perception, comprehension, reasoning, response and reaction.

Acceptance of information takes place in the human body and based on which an event experiencing occurs. Sudden and unpredictable environmental events that one experiences, require quick response and quick resolution. Experiences with timely and correct resolution lead to a safe exit from that situation, while experiences with incorrect and untimely resolution lead directly to traffic threatening. The shortness of the speed of occurrence and duration of the event determines both the essence and gravity of the course of the event, as well as its duration.

The response time of the driver depends on the subjective characteristics of the driver as well as on the conditions of objective circumstances. This is the time it takes the driver to know the danger and make the decision to stop the car or maneuver right or left. Perception is the process of receiving and evaluating information and drawing conclusions about forced information.

3 The Importance of Visual Abilities in the Driver's Timely Response Function

The basic component that is necessary for receiving external visual information, the safety of managing and maintaining direction is visibility. The field of view of a person is called the space that encompasses the view. The ability to capture the environment in the field of view is the limit to which objects are observed and occurrences beyond the point of fixation. The field of view of man consists of a horizontal and vertical field of view. The vertical field of view of a person reaches an angle of 115°, while the horizontal field of view reaches an angle of 160°–180°.

The sharpest field of view is up to 3° from the point of fixation, that is, from the symmetry. 3° to 6° is a very good field of view, 6° to 10° is a good field of view, and 10° to 20° is a satisfactory field of view. A field of vision of 20° or more is called the peripheral field of view. Objects in the peripheral field of view can be seen without clear details and colors. The width of the field of vision changes depending on the speed of movement of the vehicle, so that the field of vision is functionally narrowed with increasing speed, because the driver has to fix the road further from himself [8].

In the sharp field of view, the driver sees all the pedestrian details, the color of the clothing, estimates the speed, direction of movement, while these sharp field of view abilities are reduced if the pedestrians are distant from the line of sight. The driver's sharp field of view limited on only 6° is in large disproportion to the obligations of the driver, which are requiring spotting of pedestrians and objects throughout the field of view from the side of the vehicle and behind the vehicle (Fig. 1).

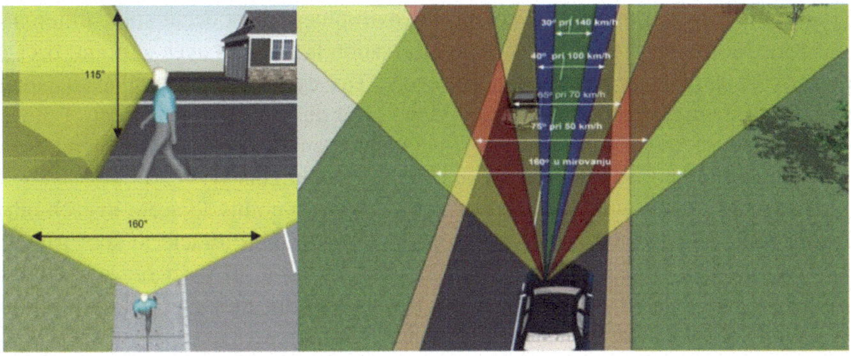

Fig. 1. Vertical and horizontal field of view, wideness offield view of drivers at different speeds [9]

Specific traffic situations and scenarios require different maneuvers from the driver which are demanding the 360° field of view. As the clear sighting as a precondition for understanding of visual stimulation, connects to the sharp field of view, there is a need and obligation for the driver to place all relevant safety and driving performance items in the sharp field of view. Drivers achieve this by body movements, head movements and eye movements. There are three separate actions in the eye movement:

- Switching view on another object, with a jump time of 0.1 (s) for movement up to 20°, and 0.15 (s) for a jump of more than 20°.
- The fixation of an object in average takes 0.15 (s) and depends on the error in stopping the view of an object, and the error is greater if the angle of view switch is greater.
- Binocular coordination is the process of coordinating the view of two eyes on the same object and the accommodation of the lens of the eye, which depends on the difference between the distance of the object the driver is looking at and the new object on which he switches his view, with an average time of 0.4 (s) (Table 1).

Table 1. Bus driver visual perception activities [10]

Activity of the view	Time in (%)
Straight ahead	78
Right ahead	4,5
Left ahead	5,4
Right	1,4
Left	1,5
Right exterior mirror	1,7
Left exterior mirror	6,4
Panel in the cabin	0,5
Towards the fall of the cabin	0,6

For the occurrence of eye movement, in environment must occur some differences or changes. More recent research shows that fixation is equal to eyes to point, because eyes are constantly moving at different levels. The eye has its natural path (spatial view) and its natural retention (temporal view). The direction of movement of the eye depends solely on the properties of the visual phenomenon. The eye is directed towards greater attraction, and it is worn by a sudden or gradual change.

Visual acuity is the ability of the eye to perceive two points adjacent to each other. A driver with lower visual acuity must come closer to the obstacle to spot it than a driver with normal visual acuity, so that he has less time to react (e.g., braking or avoiding obstacles). Dynamic visual acuity is important for motor vehicle drivers, that is, the ability to spot details on a moving object that decreases as a person ages, and is caused by the drop in oculomotor control [11]. Aging weakens the visual acuity, so that the visual acuity of sixty-year-olds is at 74%, while in eighty-year-olds the visual acuity is only at 47% of the visual acuity of a healthy young man [12].

4 The Possibilities of Using New Technologies to Improve Driver Response Efficiency in Prevention of Traffic Risk

Recently, a great deal of attention has been devoted to enhancing the ability and characteristics of drivers to reduce traffic risks in the process of traffic safety management. The most important are the visible abilities and characteristics, as well as the acquired skills, which directly depend on the process of responding to traffic hazards and risks. The exploration of drivers view to improve drivers response skills and take safety actions to avoid accidents has been intensified in recent years. This is due to the increasing focus of traffic safety policies and strategies on the advancement of human factors in the prevention of traffic risks. Therefore, due to the specificity of the research area, the development of new technologies, specified hardware components and custom software solutions, which, based on the results of the research and analysis, open the field of possibilities for improving the skills of drivers in the prevention of traffic hazards, plays a key role.

Currently eye gaze tracking systems can be classified into two groups: face-recording and eye-detection systems [13, 14], and fixed-camera systems located on subjects head which are performing eye imaging [15]. View-tracking devices and applications can be categorized as diagnostic and interactive [16]. Diagnostic applications use view data as quantitative evidence of human visual processes and attention. Interactive applications use them to interact with the user based on eye movements.

The most famous companies, that is, products for monitoring eye gaze, classified by academic research aspect by the number of publications found using Google Scholar are: TobiiPro, SMI Senso Motoric Instrument, SR Research EyeLink, ISCAN, LC Technologies, EyeTech, The Eye Tribe, Ergoneers, Smart Eye, Pupil Labs, Gazepoint.

For the purpose of experimental research and analysis of the possibility of applying new technologies to improve driver response efficiency in prevention of traffic risk, the Pupil Core platform was used for comprehensive eye gaze monitoring and gaze-based interaction [17]. It consists of customizable hardware components in the form of head glasses with a set of cameras, a software framework for eye view tracking, and a graphical user interface for data playback, visualization and analysis (Fig. 2).

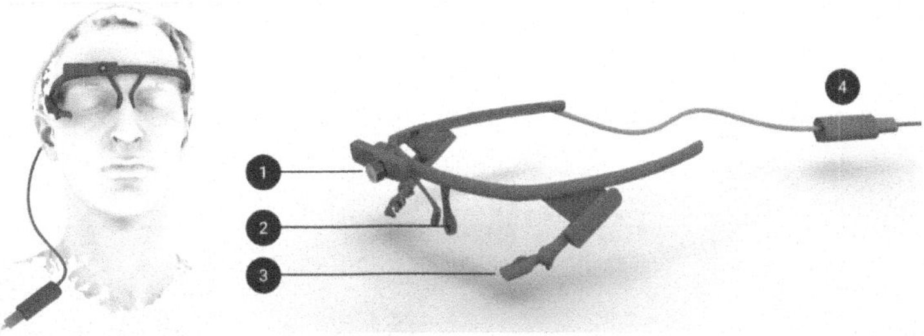

1. The camera for recording view field
2. Adjustable head / nose support elements
3. Camera for recording eye movements
4. USB connector for connecting to software and PC

Fig. 2. Pupil Core hardware components

4.1 Possibilities of Application of the System for the Research in Education of Drivers and Improving the Efficiency of Reaction on Risks in Traffic

Eye gaze tracking systems offer a wide range of options and output parameters for analysis. Investigating the driver's eye view features in real or simulated conditions provide a wide range of useful information and parameters that can be used to improve the driver's response and behavior to traffic risks. This primarily refers to novice drivers who are in the process of training, where, through the improvement of educational mechanisms, combined with practical training in real-world conditions and training using a driving simulator, would enable the formation of more efficient driver skills. Also, these training mechanisms can also be applied to conditionally experienced drivers to improve the ability and skills to respond to traffic risks.

By analyzing the characteristics of the most famous eye gaze systems and tools, great potentials have been identified for use in traffic hazard prevention processes [18]. The most significant features and capabilities of these systems, the output of which the driver's eye view can be used to significantly improve the response efficiency of novice drivers in the training and risk prevention of traffic hazards, are presented below [19–22].

Views and fixation of views represent the basic output parameters of monitoring and research. The view is a space that is covered by the eyes of the driver and depending on the frequency of data collection, the number of views within one second depends. If the set of points of view is very close in time and space, then the set of those points is a fixation, marking the period in which the eyes are directed towards an object. Driver's gaze fixes are a measure of visual attention to potential risks from the traffic environment.

View visualization using heat maps shows the general distribution of the driver's viewpoints. They appear as an overlay of color gradients in the driver's field of view.

The use of heat maps is an easy method to quickly visualize which elements from the environment attract more attention than others. It is possible to display heat maps of view visualization at different time intervals depending on the required parameters for analysis (Fig. 3).

Fig. 3. Visualization of gaze fixation [23] and gaze visualization by heat maps [19]

Analysis and metrics of interest points of area is a tool that allows you to select segments within the driver's field of view that are of interest for the driver, or that have caused a specific visual stimulus, such as the appearance of pedestrians near the roadway. Within this visual stimulus, analysis and metrics of special interest areas, such as monitoring how much the driver's gaze is focused on the pedestrian's face, body, hands or feet, as well as turning the gaze back on, are provided. This feature is of particular importance in the analysis of biomechanics and pedestrian behavior, correlating with the timely response of drivers and taking safety actions to avoid the occurrence of traffic accidents.

The time until the first fixation of the view indicates how long it takes the driver to view a particular object or object of interest such as pedestrians, cyclists, vehicles and other objects. This tool can provide information on how certain aspects of the visual scene are prioritized, as from a driver's perspective, pedestrians near the roadway should have a higher priority than adult pedestrians near the roadway.

Quantifying the time spent viewing a particular object of interest provides useful information about the driver's motivation and attention to visual stimuli from the environment. The long duration of looking at a particular segment in the field of view may indicate a high level of interest, while a shorter duration of time may indicate that other areas might be more interesting. However, further research on the emotional response to the visual scene is needed in this segment to reach more complete conclusions.

Review of the data of orientation of a view to some object of interest depending on specific driver groups such as drivers in the training phase, novice drivers or experienced drivers. This information can be useful in the process of educating drivers to highlight potential risks such as the sudden pedestrian exit onto the roadway behind obstacles and objects that prevent them from being spotted by drivers.

The fixation sequences are based on spatial and temporal information when and where the driver looked. This allows the importance and order of priorities in the driver's visual scene to be defined. There is usually a segment of central fixation bias, but the following components will be representative of what is most motivating for the driver in a traffic environment.

The number of repeated views gives an indication of how many times the driver has returned the view to a particular element in view field. This makes it possible to examine areas that have repeatedly attracted drivers. The driver can be drawn in a variety of ways, or the visual stimulus can be pleasant, confusing or frustrating. Therefore, this area requires additional testing in the driver's emotional state segment.

The first duration of view fixation provides information on how long that first fixation lasted. This information is of particular importance if used in conjunction with the duration time until first fixation of the view. If the driver has a short time to first fixation and a long duration of first fixation, then the fixation area is very attractive.

Average fixation duration is a very important piece of information that can be determined for individuals or groups of drivers. The information obtained on the average duration of fixations can be directed to additional research on why individual fixations lasted much shorter or significantly longer than average.

The presented features of the driver's gaze monitoring system can be used to advance methodological procedures for managing black spots on the road [24]. Examining the driver's view and how to respond to specific conditions at identified black spot locations using a gaze tracking system can significantly advance the process of in-depth black spot analysis in order to understand the causes of traffic accidents.

4.2 Testing and Analysis of the Driver's View in Correlation with the Driver's Response to Potential Risks from the Traffic Environment

The device and components of the Pupil Core platform were used for testing and analyzing the driver's view. Drivers with years of driving experience were tested and tested in real-world traffic conditions, and additionally in virtual conditions using a driving simulator. Testing and analysis of the driver's view was performed on one-way streets, with the right lane being rolling, while vehicles on the left lane. In addition, only driving recordings in ambient conditions were selected when there were no moving obstacles in the driver's field of view. Based on the results of the tests performed, where some slight discrepancies were recorded, some generalized conclusions were reached, which are described below.

Testing begins by entering in the described street, where at the very entrance to the street, the driver directs his gaze to the traffic lane without fixation, and then switches his gaze to the left on parked vehicles, where usually the first, second or third parked vehicles are fixed. Afterwards, the driver switches his gaze to the traffic lane in front of him, where during the journey the driver's gaze is partly directed to the right side by the roadway, or to the left towards parked vehicles. However, the greatest retention and focus of the view, over 75% of the time, is on the traffic lane, where significant fixation of the space on the traffic lane surface occurs.

Based on the recorded test results, an analysis of the driver's view in the specified conditions can be performed in correlation with the risk from the traffic environment. The orientation of the driver's gaze on the traffic lane in front of him at the start of a one-way street may be related to the effect of central fixation bias. Subsequently, shifting your gaze to parked vehicles on the left can be correlated with the risks and dangers of other vehicles that the driver would normally expect to appear in the opposite traffic lane. However, after keeping a gaze on parked vehicles and forming fixations on them, based on the information collected (passing a second or third vehicle), the driver realizes that vehicles are parked in the left traffic lane, most likely parked vehicles being processed and understood by the driver as elements without any potential risk. Thereafter, passing through a large number of parked vehicles most likely in the driver creates a sense of monotony, which, under the influence of the central fixation bias, shifts the view of the road lane ahead. Toggling right and left as you continue driving may be correlated with the specifics of the elements and areas of interest that motivate the driver to switch views, which appear in the field of view during the ride, such as the color and type of parked vehicles or objects along the road, billboards, traffic signs and etc.

Described analysis and the conducted testing are highlighting the potential risks and unexpected traffic hazards that drivers must pay particular attention to. A striking example is the sudden exit of pedestrians on the roadway behind parked vehicles or other objects near the roadway. The main feature of these risks and dangers is the very short time available for the driver to respond and to take precautionary measures to avoid pedestrian crashes. Therefore, drivers with better abilities and acquired skills of adequate access and focused gaze while driving, have greater ability to detect risks in a timely manner (sudden pedestrian exits on the roadway), or greater opportunities to avoid pedestrian collisions. The mentioned analysis and the particular example described here, are opening up a wide range of potential uses of the eye movement monitoring system for the education and training of novice drivers as well as drivers with driving experience. Thus, these eye movement monitoring systems can be used to train drivers and then use the results of the analysis to suggest and correct wrong reactions, that is, to improve the driver's ability and skills (Figs. 4 and 5).

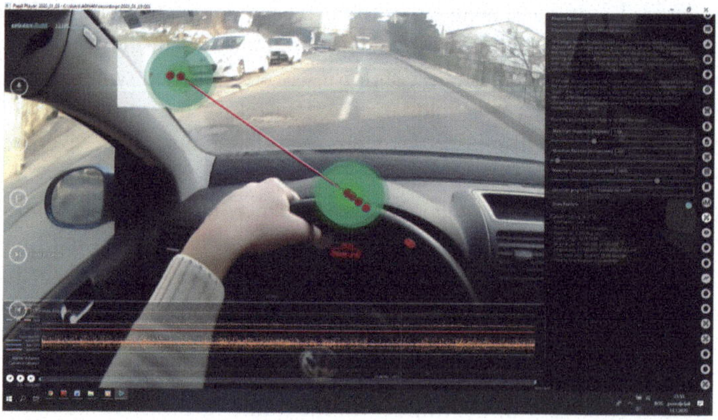

Fig. 4. Pupil Core user interface view when analyzing driver views in real-time traffic conditions

Fig. 5. Pupil Core user interface view when analyzing driver views in a virtual environment using the Carnetsoft driving simulator [25]

5 Conclusion

The paper presents the basic characteristics of new technologies for eye movement monitoring and the possibilities of application in the analysis and improvement of the efficiency of the driver's response to the risks from the traffic environment. Particular emphasis is given to the application of an eye movement monitoring and analysis system in preventive action to improve traffic safety through educational mechanisms in the training process for young novice drivers. The interaction of the driver's timely response and visual stimuli from the traffic environment has not yet been fully explored. Comprehensive analysis, in addition to the application of the driver's eye movement monitoring system, requires the integration of other systems for the analysis of parameters such as emotional state, fatigue, fear, head movements, facial expression and etc. Eye movement monitoring systems can contribute to improving traffic safety when making decisions in the process of designing and designing road elements and structures near the road, especially in terms of traffic visibility.

References

1. Drewes, H.: Eye gaze tracking for human computer interaction, March 2010
2. Underwood, G., Chapman, P., Brocklehurst, N., Underwood, J., Crundall, D.: Visual attention while driving: sequences of eye fixations made by experienced and novice drivers. Ergonomics **46**(6), 629–646 (2003)
3. Gesser-Edelsburg, A., Guttman, N.: Virtual versus 'actual' parental accompaniment of teen drivers: A qualitative study of teens' views of in-vehicle driver monitoring technologies. Transp. Res. Part F Traffic Psychol. Behav. **17**, 114–124 (2013)
4. Taylor, T., et al.: The view from the road: the contribution of on-road glance-monitoring technologies to understanding driver behavior. Accid. Anal. Prev. **58**, 175–186 (2013)

5. Wang, M., Walders, K., Gordon, M.E., Pelz, J.B., Farnand, S.: Auto-simulator preparation for research into assessing the correlation between human driving behaviors and fixation patterns. In: IS and T International Symposium on Electronic Imaging Science and Technology (2018)
6. Valdés Diaz, D.M., Ríos, B.C., Knodler, M.: Evaluation of safety enhancements in school zones with familiar and unfamiliar drivers
7. Elamrani Abou Elassad, Z., Mousannif, H., Al Moatassime, H., Karkouch, A.: The application of machine learning techniques for driving behavior analysis: a conceptual framework and a systematic literature review. Eng. Appl. Artif. Intell. **87** (2020)
8. Rotim, F.: Elementi sigurnosti cestovnog prometa: Ekspertize prometnih nezgoda. Znanstveni savjet za promet JAZU, Zagreb (1990)
9. Lindov, O.: Road Traffic Safety. Faculty of Traffic and Communication (2007)
10. Farland, R.A.M.: Psychological and psychiatric aspects of highway safety. J. Am. Med. Assoc. **163**, 233–237 (1957)
11. Davidse, R.: Assisting the Older Driver – Intersection Design and In-Car Devices to Improve the Safety of the Older Driver. Rijksununiversiteit Groningen, Leidschendam (2007)
12. Bilban, M.: Older drivers in road accidents. Ljubljana (2002)
13. Morimoto, C.H., Flickner, M.: Real-time multiple face detection using active illumination. In: Proceedings of the 4th IEEE International Conference on Automatic Face and Gesture Recognition, FG 2000, pp. 8–13 (2000)
14. Smith, P., Shah, M., Da Vitoria Lobo, N.: Monitoring head/eye motion for driver alertness with one camera. In: Proceedings of International Conference on Pattern Recognition, vol. 15, no. 4, pp. 636–642 (2000)
15. Li, D., Winfield, D., Parkhurst, D.J.: Starburst: a hybrid algorithm for video-based eye tracking combining feature-based and model-based approaches, pp. 79–79 (2006)
16. Duchowski, A.T.: A breadth-first survey of eye-tracking applications. Behav. Res. Methods Instrum. Comput. **34**(4), 455–470 (2002)
17. Kassner, M., Patera, W., Bulling, A.: Pupil: an open source platform for pervasive eye tracking and mobile gaze-based interaction. In: Adjunct Proceedings of the ACM International Joint Conference on Pervasive and Ubiquitous Computing, UbiComp 2014, pp. 1151–1160 (2014)
18. Tobii Pro Lab - Software for eye tracking and biometrics (2015)
19. Eye Tracking Pocket Guide - iMotions. https://imotions.com/guides/eye-tracking/. Accessed 12 Jan 2020
20. ISCAN® - Home. http://www.iscaninc.com/home.html. Accessed 13 Jan 2020
21. Vehicle testing kit for real world driving|ERGONEERS. https://www.ergoneers.com/en/hardware/vehicle-testing-kit/. Accessed 13 Jan 2020
22. Driver Monitoring System|Eye Tracking Technology|Smart Eye. https://smarteye.se/automotive-solutions/. Accessed: 13 Jan 2020
23. Crabb, D.P. et al.: Exploring eye movements in patients with glaucoma when viewing a driving scene. PLoS One **5**(3) (2010)
24. Lindov, O., Kiso, F., Omerhodzic, A., Begovic, M.: BSM – black spot management as methodology approach in increasing safety of road traffic. In: IMSC 7th International Maritime Science Conference, pp. 445–451 (2017)
25. Carnetsoft car driving simulator for training, assessment and research. https://cs-driving-simulator.com/. Accessed 14 Jan 2020

Predictive Model of Personalization of Services of Automated Mobility Based on the Records of User Movement in Mobile Networks

Mirza Berković(✉) and Amel Kosovac

Faculty of Traffic and Communications, University of Sarajevo, Zmaja od Bosne 8, 71 000 Sarajevo, Bosnia and Herzegovina
mirza.berkovic@gmail.com

Abstract. The aim of this paper is to investigate and develop a predictive model of personalization of automated mobility services based on historical and realtime data on the trajectory of movement of users in mobile networks. The problem is based on the fact that the concept of automated mobility enables a variant of modeling the parameters of the vehicle trajectory and personalization of the service according to the wishes and needs of users, whereby optimization and personalization can be performed both on the user side and on the side of the mobility service provider. The paper uses a research approach within the domain of intelligent transport systems based on the extraction of telecommunication activity data into the index of personalization of mobility parameters using the methodology of predictive analytics and neural inference based on neural network. The research findings establish and show a clear correlation of historical and realtime spatial data of the movement of users and their activities in the mobile network with indicators of quality of urban and sustainable mobility. The significance of the model is that only a satisfied user of a personalized automated mobility service no doubt intensely and more frequently uses a given service on the one hand, and the automated mobility system can predictively extract a set of parameter values that satisfy that user and offer the optimum variant of the service.

Keywords: Predictive model · Mobility personalization · Automated mobility

1 Introduction

Automated mobility is ushering in a new digital age of automation and connectivity that allows vehicles to talk to other systems and users inside and outside traffic. Advances in the field of artificial intelligence are opening up completely new paradigms of cooperation among road users, making a step that brings great benefits to the whole mobility system. Automated mobility is based on connecting the car to an intelligent management domain, where it is crucially dependent on data, especially on realtime data which also enables personalization of the mobility service. Innovative artificial intelligence and mapping technologies for automated driving are crucial, and high precision maps are crucial for highly automated driving. In such a system, safety and comfort are largely dependent on up-to-date data, in fact automated driving is only

I. Karabegović (Ed.): NT 2020, LNNS 128, pp. 581–595, 2020.
https://doi.org/10.1007/978-3-030-46817-0_68

possible with high precision maps. High-speed driving tickets and maps used in navigation systems generally differ in two respects. First, the precision is significantly higher. Second, the ticket material for highly automated driving consists of multiple layers. The traditional base layer for navigation is used to calculate routes A to B, including a number of roads along which it is necessary to drive. The localization layer uses a new positioning concept that provides extremely accurate map data, and is used by the automated vehicle to determine the lane position accurately. At the same time, the vehicle compares the environmental recognition with the relevant information in the localization layer, and thus can accurately determine its position relative to the road and the environment. The planning layer following the localization layer contains not only attributes such as road surface markings, traffic signs, speed limits, etc., but also 3D road geometry information, including turns and slopes. With this very detailed lane information, an automated vehicle can decide when and how to change lane. At the same time, speed limit information must be available at the same time. Only in such an information environment can vehicles choose the best proactive driving strategy. It is very important to note the changes on the roads, such as changed lane settings or new road signs, and it is essential to use feedback from the vehicle fleets of vehicles equipped with the necessary sensors.

On the other hand, the system of personalization of the services of automated mobility in the most general case implies the possibility of adjusting the characteristics of the services to the wishes and needs of a particular user. The need to enable personalization of mobility services is clearly evident in today's dynamic multimodal transport environment. At the same time, due to the nature of the work, mobile networks collect data that can be used for analyzes outside the telecommunication system, especially in the field of urban mobility and research of patterns of movement of users. Predictive analytics, on the other hand, can be used to predict what users will do and how they will behave in the future, based on what we know about them today, it's really a way of aligning data with predictive models and defining a likely outcome.

The paper will present a model for predictive personalization of automated mobility services based on the trajectory records of users' movement in mobile networks so that automated driving providers will be able to predict the behavior of system users to design a personalized service. Predictive analytics are crucial to the creation of the model, which becomes indispensable in analyzing existing data and predicting future unknown events. The aim of the paper is to prove that big data, along with analytics, can offer organizations impressive opportunities to improve the effectiveness of mobility services. Section 2 presents the research to date on the feasibility of using mobile technologies in the field of mobility personalization. Section 3 presents the key features of automated mobility. Section 4 explains the importance of personalizing mobility services. Section 5 explores the possibilities of recording trajectories of users in mobile networks. Section 6 proposes a predictive model. Section 6.3 gives the results of the study. Finally, Sect. 7 concludes and discusses possible directions for future research.

2 Related Works

Significant steps in the study of urban mobility on the basis of mobile network data have been made by Carlo Ratti [1]. Based on data on the number and duration of voice calls from base stations, it identified the zones of attraction and areas where users were staying [1]. Based on the intensity of the use of mobile phones (number of calls) and the length of the conversations, I formed a visualization of the dynamics of daily migration of users in urban agglomeration, and based on data on their mobile activities. He also noted the possibility of applying data in the analysis of the movement of passengers between cities, for monitoring the movements of the population in incident and hazardous situations, as well as for analyzing the character of migration during weekdays and weekends. With the great potential of a new data source, he concluded that the biggest shortcoming was precisely the problem of obtaining this type of data, both technically and legally for institutions outside the system. According to [2], two dimensions of mobility are defined, namely mobility as a function of frequency of movement and spatial reach. The frequency of movement represents the number of trips an individual user makes, the higher the frequency, the greater the participation in mobility. The second dimension is spatial reach, where it is worth it: the further a person travels, the greater is his participation in physical mobility. The author also defined the first mobility indicator as the number of trips referring to the number of network activities performed by the user on cells different from the previous one. According to [3], several sources of data were used for the analysis of user movement. data from billing records for mobile services, data from cellular network signaling, data from conventional traffic counters, as well as data on the movement of taxis and public transit vehicles based on records from their internal positioning systems. According to this conclusion, the fusion of data from different sources is more adequate to determine the trajectory of the mobility trajectory. The author also defined an estimate of total personal mobility, consisting of the sum of the straight distances between the areas the user visited during the day, distinguishing between personal mobility related to the vehicle (expressed in vehicle kilometers), as well as personal mobility which is related to other types of traffic. According to [4], the process of determining travel matrices is defined as the process of defining travel between source and destination, with the main emphasis being placed on the model of determining the locations where the user has stayed for an appropriate period of time, that is, by their definition. The source of the trip is defined as a location covered by one or more base stations where the user has been staying for a predefined period of time, so that the possibility of being there only in transit is excluded. They define travel as the shift between two areas of residence. In further elaboration, they also defined the distance traveled by the users, by assigning the most likely road route to the trip, using the criteria of the shortest journey. In the end, they assigned each user a so-called mobility score, consisting of the total number of trips each individual user had, the number of locations they were staying at, and the distance per trip.

According to [5], the mobile network is portrayed through cities as a living ecosystem that is constantly changing and responding to events, such as matches, concerts, traffic closures, etc., whereby data from billing records for telecommunication services is usable for monitoring the collective behaviors, perturbations and fluctuations that take place in complex urban environments. Customer records show intense activity in residential areas outside of normal business hours, as well as intense activity in business areas and downtown during business hours, that is, increased activity in the area of transportation nodes between these stages. Additionally, additional data such as the user's approach angle to the cell, a characteristic timing advance, and signal strength were used to increase localization accuracy over a procedure that uses only the base station position. According to [6], the site was realized to identify the relative activity intensity within the urban area. The area is subdivided into urban cell coverage areas, which are then relative to the number of calls in all cells during this period, and then normalized to their average over a defined period of time. The results indicated significant differences in user activity with respect to time of day as well as with days of the week. He then grouped the areas with similar intensity of activity over a period of time, thus defining the areas by purpose. According to [7], a model of prediction of user movement in urban areas has been developed. The aim of the research was to define the movement patterns of users for traffic planning purposes, as well as for monitoring the spread of infections or diseases. The conclusion was that 93% of travel was predictable based on knowledge of common patterns of user movement, which is a good starting point for estimating emissions. According to [8], the analysis of signaling data on the radio interface of the network was carried out in an attempt to determine the speed of movement of users around the network. The position of the vehicle at each exchange of the signaling message between the mobile terminal and the network (precision of approximately 150 m) has been estimated, and the speed and direction of movement of the vehicle have been determined on the basis of such locations. Compared to data from conventional traffic counters at these locations, the error was found to be 10% in the rural area and 28% in the urban area. According to [9], an analysis of data on established and received calls, incoming and outgoing short text messages and internet access was made to estimate the movement of tourists across Estonia based on the analysis of 9.2 million records. The country of origin of tourists and the residence time at individual destinations were analyzed. The result of the research is correlated with the records of movement and retention of tourists by the state tourist board, with a high correlation coefficient of 0.97. The key disadvantage here is the fact that tourists who do not use a mobile phone are not recorded, which is generally a relatively small number.

According to [10], using the base station changeover data (handover), an attempt was made to determine the frequency of traffic on the road network. The switching data served as a trigger to detect the number of terminals that "passed" through the virtual traffic counter. The number of registered mobile phones and the number of vehicles detected by traditional traffic counters were found to be highly correlated. It should be emphasized here that triggers are key elements for identifying vehicle activity, especially volatile emissions. According to [11], using telecommunication activity samples of six million unnamed country users that were collected over a month, which included telecom activity time, base station ID, and depersonalized user tag, correlations between the number of calls made and the distance traveled were determined. With peak values recorded in the morning and afternoon periods and lowest values recorded at night. According to [12], the possibility of using data from billing records for telecommunication services to identify travel matrices was considered. The conclusion is that to accurately determine travel matrices, it is necessary to analyze the data collected over a long period of time. In the absence of real data, simulated data from the mobile network were used. Based on simulated data on mobile phone usage and actual traffic count data, they concluded that the accuracy of determining the travel matrices relative to the real situation is 95%, with the simulation input (number of users) being weighted by their market shares and the total number of users was estimated based on statistics on penetration of users. The number of users is also mapped to the number of cars, with each car being assigned the same number of users as the average in the car (determined by traffic counts). According to [13], the following mobility indicators were identified: (a) the number of cells (the number of cells from which the user performed network activities over a period of time), (b) the straight distance (the straight distance traveled) as the sum of all the straight distances between the base stations from which successive activity was achieved, then (c) the greatest distance traveled as a straight distance between the two farthest base stations at which the user had his activity and (d) the mass center indicator. The mass center indicator is calculated by establishing mass centers for all cells used by the user and then calculating the straight line distance between all mass centers at which the user's activity is recorded. The value of the center of mass is the root of the mean of the squares of these distances.

According to [14], particular emphasis was placed on the ability to define travel matrices based on the analysis of a set of data from telecommunication service billing records. Analyzing data from billing records for telecommunications services determines travel matrices, which are then validated using existing travel matrices obtained by traditional methods, such as population surveys and traffic counter data. Also, since the position in which the user started to perform telecommunication activity (which does not necessarily coincide with the actual source of the trip) is taken as the source of the trip, they have introduced the term so-called. transient travel matrices, corresponding to journeys with sources and destinations registered in the billing records. Other sources of data were also used (data on trajectories of public transport vehicles, data on the movement of taxi vehicles, data from smart city public transport ticketing systems), as well as tools for microsimulation of traffic flows. The conclusion is that the use of data from communication networks is particularly suitable for generating travel matrices of extremely complex urban areas, where there is limited application of traditional methods, and where a heterogeneous pattern of land use prevails, and in urban areas where travel is numerous and often asymmetrical. According to [14], processes for determining travel matrices based on a depersonalized subset of billing records for telecommunications services have also been identified. The process of determining the source and destination of the trip is defined, as well as the appropriate time and space filters that eliminate the so-called false trips, that is, those that did not occur. Travels are then "counted" between individual areas, on the basis of which travel matrices are defined for predefined time ranges. According to [15], it is possible to combine traffic map data with smartphone location service trajectories and big data analytics to simulate real-time air pollution. Almost every step of modeling air pollution in traffic depends on traffic speed. According to these authors, the traffic map is a very efficient preprocessor for calculating traffic pollution in real time. It should be emphasized here that in addition to traffic speed, there are other factors for estimating traffic emissions.

According to [16], it is concluded that there is a possibility to calculate air pollution in traffic through mobile devices, i.e. by mapping the trajectory of a telephone. The advantage of the process is its low cost, mainly because of the data for calculating the trajectory of a mobile phone and modeling the dispersion of emissions in an area. The combination of an inexpensive and standardized mobile phone provides mobility data across the population of pollution studies. The system is based on fully automatic calculation and real-time data. Each path reflects one individual. Therefore, the opportunity to provide information to each individual opens up. Web server auto-feedback applications can be developed for citizens who want to get their own path information or want to contribute to a data set. An integrated model of air pollution assessment should be developed that takes into account all types of traffic and emission factors at a given trajectory of the user and even electric cars powered by non-renewable energy sources. According to [17], the key advantage of collecting location information from mobile phones is that it allows more people to be tracked in a cheaper and less burdensome way than using a global GPS tracker.

According to [17], existing models used to determine vehicle pollution in traffic flow have their advantages and disadvantages, and they relate to the ability to collect input data and the accuracy of the results. A more model has been developed to estimate the level of pollutant emissions from different modes of transport. A number of models link information about driving conditions, driver and vehicle behavior to environmental factors when evaluating motor vehicle emissions. It should be emphasized here that previous emission models are based on inaccurate data on the relationship between vehicle/user activity and vehicle emissions. Among the methods that partially involve user activity and behavior is the concept of a personal environmental impact report (PEIR) [18]. It is a platform that uses location data sampled from mobile phones to calculate personalized environmental impact and exposure estimates [19].

The authors [20] also suggested the use of a mobile phone with a built-in GPS receiver and an accelerometer to identify the mode of transport, including whether the individual is stationary, walking, running, cycling or motorized transport [20]. Additional research has been carried out under the Seventh Framework Program (FP7) and the Web of Environmental Observation and its Future Internet Services (ENVIROFI) applications, where a mobile-based application designed to provide relatively accurate quality data is designed air at the user's location based on GPS coordinates provided by the user's phone [21]. Also within the CitiSense project [22], a mobile air quality system was developed that allowed users to monitor air quality exposures.

3 System of Automated Mobility

An automated mobility system is defined as a set of mobile solutions that use appropriate technologies and methodologies and the availability of large amounts of data in order to establish the sustainability of transport systems. The benefits of the system include increased traffic safety, reduced individual travel costs, greater personalization of services and sustainable travel. Automated vehicles are networked, and automated mobility technology is advancing and expanding to all types of road transport, including utility vehicles and trucks. Such technologies can significantly reduce traffic congestion, greenhouse gas emissions and air pollutants and increase mobility by opening road transport to the elderly or disabled or reduced mobility. The key issue is protecting data in automated vehicles is how to guarantee computer security and protect automated vehicles from cyber attacks. Since automated vehicles share the road with vehicles operated by drivers, pedestrians and cyclists, safety measures and traffic coordination are needed. One of the challenges is the issue of responsibility. Since vehicles are not driven by humans but by technology, it must be made clear who is responsible for the accident: driver or manufacturer?

On the other hand, such vehicles should respect human dignity and freedom of choice. Laws on automated and interconnected transport should cover all modes of transport, including short sea shipping, inland navigation vessels, unmanned aerial vehicles and light rail systems. Furthermore, international standardization efforts need to be further coordinated to ensure the safety and interoperability of vehicles across borders. Data capture systems should be mandatory in automated vehicles to enhance accident investigations and address liability issues. Appropriate rules need to be quickly developed regarding data protection and ethics in the automated transport sector, as well as increasing citizens' confidence in driverless vehicles. Particular attention should be paid to the development of automated vehicles accessible to persons with reduced mobility or disability.

The elements of automation and vehicle connectivity are contained in:

- vehicles use digital technologies,
- are equipped with sensors, cameras, computers, GPS, satellite receivers and short-range radars and perform all or part of the driving tasks,
- some or all of the driving functions may be transferred to a computer system,
- are equipped with devices to communicate with other vehicles or infrastructure via the Internet,
- related technologies are complementary,
- enable predictive analytics.

On the other hand, the main trends in car connectivity are in data collection, whether it is collective intelligence, where data from multiple cars is collected, through which various services can be offered, or predictive diagnostics, where data is collected from one car. A very important issue is also the layers of protection used in cars. There are control units, that is, hardware protection, which stores all the "keys", a gateway that separates the security domain in the car and the Internet, then a firewall and security.

The open question is how self-driving cars can be integrated into an intermodal mobility system that also includes public transport and shared transport. The goal is to achieve a 4/5 SAE driving system for fully automated, driverless vehicles, including vehicle management software. That is, to develop an intelligent, ready-to-use system that can be integrated into different types and models of vehicles. Efforts to develop vehicle motion management software need not only rely on artificial intelligence and mileage collection. More intensive use of simulations and specially designed test polygons to take account of all driving situations.

4 The Problems of Personalization of Automated Mobility Services

Personalization is the selection, customization, or filtering of information, services or products for a person with respect to his or her characteristics, interests and needs, which defines the profile of that person as user or customer. It is a growing trend in the area of connectivity. Increased personalization of services has brought connectivity to the next phase in development so that devices become intelligent companions that make up everyday life.

According to [23], the concept of automated mobility consists of three key characteristics:

- mobility forecasting,
- optimization of mobility, and
- personalization of mobility.

The prediction feature is the key to adapting the system to changing environmental conditions. Real-time data must be included to anticipate online conditions so that proactive decisions can be made. Feature optimization is key to the decision-making mechanism to achieve system-level goals, such as minimizing travel time or maximizing revenue across the network. On the other hand, the characterization of personalization as a subject of research in this paper is the key to achieving benefits and attracting travelers through understanding their preferences. Of course, it is necessary to have methods for assessing and updating individual-level preferences and being able to optimize travel, provided they are consistent with preferences and maintain system-level goals. Furthermore, smart mobility solutions need to be evaluated in terms of impact at the user and system level, with simulation offering key answers.

Personalized communication between the car and its driver will also be expanded: new features connect the car to its environment, smart home and service. These features will allow for highly automated driving. A car connected to a smart home or smart city via the cloud provides measurable benefits. Connectivity turns the car into a four-wheeled assistant. With personalization of connected solutions, technologies and services are not only tailored to individual needs. Mobility as a service is not such an old idea of selling passengers a personalized package of multimodal mobility services (car sharing, public transportation, taxis, bicycles) that is paid for in a single transaction for each trip or time period, rather than separate transactions with each individual transport service provider.

Personalized mobility service is a traffic strategy that focuses on targeting users and messages to the future user. The process is carried out with the help of data analysis and digital technology. Personalized Mobility Models is a one-on-one concept, as the name suggests, fully tailored to the needs of each individual system user. In order for the concept to be successful, the mobility service provider should be able to obtain as much personal information about the current or potential mobility user as possible. Today, there is a huge sea of information about individual users. This data is then analyzed and harnessed to provide personalized, targeted services, referrals and messages designed to attract the individual based on past behavior, interests and other data [23].

The user profile is defined on the basis of data collected by recording the interaction of users with the transport system, the transactions performed, or by directly searching for preference data. Personalization is possible because of the knowledge of the virtual reflection of traffic system users, the availability of product metadata, and the availability of methods to combine these datasets to generate recommendations for that user. The essence of personalization is to try to get the urban mobility system as close as possible to each user, especially by preparing a unique offer for each user or for each user segment separately. When planning a traffic bid, the wishes of those who influence the target user, such as, for example, other family members, friends and the like, should not be neglected. Personalization should first and foremost focus on what the user wants. Companies can achieve this only if they know the user well and by using different methods and strategies [25].

Today, with the advent of big data, companies can better investigate the interests and preferences of users. The data enables companies to develop personalized services that are even more responsive to customer needs [26]. In order for companies to prepare personalized offers, the four overlapping steps that increase complexity and benefit to the company are important:

- Identification of users of mobility services,
- Differentiation of users of mobility services,
- Interaction with users of mobility services,
- Customizing users with mobility services.

It is not enough just to have user information for the success of the concept of personalized mobility. It is crucial to persuade the right target group, with the right offer at the right time [27]. Adaptation of traffic supply to the needs of the individual is becoming more significant, and with modern tools it is becoming easier. In many cases, a personalized mix service is a prerequisite for gaining new and retaining old users.

5 Identification of Movement Data in Mobile Networks

Mobile phones have become advanced communication devices, used for telephony in the narrow sense, text, image and video messaging, email exchange, internet access, document processing, photo and video recording, listening to radio and music, etc. Due to popularity and easier to use than traditional computers, mobile phones are expected to create a huge space for personalizing the services and content offered to users [28]. In addition to the general features, personalization of the service in the case of mobile networks has two specifics. First of all, it is necessary to adapt the service to the still limited software resources of the mobile terminal [29]. Another feature is the mobility of users, who need to ensure that other activities are carried out smoothly, such as moving on the street or driving public transport while the service is being performed. Research has shown that users treat mobile terminals in a highly personal way [30]. There is no user who has not "personalized" their mobile phone by setting ringtones, screen backgrounds, masks, etc.

According to [31], a system that uses information about a user's current position to provide services and location-based content is described. User position is determined by GPS receivers, infrared sensors, or WLAN sensors. Each measurement represents an ordered four (x_i, y_i, z_i, t_i), the first three members of which are the spatial coordinates of the user and the fourth instant of measurement. The MPS motion profile is a pair (mod_p (S), mod_m (S)), in which S = (p_1, p_2, …, p_n) is the measurement sequence. The first member of the movement profile represents an abstraction of locations from S, and the second an abstraction of the movement itself.

The goal of motion modeling is to provide recommendations to the user of mobility services that are of potential interest to him. This can be done either by collaborative recommending (comparing movement profiles with profiles of other users), or by prediction based on the structural features of the user's movement. Collaborative recommendations are made by recording movement sequences within one area and then defining clusters of related movements. They are then replaced by medioids, curved lines that have the least average distance from all elements of the cluster (Fig. 1).

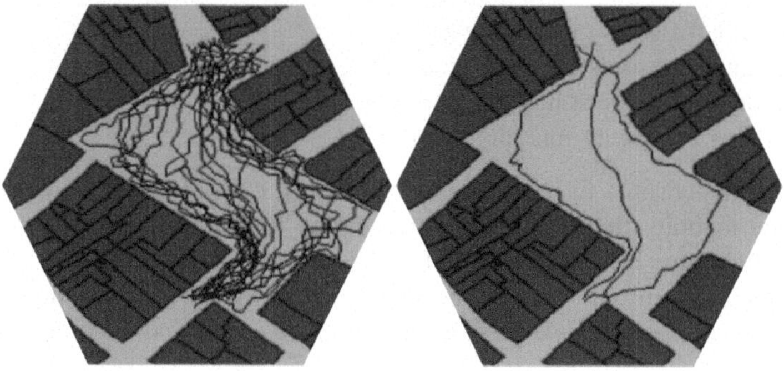

Fig. 1. Typical city sequences (left) and medioids of observed clusters (right)

The distance of the movement sequences S_1 and S_2 is defined by the expression:

$$\mathrm{dist_s}(S_1,\ S_2) = w_e \min\left(\mathrm{dist_e}(S_1,\ S_2),\ \mathrm{dist_e}(S_1,\ S_2)\right) + w_0\, \mathrm{disorder}(S_1,\ S_2) \quad (1)$$

Here we and w_0 are weight coefficients, diste is so-called. edit sequence distance, and disorder the number of positions in S_1 that are in reverse order with respect to S_2. The edit distance is taken from string comparison and corresponds to the minimum number of insertion, deletion, and replacement operations that must be performed to make the first argument equal to the second. By separating the media and comparing the movement profiles of different users within one spatial area, it is possible to identify regularities in the movement and formulate recommendations. The prediction of user movement is made by observing the sequence S. A typical example is shown in Fig. 2.

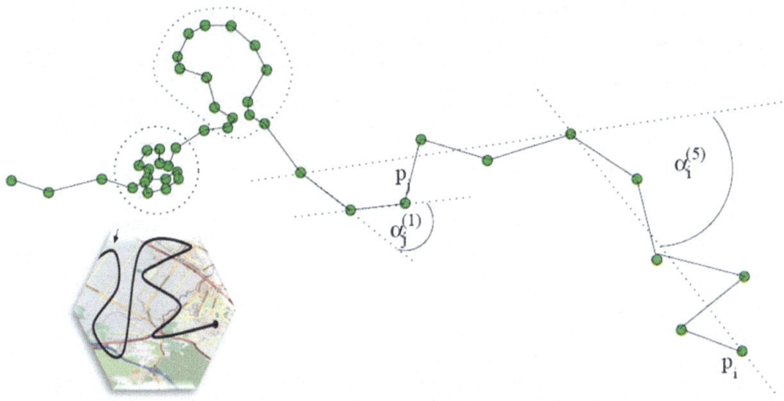

Fig. 2. Sequence of user movement

Sequence analysis *S* identifies the following data and phenomena:

- existence of loops, their duration, speed of movement, frequency of mileage, etc.,
- location retention, average duration, type of sites visited, etc.,
- diverting from the original route, distinguishing between intentional and compulsory routes (e.g. due to a street diversion).

6 Designing a Predictive Model the Personalization of Automated Mobility Services

The predictive model of personalization of automated mobility services is the output generated by the process of predictive analysis of the trajectory of movement of users. The model captures the relationships between past and future behaviors and movement trajectories revealed by the analysis [32]. Once a model has been created, it can be used to generate new predictions about future movement trajectories.

Predictive personalization models of automated mobility services represent the link between how a user behaves as a role model and some of the known user characteristics. The goal is to estimate the likelihood that a similar user from another set will behave the same way. This helps us identify patterns that indicate the performance of the trajectory of a user's movement. The proposed model can even perform calculations at the exact moment when the user makes a real-time motion.

6.1 Creating of Predictive Model

The basic process involved running one or more algorithms against a set of data from a mobile network record with known values for the dependent variable (i.e., what is being attempted). We then split the data set in half and used one set to create a path model and another set up to test the path model. If one wants to predict which user will go on a path, the algorithm references a database of users who have navigated that path over the past 12 months. Then, the result of the path model relative to the other part of the database is run to see how well it predicts which users traveled that path (Fig. 3).

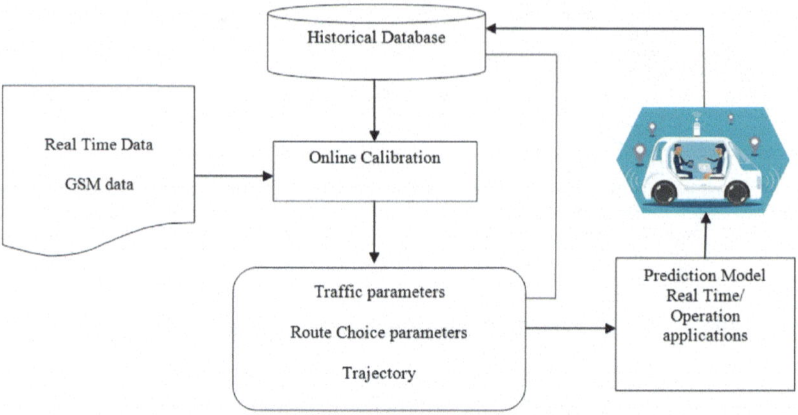

Fig. 3. Model architecture

As one might imagine, the process of requesting a path, testing and validation is iterative, ie it is repeated many times. So it was necessary to analytically identify and examine many combinations of variables to see which ones have the greatest impact. Most begin with the use of statistical tools to identify significant trends in data as well as previous analytical work internally. Also, one can talk to users who are close to the observed movement trajectory and rely on their own knowledge to join the most important variables for inclusion in the model. As a result, the list of variables in the initial version and in the final model was removed. Along the way, different algorithms need to be tested to see what works best on the record dataset. There are various ways to add new types of data or recombine existing fields to improve model accuracy. This iterative process makes it possible to create models that require intense and long-lasting use.

6.2 Benefits of Personalized Model of Automated Mobility Services

The proposed model is designed to generate predictive trajectories of users' movement trajectories and personalize mobility services with behavioral patterns. The two head benefits of this model's predictive are:

- determine the personalization of the service, who will buy the service, and what mobility service to offer them,
- determine which users are most likely to opt out of the mobility service.

Models can be particularly useful in determining which of the current users will reuse the automated mobility service and which services are most commonly used. The advantage of using models on current users is that they usually have much more important data to work with. In the model, user behavior data and trajectories are much more powerful than statistical data. In other words, it is better to predict what users will do in the future based on what they have already done in the past than base the model on who they are.

The question is what could be predicted when a large group of mobility users would respond to the new service offered to them. This is what a predictive model can, that is, determine:

- Which user is likely to use an automated mobility service that they are unlikely to use,
- Which user is at risk of leaving, and who will remain,
- Which mobility services he wants to use most.

Principles underlying the proposed model:

- Potential users of automated mobility and those existing in many segments respond in predictable ways. This predictability is vital because if everyone has unpredictable reactions to services it is difficult to define a model.
- Indications of expected behavior of service users may sometimes be recognized in their previous behavior. Behavior used in predictions is usually stored in databases in the form of transactions. Traffic data can be added from external sources. However, this does not always work. It is not always possible to predict customer behavior based on data in the database.
- The predictive model evolves from responses to previous movement scenarios. It is difficult to run and run a successful predictive model unless information has already been sent to the user to get a reaction from them. This means that one cannot just take a file with names and addresses and create a model that would identify automated mobility users who would be interested in a personalized service.

6.3 Enhancement of the Proposed Predictive Model Using Artificial Intelligence

With artificial intelligence, the predictive model can represent an incredible volume of real-time information about habits and patterns of user movement trajectories in the mobile network. This means that predictive models can consider much more information than ever before, making the result more accurate and effective.

On the other hand, artificial intelligence in just a few moments can achieve the remarkable result of evaluating the behavior patterns of a potential personalized mobility user. It estimates billions of variables in real time, while also making decisions to analyze over one million opportunities per second. When this speed is applied to predictive modeling, the result is real-time decision-making and action.

An example of a basic model of an automated mobility service that incorporates only some of the features of personalization is a smartphone-based system for evaluating individual real-time travel decisions, offering information and incentives to optimize system-wide energy performance. The system-wide incentive strategy is optimized using real-time multimodal traffic network prediction, while the application menu is tailored to each user based on his or her travel preferences and requirements. Joining these methodologies in a convenient and efficient way is crucial for smart automated mobility.

7 Conclusion

The results of the research show that it is possible to design a predictive model for the realization of personalization of mobility automation services. The research results establish and show a clear correlation of historical and realtime spatial data of the movement of users and their activities in the mobile network with indicators of quality of urban and sustainable mobility. The only problem encountered in the practical implementation of the personalization model is the mathematical description of mobility services, content and user preferences based on movement trajectories. Also, a potential danger that needs to be addressed now is the development of particular personalization algorithms. In addition, many more technical problems need to be addressed to make sure that the vehicle is fully capable of feeling and understanding its surroundings and to take appropriate human-like actions predictably. The benefits of the proposed predictive model are that it can determine who will use the automated mobility service and determine which users are most likely to opt out of the service.

References

1. Ratti, C., Sevtsuk, A., Huang, S., Pailer, R.: Mobile landscapes - Graz in real time. In: Gartner, G., Cartwright, W., Peterson, M.P. (eds.) Location Based Services and TeleCartography, pp. 433–444. Springer, Heidelberg (2007)
2. Williams, N.E., Thomas, T.A., Dunbar, M., Eagle, N., Dobra, A.: Measures of human mobility using mobile phone records enhanced with GIS data. PLoS ONE **10**(7), 1–16 (2015)
3. Calabrese, F., Di Lorenzo, G.: Estimating origin - destination flows using mobile phone location data. Cell **10**, 36–44 (2011)
4. Dash, M. et al.: CDR-To-MoVis: developing a mobility visualization system from CDR data. In: Proceedings of the International Conference on Data Engineering, pp. 1452–1455, May 2015
5. Pulselli, R.M., Pulselli, F.M., Ratti, C., Tiezzi, E.: Dissipative structures for understanding cities: resource flows and mobility patterns, pp. 271–279 (2005)
6. Reades, J., Calabrese, F., Sevtsuk, A., Ratti, C.: Cellular census - explorations in urban data collection. IEEE Pervasive Comput. **6**(3), 30–38 (2007)
7. Bengtsson, L., Lu, X., Garfield, R.: Internal population displacement in Haiti, earthquake (2010)
8. Ygnace, J.-L.: Travel time/speed estimates on the french Rhone corridor network using cellular phones as probes. Final report, Serti V Program STRIP. Inst. Natl. Rech. sur les Transp. leur Secur. (2001)
9. Ahas, R., Aasa, A., Mark, Ü., Pae, T., Kull, A.: Seasonal tourism spaces in Estonia - case study with mobile positioning data. Tour. Manage. **28**(3), 898–910 (2007)
10. Ygnace, J.-L.: Travel time/speed estimates on the french Rhone corridor network using cellular phones as probes. Final report, Serti V Program STRIP. Inst. Natl. Rech. sur les Transp. leur Secur. (2011)
11. Candia, J., González, M.C., Wang, P., Schoenharl, T., Madey, G., Barabási, A.-L.: Uncovering individual and collective human dynamics from mobile phone records. J. Phys. A: Math. Theor. **41**, 224015 (2008)

12. Caceres, N., Wideberg, J.P., Benitez, F.G.: Deriving origin–destination data from a mobile phone network. IET Intell. Transp. Syst. **1**(1), 15 (2007)
13. Alexander, L., Jiang, S., Murga, M., González, M.C.: Origin-destination trips by purpose and time of day inferred from mobile phone data. Transp. Res. Part C Emerg. Technol. **58**, 240–250 (2015)
14. Vidović, K., Lučić, A., Šarić, D.: Methodology of O-D matrix determination from telecommunication charging data records, in 36. skup o prometnim sistemima s međunarodnim sudjelovanjem Korema "Automatizacija u prometu 2016" (2016)
15. Skjetne, E., Liu, H.-Y.: Traffic maps and smartphone trajectories to model air pollution, exposure and health. J. Environ. Prot. **8**, 1372–1392 (2017)
16. Liu, H.-Y., Erik, S., Mike, K.: Mobile phone tracking: in support of modelling traffic-related air pollution contribution to individual exposure and its implications for public health impact assessment (2014)
17. Mehanović, M.: Planiranje u saobraćaju prijevozu i komunikacijama, Fakultet za saobraćaj i komunikacije. Univerzitetski udžbenik, Sarajevo (2017)
18. Dijkema, M.: Traffic related air pollution: spatial variation, health effects and mitigation measures. Ph.D thesis. Utrecht University, Institute for Risk Assessment Sciences (2011)
19. UCLA CENS: PEIR: personal environmental impact report. http://www.viewingspace.com/peir.html
20. Mun, M., Reddy, S., Shilton,K., Yau, N., Burke, J., Estrin, D., Hansen, M., Howard, E., West R., Boda, P.: PEIR, the personal environmental impact report, as a platform for participatory sensing systems research. In: Zielinski, K., Wolisz, A., Flinn, J., LaMarca, A. (eds) Proceedings of MobiSys 2009: The 7th International Conference on Mobile Systems, Applications and Services, 22–25 June 2009, Wroclaw, Poland, pp. 55–68. ACM, New York (2009)
21. Reddy, S., Mun, M., Burke, J., Estrin, D., Hansen, M., Srivastava, M.: Using mobile phones to determine transportation modes. ACM Trans. Sensor Netw. (TOSN) **6**(13), 1–13 (2010)
22. ENVIROFI: The environmental observation web and its service applications within the future internet. http://www.envirofi.eu/
23. Atasoy, B., de Azevedo, C.L., Akkinepally, A.P., Seshadri, R., Zhao, F., Abou-Zeid, M., Ben-Akiva, M.: SMART mobility via prediction, optimization and personalization (2018)
24. Galetto, M.: What is predictive marketing? Preuzeto 22 July 2018 (2018). https://www.ngdata.com/what-is-predictive-marketing/
25. Bordon, A., Pipuš, A.: Personalizirani marketing postaje temelj uspješne prodaje. Preuzeto 5 September 2018 (2014). http://www.media
26. Zhang, A.: Data Analytics, Poljska, Amazon Fulfillment (2017)
27. Bordon, A., Pipuš, A.: Personalizirani marketing postaje temelj uspješne prodaje. Preuzeto 5.9.2018 (2014). http://www.media
28. Subramanya, S.R., Yi, B.K.: Mobile content customization. IEEE Multimedia **12**(4), 103–104 (2005)
29. Lan, T.-H., Tewfik, A.H.: A resource management strategy in wireless multimedia communications – total power saving in mobile terminals with a guaranteed QoS. IEEE Trans. Multimedia **5**(2), 267–281 (2003)
30. Ichikawa, F., Chipchase, J., Grignani, R.: Where's the phone? A study of mobile phone location in public spaces. Proceedings of the IEEE Mobility Conference, Guangzhou, China (2005)
31. Bauer, M., Deru, M.: Motion-based adaptation of information services for mobile users. In: Ardissono, L., Brna, P., Mitrovic, A. (eds.) User Modeling 2005. LNCS, vol. 3538, pp. 271–276. Springer, Heidelberg (2005)
32. Finlay, S.: Predictive Analytics in 56 Minutes. CreateSpace Independent Publishing Platform, Scotts Valley (2015)

Logistics Optimization of Agricultural Products Supply to the European Union Based on Modeling by Petri Nets

Olexiy Pavlenko[1(✉)], Natalya Shramenko[2,3], and Dmitriy Muzylyov[2]

[1] Kharkiv National Automobile and Highway University, Yaroslava Mudrogo Str. 25, Kharkiv 61002, Ukraine
ttpov@ukr.net

[2] Kharkiv Petro Vasylenko National Technical University of Agriculture, Kharkiv, Ukraine
nshramenko@gmail.com, murza_1@ukr.net

[3] Ukrainian State University of Railway Transport, Kharkiv, Ukraine

Abstract. The using of digital logistics, based on intensive developments of smart technologies, into the system of product supply allows achieving optimal conditions for a certain period with the expected efficiency level. At the same time, it is necessary to consider increasing demand trend for these services in the sphere of agro-products supply from Ukraine to European Union markets. The use of "smart systems, "Internet of Things" (IoT), big data allows to quickly obtain info from hardware and other drives: technical tools installed on trucks, containers, ships, for operational monitoring and control of technological processes, as well as for making accurate forecasts. The presented supply system is distributive and parallel working with asynchronous interactions. Therefore, it is proposed to use the type of Petri nets - CPN tools to study systems behavior in enough depth and to obtain information about their most important characteristics. Models developed based on Petri nets have shown that in addition to clarity and simplicity in application, they allow considering various probabilistic factors. This fact allows us to simulate, analyze and predict works for all system participants according to promptly calculated values for each specific management object.

Keywords: Robot · Stock · Europe · Smart · Models · Petri Nets · Management · Agricultural products · Internet of Things

1 Introduction

Cargo transportations are now seen as a complex system influenced by globalization effects, the integration of different transport kinds, geographically distributed operations and expanded business models [1]. This complexity is also compounded by the need to respond in real-time to unexpected situations detected during the transportation phase (e.g. weather conditions, strikes, accidents). Thus, critical problems are timely

I. Karabegović (Ed.): NT 2020, LNNS 128, pp. 596–604, 2020.
https://doi.org/10.1007/978-3-030-46817-0_69

delivery of cargo, transportation efficiency to meet requests of a very demanding market. The ability to detect and remove all possible obstacles and risks in real-time becomes the main competence for logistics companies [2, 3].

The active use of modern information technologies has made it possible in recent years to accelerate processes of obtaining, presenting, accumulating, processing and using the information in the cargo delivery system [4]. But carriers are also difficulties and problems in creating reliable information systems in companies on this stage. The main problem is the lack of operation information gathering from stakeholders. In general, information is not consistent with reality. Most companies suffer losses due to late or unreliable operational data received at each stage of supply chains [5]. Today, attention is increasingly drawn to information flows, through which material flows are planned, managed it and controlled it [6]. Improved informativity and organization can often have a greater impact than technological innovation [7].

Ukraine is currently one of the world's leading exporters of agricultural products. The agro-industrial complex (agribusiness) is a priority sector of the Ukrainian economy and, soon, its entry strategies into European Union (EU) market are essential for future developments of Ukraine. Agricultural products were exported to the EU by 4.76 billion US dollars during periods January-September 2019, which is 106.9% more than last year [8]. This volume is 30% of total exports. The largest volume of agricultural exports (AP) is 1.06 billion US dollars (Table 1) to the Netherlands. The grain cargoes (Table 2) have largest volumes from 16 items shipped to this country.

Table 1. Agricultural exports volumes from Ukraine to ten EU countries

№	Country	Export amounts, thousands of US dollars
1	Netherlands	1062586,56
2	Spain	765642,13
3	Germany	596009,22
4	Italy	456854,32
5	Poland	382205,05
6	Belgium	342315,23
7	France	198895,59
8	Great Britain	190718,25
9	Portugal	146400,54
10	Lithuania	101443,46

Table 2. Agricultural exports volumes from Ukraine to the Netherlands

№	Products	Export amounts, thousands of US dollars	Current period percentage to previous year
1	Live animals of husbandry	42,3	184,3
2	Meat and edible subproducts	80724,8	87,0
3	Fish and crustaceans	605,4	109,6

(*continued*)

Table 2. (*continued*)

№	Products	Export amounts, thousands of US dollars	Current period percentage to previous year
4	Milk and dairy products, poultry eggs; natural honey	6680,2	60,3
5	Vegetables	1210,2	109,6
6	Edible fruits and nuts	7033,1	80,0
6	Coffee, tea	288,8	67,9
7	Grain crops	463152,2	138,8
8	Flour milling and cereal industries products	971,2	97,4
9	Seeds and fruits of oilseeds	161930,0	203,0
10	Plant materials for manufacture	127,3	114,0
11	Fats and oils of animal or vegetable origin	335210,4	190,4
12	Meat, fish products	219,8	108,9
13	Sugar and sugar confectionery products	135,1	1323,6
14	Cocoa and its products	726,8	75,5
15	Finished grain products	2133,4	140,1
16	Processed vegetable products	1394,8	31,0

Thus, effective structures of interaction between process participants must be defined in order to form optimal logistics for agricultural product supply from Ukraine to the EU. This structure must be based on smart solutions and considering also trends in increasing demand for these services, risk components, failure levels, parallelism, and glitches interaction.

2 Literature Analysis

Effective supply tracking is critical to managing global trade and logistics [9]. To cope with the complexity of growing clients' requirements, the organization and management of supply chains constantly move to cooperation, intellectuality and service orientation [10–12]. Many logistics companies in most world countries have made traceability a mandatory procedure in supply chains [13]. And flow management seems challenging for multimodal transportation chain participants [14, 15]. It is the main driving force in supply chains. What is proved by proposing an intelligent traceability system for goods movement in containers in the context of multimodal transport [16]. Some works also propose a system of tracking movement in products supply chains, which has many links and complex factors [17]. At the same time, models of studied objects are optimized by using modern information systems and technologies [18–22].

Delivery system construction of agro-products based on smart solutions takes place on the generalization of all information, setting acceptable deadlines for works performance, determining resources adequacy, providing all participants of the process with clear, accurate, specific tasks [23, 24]. Neural networks are mainly used to solve problems on optimization of supply logistics functioning and predictive models' construction [25–27]; also, Petri networks are ordinary, colored and hierarchical in its various implementations [28–31]. This study will use color Petri Nets - CPN Tools, to determine real-time object behavior and evaluate system performance.

3 Design of Model

The delivery of expensive and important cargoes requires continuous monitoring and tracking of delivery [32]. Most existing tracking methods use RFID (radio frequency identification), GPS (global positioning system), NFC (near field communication), RTLS (real-time location system) [33, 34]. Creating information systems requires systems thinking. Logistics system structure of enterprises, material flow, which providing logistics information systems are interrelated and interdependent. Therefore, in order to design the logistics of agricultural products supply, it is necessary to determine participants and levels of their interaction (Fig. 1). The main element of interaction is

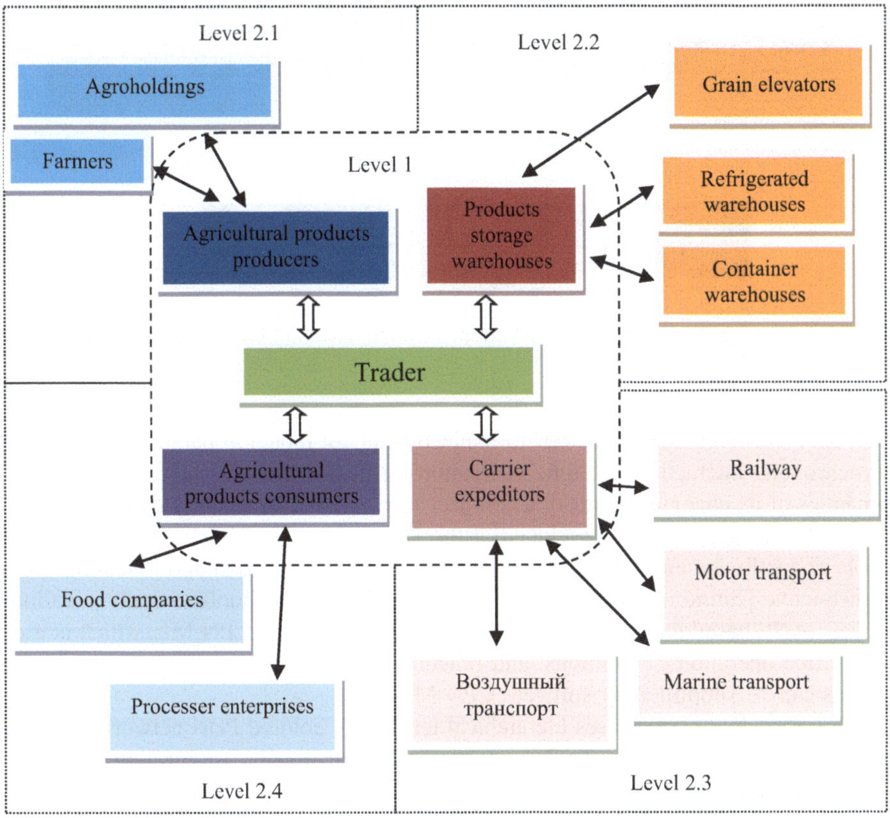

Fig. 1. Information exchange between participants of supply logistics on two levels

information exchanges. Levels (2.1–2.4) are formed based on information generalizations. This enables you to create a database for each management object. Amount of input and output data must be determined to create a database of each level 1 element. For example, traders are producers of agricultural products (Fig. 2).

Presentation of products delivery logistics should be based on objects structure descriptions that making up systems, their attributes, operations, relationships with other objects [35]. This possibility is provided by object-oriented modeling, which has unique characteristics, is a simple and concise method of modeling and competent capabilities of mathematical analysis in any activity fields, including logistics systems [36]. The object-oriented model should reflect concepts and objects of the real world that are important for systems being developed. It is object-oriented technology that combines with Petri Nets [37].

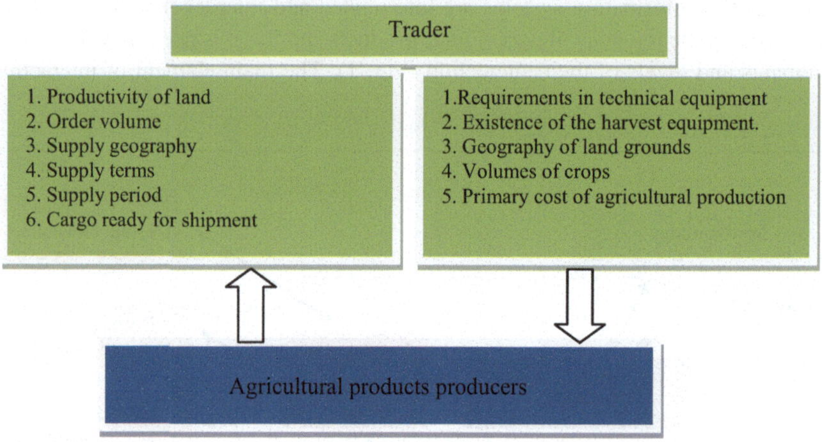

Fig. 2. Scheme of database formation at the first interaction level: trader - producer of agricultural products

Petri Net is an evident and well-formalized behavior model of parallel systems with asynchronous interactions. It reflects relationships structure of system elements and dynamics of its changing states in a compact form under the set entry conditions [38]. The tool uses the program developed at the University of Aarhus (Denmark) - CPN Tools as a software environment for the study of Petri Nets (SPs) and modeling. [39]. A distinctive feature of CPN Tools is the availability of rich tools to analyze various aspects of SP-based model performance. CPN Tools supports CPN ML, which is used to describe operations, conditions, and functions. Process of designing and analyzing models can be simplified by software CPN ML.

This modeling system uses hierarchical temporary colored Petri networks to build models. Simulation in CPN Tools is discrete and situationally, suggesting an instantaneous change in Petri network state at certain points in time, called steps. Timestamps allow simulating processes based on runtime [31]. Model for receiving information from participants in the delivery process (Fig. 3) was design based on the Fig. 1: trader – t1, agricultural products producers – t2, Agricultural products consumers – t3, Products storage warehouses – t4, Carrier expeditors – t4. Accumulation and determination of information takes place in transitions p1–p8. It is assumed to design models of information exchange between participants of supply logistics in future studies. There must also consider material flows that directly affect rates of data receipt and processing in systems. These models will allow obtaining forecast data for each control object and for the whole system in dynamics [40].

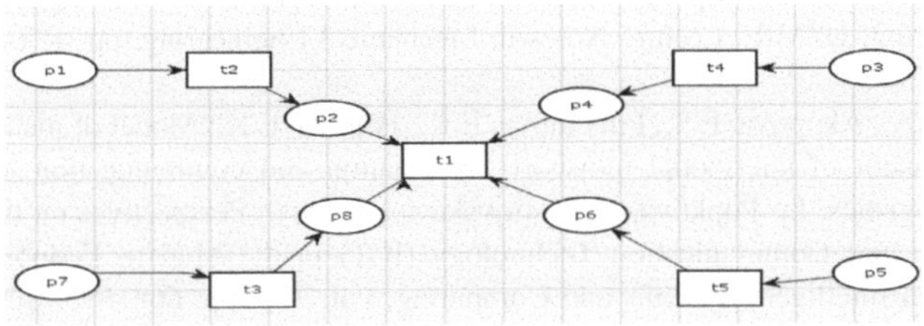

Fig. 3. Schematic representation model when trader receiving information from Level 1 process participants in software CPN Tools

4 Conclusion

The new tendency was identified in the Industry 4.0 context. This fact has made it possible to create high-level requirements for information exchange. It is SMART solutions that allow you to achieve optimal conditions for a certain period with expected efficiency levels. Such temps of technology development required to use early forecasting tools. And so that the data can be systematically processed into information to explain uncertainties and thus make more "reasonable" decisions.

Design of agricultural products supply systems based on smart solutions took place by accumulating all information, defining acceptable terms of works performance, finding resources adequacy, provision of clear, accurate, specific tasks to all participants process. This was considered in supply logistics models design based on Petri nets - CPN Tools. It has been identified that this tool will allow to simulate, analyze and predict works of all system participants according to the promptly calculated values for each specific management object.

Model's example for receiving information by a trader from participants in the process of supplying agricultural products at the first level in software CPN Tools has been developed. Thus, a colored Petri net is one of the best for information classification and its transmission, and integrated information management is an effective system of risk identification and decision-making in the supply chain of agricultural products.

References

1. Costa, R., Jardim-Goncalves, R., Figueiras, P., Forcolin, M., Jermol, M., Stevens, R.: Smart cargo for multimodal freight transport: when "Cloud" becomes "Fog". IFAC-Papers OnLine **49**(12), 121–126 (2016)
2. Gregora, T., Krajčoviča, M., Więcek, D.: Smart connected logistics. Procedia Eng. **192**, 265–270 (2017)
3. Shramenko, N.Y., Shramenko, V.O.: Mathematical model of the logistics chain for the delivery of bulk cargo by rail transport. Sci. Bull. Nat. Mining Univ. **5**(167), 136–141 (2018)
4. Peraković, D., Periša, M., Sente, R.E.: Information and Communication Technologies Within Industry 4.0 Concept. In: Ivanov, V. et al. (eds.) Advances in Design, Simulation and Manufacturing, DSMIE. Lecture Notes in Mechanical Engineering, pp. 127–134. Springer, Cham (2018)
5. Zelikov, V.A., Akopova, E.S., Pilivanova, E.K., Popova, L.K.: Model of management of the risk component of intermodal transport: information and communication technologies of transport logistics. In: Popkova, E., Ostrovskaya, V. (eds.) Perspectives on the Use of New Information and Communication Technology (ICT) in the Modern Economy. ISC 2017. Advances in Intelligent Systems and Computing, vol. 726, pp. 668–695. Springer, Cham
6. Shramenko, N.Y., Shramenko, V.O.: Optimization of technological specifications and methodology of estimating the efficiency of the bulk cargoes delivery process. Naukovyi Visnyk Natsionalnoho Hirnychoho Universytetu, vol. 3, pp. 146–151 (2019)
7. Velykodnyi, D., Pavlenko, O.: The choice of rational technology of delivery of grain cargoes in the containers in the international traffic. Int. J. Traffic Transp. Eng. **7**(2), 164–175 (2017)
8. Officialsite of Derzhkomstatu Ukrayiny. http://www.ukrstat.gov.ua/operativ/operativ2019/zd/kr_tstr/arh_kr_2019.htm
9. Hasan, H., AlHadhrami, E., AlDhaheri, A., Salah, K., Jayaraman, R.: Smart contract-based approach for efficient shipment management. Comput. Ind. Eng. **136**, 149–159 (2019)
10. Pan, S., Zhong, R.Y., Qu, T.: Smart product-service systems in interoperable logistics: design and implementation prospects. Adv. Eng. Inform. **42**, 123–132 (2019)
11. Wang, Q., Luo, R.: The mechanism and empirical study of intelligent logistics technology improving the efficiency of logistics industry - taking the "Core Area" of the silk road economic belt as example. Procedia CIRP **83**, 285–291 (2019)
12. Giustia, R., Manerbaa, D., Brunoa, G., Tadeia, R.: Synchromodal logistics: an overview of critical success factors, enabling technologies, and open research. Transp. Res. Part E **129**, 92–110 (2019)
13. Konovalenko, I., Ludwig, A.: Event processing in supply chain management – The status quo and research outlook. Comput. Ind. **105**, 229–249 (2019)
14. Shramenko, N.: Evaluation of the effectiveness of piggyback traffic in the context of creating transport and logistics clusters. Sci. Bull. Nat. Mining Univ. **6**(162), 151–155 (2017)
15. Gökkuş, Ü., Yıldırım, M.S., Aydin, M.M.: Estimation of container traffic at seaports by using several soft computing methods: a case of Turkish Seaports. Discrete Dyn. Nat. Soc. **2017**, 1–15 (2017)
16. Abed, M., Charfeddine, I., Benaissa, M., Starostka-Patyk, M.: Intelligent traceability system of containerized goods. Appl. Mech. Mater. **309**, 241–251 (2013)
17. Muzylyov, D., Kravcov, A., Karnayh, N., Berezhnaja, N., Kutiya, O.: Development of a methodology for choosing conditions of interaction between harvesting and transport complexes. Eastern Eur. J. Enterp. Technol. **2**(3), 11–21 (2016)
18. Li, F., Zhu, Y.P., Wu, H.R.: Modeling and optimization of traceability system for agriculture products supply chain. Adv. Mater. Res. **605–607**, 574–579 (2013)

19. Qu, J.H., Yao, X.S., Ying, J.L.: Agricultural products logistics operational pattern based on information center. Adv. Mater. Res. **361–363**, 1679–1683 (2012)
20. Vendrell-Herrero, F., Bustinza, O.F., Parry, G., Georgantzis, N.: Servitization, digitization and supply chain interdependency. Ind. Mark. Manag. **60**, 69–81 (2017)
21. Xue, L., Zhang, C., Lingand, H., Zhao, X.: Risk mitigation in supply chain digitization: system modularity and information technology governance. J. Manag. Inf. Syst. **30**, 325–352 (2013)
22. Aulin, V., Lyashuk, O., Pavlenko, O., Velykodnyi, D., Hrynkiv, A., Lysenko, S., Holub, D., Vovk, Y., Dzyura, V., Sokol, M.: Realization of the logistic approach in the international cargo delivery system. Commun. Sci. Lett. Univ. Zilina **21**(2), 3–12 (2019)
23. Korczakand, J., Kijewska, K.: Smart logistics in the development of smart cities. Transp. Res. Procedia **39**, 201–211 (2019)
24. Tangaand, C.S., Veelenturf, L.P.: The strategic role of logistics in the industry 4.0 era. Transp. Res. Part E **129**, 1–11 (2019)
25. Shramenko, N., Muzylyov, D.: Forecasting of overloading volumes in transport systems based on the fuzzy-neural model. In: Ivanov, V. et al. (eds) Advances in Design, Simulation and Manufacturing II. DSMIE2019. Lecture Notes in Mechanical Engineering, pp. 311–320. Springer, Cham (2020)
26. Jin, C.F., Yang, H.M., Wang, L.L.: Research on optimization and debugging simulation model of logistics center based on neural network. Appl. Mech. Mater. **37–38**, 1060–1063 (2010)
27. He, W., Lu, T., Yu, C.Q.: A novel traffic flow forecasting method based on the artificial neural networks and intelligent transportation systems data mining. Adv. Mater. Res. **842**, 708–711 (2014)
28. Shramenko, N., Pavlenko, O., Muzylyov, D.: Information and communication technology: case of using petri nets for grain delivery simulation at logistics system. In: CEUR Workshop Proceedings, vol. 2353, pp. 935–949 (2019)
29. Stetsenko, I.V., Dyfuchyn, A., Dorosh, V.I.: Petri-object simulation: sofyware package and complexity. In: Intelligent Data Acquisition and Advanced Computing Sysytems: Technology and Applications, 1 (IEEE 8th International Conference), pp. 381–385 (2015)
30. Gaied, M., M'halla, A., Lefebvre, D.: Robust control for railway transport networks based on stochastic P-timed Petri net models. Proceedings of the Institution of Mechanical Engineers, Part I: Journal of Systems and Control Engineering **233**(7), 830–846 (2019)
31. Naybour, M., Remenyte-Prescott, R., Boyd, M.J.: Reliability and efficiency evaluation of a community pharmacy dispensing process using a coloured Petri-net approach. Reliab. Eng. Syst. Safe. **182**(C), 258–268 (2019)
32. Yang, G.H., Xu, K., Li, V.O.: Hybrid cargo-level tracking system for logistics. In: Vehicular Technology Conference (VTC 2010-Spring, IEEE 71st), pp. 1–5 (2010)
33. Marinagi, C., Skourlas, C., Galiotou, E.: Advanced information technology solutions for implementing information sharing across supply chains. In: Proceedings of the 22nd Pan-Hellenic Conference on Informatics, pp. 99–102 (2018)
34. Yaqiong, L.V., Lei, T., Lee, C.K.M., Xin, T.: IoT based Omni-Channel Logistics Service in Industry 4.0. In: International Conference on Service Operations and Logistics, and Informatics (SOLI, IEEE), pp. 240–243 (2018)
35. Kamburjan, E., Hähnle, R., Schön, S.: Formal modeling and analysis of railway operations with active objects. Sci. Comput. Program. **166**, 167–193 (2018)
36. Liu, Y.P., Fu, J.H., Feng, H.Y.: A survey on the object-oriented Petri net modeling. Appl. Mech. Mater. **121–126**, 4350–4356 (2012)

37. Ghoniem, R.M., Abas, H.A., Bdair, H.A.: Three-dimensional simulation system based intelligent object-oriented paradigm for conducting physics experiments. Procedia Comput. Sci. **135**, 490–502 (2018)
38. Chinnusamy, T.R., Karthikeyan, T., Krishnan, M., Murugesan, A.: A comprehensive survey of flexible manufacturing system scheduling using Petri nets. Adv. Mater. Res. **984–985**, 111–117 (2014)
39. A tool for editing, simulating, and analyzing Colored Petri nets. http://cpntools.org/
40. Liu, L., Liu, X., Liu, G.: The risk management of perishable supply chain based on coloured Petri Net modeling. Inf. Process. Agric. **5**(1), 47–59 (2018)

Directions for Quality Assurance of Specialists Training in Logistics and Transport Spheres from a Competence Approach Perspective

Natalya Shramenko[1,2(✉)], Dmitriy Muzylyov[1(✉)], Vladyslav Shramenko[3,4(✉)], and Pavlo Mazyliuk[5(✉)]

[1] Department Transport Technology and Logistics, Kharkiv Petro Vasylenko National Technical University of Agriculture, Alchevskyh Str. 44, Kharkiv 61002, Ukraine
nshramenko@gmail.com, murza_l@ukr.net

[2] Department of Operational Work Management, Ukrainian State University of Railway Transport, Feierbakh Square 7, Kharkiv 61050, Ukraine

[3] Faculty of Mathematics and Computer Science, V. N. Karazin Kharkiv National University, Svobody Sq. 4, Kharkiv 61022, Ukraine
shramenko.vlad@gmail.com

[4] Faculty of Technological Systems and Logistics, Kharkiv Petro Vasylenko National Technical University of Agriculture, Alchevskyh Str. 44, Kharkiv 61002, Ukraine

[5] Department of Cars and Transport Technologies, Lutsk National Technical University, LvivskaStr. 75., Lutsk 43018, Ukraine
mazylyuk@ukr.net

Abstract. An experience analysis of educational systems in many countries shows that one way to update education content and educational technologies is to apply a professional approach that only complements the list of educational innovations without reducing classical approaches importance. The analysis of recent studies has been carried out. It helps to determine that a competency approach implies the need to strengthen practical orientations of education. The prerequisites for the realization of a competence approach in education are defined, due to the European integration of Ukrainian higher school, with features of modern society live and increasing requirements by employers. Professional activity features for specialists in logistics and transport spheres at modern conditions are described. These specific skills are consisting in intense intellectual activity, ability to make decisions independently and quickly, to use information and communication technologies for optimization of technological processes in transport. The directions of education to ensure the quality of training and increase the competence of specialists in the field of logistics and transport are proposed. They based on the close interaction of theory and practice, which contribute to the graduate's acquisition of the potential of competitiveness in constantly changing conditions, aimed at developing the abilities of students to solve professional problems and problems with the application of knowledge, skills, and professional experience.

I. Karabegović (Ed.): NT 2020, LNNS 128, pp. 605–611, 2020.
https://doi.org/10.1007/978-3-030-46817-0_70

Keywords: Education · Competent approach · Competence · Quality training of specialists · Skills

1 Introduction

The vocational training process of specialists in technical spheres, including in the specialty "Transport technologies," is characterized by the integration process into European systems. An analysis of experiences about educational systems shows in many countries that one way to update education contents and educational technologies is to adopt a competency approach that only complements the list of educational innovations without reducing classical approaches importance.

The main task for higher educational institutions is to ensure the high quality of students' vocational training [1] at initial stages when Ukrainian education integrates with the world educational space. That's why is necessary to modernize education contents and to view goals and results of vocational education [2].

Process of students professional education for Ukrainian transport industry, including in "Transport technologies", is characterized by integration processes into the European system, current trends of social and economic development in country, considering changing conditions at global markets [3], also, by rates acceleration of society development [4, 5] and general population informatization [6]. Employers are making increasing demands for professional personnel training, namely, availabilities of systemic knowledge and communicative qualities (soft skills); ability to self-organization; qualities to think creatively; skills to operate effectively in a global market environment. Therefore, it is necessary to introduce an innovative approach to the training process organization of specialists in transport systems, which would be considered with employers' requirements, on the one hand, and with aims of higher education, on others.

2 Literature Review

It should be noted that researchers pay much attention to problems of approaches realization in education based on competency [2]. In [7] the competent approach is considered as a type of education contents, which are not limited to a knowledge-oriented component, but involves a holistic experience of solving life problems, performing key functions, social roles, competences.

References [7–9] present many concepts interpretations "one competency", "general competencies "Sources analysis shows that all definitions proposed by scientists of these concepts are not contradictory but complement each other. Researchers prove that concepts of "one competency" and "general competence" are different, interrelated and interdependent [10], that is, one competency is realized in the general competence [11]. Scientists conclude that one of the education results is the acquisition by person competencies set necessary for activities in various spheres of public life.

The authors consider competence as an integral characteristic of a specialist, which determines his/her skills to adapt quickly to professional activity and to solve specific problems arising in various spheres of life by using knowledge, educational, professional and life experience in accordance with the developed value guidelines [12].

Competency approach is seen as the basis for education modernization in scientific publications [13], as an innovative approach to professional training organization [14], where attention is focused on education results rather than on process. Competency approach should be considered not only as a means of updating education contents but also as a mechanism of bringing it into line with modern times requirements. This definition highlights as the main thought in research [10]. The competency approach generally essence lies in practical-oriented focuses of educational processes.

The [15] considers introducing specifics competency approach into education practices and states that recent scientific research has developed a steady tendency to assert not only concepts of a competency approach but also to understand its essence since competency approach implementation is based on understanding that the progress of mankind depends primarily from individual-level development.

Scientific publications analysis showed that there is a significant discrepancy in the content and nomenclature of basic competences in various education fields [16]. This shows that studies of competency approach and determination of a stable set of key competencies need to be carried out depending on the type of university.

3 Aim and Research Tasks

The research purpose is to define education focuses to ensure training qualities of specialists in fields of logistics and transport from positions of a competency approach.

Research tasks are:

- prerequisites definition for competency-based approach realization in education;
- determination of features a professional activity for experts in fields of logistics and transport due to modern conditions;
- events definition for ensuring quality of preparation and increase specialists competence in logistics and transport spheres.

4 Activities to Enhance Competencies of Logistics and Transport Professionals and to Ensure Their Training Qualities

Summarizing the various author's visions of concepts essences "one competency" and "general competencies", it can be concluded that one competency is expected and measurable achievements, which confirm a person's abilities to perform certain actions independently, after studying corresponding knowledge system.

The concept of "general competencies" can be interpreted as educational activities results of the individual, abilities to realize acquired knowledge and skills in practical activities and various life situations. That's why it is a personal characteristic of specialists who owns a few relevant competences.

Prerequisites have been identified for realization competence approach in education sphere as analysis results:

- features of modern society development: legislation updates, informatization of all social spheres, development and introduction in practice of modern information and communication technologies, the formation of a common information space, entry into world information space;
- constantly increasing requirements of society to ensuring high-quality education at all its levels;
- necessities of mastering in graduate competitiveness potential for the modern, constantly changing social and economic conditions modern multiple-factor socio-political, market and economic, information and communication-saturated space;
- ever-increasing requirements of employers concerning presence at university graduates of system, intellectual, communicative qualities, skills to self-organization, self-updating, to self-learning ability;
- the integration processes of the Ukrainian education system into the world educational space;
- world economy globalization.

Switching to a competency approach means:

- reorientation from process to education results;
- accent shift from accumulation of standardly certain knowledge, skills on formation and development in students of ability practically to work, apply the gained knowledge in concrete situations.

It is necessary to consider professional activity features of specialists in logistical and transport spheres for modern conditions according to order to implement a competence approach in education:

- the highly intense intellectual activity of specialists caused by information environment nature in which transport experts, and complexity of problems which require operational solutions works (in minimal terms);
- purposeful use of information technologies for technological processes optimization on transport and control of vehicles condition, road and transport network and transport systems in general;
- continuous control of changes in legislation and a regulatory framework in the transport industry;
- skill to communicate - abilities to contact people and have skills to guide in an operational situation;
- independence which demands from workers of skill independently to make decisions, with own hand to fix and confirm the facts, to operate;
- self-education throughout all professional activity.

Measures to guaranty training qualities of specialists in fields of logistics and transport according to the competence approach on vocational-oriented disciplines examples of the Department of Transport Technologies and Logistics of Kharkiv Petro Vasylenko National Technical University of Agriculture:

- reflection of system approach found to display in formation of "problem" lectures where questions from classical theoretical approaches before practical application of modern scientific developments are considered [17, 18];

- theoretical courses of some disciplines are implementation basis of master's and baccalaureate qualification theses;
- revitalization of studying dialogue and educational discussion during lecture and seminar classes;
- approximation research process and design to real operating conditions of enterprises;
- real data from manufacturing and transport enterprises received in technological practice are used as a design and statistical data;
- disciplines courses use wide ranges of mathematical methods, such as the theory of chances and mathematical statistics, mass service theory, graph theory, network planning, game theory, fuzzy sets theory [19, 20];
- all opportunities are created to students for actively applied language skills of programming and computer modeling basis, abilities to use modern computer technologies [21];
- packages of application programs are provided to students at the solution of practical problems: transport task solution, shortest distance search on transport networks, delivery route formation;
- students can use possibilities of scientific and methodical and laboratory bases of the department where in video formats in the real-time is possible to study performance technologies of some specific operations and to fix it duration;
- modern educational technologies are applied in distance courses form to training skills for independent work at students;
- conditions for exchange with ideas, knowledge, abilities and scientific-practical experience between students and lecturers from different countries are created.

It is necessary to highlight characteristic results of education in competence approach implementation:

- the student - expansion of academic and professional recognitions, mobility, comparability, and compatibility of diplomas and qualifications, competitiveness in the labor market, successful socialization in society, personal development, a self-improvement possibility;
- employer - employment of experts who are almost trained capable of fast adaptations to professional activity, having skills creatively to think, trained to work actively and structurally;
- lecturer – professional growth;
- university – positive image strengthening;
- society – improvement of higher education quality at all its levels, rates growth of spiritual and general culture.

5 Conclusion

Thus, the quality improvement of specialists training for transport systems is based on the practical-orientation strengthening of educational processes and is characterized by orientation towards a competency approach.

As a result of the study, the prerequisites for competency approach realization in education are defined, due to the European integration of Ukrainian universities and increasing requirements of modern society.

Professional activity features for specialists in logistics and transport spheres at modern conditions are described. These specific skills are consisting in intense intellectual activity, ability to make decisions independently and quickly, to use information and communication technologies for optimization of technological processes in transport.

Directions of education are proposed for ensuring high training qualities and increasing competencies of specialists in logistics and transport spheres. This approach must be based on close interactions of theory and practice, aimed at developing students' abilities to solve professional tasks and problems with knowledge application, skills and professional experience.

A promising direction of studies is standards development for appropriate level formations of professional competencies at students with practical-oriented approaches introduction.

References

1. Lyashenko, O.: Quality of education as a basis for the functioning and development of modern education systems. Pedagogy Psychol. **1**(46), 5–12 (2005)
2. Kremen′, V.: New requirements for quality education. Educ. Ukraine **45–46**, 6–7 (2006)
3. Bosov, A., Khalipova, N.: Formation of separate optimization models for the analysis of transportation-logistics systems. Eastern-Eur. J. Enterp. Technol. Control Processes **3**(87), 11–20 (2017)
4. Kapskij, D.: Development of the system of road traffic safety improvement in accident seats of urban areas. Transp. Telecommun. **10**(1), 30–37 (2009)
5. Kapskij, D., Korzhova, A.: The analysis of various measures of the speed control of the traffic in the cities and settlements of the Republic of Belarus. J. Sustain. Dev. Transp. Logistics **2**(2), 26–34 (2017)
6. Matsiuk, V.: A study of the technological reliability of railway stations by an example of transit trains processing. Eastern Eur. J. Enterp. Technol. **1**(3), 18–24 (2017)
7. Nagorna, N.: Formation of students' concepts of competency and competence. Educ. Cult. **1–2**(11–12), 266–268 (2007)
8. Lokshyna, O.: European reference system as a tool for implementing competent approach in education of EU member states. Pedagogy Psychol. **1**, 131–142 (2007)
9. Rud′, M.: Competent approach in education. Bull. Lviv Univ. Series Pedagogy **21**(1), 73–82 (2006)
10. Pobirchenko, N.S.: Competency approach in high school: theoretical aspect. Educ. Pedagogical Sci. **3**(152), 24–31 (2012)
11. Ovsiyenko, L.M.: The essence of the concepts of "one competency", "general competencies", "competence approach", "quality of education" in the light of the modern educational paradigm. Sci. Bull. Donbass, **2** (2013) http://nbuv.gov.ua/j-pdf/nvd_2013_2_32.pdf
12. Zeer, E., Symanyuk, E.: Competency-based approach to the modernization of vocational education. High. Educ. Russia **4**, 22–28 (2005)

13. Zimnyaya, I.A.: Competency approach. What is its place in the system of modern approaches to the problems of education? (theoretical and methodological aspect). High. Educ. Today **4**, 20–27 (2006)
14. Zablocka, O.: Competence approach as educational innovation: a comparative analysis. Bull. Zhytomyr State Univ. Series Pedagogical Sci. **40**, 63–68 (2008)
15. Chasnikova, O.: Competent approach in education as a basis for its reform, Public education (2014). http://narodnaosvita.kiev.ua/?page_id=2607
16. Lokshyna, O.: Becoming a competence idea in European education. Pedagogical Thought **2009**, 21–30 (2009)
17. Shramenko, N.: Evaluation of the effectiveness of piggyback traffic in the context of creating transport and logistics clusters. Sci. Bull. Nat. Mining Univ. **6**(162), 151–155 (2017)
18. Shramenkoand, N.Y., Shramenko, V.O.: Mathematical model of the logistics chain for the delivery of bulk cargo by rail transport. Sci. Bull. Nat. Mining Univ. **5**(167), 136–141 (2018)
19. Shramenko, N.Y., Shramenko, V.O.: Optimization of technological specifications and methodology of estimating the efficiency of the bulk car goes delivery process. Sci. Bull. Nat. Mining Univ. **3**, 146–151 (2019)
20. Shramenko, N., Muzylyov, D.: Forecasting of overloading volumes in transport systems based on the fuzzy-neural model. In: Advances in Design, Simulation and Manufacturing II. DSMIE 2019. Lecture Notes in Mechanical Engineering, pp. 311–320. Springer, Cham (2020)
21. Shramenko, N., Pavlenko, O., Muzylyov, D.: Information and Communication Technology: Case of Using Petri Nets for Grain Delivery Simulation at Logistics System. In: CEUR Workshop Proceedings, vol. 2353, pp. 935-949 (2019)

Mathematical Model of Reverse Loading Advisability for Trucks Considering Idle Times

Dmitriy Muzylyov[1(✉)] and Natalya Shramenko[1,2]

[1] Kharkiv Petro Vasylenko National Technical University of Agriculture, Alchevskyh Str. 44, Kharkiv 61002, Ukraine
murza_1@ukr.net

[2] Ukrainian State University of Railway Transport, Feierbakh Square 7, Kharkiv 61050, Ukraine
nshramenko@gmail.com

Abstract. The article is aimed at determining the expediency of rational option choosing by reverse transportation loading during goods delivery in intercity directions. Five possible alternatives are proposed to truck returns while waiting for backloading at first step of researches. An integrated criterion is selected as a condition for rational option choosing of truck returns. It is a synergistic effect for carriers and customers of transport services. Therefore, rational option search to return vehicles with reverse loading is to choose the best values: maximum cost-effectiveness of transportation and minimum delivery period. The result is a compromise solution between carrier and customer interests. Such an approach will allow transport companies to expand their influence sphere and find new permanent clients.

Keywords: Cargo · Supply · Trucks · Logistics · Supply chain · Idle time · Reverse loading · Backloading transportation · Criterion · Synergetic

1 Introduction

The use of a synergistic approach to supply chain management will allow most transport process indicators integration, which will significantly improve transportation conditions.

A synergistic effect is called such an approach in logistics, which is achieved by a mutual strengthening of ties of one logistics system with others according to the portal "Trade Master Group" [1]. One option for a synergistic approach is to ensure the timely reverse loading of trucks. This is especially true in long-distance transport.

The procedure of transit cargo loading is used when it is necessary to minimize expenses for vehicle idle mileage at reverse direction routes. This is main essence of backloading transportation strategy. At the same time in reverse directions, it can deviate vehicle rides from route trajectory, which was in forward directions. Backloading transportation allows increasing levels using mileage and trucks cargo capacity on routes. Information on available applications for transportation can be found on the Internet through specialized logistics sites.

I. Karabegović (Ed.): NT 2020, LNNS 128, pp. 612–620, 2020.
https://doi.org/10.1007/978-3-030-46817-0_71

Carrier may have a question about advisability to wait for order in case of a one-time request for backloading transportation. After all, order presence for cargo carriage is a random value [2] in the required direction by transport companies.

Having studied the latest works on this topic [3–5], it can be concluded that in practice optimal order choosing for cargo transportation in reverse directions is made on an intuitive level and based on the professional experience of carrier's managers. At the same time, logistics operators apply many efforts to quickly search for corresponding orders [6] to minimize idle times of trucks under reverse loading expectations. As a financial criterion of cargo acceptance for transportation is the largest freight rate for a specific transportation variant, which should be not lower than the average for cargo corresponding categories [7].

It is necessary to develop a new criterion for correct determinations of idle time duration for trucks before backloading transportation orders will appear in the accompanying directions considering probabilistic approaches specified by the previous researchers [8, 9].

2 Literature Analysis

The research conducted with transport companies functioning results in Brazil [10] became one of the last innovations in choosing optimal variant backloading transportation. The authors propose to consider added cargo values when forming reverse loads. This should significantly increase transportation profitability, especially in long-distance routes. Also important is to consider environmental components of cargo supply [11, 12]. After all, a rational route will reduce emissions amount into the atmosphere [13, 14]. The study describes possible influences of the shipper on carbon emissions from deep-sea container supply chains [15]. The authors clearly argue that reverse loading choice depends not only on an economic component but also on pollution at port zone. Some works [16–18] suggest necessities to forecast transportation volumes for a week, especially during loading/unloading ships [19, 20].

Another feature of the best reverse loading option choosing is the inability to find reverse loads corresponding to cargo capacities of trucks by 100%. Sometimes batches of cargo are very small that's why questions will appear about economic advisability of delivery [21]. Therefore, carriers must correctly assess profitability of future delivery in reverse directions [22, 23]. Correct assessment of economic components is particularly important when choosing a multi-modal supply chain [24]. The relevance of problem-solving to rational option finding of reverse loading is described on the website of one transport company [25]. The main goal is to increase transportation profitability over transport cycles. This is especially true on long-distance routes, where reverse return distances of empty trucks significantly reduce incomes of transport companies [26–28].

The technology of delivering containers with agricultural goods from ports [29, 30] is a good example of choosing a rational route during backloading transportation. It is particularly important to find an optimal delivery scheme for reverse loading under resource allocation conditions [31] or for transporting bulk cargoes at metallurgical [32] and mining industries [33].

Agricultural cargo delivery options should consider delivery consistency conditions across supply chains [34].

3 Design of Mathematical Model

Theory and practice analysis of cargo delivery systems in long-distance routes shows that there is a wide variety of supply schemes. They involve different quantities of vehicles and participants in transportation. Therefore, there is a necessity to formalize schemes by similar parameters. This will allow considering more correctly the average idle time for reverse loading of trucks. All existing transport technologies were combined into five groups by results of preliminary analysis (Fig. 1).

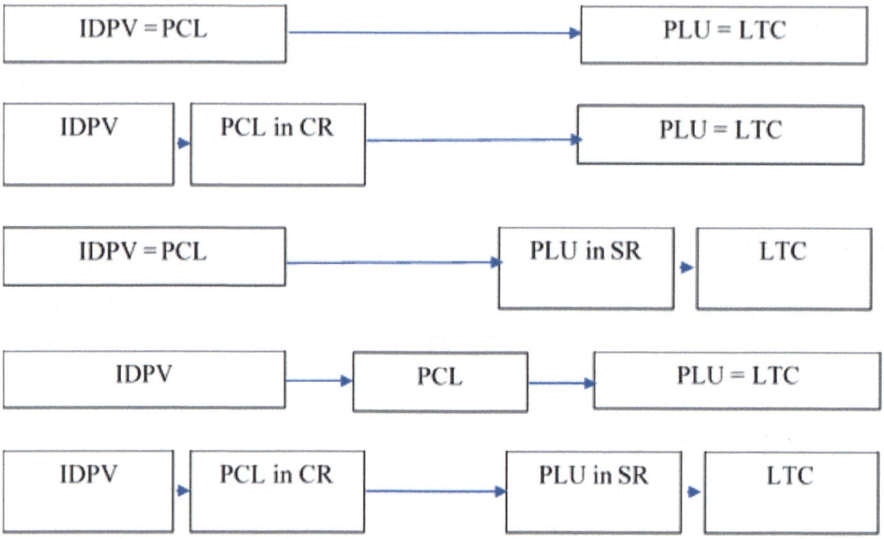

Notes: IDPV - Initial departure place of vehicles (location of trucks after direct cargo transportation); PCL– place of cargo loading; PLU - place of cargo unloading; SR - shipper's region (locality is at primary loading point in direct cases routing); CR - consignee's region (locality is at primary unloading point in direct cases routing); LTC - location of the transport company.

Fig. 1. Five types of options for backloading transportation

Any city or distribution center (DC) shall be understood as places of loading and unloading. It should be noted that the sixth option of transport and technological scheme (when possibilities of reverse loading exist at a distance of more than half-route from the previous shipper) was not considered. This is due to the economic impracticability of designing such a backloaded supply chain. This parameter depends directly on the type of transport and technological scheme according to collected information by analysis about idle time.

At the same time, it is necessary to follow the methodology of integral criteria designing, which simultaneously considers different participants' interests in supply chains [35–37]. It can be concluded based on previous studies that in classical approaches to assessing transport process efficiency, operative decisions of processes become a priority. However, based on articles on this topic in foreign sources (USA, China) it can be concluded that at current moments the transport process efficiency is not only assessed by technical parameters. The assessment of transport services efficiency comes to the fore by qualitative indicators. This means that at the current stage of transport services market development such transport companies are more competitive which are assess transport process efficiency not only from technical sides but also from service quality to potential clients. Therefore, such a parameter as an idle time for waiting reverse loads plays a key role in choosing transportation options.

The conceptual scheme of designing a mathematical model for choosing rational options of transportation at reverse loading considers idle times (submitted in Fig. 2).

Input parameters were selected to consider two conditions at model designing:

1. Lack of complementary factors.
2. Absence of mutually exclusive input values.

Compliance with these two aspects guarantees to design an adequate mathematical model. This will cause a possible error in the process simulation to be within 5%. This is full compliance with recommendations for models designing which describe transport systems function.

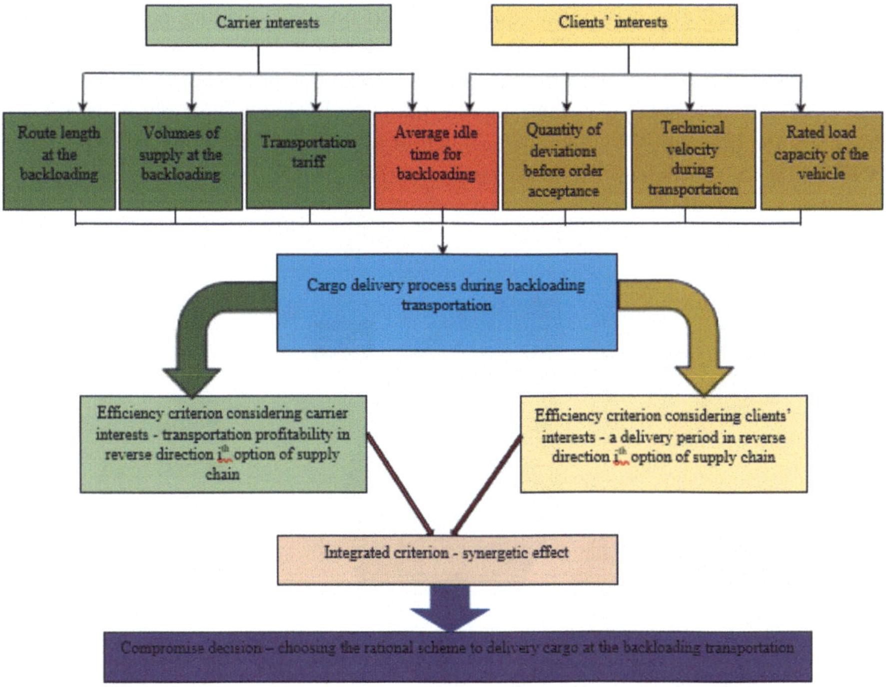

Fig. 2. Conceptual scheme to design mathematical model of backloading advisability for trucks considering idle times

This is important to consider trucks load volume when determining the idle time of waiting for backloading transportation based on Fig. 3. The order probability for reverse loading is higher than less cargo volume must be transported according to practice experience. Batches sizes reverse loading does not often correspond to cargo capacities of vehicles. Therefore, it is proposed to calculate idle times of backloading transportation for four options (not counting five - when vehicles return to the locality empty).

Route length in the reverse direction must be also considered when determining alternatives for calculating idle times of backloading transportation. Therefore, the initial alternatives matrix with sizes *n* and *k* is compiled to calculate average values of the idle time at reverse load. Where *n* is option quantity based on truck loading percentage alternatives and *k* is a variant quantity for each route length scheme.

Average values of idle time of waiting for backloading transportation is used in future calculations. Calculated as an arithmetic mean by next dependency:

$$t^{idlebl} = \frac{\sum_{i=1}^{n}\sum_{j=1}^{k} t_{ij}^{idlebl}}{n \cdot k} \tag{1}$$

where t_{ij}^{idlebl} - idle time for backloading at i-th percent of vehicle cargo capacity usage and j-th route length in reverse direction, day;

n, k - accordingly, n is option quantity based on truck loading percentage alternatives and k is a variant quantity for each route length scheme.

It is necessary to carry out formations of combinations proceeding from researches [38, 39]. Trucks not always have an opportunity to be loaded completely in the reverse directions for quantitative values determination of average idle times of expectation backloading transportation. This tendency is accurately visible from the diagram (Fig. 3).

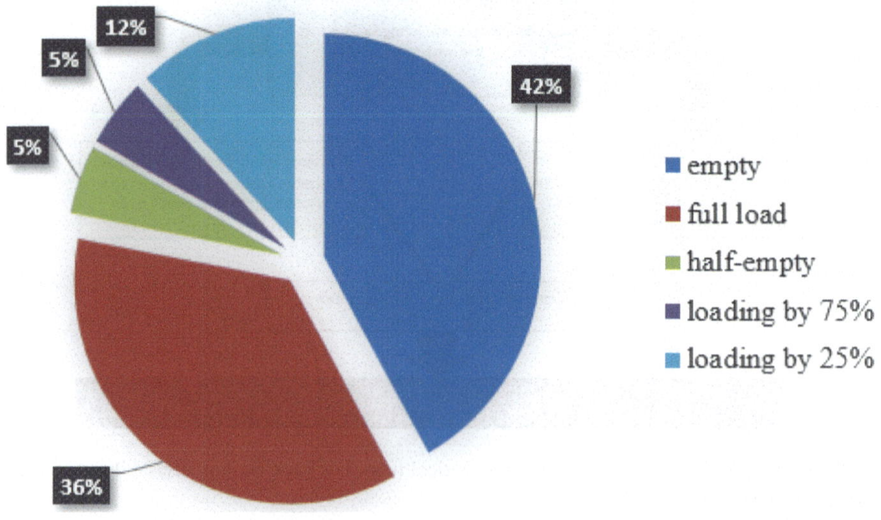

Fig. 3. Cargo loading percentage in vehicles at reverse transportation

Efficiency measuring transport services regarding quality is a very important step in efficiency assessment of transport processes in whole. However, the efficiency determination of transport processes only from quality service cannot be unequivocal to choose optimal transportation options. So, in order to choose a more reliable transportation option, it is necessary to approach efficiency assessment of transport processes both from classical aspects - technological sides, and at the same time to consider options of quality service provision. Only in a complex will these approaches yield a more accurate result.

Especially designing of integrated criteria is very important in delivery options formation in a significant competitive environment and in the non-constant nature of service demand [40].

Model of the process in question can be represented as functions 2 and 3 in mathematical form:

$$R = f\left(t^{idlebl}, L^{br}, Q^{reverse}, T\right) \tag{2}$$

where t_{ij}^{idlebl} - average idle time for backloading, days; L^{br}-route length at the backloading transportation, km (miles);$Q^{reverse}$-volumes of supply at the backloading transportation, tons;T - transportation tariff, UAH/tons*km.

$$T^{deliv} = f\left(t^{idlebl}, N^{deviat}, V^{tech}, q^{cc}\right) \tag{3}$$

where N^{deviat}-quantity of deviations before order acceptance, units (orders); V^{tech}-technical velocity during reverse transportation; km/hour; q^{cc} - rated load capacity of the vehicle according technical specifics, tons.

Target function can take the following generalized form:

$$C^{INT} = \begin{cases} f(R) \to \max \\ f(T^{deliv}) \to \min \end{cases} \tag{4}$$

where R- transportation profitability in reverse direction ith option of supply chain, %; T^{deliv} - delivery period in reverse direction ith option of supply chain, days.

It is recommended to calculate average idle times for reverse loading based on data of transport companies operate in order to supply chain rationalization, for example by using Larditrans.

Regression models will be built for each component of the criterion in future researches. Choosing of rational backloading transportation options for trucks with reverse loading is supposed to be carried out based on graphical methods.

4 Conclusion

Five possible alternatives are proposed to truck returns while waiting for backloading at first step of researches. An integrated criterion is selected as a condition for rational option choosing of truck returns. It is a synergistic effect for carriers and customers of transport services. Therefore, rational option search to return vehicles with reverse loading is to choose the best values: maximum cost-effectiveness of transportation and minimum delivery period. The result is a compromise solution between carrier and customer interests. Such an approach will allow transport companies to expand their influence sphere and find new permanent clients.

References

1. Portal of top managers of wholesale and retail trade “TradeMasterGroup”. https://trademaster.ua/articles/312534
2. Horbachov, P.F., Naglyuk, I.S., Makarichevand, O.V., Mospan, N.V.: Evaluation of the effectiveness of carriers’ strategies for servicing one-time applications for intercity cargo transportation. Automobile Transp. **37**, 61–68 (2015)
3. Torres, I., Rosete, A., Cruz, C., Verdegay, J.L.: Fuzzy constraints in the Truck and Trailer Routing Problem. In: Fourth International Workshop Proceedings, 2013, pp. 71–78 (2013)
4. Liu, P., Mu, D., Gong, D.: Eliminating overload trucking via a modal shift to achieve intercity freight sustainability: a system dynamics approach. Sustainability **9**(3), 398 (2017)
5. Liu, P., Mu, D.: Evaluating sustainability of truck weight regulations: a system dynamics view. J. Ind. Eng. Manag. **8**, 1711–1730 (2015)
6. Qu, Y., Bektasë, T., Bennell, J.: Sustainability SI: multimode multicommodity network design model for intermodal freight transportation with transfer and emission costs. Netw. Spat. Econ. **16**, 303–329 (2016)
7. McKinnon, A.C., Ge, Y.: The potential for reducing empty running by trucks: a retrospective analysis. Int. J. Phys. Distrib. Logistics Manag. **36**(5), 391–410 (2006)
8. Muzylyov, D., Shramenko, N., Shramenko, V.: Integrated business-criterion to choose a rational supply chain for perishable agricultural goods at automobile transportations. Int. J. Bus. Perform. Manag. **21**(1/2), 166–183 (2020). https://doi.org/10.1504/IJBPM.2020.10027634
9. Shramenko, N., Muzylyov, D., Shramenko, V.: Methodology of costs assessment for customer transportation service of small perishable cargoes. Int. J. Bus. Perform. Manag. **21** (1/2), 132–148 (2020). https://doi.org/10.1504/IJBPM.2020.10027632
10. Mazier, H.J., Azevedo Motta, G., Vilas Boas Pini, G.: Impactos da implementação de KPI′S Logísticosemumausinasucroenergética do interior do estado de São Paulo, XXXIX Encontro Nacional De Engenharia De Producao “Osdesafios da engenharia de produção para umagestãoinovadora da Logística e Operações” Santos, São Paulo, Brasil, 15 a 18 de outubro de 2019, pp. 1–14 (2019)
11. Davenport, J., Davenport, J.L. (eds.): The Ecology of Transportation: Managing Mobility for the Environment. Environmental Pollution. Springer, Netherlands (2006)
12. Miller, P., de Barros, A.G., Kattan, L., et al.: Public transportation and sustainability: a review. KSCE J. Civil Eng. **20**, 1076–1083 (2016)
13. Karnaukh, M., Muzylyov, D., Shramenko, N.: Technological aspects of energy optimization during operating vehicles and increase their environmental safety. In: 2nd International Scientific and Practical Conference “Energy-Optimal Technologies, Logistic and Safety on Transport” (EOT-2019). September 2019, Lviv, Ukraine. MATEC Web of Conferences, vol. 294, no. 01013, pp. 1–7 (2019)
14. Ascensão, F., Capinha, C.: Aliens on the move: transportation networks and non-native species. In: Borda-de-Água, L., Barrientos, R., Beja, P., Pereira, H. (eds.) Railway Ecology, pp. 65–80. Springer, Cham (2017)

15. McKinnon, A.: The possible influence of the shipper on carbon emissions from deep-sea container supply chains: an empirical analysis. Marit. Econ. Logist. **16**, 1–19 (2014)
16. Atalla, T., D'Errico, M.C.: Energy project financing in the GCC region: an empirical investigation. Energy Transitions **3**, 13–30 (2019)
17. Shramenko, N., Muzylyov, D.: Forecasting of Overloading volumes in transport systems based on the fuzzy-neural model. In: Advances in Design, Simulation and Manufacturing II. DSMIE 2019. Lecture Notes in Mechanical Engineering, pp. 311–320. Springer, Cham (2020)
18. Schulte, F., González, R.G., Voß, S.: Reducing port-related truck emissions: coordinated truck appointments to reduce empty truck trips. In: Corman, F., Voß, S., Negenborn, R. (eds.) Computational Logistics. ICCL 2015. Lecture Notes in Computer Science, vol 9335, pp. 495–509. Springer, Cham (2015)
19. Albjerk, N., Danielsen, T., Krey, S., Stålhane, M., Fagerholt, K.: A vessel pickup and delivery problem from the disruption management in offshore supply vessel operations. In: Paias, A., Ruthmair, M., Voß, S. (eds.) Computational Logistics. ICCL 2016. Lecture Notes in Computer Science, vol 9855, pp. 50–64. Springer, Cham (2016)
20. Shramenko, N., Pavlenko, O., Muzylyov, D.: Information and Communication Technology: Case Of Using Petri Nets For Grain Delivery Simulation At Logistics System. In: CEUR Workshop Proceedings, vol. 2353, pp. 935–949 (2019)
21. Balaji, A.N., Mukund Nilakantan, J., Nielsen, I., et al.: Solving fixed charge transportation problem with truck load constraint using metaheuristics. Ann. Oper. Res. **273**, 207–236 (2017)
22. Powell, M.: Logistics and Supply Chain Steps to Cost Savings and Efficiency (Guide), *PROTRANS*, post. 08 January 2019. http://www.protrans.com/logistics-supply-chain-steps-cost-savings-efficiency-guide
23. Grunewald, M., Volling, T., Müller, C., et al.: Multi-item single-source ordering with detailed consideration of transportation capacities. J. Bus. Econ. **88**(7–8), 971–1007 (2018)
24. Martel, A., Klibi, W.: Transportation in the supply chain. Designing Value-Creating Supply Chain Networks, pp. 161–206. Springer, Cham (2016)
25. What is Backloading? The cheapest way to move interstate, *AustateRemovals* - https://www.austate.com.au/what-is-backloading
26. Miller, T.: Distribution and transportation planning and scheduling. Hierarchical Operations and Supply Chain Planning, pp. 95–158. Springer, London (2002)
27. Carlos, M., Gallardo, A., Edo-Alcón, N., Abaso, J.R.: Influence of the municipal solid waste collection system on the time spent at a collection point: a case study. Sustainability **11**(22), 6481 (2019)
28. Fritzsche, R.: Damage to wheel bearings during car transportation on a truck. ATZ Worldwide **121**, 54–59 (2019)
29. Velykodnyi, D., Pavlenko, O.: The choice of rational technology of delivery of grain cargoes in the containers in the international traffic. Int. J. Traffic transp. Eng. **7**(2), 164–175 (2017)
30. Aulin, V., Lyashuk, O., Pavlenko, O., Velykodnyi, D., Hrynkiv, A., Lysenko, S., Holub, D., Vovk, Y., Dzyura, V., Sokol, M.: Realization of the logistic approach in the international cargo delivery system. Commun. Sci. Lett. Univ. Zilina **21**(2), 3–12 (2019)
31. He, S., Li, S., Ma, H.: Integrating fluctuations into distribution of resources in transportation networks. Eur. Phys. J. B **76**(1), 31–36 (2010)
32. Turpak, S.M., Taran, I.O., Fomin, O.V., Fomin, O.V., Tretiak, O.O.: Logistic technology to deliver raw material for metallurgical production. Sci. Bull. Nat. Mining Univ. **1**(1), 162–169 (2018)
33. Shramenko, N.Y., Shramenko, V.O.: Mathematical model of the logistics chain for the delivery of bulk cargo by rail transport. Sci. Bull. Nat. Mining Univ. **5**(167), 136–141 (2018)

34. Muzylyov, D., Kravcov, A., Karnayh, N., Berezhnaja, N., Kutiya, O.: Development of a methodology for choosing conditions of interaction between harvesting and transport complexes. Eastern Eur. J. Enterp. Technol. **2**(3), 11–21 (2016)
35. Tomlyak, S., Polyakov, A.: Ways to increase the efficiency of carriage of goods by road. Intercollegiate Collect. "SCI. NOTES", **46**, 529–537 (2014)
36. Shramenko, N.Y., Shramenko, V.O.: Optimization of technological specifications and methodology of estimating the efficiency of the bulk cargoes delivery process. Naukovyi Visnyk Natsionalnoho Hirnychoho Universytetu, **3**, 146–151 (2019)
37. Ant Logistics. https://ant-logistics.com/uk/main.html
38. Make back-loading work for you. https://collaboration.fhwa.dot.gov/dot/fhwa/pm/Lists/aReferences/Attachments/83/BackLoading.pdf
39. Shakantu, W.M., Muya, M., Tookey, J.E., Bowen, P.A.: Evaluating truck empty running in construction: a case study from Cape Town, South Africa. Australasian J. Constr. Econ. Build. **8**(2), 41–49 (2008)
40. Azemsha, S.: Statistical modeling of trucks on international routes with various strategies for the adoption of reverse loading. Transp. Telecommun. **8**(1), 53–61 (2007)

Model for Choosing Rational Technology of Containers Transshipment in Multimodal Cargo Delivery Systems

Natalya Shramenko[1,2(✉)], Dmitriy Muzylyov[1(✉)], and Vladyslav Shramenko[3,4(✉)]

[1] Department of Transport Technology and Logistics, Kharkiv Petro Vasylenko National Technical University of Agriculture, Alchevskyh Str. 44, Kharkiv 61002, Ukraine
nshramenko@gmail.com, murza_1@ukr.net

[2] Department of Operational Work Management, Ukrainian State University of Railway Transport, Feierbakh Square 7, Kharkiv 61050, Ukraine

[3] Faculty of Mathematics and Computer Science, V. N. Karazin Kharkiv National University, Svobody Sq. 4, Kharkiv 61022, Ukraine
shramenko.vlad@gmail.com

[4] Faculty of Technological Systems and Logistics, Kharkiv Petro Vasylenko National Technical University of Agriculture, Alchevskyh Str. 44, Kharkiv 61022, Ukraine

Abstract. The study's purpose is to increase grain cargo delivery efficiency in containers in multimodal railway and water transportation by choosing a rational technology for container transshipment at port terminals. Typical technologies for port terminals are container cargoes transshipment from railway carriages to ship according to the direct option technology, as well as container cargoes transshipment through warehouses. The analysis of container transportations volumes dynamics was carried out in Ukraine, including grain supply. Most relevant is the daily total cost of container cargo transshipment at port terminals as efficiency criteria. Numerical experiments were conducted for various container quantities incoming to port terminals. As a result, total costs are dependent on container quantities incoming to port for alternative transshipment technologies. The obtained graphs allow determining rational technologies of container cargoes transshipment at port terminals depending on their quantity incoming to port in railway carriages.

Keywords: Terminal · Container · Direct option · Technology · Logistics · Scheme · Cargo · Sea port · Costs transshipment · Railway carriage · Ship · Wagon · Warehouse · Supply

1 Introduction

World experience shows that one of the effective ways to develop transport systems that ensure goods delivery in accordance with basics requirements of consumers for transport services are wide applications of multimodal transport technologies [1] and

I. Karabegović (Ed.): NT 2020, LNNS 128, pp. 621–629, 2020.
https://doi.org/10.1007/978-3-030-46817-0_72

logistics principles [2] in transport process organization. Multimodal systems [3] and intermodal cargo delivery technologies are particularly relevant [4, 5].

The recent emphasis on organization [6] and efficient operations of cargo delivery systems [7] is due to reductions in trade cycle length, increases in storage costs [8] and necessities to accelerate responses to consumer demand [9]. One of the main approaches aimed at improving cargo delivery efficiency is designing scientifically based provisions on delivery systems parameters management and optimization processes of transport complex during interactions between various kinds of transport in hubs, which are seaports.

2 Literature Review

The necessity to organize coordinated [10], uninterrupted operation of related kinds of transport [11] becomes particularly important when carrying out a transport process involving several kinds of transport, which lead to processes formation and implementation inherent in each of them.

The modern form of transport processes organization is transport and technological schemes (TTS) [12], which specifically unite local physical objects included in systems [13].

TTS types can be divided into single-kind and systems including several kinds of transport [14], i.e. TTS with combine transportations [15]. Transport and technological processes are generally realized in delivery systems with participation by various kinds of transport [16]. Thus, it is multimodal transport that should be considered as transport and technological bases of delivery systems.

Efficient organization and functioning of delivery systems require detailed investigation of transport and technological processes [17], which are provided with various kinds of transport participation [18], from viewpoints of their compliance with logistics principles [19].

Multimodal transportation is organized based on terminal cargo delivery systems [20]. Thus, the terminal complex should be considered in multimodal transportation as a transport and technological basis of delivery systems [7]. The main functions of terminal systems are cargo flow concentration and separation [21], ensuring reliable goods supply in short-time periods, reducing transportation and goods storage costs [22].

Container transportations are very important in multimodal delivery systems [23]. Regularity operation of port container terminals is at the same time mandatory for creating stable and reliable intermodal transport networks [24].

The continued growth of global containerization [25] has led to deploying bigger container ships [26]. Researchers suggest that the next generation of container ships with super tonnage (about twice as powerful as today's largest container ships) will be served by an only limited number of deep-sea or offshore transfer ports [27]. This will lead to necessities change port infrastructures, with a view to developing models for assessments of international and inland container cargo flows using maritime, rail and road transport means [28]. A key feature of infrastructure investment is structuring a protected risk management strategy to address uncertainties [29].

Port is an important transport hub during supply chain formation [30] especially at organizing multimodal cargo delivery processes. Port infrastructure [31] and management structure [32] should also facilitate optimal cost-effectiveness use of port facilities. On the other side, inter-operational downtime of vehicles should be avoided with a view to rational using fleets of motor vehicles [33] and railway rolling stock [34, 35] arriving at the port for containers transshipment to ships.

Container transportations of bulk goods [36] more and more usually appear in recent decades including grain cargoes [37]. Articles [38, 39] describe and classify the main logistics processes and operations on container terminals and provide an overview of how to optimize them.

Part of containers transshipment is loaded at ports directly to ships according to direct options [40], and part is going into warehouses [41] for short-time storage. The study results [42] show that direct overload mode can significantly reduce operating costs.

Thus, this research is aimed at choosing rational technology for container cargo transshipment through port terminals during intermodal transportation by using railway and water transport.

3 Aim and Research Tasks

The research purpose is to increase delivery efficiencies of grain cargoes in containers in multimodal railway and water transportation by choosing a rational technology of containers transshipment at port terminals.

Research tasks are:

- to analyze the dynamics of container cargo transportation in Ukraine using railway and water transport, including grain delivery;
- to study status and methods to improve operational efficiencies of port terminals;
- to design a mathematical model of container cargoes transshipment at seaports to determine rational technologies of its function;
- to simulate of container cargo transshipment process at port terminals by using alternative technologies.

4 Design of Mathematical Model

Containers' exports and imports distribute through Ukrainian seaports in 2018 almost equally - 47.4% and 48.8%, respectively [43]. Ukrainian seaports overloaded 846.5 thousand TEU of container cargoes according to the results of 2018. This is 18.7% more than in 2017. The next three ports became leaders in container transshipment: Odessa (598.6 thousand TEU), Yuzhny (124.9 thousand. TEU) and Chernomorsk (122.6 thousand. TEU). Yuzhny increased transshipment of these goods in 2018 by more than 75% due to international container line Maersk transition into a port from Odessa. However, despite this Odessa managed not only to preserve container volumes but also to increase them by 15%. The global container market will grow by an average of 3.8% per year by 2023 [43] according to the United Nations Conference on Trade and Development (UNCTAD).

Among the main types of cargo exported by containers are grain, fruit, food cargoes, metal products, construction materials. Equipment, textiles, food loads, alcohol, plastic are mainly imported [44].

Ukrainian Railway has operated about 150-grain containers used for transportation grain cargoes from farm elevators to seaports since 2019. The design of these containers allows grains to be unloaded directly to ships. The transport process is greatly accelerating in this case, therefore, reduces its cost. Ukraine has launched ten new container trains, four of which are international since 2019. 1.7% of total cargo volumes are generally transported in containers in Ukrainian territory. Such service allows reducing cargo cycles which performed by rolling stock more than three times and, as a result, contributes to final costs reduction of cargo transportation up to 25% [45].

Transportation of grain goods by rail increased in 2019 by 20.1% compared to 2018 - to 40.5 million tons. 35.3 million tons of cereals were exported during 2019-30% more than in 2018 [46] according to Ukrainian Railway.

International traffic especially grain transportation was increased last year. This tendency can be seen according to the analysis results of container traffic dynamics.

Typical technologies for port terminals are container cargoes transshipment from railway carriages to ship according to the direct option technology, as well as container cargoes transshipment through warehouses. However, the disadvantage is that cargo turnover is not always rationally distributed between warehouses and overloaded volumes according to direct technology. As a result, queues of vehicles are formed while idle waiting for service, and therefore their transportation cycle time is increased.

The study object is process of container cargoes transshipment at port terminals when delivering grain cargoes using multimodal railway and water transportation. A subject is represented as a 'Black Box' (Fig. 1).

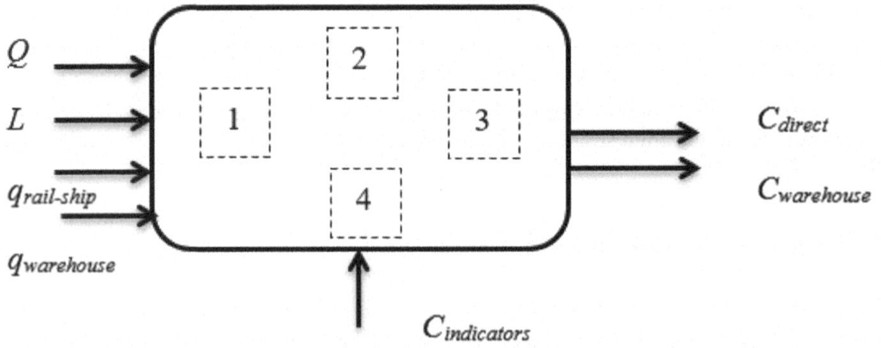

Notes: 1- transportation by railway carriage into port; 2 - wagon idle waiting for transshipment; 3 - container transshipment; 4 - a process of storing containers in a port warehouse.
Input factors: Q is containers quantity in one arrival to be transshipped at ports, units; L - mileage to port , km; $q_{rail\text{-}ship}$, $q_{warehouse}$ - processing capacity of transshipment fronts according to direct options "railway carriage - ship" and through warehouses, respectively, units.
External factors: $C_{indicators}$ are cost indexes, UAH.
Output factors: C_{direct}, $C_{warehouse}$ - total costs for container transshipment in the ports according to direct options 'railway carriage - ship' and through warehouses, respectively, UAH.

Fig. 1. 'Black Box' cybernetic model

Most relevant is the daily total cost of container cargo transshipment at port terminals as efficiency criterion.

Mathematical simulation is chosen for study, which allows describing more fully relationships between elements of researching objects. Mathematical models have been developed to describe transshipment processes according to alternative options at seaports for choosing rational technology for cargoes overloading in containers.

Mathematically model describing processes of container transshipment in the ports according to direct options ‘railway carriage - ship’ is as follows:

$$C_{direct} = f(C_{shloc},\ t_{shloc},\ C_{wag},\ L_{add},\ k,\ t_{idle}^{ship},\ n_{ship},\ q_{rail-ship},\ Q,\ C_{op},\ C_{whour},\ C_{up/un}) \\ \to \min, \tag{1}$$

where C_{shloc} - specific costs per 1 h of shunting locomotives operation at freight traffic for wagons group delivery to cargo fronts, UAH/h; t_{shloc} - operating time of shunting diesel locomotive at supply and retraction of wagons group to/from cargo front, hour; C_{wag} - unit costs per 1 wagon-kilometer from total mileages by freight fleet wagons during supply to transshipment front, UAH/wagon-km; L_{add} - additional mileage of freight fleet wagons at supply-retraction to (from) cargo fronts of transshipment according to direct options ‘railway carriage - ship’ in seaports, km; k - railway carriages quantity in one arrival to a port, wagon; t_{idle}^{ship} - idle waiting by railway carriages for transshipment to ship, hour; n_{ship} - coefficient of double operations at containers transshipment by direct option ‘railway carriage - ship’; C_{opc} - container operation cost, UAH/hour; C_{whour} - railway carriage idle hour cost, UAH/wagon-hour; $C_{up/un}$ - cost of uploading and unloading works at port terminals, UAH/container.

System of restrictions:

$$\begin{cases} t_{mileage} + t_{idle}^{ship} + t_{transsh} \le T_{port}, \\ 12 \le Q \le 60 \end{cases} \tag{2}$$

where T_{port} - time of container cargoes stay at port terminals, hour; $t_{mileage}$ - mileage time of railway carriages to (from) cargo front of transshipment according to direct option ‘wagon - ship’ in port, hour; $t_{transsh}$ - time for container transshipment according to direct option ‘railway carriages - ship’, hour.

A mathematical view of models describing processes of goods transshipment in containers through warehouses is as follows:

$$\begin{gathered} C_{warehouse} = f(C_{shloc},\ t_{shloc}^{wareh},\ C_{wag},\ L_{add}^{wareh},\ k,\ t_{idle}^{wareh}, \\ n_{wareh},\ q_{wareh},\ Q,\ C_{op},\ C_{whour},\ t_{stor}, C_{stor},\ MP_c,\ d) \to \min \end{gathered} \tag{3}$$

where t_{shloc}^{wareh} - operating time of shunting diesel locomotive at supply and retraction of wagons group to/from port warehouses, hour; L_{add}^{wareh} - additional mileage of freight fleet wagons at supply-retraction to (from) port warehouses, km; t_{idle}^{wareh} - idle waiting by railway carriages for transshipment to warehouses, hour; n_{wareh} - coefficient of double operations at containers transshipment by undirect option 'railway carriage – warehouse - ship'; t_{stor} - storage time of containers in port warehouses, hour; C_{stor} - hour cost of cargo storage in port warehouses, UAH/hour; MP_c - market prices of goods in container, UAH/container; d - percent rate on bank deposits, % (proportion).

System of restrictions:

$$\begin{cases} t_{mileage}^{wareh} + t_{idle}^{wareh} + t_{transsh}^{wareh} + t_{stor} \leq T_{port} \\ 12 \leq Q \leq 60 \\ t_{transsh}^{wareh} + t_{stor} \leq T_{warehouse} \end{cases}, \quad (4)$$

where $T_{warehouse}$ - time of container cargoes stay at warehouses, hour; $t_{mileage}^{wareh}$ - mileage time of railway carriages to (from) warehouses in a port, hour; $t_{transsh}^{wareh}$ - time for container transshipment through warehouses, hour.

5 Research Results

Simulation is carried out with conditional numerical values of sometime parameters to illustrate the validity of proposed models because ports are objects with special regimes. Numerical experiments were conducted for various container quantity incoming to port terminals. Total costs are dependent on container quantity incoming to ports for alternative transshipment technologies (Fig. 2a) as experiment results.

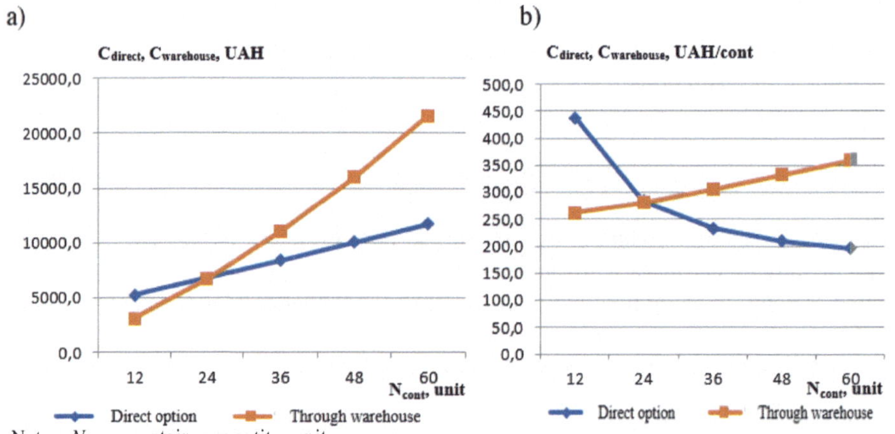

Notes: N_{cont} – container quantity, unit.

Fig. 2. Dependencies of total costs (a) and specific cost (b) at various technologies of container transshipment in port terminals

Analytical cost calculations have been similarly made for alternative container transshipment technologies when various container quantity arrive at seaports. As a result, specific costs are dependent on containers quantities incoming to port (Fig. 2b) for alternative transshipment technologies.

The graphs (Fig. 2) allow determining rational technologies of container cargo transshipment at the port terminal depending on their quantity arriving at ports by railway carriages. Rational technology is container transshipment through warehouses, as total costs of container transshipment are lower than in the case of transshipment according to direct options ‘railway carriage - ship’. This tendency is true in the case of goods transshipment in 3-ton containers up to the amount of 24 containers inclusive. Rational technology is containers overloading according to the direct option ‘railway carriage - ship’ when amounts of container transshipment are more than 24 containers (3-ton).

6 Conclusion

International traffic especially grain transportation was increased last year. This tendency can be seen according to the analysis results of container traffic dynamics.

Literature analysis showed that it is necessary to consider technological peculiarities in transport hubs where different kinds of transport interacting Technological features in hubs of interaction are not taken into account in most researches. Interests of all subject’s delivery process do not always consider during model design,

Alternative technologies for cargo container transshipment at the port terminal have been analyzed. Most relevant is the daily total cost of container cargo transshipment at port terminals as efficiency criterion.

Mathematical models have been developed to describe transshipment processes according to direct options “railway carriage - ship” and through warehouses. Restriction system is justified for mathematical models.

Dependencies of total costs on containers quantity in alternative technologies of container cargoes transshipment have been designed according to simulation results. The obtained graphs allow determining rational technologies of container cargoes transshipment at port terminals depending on their quantity incoming to port in railway carriages.

References

1. Bektas, T., Crainic, T.: A Brief Overview of Intermodal Transportation, CIRRELT (2007). https://www.cirrelt.ca/DocumentsTravail/CIRRELT-2007-03.pdf
2. Shramenko, N.Y., Shramenko, V.O.: Mathematical model of the logistics chain for the delivery of bulk cargo by rail transport. Sci. Bull. Nat. Mining Univ. **5**(167), 136–141 (2018)
3. Shyriaieva, S.V., Kravchuk, A.V.: Analysis of trends in multimodal transport of Ukraine and abroad. Manag. Project Syst. Anal. Logist. **10**, 297–302 (2012)
4. Troitskaya, N.A., Chubukov, A.B., Shilimov, M.V.: Multimodal transportation systems and intermodal technologies, Moscow, vol. 336 (2009)
5. Shramenko, N.: The methodological aspect of the study feasibility of intermodal technology of cargo delivery in international traffic. Sci. Bull. Nat. Mining Univ. **4**(160), 145–150 (2017)
6. Petrashevskyi, O.L., Kyrychenko, A.I.: Ways of improving effective management of goods delivery under multimodal transportation. Probl. Transp. Coll. Sci. Works **9**, 3–16 (2012)

7. Shramenko, N.: Evaluation of the effectiveness of piggyback traffic in the context of creating transport and logistics clusters. Sci. Bull. Nat. Mining Univ. **6**(162), 151–155 (2017)
8. Shramenko, N.: Effect of process-dependent parameters of the handling-and-storage facility operation on the cargo handling cost. Eastern-Eur. J. Enterp. Technol. **5/3**(77), 43–47 (2015)
9. Shramenko, N., Muzylyov, D.: Forecasting of overloading volumes in transport systems based on the fuzzy-neural model. In: Advances in Design, Simulation and Manufacturing II. DSMIE 2019. Lecture Notes in Mechanical Engineering, pp. 311–320. Springer, Cham (2020)
10. Sokolova, O.E.: Conceptual bases of formation of multimodal cargo transportation system. Sci. Intensive technol. **1**(21), 114–118 (2014)
11. Pidlisnyi, P.I., Braykovska, A.M.: Improving the efficiency of multimodal cargo transportation by implementing multimodal technologies. Formation market Relat. Ukraine Coll. Sci. Works, **2**(141), 67–74 (2013)
12. Zabors'kyy, L.O.: Methodical bases of organization of transport and technological processes in cargo delivery systems. Abstract of Ph.D. thesis, 20 (2008)
13. Gerami, V.D., Kolik, A.V.: Management of transport systems. In: Logistics Transportation Support: Textbook and Workshop. National Research University Higher School of Economics, vol. 508 (2014)
14. But'ko, T.V., Lomot'ko, D.V., Golovko, T.V.: Structural approach to the analysis of the port railway junction. In: Collection of Scientific Papers on the Materials of the International Scientific-Practical Conference, vol. 1, pp. 19–20 (2008)
15. Hatuev, R.A., Osintsev, N.A.: The choice of the cargo delivery systems in multimodal containers transportation. Actual Prob. Sci. Technics Educ. **1**(1), 63–66 (2015)
16. Lomot'ko, D.V.: Improvement of technology of transfer of cargo flow at interaction of railway and sea transport. In: Proceedings of the Ukrainian State Academy of Railway Transport, pp. 91–97 (2014)
17. Shramenko, N.Y., Shramenko, V.O.: Optimization of technological specifications and methodology of estimating the efficiency of the bulk cargoes delivery process. Sci. Bull. Nat. Mining Univ. **3**, 146–151 (2019)
18. Lapkina, I.O., Malaksiano, M.O.: Modelling and optimization of perishable cargo delivery system through Odesa port. Actual Prob. Econ. **177**, 353–365 (2016)
19. Mirotin, L.B., Bul'ba, A.V., Demin, V.A.: Logistics, technology, design of warehouses, transport hubs and terminals, Rostov-on-Don, Phoenix, vol. 408 (2009)
20. Shramenko, N., Pavlenko, O., Muzylyov, D.: Information and communication technology: case of using petri nets for grain delivery simulation at logistics system. In: CEUR Workshop Proceedings, vol. 2353, pp. 935–949 (2019)
21. Shramenko, N., Muzylyov, D., Shramenko, V.: Methodology of costs assessment for customer transportation service of small perishable cargoes. Int. J. Bus. Perform. Manag. **21** (1/2), 150–166 (2020). Special Issue on: TBM 2019 Transformative Business Models – Disruptive Innovation in Finance, Logistics and Tourism
22. Shramenko, N.: Methodology for evaluation of synergy effect in terminal cargo delivery system. Actual Prob. Econ. **8**(182), 439–444 (2016)
23. Rodrigue, J.-P., Slack, B.: Intermodal Transportation and Containerization. The geography of transport systems (2013). https://people.hofstra.edu/geotrans/eng/ch3en/conc3en/ch3c6en.html
24. Kurapati, S., Lukosch, H., Verbraeck, A., et al.: Improving resilience in intermodal transport operatons in seaports: a gaming approach. EURO. J. Decis. Process. **3**, 375–396 (2015)
25. Haralambides, H.: Globalization, public sector reform, and the role of ports in international supply chains. Marit. Econ. Logist. **19**, 1–51 (2017)

26. Haralambides, H.E.: Gigantism in container shipping, ports and global logistics: a time-lapse into the future. Marit. Econ. Logist. **21**(1), 1–60 (2019)
27. Ircha, M.: Serving tomorrow's mega-size containerships: The Canadian Solution. Marit. Econ. Logist. **3**, 318–332 (2018)
28. Wang, H., Nozick, L., Xu, N., Gearhart, J.: Modeling ocean, rail, and truck transportation flows to support policy analysis. Marit. Econ. Logist. **20**(3), 327–357 (2018)
29. Ho, M., Ho, K.: Risk management in large physical infrastructure investments: the context of seaport infrastructure development and investment. Marit. Econ. Logist. **8**, 140–168 (2006)
30. Deev, E.A., Kornilov, S.N.: The method for optimization multimodal container transportation. Modern Prob. Transp. Complex Russia **5/1**(6), 17–20 (2015)
31. Juhel, M.: Globalisation, privatisation and restructuring of ports. Int. J. Marit. Econ. **3**, 139–174 (2001)
32. Tyulenev, K.G.: Management of Container Transportation in Foreign Economic Activity, vol. 112. Book, Institute of Transport Problems Russian Academy of Sciences, St. Petersburg (2017)
33. Muzylyov, D., Shramenko, N., Shramenko, V.: Integrated business-criterion to choose a rational supply chain for perishable agricultural goods at automobile transportations. Int. J. Bus. Perform. Manag. **21**(1/2), 166–183 (2020)
34. Bojovic, N., Milenkovic, M.: The best rail fleet mix problem. Oper. Res. Int. J. **8**, 77–87 (2008)
35. Valkova, S.S., Stepanets, A.V., Veryutina, V.E.: The formation of complex of tasks on a management by cargo handling of railway wagons in a seaport. Sci. Tech. Bull. Volga Region **1**, 146–150 (2013)
36. Kawasaki, T., Matsuda, T.: Containerization of bulk trades: a case study of US–Asia wood pulp transport. Marit. Econ. Logist. **17**(2), 179–197 (2015)
37. Kirichenko, S.A., Lakhmetkina, N.Yu.: Multimodal container transportation of grain. World Transp. **13**(1), 116–125 (2015)
38. Steenken, D., Voß, S., Stahlbock, R.: Container terminal operation and operations research - a classification and literature review. OR Spectrum **26**(1), 3–49 (2004)
39. Stahlbock, R., Voß, S.: Operations research at container terminals: a literature update. OR Spectrum **30**, 1–52 (2008)
40. Liang, C., Hwang, H., Gen, M.: A berth allocation planning problem with direct transshipment consideration. J. Intell. Manuf. **23**, 2207–2214 (2012)
41. Panasenko, N.N., Yakovlev, P.V.: Containerization of the international transport system. Bull. Astrakhan State Techn. Univ. Series Marine Technol. **4**, 103–116 (2016)
42. Zeng, Q., Feng, Y., Chen, Z.: Optimizing berth allocation and storage space in direct transshipment operations at container terminals. Marit. Econ. Logist. **19**(3), 474–503 (2017)
43. Bystritskaya, O.: Container Cargo: Maximum over the past 10 years, Center for Transport Strategies. https://people.hofstra.edu/geotrans/eng/ch3en/conc3en/ch3c6en.html. Accessed 7 Feb 2019
44. Seaports of Ukraine in 2018 increased container handling by almost 19% - AMPU, Multimedia broadcasting platform of Ukraine "UkrInform" (2019). https://www.ukrinform.ru/rubric-economy/2624014-morporty-ukrainy-v-2018-godu-uvelicili-perevalku-kontejnerov-pocti-na-19-ampu.html
45. Ukrainian Railway launched three scheduled grain container trains from Chernomorsk-Port station, Analytical Agency "AGRICULTURE". http://www.agriagency.com.ua/news/27435.html. Accessed 25 Sept 2019
46. Ukrainian Railway increased grain transportation by 20.1% in 2019, Analytical Agency "AGRICULTURE". http://www.agriagency.com.ua/news/27570.html. Accessed 13 Jan 2020

A Review of Enabling Technologies and Solutions for IoT Based Smart Warehouse Monitoring System

Alem Čolaković(✉), Samir Čaušević, Amel Kosovac, and Ermin Muharemović

Faculty of Traffic and Communication, University of Sarajevo, Sarajevo, Bosnia and Herzegovina
alem.colakovic@gmail.com

Abstract. Warehouses are premises used to store goods or products with the key task of ensuring its security as well as maintaining its value and quality. Some unexpected events in the warehouse such as temperature fluctuations and humidity changes, and other unwanted events may destroy the value of goods and products.In such circumstances, continuous monitoring of the warehouse and condition control should be carried out. Internet of Things (IoT) enables the development of various cost-effective solutions based on modern Information and Communication Technologies (ICT) and infrastructure. IoT brings new opportunities and benefits for the development of Warehouse Management Systems (WMS). This paper put in light some possibilities of IoT applications for a smart warehouse monitoring system. We present a model of IoT system that helps to overcome some shortcomings in existing solutions for warehouse monitoring control.

Keywords: Internet of Things (IoT) · Warehouse Management System (WMS) · Monitoring system · Smart logistics · Application model

1 Introduction

A warehouse is a complex environment that uses a combination of different functions, processes, and operations to perform a basic mission that includes receiving, storing, joining, manipulating, and issuing different products. The various processes, functions, and operations that take place within the warehouse must be actively monitored for the safety, efficiency, and quality of the performance to be successful. Safety and effectiveness in the warehouse are important indicators of management and control of warehouses [1]. Storing products that are sensitive to temperature, pressure, and humidity in a warehouse requires special monitoring and control.

With the advent of Industry 4.0, space has been opened to develop new applications and potentially improve the control and management of warehouses. One of the drivers of innovation is the paradigm known as the Internet of Things (IoT). This industry requires an integrated system of different subsystems that include sensors, different hardware and software platforms, communications and other technologies. IoT

I. Karabegović (Ed.): NT 2020, LNNS 128, pp. 630–637, 2020.
https://doi.org/10.1007/978-3-030-46817-0_73

platforms can generate a large amount of data to be processed quickly with the aim of faster-informing stakeholders and to respond faster to changes that read sensors in the system. The IoT concept-based system enables physical systems to communicate with one another for the purpose of executing various processes such as sensing, actuation, communication, computation, and management [2].

The focus of this paper is a description of a warehouse monitoring and control system based on IoT technologies. After the introductory part, the importance of the process of monitoring and controlling certain sizes, such as temperature, humidity, and lighting in warehouses, is described. Section 3 proposes a solution with a detailed structure of the IoT system and a description of the required infrastructure and technologies. In addition to the specification of technologies and infrastructures that can be used to design a warehouse monitoring and control system, the paper presents an algorithm for the operation of the proposed system. The final section contains concluding considerations.

2 A Warehouse Monitoring and Control

The basic task of a warehouse is to preserve the value and quality of raw materials, semi-finished products or products stored in premises called warehouses. These materials and products can be influenced by various physical, chemical and atmospheric conditions. To develop a smart warehouse monitoring and control system, it is necessary to know for which products ant materials it is designed and the manner of organization of the storage space.

The monitoring and control system should consider the basic functions of the warehouse [3, 4]: receipt, unloading, storage, and issuing goods. Each of these functions should be separately analyzed to cover different risks of unwanted conditions such as temperature fluctuations. It is necessary to supervise and control all these key procesess. Special attention should be on storage and safekeeping procesess. It is necessary to constantly perform supervision and control [5]. There are several IoT solutions for the applied in logistics, which entail various aspects of monitoring, control, and automation [2, 5–10]. IoT solutions provide other possibilities such as the fire warning program [11] and can enhance other performance such as energy efficiency [12]. Therefore, IoT is an excellent platform providing connectivity between various technologies to enables remote monitoring, control, and automation of different environments including warehouses [2, 11, 13, 14]. To perform these tasks, different sensors and other components are required. The technology stack depends on the type of product being monitored [2, 6, 12, 15, 16].

Particular attention should be paid to products such as food, specialty paints, and coatings, a large proportion of pharmaceutical and medical products, etc. Storage of these products requires monitoring and control of various conditions such as the temperature, humidity, and illumination of the storage area. Figure 1 shows an example of warehouse zoning, labeled by function and possible locations for setting different types of sensors for monitoring of these conditions.

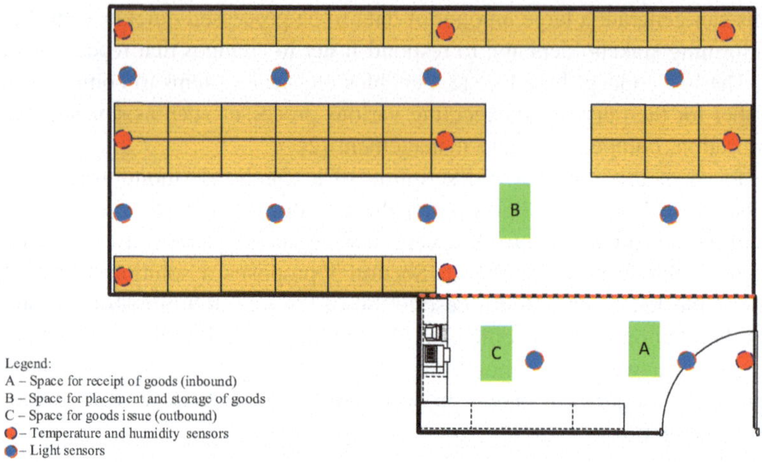

Fig. 1. An example of a warehouse with designated locations for sensors, functions and zones

The use of warehouse sectors greatly facilitates the implementation and creation of systems that monitor and record the parameters measured by the sensors.

3 IoT Based System for Warehouse Monitoring and Control

Internet of Things (IoT) is the internetworking paradigm enabled by technology stack which provides a seamless connectivity between physical and virtual objects [15]. IoT systems are based on several processes such as sensing, identification, actuation, communication, computation, and management. IoT enabling technologies should enable these processes in order to facilitate the development of intelligent services such as smart warehouse monitoring. An IoT system for warehouse monitoring and control should enable the collection and processing of data to respond to potential unwanted events and undesirable conditions (e.g. high temperature) in real-time. The goal of this system is the monitoring of these conditions and the automation of the control.

The main challenges in developing such solutions are the result of the complexity of the IoT system and the heterogeneity of technologies across all layers of the system architecture. To simplify the development of these systems and solve the problem of complexity, appropriate architectures are needed that guide the application and integration of specific technologies for the development of IoT solutions. These architectures should support multi-system integration and inter-domain interactions with full interoperability and the ability of simple management processes. This imposes the need to understand the structure of the IoT system, which emphasizes the importance of the need for a comprehensive system framework, i.e. conceptual model. For the purpose of this paper, we use the layered conceptual architecture of an IoT system that include four functional blocks [15].

3.1 IoT System Structure and Technology Stack

An IoT system for warehouse monitoring and controlling can include several applications such as monitoring the conditions where goods are stored, triggering certain actions, controlling access to premises, etc. These applications can be implemented in various scenarios such as web or mobile application. The implemented application should send instructions to perform actions in case of detection of certain anomalies such as unwanted temperature or humidity in a particular storage sector (zone). Figure 2 shows the structure models of the proposed IoT system.

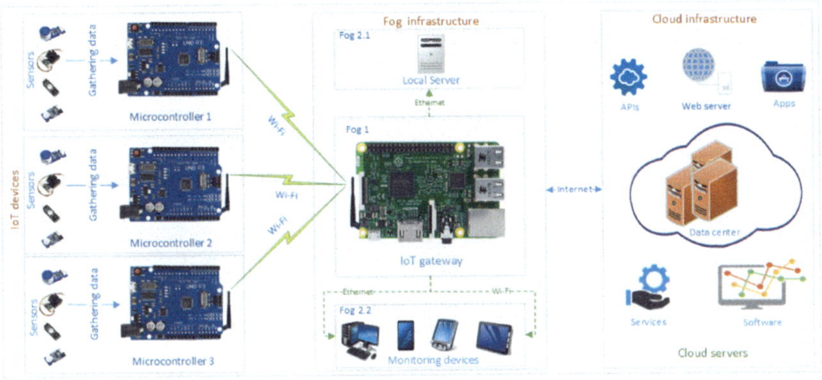

Fig. 2. IoT system structure

The specification of technologies and infrastructure can cover a large number of components depending on IoT system structure. A large number of possible technologies are available in the market for the development of the described system. For example, the technology stack and infrastructure specifications for the warehouse monitoring and control system are shown in Table 1. Sensors are used to measure temperature, humidity, and illumination. For example, experiments of using SHT10 sensor as the detector of temperature and humidity in the warehouse shows the practical value of the application [14]. They are connected to a microcontroller that performs filtering, aggregation, preprocessing and data forwarding to the IoT gateway. For example, NodeMCU (ESP8266EX MCU) is a cost-effective microcontroller that acts as an aggregator of multiple raw datasets collected from sensors. It is equipped with a Wi-Fi module for sending data to the network gateway. The gateway can be implemented using the Raspberry Pi platform because it enables to implement a program for deciding whether to forward data to a cloud server, local server, or terminals. These decisions are predefined in the program code embedded on the platform. Because of these functionalities, the Raspberry Pi can be considered as a fog layer in the system architecture. For cloud services such as advanced data processing and storage, it can be used private infrastructure or services such as Microsoft Azure, AWS, etc. A key role of this infrastructure is to analyze and to store data in the database (e.g. MongoDB), and to host a web application. This implementation requires a permanent internet connection. On the other hand, a local server can be used as alternative solution to make the system cost effective, secure and reliable, since it eliminates the need for internet connection. Mobile applications (e.g. android) allow different stakeholders (e.g. product owners) to have an insight into the warehouse conditions in real-time. In this case, terminal devices can be considered as another fog layer in the described system architecture.

Table 1. Technology and infrastructure specification

Domain	Enabling technologies and infrastructure	
Application Domain	Application	Monitoring and control of warehouse conditions
	OS and API	OS (Android for mobile devices, Apache web server, Windows server) RTOS (Raspbian for Raspberry Pi platform) API (WebGL JavaScript API, HTTP Server API, API za Raspberry Pi, Google APIs for Android)
	SDK and IDE	Web application: HTML, CSS, JavaScript, PHP Mobile app: Android SDK with APIs libraries and debugging tools Raspberry Pi: Arduino IDE, Python Programming a microcontroller: Arduino IDE
	Architecture	RESTful (IP address + application port number, e.g. 8125)
Middleware Domain	Cloud and fog platforms	Cloud: private cloud infrastructure Fog 1: Raspberry Pi 3 Model B+ Fog 2: Local Windows Server
	Data manipulation	NoSQL database (MongoDB) File format: CSV (comma-separated values)
Networking Domain	Protocol stack	Application (CoAP, HTTPS), Transport (UDP), Network (IPv4) Network interface (IEEE 802.11n)
	Network interface	WLAN: ESP8266 Wi-Fi module LAN: Ethernet 1000Base-T, Cable Cat 6, lGbps WAN: Internet
Devices Domain	Microcontroller and equipment	NodeMCU (ESP8266EX MCU), Tensilica L106 32-bit RISC processor 160 MHz, voltage transformers, batteries, resistors, etc.
	Sensors and actuators	Actuators for automatic parameter control Sensors: temperature, humidity, light, etc.

3.2 The Procedure Algorithm of IoT System

Sensors role is to measure temperature, humidity and lighting. The NodeMCU perform data aggregation collected by sensors as well as simple data filtering to determine when to forward the data to the IoT gateway. Once the data is collected and filtered, it is sent to the IoT gateway (Raspberry Pi). This component presents a central hub that acts as an aggregator of multiple raw datasets generated by sensors as well as a network gateway. A decision-making mechanism implemented on this infrastructure to make the decision to process or forward data to other system component. It converts unstructured data coming in different formats from sensor nodes into format that can be used by destination components such as monitoring terminals (e.g. mobile devices or computers in the monitoring center) and mechanisms for taking specific actions (e.g. actuators).

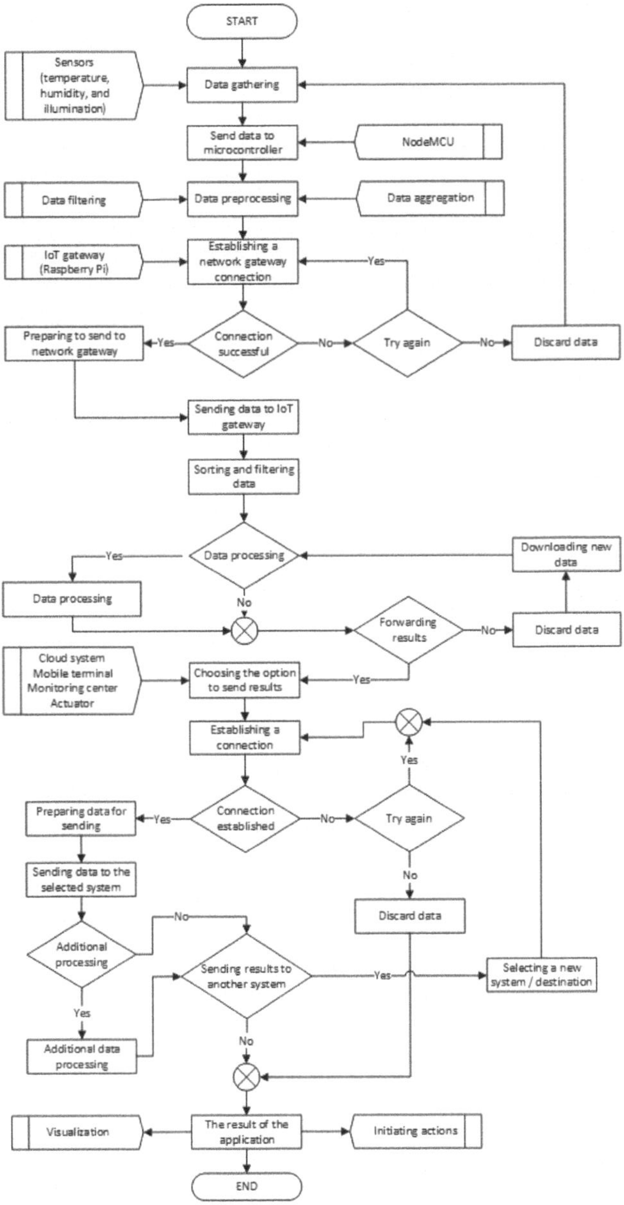

Fig. 3. IoT system operation algorithm

There are several possible system implementations that include various data processing systems. For example, the following components can be used for data processing: IoT devices, Raspberry Pi (F1), local server or terminal devices (F2), and cloud system (C). The key issue is to choose the best option for task allocation in order to achieve the desired level of technical performance including reliability, security, Quality of Service (QoS), and cost-effectiveness. This issue means system optimization by reducing system operating costs and optimizing the usage of available resources. The algorithm of the described system for the warehouses monitoring and control is shown in Fig. 3.

4 Conclusion

IoT technologies have caused a huge wave of innovation in almost every vertical market, including logistics. This paper presents one of the ideas for an application used for warehouse monitoring and control. This application is based on processes such as real-time detection and measurement of various events, as well as the processing and storage of data. To enable these functionalities, a set of different technologies such as sensors, hardware, software, cloud services, and communication technologies are required. There are many possible components with the same functionality, performance, quality, and price. These components can be interconnected in different ways. Therefore, the IoT system structure depends on the connectivity possibilities of used components and the conditions in which the system is used.

The paper presents the model of smart warehouse monitoring and control systems by using the "IoT technology stack". A systematic approach based on structural architecture blocks were used to create the proposed system specification. The main components of this system are programs for data processing and visualization of results, sensors, network gateway, fog and cloud systems for data processing and storage, actuators for automatization of certain actions in case of unwanted events, and the system power sources. The system structure and components determine how the system works. The workflow diagram was introduced to describe operations algorithm with different alternatives of integrating the technologies and components. The proposed solution is of particular interest to logistics companies providing sensitive goods storage services. Logistics operators, transport companies, customers and other stakeholders can achieve greater control over goods, reducing the potential for damage to goods and increasing the quality of the warehousing process. This solution can help companies improve the competitiveness of their business by providing value-added services.

References

1. Vrysagotis, V., Kontis, P.A.: Warehouse layout problems: types of problems and solution algorithms. J. Comput. Model. **1**(1), 131–152 (2011)
2. Halawa, F., Daoud, H., Lee, I., Li, Y., Yoon, S.W., Chung, S. H.: Introduction of a real time location system to enhance the warehouse safety and operational efficiency. Int. J. Prod. Econ. (2019)
3. Faber, N., de Koster, M.B.M., Smidts, A.: Organizing warehouse management. Int. J. Oper. Prod. Manag. **33**(9), 1230–1256 (2013)
4. Šamanović, J.: Prodaja, distribucija, logistika: teorija i praksa. Sveučilište u Splitu, Ekonomski fakultet, Split (2009)
5. Čaušević, S., Čolaković, A., Hasković, A.: The model of transport monitoring application based on Internet of Things. In: International Scientific Conference on Science and Traffic Development (ZIRP), pp. 63–72 (2018)
6. Liu, Y., Zhong, D., Cui, B., Zhong, G., Wei, Y.: Study on real-time construction quality monitoring of storehouse surfaces for RCC dams. Autom. Constr. **49**(PA), 100–112 (2015)
7. Chandra, A.A., Lee, S.R.: A method of WSN and sensor cloud system to monitor cold chain logistics as part of the IoT technology. Int. J. Multimed. Ubiquit. Eng. **9**(10), 145–152 (2014)
8. Ruan, J., Shi, Y.: Monitoring and assessing fruit freshness in IOT-based e-commerce delivery using scenario analysis and interval number approaches. Inf. Sci. **373**(October), 557–570 (2016)
9. David, D.R., Nait-Sidi-moh, A., Durand, D., Fortin, J.: Using Internet of Things technologies for a collaborative supply chain: application to tracking of pallets and containers. Proc. Comput. Sci. **56**(1), 550–557 (2015)
10. Liu, Y., Han, W., Zhang, Y., Li, L., Wang, J., Zheng, L.: An Internet-of-Things solution for food safety and quality control: a pilot project in China. J. Ind. Inf. Integr. **3**, 1–7 (2016)
11. Jiang, J., Gao, Z., Shen, H., Wang, C.: Research on the fire warning program of cotton warehousing based on IoT technology. In: 2015 International Conference on Logistics, Informatics and Service Science, LISS 2015 (2015)
12. Li, W., Kara, S.: Methodology for monitoring manufacturing environment by using Wireless Sensor Networks (WSN) and the Internet of Things (IoT). Proc. CIRP **61**, 323–328 (2017)
13. Deshmukh, P.R., Bhalerao, D.: An implementation of MQTT through the application of warehouse management system for climacteric fruits and vegetables. In: Proceedings of the 2nd International Conference on Communication and Electronics Systems, ICCES 2017, pp. 844–849 (2018)
14. Yao, Z.Q., Zhu, H.J., Du, W.H.: Design and implementation of automated warehouse monitoring system based on the internet of things. Appl. Mech. Mater. **543–547**, 1099–1102 (2014)
15. Čolaković, A., Hadžialić, M.: Internet of Things (IoT): a review of enabling technologies, challenges, and open research issues. Comput. Netw. **144**, 17–39 (2018)
16. Bolf, N.: Nadzor temperature tijekom prijevoza, skladištenja i prerade. Kemija u industriji/J. Chem. Chem. Eng. **66**(5–6), 320–321 (2017)

Theoretical Aspects of Diagnosing of Car Engine at the Time of Acceleration

O. P. Sitovskyi(✉), B. M. Dembitskyi, Y. V. Bulik, and P. V. Mazyliuk

Department of Cars and Transport Technologies, Lutsk National Technical University, Lvivska Street, 75, Lutsk 43000, Ukraine
sitovsky@ukr.net

Abstract. During the operation of the car, a question often arises of the diagnosis of the technical state of their engines. Due to the fact that specialized stands are quite expensive, not all businesses have the ability to use them. Therefore, there is a problem diagnosing the technical condition of engines by other methods. In this article the method of diagnosing the technical condition of engines in terms of time of acceleration of the car. The proposed method is simple and relatively cheap, since it involves diagnosing engines of the car's. To assess the suitability of the methodology and conducted experiments conducted evaluation of measurement uncertainty. The obtained results quite accurately characterize the state of the internal combustion engine, the maximum deviation is 8.7%. The given methodology will be useful for operating enterprises, for diagnosing motors of the cars.

Keywords: Car · Diagnostics · Engine · Power · Acceleration time · Speed · Torque

1 Introduction

The engine is one of the main units of the car. The efficiency of the entire vehicle depends on the technical condition of the engine. Today in Ukraine is quite acute a problem of effective and affordable diagnostics of technical condition of the engine separately and the transmission of the car in general. Modern cars are equipped with electronic control systems. They are quite sensitive to changes of the technical condition of the car. However, the result of natural wear of engine is increased fuel consumption, worsening traction and speed properties, reduced speed, increased emission levels. These factors certainly lead to increased operating costs, reduce the efficiency of transportation of passengers, cargo, and have a negative impact on the environment [1]. The consequence of this is the transition to alternative transportation technologies [2, 3] and a decrease in the synergistic effect of the functioning of transport systems [4].

At this stage of the development of science and technology offered quite a lot of different methods of diagnosing internal combustion engines according to the parameters of the crankcase oil, the amount of compression; the leakage of compressed air; power traction and dynamic stands; noise and vibration; for fuel consumption.

I. Karabegović (Ed.): NT 2020, LNNS 128, pp. 638–646, 2020.
https://doi.org/10.1007/978-3-030-46817-0_74

A special case is the diagnosing of the engine on the signals of the electronic control unit using motor-scanners, motor testers and other similar equipment. Research conducted by Pestrykov and Yevkarpiev [5] demonstrate that according to information received from the electronic control unit, it is possible to diagnose the technical condition of the engine for fuel consumption, capacity. It is worth noting that such diagnostic methods cannot effectively be used for a number of reasons:

- it is impossible to assess the method of diagnosing because of lack of data regarding the accuracy of calibration of measurement modules;
- the actual lack of data of producers regarding the hour cost of fuel leads to the need for a significant number of calculations.

Method of diagnosing of internal combustion engines using instantaneous angular acceleration given in [6] allows checking the technical condition of the cylinder group, however, does not give complete information about engine capacity. This method of diagnosis proposed in [7, 8], however it differs because it is based on research release rate heat of combustion. This method is proposed to apply mainly during the simulation work and research of internal combustion engines. Method of evaluating engine torque for instantaneous velocity is described in [9]. Assessment is carried out in stationary and transient modes.

The given model allows to accurately evaluate the technical condition of the internal combustion engine. Many works [10–13] are dedicated to the estimation of the engine torque, in addition there are a number of patented techniques, which can determine how the parameters of the engine as a whole and its condition [14–18]. It is worth noting sufficiently effective methods of vibro-acoustic diagnostics of internal combustion engines [19–22]. There is no doubt that all the methods of diagnosis and assessment of technical condition of internal combustion engines are efficient and accurate. However, in the operating company do not always have the possibility and feasibility to establish appropriate diagnostic facilities. Therefore, the main task in the technical diagnostics of car engines is to provide a simple to use and accurate method of estimation of its technical condition. In addition, an important issue is the possibility of collecting statistical data on the change of technical state of engine, and use the minimum amount of data of the manufacturer, because sometimes they are impossible to obtain, for example external high-speed characteristics, fuel consumption etc.

2 The Results of the Research

2.1 Theoretical Substantiation

Evaluation of the technical condition of the engine is proposed to conduct by a road test on the parameters of acceleration and coasting of the vehicle with a given speed [23, 24]. For the implementation of this method, the parameters must be put to speed characteristics of the power or the torque of the engine. It is known that [25] the equation of motion of the car is:

$$P_k = P_f + P_h + P_w + P_j, \quad (1)$$

where P_f, P_h, P_w, P_j – accordingly, the force of rolling resistance, recovery, air and disperse, N; P_k – the traction on the drive wheels, N.

Traction power to the drive wheels of the vehicle is determined [25]:

$$P_k = \frac{M_e \cdot u_k \cdot u_o \cdot \eta_{tr}}{r_k}, \quad (2)$$

where M_e – effective torque of the internal combustion engine, Nm; u_k, u_o – accordingly, the gear ratios of the gearbox and the main gear; η_m – the efficiency of the transmission; r_K – the rolling radius of the wheel, meters.

Support of the vehicle P_f, P_h, P_w can be determined from the deceleration measured during the free coasting, because in this case the car is moving and overcomes the resistance force due to the inertia:

$$P_{jslow.} = P_f + P_h + P_w. \quad (3)$$

On the other hand, according to Newton's second law the force of inertia is equal to:

$$P_{jslow.} = m_a \cdot j_{slow.}, \quad (4)$$

where $j_{slow.}$ – the instantaneous deceleration of the vehicle in coasting mode, m/s^2.

During the road test, you must use the requirements of UN Regulations No. 68 [26] in respect to the requirements for a test site and atmospheric conditions.

By converting dependencies (1–4) and taking into account the factor of the rotating masses, the dependence for determining engine torque for acceleration on the appropriate gear:

$$M_e = \frac{m_a \cdot (j_{acel.} \cdot \sigma + j_{slow.}) \cdot r_k}{u_k \cdot u_o \cdot \eta_{tr.}}, \quad (5)$$

where $j_{acel.}$ – the instantaneous acceleration of the car during acceleration to set speed, m/s^2.

It should be noted that according to (5) is received, the instantaneous acceleration, so the calculations are obtained for different values of torque. Then, the corresponding engine output will be equal to [26]:

$$N_e = \frac{M_e \cdot n_{ef}}{9550}, \ kW. \quad (6)$$

where n_{ef} – engine revolutions, min^{-1}.

In technical characteristics of the car are usually listed such parameters as: the maximum power of the engine and the speed at which it is achieved, the maximum torque of the engine and the speed at which it is achieved, and acceleration to a certain speed, usually 60, 80 or 100 km/h, depending on vehicle class. However, when determining the time of acceleration, it is necessary to carry out the overclocking-vehicle to a speed at which the engine goes to maximum power and maximum torque. Therefore, before the experiment, it is necessary to determine the vehicle speed at which the engine will be released at the maximum values, that is, that is, the speed at which acceleration will be carried out.

Determination of the reference speed acceleration is based on:

$$V_{ef} = \frac{n_{ef} \cdot \pi \cdot r_{\text{K}}}{30 \cdot u_{tr.}}, \tag{7}$$

where n_{ef} - the speed at which the engine comes on to maximum power or maximum torque, min^{-1}.

The resulting speed V_{ef} will be the normative value, which is necessary to carry out the acceleration of the vehicle. In addition, to account for the total resistance to movement necessary to carry out the measurements of the coasting the vehicle speed V_{ef}. It is clear that the smaller the ratio, the greater will be the speed of the car, and as a result, you will need to have a considerable length of the test plots for research. The results of calculations carried out for a number of passenger cars indicate that the most optimal will conduct a diagnosis of the engine on the II and III gear. So, for Audi A3 speed at II gear - 24.9 km/h on III gear - 35.4 km/h for the Citroen Berlingo, respectively, 43.4 and 66,3 km/h, for Honda Accord and 103.9, respectively, 73,9 km/h, for VW Passat respectively, 43.9 and 63,4 km/h.

Because, usually, the operating company has no data relative to the external speed characteristics, then as the control point is taken the maximum torque, and to assess the condition of the engine throughout the Rev range is to use schedule based on torque speed (rpm) of the vehicle, the construction of which will not cause difficulties. The technique of building a conditional dependence of torque from revs of the car are given below.

2.2 The Results of the Experiment

To assess the adequacy of the proposed methodology was made a series of experiments. For diagnosis, the following conditions were established and observed: road surface is smooth, dry and clean the slope in the transverse direction not more than 0.5%, in length – no more than 1%. The wind speed did not exceed 5 m/s, ambient air temperature ranged from +5 to +25 °C. To ensure the accuracy of the measurements the tests were carried out in two opposite directions.

Check of parameters was carried out with the use of modern measuring equipment, which is much cheaper compared to the cost of traction and dynamic stands, but at the same time, is sufficiently precise and allows to obtain a series of data. Device RecelogicVbox III Mini (Fig. 1), during use of the test, "Accel", fixes the time path and acceleration. Display resolution of 0.1 s. Tachometer RPM VC2 for fixing the rpm of the car. All parameters were recorded in real-time. The equipment that was used to determine and check parameters of the car calibrated by accredited bodies for conformity of assessment.

The measurements were carried out for Geely CK. It was carried out on two runs in both directions. The results of the experiment are shown in Fig. 2. The value of the

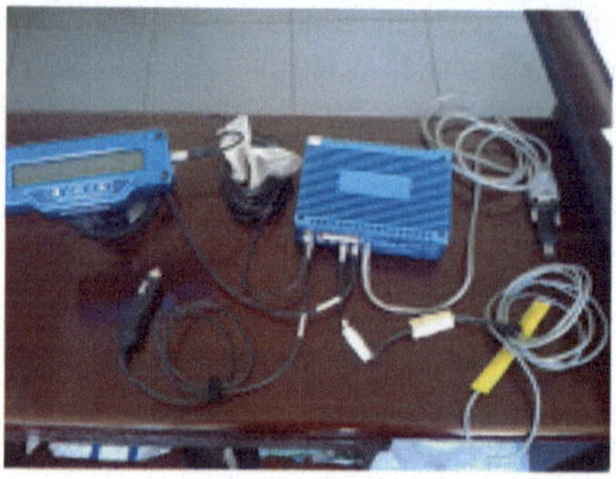

Fig. 1. RecelogicVbox III Mini

approximation of the trend line is of high value, which indicates a sufficient accuracy of the conducted experiment. The graph shown in Fig. 2 can appropriately be called the conditional characteristic torque of the vehicle as it does not match the factory data, but serves only to determine the technical condition of the car engine. For the control point of the graph is taken the maximum torque which this engine is 128 Nm, speed 3400 min^{-1}. For further application of the methodology, it is necessary to conduct validation.

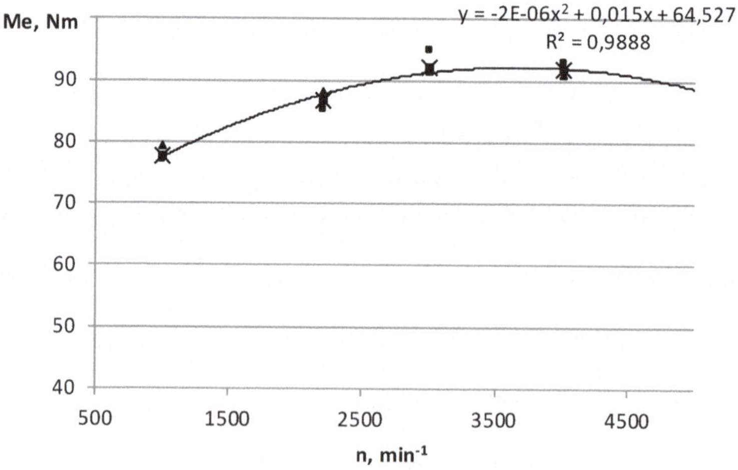

Fig. 2. The torque of the car Geely CK obtained experimentally

2.3 Validation

Evaluation of the suitability of the methodology is carried out by estimating the uncertainty of measurements. The dependence for determining the total standard uncertainty of measurement of acceleration $u_c(j_{CK})$, according to state standard DSTU-N RMG 43:2006 [27]:

$$u_c(j_{CK}) = \sqrt{u_A^2(\bar{j}_{CK}) + \frac{\bar{j}_{CK}^2}{(100\%)^2}\left[\tilde{u}^2(v)\right]}, \tag{8}$$

where u(v) – the relative standard uncertainty type B of instrument for measuring speed RecelogicVbox III Mini; $u_A(\bar{j}_{CK}) = \sqrt{\frac{1}{n(n-1)}\sum_{i=1}^{n}(j_{CKi} - \bar{j}_{CK})^2}$ – standard uncertainty of type A is conversion speed by using the registered software.

The expanded uncertainty is determined from the expression:

$$U(j_{CK}) = ku_c(j_{CK}), \tag{9}$$

where k – the coverage ratio, which is defined as the Student coefficient for a given probability (usually p = 0.95) and the effective number of degrees of freedom ν_{eff}, which is calculated according to the formula of Welch-Satterthwaite.

$$\nu_{eff} = (n-1)\left[\frac{u_c(j_{CK})}{u_A(\bar{j}_{CK})}\right]^4. \tag{10}$$

For the results of the experiment shown in Fig. 2, there were calculated and got the estimates of uncertainty listed in Table 1. It is worth noting that at this stage of the assessment of uncertainty were not taken into account a number of elements for which it is necessary to conduct additional research. These include meteorological test conditions, road surface, tire condition of the vehicle and the pressure in them.

Table 1. Results of estimation of measurement uncertainty

$n_{eф}$, min^{-1}	M_e, Nm	$U(j_{CK})$, at p = 0,95	Confidence interval	Deviations, %
1000	77,71077	2,824654	77,7 ± 2,8	7,0
2200	86,92204	2,720482	86,9 ± 2,7	6,1
3000	92,15523	4,2031	92,2 ± 4,2	8,7
4000	91,8459	2,02021	91,8 ± 2,0	4,3

2.4 Evaluation Criteria

Concerning the results of evaluation of measurement uncertainty is established that the deviation of the results is in the range 4…9%, at a confidence level of p = 0,95. Therefore, we can assume that the technical condition of the internal combustion engine is satisfactory, if after re-checking the values are within the confidence interval (see Fig. 3). Otherwise, it is necessary to reduce the interval diagnosis to monitor changes of its power characteristic.

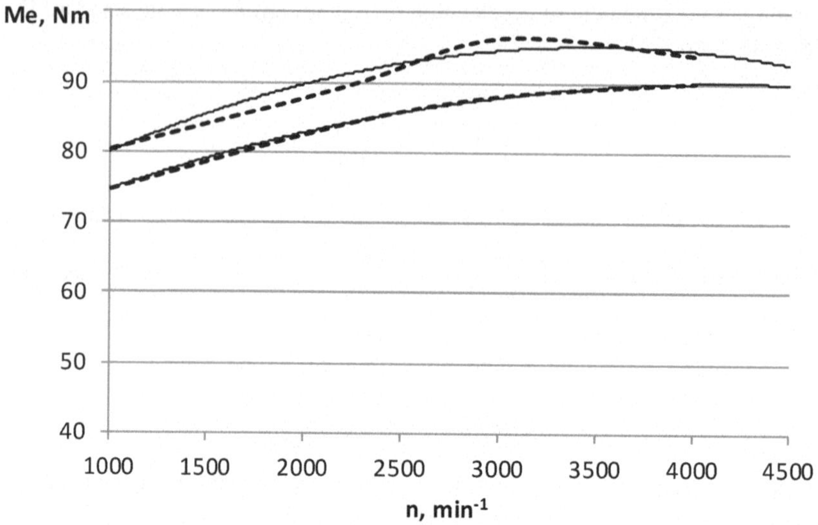

Fig. 3. The confidence interval and the criteria for the evaluation of engine torque

3 Conclusion

Analysis of previous studies indicates the absence of a simple and effective method for diagnosis of traction-speed characteristics of internal combustion engine vehicles in terms of maintenance companies. However, it can be noted that based on available plant-specific data, the most rational assessment of the technical condition of the engine can be delivered by torque. The given dependence allows calculation of the torque after determining the parameters of traction-speed properties of the vehicle in the mode of "acceleration-coasting". Evaluation of uncertainty of the results of the experiment shows the acceptable deviation of the measured values. The largest deviation is 8.7%, in the control point, which is adopted as the maximum torque set by the manufacturer. As the criterion for assessing the technical condition of the engine was taken the confidence interval calculated in the evaluation of measurement uncertainty.

Operating companies, service stations and other interested organizations, can use the proposed method of evaluation of technical condition of the engine. Timely diagnostics of the technical condition of the engine will reduce fuel consumption, costs on repair of vehicles, thereby increasing efficiency of their use.

References

1. Karnaukh, M., Muzylyov, D., Shramenko, N.: Energy-optimized technologies, energy efficiency and energy management on transport. In: 2nd International Scientific and Practical Conference "Energy-Optimal Technologies, Logistic and Safety on Transport", EOT-2019, MATEC Web of Conferences, vol. 294 (2019)

2. Shramenko, N.: The methodological aspect of the study feasibility of intermodal technology of cargo delivery in international traffic. Sci. Bull. Natl. Min. Univ. **4**(160), 145–150 (2017)
3. Shramenko, N.: Evaluation of the effectiveness of piggyback traffic in the context of creating transport and logistics clusters. Sci. Bull. Natl. Min. Univ. **6**(162), 151–155 (2017)
4. Shramenko, N.: Methodology for evaluation of synergy effect in terminal cargo delivery system. Actual Probl. Econ. **8**(182), 439–444 (2016)
5. Pestrikov, V.M., Evkarpiev, V.E.: Osobennosti diagnostiki sovremennyh avtotransportnyh sredstv. Tehniko-tehnologicheskie problemy servisa. Nauchno-tehnicheskij zhurnal Sankt-Peterburg, Sankt-Peterburgskij gosudarstvennyj ekonomicheskij universitet **4**(30), 14–20 (2014)
6. Le Khac, B.: Diagnostic internal combustion engine based on crankshaft angular acceleration (2017). https://www.researchgate.net/publication/267801621_DIAGNOSTIC_INTERNAL_COMBUSTION_ENGINE_BASED_ON_CRANKSHAFT_ANGULAR_ACCELERATION
7. Bueno, A.V., Velásquez, J.A., Milanez, L.F.: Internal combustion engine indicating measurements. In: Haq, Z. (ed.) Applied Measurement Systems. InTech (2012). http://www.intechopen.com/books/applied-measurement-systems/internal-combustion-engineindicating-measurements. ISBN 978-953-51-0103-1
8. Businaro, A., Cavina, N., Corti, E., Mancini, G., Moro, D., Ponti, F., Ravaglioli, V.: Accelerometer Based methodology for combustion parameters estimation. Energy Proc. **81**, 950–959 (2015). https://doi.org/10.1016/j.egypro.2015.12.152. http://www.sciencedirect.com/science/article/pii/S1876610215028015. ISSN 1876Accelerometer Based Methodology for Combustion Parameters Estimation-6102
9. Franco, J., Franchek, M.A., Grigoriadis, K.: Real-time brake torque estimation for internal combustion engines. Mech. Syst. Sig. **22**(2), 338–361 (2008). https://doi.org/10.1016/j.ymssp.2007.08.002. ISSN 0888-3270
10. Lee, B., Rizzoni, G., Guezennec, Y., Soliman, A., et al.: Engine control using torque estimation. SAE Technical Paper 2001-01-0995 (2001). https://doi.org/10.4271/2001-01-0995
11. Jianqiu, L., Minggao, Y., Ming, Z., Xihao, L.: Advanced torque estimation and control algorithm of diesel engines. SAE Technical Paper 2002-01-0198 (2002). https://doi.org/10.4271/2002-01-0198
12. Langman, J., Pedryc, N.: Diagnostics for low-power internal combustion engines (2017). https://www.researchgate.net/publication/267946589_DIAGNOSTICS_FOR_LOW-POWER_INTERNAL_COMBUSTION_ENGINES
13. Gawande, S.H., Navale, L.G., Nandgaonkar, M.R.: Power balancing of inline multicylinder diesel engine. Adv. Mech. Eng. **2012**, 1–9 (2012)
14. Rizzoni, G., Guezennec, Y., Soliman, A.: Engine control using torque estimation, US Patent Number US20030167118 A1, September 2003
15. Ling, R., Harris, O.G., Rushinsky, J.: Systems and methods for determining engine torque values, US Patent Number US20100153025 A1, June2010
16. Fredriksson, K., Wattwil, F., Ahlinder, S.: Method for determination of mean engine torque, US Patent Number US20090132182 A1, May 2009
17. Varoquie, B., Denert, C.: Method for determining the engine torque delivered by a multi-cylinder engine, International Application No.: PCT/EP2015/002533, 30 June 2016
18. Verdejo, J.R., Grimes, M.R., Kalweit, N.J., Mathews, D.S.: Method and control system for controlling an engine function based on crankshaft acceleration, US Patent Number US7918212 B2, April 2011

19. Delvecchio, S., Bonfiglio, P., Pompoli, F.: Vibro-acoustic condition monitoring of Internal Combustion Engines: a critical review of existing techniques. Mech. Syst. Sig. Process. **99**, 661–683 (2018). https://doi.org/10.1016/j.ymssp.2017.06.033. ISSN 0888-3270
20. Bánlaki, P., Kulcsár, Sz.: Monitoring the operation of internal combustion engines using order analysis of noise and vibration data. J. Mach. Manuf. **XLIX**, 142–144 (2009)
21. Syrbakov, A.P., Korchuganova, M.A.: Technical diagnosis of internal combustion engines in the parameters of vibration. Appl. Mech. Mater. **756**, 53–58 (2015)
22. Beresnev, A., Beresnev, M.: Vibroacoustic method of IC engine diagnostics. SAE Int. J. Engines **7**(1), 1–5 (2014). https://doi.org/10.4271/2014-01-0282
23. Dembic'kij, V.M.: Metodi diagnostuvannja tehnichnogo stanu dviguniv vnutrishn'ogo zgorjannja: problemi ta perspektivi. Tezi dopovidej IV Vseukraïns'koï naukovo-tehnichnoï konferencii̇̈ «Aktual'ni problemi konstrujuvannja, ekspluatacii̇̈ ta remontu obladnannja lisovogo kompleksu» 16–17 listopada 2017 roku, pp. 53–54. inf.-vid. viddil Luc'kogo NTU, Luc'k (2017)
24. Sitovs'kij, O.P.: Pobudova chastkovih shvidkisnih harakteristik dviguniv v ekspluatacijnih umovah. Tezi dopovidej Vseukraïns'koï naukovo-praktichnoï konferencii̇̈ «Novitni shljahi stvorennja, tehnichnoï ekspluatacii̇̈, remontu i servisu avtomobiliv», pp. 229–231. Vijs'kova akademija, Odesa (2015)
25. Tarasik, V.P.: Teorija dvizhenija avtomobilja: Uchebnik dlja vuzov. BHV-Peterburg, SPb. (2006). 478 p.
26. UN Regulation № 68 Uniform provisions concerning the approval of power-driven vehicles including pure electric vehicles with regard to the measurement of the maximum speed, Geneva (1987)
27. DSTU-N RMG 43:2006 Metrologija. Zastosuvannja «Rukovodstva po vyrazheniju neopredelennosti izmerenij»

New Technologies in the Field Energy: Renewable Energy, Power Quality, Advanced Electrical Power Systems

Analysis of Groundwater Source Heat Pump Operation with Improvement Suggestions

Damir Špago(✉), Mirna Nožić, and Safet Isić

Faculty of Mechanical Engineering, "Džemal Bijedić" University of Mostar, Sjeverni logor bb, 88104 Mostar, Bosnia and Herzegovina
damir.spago995@gmail.com

Abstract. Heat pumps are one of the most used heating and cooling solutions in buildings today, especially in buildings that fall into energy efficient categories, such as categories A and B. Their advantage, in addition to high energy efficiency, lies in the low or non CO2 emissions, and corresponds to the tendency of modern building construction. Most commonly used systems are systems which use water, air or ground as a renewable energy source. This paper will present an analysis of the influence of ambient air temperature on the operation of such a system that uses groundwater as a source, as well as measures to improve the performance or reduce the electrical energy consumption.

Keywords: Heat pumps · Energy efficiency · Solar energy – PV panels · Geothermal energy

1 Introduction

Heat pumps are devices that operate on the principle of the left-hand Carnot cycle. They operate on a very similar principle to that of refrigerators, air conditioners and others. Taking heat energy from a heat source such as groundwater in this case, heat energy is transferred to a working medium which can be a freon. The evaporated freon is then injected into a compressor, where its pressure and temperature increase. Using the heat of the evaporated freon to heat the water or directly the air in the room in this way, it is possible to obtain 4 kWh of thermal energy for 1 kWh of electricity. Many factors influence this, such as type of heat source, proper selection of the heat pump, mode of operation of the heat pump, equipment correctness, external influencing factors, etc. To describe the energy efficiency of a heat pump, in practice, the term COP - coefficient of performance is used. It represents the ratio of generated heat energy and electrical energy that is consumed.

2 Groundwater Source Heat Pumps

During winter days, the groundwater temperature remains constant at 8–14 °C, which makes this heat source very acceptable for use in such installations. The COP of these systems is relatively high and constant throughout the year. The systems are constructed in such a way that the water is taken in by submersible pumps from wells,

I. Karabegović (Ed.): NT 2020, LNNS 128, pp. 649–656, 2020.
https://doi.org/10.1007/978-3-030-46817-0_75

which are previously dug in places where the existence of groundwater is determined. It is recommended to dig 2 wells at a distance of 15 m from each other in order to avoid water mixing and to achieve a minimum water flow of 2 m^3/h, which can be achieved already at a depth of 5 m. Manufacturers mostly recommend using a plate heat exchanger in the intermediate circuit since the exchangers in the crane itself are very sensitive to the water quality of the wells (various impurities, sand, gravel, limescale, etc.). The heat transfer between the intercooler and the heat exchanger is most commonly carried out using brines (Fig. 1).

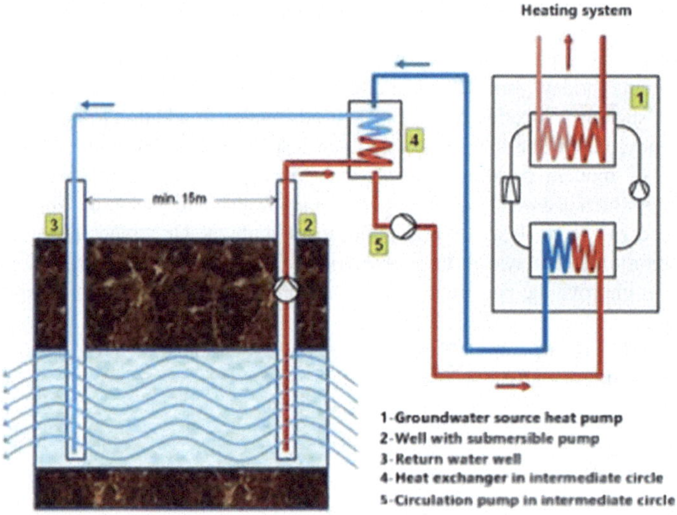

Fig. 1. Groundwater source heat pump (schematic view of working principal) [6]

In order to select the appropriate type of heat pump, in addition to the required heat source, it is necessary to determine the required amount of heat energy for heating, which is obtained by calculating the heat losses and heat gains of the building according to the appropriate standards.

3 Selected Type of Heat Pump

Total heat losses are calculated as the sum of transmission and ventilation heat losses according to standard: EN12831

$$\Phi_{\mathrm{Tot.}} = \Phi_T + \Phi_V \tag{1}$$

Where Φ_T are transmission heat losses and Φ_V are the ventilation heat losses.
The total heat losses of the building where the heat pump is located are:
Φ_T = 121,115 [kW]
The total amount of heating area of the building is 1986 m^2 [5]

Taking into account the previous calculation, which was taken from a HVAC Main Design project of the building where two water-to-water heat pumps of the following characteristics were selected:

Heat capacity = 86,3 kW (each)
Compressor type: scroll compressor (2 in each heat pump)
Refrigerant: R410 A
Nominal electricity power: 21,2 kW
Nominal COP: 4,07 [5]

The previous characteristics of the heat pumps are only for the heating operation mode, because the experimental research was done in that period.

4 Experimental Research

For the purpose of analyzing the operation of a given installation, measurements were made of some of the operating parameters such as outside air temperature, hot and return/cold water temperatures, heating energy output, flow, etc. The water temperature from the well was not measured and was taken as constant (14 °C for the measured area). Measurements were made so that the data was taken every hour using a installed calorimeter from manufacturer Siemens UH-50 and an outdoor air temperature thermometer. The operation of the heat pump was automatically controlled.

The assumption was that due to the lack of continuity in operation (interruption of work from Friday to Monday, and work from 7 h to 14 h on working days), the selected heat pump will consume more electrical energy than it would in the case of continuous work throughout the week. The impact of outside air temperature on operation of the HP was also taken in consideration.

The measurements were made within a month. The characteristic three days in the last week of measurement will be shown in detail, because of the completeness of the data. The characteristic days: Monday (07.01.2019), Wednesday (09.01.2019), Friday (11.01.2019) (Tables 1, 2 and 3) (Figs. 2, 3 and 4).

Table 1. Recorded temperatures for Monday (07.01.2019)

Outside air temp. °C		Hot water temp. °C	Cold/return water temp. °C	Heat output at measuring moment (kW)
9 h	2	44,5	39,7	82,6
10 h	3,5	43,3	38,5	81,3
11 h	5	44,4	39,5	84,1
12 h	5	42,1	37,1	84,4
13 h	5,5	42,7	39,8	50,3
14 h	7,5	42,4	39,5	50,4

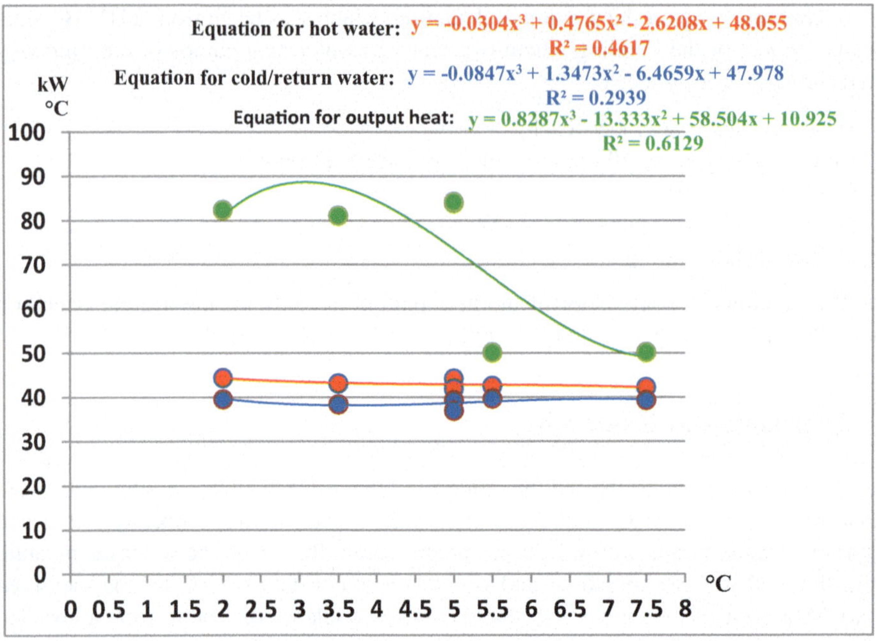

Fig. 2. Equations and graphs of measured parameters (Monday 07.01.2019)

Table 2. Recorded temperatures for Wednesday (09.01.2019)

Outside air temp. °C		Hot water temp. °C	Cold/return water temp. °C	Heat output at measuring moment (kW)
8 h	3	43,4	40,5	47,7
9 h	3	41,3	38,4	50,1
10 h	4	40	37,7	52,1
11 h	5	40,4	37,7	49,4
12 h	7	40,8	37,9	50,1
14 h	9	42	39,1	49,6

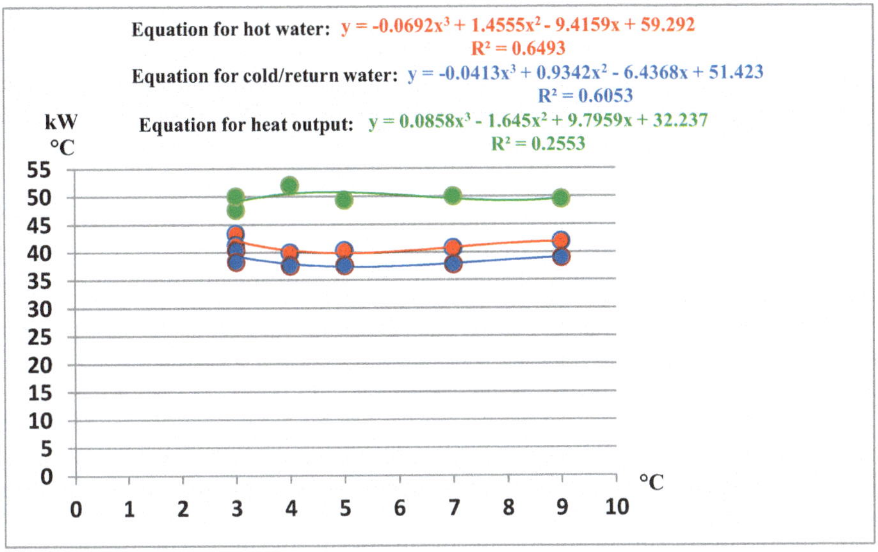

Fig. 3. Equations and graphs of measured parameters (Wednesday 09.01.2019)

Table 3. Recorded temperatures for Friday (11.01.2019)

Outside air temp. °C		Hot water temp. °C	Cold/return water temp. °C	Heat output at measuring moment (kW)
9	6	44,3	41,2	51,6
10 h	6	43,8	41,1	50,6
11 h	7	46,1	43,2	50,1
12 h	7	43,9	41,6	38,1
13 h	9	44,1	41,7	39,9
14 h	10	43,1	42,4	13,6

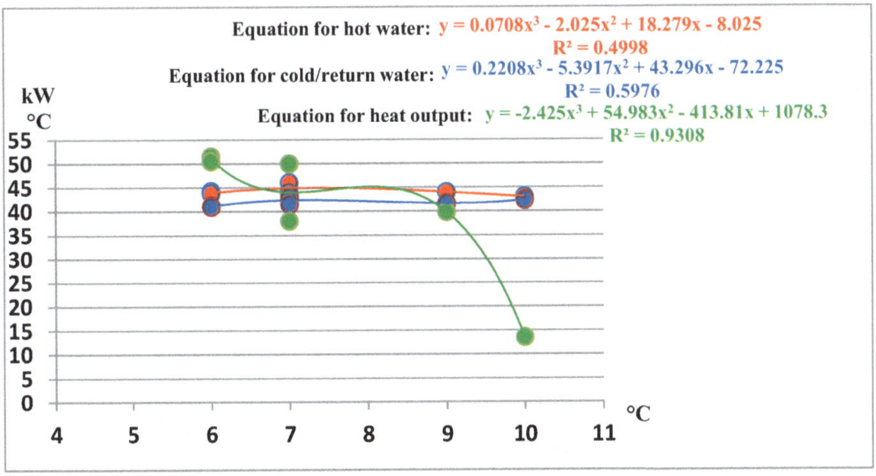

Fig. 4. Equations and graphs of measured parameters (Friday 11.01.2019)

For the parameters shown in tables above, the corresponding graphs and equations were extracted that describe this data using Microsoft Excel software. The graphs and equations were made so that the values on the ordinate (temperatures of hot and cold water which are given in °C and heat output energy in kW) were relative to the temperature of outside air which is applied on the x-axis. Using a simple equation, the COP of the heat pump at measured moments was calculated.

$$COP_{Heat} = \frac{\Delta Qh}{\Delta A} \tag{2}$$

Where ΔQh is the heat output energy, and ΔA is the electrical energy consumption of the compressor. Knowing that the heat pump operates at a constant rated electric power of 21,2 kW, the following results were obtained (Tables 4, 5 and 6):

As it can be seen from the attached analysis, using a software package (Microsoft Excel) appropriate graphs have been drawn which connect the collected data for the measured period and certain polynomial equations of the third order describing them have been obtained.

Table 4. COP results for Monday (07.01.2019)

Time [h]	Outside air temp. [°C]	Heat output [kW]	COP
9	2	82,6	3,89
10	3,5	81,3	3,83
11	5	84,1	3,97
12	5	84,4	3,98
13	5,5	50,3	2,37
14	7,5	50,4	2,37

Table 5. COP results for Wednesday (09.01.2019)

Time [h]	Outside air temp. [°C]	Heat output [kW]	COP
8	3	47,7	2,25
9	3	50,1	2,36
10	4	52,1	2,46
11	5	49,4	2,33
12	7	50,1	2,36
14	9	49,6	2,24

Table 6. COP results for Friday (11.01.2019)

Time [h]	Outside air temp. [°C]	Heat output [kW]	COP
9	6	51,6	2,43
10	6	50,6	2,38
11	7	50,1	2,36
12	7	38,1	1,79
13	9	39,9	1,88
14	10	13,6	–

5 Conclusion

From the data analysis it can be concluded that the direct impact of external air temperatures to the work of groundwater source heat pump is missing. The impact of external air temperatures on the work of heat pumps can be linked indirectly over the temperatures of the heating rooms in the building. Namely, it is obvious that temperatures of hot and cold water are quite constant during the week, but the temperature differences and therefore the heat energy output is highest at the beginning of the week and is dropping at the end of the week. Also, during the data analysis, it was observed that the heat pump during work had not once reached the projected temperature of the hot water from 47 °C, and therefore the heat pump was working constantly. All this points to us that the assumption at the beginning was correct. The heat pumps working mode during the day and the lack of work during the weekend highly affects the consumption of electrical energy and time spent in working mode. Also, the impact of the missing second heat pump is obvious. Some measures for increasing efficiency and reducing electrical energy consumption can be made, but they need to be calculated carefully so the most convenient could be chosen. Some of the suggestions are:

- Installing the missing (second heat pump, which doesn't have to be the same type as the installed one. Maybe a solar or air-air system would be suitable, and even cost less.
- Installing Photovoltaic panels so some of the electrical consumption can be reduced.
- Adding additional thermal insulation on the building, reducing that way the heat losses of the building.
- Using other sources to heat water (electrical heater, solar collectors).

References

1. Nožić, M.: Termodinamika. Univerzitet "Džemal Bijedić, Mašinski fakultet Mostar", Mostar (2011)
2. Egg, J., Howard, B.C.: Geothermal HVAC; Green Heating and Cooling. McGraw-Hill Education, New York (2011)
3. Technical catalogue of SYSTEMAIR Heat Pump
4. Technical catalogue of calorimeter SIEMENS

5. Main design project of heating and air conditioning "Faculty of Mechanical Engineering" University of "Džemal Bijedić" in Mostar
6. Špago, D.: Analiza korištenja geotermalne energije za grijanje i hlađenje visokoškolske ustanove i prijedlog za poboljšanje efikasnosti, Master Thesis, Univerzitet "Džemal Bijedić, Mašinski fakultet Mostar", Mostar (2019)

Air Staging and Reburning to Achieve Low Emissions During Co-firing Coal and Biomass

Nihad Hodzic[1(✉)], Anes Kazagic[2], and Kenan Kadic[2]

[1] Faculty of Mechanical Engineering Sarajevo, University of Sarajevo, Vilsonovo setaliste 9, 71000 Sarajevo, Bosnia and Herzegovina
hodzic@mef.unsa.ba

[2] JP Elektroprivreda BiH d.d. - Sarajevo Power Utility, Vilsonovo setaliste 15, 71000 Sarajevo, Bosnia and Herzegovina

Abstract. In this paper results of research of co-firing coals from Middle Bosnian basin with waste woody biomass are presented. Tested fuels were subjected to pulverized combustion under various temperatures and various technical and technological conditions: different mass ratio of fuel components in the mixture, different coefficient of excess air for combustion, staged air supply into the reaction zone and application of natural gas as an additional fuel (reburning technology). The results are related to the emission of flue gas components (CO_2, CO, NO_x and SO_2) with respect to the technological conditions of combustion. In addition to the valorization of the influence of the primary fuel composition and process temperature on the emission values of the flue gas components, it has been shown that the application of primary measures in the combustion chamber results in lower or higher positive effects in terms of reducing the emissions of individual components - e.g. the NO_x emission is reduced from 837 mg/m_n^3 in conventional combustion to 710 mg/m_n^3 using air staging. It is also shown that when using natural gas as additional fuel there is an additional reduction in NOx emissions, and this is proportional to the amount of natural gas - e.g. for process temperature of 1350 °C and 10% of natural gas energy share co-firing with coal, a NO_x emission reduction of more than 250 mg/m_n^3 was recorded, compared to emissions without additional fuel.

Keywords: Biomass · Coal · Combustion · Natural gas · NO_x emission · OFA air

1 Introduction

There are significant coal reserves in Bosnia and Herzegovina. The balance and exploitation reserves of coal in Bosnia and Herzegovina are about $4.5 \cdot 10^9$ t, of which about 40% refers to brown coal and about 60% to lignite [1]. However, the quality of Bosnian coal varies significantly from one mining basin to another, and even from one mine to another within the same mining basin. The basic characteristics of these coals are low heating value, high mineral mass and moisture content and poor reactivity [2]. Consequently, various technical and technological problems arise in the combustion of these coals, including instability and discontinuation of the combustion process. The

I. Karabegović (Ed.): NT 2020, LNNS 128, pp. 657–666, 2020.
https://doi.org/10.1007/978-3-030-46817-0_76

instability of the combustion process, as well as the level of emissions - especially NO_x emissions - is directly related to the way and place of introduction of combustion air. Coals from Middle Bosnian mining basin (e.g. Kakanj, Breza and Zenica mines) are predominantly delivered to Thermal Power Plant Kakanj in which it is fired in pc boilers with slag tap furnace. It is well known that this combustion technology is characterized by high temperature in the furnace, especially in the melting chamber (e.g. 1300 ÷ 1550 °C), which results in very high emissions: emission of NO_x is up to 900 mg/m_n^3, while SO_2 emission is about 8000 mg/m_n^3 due to the content of sulfur in coal mixtures of about 2.5% [3].

Bosnia and Herzegovina also has significant energy potential from waste biomass derived from agricultural production and forestry, primary and secondary wood processing. It is estimated that the total annual technical energy potential of biomass remains in BiH is more than 33 PJ, which is equivalent to more than 3 million tons of BIH lignite [4].

In principle, a significant reduction in CO_2 emissions from coal-fired power plants can be achieved by introducing waste wood biomass as a renewable energy source into regular operation, while the reduction of NO_x emissions relative to conventional combustion systems can be sufficiently achieved by the use of burners of the newer generation (Low-NO_x Burner, LNB), staged air supply (OFA), staged primary fuel intake into the furnace and using additional fuel (e.g. natural gas, reburning technology). LNB burners work on the principle of a staged combustion air supply at the single burner level - these burners result in a significant reduction in NO_x emissions and as such today they are the essential equipment of newer boilers, which usually work with supercritical or ultra-supercritical steam parameters and high flexibility is required [5]. In the paper [6], the results i.e. effects of the staged supply of combustion air on the NO_x emission (dried lignite was used as fuel), were presented. It has been shown that in this way there is a significant reduction in NO_x emission - the results are related to the investigation of the influence of the coefficient of excess air, the distribution of combustion air, and the distance of the position of introduction of the OFA air into the reaction zone relative to the burner. In the paper [7] it was stated that, with Low-NO_x burners, the NO_x emission can be reduced by a fifth, but that due to the high combustion temperature this emission is still high (1036 mg/m_n^3 in dry flue gases with 6% O_2), and for the further reduction of NO_x emissions, a zonal or staged combustion air supply (OFA) is recommended. Similar positive effects of the staged supply of combustion air on the reduction of NO_x emissions are presented in works [8] and [9] for coal combustion; as well as for co-firing coal with biomass [10–14].

By subjecting coal blends, as well as mixtures of coal and waste woody biomass, to combustion at different temperatures with the application of staged combustion air supply as primary measure, and with the introduction of a portion of the OFA air at different distances from the primary burner and with the natural gas as third fuel, it is possible to determine the appropriate response of the combustion process through the measurement of process parameters and their analysis, i.e. it is possible to come to relevant conclusions about the impact of the location of the introduction of the OFA air and the application of natural gas to the process itself, and consequently the effect on the emission of flue gas components into the environment (CO_2, CO, NO_x, and SO_2).

Based on these conclusions it is possible to quantify and sublimate the characteristics of co-firing coal with woody biomass and natural gas, including the advantages of conversion of primary energy from fuels under conditions of solid fuel pulverized combustion technology with a staged combustion air supply [15].

2 Fuel Test Matrix, Test Regimes and Laboratory Plant

Fuel Test Matrix and Test Regimes: For the purpose of laboratory investigations of the influence of the staged supply of the combustion air into the furnace as well as the application of natural gas as additional fuel while co-firing coal with waste woody biomass on the emission of flue gas components, a matrix of basic test fuels was established [15] - Table 1.

– The usual coal blend that has been fired in TPP Kakanj for the last few years - fuel label: U100. This coal blend was produced by mixing coals delivered to the coal depot of TPP Kakanj from several mines (Kakanj, Breza, Zenica, Gracanica, Livno, Nova Bila, Banovići, etc.) in approximately the same percentage as they are delivered from the mentioned mines. For laboratory testing purposes a sample of this coal mixture (coal powder) is excluded directly behind the mills during real operation of the Unit 5 in TPP Kakanj.

Table 1. Basic properties of test fuel, [15]

Fuel/label	U100	B100	U95B5	U93B7	U90B10
Proximate analysis, %					
Moisture	13,90	21,63	19,06	18,09	14,67
Ash	37,88	0,52	34,33	33,05	34,14
Volatiles	28,97	64,14	29,32	31,16	32,49
C_{fix}	19,25	13,72	17,28	18,59	18,70
Combustible	48,22	77,86	46,60	48,86	51,18
Ultimate analysis, %					
Carbon	36,62	38,92	32,12	33,36	33,25
Hydrogen	2,60	4,95	2,20	2,52	2,84
Sulfur	2,06	0,14	1,57	1,59	1,87
Nitrogen	0,72	0,17	0,81	0,75	0,67
Oxygen	10,22	33,67	9,91	10,63	12,57
Heating values, kJ/kg					
HHV	13351	15564	12651	13446	13572
LHV	12496	14081	11759	12510	12655

– Wood biomass sawdust - fuel label: B100. Sawdust is a mixture of beech and spruce formed during primary wood processing, in an approximate 50:50 weight ratio. The sawdust sample was excluded after delivery to the TE Kakanj depot.

- Mixture made of U100 coal blend and waste wood biomass B100 - fuel label: U95B5 and U93B7. These mixtures consist of 5 and 7% wood biomass respectively and are excluded behind the mills.
- Mixture made of coal U100 and waste wood biomass B100 - fuel label: U90B10. This coal mixture consists of 10% wood biomass and is formed after drying and grinding the components in a laboratory mill (Fig. 1).

Fig. 1. (a) AUCR experimental line: front view of four visible zones with temperature measurement of the reaction tube and process control; (b) pulverized fuel tank, dosing and supply; (c) worktable with measurement equipment during the test

The mentioned solid fuels (mixtures of coal and biomass) were subjected to a laboratory investigation of combustion under the conditions of staged supply of combustion air into the combustion chamber and with natural gas as additional fuel. Test regimes are defined and performed in the range of process temperatures corresponding to the temperatures from the real boiler operation in TPP Kakanj: 1350 ÷ 1450 °C. A set of these research was aimed at gaining insight into the characteristics of co-firing coal with biomass, with a particular emphasis on emissions of flue gas components, primarily NO_x. Test regimes of the co-firing with natural gas as an additional fuel are carried out with 5 and 10% of the natural gas energy share. These tests of combustion with natural gas have been assigned the following markings: UPn and UB(95)Pn, where "n" represents the energy share of gas in total energy input supplied into the combustion chamber.

3 Results of the Research: Emission of Flue Gas Components

Co-firing of Coal with Biomass: For the same combustion temperature, NO_x emission in co-firing was found to be at the same emission level as for combustion of coal without biomass, and that there is practically no change in that emissions with the change in the fraction of wood biomass in the mixture. The average NO_x emission difference depending on the mode of supply of combustion air is 250 mg/m_n^3: in the case of conventional air supply, the average NO_x emission is 942 mg/m_n^3, while in the case of a stage combustion air supply, the emission is 692 mg/m_n^3, Fig. 2 - left. On the other hand, the sulfur content decreases proportionally with the increase of fraction of wood biomass in the coal mixture. Despite the fact that the results regarding SO_2 emission are dispersed, it can still be concluded that the SO_2 emission reduces proportionally with the increase in biomass content in the mixture. Additionally, based on the results of the measurements, it can be concluded that the SO_2 emission practically does not depend on the way of combustion air supply - that emission is high and at a process temperature of 1350 °C it is approximately 5300 mg/m_n^3, Fig. 2 - left.

Total and net CO_2 emission due to the combustion of wood biomass as a renewable fuel in a coal mixture is shown in Fig. 2 - right. Increasing the fraction of wood biomass in the mixture linearly reduces the net CO_2 emission, which at 10% of biomass in the mixture is 0.233 kg/m_n^3. Somewhat more pronounced CO emissions from co-firing of coal with biomass, especially at 10% of the biomass content, compared to the combustion of coal, can be related to the granulation of wood biomass in the mixture that is not mechanically treated (not grinded in the laboratory mill) - a certain fraction of biomass particles was up to 4 mm, Fig. 2 - right.

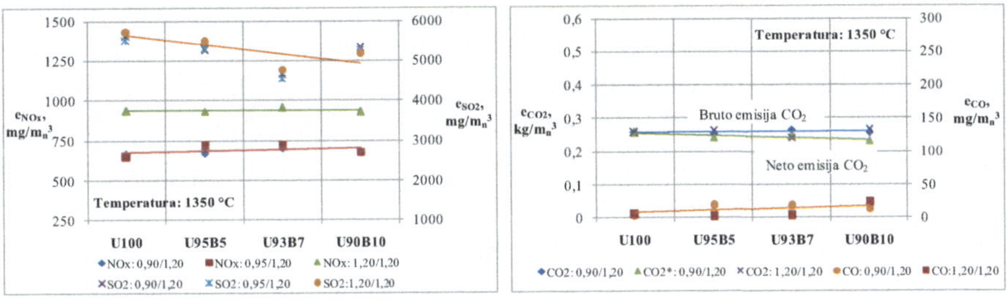

Fig. 2. Emission of NO_x and SO_2 (left) and CO_2 and CO (right) in the co-firing of coal with wood biomass at process temperature of 1350 °C

With the increase of combustion temperature, NO_x and SO_2 emissions are also increasing. For example, at temperature of 1450 °C the average NO_x emission is 800 mg/m_n^3 and the SO_2 emission is about 5200 mg/m_n^3. In these test regimes, the previously stated conclusions were confirmed that the NO_x emissions practically do not change with the increase in the fraction of wood biomass in the mixture, and that the SO_2 emission decreases proportionally by this increase of biomass fraction in the mixture.

The effect of natural gas as reburning fuel on NO_x emission when combusting the basic test solid fuels at process temperature of 1350 °C is shown in Fig. 3 - left. In addition, for given solid fuels, the NO_x emission values for the cases of 0%, 5% and 10% natural gas energy share in combustion are shown for comparison. In relation to emissions without the use of gas as additional fuel, a reduction in this emissions by more than 250 mg/m_n^3 was recorded for the case of 10% of the gas energy share. For example, the NO_x emission for the U100 coal mixture was reduced to 489 mg/m_n^3 or by about 27%, while when co-firing coal with biomass these emissions were reduced from 720 mg/m_n^3 to 480 mg/m_n^3 [15]. The efficiency of the use of natural gas in combustion of the basic fuel as a function of the process temperature is presented in Fig. 2 - right, with the results referring to energy share of the natural gas of 10% of the total energy input into the combustion chamber.

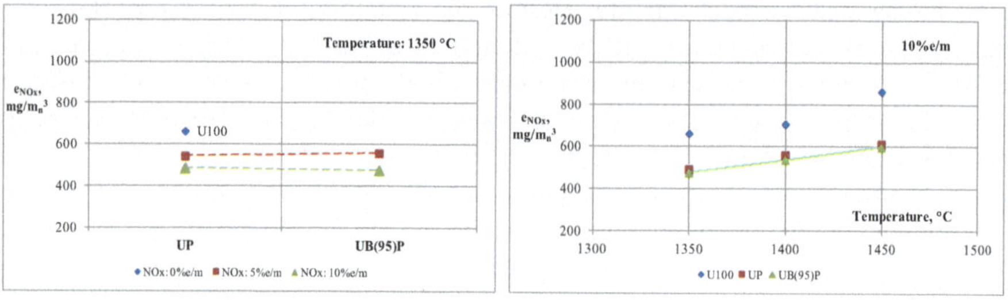

Fig. 3. NO_x emission (left) and influence of temperature on NO_x emission (right) for fuels of different composition burning with natural gas

On the previous diagram (right) NO_x emission for combustion with reburning technology with natural gas is compared with the emissions measured during combustion of the coal mixture U100 without the application of natural gas. It can be noted that NO_x emission for U100 with reburning technology with 10% gas content measured by energy (UP) and at temperature of 1450 °C is 615 mg/m$_n^3$, which is approximately 30% less than the emission without the use of natural gas. Also, when coal and biomass were co-fired under the same conditions (UB(95)P) and at temperature of 1450 °C, the emission was 602 mg/m$_n^3$, which is practically the same 30% less than the emission when combusting only U100 (866 mg/m$_n^3$) [15].

The effect of natural gas as reburning fuel on the SO_2 emission of combustion of basic test solid fuels at a process temperature of 1350 °C is shown in Fig. 4 - left. In general, for combustion under these conditions, there is a slight decrease in the SO_2 emission. For example, the SO_2 emission of reburning with 10% of natural gas is in average by 325 mg/m$_n^3$ lower than the SO_2 emission of combustion of coal blend U100. During these test regimes with natural gas, a very low level of CO emission was recorded, especially for combustion of basic fuels with smaller granulation and at a temperature above 1400 °C - the CO emission is practically negligible, below 5 mg/m$_n^3$, Fig. 4 - right.

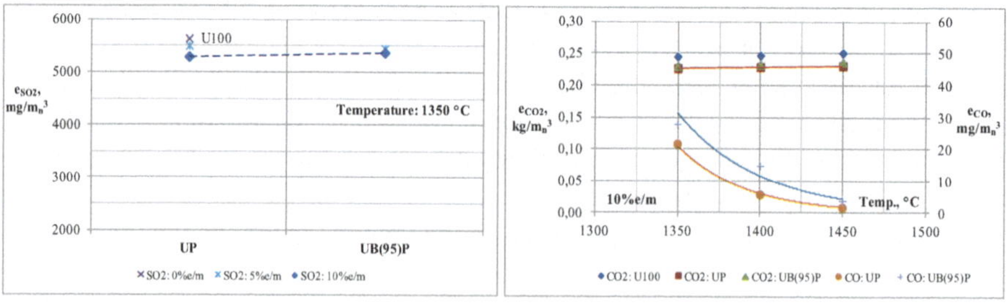

Fig. 4. Emission of SO_2 (left) and CO_2 and CO (right) for different fuel compositions with reburning technology with natural gas

The previous phenomenon can be explained by intense reaction, that is, by the combustion of natural gas in the reburning zone in which the products of incomplete combustion (such as CO) that were formed in the primary combustion zone in which the combustion air is lacking ($\lambda < 1$, the sub-stoichiometric zone) burn additionally [15]. Also, it is a known fact that the combustion of natural gas, in comparison with the combustion of liquid and especially solid fuels, is accompanied by the lowest CO_2 emissions. This was also reflected in the results of the test regimes for combustion of selected solid fuels with reburning technology with natural gas - see Fig. 5. In this case, the CO_2 emission averaged about 0.231 kg/m_n^3 and is lower by almost 0.020 kg/m_n^3 in relation on emissions from the combustion of U100 coal mixture without the application of natural gas.

Comparative Overview of the Efficiency of Primary Measures in the Combustion Chamber on the NO_x Emission: The effects of primary measures in the combustion chamber for fuels U100 and U95B5, for the process temperature range from 1350 to 1450 °C are presented in Fig. 5 - left. There is a noticeable decrease in emissions at a temperature of 1450 °C, from 1154 to 615 mg/m_n^3 or about 47% relative - these results refer to NO_x emissions for combustion of coal U100. The NO_x emission for co-firing coal with biomass (U95B5) with the application of different primary measures in the combustion chamber, including the efficiency of these measures by reducing this emissions relative to those in conventional combustion, is presented in Fig. 5 - right.

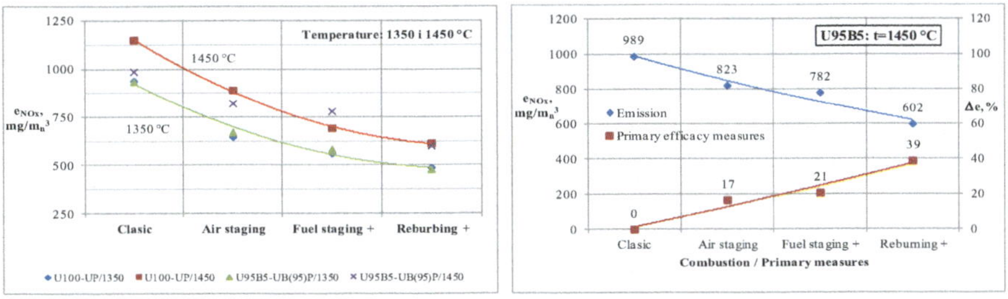

Fig. 5. Effects of application of primary measures in the combustion chamber on coal combustion in function of temperature (left) and for co-firing of coal with wood biomass at the same temperature (right)

From the results presented in the previous figure, it is concluded that the use of natural gas as an additional fuel is the most effective primary measure for the reduction of NO_x emissions. Also, in relation to other primary combustion measures, the impact of the combustion temperature on this measure is the smallest: the emission difference for process temperatures of 1450 °C and 1350 °C using reburning technology by introducing natural gas is on average 135 mg/m_n^3, whereas this difference in the application of staging combustion air as primary measure is on an average of 180 mg/m_n^3 [15].

4 Conclusion

Based on the results of the research, the following conclusions can be drawn regarding the impact of the staged air intake into the combustion chamber and the use of natural gas as additional fuel on the emission of flue gas components in the co-firing of coal with waste woody biomass:

- NO_x emissions in co-firing coal with waste woody biomass is at the emission level at combustion of coal blend. There is practically no change in this emission with the change in the fraction of woody biomass in the mixture. The average difference in NO_x emissions depending on the mode of combustion air is 250 mg/m_n^3: for the classical air intake the average emission is 942 mg/m_n^3, while for the mode with air staging it is 692 mg/m_n^3.
- The SO_2 emission is slightly reduced by increasing the fraction of woody biomass and practically does not depend on the mode of combustion air intake - the emissions are high and generally over 5000 mg/m_n^3.
- By increasing the fraction of woody biomass in the mixture, the CO_2 neet emission is proportionally reduced, which at the 10% biomass in the blend is 0.233 kg/m_n^3.
- When co-firing coal-biomass mixture with natural gas as additional fuel (UB(95) P10) at temperature of 1450 °C, the measured NO_x emission is 602 mg/m_n^3, which is about 30% less than emission with coal U100 which is 866 mg/m_n^3.
- When using the natural gas in reburning test regimes there is a slight decrease of SO_2 emission. Burning the coal U100 with 10% of natural gas, the SO_2 emission is on average about 325 mg/m_n^3 is lower than the SO_2 emission when burning coal only.
- In combustion regimes with natural gas, CO emission is very low, especially when at temperatures above 1400 °C, where CO emissions are practically negligible: below 5 mg/m_n^3. In doing so, CO_2 emissions are lower by almost 0.020 kg/m_n^3 compared to emissions when combusting U100 coal mixture without natural gas.
- By using natural gas in combustion of coal U100 at temperature of 1450 °C, the emission of NO_x decreased to 615 mg/m_n^3 or by almost 50% compared to the emission of conventional combustion: 1154 mg/m_n^3.
- When using natural gas as additional fuel, compared to other primary combustion measures, the impact of the process temperature on the NO_x emission is the smallest: the difference in emissions at temperatures of 1450 and 1350 °C is on average 135 mg/m_n^3 - the difference for the application of the combustion air staging as primary measure is on average about 180 mg/m_n^3.
- Efficiency of primary measures by staging the base fuel and using natural gas as additional fuel weakens with rising process temperatures. Efficiency of the staging basic fuel at 1450 °C is 21% compared to 32% at 1350 °C, while the efficiency of natural gas utilization for those temperatures is 39% and 46% respectively. In doing so, efficiency of the application of staging combustion air is practically the same for both observed combustion temperatures.

Acknowledgements. Part of the results presented in this paper were created during the research in the framework of the project: *Advantages of co-firing of coal with waste wood biomass under the conditions of staged air supply*, financed by the Ministry of Education and Science of the Federation of Bosnia and Herzegovina for 2018 and infrastructural assisted by the Faculty of Mechanical Engineering of University of Sarajevo.

References

1. Studija energetskog sektora u BiH, Konačni izvještaj, Modul 8 - Rudnici uglja, Konzorcij: Energetski institut Hrvoje Požar, Hrvatska; Soluziona, Španjolska; Ekonomski institut Banjaluka, BiH; Rudarski institut Tuzla, BiH (2009)
2. Uticaj kvaliteta uglja na troškove proizvodnje električne energije i cijenu uglja; Studija; Naručilac: JP Elektroprivreda BiH d.d. Sarajevo, Izvršilac: Mašinski fakultet Sarajevo (Lider) i Rudarski institut d.d. Tuzla (član), Sarajevo (2014)
3. Smajević I., Hodžić N.: Betriebserfahrungen im Kraftwerk Kakanj bei der Verbrennung der starkverschmutzungsneigenden mitelbosnischen Braun-kohle, Dresden, 25–26 Okto-bar 2000, pp. 151–160 (2000)
4. Advanced Decentralised Energy Generation Systems in Western Balkans - ADEG, Projekt FP6, National Technical University of Athens, Institut IVD Stuttgart, Fakultet Strojarstva i Brodogradnje Zagreb, Mašinski fakultet Sarajevo, Institut Vinča, IST Portugal, voditelj projekta ispred MF Sarajevo: A. Lekić, podprojekat: Biomasa, voditelj podprojekta: I. Smajević, član istraživačkog tima: N. Hodžić, i dr. (2004–2007)
5. Tsumura, T., Okazaki, H., Dernjatin, P., Savolainen, K.: Reducing the minimum load and NO_x emissions for lignite-fired boiler by applying a stable-flame concept. Appl. Energy **74** (3–4), 415–424 (2003)
6. Wang, J., Fan, W., Li, Y., Xiao, M., Wang, K., Ren, P.: The effect of air staged combustion on NO_x emissions in dried lignite combustion. Energy **37**(1), 725–736 (2012)
7. Kuang, M., Li, Z., Liu, C., Zhu, Q.: Experimental study on combustion and NO_x emissions for a down-fired supercritical boiler with multiple-injection multiple-staging technology without overfire air. Appl. Energy **106**, 254–261 (2013)
8. Kuang, M., Li, Z., Ling, Z., Zeng, X.: Evaluation of staged air and overfire air in regulating air-staging conditions within a large-scale down-fired furnace. Appl. Therm. Eng. **67**(1–2), 97–105 (2014)
9. Rozendaal, M.: Impact of coal quality on NO_x emissions from power plants. Delft University of Technology, Delft (1999)
10. Kazagić, A., Smajević, I.: Experimental investigation of ash behavior and emissions during combustion of Bosnian coal and biomass. Energy **32**(10), 2006–2016 (2007)
11. Kazagić, A., Smajević, I.: Synergy effects of co-firing of wooden biomass with Bosnian coal. Energy **34**(5), 699–707 (2009)
12. Kazagić, A., Smajević, I., Duić, N.: Selection of sustainable technologies for combustion of Bosnian coals. Therm. Sci. **14**(3), 715–727 (2010)
13. Smajević, I., Kazagić, A., Musić, M., Bečić, K., Hasanbegović, I., Sokolović, Š., Delihasa-nović, N., Skopljak, A., Hodžić, N.: Co-firing Bosnian Coals with woody biomass: experimental studies on a laboratory-scale furnace and 110 MWe power unit. Therm. Sci. **16**(3), 789–804 (2012)

14. Nareddy, S., Moyeda, D., Zhou, W., Marquez, A., Swanson, L., Duval, S.: NO_x reductions achieved through air staging in cyclone fired boilers. In: 32nd International Clearwater Coal Conference, Clearwater, Florida, USA, 01–05 June 2008 (2008)
15. Hodžić, N.: Istraživanje kosagorijevanja uglja i biomase usmjereno na smanjenje emisija primarnim mjerama u ložištu, Doktorska disertacija, Mašinski fakultet Univerziteta u Sarajevu, Sarajevo (2016)

Projects of Renewable Energy Resources: An Analytical Overview of the Windfarm's Electricity Generation on the Hillside Možura

Jovanović Jovana[1(✉)], Sun Xiaoqin[2], and Djukanović Milena[3]

[1] Faculty of Civil Engineering and Management, University Union Nikola Tesla, 11 000 Belgrade, Serbia
jocka747@gmail.com
[2] Faculty of Civil Engineering, University of Science and Technology, Changsha, China
[3] Faculty of Electrical Engineering, University of Montenegro, Podgorica, Montenegro

Abstract. Alternative energy management thrives for a long time. It covers up green-field investments, such as the construction of new production and energy facilities, distribution hubs, renewable energy technologies. With the ongoing wind energy projects, an adequate share of wind generator's farms in annual electricity production from renewable energy resources in Montenegro is estimated to be 11.4%. This paper analyses the concrete exemplar of wind generator's farm, located in hilly area of the southern part of Montenegro, called Možura: the installed power of wind generator's farm; electricity dispatch among constituent parts of wind generator's farm; electricity production within the wind generator's farm; investment analysis; the generated noise level per a wind generator of Možura wind farm and CO_2 emission reductions (savings). Additionally, attached is the prognosis of future development of renewable energy resources up to 2025th year for Montenegro. Logistic metrics, numerical data, and analytical approach are used to underpin this research. Results are analytically, narratively discussed per sections and subsections through tabular displays and graphics, which are added to complete the analysis.

Keywords: Alternative energy management · Greenfield investments · Renewable energy resources · Wind generator's farm · Generated noise level

1 Introduction

High-energy cogeneration of electricity from renewable energy resources is the alter-ego of almost every country in today's world. Whether by the cultivation of bioenergy feedstock, or laying out micro or macro hydro turbines with or without pumps, or finding optimal tilt angle of the photovoltaic module, renewable energy can be harnessed over again.

There is a multitude of literature on wind energy and its exploitation: decentralized and centralized micro grids, certain models of wind generators, and particular techniques on estimating the intensity of wind energy. As stated in the article [1], micro

I. Karabegović (Ed.): NT 2020, LNNS 128, pp. 667–679, 2020.
https://doi.org/10.1007/978-3-030-46817-0_77

grids of wind turbines required to power electronic interfaces like DC/AC or DC/AC/DC converters, to connect to the main electricity distribution network. Different types of AC wind turbine generators like (squirrel-cage induction generator, doubly-fed induction generator, and synchronous generator) were also discussed based on speed range, maintenance costs, size and weight [2]. In the article [3], Weibull distribution methods were in charge for the estimation of power generation from wind energy. Furthermore, equivalent loads of wind turbines in vast wind farms were investigated by using large Eddy Simulations [4].

In this paper was discussed the versatility of the Možura wind generator's farm, located in one poor, developing country such as Montenegro. In the last decades, the country of Montenegro is experiencing the relatively high penetration of renewable energy resource technologies and one of them is this project.

The general contributions of this manuscript can be summarized as follows:

- The high-grade utility of renewable energy resources' projects, thus the utility of the Možura wind generator's farm.
- An operational scheme of the constituent parts of the Možura wind generators' farm project.
- Electricity production system within a wind generators' farm.
- Specifics of the exposed and discussed renewable energy resource project, i.e. wind generator's farm.
- The incentive scheme and investment details of a mentioned wind generators' farm project.
- The generated noise level appraisal for an adopted height of a wind generator.
- The accomplished CO_2 emission reductions.

2 Analytical Overview of the Project of Wind Generator's Farm on Možura (Montenegro) and Its Specifications

2.1 Project Description and Its Baseline Scenario

Installation of wind turbines or any other electricity generation facilities which engage renewable energy resources, on a certain area, fortifies the electrical infrastructure of that area and contributes to electric power supply system to be carbon neutral.

For example, in Denmark wind turbines per 2017th year delivered electric power equivalent to 43.6% of overall Denmark's annual electricity consumption. The Danish government also aims to reduce the number of wind turbines installed in Denmark, from roughly 4300 to 1850 machinery (on-shore and off-shore) by 2030 [5, 6]. It is outlined in [7] that the newer onshore wind turbines' installations in Denmark, are now mostly an upgrade of old wind turbines' installations because of the lack of available mainland.

Regarding the country of Montenegro, it imports more than one-third of electricity demand, between 35% and 40%. This implies transboundary electricity import, namely the electricity import from abroad [8]. The current scenario in the transmission system of electricity in Montenegro is that electricity is mainly produced from fossil fuels, such as lignite.

The baseline scenario of the proposed project activity of wind generator's farm on Možura (Montenegro) is the production of electricity by installment of 23 new grid-connected wind generators on Možura hillside, on the Mediterranean coast. Možura hillside is located on the borders of two Mediterranean city municipalities, in the southern part of Montenegro. Each of the 23 wind generators has a nominal electric power of 2 MW, reaching the total installed wind generator's capacity of 46 MW. The type of wind generators is Gamesa G97-2.0 MW.

Project's expectations and appraisals are that the annual amount of produced electrical energy will be approximately 96.968 GWh. The produced electricity from wind generator's farm goes through a transformer station, from the 30 kV transmission line to the 110 kV transmission line. GHG emissivity factor of newly installed wind turbines is estimated at 0 tCO_2/MWh, instead of exploitation of fossil fuels in power plants, which are the main causes of greenhouse effect [9].

In Table 1, technical specifications of wind farm generators on Možura hillside are shown [9].

Table 1. Technical specifications of wind farm generators on Možura hillside, Montenegro

Total capacity (power) of the wind generator's farm	46	MW
Wind generator	G97-2.0 MW	
Estimated power per wind generator	2.0	MW
Number of wind generators	23	–
Equivalent hours of function per year	2108	Hours
Annual production of electrical energy	96.968	GWh
Planned load factor	24.06	%
The width of the transmission line	1.8	km
Transmission line voltage	30	kV
Diameter	97	m
Cover area	7390	m^2
Height of the node	78	m

With wind farm generators on Možura hillside, (Montenegro), is going to be generated approximately 96.968 GWh of electrical energy on an annual basis. This electricity output is green electricity since it is generated with clean technologies. 900 GWh of electricity corresponds approximately to one-fourth of annual electricity consumption in Montenegro, which means that 3600 GWh of electricity is the annual electricity consumption of entire Montenegro. This further means that annual electricity production of such wind farm (Možura) covers 1/37th part of annual electricity consumption of Montenegro.

Furthermore, the planned load factor of 24.06% (from Table 1) corresponds to the one-hour load factor of Možura wind farm (wind generators). Equivalent hours of wind farm (wind generators) function per year is 2108 h, which divided by 365 days give a function sum of 5.77–6 h of work of wind generators per day. That means depending on wind gust intensity during the day, daily work of wind generators is approximately

6 h. Such a low value of full load hours is a typical trait of wind generator's farm. This function sum of 5.77 h divided by one day (24 h) gives this percentage of load factor hourly (24.06%).

It can also be calculated the singular generated electricity, i.e. generated electricity per one wind generator. The electric power of one wind generator of 0.002 GW multiplied by equivalent hours of function annually 2108 h gives annually generated electricity of 4.216 GWh per one wind generator.

On Fig. 1 is given a detailed view of Možura wind farm generator, as a general review of the inner structure of a wind turbine [10].

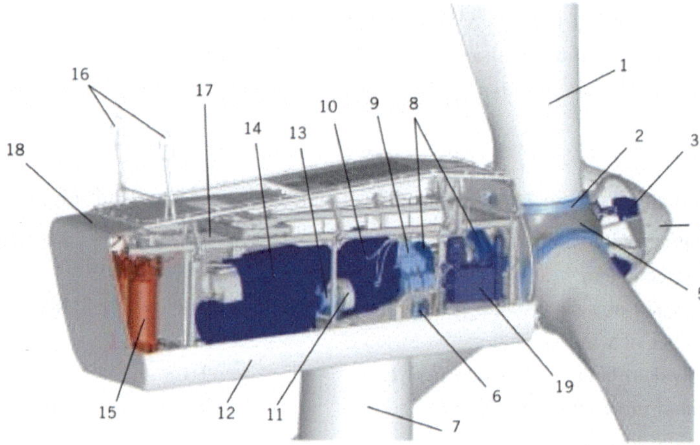

Fig. 1. Možura wind farm generator. 1. Blades of the turbine, 2. Carrier of the blades, 3. Hydraulic angle rotator, 4. Housing hub, 5. Hub, 6. Active corner rotation controller, 7. Tower, 8. Main shaft with two housings, 9. Shock absorbers, 10. Gearbox, 11. Main brake, 12. Supporting cantilever of gondola, 13. High speed shaft, 14. Double power generator, 15. Transformer, 16. Anemometer and windshield, 17. Upper controller, 18. The cabin of the gondola, 19. Hydraulic unit

The project activity of wind farm generators on Možura hillside, Montenegro is a CDM (Clean Development Mechanism) investment. CDM projects or investments earn tradable, saleable, certified emission reduction credits (CERs) that can be used by both, developed and developing countries in achieving GHG emissions reductions. CER is a market financial instrument, created as a result of an appropriate CDM project and represents a reduction in CO_2 emissions by one ton. The certified emission reduction credits (CERs) lie in the European emission trading system as a command and control environmental regulation policy. The carbon footprint tax has been formally adopted since the 1990s [11]. There are over 90% of registered CDM projects in the world, which are being accomplished in India, China, Brazil, and Mexico. With CDM project activities, the baseline emissions and achieved emission savings of greenhouse gases are determined.

This project activity is realized by ACM0002, (Version 12.1.0) consolidated methodology for grid-connected electricity generation from renewable resources.

Figure 2 shows the principle of electricity flow from the wind generator's farm on Možura to the national distribution electricity network.

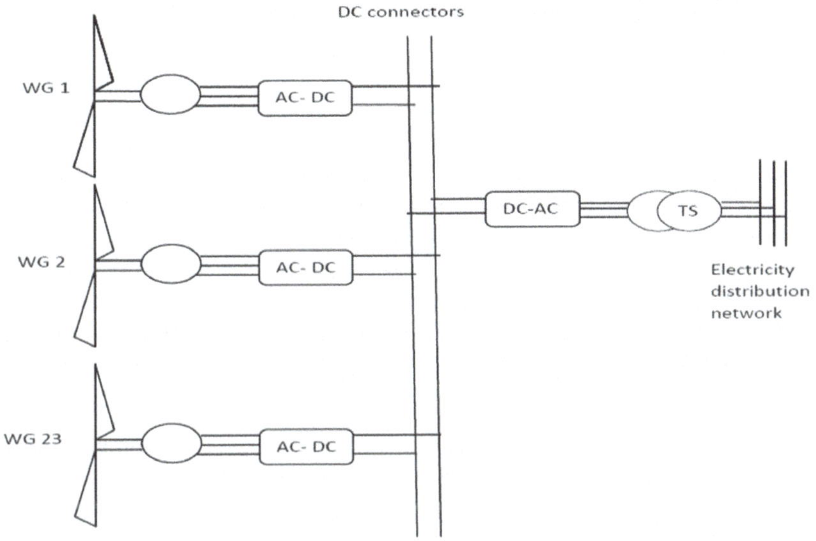

Fig. 2. Principal electrical scheme of Možura wind generators with synchronous generators [by authors] WG – wind generator, AC-DC converter, DC-AC converter, TS - Transformer

The approximate surface of wind generators' foundation is 13.8 m^2. The foundations are of quadrangular shape, with a maximal depth of 1.5 m. The foundations are designed in a way to minimize the risk of wind generator's rolling over and risk of any negative environmental impact [10].

On Figs. 3 and 4 are given construction phases of wind generators in a wind farm Možura, (Montenegro) [12].

Fig. 3. The beginning of the construction of a wind generator's tower at the construction site of the wind farm on Možura hillside (left)

Fig. 4. Several constructed wind generators at the site of the wind farm on Možura hillside (right)

2.2 Long-Term Energy Supply in Montenegro

New age electrification worldwide envisions 80% renewable energy conversion by 2030 and 100% by 2050. Such converting of renewable energy to electricity will provide around 24.3 million of new, full-time jobs, stabilize energy prices and markets, lessen power disruption and more facilitate access to the electricity. Transition to the 100% renewables, in the 2050th, across the globe, concretely to WWS model (wind, water, and solar power) will also reduce the 139 country's electricity load by 42.5% [13].

In the case of Montenegro preliminarily, when it comes to wind energy potential, the capacity installment of wind farms was estimated at 400 MW and annual production of electricity from wind energy resources to 900 GWh. With the ongoing wind energy projects, an adequate share of wind generator's farms in annual electricity production from renewable resources in Montenegro is estimated to be 11.4% [14].

From the current net importer of electricity, Montenegro plans to become the net electricity exporter after 2020th. Montenegrin energy policy has a few main strategic commitments and they refer to:

- Maintenance, revitalization, and modernization of existing and building of new infrastructure for electricity production, transmission and distribution, based on international technical standards, energy efficiency and reduction of negative impact on the environment
- Gradual reduction of dependence on electricity import: by reduction of specific, final energy consumption, enhancement of production of primary and secondary energy by using own resources, reduction of energy losses from production to ultimate consumption
- Inquiring of existing barriers for the activation of long-term energy development.

Montenegro has a possibility of accelerated exploitation of renewable energy resources, such as small hydropower plants, wind, solar energy, biomass, and waste. It was also a relatively high potential for energy savings (20%–30%), by implementing the energy efficiency measures [15].

The country is very dependent on coal, having great coal backlogs but regardless of that imports power in a huge percentage. Also, oil and oil derivates are 100% imported.

Montenegrin national target for the share of energy from renewable resources in gross final energy consumption in 2020th is predicted as 33%.

In compliance with all energy scenarios of Montenegro which have been analyzed, a significant series of electricity, approximately (1460–1570 GWh) will continue to be imported annually. Since that 3600 GWh is annual electricity consumption of entire Montenegro, it can be concluded that nearly one half of annual Montenegrin electricity consumption is imported.

Based on the prognosis of future electricity supply to Montenegro up to 2025, the import of electricity will be therefore increased until the new 2nd block of the national, thermal power plant of 225 MW power is put in operation. For example, since the energetic infrastructure of Montenegro is going to be newly built and renewed, the total potential of Montenegrin central area for wind turbines' installment is currently assessed to 100 MW.

So far, without new and ongoing RES projects, the Montenegrin electricity system makes 2 hydropower plants and 1 thermal power plant [16].

Below, the Table 2 displays the existing and future projections of generated electricity, up to 2025th, after the launch of all power plants in Montenegro [9].

Table 2. Long-term energy supply in Montenegro (GWh)

The year	Hydropower plants (without HPP Piva)	Thermal powerplants	HE Piva	Other renewable energy resources	Import	Export	Total
2010	971	1152	1065	120	1456	0	4764
2015	1927	2369	1065	296	108	783	4982
2020	1927	2290	1065	406	194	511	5371
2025	1927	2325	1065	485	250	262	5790

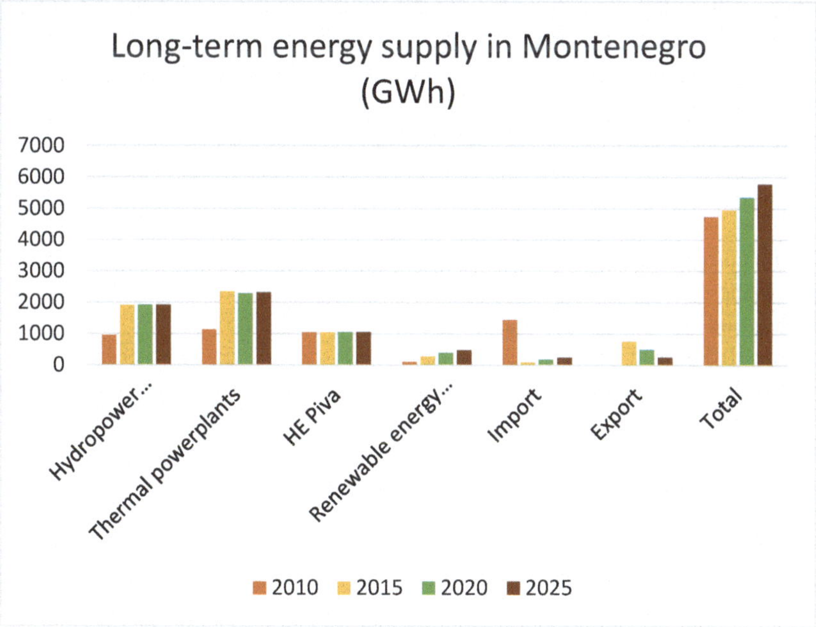

Fig. 5. Energy supply in Montenegro (GWh) through time sequences

From the statistics shown in Table 2 and corresponding graphic in Fig. 5, the overall data for these yearly sequences demonstrate the supreme dominance of electricity generation from hydropower plants, approximately 60%. For 2010th year the electricity generation from all hydropower plants (Piva and others), accounted for 42.74%, while through the other shown time sequences this percentage is significantly upgraded. It can be observed that the electricity production of hydropower plant Piva stays continuous through the timed sections, as well as the electricity production from other hydropower plants starting from 2015th.

Furthermore, the total future contribution of other renewable energy resources (solar, wind, biomass…) in electricity generation can be neglected, with the merely foreseen percentage of 8.37% in 2025th. The electricity generation from coal combustion still stands as a substantial energy resource for Montenegro, with a share of 40% through the time. On behalf of import of the energy services, the lowest import rate was in 2015th, fluctuating up to 4.3% in 2025th. From all this, it can be concluded that in Montenegro beside hydropower, the integration of other renewable energy resources is not so radical for a given time horizon. But with setting new time targets and horizons, the integration of such resources should be considered for the country of Montenegro.

2.3 Investments' Analysis of Možura Wind Farm Project

There are various financial, upfront incentive schemes to promote the use of renewable energy resources. Such incentives are state incentives and comprise direct cash incentives such as loans, grants, rebates, buy-downs, income-tax credits and deductions and industry recruitment incentives [17].

For example, the so-called Performance-Based Incentive (PBI) is based upon electricity cogeneration system's estimated or predicted future performance. The PBI one-time incentive payment is calculated by the following equation:

$$\text{PBI incentive payment} = \textit{Incentive rate} \times \textit{System rating}\ (\text{kW}) \times \textit{Design factor}$$

For this project activity, it was taken the investment tax credit (ITC), which usually passes to the tax-burdened company. Also the reference, investment analysis was conducted and as a financial indicator of this project's profitability served internal return rate (IRR).

Generally, in financial analysis of investment projects, the high-ranked forecast of expected project's cash flows, along with the appraisal of project risk is done. The financial determinants for assessing the investment projects' profitability are net present value (NPV), internal return rate (IRR), benefit-cost ratio (BCR), and measurement of project sustainability, project liquidity test during the credit payback period as well as the test of project's sensitivity.

Overall revenues from this project activity are revenues from the sales of electricity, as well as the revenues from the certified emission reduction credits (CERs). The revenues from the sales of electricity are estimated, according to the amount of electricity which will be sold to the distribution network and the unit price by which will be sold. Možura wind generator's farm will sell to the national distribution network, produced electricity by the price of 95.99 euro/MWh first 12 years, after which the electricity price will be market price [9].

The price of the generated and delivered electricity per year, from Možura wind generator's farm, can be calculated as follows:

$$96\,968 \text{ MWh } \times 95.99 \text{ euro/MWh} = 9307958.32 \text{ euro}$$

This is in case, that total annual electricity production of Možura wind generator's farm is delivered to the national electricity distribution network, in her first 12 years of exploitation.

Furthermore, the internal return rate of credit, with and without CER revenues was displayed in Table 3.

Table 3. Internal return rate of credit with and without CER revenues

Scenario	Project: internal return rate of credit
Without CER revenues	7.98%
With CER revenues	9.64%

CER revenues are revenues that originate from accomplished certified emission reduction credits, that each country has different by one ton of CO_2 [9].

2.4 The Generated Noise Level

The extent of adaptability of surroundings on the wind generators begins with the tolerance's threshold of emitted noise from the turbines. Many noise-abatement and mitigation measures are assumed by residents and others, like the relocation of residents, buffer modules per installed wind generators, ameliorated acoustic insulation in nearby households [18–20].

In the laid out project, (the wind generator's farm on Možura), the highest sound power level of wind generators is 105.8 dB (A). Based on quantifying data for a noise map model, as the most relevant wind blowing directions for creating noise, are taken north-east and south-west directions. Estimation and simulation of noise expansion in the surroundings of the wind farm are carried out by (for this occasion) specific software for the given natural conditions, spatial position, and layout of the wind generators. The calculation is done based on the known expressions for noise calculation in the conditions of free propagation of sound, without correction caused by the morphological form of relief and air absorption [10].

The estimated values of the noise per (G97-2MW) wind generator are made according to the wind speed from 3 m/s to 10 m/s (at the anemometer's height of 10 m). Influential parameters of noise are also the height of the wind generator H = 78 m and the wind speed at the height of the wind generator. The estimated noise values, according to these parameters, are shown in Table 4 [10].

Table 4. The estimated noise values per wind generator

Wind speed (m/s) at the anemometer's height of 10 m	The height of the wind generator H = 78 m	
	Wind speed (m/s) at the height H	Noise level in dB (A)
3	4.2	95.3
4	5.6	95.9
5	6.9	99.6
6	8.3	103.0
7	9.7	105.6
8	11.1	105.8
9	12.5	105.8
10	13.9	105.8

2.5 THE CO_2 Emission Reductions

Carbon footprint metering and it's pricing has been significantly promoted a strategy to lower the impact of GHG emissions on the environment. The mitigation report, originating from the Intergovernmental Panel on Climate Change (IPCC), claims that lifestyles and behaviors of consumers have an impact on energy consumption patterns and GHG emissions [21]. A growing body of other scientific literature also addresses the transformation towards a low carbon economy and society, what that entails, etc. [22–24].

Since it is mentioned, that this project is CDM (Clean Development Mechanism) project and it earns certified emission reduction credits for each ton of CO_2, it helps in achieving GHG emissions reductions. Accordingly, this paper tackles the CO_2 emission reductions, since the crediting of the project.

The crediting interval of this project activity started on 1st January 2013th. This interval is anticipated for 7 years and possibly will be repeated. For 7 years of crediting interval, the total CO_2 emissions reductions (savings) are calculated as 565 194 tons of CO_2 [9].

In Table 5 are displayed CO_2 emission reductions in tons per each year of crediting interval [9].

Table 5. CO_2 emission reductions (t) per year of crediting interval

Year	Estimation of CO_2 emission reductions in tons (t) per year
2013	80 742
2014	80 742
2015	80 742
2016	80 742
2017	80 742
2018	80 742
2019	80 742
Total estimated CO_2 emission reductions (t)	565 194
Total number of credit years	7 years

3 Conclusion

Renewable energy resources (RES), are playing a pivotal role in creating and updating new, more energy-intensive industries. In this manuscript, the significance of wind generators' farms in different domains was highlighted such as green electricity production, flattening country's energy demand's peaks, profitability formula, generated noise extent per wind generator and overall CO_2 emission reductions per crediting interval of project. Also, the forecast for renewable energy condition in Montenegro until 2025th was enclosed.

This article brings us incrementally towards the following conclusions.

- The mentioned wind generator's farm on Možura hillside, Montenegro is on-grid connected electricity system, which means it is centralized and delivers electricity to the main, national electricity distribution network. The annual electricity production of such wind farm covers 1/37 part of annual electricity consumption of entire Montenegro. The daily work of wind generators is assessed to 6 h, which corresponds to full load hours daily. The electricity output of the wind farm is green electricity since it is generated with clean technologies.
- The wind farm generators are with synchronous generators and each generator has AC-DC converter and later DC-AC converter to plug to the main, national distribution network.
- In the done forecast of renewable energy condition in Montenegro until 2025th, the electricity generation from other renewable energy sources (solar, wind, biomass) is slight, with the total percentage of 8.37%. That means that the exploitation of wind energy in Montenegro can still be neglected in the close future.
- Internal Return Rate of credit is different with and without CER revenues.
- The highest noise (sound power) level of wind generators is 105.8 dB (A) according to the considered height of the wind generator and height of the anemometer.
- With this project of the wind farm, the total CO_2 emissions reductions are calculated as 565 194 tons of CO_2 for the considered period of 7 years.

Besides above mentioned, such renewable energy projects change permanently the old-fashioned economics, cash flows, and environment and energy sectors of a country. This manuscript is particularly relevant from the point of clean energy (electricity), because it shows the potentials of a certain renewable energy resource project in a poor, developing country.

Acknowledgements. This manuscript was triggered and supported by the Energy Directory within the Federal Ministry of Economy and Electric Power Industry of Montenegro. We thank all executives of energy departments for providing us necessary information. The underlying idea of this manuscript was the research on a recently installed wind farm in Montenegro.

References

1. Hirsch, A., Parag, Y., Guerrero, J.: Microgrids: a review of technologies, key drivers, and outstanding issues. Renew. Sustain. Energy Rev. **90**, 402–411 (2018)
2. Goudarzi, N., Zhu, W.D.: A review on the development of wind turbine generators across the world. Int. J. Dyn. Control **1**, 192–202 (2013)
3. Azad, A.K., Rasul, M.G., Islam, R., Shishir, I.R.: Analysis of wind energy prospect for power generation by three Weibull distribution methods. In: Proceeding of the 7th International Conference on Applied Energy, ICAE 2015 (2015). Energy Procedia **75**, 722–727
4. Andersen, S.J., et al.: Performance and equivalent loads of wind turbines in large wind farms. J. Phys. Conf. Ser. **854** (2017). https://doi.org/10.1088/1742-6596/854/1/012001
5. https://www.cleantechnica.com
6. https://www.windpowermonthly.com
7. Wang, J., Zong, Y., et al.: A review of Danish integrated multi-energy system flexibility options for high wind power penetration. Clean Energy **1**, 23–35 (2017)
8. Ministry of Economy of Montenegro: Development of the Energy Strategy of Montenegro Until 2025, p. 14. White Book (2007). National document
9. Montenegrin Ministry of Sustainable Development and Tourism: PDD project Možura wind park. National document. www.mrt.gov.me/ResourceManager/FileDownload.aspx?rld
10. The environmental impact assessment study on the construction of the Možura wind park, with identification project number 02724391 (2012). National document
11. Borghesi, S., Montini, M.: The best (and worst) of GHG emission trading systems: comparing the EU ETS with its followers. Front. Energy Res. **4**, 1–19 (2016)
12. Figures 3 and 4 are available online at https://www.cdm.me
13. Jacobson, M.Z., et al.: 100% clean and renewable wind, water, and sunlight all-sector energy roadmaps for 139 countries of the world. Joule **1**, 108–121 (2017)
14. Vujadinovic, R.V., Tombarevic, E.M., Karadzic, U.M.: Valorization of potentials of wind energy in Montenegro. Therm. Sci. **21**(5), 1893–1903 (2017)
15. Ministry of Economy of Montenegro: Energy development strategy of Montenegro until 2030, action plan for 2016–2020 (2016)
16. Ministry of Economy of Montenegro: Energy Development Strategy Until 2025. White Book (2007)
17. Gouchoe, S., et al.: Case studies on the effectiveness of state financial incentives for renewable energy. National Renewable Energy Laboratory, NREL (2002)
18. Botelho, A., et al.: Effect of wind farm noise on local residents' decision to adopt mitigation measures. Int. J. Environ. Res. Public Health **14**, 1–20 (2017)
19. Song, K., et al.: Community survey on noise impacts induced by 2 MW wind turbines in China. J. Low Freq. Noise Vib. Active Control **35**, 279–290 (2016)
20. Van Kamp, I., Van den Berg, F.: Health effects related to wind turbine sound, including low-frequency sound and infrasound. Acoust. Aust. **46**, 31–57 (2018)
21. Edenhofer, O., et al.: Climate change: mitigation of climate change. Contribution of working group III to the fifth assessment report of the intergovernmental panel on climate change-IPCC (2014)

22. Schanes, K., Giljum, S., Hertwich, E.: Low carbon lifestyles: a framework to structure consumption strategies and options to reduce carbon footprints. J. Clean. Prod. **139**, 1033–1043 (2016)
23. Liu, H., Fan, X.: Value-added-based accounting of CO_2 emissions: a multi-regional input-output approach. Sustainability **9**, 1–18 (2017)
24. Gao, T., et al.: A comparative study of carbon footprint and assessment standards. Int. J. Low-Carbon Technol. **9**, 237–243 (2014)

Development of Hybrid System for Air-Conditioning of Almost Zero Energy Buildings

Edin Šunje(✉), Sead Pašić, Safet Isić, Emir Nezirić, and Edin Džiho

Faculty of Mechanical Engineering, University "Džemal Bijedić" in Mostar, University Campus, 88 000 Mostar, Bosnia and Herzegovina
edin.sunje@unmo.ba

Abstract. This paper presents a conceptual design of hybrid air-conditioning system which uses solar energy to heat/reheat water and generate electricity (solar collectors and photovoltaic panels), wind energy to produce electricity (small horizontal or vertical wind generator), and geothermal energy of water or soil in the building environment. The paper presents a pilot plant for hybrid air-conditioning system at the building of Faculty of Mechanical Engineering in Mostar and analyzes the efficiency and cost-effectiveness of system and its individual components.

Keywords: Zero energy buildings · Air-conditioning · Geothermal energy · Solar energy

1 Introduction

Air-conditioning of large buildings (public buildings, schools, hospitals, etc. …) is still mostly based on the combustion of solid fuels during the winter and usage of small individual air-conditioners during the summer months. The combination of large proportion of such buildings and the warming them by burning solid fuels during the winter months often causes increased air pollution in the environments. Cooling of these buildings during the summer is also not efficient, because of very high air temperatures, that commonly occur during this period. In such conditions the efficiency of these units is significantly reduced.

Apart from the other elements that define almost zero energy buildings, a special problem is production of energy for own needs [1–6]. This paper presents a conceptual design of hybrid air-conditioning system that uses solar energy to heat/reheat water and generate electricity (solar collectors and photovoltaic panels), wind energy to produce electricity (small horizontal or vertical wind generator), and geothermal energy of underground water nearby the building.

This paper presents a pilot plant for a hybrid air-conditioning system at the building of Faculty of Mechanical Engineering in Mostar (Fig. 1). This building is of a recent date and has been constructed to all building standards, so that it can be transformed and used as an energy efficient facility. To date, the Faculty has invested significant funds for procurement and installation of equipment that has made it possible to significantly

I. Karabegović (Ed.): NT 2020, LNNS 128, pp. 680–687, 2020.
https://doi.org/10.1007/978-3-030-46817-0_78

improve the energy efficiency of the facility. This primarily refers to the construction of a system for heating and air-conditioning of buildings using a heat pump that use geothermal energy of underground water, the procurement of 800 W wind turbines and 40 photovoltaic panels with a total power of 10 kW, as well as the equipment necessary for the conversion and storage of electricity (inverters and batteries). Integrating all these components into one hybrid system it is established a valuable base for further upgrades in order to achieve zero energy building of the Faculty.

Paper presents result of total electricity consumption in the last three years, in the period in which classic air-conditioning system migrated to the new hybrid system.

In this paper is also presented an analysis the possibilities of further upgrades of the system through increasing of capacity of solar thermal collectors and measuring equipment necessary for monitoring and optimization of energy production and consumption.

Fig. 1. Test facility et Faculty of Mechanical Engineering Mostar.

2 Heat Energy Production

2.1 Heat Pump

Heat pumps are commonly used for air-conditioning of private and public buildings. It rises heat from one lower level to higher heat level. As the heat source it uses air, ground and underground water energy sources. Underground water energy source is considered the most efficient. System air WQH75 has been installed at the test facility at the building of Faculty of Mechanical Engineering in Mostar. Installed pump uses underground water source as external energy source as it is shown at Fig. 2 and Fig. 3.

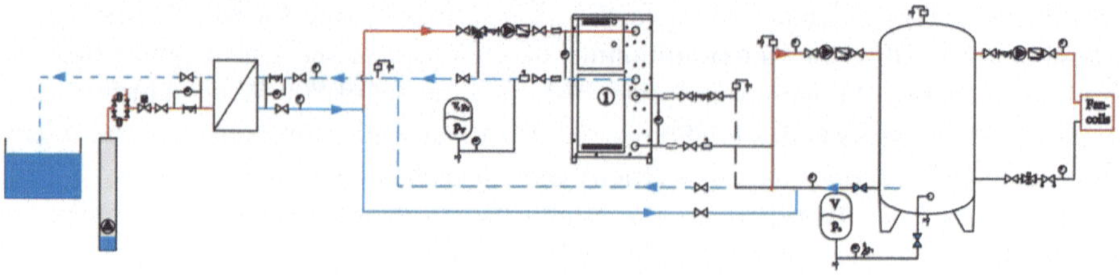

Fig. 2. Scheme of integration of heat pump into heating and air-conditioning system.

Fig. 3. Heat pump system air WQH75, 31 kW, integrated in air-conditioning system at Faculty of Mechanical Engineering in Mostar.

2.2 Thermal Collectors

Energy from the Sun has been delivered as a heat or light. Heat energy is usually used for water heating for air conditioning, sanitary water, electricity production etc. Solar thermal systems are used for heat transfer from Sun to water, while PV systems are used for electric energy production. Thermal collector with a vacuum pipe has been installed at the flat roof of the faculty building and has been used as an additional heater that helps the heat pump to rise the temperature of media in the air-conditioning system (Fig. 4).

Fig. 4. Thermal collector at the roof of Faculty building.

3 Electric Energy Production

3.1 Photovoltaic Plant

Photovoltaic system is used for electric energy production from the Sunlight. Mostar has a huge solar energy potential due to its geographical location. Photovoltaic plant has been installed at the faculty building flat roof with total installed power of 10 kW. Electricity produced from PV is used for heat pump operation, fan coils and partially for daily energy consumption (Fig. 5).

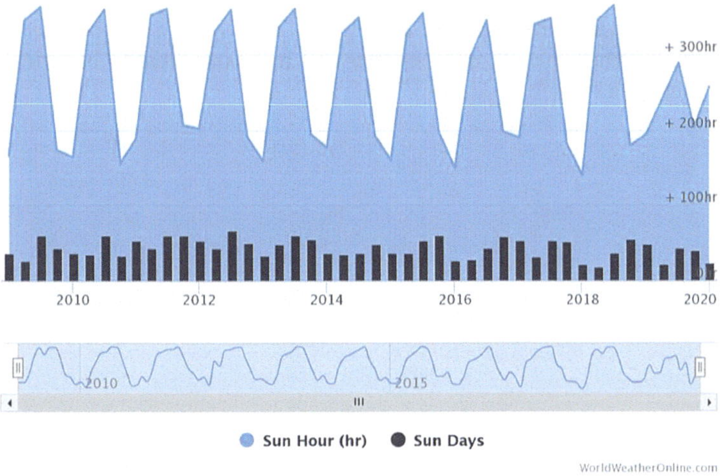

Fig. 5. Sun hour and sunny days in Mostar.

Photovoltaic plant of 10 kW installed power is mounted at roof of Faculty building. Subconstruction is adjustable to obtain optimum angle of the solar panels in different time during a year (Fig. 6).

Fig. 6. Photovoltaic plant at the building roof of Faculty of Mechanical Engineering in Mostar.

3.2 Small Wind Turbine

Beside the solar energy potential, Mostar area has also huge wind potential. Energy from the wind is also used for electric energy production. Energy produced is used for battery charging also covering energy consumption during the cloudy days and night period. Artas 500 windmill with has been installed as a part of test wind power plant. Total installed power is 900 W (Fig. 7).

Small wind turbine has its own inverter and it is connected to batteries to colect produced energy.

Fig. 7. Artas 500 windmill installed at Faculty of Mechanical Engineering in Mostar.

4 System Components Integration

All of the components for production heat and electrical energy are integrated into one hybrid system as it presented in Fig. 8. and Fig. 9.

Fig. 8. Photovoltaic panels, thermal collector and wind turbine integrated in the hybrid system.

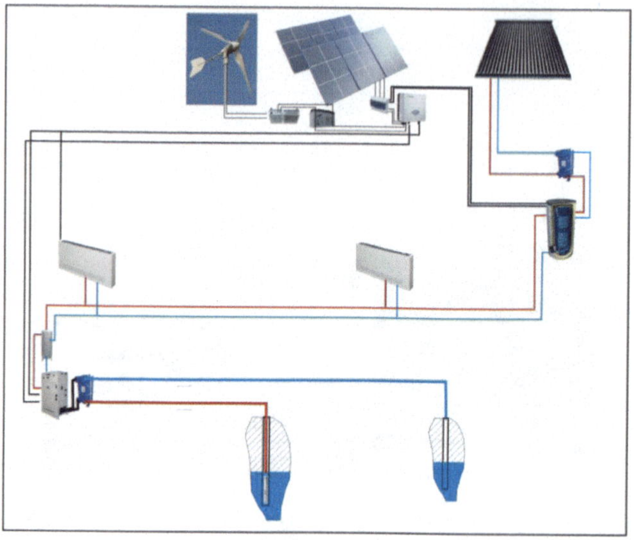

Fig. 9. Scheme of integration of components into hybrid air-conditioning system.

5 System Efficiency Analysis

The efficiency of developed system was monitored through the total electricity consumption in the period 2018–2019 and compared to the state of consumption before the system was installed in 2017. Figure 10 shows data on the total building area included in the air conditioning system in 2017–2019 and data on total energy consumption per square meter. It is evident that total consumption in 2017 for heating of approximately 900 m^2 of space is equal to the total consumption for heating and cooling of the space of approximately 1400 m^2. The disadvantage of this comparison is that hybrid system was partially commissioned in mid-2018, while the solar power plant was put into operation at the end of 2019, so results on system efficiency were not available for almost one year. Although the presented data shows that significant prerequisites for

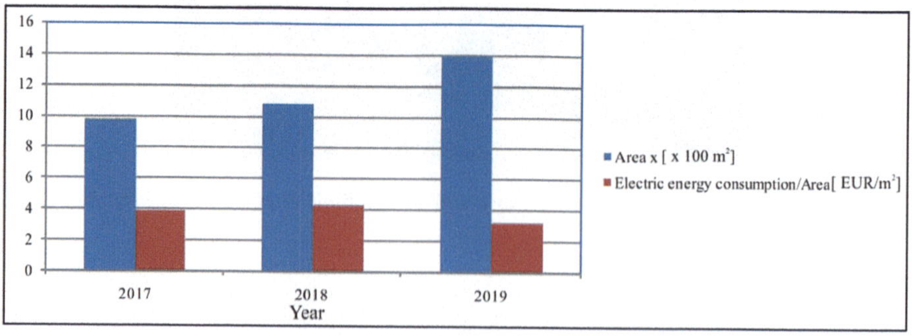

Fig. 10. Total energy consumption in period 2017–2019.

achieving zero energy building have been achieved, testing in 2020 will provide fully relevant data on system efficiency.

6 Conclusion

The development and implementation of renewable energy is one of the top priorities in the world today. However, one of the characteristics of these energy sources is their volatility; solar energy can be used during the day, while at night there is none; Wind energy can also be used only occasionally. The aim of research proposed in this project is to develop a hybrid system that would use all of the mentioned renewable energy sources in an optimal way, define the necessary capacities for energy storage, all with the aim of meeting the electricity needs of a particular object that is, achieving almost zero energy building. Optimal use of renewable energy sources: sun, wind and geothermal energy, for each specific micro location, it is possible to significantly reduce costs for electricity consumption, that is, to achieve the effect of almost zero energy building.

References

1. Voss, K., Musall, E.: Net Zero Energy Buildings – International Projects of Carbon Neutrality in Buildings, Munich (2011). ISBN 978-3-920034-80-5
2. Marszal, A.J., Heiselberg, P., Bourelle, J., Musall, E., Voss, K., Sartori, I., Napolitano, A.: Zero energy building – a review of definitions and calculation methodologies. Energy Build. **43**(4), 971–979 (2011). ISSN 0378-7788
3. Chwieduk, D.: Solar Energy in Buildings, Thermal Balance for Efficient Heating and Cooling, 1st edn. Academic Press, Cambridge (2014). ISBN 978-0-12-410514-0
4. Isabella, O., Jager, K., Smets, S.A., van Swaaij, R., Zeman, M.: Solar Energy: The Physics and Engineering of Photovoltaic Conversion. Technologies and Systems. UIT Cambridge Ltd., Cambridge (2016). ISBN 9781906860325
5. Primorac, M.: Economic analysis of photovoltaic system application. Educ. Entrep. **16**, 2 (2016)
6. Puljić, M.: A project to increase energy efficiency in a family home using solar collectors, Pula (2016)

New Technology of Arterial Pipeline Tubes Failure Prediction

Evgeny A. Deulin(✉) and Ekaterina I. Ikonnikova

Mechanical Engineering Department, MT-11, Bauman Moscow State Technical University (BMSTU), 2nd Baumanskaya st. 5, Moscow 105005, Russia
deulin@bmstu.ru

Abstract. The presented technology method relates to the distributing and cordial pipelines tubes diagnosing methods and its failure prediction and is based on Nano technological methods. It may be used for the estimation of residual resource of cordial pipelines tubes, which work in long time conditions of variable and state loads. The author's results of steel samples investigation with Secondary ION Mass Spectrometry and with Russian Electron Microscopy are presented, that help authors come to new conclusions about the results of transporting gas micro particles friction processes with the tube's wall, which leads to the steel samples destruction. It was shown that the sample microstructure forming process coincides with hydrogen molecules penetration process that influences on the steel properties according the "dry friction" theory being created

Keywords: Gas · Transporting · Cordial · Tubes · Reliability · Secondary Ion Mass Spectrometry · Failure · Prediction · Nano scale · Technology

1 Introduction

The presented technology [1] relates to the arterial pipelines tubes failure prediction and is based on method of Secondary Ion Mass Spectrometry (SIMS) analysis usage, instead of traditional ones. It may be used for the estimation of residual resource of arterial pipe lines tubes, which work in long time conditions of variable and state loads, and with joint length more 50.000 km, Fig. 1.

Let us consider traditional methods, which are being used for arterial tubes diagnosing, they are:

1 - ultra sonic diagnosing method [2] 2 - electromagnetic and acoustic(al) diagnosing method of tube material [3] 3 - electrometric method of tube material control [4]. 4 - radiographic method of tube material defects control [5] 5 - chemical, spectral analysis of cardial pipe line tube material methods [6].

All above-mentioned methods are nondestructive check methods i.e. correspond to nowadays diagnosing orthodox methods. Nevertheless, the diagnosing results interpretation is the main problem of modern diagnosing system creation. The presented method [1] is based on the SIMS analysis usage. The process of hydrogen dissolution in the result of gas flow movement into the arterial gas pipe line tubes in the near surface volume is the adequate valuation of the hydrogen dissolution as the result of so

I. Karabegović (Ed.): NT 2020, LNNS 128, pp. 688–700, 2020.
https://doi.org/10.1007/978-3-030-46817-0_79

Fig. 1. The view of arterial pipelines tube.

called "dry friction" process [7, 8]. The process of hydrogen dissolution in the result of gas flow movement into the arterial gas pipe line tubes in the near surface volume is the adequate valuation of the hydrogen dissolution as the result of well-known friction process.

In the process of long-term cardinal tubes exploitation the deterioration of plastic and viscous properties of tube metal take a place. The plastic properties of tube metal deterioration is a result of deformation based ageing and may lead to mechanical properties of the tube discrepancy to the standard ones and sometimes is a reason of arterial pipe line tube destruction, as it shown on Fig. 2.

Fig. 2. The view of arterial pipeline tubes failure (crack)

The parameters of the being tested samples were calibrated and used for the pipe line residual resource detection [1] as it is shown in Russian patent [8]. As a rule the steel viscosity detection with help of magnetic noise method leads to large errors. The reason is that the steel strength dependence as a function of magnetic noise signal usually looks as a curve with maximum in area of 2–4% plastic deformation and with minimum in area of 10–15% plastic deformation that leads to mistakes in the process of plastic (mechanic) properties of the steel being searched determination with help of magnetic noise method. The task of the method being presented is the diagnosing process of the arterial tube metal aging process, which is considered as corrosive and "hydrogen illness" of the tubes.

2 Experimental

Because the presented method is based on the SIMS analysis it was shown experimentally with help of SIMS installationы (see Fig. 3, 4) that the process of gas flow movement into the arterial gas pipe line tubes increases the dissolved hydrogen concentration 100–300 times on the depth 1–20 mm after 20 year period of the tube exploitation (tube aging). It was shown, that the results of SIMS analysis might be used for the process of arterial gas pipe line tubes destruction as the result of hydrogen illness.

Fig. 3. The view of TOF SIMS installation (cost 1.200.000 USD): The installation vacuum chamber overview (left), the sample under the ion source into ultra high vacuum chamber (right)

It is known [8], that the SIMS analysis method helps to detect the concentration of the dissoluted gases at analysis adequacy about 0,0001%, including H_2 and D_2 at the depth not more then 1–10 mic. From the other side, namely SIMS analysis method helps to predict [7] and then to discover [8] the role of so called "dry friction" in the process of hydrogen dissolution with catastrophic high level of atoms concentration. The technical result of new patent is in authenticity increasing of residual resource of cardial pipe line detection. The method consists of some successive ("know how") steps.

The advantage of the method being reported is the ability of SIMS analysis method to fulfil the detection of hydrogen concentration distribution in the sample at the depth 0.01–10 μ and at time about 2–8 h. The SIMS analysis method being described is based on the periodical information about "hydrogen illness" process and state of linear parts of arterial pipe lines.

Fig. 4. The view of CAMECA-4 [9] SIMS installation (cost 4.500.000 USD)

The possibilities and advantages of the SIMS method being invented are:

1. possibility to fulfill with the highest adequacy the dissoluted gases analysis
2. overcome the depth limitations (0,01–10 mic) of SIMS method and ensure the depth 1–30 mm.
3. to ensure nondestructive check being invented diagnosing method's parameters
4. to ensure nondestructive check with the planned period about 1–3 months.

3 Theoretical Aspects

The theoretical concept of this process is based on the idea, that the hydrogen (deuterium) dissolution process in the metal of tubes is a result of the gases flow movement inside the pipe line tubes and it is similar to the process of these gases dissolution into the metal of solid surfaces at "dry" friction process [9]. The task of this method is corrosive and "hydrogen illness" of the tubes of arterial pipelines process control and the variation of mechanical properties of a pipe line tube material forcasting. The theoretical concept of this process is based on the idea, that the hydrogen (deuterium) dissolution process in the metal of tubes is a result of the gases flow movement inside the pipe line tubes and it is similar [10] to the process of these gases dissolution into the metal of solid surfaces at "dry" friction process [7].

The measured hydrogen concentration with SIMS analysis helps us to see results of penetrated gases concentration variation and to compare it with analytical expression (1):

According 2nd Ficke equation [11] concentration of dissolved gases C(x, t): as a function of time t and distance (depth) x may be expressed:

$$C_H(x,t) = \left(C_{surfH} - C_i\right) \cdot \left[1 - erf\left(\frac{x}{2 \cdot \sqrt{D \cdot t}}\right)\right] + C_i, \quad (1)$$

where:

C_H - hydrogen concentration through the wall thickness;
C_{surfH} - hydrogen original concentration on the work surface of the pipe wall;
C_i - initial hydrogen concentration in the metal lattice;
D - diffusion constant;
x - distance from the surface of the pipe wall;
t - operating time.

According the theory [7] the initial gases concentration on the surfaces C_0, doesn't depend on the time $C(0, t) = C_0$.

The author's preposition was done that the process of hydrogen penetration from the inner gas stream into the tube material is based on the physical nature like a phenomena of gases dissolution in the materials of friction couple at traditional "dry friction process" [12]. The chemistry components of the tube steel are shown in Table 1.

Table 1. The chemistry components in the samples made of the tube steel

Steel	Elements, %							
	C	Mn	Si	Nb	V	S	P	Mo
						Less then		
10Г2ФБ (Russian type)	0,09–0,12	1,55–1,75	0,15–0,35	0,02–0,04	0,09–0,12	0,006	0,020	–

For "Traditional Specialists" working with gases in atmosphere it seems unbelievable, that atmospheric hydrogen of may be considered as a source C_{0H} in friction process, because the partial hydrogen pressure in atmosphere is eight time less the partial pressure of helium.

Nevertheless, according the theory [7] this hydrogen concentration in the first atom's monolayer, (on the surfaces) is equal $C_{0H} = 10^{23} - 10^{24}$ at/sm^3, that 10^4–10^7 times more, then traditional hydrogen concentration in steel $C_H = 10^{17} - 10^{19}$ at/sm^3 measured in samples lattice steel. We know, that the thickness of this monolayer is equal to 3×10^{-7} μ and for traditional Specialists it seems too small for the hydrogen concentration increasing in the steel lattice.

4 Methodology Aspects

An "in-situ" method of study of H and D dissolution in the pipe walls is discussed taking into account the significance of the gas-main pipe life cycle diagnosis problem. SIMS method provides opportunities that can be utilized in the proposed device.

It is well known [7, 9], that SIMS analysis allows taking measurements of concentration of dissolved gases including H_2 and D_2 with the accuracy of about 0.0001%, at a depth down to 1 … 10 μm. This method help to authors before to predict [8] and then to discover [10] the role of dry friction in the process of hydrogen dissolution that raises the hydrogen concentration in steel up to catastrophically high level. The aims of the described method are following:

(1) Application of the high-accuracy SIMS method to analysis of the dissolved gases [6];
(2) Overcoming 1 … 10 μm limitation of depth of the SIMS method and ensuring 1 … 30 mm depth SIMS analysis of a pipe wall [8];
(3) Nondestructive inspection usage based on the invented diagnostic technique;
(4) The reliability of scheduled nondestructive inspections of a gas-main pipeline increase by means of invented method usage every 1–3 months;
(5) Creating an opportunity for a nondestructive inspection of gas-main pipelines in the zones, where the pipe condition is analyzed with pipe wall samples being cut from a typical pipeline being operated at the same conditions, see Fig. 3. The measured results in a such zones Fig. 4 had to correspond to tube state and must be compared with the results of routine analysis;
(6) Conversion of the qualitative results of SIMS analysis into quantities' values.

The following technique of the SIMS diagnostic method was designated to meet the above listed requirements:

(1) a representative sample of the pipe is chosen to ensure reliability of SIMS analysis of the pipe wall material;
2) The original sample, Fig. 3 must be cited into "elements" for SIMS analysis, see Fig. 4, 5. The proposed layout takes into account difference between the depth of SIMS analysis (x_1 = 1 … 10 μm) and the thickness of the pipe wall (x_2 = 1 … 30 mm);

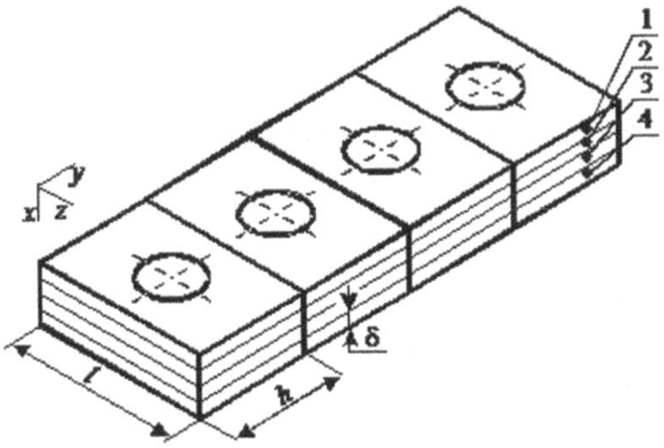

Fig. 5. The view of the initial sample, made of arterial pipe line tube and the view of cutting lines it into "elements" for SIMS analysis.

(3) The results of SIMS analysis are converted from qualitative form into quantities' values by means of ion implantation of H and D in analyzed samples;
(4) The analyses are implemented periodically to accumulate empirical data about diffusion of hydrogen into the gas-main pipeline pipe wall material.

Elements increase starting numbers from the outer surface of the pipe. The sample sizes: l = 24 mm, wideness h = 10 mm, thickness δ = 3 mm.

Below on Fig. 6 we may see layout for the original sample (made of arterial pipe line tube) cutting into "elements" for SIMS analysis (Fig. 7).

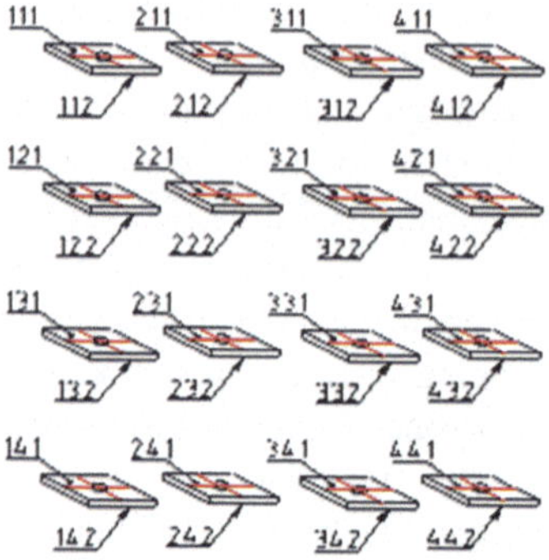

Fig. 6. The cheme of the sample's elements indexation before SIMS analysis

Fig. 7. The view of sample's elements, prepared for SIMS analysis

For Specialists working with gases in atmosphere it seems unbelievable, that atmospheric hydrogen may be considered as a source atomic concentration C_{0H} increasing in friction process, because the its partial pressure in atmosphere is 8 time less then the partial pressure of helium.

Nevertheless, according the theory [7] this hydrogen concentration in the first atom's monolayer, (on the surfaces) is equal $C_{0H} = 10^{23} - 10^{24}$ at/sm^3, that 10^4–10^7 times more, then traditional hydrogen concentration in steel $C_H = 10^{17} - 10^{19}$ at/sm^3 measured in samples made of steel. We know, that the thickness of this monolayer is equal to 3×10^{-7} μ and for traditional "specialists" it seems too small for the hydrogen concentration increasing in the steel samples.

Figure 8 shows the quantities results of the hydrogen concentration distributions in the "elements" of the sample at different distances x, from inner tube's wall that symbols correspond to Fig. 3.

All graphs of the C_H concentration distributions at the depth x = 0–0,5 mic. (left parts of the graphs of Fig. 8) show the concentration rise near x = 0. The reason of this rise is friction (cutting) process of "original sample" into "elements", and the intensity of hydrogen dissolution at cutting process is rather more, then at friction of gas flow micro particles with the tube wall. The right parts of the same graphs correspond to the hydrogen concentration after it penetration into the tube material at service period on the distance x = 0,1–20 mm from the inner surface of the pipe line tube.

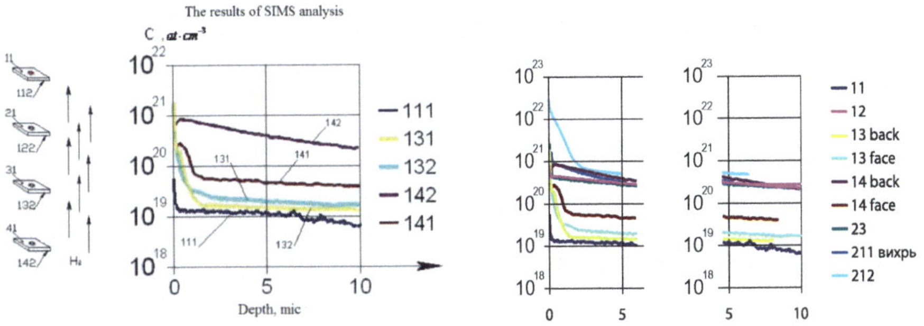

Fig. 8. The quantities results of the hydrogen concentration distributions in the "elements" of the sample at different distances x, from inner tube's wall which symbols correspond to Fig. 5 and 6 (left). Right diagrams present the left graphs being divided into two parts. The left parts of these graphs correspond to the hydrogen-increased concentration because of tube cutting process into the samples. The right parts of the same graphs correspond to the hydrogen concentration after its penetration during many year long service period

Figure 9 shows the theoretically calculated hydrogen concentration distribution in the tube wall according Eq. (1) and according the right parts of the diagrams Fig. 8.

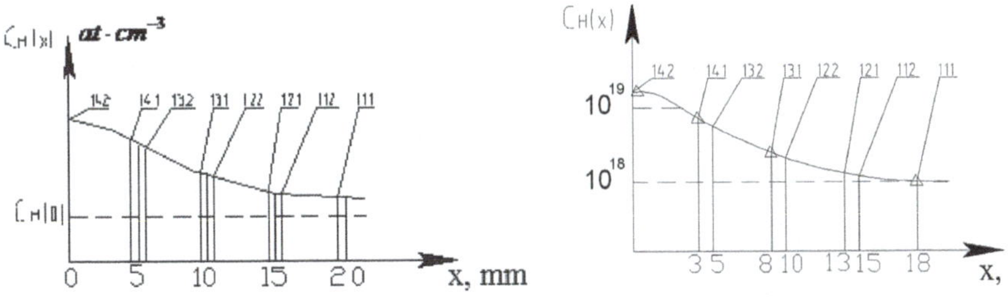

Fig. 9. The left graph shows theoretically calculated hydrogen concentration distribution. Right graph shows experimental results. The numbers correspond to Fig. 5, 6 samples dividing

The opportunities provided with the method of SIMS analysis are utilized in the device designed by authors to clarify the characteristic properties of mechanically stimulated dissolution process and to facilitate further research and development of a gas-main pipeline pipe protection against hydrogen dissolution and "hydrogen illness" of oil and gas pipelines.

5 Experimental Results in Atmosphere and in Vacuum

All methods of pipe-line tube failure prediction, mentioned above (in "Introduction") are considered as nondestructive diagnostic techniques, and meet the requirements to standard diagnostic method. The gas flow allows making an adequate estimate of hydrogen dissolution in a material of pipeline wall. Hydrogen dissolution in the wall of the gas pipeline is caused by the gas traveling near the surface of the pipe wall show [12, 13] that results of SIMS analysis can be used to predict a gas-main pipeline destruction caused by hydrogen illness and consider SIMS analysis as nondestructive diagnostic techniques.

The author's results of steel being prepared (by rolling methods) for arterial pipe line's tubes were done with SIMS analysis also as with Electron Microscopy (УЭМВ100К) installation. They help us come to non-visible conclusions about the friction processes which leads to the samples (Russian made steels 40, 09Г2ФБ, ШХ 40, Т15К6) destruction at traditional friction, cutting, rolling and friction processes. It was shown that the sample microstructure forming process coincided with hydrogen molecules forming process influences on the steel properties according the "dry friction" theory being created. The process of mechanically stimulated hydrogen dissolution was predicted in 1997 [14]. It was shown with SIMS analysis [7, 10], that the concentration of dissolved hydrogen increases many times at a depth of 1… 5 microns even after some seconds of friction process on nano scale level. SIMS analysis method was used in our experiments because it helps to detect the concentration of different gases at analysis adequacy about 0,0001%, including H_2 and D_2 at the depth not more 10–15 μ. The main results of this method usage are presented below. The idea of the experiment was to receive direct and indirect results of new hydrogen and its isotopes atoms appearing at friction process. The metal cutting process was considered as a "dry friction" process at ultimate friction parameters. The tritium atoms were used for the research because only this hydrogen isotope is radioactive and its appearing in the samples may be measured as with SIMS analysis (direct method) also as with gamma meter (indirect method). The cutting (friction) process parameters were: Friction (cutting) speed V = 72,2 m/min, cutting step S_0 = 0.2 mm/rotation, depth of cutting, t = 1 mm, being cited material (Russian Steel 40 type), speed of rotation N = 315 min^{-1}. The Gamma Radiation being measured at cutting process was compared with background radiation. The same comparison was done for X-rays Background and measured radiations, and the results we can see in Table 2.

Table 2. The figures of the radiation from the samples before and after friction (cutting) process

(A) Gamma-rays background 10^{-3} Sv/hour	Gamma Radiation at cutting 10^{-3} Sv/hour		
0.10	0.19	0.11	0.16
(B) X-rays Background radiation, 10^{-3} Sv/hour	X-rays Radiation at cutting, 10^{-3} Sv/hour		
0.11	0.13		

The indirect results show, that the friction process is a reason of Gamma and x-Rays Radiations appearing, as it coincides with the theoretical aspects [1, 11] of atoms Tritium appearing in solids of friction pair. We can see, that the cutting (rough friction) process increases the Gamma- rays background radiation in 1,1–1,9 times and we can see also, that the same process increases X-rays Background radiation in 1,2 times. The direct results of atoms Tritium appearing were received with SIMS analysis (TOF SIMS-5 installation) with the descried above samples and it show [7] that atoms Tritium concentration increases in the cutting instrument from $C_D = 5.10^{19}$ at/sm^3 till $C_D = 1.10^{20}$ at/sm^3 i.e. two times.

6 The Tube's Sheet Rolling Influence on Steel Properties

The main task of this experiment was to estimate the sheets rolling process influence on sheet's metal mechanical properties, being prepared for arterial pipe tube manufacturing. The initial results were received with the SIMS analysis and shows [3], that the process of mechanic elements hydrogen illness and their destruction are the result of hydrogen dissolution at friction in rolling process. It was shown above, the SIMS analysis method was chosen because it helps to detect the concentration of different gases at analysis adequacy about 0,0001%, including H_2 and D_2 at the depth not more then 1–10 mic.

The main experiment was done in frames of converter steel melt on Azov Steel Co plant (Ukraine). The samples cited from steel sheet, heated till 1100° C and 1250°°C, and then were cooled by room air. The experiments which were done by Ikonnikova [15, 16] with the electron microscope (УЭМВ100К) show, that the plastic properties of the metal deterioration as a result of deformation may be seen as the results of sulfide inclusions and its hydrogen environments transformation at sheets rolling process we can see on Fig. 10 and Fig. 11, where the form of sulfide particles (black, Fig. 10). May be explained as a result of hydrogen diffusion transforms into "roundish" form particles (Fig. 11), and where the hydrogen concentration becomes less and the sheet's mechanical properties become better.

Fig. 10. Sheet for the tube made of steel, after 1st rolling before heating (scale ×3200, Ikonnikova's result, AzovSteel Co)

Fig. 11. The same sheet of steel after 1250° C heating (scale ×4000, Ikonnikova's result, Azov Steel Co)

It is well known that sulfide particles are the source of hydrogen atoms accumulation. It is well known also, [16] that namely sulfide particles are the reason of hydrogen illness and the cracks appearing in the sheets being rolling. The mentioned above process of hydrogen atoms diffusion from the "limited sources" of hydrogen may be seen after comparison micro particles form (and color) on Fig. 10 and 11 because roundish form and bright color of the same micro particles shows that steel mechanical properties become better.

The process of hydrogen atoms diffusion from the "limited sources" was done with the D_2 atoms SIMS analysis [17]. In the role of "limited sources" was the sample surface being covered with one monolayer (0,35 more correct figure) of heavy water.

Figure 12 shows the deuterium atoms concentration distribution in the sample covered with the deuterium (heavy water initially) one monolayer, after 30 s of friction (d1 - long graph for the friction zone of the sample, d2, d3 - near the friction zone) We can see high deuterium initial concentration distribution (d2, d3) decreasing as result of deuterium penetration from the surface into the sample depth about 5 microns, and these graphs show that deuterium initial concentration distribution (d2, d3) increasing about 25 times at the depth 0.3–4.5 μ from the surface according Sieverts low (without friction). We can see, that near the surface ("limited sources") where the initial deuterium atoms concentration was $C_{0D} = 10^{22}$ at/sm^3 and this concentration decreases till $C_{0D} = 10^{18}$ at/sm^3 (see graph d1) that shows, that the hydrogen illness was removed, and our patient (steel sheet) comes to "good health". From the another side the simple calculation of hydrogen atoms concentration increasing near the surfaces after the rolling friction action (after sheets rolling) shows, that the initial hydrogen atoms concentration in sheet steel $C_H = 10^{17} - 10^{19}$ at/sm^3 increases till $C_{0H} = 10^{23}$ at/sm^3. at the depth 0,1–3,5 mm after 3–5 min rolling process. It shows, that the "hydrogen illness" comes inside the steel sheet from the surfaces.

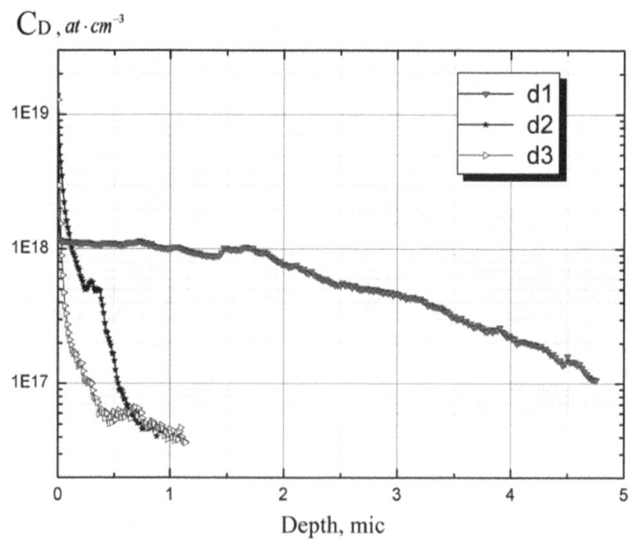

Fig. 12. Deuterium atoms concentration distribution in the sample covered with the deuterium (heavy water initially) one monolayer, after 30 s of friction (d1- long graph for the friction zone of the sample, d2, d3-near the friction zone)

7 Conclusion

We may see that the pipe tubes are one of the key components of oil-and-gas infrastructure. Stress-corrosion is one of the most considerable modes of failure. Such corrosion takes place in railway transport as well.

It was shown, that metal is ageing during a long operation period outer and inner pipe wall surfaces corrode and all this leads to pipe breakdown with heavy ecological and economical costs of pipe unreliability.

It was shown also, that hydrogenation or "hydrogen illness" revealing itself goes in pipe material when pipelining hydrocarbons requires further study and perfection of diagnostic techniques.

Modern diagnostic systems are perfect from the technical point of view.

The requirement to provide a non-destructive inspection is an important criterion applied to diagnostic systems.

References

1. Patent of RF № 2391601 method of residual resource of cordial pipeline detection, 21 July 2008
2. Aleshin, N.P., et al.: Acoustic control methods/non-destructive control, 283 p. High School, Moscow (1991). (in Russian)
3. New methods of steel materials mechanical properties control, 439 p. Mir, Moscow (1972). (in Russian)
4. Method of underground communications electro-meter control, 40 p. ВНИИСТ-ВНИИГАЗ, Moscow (2000). (in Russian)
5. Non destructive control welded joints/radio graphs control method/Russian standard: ГОСТ 7512-82. Russian standard edition (1983). (in Russian)
6. New methods strength determination of fragile destruction of materials (Новые методы оценки сопротивления материалов хрупкому разрушению), 439 p. Мир, Moscow (1972). (in Russian)
7. Deulin, E.A., Basmanov, V.S.: Vacuum Mechanics of Nuclear and Space Technique, 388 p. Telecom, Moscow (2019)
8. Deulin, E.A., et al.: Mechanics and Physics of Precise Vacuum Mechanisms, 234 p. Springer, Cham (2010)
9. Advertising materials of CAMECA: "Science @ Metrology Solutions" IMS 7f "High performance universal magnetic sector SIMS". www.cameca.com
10. Deulin, E.A., Mashurov, S.S., Mirzoev, A.M.: Research of the reasons of hydrogen illness of arterial pipe line tubes and vacuum protection methods development. In: Conversion in Machine Building, vol. 3, no. 82, pp. 32–36 (2007). (in Russian)
11. Sieverts, A., Zapf, G., Moritz, H.: Zs. Phys. Chem. A **183**, 19, 32–36, 268 (1938/1939)
12. Deulin, E.A., Goncharov, S.A., de Segovia, J.L., Nevshoupa, R.A.: Mechanically stimulated solution of adsorbed hydrogen and deuterium in steel. Surf. Interface Anal. **30**, 635–637 (2000)
13. Deulin, E.A., Mashurov, S.S., Mirzoev, A.M.: Research of the reasons of hydrogen illness of arterial pipe line tubes. In: Conversion in Machine Building, vol. 3, no. 82, pp. 32–36 (2007). (in Russian)

14. Deulin, E.A.: Exchange of gases at friction in vacuum. In: ECASIA 1997, pp. 1170–1175. Wiley, November 1997
15. Ikonnikova, E.I., et al.: Dislocation structure of "dressed steel" (Russian type 08КП) after artificial and natural long deformation processes. Metal Thermo Working. Industrial Book № 5, pp. 19–120. Metallurgy, Moscow (1977). (in Russian)
16. Ikonnikova, E.I., et al.: Rolling steel pressing range influence on re-crystallization could rolled sheet steel (Russian type 08КП). Sheet Working Industry. Industrial Book № 3, pp. 50–56. Metallurgy, Moscow (1974). (in Russian)
17. Deulin, E.A., Konovalov, V.V., Trifonov, A.J., et al.: Mechanically stimulated heavy water (D_2O) dissociation and deuterium dissolution in stainless steel. Vacuum Technique and Technology, vol. 13, No. 3, pp. 17–22 (2003). (in Russian)

Functional Modeling of the Means for Heat Consumption Monitoring During Its Design Using the Information

Viktoriia Antypenko(✉), Ruslan Okopnyu, Viktor Nenia, Anna Marchenko, and Bohdan Antypenko

Faculty of Electronics and Information Technologies, Sumy State University, Sumy, Ukraine
viktoriait@ukr.net

Abstract. This article presents a complex method of functional modeling for technical objects with different levels of detail description. During decomposing, the main function of a technical object is represented by simpler functions which are performed sequentially and in parallel. Several types of functions formalized description such as graphical, tabular and analytic one are presented. General descriptions of IDEF0 functional modeling methodology have been inherited and partially modified and supplemented according to the requirements of engineering practice. As a result, it allows getting quantification of technical objects. Thus, a well-grounded complex method of functional modeling for technical objects is developed. It has been tested practically on the methodology for designing and controlling the process of heat consumption monitoring in the apartment building.

Keywords: Formal description · Information technology · Regulator · Controller · Weather dependent control · Remote control · Monitoring system · Heating system · Saving energy resources · Energy efficiency

1 Introduction

Nowadays, the transition to the design and manufacturing of technical objects as "smart" objects has become an important direction of technology development. Such technical objects include both means of performing the basic functional purpose and computers (controllers), as well as software and information support. They recognize changes of both the external environment and their own state, effects on themselves and change the parameters of their functioning accordingly. This direction of the technology development is supported by the INCOSE International Systems Engineers Council and envisaged by the Industry 4.0 strategy of Industrial Development.

2 Related Work

The problem of rational use of thermal energy has been determined as urgent all over the world. For example, in the US, houses consume 40% of all energy and about 48% of it is used to maintain comfortable climatic conditions [1]. This makes it important to

I. Karabegović (Ed.): NT 2020, LNNS 128, pp. 701–708, 2020.
https://doi.org/10.1007/978-3-030-46817-0_80

develop means of monitoring and regulating the thermal energy consumption and, consequently, energy carriers for its generation.

In addition to the economic factor, safety requirements are also established to the heat generation means located in residential buildings. Solving safety issues in a technical sense automatically resolves human safety issues [2]. Developing means should prevent the operation of heat generation equipment in abnormal modes.

The used indoor climate control means currently are based on static models of thermal comfort [3]. In this case, residents are considered as passive elements of the thermal environment of the house. It is used an assumption that people have a fairly constant requirements range of their comfort. To meet these requirements, it is necessary to have the possibility to implement different configurations of heat distribution in different places of the house [2].

The development of technical means for human use is a deliberate process, which is intended to involve the use of modern technologies, equipment and materials. In the area of monitoring the state of the distributed environment, the use of wireless networks and objects connected to the Internet is relevant today [4]. This trend extends from the manufacturing environment to the individual accommodation. The effective application of these technologies requires the integration of finished components into a single entity that has the necessary functionality.

Another trend of today is to create "smart" objects, especially those that work with people or work for people in changing environments. For this subject area, it is advisable to carry out a logical analysis of the following issues [5]: determining the operation of equipment in an abnormal emergency mode and its switching off with warning a person who makes the final decision, determining the deviation from the required mode according to these circumstances and automatic return of the equipment operation and equipment itself to the proper mode of operation. Monitoring of state must be performed on a continuous basis with the selected discrete time step. Activation of control must be performed if necessary [6].

A typical approach to the design of monitoring means will be considered using the example of works [7–9], where the methods and techniques of the already considered works [1, 2, 4–6] are repeated. At first, a general block diagram is proposed, and then components or modules implemented by specific technologies of concrete sizes are described. The purpose of the used components is indicated but the functions performed by them are not stated or used. Both existing developments (LabView, Proteus, MatLab) and owned ones written in Micro-C or Visual Basic languages are used as software. Under these conditions, it is impossible to apply a model-based approach to design, which is the basis for creating "smart" or otherwise "cyber-physical" objects.

The work [10] takes a special place. It describes a monitoring means based on a subject area model. Preliminarily the model is presented in analytical form. Then it is integrated into the controller model and used when setting the operating mode by the regulator. The condition for the effective application of such solution is given in [10].

Based on the performed analysis, the aim of this paper to develop such method for functional modelling of technical objects that would be used at all stages of their life cycle. To achieve it, the following tasks must be performed: analysing existing methods of describing the technical objects functioning; developing a new method or choosing and improving an existing one for implementation the possibility of formal analysis of a technical object; testing the proposed method practically on a tool for monitoring the heat consumption in a residential building.

3 Research Methods

The methodology of the research is based on the use of the following ones:

- the system approach and system analysis in the development of conceptual solutions;
- a process theory in the modeling the functioning of the proposed means;
- the theory of fuzzy sets within a considering the factors determined by qualitative indicators;
- the hierarchical theory for the decomposing the functions of the system, the description of which consists from the circuit theory in the practical implementation of the proposed models.

4 Results

4.1 Conceptual Model of the Information Technology for Heat Consumption Monitoring

Relying on the achievements of modern information technologies in general and achievements in the fields of telecommunications, hardware and software, and taking into account the persistent trends of the directions of their development and application, there is an approach, which considers the use of the Internet and wireless collecting the information from temperature sensors that are located in the boiler, different rooms of a building and exterior.

Various wired and wireless communication modules have been added to implement the possibility for remote controlling of the device operation. Data on all measured parameters of the heating system operation are transmitted to the monitoring system server through one of these channels every minute. These data are needed to further analyze the effectiveness of the regulatory process.

4.2 Functional Modeling of Information Technology for the Indoor Temperature Regulation

The maintaining a comfortable or technologically sufficient temperature in the building under any climatic conditions of the heating period is the main function of the system *F0*.

The main function is further decomposed. It is presented by the Eq. (1):

$$[F0] = [F1] + [F2] + if(c1, [F3], null) + if(c2, [F4], null), \tag{1}$$

where [] – square brackets mean that the function is complex;
() – round brackets are used to group functions into units;
F1 – a function of the maintaining set temperature in the room;
F2 – a function of the saving information about the system current state which is served;
F3 – a function of the information analysis for the specified period;
F4 – a function of the improvement the regulation algorithm taking into account the operation results.
c1, *c2* – appropriate conditions.

Functions *F1* and *F2* are performed regularly after the set time interval and functions *F3* and *F4* are performed only when the conditions *c1* and *c2* are fulfilled respectively.

The functions used in (1) do not have the traditional mathematical content, but reflect the functional opportunities that will be implemented by information technologies during the performance of the processing information processes.

The operator (+), which is used in the Eq. (1), means the fact that the functions are performed exclusively consecutively and their performance in parallel is forbidden. Respectively, the next operator (*) means that the functions, which are associated by it, can be performed at the same time or asynchronously with any values of individual phase shifts within the time of the most long-term process.

The implementation of the first function, namely *F1*, with the help of information technology, is considered. For its decomposition, a typical chain of operations, which the information technology contains, is used. They are a function of gathering the information from indoor sensors *F1a*, a function of its transmitting through the communication channels *F1b*, a function of its processing *F1c*, a function of its using for the regulating the outdoor sensors *F1d* and a function of information storing *F1e*. The decomposition of the function *F1* is presented by the Eq. (2):

$$[F1] = [F1a] + [F1b] + [F1c] + [F1d] + [F1e]. \tag{2}$$

The function *F1a* is implemented by three functions represented by the Eq. (3):

$$[F1a] = [F1a0] + [F1a1] + [F1a2], \tag{3}$$

where *F1a0* – a function of temperature determination in the source of thermal energy – inside it;
F1a1 – a function of determination the temperature of external environment;
F1a2 – a function of determination the temperature in different parts of the building.

The function *F1a0* is determined in the same way and presented by the Eq. (4):

$$[F1a0] = F1a0a + F1a0u + F1a0c, \tag{4}$$

where $F1a0a$ – a function of temperature determination inside the thermal energy source;
$F1a0b$ – a function of temperature determination at the input of the thermal energy source;
$F1a0c$ – a function of determination the temperature at the outlet from the thermal energy source.

The hierarchical decomposition of the remaining functions is performed by the same defined rule.

4.3 Structural Model of the Controller

On the basis of the proposed functions hierarchy, the following structural model of the device for the temperature regulation in the building is presented in Fig. 1. The industrial microcontroller 7 is its basis. The elements represented in Fig. 1 perform the following functions. The microcontroller 7 receives the information from the temperature sensors from the multiplexer 5. The last one performs the function $F1a$ according to the Eq. (3) and also defines all its components.

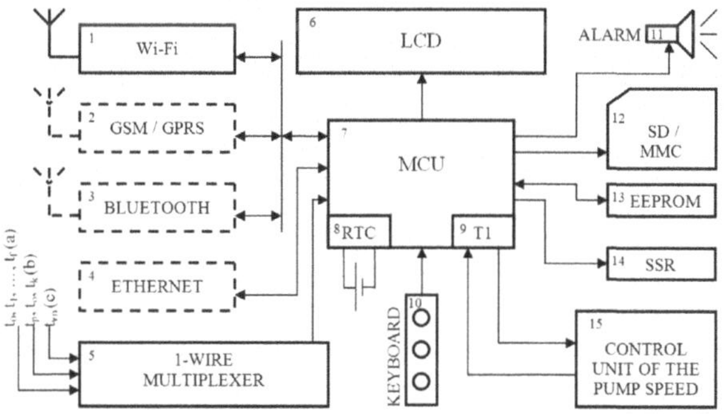

Fig. 1. Structural model of the controller

A list of these units and their functions is the following:

1. Unit 1. Wi-Fi Data Transfer Module. Used to connect a device with the monitoring system website through the Internet by the standard IEEE 802.11 b/g/n.
2. Unit 2. GSM/GPRS Data Transfer Module. Used to connect a device with the monitoring system website through the Internet by the standard 2G.
3. Unit 3. Bluetooth Data Transfer Module. Used to provide the access to manage the device directly using a tablet or mobile phone through a specialized application. The standard IEEE 802.15.1 is used.
4. Unit 4. Ethernet Data Transfer Module. Used to connect a device with the monitoring system website through the Internet by the standard IEEE 802.3 (10BASE-T). The most commonly used in the industry, unlike wireless data transfer environments.

5. Unit 5. 1-Wire multiplexer. It is needed to switch different wire segments and to increase the reliability of data transmission through the wire by reducing the line impedance. There can be room temperature sensors (*a*), temperature sensors of the boiler (*b*) and a sensor of the outdoor environment temperature (*c*) on the separate segments of the multiplexer. This unit implemented the function *F1a0*which consists of the simplest functions. Other functions are implemented in the same way.
6. Unit 6. Display. It is able to display 16 characters on 2 rows and used to indicate the current state of the controller and to navigate through the menu items.
7. Unit 7. MCU (Microcontroller Unit). It is a single-chip microcontroller which has the ARM architecture and the Cortex-M3 core. Its model is STM32F103. MCU is the main control unit of the device.
8. Unit 8. RTC (Real Time Clock). It is a part of the microcontroller and requires an external battery to keep the clock running when the main power source is turned off.
9. Unit 9. T1. T1 is a timer and also is a part of the microcontroller. It controls the delay of switching on the half-period of the AC voltage variable to regulate the power transmitted to the pump engine of the heat delivery by hardware.
10. Unit 10. Keyboard. It consists of 3 separate buttons available on the front device panel and used for local manage of the device.
11. Unit 11. Alarm. This buzzer is used to alert the occupants or staff about the malfunction of the boiler.
12. Unit 12. SD/MMC (Secure Digital Memory Card/MultiMedia Card). SD/MMC is a memory card. It is used as a reserve unit of flash ROM (Read Only Memory) for storing statistic data about the operation modes of the heating system. At problems emergence in the website work it can be used to import data and also there is an opportunity to analyze files using a spreadsheet like Excel.
13. Unit 13. EEPROM (Electrically Erasable Programmable Read-Only Memory). It is a type of non-volatile memory of 8kB with access to each byte. It is used to store device settings.
14. Unit 14. SSR (Solid-state relay). It has one group of normally open switches, used to control the operation of a fuel boiler with its own electronic system of automatics.
15. Unit 15. Control unit of pump speed. It is a power part of the circuit for controlling the shaft rotational speed of the heat delivery pump.

4.4 Hardware

The control unit of the proposed heating system is constructed on the basis of modern architecture of single-crystal microcontrollers Cortex-M3 namely STM32F103. The microcontroller contains a sufficient number of peripheral devices such as timers, controllers of digital consecutive buses I2C which are used, for example, to connect an EEPROM memory chip for saving the settings, the SPI which is used to connect an MMC/SD flash memory card for saving statistical data about heating and boiler systems working, internal and external temperatures. For manual manage of the heating system operating modes there is LSD displaying the current status of the control object, temperatures of the external environment, temperature to which the regulator has to aim, indoor temperature, current time, boiler operation mode (heating, cooling, maintenance of the set temperature), the state of the data transmission interface and signal level for wireless data transmission systems in the device.

At the hardware level, an alarm system built on two internal timers of the STM32F103 microcontroller was added to the design. The conditions for the siren activation are set by means of fuzzy logic model.

4.5 Software

The control program for the microcontroller is written in a high level language C++ in the integrated development environment IAR-EWARM for ARM Cortex-M. The functions of the developed program give the opportunity to set an algorithm for the process of regulating the operation of the heating system using the Fuzzy Logic methodology in the visual environment of the construction and adjustment of the fuzzyTECH model [11]. The fuzzyTech program is designed to solve various fuzzy simulation tasks in difficult-to-evaluate and difficult-to-control parameters. The main distinctive feature of fuzzyTECH is setup the Serial Link Debug Mode. This mode allows to connect a project developed by fuzzyTECH with another system (industrial controller, etc.) by means of the consecutive communication interface and conduct a remote setup of the project in real time using all visual means of the environment. After successful setting up of the fuzzy output model, the environment allows to generate a code in the programming language C, which implements a functional fuzzy model. It is required to connect this code to the main IAR-EWARM project and compile the binary firmware file for its downloading to the microcontroller.

5 Conclusion

The hardware and software of the automatic system of heating regulation have been developed on the basis of a single-chip microcontroller with the opportunity of remote control of its work. This solution allows simply modifying and/or adjusting the controller behavior model at the controlling the operating modes of the heating system using a model based on fuzzy logic and is created in the fuzzyTECH visual environment. The approach of the adaptive fuzzy output of the controlling influence of the temperature regulator has demonstrated its ability to maintain the temperature in the room accurately.

The design of the regulatory device has been performed on the basis of the proposed hierarchical decomposition by performing the functions of an arbitrary structure. It is convenient for its algorithmic implementation in the form of information technology.

The developed hardware and software complex has shown a high enough accuracy of maintaining the temperature in the room, with the magnitude of the regulation error of 1.43% [12]. This has been achieved due to the smooth change of engine rotations of the circulation pump in the system of quantitative regulation of heat consumption.

References

1. Abdallah, M., Clevenger, C., Golparvar-Fard, M.: Developing a thermal comfort report card for building. Procedia Eng. **118**, 675–682 (2015)
2. Babiarz, B.: Aspects of heat supply security management using elements of decision theory. Energies **11**(10), 2764 (2018)
3. Youssef, A., Caballero, N., Aerts, J.M.: Dynamic model-based monitoring of human thermal comfort for real-time and adaptive control applications. Biomed. J. Sci. Tech. Res. **19**(4), 14526–14532 (2019)
4. Li, W., Kara, S.: Methodology for monitoring manufacturing environment by using wireless sensor networks (WSN) and the Internet of Things (IoT). Procedia CIRP **61**, 323–328 (2017)
5. Rangaswamy, T., Prakash, S.: Smart logic system for interlocks and protections of boilers in a thermal power plant. Int. J. Adv. Res. Electr. Eng. **1**(1), 8–14 (2018)
6. Guo, L.: Design of an intelligent heat supply monitoring system based on android system. In: 2017 International Conference Advanced Engineering and Technology Research (AETR 2017), pp. 138–143. Atlantis Press, March 2018
7. Noh, S.K., Kim, K.S., Ji, Y.K.: Design of a room monitoring system for wireless sensor networks. Int. J. Distrib. Sens. Netw. **9**(8) (2013). https://doi.org/10.1155/2013/189840
8. Khairi, N.A., Jambek, A.B., Boon, T.W., Hashim, U.: Design and analysis of a wireless temperature monitoring system. In: RSM 2013 IEEE Regional Symposium on Micro and Nanoelectronics, pp. 105–108. IEEE, September 2013
9. Zungeru, A.M., Mangwala, M., Chuma, J., Gaebolae, B., Basutli, B.: Design and simulation of an automatic room heater control system. Heliyon **4**(6) (2018). https://doi.org/10.1016/j.heliyon.2018.e00655
10. Stepanets, O., Andrasovych, A.: An algorithm of regulating an energy-efficient hot water system with the object model in the controller. Восточно-Европейский журнал передовых технологий **4**(8), 11–18 (2016)
11. Leonenkov, A.: Nechetkoe modelirovanie v srede MATLAB i fuzzyTECH. BHV-Peterburg (2005)
12. Okopnyu, R., Nenia, V., Zakharchenko, V.: Simistorne regulyuvannya obertiv odnofaznogo dviguna vidcentrovanogo cirkulyacijnogo nasosu u sistemi opalennya budivli. Visnik Harkivskogo nacionalnogo tehnichnogo universitetu silskogo gospodarstva imeni Petra Vasilenka (196), 69–70 (2018)

Renewable Energy Sources in Construction of Energy Efficient Residential Buildings

Miron Torlo[1(✉)], Ismar Kreso[2], and Šunje Edin[1]

[1] Faculty of Mechanical Engineering, University "Džemal Bijedić" in Mostar, University Campus, 88 000 Mostar, Bosnia and Herzegovina
miron.torlo@gmail.com

[2] "SMED Engineering BH", Mostar, Bosnia and Herzegovina

Abstract. Energy consumption is increasing every day, which is why renewable energy is increasingly being offered as one of the most important solutions in the fight against climate change and the energy crisis. Exactly one of these forms is solar energy. Solar radiation is directly converted to heat or electricity. The conversion is done by different devices, some of which are photovoltaic cells for generating electricity.

This paper shows how to increase energy efficiency in residential buildings by using photovoltaic cells. The materials for making them are described, and the principles of operation of photovoltaic cells are explained. One of the new structural solutions has been arranged, the method of their mounting is shown, and the technical and economic analysis of these solutions is given.

Keywords: Energy efficiency · Sunlight · Photovoltaic cells

1 Introduction

Renewable energy sources are a long-term viable option for energy supply. In this paper, special attention is given to solar energy, how to increase energy efficiency in residential buildings by using solar photovoltaic cells. Solar energy is a renewable energy source, it is also a very clean source of energy, so there are no harmful emissions and pollution from the use of photovoltaic cells after installation.

Unlike non-renewable ones, renewable energy sources are characterized by variable energy production because the sources themselves are such by their nature, the wind blows at different speeds, the sun does not radiate with the same intensity.

Renewables have come a long way from minimizing, or rather, labeling with alternative epithets to the present day, when they represent one of the fastest growing industries.

Solar technology is widely characterized as passive solar or active solar technology, depending on how sunlight is collected, converted and distributed. Active solar techniques include the application of photovoltaic cells to convert sunlight into usable products.

I. Karabegović (Ed.): NT 2020, LNNS 128, pp. 709–719, 2020.
https://doi.org/10.1007/978-3-030-46817-0_81

One of the technical solutions that can be marketed for today's electricity production is integrated roofs with photovoltaic cells. These solutions are very similar to classic solar panels. However, in technical terms, they are more advanced than conventional solar panels. These types of panels serve as a roof covering, insulation and solar panel with their structural design and therefore fulfill all the needs of today's users.

2 Integral Solution of Tin Roofing with Photovoltaic Cells

Today, solar panels are most often mounted on the roofs of buildings, for example in Germany in 2018, more than 8000 MW of solar panels were installed on roofs. These energy efficiency solutions are certainly good, but there are drawbacks to the extra weight on the roofs and certainly aesthetic issues (Fig. 1).

Fig. 1. Roof with solar panels.

An essential feature of this solution is the significant additional mass on the roof that must be considered. This paper presents a different approach to this problem. It is integrated roof with photovoltaic cells. There are already solutions in the world that do not have additional structures that burden the existing roof structure. These are 3 in 1 solutions (cover, insulation and solar cells). Tesla uses shingles for the 3 in 1 cover in Fig. 2 [7].

Fig. 2. Tesla's roof 3 in 1.

Both in the world and in Bosnia and Herzegovina there is an increasing trend of electricity generation through solar panels. This area has been addressed under local law through the Renewable Energy Act. This law also provides for a significant incentive price for the construction of micro power plants with an installed capacity of up to 23 kW. Because of this, the production of energy using roof surfaces is economically very cost effective and attractive to households.

All of above also led us to define a 3 in 1 technical solution with a combination of existing products on the market.

In our market the most commonly used roofs (tile, shingles and sheet metal).

The trapezoidal sheet was used as a choice. Since the standard dimensions of the trapezoidal sheet do not fit our technical solution, it is necessary to define a tool for plastic deformation of the sheet thickness of 0.5 to 0.6 mm, so that the area of the individual plate is 1 m^2 [4].

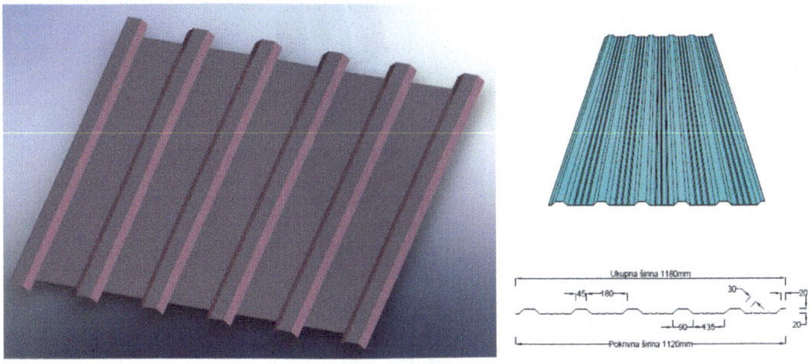

Fig. 3. Trapezoidal sheet.

The layers that make up the solar panel are shown in Fig. 4 [2].

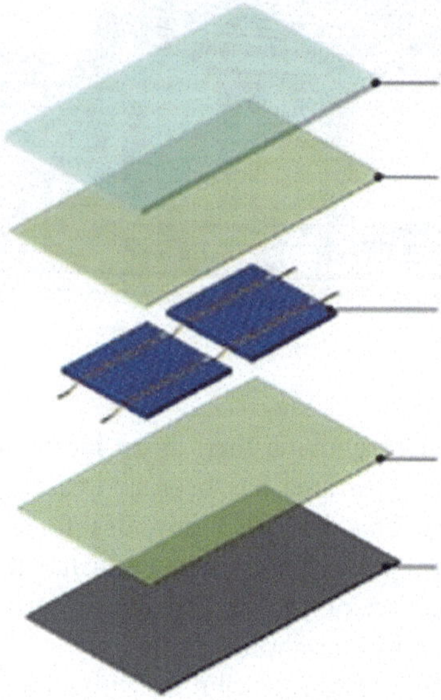

Fig. 4. Parts of solar panel

Figure 5 shows a standard solar panel with the difference that they are made in the form of strips whose width corresponds to the width of the solar cell or the distance between two trapezoids on the trapezoidal sheet [2].

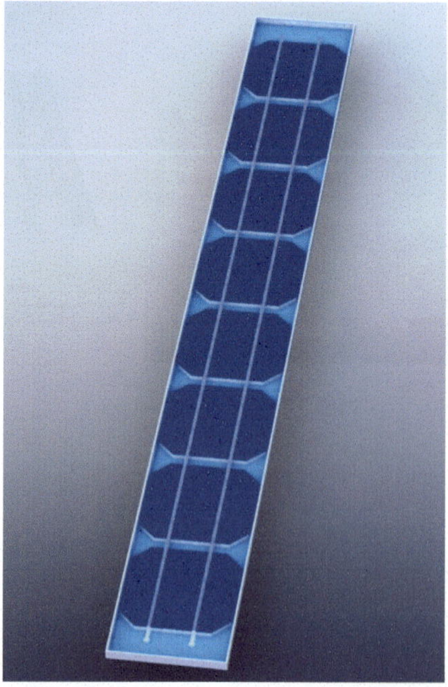

Fig. 5. Solar panel.

Components listed above are mounted in the solar panel shown in Fig. 5.

Such plate is mounted in an aluminum frame, which is also standard, shown in Fig. 6.

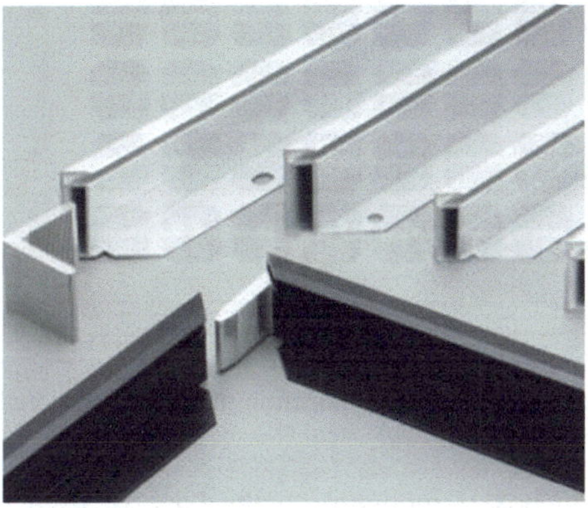

Fig. 6. Standard aluminium frame

Since the trapezoidal sheet is made in the modules shown in the Fig. 3, there are 5 spaces between them for the installation of such solar panels, as can be seen in Fig. 7.

Fig. 7. Connection of subassemblies into trapezoidal sheet.

The solar panels that are mounted on the tin panel are connected into one electrical unit (serial connection) as shown in Fig. 8. such that the power of one panel outputs 140 W, a voltage of 24 V and a current of 5.83 Amps [3].

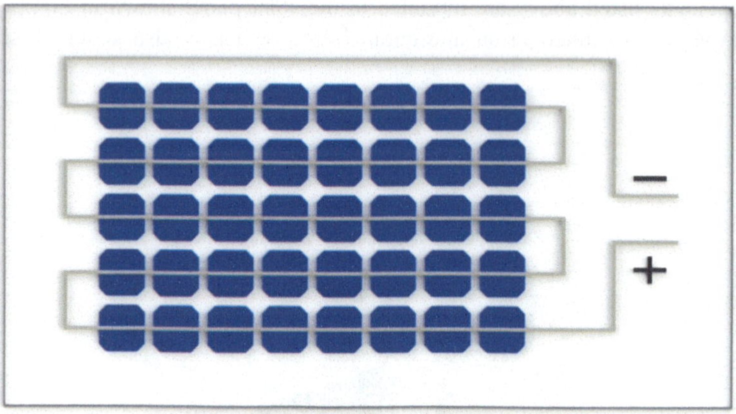

Fig. 8. Scheme of connection of photovoltaic cells.

A standard junction box is mounted on the back of each individual trapezoidal panel to be connected to the next panel, as shown in Fig. 9.

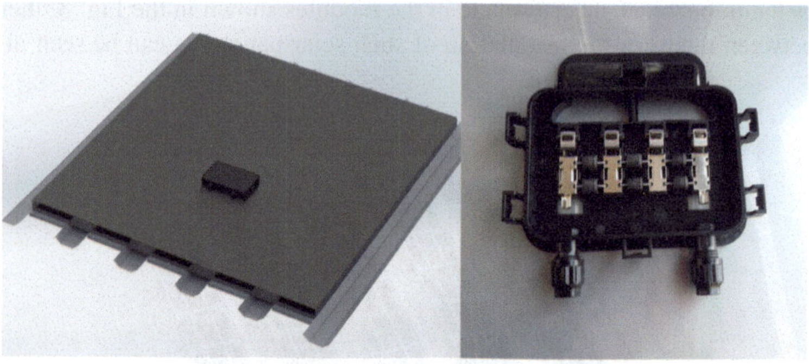

Fig. 9. Junction box on the back of the trapezoidal panel.

The interconnection of the trapezoidal panels is usually done with the cables at the ends of which are the MC4/MC5 connectors, shown in Fig. 10.

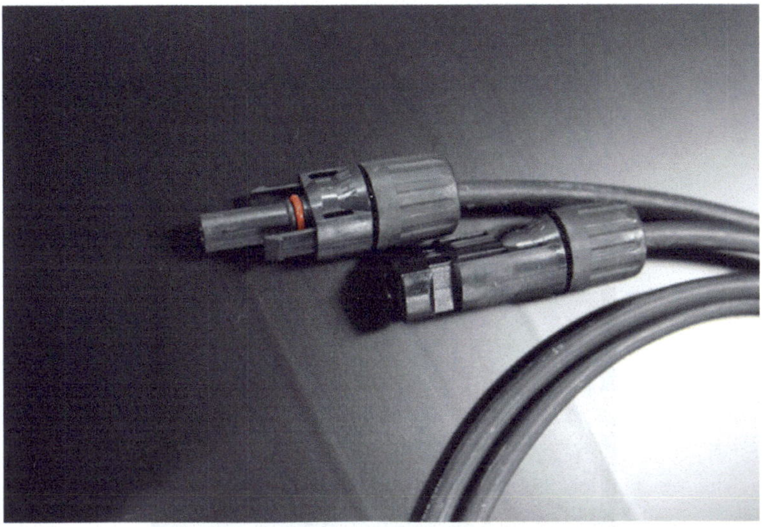

Fig. 10. MC4/MC5 connectors.

The panels are then mounted on a standard timber roof structure in such a way as to create an integral roof with hydro and thermal insulation. The following figure shows a standard wooden roof structure (Fig. 11).

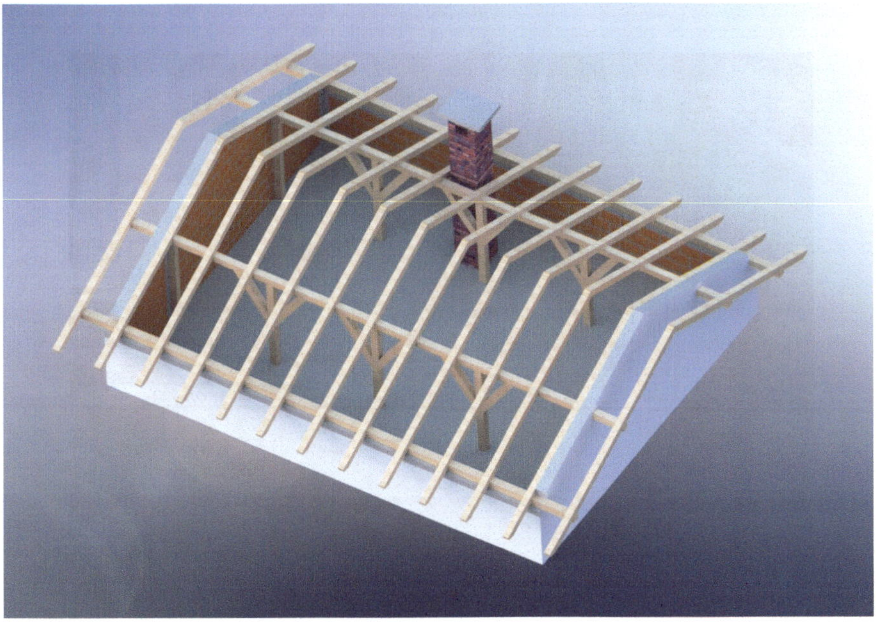

Fig. 11. Standard wooden roof structure.

This construction is suitable for roofs with two slopes, north-south orientation, where trapezoidal sheets with solar cells are placed on the south side. The connection of trapezoidal sheets to the roof structure is done with standard screws, shown in the Fig. 12.

Fig. 12. Trapezoidal sheet fastening screw for wooden construction.

By installing thermal insulation we get a 3 in 1 system and get an energy efficient object [8] (Fig. 13).

Fig. 13. Installation of thermal insulation.

3 Comparison and Analysis of Classical and Integrated Roof Coverings (Technical and Economical)

In the following table we will present a comparison and analysis of classic and integrated roofing from a technical and economical point of view (Fig. 14).

Fig. 14. Classic and integrated solar panels.

3.1 Technical Analysis and Comparison

In the following table are summarized analysis and comparison of the technical aspects of classic and integrated solution (Table 1).

Table 1. Comparison of classic and integrated panels

Classic	Integrated
Highly developed and often applicable	They are in the development phase
Suitable for existing roof structures	Suitable for new constructions
Additional structural load	Less structural load
The roof consists of two covering bodies (tile and panels)	3 in 1 roof (panels are also cover, insulation and electricity generator)
Simpler technical design	More complex technical design
They require less training for the assembly workers	They require the training and precision of the workers when assembling
The aesthetic appearance of the building or the roof is disturbed	Aesthetically, it makes a slight difference between cover and panel
Harder maintenance	Easier maintenance

3.2 Economical Analysis and Comparison

The cost of small photovoltaic systems is significantly higher than the cost of conventional technologies but has a downward trend. Also, photovoltaic systems occupy a significant space, so the trend of installing them on roof surfaces is increasingly interesting. The advantage of these systems is that electricity production is at the point of consumption. Therefore, the production price corresponds to the sales price. Also, due to the positive legislation, the price of green energy is subsidized by the state, so we have a direct economic benefit due to the reduction of fossil fuel imports and a great environmental benefit. For the technical solution presented in this paper, an orientation analysis of the current market prices for the production of an integrated solar photovoltaic panel was made. This analysis is given in Table 2 [5].

Table 2. Price and material characteristics of photovoltaic trapezoidal panel.

	The name of the material	Characteristics	Price
1.	Trapezoidal sheet	1 m^2	10 km
2.	Solar cell 40 pieces	3.5 W	14 km
3.	Tempered glass 3.5 mm	0.78 m^2	6.8 km
4.	EVA film	1.55 m^2	3.2 km
5.	Aluminum frame	12.5 m	16 km
6.	Waterproof sheet 4 mm	0.78 m^2	6.3 km
7.	Photovoltaic cell connection wire	12 m	3.5 km
8.	Junction box	1 komad	8.2 km
9.	Sheet metal box	2.3 m^2	5 km
Total			73 km

The table shows that the price of the solar panel itself is approx. 63 km for 140 W power. the price of a classic solar panel is approx. 85 km for 140 W power, considering that the price of the panel presented in the table does not show the labor price and the trade margin. So the price of a classic panel and an integrated panel is about the same as the price of a classic panel increases for additional construction, which is a direct economic saving.

4 Conclusion

Integrated roofs with photovoltaic cells, whose construction is presented and explained in the paper, are certainly one of the solutions that will make it possible to obtain electricity from renewable energy sources, which are becoming more and more prevalent in today's world. Sun energy is one of the most useful ways to get electricity from renewable energy sources.

These are the reasons that have led to the need to find new solutions for obtaining electricity especially on residential premises that occupy a significant part of the space. There are already constructive solutions in the world that generate electricity from solar energy.

However, integrated solar panels offer better technical solutions in terms of structural design. The novelty of such constructive solutions also relates to the aesthetic appearance and functionality of the roof on which they are installed. This is one very important component affecting the acceptability of a new solution in the world, and subsequently their increasing use and distribution. In addition to the aesthetic appearance, roofs with integrated solar panels also meet technical and economical aspects. From a technical point of view, they combine the three functions of the roof, which puts them among the most ideal solutions.

The persistence of such a solution represents a significant advance in the world of mechanical engineering. Integrated solar panels are on track to meet all the conditions required by today's standards.

References

1. Federal Ministry of Energy of Mining and Industry: Law on the Use of Renewable Energy Sources and Efficient Cogeneration http://www.fmeri.gov.ba/zakon-o-obnovljivim-izvorima-energije-i-kogeneracije.aspx
2. Lambić, M.: Solar technology - thermal and photovoltaic systems. AGM BOOK (2013)
3. Majdančić, L.J.: Photovoltaic systems [manual]. Ruđera Boškovića Technical School in Zagreb (2010)
4. Nožić, M.: Manufacturing technologies: (deformation treatments). Džemal Bijedić University, Faculty of Mechanical Engineering (2007)
5. Primorac, M.: Economic analysis of photovoltaic system application. Educ. Entrepreneurship **16**, 2 (2016)
6. Puljić, M.: A project to increase energy efficiency in a family home using solar collectors, Pula (2016)
7. Tesla. https://www.tesla.com/solarroof
8. Knauf roof insulation. https://pim.knaufinsulation.com/files/download/kosi-krovovi-potkrovlje-prospekt-2018-07.pdf

Computer Studies of the Tightness of the Drill String Connector Depending on the Profile of Its Tapered Thread

Oleh Onysko(✉), Lubomyr Borushchak, Volodymyr Kopei, Tetiana Lukan, Iulia Medvid, and Viktor Vryukalo

Computerized Machine Building Department, Ivano-Frankivsk National Technical University of Oil and Gas, Karpatska, 15, Ivano-Frankivsk 76000, Ukraine
onysko.oleg@gmail.com

Abstract. Drill connector is named as tool-joint play an important role in ensuring the impermeability of the drill string. The tightness closure of the ends of the box and the pin depends on the contact pressure between the ends of the box and the pin of the on one hand and the pressure of the drilling fluid inside of the pipe on the other hand. In common tool-joint, the contact pressure must exceed the pressure of the drilling fluid inside the drill string. Computer studies conducted by the authors in the SOLIDWORKS Float Simulation show the reduce of the effect of the drilling fluid pressure on the amount of contact pressure between the ends of the box and the pin in the process of initial depressurization.

Keywords: Drill tool-joint · Mud fluid · Tapered thread · Hydraulic simulation · Thread surfaces

1 Introduction

Tightness is one of the main operational indicators of the drill string. This indicator is both an influential factor for the environment preservation and for the loss of hydraulic energy during drilling [1]. Especially hydrosphere as a part of the multicomponent dynamic system that is influenced by oil and gas drilling engineering is depended on it [2]. In addition the process of drilling oil and gas wells is the source of air pollution through drilling mud evaporation containing hazardous chemical substances [3]. Through the cavity of the drill string under a significant pressure up to 10–20 MPa drill wash solution is pumped. This solution is the main source of water and air pollution during drilling. Therefore, avoiding its losses due to the leakage from the drill string is an urgent task. The drill string consists mainly of drill pipes. They are joined by means of special connectors. Its are called tool joint, which consist of box 1 and pin 2 (Fig. 1).

I. Karabegović (Ed.): NT 2020, LNNS 128, pp. 720–729, 2020.
https://doi.org/10.1007/978-3-030-46817-0_82

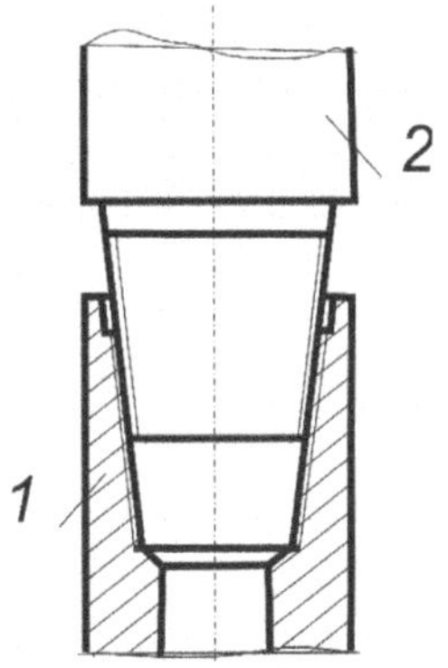

Fig. 1. Scheme of drill pipe tool-joint

There is special tool-joint tapered thread during drilling. Therefore, avoiding its losses due to the leakage from the drill string is an urgent task.

2 Literature Review

When the pin (2) and the box (1) are screwed together, a conical screw channel (3) is formed between the pin crest and the box root and vice versa (Fig. 2).

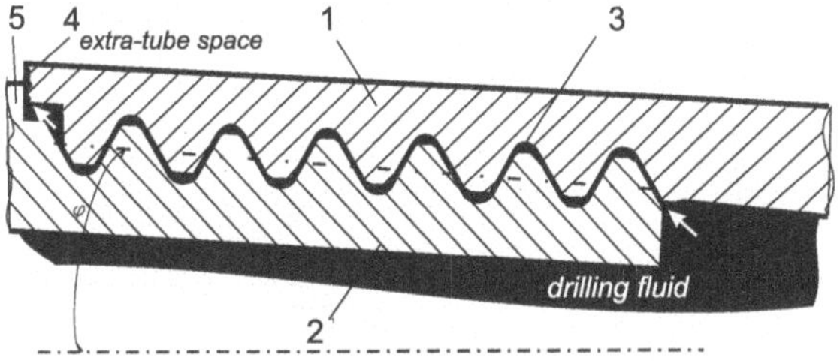

Fig. 2. Conical screw channel formed between the pin crest and the box root and vice versa

The flow characteristics of the mud in the channel depend on its transverse dimensions. If there is depressurization between the box end (4) and the pin end (5), this channel, due to the small geometric dimensions and features of drilling fluids, can counteract fluid flow [4]. Source [4] shows an example of a computer study of channel velocity distribution using a Float Simulation application. However, the studies carried out are only relevant to size of the tool-joint tapered thread 2 7/8 REG, which corresponds to profile shape I, which is not commonly used in drilling practice. The existing API7 standard provides for other shapes - II, III, IV, V. Due to the greater wear resistance of tapered thread on the use of form IV standard prefers it. In addition, there are no studies on channel pressure distribution in source [4]. Papers [5, 6] present theoretical and computer studies of premium class threads for the connection of oil and gas pipes. But the work is aimed at investigating of the contact pressures and deformations in the threads and does not show study of the tightness of these connections. Articles [7, 8] shows computer studies of oil and gas drill pipes tool joint tapered thread. But it deal with investigation of stress between thread surfaces only. Source [9, 10] presents the ways to improve the reliability and tightness of casing strings but not drill strings. Article [11] shows the manufacturing of the tapered thread with the small geometric dimensions channel and don't deal with investigation of the tightness of the tool-joint tapered thread.

3 Research Methodology

3D model of the channel that created by screwing the pin and the box together is one solid piece. For simulation purposes, the channel begins and ends with straight sections of the same section. And they are accessed through the channels of circular cross-section, which in the hydraulic simulation will not participate (Fig. 3). According to form IV is used tapered thread NC31. Its parameter are: taper 1:6, pitch 6,35 mm, profile high h = 2,64 mm, end major diameter d_3 = 71,23 mm, length 70 mm.

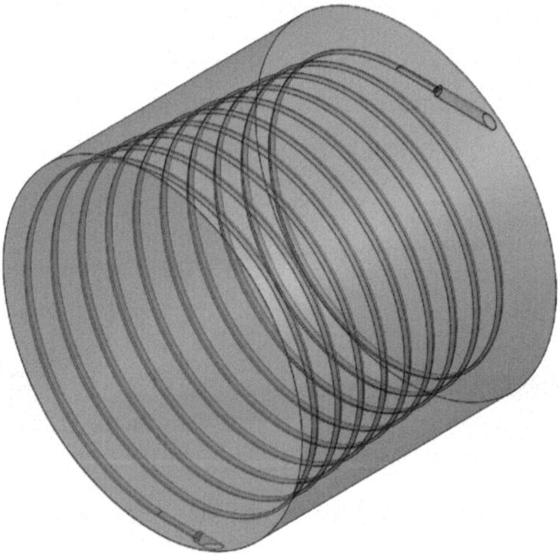

Fig. 3. 3D Model of the conical screw channel formed between the pin crest and the box root and vice versa

Input and output are continued in straight sections so SOLIDWORKS FlowSimulation can create LID1 and LID2 plugs. They are needed to limit the volume for which the simulation will take place, as well as to impose boundary conditions (Fig. 4).

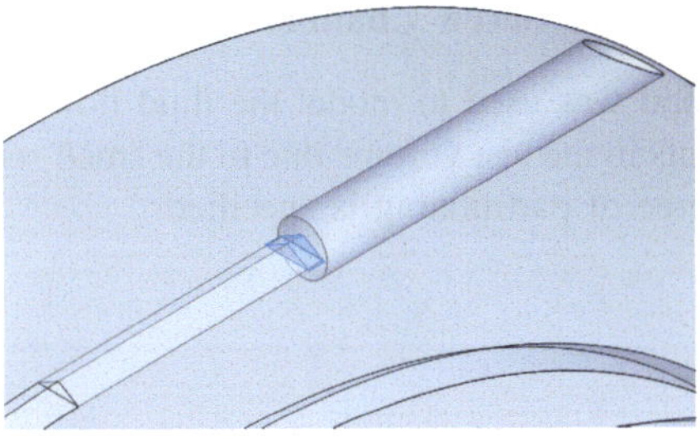

Fig. 4. Channel plug (highlighted in blue)

3.1 The Model of the Hydraulic Fluid

Specified input data for the FlowSimulation program using:

(1) the type of analysis is chosen internally because the working fluid is limited by the walls of the solid. The heat exchange in this model is not taken into account as it is believed that the body is cooled by the flow of solution through the central opening of the column.
(2) physical characteristics of the mud: density 1200 kg/m^3, initial kinematic viscosity of 10 cSt. The viscosity characteristics were chosen according to the Herschel-Bulkley model.
(3) the subarea of the working environment is established.
(4) boundary conditions of the experiment: pressure at the inlet equal to 10 MPa and pressure of the external environment (at the outlet) equal to atmospheric - 0,1 MPa.

So Drilling fluid parameters for SOLIDWORKS FlowSimulation medium are shown in Table 1.

Table 1. Drilling fluid parameters in SOLIDWORKS FlowSimulation medium

Type	Non-Newtonian
Name	Drilling Pulp
Density	1200.00 kg/m^3
Specific heat	1000.0 J/(kg*K)
Thermal conductivity	0.2500 W/(m*K)
Viscosity	Herschel-Bulkley model
Consistency coefficient	0.5000 Pa*s
Yield stress	20.00 Pa
Power-law index	0.3

3.2 The FEM Model of the Screw Channel

The finite element method was used to model the fluid flow in the channel. Configuration the grid of elements in the test volume due to the small section dimensions of the channel needs high degree of partitioning is specified.

4 Results

4.1 Geometric Models of the Cross Section of the Screw Channel

Figure 5 shows a cross-section of the screw channel in accordance with the standard API 7 form IV. Values a = 1,65 mm, h_1-h = 0,46 mm.

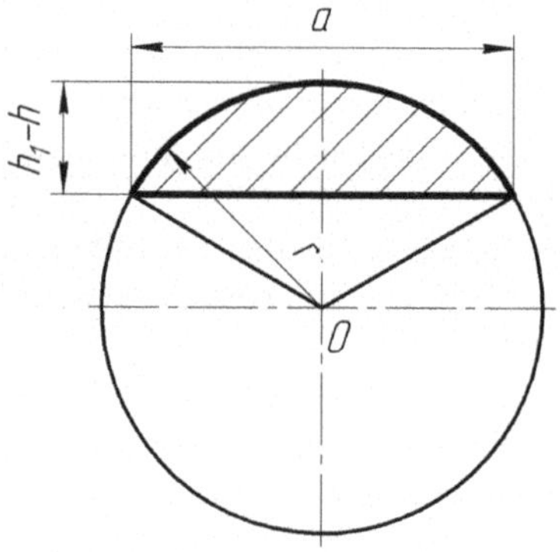

Fig. 5. Cross-section of the screw channel in accordance with the standard API 7 form IV.

Figure 6 shows a reduced (highlighted in gray) cross-section of the screw channel of form IV. Values a = 1,65 mm, h_1-h = 0,46 mm, a_1 = 1 mm, |FB| = 0,134 mm in accordance to methodology [4].

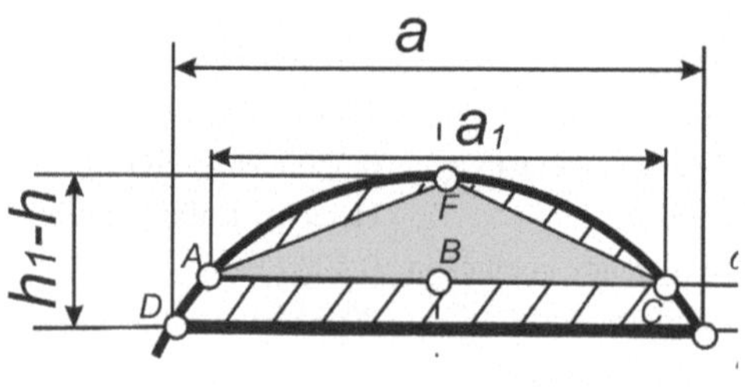

Fig. 6. Cross-section of the reduced cross-section of the screw channel of form IV (highlighted in gray)

4.2 Pressure Drop Models

The obtained data show a pressure drop in the channel of standard sizes from 10 MPa to the level of the environment without differences in the cross section. Standard Channel Pressure chart (Input Pressure 10 MPa) is shown in Fig. 7.

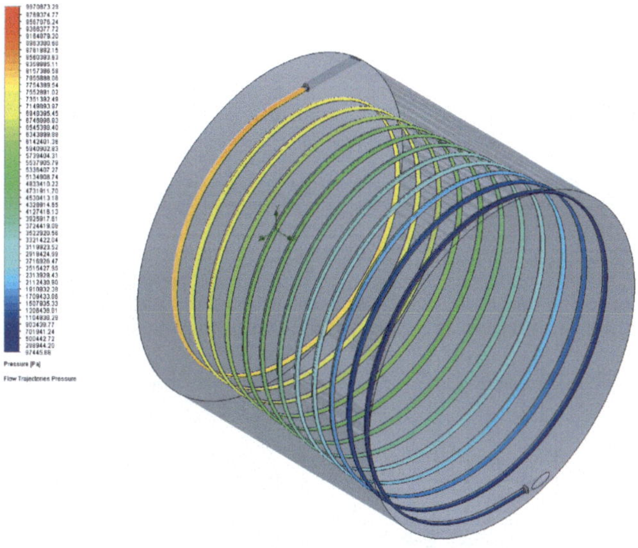

Fig. 7. Standard Channel Pressure chart (Input Pressure 10 MPa)

Reduced Channel Pressure chart (Input Pressure 10 MPa) is shown in Fig. 8.

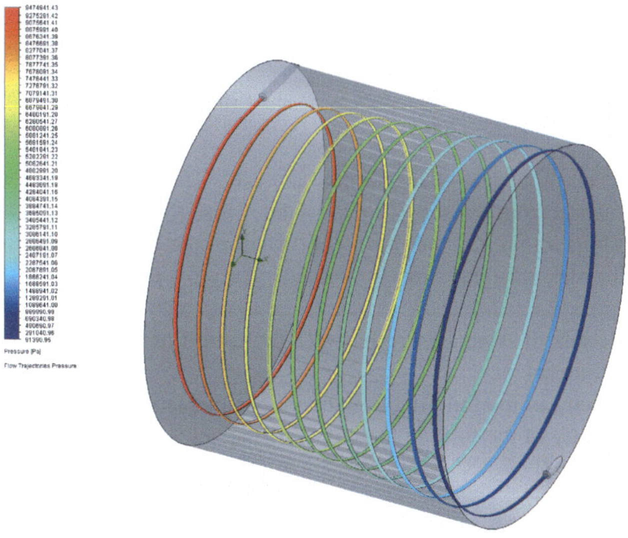

Fig. 8. Reduced Channel Pressure chart (Input Pressure 10 MPa)

4.3 Models of the Distribution of Fluid Speed in the Screw Channel

3 parts of the plot were selected for analysis. The first one is 360° after the beginning (transition from the straight part to the spiral) (Fig. 9, No. 1), the second one in the middle (Fig. 9, No. 2), the third - 360° before the end (the transition from the spiral part to the straight line) (Fig. 9, No. 3).

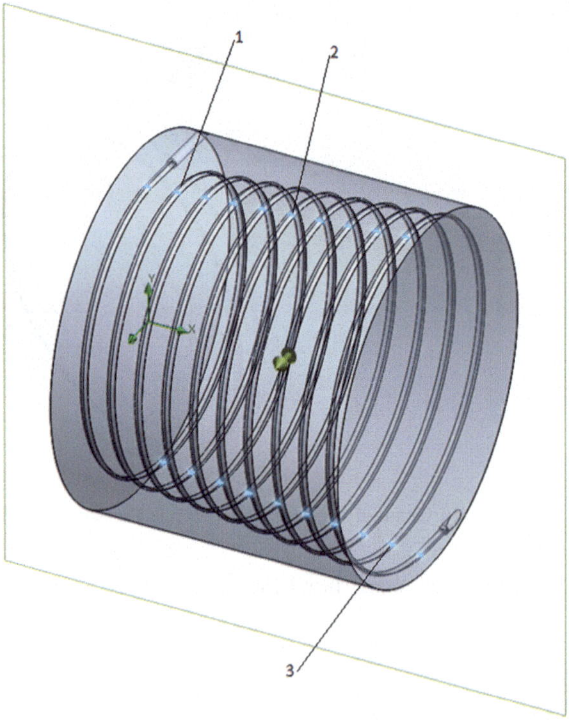

Fig. 9. Reduced Channel Pressure chart (Input Pressure 10 MPa)

3 parts of the plot were selected for analysis. The first one is 360° after the beginning (transition from the straight part to the spiral) (Fig. 9, No. 1), the second one in the middle (Fig. 9, No. 2), the third - 360° before the end (the transition from the spiral part to the straight line) (Fig. 9, No. 3).

4.3.1 The Models of Average Velocity Distribution of Fluid Speed in the Screw Channel

The average velocity of the fluid in the standard channel is from 50 to 55 m/s (Fig. 10). The average velocity of the fluid in the reduced channel is from 26 to 31 m/s (Fig. 11).

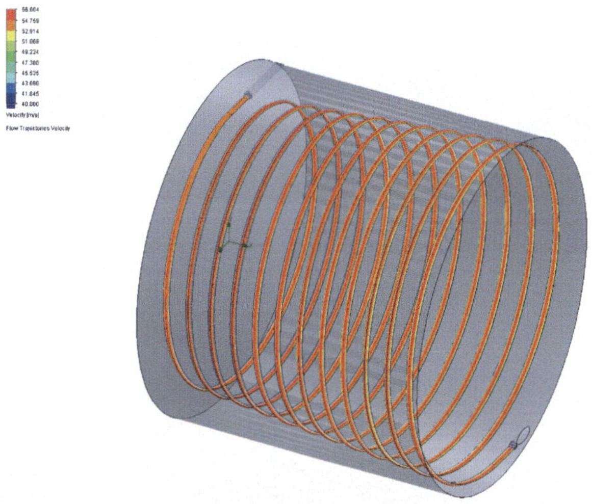

Fig. 10. Average velocity of the fluid in the standard channel

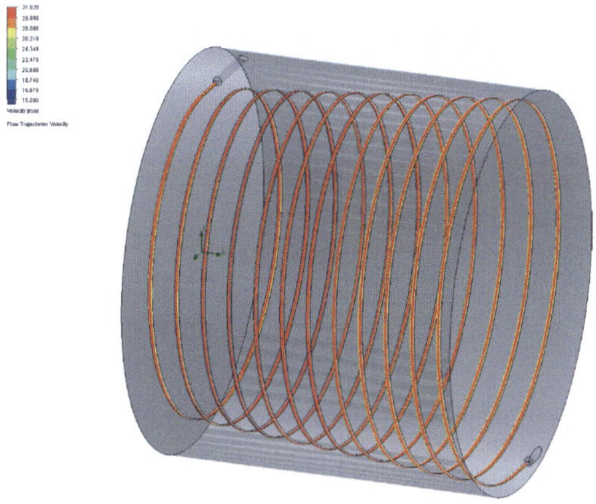

Fig. 11. Average velocity of the fluid in the reduced channel

4.3.2 The Models of Velocity Distribution of Fluid Speed in the Reduced Screw Channel

Figures 12, 13, 14 show the plot of the transverse velocity distribution in the reduced screw channel. The velocity distribution in the first thread (Fig. 12) is from 27 m/s to 30 m/s, the average section of the channel (Fig. 13) - from 23 m/s to 31 m/s, in the last turn (Fig. 14) - from 22 m/s to 31 m/s. The pattern of concentration of higher velocities in the center of the flow is closer to the end of the channel.

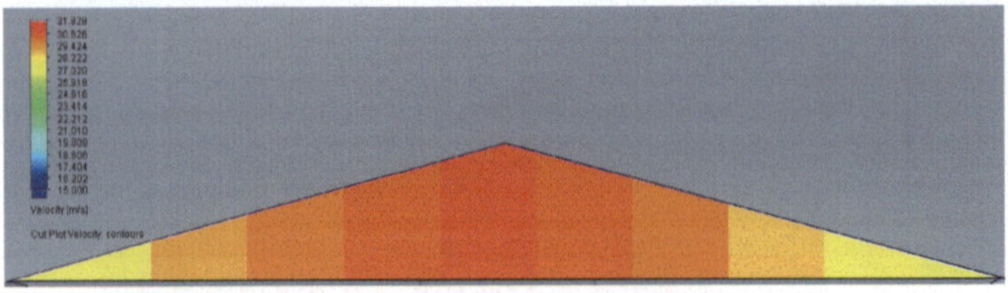

Fig. 12. Plot of the velocity distribution in the cross section of the helical channel on the first thread.

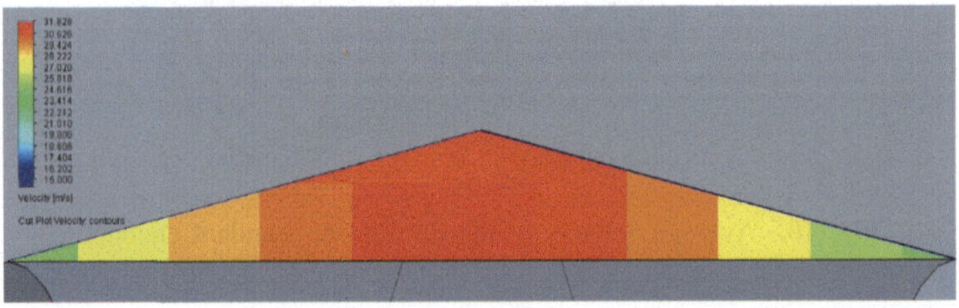

Fig. 13. Plot of the velocity distribution in the cross section of the helical channel on the middle thread.

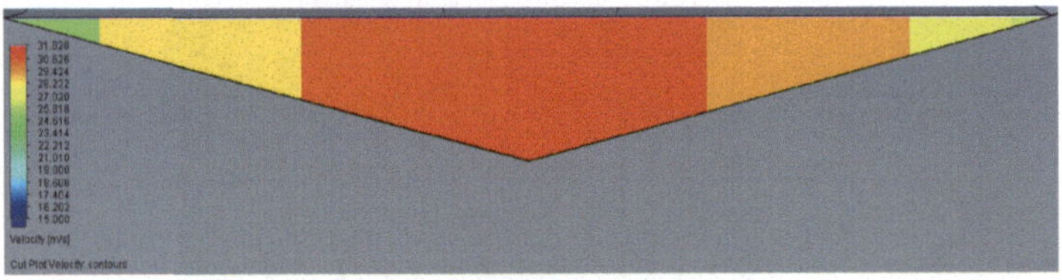

Fig. 14. Plot of the velocity distribution in the cross section of the helical channel on the thread before the end of channel.

5 Conclusions

1. The results showed that reducing of the section height from 0.46 mm to 0.134 mm and the width of the base from 1.6 mm to 1 mm led to a decrease in the speed of the drilling fluid in it 1.5–2 times.
2. The higher speed drilling fluid tends to concentrate in the center of the channel cross section.
3. Basing on the points 1.2 we can conclude that the drilling fluid leakage is reduced in the case the partially permeability of tool-joint if the gaps between the roots of the pin and crest of the box of the thread and vice versa reduces.

References

1. Chudyk, I.I.: Dovtrathidravlichnoienerhiipidchaspromyvanniasverdlovyny. Rozvidkatarozrobkanaftovykh i hazovykhrodovyshch **2**(31), 43–47 (2009). http://elar.nung.edu.ua/bitstream/123456789/4288/1/426p.pdf
2. Mandryk, O.M., Arkhypova, L.M., Pukish, A.V., Zelmanovych, A., Yakovlyuk, Kh.: Theoretical and methodological foundations of sustainable development of Geosystems. In: IOP Conference Series: Materials Science and Engineering. International Conference on Innovative Ideas in Science (IIS 2016), Baia Mare, Romania, 10–11 November. International Databases Web of Science, vol. 200, p. 012018. IOP Publishing (2017). http://iopscience.iop.org/article/10.1088/1757899X/200/1/012018
3. Shkitsa, L., Yatsyshyn, T., Lyakh, M., Sydorenko, O.: Means of atmospheric air pollution reduction during drilling wells. IOP Conf. Ser.: Mater. Sci. Eng. **144** (2016). https://doi.org/10.1088/1757-899X/144/1/012009
4. Borushchak, L., Borushchak, S., Onysko, O.: Influence of the technological gap value of the tool-joint tapered thread on the drilling mud flow rate in its screw coupling. Ukrainian J. Mech. Eng. Mater. Sci **3**(2), 24–31 (2017)
5. Xu, H., Shi, T., Zhang, Z., Shi, B.: Loading and contact stress analysis on the thread teeth in tubing and casing premium threaded connection. Math. Probl. Eng. **2014,** 11. https://www.hindawi.com/journals/mpe/2014/287076/. Article Id 287076
6. Sugino, M., Nakamura, K., Yamaguchi, S., Daly, D., Briquet, G., Verger, E.: Development of an innovative high-performance premium threaded connection for OCTG. In: Proceedings of the Offshore Technology Conference (OTC 2010), pp. 1883–1891, May 2010
7. Wang, Y., Xia, B., Wang, Z., Chai, C.: Model of a new joint thread for a drilling tool and its stress analysis used in a slim borehole. Mech. Sci. **7**, 189–200 (2016)
8. Luo, S., Wu, S.: Effect of stress distribution on the tool joint failure of internal and external upset drill pipes. Mater. Des. (1980–2015) **52**, 308–314 (2015)
9. Vasylyshyn, V.: Ways to improve the reliability and tightness of casing strings. Int. Sci. J. Internauka **2**(24), 134–135 (2017). 1т
10. Saruev, A.L., Saruev, L.A., Vasenin, S.S.: Drill pipe threaded nipple connection design development. In: IOP Conference Series: Earth and Environmental Science, vol. 27, p. 012056 (2015). https://doi.org/10.1088/1755-1315/27/1/012056
11. Kopei, V., Onysko, O., Panchuk, V.: The application of the uncorrected tool with a negative rake angle for tapered thread turning. In: Conference Proceedings DSME 2019. 2-th International Advances in Design, Simulation and Manufacturing II The Innovation Change. Book of Abstracts-2019. Lecture Notes is Mechanical Engineering, pp. 149–158. https://doi.org/10.1007/978-3-030-22365-6_15

Trend Analysis of Production and Distribution of Wing Pumps: A Case Study of Firm PPT-Trstenik (Serbia)

Milutin Živković[1(✉)], Predrag Dašić[1], Milan Radosavljević[2], and Maja Anđelković[3]

[1] High Technical Mechanical School of Professional Studies, 37240 Trstenik, Serbia
milutinzivkovicts@gmail.com
[2] Faculty of Business Studies and Law (FPSP), 11070 Novi Beograd, Serbia
[3] Faculty of Information Technology and Engineering (FITI), 11070 Novi Beograd, Serbia

Abstract. Considering the current state of production and marketing of hydraulic devices and systems in Serbia and world-renowned manufacturers, significant indicators and trends can be observed. Modern structural and production solutions for hydraulic devices are based on the application of new technologies and the introduction of new production capacities, processes and materials in the process of making or revitalizing existing devices and assemblies. As a significant activity of mechanical engineering, the hydraulics industry is also a reference of the level of development of the national economy of a country and the basis for profit. This paper presents the trend analysis and statistical modelling of production and distribution of wing pumps of the PPT firm from Trstenik (Serbia). Also, the reasons for the loss of the market and the techno-economic indicators of the production and marketing of the once famous manufacturer of PPT Petoletka wing pumps from Trstenik (Serbia) are given. All the problems and reasons for losing the market can be observed on this problem, although the situation with the production and placement of other devices is no better.

Keywords: Statistical analysis · New technologies · Production · Distribution · Wing Pump (WP)

1 Introduction

The twenty-first century is called by the experts and professional public a century of knowledge [1–3]. Modern scientific disciplines and technologies have therefore become the most important drivers of social development, in the true sense of the word. Therefore, it can be said that the development of new products, processes and services will be based on scientific principles, which implies the development and application of new technologies, as these would be the main drivers of economic progress. New jobs will be created in the modern and high technology sector by implementing knowledge in modern products as well as in the services sector [1]. Technological development,

I. Karabegović (Ed.): NT 2020, LNNS 128, pp. 730–740, 2020.
https://doi.org/10.1007/978-3-030-46817-0_83

new trends in production, process automation as well as complete systems will influence significant changes in the number of employees in certain branches of industry. With modern technologies and products, the costs of direct live work (machine time) will decrease, and the tendency of increasing the participation of intellectual-indirect work will decrease: development time and appearance on the market, preparatory times, design time, design innovations and increasingly significant implementation of the new knowledge etc. In this sense, all experts attach great importance to the introduction of new energy sources, and also pay considerable attention to environmental issues. The constant question is, how to survive in a demanding market? An example is the case study relating to the production of wing pumps by PPT from Trstenik (Serbia). Namely, for the last 40 years they have been successfully manufactured in this once famous factory of hydraulic devices and systems. The authors of the paper, using commercial results, will point out the reasons and problems in the production and placement of these pumps [4–7] and present the trend analysis and statistical modelling of the production and distribution of wing pumps of a PPT firm from Trstenik (Serbia).

2 Trend of the Development and Production of Pumps in the World

The experience of industrialized countries shows that knowledge directly influences the development of products used in the application of new materials, technologies and production techniques [1–3, 8]. It is also the basis for successful overall social development and is directly linked to education, especially higher education. That is why colleges and institutes are the right place to spread new knowledge and visions of the future. However, only a developed society cares specifically for knowledge and education. Therefore, Alvin Toffler believes that if we take technology as a great engine, we can understand knowledge as its fuel [9]. Increasingly, experts can be heard talking about the so-called "Industry 4.0" and the ways in which this phenomenon affects the life of the modern man. It may be more accurate to say that this is the fourth phase of the industrial revolution and the impact of modern technologies on industrial production. If the first phase was the transition from craft to factory production, the second introduction of steel and electrification, the third transformation into digital technology in the second half of the last century - the fourth industrialization refers to the introduction of modern telecommunications, the processing and analysis of huge amounts of data, and the increasing impact the so-called Internet of Things (IoT) [3] to classic industries. Industry 4.0 brings us the so-called "smart factory", which has expressed the involvement of digitalization, automation and artificial intelligence (AI), all with the aim of optimizing the procurement, production and marketing of products that should make sense of the word, to be of better quality than before. In a word, it is about the introduction of information technology (IT) into industrial production, solutions such as connected devices, internet communication and artificial intelligence. One of the main problems when it comes to introducing new technological solutions is the possibility of losing a job, i.e. fear that machines will replace us. Yes, robots will certainly reduce the number of direct manufacturing employees, but new jobs and new professions will emerge that will require human creativity and the ability to decide on things that are not based on pure mathematics. All actors are required to work harder and to educate themselves continuously, because that is the nature of the time in which we live. The industry is always based on reliable and proven solutions, but there is a belief that it is difficult and slow to change. On the other hand, the previous three industrial revolutions have faced similar challenges, so the human race has been able to cope with them and move forward.

The market for the production and application of hydraulic systems and devices is in direct proportion to the level of technical, technological and general economic development of each country [5–7]. Their modern development is based on increasingly demanding modes of operation, both in terms of quality and reliability, and in view of all their advantages. Also, energy efficiency improvements are taken into account, i.e. an increase in overall efficiency as well as a lifetime of the device from a marked increase in operational reliability. Improved environmental performance, noise and vibration reduction through optimization of operating modes and ergonomic management performance have also been neglected, leading to reduced production and maintenance costs. These are the basic parameters of designing and selecting the optimal solution for a hydraulic system. In addition to the basic advantages of a hydraulic device, there are also disadvantages in the power transformation process.

Among the most influential are the high technical and technological complexity, which requires specific production conditions and technologies (procedures, treatments, materials), starting with geometric complexity and accuracy, types and quality of materials, topography of treated surfaces and significantly increased wear. All in all, the disadvantages of these devices are far below the listed advantages of the devices and can minimize them and reduce them to acceptable conditions of exploitation, maintenance and above all by increasing the quality of production with the introduction of new technologies. Figure 1 shows the historical course of development of hydraulic pumps/motors, and this also applies to the development of other hydraulic devices [10].

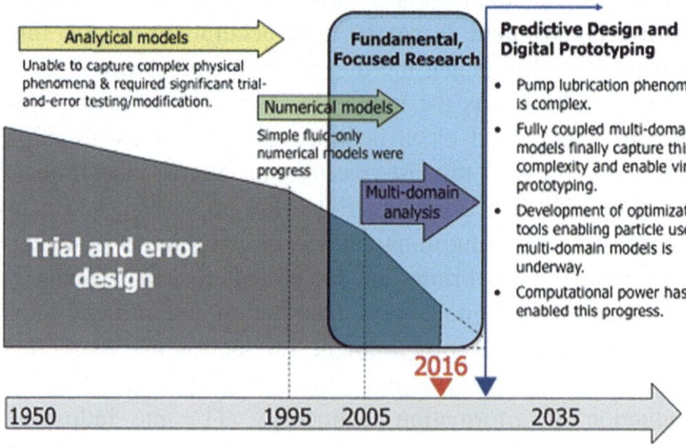

Fig. 1. New directions in pump design [10]

Because of all this, there are more methods and methodologies for designing components of technical systems [11–17] and determining their reliability [18].

3 Materials and Methods

The trend analysis was performed for the production and placement of wing pumps manufactured by PPT from Trstenik (Serbia), licensed by ZF Group from Friedrichshafen (Germany). Figure 2 shows its appearance and components [4], which are primarily used in power steering systems (PSS). Table 1 shows the machines on which its vital parts are made, with the year it was purchased and included in the manufacturing process [4].

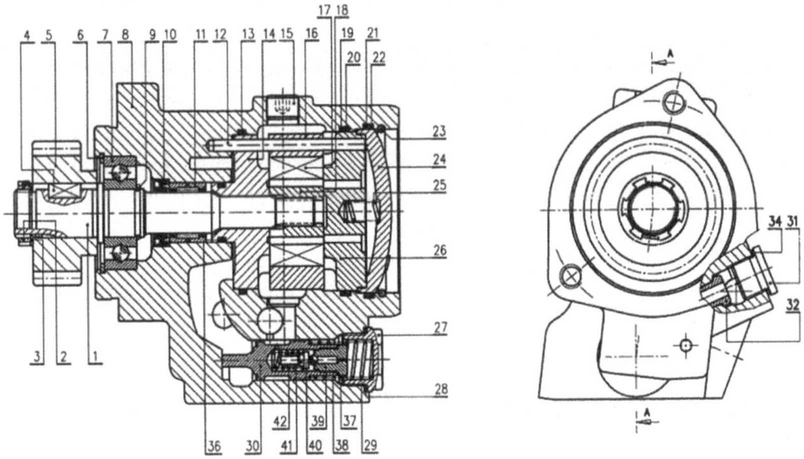

Fig. 2. Appearance of a wing pump produced in firm PPT from Trstenik (Serbia)

Table 1. Machine tools involved in wing pump production and years of supply

No.	Machine type and designation		Years of supply
1.1	Lathes	PIMET 160 (revolver lathe)	1977
1.2		FAR 250	1976
1.3		PA–B22/750	1974
1.4		NE–480 (NC)	1976
2.1	Milling machines	HITAČI SEIKI VK 5	1988
2.2		ALG 200	1977
2.3		KOEPFER 151 (tooth making)	1977

(*continued*)

Table 1. (*continued*)

No.	Machine type and designation		Years of supply
3.1	Grinders	BRD 60 (straight grinding)	1983
3.2		B15U/700	1988
3.3		D7B (straight grinding)	1977
3.4		PROFIMAT 412 (straight grinding)	1988
3.5		KS–2–8 (copier grinder)	1977
3.6		CNC S40–4 (round grinding)	1990
4.1	Drills	BD 4	1974
4.2		G 25	1965
5	Purging machines	16 DEH-2000	1978
6	Testing device	purpose designed	1982

Data on values of production and distribution of wing pumps of the PPT firm from Trstenik (Serbia) has been retrieved from the documentation [4] and papers [5–7], with a certain calculations from the part of the authors. For the trend analysis, we used the following parameters: annual growth rate (AGR), average annual growth rate (AAGR), compound annual growth rate (CAGR) and cumulative growth index (CGI), which is described in the papers [19–22].

For the calculation of descriptive statistics parameters and approximation of data about production and sale of wing pumps we used software package CoPRMTV, which is described in the papers [19–22] and software system MS-Excel.

4 Results and Discussions

In Table 2 data are given about production and sold of wing pumps (pieces) in firm PPT-Trstenik (Serbia) for the period 1994–2017 [4–7], with the calculation of trend analysis parameters (AGR and CGI) by the authors.

Table 2. Overview of produced and sold of wing pumps (pieces) in firm PPT-Trstenik (Serbia) for the period 1994–2017

Year	Production of wing pumps	Cumulative	AGR	CGI	Sold wing pumps	Cumulative	AGR	CGI
1994	2670	2670	–	100.00	2482	2482	–	100.00
1995	5722	8392	114.31	214.31	5566	8048	124.25	224.25
1996	5850	14242	2.24	219.10	5275	13323	−5.23	212.53
1997	10687	24929	82.68	400.26	10935	24258	107.30	440.57
1998	13946	38875	30.49	522.32	14667	38925	34.13	590.93
1999	17922	56797	28.51	671.24	16000	54925	9.09	644.64
2000	22148	78945	23.58	829.51	20500	75425	28.13	825.95
2001	27200	106145	22.81	1018.73	27200	102625	32.68	1095.89
2002	29906	136051	9.95	1120.07	29846	132471	9.73	1202.50

(*continued*)

Table 2. (*continued*)

Year	Production of wing pumps	Cumulative	AGR	CGI	Sold wing pumps	Cumulative	AGR	CGI
2003	22650	158701	−24.26	848.31	23328	155799	−21.84	939.89
2004	27271	185972	20.40	1021.39	27254	183053	16.83	1098.07
2005	14344	200316	−47.40	537.23	14279	197332	−47.61	575.30
2006	7445	207761	−48.10	278.84	7445	204777	−47.86	299.96
2007	11184	218945	50.22	418.88	11038	215815	48.26	444.72
2008	2430	221375	−78.27	91.01	2430	218245	−77.99	97.90
2009	1403	222778	−42.26	52.55	1403	219648	−42.26	56.53
2010	836	223614	−40.41	31.31	795	220443	−43.34	32.03
2011	943	224557	12.80	35.32	918	221361	15.47	36.99
2012	670	225227	−28.95	25.09	663	222024	−27.78	26.71
2013	610	225837	−8.96	22.85	606	222630	−8.60	24.42
2014	495	226332	−18.85	18.54	501	223131	−17.33	20.19
2015	865	227197	74.75	32.40	856	223987	70.86	34.49
2016	595	227792	−31.21	22.28	582	224569	−32.01	23.45
2017	1100	228892	84.87	41.20	1096	225665	88.32	44.16

Figure 3 shows the trend of production and sold of wing pumps (pieces) in the firm PPT-Trstenik (Serbia) for the period 1994–2017.

Highest growth of AGR rate for production of wing pumps (PWP) (pieces) in firm PPT-Trstenik (Serbia) was by 114.31 in 1995 and then 84.87 in 2017, 82.68 in 1997 and etc., while the highest fall was by −78.27 in 2008 and then −48.10 in 2006, −47.40 in 2005 and etc. (Table 2 and Fig. 3). Production of wing pumps (pieces) has a decreasing trend with CGI = 41.20%, AAGR = 8.21% and CAGR = −3.78% per year in 2017 compared to 1994.

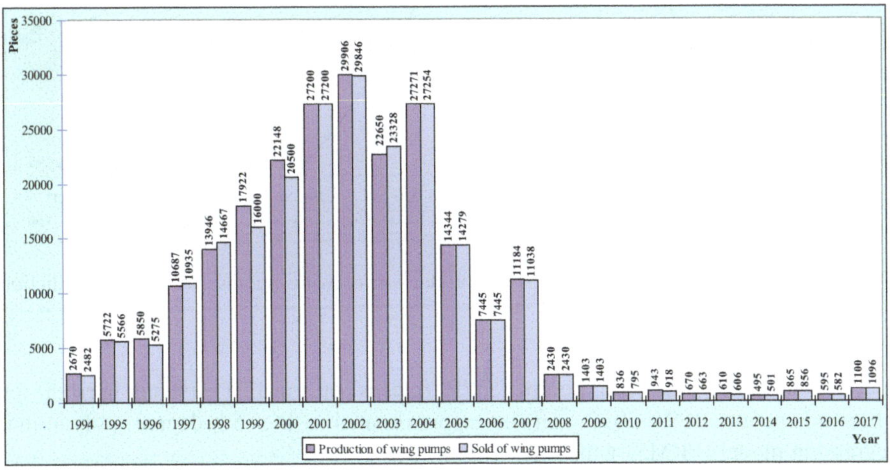

Fig. 3. Trend analysis of production and sold of wing pumps (pieces) in firm PPT-Trstenik (Serbia) for the period 1994–2017

Highest growth of AGR rate for sold wing pumps (SWP) (pieces) in firm PPT-Trstenik (Serbia) was by 124.25 in 1995 and then 107.30 in 1997, 88.32 in 2017 and etc., while the highest fall was by −77.99 in 2008 and then −47.86 in 2006, −47.61 in 2005 and etc. (Table 2 and Fig. 3). Production of wing pumps (pieces) has a decreasing trend with CGI = 44.16%, AAGR = 9.27% and CAGR = −3.49% per year in 2017 compared to 1994.

The data about production of wing pumps (PWP) (pieces) in firm PPT-Trstenik (Serbia) for the observed period 1994–2017 can be approximated using 5th-degree polynomial regression model (PRM5) which has the form (Fig. 4):

$$\begin{aligned} PWP = {} & 11238112211.064 \cdot 10^6 - 28015352811465.5 \cdot y + 27935498329.3166 \cdot y^2 \\ & - 13927858.92872 \cdot y^3 + 3472.00152 \cdot y^4 - 0.3462 \cdot y^5 \end{aligned} \tag{1}$$

where:
y - year and
PWP - production of wing pumps (pieces).

Performance PRM5 (Eq. 2) are: coefficient of correlation is R = 0.95420, coefficient of determination is $R^2 = 0.93105$ and adjustment coefficient of determination is $AdjR^2 = 0.91190$.

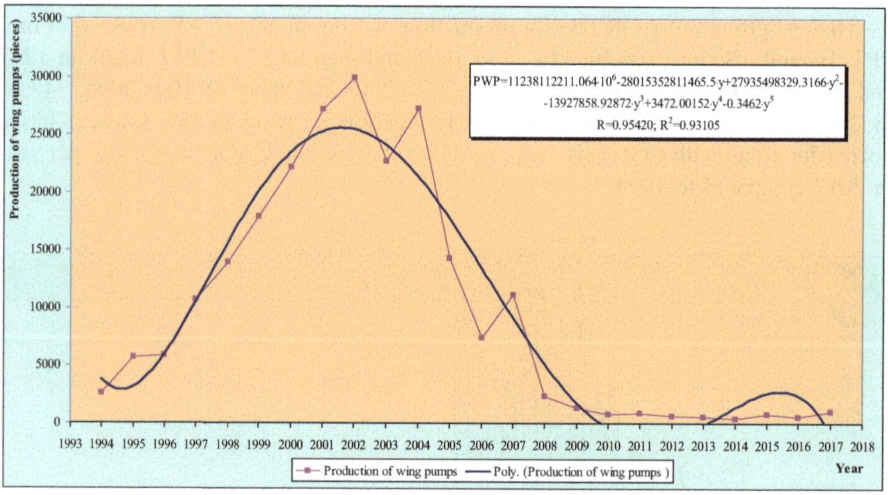

Fig. 4. Graphical representation of data of production of wing pumps (PWP) and prediction value of PRM5

The data about sold wing pumps (SWP) (pieces) in firm PPT-Trstenik (Serbia) for observed period 1994–2017 can be approximated using 5th-degree polynomial regression model (PRM5) which has the form (Fig. 5):

$$\begin{aligned} SWP = {} & 11183462208.2535 \cdot 10^6 - 27878618079153 \cdot y + 27798656712.7476 \cdot y^2 \\ & - 13859386.09362 \cdot y^3 + 3454.87066 \cdot y^4 - 0.34449 \cdot y^5 \end{aligned} \tag{2}$$

where:
y - year and
SWP - sold of wing pumps (pieces).

Performance PRM5 (Eq. 2) are: coefficient of correlation is R = 0.96113, coefficient of determination is $R^2 = 0.92377$ and adjustment coefficient of determination is $AdjR^2 = 0.90260$.

As stated above, technical and technological, functionally geometric and structural performance of the device, as well as production conditions (technological level): fabrication, control, assembly and testing make hydraulic devices, their vital assemblies and constituents complex, in comparison to devices in other fields of mechanical engineering. The focus in the production of these devices, in Prva Petoletka from Trstenik, is on licensed cooperation realized until 1980 with leading West manufacturers. They have been successfully manufactured for almost 40 years, which exceeds the existing domestic market needs and the needs of our country in terms of production potential and range. For these reasons, as well as the lack of competitiveness to successful manufacturers, it is in evident problems. They are reflected in the evident technological setbacks and the operating and functional parameters of the manufactured devices are lower, compared to the modern devices of the leading manufacturers. They relate to design, price, better technical characteristics resulting from the application of new materials and technologies. It is the technological lag as evident in the production, which basically consists of conventional processes, materials and treatments.

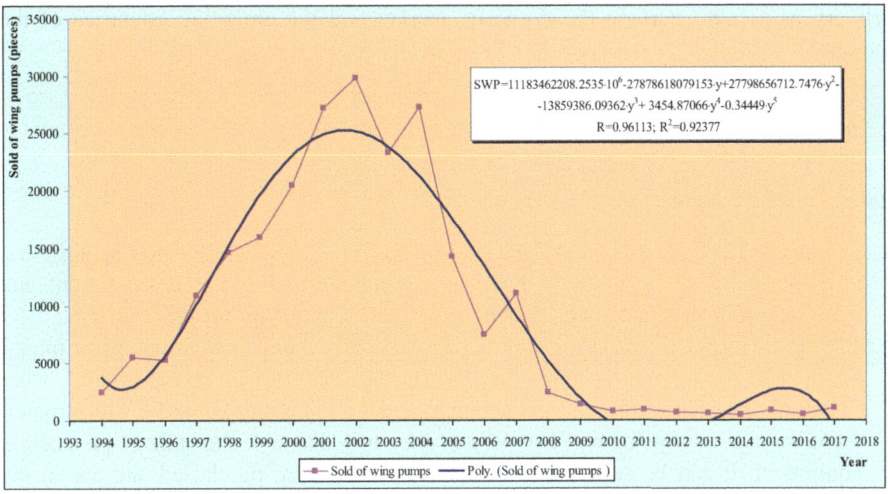

Fig. 5. Graphical representation of data of production of wing pumps (SWP) and prediction value of PRM5

In particular, significant structural and technological differences between the existing devices for the transmission and transformation of energy and power (pumps, hydromotors) with respect to the latest generation of leading European and world manufacturers should be noted: "LINDE", "ZF", "BOCH REXROT", "DAMFOS", "POCLAIN" and etc. It is also interesting to note that in 1984 a study was done on the site of PPT Trstenik (by the number of devices in the production program) produced in Trstenik. It was about the production of pumps/motors ((gears (external and internal toothing)), wing, piston (axial, regular, radial, all types of cylinders, manifolds and valves of flow and pressure), auxiliary devices (filters, fillers, vents, coolers, rheostats, accumulators, fittings, flexible hoses, gaskets, etc. in relation to world competition. Based on the analysis of 140 manufacturers of hydraulic equipment, then the most prominent world manufacturer Rexroth produced 38 types of devices and PPT from Trstenik produced 42 types of devices. The rapid development and production of these devices is a consequence of the accelerated at that time evident development of the then great Yugoslavia. It is certain that not all of these devices were mass-produced, but there was design documentation for them and all prototype research and testing was carried out. All this was confirmed by the quality reports, and, for the stage of development at the time of the then world-wide factory. These were the competing products.

The problem of procurement of starting materials, semi-finished products, finished goods, technological obsolescence of production, measuring and control equipment as well as compliance with the delivery time forces the manufacturer to take into account the dynamics of procurement in order to respect the delivery time of the finished devices. Production is now realized by engaging a smaller number of people, according to production needs, and Table 2 gives the business results. According to the available data, the export of these pumps was reduced to delivery to KAMAZ company as well as a spare part for the ongoing maintenance of the PS system of mobile machines, i.e. delivery as a spare part on the domestic market and the emerging countries of the former Yugoslavia.

5 Conclusion

Based on the foregoing it can be concluded:

- Competitiveness is associated with investment in research and development, both o new products, by innovation of existing products, and in the development and deployment of new machines, technologies and manufacturing processes;
- The best example is shown in the case o the wing pump, which has been manufactured by PPT from Trstenik (Serbia) in the quantities given in Table 2, from which all the reasons for market loss and technological lag behind leading manufacturers can be considered. This is best shown in Table 1, the state of production equipment. It can be seen that the machines have been purchased and are in the production process when batch production started;
- The production of other devices in PPT-Trstenik (Serbia) is similar.

References

1. Dašić, P., Milosavljević, M.: Nano-materials and nano-technologies: a new era of digital devices. In: Proceedings of the 3rd International Scientific Conference Agribusiness MAK-2016: "Western Balkans and Europe", Kopaonik, Srbija, 29–30 January 2016, pp. 1–40. Etno Centar GEGULA, Kraljevo (2016). (in Serbia)
2. Kereluik, K., Mishra, P., Fahnoe, C., Terry, L.: What knowledge is of most worth: teacher knowledge for 21st century learning. J. Digit. Learn. Teach. Educ. **29**(4), 127–140 (2013)
3. Bradley, D., Russell, D., Ferguson, I., Isaacs, J., MacLeod, A., White, R.: The Internet of Things – the future or the end of mechatronics. Mechatronics **27**, 57–74 (2015)
4. Commercially technical documentation for pumps in group 3115 and ZPB (higher pressures). PPT-Hidraulika a.d. (2018). (in Serbian)
5. Živković, M.: Study on improving the design characteristics of the wing pumps. Firm PPT, Trstenik, Serbia, p. 35 (2017). (in Serbian)
6. Živković, M., Dašić, P., Petrović, Z.: Influence of new technologies on higher energy efficiency of hydrostatic devices and systems. Lecture Notes in Networks and Systems (LNNS), vol. 42, pp. 386–396 (2019)
7. Živković, M., Dašić, P., Pravdić, P.: Analysis of the comparative advantages of gear pumps indicators. Lecture Notes in Networks and Systems (LNNS), vol. 76, pp. 158–164 (2020)
8. Lonnie, J.L., Lanke, E., Alles, P.: Estimating the impact (energy, emission and economics) of the US fluid power industry. Report to US Department of Energy DE-AC05-00OR22725 and ORNL/TM-2011/14. Oak Ridge National Laboratory and the National Fluid Power Association, Oak Ridge, Tennessee, USA, p. 45 (2012)
9. Toffler, A.: The Third Wave (Book Review), vol. I, no. 2, pp. 115–143. Journal of Managerial Sciences
10. Ivantysynova, M.: Technologies and innovations for hydraulic pumps. In: Proceedings of the 10th International Fluid Power Conference, Dresden, Germany, Group 3 - Pumps, March 2016, pp. 135–138 (2016)
11. Dašić, P.: Response Surface Methodology: Selected Scientific-Professional Papers. SaTCIP Publisher Ltd., Vrnjačka Banja (2019). 312 p. (in Serbian)
12. Formato, A., Guida, D., Ianniello, D., Villecco, F., Lenza, T.L., Pellegrino, A.: Design of delivery valve for hydraulic pumps. Machines **6**(4), 44 (2018)
13. Formato, G., Romano, R., Formato, A., Sorvari, J., Koiranen, T., Pellegrino, A., Villecco, F.: Fluid-structure interaction modelling applied to peristaltic pump flow simulations. Machines **7**(3), 50 (2019)
14. Pappalardo, C.M., Guida, D.: Dynamic analysis of planar rigid multibody systems modelled using natural absolute coordinates. Appl. Comput. Mech. **12**(1), 73–110 (2018)
15. Pappalardo, C.M., Guida, D.: On the computational methods for the dynamic analysis of rigid multibody mechanical systems. Machines **6**(2), 20 (2018)
16. Pappalardo, C.M., Guida, D.: Use of the adjoint method in the optimal control problem for the mechanical vibrations of nonlinear systems. Machines **6**(2), 19 (2018)
17. Zhou, X., Liang, X.M., Zhao, G., Su, Y., Wang, Y.: A new computational fluid dynamics method for in-depth investigation of flow dynamics in roller pump systems. Artif. Organs **38**(7), E106–E117 (2014)
18. Dašić, P.: Reliability of Technical Systems: Selected Scientific-Professional Papers. SaTCIP Publisher Ltd., Vrnjačka Banja (2019). 308 p. (in Serbian)
19. Dašić, P.: Application of polynomial regression models for approximation of time series. J. Econ. Manage. Based New Technol. **1**(2), 81–160 (2012)

20. Dašić, P.: Statistical analysis and modelling of global competitiveness index (GCI) of Serbia. J. Res. Dev. Mech. Ind. **10**(4) (2018)
21. Tošović, R., Dašić, P., Ristović, I.: Sustainable use of metallic mineral resources of Serbia from an environmental perspective. Environ. Eng. Manage. J. **15**(9), 2075–2084 (2016)
22. Turmanidze, R., Dašić, P., Popkhadze, G.: Analysis of some techno-economical indicators of Georgia. J. Res. Dev. Mech. Ind. **11**(4) (2019)

New Technologies in Agriculture, Ecology, Chemical Processes

Carbon Based Aerogels and Xerogels for Removing of Toxic Organic Compounds

Amra Bratovcic[1(✉)] and Irena Petrinic[2]

[1] Department of Physical Chemistry and Electrochemistry, Faculty of Technology, University of Tuzla, Tuzla, Bosnia and Herzegovina
amra.bratovcic@untz.ba

[2] Faculty of Chemistry and Chemical Engineering, University of Maribor, Maribor, Slovenia

Abstract. In recent years with the rapidly growing industries and population, water pollution has emerged as major challenge for scientific community that demands an intense and real-world solution. Carbon based aerogels and xerogels are exceptional classes of porous materials with a number of excellent physicochemical properties such as low density, high porosity, high surface area and adjustable surface chemistry. These materials possess an extraordinary adsorption capacity for the removal of toxic organic compounds from the environment. This article aims to give an overview regarding the capacity adsorption of different organic solvents, endocrine disrupting compounds, painkillers, antibiotics and dyes on the carbon aerogels and xerogels in water solutions.

Keywords: Carbon aerogel · Carbon xerogel · Adsorption · Organic compounds

1 Introduction

In the last twenty years, several parts of the world have witnessed the enrichment of toxic organic chemicals in the water which could seriously damage human health. The textile dyes in the wastewater are mostly composed of toxic organic materials, while the oil spills contain highly toxic aromatic hydrocarbons which could be decomposed by microorganisms via consuming oxygen in the water and lead to the death of organisms and thus destroys the ecological balance of the water body [1, 2].

Wastewater treatment technologies have become very important due to the dangers and health risks posed from water pollution. Several techniques have been adopted to remove toxic organic compounds from water such as membrane filtrations [3], adsorption [4], biological treatment [5], and chemical treatment [6]. Among various methods employed to address the issues of toxic organic compounds, adsorption by porous materials was considered as one of the most efficient techniques without any secondary pollution. It often allows to recycle or reuse of adsorbents for several cycles.

Based on the numerous findings in relation to aerogel properties, the material has attracted a tremendous attention of many researchers. Aerogel is a general terminology referring to any material derived from organic, inorganic or hybrid molecular precursors that are normally prepared by a sol-gel process and an appropriate drying

I. Karabegović (Ed.): NT 2020, LNNS 128, pp. 743–749, 2020.
https://doi.org/10.1007/978-3-030-46817-0_84

technology with which the three-dimensional and the highly porous network is conserved. Almost all aerogels are prepared by a wet chemical synthesis approach, a sol-gel reaction, but through different starting precursors, operating and provision requirements [7]. It is well known that the sol-gel process allows a good control of composition, texture and structural characteristics of the final products.

When gels are prepared by hydrolysis and condensation of metal or semimetal alkoxides, the pore liquid mainly consists of water and/or alcohols. When the pore liquid is replaced by air without decisively altering the network structure or the volume of the gel body, aerogels are obtained (or cryogels when the pore liquid is removed by freeze-drying). A xerogel is formed upon conventional drying of the wet gels, that is, by increase in temperature or decrease in pressure with concomitant large shrinkage (and mostly destruction) of the initially uniform gel body [7]. Organic xerogels which are made by removing the solvent from the gel structure by drying conventionally with nitrogen or air under normal conditions [8]. The capillary pressure of the liquid inside a pore leads to the collapse of the gel-network during oven drying, which results in a dense polymeric structure called xerogel [9]. This structure is carbonized in an inert atmosphere (pyrolysis) to produce the corresponding carbogel.

The aim of this paper is to explain the difference between carbon aerogel and xerogel as well as to compare their adsorption capacity for removing of toxic organic compounds.

1.1 Methods of Drying

The method of drying the aqueous organic gel is one of the key factors for the synthesis of a mesoporous carbon. Indeed, it is necessary to prevent the collapse of the 3D network which may occur as result of the strong capillary forces of water, and this may be accomplished by solvent exchange. Drying of the gel is the final and most critical step in the production process of aerogels. Drying is ruled by capillary pressure except for supercritical drying and freeze-drying [10], causing consequent shrinking and possible fracture due to the small pore sizes and the resulting capillary tension. As a result, two different drying processes are most often used: ambient pressure drying [10–14], here capillary tension cannot be avoided, and supercritical drying (using CO_2, acetone, or ethanol) [15–17] where the pore liquid is removed above the critical temperature and pressure to avoid capillary tension.

The liquid phase is replaced with air using either freeze-drying or supercritical carbon dioxide (CO_2) drying techniques. The latter preserves the original structure of the gel, which leads to the creation of large surface areas.

Liang and co-workers carried out a comparison of the basic physical properties between a resorcinol - formaldehyde (RF) aerogel from supercritical acetone drying and an RF aerogel from supercritical CO_2 drying. They found that the shrinkage and density of the RF aerogel from supercritical acetone drying are larger than those of the RF aerogel with supercritical CO_2 drying. The experimental use of an initial nitrogen pressure shows that the gel texture can be preserved using supercritical acetone drying [18].

Aerogels can be obtained in various sizes and shapes, ranging from beads to disks to monoliths, by the removal of the liquid from a corresponding gel [19].

Freeze-drying is a commonly used operation which can remove water at a low temperature under vacuum where the action of capillary forces is avoided so that the materials' microstructure can be largely preserved, leading to a high specific surface area of the resultant aerogel. Additionally, in the following pyrolysis process, the removal of non-carbon atoms (O, H) in the form of gases will create numerous micropores on the carbon fibers [20]. This will further increase the specific surface area of carbon aerogel (CA) and lower its density, giving rise to an improved toxic organic solution absorption capacity [21].

2 Removing of Toxic Organic Compounds by Adsorption

2.1 Toxic Organic Compounds

Organic pollutants include pesticides, fertilizers, hydrocarbons, phenols, plasticizers, biphenyls, detergents, oils, greases, pharmaceuticals, proteins and carbohydrates. Among all toxic organic compounds, aromatic compounds, pesticides and herbicides, dyes [22] and plastics [23] in these studies were considered. The author has highlight that many organic compounds are toxic and have endocrine disrupting properties. These organic compounds when appear in wastewater becomes a serious hazard to ecological and human health due to increasing sewage discharge generated from industrial manufacture and agriculture irrigation like dyes, pesticides, plasticizers.

In the class of aromatic compounds, polycyclic aromatic hydrocarbons (PAHs) are mostly colourless, white, or pale yellow solids. They are a ubiquitous group of several hundred chemically related compounds, environmentally persistent with various structures and varied toxicity. PAHs are commonly detected in air, soil, and water. Bisphenol A (BPA), estradiol, and estrone are typical emerging pollutants commonly found in water resources and effluents from wastewater treatment plants.

Monochlorophenols and dichlorophenols serve as intermediates in the production of pesticides and are also used as additives to inhibit microbial growth in a wide array of products such as adhesives, oils, textiles, as well as pharmaceutical products.

More than 50% of commercial dyes and pigments belong to azo compounds which are most used in the textile industry for the dyeing of cotton, paper, silk, leather, wool, and dyeing of poly-amide, acrylic, polyolefin, polyester, viscose and cellulose acetate fibers.

Polyethylene terephthalate (PET), low-density polyethylene (LDPE), high-density polyethylene (HDPE), polyvinyl chloride (PVC), polypropylene (PP) and polystyrene (PS) are the most widespread and studied plastic materials. Other plastics include acrylic, nylon, polycarbonate, polyurethanes, polylactic acid, and other biodegradable plastics.

2.2 Adsorption of Organics by Carbon Aerogels

Capability of all solid substances to attract to their surfaces molecules of gases or solutions with which they are in contact is called adsorption.

A common and effective method for the removal of various contaminants from water is physical adsorption, which involves competing attractive and repulsive forces between the adsorbate (contaminant) and adsorbent molecules. Aerogels can easily meet the adsorption requirements and their application in the removal of the aforementioned toxic organic pollutants is very promising. Carbon aerogels are mostly obtained by pyrolysis, at temperatures above 500 °C, of organic aerogels. Carbon aerogels are mesoporous materials with low density, high surface area, and a porous interconnected network, which are ideal properties for an effective adsorbent. The most extensively studied organic aerogels are the resorcinol-formaldehyde and melamine-formaldehyde aerogels. Organic aerogels have some common properties despite the structural differences originating from different precursors used during preparation. Specifically, they have pores with small diameters, high specific surface areas (400–1000 $m^2\ g^{-1}$), and are present as solid matrices, which are formed from either connected colloidal particles or polymeric chains of about 10 nm in diameter. RF aerogels are stiffer and stronger than silica aerogels.

Yuan et al. [24], have prepared ultra-lightweight and highly porous carbon aerogelsfrom the cheap and abundant bamboo pulp fibers via freeze-drying followed by pyrolysis. Prepared carbon aerogel (CA) showed an ultra-low density (5.65 mg cm^{-3}), high hydrophobicity (water contact angle of 135.9°) and a large specific surface area (379.39 $m^2\ g^{-1}$) as well as great mechanical properties. The absorption capacities of CA for organic solvents/oils were extraordinary (50–150 g/g). Particularly, its absorption on organic solvents was superior to many other bio-based CAs. The reusability of CA was also found impressive. For over five absorption–desorption cycles, the CA still showed excellent absorption behaviours on organic solvents and oils.

Fang and co-workers [25] have synthetized carbon aerogels from organic gels using a resorcinol–formaldehyde polycondensation reaction as an alternative adsorbent to anthracite coal for more efficient removal of hexavalent chromium and estrone. Prepared carbon aerogel had a surface area of 980 m^2/g, pore volume 1.1 cm^3/g and average pore diameter 4.6. For high concentrations of contaminants in water, carbon aerogel removed 95% or more of the hexavalent chromium and virtually all of the estrone. In both cases, carbon aerogel significantly outperformed anthracite coal in the removal of these contaminants from water.

Table 1 summarizes some of the results obtained from the removal of various organics from water through adsorption onto carbon aerogels.

Table 1. Removal of various organics from water using carbon aerogels

Aerogel type	Source	Method of drying	Removed organic/sorption capacity/%removal	Ref.
Carbon aerogel	Bamboo pulp fibers	Freeze-drying	Organic solvents, oils/50–150 g/g	[24]
Carbon aerogel	R-F	Supercritically dried with CO_2	Endocrine-disrupting compounds (17α-Ethinyloestradiol (EE2) and 17β-Estradiol (E2)/>99.9)	[25]
Microporous monolithic carbon aerogels	R-F	Supercritically dried with CO_2	Toluen/1180 mg g^{-1}	[26]

2.3 Adsorption of Organics by Carbon Xerogels

There are not many studies focused on the elimination of emerging contaminants by synthesized materials with extended mesoporosity such as carbon xerogels. In this sense, Álvarez et al. [27] prepared carbon xerogels by polycondensation of resorcinol with formaldehyde and applied as adsorbents in the removal of two emerging contaminants, caffeine and diclofenac, from aqueous solutions.

Carabineiro et al. [28] have studied the adsorption capacity of ciprofloxacin (CPX) which is a synthetic antibioticon three types of carbon-based materials: activated carbon, carbon nanotubes and carbon xerogel. Adsorption capacities ranging from approximately 112 to 231 mgCPX/gC^{-1} were obtained (for carbon xerogel and activated carbon, respectively).

Figueiredo et al. [29] have studied the adsorption of textile dyes (Reactive Red 241 and Acid Blue 113) onto carbon materials with extended mesoporosity.

Table 2 summarizes some of the results obtained from the removal of various organics from water through adsorption onto carbon xerogels.

Table 2. Removal of various organics from water using carbon xerogels

Xerogel type	Source/surface modification	Method of drying	Removed organic/sorption capacity/%removal	Ref.
Carbon xerogel	R-F 1. urea solution 2. sulfuric acid	In the oven 60 °C–150 °C (20 °C day^{-1}). Calcination under N_2 flow	1. caffeine/182.5 mg g^{-1} 2. diclofenac/80.0 mg g^{-1}	[27]
Carbon xerogel	R-F	In the oven 65 °C–150 °C (20 °C day^{-1}). Calcination under N_2 flow	Ciprofloxacin (CPX)/112–231 mgCPX/g	[28]
Carbon xerogel	R-F	In the oven 60 °C–120 °C (20 °C day^{-1}). Calcination under N_2 flow	Reactive Red 241/mg g^{-1} at pH = 12; 1.14 ± 0.12 at pH = 7; 1.47 ± 0.09 at pH = 2; 1.79 ± 0.15 Acid Blue 113/mg g^{-1} at pH = 7; 1.56 ± 0.09	[29]

3 Conclusion

This study show that carbon aerogels and xerogels are promising materials for removing of toxic organic compounds because of their unique and adjustable physical properties such as low density, extremely high specific surface area and porosity. In most of these studies, carbon-based aerogels and xerogels have been found to be very good sorbents materials having high sorption capacity towards the organic solvents, dyes, pesticides, antibiotics, painkillers and many other tested organic components. Also, the regeneration capability of carbon aerogels and xerogels were also promising for an economically and large scale production point of view. Nonetheless, the present work demonstrates that there is a great opportunity for carbon aerogels and xerogels environmental applications because of the extraordinary adsorption properties derived from their chemistry and porous architecture.

References

1. Nematollahzadeh, A., Shojaei, A., Karimi, M.: Chemically modified organic/inorganic nanoporous composite particles for the adsorption of reactive black 5 from aqueous solution. React. Funct. Polym. **86**(26), 7–15 (2015)
2. Nalan, O.S.K., Aslı, C., Tamer, U., Turgay, T.: Microalgae immobilized by nanofibrous web for removal of reactive dyes from wastewater. Ind. Eng. Chem. Res. **54**(21), 5802–5809 (2015)
3. Jian, W., Kitanaka, A., Nishijima, W., Baes, A.U., Okada, M.: Removal of oil pollutants in seawater as pretreatment of reverse osmosis desalination process. Water Res. **33**(8), 1857–1863 (1999)
4. Ayotamuno, M.J., Kogbara, R.B., Ogaji, S.O.T., Probert, S.D.: Petroleum contaminated ground-water: remediation using activated carbon. Appl. Energy **83**(11), 1258–1264 (2006)
5. Pasila, A.: A biological oil adsorption filter. Mar. Pollut. Bull. **49**(11–12), 1006–1012 (2004)
6. Hartmann, M., Kullmann, S., Keller, H.: Wastewater treatment with heterogeneous fentontype catalysts based on porous materials. J. Mater. Chem. **20**(41), 9002–9017 (2010)
7. Hüsing, N., Schubert, U.: Aerogels-airy materials: chemistry, structure, and properties. Angew. Chem. Int. Ed. **37**, 22–45 (1998)
8. Al-Muhtaseb, S.A., Ritter, J.A.: Preparation and properties of resorcinol– formaldehyde organic and carbon gels. Adv. Mater. **15**(2), 101–114 (2003)
9. Pekala, R.W., Alviso, C.T.: Carbon aerogels and xerogels. In: Renschler, C.L., et al. (eds.) Novel forms of Carbon, MRS Symposium Proceedings, vol. 270, pp. 3–14 (1992)
10. Chen, H.B., Hollinger, E., Wang, Y.Z., Schiraldi, D.A.: Facile fabrication of poly(vinyl alcohol) gels and derivative aerogels. Polymer **55**(1), 380–384 (2014)
11. Guo, K., Song, H.H., Chen, X.H., Du, X., Zhong, L.: Graphene oxide as an anti-shrinkage additive for resorcinol–formaldehyde composite aerogels. Phys. Chem. Chem. Phys. **16**(23), 11603–11608 (2014)
12. Liu, L., Yang, J., Meng, Q.H.: The preparation and characterization graphene-cross-linked phenol–formaldehyde hybrid carbon xerogels. J. Sol-Gel. Sci. Technol. **67**(2), 304–311 (2013)
13. Reuß, M., Ratke, L.: RF-aerogels catalyzed by ammonium carbonate. J. Sol-Gel. Sci. Technol. **53**(1), 85–92 (2010)

14. Zhu, Y.D., Hu, H.Q., Li, W.C., Zhang, X.Y.: Cresol-formaldehyde based carbon aerogel as electrode material for electrochemical capacitor. J. Power Sources **162**(1), 738–742 (2006)
15. Tamon, H., Ishizaka, H., Araki, T., Okazaki, M.: Control of mesoporous structure of organic and carbon aerogels. Carbon **36**(9), 1257–1262 (1998)
16. Pekala, R.W.: Organic aerogels from the polycondensation of resorcinol with formaldehyde. J. Mater. Sci. **24**(9), 3221–3227 (1989)
17. Qin, G.T., Wei, W., Guo, S.C.: Semi-continuous drying of RF gels with supercritical acetone. Carbon **41**(4), 851–853 (2003)
18. Liang, C.H., Sha, G.Y., Guo, S.C.: Resorcinol-formaldehyde aerogels prepared by supercritical acetone drying. J. Non-Cryst. Solids **271**(1–2), 167–170 (2000)
19. García-González, C.A., Jin, M., Gerth, J., Alvarez-Lorenzo, C., Smirnova, I.: Polysaccharide-based aerogel microspheres for oral drug delivery. Carbohyd. Polym. **117**, 797–806 (2015)
20. Dumanlı, A.G., Windle, A.H.: Carbon fibres from cellulosic precursors: a review. J. Mater. Sci. **47**(10), 4236–4250 (2012)
21. Liu, H., Cao, C., Wei, F., Huang, P., Sun, Y., Jiang, L., Song, W.: Flexible macroporous carbon nanofiber film with high oil adsorption capacity. J. Mater. Chem. A **2**(10), 3557–3562 (2014)
22. Bratovcic, A.: Photocatalytic degradation of organic compounds in wastewaters. Technologica Acta **11**(2), 17–23 (2019)
23. Bratovcic, A.: Degradation of micro- and nano-plastics by photocatalytic methods. J. Nanosci. Nanotechnol. Appl. **3**, 206 (2019)
24. Yuan, W., Zhang, X., Zhao, J., Li, Q., Ao, C., Xia, T., Zhang, W., Lu, C.: Ultra-lightweight and highly porous carbon aerogels from bamboo pulp fibers as an effective sorbent for water treatment. Results Phys. **7**, 2919–2924 (2017)
25. Fang, J., Solis, J., Lan, E., Dunn, B.: Carbon aerogels for wastewater treatment. In: Aparicio, M., Jitianu, A., Klein, L. (eds.) Sol-Gel Processing for Conventional and Alternative Energy, Advances in Sol-Gel Derived Materials and Technologies, pp. 201–216. Springer, Boston (2012)
26. Maldonado-Hodar, F.J., Moreno-Castilla, C., Carrasco-Marin, F., Perez-Cadenas, A.F.: Reversible toluene adsorption on monolithic carbon aerogels. J. Hazard. Mater. **148**(3), 548–552 (2007)
27. Álvarez, S., Ribeiro, R.S., Gomes, H.T., Sotelo, J.L., García, J.: Synthesis of carbon xerogels and their application in adsorption studies of caffeine and diclofenac as emerging contaminants. Chem. Eng. Res. Des. **95**, 229–238 (2015)
28. Carabineiro, S.A.C., Thavorn-amornsri, T., Pereira, M.F.R., Serp, P., Figueiredo, J.L.: Comparison between activated carbon, carbon xerogel and carbon nanotubes for the adsorption of the antibiotic ciprofloxacin. Catal. Today **186**(1), 29–34 (2012)
29. Figueiredo, J.L., Sousa, J.P.S., Orge, C.A., Pereira, M.F.R., Órfão, J.J.M.: Adsorption of dyes on carbon xerogels and templated carbons: influence of surface chemistry. Adsorption **17**, 431–441 (2011)

Supercritical Fluid Extraction of Lycopene and Omega-3

Maria Margotta[1] and Marco Claudio De Simone[2(✉)]

[1] MEID4 Academic Spin-Off of the University of Salerno, Via Giovanni Paolo II, 132, 84084 Fisciano, SA, Italy
[2] Department of Industrial Engineering, University of Salerno, Via Giovanni Paolo II, 132, 84084 Fisciano, SA, Italy
mdesimone@unisa.it

Abstract. An ideal extraction method should be rapid, produce a quantitative recovery without degradation and the extracts should be easily separated from the solvent. Supercritical fluid technology offers features that overcome many limitations of conventional extraction methods. There are many advantages to their application, especially in the food sector (for example the extraction of lycopene from tomato skins or the extraction of fish oil rich in omega-3 from fish by-products); the extraction method influences the yield and quality of the product obtained, and also allows to operate with moderate temperatures and with shorter timescales. The processes involving SCF are sustainable, environmentally friendly and cost efficient, and offer the possibility of obtaining new products. The disadvantage, however, lies in the fact that it is necessary to use expensive equipment and to perfect the particular phases of the process to obtain a specific product.

Keywords: Supercritical fluids extraction · Supercritical carbon dioxide · Supercritical fluids fractionation · Lycopene · Omega-3 Fatty acid

1 Introduction

The application of supercritical fluids as an alternative to conventional processes of precipitation, separation and extraction it is a field of research in continuous innovation.

The use of supercritical CO_2 offers innumerable advantages, including moderate operating temperature, density and solvent power that can be easily adjusted thanks to a minimum change in temperature or pressure, flammability and low cost.

Nowadays, SC-CO_2 is used for large-scale extractions in the food industry, as an alternative to traditional separation systems, such as fractional distillation, steam current extraction, extraction with organic solvents.

A large application of supercritical fluids occurs in the agro-industrial field, above all for the recovery of nutrients from processing waste. In fact, in the last few years the concept of "zero waste" has become increasingly aware: waste can be reused within the food production chain. Many residues, in fact, are rich in bioactive species [1].

The aim of this work is to illustrate how supercritical CO_2 is able to extract lycopene from tomato production waste and omega-3 rich oil from fish by-products.

I. Karabegović (Ed.): NT 2020, LNNS 128, pp. 750–758, 2020.
https://doi.org/10.1007/978-3-030-46817-0_85

2 Supercritical Fluids Properties

When the substance is above its critical temperature and pressure, the fluid is called supercritical. This fluid has properties similar to gas and liquid: it is similar to gas because it is a compressible fluid and it is similar to liquid for the comparable density value [2].

It is possible to show the dependence of solubility (δ) with density (ρ) using the Hildebrand formula (1):

$$\delta = 1.25\ P_c^{0.5}\left(\frac{\rho_r}{\rho_{r,liq}}\right) \tag{1}$$

where δ is a solubility of CO_2 in $(\text{cal/cc})^{0.5}$, P_c is a critical Pressure in bar, ρ_r is the reduced density of the SCF solvent or liquid and $\rho_{r,liq}$ is the reduced density of the liquid at its normal boing point, a value of about 2.66.

With this formula (1) the density is seen as a property that measures the solvent power of the supercritical fluid. With small variations of pressure, the fluid changes its density and, consequently, its solvent power.

Other important properties are the viscosity and the diffusion coefficient: the speed with which a fluid is able to interact with a specific solute depends on them. The transport properties for a supercritical fluid are particularly favorable, since they are closer to the properties of the gaseous state rather than to those of the liquid state.

Extraction with a supercritical fluid of comparable solvent power is therefore faster, a further reason why the use of supercritical fluids in different industrial sectors continues to be tested more and more.

The advantages of extraction with supercritical fluids (compared to extraction with liquids) are the speed of the operation, given the low viscosity and high diffusivity of SCFs, the high commercial value of the extracted product and purity, and the disengagement from environmental limitations.

The solvent capacity of a supercritical fluid is comparable to a liquid of equal polarity but, unlike liquids, it is able to penetrate, like a gas, inside the solid matrix. As regards the extractive selectivity of the supercritical fluid, the fractionation of liquids and solids can be obtained if the constituents of the matrix to be fractionated differ in volatility (the compounds with higher vapor tension are more soluble), mass (the pressure necessary for the extraction increases with the molecular weight of the compounds) and polarity (for example, in a non-polar supercritical fluid such as CO_2 the solubilization of apolar compounds is favored).

2.1 Characteristics of Supercritical CO_2

The most used supercritical fluid is carbon dioxide: it is inexpensive, chemically inert, non-toxic, non-flammable and readily available at high purity and low cost. Furthermore, the critical point of the CO_2 is easily reachable (critical temperature 31.1 °C and critical pressure 73.8 bar) which allows using the fluid at average temperatures (40–60 °C) without leaving harmful organic residues [3]. The choice of the type of fluid to be used in a specific process depends on technological, economic and environmental impact considerations: it must have relatively low critical temperature and pressure values, it must also be easily separable from other materials present at the end of the process, it must not be flammable or toxic, and have a low cost [4].

In Table 1 there are some fluids that can be used in supercritical conditions.

Table 1. Possible supercritical solvents and related critical temperature and pressure parameters.

Fluid	T_c [°C]	P_c [bar]
Carbon Dioxide	31.1	73.8
Difluoromethane	111.8	39.0
Trifluoromethane	28.8	39.0
Ethane	32.2	48.8
Ethylene	9.3	50.4
Propane	96.7	42.5
Propylene	91.8	45.0
H_2O	374.2	220.5
N_2O	36.5	71.7

3 Supercritical CO_2 Extraction of Lycopene from Tomatoes in the Presence of Vegetable Oil as a Co-solvent

Lycopene ($C_{40}H_{56}$) is the carotenoid pigment most present in tomatoes (it represents about 85% of the total carotenoids) and gives it its characteristic red color. Its presence varies considerably depending on the variety, degree of ripeness, environmental conditions and cultivation techniques. In mature tomatoes typical lycopene values of 50 mg/kg are found, but in some varieties this level can even exceed 70 mg/kg.

The main problem in obtaining lycopene is its solubility and stability: it is insoluble in water but soluble in highly toxic organic solvents, such as benzene, chloroform, dichloromethane and decomposes easily over time. In order to overcome these difficulties and obtain lycopene without traces of organic solvents, the scientists considered the use of the extraction technique with supercritical fluids (SFE) [5].

Experiments conducted with and without co-solvent at pressures and temperatures between 335 and 450 bar and 45–70 °C, respectively, have shown that the quantity of lycopene that can be extracted, depends on the experimental conditions. Furthermore, the maximum quantity of lycopene extractable from the dried tomato (6% humidity, average grain size about 1 mm), at 450 bar and 66 °C in the presence of co-solvent and using a flow rate about 20 kg of CO_2/h, it was 60% [6].

The presence of the co-solvent improves the yields of the lycopene extract and has a beneficial role in pigment stability.

Various vegetable oils have been tested (almond oil, peanuts, hazelnuts and sunflower seeds) but only in the case of hazelnut oil (cheaper and its low acidity prevents the degradation of lycopene during extraction) have been obtained higher extraction yields. A greater quantity of oil provides a more diluted extract of lycopene and for this reason the average value of 10% was chosen.

Since there is a high degree of humidity in tomatoes, the raw material was initially dried and then ground to obtain significant yields of lycopene. With smaller particle sizes, the ratio surface/volume of the sample increases and therefore the contact between the solid phase and the liquid phase increases. For this reason, an average particle size of 1 mm was chosen.

Extraction with SC-CO_2 was performed with the apparatus shown Fig. 1, equipped with an AO1 extractor (10 l) and three separators S1 (1.5 l), S2 (1.5 l) and S3 (0.3 l) in series. The CO_2, pressurized to the desired level by the P1 pump (Milton Roy), flows through a heating coil (E2) before reaching the AO1 extraction bed.

Then, leaving AO1, the fluid subsequently flows towards the separator S1 through a cooling coil (E3) and then towards the separator S2 passing through a micro-measurement valve (VL1) with consequent reduction of the pressure and, therefore, reduction of the density of CO_2. The solute that precipitates from the stock solution can be collected in S2. The fluid subsequently arrives in the separator S3 through the second micrometric valve (VL2) with a reduction of pressure up to its final value and the solute is then separated and collected in S3. The CO_2 that comes out of the S3 separator can be recycled by passing through the F1 filter and the E1 condenser. The S1 separator operates with constant pressure; the variation of the temperature reduces the solubility of the extract, producing the separation and its precipitation. In the separators S2 and S3 the solubility decreases with the variation of the pressure.

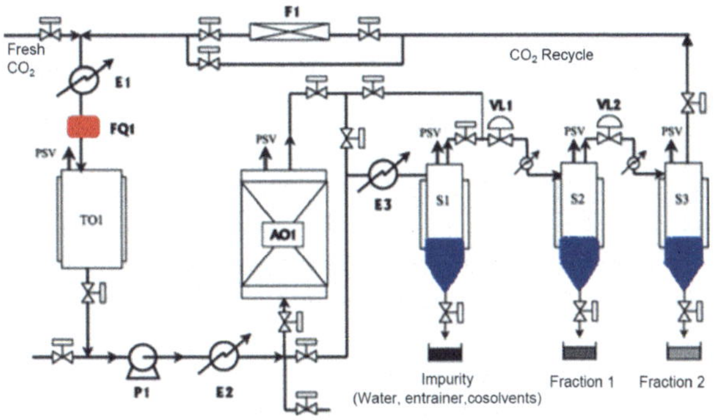

Fig. 1. SFE apparatus

AO1, extractor; S1-3, separators; TO1, CO_2 supply tank; F1, filter; P1, pump; E1, condenser; E2, heating battery; E3, cooling coil; VL1–2, micrometric valves; FQ1, flow meter; PVS, safety valves.

Several experiments were performed varying the Pressure and Temperature and it was noted that:

- An increase in the Pressure, working with a constant Temperature, increases the density of the solvent and therefore the solubility of the lycopene and at higher

densities the molecular interactions between the solvent and the solute are enhanced, resulting in a greater dissolution of the solute.
- An increase in temperature, working at constant Pressure, reduces the density of the solvent and, consequently, reduces the solubility of lycopene but promotes the transport of the solute from the matrix to the solvent.

The best operating conditions experimentally achieved are the following:

- P = 450 bar
- T = 65–70 °C
- CO_2 mass flow rate = 18–20 kg/h
- Average grain size of the material = approximately 1 mm
- Quantity of hazelnut oil = 10% w/w.

The optimization of the parameters in the extraction with SC-CO_2 as temperature, pressure, flow rate, extraction time, as well as the moisture content and particle size, leads to higher yields of lycopene.

The presence of the co-solvent probably promotes better transport and a better solubility of the pigment from the solid phase to the supercritical phase and allows it to avoid the degradation of lycopene and keep the pigment stable over time.

4 Extraction of Fish Oil Rich in Omega-3 from Fish by-Products

Fish oil is the most generous food source of omega-3, in particular, of eicosapentaenoic acid (EPA) and docosahexaenoic acid (DHA). They are polyunsaturated fatty acids that exert a strong positive influence on human health.

Fishes, like men, are not able to directly synthesize omega-3 s but accumulate them in their tissues by absorbing them from microalgae or from the animals they eat.

The synthesis of DHA occurs in phytoplankton and animals, but not in plants: DHA is absent from foods of plant origin, including vegetable and oil fats, cereals, nuts, and seeds. However, it is present in the lipids of animal tissues, with the richest source of fatty fish. Meat and milk from ruminants also contain small amounts of DHA.

The production of fish oil takes place mainly through the so-called “wet reduction process”: whole fish or fish by-products (mainly viscera, heads and spines) are treated by grinding, heating and centrifuging in one or two stages.

This process makes it possible to obtain high volumes of raw fish oil, although in order to make the oil suitable for edible purposes, subsequent refining phases are necessary.

Four different extraction methods were evaluated to obtain fish oil rich in omega-3:

- cold extraction (CE) or centrifugation
- wet reduction (WR)
- enzymatic extraction (EE)
- supercritical fluid extraction (SFE).

The quantity of raw material used in each extraction method was about 100 g [7] (Fig. 2).

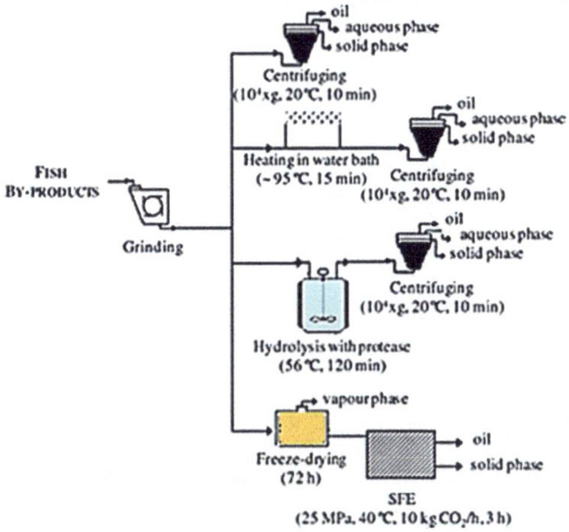

Fig. 2. Scheme of the different fish oil extraction procedures

In the cold extraction, in the wet reduction and in the enzymatic extraction, the fish waste was previously thawed at room temperature for 12 h and the co-extracted water together with the oil was removed by centrifugation [8].

For enzymatic extraction the enzyme used was a food-grade protease, Alcalase 2.4 L (bacterial protease from Bacillus licheniformis).

For the SFE, it was performed in a semi-pilot plant; the raw material used for this extraction method was previously freeze-dried (lyophilized).

The quality of the oil obtained with the different extraction methods was evaluated by determining several parameters:

- moisture content and volatile matter
- neutral lipid composition
- fatty acid profile
- acidity value
- peroxide value
- traces of metals

Extraction with SC-CO_2 allows obtaining fish oil with a high quantity of omega-3 fatty acids, using a non-oxidizing atmosphere and mild temperatures, which prevent the oxidation of polyunsaturated fatty acids.

Several extraction tests were carried out in different fish by-products (off cuts from hake, orange roughly, salmon and livers from jumbo squid) and it was observed that, at the beginning of the process, the extracted oil had a linear dependence with the amount of SC-CO_2 flowing through the extractor, which may indicate that the internal mass transfer is negligible and that the process is controlled by the solubility of the oil in SC-

CO_2 or that the internal mass transfer is constant and the extraction rate depends on the internal structure of the solid matrix.

The values of the slopes of the extraction curves estimated at time zero are different for each fish product examined, which can be attributed not only to the different internal structure, but also to the different solubility of fish oil in SC-CO_2 (Fig. 3).

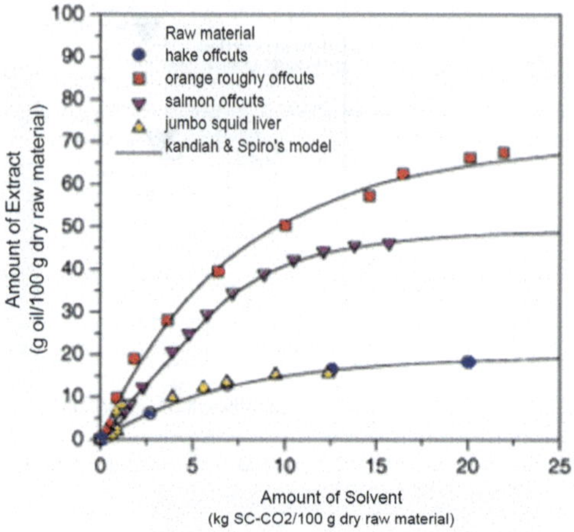

Fig. 3. Extraction curves obtained for SFE of oil from different fish by-products.

The different extraction curves obtained, correctly followed the empirical model of Kandiah and Spiro (1990): the process is controlled by two diffusion stages depending on the quantity of oil accessible to the SC-CO_2. As can be seen, at the beginning the most accessible quantity of oil is high since there is low internal resistance to mass transfer [8].

5 Results

The extraction time using supercritical CO_2 is shorter than the extraction time using common solvent, since CO_2 has a strong permeability and, moreover, the first advantage of this technique is that, at the end of the extraction process, the carbon dioxide is brought to room temperature and atmospheric pressure and consequently gasifies, leaving the substances extracted from the solid matrix [9, 10]. The extraction method with supercritical fluids is very advantageous and ecological and allows the valorization of waste products as a derived raw material or as another product. The extraction of fish oil by wet pressing is the most commonly used method for industrial-scale production and it is basically performed in four phases: cooking of the fish, pressing, decanting and centrifugation. Another conventional procedure is solvent extraction: the main limitation of this technique is that it requires a relatively dry sample that is destroyed, and also takes a long time, in addition to generating large amounts of residual solvent. After fish oils are extracted, they require a purification process to achieve quality characteristics that make them acceptable for human and animal consumption, as they contain insoluble impurities [11].

6 Conclusions

The use of SC-CO_2 in the extraction of lycopene from tomato skins has produced numerous advantages especially in the higher yields obtained, in the selectivity of the extraction and in obtaining lycopene without traces of organic solvents.

In fact, the extraction techniques commonly used for the extraction of lycopene from vegetable matrices use organic solvent, which, however, has two drawbacks: the first is that it is not specific for lycopene, but extracts many other pigments (carotenes, xanthophylls) with the consequence that the purification process of lycopene becomes long and expensive and in general does not lead to having lycopene of high degree of purity; the second contraindication is that the raw material used, becomes a special waste that will have to be incinerated with considerable cost increase and cannot be recovered in any way. The organic solvent that causes the least problems is ethanol, but any solvent extraction requires a subsequent stage of purification. The traditional refining process includes various phases, such as degreasing, neutralization, bleaching, deodorization and, in some cases, winterization. Despite the need for a high investment cost, the supercritical fluid extraction of essential oils requires lower processing costs and downstream processes that make this process competitive with regards to steam distillation. In the industrial field the application of supercritical fluids in the extraction of omega-3 should be extended to the whole process: extraction, fractionation, concentration of omega-3, in order to obtain omega-3 concentrates used in functional foods and in the pharmaceutical field.

References

1. Levelt Sengers, J.M.H., Erdogan, K.: Supercritical Fluids: Fundamentals for Application. Kluwer Academic Publishers, Berlin (1994)
2. Hitchen, S.M., Dean, J.R.: Properties of supercritical fluids. In: Applications of Supercritical Fluids in Industrial Analysis, pp. 1–11. Springer, Dordrecht (1993)
3. Reverchon, E., De Marco, I.: Supercritical fluid extraction and fractionation of natural matter. J. Supercrit. Fluids **38**(2), 146–166 (2006)
4. Guida, R., De Simone, M.C., Dašić, P., Guida, D.: Modeling techniques for kinematic analysis of a six-axis robotic arm. In: IOP Conference Series: Materials Science and Engineering, vol. 568, no. 1, p. 012115. IOP Publishing (2019)
5. Choksi, P.M., Joshi, V.Y.: A review on lycopene-extraction, purification, stability and applications. Int. J. Food Prop. **10**(2), 289–298 (2007)
6. Vasapollo, G., Longo, L., Rescio, L., Ciurlia, L.: Innovative supercritical CO_2 extraction of lycopene from tomato in the presence of vegetable oil as co-solvent. J. Supercrit. Fluids **29** (1–2), 87–96 (2004)

7. Rubio-Rodríguez, N., Sara, M., Beltrán, S., Jaime, I., Sanz, M.T., Rovira, J.: Supercritical fluid extraction of fish oil from fish by-products: a comparison with other extraction methods. J. Food Eng. **109**(2), 238–248 (2012)
8. Bonilla-Méndez, J.R., Hoyos-Concha, J.L.: Methods of extraction refining andconcentration of fish oil as a source of omega-3 fatty acids. Corpoica Cienciay Tecnología Agropecuaria 19 (3), 645–668 (2018)
9. Boguniewicz-Zablocka, J., Lukasiewicz, E., Guida, D.: Analysis of the sustainable use of geothermal waters and future development possibilities-A case study from the Opole region, Poland. Sustainability **11**(23), 6730 (2019)
10. Formato, A., Guida, D., Ianniello, D., Villecco, F., Lenza, T., Pellegrino, A.: Design of delivery valve for hydraulic pumps. Machines **6**(4), 44 (2018)
11. Bonilla-Méndez, J.R., Hoyos-Concha, J.L.: Methods of extraction refining and concentration of fish oil as a source of omega-3 fatty acids. Corpoica Ciencia y Tecnología Agropecuaria **19**(3), 645–668 (2018)

Assessment of Cancerogenic Health Risk of as, Cd, Pb and Ni from Tobacco Smoke

Alma Leto(✉), Aida Šukalić, and Alma Mičijević

Džemal Bijedić University of Mostar, 88 104 Mostar, Bosnia and Herzegovina
alma.leto@unmo.ba

Abstract. The aim of this study was to determine the cancerogenic risk to human health of As, Cd, Pb, and Ni intake via cigarette tobacco smoke, made primarily from indigenous Herzegovina tobacco varieties. Since such research has never been done on these Herzegovina varieties VH, VH32, and Ravnjak, they radically compared the possibilities next to them, examining the cultivar Virginia and Berley, also produced in BiH. Samples were taken from 16 localities, namely, all localities in BiH today. Samples were used per insertion, from medium insertions, such as the highest quality, handmade cigarettes that were later dropped on smokers, and the smoke condensate tested for Cd, Pb, As, and Ni content. A tobacco smoke cancerogenic risk assessment was made after setting up the CSF by the US EPA, assuming that 1 pack of cigarettes is smoked daily. CDI - chronic daily dose for 4 heavy metals and ILCR for each type of tobacco was determined. CDI Calculated risk had cancer over the life span of As, Cd, Pb, and Ni ranging from 1.17E−4 for Virginia to 1.48E−4 for Ravnjak, which did not reach the common benchmark of 10^{-6}–10^{-4} is used by the US EPA to protect public health.

Keywords: Tobacco · Smoke · As · Cd · Pb · Ni · Carcinogenic risk

1 Introduction

In Bosnia and Herzegovina, the incidence of smoking is much higher than in most European countries. Every third BiH resident is a daily smoker, which is almost half the European average. Of particular concern is the fact that the incidence of smoking among young people in our country is alarming. A European survey on the use of tobacco, alcohol, and drugs among young people aged 13–15 found that as many as 14.5% of them consume tobacco or other tobacco. The figures are even more serious when it comes to secondhand smoke, i.e. exposure of children to tobacco smoke in public space, so this percentage goes up to an alarming 85%.

Cigarette smoking is a major cause of various diseases, such as cancer of the mouth and nasal cavity, pharynx, bronchus, and lungs, paranasal sinuses, larynx, pancreas, bladder, stomach, cellular carcinoma of the cervix, myeloid leukemia, etc. Cigarette smoke is known to contain about 4000 different chemicals that enter the body after smoke inhalation and is transmitted through blood to specific organs. In cigarette smoke, the highest content is nicotine, carbon monoxide, polycyclic aromatic hydrocarbons, nitrogen, hydrogen cyanide, carbon dioxide, oxygen, and metals.

I. Karabegović (Ed.): NT 2020, LNNS 128, pp. 759–769, 2020.
https://doi.org/10.1007/978-3-030-46817-0_86

According to the World Health Organization, one person dies every 8 s in the world, causing tobacco damage. 1.3 billion people in the world consume tobacco products, and about 5 million people die from smoking-related diseases annually. If current trends and bad habits continue, the death toll could reach 10 million by 2030. Tobacco consumption is one of the main causes of preventable death, so smoking cessation is a very important decision to preserve one's health. Research indicates that for each year of smoking, after the age of 35, three months of life are lost.

The harmful effects of smoking such as the use of tobacco and related products and exposure to tobacco smoke, usually occur only after several (tens) years of "smoking experience".

This paper deals with 5 types of tobacco VH, VH32, Berley, Ravnjak, and Virginia with 16 locations in Bosnia and Herzegovina. Smoking on a Borgwaldt smoking machine according to ISO 3308: 2000, a standard analytical smoking machine for 20 cigarettes of each variety, under standard conditions, calculated the concentration of As, Cd, and Pb in tobacco smoke. Subsequently, a risk assessment of 3 cancerogenic heavy metals from tobacco smoke was performed. The tobacco smoke cancerogenic risk estimate was calculated according to the US Environmental Protection Agency's (US EPA) CSF assuming that 1 pack of cigarettes is smoked daily. The total number of smoking years (presented here) is one of the common measures of total smoking exposure. (1 pack per day = 365 packets per year = 7,300 cigarettes.) Z is a measure directly related to the calculation of CDI values. The basic elements of chemical hazard risk assessment include four steps: hazard identification, hazard characterization, exposure assessment, and risk characterization. The risk assessment methodology for chemical contaminants is defined by WHO (World Health Organization), US EPA (Environmental Protection Agency) and EFSA (European Food Safety Authority) [1].

Heavy metal toxicity

Cadmium

Cadmium poisoning can be acute and chronic. Acute poisoning is caused by the inhalation of vapors or particles of cadmium salts (oxide, chloride, sulfide, sulfate, carbonate, and acetate) [2]. Chronic occupational poisoning can result from prolonged exposure to cadmium by inhalation or oral route, and systemic exposure to cadmium leads to increased calcium excretion [3].

In 1987, the International Organization for Research on Cancer (IARC) classified cadmium and its components as probable carcinogens (Group 2A) based on occupational exposure studies. More recent data suggest that human exposure in the general population is statistically associated with an increased risk of endometrial, bladder, and breast cancer. In 1993, IARC classified cadmium and its components in Group 1 as human carcinogens based on evidence of human lung cancer resulting from occupational inhalation exposure and based on animal studies [4]. One of the consequences of over-entering this element is the attachment to the so-called. zinc finger proteins that directly result from changes in the transcription of ribonucleic acid, i.e., the genetic code. Cadmium also prevents the production of antibodies and reduces the secretion of insulin in the human body. Of all plants, tobacco is the largest cadmium assimilator in the soil. In most plants, its concentration decreases from root to leaf, fruit to seed. Tobacco is most commonly found in tobacco, i.e. in that part of the plant used as a commercial product. Although, for example, tomato and soil tobacco accumulate almost the same amount of cadmium, it is up to seven times more in tobacco leaf than in tomato [5]. The usual amounts of cadmium in the tobacco leaf are 0.5 to 0.7 micrograms per gram and for cigarettes 1 to 4 micrograms per gram of tobacco. The content in the smoke condensate of cigarettes without filters is greater than 0.1 micrograms per cigarette, i.e. the transition to smoke condensate is from 15–20%. In the smoke condensate of cigarettes with a filter, the cadmium content does not exceed 0.1 micrograms per cigarette with a transition rate of 2–12% [6].

Arsenic

Arsenic (As) is a xenobiotic that can be found in the Earth's crust and is widely distributed throughout the environment by air, water, and earth. According to the IARC classification, arsenic and its compounds are carcinogens for humans [7]. It occurs in four oxidation states as As (0) (arsenic), As (−III) (arsine), As (+III) (arsenite), As (V) (arsenate) [8]. When building chemical bonds with oxygen, chlorine, and sulfur, arsenic produces very toxic inorganic compounds. This form of arsenic is found in industry, drinking water and plants as a result of irrigation with such water. Because of its high toxicity and slower excretion, it is of great health importance. Prolonged exposure to inorganic arsenic due to its cumulative effect can lead to chronic poisoning [9]. The trivalent forms of arsenic are thought to react with -SH groups in vital cellular enzymes, inhibiting reactions in the Krebs cycle and preventing the breakdown of carbohydrates, fats, and proteins. This affects the oxidation of cells and the paralytic effect of arsenic on the smooth musculature of blood vessels whereby arsenic is classified as vascular poisons. An excessive amount of arsenic in the human body can cause various acute and chronic diseases, often with fatal outcomes. The most common diseases caused by a large amount of arsenic are diseases of the respiratory tract, digestive tract, liver (so-called as-cirrhosis), blood flow, skin diseases, conjunctivitis. The mechanism of toxicity is reflected in the inhibition of the enzymes with the thiol (SH) group, the general hypo methylation of deoxyribonucleic acid and the hyper methylation of tumor suppressor genes resulting in accelerated cell division and oxidative stress. Arsenic compounds with an oxidation state of +3 are much more toxic than those in which the oxidation state of arsenic is +5. After absorption, arsenic compounds are rapidly distributed to all organs, and especially to the kidneys, liver, and spleen. Arsenic bound to organic compounds is excreted more rapidly from the human body than those compounds where arsenic is bound to inorganic compounds [10]. The absorption of arsenic into tobacco depends most on soil texture and iron content [11]. Thus, the absorption is higher on larger granulate soils and with increased phosphorus fertilizer intake, and at higher iron content the absorption is less. As a result of the use of arsenic-based insecticides until 1960, tobacco contained a significantly higher amount of this element (7.0–51 micrograms per gram of tobacco). In the sixties, it was in the range of 3–12 micrograms per gram of tobacco, and in the seventies from 0.005–1.0 micrograms per gram of tobacco [6]. Tobacco from these areas in the 1980s contained 0.1–0.8 micrograms per gram of tobacco [12]. By burning tobacco, some of the arsenic goes to the smoke condensate, and depending on the construction of the cigarette, the conversion coefficient of arsenic from tobacco to raw smoke condensate is about 10% [6].

Nickel

The role of nickel in plants has not been well studied, but it is still considered to be a necessary element in the metabolism of higher plants [13]. On the other hand, according to Uren 1992, and according to Bogdanovic, D. 2007 [14], the toxicity of nickel to plants is reflected in chlorosis, cessation of root system growth and, sometimes, intercostal necrosis. It is also known that the toxic dose for this element is eight times higher than zinc. The plants adapt it in the Ni +2 ion form and the form of chelate. It accumulates most in leaves and seeds. Its contents are very different not only for different types of plants and growing conditions but also for different varieties. There are plant species with a high ability to accumulate this element (in the list of some plants it can be as high as 8000 ppm) [15]. It is found in food of plant origin, fish, shellfish, drinking water (which is likely to be reached through water pipes), and absorption from water is 40 times greater than absorption from food. In humans, it harms the immune system, causes contact dermatitis, lung fibrosis, cardiovascular and renal diseases, and respiratory tract cancer. The mechanism of toxicity is reflected in the fight for binding sites in tissues with magnesium, calcium, iron, manganese, which directly causes oxidative stress. The average daily intake of this element is 100 micrograms, while allergic reactions can occur as early as 200 micrograms per day, with a tolerable dose of up to 400 micrograms per day. Exposure of people to high levels of nickel can cause many pathological effects, such as allergic contact dermatitis, lung fibrosis, cardiovascular and renal diseases, and respiratory tract cancer. The different forms of nickel differ in toxicity and cancer-causing potential. Studies have shown that the effects that nickel may cause in an organism depend on the solubility of its compounds. Soluble nickel compounds (nickel chloride, sulfate, and carbonate) are considered more toxic than insoluble compounds, while insoluble compounds (nickel sulfide and oxide) are considered to have higher carcinogenicity. The International Agency for Research on Cancer [4] classified metal nickel in group 2B (a group of substances that can be carcinogenic to humans) and nickel compounds in group 1 (a group of substances that are carcinogenic in humans). The U.S. Environmental Protection Agency (U.S. EPA) has classified dust formed during nickel processing and nickel sub sulfide into Group A (a group of substances that are carcinogenic to humans). Other nickel compounds are not classified by the EPA.

Lead

The major routes of lead uptake are the respiratory tract, the skin and the digestive tract [16]. Once ingested, lead exerts toxic effects on the liver, kidneys, and brain. As a cumulative metal, lead is competitive with essential metals (iron, calcium, copper, and zinc) for their many functions in the body, especially those related to the presence of free -SH groups in parts of protein and enzyme biomolecules. The presence of this metal is attributed, in some cases, to the onset of hypertension, cardiac arrhythmias, malignant changes in the digestive tract, lungs and kidneys [17]. Man is exposed to the toxic effects of lead by ingesting food and water that are contaminated with lead, but also by inhaling soil particles, dust and old dyes. Lead is rapidly absorbed into the bloodstream and adversely affects the central nervous, cardiovascular and immune systems, as well as the kidneys. It causes the destruction of red blood cells and impairs the synthesis of heme, resulting in various forms of anemia [18]. It acts not only on protein synthesis but also on ribosome synthesis [19] as the most potent inhibitor of globin synthesis. Although lead in inorganic compounds is more dangerous than inorganic compounds, this activity applies to both of these forms. Of the organic compounds, the most dangerous and well-known are tetra methyl lead and tetra methyl lead, which are used as anti-detonators in gasoline, so today it is sought to be replaced by some other organic compounds, and where possible completely discarded. The mechanism of toxic action is reflected in the formation of coordinative bonds with amino (NH2) and thio (SH) groups of amino acid residues that remain by metabolic degradation of proteins instead of zinc ions, which directly leads to a change in structure. Temporary tolerable weekly entry for this element per kilogram of body weight, according to the World Health Organization, and in the territory of the European Union countries, it is 14% of PTWI. At the time of use of lead-arsenic as a protective agent, more damage was seen on plants from sandy soil and those containing less iron. Lead concentrations above 5 micrograms per gram of tobacco do not cause major damages to plants. In the various tobaccos tested, the lead content ranges from 0.00 to 200 micrograms per gram of tobacco, while the concentrations in commercial tobacco cigarettes range from 1.60 to 22.75 micrograms per gram of dry matter. Lead transfer to flue condensate is in the range of 0.37–1.20 micrograms per cigarette, i.e. 2–16% [6].

2 Methodology

Sixteen representative localities were selected in the entire territory of Bosnia and Herzegovina. Territories ranged from the areas where tobacco is produced in our country today, namely the area of Central Bosnia Posavina and the territory of Herzegovina. When selecting the site, care was taken to cover different types of land, i.e. all types of land cultivated with tobacco. The survey also covered all the varieties, that is, the types of tobacco grown in our country today. Three varieties of Herzegovinian type and one variety of Virginia type and burley were used as the research material. Of the Herzegovina varieties, Ravnjak 108, VH32 and Greater Herzegovina (VH) were used in the study, in the Virginia flue-cured tobacco, testing was performed on the DH 17 variety and in the Berley tobacco on the Podravac variety. Tobacco samples were collected from each site (lower, middle and upper), dried and transferred to laboratories. Dried tobacco samples from all sixteen sites, after the harvest, were prepared at the premises of the Tobacco Factory in Mostar (cut and moistened to a moisture content of 18–20% because this is the optimum moisture for handling tobacco) and stored in plastic boxes. For the analysis of tobacco smoke, medium picking samples of all tobacco samples were taken as the highest quality and most representative. From each sample, 40 cigarettes were made by hand, meaning a total of 640 cigarettes, each of which had the prescribed weight, ranging from 0.75 to 0.80 grams. Cigarette making was done in the laboratory of the Faculty of Agriculture in Mostar. Out of these 40 cigarettes of each medium-sized sample, 20 were selected, which most met the prescribed standards in terms of weight, appearance, and filling and were smoked on a smoking machine at the Tobacco Institute in Zagreb.

Chronic Daily Dose (CDI) for One Pack of Cigarettes per Day
In elaborating a general approach taken by others [19], for any toxicant, the immediate application of risk assessment methods is calculated as follows:

$$CDI_{\text{i}}\left(\frac{mg}{kg}d\right) = \frac{A^{i}x\,10^{-3}mg/\mu gxSRx365d/godxED}{BWxAT}$$

CDI¡ (mg/kg-d) - Average (chronic) rate of daily intake of a toxicant by smoking when the total smoking dose is 1 pack per day when the dose averages over 70 years of age

Ai (μg/cigarette) - Yield of toxic substance in cigarette smoke
SR (cigarettes/d) average smoking rate;
ED (years), duration of exposure to smoking;
BW (kg), average weight;
AT (days), the average duration for smoking doses.

Following the others [20], the assumption is made that 100% of the toxicant is deposited in the respiratory tract, which would be typical of any conservative calculation of the risk assessment. The smoking dose z (year-packs) can be achieved as follows: For z = 1 pack, one box (20 cigarettes) can be smoked daily for 1 year. The total amount of cigarettes smoked is given by SRx365d/godxED, so that:

$$z = \frac{SRx365d/godxED}{7.300cigarets/pack - annually}$$

Substituting z in the equation we obtain:

$$\text{CDI}_{\text{i}}\left(\frac{\text{mg}}{\text{kg}}\text{d}\right) = \frac{\text{A}^{\text{i}}\text{x}\,10^{-3}\text{mg}/\mu\text{gx}7.300}{\text{BWxAT}}$$

ILCR - An Incremental Lifetime Risk of Cancer
Incremental cancer lifetime risk of ILCR is calculated as an increase in the risk assessment for carcinogenic or potentially carcinogenic substances per dose of 20 cigarettes (1 pack) per day:

$$ILCR = CDIxCSF$$

in which CSF (carcinogenic factor for calculation) is expressed in units of mg/kg per day, these values are listed in Table 1.

According to the U.S. The EPA value for adult body weight (BW) is 70 kg and the average time (AT) is 70 years.

Table 1. Cancer slope factor CSF for arsenic, cadmium and lead

Toxicant	CSF (mg/kg-d)
Arsenic	15*
Cadmium	6,3*
Lead	0,042#
Nickel	0,91#

**Integrated Risk Information System, U.S. Environmental Protection Agency.*
#Office of Environmental Health Hazard Assessment, California Environmental Protection Agency. Sept. 2004 Cancer Potency List.

3 Results and Discussion

From Herzegovina, tobacco varieties of Herzegovina (VH, VH32 and Ravnjak) were taken from the following locations: PopovoPolje, Ljubinje, DonjiMamići, GornjiMamići, Stolac-Pesjevac, Stolac-Lokve. Samples of Virginia and Berlej tobacco from the following localities were taken from the BosanskaPosavina area: Bok-1, Bok-2, Donja Mahala-1, Donja Mahala-2, Pelagicevo-1, Pelagicevo-2, Pelagicevo-Orasje, Slatina-1, Slatina-2, Brcko. The concentrations of heavy metals in tobacco smoke are shown in Table 2.

The results of an analysis in a study by [21] suggest that the content of arsenic in tobacco and cigarettes is below 0.01 mg/L, which is the level of the detection limit of the AAS machine. The average content of lead in tobacco and cigarettes is 5.823 mg/kg with a maximum of 8.3 mg/kg and at least 1.53 mg/kg. The mean content of cadmium was 1.8 mg/kg and nickel was 10.74 mg/kg. All results obtained are following the permissible daily doses recommended by WHO/FAO as In our study, the concentration of Cd (μg/per cigarette) in tobacco smoke ranged from 0.036 μg (Virginia) to 0.055 μg (Berley). The concentration of As (μg/per cigarette) in tobacco smoke ranged from 0.0093 μg μg (Ravnjak and VH32) to 0.0107 μg (Virginia). The concentration of Pb (μg/per cigarette) in tobacco smoke ranged from 0.045 μg (Virginia) to 0.568 μg (Ravnjak). The concentration of Ni (μg/per cigarette) in tobacco smoke ranged from 0.211 μg (VH) to 0.2865 In their work, [22] investigated the content of heavy metals in 80 tobacco samples in the Republic of Serbia. The mean lead content in tobacco was 0.93 μg/g (range 0.02–8.56 μg/g) and arsenic 0.15 μg/g (range < 0.02–2.04 μg/g). The mean lead content of cigarettes was 1.26 μg/g (range 0.02–6.72 μg/g) and arsenic 0.11 μg/g (range < 0.02–0.71 μg/g). [23] cited a mean ILCR of cadmium for smokers 1.07E−03, which means that 1 person is likely to become ill with 934 cancer exposed to tobacco smoke Cd, which exceeds the common reference value of 10^{-6}–10^{-4} for US public health protection EPA (32). μg (Ravnjak). well as in accordance with our study. In our work, the ILCR values for cadmium ranged from 9.77E−5 to 1.02E−4, the ILRC for arsenic from 5.76E−06 to 6.29E−06 and for nickel from 1.04E−05 to 8. 54E−06 which is in line with the benchmarks used by the US EPA, while for lead ILRC is 7.90E−8 to 8.82E−8 which is slightly lower than the reference values and presents a low risk of cancer. The total ILRC ranged from 1.03E−4 to 1.38E−4. The Chinese brands tested [24] averaged 0.55 μg/g Cr, 0.78 μg/g As, 3.24 μg/g Cd, and 2.54 μg/g Pb (Table 3).

Table 2. Locations, tobacco varieties and concentrations of heavy metals in tobacco smoke

Locality	Sort/Type	Conc. Cd (μg/1 cigarette) in smoke	Conc. As (μg/1 cigarette) in smoke	Conc. Pb (μg/1 cigarette) in smoke	Conc. Ni (μg/1 cigarette) in smoke
Popovo Polje	VH	0,399	0,0096	0,0435	0,211
Ljubinje	VH	0,405	0,0105	0,0486	0,236
Gornji Mamići	Ravnjak	0,568	0,0093	0,0530	0,2741
Donji Mamići	Ravnjak	0,465	0,0056	0,0499	0,2865
Lokve	VH32	0,388	0,0093	0,0491	0,226
Pješivac	VH32	0,372	0,0095	0,0498	0,2105
Slatina-2	Burley	0,560	0,0094	0,0495	0,2535
Bok-2	Burley	0,530	0,0110	0,0575	0,2485
Donja Mahala-1	Burley	0,49	0,0100	0,0437	0,228
Pelagićevo1	Burley	0,421	0,0098	0,055	0,244
Pelagićevo-Orašje	Burley	0,461	0,0111	0,0356	0,2605
Pelagićevo2	Virginia	0,367	0,0095	0,0511	0,2015
Brčko	Virginia	0,045	0,0099	0,0449	0,209
Slatina-1	Virginia	0,521	0,0095	0,0495	0,263
Donja Mahala-2	Virginia	0,441	0,0098	0,036	0,221
Bok-1	Virginia	0,514	0,0107	0,0580	0,2555

Table 3. Average concentration and values of CDI and ILCR for As, Cd, Pb and Ni in 5 tobacco varieties

	Tox.	Var.					
		VH	Ravnjak	VH32	Burley	Virginia	Total
Concentration of C (μg/1 cigarette) in smoke	As	0,01005	0,00745	0,0094	0,01026	0,00988	
	Cd	0,402	0,5165	0,38	0,4924	0,3904	
	Pb	0,04605	0,05145	0,04945	0,04826	0,0479	
	Ni	0,2235	0,2803	0,21825	0,2469	0,23	
Chronic daily dose of CDI	As	4,10E−07	3,04E−07	3,84E−07	4,19E−07	4,07E−07	
	Cd	1,64E−05	2,11E−05	1,55E−05	2,01E−05	1,59E−05	
	Pb	1,88E−06	2,10E−06	2,02E−06	1,97E−06	1,96E−06	
	Ni	9,12E−06	1,14E−05	8,91E−06	1,01E−05	9,39E−06	
Incremental cancer lifetime risk of ILCR	As	6,15E−06	4,56E−06	5,76E−06	6,29E−06	6,11E−06	**1,18E−04**
	Cd	1,03E−04	1,33E−04	9,77E−05	1,27E−04	1,02E−04	**1,48E−04**
	Pb	7,90E−08	8,82E−08	8,48E−08	8,27E−08	8,06E−08	**1,12E−04**
	Ni	8,30E−06	1,04E−05	8,11E−06	9,19E−06	8,54E−06	**1,17E−04**

Results of the [25], the average values of all toxins tested were well above the permitted limits. The estimated lifetime risk of cadmium and lead cancer is 1 lac (100000) to 10 lac (1000000) times the minimum 10E−4 (0.00001) to 10E−6 (0.000001), which is the 'target range' for potentially hazardous substances, according to US Pat EPA. Similarly, arsenic levels ranged from 0.15 to 14.04 mg/g, averaging 1.25 mg/g. The estimated average bioavailable concentration of arsenic is 0.125–0.25 mg/g, which is higher than the allowable standard of 0.01 mg/g. In our work, the average concentrations were 0.44 μg/1 Cd cigarettes, 0.009 μg/1 cigarettes Aces and 0.048 μg/1 of Pb cigarettes. During April 2015, the Federal Inspection Directorate increased the inspection of illegal tobacco sales in April 2015 and seized a total of 513.3 kg of illegally traded cut tobacco and 936 boxes of cigarettes from different manufacturers that did not have excise duty mark. To obtain reliable and unbiased results, five samples were excluded by random sampling and sent to an accredited laboratory at ASL Analytic Service Laboratory GmbH, Hamburg, Germany for analysis of foreign matter content such as pesticides and heavy metals. The results of the analyzes were as follows:

- In all five samples the lead concentration exceeded the maximum allowable concentration on average four times;
- The concentrations of cadmium in the four samples tested are five times higher than the maximum allowable concentration, while in one sample the concentration of this heavy metal is even thirty times higher than the allowed concentration;
- In the five analyzed samples, the concentrations of arsenic are on average one and a half times higher than the allowed maximum values;
- The nickel concentration in the five analyzed samples is up to fifty times higher than the prescribed maximum allowed concentration;
- In only one test sample did the tobacco concentration of mercury exceed the maximum allowable concentration ten times.

In our work, all samples were in the reference values for all tested heavy metals.

4 Conclusion

The Risk Assessment Framework is a useful tool to get you started on considering the specific aspects of the risk of tobacco smoke-specific cancers. This framework makes it possible to link the levels of carcinogens in cigarette smoke and the perceived health risks of cigarette smoking made from typical tobacco varieties produced in BiH.

As the number of illegal black markets for tobacco and tobacco products has increased in Bosnia and Herzegovina due to the increase in cigarettes, this paper seeks to point out that cut tobacco that is illegally sold in makeshift packaging (unmarked boxes - so-called shirts, nylon bags) without a label Manufacturers, origin and quality pose a great risk to consumer users, which is reflected in several segments:

- its quality is not controlled, neither as raw tobacco in the leaf or as cut tobacco;
- raw tobacco in the leaf has not undergone the fermentation phase (subjected to high temperatures);

- Moisturizing, cutting and packaging of cut tobacco is done in makeshift warehouses in unhygienic conditions (cut tobacco is highly susceptible and can absorb bad odors and mold);
- cut tobacco is not subject to any control in terms of the content of nicotine, tar, carbon monoxide, pesticides, heavy metals, microorganisms, and other harmful additives found on tobacco leaf and other harmful ingredients that may be found in uncontrolled cultivation and cut tobacco;
- inadequate packaging often leads to the formation of tobacco molds;
- during the technological process, such tobacco did not undergo a phase of exposure to high temperatures, evaporation, and removal of impurities, unlike the technological process that takes place under controlled conditions in tobacco factories.

References

1. Dorne, J.L., Kass, G.E., Bordajandi, L.R., Amzal, B., Bertelsen, U., Castoldi, A.F., Heppner, C., Eskola, M., Fabiansson, S., Ferrari, P., Scaravelli, E., Dogliotti, E., Fuerst, P., Boobis, A. R., Verger, P.: Human risk assessment of heavy metals: principles and applications. Met. Ions Life Sci. **8**, 27–60 (2011)
2. Wentz, P.W.: Chelation Therapy: Conventional Treatments. Advance for Administrators of the Laboratory, Lab. Corp, Burlington (2000)
3. Godt, J., Scheidig, F., Grosse-Siestrup, C., Esche, V., Brandenburg, P., Reich, A., Groneberg, D.A.: The toxicity of cadmium and resulting hazards for human health. J. Occup. Med. Toxicol. **1**, 22 (2006)
4. International Agency for Research on Cancer: Agents reviewed by the IARC monographs, vol. 1–99. IARC (2008)
5. Clark, B.R., Brennan, E.: Tob. Sci. **27** (1983)
6. Šorak–Pokrajac, M.: Istraživanje promjena toksičnoh tvari u dimu cigareta, Doktorska disertacija, Zagreb (1991)
7. IARC, International Agency for Research on Cancer: A review of humcarcinogens. Part C: Arsenic, metals, fibres and dust. WHO, Lyon (2012)
8. Mandal, B.K., Suzuki, K.T.: Arsenic round the world: a review. Talanta **58**, 201–235 (2002)
9. WHO: Background document for development of WHO Guidelines for drinking-water quality. Arsenic in drinking water, pp. 1–24 (2011)
10. Clayton, G.D., Clayton, F.E.: Am. J. Physail 201, **63**, 927 (1961)
11. Small Jr., H.G., McCants, C.B.: Agron. J. **54**, 129–133 (1962)
12. Dermelj, M., Ravnik, V., Byrne, A.R., Vaskelj, A.: Microchim, acta I (Wien), pp. 261–270 (1978)
13. Mengel, K., Kirkby, E.A.: Principles of Plant Nutrition, 4th edn. Inter. Potach Inst., Bern (1987)
14. Bogdanović, D.: Izvori zagađenja zemljišta niklom. Letopis naučnih radova **31**(1), 21–28 (2007)
15. Šaćiragić, B.: Agrohemija, Univerzitetska knjiga, Sarajevo, pp 118–119 (2000)
16. Fischbein, A.: Occupational and environmental lead exposure. In: Rom, W.N. (ed.) Environmental and Occupational Medicine, vol. 2, pp. 735–758. Little Brown, Boston (1992)

17. Counis, M.F.: L-DNase II, a molecule that links proteases and endonucleases in apoptosis, derives from the ubiquitous serpin leukocyte elastase inhibitor. Mol. Cell. Biol. **18**, 3612–3619 (1998)
18. Valldron, H.A.: Brit. J. Med. **23**, 83 (1966)
19. Ferm, V.H.: Experientia **25**, 56 (1969)
20. U.S. Environmental Protection Agency: Risk assessment guidance for superfund. In: Vorhees, D.J., Dodson, R.E. (ed.) Estimating Risk To Cigarette Smokers from Smoke Constituents in Proposed "Testing and Reporting of Constituents of Cigarette Smoke" Regulations. Menzie-Cura & Associates, Inc., Boston (1999)
21. Sebiawu, G.E., Mensah, N.J., Ayiah-Mensah, F.: Analysis of heavy metals content of tobacco and cigarettes sold in Wa Municipality of Upper West Region, Ghana. Chem. Process Eng. Res. **25**, 24–33 (2014)
22. Lazarević, K., Nikolić, D., Stošić, L., Milutinović, S., Videnović, J., Bogdanović, D.: Determination of lead and arsenic in tobacco and cigarettes: an important issue of public health. Cent. Eur. J. Public Health **20**(1), 62–66 (2012)
23. Wu, H., Liao, Q., Chillrud, S.N., Yang, Q., Huang, L., Bi, J., Yan, B.: Environmental exposure to cadmium: health risk assessment and its associations with hypertension and impaired kidney function. Sci. Rep. (2016). https://www.ncbi.nlm.nih.gov/pmc/articles/PMC4944124/
24. O'Connor, R.J., Li, Q., Stephens, W.E., Hammond, D., Elton-Marshall, T., Cummings, K. M., Giovino, G.A., Fong, G.T.: Cigarettes sold in China: design, emissions and metals. Tob. Control (2010). https://tobaccocontrol.bmj.com/content/19/Suppl_2/i47
25. Muhammad Saeed, Z., Muhammad, N., Khan, S.A., Gul, F., Khuda, F., Humayun, M., Khan, H.: Assessment of potential toxicity of a smokeless tobacco product (naswar) available on the Pakistani market. Tob. Control (2011). https://tobaccocontrol.bmj.com/content/21/4/396

Application of Eco-Compatibile Technology – Phytoremediation – Case Study with Phytoaccumulator *Plantago lanceolata*

Fatima Muhamedagić[1(✉)] and Mehmed Cero[2]

[1] Biotechnical Faculty, University of Bihac,
Luke Marjanovica b.b, 77000 Bihac, Bosnia and Herzegovina
fatima.muhamedagic@gmail.com
[2] Federal Ministry of the Environment and Tourism,
Sarajevo, Bosnia and Herzegovina

Abstract. Phytoremediation is known as eco-compatible technologies in an environmentally friendly technology segment. This paper presents the results of phytoremediation using wild facultative natural phytoaccumulation of the natural habitat of *Plantago lanceolata.* The analyzed included 18 samples from three locations. All locations are located in Una National Park, namely: the first location – Kalati, second location – Cukovi and third location – Rajinovac. The AAS method was used to determine the total content (mg/kg) of the toxic element in soil and plant samples. The results are presented in the values of the phytoaccumulation factors – PF. According to the obtained PF factor values, *Plantago lanceolata* conducted a moderate phytoaccumulation of cadmium and arsenic, while for nickel conducted to be a low phytoaccumulation.

Keywords: Eco-compatibile technologies · Phytoremediation · *Plantago lanceolata* · Phytoaccumulation Factor (PF)

1 Introduction

With the industrialization and urbanization of developing countries and the increasing demands of mankind, the consumption of heavy metal elements has been growing enormously. As a result, pollution by heavy metals has become an increasingly serious issue, which has raised public awareness. Unlike organic pollutants, which can be entirely, or partly degraded by the action of various (micro)organisms or detoxified by the action of plants, heavy metals are non-degradable and can only be transformed into forms with altered toxicity and/or mobility/bioavailability [1]. The persistence of heavy metals in environmental matrices, especially soils, sediments and water bodies, is a result of the elemental character of heavy metals [2]. Pollution by heavy metals, therefore, represents a long-term threat for the environment and human health, not only in industrial and post-industrial areas.

Pollution by heavy metals (HM) represents a serious threat for both the environment and human health. Due to their elemental character, heavy metals cannot be chemically degraded, and their detoxification in the environment mostly resides either in stabilization in situ or in their removal from the matrix, e.g., soil. Clean-up

I. Karabegović (Ed.): NT 2020, LNNS 128, pp. 770–778, 2020.
https://doi.org/10.1007/978-3-030-46817-0_87

technologies have been developed for the removal of heavy metals [3], but often these are expensive or have some environmentally deleterious consequences. Soil washing techniques, both in situ and ex-situ, aim to remove metals from the soil but have detrimental effects on their physical and biological properties and result in a waste product with a high metal content, which itself requires disposal. Phytoremediation is the use of plants to reduce the concentrations or toxic effects of contaminants in the environments [4]. Furthermore, it has been used for cleaning up the environment with high success [5].

2 Phytoremediation

Phytoremediation is the name given to a set of technologies that use different plants as a containment, destruction, or an extraction technique. This technology has received attention lately as an advanced, cost-effective alternative to the more established treatment methods used at hazardous waste sites [6–8]. Phytoremediation is an emerging technology that uses various plants to degrade, extract, contain, or immobilize contaminants, including metals, pesticides, hydrocarbons, and chlorinated solvents from soil and water. Phytoremediation, also called green remediation, botano-remediation, agroremediation, or vegetative remediation is considered a publicly appealing (green) remediation technology that uses vegetation and associated microbiota, soil amendments and agronomic techniques to remove, contain, or render the heavy metals harmless in the soil [9, 10]. Phytoremediation is a novel, less expensive, efficient, environment and eco-friendly remediation strategy with good public acceptance [11–13].

Phytoremediation is the lower capital costs, aesthetic benefits, minimization of leaching of contaminants and soil stabilization. The operational cost of phytoremediation is also substantially less and involves mainly fertilization and watering for maintaining plant growth. Phytoremediation can be classified into different application such as phytoextraction, phytodegradation, phytostabilization, phytofiltration or rhizofiltration and phytovolatilization Fig. 1.

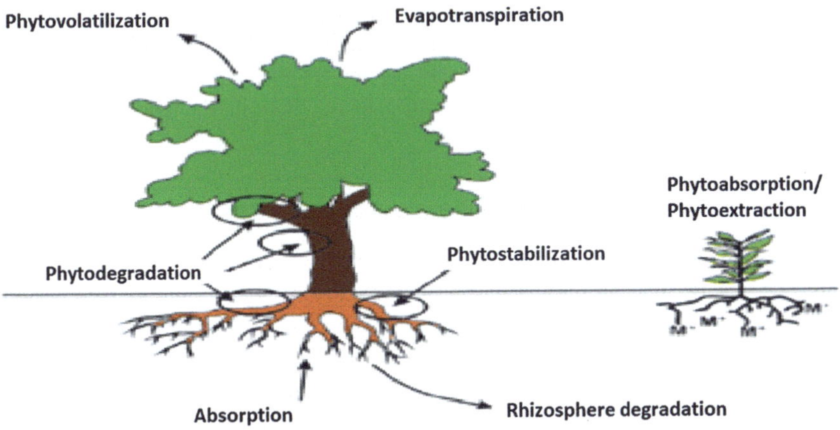

Fig. 1. Process of plant absorption and degradation of absorbet particles

National park "Una" unique natural complex in the southeastern part of Europe due to its location on the border of three climate areas, karst formations and hydrography, which are, together with a basin of Korana and Krka, unique in the European range. Waterfalls and rapids of "Una" represent a remarkable heritage, and the entire upper flow is characterized by rich biodiversity. Due to its sensitivity as highly sensitive ecosystem National park "Una" was taken as the place for a performance of a case study of phytoremediation with a plant *Plantago lanceolata* L. The main objective of our experiment was to investigate whether promoting plant growth by organic matter additions increases the uptake of heavy metals.

This paper presents results of the field research, sampling and laboratory testing. For the purpose of this paper the three pilot sites were taken: site Kalati, site Cukovi and site Rajnovac Fig. 2.

Fig. 2. Pilot sites (a) Kalati, (b) site Cukovi and (c) site Rajnovac (National park "Una")

After the field observance of the National park "Una", the three sites were taken as pilot sites. These areas were previously being determined as the places of the high presence of the heavy metals (Cd, Ni and As). Preparation of soil samples was carried out from the side of researchers and examined in a laboratory. After the first vegetation season was near the end, the vegetation was taken to the examination along with a soil samples. Method for determination of the concentration of heavy metals collected plant material was: Atomic absorption spectrometry – AAS.

Data retrieved were analyzed based on the value of phytoaccumulation factor (PF factor) which was calculated from the ratio of the results of the mean value of total content of cadmium (Cd), nickel (Ni) and arsenic (As) in the soil and the mean concentration in plant material.

2.1 Calculation Phytoaccumulation Factor

The Phytoremediation factors are the ratio of the concentration of heavy metals in plants and in soils. It is an indicator of a plant's capacity to accumulate heavy metals. The Phytoremediation factors were calculated using Eq. 1: The formula for phytoremediation factor [14].

$$PF = \frac{C_{plant}}{C_{soil}} \tag{1}$$

Where is:

C_{plant} – the concentration of metals in the plant (mg/kg)
C_{soil} – the concentration of metal in the soil (mg/kg)

After obtaining Phytoremediation factor for the plant *Plantago lanceolata,* the following factor will be compared in accordance with the values given in the Table 1.

Table 1. Class of plant phytoaccumulator range

Value phytoaccumulation factor	Range
High phytoaccumulator	Phytoaccumulation factor between 1–10
Moderate phytoaccumulator	Phytoaccumulation factor between 0.1 to 1.0
Low phytoaccumulator	Phytoaccumulation factor between 0.01 to 0.1
Non phytoaccumulator	Phytoaccumulation factor between 0.01

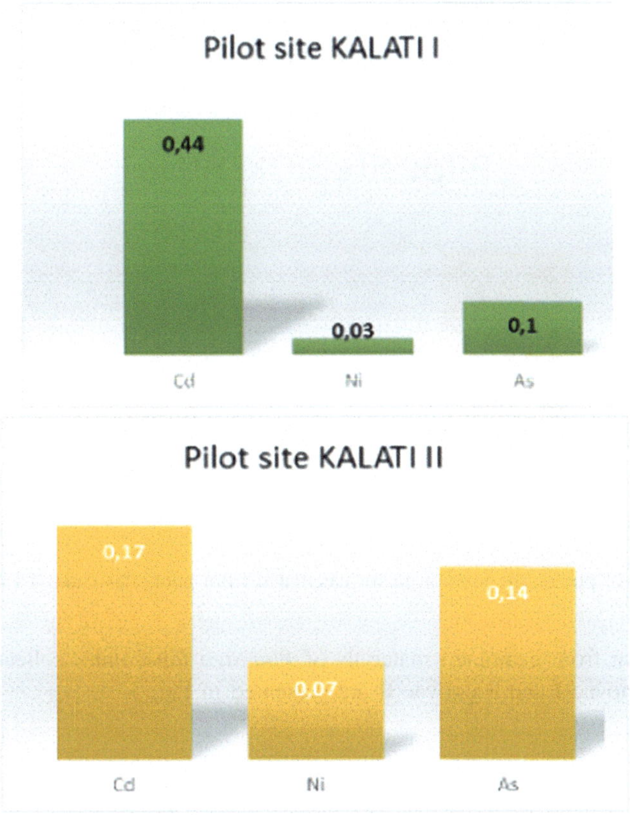

Fig. 3. Value of phytoaccumulation factor calculated from pilot sites Kalati I and Kalati II

Results of this research show that levels of Phytoaccumulation factor, relative to the toxic elements analyzed (Cd, As, Ni) showed that the *Plantago lanceolata L.* phytoaccumulator had generally moderate values for cadmium and arsenic while for nickel low PF values.

Calculation from examined materials of *Plantago lanceolata,* collected from the location Kalati I and Kalati II are presented in Fig. 3.

Calculation from examined materials of *Plantago lanceolata,* collected from the location Cukovi I and Cukovi II are presented in Fig. 4.

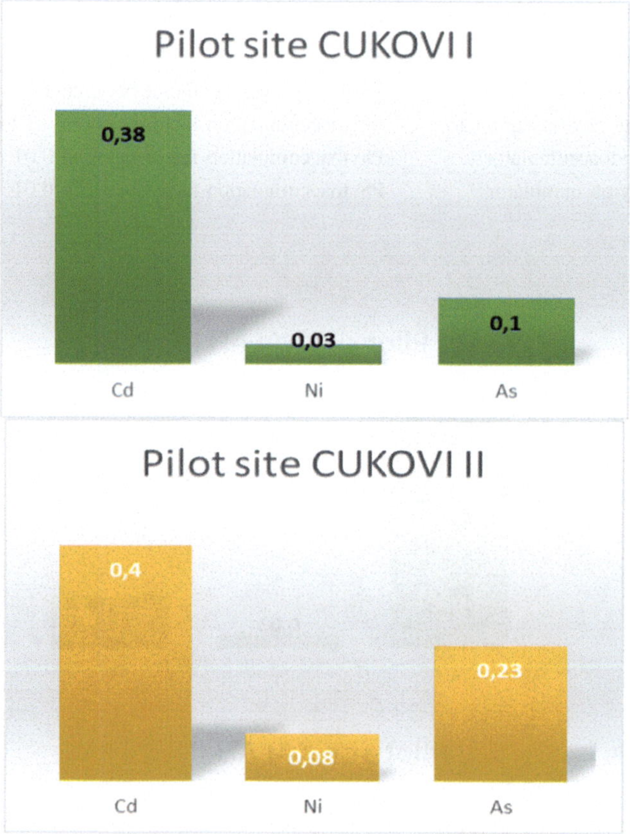

Fig. 4. Value of phytoaccumulation factor calculated from pilot sites Cukovi I and Cukovi II

Calculation from examined materials of *Plantago lanceolata*, collected from the pilot site Rajnovac I and Rajnovac II are presented in Fig. 5.

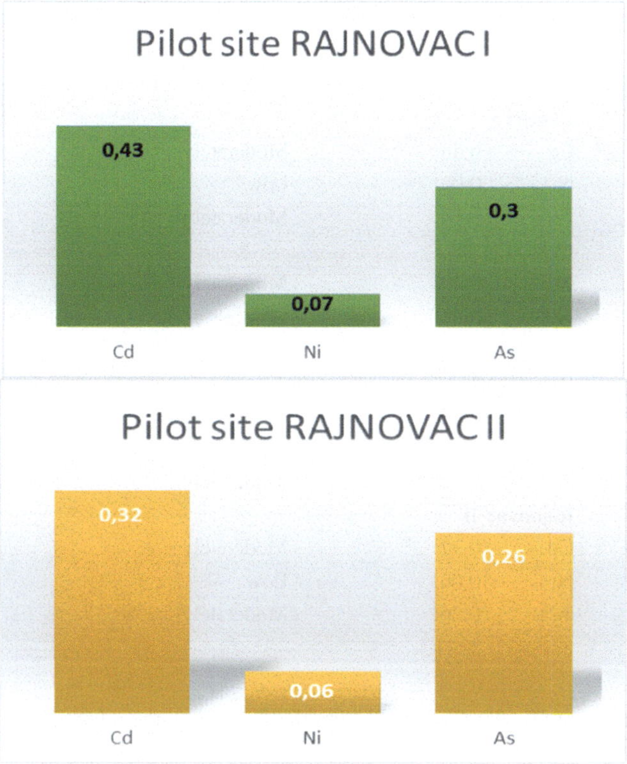

Fig. 5. Value of phytoaccumulation factor calculated from pilot sites Rajnovac I and Rajnovac II

The data retrieved from upon the calculation was then compared in accordance with a Table 1 and results are presented in Table 2.

Table 2. Comparison of the PF-factor obtained from pilot sites with a class of plant phytoaccumulator range

Element	Value of PF factor	Phytoaccumulator *(Plantago lanceolata L.)*
Kalati I		
Cd	0,44	Moderately
Ni	0,03	Low
As	0,10	Moderately
Kalati II		
Cd	0,17	Moderately
Ni	0,07	Low
As	0,14	Moderately

(*continued*)

Table 2. (*continued*)

Element	Value of PF factor	Phytoaccumulator (*Plantago lanceolata L.*)
Ćukovi I		
Cd	0,38	Moderately
Ni	0,03	Low
As	0,10	Moderately
Ćukovi II		
Cd	0,40	Moderately
Ni	0,08	Low
As	0,23	Moderately
Rajnovac I		
Cd	0,43	Moderately
Ni	0,07	Low
As	0,30	Moderately
Rajnovac II		
Cd	0,32	Moderately
Ni	0,06	Low
As	0,26	Moderately

This research application of phytoremediation using the phytoaccumulator *Plantago lanceolata L.* showed the following results for PF factors:

- The first locality of Kalati I and Kalati II
  - moderate PF values for cadmium 0.17–0.44
  - moderate PF values for arsenic 0.10–0.14
  - low PF values for nickel 0.03–0.07
- Second locality of Cukovi I and Cukovi II
  - moderate PF values for cadmium 0.38 to 0.40
  - moderate PF values for arsenic 0.10–0.23
  - low PF values for nickel 0.03 - 0.08
- Third locality Rajinovac I and Rajinovac II
  - moderate PF values for cadmium 0.32–0.43
  - moderate PF values for arsenic 0.26–0.30
  - low PF values for nickel 0.06–0.07.

3 Conclusions

Heavy metals are one of the most critical threats to the soil and water resources, as well as to human health. Phytoremediation, in a contest of this research, has been proven to be a very successful, inexpensive and simple green technology. The results presented in this research demonstrate that plant *Plantago lanceolata L.* is an efficient tool for bioremediation of high metal-contaminated sites. These plants demonstrate high growth potential with high tolerance to heavy metal contaminated areas, acting as cover crops against the direct impact of the rainfall in the soil surface, reducing soil and water losses by surface runoff with consequent contamination of the environments. Furthermore, high macro and micronutrients were accumulated in the biomass, showing an important role in nutrient cycling.

In the application of Phytoremediation with a *Plantago lanceolata L.* for cadmium and arsenic, the results show moderate phytoaccumulation at all three pilot sites (Phyto accumulation factor: for cadmium 0.17–0.44, for arsenic 0.10–0.30). For the nickel element, *Plantago lanceolata L.* showed low phytoaccumulation at all three study sites (Phyto accumulation factor 0.03–0.08).

Choosing the right plant species as a phytoaccumulator is one of the first steps that lead to successful desired goals. In this case, the *Plantago lanceolata L.* was proven to be natural phytoaccumulator and represents a good choice because it does not damage the natural environment or the other vegetation. The ability of *Plantago lanceolata* L. to accumulate heavy metals, based on the outcomes of the present research, it was concluded that *Plantago lanceolata L.* is capable of accumulating all selected heavy metals from soil.

Further development and refinement of the application of phytoremediation and its processes would require moving towards commercialization and wider application in BiH, especially in the areas of contaminated (industrial, urban and similar) zones.

Owing to that examined metals, their toxicity and possible bioaccumulation, these compounds should be subject to mandatory monitoring. Governments should promote harmonized data collection, research, legislation and regulations, and consider the use of indicators. Assessments determining the chemical concentration scenario and the use of biomarkers should provide useful data to set standards and guideline values designed to protect human and environmental health from heavy metal contaminants.

From the point of view of ecology, this green technology, through further refinement, could in the future take the leading role in the process of purification and rehabilitation of environment (soil, water, air).

References

1. Silver, S., Phung, L.T.: Genes and enzymes involved in bacterial oxidation and reduction of inorganic arsenic. Appl. Environ. Microbiol. **71**, 599–608 (2005). https://doi.org/10.1128/AEM.71.2.599-608.2005
2. Chapman, P.M., Thornton, I., Persoone, G., Janssen, C., Godtfredsen, K., Z'Graggen, M.N.: International harmonization related to persistence and bioavailability. Hum. Ecol. Risk Assess. Int. J. **2**, 393–404 (1996). https://doi.org/10.1080/10807039609383618
3. Luo, Y.M.: Current research and development in soil remediation technologies. Prog. Chem. **21**, 558–565 (2009)
4. Ali, H., Khan, E., Sajad, M.A.: Phytoremediation of heavy metals—concepts and applications. Chemosphere **91**(7), 869–881 (2013). https://doi.org/10.1016/j.chemosphere.2013.01.075. Pmid:23466085
5. Pandey, V.C.: Suitability of *Ricinus communis* L. cultivation for phytoremediation of fly ash disposal sites. Ecol. Eng. **57**, 336–341 (2013). https://doi.org/10.1016/j.ecoleng.2013.04.054

6. Pulford, I.D., Watson, C.: Phytoremediation of heavymetal-contaminated land by trees: a review. Environ. Int. **29**, 529–540 (2003)
7. Jadia, C.D., Fulekar, M.H.: Phytoremediation of heavy metals: recent techniques. Afr. J. Biotechnol. **8**(6), 921–928 (2009)
8. Zhang, B.Y., Zheng, J.S., Sharp, R.G.: Phytoremediation in engineered wetlands: mechanisms and applications. Procedia Environ. **2**, 1315–1325 (2010)
9. Cunningham, S.D., Ow, D.W.: Promise and prospects of phytoremediation. Plant Physiol. **110**, 715–719 (1996)
10. Helmisaari, H.S., Salema, M., Derome, J., Kiikkila, O., Uhlig, C., Nieminen, T.M.: Remediation of heavy metal-contaminated forest soil using recycled organic matter and native woody plants. J. Environ. Qual. **36**, 1145–1153 (2007)
11. Singh, A., Eapen, S., Fulekar, M.H.: Potential of Medicago sativa for uptake of cadmium from contaminated environment. Rom. Biotechnol. Lett. **14**, 4164–4169 (2009)
12. Saier, M.H., Trevors, J.T.: Phytoremediation: water, air and soil pollution, S61–S63 (2010)
13. Revathi, K., Harbabu, T.E., Sudha, P.N.: Phytoremediation of chromium contaminated soil using sorghum plant. Int. J. Environ. Sci. **2**(2), 417–428 (2011)
14. Zgorelec, Ž.: Phytoacumulation of metals and metalloids from soil polluted by coal ash. Doctoral dissertations, Faculty of Agriculture, University of Zagreb, Zagreb, p. 105 (2009)

Human Risk Assessment Based on the Content of Inorganic and Organic Pollutants in Sarajevo's Playgrounds

Aida Šapčanin[1(✉)], Ekrem Pehlić[2], Mirsada Salihović[1], and Alisa Smajović[1]

[1] Faculty of Pharmacy, University of Sarajevo, Zmaja od Bosne 8, 71 000 Sarajevo, Bosnia and Herzegovina
ida@bih.net.ba

[2] Faculty of Health Studies, University of Bihać, Bihać, Bosnia and Herzegovina

Abstract. Soil is an important part of ecosystem actively involved in the natural cycle of substances, so it is necessary to investigate the content of pollutants and to establish preventive activities on this basis in order to reduce soil pollution, and thus preserving the flora and fauna, as well as human health. The main objective of this study was the human risk assessment based on the content of inorganic and organic polutants in the soil of children playgrounds. The contents of inorganic pollutants such as Cd, Pb, Cr, Ni, Cu, Zn, Co, Se and As, and organic pollutants such as naphtalene, fluoranthene and benzo(a)pyrene have been measured in Sarajevo's playgrounds soil and the HQ and HI have been calculated. Risk characterization relevant for the present investigation comprises calculations of non-carcinogenic risk for ingestion and inhalation of soil. Our results suggest that children are at higher risks than the adults due to their contact with potentially polluted soil.

Keywords: Soil · Inorganic pollutants · Organic polutants · Playgrounds · Health risk assessment

1 Introduction

The risk assessment is a multistep procedure that comprises the following: data collection, exposure assessment, toxicity assessment and risk characterisation. A non-cancerogenic impact of particular elements is assessed by calculating the hazard coefficient (HQ) and hazard Index (HI) [1–3]. The commonly used chemical risk assessment paradigm entails identification and characterisation of a chemical hazard, which is then compared with an estimate of human exposure to the chemical, in order to assess the risk posed. Risk assessment becomes more complex when it is necessary to assess mixtures of similar chemicals, such as the PAHs since the hazard, mechanism of toxicity and potency may vary between chemicals. Despite the numerous studies attempting to quantify exposure factors relevant to a risk assessment for children during playing activities, there is a significant variability in their numerical results reflecting the difficulties involved [3–7]. Besides, some of these factors, like exposure frequency, cannot be directly extrapolated from one scientific study to another since playing habits

I. Karabegović (Ed.): NT 2020, LNNS 128, pp. 779–785, 2020.
https://doi.org/10.1007/978-3-030-46817-0_88

and time spent outdoors may differ basically from one geographical area to another. The uncertainty in the estimates of the degree of exposure and in toxicity data for different elements in their different chemical forms confers, in turn, a significant degree of uncertainty to the risk analysis, but should not invalidate the potential of risk estimation to identify the contaminants of most concern and the most relevant routes of exposure.

The goal of this study was to assess health risks posed to children and adults based on the contents of inorganic pollutants in the soil of Sarajevo urban area where Cd, Pb, Cr, Ni, Cu, Zn, Co, Fe, Se and As, [1] and organic pollutants such as naphtalene, fluoranthene and benzo(a)pyrene were identified as potential hazardous agents in the soil, especially at playgrounds. Successfully conducted risk assessment does not mean improvement at the same time if this result is not implemented, that is, if activities are taken to solve the problem and remove the risk or significantly reduce the risk. This goes in support of the aforementioned, that the goal of health risk assessment is actually to detect its existence and the extent to which it exists and then, if necessary, to plan activities for its elimination. Risk management is a process of identifying, evaluating, selecting and implementing activities that will completely eliminate the health risk for a population exposed to a harmful substance from the environment and, if this is not possible, at least significantly reduce it.

2 Materials and Methods

Soil samples for the determination of inorganic pollutants were prepared by microwave assisted acid digestion and determined by using an atomic absorption spectrophotometer [8, 9]. Furthermore, the potential health risk assessment calculated for a lifetime of exposure (ingestion and inhalation), based on USEPA model [4–6, 10, 11], was determined as the non-carcinogenic risk. The value of total hazard index for both children and adults was calculated for maximum, minimum and average concentration of pollutants [1, 2].

The non-carcinogenic effects of inorganic and organic pollutants were assessed by using Eqs. (1 and 2):

$$ADD_{ing} = C_{soil} \times \frac{IngR \times EF \times ED}{BW \times AT} \times 10^{-6} \tag{1}$$

$$ADD_{inh} = C_{soil} \times \frac{InhR \times EF \times ED}{PEF \times BW \times AT} \tag{2}$$

ADD_{ing} - Average daily doses (mg/kg day) via ingestion
ADD_{inh} - Average daily doses (mg/kg day) via inhalation

The naphtalene, fluoranthene and benzo(a)pyrene were analyzed by using a Shimadzu single quadrupole GCMSQP2010 Ultra gas chromatograph-mass spectrometer according to EPA protocols. Maximum allowable concentration for organic pollutants was estimated by using the EPA regulations [9] (Table 1).

Table 1. Definition and reference values of some health risk assessment parameters used for pollutants in the soil samples [4–6, 10, 11]

Factor	Definition	Unit	Value	
			Children	Adults
C_{soil}	Pollutant content in soil	mg/kg		
IngR	Ingestion rate of soil	mg/day	200	100
EF	Exposure frequency	days/year	350	350
ED	Exposure duration	years	6	20
BW	Body weight of the exposed individual	kg	15	80
AT	Average time	days	365ED	365ED
InhR	Inhalation rate of soil	m^3/day	7,63	12,8
PEF	Particle emission factor	m^3/day	$1,36 \times 10^9$	$1,36 \times 10^9$

2.1 A Non-carcinogenic Risk Assessment

Risk assessment is a function of hazard and exposure and is defined as the process of estimating the expectation of occurrence of an event and the probable consequence of negative health effects on human exposures to environmental hazards over a specified time period [1, 2]. Determination of hazard elementary aims to investigate toxicants that are present at any given location, their exact content, and spatial distribution. In the study area, inorganic pollutants like: Cd, Pb, Cr, Ni, Cu, Zn, Co, Se and As and organic pollutants such as naphtalene, fluoranthene and benzo(a)pyrene were identified as possible hazards for the community. Non-cancerogenic activities of pollutants were assessed based on calculations of the hazard coefficient (3) and the hazard index (4).

$$HQ = \frac{ADD}{RfD} \tag{3}$$

HQ – Hazard coefficient
RfD – Chronic reference dose (mg/kg day) of a specific pollutant

$$HI = \sum_{k=1}^{n} HQ_k = \sum_{k=1}^{n} \frac{ADD_k}{RfD_k} \tag{4}$$

HI – Hazard Index.

3 Results and Discussion

Due to the children's exposure to urban soil, non-cancerogenic hazardous index (HI) for children was estimated using 95th percentile values of total metal concentration. The results for non-carcinogenic children health risk, based on metal concentrations (minimal and maximum) in playgrounds and exposure by two different routes were calculated [1, 2, 4–6, 10, 11]. The hazard coefficients based on organic pollutants for children and adults are shown in Tables 2 and 3.

Table 2. The hazard coefficient based on organic pollutants for children

Organic pollutant	Ingestion		
	RfD_{ing}	ADD_{ing}	HQ
Naphtalene	$2{,}00 \times 10^{-2}$	$3{,}9635 \times 10^{-7}$	$1{,}98 \times 10^{-5}$
Fluoranthene	$4{,}00 \times 10^{-2}$	$4{,}5516 \times 10^{-6}$	$1{,}14 \times 10^{-4}$
Benzo(a)pyrene	$3{,}00 \times 10^{-4}$	$2{,}6402 \times 10^{-6}$	$8{,}80 \times 10^{-3}$

Table 3. The hazard coefficient based on organic pollutants for adults

Organic pollutant	Ingestion		
	RfD_{ing}	ADD_{ing}	HQ
Naphtalene	$2{,}00 \times 10^{-2}$	$3{,}7157 \times 10^{-8}$	$1{,}86 \times 10^{-6}$
Fluoranthene	$4{,}00 \times 10^{-2}$	$4{,}2671 \times 10^{-7}$	$1{,}07 \times 10^{-5}$
Benzo(a)pyrene	$3{,}00 \times 10^{-4}$	$2{,}4751 \times 10^{-7}$	$8{,}25 \times 10^{-4}$

Calculations of the non-carcinogenic hazard index (HI) based on the content of investigated inorganic pollutants are shown in Fig. 1.

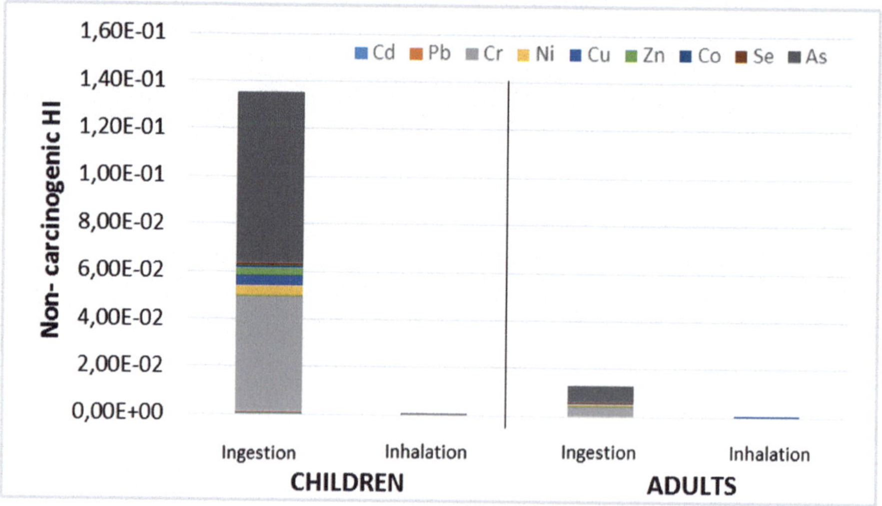

Fig. 1. The non-carcinogenic hazard index (HI) based on the content of investigated inorganic pollutants

Calculations have shown that values for HQ and HI obtained from children's playgrounds are lower than the standard values equal to 1, so that they do not pose a direct risk for children. The highest values were for Pb and Cr and the result was similar to the results obtained in Mugoša et al., study (2015). The results for HI for Pb from all investigated locations were very close to the upper limit of the safe level and it seems that they pose a long term risk. [12–14]. The highest values of HI are obtained if taken orally, and the lowest if inhaled. In comparison to other investigations, for example, exposure to Pb in the urban soil in Podgorica may also pose a health threat, especially to young children [14]. Because of its negative effects on the children's central nervous system, monitoring of Pb content in soil is of great importance.

Calculations of the non-carcinogenic hazard index (HI) based on the content of investigated organic pollutants are shown in Fig. 2.

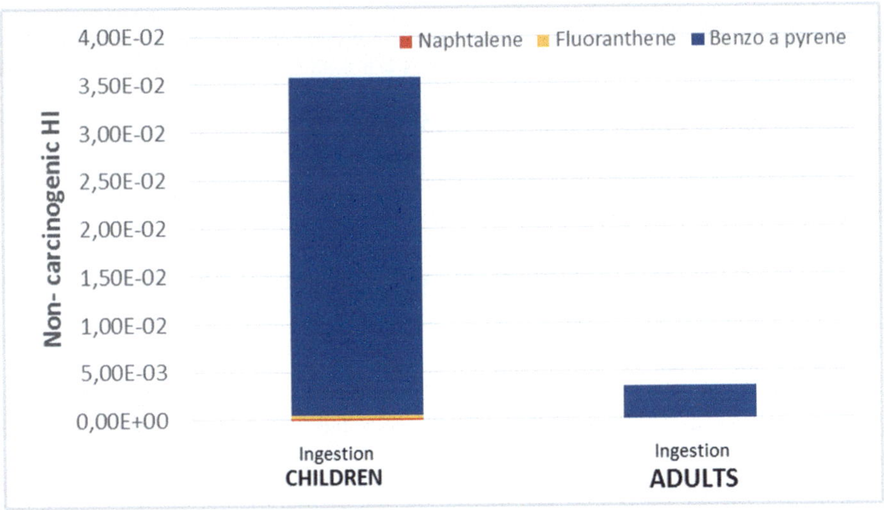

Fig. 2. The non-carcinogenic hazard index (HI) based on the content of investigated organic pollutants

Comparing HI values for both children and adults, it can be concluded that children are at much higher risk of being exposed to non-cancerogenic risks caused by exposure to organic toxicants elements from the soil.

These figures of risk should be interpreted logically: firstly, because the estimates of exposure factors and the toxicity data are affected by a high degree of concern that is conveyed to the numerical outcome of the risk model. And secondly because other opportunities of potential exposure to different toxicants (i.e., dietary intake, inhalation of urban aerosol, indoor exposure to dust particles at urban area) which have not been assessed in this study, should elevate the overall figure of risk. The risk assessment has proven to be a very useful tool to reveal the true meaning and relevance of the content of inorganic and organic pollutants found in urban area. Considering toxicity and level of exposure, risk assessment allows the identification of the pollutants and pathways of most concern in a situation of urban environmental exposure. In the case of Sarajevo's playgrounds, the highest risk experienced by children exposed to trace elements in the soil of playgrounds increase from the ingestion of soil particles during the playing activities.

4 Conclusion

Calculated HQ and HI of inorganic and organic pollutants were lower than the safe limit of 1, indicating that there is no direct risk to human health from inorganic neither organic pollutants in the investigated area; however, these levels should be monitored in a long-term perspective. Our results suggest that children are at higher risks than the adults due to their contact with potentially polluted soil.

Acknowledgements. This work was supported by the Federal Ministry of Education and Science in Bosnia and Herzegovina and was carried out within the framework of the project „Assessment of health risk caused by heavy metal contamination of the soil at children's playgrounds in Sarajevo" (Grant no. 0101-7552-17/15, dated 14.12.2016).

References

1. Sapcanin, A., Smajovic, A., Pehlic, E., Salihovic, M., Jancan, G.: Health risk assessments based on the contents of heavy metals in Sarajevo urban soil. In: Karabegovic (eds.) New Technologies, Development and Application II. NT 2019. Lecture Notes in Networks and Systems, vol. 69, pp. 595–603. Springer (2020)
2. Kamunda, C., Mathuthu, M., Madhuku, M.: Health risk assessment of heavy metals in soils from witwatersrand gold mining Basin, South Africa. Int. J. Environ. Res. Public Health **13** (7), 663 (2016)
3. Ružić, I.: Procjena rizika po zdravlje čovjeka i kvalitetu okoliša. Hrvatske vode: časopis za vodno gospodarstvo **6**(22), 43–57 (1998)
4. USEPA: Risk assessment guidance for superfund: human health evaluation manual. EPA/540/1-89/002, vol. 1. Office of Solid Waste and Emergency Response (1989)
5. USEPA: US environmental protection agency, resident soil table June 2017; dostupno na (2017). https://semspub.epa.gov/work/03/2245085.pdf
6. USEPA: Supplemental guidance for developing soil screening levels for superfund sites. OSWER 9355.4-24. Office of Solid Waste and Emergency Response (2001)
7. Salmani-Ghabeshi, S., Palomo-Marin, M.R., Bernalte, E., Rueda-Holgado, F., MiroRodriguez, C., Cereceda-Balic, F., Fadic, X., Vidal, V., Funes, M., Pinilla-Gil, E.: Spatial gradient of human health risk from exposure to trace elements and radioactive pollutants in soils at the Puchuncavi-Ventanas industrial complex, Chile. Environ. Pollut. **218**, 322–330 (2016)
8. Sapcanin, A., Cakal, M., Jacimovic, Z., Pehlic, E., Jancan, G.: Soil pollution fingerprints of children playgrounds in Sarajevo city, Bosnia and Herzegovina. Environ. Sci. Pollut. Res. **24** (12), 10949–10954 (2017)

9. Sapcanin, A., Cakal, M., Imamovic, B., Salihovic, M., Pehlic, E., Jacimovic, Z., Jancan, G.: Herbicide and pesticide occurrence in soils of children playgrounds in Sarajevo, Bosnia and Herzegovina. Environ. Monit. Assess **188**(8), 450–456 (2016)
10. USEPA: Risk assessment guidance for superfund Volume I: human health evaluation manual (Part A); Office of Emergency and Remedial Response, Washington, DC, USA (2004)
11. USEPA: Human health evaluation manual, supplemental guidance: update of standard default exposure factors. OSWER Directive 9200.1-120. Office of Solid Waste and Emergency Response (2014). https://rais.ornl.gov/documents/OSWER_Directive_9200.1120_ExposureFactors_corrected2.pdf
12. De Miguel, E., Iribarren, E., Chacon, I., Ordonez, E., Charlesworth, S.A.: Riskbased evaluation of the exposure of children to trace elements in playgrounds in Madrid (Spain). Chemosphere **66**, 505–513 (2007)
13. Shi, G., Chen, Z., Bi, C., Wang, L., Teng, J., Li, Y., Xu, S.: A comparative study of health risk of potentially toxic metals in urban and suburban road dust in the most populated city of China. Atmos. Environ. **45**, 764–771 (2011)
14. Mugoša, B., Đurović, D., Nedović-Vuković, M., Barjaktarović-Labović, S., Vrvić, M.: Assessment of ecological risk of heavy metal contamination in coastal municipalities of Montenegro. Int. J. Environ. Res. Public Health **13**(4), 393–408 (2016)

Influence of Initial Alcohol Content in Distillate on Distribution of Methanol and Higher Alcohols During Redistillations

Nudžejma Jamaković(✉) and Nermina Spaho

Faculty of Agriculture and Food Science, University of Sarajevo, Zmaja od Bosne 8, 71000 Sarajevo, Bosnia and Herzegovina
nudzejma.sahmatra@gmail.com

Abstract. Quality of strong alcoholic beverage depend on quantity and quality of volatile compounds that can exhibit both positive (esters, higher alcohols) and negative effects (methanol). In order to improve the sensory quality of the products but also to improve the quality from a toxicological point of view, the aim is to minimize the effect of less desirable components, and highlight the effect of more desirable components. When will some compound distil, it depends on its boiling point, but also its solubility in alcohol or water as the main carriers of all volatile compounds. Therefore, the content of alcohol in initial mixture is very important for distribution of other volatile compounds during distillation. The aim of this paper is to examine how the amount of alcohol in initial distillate influence on evaporation dynamics of methanol and higher alcohol during redistillations. The evaporation dynamics will be calculated by separating the fractions during the redistillation of two initial distillates with 20% vol. and 60% vol. of alcohol. Content of methanol and higher alcohol was determined by spectrophotometric method. The results showed that in the low-alcoholic mixtures significantly higher content of higher alcohols are separated in the first fraction (3041,01), and least of all in last fraction (575,18 mg/l a. a.). In the same time the content of methanol was higher in last fraction (0,4943 vol.% a. a.), then in the first fraction (0,1665 vol.% a. a.). Opposite dynamics of those compounds were evident in redistillation of high-alcoholic mixtures.

Keywords: Methanol · Higher alcohols · Distillation

1 Introduction

Strong alcoholic drink consists a huge number of volatile compounds that can manifest their positive but and negative effects too, which again depends on a number of factors. During the process of strong alcoholic drink production, the raw material passes through different stages of change, resulting in new different aromatic compounds. At the end, the aroma compounds of spirits are made upto four aromas: the primary aroma compounds that are originate from the fruit itself, the secondary aroma compounds that are produced during the fermentation, the tertiary aroma compounds that are formed during the distillation and the quaternary aroma compounds that are formed during the

I. Karabegović (Ed.): NT 2020, LNNS 128, pp. 786–794, 2020.
https://doi.org/10.1007/978-3-030-46817-0_89

spirits maturation [1–5]. Since the volatile components are concentrated during the distillation process, it is clear that the main carriers of all the volatile components are ethanol and water. The aim of distillation is to concentrate ethanol and all positive aroma compounds. So, all volatile compounds can be classified into a group of compounds that are more volatile than ethanol and thus significantly evaporate before ethanol, and in a group of compounds that are less volatile than ethanol and more intensively distillate after ethanol [3, 5–7].

There are a large number of factors that will influence on the volatility and distribution of individual compounds during distillation. Among them, the most important factor is boiling point, but also the solubility of a particular volatile compounds in water or ethanol, as the main carriers of all volatile components during distillation. Thus, Leutate [8] separated all the compounds which evaporated during distillation into five groups according to their boiling point and their partially or completely solubility in alcohol or water. Thanks to different boiling point of compounds and their different solubility, the volatile compounds can be concentrated in different distillation fractions. The main reason for the distillation cutting is separation of fractions or cuts with higher concentration of unwanted compounds. Thus, mainly three cuts are separated namely head, hart and tail. The head and tail consist a higher concentration of undesirable and harmful compounds. Undesirable or unwanted compounds are those that have an unpleasant aroma and sharp and stinging mouth feeling. Sometimes one compounds in low concentration is desirable but becomes undesirable at high amount [5]. For the human health it is more important to know what is content of harmful compounds in the spirits. One of them is methanol. Since it is toxic to humans, the level of methanol is regulated by EU Regulations (No. 110/2008) [10]. These Regulations determines the maximum concentration of methanol of fruit spirits to 1000 g/hl a. a and it should not exceed to 1200 g/hl a. a. for various fruit spirits. In the fruit spirits the content of methanol is often higher in ratio to all other compounds [1, 9, 11, 12]. Methanol intensively appears during alcohol fermentation by demethoxylation of esterified methoxy groups in pectin [5] thus, its presence determines the authenticity of spirits, and its concentration levels are an indicator of the level of safety for consumption [13].

However, some authors state the higher alcohols as quantitatively the largest group of aroma compounds in spirts [14, 15]. They are one of the most important groups of aromatic compounds because they produce esters with acids. The higher alcohols are responsible for the nice fruit and flower odour but just if they are present in low quantity. If they are present in larger quantities then they are very unpleasant and even toxic.

The goal of distillation is to remove the alcohol from methanol as much as possible while retaining a certain amount of higher alcohol and other pleasant compounds. The concentration of methanol and higher alcohols depends on the raw material [9, 16], then the fermentation conditions [1, 17], and the distillation techniques used [4, 16–19]. Its behaviour will be different depending on alcohol strength in the initial mixture [20].

The aim of this paper is to determine whether the concentration of 20% vol. and 60% vol. of alcohol in the initial mixture is sufficient to exhibit the opposite evaporation dynamics of methanol and total higher alcohols during distillation.

2 Material and Working Methods

In this research distillate with 33% vol. of alcohol was used. The distillate was divided into two parts, the first part being diluted to a strength of 20% vol. alcohol, and the second part is amplified with absolute alcohol at a strength of 60% vol. This means that two different samples of distillates were used, with different initial alcohol in the mixture. A volume of 100 ml of each sample was taken and placed on distillation on a laboratory glass distillation apparatus. During the distillation of initial alcohol-water mixture, the fractions of 15 ml were separated. During distillation of initial mixture with 20% vol. alcohol 4 fractions was separated, and 6 fractions were separated during distillation of initial mixture with 60% vol. alcohol. The number of separated fractions is determined on the basis of alcohol in fractions, and the final fraction (tail) had less than 5% vol. alcohol (Fig. 1). During distillation of lower-alcohol distillate, this was already achieved in the fourth fraction, and the alcohol concentration down upon fell below 5% vol. in the sixth fraction during distillation of higher-alcoholic distillate. The experiment was repeated three times, which means that there was a total of 12 fractions of lower-alcohol distillates and 18 fractions of higher-alcohol distillate for the analysis.

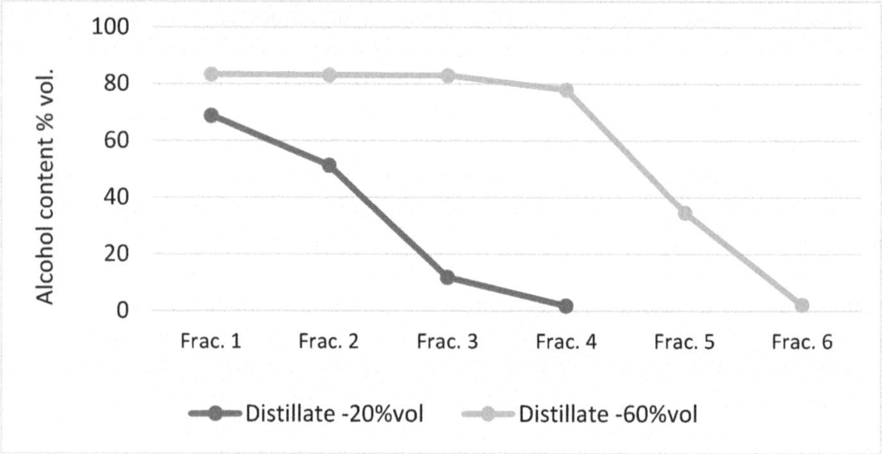

Fig. 1. The content of alcohol in separated fractions during the distillation of lower alcohol (20% vol.) and higher alcohol (60% vol.) distillate (n = 3)

The content of methanol and total higher alcohols was determinate in the obtained fractions. These compounds were determined by the spectrophotometric method OIV MA-AS312-03B [21], and the results of the concentration of methanol and higher alcohols were shown as an average of three repetitions. One-way analysis of variance (ANOVA) was performed to detect whether significant differences ($p < 0.05$) existed between mean concentration of tested compounds in analysed fractions depending on initial content of alcohol in mixture, distillation technique used and aging time. ANOVA followed by LSD test to verify statistical differences at the 0.05 significance level.

3 Results of Work and Discussion

Results of average content of total higher alcohols and methanol in the fractions of lower and higher alcohol distillate mixture were shown in Figs. 2 and 3 as well as Figs. 5 and 6 respectively. The distribution of methanol during distillation of lower alcohol mixture was presented in Fig. 2. As it can be seen in Fig. 2, significant differences were observed among the content of methanol among the fractions. The concentration of methanol increased during distillation thus the significantly the lowest content was in the first fraction and the highest content was in the third fraction. At the end of distillation, concentration of methanol decreased. The similar results were published by other authors [22–24]. As methanol is better soluble in water than in ethanol it is obviously that in low alcohol mixture methanol rather evaporated with water than with ethanol.

Distribution of methanol during distillation of higher alcohol mixture is completely different in ratio to the distribution of methanol during distillation of lower alcohol mixture (Fig. 3).

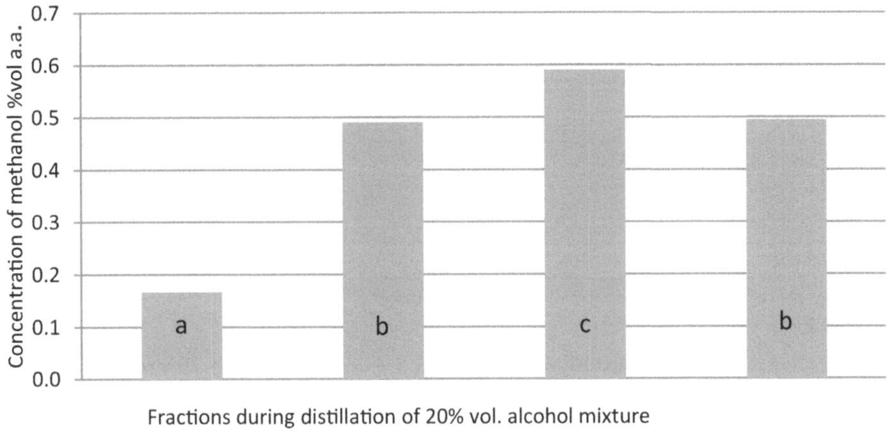

Fig. 2. The content of methanol (n = 3) in different fractions of 20% vol. alcohol with the results of the LSD test, where different letters on the columns show statistically significant differences for $p < 0.0$

As it can be seen in Fig. 3, the distribution of methanol during distillation of higher alcohol mixture was different in compere to its distribution during distillation of low alcohol content mixture. The highest concentration of methanol was at the beginning of the distillation, which is expected considering its lower boiling point than ethanol. In the heart fraction, its concentration was significantly the lowest. At the end of distillations content of methanol significantly increased. Similar results were published by Balcerek et al. [20], Douady et al. [25] and Gaymon [26].

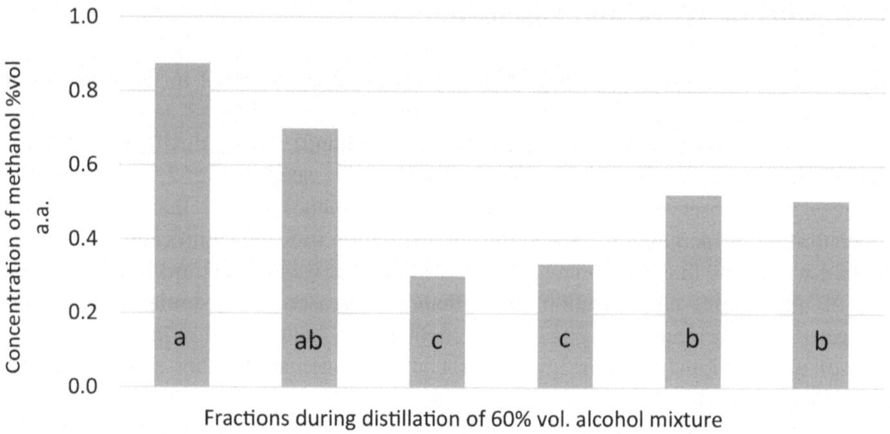

Fig. 3. The content of methanol (n = 3) in different fractions of 60% vol. alcohol with the results of the LSD test, where different letters on the columns show statistically significant differences for $p < 0.05$

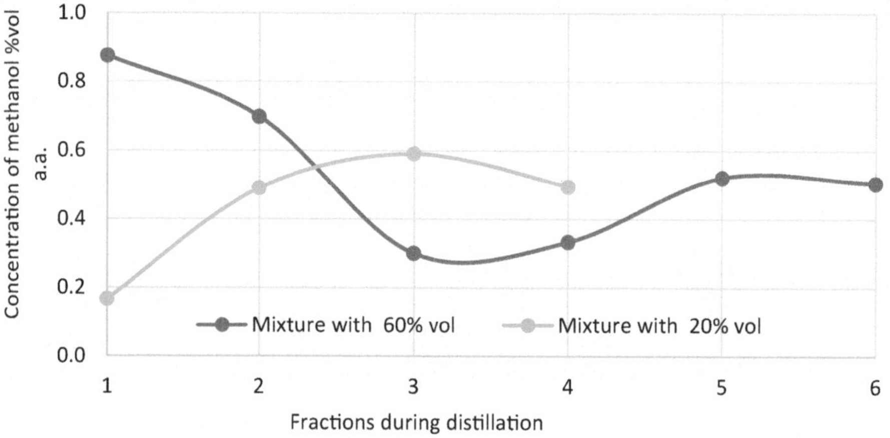

Fig. 4. A comparative overview of the distribution of methanol during the distillation of a lower and higher alcohol mixture

According to a comparative overview of methanol distribution during the distillation of low-alcohol and high-alcohol distillates (Fig. 4), generally more methanol in the central fractions was in the distillation of the low-alcoholic mixture, while the heart fraction is better "purified" than methanol if a strong alcoholic mixture is placed on the distillation.

As it can be seen in Fig. 5, the highest content of total higher alcohols recorded at the beginning of the distillation in the first fraction. During distillation time, the content of the higher alcohols decreased steadily. These differences in all fractions are statistically significant for $p < 0.05$. Although the boiling point of higher alcohols is above 100 °C (the most abundant higher alcohols: iso-butanol 108.4 °C and the iso-amyl 132 °C), they were not concentrated at the end of distillation when evaporating lower volatile compounds. However, as they are well soluble in ethanol and poor in water, they will rather follow dynamic of ethanol evaporating in distillation of lower alcohol mixture. Obviously, the initial mixture of 20% vol. alcohol is insufficiently strong to dissolve higher alcohols. These results are in accordance with the results of other authors [4, 8, 27–30]. Thus, methanol and higher alcohols during distillation of low alcohol mixtures behave quite differently.

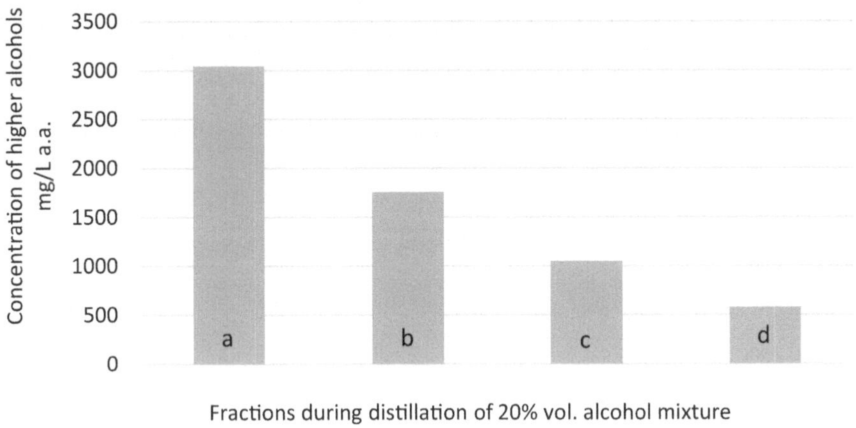

Fig. 5. The content of higher alcohols (n = 3) in different fractions of 20% vol. alcohol with the results of the LSD test, where different letters on the columns show statistically significant differences for $p < 0.05$

According to the results shown in the Fig. 6 it can be seen that the distribution of higher alcohols during the distillation of the higher alcohol mixture is completely different in relation to their distribution during the distillation of the lower alcohol mixture (Fig. 5). Their quantity in the fractions is more balanced, at the beginning it is 565 mg/l a. a., it drops to about 400 mg/l. a. a., in order to achieve a maximum value in the middle of distillation (>750 mg/l. a. a.). At the end of the distillation, the distribution of higher alcohols is almost identical to the distribution at the beginning of distillation. Since in the group of total higher alcohols there are a large number of individual alcohols with different boiling points, it is obvious that they distillate towards their boiling points without significant concentration at any of the fractions. Similar results were reported by Balcerek et al. [20] and Gaymon [26].

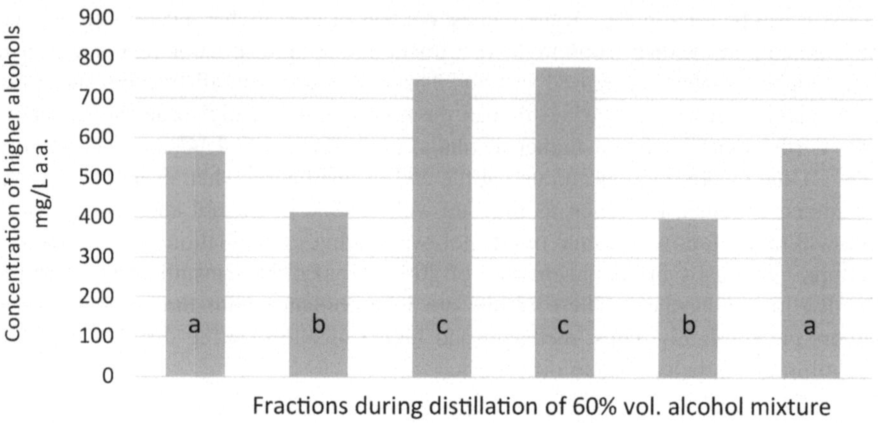

Fig. 6. The content of higher alcohols (n = 3) in different fractions of 60% vol. alcohol with the results of the LSD test, where different letters on the columns show statistically significant differences for $p < 0.05$

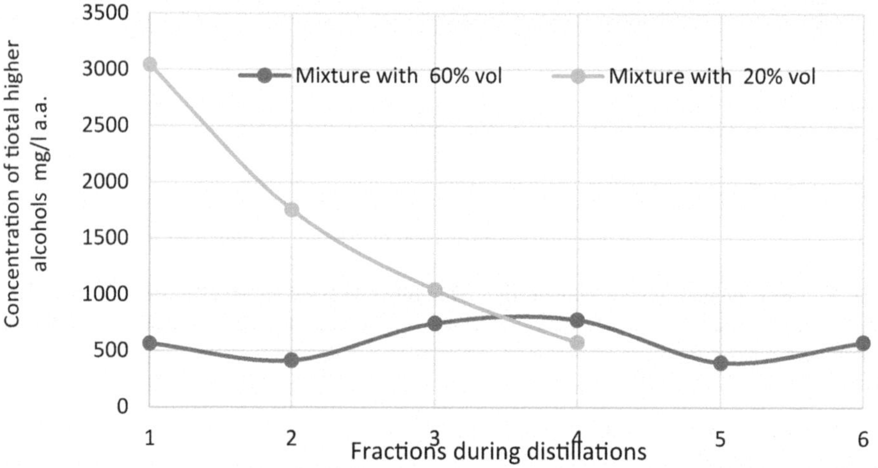

Fig. 7. A comparative overview of the distribution of higher alcohols during the distillation of lower and higher alcohol mixture

At the Fig. 7. it can be seen the distribution of content of total higher alcohols shows distinct decreasing during distillation of lower alcohol mixture. In general, these fractions had a higher content of total higher alcohols than fractions obtained during distillation of higher alcohol mixture.

4 Conclusion

According to the results of this study, it can be concluded that different content of alcohol in alcohol-water mixture affected on the distribution of methanol and total higher alcohol during distillation. Distribution of total higher alcohols and methanol is very uniform in the distillation of the low-alcoholic mixture but in the opposite direction. Alcohol strength up to 20% vol. in the distilling mixture, enhances concentration of methanol in the last/tail fractions, contrary, concentration of total higher alcohol is higher at the beginning of distillation. The opposite behavior of those two compounds occurred during distillation of higher alcohol mixture.

Conflicts of Interest. The authors declare no conflict of interest.

References

1. Nikićević, N., Tesevic, V.: Possibilities for methanol content reduction in plum brandy. J. Agric. Sci. **50**, 49–60 (2005)
2. Nikićević, N., Tešević, V.: Production of top quality fruit brandy. University of Belgrade, Belgrade, pp. 101, 102, 105 (2010)
3. Sliwinska, M., Wisniewska, P., Dymerski, T., Wardencki, W., Namiesnik, J.: The flavour of fruit spirits and fruit liqueurs: a review. Flavour Fragr. J. **30**, 197–207 (2015)
4. Spaho, N., Dürr, P., Grba, S., Velagić-Habul, E., Blesić, M.: Effects of distillation cut on the distribution of higher alcohols and esters in brandy produced from three plum varieties. J. Inst. Brew. **19**, 48–56 (2013)
5. Spaho, N.: Distillation techniques in the fruit spirits production. In: Mendes, M. (ed.) Distillation - Innovative Applications and Modeling, pp. 129–152. InTech (2017). https://www.intechopen.com/books/distillation-innovative-applications-and-modeling/distillation-techniques-in-the-fruit-spirits-production
6. Scholten, G., Kacprowski, M.: Qualitatssteigerung durch destillat behandulg. Kleinbrennerei. **10**, 11–15 (1998)
7. Claus, M.J., Berglund, K.A.: Fruit brandy production by batch column distillation with reflux. J. Food Process. Eng. **28**, 53–67 (2005)
8. Léauté, R.: Distillation in alambic. Am. J. Enol. Vitic. **41**, 90–103 (1990)
9. Popović, B., Gavrilović-Damnjanović, J., Mitrović, O., Ogašanović, D., Nikićević, N., Tešević, V.: Major volatile components and sensory characteristics of plum brandies produced from plum cultivars developed in Cacak. Acta Hort. **825**, 575–581 (2009)
10. EEC, Council Regulation: Laying down general rule on the definition, description and presentation of spirit drinks. No. 110/2008 Official Journal of the European Union, L 39/16 (2008)
11. Ruso Coldea, T.E., Socaciu, C., Parv, M., Vodnar, D.: Gas-chromatographic analysis of major volatile compounds found in traditional fruit brandies from Transylvania, Romania. Not. Bot. Horti. Agrobo. **39**(2), 109–116 (2011)
12. Soufleros, E.H., Mygdalia, A.S., Natskouli, P.: Characterization and safety evaluation of the traditional Greek fruit distillate "Mouro" by flavor compounds and mineral analysis. Food Chem. **86**, 625–636 (2004)
13. Dimitrov, D., Yoncheva, T., Haygarov, V.: Methanol content in grape and fruit brandies: a major indicator for authenticity and safety. Ukrainian Food J. **5**(2), 237–247 (2016)

14. Cortés, S.M., Gil, M.L., Fernandéz, E.: The influence of redistillation in the distribution of volatile components of Marc spirit (Aguardiente) and its repercussion on the aromatic quality. Sci. Aliment. **22**, 265–275 (2002)
15. Roudrigez Madrera, R., Valles, B.S.: Determination of volatile compounds in cider spirits by Gas chromatography with Direct Injection. J. Chromatogr. Sci. **45**, 428–434 (2007)
16. Faletar, J., Blesic, M., Smajic, M., Begic-Akagic, A., Alihodzic, A., Spaho, N.: Dynamics of evaporation of the certain volatiles during plum brandy distillation. In: Proceedings of the 24th International Scientific-Expert-Conference of Agriculture and Food Industry, Sarajevo, Bosnia and Herzegovina, 25–28 September 2013, pp. 204–208 (2013)
17. Jamaković, N.: The influence of the fermentation type on the quality parameters of the cherry brandy. University of Sarajevo, Faculty of Agriculture and Food Science, Sarajevo, p. 25 (2013)
18. Matias-Guiu, P., Rodríguez-Bencomo, J.J., Orriols, I., Pérez-Correa, J.R., López, F.: Floral aroma improvement of Muscat spirits by packed column distillation with variable internal reflux. Food Chem. **213**, 40–48 (2016)
19. Sacher, J., García-Llobodanin, L., López, F., Segura, H., Pérez-Correa, J.R.: Dynamic modeling and simulation of an alembic pear wine distillation. Food Bioproducts Process. **91**, 447–456 (2013)
20. Balcerek, M., Pielech-Przybylska, K., Patelski, P., Dziekońska-Kubczak, U., Strak, E.: The effect of distillation conditions and alcohol content in "heart" fractions on the concentration of aroma volatiles and undesirable compounds in plum brandies. J. Inst. Brew. **123**(3), 452–463 (2017)
21. Method OIV-MA-AS312-03B Compendium of International Methods of Analysis-OIV Methanol OIV-MA-AS312-03B: R2009. Resolution Oeno 377/2009
22. Cortés, S., Gil, M.L., Fernandéz, E.: Volatile composition of traditional and industrial Orujo spirits. Food Control **16**, 383–388 (2005)
23. Rodriguez Madrera, R., Mangas Alonso, J.J.M.: Obtention of cider spirit by distillation in a rectification column. Alimentaria **277**, 89–93 (1996)
24. Spaho, N., Alihodžić, A., Begić-Akagić, A., Blesić, M.: Content of methanol in the apple pomace distillates. Works of the Faculty of Agricultural and Food Sciences University of Sarajevo, LV, pp. 201–211 (2010)
25. Douady, A., Puentes, C., Awad, P., Martine Esteban-Decloux, M.: Batch distillation of spirits: experimental study and simulation of the behaviour of volatile aroma compounds. J. Inst. Brew. (2019). https://doi.org/10.1002/jib.560
26. Guymon, J.F.: Chemical Aspects of Distilling Wines into Brandy. University of California, Davis, 1 November 1974. https://doi.org/10.1021/ba-1974-0137. http://pubs.acs.org, 2019. godine
27. Hernandez-Gomez, L.F., Ubeda-Iranzo, J., Garcia-Romero, E., Briones-Perez, A.: Comparative production of different melon distillates: chemical and sensory analyses. Food Chem. **90**, 115–125 (2005)
28. Rodríguez Madrera, R., Mangas Alonso, J.J.: Distribution of the principal minor volatiles during cider distillation in "alquitara". Acta Aliment. **40**, 262–269 (2010)
29. Rodríguez-Bencomo, J.J., Pérez-Correa, J.R., Orriols, I., López, F.: Spirit distillation strategies for aroma improvement using column reflux. Food Bioprocess Technol. **9**(11), 1885–1892 (2016)
30. Zhao, Y., Tian, T., Li, J., Zhang, B., Yu, Y., Wang, Y., Niu, H.: Variations in main flavor compounds of freshly distilled brandy during the second distillation. Int. J. Food Eng. **10**, 809–820 (2014)

Comparative Composition of Spreča River Fauna (Bosnia and Herzegovina)

Isat Skenderović(✉), Avdul Adrović, Edina Hajdarević, and Alen Bajrić

Faculty of Natural Sciences and Mathematics, University of Tuzla, Univerzitetska 4., 75000 Tuzla, Bosnia and Herzegovina
isat.skenderovic@untz.ba

Abstract. Hydrobiological studies of the macroinvertebrates of the water catchment area of the river Spreča during 2019 revealed the presence of 60 taxa classified into 5 classes and 11 groups. Qualitative-quantitative analysis of the studied watercourses revealed different presence of individuals and taxa. The highest diversity was found in the samples of the Oskova River - 36, Spreča River - 28, while the lowest diversity was found in the Gostelja River (25). The results of the analysis showed the dominant participation of the Insecta class with 6 groups. Other established classes are represented by a smaller number of groups. The results of the diversity index indicate the heterogeneous composition of the macroinvertebrate communities. The highest values of the Shannon - Weaver Diversity Index were found on the Oskova River - Mačkovac locality. Cluster analysis showed that, in the presence of a number of taxa, the most similar localities on the Gostelja River are Križaljka and Stupari (L7 and L9) with a similarity percentage of 73.74%. On all other sites, the percentage of similarities was much smaller. Ichthyological analysis of the Spreče River basins revealed the presence of 24 fish species from 6 families.

Keywords: Fauna · Biodiversity · Spreča

1 Introduction

In this study the results of qualitative - quantitative analysis of macroinvertebrate composition and ichthyopopulation of the water catchment area of the river Spreča are presented. Streams and rivers in Bosnia and Herzegovina are characterized by high levels of biodiversity in both the plant and animal components. Macrophytes, benthic macroinvertebrates and fishes are used in evaluating of the water quality. Macroinvertebrates of zoobenthos are invertebrate organisms from the bottom of streams and rivers, with body sizes over 0.5 mm. Macroinvertebrates are dominant and diverse in the benthic river community. Fish communities are used as significant indicators of anthropogenic impacts on the diversity of ichthyopopulations in aquatic ecosystems. Spreča river is 130 km long. Its catchment area is 1945 km^2 and belongs to the larger rivers of Bosnia and Herzegovina. On the way from the source of the river Spreča to the estuary of Bosnia, the river receives many tributaries. The total length of the Gostelje, from Stupari to the mouth of Oskova, is about 28 km. The catchment area of Gostelje

I. Karabegović (Ed.): NT 2020, LNNS 128, pp. 795–802, 2020.
https://doi.org/10.1007/978-3-030-46817-0_90

is 185 km^2. The Oskova River belongs to the hydrographic system of the Spreča River and it is one of the most important tributaries [1]. The aim of the paper was to determine the degree of species diversity of fishes and macroinvertebrates at defined localities on the longitudinal profile of the studied watercourses during 2019, and to determine the degree of heterogeneity of ichthyofauna and macroinvertebrates based on the results.

2 Materials and Methods

Macroinvertebrates and fish sampling at defined localities of the water catchment area of the river Spreča were carried out during June and October 2019. Fauna sampling was carried out in a total of 9 localities. On the Spreča River: Spreča estuary into lake Modrac - locality 1 (L1); Strašanj - locality 2 (L2); Ciljuge - locality 3 (L3). On the river Oskova: Višća - locality 4 (L4); Oskova - locality 5 and Mačkovac - locality 6 (L6). Samples were taken from the Gostelja river from three localities: Križaljka (Živinice) - locality 7 (L7), Podgajevi - locality 8 (L8) and Stupari - locality 9 (L9). The research was carried out within the project "Ecological valorization of the biodiversity of the fauna of aquatic ecosystems of the Spreča River Basin", which was supported by the Environmental Protection Fund of the Federation of Bosnia and Herzegovina. The fishing was made by electroagregate "SAMUS MS 725" 650 W. Macroinvertebrates zoobenthos sampling was carried out with "*kick sampling*" method [2]. Determination of sampled fish and macroinvertebrates was performed up to species level according to the available keys for determination [3–7]. The diversity of fish and macroinvertebrate communities is represented by the Shannon-Weaver Diversity Index [8]. The cluster analysis was performed based on the Bray - Curtis similarity index [9].

3 Results and Discussion

3.1 Macroinvertebrate Diversity of the Spreča River Watershed

The analysis of macrozoobenthos of the investigated rivers of the water catchment area of the river Spreča revealed the presence of a total of 1,100 individuals, which were classified into five classes (11 groups), or 60 taxons. Representatives of the Insecta - Hexapoda class with 6 groups are the most dominant participants. Other established classes are represented by a smaller number of groups. The largest number of taxa 49 (81.6%) belongs to the Insecta class, six to the Gastropoda group, three taxa to the Oligochaeta group and one taxon to the Turbellaria and Crustacea groups. The biodiversity of taxa of stated macroinvertebrate groups varied among the studied rivers. According to the number of taxa, zoobenthos specimens from the Oskova river are distinguished with 36 taxa present, then with 28 taxa from the Spreča river, while the number of identified taxa from Gostelja river is only 25. The analysis of the results of zoobenthos composition on the longitudinal profile of the Spreča river shows a change in taxa distribution. The studied part of this fluid is dominated by representatives of Gastropoda, Trichoptera, Diptera, Plecoptera and Ephemeroptera. Representatives of

Table 1. Qualitative - quantitative composition of macroinvertebrates in samples of the water catchment area of the river Spreča

Taxon	River	Spreča			Oskova			Gostelja		
	Locality	L1	L2	L3	L4	L5	L6	L7	L8	L9
Gastropoda	*Lymnaea truncatula*	4		21				12	8	11
	Lymnaea peregra	3	2	3						
	Theodoxus fluviatilis		21	32					3	
	Holandriana holandrii	3	5	5						
	Ancylus fluviatilis				6		11			
	Physa fontinalis					3	6			
Turbellaria	*Dugesia gonocephala*		3					3		4
Crustacea	*Gammarus balcanicus*		32	23	4	5		5	5	7
Oligochaeta	*Tubifex tubifex*	9		3	4	5				
	Nais pardalis	4		4	5					
	Stylodrilus heringianus						3			
Ephemeroptera	*Baetis rhodani*		5		2		12			
	Baetis scambus					4		3		
	Baetis lutheri						6	21	6	13
	Baetis fluvistriga								5	
	Heptagenia quadrilineatus	4	3	4		3	7			
	Heptagenia sulphurea						6	5	6	5
	Epeorus assimilis		2	1						2
	Ecdyonurus venosus				4	5				
	Ephemera sp.				6					
	Ephemera hellenica							6		4
	Ephemera danica								2	
	Cloeon simile						15			
	Caenis robusta						32			
	Procloeon fragile						2			
	Rhithrogena carpatoalpina							5		
Plecoptera	*Leuctra hippopus*	2	6	8		2	3			
	Leuctra fusca		7							
	Leuctra sp.					7				
	Perla sp.					3				
	Perla marginata				2	12				
	Perla burmeisteriana				3	21	23			
	Perla illiesi			2			2			
	Dinocras megacephala				2	3		3		3
	Nemoura cinerea				3		3	3	4	
	Brachyptera seticornis						4	4	3	3
	Protonemoura auberti							2	3	6
	Perlodes microcephalus							3		
	Isoperla tripartita tripartite									3

(*continued*)

Table 1. (*continued*)

Taxon	River	Spreča			Oskova			Gostelja		
	Locality	L1	L2	L3	L4	L5	L6	L7	L8	L9
Odonata	*Gomphus vulgatissimus*		37		11		37		3	
	Anax imperator		5							
Trichoptera	*Athripsodes aterrimus*	2		1						
	Hydropsyche angustipennis	3	17		3	4	6			
	Hydropsyche siltalaii		66		23	17	18			
	Hydropsyche instabilis	5					5	13	3	17
	Hydropsyche incognita					2				
	Hydropsyche sp.					2				
	Odontocerum albicorne		4							
	Limnephilus bipunctatus						3			
	Rhyacophila nubila						3		2	
	Rhyacophila dorsalis	3		21						
	Philopotamus montanus								11	
Diptera	*Tipula lateralis*			43						
	Tipula saginata		9	3				4		
	Chironomus thummi	23	29		22	33				
	Simulium sp.	6		7						
	Pedicia imaculata							4		5
	Tabanus maculicornis								5	
Coleoptera	*Elmis aenea*	4	3	3			2			
	Gyrinus natator	3	3			2	2			

Odonata, Coleoptera and Oligochaeta are represented by two taxa each. Other established macroinvertebrate groups are represented by one species.

Community macroinvertebrates of benthos of river Oskova on the longitudinal profile of investigated period is composed of the largest number of established taxa. In terms of biodiversity, an important place belongs to the Ephemeroptera group, represented by 104 individuals classified into 10 taxa. Of the nine established species of the order Plecoptera, the species *Perla burmeisteriana* occurs in a significantly higher number than the other established species.

The bottom fauna of the Gostelja consists of 8 groups, with 25 determined taxa. Significant participation in the fauna of the bottom of the Gostelja river belongs to aquatic insects from the order Ephemeroptera, Plecoptera, Diptera and Trichoptera. Earlier studies of the Spreče river and Konjuh Mountain springs have revealed the dominant presence of Ephemeroptera and Trichoptera [10]. The presence of other taxa in the investigated rivers is shown in Table 1.

Table 2. Percentage of locality similarities of the studied watercourses based on the presence of macrozoobenthos communities

Similarity Matrix	L1	L2	L3	L4	L5	L6	L7	L8	L9
L1	*	24.93	27.48	37.08	36.02	12.46	10.34	9.52	11.18
L2	*	*	30.25	36.21	31.63	32.34	6.76	6.71	7.02
L3	*	*	*	7.75	8.2	5.57	14.29	12.65	14.23
L4	*	*	*	*	52.36	29.58	9.18	11.83	6.56
L5	*	*	*	*	*	30.23	9.61	4.95	7.41
L6	*	*	*	*	*	*	14.98	18.57	12.93
L7	*	*	*	*	*	*	*	42.42	73.74
L8	*	*	*	*	*	*	*	*	43.42
L9	*	*	*	*	*	*	*	*	*

Table 2 shows the locality similarities, i.e. the percentage of their similarity based on the presence of macroinvertebrate communities. It is observed that the most similar localities are Križaljka and Stupari (L7 and L9) on the Gostelja river with a similarity percentage of 73.74%. The other studied localities showed lower percentage of similarities. The results of the macroinvertebrate communities diversity by study localities are presented in Fig. 1.

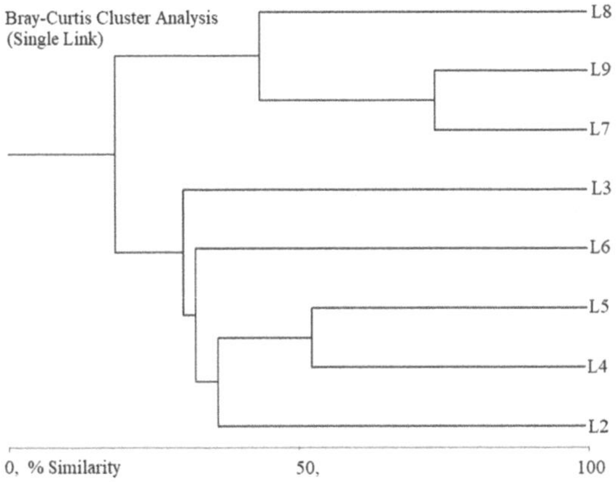

Fig. 1. Presentation of macrozoobenthos cluster analysis of the studied streams and rivers

The results of the Shannon - Weaver diversity index indicate on the heterogeneous composition of the macroinvertebrate communities of the studied streams and rivers. The highest values of the diversity index were found at locality 6 (Mačkovac) at Oskova. Slightly smaller values of diversity were registered at other surveyed localities of the water catchment area of the river Spreča.

3.2 Diversity of Ichthyopopulations of the Water Catchment of the River Spreča

Ichthyological studies have shown the presence of 24 fish species in the analyzed localities of the water catchment of the river Spreča. Based on the fish composition analysis, from three localities on the river Spreča, the presence of 14 fish species was determined. Fishes from three different families were found: Cyprinidae, Percidae and Esocidae. Results are shown in Table 3.

Table 3. Composition of fish of the water catchment of the river Spreča

River	Family	Spreča			Oskova			Gostelja		
Fish species		L1	L2	L3	L4	L5	L6	L7	L8	L9
Abramis brama	Cyprinidae	5		10						
Abramis sapa								7		
Alburnus bipunctatus		8	43	87		7	6	15	16	43
Alburnus alburnus		17	2	3		9	8	3		
Scardinus erythrophthalmus		13								
Chondrostoma nasus			5	3						
Squalius cephalus		7	37	59	4	11	8	65	29	
Rutilus rutilus		12	32	37				6	3	
Tinca tinca								6		
Phoxinus phoxinus							4	5	6	97
Gobio gobio								23	5	
Gobio obtusirostris			3	7				24	8	54
Rhodeus sericeus			2					14		
Barbus barbus			3					25	11	3
Barbus balcanicus			19	63	5	4		33	21	7
Perca fluviatilis	Percidae	6		6				12		
Acerina cernua			4							
Esox lucius	Esocidae	8								
Cobitis taenia	Cobitidae							8		
Cobitis aurata									3	5
Cobitis elongatoides								4		
Sabanejewia balcanica								7		
Cottus gobio	Cottidae									12
Salmo truta m. fario	Salmonidae									2

Cluster analysis of ichthyopopulations was carried out based on the Bray - Curtis similarity index. By cluster analysis, we attempted to determine the degree of similarity or at least the mutual differences between ichthyocenosis at different sites of the studied watercourses. According to the fish species diversity, the most similar localities are L5 - Oskova and L6 - Mačkovac on the Oskova river with a similarity percentage of 77.19%

(Table 4). Also, the samples obtained at locations L2 and L3 show the high percentage of similarity (65.41%). All other samples show a lower percentage of similarity.

Table 4. Percentage similarity of localities based on the diversity of fish species

Similarity									
Matrix	L1	L2	L3	L4	L5	L6	L7	L8	L9
L1	*	25.66	23.36	9.41	42.99	41.18	18.02	20.22	5.35
L2	*	*	65.41	11.32	26.52	18.18	42.75	57.94	30.03
L3	*	*	*	6.34	16.34	11.29	48.49	40.32	22.89
L4	*	*	*	*	40	22.86	6.77	16.22	4.31
L5	*	*	*	*	*	77.19	17.36	33.08	8.66
L6	*	*	*	*	*	*	14.84	28.12	8.03
L7	*	*	*	*	*	*	*	54.04	22.5
L8	*	*	*	*	*	*	*	*	26.46
L9	*	*	*	*	*	*	*	*	*

The results of the ichthyopopulations diversity between different localities were presented in Fig. 2. The highest levels of similarity are found in fish samples from L5 and L6 localities.

Based on the results of the Shannon - Weaver's index of diversity it can be concluded that there is quite difference between analysed localities, indicating the heterogeneous composition of the ichthyopopulation communities. The highest value of the diversity index was found at locality L7. Slightly smaller values of diversity were registered at the remaining two localities of the Gostelja river and Spreča river. The lowest values of the diversity index were registered at the localities of the Oskova river.

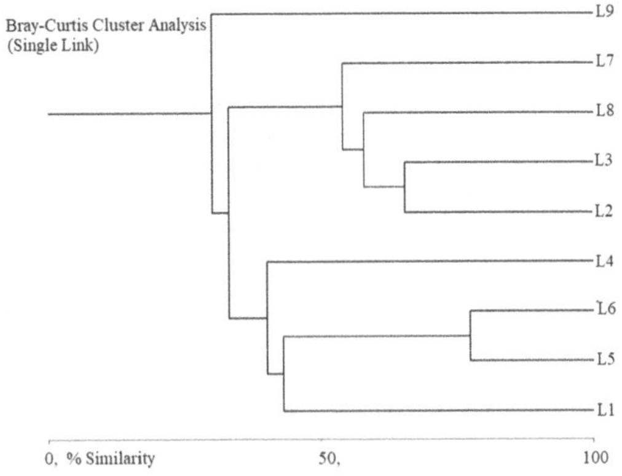

Fig. 2. Display of cluster analysis of samples of fishes

4 Conclusion

Hydrobiological studies of macroinvertebrates and fishes of the water catchment area of the river Spreča presents high degree of biodiversity, which indicate the favorable life conditions. Analysis of the macrozoobenthos in investigated rivers revealed the presence of 60 taxa from 5 classes. The results of the analysis showed the dominant participation of the Insecta class, with regard to other established classes. The Shannon - Weaver index of diversity and Cluster Analysis found that the most favorable living conditions prevailed in headwaters of Gostelja River, while in other localities are getting worse. Also, the high specific diversity of fishes in the Gostelja River indicates favorable environmental factors. Smaller values of fish diversity were registered at the Spreča and Oskova localities. The richness of fauna indicates that it is necessary to make extensive inventories of ichthyofauna and macroinvertebrates in the aim of the biodiversity conservation.

References

1. Skenderović, I., Adrović, A., Hajdarević, E., Bajrić, A.: Zoobenthos of macroinvertebrates in some streams of the water catchment area of lake Modrac (Bosnia and Herzegovina). In: 10th International Scientific Conference "Science and Higher Education in Function of Sustainable Development" Užice, Serbia (2017)
2. Dall, P.C., Friberg, N., Lindegard, C., Toman, M.J.: A practical guide of biological assessment of stream water quality. Biological Assessment of Stream Water Quality. University of Ljubljana, Ljubljana (1995)
3. Kerovec, M.: Priručnik za upoznavanje beskralješnjaka naših potoka i rijeka. SNL, Zagreb (1986)
4. Quigley, M.: Invertebrates of Streams and Rivers: A key to Identification. Edward Arnold Pub. UK (1977)
5. Sansoni, G.: Atlante per il riconoscimento dei macroinvertebrati dei corsi d'acqua italiani. Provincia autonoma di Trento. Centro italiano Studi di Biologia Ambientale (1988)
6. Waringer, J., Graf, W.: Atlas der Österreichischen Köcherfliegenlarven. Facultas Univeritätsverlag, Wien (1997)
7. Sofradžija, A.: Slatkovodne ribe Bosne i Hercegovine. Vijeće Kongresa bošnjačkih intelektualaca Sarajevo (2009)
8. Shannon, C.E., Weaver, W.: The Mathematical Theory of Communication. The University of Illinois, Illinois (1949)
9. Bray, J.R., Curtis, J.T.: An ordination of upland forest communities of southern Wisconsin. Ecol. Monogr. **27**, 325–349 (1957)
10. Skenderović, I., Adrović, A., Hajdarević, E., Bajrić, A.: Communities of aquatic macroinvertebrates from Konjuh Mountain headwater streams. In: International Conference "New Technologies, Development and Applications", pp. 587–594. Springer, Cham (2019)

A Two-Stage Multi Criteria Decision Making in Crop-Livestock Farming

Grujica Vico[1(✉)], Radomir Bodiroga[1], and Savo Stupar[2]

[1] Faculty of Agriculture, University of East Sarajevo,
Vuka Karadžića 30, 71123 East Sarajevo, Bosnia and Herzegovina
vicogrujica@yahoo.com

[2] School of Economics and Business, University of Sarajevo,
Sarajevo, Bosnia and Herzegovina

Abstract. Decision making process is one of the key stage for the successful farm management. In the recent period, the problems are more complex with new issues, which are coming from the climate change, sustainability, circular agriculture, etc. This research aims to show two-stage approach to solving decision making problem in mixed crop-livestock farming. In the first stage, the problem can be treated as well structured problem under strong constraints, and then it could be solved as MADM problem in the second stage. Linear programming (LP) used in the first stage to solve four LP solutions respecting four objectives. In the second phase, two new criteria were set up, and then the new multi-attribute decision making (MADM) problem created and solved by TOPSIS method. The results show that combining two different decision making methods can help farm managers to solve complex problems respecting a numerous both quantitative and qualitative criteria.

Keywords: Linear programming · Optimization · TOPSIS, model

1 Introduction

Animal production from the aspect of management implies a complex production system that requires the achievement of a number of goals. These are economic, environmental and social goals, and there have been lately more and more current issues concerning the ecological sustainability of production and animal welfare. Climate changes open up new questions that the animal production manager needs to answer. Their influence, among other things, will be reflected through the quantity and reliability of forage production; forage quality; thermal stress on livestock; water demands for both animal needs and for growing forage; pest, disease and weed challenges; land degradation processes; and various social and economic aspects including trade [1]. New circumstances are bringing new criteria for the sustainable livestock production like: demand and supply of consumable livestock products, land area utilized for agriculture, variability of production, net annual soil losses, nutrient balances and losses etc. [2]. The newest dairy industry is characterized by diverse production and marketing systems [3]. The aforementioned imposes the need to pay more attention to the planning and decision-making processes in livestock production management. For more than 50

I. Karabegović (Ed.): NT 2020, LNNS 128, pp. 803–810, 2020.
https://doi.org/10.1007/978-3-030-46817-0_91

years, various methods of operational research have been used to manage agricultural resources [4]. Recent years, more and more publications have conducted which show different problems that solved with Multi criteria decision making (MCDM) methods. Some of them use Objective Decision Making (MODM) techniques such as weighted goal programming used for optimization of rations for sport horses [5]. Weighted goal programming with a penalty function also used for the optimization of a daily dairy cow ration [6]. A multi-goal mathematical approach for the optimization of crop planning on organic farms has been developed to help farmers to obtain answers regarding four main issues in crop rotation planning; (a) cropping strategy; (b) potential hiring of extra resources; (c) quantity of harvested crops and (d) quantity of fertilizer being saved in the second year of planning [7]. A hybrid and multi-attribute approach based on combination AHP-TOPSIS MADM techniques used to select one tractor between six alternatives with six attributes [8]. In numerous research TOPSIS used to evaluate macroeconomic measures on different aspect of small farms. It used to quantify the effects of the CAP direct payments on the socioeconomic sustainability of small farms in Lithuania [9]. Multi stage approach used to analyze two vegetable production systems with different cultures as alternatives with purpose to find the most appropriate variants [10]. Four MADM methods includes AHP, TOPSIS, ORESTE and WSM used to evaluate nine alternatives in vegetable production according to twelve criteria. There were 3 main criteria and 9 complementary criteria [11].

The aim of this work is to find the best production structure which includes both, animal and crop production, by using two stage DM process, which includes creating whole-farm LP model in the first stage and ranking solutions with new criteria, using TOPSIS method, in the second stage.

2 Materials and Methods

An LP model has created to represent whole-farm production system. Accepted approach involves two stage of research. LP technique was applied Ii the first stage, while TOPSIS was used in the second stage. The data were collected from the farm management that in the further course of the research served as the basis for defining the activities in the model, and technical coefficients and constraints. Using pre-prepared set of patterns, the data about general conditions of livestock production such as: labor, land, buildings, machinery and equipment for animal husbandry, herd structure, livestock turnover, production and distribution of cow's milk, meal structure, capacity and volume crop were collected. For each enterprises gross margin calculation were created as difference between total production value and total variable cost. The four kinds of crop are involved in the model: maize silage, alfalfa hay, maize for grain and barley. The model was developed according to the next assumptions for livestock production: fertility rate - 90%; average weight of caves after calving - 50 kg; average weight of calves (120 days) - 140 kg; age of the first insemination - 16 months; average age of the first calving - 25,5 months; successful rate of the first insemination - 85%; number of exploating lactations - 5. There are 45 ha arable land, barn capacity is 60 daiiry cows.

Four objectives are used in LP model: Maximization gross margin (A1); Maximizitaion arable land usage (A2); Maximization land area under legume (A3); Minimization labour force usage (A4).

Solving the models were performed by using the Solver tool, which is the addition to the MS Excel. This tool enables an optimal solution, with a set of reports required for post optimum analysis. The procedure of solving problems through multi-criteria ranking by applying the TOPSIS method via six steps are given in [12].

3 Results and Discussion

Agricultural planning problems, involve multiple objectives such as maximizing crop production, maximizing overall profit, minimizing labor costs, water requirements and others [13]. Methods of modeling and linear programming were used in studies that deal with the formulation of meals for different categories of cattle. In several published papers, these methods were used to formulate meal rations [14, 15] for fattening bulls [16, 17], and for combined-type races [18, 19]. The method of linear programming is being used continuously in scientific research on the optimization of livestock farms. By using this method, a number of authors successfully dealt with problems in the management of livestock production [20–25].

On the other hand, in order to solve complex problems in the field of agronomy, the use of MCDM techniques has become more and more in use recently. The general characteristics of each multi-criteria problem from the standpoint of the decision-making process are: more criteria for decision-making, more different alternatives for choice, the role of the decision-maker and the process of selecting one final solution. MCDM can be classified into: MODM, MADM and Combination of MODM and MADM [26].

3.1 Mathematical Model

Objective function:

$$(\max)f = \sum_{i=1}^{p}\sum_{j=1}^{q} c_{ij}x_{ij} \tag{1}$$

Constraints:

$$\sum_{i=1}^{p}\sum_{j=1}^{q} a_{ijkl}x_{ij} \leq \; = \; \geq u_{kl} \; \mathrm{k} = 1, 2, \ldots, \mathrm{r} \; \mathrm{l} = 1, 2, \ldots, \mathrm{s} \tag{2}$$

Conditions of non-negativity:

$$\sum_{i=1}^{p}\sum_{j=1}^{q} a_{ijkl}x_{ij} \leq \; = \; \geq u_{kl} \; \mathrm{k} = 1, 2, \ldots, \mathrm{r} \; \mathrm{l} = 1, 2, \ldots, \mathrm{s} \tag{3}$$

Indexes: p – number of group activities; q – number of activities in a group; r – number of group limitations; s – number of limitations in a group

Activities: x_{ij}; i = 1, 2, …, p j = 1, 2, …, q

Constraints: u_{kl}; k = 1, 2, …, r l = 1, 2, …, s

Coefficients in a objective function: c_{ij}; i = 1, 2, …, p j = 1, 2, …, q

Coefficients in constraints: a_{ijkl} - the quantity of jth- activities within a ith- group of activities l^{th}- constraints and r^{th}-group of limitations

Groups of activities: lines of farm production i = 1 j = 1; fodder production at a farm i = 2 j = 1, 2, …, 8; dry substance from the concentrated part of the meal i = 3 j = 1, 2; purchased fodder i = 4 j = 1, …, 19; other purchased inputs i = 5 j = 1, …, 7; other variable expenses i = 6 j = 1; labor i = 7 j = 12; finished products i = 8 j = 5.

Groups of constraints capacities k = 1 l = 1, 2; crop rotation k = 2 l = 1; balance of nutritive k = 3 l = 1, …, 13; minimal I maximal components quantity in a meal k = 4 l = 1, …, 12; balance of other inputs k = 5 l = 1, …, 8; balance of labor k = 6 l = 1, …, 12; available labor k = 7 l = 1, …, 12; balance of mechanization k = 8 l = 1, …, 4; balance of finished products k = 9 l = 1, …, 5.

3.2 LP Model Solutions

A set of quantitative useful information to farm managers can be get by solving this constructed model. Among other things, information relating to: the scope and structure of production on the farm, the amount of total income, variable costs and margins of cover, structure and scope of the required inputs for crop and livestock production, scope and dynamics of engagement available resources, etc. The individual data directly read from the optimal solution, while the other required smaller or larger additional calculations.

The solving procedure is resulting with four solutions. For this approach, just four indicators for each one are shown. They are corresponding with objectives in each one LP model (Table 1).

Table 1. Indicator values per alternatives

Indicators	Alternatives			
	A1	A2	A3	A4
Gross margin (EUR)	83.335,46	82.394,18	81.697,80	74.053,98
Arable land usage (ha)	44,01	44,27	43,31	31,03
Land area under legume (ha)	13,61	15,38	16,00	12,78
Total working hours of labor force (hrs)	838,86	837,70	828,02	803,91

Considering objective in the first model (A1), where is the maximization of the Gross margin set as the objective in the function, it has the highest value (83,335.46 EUR). Follow this logic, A2 has the highest values in the second row, A3 in the third row and A4 has the lowest value for the "Total hours of labor force" (Table 1).

3.3 Ranking Alternatives by TOPSIS

A two-stage DM approach with LP and SAW used for ranking production structures in crop production. In the first stage a three LP problems solved simultaneous, while SAW method used for the selection the best alternative (solution) in the second stage [27]. Except of the criteria used in the first stage, two new criteria have been added. The set of criteria has given in the Table 2.

Table 2. Set of criteria

Criteria	Mark	Unit	To be max	To be min	Weight
Gross margin	C1	EUR	+		0,300
Arable land usage	C2	Ha	+		0,150
Land area under legume	C3	Ha	+		0,150
Total working hours of labor force	C4	Hour		–	0,150
Minimization of labor force variation	C5	CV (%)		–	0,125
Minimization of mechanization variation	C6	CV (%)		–	0,125

After the defining set of criteria and their weight coefficients, the row data for both, alternatives and criteria are shown in Table 3.

Table 3. The criteria values for the alternatives

Criteria	Alternatives			
	A1	A2	A3	A4
C1	83,335.46	82,394.18	81,697.80	74,053.98
C2	44.01	44.27	43.31	31.03
C3	13.61	15.38	16.00	12.78
C4	838.86	837.70	828.02	803.91
C5	16.19	15.68	14.33	12.99
C6	52.07	51.22	51.02	45.64

The following TOPSIS procedure includes: calculate the normalized decision matrix; calculate the weighted normalized decision matrix; determine the ideal and negative-ideal solutions; through n-dimensional Euclidean distance calculate the separation of each alternative from the ideal solution and calculate the relative closeness to the ideal solution. The summarized results with relative closeness to the ideal solution for each one alternatives and their rank, are shown in the Table 4.

Table 4. Ranking alternatives by TOPSIS

Alternative	Relative closeness	Rank
A1	0.6779	3
A2	0.7919	2
A3	0.8201	1
A4	0.2098	4

Respecting the seven includes criteria, alternative A1 has been shown as the best choice. It is followed by alternative A2, A1 and finally, A4 ranked on the fourth rank.

4 Conclusions

Decision making and problem solving are the crucial for the good management processes in the bot livestock and crop production. There are decision making problems where a numerous criteria should be respected. There is not a way to set some of them in objective function of an LP model. Those criteria could be added to the LP solutions, and then the new MADM problem will be created and solved. This study shows a two stage approach to solve complex decision making problem in crop-livestock farming. It includes solving four LP problems in the first stage and then setting new criteria and finally solving the new MADM problem by using TOPSIS procedure. The results of this study showed that the combining two DM techniques can help farm managers to solve more complex problems, respecting different criteria and finally they will be able to find as best as possible solution.

References

1. Howden, S.M., Crimp, S.J., Stokes, C.J.: Climate change and Australian livestock systems: impacts, research and policy issues. Aust. J. Exp. Agric. **48**(7), 780–788 (2008)
2. De Wit, J., Oldenbroek, J.K., Van Keulen, H., Zwart, D.: Criteria for sustainable livestock production: a proposal for implementation. Agric. Ecosyst. Environ. **53**, 219–229 (1995)
3. Kariuki, C.M., Van Arendonk, J.A.M., Kahi, A.K., Komen, H.: Multiple criteria decision-making process to derive consensus desired genetic gains for a dairy cattle breeding objective for diverse production systems. J. Dairy Sci. **100**(6), 4671–4682 (2017)
4. Sharma, D.K., Jana, R.K., Gaur, A.: Fuzzy goal programming for agricultural land allocation problems. Yugoslav J. Oper. Res. **17**, 31–42 (2007)
5. Prišenk, J., Pažek, K., Rozman, Č., Turk, J., Janžekovič, M., Borec, A.: Application of weighted goal programming in the optimization of rations for sport horses. J. Anim. Feed Sci. **22**, 335–341 (2013)
6. Žgajnar, J., Juvančič, L., Kavčič, S.: Combination of linear and weighted goal programming with penalty function in optimisation of a daily dairy cow ration. Agric. Econ. (AGRIECON) **55**(10), 492–500 (2009)
7. Prišenk, J., Turk, J.: A multi-goal mathematical approach for the optimization of crop planning on organic farms: A slovenian case study. Pak. J. Agric. Sci. **52**(4), 971–979 (2015)

8. Garcia-Alcaraz, J.L., Maldonado-Macias, A.A., Hernandez-Arellano, J.L., Blanco-Fernandez, J., Jimenez-Macias, E., Saenz-Diez Muro, J.C.: Agricultural tractor selection: a hybrid and multi-attribute approach. Sustainability **8**, 157 (2016)
9. Volkov, A., Balezentis, T., Morkunas, M., Streimikiene, D.: Who benefits from CAP? The way the direct payments system impacts socioeconomic sustainability of small farms. Sustainability **11**(7), 2112 (2019)
10. Pozderec, S., Bavec, M., Rozman, Č., Vinčenc, J.: Multi-criteria assessment of vegetable production business alternatives. Organizacija **48**(3), 203–213 (2015)
11. Matějček, M., Brožová, H.: Multiple attributes analysis of vegetable production. In: MCBANTA 2011, Proceedings of the 12th WSEAS International Conference on Mathematics and Computers in Biology, Business and Acoustics, Romania, 11–13 April, pp. 27–32 (2011)
12. Opricovic, S., Tzeng, G.H.: Compromise solution by MCDM methods: a comparative analysis of VIKOR and TOPSIS. Eur. J. Oper. Res. **156**, 445–455 (2004)
13. Weintraub, A., Romero, C.: Operations research models and the management of agricultural and forestry resources: a review and comparison. INFORMS J. Appl. Anal. **36**(5), 383–482 (2006)
14. O'Connor, J.D., Sniffen, C.J., Fox, D.G., Milligan, R.A.: Least cost dairy cattle ration formulation model based on the degradable protein system. J. Dairy Sci. **72**, 2733–2745 (1989)
15. Reyes, A.A., Blake, R.W., Shumway, C.R., Long, J.T.: Multistage optimization model for dairy production. J. Dairy Sci. **64**, 2003–2016 (1981)
16. Glen, J.J.: A parametric programming method for beef cattle ration formulation. J. Oper. Res. Soc. **31**, 689–698 (1980)
17. Crabtree, J.R.: Interactive formulation system for cattle diets. Agric. Syst. **8**, 291–308 (1982)
18. Nicholson, C.F., Lee, D.R., Boisvert, R.N., Blake, R.W., Urbina, C.I.: An optimization model of the dual-purpose cattle production system in the humid lowlands of Venezuela. Agric. Syst. **46**, 311–334 (1994)
19. Nicholson, C.F., Blake, R.W., Urbina, C.I., Lee, D.R., Fox, D.G., Van Soest, P.J.: Economic comparison of nutritional management strategies for Venezuelan dual-purpose cattle systems. J. Anim. Sci. **72**, 1680–1696 (1994)
20. Olson, T.A., Willham, R.L., Boehlje, M.D.: Linear programming as a tool for planning beef cattle breeding experimentation. J. Anim. Sci. **51**, 847–859 (1980)
21. Costa, F.P., Rehman, T.: Unravelling the rationale of 'overgrazing' and stocking rates in the beef production systems of central Brazil using a bi-criteria compromise programming model. Agric. Syst. **83**, 277–295 (2005)
22. Veysset, P., Bebin, D., Lherm, M.: Adaptation of Agenda 2000 (CAP reform) and optimisation of the farming system of French suckler cattle farms in Charolaise area: a model-based study. Agric. Syst. **83**, 179–202 (2005)
23. Crosson, P., O'Kiely, P., O'Mara, F.P., Wallace, M.: The development of a mathematical model to investigate Irish beef production systems. Agric. Syst. **89**, 349–370 (2006)
24. Havlik, P., Jacquet, F., Boisson, J.M., Hejduk, S., Vesely, P.: Mathematical programming models for agri-environmental policy analysis: a case study from the White Carpathians. Agric. Econ. (AGRIECON) **52**, 51–66 (2006)
25. Oltjen, J.W., Ahmadi, A.: TAURUS beef cattle ration formulation and evaluation software: windows version. In: Proceedings of 4th World Conference on Computers in Agriculture and Natural Resources, 23–25 July, Orlando, Florida, USA, pp. 09–413 (2006)

26. Kumar, A., Sah, B., Singh, R.A., Deng, Y., Xe, X., Kumar, P., Bansal, R.C.: A review of multi criteria decision making (MCDM) towards sustainable renewable energy development. Renew. Sustain. Energy Rev. **69**, 596–609 (2017)
27. Vico, G., Prodanović, R., Bodiroga, R.: A two-stage approach for multiple criteria decision making in plant production. In: Proceedings of the 22th International Symposium on Biotechnology, 10–11 March Čačak, Serbia, pp. 413–418 (2017)

New Technologies in Civil Engineering, Architecture, Quality Control

Proposed New Quality Control Plan Procedure for Road Bridges

Naida Ademović(✉)

Faculty of Civil Engineering, University of Sarajevo,
Patriotskelige 30, 71 000 Sarajevo, Bosnia and Herzegovina
naidadem@yahoo.com, naidadem@gmail.com

Abstract. In order to speed up the formulation of the guidelines for implementation of the performance-based bridge assessment a COST Action TU 1406 was established. The action gathered experts for the scientific field and practitioners that worked together for four years from 2014 to 2019. One of the main outputs of this Cost Action was the development of the procedure for the Quality Control Plan (QCP) for the road bridges. The starting point for development were results obtained by three working groups (WGs) obtained from a wide-ranging and exhaustive work. The result of the WG1 was a list of applicable Performance Indicators (PI), while WG2 gave clear instructions on Performance Goals (PG). WG3 used these results as the basis for the quality control plan proposal. Incorporating all this information was done by WG4 and a new procedure was developed for the implementation of developed guidelines, done by WG5, for the preparation of the QCP. The idea was to produce a unique method that would be applicable to different types of bridges (frame, girder, arch, and truss) at the European level regardless of the specific formulations of the QCP in each country. This paper will show the basic steps of the new procedure and its specific features. This is a qualitative procedure based on expert judgment. In order to move to a quantitative approach much work needs to be done.

Keywords: Road bridges · Quality Control Plan · Qualitative analysis · COST TU 1406 · Key performance indicators

1 Introduction

Standard ISO 2394 [1] summaries the fundamental requirements that all structures, as well as their integral elements, have to fulfill, from the concept, design, construction and finally to the maintenance. This all with the goal that the structures are suitable for their usage within the frame of their design working life, adequate economic limits, and expectations. The design codes of most of the counties are based on this standard.

The basic constituent of the worldwide economy is freight transport, which can be defined as the ability to move goods from its origins to certain destinations. In the past decades, freight transport has increased rapidly and it is predicted that the high growth rates will continue in the coming decades [2]. Based on the current demand pathway

I. Karabegović (Ed.): NT 2020, LNNS 128, pp. 813–820, 2020.
https://doi.org/10.1007/978-3-030-46817-0_92

the global freight demand will triple between 2015 and 2050. The freight transport is continuing and it is foreseen to rise by around 40% by 2030 and by little over 80% by 2050 as per the European Commission's data. Of the 108 trillion tonne-kilometers (t-km) transported worldwide in 2015, 70% traveled by sea, 18% by road, 9% by rail and 2% by inland waterway [3]. European road freight transport for a one-year period from 2016 to 2017 increased by 4.5% in terms of t-km and for a period of five years (2013–2017) it had the highest increase of 11.8% [2], as indicated in Fig. 1. Due to the increase of freight transportation and freight weight, it is essential to assess the state of the existing road bridges for safety and security reasons. Additionally, a negative impact is reflected by an enormous decrease in funds assigned for road maintenance from 2008. All this will be reflected on the society and manifested through pollution, higher noise, congestion which will then affect the economy. The health and comfort of the population will be directly impacted by such a scenario.

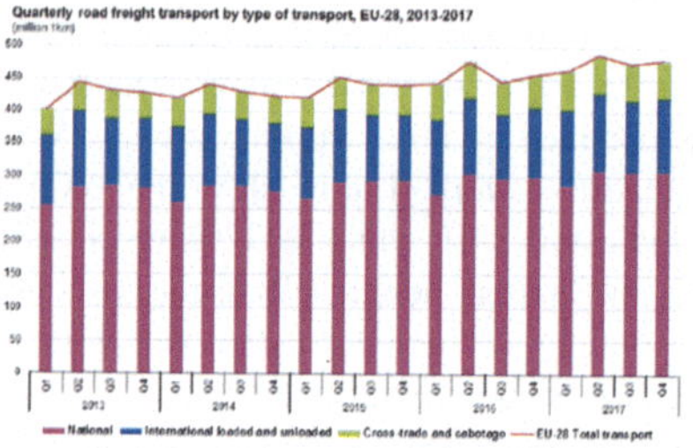

Fig. 1. Quarterly road freight transport by type of transport, EU-28, 2013–2017 (million tonne-kilometres) [2].

One of the objectives of the COST Action TU1406 was to gather together researchers and engineers in practice to formulate and haste up the preparation and creation of the guidelines for implementing the performance-based bridge assessment [4]. A well-established management bridge system is one of the prerequisites for the performance of adequate maintenance measures. This Action proposed a Quality Control Plan which is based on the identified Key PIs (Reliability, Availability, Safety and Cost, which are obtained from performance indicators (PI) being correlated by the identified damage processes [5].

2 The Methodology Developed in COST TU1406

Within the COST Action TU1406, a methodology for bridge assessment was proposed. The first step was the identification of the damage type/process (Fig. 2). Damage of a bridge element is a physical disruption or a change in its condition, caused by external actions, such that some aspect of, either the current or future performance of the component (and perhaps consecutively a complete structure) is impaired. Performance indicator, as defined in the fib Model Code 2010 is a superior term of a bridge characteristic, which indicates the condition of a bridge. It can be expressed in the form of a dimensional performance parameter or as a dimensionless performance index. The clustering of the performance indicators was done in a way to enable easier identification of methods and procedures for disclosure and quantification of performance indicators. Additionally, it was necessary to identify levels of their contribution to a certain structural performance goal defined as a type of structure's property (behavior) that is required based on the assessment of different performance indicators.

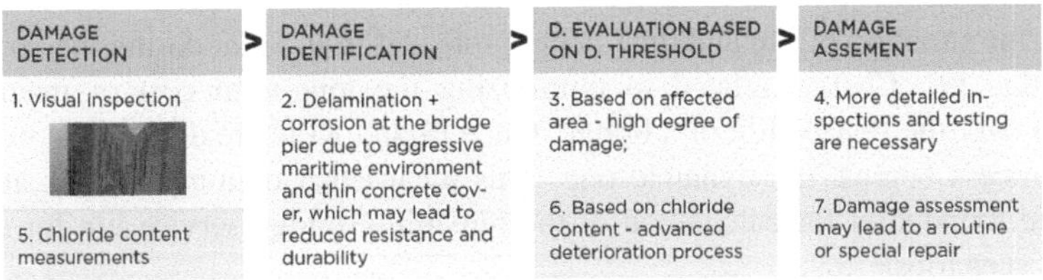

Fig. 2. Damage assessment implies the detection of damages as well as their identification and evaluation [6].

After the clustering procedure, 386 different performance indicators were defined. Performance Indicators (PI) or in particular Key Performance Indicators KPIs make it possible to define a set of objectives aimed to establish QC plans which ensure desired bridge quality service. KPIs are determined from a number of Performance Indicators, PIs collected at an operational level. Key performance indicators can be various, from Reliability (R), Availability (A), Maintainability (M), Safety (S), Security (Se), Health (H), Politics (P), Inspection (I), Durability (D), Cost (C) and Environment (E). In this case, four KPIs were chosen and elaborated, e.g. Availability, Reliability, Safety and Cost.

The methodology is composed of static and dynamic tasks and passes several levels. First of all, it is necessary to obtain as much as possible information about the bridge, from existing data and in accordance with that form an ID of the bridge. Levels from two to seven are of static nature and they can be conducted in the office (green color) or on the site (orange color). After this, dynamic tasks are foreseen, from twelve to fifteen (blue color), as these processes are changeable with respect to the envisaged scenarios. This is clearly presented in Fig. 3 [7]. Different methods will provide different acquisition procedures; however, the results should be presented in the same form enabling comparisons and further usage [8]. Once the data on the bridge level is obtained this data can be further incorporated to the network level as the importance of the bridge on the network level is essential in the analysis of management strategies and thus upgrade the decision process in the bridge management system.

Identifying possible failure modes being dependent on the bridge typology is essential as it will reveal locations in the necessity of overall and detailed inspection. The bridge failure probability is obtained from the failure mode analysis which has a straightforward relationship with the Reliability (R). The first step would be the determination of the "current" reliability (structural safety and serviceability) of the bridge. This means assessment of the undamaged ("virgin") bridge taking into account all appropriate load cases. For existing structures, the issue which always arises is the uncertain knowledge about built-in material characteristics and applicable loads and load combinations in the design process. Due to these unknowns, the other scenario would be to conduct on-site investigations and identify the damage processes, select PIs and make connections to the adequate KPIs, being related to the demands defined by the owners. It is important to define the KPIs. Reliability is defined as the probability that a bridge will be fit for purpose during its service life. It is the complement to the probability of structural failure (safety), operational failure (serviceability) or any other failure mode. Availability is the proportion of time a bridge is open for service. Safety is the situation of life and limb being protected from harm during the service life of a bridge [9]. Cost is related to minimizing the long-term cost of maintenance activities over the service life of a bridge. Once PI and KPIs are defined the next stage of the process begins, the dynamic one. This means that defining the PIs and KPIs forms the foundation for calculation of the remaining bridge service life for different foreseen scenarios.

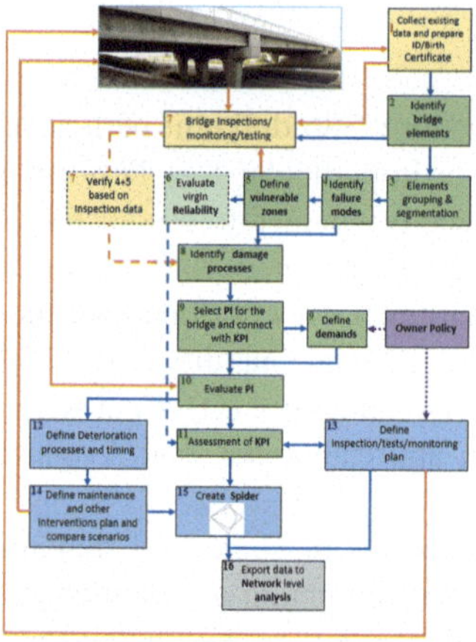

Fig. 3. Case study preparation [7].

The knowledge of bridge deterioration rates is essential for cost-effective asset management and long-range transportation planning. Among different elements which have an impact on the bridge conditions, it was confirmed that age [10, 11], and material [12] are essential. Structural characteristics of the bridge and traffic volume are the next two contributors. The value of these KPIs is more important today then e.g. in one, two or 10 years. Thus, it is regarded that the interventions in the short term are more expensive but the benefits are also more valuable. It was necessary to normalize all KPIs in the scale from 1 to 5 (1 the best to 5 the worst condition) and the Net Present Value (NPV) can be used (see Fig. 4).

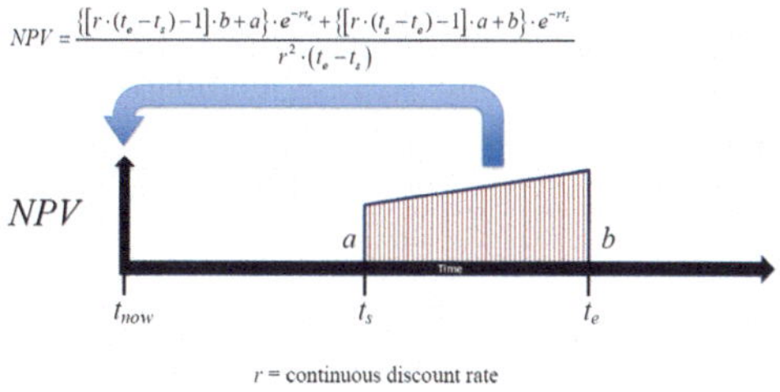

Fig. 4. Evaluation of the net present value for time-dependent KPIs case study preparation [13].

This was done in order to reduce the KPIs to the same scale as for any time instance, the normalization is performed i.e. the NPV is divided with the NPV calculated if all KPIs were equal to one over the whole investigation period. These values can be regarded as "average" long term KPIs and on the basis of this, a decision will be made regarding the adequate and appropriate maintenance strategy with the application of methods developed in [5].

3 Relating Different Maintenance Periods

Once this is done the planning of different maintenance and intervention plans with target reliability and safety over a relevant period time based on the bridge age are elaborated. Generally, two maintenance scenarios are recommended: referenced and preventive. The referenced scenario represents a scenario when there are no available funds for necessary maintenance/repair activities and the bridge will be critically deteriorated and with traffic restrictions on year 100. Preventive scenario foresees a scenario with 100-year service life, according to the normal service life of bridges, for which there will be enough funds for undertaking needed mandatory inspections and maintenance/repair actions.

For the two possible scenarios related cost estimates will be conducted for the scenarios QCP and the function defining the decrease of the KPIs will be formulated over the selected time period. As a result, time-dependent graphs for the KPIs will be constructed (Fig. 5).

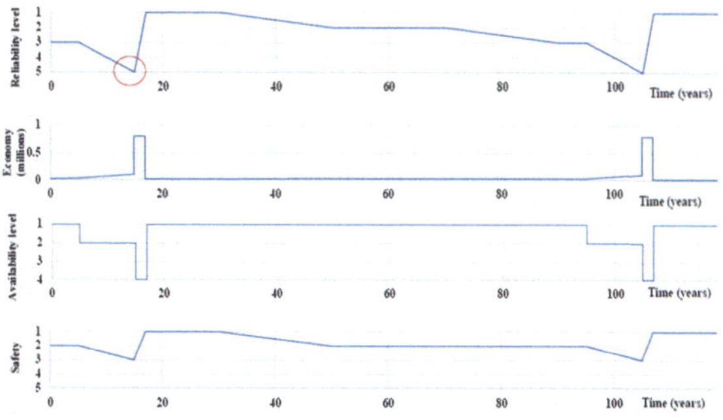

Fig. 5. Normalized KPIs over time [13]

A convenient way of presenting the KPIs is by a "Spider net diagram" (Fig. 6). Here, each axis is associated with one KPI. A comparison of the two considered approaches is based on the area of each scenario.

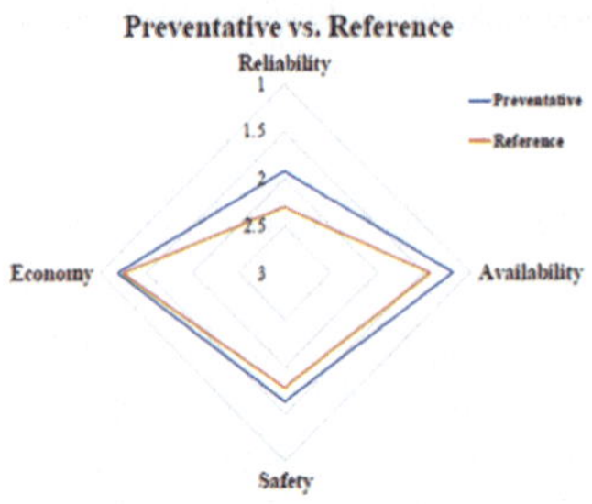

Fig. 6. Spider diagram [13]

For better visualization and in order to be able to see the changes over the time duration, the time axis is added orthogonally on the plane of the diagram. In this way, a "performance tube" is created [13] (Fig. 7).

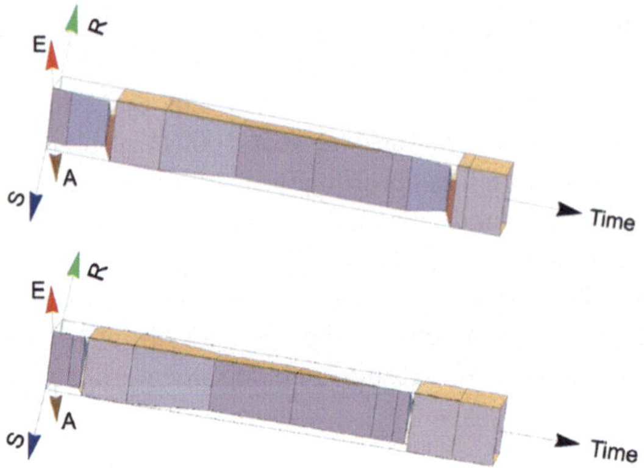

Fig. 7. Spider diagram in time [13]

4 Practical Application

Using this procedure nine case studies were prepared. The time and work on the site which was necessary for the preparation of the QCP for all the bridges were within the range of the common practice for all the involved countries. For accurate and elaborate identification of the vulnerability zones and the failure mechanics experienced and highly educated bridge engineers are needed to conduct the inspection analysis. Once done it should be stored in the bridge management system for further use in the future by the adequate personnel.

Through the elaborated examples it has been demonstrated that the proposed methodology is flexible and depending on the level of complexity the methodology can be applied in full or partially. Additionally, depending on the level of knowledge the theoretical models range from a simple one with the application of the "expert judgment", to the usage of "β" value or factor of safety (FoS) reduction calculations, to various deterioration models. The level of detail depends on the bridge owner and his requirements.

5 Conclusion

A new methodology for the preparation of the QCP for road bridges has been presented which can be used for better decision-making regarding bridge management. Thought the COST Action TU 1406 a guideline for standardization and development of quality control plans for roadway bridges has been proposed independent on local bridge regulations in each specific county. The process is based on defining and identifying performance indicators, performance goals and key performance indicators. A multi-criteria decision-making (MCDM) approach can thoroughly associate the inputs with cost-benefit models to rank existing decision options regarding bridges at the component, system or network levels. Further developments are seen in the standardization of the assessment procedures, collection of accompanying PIs and their connection with the relevant KPIs, as well as the introduction of new KPIs and the application of more robust maintenance tools in the practice.

References

1. International Organization for Standardization-ISO (2015) General principles on reliability of structures, ISO 2394 (2015)
2. IEA. Transportation projections in OECD regions, Detailed report, International Energy Agency, Paris (2002)
3. OECD. Chapter 1. How transport demand will change by 2050 (2019). https://www.oecd-ilibrary.org/sites/57f51e1e-en/index.html?itemId=/content/component/57f51e1e-en
4. European Cooperation in Science and Technology. Memorandum of Understanding for COST Action TU1406, COST (2014)
5. Stipanovic, I., Chatzi, E., Limonggelli, M., Gavin, K., Allah Bukhsh, Z., SkaricPalic, S., Xenidis, Y., Imam, B., Anzlin, A., Zanini, M., Klanker, G., Hoj, N., Ademovic, N.: COST Action TU 1406, WG2 Technical report – Performance Goals for Roadway Bridges (2017)
6. Strauss, A., MandićIvanković, A.: COST Action TU1406, WG1 Technical report – Performance Indicators for Roadway Bridges (2016)
7. Kedar, A., Sein, S., Panetsos, P., Ademović, N., Duke, A., Ryjacek, P.: COST TU 1406, WG4 Technical report – Preparation of a Case Study (2019)
8. Rücker, W., Hille, F., Rohrmann, R.: Guideline for the assessment of existing structures. SAMCO Final Report (2006)

9. Kedar, A., Sein, S., Ademović, N., Panetsos, P., Sander, S., Ademovic, N., Ryjacek, P., Duke, A., Matos, C.J.: Developing case studies for implementing COST TU1406 quality control plan procedure for typical highway bridges. In: IABSE Symposium, Guimaraes 2019: Towards a Resilient Built Environment Risk and Asset Management – Report Guimarães, Portugal, pp. 531–539 (2019)
10. Jiang, M., Sinha, K.C.: Bridge Service Life Prediction Model Using the Markov Chain. Transportation Research Records 1223. Transportation Research Board, Washington, D.C. (1989)
11. Kallen, M.J., Van Noortwijk, J.M.: Statistical inference for Markov deterioration models of bridge conditions in the Netherlands. In: Cruz, P.J., Frangopol, D.M., Neves, L.C. (eds.) Bridge Maintenance, Safety, Management, Life-Cycle Performance and Cost: Proceedings of the Third International Conference on Bridge Maintenance, Safety and Management, pp. 535–536. Taylor & Francis, London (2006)
12. Dunker, K.F., Rabbat, B.G.: Highway bridge type and performance patterns. J. Perform. Constr. Facil. **4**(3), 161–173 (1990)
13. Hajdin, R., Kušar, M., Mašović, S., Linneberg, P., Amado, J., Tanasić, N., Ademović, N., Costa, C., Marić, M., Almeida, J., Galvão, N., Sein, S., Zanini, M., Pendergast, L.: COST Action, TU 1406, WG3 Technical report Establishment of a Quality Control Plan (2018). ISBN: 978-86-7518-200-9

Problem of Characteristic Snow Load in the Eastern Part of Bosnia and Herzegovina

Rašid Hadžović[1(✉)] and Bakir Krajinović[2]

[1] Faculty of Civil Engineering, "Džemal Bijedić" University of Mostar, 88 104 Mostar, Bosnia and Herzegovina
rasid.hadzovic@unmo.ba

[2] Federal Hydrometeorological Service, Sarajevo, Bosnia and Herzegovina
bakir.krajinovic@fhmzbih.gov.ba

Abstract. In and through this paper we analyzed the characteristic snow load in the eastern part of Bosnia and Herzegovina. We cross-referenced the data from the map of characteristic snow load map with the data of snow depth in the same area during the period 1970–1980 as well as the maximum snow depth data from the January of 1976. The values of snow load are increased in respect to the period prior to defining the map of load in 2018, raising certain doubts with several design engineers. This paper demonstrated that the assigned values of characteristic snow load are realistic and applicable in construction projecting.

Keywords: Characteristic snow load · Snow load map · Eastern part of Bosnia and Herzegovina

1 Introduction

Characteristic snow load in Bosnia and Herzegovina is defined by the map of snow load in Bosnia and Herzegovina 2018, as part of the project implemented by the Institute for standardization of Bosnia and Herzegovina, with the support of the Czech Development Agency. The working group successfully completed the draft/proposal of both electronic and hard copy version of the snow load map for Bosnia and Herzegovina, adopted by the Technical Committee BAS/TC 58 of Working Group WG1 [3]. This map can be found online [6]. In the interactive map, by inputting the geographical coordinates of the object, or by directly clicking on the location of the object, one click automatically gets access to the data – the value of the characteristic snow load on the ground, according to the Eurocode 1 regulations. The values defined in the map are derived from the calculations for the period 1961–1990, from the main measurement locations in Bosnia and Herzegovina, whereas the values for water column were derived from melting of snow, also significant for calculating the snow load. On the measuring sites located at high altitudes there are no measurements for the height of the water column, due to the high amounts of snow (ex. Bjelašnica measures up until 3 m of snow depth). The eastern part of Bosnia and Herzegovina is covered with measuring site in Goražde, Foča and other sites where the height of the water column is not measured, Čemerno site – where there are defective measurements on the height of the

I. Karabegović (Ed.): NT 2020, LNNS 128, pp. 821–828, 2020.
https://doi.org/10.1007/978-3-030-46817-0_93

water column and Kalinovik where the data on the height of the water column is homogenous. In order to prove the reliability of the map, in the eastern area of B&H, in the populated areas, and on lower altitude sites we conducted the analysis on the maximum snow depth, newly formed snow cover, as well as the maximum and minimum temperature, using the methods of correlation and analysis.

2 Snow Data

2.1 Snow Data Introduction

In accordance with the recommendations of the World Metrology Organization (WMO) for statistical processing and obtaining quality budget results, the minimum period must be 30 years. The data provided for the analysis from the Institute are: snow depth, sampling height and water content for winter period 1961–1991 for all major measuring points [1] and also in the paper is analyzed period 1970–1980., and January 1976. of East part of Bosnia and Herzegovina.

2.1.1 Heights of Snow

For the calculation of snow load it is necessary to have measured heights of snow calculated as monthly maximum. We have 5 measuring places with measured data of snow depths in the eastern part of Bosnia and Herzegovina. As a rule, the depth of the snow increases with the rise in altitude, so that the depth of the snow should be bigger in the mountain parts than the lowlands.

2.1.2 Height of the Water in the Snow (VSV)

The height of the water column in the snow is derived through the snow melting according to the regulations. It is necessary in order to achieve the snow density, as the basis for the calculation of the snow load. The height of the water column in the snow is measured only on the 15 main measuring sites in Bosnia and Herzegovina, so it is very difficult to determine the snow load where there is no data about the height of the water column in the snow.

2.1.3 Snow Density

Snow load *(kN/m²)* is calculated as snow height multiplied by snow density. Snow density is a key element in determining the snow load and is expressed in *(kN/m³)*. The density of snow is determined based on the existing data from the amount of water in the snow. The snow density increases with the duration of the snow cover, and depends on the location of the building, the climate and the altitude.

According to the originally measured heights of the content VSVu *(mm/cm)*, the density of snow is calculated in *(kN/m³)* [1, 2]:

$$\rho_{snijeg} = VSV\frac{mm}{cm} \times \rho_{voda}\frac{g}{cm^3} \times 10^{-2}\left[\frac{kN}{m^3}\right] \quad (1)$$

2.2 Characteristic Snow Load

Map of the characteristic snow load on the ground show as calculated values by Gumbel's distribution for return period of 50 years based on maximum meteorological snow load.

Characteristic snow load is random variable necessary for probabilistic calculations.

Gumbel's distribution following formula (2) is applied [1]:

$$s_K = \widehat{X} + \frac{1}{a}[\ln n - \ln(-\ln p)] \tag{2}$$

- Characteristic snow load on the ground S_k
- modus

$$\widehat{X} = \bar{X} - \frac{c}{a} \tag{3}$$

- coefficient of dropout

$$a = \frac{1}{\sigma} \cdot \frac{\pi}{\sqrt{6}} \tag{4}$$

- number of analyzed years $n = 50$
- Fractional value of the action $p = 0, 95$
- The mean value of the meteorological snow load $\bar{X}$
- standard deviation of meteorological snow load σ
- Euler's constant $C = 0, 577$

Calculation of the characteristic snow load on the ground, for each measurement point, was completed and the data were entered into the numerical model for the development of map of snow load in Bosnia and Herzegovina.

A digital map of snow load is defined for measuring place Goražde, calculated by Kepen climate classification index, followed by the analysis of snow loads from the neighboring places with measured height of the water in snow and the measuring points in the border areas with Serbia, Montenegro and Croatia. The Digital Land Model (DMT) Korsten as the first substrate for all mapping has a very high resolution of 200 m. This resolution has been selected because of the project's requirements, because all displays are entered on a digital map whose image goes to the level of the building itself. The method used for spatial distribution is Universal Kriging, and the System for Automated Geoscientific Analysis, Version: 2.1.1.

As it is visible in the printed map above (Fig. 1), each color represents one zone, so the minimum load value is <0.75 kN/m^2, and the maximum >7.50 kN/m^2. Due to the large differences in altitude, the snow load is given in 9 different zones with a value difference of 1 kN/m^{2}in order to minimize dispersion of the result [2].

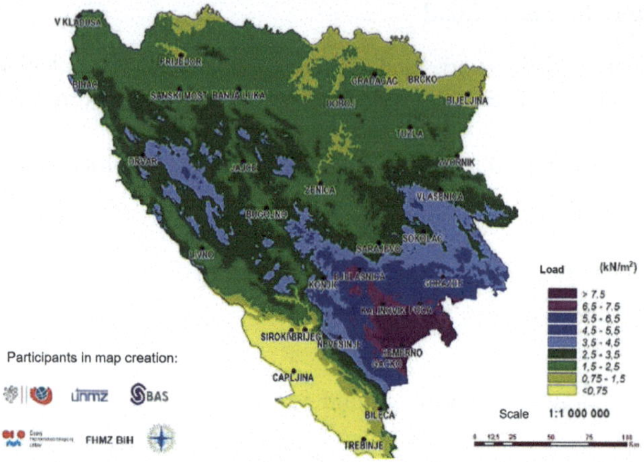

Fig. 1. Map of characteristic snow load on the ground of Bosnia and Herzegovina [2]

2.3 Snowload in the Eastern Part of Bosnia and Herzegovina

The biggest problem was the 9th zone due to the large amounts of snow and mountain massifs in the east of Bosnia and Herzegovina. If a digital map is not available, the map thus prepared can be used except in zone 9, where there is a possibility of a greater amount of snow than 7.5 kN/m^2. It is recommended that the upper values are taken in the zones, and that additional effort is being made in the zone 9 to open the web site since the values can be higher than 10 kN/m^2 [6].

Calculated snow load for Kalinovik for period 1961–1991 is 6,79 kN/m^2, but there is not enough data for Foča and Goražde to calculate snow load.

3 Analysis of Snow Cover in Eastern Part of Bosnia and Herzegovina

The area of Eastern Bosnia and Herzegovina, especially the river Drina valley from Foča to Goražde has significantly higher values of snow load in comparison to the values estimated by the standards created prior to 1991. This difference in values caused raising suspicions with several design engineers in reliability of the map of snow load in Bosnia and Herzegovina, adopted by the Technical Committee TC/BAS 58 WG 1. Since the problem with lack of quality network of measuring sites with the option of measuring of snow for acquiring data for the water content and snow density has already been stated, the mapping of calculated snow load in these areas became problematic. The above mentioned adopted map was created on the basis of data – measured values of snow density and water ratio in the snow cover on the basis of which the load distributed to 50 years is calculated.

Furthermore, we used the basis of Köppen climate classification, as well as the maximum snow depth for all the available measuring stations, that are larger in number than the stations measuring the water ratio in the snow and the snow density [4, 5]. The results are displayed in the Fig. 1.

In order to demonstrate the reliability of the map in the east of Bosnia, especially the river Drina valley – from Foča to Goražde, we conducted an additional analysis of this area. Since measuring sites Goražde and Foča do not have the possibility of measuring neither the water ratio in the snow, nor the density of the snow cover, and therefore there is no data about the snow load, we used the data of maximum snow depth for proving the significant correlation.

Regarding the correlation, 10 characteristic maximum values of snow cover height, for Foča the values fluctuate and in the case of Goražde between 0,58 and 0,98. Although this is the case of characteristic meteorological value parameter, these correlations are significantly high, especially for five years with maximum height.

We analyzed the data from January 1976 thoroughly, because at that period of time the highest value of snow cover was measured in these two towns, and for correlation we used the station Kalinovik, since it has the most accurate data in this area.

Correlation for maximum temperatures fluctuates between 0,9 for Goražde and Foča (Fig. 2), 0,4 for Goražde and Kalinovik (Table 1), and 0,6 for Foča and Kalinovik (Table 2). Minimum temperatures have far better correlation, fluctuating between 0,9 for Goražde and Foča, 0,8 for Goražde and Kalinovik and 0,8 for Foča and Kalinovik.

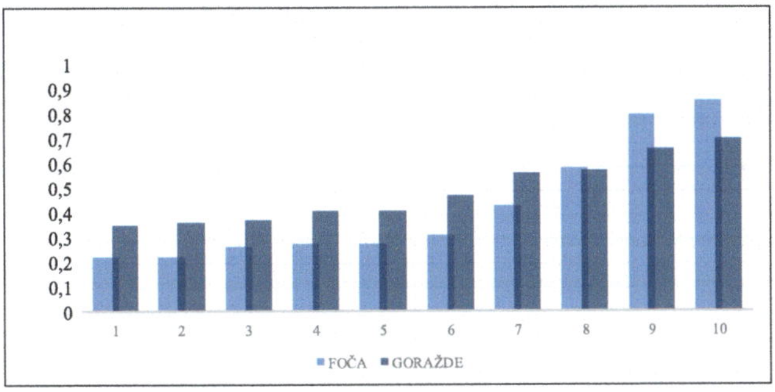

Fig. 2. Ten maximum values of snow cover for meteorological stations Foča and Goražde (m)

Table 1. Values of correlation for Foča and Kalinovik for daily values of snow depth in 10 years with maximum values of snow depth in m

Years	1970	1981	1983	1973	1989	1987	1980	1971	1985	1976	Mean corr. coeff.
Place	12	2	12	11	11	1	1	1	1	1	
	0,87	0,04	0,82	0,85	-0,26	0,73	0,59	0,88	0,92	0,97	0,64
FOČA	0,23	0,23	0,27	0,28	0,32	0,32	0,44	0,60	0,82	0,88	
KALINOVIK	0,71	0,25	0,44	0,99	0,91	0,66	0,91	0,99	0,82	0,79	

Table 2. Values of correlation for Goražde and Kalinovik for daily values of snow depth in 10 years with maximum values of snow depth in m

Years	1968	1962	1966	1971	1980	1969	1967	1985	1963	1976	Mean corr. coeff.
Place	1	12	1	1	1	12	1	1	1	1	
	0,81	0,69	0,89	0,95	0,58	0,89	0,97	0,89	0,95	0,98	0,86
GORAŽDE	0,36	0,37	0,38	0,42	0,42	0,48	0,58	0,59	0,68	0,72	
KALINOVIK	0,71	0,25	0,44	0,99	0,91	0,66	0,91	0,99	0,82	0,79	

Maximum value of snow cover for January 1976. is significant - 0,9 for all the values which is very important for proving our thesis. Maximum value of snow cover was recorded on January 29th in Goražde – 0,72 m (Fig. 4), Foča on January 28th – 0,88 m (Fig. 3), and also Kalinovik on January 28th – 0,79 m of snow cover (Fig. 5).

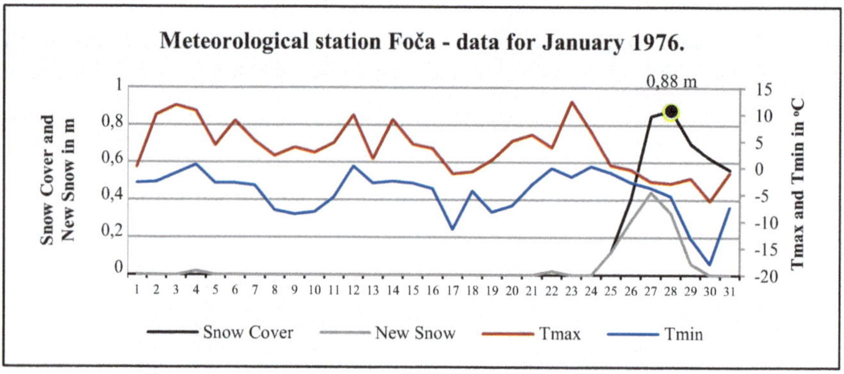

Fig. 3. Maximum snow cover and new snow in relation with maximum and minimum temperature on meteorological station Foča

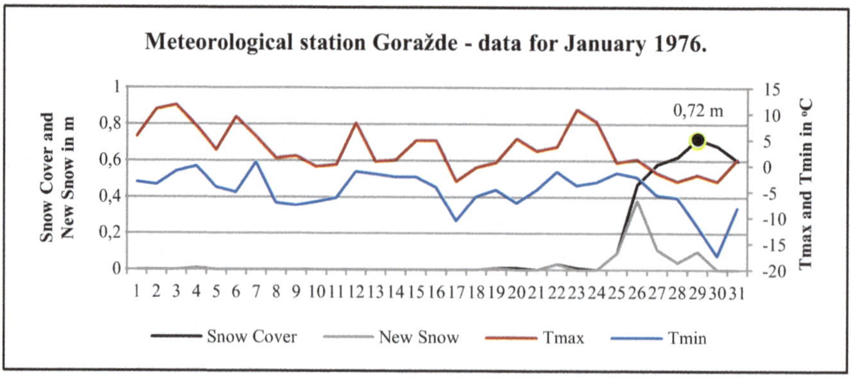

Fig. 4. Maximum snow cover and new snow in relation with maximum and minimum temperature on meteorological station Goražde

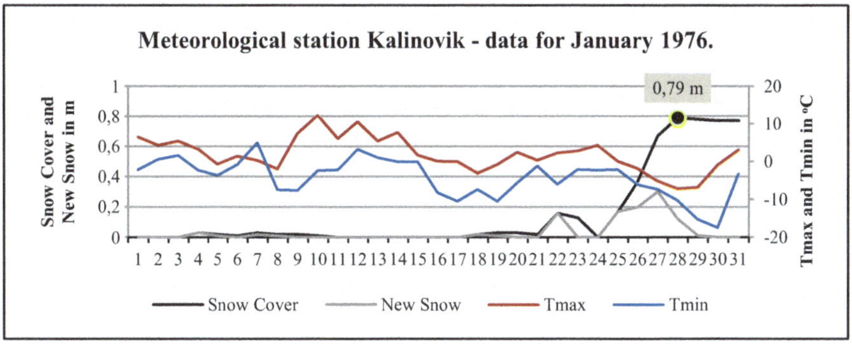

Fig. 5. Maximum snow cover and new snow in relation with maximum and minimum temperature on meteorological station Kalinovik

Since the correlation between these parameters for the examined stations is high for January 1976, and the difference between the values of snow cover of depth is small – the difference between Foča (0,88 m) and Goražde (0,72 m) is 0,16 m, and the maximum value for Kalinovik is 0,79 m, so it also fits into this amplitude. It can be concluded that the snow load for Goražde and Foča is approximately equal to the one measured in Kalinovik (Table 3).

Table 3. Basic data about the examined meteorological stations

Station name	Latitude	Longitude	Altitude m	Difference in spatial distance km	Difference in sea level m
Goražde	43° 40′	18° 59′	345	G-F 25 G-K 47	G-F –50 G-K -728
Foča	43° 30′	18° 48′	395	F-K 27	F-K -678
Kalinovik	43° 31′	18° 27′	1073	–	–

4 Conclusion

On the basis of presented values of snow depth and completed calculations in snow load at Eastern part of Bosnia and Herzegovina, it can be concluded that the values given in the map of characteristic snow load are satisfactory, as this paper ascertained through demonstration of correlation between maximum snow depths.

Climate change, among other factors, manifest through increase in extreme values of snowfall. This is a known fact since 1999. At 2012 was the biggest snow falls in Bosnia and Herzegovina and in it is great possibility in the future to snow falls be bigger. Hence, there have been recorded maximum values in snow cover, causing numerous troubles to traffic, collapsing of metal and wooden constructions, as well as numerous material damages and expenses, all in relation to the snowfall and the snow load.

References

1. Hadžović, R., Peroš, B.: Pouzdanost konstrukcija dominantno opterećenih snijegom u Bosni i Hercegovini, Građevinski fakultet Univerziteta "Džemal Bijedić" u Mostaru, Mostar 2016, knjiga
2. Hadžović, R., Krajinović, B.: Map of characteristic snow loads on the ground of Bosnia and Herzegovina. In: 5th International Conference on New Technologies, Development and Application, NT-2019, Sarajevo, Bosnia and Herzegovina, 27–29 June 2019
3. Group of authors: PriručnikzaEurokodove. http://eurokodovi.ba/handbook/?lang=bs
4. Jonas, T., Marty, C., Magnusson, J.: Estimating the snow water equivalent from snow depth measurements in the Swiss Alps. J. Hydrol. **378**, 161–167 (2009)
5. Krajinovic, B., et al.: Atlas klimeFederacijeBosne i Hercegovine, Federalnihidrometeo-rološkizavod, BiH, Sarajevo (2018)
6. Interaktivna karta opterećenja snijegom Bosne i Hercegovine. http://eurokodovi.ba/snijeg/

Influence of Wind Load to the Billboards at Mostar Valley

Rašid Hadžović[1(✉)] and Osman Tibo[2]

[1] Faculty of Civil Engineering, "Džemal Bijedić" University of Mostar, USRC "Mithat Hujdur Hujka", 88 104 Mostar, Bosnia and Herzegovina
rasid.hadzovic@unmo.ba
[2] Babić doo Jablanica, Jablanica, Bosnia and Herzegovina

Abstract. The aim of this paper is to analyze the wind load in the Mostar valley since wind speeds are up to 240 km/h. The values of bending and resistance moments relative to the surface of the billboards were calculated and the necessary equations were obtained. For 9 different cases of size of billboards, the values of the bending moments and resistance moments over to increasing wind speed were calculated and equations were obtained describing the coefficients "a" and "c" needed to calculate Mmax and Wpot. Finally, two examples were made as evidence of the analysis.

Keywords: Billboards · Wind speed · Wind load · Bending moment · Resistance moment · Mostar valley

1 Introduction

Through this assignment we conducted the research of influence of wind load on the billboards and the reasons for their crashing. Billboards have been increased in size through linear increase in width and height by 25 cm, followed by determining the variations of wind load, depending on these two dimensions, as well as the bending moment, resistance moment and the cross-sectional area surface. Billboard profiles are zinc coated (galvanized) HOP profiles, to which the aluminum billboard sheets will be attached.

The wind load analysis was conducted for different billboard dimensions (from $L = 4$ m and $H_2 = 2$ m, to $L = 6$ m and $H_2 = 4$ m) and wind speed assigned to Mostar – $V_0 = 35$ m/s. The next step was to define the cases of billboard surface, for which we determined the necessary profile dimensions and tested the main normal stresses and the deflection of the billboard top, by increasing the wind speed from 35 m/s to 50 m/s. Simultaneously, we monitored the behavior of aluminum billboard sheet and demonstrated it using diagrams. For each created diagram, mathematical function (curve) was developed. Those curves are being used for detecting all parameters presented by certain diagram for different values of billboard area and wind speed. We established the connection between the billboard area and the wind speed, demonstrated it via diagrams and illustrated the examples of all the possible variations, in order to be able to determine the adequate profile for each billboard area and wind speed.

I. Karabegović (Ed.): NT 2020, LNNS 128, pp. 829–838, 2020.
https://doi.org/10.1007/978-3-030-46817-0_94

Since the equations result in lesser values than the actual ones, and in order to avoid any possibility of mistake, the resulting values need to be multiplied by the correction coefficient $k = 1{,}02$. Static system of billboards is a cantilever beam, and the load distribution on profiles is symmetrical.

Dimensioning of profile is completed through the allowed stresses for the used materials (160 MPa), followed by the examination of the normal stresses for the selected profiles and the load for which they were selected.

After detailed calculations and dimensioning of the observed billboards, we took into consideration two different examples as a proof of analysis (Fig. 1).

Fig. 1. Sample of destroyed billboard at south part of Mostar

2 Billboards Calculation Under Wind Load in Mostar Valley

2.1 Wind Load Calculation

Wind load is calculated by Eurocode 1 norm [1, 2] (Fig. 2).

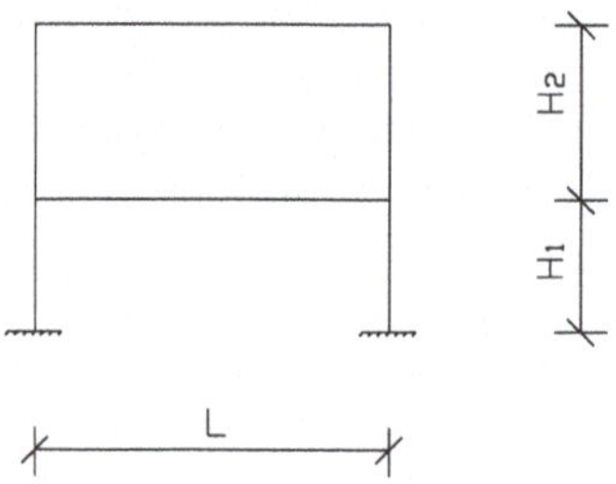

Fig. 2. Sample of billboard

V_0 – wind speed
V_{ref} – reference wind speed
q_{ref} – reference pressure of mean wind speed
H – sea level (Mostar at 55 m)
C_f – wind force coefficient
$V_0 = 35$ m/s basic wind speed for Mostar

$$V_{ref} = V_o * (1 + 0{,}001 * H) \tag{1}$$

$$q_{ref} = \frac{P}{2} * C_f * V_{ref}^2 \tag{2}$$

C_e – Exposure coefficient for II category of terrain

$$\begin{gathered} W_e = q_{ref} * C_e * C_{pe} * C_f \\ C_f = 1{,}8 \end{gathered} \tag{3}$$

$$q = W_e * \frac{L}{2} \tag{4}$$

2.2 Dimensioning of Billboard Profile

Dimensioning of billboard profile is made by JUS standards [3].

M_{max} – maximum bending moment
δ – normal stress in the cross section
W – the resisting moment of the cross section (Fig. 3)

$$\Delta = \frac{M}{W} \tag{5}$$

$$W_{pot} \geq \frac{M}{W} \tag{6}$$

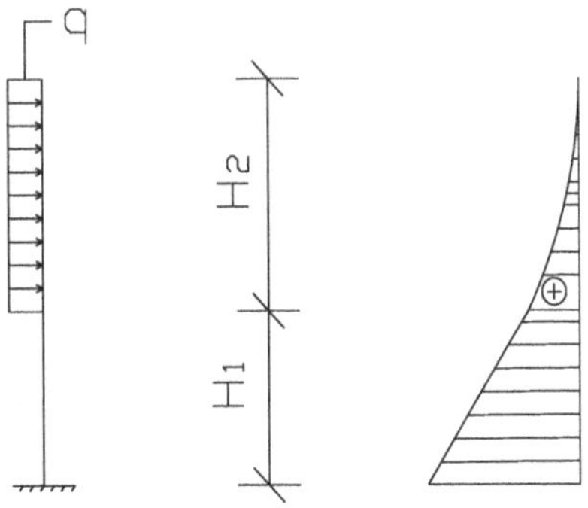

Fig. 3. Static system and diagram of bending moments [4]

3 Metal Sheets Calculation

The billboards will be comprised from different metal sheets arranged and merged in order to create a single surface. Geometric characteristics are taken from the crashed billboard (Fig. 4 and Table 1).

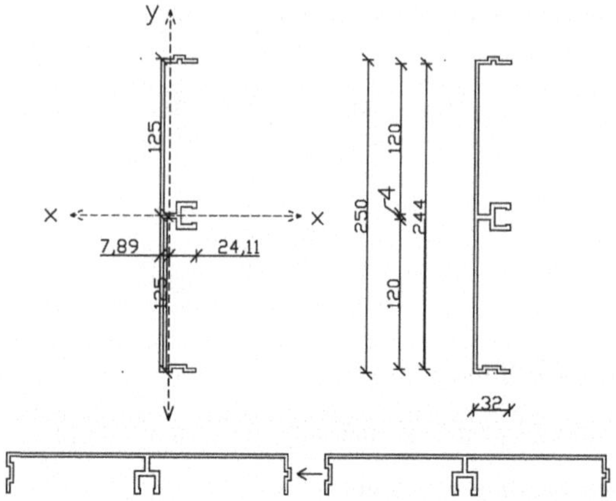

Fig. 4. Cross section and type of connection (cm)

After the profile parameters have been adopted, calculations made for every billboard case I–IX for wind speed $V_0 = 35$ m/s to get:

Equation of maximum bending moments of profile in relation to the billboard area:

$$M_{max} = 0{,}0794x^2 + 2{,}0385x - 1{,}6656; \\ R^2 = 0{,}9988 \text{ (coefficient of determination)}. \tag{7}$$

Equation of necessary resistance moment of cross-section profiles in relation to the billboard area:

$$W_{pot} = 0{,}4982x^2 + 12{,}682x - 9{,}9933; \\ R^2 = 0{,}9989 \text{ (coefficient of determination)}. \tag{8}$$

Table 1. Parameters for different size area of billboards (V_0 = 35 m/s), E = 210000 MPa

Parameter	I	II	III	IV	V	VI	VII	VIII	IX
H_1 (m)	1,50	1,50	1,50	1,50	1,50	1,50	1,50	1,50	1,50
H_2 (m)	2,00	2,25	2,50	2,75	3,00	3,25	3,5	3,75	4,00
L (m)	4,00	4,25	4,50	4,75	5,00	5,25	5,5	5,75	6,00
C_e	1,75	1,75	1,75	1,75	1,75	1,87	1,89	1,90	1,91
C_{pe}	0,75	0,75	0,75	0,75	0,75	0,75	0,75	0,75	0,75
W_e (kN/m^2)	2,01	2,01	2,01	2,01	2,01	2,15	2,17	2,18	2,19
I_z (cm^4)	920,3	1153,9	1442,9	2410	2832,7	3621,5	4337,5	5134,3	7315,5
t (cm)	1,2	1,6	1,6	1	1,2	1,6	2	2,5	1,6
q (KN/m)	4,02	4,278	4,529	4,78	5,03	5,64	5,97	6,28	6,5
M_{max} (KNm)	20,13	25,26	31,14	37,80	45,29	57,35	68,01	79,53	92,2
W_{pot} (cm^3)	125,8	157,9	194,6	236,2	283,1	358,4	425,0	497,0	576,7
Adopted	**140 × 140 × 6**	**140 × 140 × 8**	**150 × 150 × 8**	**200 × 200 × 5**	**200 × 200 × 6**	**200 × 200 × 8**	**200 × 200 × 10**	**200 × 200 × 12,5**	**250 × 250 × 8**

4 Structural Analysis of Billboards Cases I–IX for Wind Speed Increasing

Calculations made for every case of billboard I–IX over increasing wind speed $V_0 = 35–50$ m/s and got equations for: bending moments of profile over increasing of wind speed (Table 2), necessary resistance moments of profile over increasing of wind speed (Table 3) [5]. Values from Tables 2 and 3 are necessary to establish new functions for describing parameters "a" and "c" for calculation of M_{max} and W_{pot}.

This connection is only possible for billboards where L = H2 + 2 is most commonly used.

4.1 Equations from Diagrams for Different Billboards Area M_{max} and Increasing Wind Speed V_0 = 35–50 m/s

Table 2. Parameters from 4 to get equations by least squares method to get M_{max}

x (m^2)	a	c
8	1,3967	29,247
9,56	1,7533	36,715
11,25	2,1609	45,251
13,06	2,6901	58,169
15	3,5129	80,981
17,06	3,9796	83,338
19,25	4,7194	98,829
21,56	5,5187	115,57
24	6,4035	134,1

Equation to get value of "a" for calculation of M_{max}:

$$a = 0{,}0052x^2 + 0{,}1496x - 0{,}1622$$
$$R^2 = 0{,}999 \text{ (coefficient of determination)} \tag{9}$$

Equation to get value of "c" for calculation of M_{max}:

$$c = 0{,}1026x^2 + 3{,}3404x - 4{,}5134$$
$$R^2 = 0{,}9987 \text{ (coefficient of determination)} \tag{10}$$

4.2 Calculation of W_{pot} over Diagrams for Different Billboards Area and Increasing Wind Speed V_0 = 35–50 m/s

Table 3. Parameters from 4 to get equations by least squares method to get W_{pot}

x (m^2)	a	c
8	8,7304	182,82
9,56	10,958	229,47
11,25	13,506	282,82
13,06	16,813	363,56
15	21,955	506,13
17,06	24,873	520,86
19,25	29,496	617,68
21,56	34,492	722,29
24	40,022	838,11

Equation to get value of "a" for calculation of Wpot

$$\begin{aligned} a &= 0{,}0327x^2 + 0{,}9349x - 1{,}0112 \\ R^2 &= 0{,}9997 \text{ (coefficient of determination)} \end{aligned} \tag{11}$$

Equation to get value of "c" for calculation of W_{pot}

$$\begin{aligned} c &= 0{,}6413x^2 + 20{,}875x - 28{,}172 \\ R^2 &= 0{,}9987 \text{ (coefficient of determination)} \end{aligned} \tag{12}$$

4.3 Examples of Using Equations for Calculation M_{max} and W_{pot}

Example 1:

Sample way 1:

Making calculation for the billboard of dimensions H_2 = 4,5 m and L = 6,5 m to get M_{max} and W_{pot} for P = 29,25 m^2 and wind speed V_0 = 35 m/s through (7) and (8):

$$M_{max} = 0{,}0794 * P^2 + 2{,}0385 * P - 1{,}6656$$

$$\mathbf{M_{max} = 125,90\,KMm}$$

$$W_{pot} = 0{,}4982 * P^2 + 12{,}682 * P - 9{,}9933$$

$$\mathbf{W_{pot} = 787,19\,cm^3}$$

Sample way 2:

Make calculation for the billboard with dimensions H_2 = 4,5 m and L = 6,5 m to get M_{max} and W_{pot} using values from Tables 2 and 3.

Calculation of bending moment M_{max}:

To get "a" from Eq. (9) for P = 29,25 m^2:

$$a = 0{,}0052x^2 + 0{,}1496x - 0{,}1622 = 8{,}66$$

To get "c" from Eq. (10) for P = 29,25 m^2:

$$c = 0{,}1026x^2 + 3{,}3404x - 4{,}5134 = 180{,}97$$

Equation to get bending moments of billboard H_1 = 4,5 m and L = 6,5 m and wind speed V_0 = 35 m/s:

$$M_{max} = a * V_0 - c = (8{,}66 * 35 - 180{,}97) * 1{,}02$$
$$\mathbf{M_{max} = 124,57\,KNm}$$

Calculation of resistance moment W_{pot}:

$$W_{pot} = a * V_0 - c \tag{13}$$

To get "a" from Eq. (11) for P = 29,25 m^2:

$$a = 0{,}0327 * P^2 + 0{,}9349 * P - 1{,}0112 = 54{,}31$$

To get "c" from Eq. (12) for P = 29,25 m^2:

$$c = 0{,}6413 * P^2 + 20{,}875 * P - 28{,}172 = 1131{,}09$$

Calculation of resistance moment W_{pot} (13):

$$W_{pot} = 54{,}31 * V_0 - 1131{,}09$$
$$W_{pot} = (54{,}31 * 35 - 1131{,}09) * 1{,}02$$
$$\mathbf{W_{pot} = 785,15\,cm^3}$$

Example 2:

Calculate dimensions of profile for billboard H_2 = 3 m and L = 5 m and for wind speed V_0 = 60 m/s:

H_1 = 1,5 m;
H_2 = 3 m;
L = 5 m;
V_0 = 60 m/s

Bending moment calculation from Table 2 equation for area P = 15 m^2:

$$M_{max} = 3{,}5129 * V_0 - 80{,}981 = \mathbf{132,39\,KNm}$$

Resistance moment calculation from Table 3 equation for area P = 15 m^2:

$$W_{pot} = 21{,}955 * V_0 - 506{,}13 = \mathbf{827,39\,cm^3}$$

USVOJENO: HOP 250 × 250 × 12.5; Iz = 10589.61 cm^4; Wz = 847,17 cm^3

$$\delta = \frac{132{,}39 * 10^6}{847170} = 157{,}07\,\text{MPa} < \delta_{dop} = 160\,\text{MPa}$$

Control of calculation:

$$V_{ref} = V_o * (1 + 0{,}001 * H) = 60 * (1 + 0{,}001 * 55) = 63{,}3\,\text{m/s}$$

$$q_{ref} = \frac{P}{2} * V_{ref}^2 = \frac{1{,}25}{2000} * 63{,}33^2 = 2{,}50\,\text{kN/m}^2$$

$$W_e = q_{ref} * C_e * C_{pe} * C_f = 2{,}5 * 0{,}75 * 1{,}75 * 1{,}8 = \mathbf{5,92\,KN/m^2}$$

$$q = W_e * \frac{L}{2} = 5{,}92 * \frac{5}{2} = 14{,}80\,\text{KN/m}$$

$$M_{max} = 14{,}80 * 3 * (1{,}5 + \frac{3}{2}) = \mathbf{133,20\,KNm}$$

$$W_{pot} \geq \frac{M}{\delta dop} = \frac{133{,}20 * 10^3}{160} \geq 832{,}5\,\text{cm}^3$$

USVOJENO: HOP 250 × 250 × 12.5; Iz = 10589,61 cm^4; Wz = 847,17 cm^3

$$\delta = \frac{133{,}20 * 10^6}{847170} = 157{,}07\,\text{MPa} < \delta_{dop} = 160\,\text{MPa}$$

5 Conclusion

In the area of Mostar, more precisely, the southern part from Tekija to the airport, there is occurrence of extreme wind speed (the wind blowing from the north - “Bora wind”) during winter months. The wind reaches the speed up to 240 km/h, so it is necessary to pay close attention when designing and installing the billboards. We witnessed numerous cases of these billboards being crashed and destroyed after the wind reaches high speeds.

Analyzing the wind load and adopting the necessary dimensions of the profiles for different billboard areas and different wind speeds, as well as the comparison with the profiles of the billboards destroyed by the wind, we concluded that the reason for the destruction of these billboards was poorly executed dimensioning. Weak cross-sectional billboard profiles were the result of either bad calculation, or consideration of only low wind speeds.

Acquired equations can be used for manufacture of different area billboards for the region of Mostar, as well as to define the necessary profile for the given billboard area with simple insertion of the necessary values into the given equations.

References

1. Priručnik za Eurokodove (2018). http://eurokodovi.ba/handbook/?lang=bs. Sarajevo
2. Rajčić, V., Mandić, A., Čizmar, D.: Djelovanje snijega i vjetra na nosive konstrukcije. Zagreb (2009)
3. Pakrov, A.: Zbirka jugoslovenskih pravilnika i standarda za građevinske konstrukcije. Beograd (1995)
4. Pavlović, Z., Pavlović, S., Petrović, M.: Priručnik za noseće čelične konstrukcije. Beograd (1987)
5. Verbič, B.: Otpornost materijala. Sarajevo (1989)

Presentational Advances in User-Based Design Processes in Architecture and Civil Engineering: Virtual Reality Dimension

Amer Alić[1], Adnan Novalić[1], Maja Popovac[2], and Ahmed El Sayed[1](✉)

[1] International Burch University, Ilidža, Sarajevo, Bosnia and Herzegovina
ahmed.elsayed@ibu.edu.ba
[2] Faculty of Civil Engineering, University "DžemalBijedić", Mostar, Bosnia and Herzegovina

Abstract. Virtual Reality has been acquired in certain fields where the interaction is important such as the Architectural-Engineering-Construction field. This paper aimed to investigate the impact of using Virtual reality in Designing Process on improving the communication between Designers and Users. Using Qualitative research methodology, the findings agree that the major disadvantage is difficult software workflow for specific industries. It has been concluded that the increased interest in VR and immersive environments is allowing better understanding and identification of problems and is offering more accurate visualization and better project communication.

Keywords: Virtual Reality · Virtual World · Visualization · Architectural and Civil design

1 Introduction

Silvestri, Motro, Maurin, & Dresp-Langley pointed out that drawings generated by computer and virtual world of data have been replaced the manual/hand drawings and model making in a very short period of time [1].

"Digital design" or "digital architecture" has been used by Oxman to introduce the contemporary architectural practices that are supported by these innovative platforms for devising the project, its production, and construction [2].

Keene stated that that architecture is product of people, a sort of artifact that is consisted of experiences in order to trigger, intrigue, impress and entertain [3], and which reverberate with them and their being [4]. As a psychological need, a social condition interrelated with spiritual attributes for humans it implies countless possible values and meanings [5].

The definition of Architecture differs due to the variety of individual experiences of it. Bianco conducted qualitative research during the conference on architecture practitioners and based on results following definitions were concluded: Professionals defined Architecture as an art and as an applied science which fulfills the natural environment. It is the synchronization of functionality and aesthetical values of space

I. Karabegović (Ed.): NT 2020, LNNS 128, pp. 839–852, 2020.
https://doi.org/10.1007/978-3-030-46817-0_95

that contribute to the quality of life of users. Students of architecture defined it as the creation of sufficient space that is more than just physical - a set of sensory aspects of space that are supposed to fulfill individual needs of users and solves their habitation challenges Finally, academics define architecture as a space that corresponds to the living requirements that are emerged from interrelation of environment and human activities [6].

2 Problem Statement

Ibrahim and Rahimian stated that the detrimental effect on designers' reasoning procedures is being caused by the majority of current CAD tools, obstructing their cognitive activities [7].

The use of conventional Computer-Aided Design (CAD) software lacks of flexibilities and adaptabilities with the intuitive conceptual intentions of designers [8], and, as such causes struggle and frustration during the use. It has been pointed out that main two negative influences on the creativity and problem-solving potential of designers is the current demanding and hardly usable user interfaces of CAD applications [8–10].

As indicated by Cross, designers need to have a medium which enables expression and reflection on ideas [11]. Although the most of the 3D visualization software demand the a higher degree and longer period for user to achieve the desired shapes, forms and textures that architectural designers configure, it is not simply possible for all designers to reach the advanced degree of their skills [12].

It has been argued that the obtained skill is rarely sufficient since it is detected that even the experienced designers have difficulties in using and comprehending their design capabilities [13]. Consequently, Kwon, Choi, Lee, and Chai claimed that described pre-thought approach disturbs the potential of the design processes and the teamwork that usually goes along with it. For novice designers, it could, therefore, lead to miscommunication [8].

Jaffar, Tharim, and Shuib [14]; and Mitkus and Mitkus [15] identifies ineffective communication as a source of conflict in construction projects, in the form of poor documentation and lack of traceability [16]. Around 70% to 80% of production overheads are being determined in the early stages of problem finding and analysis [17]. While traditional drawing methods such as plan, section, and elevation is used successfully for centuries, they remain bound to a flat surface distancing the viewer from the idea of an environment [18], and they are equally limited in terms of their usefulness in the conceptual design phase [7]. Professional consultations on design with clients who engage but have insufficient understanding may provide an area for improvement.

According to Saleh and Nassar, the participation of users presents an important issue that should be considered so that architects can meet the customers' needs and improve the whole decision-making process [19]. The involvement of end-users has been usually minimized and eliminated due to the lack of time, resources and increasing numbers of participants in the design and construction phases of AEC projects [20, 21]. Carvajal and Shen et al. highlighted the need for effectiveness in design communication to future occupants despite the difficulty in understanding by those who are not involved in the process and despite the problems related to users who are not primarily trained for such activities and, in accordance to it they are incapable to comprehend and visualize their own intentions that are supposed to bring the solution to certain spatial problem [22–25]. Majority of CAD software does not allow an intuitive evaluation and management of 3D models for users without an education about the used software (CAD) [26].

3 Virtual Reality (VR)

Sherman and Craig in 2018 stated that Virtual reality is an instrument consisted of interactive computer simulations that perceives the users' actions and state of the body which replaces or increases the intensity of sent information to various human senses which involves users directly into virtual space and its properties [27]. Dionisio, Burns III, and Gilbert in 2013 described the virtual reality as simulations of three-dimensional volumes or environments emerged from computer that seem physically real, direct, touchable [28]. Burdea and Coiffet in 2003 defined VR as a technology that inserts the aspects of dipping (immersion) of users' sensors into three-dimensional virtual models which allow opportunity for users to experience what has been created, which is not possible with the usual forms of representation [29]. Steuer stated that VR is a human experience made possible by the feeling of real presence [30].

Sherman and Craig defined immersion as a sensational act of being present in a surroundings that may be the product of pure mental state (ultimate goal of most tool creators is a mental immersion), or might be achieved through physical tools, in which the consisting character of Virtual Reality is immersion [27].

Presence is the illusion of experience of an artificially configured environment upon the perception of it – result of interpretation of outer stimulants which are recorded a result of the interpretation of external stimuli captured by our senses [31]. The sense of presence originates from perceiving oneself inside the environment [32]: a complex "awareness phenomenon" [33]. The possibility to experience sensations and movement has been given in the VR environment [34]. According to Slater immersion represents the objective degree of credibility of the sensor information that VR provides to the user. A greater sense of presence is being provoked by a higher degree of immersion [35].

Sherman and Craig in 2018 defined a virtual environment as the purport of assigned medium that exists mainly in the mindset of its user, or it could be shared with more than one user [27]. The authors stated that the virtual world is a description of a collection of objects in space and the rules governing those objects. When seen through a system that integrates objects and interactions in an interactive physical presentation, you get a virtual experience of reality. The authors also defined the virtual world as an imaginary space often manifested through a medium [27]. Nowadays the term Virtual Worlds (VWs) is used to refer to Desktop Multi-User Persistent 3D Environments [36].

Schnabel and Kvan in 2003 noted that designer tends to work interactively in 3rd dimension in which every artificially configured element is a point of experience aligned with the interaction and movement achievable in the real-world reality [37]. As pointed out by Warwick, Gray, and Roberts in 1993, creating an enriched virtual environment has been allowed by interacting with digital objects [38]. Sherman and

Craig described one form of interactivity as the skill that affects a computer-based environment or as the possibility to change one's point of view and observation within a that environment and that it may be connected with the ability of the users within the previously set environmental frame to physically move within the world in order to obtain more suitable opportunity for the new point of view throughout the numerous movements of their head [27].

3.1 The Hardware

VR devices can be subdivided into input and output devices, followed by subtypes within each category. Peripheral input and output of devices can be applied to achieve interaction with virtual environments in a more intuitive way. The methods used in these cases include the positioning of several tracking devices, allowing head and eye movements of users to be followed and supervised, which includes numerous control devices and visual, aural and haptic inputs gathered information. Navigation and manipulation of objects within the virtual world have been allowed by handheld controllers [39]. Tracking systems interlaced with human gesture recognition algorithms made possible casual and regular body movements to be translated into techniques of functional interaction [40]. According to Brewster what allows users to utilize their sense of touch which makes possible for them to feel their sense of touch is the haptic technology that constitutes a wide part of VR [41]. Laycock and Day indicated that haptic devices provide strong feedback in understanding of how objects in a virtual world physically interact, which prove its interaction enhancement aspects [42]. It's possible to simulate a 3D sound with more complex simulations, so you can play sounds on your headphones coming from certain places in 3D space. Output units for simulating odor, heat, wind, etc. are also available, which further increases the user experience in the virtual world.

3.2 CAVE (Cave Automatic Virtual Environment) and HMD (Head Mounted Display) Systems

A virtual reality system (CAVE) was developed for the first time in 1992 where, unlike other systems, stereoscopic images were projected onto the wall and the user viewed them with 3D lenses resulting in a wider field of view than other systems. The CAVE is usually a 10′ by 10′ by 10′ cubic room that sits in a larger darkened room, the cube shape is used as an approximation of a sphere [43].

The working principle of CAVE is that visual images are generated by projectors outside of CAVE and controlled by users in CAVE through physical movement [44]. The motion tracking system captures in real-time the user's position, while the images appear on the stereoscopic LCD screen. Computers generate a pair of images quickly, using data for movement tracking. The glasses are synced with projection systems so that each eye only sees the right picture. The CAVE has been being increasingly used and explored in many projects and application domains. It can be seen used in military and training applications, education, medicine, scientific visualizations, and many other areas of human activity [44].

The advantage of the system is its high levels of immersion and the ability to involve more people in the environment while it's an inconvenience is the high price.

HMD is the basic virtual reality system equipment worn on the head. It can be a helmet, visor or VR glass which contains the user's two screens as a key component, allowing a stereoscopic 3D view to be shown to the person. This is achieved when the same image is displayed on each screen but shifted to the same distance from the user's eyes that imitate the human world's visual perception. Rolland and Cakmakci listed image resolution and field of view dimensions as the most important HMD device parameters [45].

Rolland and Hua described that each HMD device has a basic structure of four parts [46]:

- Image source - Used to generate optically displayed image data. Advancements in technology have enabled the creation of efficient, small, lightweight display sources to meet perceptive intensity and resolution needs.
- Optical system - Used to transmit information about the image source to human eyes. The optics used with an HMD device consists typically of several optical elements (mostly lenses) with a splitting optic beam. The initial HMD models, developed for visual purposes, were monocular and contained a single optical beam divider in front of the eyes, and binocular designs became dominant with the development of small displays.
- Mounting platform - A headband, visor, goggles or helmet, used to attach the device to the user's head. In addition to being used as a connection between the user and the technology, the device should remain stable and ergonomic so that the necessary alignment between the vision and HMD optics is maintained.
- Device tracking - Optional if the HMD is used only for the display of information or visual content. However, it is often necessary to supply an external image with a sensor or synthetic database. The image displayed to the user as well as the current location and orientation of the user's head should be maintained up to date when this type of optical content is displayed.

3.3 The Software

Komura, Lau, Lin, Majumder, Manocha, and Xu in stated that the development of software for the real VR experience, an experienced programmer usually has to do all the necessary activities to create a VR system together by using software libraries [47]. The software provides programs or specific commands and instructions that tell the computer how certain tasks can be performed, i.e. the software is a way to write an algorithm understandable for the computer [47].

The most important VR experience is acquired by the use of game engines, which is primarily a tool for creating interactive 3D games in real-time and has the resources to create large-scale virtual environment simulations, rendering, networking, etc. [48]. They provide powerful 3D rendering and representations of physical laws [49].

4 Application of VR in Architectural and Engineering Design Practices

During the past two decades, the Architecture, Engineering, Construction and Facility Management (AEC/FM) has been witnessing a steady increase of interest in Virtual Reality (VR) for improving existing work processes [50]. Amongst many opportunities in the usage of VR, it has been evidenced that immersive environment can provide terms for better spatial understanding of virtual prototypes [51] in comparison to 2D and non-immersive 3D representations [37] allowing the identification of design issues that would be hard to identify otherwise [52], while predicting human-building interactions giving the researchers and designer reliable user behavior data [53]. Furthermore, the use of an immersive system enables cost efficiency through discarding physical mock-ups and faster decision-making processes [3]. By adopting such systems safety training can be benefited [54] due to the promotion of high-level involvement that optimizes the learning process [55]. VR tackles a variety of problems in design, construction [56], project design and configuration [57], architectural structures [58], safety on construction site [54], coordination of construction activities [59], management and maintenance [60] and all built environment (real estate) [61]. As a result of using VR, the entire AEC/FM workflow can be improved [56], mentioning the benefits of VR in following segments.

4.1 Design Process

According to Kan and Gero [62] design (as a noun) has been defined as a series of decisions which expose the relationship of shape and forms, materialization, and efficiency and the design activities have been clustered as brainstorming (creative thinking) and understanding (deductive knowing) [63], free-hand/manual drawing and sketch in and reciprocal actions of influences [64], social construction of design solutions and designing-by-making [65]. As stated by Noël, Nguyen, Ba, and Sadeghi design has been described as a cognitive process (brainstorming process) where reliable data/info must be shared with collaborators for more efficient decision-making [66]. It has been reported by Lawson and Dorst [67] that switching between analysis and creativity, between "problem" and "solution" without any effort occurs while steeped deeply in design activity that can be supported with VR in many design activities such as architecture and organizational or interior space design [68]; virtual design studios [69]; and collaborative learning in design education [70]. During the design process of an idea (Idea development = design process), Immersive virtual environments are influential tools for building 3D models in order to analyze, process and forward where necessary all issues before the start of the phases of construction [3]. The understanding of the architectural design and on the overall quality has been impacted by the method used for representation through the process [71]. According to [72], the designers have been pushed to better perceive space by the manipulation of virtual environments during the design process taking the customer requirements and the functions of the building into consideration during the design process. Carvajal stated that occupant satisfaction, comfort, and productivity depend on the decisions made in a building's design phase [22].

4.2 Visualization

Visualization is crucial for architecture. Proper visualization of objects and spatial data gives opportunities for new perspectives and thus alternative views on the world constructed with spatial data. According to Herbert and Chen [73], the advancement of technology and computerized visualizations over the past several decades marked a clear shift from using 2D methods to using 3D visualizations, which in comparison with 2D visualizations, have been considered more capable for communication between user groups [22]. Therefore, as a new emerging technology that can provide an immersive environment, VR is a combination of the need to visualize complex models, spatial relationships, and rich human-computer interaction capabilities. For the architect, the process of visualizing and immersing into design while achieving much clearer understanding of both qualitative and quantitative nature of the space is being enabled with the help of VR [59], creating the sensation of being "inside" the visualization resulting in greater spatial understanding for the users [74]. Sharing the same virtual world through interactions can contribute to collaboration with team members, although physically, in other locations [75]. It has been suggested that visualizations can enhance public participation and understanding by allowing people to meet with professionals about proposed projects [76] and by enabling them to see a realistic image of the final product. As stated by Ribarsky, Bolter, Den Bosch, and Van Teylingen in 1994 amongst scientific and business applications in architecture, the positive impact on daily working practices is also considered allowing greater efficiency with less effort than traditional visualization techniques [77].

4.3 Collaboration and Communication

Architecture is primarily a visual profession and experts who develop architectural plans are trained to read and recognize various architectural problems visually, while Florio in 2011 stated that "experiential-spatial" human experience in the real world can contribute to the development of buildings that really match end-users demands – technical, functional, and symbolic [78].

Until VR, architects had to present their design ideas utilizing different proportional models and drawings. It has been affirmed by Beurden et al. in 2006 that the use of virtual models, compared to the use of technical drawings, can contribute to greater efficiency in project communication. According to the authors, virtual models have a greater connection with the way that a human being perceives real objects, facilitating the identification and understanding of the conceived object [79].

The use of VR technology for collaboration in various stages considers to be the focus of many research studies in the last couple of decades [80] and according to Regenbrecht and Donath [81], VW provides a platform for designing novel collaborative design environments. Addressing the point, the authors define VR as "the component of communication which takes place in a computer-generated space and embeds humans as an integral part of the system" [81].

One-to-one scale and immersive virtual mock-ups provide a better understanding of a project to end-users and stakeholders resulting in a better sense of realism and improved communication [82].

Presenting and comparing the design concepts with plausibility for customers to see the results, makes collaboration and decision making much easier before the construction process starts [51] and according to Frost and Warren [83], VR will play a significant role in the architecture of the future, which will remove barriers to communication and bridge the gap between client vision and architectural output.

4.4 Education and Training

Gant [84], claimed that interaction and collaborative forms are more completely than classical models of education, due to a capacity of short-term retention of 20% of all information that we hear, 40% of the information that we see and hear, and 75% of the information that we see and participate in. Sheridan [85], stated that immersion and imagination from various senses could be enhanced by the technology. Following the improvement in students' cognitive learning process from one-way infusion to bidirectional interaction [70].

Seeing their instructor designing or by experiencing collaborative design with them, proved to be helpful for students [86], while discussing design issues which drive the evolution of the students' designs [87]. In particular, the perception and comprehension of the third dimension of forms, design components, and structural elements, indicate the suitability of VR in design education. People benefit from these potentials, which subsequently enhances learning/education processes, as creative and innovative potential arises [88].

Ku and Mahabaleshwarkar [89] affirmed that using VR contributes to safety training and developing a better understanding of construction processes for students.

4.5 Virtual Prototyping and Walkthrough

"Virtual prototyping (VP) is a relatively new technology that involves the use of virtual reality (VR) and other computer technologies to create digital prototypes" [90]. These prototypes empower designers to explore and express their imagination with greater ease [91], generating walkthrough models to test the design with the clients in a more direct manner [92].

Burdea and Coiffet stated that adding the dimensions of immersion and interactivity to three-dimensional computer-generated models and offers an exploration that is not viable with the traditional form of representation, allowing the walk around designed space and thus experience the functionality of each room while the controller can be used by architects to change the look of a space allowing the furniture, materialization, and colors to be decorated in a manner that is suitable until a successful result is achieved [29].

This method enables architects to explore, evaluate and communicate their project concepts in a holistic manner.

5 Conclusion

This review focused on studies addressing VR as technology and its use and deployment characteristics. Literature was obtained from peer-reviewed journal articles, conference papers, theses, and books. Majority of the literature dated from late 20 century and up to date.

Much of this research aimed to inform about architectural problems, the characteristics of VR, plausible capabilities and solutions by introducing VR into everyday life.

The information gathered provides a good understanding of the challenges and issues related to the topic and related questions: What new possibilities does VR provide on interacting and exploring projects within an immersive virtual environment? What difference does it make experiencing immersive and one to one scale environment? During the design phase, when the client gets involved and what impact will this make on architecture?

The summary of key information is presented through problem identification, description of Virtual Reality and examples of applications in architecture.

In the introduction of the literature review, it was pointed out how architecture is marked by the progression of many fields referring to advances in technologies. Later in the review, many authors presented various definitions of architecture while Bianco presented that academics consider architecture as a functional space fit for human needs developed through a dialogue between human activity and the environment [6].

It has been noted that advancements in IT and architecture had a major influence on the situation. For the problems in the design part of the architecture, authors pointed out that using conventional CAD programs has many negative consequences on designer's creativity and process mentioning that designers need to have a medium that enables expression and reflection on ideas.

For the Communication problems, related literature highlighted an ineffective communication as a potential source of conflict, because plans, sections, and elevation remain presented as 2D drawings, distancing the viewer from the complete understanding. The research presented also an important issue in the involvement of end-users without CAD or computer knowledge, later they highlighted the need for effectiveness in design communication for future occupants.

Literature related to Virtual Reality offered an overview of new technology and how it acquired a relevant role in many fields based on interaction. The authors defined VR as a computer-generated simulation of 3D environment that enables the sense of immersion and interactivity. Furthermore, the authors defined immersion as a sensation of being involved with the environment and listed several types, mostly mental and physical. It was pointed out how slow advancement in IT during the last century marked VR as expensive and not easy to use technology, until the last decade when technology caught up with VR as an idea. Research on AR described a technology with some similarities to VR while working on the different principles it still allows immersion with content adding the virtuality onto the real world.

The literature highlighted the increase of interest in VR and immersive environments allowing better understanding and identification of hard to find problems, describing the design as a complex process in which immersive virtual environments proved as effective tools allowing designers to perceive space better than before. It has been explained that advancement of technology-enabled a transition from 2D methods of drawing and presentation to 3D enabling more accurate visualization, and it has been stated that a better project communication can be achieved with the use of virtual models providing the immersive and one to one scale examination, thus will be a significant part of future architecture workflow. The review highlighted that most memorized information comes from participation and discussing the project in VR with the professor enhances the evolution of student's design and improving the understanding of the third dimension and form by allowing the walking and exploration.

On the positive and negative results, the findings largely agree that recent developments in VR allow new opportunities in many industries, while the major disadvantage is difficult software workflow for specific industries. An extensive review of the literature revealed areas for improvement and collected information is expected to help all the individuals as well as the companies to understand more the potential of VR technology and the importance of implementing with design and presentation of a project. Following the literature review, major changes can be predicted with the everyday use of VR technology.

References

1. Silvestri, C., Motro, R., Maurin, B., Dresp-Langley, B.: Visual spatial learning of complex object morphologies through the interaction with virtual and real-world data. Des. Stud. **31** (4), 363–381 (2010)
2. Oxman, R.: Theory and design in the first digital age. Des. Stud. **27**(3), 229–265 (2006)
3. Majumdar, T., Fischer, M.A., Schwegler, B.R.: Conceptual design review with a virtual reality mock-up model. In: Joint International Conference on Computing and Decision Making in Civil and Building Engineering (2006)
4. Greenberg, S.: The vital museum. In: Reshaping Museum Space. Architecture, Design, Exhibitions, p. 241. Routledge, Taylor & Francis (2005)
5. Keene, S.: Museums and the interpretation of visual culture. Mater. Relig. **1**(2), 281–283 (2005)
6. Bianco, L.: Defining architecture: a qualitative research. Ann. Univ. Architect. Civil Eng. Geodesy Sofia **51**(1), 59–66 (2018)
7. Ibrahim, R., Rahimian, F.P.: Comparison of CAD and manual sketching tools for teaching architectural design. Autom. Constr. **19**(8), 978–987 (2010)
8. Kwon, J.H., Choi, H.W., Lee, J.I., Chai, Y.H.: Free-hand stroke based NURBS surface for sketching and deforming 3D contents. In: Pacific-Rim Conference on Multimedia, Berlin, Heidelberg (2005)
9. Bilda, Z., Demirkan, H.: An insight on designers' sketching activities in traditional versus digital media. Des. Stud. **24**(1), 27–50 (2003)
10. Rahimian, F.P., Ibrahim, R., Baharudin, M.N.: Using IT/ICT as a new medium toward implementation of interactive architectural communication cultures. In: International Symposium on Information Technology, vol. 3 (2008)
11. Cross, N.: Designerly ways of knowing. Des. Stud. **3**(4), 221–227 (1982)
12. Levet, F., Granier, X., Schlick, C.: 3D sketching with profile curves. In: International Symposium on Smart Graphics, Berlin, Heidelberg (2006)

13. Suwa, M., Purcell, T., Gero, J.: Macroscopic analysis of design processes based on a scheme for coding designers' cognitive actions. Des. Stud. **19**(4), 455–483 (1998)
14. Jaffar, N., Tharim, A.A., Shuib, M.N.: Factors of conflict in construction industry: a literature review. Procedia Eng. **20**, 193–202 (2011)
15. Mitkus, S., Mitkus, T.: Causes of conflicts in a construction industry: a communicational approach. Procedia-Soc. Behav. Sci. **110**(15), 777–786 (2014)
16. Lester, A.: Project Management, Planning and Control: Managing Engineering, Construction and Manufacturing Projects to PMI, APM and BSI Standards, 7th edn, p. 650. Elsevier Science, Amsterdam (2017)
17. Schütze, M., Sachse, P., Römer, A.: Support value of sketching in the design process. Res. Eng. Des. **14**(2), 89–97 (2003)
18. Lockard, W.K.: Design Drawing, 2nd edn, p. 280. Crisp Publications, Menlo Park (1993)
19. Saleh, A., Nassar, U.: The use of new visualization Tools to Enhance Public Participation in Riverfronts' Landscape regeneration. In: 4th Urbenviron International Congress on Environmental Planning and Management. Green Cities: a Path to Sustainability (2011)
20. Bullinger, H.J., Bauer, W., Wenzel, G., Blach, R.: Towards user centred design (UCD) in architecture based on immersive virtual environments. Comput. Ind. **61**(4), 372–379 (2010)
21. Oijevaar, K., Jovanovic, M., Den Otter, A.F.: User involvement in the design process of multifunctional buildings. In: Changing Roles: New Roles and New Challenges (CR09), Delft (2009)
22. Carvajal, A.: The use of visualization technologies for the integration of building end-user perspectives, a managerial approach. In: Computing in Civil Engineering (2005)
23. Liu, H., Singh, G., Lu, M., Bouferguene, A., Al-Hussein, M.: BIM-based automated design and planning for boarding of light-frame residential buildings. Autom. Constr. **89**, 235–249 (2018)
24. Shen, W., Zhang, X., Shen, G.Q., Fernando, T.: The user pre-occupancy evaluation method in designer–client communication in early design stage: a case study. Autom. Constr. **32**, 112–124 (2013)
25. Skov, M.B., Kjeldskov, J., Paay, J., Husted, N., Nørskov, J., Pedersen, K.: Designing on-site: facilitating participatory contextual architecture with mobile phones. Pervasive Mob. Comput. **9**(2), 216–227 (2013)
26. Naef, M., Payne, J.: AutoevalmkII-interaction design for a VR design review system. In: 2007 IEEE Symposium on 3D User Interfaces (2007)
27. Sherman, W.R., Craig, A.B.: Understanding Virtual Reality: Interface, Application, and Design, 2nd edn, p. 938. Elsevier Science, Amsterdam (2018)
28. Dionisio, J.D., Burns III, W.G., Gilbert, R.: 3D virtual worlds and the metaverse: current status and future possibilities. ACM Comput. Surv. (CSUR) **45**(3), 1–38 (2013)
29. Burdea, G.C., Coiffet, P.: Virtual Reality Technology, 2nd edn, p. 464. Wiley, Hoboken (2003)
30. Steuer, J.: Defining virtual reality: dimensions determining telepresence. J. Commun. **42**(4), 73–93 (1992)
31. Gifford, R.: Environmental Psychology: Principles and Practice, p. 372. Optimal Books, Colville (2007)
32. Wann, J., Mon-Williams, M.: What does virtual reality NEED?: human factors issues in the design of three-dimensional computer environments. Int. J. Hum. Comput. Stud. **44**(6), 829–847 (1996)
33. Witmer, B.G., Singer, M.J.: Measuring presence in virtual environments: a presence questionnaire. Presence **7**(3), 225–240 (1998)
34. Fuchs, P., Moreau, G., Guitton, P.: Virtual Reality: Concepts and Technologies. CRC Press, Paris (2011)

35. Slater, M.: A note on presence terminology. Presence Connect **3**(3), 1–5 (2003)
36. Bartle, R.A.: Designing Virtual Worlds, p. 741. New Riders, Indianapolis (2004)
37. Schnabel, M.A., Kvan, T.: Spatial understanding in immersive virtual environments. Int. J. Architect. Comput. **1**(4), 435–448 (2003)
38. Warwick, K., Gray, J., Roberts, D.: Virtual Reality in Engineering. Institution of Electrical Engineers, Michigan (1993)
39. Bowman, D.A., Coquillart, S., Froehlich, B., Hirose, M., Kitamura, Y., Kiyokawa, K., Stuerzlinger, W.: 3D user interfaces: new directions and perspectives. IEEE Comput. Graph. Appl. **28**(6), 20–36 (2008)
40. Mitra, S., Acharya, T.: Gesture recognition: a survey. IEEE Trans. Syst. Man Cybern. Part C (Appl. Rev.) **37**(3), 311–324 (2007)
41. Brewster, S.A.: The impact of haptic 'Touching' technology on cultural applications. In: Proceedings of EVA 2001, Glasgow (2001)
42. Laycock, S.D., Day, A.M.: A survey of haptic rendering techniques. In: Computer Graphics Forum, Oxford (2007)
43. Cruz-Neira, C., Sandin, D.J., DeFanti, T.A., Kenyon, R.V., Hart, J.C.: The CAVE: audio visual experience automatic virtual environment. Commun. ACM **35**(6), 64–73 (1992)
44. Muhanna, M.A.: Virtual reality and the CAVE: taxonomy, interaction challenges and research directions. J. King Saud Univ.-Comput. Inf. Sci. **27**(3), 344–361 (2015)
45. Rolland, J.P., Cakmakci, O.: The past, present, and future of head-mounted display designs. In: Optical Design and Testing II (2005)
46. Rolland, J.P., Hua, H.: Head-mounted display systems. Encycl. Opt. Eng. **2** (2005)
47. Komura, T., Lau, R.W., Lin, M.C., Majumder, A., Manocha, D., Xu, W.W.: Virtual reality software and technology. IEEE Comput. Graph. Appl. **35**(5), 20–21 (2015)
48. Trenholme, D., Smith, S.P.: Computer game engines for developing first-person virtual environments. Virtual Reality **12**(3), 181–187 (2008)
49. Mól, A.C., Jorge, C.A., Couto, P.M.: Using a game engine for VR simulations in evacuation planning. IEEE Comput. Graph. Appl. **28**(3), 6–12 (2008)
50. Wang, X., Schnabel, M.A.: Mixed Reality in Architecture, Design, and Construction, p. 274. Springer, Heidelberg (2008)
51. Berg, L.P., Vance, J.M.: An industry case study: investigating early design decision making in virtual reality. J. Comput. Inf. Sci. Eng. **17**(1), 45–67 (2017)
52. Okeil, A.: Hybrid design environments: immersive and non-immersive architectural design. J. Inf. Technol. Constr. (ITcon) **15**(16), 202–216 (2010)
53. Kuliga, S.F., Thrash, T., Dalton, R.C., Hölscher, C.: Virtual reality as an empirical research tool—exploring user experience in a real building and a corresponding virtual model. Comput. Environ. Urban Syst. **54**, 363–375 (2015)
54. Sacks, R., Perlman, A., Barak, R.: Construction safety training using immersive virtual reality. Constr. Manag. Econ. **31**(9), 1005–1017 (2013)
55. Faas, D., Bao, Q., Frey, D.D., Yang, M.C.: The influence of immersion and presence in early stage engineering designing and building. AI EDAM **28**(2), 139–151 (2014)
56. Messner, J.I., Yerrapathruni, S.C., Baratta, A.J., Whisker, V.E.: Using virtual reality to improve construction engineering education. In: American Society for Engineering Education Annual Conference & Exposition (2003)
57. Du, J., Shi, Y., Mei, C., Quarles, J., Yan, W.: Communication by interaction: a multiplayer VR environment for building walkthroughs. In: Construction Research Congress 2016 (2016)
58. Kaufmann, H., Schmalstieg, D., Wagner, M.: Construct3D: a virtual reality application for mathematics and geometry education. Educ. Inf. Technol. **5**(4), 263–276 (2000)

59. Bouchlaghem, D., Shang, H., Whyte, J., Ganah, A.: Visualisation in architecture, engineering and construction (AEC). Autom. Constr. **14**(3), 287–295 (2005)
60. Shi, Y., Du, J., Lavy, S., Zhao, D.: A multiuser shared virtual environment for facility management. Procedia Eng. **145**, 120–127 (2016)
61. Chen, S.E.: Quicktime VR: an image-based approach to virtual environment navigation. In: Proceedings of the 22nd Annual Conference on Computer Graphics and Interactive Techniques (1995)
62. Kan, J.W., Gero, J.S.: Quantitative Methods for Studying Design Protocols. Springer, Dordrecht (2017)
63. Cross, N.: Designerly Ways of Knowing. Springer, Milton Keynes (2006)
64. Lawson, B.: How Designers Think, 4th edn, p. 336. Routledge, Abingdon (2006)
65. Jones, J.C.: Design Methods: Seeds of Human Futures. Wiley, Hoboken (1986)
66. Noël, F., Nguyen, A., Ba, N., Sadeghi, S.: Qualitative comparison of 2D and 3D perception for information sharing dedicated to manufactured product design. In: 2012 IEEE 3rd International Conference on Cognitive Infocommunications (CogInfoCom) (2012)
67. Lawson, B., Dorst, K.: Design Expertise, p. 322. Routledge, New York (2013)
68. Mobach, M.P.: Do virtual worlds create better real worlds? Virtual Reality **12**(3), 163–179 (2008)
69. Bessière, K., Ellis, J.B., Kellogg, W.A.: Acquiring a professional "second" life problems and prospects for the use of virtual worlds in business. In: Extended Abstracts on Human Factors in Computing Systems, CHI 2009 (2009)
70. D'Souza, N., Yoon, S.Y., Islam, Z.: Understanding design skills of the Generation Y: an exploration through the VR-KiDS project. Des. Stud. **32**(2), 180–209 (2011)
71. Bassanino, M., Wu, K.C., Yao, J., Khosrowshahi, F., Fernando, T., Skjærbæk, J.: The impact of immersive virtual reality on visualisation for a design review in construction. In: 2010 14th International Conference Information Visualisation (2010)
72. Schnabel, M.A., Kvan, T., Kruijff, E., Donath, D.: The first virtual environment design studio. In: Proceedings of the 19th ECAADE-Conference, Helsinki, Finland (2001)
73. Herbert, G., Chen, X.: A comparison of usefulness of 2D and 3D representations of urban planning. Cartography Geogr. Inf. Sci. **42**(1), 22–32 (2015)
74. Bowman, D.A., Ray, A.A., Gutierrez, M.S., Mauldon, M., Dove, J.E., Westman, E., Setareh, M.: Engineering in three dimensions: immersive virtual environments, interactivity, and 3D user interfaces for engineering applications. In: Geotechnical Engineering in the Information Technology Age, GeoCongress 2006 (2006)
75. Seng, W.K., Palaniappan, S., Yahaya, N.A.: A framework for collaborative graphical based design environments. In: 6th International Workshop on Database and Expert Systems Applications (DEXA 2005) (2005)
76. Bilge, G., Hehl-Lange, S., Lange, E.: The use of mobile devices in participatory decision-making. JoDLA–J. Digit. Landscape Archit. **I**(1), 234–242 (2016)
77. Ribarsky, W., Bolter, J., Den Bosch, A.O., Van Teylingen, R.: Visualization and analysis using virtual reality. IEEE Comput. Graph. Appl. **14**(1), 10–12 (1994)
78. Florio, W.: An analysis of the design process under the cognition theory: seven difficulties in the atelier. Arquiteturarevista **7**(2), 161 (2011)
79. Borsboom-van Beurden, J.A., Van Lammeren, R.J., Hoogerwerf, T., Bouwman, A.A.: Linking land use modelling and 3D visualisation. In: Innovations in Design & Decision Support Systems in Architecture and Urban Planning, Dordrecht (2006)
80. Benford, S., Greenhalgh, C., Rodden, T., Pycock, J.: Collaborative virtual environments. Commun. ACM **44**(7), 79–85 (2001)

81. Regenbrecht, H., Donath, D.: Architectural education and virtual reality aided design (VRAD). In: Designing Digital Space-An Architects Guide to Virtual Reality, pp. 155–176 (1997)
82. Oerter, J., Suddarth, W., Morhardt, M., Gehringer, J., McGinnis, M.L., Shockley, J., Baysa, A.: A system architecture and simulation environment for building information modeling in virtual worlds. J. Defense Model. Simul. **11**(3), 205–210 (2014)
83. Frost, P., Warren, P.: Virtual reality used in a collaborative architectural design process. In: 2000 IEEE Conference on Information Visualization. An International Conference on Computer Visualization and Graphics (2000)
84. Gant, P.A.: Maximizing multimedia for training purposes. The Technology Source (1998)
85. Sheridan, T.B.: Interaction, imagination and immersion some research needs. In: Proceedings of the ACM Symposium on Virtual Reality Software and Technology (2000)
86. Dorta, T., Kinayoglu, G., Boudhraâ, S.: A new representational ecosystem for design teaching in the studio. Des. Stud. **47**, 164–186 (2016)
87. Goldschmidt, G., Hochman, H., Dafni, I.: The design studio "crit": teacher–student communication. AI Edam **24**(3), 285–302 (2010)
88. Achten, H.H., Van Leeuwen, J.P.: Feature-based high level design tools. In: Computers in Building, Boston (1999)
89. Ku, K., Mahabaleshwarkar, P.S.: Building interactive modeling for construction education in virtual worlds. J. Inf. Technol. Constr. (ITcon) **16**(13), 189–208 (2011)
90. Gowda, S., Jayaram, S., Jayaram, U.: Architectures for internet-based collaborative virtual prototyping. In: The 1999 ASME Design Technical Conference and Computers in Engineering Conference, DETC99/CIE-9040, Las Vegas, Nevada (1999)
91. Schnabel, M.A., Wang, X., Hartmut Seichter, T.K.: Touching the untouchables: virtual-, augmented-and reality. In: The Association for Computer-Aided Architectural Design Research in Asia: Beyond Computer-Aided Design, CAADRIA 2008 (2008)
92. Ormerod, M., Aouad, G.: The need for matching visualisation techniques to client understanding in the UK construction industry. In: Proceedings of the 1997 IEEE Conference on Information Visualization (Cat. No. 97TB100165) (1997)

Using the Semi-professional UAV System in Surveying the Medium Size Area of Complex Urban Surface

Jasmin Ćatić, Admir Mulahusić(✉), Nedim Tuno, and Jusuf Topoljak

Faculty of Civil Engineering, University of Sarajevo, Ul. Patriotske lige 30, 71000 Sarajevo, Bosnia and Herzegovina
admir_mulahusic@gf.unsa.ba

Abstract. The rapid development of Unmanned Aerial Vehicles (UAV) throughout the 21st century, and the advantages of using such systems for photogrammetric measurements, has made photogrammetry a generally more acceptable and promising branch of geodesy. The prices of professional unmanned aerial vehicles have become more affordable lately, making UAV photogrammetry more economical than traditional aerial photogrammetry. The process of obtaining the final results – 3D model of captured object, Digital orthophoto, Digital Surface or Terrain model (DSM or DTM), is much easier by using this method. It is also possible to apply this method in the urban environments. Combining photogrammetry and Global Navigation Satellite Systems is also a rising trend nowadays. Real Time Kinematic (RTK) method is used for georeferencing and accuracy analysis. Using the semi-professional UAV, a high accurate surface or terrain models and Digital orthophoto could be achieved. Paper presents the use of such UAV (DJI Phantom 4 Pro) for obtaining Digital orthophoto and DSM of the area around the Faculty of Civil Engineering - University of Sarajevo, which is a medium size area of complex urban surface. Achieved 3D accuracy was 7,5 cm.

Keywords: UAV · DSM · Digital orthophoto · RTK · Photogrammetry · Phantom · DJI

1 Introduction

Photogrammetry is the art and science of determining the position and shape of objects from photographs and it has traditionally been the most appropriate method of collecting geometric data on complex structures. The main use of photogrammetry is in the production of maps, either as line maps or as orthophoto maps [1, 2]. Classical aerial photogrammetry, as a geodetic survey technique, is often unprofitable in cases of mapping the smaller areas. The most important characteristic of aerial photogrammetry is the realization of geodetic survey when measurement camera is not on the ground. Carriers of the aerial photogrammetry cameras were balloons, zeppelins, airplanes, and today's very successful carriers are Unmanned Aerial Vehicles (UAV). Unmanned aerial vehicles are vehicles capable of performing continuous flight without pilots [3].

I. Karabegović (Ed.): NT 2020, LNNS 128, pp. 853–860, 2020.
https://doi.org/10.1007/978-3-030-46817-0_96

The use of unmanned aerial vehicles in surveying is still a poorly researched area. The most common applications of unmanned aerial vehicles in the field of geodesy are determination of 3D model of various objects, Digital Terrain Model (DTM), Digital Surface Model (DSM) and Digital Ortho photo plan.

It is very difficult to accurately specify the use of unmanned aerial vehicles in surveying, but one that stands out the most is the design of surveying surfaces that can be used in many fields. This research examines the possibility of using semi-professional unmanned aerial vehicle for the purpose of surveying the medium size area of complex urban surface, and presents accuracy that can be achieved.

2 Semi-professional Unmanned Aerial Vehicle DJI Phantom 4 Pro

In this research, the semi-professional UAV DJI Phantom 4 Pro (Fig. 1) was used for surveying an urban surface area around the Faculty of Civil Engineering in Sarajevo. Phantom 4 Pro (P4P) weights 1,388 kg, and has one-inch 20 MP CMOS sensor (Table 1).

Fig. 1. Phantom 4 Pro [4]

Table 1. Technical specifications and operational conditions of the DJI Phantom 4 Pro [4]

Category	Characteristics
Camera sensor	One-inch CMOS
Camera field of view	84°
Time of flight	Approximately 30 min
Maximum speed	72 kph
Camera resolution	20 MP
Diagonal size	350 mm
Weight	1,388 kg
GNSS	GPS and GLONASS
Obstacle avoidance	Forward, backward and downward

DJI Phantom 4 Pro is compatible with several applications for flight planning and flight realization. Flight planning can be done in various applications like Google maps, Pix4D, Dronedeploy, etc. Mission upload and flight realization can also be done in various applications like Pix4D, 3D survey, DJI GO 4, etc. In this work, Pix4D is used for mission planning and for flight realization.

3 Methodology and Data

Position of the test site is in the center of Sarajevo, and it presents a wider area of Faculty of Civil Engineering in Sarajevo. Approximate area size is 7,6 ha and it is shown in Fig. 2.

Fig. 2. Digital orthophoto

Approximate positions of Ground Control Points (GCP) are defined, and by field reconnaissance its definite positions are determined and marked on the ground. Photo signalization is done by orange paint in the shape of X. Seven GCP points were used for georeferencing a model in the State coordinate system of Bosnia and Herzegovina.

GCP coordinates were determined by the GNSS RTK (Real Time Kinematic) method using a Trimble R8 instrument.

After measuring and obtaining coordinates of GCP points, next was mission planning and flight realization. As already mentioned, mission was planned in Pix4D application, and created according to the parameters in Table 2. After the mission was created, it was uploaded to UAV and the flight was done automatically throughout the application. Photos were stored on the SD card in UAV.

Table 2. Mission parameters (Pix4D)

Parameter	Value
Area dimensions	241 × 305 m
Overlap	80%
Camera angle	90° (ortho)
Flight altitude	80 m
Flight lines length	1511 m
Time of Flight	9 min
Calculated number of images	135
Area type	Polygon

Calculated ground resolution (Ground Sample Distance – GSD) was 2,19 cm/pix. The ground sample distance is the real-world size of a pixel in images which generally sets a physical limit on the detail of an aerial survey [5].

After finishing geodetic survey and obtaining photos of the field area, next was processing. Processing was done by software Agisoft Metashape Pro. Agisoft Metashape is a stand-alone software product that performs photogrammetric processing of digital images and generates 3D spatial data [6]. Camera calibration was done in software and the corrected parameters were used in further processing, which was done in high quality. Final results are Digital Surface Model (DSM) and Digital orthophoto.

It is important to make a difference between Digital Surface Model and Digital Terrain Model. The Digital Terrain Model is an ordered set of sampled data points that represent the spatial distribution of various types of information on the terrain (both topographic and non-topographic), e.g. elevation, slope, slope form, rivers, ridge lines, break lines, etc. It usually represents the elevation of "bare Earth", i.e. the shape of terrain without any objects on it [7]. The Digital Surface Model (DSM) is the height of the reflecting surfaces of trees, buildings and other features elevated above "bare Earth's" surface. The DSM represents the Earth's surface and includes all objects on it [8].

Complete work and processing procedure with time spent is shown in Table 3.

Table 3. Complete work procedure with time spent on each segment

Work/procedure segment	Time spent
Measuring GCP	0,5 h
Flight/mission plan	0,5 h
Georeferencing	0,5 h
Camera calibration and photo alignment	0,5 h
Dense point cloud generation	1 h
DSM and DOF generation	0,1 h
Accuracy rating	1 h
Total	4,1 h

4 Results and Discussion

Final results of this research were DSM and Digital orthophoto of the medium size complex urban area around the Faculty of Civil Engineering in Sarajevo. Resolution of Digital orthophoto is 2,15 cm/pix, and the resolution of DSM is 4,3 cm/pix. Dense point cloud, as an intermediate product, counted approximately 71 million points. Results can be exported from Agisoft Metashape software and used in any other software which supports exported formats (tiff, jpeg). Parts of DSM and Digital orthophoto are shown in Fig. 2 and Fig. 3, respectively.

4.1 Accuracy Analysis

In evaluating the scientific, professional and practical value of surveying products, the most important indicator of their quality is geometric (positional) accuracy [9, 10]. Accuracy analysis of the Digital orthophoto was done by using the coordinates of 16 points which were determined additionally, by tachymetric measurements. Coordinates of detail points, like manhole and drain covers, were used for coordinate comparison.

The most important factors for the accuracy analysis of digital orthophoto plan are: flight height and characteristics of measurement of camera, pixel size, the accuracy of DSM and camera position.

Coordinates from digital plan (details on a digital plan) were compared with coordinates of the same details on the ortophoto plan (DSM is in the background for determining heights) and the differences were calculated (Fig. 4 and Fig. 5).

The maximum difference in the coordinates along the y-axis is 5,3 cm, along the x-axis is 6 cm and the maximum difference in height is 9 cm. Based on these differences, root mean square 2D error, root mean square height error and root mean square 3D error were calculated:

- root mean square 2D error: 4,5 cm,
- root mean square height error: 6,0 cm,
- root mean square 3D error: 7,5 cm.

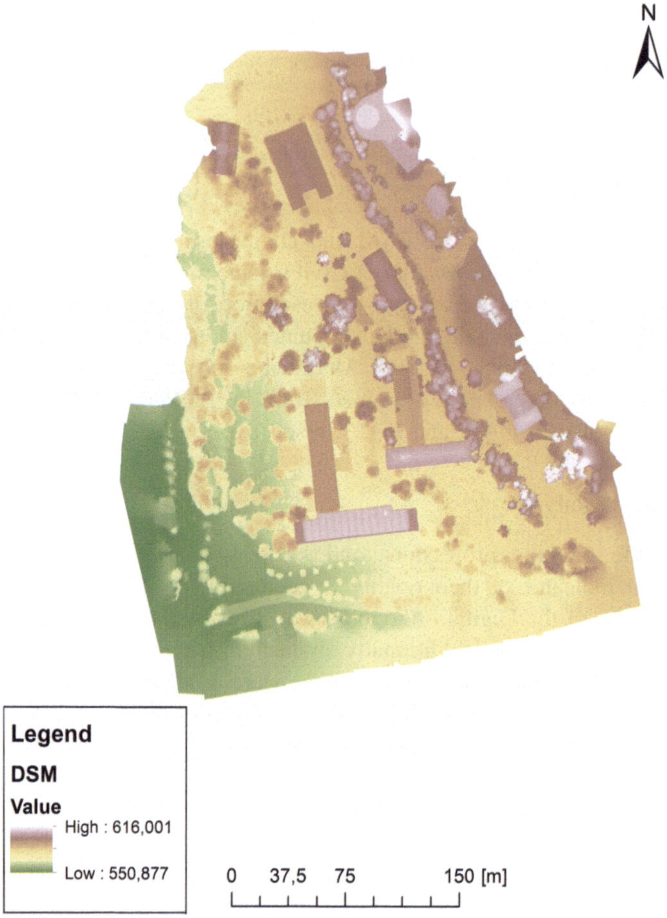

Fig. 3. Digital Surface Model (DSM)

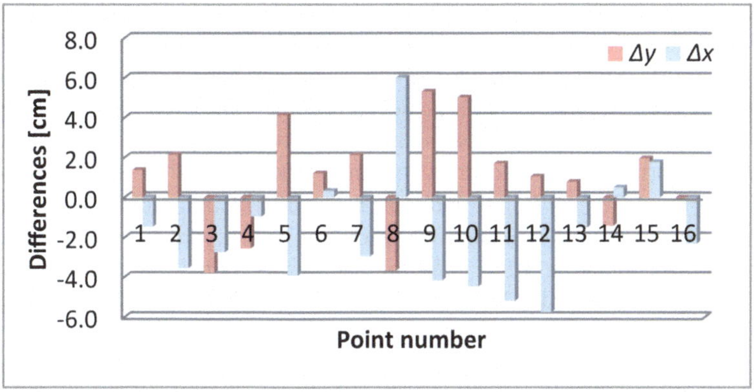

Fig. 4. Differences of *y* and *x* coordinates

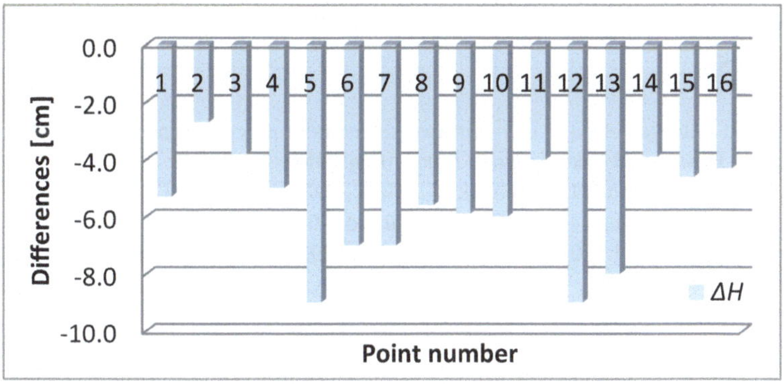

Fig. 5. Height differences

5 Conclusion

In this paper, a procedure of obtaining a Digital orthophoto plan and Digital Surface Model is described, and the accuracy analysis was done. Testing site was a medium size area of complex urban surface around the Faculty of Civil Engineering in Sarajevo. Resolution of Digital orthophoto plan was 2,15 cm/pix, and the resolution of DSM was 4,3 cm/pix. According to Ground Sample Distance (GSD), which was 2,19 cm/pix, achieved meaningful resolutions of orthophoto and DSM aren't better than GSD, considering significant digits (millimeters).

An accuracy assessment was performed for Digital orthophoto and DSM. For this purpose, coordinates of 16 points were determined from a digital plan of the area, and those were compared to corresponding 16 points measured on Digital orthophoto and DSM. Points were some terrain details, like drains, manholes, etc. Coordinate differences were calculated and also a height differences. Biggest coordinate difference for *y*-axis is 5 cm, for *x*-axis 6 cm and biggest height difference is 9 cm. According to these differences, root mean square 2D (*yx*) error, root mean square height (*H*) error and root mean square 3D error were calculated. 2D RMSE is 4,5 cm, height RMSE is 6,0 cm and 3D RMSE is 7,5 cm. Used method and parameters (flight altitude, number of GCPs and geometry) along with achieved accuracy are good enough for geodetic purposes, cadastral, planning or even some engineering purposes where high accuracy isn't quite important. Better accuracy of final products can be achieved by larger number of Ground Control Points and their better distribution around the survey area. Flying with lower altitude will improve resolution and accuracy. For example, if the flying altitude for Phantom 4 Pro is 50 m, resulting GSD will be 1,37 cm/pix.

The semi-professional UAV can be used in many different branches of geodesy (cadastral, planning, background surface base, volume calculations, etc.). It is possible to achieve high accuracy for a short period of time spent in the field, which confirms the cost-effectiveness of the method. One of the greatest advantages of UAVs is obtaining Digital orthophoto in a very short period of time. Using UAVs in urban area can produce Digital orthophoto or DSM with a quite satisfying accuracy.

References

1. Kraus, K.: Fotogrametrija 1. dio - Osnove i standardni procesi, Synopsis, Zagreb, Croatia – Sarajevo, Bosnia and Herzegovina (2006). (translation A. Mulahusić)
2. Mulahusić, A., Tuno, N., Topoljak, J., Balić, Dž.: Izdelava 3D-modela kompleksnega kulturno zgodovinskega spomenika z uporabo digitalne fotogrametrične postaje. Geodetski vestnik, **60**(1), 28–41 (2016). http://doi.org/10.15292/geodetski-vestnik.2016.01.28-41
3. Mulahusić, A., Tuno, N., Topoljak, J., Čengić, F., Kurtović, S.Y.: Collecting geospatial data using unmanned aerial photogrammetric system. In: Avdaković, S., Mujčić, A., Mujezinović, A., Uzunović, T., Volić, I. (eds.) IAT 2019. Lecture Notes in Networks and Systems, vol. 83, pp. 563–575. Springer, Cham (2020). https://doi.org/10.1007/978-3-030-24986-1_45
4. DJI Official website (2019). https://www.dji.com/phantom-4-pro/info
5. Hughes, A., Teuten, E., Starnes, T.: Drones for GIS – Best Practice (2018). https://www.wildlabs.net/sites/default/files/community/files/drones_for_gis_-_best_practice_1.3_0.pdf
6. Agisoft Metashape Official website (2019). https://www.agisoft.com/
7. Kokalj, Ž. Hesse, R., Mulahusić, A.: Vizuelizacija rasterskih podataka laserskog skeniranja iz zraka. Institut za antropološke in prostorske študije – Založba ZRC, Ljubljana, Slovenia (2018). https://doi.org/10.3986/9789610501008
8. Singh, S.: Confused Between DEM, DTM and DSM (2019). http://www.gisresources.com/confused-dem-dtm-dsm/
9. Tuno, N., Mulahusić, A., Kogoj, D.: Vrednotenje modelov transformacije geodetskih načrtov. Geodetski vestnik **59**(4), 736–751 (2015). https://doi.org/10.15292/geodetski-vestnik.2015.04.736-751
10. Mulahusić, A., Tuno, N., Topoljak, J., Balić, Dž, Hadžiosmanović, E., Stanić, S., Hajdar, A.: Primjena fotogrametrije i laserskog skeniranja kod zaštite spomenika kulturno historijske baštine. Geodetski glasnik **44**, 34–57 (2013)

Wood as a Building Material – Renewable Source to Preserve the Environment and Health in B&H

Sanela Klarić(✉) and Sarina Šabotić

Faculty of Architecture, International Burch University, Francuske revolucije bb, 71 000 Sarajevo, Bosnia and Herzegovina
sanela.klaric@ibu.edu.ba

Abstract. The building sector, which contributes towards Greenhouse gas (GHG) emissions with an amount of approximately one third, has the largest potential to cut energy consumption, as well as current pollution, through the use of local, natural, renewable and organic materials. The subject of the present paper is the comparison of calculated amounts of the CO_2 emissions, energy demand for heating, as well as Global Warming Potential (GWP), which was conducted through the comparative analysis of reconstructions of schools in Bosnia and Herzegovina (B&H) and the on-going EU "Renew School" project. The final results show that B&H is significantly trailing the European Union (EU) energy efficiency standards for buildings when focused on energy demand for heating, and Global Warming Potential. In contrast, return to B&H's traditional wood products industry as the primary source of construction material would yield many possible benefits including social-economic development, reduction of CO_2 emissions, and improving the environment through the reduction of construction waste.

Keywords: Wood · Energy efficiency · CO_2 reduction · New technologies in school renovations

1 Introduction

It is estimated that the building sector consumes up to 40% of global energy and is responsible for up to 40% of the current CO_2 emissions. However, compared to the other major emitting sectors, the building sector has the largest potential to cut energy consumption during the building life-span (existing and new) by an estimated 30% to 80% [1]. Wood is one of the oldest construction materials with excellent properties and, as one of the most important renewable materials, available in many local regions, also thanks to the constant technical improvements, has been 'rediscovered' in the construction industry after dereliction during the age of industrialization [2]. With an extraordinarily pure life cyclewood accumulates carbon dioxide (CO_2) from the atmosphere and, only during the burning process, emits the same amount of CO_2, representing the carbon dioxide neutral material. "Wood products can store CO_2 for centuries" [3]. It is estimated that people spend on average about 90% of their time indoors, despite the fact that spending time in nature is essential for health and

I. Karabegović (Ed.): NT 2020, LNNS 128, pp. 861–868, 2020.
https://doi.org/10.1007/978-3-030-46817-0_97

wellbeing. Moreover, the recovery time spent in the hospital can be reduced with the use of wood [4]. Insulation materials made from natural and renewable resources, in order to preserve the environment, represent the main demand in today's architecture design [5].

Impact on the environment by the emission of CO_2 of 'usual' or conventional thermal insulation, such as Styrofoam (EPS) or polyurethane instead of natural materials, is clearly higher. It is estimated that insulation such as EPS or polyurethane emits 7 kg CO_2-Eq/kg on average, while insulation made from natural materials emits up to 98% less [6]. At the end of the life cycle of EPS and XPS, products will become construction waste and will persist within the environment. Furthermore, EPS and XPS products are, even today, still used in the construction industry in most countries, including B&H [7].

1.1 EU Goals and Activities and the "Renew School" Project in the EU

Preventing dangerous climate change is a key priority for the European Union. Key EU targets for 2030: at least 40% cut in **greenhouse gas emissions** compared with 1990; at least 27% of total energy consumption from **renewable energy**; at least a 27% increase in **energy efficiency**. By 2050, the EU aims to cut its emissions substantially – by 80–95% compared to 1990 levels as part of the efforts required by developed countries as a group [8]. Lead by Strategic goals, all activities in the EU are focused on finding the best solution in the building sector and materials resources. Benefits of wood as renovation material is recognized by the "Renew School" project, to improve conditions in schools as places of major public interest, respectively to convert the present school buildings into the Nearly Zero Energy Buildings (nZEB) standards. The program is Co-funded by the Intelligent Energy Europe Program of the European Union, with many partners in the EU countries [9].

1.2 Bosnia and Herzegovina Toward EU Membership; Potentials, Resources and Current Practice in the Renovation of Schools

Bosnia and Herzegovina was identified as a potential candidate for EU membership during the Thessaloniki European Council summit in June 2003. 1st June 2015 was the date when the SAA (Stabilization and Association Agreements) for B&H entered into full force and brought increment of obligations and scope of reforms. The next step is B&H's entering into the EU as an equal member with all benefits and obligations that EU membership includes [10]. The Energy Community Treaty, which was signed on October 25th, 2005, and came into effect on July 1st, 2006, has goals to create a stable and single regulatory framework and market space in order to ensure reliable energy supply in the electricity and gas sectors. B&H was one of the eight contracting parties of the Treaty Establishing the Energy Community [11]. In 2013, in the Republic of Srpska, Law on Energy Efficiency, Law on Spatial Planning and Construction, as well as the Energy Efficiency Action Plan of the Republic of Srpska until 2018, had been adopted. However, in the Federation of B&H, a draft on the Energy Efficiency Law is prepared and its adoption is expected to happen soon. Also, the Energy Efficiency Action Plan for the Federation of B&H is at the preparation stage [12].

B&H is a country with a long tradition in the wood processing industry. The forestry and timber industry was founded in the middle of the 19th century, after hundreds of years of development [13]. In a period before 1992, the timber industry of B&H participated with 10% in gross national income of former Yugoslavia, accounting for 11% in total B&H's exports. Today, the wood sector is one of the three sectors with a positive commercial balance, with 11,16% in total of B&H's exports [14]. According to the official data by FAO (Food and Agriculture Organization of the United Nations), 63% of the B&H's territory is covered by forests with an estimated increment of 10% [15].

B&H has suffered enormous devastation through war horrors in the past years, especially in cities. The restoration went in the way to provide essential conditions with minimal costs. Besides this fact, most of the schools in B&H were built in the period between 1950 and 1980 with the necessity for renovation. In May 2014, B&H was exposed to the worst floods during the past 120 years [16]. Renovations of schools in B&H have performed thanks to the funds of the UNDP, Environmental Protection Fund of the Federation of B&H and Environmental Protection, Energy Efficiency Fund of the Republic of Srpska, along with Cantons of B&H as partners with the aim to convert the present school buildings into energy-efficient buildings through renovations. Thanks to the Norwegian People's Aid, additionally 36 schools and kindergartens in B&H were renovated and erected as new buildings by 2013 [17]. One of the schools renovated under the UNDP program is taken for further investigation through this paper.

2 Methodology

In order to achieve the objectives of this paper, correspondingly to provide a high level of relevant data, the chosen methodology is the comparison of the building envelope of two schools. Respectively, a comparison of the building envelope used in the renovation of school as a part of the "Renew School" project in the EU, with the building envelope used in the renovation of a school in B&H. The results are obtained through the comparison of characteristics of the envelope of school buildings in terms of the heat transfer coefficient (U value), energy demand for heating and Global Warming Potential (GWP).

Observed schools are located in the Austrian Municipality of Arnoldstein and in B&H's Municipality of Žepče. Both municipalities have similar climate characteristics, which were conducted using the site 'Climate-data.org' for the Municipality of Arnoldstein and Municipality of Doboj, as one of the closest to the Municipality of Žepče. The municipality of Arnoldstein's climate is classified as warm and temperate, with temperature averages of 8.1 °C. Municipality of Doboj has a mild and generally warm and temperate climate, with an average annual temperature of 11.2 °C [18]. The official site of the Austrian state catalog for construction equipment and materials has been used to obtain information about the characteristics of used materials, their heat transfer coefficient (U value) and Global Warming Potential (GWP) [19].

2.1 The Description of the Construction – the "Renew School" Project in the EU

For the present paper, one of the schools from the "Renew School" project, which was built in the 1960s and renovated after 2009, was elected. Table 1 presents all layers of the wall and roof – existing as well as new layers, which were added in the process of renovation, as well as the type of used windows in the renovation process.

Table 1. The description of the construction – the "Renew School" project in the EU

	Outside wall		Roof	
	Layer description	Thickness d (cm)	Layer description	Thickness d (cm)
Existing layers	1. Inside plaster	2,00	1. Inside plaster	2,00
	2. Concrete structure	25,00	2. Concrete structure	25,00
	3. Outside plaster	2,50	3. Outside plaster	2,00
	4. EPS and plaster	8,00	4. EPS and plaster	11,00
			5. Concrete structure	6,00
New layers	5. Double T beam and blown cellulose flakes	30,00	6. Sealing tape EKV5	0,35
	6. Medium-density wood panel	1,60	7. Timber frames	
	7. Rear ventilation bet, upright wood lathes	12,00	8. Double T beam and blown cellulose flakes	50,00
	8. Façade layer	0,80	9. Medium-density wood panel	1,60
			10. Rear ventilated and inclined layer	2,00–14,00
			11. Waterproof layer	1,80
New windows			Wooden triple glazed windows	

As can be seen in the previous table, added layers are made from natural materials like wood panels, blown cellulose flakes, and wooden windows.

2.2 The Description of the Construction – the "Conventional Façade" in B&H

One of the renovated schools in B&H was elected in order to show characteristics of the "Conventional façade" in B&H. The school was built in the 1970s and is in the process of renovation. Table 2 presents construction details of the wall and roof – existing and new layers which are added as part of the renovation process and, the type of incorporated windows during the process of renovation.

In contrary to the "Renew School" project, the description of the construction of the "Conventional façade" shows that used materials for renovation are unnatural, non-ecological and non-renewable.

Table 2. The description of the construction – the "Conventional façade" in B&H

	Outside wall		Roof	
	Layer description	Thickness d (cm)	Layer description	Thickness d (cm)
Existing layers	1. Inside plaster	2,00	1. Inside plaster	2,00
	2. Brick	36,00	2. Wood boards	2,00
	3. Outside plaster	1,00	3. Concrete structure	15,00–40,00
	4. Old façade layer	0,05		
New layers	5. Insulation - EPS	10,00	4. Insulation - MV	15,00
	6. Façade layer	0,80	5. Cement screed	5,00
New windows			PVC double glazed windows	

3 Results

In the present paper, Global Warming Potential is calculated for the new, added layers as a part of renovations. GWP 100a is specified in kg-CO_2 equivalents and includes both the contribution to Global Warming in terms of greenhouse gas emissions and the quantities of carbon dioxide stored in the biomass. GWP calculations were conducted using the official site of the Austrian state catalog for construction equipment – baubook [19].

Results presented in Table 3 indicate to negative contribution to Global Warming.

Table 3. GWP – the "Renew School" project in the EU

	Outside wall		Roof	
	Layer description	GWP 100a (kg CO_2/m^2)	Layer description	GWP 100a (kg CO_2/m^2)
New layers	5. Double T beam and blown cellulose flakes	**−1,25**	6. Sealing tape EKV5	**1,58**
	6. Medium-density wood panel	**−1,00**	7. Timber frames	**−1,44**
	7. Rear ventilation bet, upright wood lathes	**0,00**	8. Double T beam and blown cellulose flakes	**−0,88**
	8. Façade layer	**−1,00**	9. Medium-density wood panel	**−1,00**
			10. Rear ventilated and inclined layer	**0,00**
			11. Waterproof layer	**0,64**
New windows				**−17,81**
Total				**−22,16**

The renovation made by using the "Conventional façade" contributes to Global Warming with a significant amount, harming climate and nature in B&H (Table 4).

Table 4. GWP – the "Conventional façade" in B&H

	Outside wall		Roof	
	Layer description	GWP 100a (kg CO_2/m^2)	Layer description	GWP 100a (kg CO_2/m^2)
New layers	5. Insulation - EPS	**4,17**	4. Insulation - MV	**4,17**
	6. Façade layer	**0,16**	5. Cement screed	**0,17**
New windows				**151,00**
Total				**159,66**

Conducted analyses presented in Table 5 indicate that the "Renew School" project renovation, in contrary to the "Conventional façade" renovation, brought achievement of lower values of heat transfer coefficient (U value) and energy demand for heating. Also, differences between achieved contributions to Global Warming Potential (GWP) are incomparable.

Table 5. Comparative analysis – the "Renew School" project in the EU vs the "Conventional façade" in B&H

		The "Renew School" project in the EU	The "Conventional façade" in B&H
Heat transfer coefficient **U-value** (W/m^2K)	Outside wall	**<0,10**	**0,31**
	Roof	**<0,06**	**0,24**
	Windows	**<0,80**	**1,20**
Energy demand for heating (kWh/m^2a)		**12**	**42**
Global Warming Potential (GWP) 100a (kg CO_2/m^2)		**−22,16**	**159,66**

4 Discussion

Starting with the Kyoto protocol from 1997, as an international agreement linked to the United Nations Framework Convention on Climate Change, EU countries are constantly pursuing the most convenient ways to preserve a healthy environment, developing new technologies and materials, which led to the completely expelling some of the materials from use. Renovations of schools in Europe are developed with the idea to create comfortable conditions for teachers and their pupils, with an aim to convert the present school buildings into the Nearly Zero Energy Buildings (nZEB) standard and to conceive the decrement of the everyday maintenance costs of the buildings.

5 Conclusion

B&H is at the beginning of worldwide actions and demands for the preservation of a healthy environment. Unfortunately, the building sector in B&H is focused on using well-known conventional materials with a short lifetime, which will cause an indispensable need for a new renovation in the future years with an enormous amount of construction waste. Moreover, most of the materials are imported, contributing to the increment of GWP by transportation. With the renovation of an observed school in B&H, a goal to convert school building into an energy-efficient building has been achieved: the energy demand for heating decreased by over 60% and the heat transfer coefficient (U value) was decreased by adding new layers, resulting in lower running costs. Unfortunately, the ventilation and recuperation systems were not installed and indoor air quality after renovation is questionable, moreover, it is not measured.

The observed school from the "Renew School" project won the ISOVER Energy Efficiency Award in 2011. The school is an example of the speed of intervention, it was built on-site for a time of just 2 months. Renovation enabled work for local companies using local natural resources and brought excellent environmental comfort given by the wood and natural materials used in construction. The renovated school has high sound insulation and low running costs due to the high energy efficiency achieved. The school has installed a central mechanical ventilation system with 80% heat and moisture recovery. The concentration of the CO_2 in the classrooms, which is constantly measured and assured to be below 1400 ppm, represents the base for control of the ventilation system. A good practice of the "Renew School" project, which was presented and analyzed in this paper, seems to be a promising approach with numerous advantages if this practice would be adopted in B&H as well.

Despite the fact that numerous schools in B&H are renovated, the number of non-renovated schools is significant. Taking the present time for a holistic strategic approach could be seized as a turning point from which all future renovations, as well as new buildings, will be made with natural, renewable and ecological materials can have tremendous effects not just for the preservation of a healthy environment but also have grade influence on developing of the local economy and potential to restore wood sector and timber industry in B&H.

References

1. United Nations Environment Program (UNEP). Buildings and Climate Change – A summary for Decision Makers (2009). http://www.unep.org/sbci/pdfs/SBCI-BCCSummary.pdf. Accessed 2016
2. Jeska, S., Pascha, K.S.: Emergent Timber Technologies: Materials, Structures, Engineering, Projects, 1st edn. Birkhäuser, Berlin (2014)

3. Kuzman, M.: Wood in Contemporary Slovenian Architecture 2010–2015. University of Ljubljana, Ljubljana, Slovenia (2015)
4. Planet Ark's Make It Wood program: Wood - Housing, Health, Humanity (2014). http://makeitwood.org/documents/doc-1253-wood–housing–health–humanity-report-2015-03-00-final.pdf
5. Klarić, S., Hujdur, Z., Aybek, I., Dautović, A.: Advantages of natural renewable materials in the building sector. In: The 3rd International Conference "The Importance of Place", pp. 445–463. CICOPBH Press, Sarajevo (2015)
6. Zabalza Bribián, I., Valero Capilla, A., Aranda Usón, A.: Life cycle assessment of building materials: comparative analysis of energy and environmental impacts and evaluation of the eco-efficiency improvement potential. Build. Environ. **46**(5), 1133–1140 (2011)
7. Nie, Z., Yang, Z., Fang, Y., Yang, Y., Tang, Z., Wang, X., Die, Q., Gao, X., Zhang, F., Wang, Q., Huang, Q.: Environmental risks of HBCDD from construction and demolition waste: a contemporary and future issue. Environ. Sci. Pollut. Res. **22**(21), 17249–17252 (2015)
8. European Commission: Energy strategy | Energy (n.d.). https://ec.europa.eu/energy/en/topics/energy-strategy-and-energy-union. Accessed 2016
9. Renew School: Renew School Project (n.d.). http://www.renew-school.eu/en/home/. Accessed 2016
10. Delegation of the European Union to Bosnia and Herzegovina, & European Union Special Representative in Bosnia and Herzegovina. Stabilization and Association Agreements (n.d.). http://europa.ba/?page_id=490. Accessed 2016
11. The State Electricity Regulatory Commission, B&H: Treaty establishing the Energy Community (2016). http://www.derk.ba/DocumentsPDFs/Energy%20Community%20Treaty.pdf
12. Pita, M.: Structure of Energy Efficiency Sector in Bosnia and Herzegovina. Presented at the Second Energy Summit in B&H, Neum, B&H (2016). https://www.usaideia.ba/en/news-second-energy-summit-in-bosnia-and-herzegovina/
13. Klarić, S.: Sustainable housing (wood, sheep's wool, and straw: challenges and potentials of traditional natural materials). International Burch University, Sarajevo (2015). (In Bosnian)
14. Foreign Trade Chamber of Bosnia and Herzegovina, & Bosnia and Herzegovina Export Promotion Agency: Export Strategy of the wood sector (2016). http://komorabih.ba/wp-content/uploads/2013/06/bhepa-izvozna-strategija-drvnog-sektora.pdf. (In Bosnian)
15. Food and Agriculture Organization of the United Nation (FAO): The Forest Sector in Bosnia and Herzegovina. (2015). http://www.fao.org/publications/card/en/c/8f594e6b-5d7e-42c1-85db-534e8745b3f9/
16. UNDP in Bosnia and Herzegovina: EU Floods Recovery Program (2016). https://europa.ba/wp-content/uploads/2016/02/EUFRP_Monografija-bh-web.pdf. (In Bosnian)
17. Lisica, D.: History of the Norwegian People's Aid in Bosnia and Herzegovina, 10 January 2013. http://www.npa-bosnia.org/1_pdf/povijest-npa-bih-hrvatski.pdf. Accessed 2016. (In Bosnian)
18. Climate – data.org. Arnoldstein Climate, &Doboj Climate. (n.d). https://en.climate-data.org/europe/austria/carinthia/arnoldstein-158018/, https://en.climate-data.org/europe/bosnia-and-herzegovina/doboj/doboj-25705/. Accessed 2017
19. Schmidinger, P.: Baubook: kologische Bauprodukte (n.d.). https://www.baubook.info/. Accessed 2016

The Use of Green Materials in Architectural and Civil Design and the Success of Construction Projects: Case Study of Bosnia and Herzegovina

Yousef Zaarir[1], Fuad Ćatović[2], Adnan Novalic[1], and Ahmed El Sayed[1](✉)

[1] Faculty of Engineering and Natural Sciences, International Burch University, Ul. Francuskerevolucijebb, 71210 Ilidža, Sarajevo, Bosnia and Herzegovina
ahmed.elsayed@ibu.edu.ba

[2] Faculty of Civil Engineering, University "DžemalBijedić", Mostar, Bosnia and Herzegovina

Abstract. This paper focused on determining the effect of the design phase on the construction project success and to investigate the awareness of construction project participants about the values and benefits gained by using green materials and buildings. Results gained by using qualitative research method demonstrated that the design has a significant impact on different phases of the construction project. It has been concluded that the design has a direct effect on the cost, technology, scope, and schedule of the project.

Keywords: Architectural and civil design · Construction project management and cost · Construction project success · Green materials

1 Introduction

The building design ensures background information for construction which includes the users' requirements, needs, safety, public satisfaction, and welfare. In developing countries, such as Bosnia and Herzegovina, the cost of a building is generally the dominant and primary concern within the construction project design and implementation. For instance, if the building design exceeds the budget of the owner, the owner may decide to withdraw from the project. Some building expenses such as insurance are generally proportional to the initial cost and owners/investors try to keep this cost as low as possible. In this manner, when designers opt for green buildings or materials, mainly due to the increase initial cost compared to conventional materials and buildings, they face resistance and rejection from the investors and contractors. However, this might always not be the case. Other costs should be well-considered when designing a building to avoid additional expenditures in maintaining and repairing buildings in the following years [1]. Compared to the other industries, the construction industry has received the most attention from the public [2]. The analysis of project success, including major factors and determinants, are inevitable approaches to improve the project sufficiency [3]. The successfulness of the construction project is

I. Karabegović (Ed.): NT 2020, LNNS 128, pp. 869–880, 2020.
https://doi.org/10.1007/978-3-030-46817-0_98

often measured by the technical performance specifications and tasks, along with a high-level satisfaction of both the users and organizers that concerns the project outcome. Besides, project management success is determined by the successful monitoring and implementation of the cost, time, and quality objectives. Overall, a successful construction project must finally meet key project objectives [4].

Due to the development of construction technology engineers mostly supervise the building phases by neglecting the role and significance of designers. Moreover, the clients and investors began to rely on contractors even for design purposes. These unfortunate practices prevail, particularly in the B&H construction industry. However, the development of the drawing software technology and computer programs began to change these perceptions about the architects and designers. In today's world, people perceive architects and civil designers as designers who are experts in using drawing programs. This research aimed to show the effects of architectural and civil design on the construction project success and to investigate the role of the architect on the success of the project construction. This research has primarily objective to raise the awareness of architects and civil designers about the importance of their designs for the success of the construction project.

2 Literature Background

2.1 Construction Project Success and the Iron Triangle

Project success has been defined divergently. In the 1960s, success was purely practical, and the project was considered to be a success only when the outcome is running correctly. The 1980s literature introduced what today is called "conventional" success criteria (cost, time, quality), without including any external [5]. Although the researchers have put a great effort into having a specific definition, the existing definitions for construction project success vary and even contradict. However, the traditional way of defining project success "the level of congruity between the predetermined goals and anticipations of the project with the actual project outcome, or simply, a project is successful when all the planned and expected results are achieved" [6].

Most of the research studies have indicated the cost (budget), time, quality and management as the main Critical Success Factors (CSFs) in project success. Thereby, when the construction project is completed within the desired time, budget and quality, the project is successful. Therefore, Critical Success Factors (CSFs) can be used as guide factors to predict the probability of project success in the future [7].

The project managers commonly use the iron triangle, which is the traditional way to measure the time, cost and quality of the project. However, the iron triangle approach in assessing project success has been modified. Thus, the researchers began to examine different aspects and new dimensions of project success. Besides, since project success is a multidimensional construct, different owners/stakeholders began to perceive it differently. Therefore, customer satisfaction and client relationships have also become the main part of the project success assignment. In this regard, the customer is the most significant and crucial of these stakeholders. Some experts have argued that "customer satisfaction is a critical aspect of project success" [8]. The new definition of the project success should contain the next success elements or determinates: the allocated time period, the cost planned in accordance with the budget, the proper performance or specification level, the acceptance by the customer/user, minimum or mutually agreed upon scope changes, the main workflow of the organization, and the corporate culture [9].

The most apparent element of the construction project is time. However, within a project management perspective time comprises the time from the beginning of the project to the end and the building's appreciation during its occupational life. Similarly, the stages before the construction should also be well-considered as they include the stages before the construction: Establishing the Need, Options Appraisal, Strategic Briefing, Outline Proposals, Detailed Proposals, Final Proposals, Production Information, Tender Documentation, and Tender Action. These stages before the construction ended by a contract agreement by a successful tenderer. Thus, pre-construction stages have to be determined by the project manager along with the construction time and total occupation of the building [10].

The cost as another major factor seems to be dominated by the capital cost. Yet, the design teams have recently considered the entire spectrum of other costs of a project such as: Acquisition costs, life-cycle costs, operational costs, and demolition costs. The cost planners often neglect the design team recommendations and the consequences of their choices in the occupational life cycle of the project [10].

Because of the actual completion of the project, it is very hard to define the quality factor. For example, quality can refer to all or some of the following criteria: Function, Performance, Aesthetics, Technical, and Environmental. Designers carry a crucial task of integrating a whole range of quality requirements into a project, whereby they combine them to supply the clients, customers, and users [10].

The relationships between project factors can generally be summarized as follows: Time generally increases in demand of higher quality; Increases in time usually result in higher costs; The cost usually increases with greater quality [10].

2.2 Architectural and Civil Design and Their Impact on the Project Success Elements

According to Gary and Hughes (2007), the design is a creative and personal activity, whereby it is crucial to understand the designer's goals and priorities. Their solid intellectual activity can only be seen when the design is completed. The aim should be to simplify and minimize building services. Complex services are likely to increase both capital and maintenance costs along with reducing user satisfaction due to a lack of understanding and control [11].

Song and Knaap (2003) investigated real estate residentials price in so-called "New Urbanist" neighborhoods which was the trend in suburban plot design. These authors deduced that the difference in value between traditional and New Urbanist is assessable along with capitalized residential property values. The buyers then were found progressively willing to pay a premium for more and smaller blocks, considering adjacency to light rail stations and better pedestrian structure. They also discovered the premium more than compensated for the price reduction because New Urbanism was smaller.

The Project Management Institute describes the costs of resources required for the completion of the project activities. Estimating the magnitude of capital expense is the key task in the first stage of the investment process, that is, a conceptual stage. At an early stage, the customer has a general concept of the project, described only in the form of a functional and useful program or conceptual design. Therefore, only studies on costs based on general information should be defined by significant price distribution. Cost estimates for the customer's needs are prepared and outlined to give the rough total cost of the construction project. The calculation results, done by different surveyors and price bases, are also remarkably affected by cost estimation [12].

Kent Holland (2001) highlighted problems that were developed after the approval of the document developed by the American Institute of Architects and called "Standard agreement between the architect and the owner/B141-1997". The most problematic expression of this new and distressed sentence is stated in the last twelve words, which state that "*the owner must cooperate with the architect to make such adjustments*." The problems brought about by this sentence are as following: There is no limit to the "corrections" required to make the architect's cost estimate in accordance with the owner's budget. This is a completely open-ended necessity; As a result of the statement number one, the amount of redesign work required to correct the architect's estimate is similarly open-ended, despite the owner's evolving program being incompatible with the owner's budget constraints; and The architect is obliged to provide redesign services for free, although there may be an increase in the estimate of work costs, due to factors not entirely related to the architect's services [13].

Estimating costs at different phases of project progress help control over the budget. This alerts the customer and the project team before they encounter a larger budget problem. Architects play an important role in visualizing the entire structure from the top-down, emerging with new structure and ideas. Such a thing causes difficulty regarding their application without engineers helping and understanding construction practicality [14].

Oftentimes, the clients do not receive the building they want, mostly because they do not know how to demand it and the architect or the other team members think that the building they will design should look differently. Customers hope the project will be successful and providers will provide an authoritative service. If these expectations are not met, they will not be satisfied [15].

When adequate details are processed, an estimate can be based on a more accurate calculation of the material cost and the labor required for the project. The main methods are quantity and cost and on-site unit costs. These methods are more precise than the volume or space techniques previously used in the architect's performance, but even these methods cannot take into account all the possibilities that go into creating a construction proposal [16].

The specified materials and the details of the proposed construction have a significant impact on the project cost. Buildings with high quality and/or innovative features are always more expensive than fully functional ones. The choice stands on how much the customer wants to pay. However, the materials must conform to their use; over-specification is very wasteful. The choice of material, and therefore cost, can be affected by factors rather than aesthetics. For instance, fast-track construction projects may use a steel frame preferred to an on-site or pre-reinforced concrete frame to decrease overall program times. Even though using the concrete as an option may be cheaper, the shorter program attained by using steel can balance the initial cost issue. Such technical decisions are taken for each building element, which has a direct impact on the final cost. Where significant repetitions can be succeeded, it can be economical to prefab some structural elements or standardize various parts and fittings [15]. When determining materials, the following should be taken into account: effects resulting from the extraction of resources; effects during production; effects on transport; effects on use in structures and effects on final destruction [17].

The initial cost, called the acquisition cost or development cost, also covers the entire cost of building creation or renovation, such as the cost of buying land/building, the professional consultants fee, the cost of materials that come to terms with the completed building, and the cost of putting it all together. The running cost or operation cost affects the use cost, which is set by the decisions made at the briefing stage and the decisions made during the design and compounded phases. There is a rare third cost, which is at the same time the demolition and material recovery cost. In this regard, the customer could have sold the building long before recycling, and in part because such costs have traditionally been related to the initial cost of future development [18].

According to [19] it is estimated that 33% of the waste on-site was caused by the architects' failure to apply waste reduction measures during the design stages. The production process of construction waste through design is complicated as buildings contain a wide range of materials and products; and various customers, have contributed directly or indirectly to the resulting waste. Yet, there is a general agreement in the literature that design changes occur while construction proceeds are the main origins of construction waste production. Osmani (2008) stated that origins and causes of construction waste arising from the design phase are: Design Changes, Design and detailing complexity, design and construction detail errors, unclear/inconvenient specification, and poor coordination and commutation with other stakeholders [20].

A general rule asserts the unit costs will be cheaper in the line with building shapes being simpler [21]. Many studies state that the building shape has an impact both on construction costs and expenditures estimated for the entire building life, including usage and energy costs. The shape choice is especially important in green buildings. Changing design parameters like shape and orientation creates the case for the designed building consuming 40% less energy [22]. The sustainability concept, implemented in building construction, is calculated to present maximum efficacy and decrease costs financially. There is significant evidence that many private and public organizations have determined building-related investments, depending on initial construction cost calculations. These determinants demand the selection of construction structures, building materials, and building facilities that are often used in conjunction with investment failures with inadequate economic control [23].

2.3 Sustainable Construction

According to Asad and Khalfan (2007) sustainable structures are designed in accordance with high environmental standards. Thus, minimizing energy requirements, reducing water consumption, using materials with less environmental effect, minimizing waste, protecting/improving the natural environment and protecting human well-being. Comprehensibly, sustainable construction covers three main areas: the industry's economic, environmental and social performance [24]. A green building is, in terms of price and expenses, 3–8% more expensive than traditional buildings. Yet, these costs are retrieved in two to three years into savings in maintenance costs. Associated with a serious decrease in operational costs, the total cost of ownership of the green building is always less than the traditional buildings [25]. The suitability of reused or recycled materials notably affect the design process as one cannot design without the knowledge of the accessibility of certain recycled materials, but likewise, without the beginning design being considered, one is now aware of what materials to seek [26]. According to Veronica and Munn (2004) the markets for reused materials are not enough. If one cannot sell recycled materials, they must organize somewhere to store them. As nobody prefers to purchase reused materials, they will consequently start storing rubbish.

2.4 Green Materials and Buildings

According to Kats (2003) the benefits provided by the Green Buildings are various and could be estimated up to 500 USD per square meter. Such financial benefits are achieved by the lower energy usage, the lower amount of waste and water costs, lower ratio of emissions and environmental costs, and lower amount of operational and maintenance costs. Of course, there are other benefits that are not financial type, such as the productivity and health benefits [27].

Kats (2003) indicated the negative effects of the poor indoor conditions in commercial buildings on the health and productivity of workers/visitors, and the estimated cost of such impact is assumed up to billions of USD per year in the United States of America. This would not be a problem if people are not spending about 90% of their time indoor. According to the estimations made by Kats (2003), the average annual cost of energy in Massachusetts for traditional buildings is about 20 USD per meter square, while for Green building it is close to 12 USD per meter square, which is another financial benefit of the usage of Green Buildings [27].

In a study conducted in the USA by Ahn and Pearce (2007), results proved that companies working in the construction industry are considering the green construction to take the lead in the construction field future, but believe the increased initial costs is the main reason why it is still underestimated and unconsidered by designers, investors, and contractors [28]. They have also concluded that the movement of Green Building will change the Construction industry and market, and therefore, the material manufacturers should lower the prices of green materials, and the investors should be aware of the financial benefits of using the green materials and building in reducing the maintenance costs and living costs [28].

Tam, Hao, and Zeng (2012) used questionnaire to explore factors affecting green building implementation for the Hong Kong Construction Industry. They concluded that the cost of designing, production of materials, material transportation, and construction of green buildings are higher than the conventional buildings, which makes investors opt for conventional buildings. However, the Industry emphasized that implementing green buildings would help the improvement of both economic and social issues in Hong Kong [29].

In a study done in New Zealand in 2013, Rehm and Ade mad an in-depth investigation on the actual cost difference between green buildings and conventional buildings. They compared the actual cost of 17 green buildings in New Zealand with the modeled costs (Designed cost). They found that, for big-sized buildings, the actual cost is lower than the modeled one, and there is no significant difference between actual and modeled costs of the buildings [30].

3 Methodology

The aim of this paper is the determine the impact of the Architectural and Civil Design on the construction project success elements. The next research questions were set by authors in order to confirm the model developed in this paper: Could the A&C design affect the optimization of the workload of other participants in the project? Are designers aware of their responsibility in the reduction of construction waste? Is sustainability in building design considered as important factor in Bosnia and Herzegovina? Are Designers in Bosnia and Herzegovina familiar with the impact of their design on the construction project success? And is the decision on the construction materials a responsibility of the designing team or of the investor in Bosnia and Herzegovina?

The population selected to test the main research questions and hypotheses are engineers and investors who work in the construction industry in Bosnia and Herzegovina. This category includes the main engineers responsible for the implementation of construction projects. It is important to highlight that managers and owners of companies are not excluded from the sample targeted in case they are not engineers. The sample included well-known construction and supervising companies and building construction investitures taking the size of the company and the number of workers they employ into consideration. The selection of companies included in the research was based on several steps. The first step was to investigate which companies and designing offices were active on projects in the last five years. Companies selected for this research are well known in the field of the construction industry and they are visible in the job market of Bosnia and Herzegovina.

The qualitative research methodology in this thesis will be based on in-depth interviews that will be run with two real estate investment companies, two supervisors, and three construction companies. The conversation between two persons taking into consideration a topic of common interest based on a system of questions and answers is defined as an interview [31]. The difference between conversations and interviews depends on the structure and the aim of interviews [32]. The contrast among discussions and meetings relies upon the structure and the point of meetings. Qualitative research gives and opens the method for accomplishing new information and finds new actualities by adding a clarification to existing speculations or new theories picked up from the genuine experience of the interviewees. As an apparatus of qualitative research methodology, in-depth interviews are considered as a real part of the most utilized ones, because of their open-ended qualities and revelation situated nature, which enables the specialist to discover and investigate considerations of the interviewee about the examined subject [33].

Total of ten interview questions were prepared by authors and used during the interviews. The first question aimed the importance of cost as a factor in construction projects success. The second one investigated the relationship between the designing team. Third and fourth questions were related to the impact of the design on the building value and whether investors appreciate and understand the importance of the Designers. Fifth and sixth questions were related to impact of materials and building shape on the elements of construction project success. Seventh question was related to awareness of designers about their role and importance for the success of projects. Eighth question investigated whether designers are considering the building sustainability aspects in Bosnia while designing. The ninth question looked for the awareness and carness of designers about the construction project success. The final question investigated whether Designers are always included in the materials' selection process or it is done by investors.

4 Results and Discussion

Total of 7 different types of companies working in the construction industry (3 construction companies, two supervisors, two real estate investment companies) based in Bosnia and Herzegovina were invited to participate in the in-depth interview explained in the previews chapter. With a total of 10 main questions and follow up questions, the duration of the interviews varied from 30 to 45 min.

Summarized response to the research question was as follows:

- RQ1: Could architectural and civil design affect the optimization of the workload of other participants in the project?
  The architectural and civil design could produce a harder or easier task for the rest of the work team who work on the rest of the phases. Most design phases in the project follows the architectural and civil solutions. The architect and the civil designer should care about the following project phases.
- RQ2: Are architects and civil designers aware of their responsibility on the reduction of construction waste?
  Architects and civil designers are not fully aware of their responsibility in reduction and production of the waste of construction materials. They should try to learn this phase from the construction site workers and site engineers. By improving the communication between designers and contractors or site engineers, the construction waste would be reduced.
- RQ3: Is sustainability in building design considered in Bosnia and Herzegovina by designers and contractors?
  Sustainability is not significantly considered in B&H and is not counted among important issues. Some architects and civil designers try to apply this concept in their designs but mostly it has a collision with the construction cost of the project, and therefore it is generally excluded by investors or customers.
- RQ4: Are architects and civil designers in Bosnia and Herzegovina familiar with the impact of architectural and civil designs on the construction project success?

Most of the Bosnian architects and civil designers are not aware of their impact on the success of the construction project. Mostly they deal with the project as a design case, not a successful project.

- RQ5: Does the responsibility of choosing the construction materials is a Designing team decision taken by the project designers?
  The architects and civil designers should choose construction materials of the building, but generally, architects and civil designers are taking into consideration the client wishes and the budget limit of the construction project. In addition, designing team need to understand the aim of the construction project or the targeted market of the project up to the investor's view.

Results obtained by this research were compared with the literature findings, and below are discussed the findings:

- The cost factor is a main and effective factor for the construction project's success.
- The conceptual design affects other designers and engineers who work after on the project. It could make it easier or harder.
- Architectural and civil design has a direct and effective impact on the value of the construction project. Sailable or rentable area.
- The cost estimation made by the architectural civil design has a main role in the investment decision.
- Building materials chosen by the designing team have a significant impact on the cost of the construction project.
- Building shape or the form of the building has a strong relationship with the cost of the construction project, and therefore it impacts significantly the success of the construction project.
- Building orientation is an important aspect for the future users of the building, it could cause an additional cost for maintenance and living condition inside the building, also it could harm the value of the building itself.
- In the design phase, the design team must focus on how to reduce the construction waste of building materials. With more experience in construction site and building standards and operations. This could reduce the construction cost than could make the project more successful.
- Sustainable design is something which many designers try to implement in their designs, but on the other hand, in general cases, it may increase the cost and the time of the construction project. Using recycled material also could increase the cost and time of the construction project.
- Green materials and buildings are the future of the construction industry, as many developed countries are setting regulations to encourage designers, investors, and contractors using the green materials and building concepts. In Bosnia and Herzegovina, Investors are affecting the designer's choice of materials due to the initial costs of construction.
- Architectural and civil design should be done with good contact and understanding between the owner or the investiture of the project. And here is the role of the design team to understand what the owner's needs are. This could help the project to be more successful from the case of time, cost, and quality.

5 Conclusion

This paper aimed to investigate the impact of the Architectural and Civil designs on the success of construction projects in Bosnia and Herzegovina. Using the Qualitative research method, based on in-depth interviews and qualitative analysis of literature that dealt with this topic, this paper proved the strong relationship between these terms. The construction costs are initially estimated by the designing team and have direct impact on the investors' decision on the project and on the project success. In developing countries, the construction project cost effect on the project success in more significant than the other two effects (Time and Quality).

Depending on the architectural solution and design, all other design team members are affected. Architectural design sets the structural system and defines the needed structural and service system solutions. Therefore, the impact of the architectural design on the other design phases, and later the construction project success, should be considered as highly significant.

The architectural and civil design has impact on the marketability of the construction. Depending on the shape, orientation, construction solution, structural solution, and building service systems, the value of the construction is increased or decreased.

The choice of materials should be only assigned to the design team without interference by the investor or the contractor. The design team should consider the effect of the materials on the project cost and present their findings to the investor in early stages in order to make a decision. Sustainability, sustainable constructions, and sustainable materials should be included in the analysis, even if the initial cost is higher, as the maintaining costs are reduces, and therefore the building value is higher. Design team should be aware of the production of construction waste and to develop a design that will reduce the production of the waste by using recycled materials or re-using the construction waste for other purposes.

The role of design team, led by the architect, should be clear to all project participants. Otherwise, the success of the project will remain questionable, and a lot of important elements will remain ignored or unconsidered by other participants.

References

1. Merritt, F.S., Ricketts, J.T.: Building Design and Construction Handbook, 6th edn, p. 1600. McGraw Hill Professional, New York (2000)
2. Han, W.S.: Reviewing the notions of construction project success. Int. J. Bus. Manag. **7**, 90–101 (2012)
3. Chan, A.P., Ada, P.C., Scott, D.: Factors affecting the success of a construction project. J. Constr. Eng. Manag. **130**(1), 153–155 (2004)
4. Al-Zahrani, J.I.: The Impact of Contractors' Attributes. The University of Manchester, Manchester (2013)
5. Khosravi, S., Afshari, H.: A success measurement model for construction projects. In: International Conference on Financial Management and Economics IPEDR, Singapore, vol. 11 (2011)

6. Zanjirchi, S.M., Moradi, M.: Construction project success analysis from stakeholders' theory perspective. Afr. J. Bus. Manage. **6**(15), 5218 (2012)
7. Ramlee, N., Tammy, N.J., Raja Mohd Noor, R.N., Ainun Musir, A., Abdul Karim, N., Chan, H.B., Mohd Nasir, S.R.: Critical success factors for construction project. In: AIP Conference Proceedings, Shah Alam, Malaysia, vol. 1774, no. 1 (2016)
8. Aggor, K.K.: Relationship Between Budget and Project Success Factors in the Ghanaian Building Construction Sector. Walden University, Minneapolis (2017)
9. Karzner, H.: Project Management, p. 72. Wiley, Hoboken (2013)
10. Smith, J., Jaggar, D.: Building Cost Planning for the Design Team, pp. 57–60. Elsevier Ltd., Liverpool (2007)
11. Baden-Powell, C.: Architect's Pocket Book. Elsevier Ltd., Hungary (2008)
12. Lesniak, A., Plebankiewicz, E., Zima, K.: Cost calculation of building structures and building works in Polish conditions. Eng. Manag. Res. **1**(2), 72 (2012)
13. Holland, K.: Role of Architect in Controlling Construction Costs (2001). https://www.constructionrisk.com/2011/02/role-of-architect-in-controlling-construction-costs/
14. Mishra, G.: "The constructor" (2018). https://theconstructor.org/construction/architects-roles-responsibilities-construction/18511/
15. Cunningham, T.: Factors Affecting The Cost of Building Work - An Overview, pp. 1–21. Dublin Institute of Technology (2013)
16. Sweet, J., Sweet, L.: Architectural cost predictions: a legal and institutional analysis. Calif. Law Rev. **56**, 996 (1968)
17. Marques, F.M., Salgado, M.S.: The building material selection importance at the building design process for its sustainability. In: CIB World Building Congress Construction for Development, Cape Town, vol. 14 (2007)
18. Emmitt, S., Yeomans, D.T.: Specifying Buildings: A Design Management Perspective. Elsevier, Butterworth-Heinemann, Amsterdam (2008)
19. Innes, S.: Developing tools for designing out waste pre-site and on-site. In: Proceedings of Minimising Construction Waste Conference: Developing Resource Efficiency and Waste Minimisation in Design and Construction, London, UK (2004)
20. Osmani, M., Glass, J., Price, A.D.: Architects' perspectives on construction waste reduction by design. Waste Manag. **28**(7), 1147–1158 (2008)
21. Seeley, I.H.: Building Economics: Appraisal and Control of Building Design Cost and Efficiency. Macmillan International Higher Education, London (1996)
22. Swaffield, L.M., Pasquire, C.L.: A critique of mechanical and electrical services cost planning: existing methods and published information. J. Financial Manag. Prop. Constr. **1**(3), 23–41 (1996)
23. Akadiri, P.O., Chinyio, E.A., Olomolaiye, P.O.: Design of a sustainable building: a conceptual framework for implementing sustainability in the building sector. Buildings **2**(2), 126–152 (2012)
24. Saravanan, V.K.: Cost effective and sustainable practices for piling construction in the UAE. Heriot-Watt University, Edinburgh, Scotland (2011)
25. Khosla, S., Singh, S.K.: Energy efficient buildings. Int. J. Civ. Eng. Res. **5**(4), 361–366 (2014)
26. Munn, S., Soebarto, V.: The issues of using recycled materials in architecture. In: The 38th International Conference of Architectural Science Association ANZAScA "Contexts of architecture", Launceston, Tasmania (2004)
27. Kats, G.: Green building costs and financial benefits. Massachusetts Technology Collaborative, Boston, MA (2003)
28. Ahn, Y.H., Pearce, A.R.: Green construction: contractor experiences, expectations, and perceptions. J. Green Build. **3**(2), 106–122 (2007)

29. Tam, V.W.Y., Hao, J.L., Zeng, S.X.: What affects implementation of green buildings? An empirical study in Hong Kong. Int. J. Strateg. Prop. Manag. **16**(2), 115–125 (2012)
30. Rehm, M., Ade, R.: Construction costs comparison between 'green' and conventional office buildings. Build. Res. Inf. **41**(2), 198–208 (2013)
31. Kvale, S., Brinkmann, S.: Interviews: Learning the Craft of Qualitative Research Interviewing. Sage, Thousand Oaks (2009)
32. Tracy, S.J.: Qualitative Research Methods: Collecting Evidence, Crafting Analysis, Communicating Impact. Wiley, Oxford (2013)
33. Guion, L.A., Diehl, D.C., McDonald, D.: Conducting an in-depth interview. McCarty Hall, University of Florida Cooperative Extension Service, Institute of Food and Agricultural Sciences, EDIS, Florida, FL (2001)

Towards Innovative Solar Energy Applications: New Urban Furniture

Marija Bojovic(✉)

Faculty of Architecture, University of Montenegro, Džordža Vašingtona bb, Podgorica, Montenegro
marijab@ucg.ac.me, marijabojovic.office@gmail.com

Abstract. As every society produces its own space, in the age of sustainability, it is inevitable to rethink the ways we live and use our cities and our resources. Smart cities are an emerging topic - this next generation of urban development requires technologically advanced, socially equitable and last, but not least, well-designed urban environments. At this crucial moment of rethinking our space and cities we are given the opportunity to redesign them as more flexible, adaptable and, and citizen-focused. This paper analyses two such examples of well-designed, innovative implementation of solar energy into urban settings.

Keywords: Applied solar energy · Smart city · Montenegrin solar bench · Flexible urban furniture · Participative design

1 Introduction

Every society, and hence every mode of production, produces its own space [1]. Therefore, in the age of sustainability, it is inevitable to rethink the ways we live and use our cities and our resources. As 21st century comes with the new challenges, the importance of the role of renewable energy sources is crucial. Smart cities are an emerging topic and opportunity for creating new urban scenery and more sustainable everyday life. This next generation of urban development requires technologically advanced, socially equitable and last, but not least, well-designed urban environments. However, we finally live in technologically empowered era, having all the advancements to translate all our utopias into reality.

At this great moment of rethinking our space and cities we are given the opportunity to redesign them, as well, as more flexible, adaptable and user-friendly, in order that they can become true, open, works of art. Concept of "openness" enables long life and historical usefulness, while insisting on the element of multiplicity, plurality, or polysemy of work (of art), and the emphasis on the role of the user, which is to become an active participant in the very process of creation of its own space [2].

Although the concept of sustainability covers environmental, cultural, economic and social sustainability of human beings, this particular article focuses on environmental issues and innovative applications of solar energy into new generation of urban furniture. With the real assumption that the average solar insolation in Montenegro is 1.450 kWh/m^2 per year, a theoretical potential of solar radiation may be estimated at about 20 PWh a year [3] we can state that the country has a large solar potential.

I. Karabegović (Ed.): NT 2020, LNNS 128, pp. 881–887, 2020.
https://doi.org/10.1007/978-3-030-46817-0_99

Additionally, very small size of the country is an advantage - Montenegro could become a cluster dedicated to implementation of smart city principles, promoting sustainable urban living of the near future.

This paper further analyses two examples of well-designed and innovative implementation of solar energy into urban settings - solar pavement and first Montenegrin Solar Bench.

2 Example of Well-Designed Solar Energy-Integrated Urban Solution

As this article argues importance of innovative solutions being greatly designed as well, one of the very interesting examples of good design and its great implementation is "Platio[1] solar pavement". This is innovative, energy-producing, anti-slippery, and space-saving modular landscape design element, made of recycled plastic with integrated solar panels within (Fig. 1).

Fig. 1. Platio solar pavement module. URL Source: www.platio.cc

The frame of the element is made in recycled plastic and it comes with modular connections, enabling easier installation. Solar cells are built in the surface of the pavement module, while the electrical wiring is built in the unit. The very surface of the piece is anti-slippery and made of glass and debris. The module is highly resistant to load, impact, scratches and weather conditions, while providing vandalism resilience as well [4]. This basic element of the solar pavement is upgradable with all the necessary smart functions, such as LEDs, variety of sensors, and even heating.

[1] Platio is innovative company specialised in developing durable, high-performance solar surfaces and the complementary infrastructure. This project has received funding from the European Union's Horizon 2020 research and innovation programme.

2.1 Application

Aesthetics of the smart solar piece is very clean and simple, therefore making this ready-to-use solution easy to integrate in the overall design of the public, semi-public, and private open spaces, as well as the roof surfaces, without reducing valuable space. The installation of the solar pavement is fast and easy to execute - due to modular dimensions and connections, the process is almost similar to conventional tiling (Fig. 2).

Fig. 2. Platio solar pavement module application in a public space. URL Source: www.platio.cc

3 Montenegrin Smart City Experience: Solar Bench

Montenegrin experience with smart cities is rather new. However, the platform is developing and growing, bringing together young scientist, engineers and, architect and designers, as well. One of these products - solar urban furniture - has already found its place on the market.

First Montenegrin solar bench has been designed and produced in Podgorica, Montenegro, in 2018, by me. It is sophisticated, self-sustainable smart solution and presents new generation of urban furniture. The basic model of the bench provides remote access to usb ports for charging mobile devices and Wi-Fi infrastructure for hotspot implementation in an urban setting. Green energy is harvested via solar panel, integrated into the seating area. This smart piece is implemented in urban settings as well as in the city parks, avoiding partial shading if possible [5]. The first version of the bench is made in painted steel and plexiglass (Fig. 3).

Fig. 3. First Montenegrin solar bench

3.1 Designing the Levels of Flexibility

The design of the bench aimed to achieve flexibility and adaptability within a diverse range of aesthetic and contextual requirements. There are two levels of flexibility of this piece of urban furniture. The vertical supports are designed as "detached" from the rest of the structure, therefore altering positions of these vertical modules the spatial formation of the bench changes significantly, enabling various end results (Fig. 4).

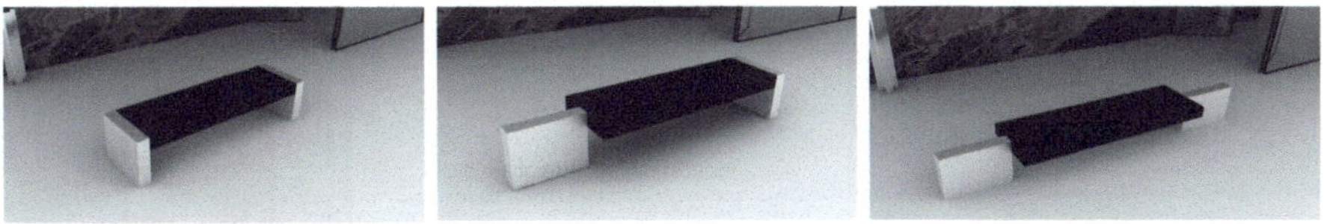

Fig. 4. First level of flexibility

Second level of flexibility addresses finishing materials and immense range of colours - the piece can be customised to fit in numerous environments. By offering great level of flexibility within this solar furniture the designer aimed to avoid usual endless repetition of the same model. As the bench is designed as an *open work*, offering not one but numerous visual outputs - it allows freedom of choice of the ordering party as a user - to actively participate in creating the final result, according to their needs and preferences (Fig. 5).

Fig. 5. Second level of flexibility, regarding finishing materials and colours

3.2 The System

Engineering components of the first Montenegrin solar bench are simple yet highly performative. The solar panel is integrated into the seating area, therefore having adequate exposure to the Sun. The wiring, connecting panel to the stationary battery, is inside the body of the bench. The battery is hidden in one of the vertical supports. The system is upgradable and features that can be added to the basic version are ambient lighting, wireless charger for mobile devices, wide range of sensors and, cloud platform delivering analytical data for the specific geo-location (Fig. 6).

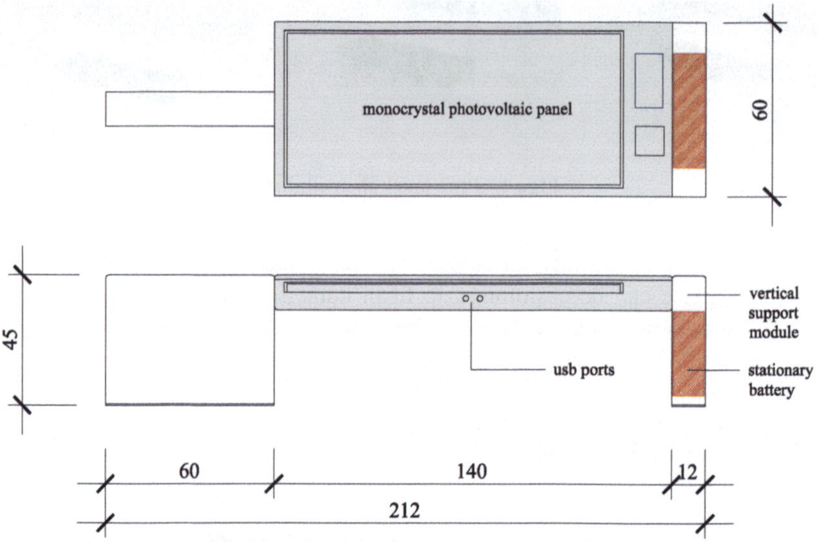

Fig. 6. First Montenegrin solar bench – the system

4 Conclusion

Smart city, as a new urban paradigm, is great opportunity to rethink our cities - the way we plan, design, build and, use them. As this new generation of urban development requires new design approach, it provides us with great new challenge to design flexible and adaptable cities - changeable due to different needs of different users over time. Therefore urban furniture can be first step in the direction of well-designed, flexible urban solutions, encouraging active participation of their users in the very process of the creation. First Montenegrin smart bench is the example of such progressive approach to designing urban furniture. However, this new urban paradigm requires interdisciplinary approach and powerful synergies of various creative parties - engineers and scientist, urban planners, architects and designers, urban sociologists and anthropologists, ensuring holistic and thorough approach to this very complex matter.

Acknowledgement. First Montenegrin Solar Bench development project is part of the "Creative Montenegro" initiative of Montenegrin Ministry of Culture, and is further recognised by Montenegrin Telekom, therefore being integral piece of their smart city platform. First Montenegrin Solar bench is 2019. BIG See "Product design" winner [6].

References

1. Lefebvre, H.: Production of Space. Basil Blackwell Ltd., Oxford (1991)
2. Eco, U.: Open work. Harvard University Press, Cambridge, Massachusetts, US

3. Report on Situation in the Energy Sector of Montenegro in 2016 (1989). http://regagen.co.me/cms/public/image/uploads/REPORT_ON_THE_SITUATION_IN_TH_ENERGY_SECTOR_OF_MONTENEGRO.pdf
4. https://platiosolar.com/smartcities#building
5. Ramadan, J.M., Goma, M.R., Al-Dhaifallah, M., Rezk, H.: Environmental impacts on the performance of solar photovoltaic systems. Sustainability **12**(2), 608 (2020). https://doi.org/10.3390/su12020608
6. http://bigsee.eu/solar-bench-by-marija-bojovic-architect-for-urb-a-from-montenegro/

Evaluation of Factors Affecting the Process of Decision Making in Construction Site

Almedina Mustafić[1(✉)], Suad Špago[2], Adnan Novalić[1], and Ahmed El Sayed[1]

[1] Faculty of Engineering and Natural Sciences, International Burch University, Ul.Francuskerevolucijebb, 71210 Ilidža, Sarajevo, Bosnia and Herzegovina
almedina_trbalic96@hotmail.com
[2] Faculty of Civil Engineering, University "DžemalBijedić", Mostar, Bosnia and Herzegovina

Abstract. This paper aimed to explore factors that may have significant impact on the decision-making process for engineering at construction site, to analyze the existing decision-making methods for the construction industry worldwide, to gather the factors affecting the decision-making process at construction site, and to list these factors according to their impact and frequency of occurring. Results gained by using qualitative research methods proved that Construction site layout planning, supplier selection process, and materials management practices at the site are the most frequent factors influencing the decision-making process at the site.

Keywords: Decision-making process · Effective models · Construction site · Project management · Decision-making support system · Layout planning · Materials management

1 Introduction

The construction site is a unique environment due to its specific requirements, which includes: the facility design, the organization of the project, the location of production facilities, and the internal traffic plan [1]. Decision making links basic strategic moves to management decisions and routine operational decisions. According to Peter F. Drucker, it is all about making choices and compromises to meet business goals, and decision-making can be defined as the choice of alternatives whose application subsequently leads to the achievement of business goals. As he stated, effective decision is the product of a schematic process with clearly defined elements, performed according to previously planned steps. Quality of decision making becomes a key link in the value chain [2].

Project management is a set of activities, skills and tools, organized into groups, that are necessary to achieve project goals. It can be defined as a continuous stream of project objectives within a specified time, cost, desired performance level, while effectively and efficiently using the resources provided [3].

James G. March explained that essential nature, contracts and assertions of morality define every individual everywhere on earth. Only depends on how and which strategy

I. Karabegović (Ed.): NT 2020, LNNS 128, pp. 888–896, 2020.
https://doi.org/10.1007/978-3-030-46817-0_100

will every decision-maker develop for himself, among his business surrounding. In his analysis, he stated that incentives and internalization of identities is best explained with an example that more experienced, successful decision-makers become socially more reliable in their decision-making than the inexperienced, unsuccessful decision-makers [4].

When it comes to the decision-making system exclusively in the construction industry, the decision-making process is, in essence, more complicated because of certain conditions specific to the construction industry. Due to the variety of decision-making situations and the fact that usually one person oversees decision-making from the beginning to the end of the project, it is recommended to use different decision support methods [5].

Considering that any decision can be decisive in its success from managerial aspect, great responsibility is placed on the staff supervising the construction site. Every decision must be justified, judged by its urgency and importance, since it can cost a lot of time, costs and relationships. Companies usually have a system or rules in place with prior information about what activities to take, so some of these actions do not require complex decision-making tools at the construction site. But some of them do require experience as a backup. If the decision sought falls outside the responsibility of the person concerned, then advice should be sought from colleagues. Complex problems involving multiple parameters can go beyond great experience, and decision-making tools and techniques must be incorporated to ensure that the most effective solution is selected [6].

Managing all projects, including construction, implies making a series of decisions that, as finally adopted, will have an impact on the appearance of the building for years to come, at a time when the raised structure will be used. The complexity of running a construction project can be reflected in the fact that dilemmas can arise from the stage of selection and organization of the plot, through the stage of document preparation and design, selection of materials and construction technology, selection of supervision and the quality or risk of the project. Both literature and practical experience advise the use of several methods based on analytical and computational procedures to facilitate the selection of the right decisions. Different methods have different advantages and disadvantages, it is important to be sufficiently instructed and choose the right one according to the situation being handled [5].

The target of new researches is focused on the obstacles facing the efficient use of Information and Communication Technologies in the process of construction at sites. An example for such is the use of the mobile technologies and RFIDs that connected the offices with building sites. In integrated project management, researches have highlighted the use of nD-modelling and construction simulations in supporting the planning of construction sites and the quality control systems [7, 8].

The new technological advanced brought the 4D visualization models which connected the 3D existing geometrical models with the daily updated scheduling data. Such advancement improved the ability of monitoring the progress at the construction site and brought an assistance during the process of decision-making. Once the elements of construction site are updated timely to the 4D visualization model, the construction site manager/engineer can find solution for dynamic issues in accordance to the real-time situation, and therefore the decision made are more accurate and precise.

Additionally, such decisions could be made faster, which makes the process of decision-making more efficient [9].

Decision support systems are computer-based systems that assist business and organizational in complex decision-making environment. They manage and process structural data to support rather than replace the decision-making process. They act as an aid, especially when it comes to complex situations, but choosing the ultimate solution is always a human choice [10].

The success of an organization as well as a project itself depends on the actors of well-developed interpersonal skills operating in different fields in this hierarchy, to meet different leadership needs. Construction companies have the benefit of hiring individuals with highly developed interpersonal skills, who can easily cope with various forms of communication, when dealing with stakeholders from other levels in the hierarchical system. The use of these skills is influenced by the positive or negative outcome of the project. Successful management is therefore conceived as possessing the imagination, the inter-intelligence, the initiative and the ability to inspire a subordinate [11].

2 Methodology

Qualitative research method is used in this paper, with the aim to sort the factors affecting the decision-making process in construction sites. Total of 13 case studies based in different countries worldwide are analyzed. Factors given in case studies are processed, evaluated, and listed in accordance to their impact on the decision-making process. For each analyzed factor from the case studies, a weight number is given depending on the impact of the factor given in the case study. The weight number is increased every time when the same factor occurs in other case study. The final rank list of factors will be made depending on the two up mentioned criteria: Impact of the factor on the decision-making process in construction site in the case study, and the frequency of occurring in other case studies.

3 Case Studies

According to Xin Ning, Ka-Chi Lam and Mike Chun-Kit Lam, several factors affect the decision-making process at construction site. In their study, they have listed them out as given below [12]:

- Location of the construction site;
- Good space utilization and configuration is one of the primary responsibilities of the construction manager of planning the utilization and configuration within the site;
- Ease of expansion which must be reflected without interference with the environment and environmental disturbance;
- Efficient movement of materials which must be provided within the construction site;

- Tie-in with external transportation generally transport processes constitute integrated activities and links to the movement of materials and passengers;
- Effective movement of personnel where the movement within the site must be ensured;
- Easy supervision and control, the builder shall ensure that the building is designed and constructed in accordance with building regulations and the permit granted for the project which must control the supervision and controller. Since material suppliers to the construction site have a direct impact on the dynamics of the work, it is necessary to pay attention to a special study of suppliers when choosing the one with which to cooperate during the construction of the project.

Ivan Marović and Tomaš Hanak stated that construction projects are, in general case, managed according to the level of detail considered while analyzing and making the decision. The project objectives should be always taken in consideration while making the decision analysis, and while evaluating the impact of the decision on the construction process. They claimed that all decisions made are based on set of assumptions, and therefore the decision-maker is always anticipating potential consequences on the project success. If the project is run entirely by an individual, a critical factor of success should be developed. In the study published in 2017 by Marović and Hanak, the introduced a new model that is used for evaluating the factor affecting the decision-making at sites, and which included the site location analysis. The model was based on both AHP (to define the Hierarchy of the decision-making design) and PROMETHEE (to analyze the impact of the location on the investment). The model proposed by Marović and Hanak brought improvement to the decision-making consistency, as it offered a systematic way to overcome problems occurring due to the poor level of information related to the case. Authors of the model highlighted that stakeholders included in analyzed projects were highly satisfied, as all decision were made carefully and precisely, and considered wide range of factors. They stated that the model is easy adoptable for other projects, locations, and environments. If decision-maker follows the model procedure, the results will be more precise and will be satisfying, claimed by the authors [13].

The location of the site where the construction is planned, according to Jajac, Bilić, and Ajduk, has a major impact on the organization of the work area, the location and design of the facility, construction and layout and the course of construction of the facility. Consequently, they found it easy to conclude that making decisions about the location of the construction is a complex and multicriteria problem. In the research done by Croatian students from the University of Split, it is represented a method for decision-support by choosing the right construction site. They have identified three groups of adequate stakeholders: Construction Experts, Socio-economic experts, and representatives of investors [14].

Construction site layout planning is lauded as one of the critical factors influencing decision making process on construction site. X. Ning, K.C. Lam, M.C.K. Lam presented in the study published in 2011 a system for making construction site decisions about the process of planning the site layout (CLSP – construction layout site planning). The processes made by the system are starting by problem identification and then followed by finding opportunities, solution development, and ends with selecting the option that has the highest effectiveness ratio and implementing it. The system take in consideration all boundaries of the layouts, as well as all other surrounding internal and external factors. The second phase, according to their study, is the identification of the object and the availability of the remaining surface (in some cases unequal surfaces) that is dynamic during construction. The third phase consists of drawing up alternative layout plans for the construction sites that meet the layout and limitation goals, and selecting and complementing the most suitable ones [12]. The system was tested in a case study considering a building of residential type and verified by authors in 2011. They found that, by using the CLSP system, the ratio of occurrence of accidents is dropped for 8.5%, and the costs of handling dropped for 9.1% [12].

The planning process of construction site should include and consider the temporary buildings nature and type (size and location) [15]. Larger projects involve a lot of workforce, subcontractors, construction technology is at a higher level, and if there is no effective workspace planning, it results in increased costs and time wasting [15]. At the same time, the very big role is on the aspect of safety. Considering all these aspects, dynamic layout planning of the site is presented by Dhanure and Pathak with a focus on transport costs, relocation cost, constraint violation cost. Authors opted for a investigating the effectiveness of the proposed model in Pune City in India, where they implemented the model and two types of construction site layouts are presented. The model is based on four sections; Decision variables (position and location of all temporary objects at construction site); Geometric constraints (preventing any physical overlap of any building at the same stage of construction); Objective function (maximizes total construction costs: transport, relocation, constraint violation); and Safety consideration. Dhanure and Pathak results are useful for the process in identification of temporary buildings' optimal positions and for the determination of their orientation. The model developed also is useful when an analysis for site layout decisions' costs is done [15].

Materials movement is an important consideration for an effective project management planning in construction sites. That is why Virtual Construction Materials as a Decision-Making Support for Material Routing was the focus of the research done by J.L. Yang and L. Mahadjoubi. The VCMR is a novel tool to aid site managers and planners to select the best movement of materials in construction site. It uses advanced GIS-fuzzy TECH system to determine certainties and uncertainties during the selection of materials routing. The tool developed by Yang and Mahadjoubi takes into consideration site layout, site access, temporary storage and new building blocks to choose the suitable materials' routing during the construction stage. The research has demonstrated that an appropriate decision support system can reduce uncertainty and help managers to make informed decisions to shift materials in construction sites. Although it is still at its early stage, its preliminary results are promising [16].

Electronic procurement can be defined as a way of delivering goods electronically, most often via the Internet [17]. The main process in deciding on e-procurement is, according to Kar, the process of selecting suppliers. The study by Arpan Kumar Khan adopted criteria for the selection of suppliers that have already been empirically established, and no new methods have been presented yet. A model prepared using Fuzzy logic and considers the criteria and constraints mentioned above was developed by author. Furthermore, the model was updated using neural networks to become a multi-objective decision-making tool. The firm using the model are able to sort the suppliers depending on their suitability. Kar in his paper proposed a classification scheme that uses incomprehensible neural networks to classify suppliers in a scenario of e-procurement. Kar proposed a methodology that focused on the impact and weight of each criterion and the relative impact of all items and their weight factor during the process of classifying the suppliers [17].

As well as the supplier, choosing the best subcontractor in a construction project is a vital process. Incorrect selection can, according to Mohemad, Hamdan, Othman, and Noor, lead to problems in the progress of the business, which includes poor quality of work and delay in delivery. Quantitative research methodology, based on a survey that was conducted by the authors in 2010, with 29 experts (in the role of prime contractor) was used to determine the evaluation of each factor influencing the selection of subcontractors. Surveys conducted proved that there are twelve factors to be considered as important, and they are: Flexibility and cooperation when resolving delays; Reputation; then Failure to comply with the quality specifications; Quality; Suppliers incompetency to deliver materials on time; Failure to complete contract; Physical resources; Tender price; Contractor's difficulty in reimbursement; Flexibility in critical activities; and Safety consciousness on the job site [10]. Making the right decision during the tender process strongly influences, according to Mohemad, Hamdan, Othman, and Noor, the successful implementation of the project. In the study, they stated that tender participation is a complex process and often involves the coordination of many tasks and individuals with different priorities. Decision Support Systems (DSS) play a major role in ordering to ensure confidentiality, greater transparency and healthy competition in tendering [10]. Traditional tendering processes are complex and fragmented. The research done by Mohemad, Hamdan, Othman, and Noor contains the general practice of tendering worldwide (US, Europe, Middle East, Asia). The advantage of decision support has attracted many researchers to implement the tendering system. The ontological extraction framework was proposed, for system of decision-support with the objective of improving the tender evaluation process [10].

It is important to choose an adequate method of construction phase in order to achieve high level of efficiency and productivity, said Ferrada, Seprell, and Skinbniewski. However, they consider that there are cases were the process of selecting the construction method is not done carefully. Qualitative research methodology was used by authors and based on semi-structured interviews included total of three contracting companies. Interviews had an objective to explore the knowledge associated with it and do they have systematic approach. Prototype - a system of knowledge of construction methods was created and tested with experts in the construction industry. They concluded that, by using a systematic approach in the process of selecting the construction method, the dependence on the individual knowledge of project managers is reduced and the project success ratio is improved [18].

According to Stefan Olander and Anne Landin, there is a natural tendency for project stakeholders to influence decision making when developing a project. The aim of their research was to show the factors influencing the stakeholder management process positively or negatively from the perspective of project implementation. Studies of two railway projects in Sweden were compared to analyze these factors. The results showed that the outcomes of stakeholders' management process depend on how the project manager presented the advantages and disadvantages of a construction project. The challenges of a project manager are to implement the project in such a way as to reduce the impact of negative factors while increasing the benefit of stakeholders [19].

During project execution, there are stakeholders who have an indirect influence (e.g., decision makers in project selection) or direct influence (e.g. project team members). As they may have a positive impact on the achievement of the project's objectives, 100 surveys were conducted by Bizon-Gorecka and Gorecki, in the Kuyawian-Pomeranian region, with construction process participants, project managers. The results showed that the investor plays a very important role in the construction investment project. In the opinion of the respondents, investor decisions have the greatest influence on the execution of construction investments; 33% said this. At the time, 26% of respondents claimed that developing the right design was also very important, making the designer of the high influence. 17% of the respondents believed all key aspects of investment success are defined in the construction law and that this is a key determinant of construction investment. Slightly less, 15% of respondents said contractor is making decisions. The smallest number of respondents, namely 9%, chose the answer: others [20].

The quality of design and documentation has, as Tilley, Mcfallan, and Tucker stated, a significant effect on the construction projects regarding the cost of construction, time and work efficiency, and general performance. Due to the Design law level of quality and the project documentation low precise level inefficiency of construction processes is recorded by authors. Such issues have a direct significant impact on project delays, inefficient use of workforce, equipment, materials, and machineries [21]. According to the authors findings, all projects' participants agreed about the issues related to the poor design documentation, and that it is impacting significantly their performances at the construction sites in Australia. They Highlighted that, the most significant negative impact of poor design documentation is on creating additional costs due to delays and interventions that aims to clarify inadequate, impractical, conflicting and unambiguous design documentation and specifications [21].

The research done by Ljevo and Vukomanovicin Bosnia and Herzegovina proved that construction sector has not yet recognized the scientific areas of project management and does not use them in day-to-day processes. Having in mind that construction companies live from the sales and production or provision in the invested projects, Ljevo and Vukomanović focused on the influence of project management on the quality of the final product. The research was completed through surveys and interviews with the participation of construction project investors, engineers and architects (all of them had more than 10 years of experience in the field). Approximately 75 projects were tested. They have found that companies within construction industry in Bosnia and Herzegovina implement the system of phases and sub-phases where they find it more suitable and easier for them, where "successful" project is defined as one that is completed until due-date, into the financial frames and has "satisfactory quality". As they stated, although the legislation does not cover this area, project management methods and techniques are available to participants in the construction industry in B&H and it is up to them to decide on their application [22].

4 Findings and Conclusions

After reviewing case studies and factors affecting the decision-making process at construction sites worldwide, the factors are sorted according to their influence and frequency of occurrence below:

- Construction site layout planning involves identifying, positioning and determining the size of temporary objects on a construction site. Larger projects involve a lot of workforce, subcontractors, construction technology is at a higher level, and if there is no effective workspace planning, it results in increased costs and time wasting.
- Supplier selection process showed as an important link in organization of managing construction project but also as a time-wasting activity.
- Materials management practices at construction site is an important consideration for an effective project management planning in construction sites.
- Subcontractor selection is a vital process in construction project management. Incorrect selection can lead to problems in the progress of the business, which includes poor quality of work and delay in delivery.
- The tender process - Making the right decision during the tender process strongly influences the successful implementation of the project. Tender participation is a complex process and often involves the coordination of many tasks and individuals with different priorities.
- According to case studies analyzed, the outcomes of stakeholder's management process depends on how the project manager presented the advantages and disadvantages of a construction project. The challenges of a project manager are to implement the project in such a way as to reduce the impact of negative factors while increasing the benefit of stakeholders.
- Site location has a great influence on the management of a construction project. Choosing the right location for a construction project is reflected in good access, close to target facilities, represented climate conditions, the possibility of connection to water, electricity and TV supply and many others.
- Appropriate choice of construction methods to be used in the execution of a construction project is an important determinant of high productivity, but sometimes this selection process is performed without the care.
- The level of Design documentation and drawing quality has a significant impact on the effectiveness of construction project, as well as on the decision-making process at site. If the quality is not adequate, the decision maker is forced to intervene in order to clarify details and missing (or inadequate) parts of the design, which increases the cost of the project due to the loss of time and the changes in decision depending on the methods used for the construction.

References

1. Scherer, R.J., Schapke, S.E.: A distributed multi-model-based management information system for simulation and decision-making on construction projects. Adv. Eng. Inform. **25** (4), 582–599 (2011)

2. Drucker, P.F.: The Effective Decision. Harvard University, Graduate School of Business Administration (1967)
3. Kerzner, H.: Project Management: A Systems Approach to Planning, Scheduling and Controlling, 10th edn. Wiley, Hoboken (2017)
4. March, J.G.: A Primer on Decision Making: How Decisions Happen. Simon & Schuster, New York (1994)
5. Szafranko, E.: Decision problems in management of construction projects. IOP Conf. Ser. Mater. Sci. Eng. **251**, 012048 (2017)
6. Poon, S., Price, A.D.F.: Decisions Made on Construction Sites. In: 15th Annual ARCOM Conference, Liverpool (1999)
7. Aquad, G., Lee, A., Wu, S.: Constructing the Future: nD Modelling. Taylor & Francis, London (2007)
8. Hartmann, T., Gao, J., Fischer, M.: Areas of application for 3D and 4D models on construction projects. J. Constr. Eng. Manag. **136**(8), 931–934 (2008)
9. Chau, K.W., Anson, M., Zhang, J.P.: Four-dimensional visualization of construction scheduling and site utilization. J. Constr. Eng. Manag. **130**(4), 598–606 (2004)
10. Mohemad, R., Hamdan, A.R., Othman, Z.A., Noor, N.M.: Decision support systems (DSS) in construction tendering processes. Int. J. Comput. Sci. Issues (IJCSI) **7**(2), 35–45 (2010)
11. Yung, Y., Jeong, M.G., Mills, T.: Identifying the preferred leadership style for managerial position of construction management. Int. J. Constr. Eng. Manag. **3**(2), 47–56 (2014)
12. Ning, X., Lam, K.-C., Lam, M.C.-K.: A decision-making system for construction site layout planning. Autom. Constr. **20**(4), 459–473 (2011)
13. Hanak, T., Marović, I.: Selection of adequate site location during early stages of construction project management: a multi-criteria decision analysis approach. IOP Conf. Ser. Mater. Sci. Eng. **251**, 012044 (2017)
14. Jajac, N., Bilić, I., Ajduk, A.: Decision support concept to management of construction projects-problem of construction site selection. Croatian Oper. Res. Rev. **4**(1), 235–246 (2013)
15. Dhanure, Y., Pathak, S.: Optimization of site layout planning for multiple construction stages with safety consideration and requirements. Int. Res. J. Eng. Technol. **3**(5), 3173–3176 (2016)
16. Mahdjoubi, L., Yang, J.: An intelligent materials routing system on complex construction sites. J. Logistics Inf. Manag. **14**(5/6), 337–344 (2001)
17. Kar, A.K.: Modeling of supplier selection in e-procurement as a multi-criteria decision making problem. Sprouts Working Pap. Inf. Syst. **9**(40), 9–40 (2009)
18. Ferrada, X., Serpell, A., Skibniewski, M.: Selection of construction methods: a knowledge-based approach. Sci. World J. **2013**, 23 (2013)
19. Olander, S., Landin, A.: A comparative study of factors affecting the external stakeholder management process. Constr. Manag. Econ. **26**(6), 553–561 (2008)
20. Bizon-Gorecka, J., Gorecki, J.: Influence of selected stakeholders of construction investment projects on the course of project. IOP Conf. Ser. Mater. Sci. Eng. **245**, 072018 (2017)
21. Tilley, P., Mcfallan, S., Tucker, S.: Design and documentation quality and its impact on the construction process. In: Customer Satisfaction: A Focus for Research and Practice in Construction. CIB W55 & W65 Joint Triennial Symposium, Cape Town (1999)
22. Ljevo, Ž., Vukomanović, M.: Characteristic project management model found in construction companies of Bosnia and Herzegovina. Technical Gazette **20**(3651), 689–696 (2013)

Economics, E-Business, Entrepreneurships

Techno-Entrepreneurship: Interdisciplinary Curricula Challenges for 4IR in Bosnia and Herzegovina

Munira Šestić[1(✉)], Elvir Čizmić[1], Mirha Bičo Ćar[1], and Dženana Hodžić[2]

[1] School of Economics and Business, University of Sarajevo, Trg oslobođenja 1, 71000 Sarajevo, Bosnia and Herzegovina
munira.sestic@efsa.unsa.ba
[2] Committee on Education, Science, Culture and Sport, Parliament of the Federation of Bosnia and Herzegovina, Sarajevo, Bosnia and Herzegovina

Abstract. Techno-entrepreneurship includes both entrepreneurial and intrapreneurial activities in which entrepreneur recognises ideas related to innovation and technology development, finding ways to its market implementation and verification in technology-intensive environments. Therefore academic entrepreneurs in Bosnia and Herzegovina are facing the urge to create curricula that would enable potential future techno-entrepreneurs to acquire interdisciplinary knowledge and skills that would enable them to market their ideas more effectively in the context of the Fourth Industrial Revolution (4IR). In this article we analyse the presence of the subject of Techno-entrepreneurship in Economic and Non-economic studies at the universities of Bosnia and Herzegovina and widely suggest guidelines and recommendations for creating its interdisciplinary curricula for the aforementioned studies following contemporary world academic practices. The aim is to domesticate the term of techno-entrepreneurship in the educational offer of the universities in Bosnia and Herzegovina, responding to the demands of modern educational requirements and rhythms brought by the Fourth Industrial Revolution.

Keywords: Techno-entrepreneurship · The Fourth Industrial Revolution · Curricula · Interdisciplinarity · Bosnia and Herzegovina

1 Introduction

Ever since it was first made public at the Hannover Messe in 2011, the concept of the Fourth Industrial Revolution (4IR) - Industry 4.0 has been promoted as a new German industry development strategy. The term revolution refers to a number of major changes in technology, processes and working modes that are significantly affecting people's lives and changing both social and economic relations and consequently the education system as well. Nowadays the term 4IR is accepted and used as the global concept of digital transformations of the industry and itindicates new industrial future of the humanity.

I. Karabegović (Ed.): NT 2020, LNNS 128, pp. 899–905, 2020.
https://doi.org/10.1007/978-3-030-46817-0_101

Due to the challenges of the 4IR context, which is characterised by globalization, time compression and continuously growing technological complexity of innovations, companies have to base their business competitiveness on techno-entrepreneurship and technology innovation [1]. Techno-entrepreneurs are the basis of the future development, so they require interdisciplinary knowledge and skills that would enable them to market their ideas more effectively, finding ways to its market implementation and verification in technology-intensive environments.

However, it should be noted that 4IR in today's world it mainly characterizes the reality of advanced and developed economies, while it is a still only a vision of a desired future and progress for the rest of the world. Therefore, it is the task of academic entrepreneurs in less developed countries to create the ground for the creation of new educational contents which will enable these countries, their economies and their human resources "to walk together" with more developed part of the world into the future brought by the 4IR.

2 Techno-Entrepreneurship

Techno-entrepreneurship is an entrepreneurship based on transformative new technology that emerges in the form of innovation or scientific discovery - it can be applied to a new product or service, thus significantly changing both the country and the region and the ultimately the world [2–5]. Techno-entrepreneurial activities can take place, not just in new emerging ventures, but also in existing companies [6, 7].

No matter which term is used, technology (based) entrepreneurship or techno-entrepreneurship [8] is recognised as one of the most important sources of new value creation and development [7, 9]. The most advanced economies have been encouraging the formation and growth of technology - based businesses [10, 11] and their policy makers recognised these firms as the most dynamic part of their entrepreneurial systems [12, 13] in the future to come.

2.1 Upgrading Triple to Quadruple Helix Concept

Academic entrepreneurship is integrated into economic development mission of the universities [14]. The Triple Helix Model is a term that represents the relationship between universities, industry and government for new knowledge production, transfer and application [15–22]. It is a base for new knowledge development and overall economic development through new institutional and social frameworks in the 4IR world. The University is a central concept to the Triple Helix Model with the capacity to provide students with new ideas, skills and entrepreneurial talent. Students should be taught to put their knowledge in use as well as thought to create new knowledge. Students can also be inter-disciplinary trained (and encouraged) to become entrepreneurs and start-up founders, contributing to economic growth and job creation.

The EU Entrepreneurship Action Plan 2020 [23] insists on entrepreneurial education and reaching out to the groups of students whose entrepreneurial potential has not been encouraged or were encouraged to a small extent (such as engineers) and the creation of an environment that encourages entrepreneurship in educational institutions (environment of entrepreneurial incentives for students). Scientists and engineers cannot just rely on their technological knowledge - it is also important for them to understand how technology can be brought successfully to the market through commercialization [24–27] (Fig. 1).

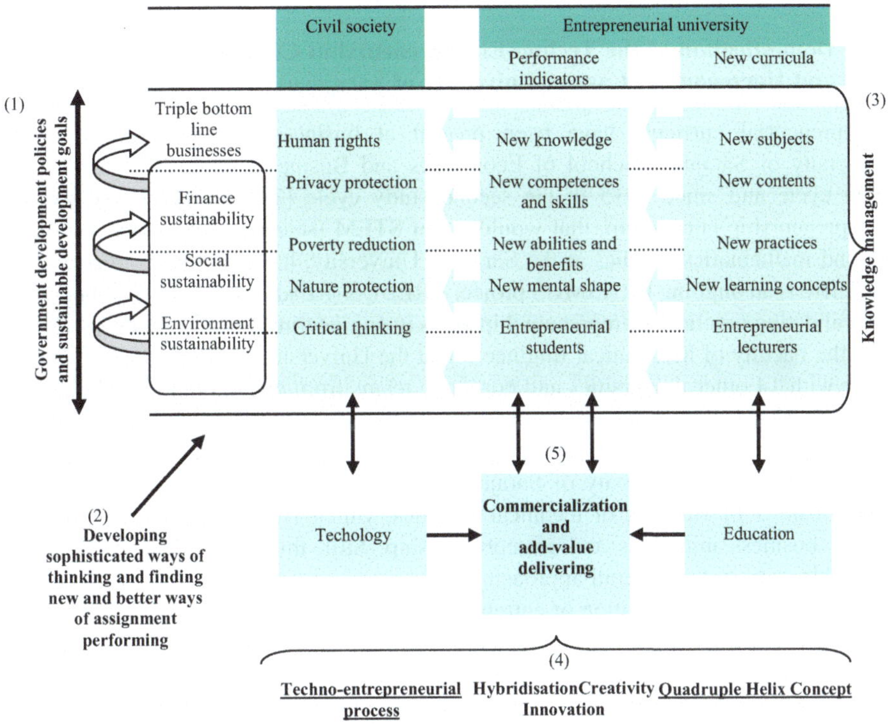

Fig. 1. Techno-entrepreneurial process and Quadruple Helix Concept [adapted and upgraded from 27]

Therefore intensive changes of the 4IR impel academic entrepreneurs to speak today not only about the Triple Helix Model but about the Quadruple Helix System, which besides others elements includes the civil society and social entrepreneurship as well. It means that the Quadruple Helix System and techno-entrepreneurial process should be understood as a hybrid framework of institutional and business development where higher education institutions must play highly important role. Therefore university professors must be simultaneously: teachers, scientists, project managers, coaches, entrepreneurs, researchers, leaders, motivators and inspirers, so they could generate new multidisciplinary curricula solutions to shape future young people that will undertake all societal, economic and technological changes with respecting total quality management, triple bottom line and sustainable development's goal requirements.

The earlier model could be used as a platform to understand the new big picture of the future development. The Quadruple Helix Model is taking into consideration a new development platform of the modern society's needs and has the potential to fulfill requirements of all stakeholders in the global context. This can be used as a starting point for a practical way in which the nation can develop new development strategies for the 4IR and speed up the process itself.

2.2 Domestication of the Techno-Entrepreneurship Curricula in Bosnia and Herzegovina (Case of University of Sarajevo)

Entrepreneurial curricula have been present at business oriented studies of the University of Sarajevo (School of Economics and Business) since 2002 in the first study cycle and since 2015 in the second study cycle [28]. However, the techno-entrepreneurship curriculum, that would target STEM (science, technology, engineering and mathematics) studies at the Sarajevo University, has not been confirmed.

There is an ongoing ERASMUS project – REBUS (REady for BUSiness-Integrating and validating practical entrepreneurship skills in engineering and ICT studies) [29, 30] with the Faculty of Mechanical Engineering at the University of Sarajevo participating along with 14 other universities and corporate teams from eight countries. The project goals are to develop innovative, entrepreneurial oriented training modules with the intention of their integration into existing curricula of the Faculty of Mechanical Engineering at the University of Sarajevo. The overall project aim is to change the consciousness of students of technical faculties, which is recognised as resistant to private business initiatives and entrepreneurship. Still, this is an isolated and commendable case, not an overall approach.

Aware of the long tradition of entrepreneurial education at the Sarajevo University, there is a question of the possibilities/opportunities for developing of an interdisciplinary approach that would take advantage of the basis and tradition of entrepreneurial education at the School of Economics and Business and the contemporary needs for incorporating the techno-entrepreneurship curricula into STEM study offerings at the University of Sarajevo.

Developing and incorporating the techno-entrepreneurship curricula into existing STEM studies at the University of Sarajevo would provide the first step for domestication of the term of techno-entrepreneurship itself and would create the basis for the continued activities of academic entrepreneurs in their economic development mission direction in Bosnia and Herzegovina.

2.3 Contextualisation of the Techno-Entrepreneurship Curricula in Bosnia and Herzegovina

Yet there are serious challenges for the creators and implementors of the future techno-entrepreneurship curricula in Bosnia and Herzegovina. Challenges primarily relate to the fact that Bosnia and Herzegovina is a transitional and post-conflict environment and that this dual context has a significant impact on the success of entrepreneurial ventures [31]. Furthermore, relevant international studies [32–34] shows that state administration of Bosnia and Herzegovina does not provide adequate and sufficient infrastructural support to either current or future entrepreneurs.

It is necessary to incorporate this knowledge into techno-entrepreneurship curricula and contextualise them for optimal results. Copying and implementing academic practices from contexts that are not imminent or by indicators similar to Bosnia and Herzegovina will have no guarantee that the expected results will be achieved. Academic entrepreneurship and university support is the core to provide clear direction, framework, policy, process, system, promotion and assessment of performance for techno-entrepreneurship [35] for needed university education transformation in Bosnia and Herzegovina for the 4IR context requirements.

3 Conclusion

The article tried to identify the challenges of techno-entrepreneurial curricula in Bosnia and Herzegovina in the future to come, but also to point out possible ways to overcome them. The goal is involvement of Bosnia and Herzegovina in global educational and developmental trends. The challenges of the initial and systematic introduction of the techno-entrepreneurial curricula need to be addressed by following the Triple and Quadruple Helix Model that identifies key stakeholders in the process and directs their roles for successful implementation. In this systematic and comprehensive manner, it is possible to ensure both increased competitiveness and development opportunities of the national economy through techno-entrepreneurship in the context of the 4IR.

There are challenges regarding the need for the interdisciplinarity of techno-entrepreneurial curricula. One should bear in mind the existence of the entrepreneurial education tradition at the University of Sarajevo. The experiences and contributions of business studies can best be utilized through an interdisciplinary co-working approach with STEM studies - bringing together all existing and future teaching, project and research experiences in the field of entrepreneurship.

Finally, the contextualisation of the techno-entrepreneurship curricula is necessary to achieve better results relevant to the context of Bosnia and Herzegovina. Only through the contextualisation of techno-entrepreneurial curricula it would be possible to provide an additional effect through the context feedback - in terms of the continuous improvement of entrepreneurial infrastructure and support. By identifying and addressing the challenges of techno-entrepreneurial curricula we could expect optimal results through adequate response to the modern educational requirements and rhythms brought by the Fourth Industrial Revolution.

References

1. Narayanan, V.K.: Managing Technology and Innovation for Competitive Advantage. Prentice Hall, Upper Saddle River (2001)
2. Petti, C. (ed.): Cases in the Technological Entrepreneurship – Converting Ideas Into Value. Edward Elgar Publishing, North-Hampton (2009)

3. Bailetti, T.: Technology entrepreneurship: overview, definition and distinctive aspects. Technol. Innov. Manag. Rev. **2**(2), 5–12 (2012)
4. Byers, T.H., Dorf, R.C., Nelson, A.J.: Technology Ventures: From Idea to Enterprise, 4th edn. McGraw Hill Education, New York (2014)
5. Badzinska, E.: The concept of technological entrepreneurship: the example of business implementation. Entrepreneurial Bus. Econ. Rev. **4**(3), 57–72 (2016)
6. Ratinho, T., Harms, R., Walsh, S.: Structuring the technology entrepreneurship publication landscape: making sense out of chaos. Technol. Forecast. Soc. Chang. **100**, 168–175 (2015)
7. Runge, W.: Technology entrepreneurship, a treatise on entrepreneurs and entrepreneurship for and in technology ventures. Karlsruher Institut für Technologie (KIT), vol. 2, Karlsruhe, Germany (2014)
8. Petti, C. (ed.): Technological Entrepreneurship in China: How Does it Work?. Edward Elgar Publishing, North-Hampton (2012)
9. Blanco, S.: How techno-entrepreneurs build a potentially exciting future? In: Therin, F. (ed.) Handbook on Research on Techno-Entrepreneurship, 1st edn. Edward Elgar, UK, USA, pp. 3–24 (2007)
10. Autio, E.: 'Atomistic' and 'systemic' approaches to research on new, technology-based firms: a literature review. Small Bus. Econ. **9**(3), 195–209 (1997)
11. World Economic Forum: Global Entrepreneurship and the Successful Growth Strategies of Early-Stage Companies, A World Economic Forum Report in collaboration with Stanford University, Graduate School of Business, SPRIE and STVP, World Economic Forum, Geneva (2011)
12. OECD: Enhancing the Competitiveness of SMEs Through Innovation: Strategies and Policies. Organisational for Economic Co-operation and Development (OECD), Bologna, Italy (2000)
13. Mason, C., Brown, R.: Entrepreneurial Ecosystems and Growth-Oriented Entrepreneurship, Background paper, OECD LEED Programme, Hague, Netherlands (2014)
14. Siegel, D.S., Wright, M.: Academic entrepreneurship: time for a rethink? Enterprise Research Centre, Research Paper No. 32, Coventry, Birmingham, UK (2015)
15. Etzkowitz, H.: Technology transfer: the second academic revolution. Technol. Access Rep. **6**, 7–9 (1993)
16. Etzkowitz, H., Leydesdorff, L.: The Triple Helix: University - Industry - Government Relations: A Laboratory for Knowledge-Based Economic Development. EASST Rev. **14**, 14–19 (1995)
17. Etzkowitz, H., Leydesdorff, L.: The endless transition: a "triple helix" of University Industry-Government Relations. Minerva **36**, 203–208 (1998)
18. Etzkowitz, H., Leydesdorff, L.: The dynamics of innovation: from national systems and "Mode 2" to a Triple Helix of University-Industry-Government Relations. Res. Policy **29**, 109–123 (2000)
19. Leydesdorff, L., Etzkowitz, H.: Emergence of a Triple Helix of University-Industry Government Relations. Sci. Public Policy **23**, 279–286 (1996)
20. Leydesdorff, L., Etzkowitz, H.: The Triple Helix as a model for innovation studies. Sci. Public Policy **25**, 195–203 (1998)
21. Etzkowitz, H.: MIT and the Rise of Entrepreneurial Science. Routledge, London (2002)
22. Rangaa, M., Etzkowitz, H.: Triple Helix Systems: an analytical framework for innovation policy and practice in the knowledge society. Ind. High. Educ. **27**(4), 237–262 (2013)
23. European Commission: EU Entrepreneurship Action Plan 2020 (2013). https://ec.europa.eu/growth/smes/promoting-entrepreneurship/action-plan_en. Accessed 04 Feb 2019

24. Barr, S.H., Baker, T., Markham, S.K., Kingon, A.I.: Bridging the valley of death: lessons learned from 14 years of commercialization of technology education. Acad. Manag. Learn. Educ. **8**(3), 370–388 (2009)
25. Zappe, S.E., Hochstedt, K., Kisenwether, E., Shartrand, A.: Teaching to innovate: beliefs and perceptions of instructors who teach entrepreneurship to engineering students. Int. J. Eng. Educ. **29**(1), 45–62 (2013)
26. Lynch, M.P.J., Kamovich, U., Longva, K.K., Steinert, M.: Combining technology and entrepreneurial education through design thinking: students' reflections on the learning process, Technology Forecasting and Social Change (2019). https://doi.org/10.1016/j.techfore.2019.06.015
27. Čizmić, E., Trgo, A.: Savremeni menadžerski koncepti-contemporary managerial concepts, Faculty/School of Economics and Business Sarajevo, Sarajevo University (2010)
28. Bičo Ćar, M., Šestić, M., Softić, S., Stupar, S.: Entrepreneurial education at universities in Bosnia and Herzegovina. Obrazovanje za poduzetništvo – E4E: znanstveno stručni časopis o obrazovanju za poduzetništvo **9**(2), 101–118 (2019)
29. REBUS - Ready for BUSiness – Integrating and validating practical entrepreneurship skills in engineering and ICT. https://www.rebusproject.net/index.php. Accessed 12 Dec 2019
30. Univerzitet u Sarajevu: Ready for BUSiness – Integrating and validating practical entrepreneurship skills in engineering and ICT (2015). https://www.unsa.ba/projekti/rebus-ready-business-integrating-and-validating-practical-entrepreneurship-skills. Accessed 12 Dec 2019
31. Čizmić, E., Šestić, M., Softić, S.: Ključni faktori uspješnosti ženskog preduzetništva u kontekstu tranzicijskog i post-konfliktnog okruženja u Bosni i Hercegovini. Zbornik radova/Sarajevo Bus. Econ. Rev. **37** (2019, accepted, print preparation)
32. Global Entrepreneurship Monitor: World Report 2016/17 (2017). https://www.gemconsortium.org/report/gem-2016-2017-global-report. Accessed 10 May 2018
33. World Bank: Doing Business 2020: Comparing Business Regulation in 190 Economies - Economy Profile of Bosnia Herzegovina (English). Doing Business 2020. World Bank Group, Washington, D.C. (2019)
34. Schwab, K. (ed.): World Economic Forum: The Global Competitiveness Report 2019 (2019). http://www3.weforum.org/docs/WEF_TheGlobalCompetitivenessReport2019.pdf. Accessed 08 Dec 2019
35. Yusof, M., Siddiq, M.S., Nor, L.M.: The role of academicians in Technology Entrepreneurship. In: Therin, F. (ed.) Handbook on Research on Techno-Entrepreneurship, 2nd edn., Edward Elgar, UK, USA, pp. 195–210 (2013)

Mechanism of Stimulate the Growth of Highly Competitive Technology Business

Yuriy A. Doroshenko[1(✉)], Ludmila A. Minaeva[1], Irina V. Somina[1], and Nataliia N. Glagoleva[2]

[1] Belgorod State Technological University named after V.G. Shukhov (BSTU), Ulitsa Kostyukova 46, 308012 Belgorod, Russia
549709@mail.ru

[2] Belgorod University of Cooperation, Economics and Law, 308012 Belgorod, Russia

Abstract. The article discusses the methodological approaches to the definition of the concepts of "high-tech business", "high technology", "information technology" and others. The main components of a highly competitive technological business are revealed: high technology, science output, and the share of people employed in R&D. The influence of investments in human capital on the development of high-tech business is shown. The main problems were revealed: the lack of highly qualified personnel and insufficient motivation of employees, lack of activity in the implementation of information technologies in production processes; insufficient efficiency of implementation of state programs in the field of science and innovations, etc. Recommendations are given to stimulate the growth of a highly competitive technology business.

Keywords: High-tech business · High technology · Information technology (IT) · Competitive technological business

1 Introduction

The operation of the enterprise is carried out by using various technologies with the help of which products or services are created that can satisfy the needs of customers. At the same time to meet the new needs of new technologies are needed. Russia, participating in global world processes, is focused on the innovative development of its economy, a departure from its resource component. In accordance with the Innovation Development Strategy, the main elements of the innovation system are formed. Measures taken to develop high-tech institutions, the venture capital market, and intellectual property are crucial for stimulating the growth of the country's economy.

One of the step of the implementation of the course on the innovative development of the country was the Strategy of the scientific and technological development of the Russian Federation. Its necessity is due to a number of external and internal factors. Internal factors are basically based on the impossibility of economic growth due to the exploitation of raw materials. External factors point to global challenges: active development in the leading countries of the latest technologies, increased anthropogenic pressures on the environment, which affects the socio-economic situation in

I. Karabegović (Ed.): NT 2020, LNNS 128, pp. 906–915, 2020.
https://doi.org/10.1007/978-3-030-46817-0_102

the world. In addition, international instability, conflict between countries makes it necessary to pay special attention to national security.

State policy aimed to enhancing scientific and technological development does not lead to quick and obvious results. This is largely due to the fact that the fundamental discoveries and their commercialization are perspective-oriented and have a certain time lag. Nevertheless, constant investments, stimulating the realization of people's creative abilities and increasing production, become the foundation of knowledge, innovation, contribute to the development of the global economy.

For the successful functioning of Russian enterprises in the conditions of tough global competition, the primary role is taken by innovations, on the basis of which the preventive development of the intellectual, technical and technological components of the enterprise spheres is ensured. Innovation activity provides an opportunity to improve all areas of business and, as a rule, is the basis for the accumulation of intellectual capital.

In today's rapidly changing world, the engine of global economic development is becoming a non-standard business idea, an innovative and high-tech business. According to experts, "at the moment the volume of the sector of high-tech products in monetary terms is about 3 trillion. dollars. By 2020, this indicator is expected to grow by at least a whole order of magnitude, which indicates a high investment attractiveness and significant development potential of this sector" [1].

Innovation and high-tech business in the world is developing in different ways. Japan, China, and South Korea are considered innovative giants. Recently, Singapore, Indonesia, the Philippines, Malaysia and Vietnam have been actively moving towards the indicators of these countries. These countries are constantly improving their innovation ecosystems, improving their results in the field of education, research and development (R&D), increasing productivity, including due to the release of high-tech products.

Nowadays Russia's competitiveness depends on its ability to accelerate the development of high-tech industries. Moreover, acceleration should be carried out both in the direction of stimulating the consumption of high-tech products by Russian enterprises, and in the direction of stimulating the production of innovative products and services.

The situation in the high-tech sector of the Russian economy is characterized by the presence of individual measures for the development of innovations, but they are not connected into a single system, there is no coordination and motion vectors, which does not allow for a synergistic effect. The mechanism of socio-economic, structural, organizational, fiscal, informational, advisory, administrative support and growth promotion of high-tech companies has not been fully developed. The "open innovation" model, based on: "the use of intellectual capital, technology flows, knowledge, close cooperation of partners with each other, staff qualification and mobility, flexibility of doing business, cooperation with external counterparties, etc., is formally implemented, but its results are not obvious [2].

2 Materials and Methods

The meaning of "high technology" appeared relatively recently - in the late 80s of the XX century. Initially, high technology correlated with some of the newest industries, in which the unit costs of research and development were quite large. Subsequently, the concept of "high technology" was added by several similar in its essence terms: "high technologies", "high-tech industries", "high-tech industries", "innovative companies". Legislatively, the concept of "high technology" was identified in the consolidated analytical report of the Accounts Chamber of the Russian Federation "The main problems and conditions for the effective reproduction of the domestic scientific potential" in 2003 [3].

In relation to organizations, high technologies are interpreted more broadly and extend both to the industries producing products and the service industries. At its core, high technologies are part of all sectors of the economy and tend to actively develop as an independent sub-industry.

A.A. Laptev and other scientists refer the industry and companies that operate in it to high-tech if they meet certain criteria. First, it is the level of knowledge-intensiveness, i.e. the share of research and development costs in the total costs of production and sales of products (services). If this share exceeds 8.5%, the company can be considered as having a high level of knowledge intensity. Such companies are leading [3]. The share of costs corresponding to 2.5% indicates the average level of knowledge-intensiveness. If the cost of research and development is 0.5% - the level of knowledge-intensiveness of such an enterprise is considered low. This classification of levels of knowledge-intensiveness is not universal. In such developed countries as Germany, the level of high technology intensity includes enterprises in which the share of R&D costs exceeds 8.5% of the volume of output (services).

In the scientific literature there is an opinion that the assignment of the level of expenditure on research and product development, although an obligatory, but insufficient criterion, since does not take into account human capital. A.A. Shpolyanskaya, emphasizing the importance of human capital in high-tech companies, believes that R&D costs will be useless if attention is not paid to knowledge, experience, and creative abilities of the staff [4]. Secondly, the company's high-tech is characterized by scientific expertise. Knowledge quotation indicates the ratio of sales of high-tech products to the volume of R&D expenditures in a certain period of time. Evaluation of the efficiency of science output is determined by the relative growth in sales of new high-tech products in comparison with the growth of the entire knowledge-intensive market. The scope of the market also includes products that were previously developed but not yet sold. Nowadays in developed countries, the industry is divided into four categories:

- extraction and primary processing of raw materials;
- heavy (material-intensive/labor-intensive industry);
- high-tech industries (a high proportion of R&D costs, but relatively low material consumption and labor intensity);
- The industry of soft technologies (software development, consulting, education, integration systems, etc.).

Thirdly, high-tech companies differ from others in the industry by the proportion of people employed in R&D. The share of those employed in R&D gives a characteristic of the level of technology development. For the industry, a high technology index and an innovation index are calculated. Calculations are conducted annually in order to identify the dynamics of technology development.

The high-tech index shows the share of high-tech companies in the region in the volume of national production in industries, as well as the share of high-tech enterprises in the GRP of a certain region in relation to the share of high-tech industries in the country's GDP.

The innovation index reflects the number of registered patents per capita. The index shows the potential commercialization of achievements in research activities, the direction of technology development.

Sometimes the meaning of "high-tech company" is associated with the abbreviation of high-tech (high-tech), as well as with such terms as "breakthrough technologies", "key competencies", "strategically important technologies". In any case, these definitions reflect the innovative focus of business organization. A distinctive feature of high-tech companies is their versatile orientation, i.e. in addition to the interdisciplinary nature of their activities, they have specific unique features.

The high-tech companies, in the opinion of I.V. Tatarintseva and A. Vasin, are reflected in the novelty of the products, the technological complexity of their production, the use of the latest technologies and the relatively short life cycle [5].

V.S. Vostrikov, seeing in high-tech companies a factor in the development of the domestic economy, points to such a property as the rapid change in products and technologies that stimulate research and generate the cumulative effect of innovation [2]. A high-tech company is a specific business organization based on the pooling of resources with the goal of developing, producing and commercializing a multifunctional, knowledge-intensive, containing product innovation with an optimal combination of economic benefits and risks. Companies which using high technology operate in various sectors of the Russian economy. In accordance with the Order of the Government of the Russian Federation of 2009. and have not lost their relevance in 2019. The high-tech sectors of the economy include: "the aviation and rocket and space industry, shipbuilding, radioelectronic industry, nuclear power industry, power engineering, information and communication technologies" [6]. These industries have certain competitive advantages and their development is the basis of long-term federal government programs. These sectors of the economy, as participants in international competition, cannot develop outside the attraction of foreign partners, investments, technologies. However, at the state level, the goal is to create independent national enterprises to support the domestic high-tech business and to bring high-tech products to the external market.

3 Results and Discussion

The political decision on the transformation of the Russian economy from energy to innovation is at the core of the Strategy for Russia's Socio-Economic Development up to 2020. Important tasks have been set: an increase in the number of organizations implementing technological innovations, the creation of conditions for the active implementation of information technologies (IT) both in the economy and in the social sphere. The solution of these tasks is due to the adoption of a set of measures to increase the level of investment in human capital, build a national innovation system, support new competitive sectors in high-tech areas, etc. [7].

The necessary to increase investment in human capital proves the increased demand for highly qualified specialists with higher education and specialists of average qualification. Requirements not only for highly specialized knowledge, but also for the presence of competences in natural, economic, and technical fields, are imposed on professionals serving the creation and production of high-tech products.

The development of high-tech business is hampered by the current shortage of personnel of all skill levels in technical areas. The situation has arisen that young cadres are not focused on technical professions, and cadres of the former vocational training are aging. This problem is typical not only for Russia, but also for the countries of Central and Eastern Europe. Even in the presence of a full-fledged material and financial base, the lack of personnel with the necessary qualifications does not allow for technological renewal of production, introduction of product innovations.

The solution of this problem is impossible without the active participation of the state. Innovative development of the economy should be based on the national doctrine, which is based on guidelines for the formation of human resources with a modern level and quality of education and focus on the requirements of high-tech industries. Transformation of approaches and methods of training specialists involves their involvement in innovation, the use of information and experience in the innovation sphere, the exchange of knowledge, as well as the formation of continuity in research [8].

Important for a high-tech business is the creation of an adequate intra-corporate mechanism for motivating employees to the required quality level. The use of the innovative potential of workers is realized with the help of various instruments of material and moral encouragement. Moreover, not only innovation itself is encouraged, but also innovation proses, which is associated with the commercialization of the results of intellectual inquiry. For employees of high-tech companies, along with material incentives are important: awareness of their importance in the company, involvement in the results of its activities, career opportunities.

Russia has a high innovative potential, but it is not fully appreciated and is not used enough by business. To build a full-fledged innovation system, it is necessary to use the research potential not only in big business, but also on average and small. The first priority should be to change the attitude to intellectual property, to patenting innovation, disseminating information and experience.

In the economies of developed countries, startups in small business are successfully operating. Moreover, startups for sale to large businesses are characterized by low costs for the creation of mobile and Internet offers. Spending risk capital in the amount of not more than 1 million dollars, the startup is subsequently sold to large companies for 5–50 million dollars. Large business, founding and developing its own start-ups, improves and creates new versions of the basic product, i.e. uses the so-called stabilizing innovation. Revolutionary innovations are created in the field of high technologies and, as a rule, are located in large technological centers. In the United States, this is the Silicon Valley, in Finland, this is the city of Helsinki, and in Russia, Skolkovo. As a result, the owners of startups (if successful) receive a business with sales of hundreds of millions of dollars.

Creation of a national innovation system should take into account the experience of foreign countries in identifying and supporting drivers for the development of high-tech sectors. For example, in Finland, which is one of the leaders in creating effective ecosystems, most start-ups are concentrated in the metropolitan area and are actively supported by the state in direct interaction with higher education institutions. In addition, the Finnish high-tech business is focused on the global market.

A start-up ecosystem is formed "at the expense of specialists, start-ups at various stages of their development and physically represented and/or virtual organizations interacting as a single system for creating new start-up companies" [1]. The adequate support of start-ups in Finland from the state allows them to grow the number is not only in the field of information technology, but also in the field of health - the most growing market for high-tech products. An integrated approach to support and stimulate high-tech business can be provided if there is a platform - information technologies that will function in all areas of the economy. IT plays a dual role: on the one hand, it is an independent industry, elements of which are included in all spheres of economic development and society, on the other hand, it is a tool for modernization for all industries [9].

The IT industry is not only the production of computer hardware and software, but also the active use of the Internet, the development and implementation of information security systems, automation of enterprise management systems, management of interaction with suppliers and consumers of the company's products.

The active implementation of IT in the national economy will contribute to:

- increase in the number of highly qualified personnel;
- elaboration of legislation in terms of securing copyright, development of patent law;
- a set of government support measures (subsidies, tax incentives, conditions for the participation of private investment, etc.).

Nowadays the IT sector is undergoing changes due to the fact that digitization is "directed" to the area of material production. Along with the solution of production management issues, the problems of using IT processes in the production itself become topical. In turn, "the dramatic expansion of the IT sphere opens up new opportunities for the IT business - in the form of rapid advancement in various industrial sectors, relying on serious investments from leading economic players" [10].

The improvement of various technological processes is facilitated by the use of system IT integrators, whose functions include: selecting, installing and maintaining the latest equipment, developing software to improve business [11]. According to M. Kozhevnikov, - the general director of Open Technologies in the IT industry is actively developing systems integration, which includes "solving a wide range of tasks: building an IT infrastructure, implementing ERP and CRM systems, service support and much more" [12]. There is a tendency of convergence of system integrators and consultants towards solving business problems of customers, offering them alternatives to improve business based on IT knowledge. System integrators offer management consulting services.

Moreover, as the IT company develops, more and more attention is paid to, because their use increases the manageability of the company and its transparency. If IT becomes a natural phenomenon, they are outsourced. The degree of importance of IT in different industries varies. In telecommunications IT companies occupy an important place, because their activities are related to technology; for raw material companies, the importance of IT is relatively small due to their reliance on natural resources.

The main factors promoting the growth of companies based on integrators of highly competitive technology businesses are:

- availability of a sustainable consumer base, long-term partnership of system integrators with customers;
- creation of conditions for continuous development, bringing the previously unrepresented new technologies to the market by the system integrator;
- investing in personnel, raising the qualification potential of engineers and managers [2].

Changes in the geopolitical situation and the acceleration of import substitution necessitated the support of new high-tech sectors of the economy, in the audit of the factors of ineffectiveness of the implementation of government programs in the field of science and innovation. Currently, the main problems of high-tech business are systemic in nature, are in the financial, legislative, personnel, organizational areas. Certain measures are being taken, but their effectiveness is insignificant. The decisions made are often of a formal nature and incline to the imitation control model.

According to V.S. Vostrikov, stimulating the growth of high-tech business should be based on the creation of a single body of state management of the scientific and technical complex of the country. The bottom-up initiative, when high-tech companies "develop and offer products that, according to their calculations, are promising for the market and important for the country, should at the same time be directed from above based on, for example, the national roadmap, which brings together setting a clear vision for innovative counterparties of the prospects of civilian high-tech production" [12].

One of the limiting factors for the development of high-tech business is the limited demand for high-tech products. In recent years, the state has paid more attention to supporting innovation, whereas it has not fully stimulated the need for domestic innovative products. This has led to a bias in the "demand-demand" system. Often, without having information about domestic products, the buyer acquires the goods of "promoted" foreign brands.

Independent promotion of high-tech products associated with significant financial resources. In addition, there are not always information resources, skills and experience in introducing new products to the market. The situation is aggravated by the foreign marketing intervention of the installation that has been established over many years, that domestic products are outdated, undervalued, etc.

Partly to mitigate the problem in a short time may be the creation of a regional, and then a federal portal - a "virtual fair" of innovative proposals. This step will allow creating a national bank of domestic high-tech resources, enabling access to domestic achievements, using it by potential innovators and investors with various opportunities.

The bank of high-tech resources, as a universal platform, will contribute to the creation of cooperation links of all interested parties and companies.

In the current economic conditions high-tech business is experiencing certain difficulties of organizational and psychological order, they are associated with self-organization and cooperation. The reasons are:

- lack of trust in partners;
- inefficient internal management;
- weak competition in many segments of the domestic market;
- lack of business experience based on new models, etc.

According to Russian business researchers, the majority of companies (65.4%) rely on their own strength or on those partners with whom they developed an innovative product; only a third of organizations actively cooperate with other companies [13].

The development of a more coherent science and technology cooperation will be promoted by a special federal project, the purpose of which is to create a network of high-tech research and production alliances with government support of contacts between all sectors of high-tech products.

Stimulating the growth of high-tech business is impossible without solving the personnel issue. The saturation of high-tech industries with the necessary specialists should be carried out through the development and implementation of a targeted training program (and retraining) of specialists. Particular attention should be paid to the training of so-called high-tech producers, the distinguishing characteristics of which are: initiative, readiness for complex coordination activities, the availability of a wide range of competencies, the ability to create a synergistic effect through a combination of intellectual resources. An important element in attracting highly qualified personnel to a high-tech business is the motivation for the transfer of specialists from foreign companies operating in Russia. As a rule, foreign companies carry out a rigorous selection of specialists, paying attention to their professional data and creative abilities. Such specialists know foreign languages, have the appropriate education, experience in teams, work skills in knowledge-intensive industries, have knowledge of the law. In addition, the perception of the cultural characteristics of a foreign company and partner companies replenishes the arsenal of elements of the corporate culture of high-tech business.

4 Conclusion

The development of high-tech business in recent years is due to the implementation of government programs that are aimed at stimulating promising areas for the introduction of new technologies. In particular, these are transport, space systems, nanotechnologies, nuclear energy, etc. Companies operating in promising industries have tax benefits - reducing the taxable base for profits by 1.5 times the actual R&D costs. The mechanism of supporting and stimulating the development of high-tech enterprises requires, on the one hand, increasing the impact of state support measures, and on the other, continuing the sustainable functioning of companies through diversification aimed at the production of innovative products.

The strategic innovation development of high-tech companies largely depends on the efficiency of their financial and economic activities and the effectiveness of the management system. The high costs of R&D and the availability of highly qualified personnel are mandatory, but insufficient elements of the development of high-tech enterprises. Common attribute attributes of the further effective functioning of Russian innovative companies are:

- knowledge intensity and high significance of R&D results throughout the all-production cycle, which makes it possible to classify R&D as the main internal operational processes of high-tech enterprises;
- highly qualified personnel (intellectual) potential;
- a significant proportion of intangible assets (intellectual property) in assets;
- a powerful industrial and technological platform that allows you to create and implement developments and innovations (scientific and technological or to the potential of commercialization);
- high added value of the final product;
- high risks and investments of the innovation process [14].

The motivation of high-tech companies to increase innovation should be based on long-term motivation. It is advisable:

- in order to develop exports of high-tech products, it is necessary to conduct an audit of measures to support R&D and high-tech exports in terms of simplicity and ease of use by large, medium and small businesses;
- stimulate the growth of the number of purchases through life cycle contracts, which will improve the quality of service and after-sales service;
- in order to create mechanisms for stimulating technological development of industries and intersectoral complexes, create, on the basis of expert councils, working groups with the participation of the state and large business to develop and implement sectoral and intersectoral programs for the joint development of promising technologies and eliminate gaps in production chains.

In order to achieve global technological leadership, it is necessary to develop and implement a strategy for the development of the digital economy.

Acknowledgements. The work was prepared during implementation of project No. 26.9642.2017/8.9 within the framework of the State task of the Ministry of Education and Science of Russia.

References

1. Kostin, K.B.: Drivers for the development of high-tech startups on the example of Finland. Russ. J. Entrepreneurship **18**, 2705–2718 (2017)
2. Vostrikov, V.S.: High-tech companies as a factor in the strategic development of the domestic economy. Creative Econ. **9**(4), 473–484 (2015)
3. Laptev, A.A.: The concept of "high-tech company" in modern macroeconomic theory. Innovations **7**(105), 35–41 (2007)

4. Shpolyanskaya, A.A.: High-tech industries: definitions and development conditions. Young Sci. **22**(102), 520 (2015)
5. Tatarintseva, I.V., Vasin, A.V., Tatarintsev, V.A.: The model of information support for the sales strategy of an innovative product. Bull. Bryansk State Tech. Univ. **4**, 76–83 (2008)
6. Order of the Government of the Russian Federation. No. 1662 from 17-11-2008 (editors from 08-08-2009). (Actual in 2018 year). http://www.zakonprost.ru/content/base/part/593361/
7. Gruzdov, V.V., Shakhmalov, F.I.: Innovative development of integrated corporate structures with state participation in modern conditions. Monograph, 448 p. MIREA, Russian Technical University, Moscow (2018)
8. Kraft, J., Zaitsev, A.V.: Improving the functioning of a high-tech enterprise in a limited labor market. Creative Econ. **11**(5), 625–636 (2017)
9. Ashmyanskay, I.S.: Information technology sector as a platform for building in innovation economy. In: Proceedings of the International Scientific Conference "Actual Issues of Economics and Management" (Moscow, April 2011 year). T.I. Moscow, RIOR, pp. 17–20 (2011)
10. Kolesov, A.: "Numeral" aimed at the market. itWeek **3**(946) (2019)
11. IT integrator - Benefit of using. http://hitech-news.ru/news/it-integrator-preimushhestva-ispolzovaniy/
12. Business of the system integrator – «Anticipation game». https://www.ot.ru/news_events/press/biznes_sistemnogo_integratora_igra_na_operezhenie/?mobile=Y
13. Innivative activity of subjects of the innovation process: Monitoring: Newsletter NIU «VSE» and Institute for Statistical Studies and Economic Knowledge, No. 3 (2015)
14. Verevka, T.V.: Assessment of the effectiveness of economic and financial activities and the effectiveness of enterprise high-tech management. Russ. J. Entrepreneurship **19**(2), 445–456 (2018)

Methodology of the Formation of a Comprehensive Support Mechanism of Innovation and Investment Development in the Region

Yuriy A. Doroshenko(✉), Irina O. Malykhina, and Irina V. Somina

Belgorod State Technological University named after V.G. Shukhov (BSTU), Ulitsa Kostyukova 46, 308012 Belgorod, Russia
549709@mail.ru

Abstract. Nowadays the situation of the national economy is characterized by the need to carry out significant, and in some industries - radical modernization processes, management reform, the formation of new economic relations based on innovation. Problems of perspective development of regions as subjects of the national economy, as well as diversification structuring of regional economies, require the formation of an integrated mechanism to support the impulses of their development. At the present time the traditional factors of economic growth of the Russian economy are almost completely exhausted. In this regard, the problem of finding new sources of development of domestic economic systems at all levels is particularly relevant. However, the key condition for the successful development of Russia in the long term is the modernization of all sectors of the economy, especially the industrial sector, which cannot be done without actively applying the results of innovation. Thus, the formation of a comprehensive mechanism to support drivers of innovation and investment development of regions is a priority to ensure the economic development of the Russian economy.

Keywords: National economy · Economic growth · Innovation and investment development · Modernization processes

1 Introduction

Nowadays while determining the prospects for the development of a region, the question of whether or not its current strategic choice will work in the prospect under consideration takes a significant place. It is about clarifying the principal circumstances, whether in principle, and if so, how effectively the established list of priority functions implemented by the region will be able to ensure the achievement of the intended goals of its social development for a strategic perspective. Solving this kind of tasks can be quite feasible, however, using the method of forming and activating the impulses of economic development in the region through the intensification of innovation and investment activities.

I. Karabegović (Ed.): NT 2020, LNNS 128, pp. 916–923, 2020.
https://doi.org/10.1007/978-3-030-46817-0_103

The problem of increasing the effectiveness of the investment activity of the regional economy is not new and is of interest to many scientific researchers in order to develop an investment strategy for the country's territorial and economic systems. Less attention is paid to improving the "quality" of investments, i.e. not a quantitative, but a qualitative component, which proportionally increases with the growth of innovative projects implemented at the expense of the financial resources of the regional investment portfolio and, on the contrary, leads to the degradation of the basic sectors of the economy as a result of their decline. The hypothesis of a relationship between the structure of a regional investment portfolio, innovation activity and sustainable economic growth is confirmed by the experience of economically developed countries [1]. Thus, the development and competitiveness of a region directly depends on its ability to quickly respond to changing external factors, create favorable conditions for doing business, and generate ideas and projects for its development. Regions that attract investments for projects, especially innovative ones, eventually become capable of concentrating value chains on their territory, providing the economic basis for sustainable development. To manage innovation, many economies and regions are developing: framework for analyzing the competitive implications of innovation [2], national innovation system (NIS), regional innovation system (RIS), global innovation system (GIS) and etc. [3–6].

Some of the scientists say [7] that the regions should be the main object of application of the theory of sustainable development, since the sustainable development of the country depends on the degree of sustainability of their development. Russian regional systems are geographically stable and historically established conglomerations with distinct individual characteristics. The geographical size and scope of activities of domestic regional economic systems allow them to independently position themselves in the external environment, taking into account the experience of combining market transformations and government management of socio-economic processes.

Presently the necessary to provide real support to individual regions with a high concentration of research, development, innovation infrastructure, production and their relationship with other subjects of the Russian Federation in terms of transfer of technologies, products and services is recognized by all participants of the innovation process. It should be noted, such support is necessary because the share of innovative products in the gross regional ratio ranges from 10 to 20%. The analysis of the constituent entities of the Russian Federation in terms of the share of innovative products in the GRP allowed us to identify the top 15 leaders in this category, where the top ten include regions with a pronounced specialization in technological areas, enshrined in regional strategic documents: the Vologda region, the Republic of Tatarstan, and the Samara Region, Republic of Mordovia, Nizhny Novgorod Region, etc. [8].

However, if we will compare the first 10 regions in terms of the costs of technological innovation, they are strong and medium-strong innovators (regions with a high concentration of human, financial capital and research infrastructure): Moscow, Moscow Region, St. Petersburg, Tyumen, Samara Regions, etc. The analysis of statistical indicators of innovative activity of the regions determined that only in 12% of regions the share of innovative products in the GRP is more than 10%, which is a very low level [8].

At the present conditions of economic and geopolitical development the government does not have sufficiently effective and efficient ways to support the regions. The regional authorities themselves, who are interested in boosting investment activity and introducing innovations in the manufacturing sector, should create favorable conditions for the inflow of investments and the formation of a comprehensive mechanism to support drivers of innovation and investment development. One of the tools for managing innovation activity is timely and high-quality monitoring, which allows you to quickly make decisions that prevent the failure to implement projects of a full innovation cycle. As indicators of innovative development of the regions are key performance indicators, in particular, the security of the priority for the industry sectors results of intellectual activity.

The problems of innovative development of regions are constantly in the center of attention of representatives of science and business. This is due to the fact that in the process of achieving the maximum competitiveness of the state, the output of domestic products to international markets, increasing the export and import of technology, the innovation system of the regions is given a priority role. Thus, the provision of conditions for the development of high-tech production, the activation of human resource potential, the formation and promotion of a driver system for the innovation and investment development of the region is an important tool for ensuring the economic growth of the country. Problems and prospects of innovation and investment development of different regions is given in papers [9–15].

2 Materials and Methods

The methodological basis of this research was formed by general scientific methods: analysis and synthesis, synthesis, system approach, deduction method. The principles of systemic and integrated approaches are applied. General scientific methods of cognition were also used; methods of analogy and comparison.

We have to note that the most important methodological aspects of the investment and innovation processes of the regional economic system were proposed by foreign and domestic scientists J. Schumpeter, R. Cantillon, P. Drucker, A. Marshall, G. Markowitz, N.D. Kondratyev, S.Yu. Glazyev, I.G. Andreeva, R. Cantillon, A. Marshall, V.Ya. Gorfinkel, M.A. Bendikova, O.A. Romanova and others. Regularity and trends of the region's innovative development are presented in the works of M. Miller, F. Modigliani, S. Ross, B. Terborg, A. G. Granberg, Yu. A. Doroshenko, I. V. Somina, D.S. Lvova, A.P. Egorshina, L.I. Abalkina and others.

The conceptual aspects of the formation of new industrialization as a separate type of social transformations based on innovations were first described in the articles of J. Galbraith and were fully reflected in the works of scientists A.A. Bogdanova, A.V. Zvereva, J. Van Dein, N.D. Kondratyeva, B.N. Kuzyka, G. Mensha, S. Kuznets, E. Toffrer, H. Freeman, Yu.V. Yakovets.

3 Results and Discussion

While study the methodological approaches in domestic and foreign science to the formation of drivers of innovative and investment development of the region, it is necessary to clarify what we mean by the economic category "driver". Thus, a driver is a subject of economic relations, an object of economic impact, a method of organization or a management method that affects the economic system and leads to streamlining of its structure, intensification of the growth processes and development of the potential of subsystems, contributing to an increase in the efficiency of the system and the achievement of economic, social, scientific -technical, technological, environmental and other type of effect. Meanwhile, the drivers of innovation and investment development, on the one hand, create conditions for the commercialization of innovations in this territorial segment, on the other hand, stimulate investment inflows to implement projects related, as a rule, using these innovations and the latest technologies [16].

Have to say, they help to the formation of favorable conditions for this development. The concept of economic development is an independent sustainable system operating in conditions of optimal configuration. Drivers can be of various origins. Nowadays interest in innovative and economic development is the strategic and highest priority stimulating development of high-tech industries and the knowledge-intensive sector, which in turn are the mechanism for creating a new economy — an innovative, founded on the newest knowledge and technology. The most important function of development is to build a self-contained system of vertical and horizontal communications, propagating and propagating impulses [16].

Some researchers view the region as a "point of economic growth," because the concept of "growth points" is based on a strategy of polarized regional development, and its goal is to form centers of industry that can have a stimulating effect on the development of adjacent territories.

Thus, T.O. Lascheva thinks that the "point of economic growth" suggests understanding the economic (managing) entity, industry or activity, which as a result of activation can ensure diversification and rationalization of the structure of the regional economy, stimulate the emergence and development of its new elements and help improve the quality of life [17].

As the results of the revitalization of points of economic development included in the elemental composition of the strategic choice of a region, one can imagine [17]:

- positive transformation of the regional economy structure;
- expanding the range of sources of financing investment processes by increasing the investment attractiveness of the region;
- ensuring the revenue part of the consolidated budget sufficient to finance the social sphere of the region;
- increasing the economic activity of the population;
- assistance in strengthening and development of inter-regional and foreign economic relations of the region.

Scientists include in the concept of "points of economic growth" a set of four elements [18]:

- territories (as constituent countries in general or regions within it);
- industries (more promising and less costly for a specific territory);
- enterprises (basic for industries or just stably functioning);
- programs (aimed at economic stabilization).

Scientific literature shows the following approach to determining the possibility of attributing to the drivers of development of a regional economic system [17]:

- employment promotion;
- promoting the employment of certain groups of the population;
- promoting the growth of living standards;
- promoting revenue growth and reducing budget expenditures;
- promoting the growth of related industries;
- promoting the sustainability and structural diversification of the local economy;
- environmental impact;
- assistance to the formation of the chosen image and perspective specialization;
- promoting overall economic growth.

Thus, the formation of a comprehensive mechanism to support drivers of innovation and investment development in the region is a complex and multifactorial task. However, it should be pursued that the number of innovations has a steady upward trend. Table 1 and Fig. 1 shows the values for Global Innovation Index (GII) for post-soviet countries (former USSR states) in 2019. The Turkmenistan and Uzbekistan are not included in the GII rankings for 2019. Source about GII are from [19], with data about population from Google and calculations rank [%] and quarter by the authors.

Table 1. Values for Global Innovation Index (GII) for post-soviet countries in 2019

Country	Capital city	Population	Score	Rank [position]	Rank [%]	Quarter
Armenia	Yerevan	2967000	33.98	64/129	49.61	Q2
Azerbaijan	Baku	8238700	30.21	84/129	65.12	Q3
Belarus	Minsk	9648500	32.07	72/129	55.81	Q3
Estonia	Tallinn	1299400	49.97	24/129	18.60	Q1
Georgia	Tbilisi	4615800	36.98	48/129	37.21	Q2
Kazakhstan	Astana	15399400	31.03	79/129	61.24	Q3
Kyrgyzstan	Bishkek	5431700	28.38	90/129	69.77	Q3
Latvia	Riga	2231500	43.23	34/129	26.36	Q2
Lithuania	Vilnius	3555200	41.46	38/129	29.46	Q2
Moldova	Chişinău	4320700	35.52	58/129	44.96	Q2
Russia	Moscow	140041200	37.62	46/129	35.66	Q2
Tajikistan	Dushanbe	7349100	26.43	100/129	77.52	Q3
Turkmenistan	Ashgabat	4884900	–	–	–	
Ukraine	Kiev	45700400	37.40	47/129	36.43	Q2
Uzbekistan	Tashkent	27606000	–	–	–	

We have to say that the effectiveness of the cluster principle of forming an innovation system (according to the experience of the Republic of Korea), in particular, the infrastructure of innovation activity, the effectiveness of which contributes to enhancing the drivers of economic development of economic entities of various levels and complexity of the organization. The formation of clusters is expressed through the merger of companies grouped around a single center by industry or science and technology.

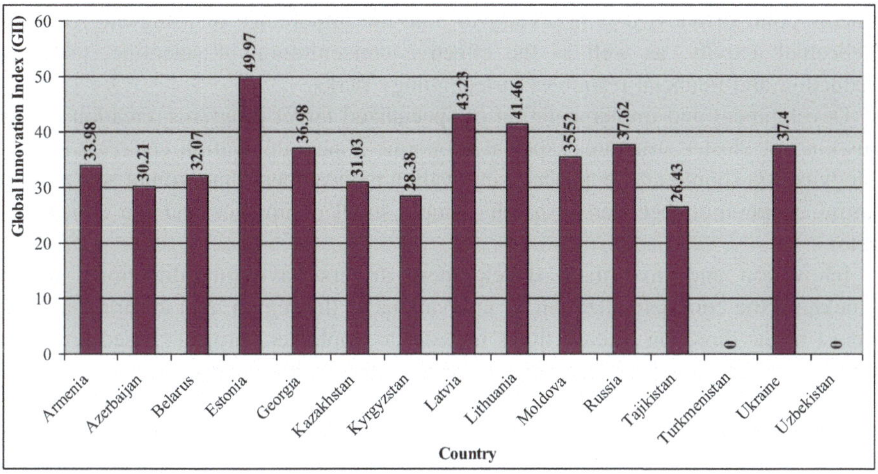

Fig. 1. Graphic representation of values for Global Innovation Index (GII) for post-soviet countries in 2019

Focusing on the successful experience of innovatively active countries, to which the Republic of Korea, Singapore and other countries that have recently become economically developed countries deservedly can be attributed, it is necessary to build an innovation and investment strategy for the economy, adapting successful practices in the Russian context. In the Republic of Korea in 2013, the Innopolis Technopark (Innopolis) was recognized as a special institute engaged in the professionalization of technology commercialization through cooperation of business, science and education [8]. The authorities of the country identified individual regions that were the core for the development of a certain type of industry in the country. Scientific, technical and innovative development of the Republic of Korea was carried out on the basis of the strategy of the territorial-branch organization of the economy, which contributed to the creation of new sources of economic growth. The basis was the introduction of large national companies in the infrastructure of the "creative economy".

Scientists believe that an effective tool for the formation and development of the Russian "creative economy" will be the intensification of innovation and investment activities through the implementation of a comprehensive support mechanism for innovation and investment drivers through the provision of continuous technology transfer and the development of high-tech business and high-tech production in industrial sectors of the economy and expansion of international scientific and technological cooperation.

4 Conclusion

From the methodology of the formation of mechanisms to support drivers or impulses of innovation and investment development of the economy of regional economic systems point of wiev, it is necessary to note the importance of using the results of intellectual activity, as well as the effective concentration of scientific, technical, production and financial resources in technology parks.

Development and implementation of specialized target programs, creation of various kind of cluster structures (special economic zones, innovation centres and technology parks) should create a special innovation environment that promotes active and fruitful cooperation between research groups, small companies and the educational sector.

Innovation and investment development drivers have multidirectional action, stimulating the commercialization of innovations in the region, and ensuring investor interest in the direction of cash flows in order to implement projects based on manufacturing, which based on innovations and the latest technologies.

Acknowledgements. The work was prepared during implementation of project No 26.9642.2017/8.9 within the framework of the State task of the Ministry of Education and Science of Russia.

References

1. Amadaev, A.A., Vasilenko, A.A.: Innovation and investment way of development of the regional economy as a basis for ensuring sustainable economic growth. New Technol. **4**, 27–34 (2015)
2. Abernathy, W.J., Clark, K.B.: Innovation: mapping the winds of creative destruction. Res. Policy **14**, 3–22 (1985)
3. Binz, C., Truffer, B.: Global innovation systems – a conceptual framework for innovation dynamics in transnational contexts. Res. Policy **46**(7), 1284–1298 (2017)
4. Doz, Y.L., Wilson, K.: Managing global innovation: Frameworks for integrating capabilities around the world. Harvard Business Press, Boston (2012)
5. Lee, A., Mudambi, R., Cano-Kollmann, M.: An analysis of Japan's connectivity to the global innovation system. Multinational Bus. Rev. **24**(4), 399–423 (2016)
6. Lee, S., Lee, H., Lee, C.: Open innovation at the national level: Towards a global innovation system. Technol. Forecast. Soc. Change 151 (2020). art. no. 119842
7. Korableva, A.A., Karpov, V.V., Kapoguzov, E.A.: Correlation between territorial innovative clusters and sustainable development of a region. Actual Probl. Econ. Law **12**(4), 800–815 (2018). (in Russian)
8. Ilina, I.E., Zharova, E.N., Agamirova, E.V., Kamenskyi, A.S.: Innovative development of Russian regions. Regionology **26**(2), 230–255 (2018). (in Russian)

9. Anđelković-Pešić, M., Anđelković, A., Dašić, P.: The theory of constraints as a basis for production process improvement model. Actual Probl. Econ. (APE) **14**(10), 251–260 (2013)
10. Bezrukova, T.L., Kuksova, I.V., Korda, N.I., Kirillova, S.S.: Formation of target programs of innovation and investment development of regions. In: Lecture Notes in Networks and Systems, vol. 57, pp. 461–472 (2019)
11. Boris, O.A., Timoshenko, P.N., Tadtaev, D.M., Rebiy, E.Y.: Integrated innovation and investment policy as the factor of a balanced development of the new economy. Qual. Access Success **19**(S2), 108–113 (2018)
12. Izmailova, M.A., Reshetova, T.Y., Rukina, I.M., Seifullaeva, M.E., Yunusov, I.A.: Problems and prospects of innovative and investment development of modern Russia. Int. J. Econ. Financ. Issues **6**(2), 95–102 (2016)
13. Kolmakova, E.M., Kolmakova, E.M., Kolmakova, I.D.: Innovation and investment potential of regions as a vector for their development in the 21st century. In: Lecture Notes in Networks and Systems, vol. 91, pp. 534–543 (2020)
14. Kuzkin, Y., Cherkashyna, T., Nebaba, N., Kuchmacz, B.: Economic growth of the country and national intellectual capital (evidence from the post-socialist countries of the central and eastern Europe). Probl. Perspect. Manag. **17**(1), 348–359 (2019)
15. Minaeva, E., Lastochkina, V., Gusev, V., Fadeev, A., Manukhina, L.: Formation of the strategy of management of innovation and investment activity of the enterprise. MATEC Web Conf. **193** (2018), art. no. 05082
16. Doroshenko, Yu.A, Somina, I.V., Malyhina, I.O.: Studying the mechanism of infrastructure support of high-tech business as an integrator of innovation-investment development. Rev. ESPACIOS **39**(47), 24 (2018). (in Russian)
17. Lasheva, T.O.: Opportunities for diversification of the regional economy structure based on the method of forming the poles of economic development. Sci. Notes St. Petersburg Acad. Manag. Econ. **2** (24), 33–41 (2009). (in Russian)
18. Gutman, G.V., Miroedov, A.A., & Fedin, S.V.: Regional Economy Management. Finance and Statistics, Moscow, Russia (2001). 295 p. (in Russian)
19. Dutta, S., Lanvin, B., Wunsch-Vincent, S. (eds.): The Global Innovation Index 2019: Creating healthy lives - The future of medical innovation. World Intellectual Property Organization (WIPO), Geneva, Switzerland (2019). 399 p.

Global Gender Gap Index: Is It Time to Measure Technology Access Gap Also?

Munira Šestić[1(✉)], Zijada Rahimić[1], Mirha Bičo Ćar[1], and Dženana Hodžić[2]

[1] School of Economics and Business, University of Sarajevo, Trgoslobođenja 1, 71000 Sarajevo, Bosnia and Herzegovina
munira.sestic@efsa.unsa.ba

[2] Committee on Education, Science, Culture and Sport, Parliament of the Federation of Bosnia and Herzegovina, Sarajevo, Bosnia and Herzegovina

Abstract. The Gender Gap Index was introduced by the World Economic Forum (WEF) in 2006 as "A framework for capturing the magnitude of gender-based disparities and tracking their progress over time", based on the idea that the equal contribution of women and men is critical for societal and economic development. The Gender Gap Index is presented yearly in a WEF report and measures the relative gap between women and men across four key areas: health, education, economy and politics - 149 countries are covered by the 2018 report. However, the context of the Fourth Industrial Revolution (4IR) brings unprecedented opportunities as well as new challenges. In this article we are questioning whether it is time to include another key monitoring area important for the context of 4IR? The question of technology access gap arises in the form of horizontal segregation not only in terms of choosing the field of education and job participation, but also in disparities in the use and application of new technologies and the choice of dominant business field when starting a business. Technology access is essential in order to take full advantage of new technologies and therefore gender-based disparity in this area is important to measure and track, for both academic researchers and policy makers.

Keywords: World Economic Forum · Gender Gap Index · Context of the Fourth Industrial Revolution

1 Introduction

The context of the Fourth Industrial Revolution (4IR) offers unlimited challenges for all. The future will bring many opportunities and demands and if they are ignored, stagnation is inevitable. The possibility of taking full advantage of new technologies places a need to utilize all our human capacities: learning, creativity, inquisitiveness and enterprising. The equal contribution of women and men in 4IR processes can obtain balanced economic and societal development and furthermore increase the chances for global broad-based progress. Overall, humanity cannot afford to give up the talent, knowledge, skills and ideas of half of its population, since "new technologies are

I. Karabegović (Ed.): NT 2020, LNNS 128, pp. 924–929, 2020.
https://doi.org/10.1007/978-3-030-46817-0_104

primarily human-produced" [1]. Without that, world cannot fully count on the prosperous future that new technologies can bring – "future that works for everyone" [1].

As of 2006, the World Economic Forum, through its Global Gender Gap Index, measures the relative gap between women and men across four key areas: health, education, economy and politics. It provides a comprehensive set of data and offers the participating countries a base to set improvement priorities within their own economic, political and cultural contexts. However, keeping in mind the focus on new technologies, the question arises for the Global Gender Gap Index: is it time to measure another key monitoring area important for the context of 4IR? In this article, we will try to justify the idea that technology access gap should be measured as well, since technology access is essential in order to take full advantage of new technologies. The gender-based disparity in this area is important to measure and track and its measurement would bring benefits and prosperity for society and economy.

2 Technology Access Gender Gap as a Consequence of Horizontal Segregation

Technology access is essential in order to facilitate full advantage of new technologies and therefore gender-based disparity in this area is important to focus on. The question of technology access gender gap arises in the form of horizontal segregation not only in terms of choosing the field of education and job participation, but also it is witnessed through disparities in the use and application of new technologies and the choice of dominant business field when starting a business.

So far, the issue of the technology access gender gap has been approached by researches with different aspects and from the different stand of time dimensions. We believe that systematic, interactive and continuous measurement and tracking of different aspects of technology access gender gap, as successfully done by the Gender Gap Index in areas of health, education, economy and politics, would give the comprehensiveness required by the context of 4IR.

The essence of the fourth industrial revolution is inextricably linked to the term of STEM (science, technology, engineering, mathematics) and this direct and dependent relationship points in which direction education, jobs, knowledge/skills, application and businesses need to be geared to achieve developmental goals. With present fears of how 4IR will affect women and their place and role in society in the future [1, 2]. WEF should remain on the verge of seeking an "all-inclusive" answer in the context of 4IR: will gender gap increase or decrease in the future to come.

2.1 Choice of Educational Area

Horizontal gender segregation in the educational orientation has been documented for decades [3, 4]. According to the UNESCO report [5] only 35% of STEM students in higher education globally are women. Obvious differences are recognized within certain STEM disciplines and for example only 3% of female students in higher education choose information and communication technologies (ICT) studies.

According to NCES data [6], the number of female students in the USA who gained STEM degree/certificate has steadily risen over the most recent 8 year period but the number of male students has also increased in the same period. Regardless of the positive trend of more women choosing STEM subjects in the college, the number of men choosing STEM subjects in college is rising much faster than the number of women. Because of that, the gender gap in STEM does not only remain, but is actually growing - in the USA the percentage of women with STEM degrees has dropped from 25% to 24%. If the trend continues, the number is going to be even lower in the future according to the numbers shown above.

Women's education choice has an impact on success of their possible future business activities and the impact of education may vary by sector with more pronounced impact identified in technology-intensive industries [7–9].

Gender imbalances in educational choices could have possible explanation in female academic strengths more than in their weaknesses [10]. Still most of the studies succeeded in mapping facts and trends, but they provide little information on the more subjective aspects of women's educational choices [11, 12].

2.2 Professional (Career) Choice

"On Facebook and Google, women employees are the 1/3 of the company's payroll, but only 16% of Facebook's technical posts and 18% on Google are owned by female engineers and only 10% on Twitter. These numbers are the norm in the whole industry - in Europe, only 7% of engineering positions are held by women" [13]. Gender professional segregation is mostly based on the belief that there are certain "male" and "female" professions [14] and women in general are dominantly educated in the social sciences [15, 16].

Different gender paths in professional choices have been documented by the WEF: its reports state that on average men are underrepresented in the fields of education, health and welfare whereas women are underrepresented in the STEM fields. New technology jobs are still majorly underrepresented by women. Despite the fact that women constitute almost 50% of the labour market, there are only 28% of women in STEM sectors as opposed to 72% of men [2]. Furthermore, very rarely are board members in the information technology industry female - pointing to vertical segregation as well. Still women on these corporate boards are more likely to be more qualified than men. About 16% of all female board members are likely to have prior professional technology experience compared to only 9% of their male colleagues. [17].

Gender differences in the relation between future career perspectives and career aspirations (with emphasis on female under-representation in STEM fields) has been mostly studied through future career perspectives and lifestyle preferences in general [18–20]. Still there are concerns that career choices are connected to academic choices made at early life age [21]. Future studies are noted to examine how social environment affects women's STEM choices [22].

Still, just driving more women for STEM carriers would not be sufficient, since more than 44% of female STEM students in China reported "gender discrimination" in the job market [23]. It can be concluded that without direct and focused institutional support on universities and the labour market, these STEM driving efforts for women may have suboptimal outputs.

2.3 Entrepreneurial Choice of Business Sector

When comparing female and male entrepreneurs, there are recognized differences in their field of education and further to this, choice of business sector for their start-ups and businesses to run [14]: female entrepreneurs are less frequently represented in the so-called fast-growing branches of advanced technology [14]. One possible reason for that can be that women are less educated in the technology and do not choose technology sector very often to start and run their business [16]. As mentioned before, most of female entrepreneurs are educated in the social sciences [15, 16] and their businesses are dominantly established and run in this part of the economy system. It can be said that the gender gap is being most pronounced in the information and communications technology (ICT) sector [24].

A recent World Bank Policy Research Paper [25] confirms that women tend to sort into different sectors than men and these sectors (dominated by women) tend to be less profitable. The paper (Idem) finds that some of the business earning gaps can be explained by the business sector choice made by female entrepreneurs: female-owned businesses in male-dominated sectors make significantly higher profits than those in traditionally female sectors. The report finds computer programming/consultancy, telecommunications and information service activities as the sectors recognized as male-dominated with more than 70% businesses in these sectors owned by man globally.

Further to these findings, the European Commission says that more women in information and communications technology (ICT), as a STEM subfield, could increase the European Union's GDP by nine billion Euros per year [26]. These facts should be grave concerns for governments that they "could fail to capture the benefits of new technologies" [1]. These concerns require more long-term and pro-active solutions, collaborative strategies and creativity for all: both individuals and societies.

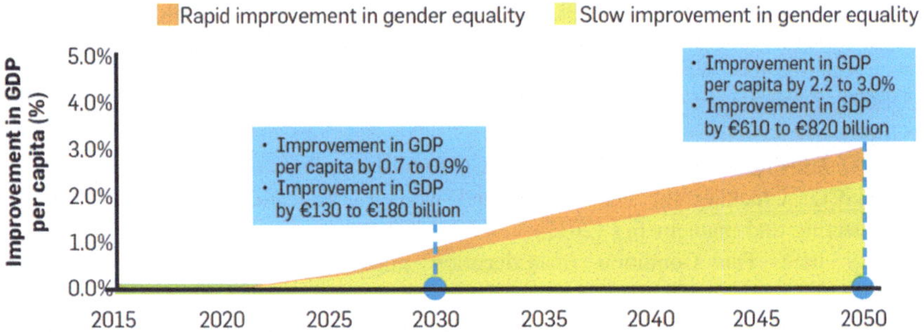

Fig. 1. Economic benefits of gender equality in the EU: how gender equality in STEM education leads to economic growth [27]

Yet with a goal of achieving success in any business model, there is a need for combination of both female and male peculiarities that have valuable entrepreneurship characteristics, regardless of gender [28].

3 Conclusion

Humanity in the context of 4IR needs the resources and diversity of all members of the society. Balanced global development in economic and societal aspects is not possible unless all individuals make their full contribution. Measuring and monitoring the technology access gender gap is a very important issue that requires a comprehensive approach in the manner in which the WEF has done it through the Global Gender Gap Index within other key areas (health, education, economy and politics) over the past 15 years.

Gender-based disparity in technology aspect is important to measure and track in overall index manner since we believe such comprehensive, interactive and continuous interest would bring dual benefits: for academic researchers, for more focused researches - with multiple questions starting with how and why, as well as policy makers – with challenges of overcoming recorded technology access gender gap and streaming to achieve optimal results for driving efforts to close or decrease the gap.

Partially determined and insufficiently explored present gender gaps would surely worsen future gender gaps. This is especially important for the issue of technology, since we are now with no doubt, experiencing most of the STEM being developed without diverse talent and with limited inclusive capacity. In the future shaped by 4IR, possible low technology integration of women will result in significant missed opportunity - humanity cannot afford to deprive itself of women's talents and skills in a field(s) where implementation of women's talents and skills are already scarce. So, the answer to the title question should be – YES it is a time.

References

1. Schwab, K.: The Fourth Industrial Revolution. Crown Publishing Group, New York (2017)
2. WEF - World Economic Forum: The Global Gender Gap Report (2018). http://www3.weforum.org/docs/WEF_GGGR_2018.pdf. Accessed 22 Dec 2019
3. OECD: Evolution of Student Interest in Science and Technology Studies, Policy Report (2006)
4. OECD: Closing the Gap - Act Now. OECD Publishing, Paris (2012)
5. UNESCO: Cracking the code: girls' and women's education in science, technology, engineering and mathematics (2017)
6. NCES IES: The Condition of Education https://nces.ed.gov/pubs2015/2015144.pdf. Accessed 15 Dec 2019 (2015)
7. Smallbone, D., Wyer, P.: Growth and development in the small firm. In: Carter, S., Jones-Evans, D. (eds.) Enterprise and Small Business, pp. 409–433. Financial Times/Prentice Hall, Harlow (2006)
8. Demartini, P.: Innovative female-led startups: do women in business underperform? Adm. Sci. MDPI, Open Access J. **8**(4), 1–15 (2018)

9. Ozkazanc-Pan, B., Muntean, C.S.: Networking towards (in)equality: Women entrepreneurs in technology. Gend. Work Organ. **25**(4), 379–400 (2018)
10. Breda, T., Napp, C.: Girls' comparative advantage in reading can largely explain the gender gap in math-related fields. PNAS **116**(31), 435–440 (2019)
11. Reinking, A., Martin, B.: The gender gap in STEM fields: theories, movements ideas to engage girls in stem. J. New Approaches Educ. Res. **7**(2), 148–153 (2018)
12. Kang, J., Hense, J., Scheersoi, A., Keinonen, T.: Gender study on the relationships between science interest and future career perspectives. Int. J. Sci. Educ. **41**(1), 88–101 (2018)
13. Fish, G.: Women in Technology: Ways to Close the Gender Gap (2016). www.techchange.org
14. Lee, I.H., Marvel, R.M.: Revisiting the entrepreneur gender-performance relationship: a firm perspective, special issue: small business, innovation, and entrepreneurship. Small Bus. Econ. **42**(4), 769–786 (2014)
15. Bowen, D., Hirsch, D.: The female entrepreneur: a career development perspective. Acad. Manag. Rev. **11**(2), 393–407 (1986)
16. Klapper, L.F., Parker, S.C.: Gender and the business environment for new firm creation. World Bank Res. Observer **26**(2), 237–257 (2011)
17. Accenture Research: Tech Experience: Women's Stepping Stone to the Corporate Boardroom? http://accenture.com/_acnmedia/pdf-29/accenture-tech-experience-womens-stepping-stone-corporate-boardroom.pdf. Accessed 22 Mar 2019 (2016)
18. Diekman, B., Brown, E.R., Johnston, A.M., Clark, E.K.: Seeking congruity between goals and roles: a new look at why women opt out of science, technology, engineering, and mathematics careers. Psychol. Sci. **21**(8), 1051–1057 (2010)
19. Sadler, P.M., Sonnert, G., Hazari, Z., Tai, R.: Stability and volatility of STEM career interest in high school: a gender study. Sci. Educ. **96**(3), 411–427 (2012)
20. Su, R., Rounds, J.: All STEM fields are not created equal: People and things interests explain gender disparities across STEM fields. Front. Psychol. **6**, 189 (2015)
21. Card, D., Payne, A.A.: High School Choices and the Gender Gap in STEM, NBER Working Paper No. 23769 (2017)
22. Mouganie, P., Yaojing, W.: High-Performing Peers and Female STEM Choices in School, Institute of Labor Economics (IZA), Bonn, *Discussion Paper No. 12455* National Center for Educational Statistics: STEM education by Gender, *NCES 2015-075,* USA (2019)
23. Kuhn, P., Shen, K.: Gender discrimination in job ads: evidence from China. Q. J. Econ. **128**(1), 287–336 (2013)
24. Kelley, J., Singer, S., Herrington, M.D.: Global Entrepreneurship Monitor 2016, Global Report, Global Entrepreneurship Research Association, London Business School (2016)
25. Goldstein, M., Gonzalez Martinez, P., Papineni, S.: Tackling the Global Profitarchy: Gender and the Choice of Business Sector, World Bank Group, Policy Research Working Paper no. 8865 (2019)
26. Fatourou, P., Papageorgiou, Y., Petousi, V.: Women are needed in STEM. Eur. Polices Incentives **62**(4), 52–72 (2019)
27. EIGE: Economic benefits of gender equality in the EU: How gender equality in STEM education leads to economic growth, Publications Office of the European Union (2017). https://doi.org/10.2839/652355
28. Pavlidou Koutroumpis, O.: The Role of Women in 4th Industrial Revolution, EUW Legal and Human Commision, European Institute for Gender Equality (2019). www.eige.europa.eu. Accessed 04 Feb 2019

Statistical Analysis of E-Government Development Index (EGDI) of Georgia

Raul Turmanidze[1], Predrag Dašić[2(✉)], and Giorgi Popkhadze[1]

[1] Faculty of Transportation and Mechanical Engineering, Georgian Technical University (GTU), Tbilisi, Georgia
[2] High Technical Mechanical School of Professional Studies, Str. Radoja Krstića 19, 37240 Trstenik, Serbia
dasicp58@gmail.com

Abstract. EGDI (e-Government Development Index) is used for measure the development of e-Gov in different regions and countries at the national level. The EGDI is a composite index based on the weighted average of following three indices: online service index (OSI), telecommunications infrastructure index (TII) and human capital index (HCI). The EGDI index is based on data collected and publication of United Nations E-Government Survey (UN-EGS) and analyzes the ranking of the 193 UN member states, from 2003. In the paper is given statistical and trend analysis of Georgia's EGDI index for the period 2003–2018. Based on statistical and trend analysis, the EGDI index in Georgia for the period 2003–2018 had an increase from 96.33% (CGI = 196.33 and CAGR = 4.60% per year in 2018 compared to 2003). The highest growth had is the TII index from 369.83% (CGI = 469.83% and CAGR = 10.87% per year in 2018 compared to 2003), followed by the OSI index of 180.45% (CGI = 280.45% and CAGR = 13.76% per year in 2018 compared to 2010), while the HCI index had a decrease of −6.37% (CGI = 93.63% and CAGR = −0.44% per year in 2018 compared to 2003).

Keywords: Development indices (DI) · e-Government development index (EGDI) · Online service index (OSI) · Telecommunications infrastructure index (TII) · Human capital index (HCI)

1 Introduction

The e-Gov or E-Government (Electronic Government) can be defined as the use of information and communication technologies (ICT) to enable citizens and businesses to interact and conduct business with government agencies via various electronic media such as: telephone touch pad, fax, smart cards (SC), self-service kiosks, e-mail, Internet, electronic data interchange (EDI), electronic data interchange for administration, commerce and transport (EDIFACT), interactive electronic data interchange (i-EDI) protocols, and etc. [1–4]. In papers [5–14], the state of e-Gov application in different regions and countries of the world have been analyzed.

For measure the development of e-Gov in different regions and countries at the national level is used EGDI (e-Government Development Index). The EGDI is a composite index based on the weighted average of following three indices: Online

I. Karabegović (Ed.): NT 2020, LNNS 128, pp. 930–938, 2020.
https://doi.org/10.1007/978-3-030-46817-0_105

Service Index (OSI), Telecommunications Infrastructure Index (TII) and Human Capital Index (HCI) [15–17]. The EGDI index is based on data collected and publication of United Nations E-Government Survey (UN-EGS) and analyzes the ranking of the 193 United Nations (UN) member states, from 2003.

The value of the EGDI index ranges from 0 to 1, and is divided into four quartiles: Low with a abbreviation LEGDI (EGDI values in the range of 0.00 to 0.25), Middle with a abbreviation MEGDI (EGDI values in the range of 0.25 to 0.50), High with a abbreviation HEGDI (EGDI values in the range of 0.50 to 0.75) and Very-High with a abbreviation VHEGDI (EGDI values in the range of 0.75 to 1.00) [16, 17].

2 Data and Methods

Data on values of EGDI index for Georgia has been retrieved from EGDI reports for period 2003–2018 [15, 16], with a certain calculations from the part of the authors. For the trend analysis, we used the following parameters: G (Growth), AG (Annual Growth), CI (Chain Index), AGR (Annual Growth Rate), CAGR (Compound Annual Growth Rate) and CGI (Cumulative Growth Index), described in the papers [18–22].

For the calculation of descriptive statistics parameters we used software package CoPRMTV and software system MS-Excel.

3 Results and Discusions

On Table 1 are given data about EGDI index of Gergia for the period 2003–2018 [15–17], with calculation by the authors for: position [%], Q (quartile) and trend analysis parameters (G, CI, CAGR and CGI).

Table 1. Data about e-government development index (EGDI) of Gergia for the period 2003–2018

Year	Score	Rank	Position [%]	Q	G	CI	CAGR	CGI
2003	0.3511	99/191	51.83	Q3	–	–	–	100.00
2004	0.3784	94/191	49.21	Q2	0.0273	107.78	7.78	107.78
2005	0.4034	83/191	43.46	Q2	0.0250	106.61	6.61	114.90
2008	0.4598	90/192	46.88	Q2	0.0564	113.98	4.46	130.96
2010	0.4248	100/193	54.64	Q3	−0.0350	92.39	−3.88	120.99
2012	0.5563	72/190	37.89	Q2	0.1315	130.96	14.44	158.44
2014	0.6047	56/193	29.02	Q2	0.0484	108.70	4.26	172.23
2016	0.6108	60/193	31.09	Q2	0.0061	101.01	0.50	173.97
2018	0.6893	60/193	31.09	Q2	0.0785	112.85	6.23	196.33

Figure 1 shows the trend of the EGDI index and with compound annual growth rate (CAGR) in Georgia for the period 2003–2018. According to data from 2003 to 2018, according to the EGDI index, Georgia was largely in the MEGDI and HEGDI category.

Highest growth of CAGR rate was by 14.44 in 2012, while the highest fall was by −3.88 in 2010 (Table 1 and Fig. 1). EGDI index has a growing trend with CGI = 196.33 and CAGR = 4.60% per year in 2018 compared to 2003.

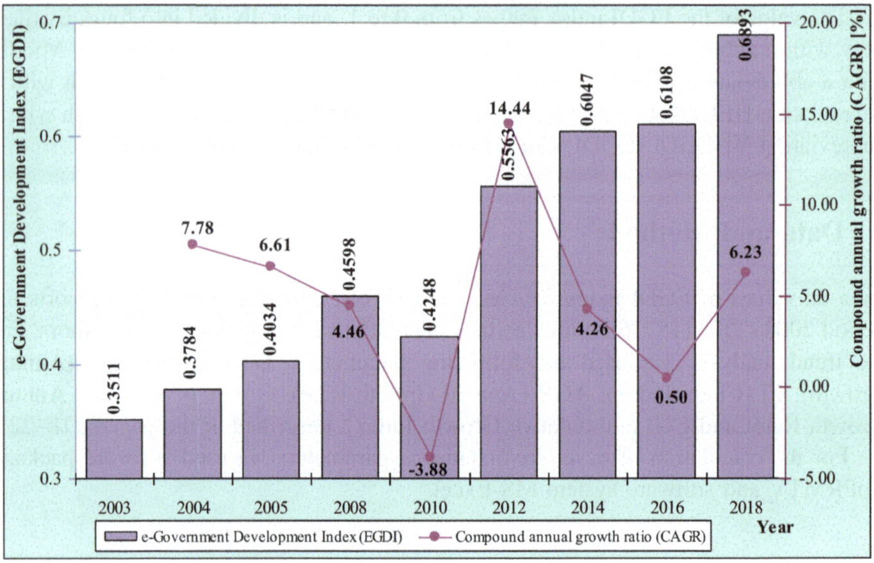

Fig. 1. Trend analysis of EGDI index with compound annual growth rate (CAGR) of Gergia for the period 2003–2018

On Table 2 are given data about Web measure index (WMI) and online service index (OSI) of Gergia for the period 2003–2018 [15, 16], with calculation by the authors for trend analysis parameters: G, CI, CAGR and CGI.

Table 2. Data about Web measure index (WMI) and online service index (OSI) of Gergia for the period 2003–2018

Year	Score	G	CI	CAGR	CGI
Web measure index (WMI) was used for the period 2003–2008					
2003	0.0480	–	–	–	100.00
2004	0.1470	0.0990	306.25	206.25	306.25
2005	0.2115	0.0645	143.88	43.88	440.63
2008	0.3545	0.1430	167.61	18.79	738.54
Online service index (OSI) was used for the period 2010–2018					
2010	0.2476	–	–	–	100.00
2012	0.6013	0.3537	242.85	55.84	242.85
2014	0.5984	−0.0029	99.52	−0.24	241.68
2016	0.6377	0.0393	106.57	3.23	257.55
2018	0.6944	0.0567	108.89	4.35	280.45

Figure 2 shows the trend of the OSI index and with compound annual growth rate (CAGR) in Georgia for the period 2010–2018. Highest growth of CAGR rate was by 55.84 in 2012, while the highest fall was by −0.24 in 2014 (Table 2 and Fig. 2). OSI index has a growing trend with CGI = 280.45 and CAGR = 13.76% per year in 2018 compared to 2010.

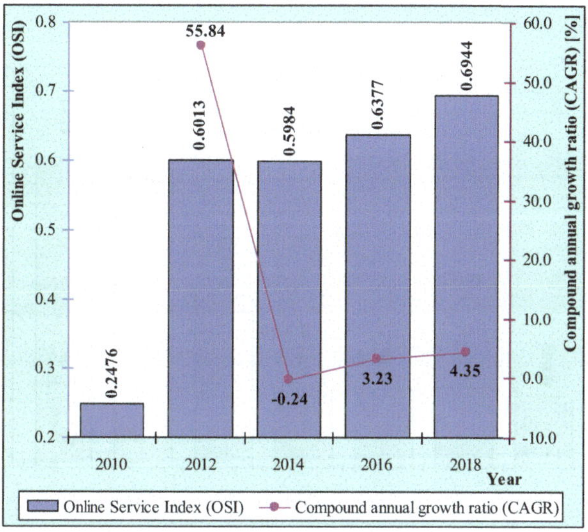

Fig. 2. Trend analysis of OSI index with compound annual growth rate (CAGR) of Gergia for the period 2010–2018

On Table 3 are given data about telecommunications infrastructure index (TII) for Gergia for the period 2003–2018 [15, 16], with calculation by the authors for trend analysis parameters: G, CI, CAGR and CGI.

Table 3. Data about telecommunications infrastructure index (TII) of Gergia for the period 2003–2018

Year	Score	G	CI	CAGR	CGI
2003	0.1150	–	–	–	100.00
2004	0.0980	−0.02	85.22	−14.78	85.22
2005	0.1086	0.01	110.82	10.82	94.43
2008	0.1072	0.00	98.71	−0.43	93.22
2010	0.1164	0.01	108.58	4.20	101.22
2012	0.2328	0.12	200.00	41.42	202.43
2014	0.4261	0.19	183.03	35.29	370.52
2016	0.4184	−0.01	98.19	−0.91	363.83
2018	0.5403	0.12	129.13	13.64	469.83

Figure 3 shows the trend of the TII index and with compound annual growth rate (CAGR) in Georgia for the period 2010–2018. Highest growth of CAGR rate was by 41.42 in 2012, while the highest fall was by −14.78 in 2004 (Table 3 and Fig. 3). TII index has a growing trend with CGI = 469.83% and CAGR = 10.87% per year in 2018 compared to 2003.

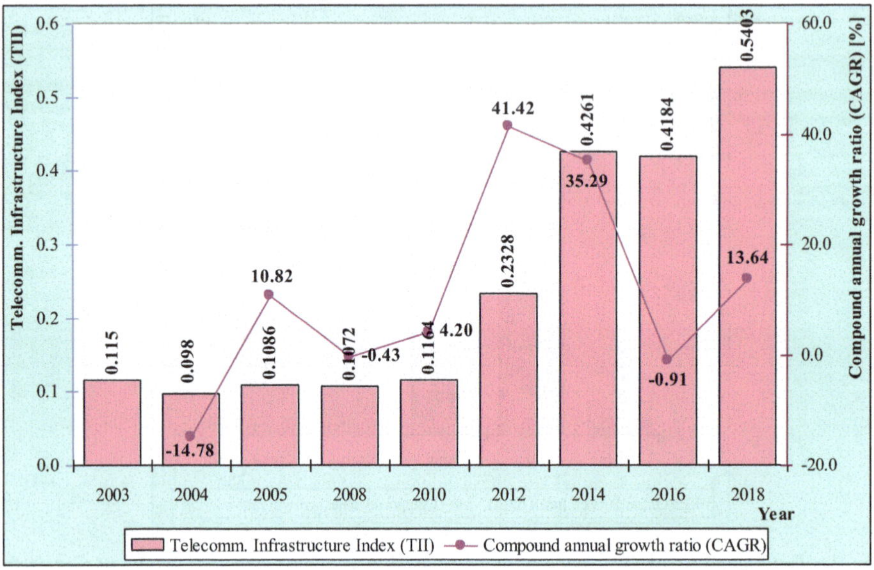

Fig. 3. Trend analysis of TII index with compound annual growth rate (CAGR) of Gergia for the period 2003–2018

On Table 4 are given data about human capital index (HCI) of Gergia for the period 2003–2018 [15, 16], with calculation by the authors for trend analysis parameters: G, CI, CAGR and CGI.

Table 4. Data about human capital index (HCI) of Gergia for the period 2003–2018

Year	Score	G	CI	CAGR	CGI
2003	0.8900	–	–	–	100.00
2004	0.8900	0.00	100.00	0.00	100.00
2005	0.8900	0.00	100.00	0.00	100.00
2008	0.9210	0.03	103.48	1.15	103.48
2010	0.9156	−0.01	99.41	−0.29	102.88
2012	0.8348	−0.08	91.18	−4.51	93.80
2014	0.7895	−0.05	94.57	−2.75	88.71
2016	0.7763	−0.01	98.33	−0.84	87.22
2018	0.8333	0.06	107.34	3.61	93.63

Figure 4 shows the trend of the HCI index and with compound annual growth rate (CAGR) in Georgia for the period 2010–2018. Highest growth of CAGR rate was by 3.61 in 2018, while the highest fall was by −4.51 in 2012 (Table 4 and Fig. 4). HCI index has a decrease trend with CGI = 93.63% and CAGR = −0.44% per year in 2018 compared to 2003.

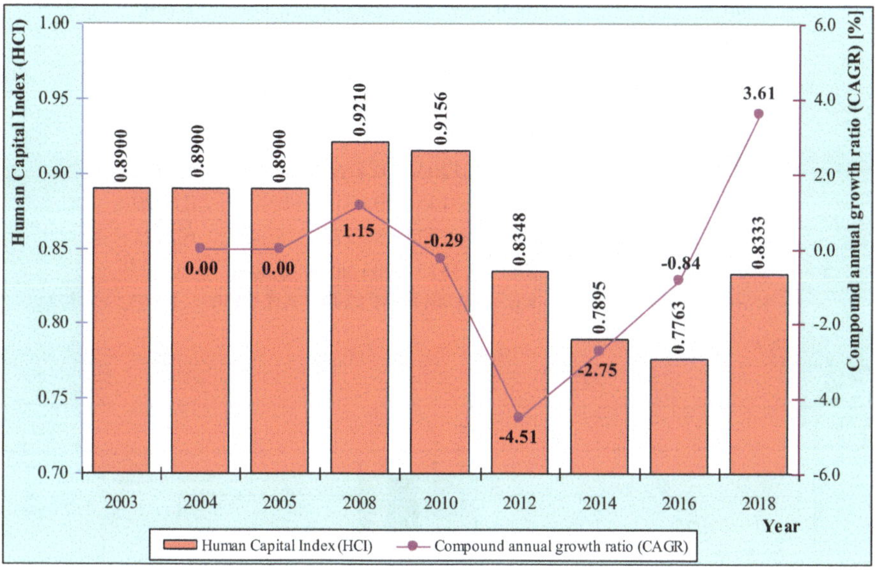

Fig. 4. Trend analysis of HCI index with compound annual growth rate (CAGR) of Gergia for the period 2003–2018

Descriptive statistics of EGDI index for Gergia for period 2003–2018 is given in Table 5. The data about EGDI index changed in intervals from 0.3511–0.6893.

On Fig. 5 is given of indicators of central tendency for EGDI, OSI, TII and HCI indices of Gergia for the period 2003–2018.

The lowest value of EGDI index in Gergia was in 2003 (0.3511) and the highest in 2018 (0.6893), with AM = 0.4976, Med = 0.4598, SD = 0.1203 and CV = 24.1724 (Table 5 and Fig. 5).

The lowest value of OSI index in Gergia was in 2010 (0.2476) and the highest in 2018 (0.6944), with AM = 0.5559, Med = 0.6013, SD = 0.1766 and CV = 31.7745 (Table 5 and Fig. 5).

The lowest value of TII index in Gergia was in 2004 (0.0980) and the highest in 2018 (0.5403), with AM = 0.2403, Med = 0.1164, SD = 0.1742 and CV = 72.4735 (Table 5 and Fig. 5).

The lowest value of HCI index in Gergia was in 2016 (0.7763) and the highest in 2008 (0.9210), while in the rest of the period the values for the HCI index varied around arithmentic mean (AM = 0.8601) (Table 5 and Fig. 4).

Table 5. Descriptive statistics for EGDI, OSI, TII and HCI of Gergia for the period 2003–2018

Measures	Sign	EGDI	OSI	TII	HCI
Sample size	N	9	5	9	9
Minimum	Min	0.3511	0.2476	0.0980	0.7763
Maximum	Max	0.6893	0.6944	0.5403	0.9210
Range (Max-Min)	R	0.3382	0.4468	0.4423	0.1447
Total	Sum	4.4786	2.7794	2.1628	7.7405
Arithmetic mean	AM	0.4976	0.5559	0.2403	0.8601
Geometric mean	GM	0.4849	0.5239	0.1913	0.8586
Harmonic mean	HM	0.4727	0.4816	0.1581	0.8570
Median	Med	0.4598	0.6013	0.1164	0.8900
Std. deviation	SD	0.1203	0.1766	0.1742	0.0534
Variance	Var	0.0145	0.0312	0.0303	0.0029
Coefficient of variation	CV	24.1724	31.7745	72.4735	6.2121
Skewness	Sk	0.3513	−1.9601	0.8415	−0.5224
Kurtosis	Ku	−1.4634	4.1399	−1.1350	−1.2632

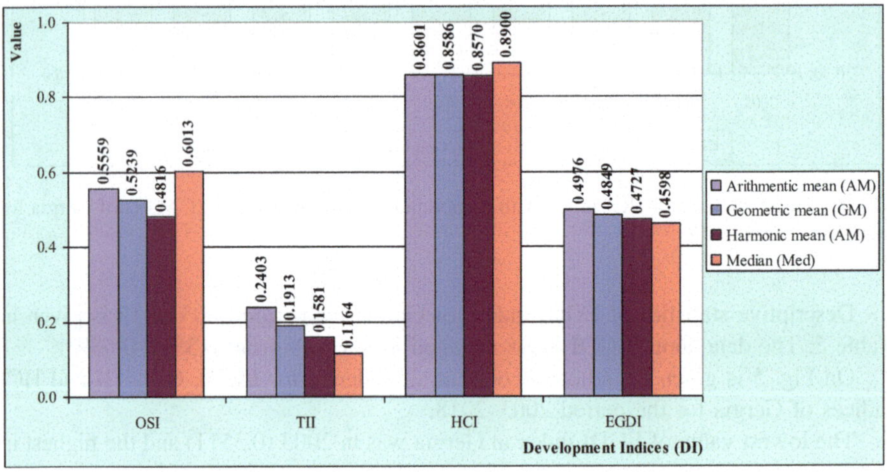

Fig. 5. Graphical representation of indicators of central tendency for EGDI, OSI, TII and HCI indices of Gergia for the period 2003–2018

4 Conclusion

Based on statistical and trend analysis, the EGDI index in Georgia for the period 2003–2018 had an increase with CGI = 196.33 and CAGR = 4.60% per year in 2018 compared to 2003. The highest growth trend have: TII index with CGI = 469.83% and CAGR = 10.87% per year in 2018 compared to 2003 and OSI index with CGI = 280.45 and CAGR = 13.76% per year in 2018 compared to 2003. The decrease trend there are HCI index with CGI = 93.63% and CAGR = −0.44% per year in 2018 compared to 2003.

According to the EGDI index, for the observed period 2003–2018, Gergia was in quartile Q2, except in 2003 and 2010 when it was in Q3.

For OSI and HCI indices the arithmetic mean (AM) is smaller than the median (Med), while for the EGDI and TII indices the arithmetic mean (AM) is greater than the median (Med).

Acknowledgment. This work was supported by Shota Rustaveli National Science Foundation (SRNSF) [PHDF-19-2224, Improving the efficiency of mechatronic systems in order to ensure the reform of "Industry-4.0"].

References

1. Almarabeh, T., AbuAli, A.: A general framework for e-government: definition maturity challenges, opportunities, and success. Eur. J. Sci. Res. **39**(1), 29–42 (2010)
2. Silcock, R.: What is e-government. Parliamentary Aff. **54**(1), 88–101 (2001)
3. Twizeyimana, J.D., Andersson, A.: The public value of e-government - a literature review. Gov. Inf. Q. **36**(2), 167–178 (2019)
4. Zouridis, S., Thaens, M.: E-government: towards a public administration approach. Asian J. Public Adm. **25**(2), 159–183 (2003)
5. Adeyemo, A.B.: E-government implementation in Nigeria: an assessment of Nigeria's global e-gov ranking. J. Internet Inf. Syst. **2**(1), 11–19 (2011)
6. Agboh, D.K.: An assessment of Ghana's global E-government UN ranking. J. Technol. Res. **8**, 1–17 (2019). Article no. 172605
7. Alkhwaldi, A., Kamala, M., Qahwaji, R.: Analysis of cloud-based e-government services acceptance in Jordan: challenges and barriers. J. Internet Technol. Secur. Trans. (JITST) **6**(2), 556–568 (2018)
8. Basu, S.: E-government and developing countries: an overview. Int. Rev. Law Comput. Technol. **18**(1), 109–132 (2004)
9. Lin, F., Fofanah, S.S., Liang, D.: Assessing citizen adoption of e-government initiatives in Gambia: a validation of the technology acceptance model in information systems success. Gov. Inf. Q. **28**(2), 271–279 (2011)
10. Qader, N.N., Kheder, M.Q.: Challenges and factors affecting the implementation of e-government in Iraq. J. Univ. Hum. Dev. **2**(3), 476–481 (2016)
11. Qaisar, N., Ali Khan, H.G.: E-government challenges in public sector: a case study of Pakistan. Int. J. Comput. Sci. Issues (IJCSI) **7**(5), 310–317 (2010)
12. Raman, M., Kaliannan, M., Yu, C.: E-business and e-government: Issues and challenges in Malaysia. Inf. Technol. J. **6**(3), 428–434 (2007)
13. Ramli, R.M.: E-government implementation challenges in Malaysia and South Korea: a comparative study. Electron. J. Inf. Syst. Dev. Countries (EJISDC) **80**(7), 1–26 (2017)
14. Williams, M.D.: E-government adoption in Europe at regional level. Transf. Gov. People Process Policy **2**(1), 47–59 (2008)
15. United Nations e-Government Survey 2008: From e-Government to connected governance. United Nations (UN), New York (New York - USA) (2008). 270 pp
16. United Nations e-Government Survey 2018: Gearing e-government to support transformation towards sustainable and resilient societies. United Nations (UN), New York (New York - USA) (2018). 270 pp

17. Turmanidze, R., Dašić, P., Popkhadze, G.: Digital infrastructure in Georgia as a condition for successful application Industry 4.0. Int. Sci. J. "Industry 4.0" **5**(1), 3–6 (2020)
18. Dašić, P.: Application of polynomial regression models for approximation of time series. J. Econ. Manag. Based New Technol. **1**(2), 81–160 (2012)
19. Dašić, P.: Statistical analysis and modeling of global competitiveness index (GCI) of Serbia. J. Res. Dev. Mech. Ind. **10**(4) (2018)
20. Stanivuk, T., Dašić, P., Aščić, A.: Approximation of global competitiveness index (GCI) for Croatia using polynomial regression model. In: Proceedings of the 7th International Conference "Economics and Management-Based on New Technologies" (EMoNT 2017), Vrnjačka Banja, Serbia, 15–18 June 2017, pp. 22–31. SaTCIP Publisher Ltd., Vrnjačka Banja (2017)
21. Tošović, R., Dašić, P., Ristović, I.: Sustainable use of metallic mineral resources of Serbia from an environmental perspective. Environ. Eng. Manag. J. **15**(9), 2075–2084 (2016)
22. Turmanidze, R., Dašić, P., Popkhadze, G.: Analysis of some techno-economical indicators of Georgia. J. Res. Dev. Mech. Ind. **11**(4) (2019)

Valorization of Montenegrin Lighthouses as Destination Icons Through Different Forms of Multimedia Arts

Anđela Jakšič-Stojanović[1(✉)] and Neven Šerić[2]

[1] University Donja Gorica, Podgorica, Montenegro
andjela.jaksic@udg.edu.me
[2] Faculty of Economics, University of Split, Split, Croatia

Abstract. In the last decades, many destinations all around the world put a lot of efforts in adequate preservation, valorization and promotion of lighthouse heritage. There are spectacular examples of the lighthouses all over the world that became destinations' icons- it is well known that many lighthouses and maritime structures became idyllic settings for weddings, celebrations, receptions, educational programs etc., while some became famous because famous museums, interesting tourism tours, extremely valuable design or idyllic surrounding.

At the moment in Montenegro there are more than forty-five lighthouses and some of them have significant importance from historical, cultural or some other point of view. But, unfortunately, they are not listed in the register of cultural and historical monuments, nor protected by any law, so their fate in the era of modernization, globalization, rapid foreign investments etc. is quite uncertain. Because of that fact, it is necessary to create completely new marketing-management approach which is based on adequate preservation, protection, valorization and promotion of lighthouses and their historical, cultural and architectural values. Our concept is based on the usage of different forms of multimedia arts and modern ICT technologies which represents completely new approach in promotion of this segment of cultural heritage- not only in Montenegro, but in the world as well.

Keywords: Lighthouses · Contemporary art · ICT technologies · Tourism · Valorization · Promotion

1 Introduction

Destination icons as symbols of tourist destination and country itself are a very powerful tool for attracting tourists and they represent attractive factors of destination itself [1]. Some of the most representative examples on global level are the Eiffel Tower in Paris, the Statue of Liberty in New York, the statue of Christ the Redeemer and the Maracana Stadium in Rio de Janeiro, double- decker in London and many others and their attractiveness may even override the attractiveness of the image of the destination itself [2]. A number of destination icons contribute to the attractiveness of the destination itself [3] while the higher degree of attractiveness of the destination contributes to its competitiveness on the global tourism market [4].

I. Karabegović (Ed.): NT 2020, LNNS 128, pp. 939–946, 2020.
https://doi.org/10.1007/978-3-030-46817-0_106

Destination icons may be natural or built attractions that are recognized by tourists as attractions and they usually represent one of the main reasons for a visit to particular destination [5]. They are very often used in tourism practice to differentiate a particular destination or tourist area from competitors which is extremely important having on mind the fact that identification of the points which may become new tourist icons becomes an imperative for all stakeholders in tourism industry [6].

In the last decades, the interest for valorization of lighthouse heritage from historical, archeological, cultural, tourist and many other points of view is continuously increasing and many governments, national, regional and local authorities and institutions, teams of experts and researchers all over the world have put a lot of efforts in order to preserve lighthouses and valorize and promote them as particular tourist attractions. According to the data from Lighthouses directory, there are more than 21700 lighthouses all around the world nowadays and many of them represent destination icons and the main reason of visit of particular destination [7]. In that sense, it is interesting to mention one good example from region- Croatian project "Stone lights" introduced new criteria of the relationship between man and the environment which significantly improved the quality of Croatian tourist offer and its position on global tourism market, differentiated it from its main competitors as well as set up new standards in destination tourism which should be reached in future [8, 9].

Having on mind that Croatia and Montenegro were parts of Yugoslavia for many decades and have a lot of in common regarding history, tradition, culture etc., as well as the fact that most of the lighthouses in both countries belong to the same cultural heritage built by Austro- Hungarians, it is not difficult to conclude that the Croatian practice regarding lighthouses could be successfully introduced in Montenegro as well. But the adoption of others experiences, no matter how successful and impressive are they, is not the idea that should be followed in long term [10]. In other words, completely new approach should be introduced which should be based on promotion of lighthouses as historic and cultural heritage through different forms of ICT technologies and multimedia arts.

2 Valorization of Montenegrin Lighthouse Heritage

There are not many available dates about lighthouses in Montenegro nor there is any institution that has updated base of lighthouses with some basic information about them such as name and location, position, characteristics, height, range, structure etc. For most of the lighthouses, the names of the constructors and the year of construction are not known as well. There are only few available photos of Montenegrin lighthouses- mostly made by tourists, usually of low quality and poor resolution. They do not belong to the register of cultural and historical monuments, they are not recognized as cultural heritage, there are scarce archeological data about their historical and cultural values, there is limited number of fairy tales and legends regarding their construction and interesting facts from the past. It is also important to mention that, unfortunately, lighthouses in Montenegro are not protected by any law so some of them are already demolished and destroyed, some of them are replaced by the new ones, some are constantly imposed to vandals etc.

That was the main reason for realization of the first research regarding Montenegrin lighthouse heritage that lasted for twelve months. The initial phase of this research started from collection of materials. Because of the fact that most institutions in Montenegro do not have relevant data regarding lighthouses, most of the data in this phase were collected from very scarce archive materials and from conversation with historians, art historians, curators etc. In next phase, data were additionally collected by field research and visit to every particular lighthouse in Montenegro that is presented on nautical maps (45 lighthouses exactly). The result of the research is creation of first Montenegrin lighthouses data base in Montenegrin history that consist professional photo base of lighthouses as well.

2.1 The Implementation of Multimedia Arts in Promotion of Lighthouse Heritage

The results of the research highlighted that in Montenegro there are some beautiful examples of lighthouses which have extreme potential for tourist, historical, cultural and other kind of valorization. But, unfortunately, the results of the research confirmed that lighthouses are not adequately (or to be more precise- not at all) preserved, protected, valorized and promoted and clearly indicated that their fate in the era of modernization is quite uncertain. In order to protect the lighthouses from future decline, it is first necessary to create a strategy of their revitalization, protection and valorization and carry out a set of measures and activities in order to position them as destinations icons. One of the possible approaches could be promotion of lighthouses and their historic, cultural, architectural etc. values through implementation of different forms of ICT technologies and multimedia arts which represents completely new approach in promotion of this segment of cultural heritage- not only in Montenegro, but in the world as well.

One on the most interesting places for realization may be island Mamula with spectacular fortress and beautiful lighthouse. This fortress served as one of the most cruel and terrifying prison in this part of Europe in the past and it was known as "YugoslavAuschwitz". The initial phase in valorization of this island would be the restauration of the for tress, establishment of historical museum as its integral part as well as organization of guided tours which would enable creation of completely new authentic tourists' experience. Let's just mention Alcatraz, the Birdman, Old Melbourne Gaol, Robben Island, Kilmainham Gaol etc. which attract great attention of tourists exactly because of that authentic, special and unusual experience. The implementation of ICT technologies could not only make the lighthouse the integral part of tourist offers, but also contribute to its attractiveness and uniqueness on tourism market.

In other words, Mamula with all its attractions in the same time may become spectacular *cinema/theatre on the water* and center of multimedia art festival which may rank Montenegro among top destinations in this part of Europe. By using different multimedia art techniques such as video mapping, the seaside would be transformed in fabulous water screen projector and the island with all its attractions such as fortress and lighthouse would become beautiful impressive scene and ambient which could attract a lot of different target groups into the destination itself. Lighthouse could also be used a place from which the projection should be realized. One of the interesting events would be the promotion of film Campo Mamula, Yugoslav black and white cult film, made in 1959 which describes a life of prisoners during the captivity, their struggle for survival etc. which could significantly contribute to authenticity of tourists' experience (Fig. 1).

Fig. 1. Scene from a film Campo Mamula presented on water screen

The water screen could also be used for realization of many different other events such as concerts of classic music, festivals of electronic music, festivals of lights, art colonies, festivals of multimedia art, multimedia opera spectacles etc. Authentic souvenirs designed by artists would be also created in order to remind tourists on spectacular experience from island Mamula (sculptures of lighthouse with water and light effects, fortress, reproduction of stone with message written at the entrance to the prison etc.).

Another example of valorization would be to use lighthouses for different projections of multimedia art works on different surfaces like stone, rocks, water etc. It would be very interesting for realization of *different multimedia projects and festivals* in which artists use text, images, lights, effects etc. in order to promote some idea (Fig. 2).

Fig. 2. Multimedia projection of a piece of art on stone and rocks

The multimedia festivals would be organized every year on different place in order to promote cultural, historical and architectural values of different lighthouses. For example, when talking about lighthouses in Ulcinj, it is important to point out there are a lot of historical and archeologic data, as well as legends and fairytales from the past that are connected with this town and that are still not adequately used and promoted in creation of tourist offer. One of them refers to history of Ulcinj as pirates' nest while other refers to slavery of Spanish writer Miguel de Cervantes. These elements could be very inspiring for multimedia artist as well as for the audience.

It could also be very attractive to make connection between existing festival and lighthouse heritage. In that sense, it would be interesting to realize project *Sea Dance-Disco Balls* which includes realization of festival Sea dance, as the best European medium-sized festival in ambient of different lighthouses which gathers numerous fans from more than 50 countries from all over the world, and engaged more than 100 famous international music stars and world-renowned artists from its establishment. It is not difficult to conclude that its popularity could significantly contribute to the positioning on lighthouses as destination icons and on the other side the lighthouses themselves and different stunning locations would have provided added value to the festival. Every year the festival would be organized on another place and different lighthouse would be promoted. For the purpose of realization of festival special disco balls and other requisites would be created by artists from different parts of the world and they would be used as a trade mark of festival. The disco balls should be placed on the lighthouse, as well as on the surface of the sea or sand. The lighthouse with shiny glossy disco balls would be used as a brilliant background and scenery for the festival and create magnificent impression based on the harmony between natural landscape, masterpiece of human and artistic work, the motive of lighthouse with disco balls would be a symbol of the festival and would be promoted through different media channels (Fig. 3).

Fig. 3. Sea Dance- Disco Ball- presentation on Tunja Lighthouse

Another interesting concept would be the realization of *art colony - project Put on the lighthouse*. The realization of this project includes different traditional or multimedia interventions on lighthouses by artists from all around the world. One approach could be based on traditional land-art techniques by putting on the lighthouse in different cloths made of natural materials such as leaves, sand, flowers, grass, stones etc. in a way that doesn't have negative effects on their appearance, construction, structure, materials etc., while other would be realized by implementation of modern ICT technologies and different multimedia forms. For example, it would be interesting to see the lighthouses dressed up in national costumes and traditional clothing of different parts of Montenegro or in traditional clothing of ex Yugoslavian countries or in national costumes of the countries where the artists come from. In that sense, by choosing inspiring themes and bringing artists from all over the world, the lighthouses would be used and promoted in one completely new way which would make themselves and the destination itself the epicenter of artistic and cultural entertainment in this part of Europe. Special guided tours would be organized by trip in order to present tourists masterpieces of multimedia art realized in nature (Fig. 4).

Fig. 4. Project put on the lighthouse, Turski Rt, Tivat

These are only some of the examples of the artistic projects that could be realized with lighthouses. It is also possible to realize various interesting projects regarding the artistic installations by traditional or multimedia art forms (Figs. 5 and 6).

Fig. 5. Project Lilihop, installation

Fig. 6. Artistic transformation

3 Conclusion

The concept *The Art of the Lighthouses* includes the realization of different multimedia art projects by using various contemporary art techniques which will present lighthouses in completely new, attractive, impressive way and make them some kind of destination icons. By combining these multimedia art techniques with the elements of event management, completely new, authentic tourist products would be created which will contribute to positioning of Montenegro as high-quality tourist destination with competitive advantage on the market and creating completely new image of the destination itself.

References

1. Križman Pavlović, D.: Marketing turističkedestinacije. Pula, Mikrorad (2008)
2. Morrison, A.M.: Marketing and Managing Tourism Destinations. Routledge, London (2013)
3. Kušen, E.: Turizam i prostor - klasifikacija turistickih atrakcija, Prostor, znanstveni casopis za arhitekturu i urbanizam **9**(1), 1–12 (2002)
4. Becken, S.: The role of tourist icons for sustainable tourism. J. Vacation Mark. **11**(1), 21–30 (2005)
5. Ramukumba, T.: The role of tourist icons for a destination. Afr. J. Hospitality Tourism Leisure **3**(2), 1–12 (2014)
6. Pearce, P., Morrison, A., Moscardo, G.: Individuals as tourist icons: a developmental and marketing analysis. J. Travel Res. **10**, 63–85 (2003)
7. The Lighthouse Directory (2020). https://www.ibiblio.org/lighthouse/. Accessed 4 Jan 2020)
8. Šerić, N.: Kamena svjetla, priče i legende o jadranskim svjetionicima, o mjestima na kojima su izgrađeni i o njihovom podmorju. Marjan Tisak, Split (2004)
9. Šeric, N.: Ekološki turisticki projekt Kamena Svjetla – cuvari lanterni povijesnih vrijednosti. Jadranska Meteorologija review **48**, 74–79 (2003)
10. Jakšić- Stojanović, A., Šerić, N.: Montenegrin Lighthouses as Destination Icons. University Mediterranean, Podgorica (2019)

Measuring Youth Attitudes on Materialism as a Consequences of a Consumer Society

Emir Kurtović(✉), Savo Stupar, Naida Jažić-Asotić, and Mirha Bičo Ćar

School of Economics and Business, University of Sarajevo, Trg oslobodjenja - Alija Izetbegovic 1, 71000 Sarajevo, Bosnia and Herzegovina
emir.kurtovic@efsa.unsa.ba

Abstract. This paper deals with the analysis of young people's attitudes towards materialism, viewed through three key dimensions: possession, selfishness and envy. In order to examine the attitudes of young people in Bosnia and Herzegovina, an empirical study was conducted on a sample of 206 respondents, aged 20 to 30 years, with an active residence in the country. The results of the research show that, contrary to our basic assumption, young people in BiH are still not strongly influenced by materialism, that is, materialism in young people is not related to their life values, i.e. the perception of their value towards important things in life. It has also been observed that the number of years of youth does not increase the propensity for materialistic values, and that the increased religiosity of young people leads to a decrease in the propensity for materialism. Taking into account all the limitations of the conducted research, the paper concludes that materialism, as a kind of social and cultural phenomenon that is leading in shaping the contemporary global image of society, is not a dominant form in Bosnian society and does not affect the behavior of young people.

Keywords: Youth · Materialism · Consumerism · Values

1 Introduction

Although we usually focus on the benefits that modern technology and modern society offer, in parallel with the processes of improvement provided by new technologies, modern innovations and technologies bring (directly and indirectly) some negative changes too. Among the undesirable changes are certainly those at the sociological level driven by hyper-production and indirectly by hyper-consumption. The emergence of consumerism and the materialism associated with it are exactly the changes that are becoming more and more prevalent in modern society, especially among young people.

The study of materialism has a long and varied history in various scientific disciplines, especially in the fields of philosophy, sociology, psychology, anthropology, etc. So we can say that materialism is not so new concept or topic for discussion and research in the sphere of marketing, especially consumer behavior. Nevertheless, marketers rely heavily on the materialistic nature of consumers to entice them to buy their goods or services. Although there is a great deal of research on this subject, we

I. Karabegović (Ed.): NT 2020, LNNS 128, pp. 947–954, 2020.
https://doi.org/10.1007/978-3-030-46817-0_107

can say that so far there is no absolute definition of materialism. According to some authors, materialism is a sign of a prosperous and secure society, while others consider it a negative consequence of a consumer-oriented society. When analyzing the available literature, it is generally observed that a negative attitude towards materialism dominates, while there are those authors who consider that materialism does not necessarily have to be either a positive or a negative phenomenon [1]. In addition, research by some authors confirms that they find a negative correlation between measuring envy, selfishness and possessiveness, and measuring satisfaction and happiness in life [2]. Measuring young people's attitudes towards materialism as a consequence of a consumer society is important, since it is young people who are most vulnerable to social influences, and exposed to different social pressures that are marketed to them through the media, the Internet and through various social groups to which they either belong or aspire to belong. The topic is significant because young people are the driving force of society, and it needs to be explored how much materialism is a phenomenon among young people in BiH society, and how much it affects their awareness and values that they aspire to in life?

2 Materialism as a Feature of the Consumer Society

There are several approaches to the conceptualization of materialism in the literature: materialism as a characteristic of personality, materialism as an indicator of value, materialism as a cycle, materialism as a sociological phenomenon, and materialism as attachment. Materialism can also be observed on an individual and social level. While at the individual level, materialism is perceived through individual own needs, goals, and self-esteem, the social approach implies that materialism is accepted as a value system in a particular society. It is also often pointed out that materialism is in fact the importance that the consumer attaches to worldly values [2], not to those of a spiritual character. Materialism can also be seen as a lifestyle where happiness is achieved through material objects, wealth or status [3]. In fact, the antisocial aspect of materialism threatens the well-being of individuals and conflicts with socially-oriented values, such as a sense of family spirit and religiosity. Research has shown that materialism is negatively correlated with high ethical standards of a consumer [4]. In addition, research that has focused on the impact of religion on consumer materialism shows that the positive effects of possessing on happiness (which is one of the dimensions of materialism) cannot be determined with certainty [5]. On the contrary, twenty years of research have consistently shown that the moment a need is met, the rise in wealth is short-lived satisfaction that has minimal impact on happiness [6]. A satisfied need is the beginning of the desire to meet new needs and there is no end to achieving satisfaction when there is always a desire for a new one [7]. Opinions have been, and will remain, divided on whether materialism is a positive or negative phenomenon of society nowdays, as has been confirmed by previous studies. In this context, the question is whether materialism is essentially an egoistic phenomenon that opposes altruism and other prosocial behaviors, does materialism affect interpersonal relationships, is the gift of material an adequate and correct substitute for a more direct expression of love between people? One of the problems that emerges in materialism research is the relationships and contributions of maintaining a positive self-identity [8]. The findings of some authors indicate that possession has therapeutic effect and value for the elderly and the mentally ill, as well as that materialism can lead to dissatisfaction in life, when materialists evaluate their standard of living using unrealistic (ideal) expectations, which increases the possibility of evaluating its standard of living as negative [9].

Excessive consumption, which goes beyond meeting the basic needs, results in another phenomenon of consumerism. It is a culture that is essentially aimed at promoting the over-acquisition of tangible goods or services through the sales process and which is based on materialism. Although these are essentially complementary terms, materialism could nevertheless be seen as a "value framework" of contemporary consumerist theory at the general level, that is, an incentive for consumer behavior on the individual [10]. It is certain that in such circumstances, the role of marketing as a business philosophy that focuses on satisfying the needs and desires of consumers while generating profits, faces the challenge of preventing unnecessary sales and consumption based on consumer lending, where even marketing guru Philip Kotler poses the dilemma that is it "the task of marketing to sell materialism and consumption" [4, 11]. A growing number of recent research is elaborating on the connection between materialism and external migration, especially of the young population leaving the country of origin for economic and social reasons [12].

3 Materialism and Youth

The young people are hit hard by the effects of materialism because they live in an environment that reinforces the importance of owning things, especially expensive things. Different cultures show different levels of acceptance of materialism. What can be taken as a fact is that materialism is more affirmed and accepted in the US than other developed countries of the world, although in recent times this value system has been increasingly accepted in China too [13]. There are numerous studies related to research on materialism and youth. The results are different and very difficult to generalize. Thus, Ger and Belk [14] have shown that Romanian students show the greatest propensity for materialism values, in a sample of twelve countries, in their well-known cross-cultural research. They are followed by students from the US, New Zealand, Ukraine, Germany, and Turkey. Students from Sweden, France, the United Kingdom, India, Thailand and Israel show the lowest levels of materialism. The study concluded that variables such as dramatic cultural or social changes could lead to an increase in materialism. They also concluded that materialism, as a phenomenon, occupies a significant place in both Eastern and Western cultures, in developed and underdeveloped economies, and in countries that prefer both individual and collectivist values. Another study was conducted among students from Mexico, China, and the US, where their propensity for materialism was measured. The results showed that Mexican students are the least prone to materialism, while those in China have a very high preference [15]. Materialism research has also been conducted with master students and their parents in France and the US, and it has been shown that students in both countries are definitely more materialistic than the elderly. The same research shows that US adolescents are materialists influenced by advertising and peer pressure toward buying the latest products. There is a phenomenon in China called the "little emperor". It is a child who gets everything he wants from his parents and grandparents. According to materialism researchers, this kind of policy has made a generation of spoiled, demanding and materialistic children [16]. Unfortunately, the only thing that all researchers agree on is that materialists are less happy and content in life than non-materialists. Of particular interest are the findings of longitudinal studies of the levels of materialism in the United States over 30 years (1976–2007), which indicate a significant increase in materialism among high school students since the mid-1970s, with a peak among Generation X [17]. In addition, the results show that young people whose growing up are marked by social instability (e.g. unemployment) and social exclusion (e.g. an increase in the number of divorced parents), in adulthood, are particularly receptive to materialistic values.

When it comes to research on youth and materialism in our region, a relatively small number of studies are noted. Some studies treat the problem of materialism as a cause of emigration intentions (Example of Croatia), which is the most acute demographic and economic problem in our country too. The authors associated poor economic conditions and materialistic values with the tendency to emigrate from Croatia. The results showed that Croats are not highly oriented towards materialism, and that materialism varies more among individuals than among different status groups [12]. Similar findings are found in the study of materialism, status and conspicuous consumption among students in Croatia, where it is stated that students are materially oriented below average and are not prone to status and conspicuous consumption [18]. Buying well-known brands, customers are prone to status consumption, while showing or publicly using products is a conspicuous consumption. When it comes to Bosnia and Herzegovina, we can say that this paper is one of the first on this topic.

4 Methodology

In accordance with the aim of the research, i.e. the attempt to determine whether young people have a positive attitude (degree of acceptance) towards materialism, the following hypotheses were defined:

H0: "Young people have a positive attitude towards materialism."
H1: "Materialism is related to the values that young people strive for in life."
H2: "With the increase in years of life within the age group under the study, increases the tendency to materialism manifested through possession, indecency and envy."
H3: "With increasing degree of religiosity within the examined age group decreases the tendency towards materialism, manifested through possession, indecency and envy."

The survey was conducted through the social network Facebook, a list of respondents' private e-mails and face-to-face. In the first two cases, an online questionnaire was used, while in the case of personal interviewing, a printed version of the questionnaire was used. As a basic research instrument, a questionnaire consisting of a total of 56 questions was grouped into 7 specific units (groups). The first three groups of questions relate to measuring the degree of propensity to materialism of respondents, i.e. its dimensions: possessiveness, inviolability and envy [2]. In this case, 24 questions were used (standardized interval/Likert scale with 5 levels of agreement). The fourth group of questions (6 in total) concerns the measurement of respondents' (young people's) attitudes towards materialism (i.e. the degree of acceptance). This scale was developed by Moschis and Churchill (1978) [19]. The fifth group of questions relates to life values (intrinsic individual values, extrinsic values, intrinsic interpersonal values) - 9 questions in total. This scale is known as the "List of values" - the LOV developed by Kahle [20]. The sixth group of questions relates to the religiosity of the respondents (standardized interval/Likert scale with 5 levels of frequency). This scale is also called the Centrality of Religiosity Scale (CRS) and was developed by Huber [21]. Finally, questions related to the demographic characteristics of the respondents (gender, age, education, socio-professional status, job satisfaction and monthly income) were also used. The research was conceived and realized as an empirical (quantitative) research.

The sample included 206 young respondents (individuals aged 20 to 30 years) with an active residence in BiH (sampling method). The questionnaire was filled in anonymously, which enabled the objectivity of the testimonies of the respondents and thus the data collected. The respondents were acquainted with the purpose of data collection during the survey, and the data obtained from this survey were processed in the SPSS statistical program. Descriptive analysis, reliability analysis (Cronbach's alpha), Kolgomor-Smirnov test and correlation analysis were used in the presentation of the research findings. Most of the respondents in the sample were women (137 or 66.5%). 67.7% of respondents were between 27 and 30 years old, which is the most dominant segment, while 13.6% of respondents belonged to the group of 20 to 23 years. The sample was dominated by respondents with a university degree (119 or 57.8%). 19.9% of respondents had completed high school, while 28 or 13.6% had a master's degree and 7% had a doctorate. Most respondents were employed (128 or 62.1%), while the number of students and unemployed persons was almost equal (19.4% and 18.4%). The largest number of respondents in the sample has monthly incomes of less than 500 KM (33.5%), followed by those from 1000 to 1500 KM (27.2%), and the smallest number of respondents is those who have incomes between 2.001 and 2.500 KM. Table 1 presents the results of descriptive statistics relating to the

Table 1. Descriptive statistics

Scale name	Arithmetic mean	Cronbach Alfa	Standard deviation
1. Propensity for materialism (1–5) (1.1) possession (1.2) indecency (1.3) envy	2.85 (1.1) 3.64 (1.2) 2.31 (1.3) 2.61	$\alpha = 0.715$ $\alpha_{1.1} = 0.760$ $\alpha_{1.2} = 0.765$ $\alpha_{1.3} = 0.753$	0.296 (1.1) 0.471 (1.2) 0.492 (1.3) 0.503
2. Degree of acceptance of materialism (1–5)	2.47	$\alpha = 0.726$	0.669
3. The scale of values (1–9) (3.1) Intrinsic individual values (3.2) Extrinsic values (3.3) interpersonal values	 (3.1) 7.68 (3.2) 7.66 (3.3) 7.46	 $\alpha_{3.1} = 0.729$ $\alpha_{3.2} = 0.849$ $\alpha_{3.3} = 0.766$	 (3.1) 1.040 (3.2) 1.267 (3.3) 1.240
Religiosity (1 do 5)	3.11	$\alpha = 0.837$	0.836

measurement of propensity for materialism, the degree of acceptance of materialism, the degree of value and religiosity. Based on the presented results related to reliability measurement, all scales show a high degree of internal homogeneity, that is, reliability because the Cronbach Alpha indicator is greater than 0,7.

The results obtained indicate that the respondents have a low level of acceptance of materialism (the overall average on the 6 questions on the scale is 2.47). When it comes to the propensity for materialism, we can conclude that young people have a low propensity for materialism (the total average of 24 questions on the scale is 2.85). Regarding the individual dimensions of materialism, the respondents (young people) had the highest degree of agreement in the case of possession (total average of 9 questions on the scale is 3.64), then envy (total average of 8 questions on the scale is 2.61), and indecision. (the overall average for the 7 questions on the scale is 2.31). The results obtained indicate that the respondents (young people) have the most pronounced tendency towards intrinsic individual values (the total average on 4 questions from the scale is 7.68). In terms of extrinsic values, the agreement rate of the respondents (young people) is relatively high (the overall average on the 3 questions on the scale is 7.66). In the case of intrinsic interpersonal values, there is also a high degree of agreement (the overall average for the 2 questions on the scale is 7.46). Regarding the degree of religiosity, the results obtained indicate that the respondents have a pronounced degree of religiosity (the overall average for the 7 questions on the scale is 3.11).

5 Results of the Research

In the case of the main hypothesis, the Kolmogorov-Smirnov test was used in the analysis to determine the extent to which the subjects were inclined (deviation to answers greater than 3) or not toward accepting materialism (deviation to answers less than 3). The results obtained are presented in tables below (Table 2). Materialism has been tested through several variables such as the happiness is bought by money, the desire to own expensive things, judging other people based on the things at their disposal, buying things to impress other people, the importance of money when choosing a job. The presented results showed that young people in B&H do not attach much importance to these variables (which averaged a grade of 2.47), while applying the above statistical test confirmed that young people do not have a positive attitude towards materialism, thus rejecting the basic hypothesis.

Table 2. Young people's attitudes towards materialism

Kolgomorov-Smirnov test – single T test						
Variable	N	Mean	SD	t	df	P (sign.)
Attitudes on materialism	206	2,47	0,699	−11,289	205	p = 0,000

When it comes to hypothesis H1, four variables are included in the analysis: degree of materialism, intrinsic individual values, extrinsic values, and intrinsic interpersonal values. Correlation analysis found that materialism in young people in B&H does not affect values in life, the perception of their value in relation to important things in life, thus rejecting this hypothesis. Specifically, between materialism and intrinsic interpersonal values, there was a statistically significant negative association ($R = -0.164$, $p = 0.018 < 0.05$). In the case of the correlation between materialism and the other two dimensions of intrinsic and extrinsic values, no statistically significant correlation was observed. Thus, in the case of intrinsic individual values, the results are ($R = -0.121$, $p = 0.083 > 0.05$) and extrinsic values ($R = -0.105$, $p = 0.134 > 0.05$). In H2 testing, five variables were used: age, materialism acceptance rate, possession, nonresponsiveness, and envy. Correlation analysis showed that between: the degree of materialism, possession, lack of respect and envy, on the one hand, and age, on the other, there was almost no correlation, so this hypothesis was rejected. The following illustrates the results of the Pearson correlation, which indicate that there is no statistically significant association between young people's age on one hand and materialism ($R = -0.036$, $p = 0.607 > 0.05$) on the other. There is also no statistically significant correlation between age and possession ($R = -0.039$, $p = 0.580 > 0.05$), indecency ($R = 0.009$, $p = 0.902 > 0.05$), and envy ($R = -0.036$, $p = 0.609 > 0.05$). Finally, research has tested the extent to which religiosity reduces the propensity for materialism. In this case, too, five variables were used, one of them being religiosity and the other the degree of materialism, possession, lack of respect and envy. The Pearson correlation results indicate that there is a statistically significant negative association between religiosity on the one hand, and degree of materialism ($R = -0.218$, $p = 0.002 < 0.05$), possession ($R = -0.187$, $p = 0.007 < 0.05$), and envy in young people ($R = -0.206$, $p = 0.003 < 0.05$), on the other hand. In the case of indecency ($R = -0.039$, $p = 0.581 > 0.05$), no statistically significant association was observed. Based on the aforementioned results of the research, we can conclude that as the degree of religiosity within the examined age group increases, the tendency towards materialism expressed through the dimensions of possession and envy decreases.

6 Conclusion

The findings of empirical research confirm that young people in B&H are still not strongly influenced by materialism, which is in line with research by other authors who have confirmed a lower level of materialism in less developed countries. Religion has also been shown to have a significant influence on attitudes toward materialism in young people. It is also important to point out some research limitations that primarily relate to a relatively small sample. Although a sample of respondents from all over Bosnia and Herzegovina was used, we believe that the size and structure of the sample is still insufficient in terms of reaching general conclusions. If the sample included more respondents with different characteristics, it would be possible to obtain different research results.

References

1. Csikszentmihalyi, M., Rochberg-Halton, E.: The Meaning of Things: Domestic Symbols and the Self. Cambidge University Press, Cambidge (1981)
2. Belk, W.R.: Materialism: trait aspects of living in the material world. J. Consum. Res. **12**, 265–280 (1985)
3. Watson, D., Tellegen, A., Clark, L.: Development and validation of brief measures of positive and negative affect: the PANAS scales. J. Pers. Soc. Psychol. **54**(6), 1063–1070 (1988)
4. Muncy, J., Eastman, J.: Materialism and consumer ethics: an exploratory study. J. Bus. Ethics **17**(2), 137–145 (1998)
5. Pace, S.: Does religion affect the materialism of consumers? An empirical investigation of buddhist ethics and the resistance of the self. J. Bus. Ethics **112**(1), 25–46 (2012)
6. Ahuvia, A.: If money doesn't make us happy, why do we act as if it does? J. Econ. Psychol. **29**(4), 491–507 (2008)
7. Cova, B., Kozinets, R., Shankar, A.: Consumer Tribes. Routledge, Oxford (2007)
8. Belk, W.R., Tumbat, G.: The cult of Macintosh. Consum. Mark. Cult. **8**(3), 205–217 (2005)
9. Sirgy, M.J., Gurel-Atay, E., Webb, D., Čičić, M., Husić-Mehmedović, M., Ekici, A., Herrmann, A., Hegazy, I., Lee, D., Johar, J.: Is materialism all that bad? Effects on satisfaction with material life, life satisfaction, and economic motivation. Soc. Indic. Res. **110**(1), 349–366 (2013)
10. Hromadžić, H.: Konzumerizam: potreba, životni stil, ideologija. Naklada Jesenski i Turk, Zagreb (2008)
11. Kotler P.: The Relationship between marketing and Capitalism. Mark. J. (2016). http://www.marketingjournal.org/the-relationship-between-marketing-and-capitalism-phil-kotler/
12. Peračković, K., Rihtar, S.: Materijalizam kao društvena vrijednost i poticaj namjerama iseljavanja iz Hrvatske. Migracije i etničke teme, godina **32**, 295–317 (2016)
13. Podoshen, J.S., Li, L., Zhang, J.: Materialism and conspicuous consumption in China: a cross-cultural examination. Int. J. Consum. Stud. **35**, 17–25 (2011)
14. Ger, G., Russell, W.B.: Cross-cultural differences in materialism. J. Econ. Psychol. **17**(1), 55–77 (1996)
15. Parker, R.S., Haytko, D., Hermans, C.:The perception of materialism in a global market: a comparison of younger Chinese and United States. J. Int. Bus. Cult. Stud. **3**, 399–411 (2010)
16. Shao, T., Herbig, P.: Marketing implications of china's little emperors. Rev. Bus. **16**(1), 16–20 (1994)
17. Twenge, G.M., Kasser, T.: Generational changes in materialism and work centrality, 1976-2007: associations with temporal changes in societal insecurity and materialistic role modeling. Pers. Soc. Psychol. Bull. **39**(7), 883–897 (2013)
18. Nefat, A., Benazić, N.: Materijalizam, statusna i upadljiva potrošnja studenata u Hrvatskoj. Ekonomski pregled **65**(3), 241–259 (2014). Sveučilište Jurja Dobrile u Puli, Fakultet ekonomije i turizma "Dr Mijo Mirković"
19. Moschis, G.P., Churchill, G.A.: Consumer socialization: a theoretical and empirical analysis. J. Mark. Res. **15**, 599–609 (1978)
20. Kahle, L.: Social Values and Social Change: Adaption to Life in America, p. 1361. Praeger, New York (1983)
21. Huber, S., Huber, O.W.: The centrality of religiosity scale (CRS). Religions **3**, 710–724 (2012)

"What the Engineers Want" – Job Expectations of the Employees in IT Industry

Muamer Bezdrob[1(✉)] and Mirha Bičo Ćar[2]

[1] Faculty of Economics, American University in Bosnia and Herzegovina, Sarajevo, Bosnia and Herzegovina
mbezdrob@ping.ba
[2] School of Economics and Business, University of Sarajevo, Trg oslobođenja 1, 71 000 Sarajevo, Bosnia and Herzegovina
mirha.car@efsa.unsa.ba

Abstract. This research is aimed to analyze worker expectations of the workplace and work environment. In particular, it explores the difference in job expectations between the technical and non-technical staff within IT industry in the F BiH. To get a deeper insight into research topics, a questionnaire was developed and had been continuously sent to all eligible employees of a single IT company each year from 2008 to 2018. The research results show that the main difference between the technical and non-technical staff within the IT industry is related to self-actualization and professional advancement. At the same time, there is no difference between the two groups of employees when it comes to the general characteristics of the work environment and the competitiveness of the job environment. These findings point to individualism as a particular personality trait that makes the difference between technical and non-technical employees in the IT industry.

Keywords: Job expectations · Job environment · IT industry · Engineers

1 Introduction

In a time of globally relentless competition for talents and a superior workforce, which is especially inherent to the high-tech industries, employee attraction and retention is one of the critical business issues. Since "knowledge workers" from high-tech industries, such as IT, are incredibly mobile and show a strong inclination toward individualism, creating a working environment that would meet their job expectations becomes of great importance for the companies' managers [1].

The purpose of this research is to adopt the existing knowledge base and to further explore and analyze certain factors that underlie the employees' job expectations. Thus the first research question of this study is: *What are the main factors that determine employees' job expectations.*

Additionally, the research primarily explores the difference in job expectations between the technical and non-technical staff that work within the IT industry. This fact leads to the second research question of this study: *What are the significant differences in job expectations between technical and non-technical staff.*

I. Karabegović (Ed.): NT 2020, LNNS 128, pp. 955–963, 2020.
https://doi.org/10.1007/978-3-030-46817-0_108

To further investigate these issues, relevant research was designed, and appropriate research model was proposed. The research model was subsequently tested using the data gathered through the survey conducted within a single IT company from the F BiH.

2 Literature Review

Two very important and widely exploited concepts within social sciences and, especially, within the field of organizational behavior are the employees' job satisfaction concept and the job motivation concept. The job satisfaction – *a positive feeling about a job*, is, arguably, the best single predictor of employees' performance, turnover, absenteeism, and withdrawal behaviors. The job motivation – *worker's intensity, direction, and persistence of effort toward attaining a goal*, is another powerful predictor of employee productivity, absenteeism, and turnover [2]. It is generally accepted that satisfied and well-motivated employees are more effective, engaged, and perform on a higher level [3]. Furthermore, those employees are more connected to the organizational goals and will opt for a longer stay at their companies [1].

A very important constituent of both job satisfaction and motivation concepts is the employees' job expectations. In accordance with Herzberg's two-factor theory [4, 5], in order just not to be dissatisfied, employees expect that so-called hygiene factors (wages, job security, working conditions, ...) are, at least, adequate. At the same time, to be motivated and, consequently, satisfied, employees look for intrinsically rewarding job characteristics – personal growth, recognition, and achievement. Employees' motivation, as the Vroom's expectancy theory asserts, depends on the strength of their expectation of a particular outcome and attractiveness of such outcome [2].

Finally, the met expectations hypothesis [6] states that when an employee's expectations are not met, the odds he/she will leave the organization increases [7]. Since its inception, many studies had been exploring the met expectations thesis, finding positive correlations between met expectations and job satisfaction, organizational commitment, job survival, and job performance [8].

Considering all mentioned theories and numerous research works that explore them (e.g., [9–12]), job satisfaction and, congruently, met job expectations are positively correlated with employees retention and commitment. Hence, sound human resources management (HRM) practices and programs should be in place to improve employees' job satisfaction and motivation [13]. Such a need is even more emphasized in modern and high tech industries, like the IT industry, which employs knowledge workers as their most valuable resources [1].

2.1 Research Model and Hypotheses

This study is aimed to analyze employees' expectations of the workplace and work environment. Its primary concern is to explore the difference in job expectations between the technical and non-technical staff that work within the IT industry. Some recent studies (e.g., [1, 14]) show that knowledge workers for high-tech industries primarily seek personal development. Besides, the work environment in high-tech industries is characterized by high internal competitiveness [15], while the adequacy of the general work environment is something that knowledge workers take for granted [4, 5]. All these leads to the following study hypotheses, which define the research model:

H_1: *Self-actualization and career advancement are more important to technical (engineering) staff than to non-technical (non-engineering) staff.*

H2: A competitive work environment is more important to technical (engineering) staff than to non-technical (non-engineering) staff.

H3: The stability and maturity of the work environment are more important to technical (engineering) staff than to non-technical (non-engineering) staff

3 Data and Methodology

The research was conducted within a single IT company ("*The Company*") from the F BiH, which employs around 60 full-time staff members (10-years average amounts to 56 employees), most of them highly qualified in the fields of IT, with an average length of service of nine years.

To test the research hypotheses, an appropriate survey questionnaire was developed and sent to all eligible employees (those who were employed in "*The Company*" for at least five consecutive years, were not members of the company's management, and did not occupy positions in ancillary services) each year during the period from the 2008 to the 2018. A total of 471 responses to the surveys were received, out of which 352 (or 74.73%) were given by technical staff, and 119 (or 25.27%) by non-technical staff. These responses were received from 85 persons (67 technical and 18 non-technical staff). In order to preserve data independence, the responses were averaged over different age periods for each respondent, thus reducing the data set to 115 cases.

3.1 Research Design and Measures

The main concern of this study is the employees' job expectations or, more specifically, the difference in job expectations between technical and non-technical staff that work within the IT industry. To further explore this topic, the corresponding research design was prepared (Fig. 1). In order to conduct appropriate testing, the series of several ANOVA was employed. There are six sets of job expectations measures (Fig. 1), which are designated with $SAaA_{Gi}$, IC_{Gi}, and WEM_{Gi} ($i = 1, 2$). Each measure-set relates to the single data-cell of the research design vector, where the vector dimension is defined by the independent variable "*Employees Profession*" (*G*). The independent variable *G* designates different groups of employees, which are differentiated by the employees' professional background – technical staff (allocated to the group G_1) and non-technical staff (allocated to the group G_2).

The dependent variables – "*Self Actualization and Advancement*" (*SAaA*), "*Internal Competitiveness*" (*IC*), and "*Work Environment Maturity*" (*WEM*), represent different facets of the employees' job expectations. Both dependent and independent variables were measured using data from the survey, while the measurement spans a period from the year 2008 to the year 2018.

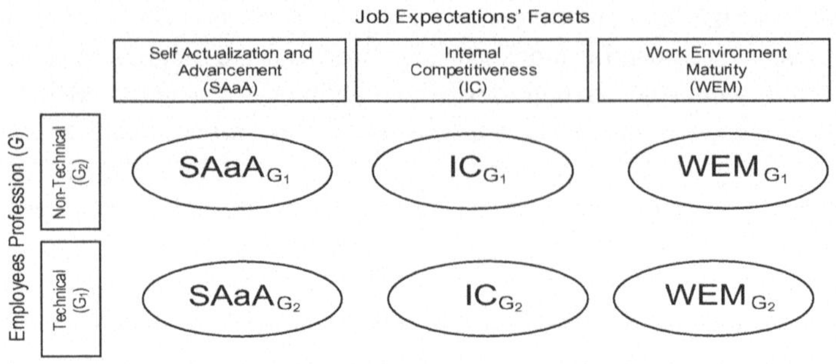

Fig. 1. Layout of the research design

3.1.1 Dependent Variables

As already stated, a questionnaire with eight questions was devised to measure various aspects of employees' job expectations. Each question was related to specific job traits in the form of a statement followed by the seven-point Likert scale. After the data were collected, the exploratory factor analysis was conducted to discover what concepts (latent variables) contribute to the employees' job expectations.

The principal component analysis (PCA) with "*direct oblimin*" oblique rotation was used as an extraction method. The sample size of 115 items provides for a stable factor solution, while the Kaiser–Meyer–Olkin measure ($KMO = 0.74$) verified the sampling adequacy for the analysis. Also, KMO values for all individual items were equal or above the acceptable limit of 0.5. Bartlett's test of sphericity [χ^2 (28) = 327.06, $p < 0.001$] indicated that correlations between items were sufficiently large for PCA.

The main analysis revealed three components that had eigenvalues greater or equal to Kaiser's criterion of 1, and that together have explained 74.2% of the variance. The extracted components, along with corresponding loadings after rotation and Cronbach's α, are presented in Fig. 2. These three components, treated as dependent variables in further analysis, are:

(a) *Self Actualization and Advancement (SAaA)* – represents employees' expectations that are related to self-actualization and professional advancement. It is measured by the following four indicators:

- *"Creative Environment"* (X_1) – measures the importance of the work environment creativity potential.
- *"Challenging Work"* (X_2) – measures the importance of the challenge that work poses to the employee.
- *"Professional Advancement"* (X_3) – measures the importance of professional advancement opportunities.

(b) *Internal Competitiveness (IC)* – represents employees' expectations related to the competitiveness of the work environment, and is measured by the following indicators:

- *"Promotion Opportunities"* (X_4) – measures the importance of potential advancement within the managerial hierarchy.

- *"Competing with Colleagues"* (X_5) – measures how important and stimulative it is for an employee to compete with colleagues professionally.

(c) *Work Environment Maturity (WEM)* – represents employees' expectations related to the self-actualization and professional advancement, and is measured by the following indicators:

- *"Employees Empowerment" (X_6)* – an indicator that measures how important it is for an employee to have clearly defined powers and responsibilities.
- *"Atmosphere of Respect"* (X_7) – an indicator that measures the importance of an atmosphere of appreciation and trust in the work environment.
- *"Organization of Work"* (X_8) – an indicator that measures the importance of work organization.

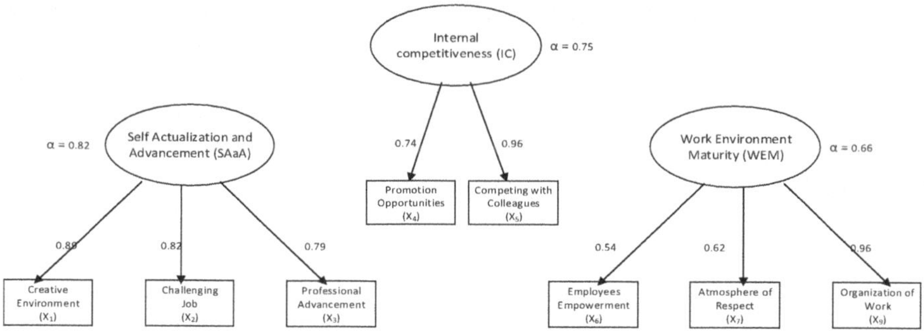

Fig. 2. The extracted components (factors) and corresponding loadings

3.2 Results

Table 1 contains the means and standard deviations of all model dependent variables for all groups of independent variable *G*. To test the differences between the defined groups of employees, series of three ANOVA tests was employed to examine three facets of the employees' job expectations.

Table 1. Descriptive statistics of dependent variables for groups of *G*

Indicator	Group of *G*	*N*	Mean	Std. deviation
SAaS	Group 1	91	0.274	0.785
Self-Actualization and Advancement	Group 2	24	−1.040	1.059
IC	Group 1	91	0.062	1.025
Internal Competitiveness	Group 2	24	−0.234	0.881
WEM	Group 1	91	0.035	0.997
Work Environment Maturity	Group 2	24	−0.133	1.022

3.2.1 Assumptions

Assumptions for ANOVA are independence, normality within groups, and homogeneity of variances in each experimental condition. There were no missing data and no significant outliers in the sample, and there was the only deviation from a normal distribution for the second group of *IC* variable. Although ANOVA is a rather robust test, this slight violation can be further remedied by decreasing the p-value while testing the significance of ANOVA statistics. Levene's test was used to assess the assumption of homogeneity of variance. This assumption was met for the variables *IC* and *WEM* ($p > 0.05$), while for *SAaS* it was violated ($p < 0.05$). To correct this violation Welch's *F* was used.

3.2.2 The ANOVA Model Estimation

The next step in the ANOVA procedure was to assess whether there exist significant differences for all three dependent variables related to the job expectations across the two groups of employees. Univariate ANOVA tests for all three dependent variables show that only variable *SAaS* (*Self Actualization and Advancement*) has a significant main effect ($p < 0.001$), while variables *IC* (*Internal Competitiveness*) and *WEM* (*Work Environment Maturity*) have a non-significant main effect (Table 2). Since the assumption of homogeneity of variance for the variable *SAaS* was not met, the robust test of equality of means (Welch's *F*) was used. This test also showed that this variable has a significant main effect [$F(1, 29.974) = 32.259$, $p < 0.001$].

Table 2. Univariate tests for group differences in dependent variables

Source	Variable	Σ of sq.	df	Mean Sq.	F	Sig.
Betw. groups	*SAaS – Self Actualization and Advancement*	32.79	1	32.79	45.63	0.000
	IC – Internal Competitiveness	1.66	1	1.66	1.67	0.199
	WEM – Work Environment Maturity	0.53	1	0.53	0.53	0.468
Within groups	*SAaS – Self Actualization and Advancement*	81.21	113	0.72		
	IC – Internal Competitiveness	112.34	113	0.99		
	WEM – Work Environment Maturity	113.47	113	1.00		

3.3 Interpretation of the Results and Discussion

The ANOVA analysis showed that there was a significant main effect for variable *SAaS* [$F(1, 113) = 45.632$, $p < 0.001$], and non-significant for variables *IC* [$F(1, 113) = 1.666$, $p = 0.199$] and *WEM* [$F(1, 113) = 0.531$, $p = 0.468$]. These results show that the difference in job expectations between the two groups of employees is primarily indicated by the variable *SAaS* (*Self Actualization and Advancement*), while the significant difference for the other two variables does not exist. The tests' outcomes fully support the first hypothesis of this research, while, at the same time, they do not provide support for the second and third research hypotheses.

Strong support for the first research hypothesis points toward the main difference between the technical and non-technical staff within the IT sector. This finding is well aligned with the existing literature [e.g., 1, 14] – the experts from high-tech industries, such as IT, are strongly driven by intrinsic motivators and look primarily for professional challenges and personal development. The fact that the third research hypothesis was not supported had been, to a great extent, expected. More specifically, when it comes to the general characteristics of the work environment, there are no significant differences between the two groups of employees. For the majority of people, it is common that they want a stable, well organized, and respectful work surrounding. Based on Herzberg's two-factor theory [4, 5], many studies confirm that extrinsic factors (working relationships, salary, stability and security, and similar) are necessary precondition for workers job satisfaction (or, rather, *no* job dissatisfaction) and, consequently, their job expectations, so this finding is in a full concordance with the existing theories.

A very interesting finding of the research is the lack of a significant difference between two groups of employees regarding the expectations about the competitiveness of the job environment. Since the "knowledge workers" with technical background show a strong inclination toward individualism [14], it would be intuitively expected that they seek more competitive work environments than their non-technical counterparts. Indeed, many high-tech companies use internal competition to stimulate productivity, quality, and innovation [15], although one can find quite the opposite economic models [16]. Such mixed standings regarding the internal competition were well elaborated by Birkinshaw [17]. This research finding is, nevertheless, aligned with those resources that advocate against the internal competition, which is especially characteristic for SMEs [1].

4 Conclusion

The research results show that the main difference between the technical and non-technical staff within the IT industry is related to self-actualization and professional advancement. Namely, the experts from the IT industry look primarily for professional challenges and personal development. At the same time, there is no difference between the two groups of employees when it comes to the general characteristics of the work environment. Surprisingly, the obtained results reveal no significant difference between the two groups of employees regarding the expectations about the competitiveness of the job environment. Nevertheless, the obtained research results are highly congruent with the existing theory on the topic.

Combining all research findings, one can conclude that individualism is just that particular personal trait that makes the difference between technical and non-technical employees in the IT industry. The main implication of the research, however, is rather practical. Namely, to retain the most wanted and valuable employees, managers of IT companies should make and maintain a highly creative and challenging working environment, where their engineers would have an opportunity for constant professional development and advancement. Furthermore, meeting their job expectations would lead to an increase in their job satisfaction, and, consequently, in their productivity and eagerness to attain the organizational goals.

Several limitations apply to this research. In order to make research feasible, a limited set of observed variables was used. The direct undesirable consequence of that is a lack of more elaborate latent variables that would enable a finer representation of the concept of employees' job expectations. Therefore, a recommendation for some future research would be to seek out new observed variables and, possibly, new factors, which would improve research model reliability and explanatory power.

Another significant research limitation is that the data analysis was conducted on a single sample, collected from one company, so the obtained results have a weak generalization potential. In order to remedy this limitation, future research may use the same (or similar) research design applied to different datasets.

References

1. Park, C., McQuaid, R., Lee, J., Seungjin Kim, S., Lee, I.: The impact of job retention on continuous growth of engineering and informational technology SMEs in South Korea. Sustainability **11**(18), 5005 (2019)
2. Robbins, S.P., Judge, T.A.: Organizational Behavior, 17th edn. Pearson Education Limited, Harlow (2017)
3. Griffin, R.W., Moorhead, G.: Organizational Behavior: Managing People and Organizations, 11th edn. Cengage Learning, South-Western (2014)
4. Herzberg, F., Mausner, B., Snyderman Bloch, B.: The Motivation to Work, 2nd edn. Wiley, New York (1959)
5. Herzberg, F.: One more time: how do you motivate employees? Harv. Bus. Rev. **40**(1), 53–62 (1968)
6. Porter, L.W., Steers, R.M.: Organizational, work, and personal factors in employee turnover and absenteeism. Psychol. Bull. **80**(2), 151–176 (1973)
7. Irving, P.G., Montes, S.D.: Met expectations: the effects of expected and delivered inducements on employee satisfaction. J. Occup. Organ. Psychol. **82**(2), 431–451 (2009)
8. Wanous, J.P., Poland, T.D., Premack, S.L., Davis, K.S.: The effects of met expectations on newcomer attitudes and behaviors: a review and meta-analysis. J. Appl. Psychol. **77**(3), 288–297 (1992)
9. Currivan, D.B.: The causal order of job satisfaction and organizational commitment in models of employee turnover. Hum. Resour. Manag. Rev. **9**(4), 495–524 (1999)
10. Dalkrani, M., Dimitriadis, E.: The effect of job satisfaction on employee commitment. Int. J. Bus. Econ. Sci. Appl. Res. **11**(3), 16–23 (2018)
11. Djoemadi, F.R., Setiawan, M., Noermijati, N., Irawanto, D.W.: The effect of work satisfaction on employee engagement. Pol. J. Manag. Stud. **19**(2), 101–110 (2019)
12. Judge, T.A., Thoresen, C.J., Bono, J.E., Patton, G.K.: The job satisfaction-job performance relationship: a qualitative and quantitative review. Psychol. Bull. **127**(3), 376–407 (2001)
13. Bhatnagar, J.: Talent management strategy of employee engagement in Indian ITES employees: key to retention. Empl. Relat. **29**(6), 640–663 (2007)
14. Sutherland, M., Jordaan, W.: Factors affecting the retention of knowledge workers. SA J. Hum. Resour. Manag. **2**(2), 55–64 (2004)

15. Eisenhardt, K.M., Galunic, C.: Coevolving: at last, a way to make synergies work. Harv. Bus. Rev. **78**(1), 91–101 (2000)
16. Brown, J.: Quitters never win: the (adverse) incentive effects of competing with superstars. J. Polit. Econ. **119**(5), 982–1013 (2011)
17. Birkinshaw, J.: Strategies for managing internal competition. Calif. Manag. Rev. **44**(1), 21–38 (2001)

Risk Management as a Part of the Business Process in Corporate Firms

Haralambie Vochitoiu(✉), Florin Vedinas, Olga Miclea, and Camelia Lavinia Unguras

University of Petrosani, University Street, No. 20, Petrosani, Romania
hvochitoiu@gmail.com

Abstract. Risk management is the process that allows business managers to balance operational and economic costs of protective measures and achieve gains in mission capability by protecting business processes that support the business objectives or mission of the enterprise. Risk management, however, is not restricted to the information technology and security realm. This is a business process that assists management in meeting its fiduciary duty to protect the assets of the organization.

Keywords: Risk · Business manager · Information security

1 Introduction

Risk management is the total process used to identify, control, and minimize the impact of uncertain events. The objective of the risk management program is to reduce the risk of performing some activity or function to an acceptable level and obtain senior management approval. Risk management is made up of four distinct processes: risk analysis, risk assessment, risk mitigation, vulnerability assessment and controls evaluation. Senior management must ensure that the enterprise has the capabilities needed to accomplish its mission or business objectives. However, the senior management of a department, business unit, group, or other such entity is considered to be the *functional owner* of the enterprise's assets and it is senior management's fiduciary duty to act in the best interest of the enterprise to implement reasonable and prudent safeguards and controls. Risk management uses the following terms [1–3]:

Risk management–The total cost to identify, control, and minimize the impact of uncertain events. The objective of risk management is to reduce risk to an acceptable level.

Risk analysis–A technique to identify and assess factors that may jeopardize the success of a project or achieving a goal. This technique also helps define preventive measures to reduce the probability of these factors from occurring and identify countermeasures to successfully deal with these constraints when they develop.

Risk assessment–The computation of risk. Risk is a threat that exploits some vulnerability that could cause harm to an asset. The risk algorithm computers the risk as a

H. Vochitoiu—Student Doctoral School, University of Petrosani,University Street, no.20, Petrosani, Romania

I. Karabegović (Ed.): NT 2020, LNNS 128, pp. 964–972, 2020.
https://doi.org/10.1007/978-3-030-46817-0_109

function of the assets, threats, and vulnerabilities. One instance of a risk within a system is represented by the formula: *Risk = Assets Threat × Vulnerability*. The total risk for a network equates to the sum of all the risk instances.

Risk mitigation–The process in which an organization implements controls and safeguards to prevent identified risks from ever occurring, while at the same time implementing a means of recovery should the risk become a reality in spite of all efforts.

Vulnerability assessment and controls evaluation–Systematic examination of a critical infrastructure, the interconnected system on which it relies, its information, or product to determine the adequacy of security measures, identify security deficiencies, evaluate security alternatives, and verify the adequacy of such measures after implementation.

2 Risk Management

System development life cycles structured to meet the needs of the information technology organization, and there-fore anything associated with the IT process. Risk management is a business process and all business decisions should have a business development life cycle. This system allows for those elements that make up information technology development, but also takes into account normal business decisions [4–7].

2.1 Business System Development Life Cycle

Effective risk management must be totally integrated into the organization'system development life cycle, and includes the following key concepts: *Analysis*, *Design*, *Construction*, *Test* and *Maintenance* (Fig. 1). As the diagram points out, risk analysis is mapped throughout the *System Development Life Cycle* for the first time risk analysis needs to be done is when there is a discussion on whether a new system or application of business process is required (Table 1) [4, 6].

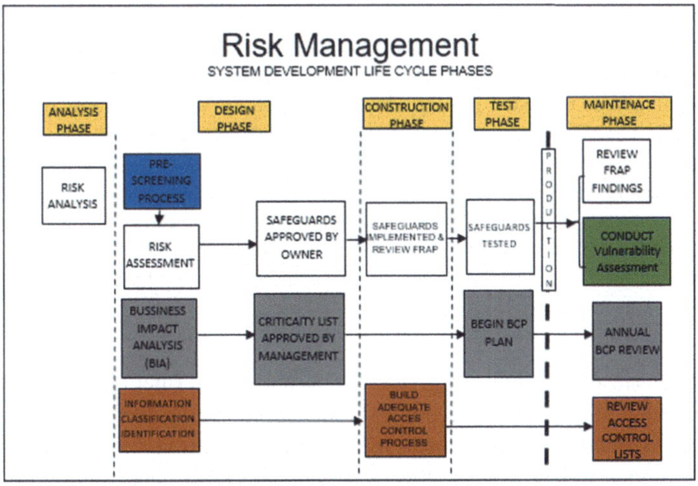

Fig. 1. System development life cycle

Table 1. System development life cycle versus risk management

Phases	Risk management activities
Analysis—The need for a new system, application, or process and its scope is documented	***Analysis***—Identified risks are used to support the development of system requirements, including security needs
Design—The system or process is designed and requirements are gathered	***Design***—Security needs lead to architecture and design trade-offs
Development—The system or process is purchased, developed, or otherwise constructed	***Development***—The security controls and safeguards are created or implemented as part of the development process
Test—System security features should be configured, enabled, tested, and verified	***Test***—Safeguards and controls are tested to ensure that decisions regarding risks identified are reduced to acceptable levels prior to movement to production
Maintenance—When changes and updates are made to the system, the changes to hardware and software are noted and the risk analysis process is revised	***Maintenance***—Controls and safeguards are re-examined when changes or updates occur or on regularly scheduled intervals

2.2 Employee Roles and Responsibilities

Risk management is a management responsibility. To be successful, the risk management process must be supported by senior management and the concept of ownership of assets established. This concept is typically presented to the organization through the *asset or information classification policy*. Sample language for this portion of the policy might be similar to that shown in Table 2 [5, 6].

Table 2. Employee responsibilities example

Owner–The company manager of a business unit or office where the information is created, or the primary user of the information. Owners are responsible for: Identifying the classification level of all corporate information within their organizational unit; Defining and implementing appropriate safeguards to ensure the confidentiality, integrity, and availability of the information resource; Monitoring safeguards to ensure their compliance and reporting situations of noncompliance; Authorizing access to those who have a business need for the information; Removing access from those who no longer have a business need for the information
Custodian–Employees designated by the owner to be responsible for protecting information by maintaining safeguards established by the owner
User–Employees authorized by the owner to access information and use the safeguards established by the owner

Employees are responsible for protecting corporate information from unauthorized access, modification, destruction, or disclosure, whether accidental or international. However, to facilitate the protection of corporate information, employee responsibilities have been established at these levels: *owner*, *custodian*, and *user*.

Employees have different roles, and these roles support the activities of the other roles and responsibilities (Table 3). These are only examples of the types of roles and responsibilities found in a typical organization. It will be necessary to ensure that all employees have as part of their job description the concept of support of information security. The organization's *information classification* policy will also establish the key concepts of *owner, custodian*, and *user*. For key position such as Corporate Information Security Officer (CISO) and Information Security Administrator (ISA), in addition to a job description, it will be necessary to create and publish a charter statement for security management, and audit might include the language shown in Table 3.

Table 3. Typicalrolesfor employees

Typical role	Risk management responsibility
Senior management	Under the standard of due car, senior management is charged with the ultimate responsibility for meeting business objectives or mission requirements. Senior management must ensure that necessary resources are effectively applied to develop the capabilities to meet the mission requirements. It must incorporate the results of the risk analysis process into the decision-making process
Chief information security officer (CISO)	This position was at one time called the chief information officer (CIO), but many organizations simply named their head of IT to this position, and therefore, the CIO has become the position known as the CIO The CISO is responsible for the organization's planning, budgeting, and performance, including its information security components. Decisions made in this area should be based on an affective risk management program
Resource owners	These are the business unit managers assigned as functional owners of organization assets; they are responsible for ensuring the proper controls are in place to address integrity, confidentiality, and availability of the information resources that they are assigned ownership. The term *owner* must be established in the asset classification policy The managers (a.k.a. owners) are the individuals with the authority and responsibility for making cost-benefit decisions essential to ensure accomplishment of organization mission objectives. This involvement in the risk management process enables the selection of business-orientated controls. The charge of being an owner supports the objective of fulfilling the fiduciary responsibility of management to protect the assets of the enterprise
Information security professional	The security program manager is responsible for the organization's security programs, including risk management. The ISA has changed its designation because the designation *officer* is normally restricted to senior executives. The officers can be held personally liable if internal controls are not adequate

Another important element found in most enterprise wide policy documents is a section on organizational responsibilities. A typical character statement for security, management, and audit might include the language shown in Table 4.

Table 4. Sample mission statements

Organization management	Officers and directors of the organization manage the assets of the organization in a prudent and diligent manner in accordance with the best interests of the customers, clients, employee, shareholders, regulatory agencies, and any other persons they represent
Audit	Audit assesses the adequacy of and compliance with management, operating, and financial controls, as well as the administrative and operational effectiveness of organizational units
Information security	Information security directs and supports the company and affiliated organizations in the protection of their information assets from intentional or unintentional disclosure, modification, destruction, or denial through the implementation of appropriate information security and business resumption planning policies, procedures, and guidelines
Regulatory affairs	Regulatory affairs represent the organization's interests by initiating strategies and affecting enterprise relationships with the executive and legislative branches or the government authority, as well as with the independent agencies. Regulatory affairs provide the organization with development, interpretation, and implementation of national legislation and regulations

When implementing risk management, it will be necessary to view this process as part of the ongoing information security life cycle (Fig. 2).

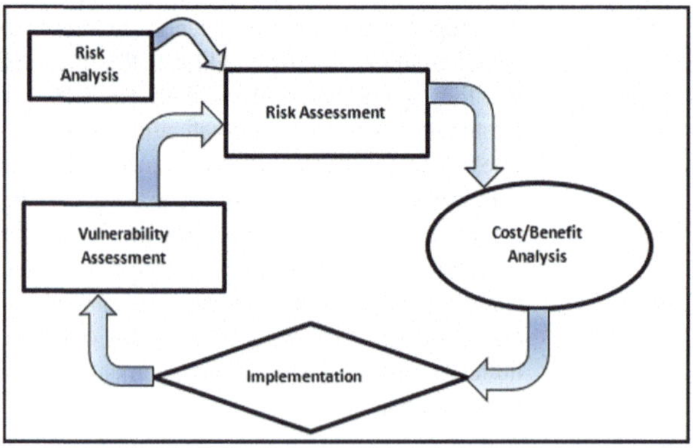

Fig. 2. Information security life cycle

A risk analysis should be conducted whenever money or resources are to be spent. Before starting a task, project, or development cycle, an enterprise should conduct an analysis of the need for the project.

2.3 Risk Assessment

Risk assessment is the second process in the risk management life cycle. Organizations use risk assessment to determine what threats exist to a specific asset and the associated risk level of the threat. As we examine the risk assessment portion of the risk management process, we will discuss six steps that will provide us with the three deliverables we need [6, 7].

(a) ***Asset definition:*** To be successfully, this first step in the risk assessment process must be as through as possible. It will be difficult to conduct an accurate risk assessment if all of the team members do not have the same vision of what is to be reviewed.

(b) ***Threat identification:*** We define a threat as an undesirable event that could impact the business objectives or mission of the business unit or enterprise. Some threats occur when existing controls that either were implemented incorrectly or have passed their usefulness now provide a weakness or threat to the infrastructure that can be exploited to circumvent the intended behaviour of the control.

(c) ***Determine probability of occurrence:*** Once a list of threats has been finalized and the team has agreed on the definitions of each threat, then will be necessary to determine how likely that threat is to occur. The risk management team will want to derive an overall likelihood that indicates the probability that a potential threat may be exercised against the risk assessment asset under review (Fig. 3):

		IMPACT		
		High	**Medium**	**Low**
PROBABILITY	**High**	**A**	**B**	**C**
	Medium	**B**	**B**	**C**
	Low	**C**	**C**	**D**

Fig. 3. Matrix of the probability parameter (*A-Corrective action must be implemented; B-Corrective action should be implemented; C-Require monitor; D-No action required at this time*)

Where: *High probability*–very likely that the threat will occur within the next year; *Medium probability*–possible that the threat may occur during the next year; *Low probability*–Highly unlikely that the threat will occur during the next year.

(d) ***Determine the impact of the threat:*** Once we have determined the probability of a threat occurring, it will be necessary to determine the impact that the threat will have on the organization (Fig. 4).

		IMPACT		
		High	Medium	Low
PROBABILITY	High	High	High	Medium
	Medium	High	High	Medium
	Low	Medium	Medium	Low

Fig. 4. Matrix of the impact parameter (*High-Corrective action must be implemented; Medium-Corrective action should be implemented; Low-No action required at this time*)

Where: *High impact*–Shutdown of critical business unit that leads to a significant loss of business, corporate image, or profit; *Medium impact*–Short interruption of critical process or system that results in a limited financial loss to a single business unit; *Low impact*–Interruption with no financial loss.

(e) ***Determine the level of the risk:*** Once a list of threats has been finalized, it will be necessary to determine how likely that threat is to occur. Once the probability that a threat might occur has been determined, the impact that the threat will have on the organisation must be addressed (Fig. 5).

		IMPACT		
		High	Medium	Low
PROBABILITY	High	High (H)	High (H)	Moderate High (MH)
	Medium	High (H)	Moderate High (MH)	Moderate Low (ML)
	Low	Moderate High (MH)	Moderate Low (ML)	Low (L)

Fig. 5. Matrix of the risk level (*High-Corrective action must be implemented; Moderate High-Corrective action should be implemented; Moderate Low-Continue to Monitor; Low-No action required at this time*)

The output from the risk analysis will lead the organization to identify controls that should reduce the level of threat occurence (Table 5).

Table 5. Threat table by threat source

Threat	Applicable Yes/No	Probability *1=Low 2=Medium 3=High*	Impact *1=Low 2=Medium 3=High*	Risk level	Control selected	New risk level
Natural (storm, landslide, floods ...)	Yes	2	1	**ML**	*Continue to Monitor*	**L**
Human Accidental (fire, explosion, human error ...)	Yes	2	3	**H**	*Corrective action must be implemented*	**MH**
Human Deliberate (terrorism, sabotage, arson, vandalism ...)	Yes	2	3	**H**		**MH**
Environmental Toxic contamination, telecommunications failure, temperature inadequacy ...)	Yes	2	2	**MH**	*Corrective action should be implemented*	**ML**

(e) ***Risk mitigation:*** Risk mitigation is systematic methodology used by senior management to reduce organizational risk. Once the risk assessment has been conducted (threats identified, risk level established, control chosen) management can use various risk mitigation techniques to complete the process.

3 Conclusion

The risk management process is a business process that supports management in its decision making. It allows the management owners of the assets to perform their fiduciary responsibility of protecting the assets of the enterprise in a reasonable and prudent manner. The purpose of risk management is to analyse the business risk of a process, application, system, or other asset to determine the most prudent method for safe operation.

We do not want, nor can we use, a control mechanism that reduces risk to zero. A security program that has as its goal 100% security will cause the organization to have 0% productivity. The risk assessment process has two key objectives: to implement reasonable and prudent controls and to document management's due diligence.

References

1. Atkinson, R.: Security Architecture for the Internet Protocol, RFC 1825, Naval Research Laboratory (1995)
2. Lisa Bryson, C.: Protect Your Boss and Your Job: Due Care in Information Security, Computer Security Alert 146, San Francisco (1995)
3. John Carrol, M.: Computer Security, 3rd edn. Butterworth-Heinemann Publisher Ltd., Woburn (1996)
4. Robert Glass, I.: Building Quality Software. PrenticeHall, EnglewoodCliffs (1992)
5. Michel Kabay, E.: The NCSA Guide to Enterprise Security. McGraw-Hill Computer Communications Series, New York (1992)
6. Thomas Peltier, R.: Information Security Risk Analisys. Auerbach Publications, New York (2001)
7. Killmeyer, T.J.: Information Security Architecture. Auerbach Publications, New York (2001)

The Application of PROMETHEE in Choosing the Best Promotion Service

Lamija Šćeta[1(✉)], Adela Delalić[1], and Halid Skaka[2]

[1] School of Economics and Business, University of Sarajevo, Trg oslobođenja – Alija Izetbegović 1, 71000 Sarajevo, Bosnia and Herzegovina
lamija.sceta@efsa.unsa.ba
[2] SKYMedia, Sarajevo, Bosnia and Herzegovina

Abstract. One of the common problems faced by companies of various types is the choice of the type of promotion that will present their products or services to a targeted group of consumers. Numerous elements influence this choice, some of which are budget, availability, sustainability, visibility, flexibility, innovation. This problem is well suited for the application of multicriteria decision making, and will thus serve as a case study in this paper.

The problem discussed in the case study is the selection of a promotion service that is adequate for companies in Sarajevo Canton. The results of this research may be of particular importance to marketing agencies when designing packages of promotion services offered to companies in Bosnia and Herzegovina.

The PROMETHEE is very useful and adequate in the multicriteria decision making process in economics and management. For the implementation of this method we used the Visual PROMETHEE software tool, which enables a wider and faster application of the method.

Keywords: Multicriteria decision making · PROMETHEE · Promotion services

1 Introduction

This paper is aimed to provide a brief overview of multicriteria methods in decision making with the special attention to the PROMETHEE (Preference Ranking Organizational Method for Enrichment Evaluation). Different versions of this method, that enable partial or complete ranking of alternatives for decision makers, have been developed. The case study analysis, presented in the last chapter of the paper, addresses choosing of adequate promotion service for Sarajevo Canton companies, by using software tool for solving multicriteria decision problems.

The main objective of this research is to give recommendation to marketing agencies which packages of promotion services they should develop. Some classic and the most commonly used promotion services and their features were used. The data were collected through a survey of potential clients of marketing agencies, and an analysis and discussion of the results are obtained.

I. Karabegović (Ed.): NT 2020, LNNS 128, pp. 973–980, 2020.
https://doi.org/10.1007/978-3-030-46817-0_110

2 Basic Elements of Decision Making Theory

The decision making process is present in our daily lives as well as in our business environment. Some decisions are made quickly and easily without being aware of the decision making process, while some situations require detailed evaluation of multiple alternatives and their consequences in order to determine the best option.

The solving of decision problem means choosing the best possible alternative according to specified goal. The alternatives are presented as various options available in decision process. By choosing of certain alternative, system will be transformed from one to another of possible states. In the decision making process, the consequences of choosing a particular alternative in terms of level to which the defined goal is met, are especially emphasized. Criteria for decision making problems are the factors used to evaluate possible alternatives in accordance with the goal set. The criteria determine the extent to which each of the existing alternatives contributes to the goal. Multicriteria decision making addresses the case of multiple criteria, which are often opposed. The decision is made when the process ends with choosing one of the possible alternatives.

The manager's tasks, in decision making process, are to develop detailed insight to decisions consequences through their quantification. Very often, decisions are related or exclude each other or even impose conditions or restriction on some other decisions. These situations lead to situation where choosing of alternative need to be made through several stages. Each stage contains a variety of available alternatives which transform the system into the one of possible states, so the complete process can be described by the decision tree. The solving of decision problem comes down to determining the best branch in the constructed decision tree. Three approaches to classification, based on mathematical models, are particularly significant: classification of decisions related to strategic planning [1], the structured form that classifies decisions into programmed and unprogrammed ones [2], and the third classification which is based on the certainty of alternatives' consequences, that classify decisions into those with certain outcomes, decisions at risk and decisions in uncertainty.

An important element in the decision making process is the answer to the question "To what extent the goal is fulfilled?" This leads to the need to quantification of elements of the decision process. Some simple situation can be treated by the use of certain standardized measures while, in the more complex situation, relative measures can be used in order to compare contribution of alternatives to the goal set.

3 Multicriteria Decision Making

In situations with predefined set of available alternatives, different models and methods are used to deal multicriteria decision making problems. In the process of evaluation of alternatives, it is necessary to consider broader picture determined by the set of given criteria, rather than focus on the only one criterion. Separate evaluation and alternative ranking based on individual criterion can lead to a significant deviation from the rank of alternatives based on taking all criteria into account. Also, it is necessary to eliminate the impact of different scales in evaluation of alternatives and to introduce a precise procedure for determining adequate weighting factors for the observed criteria.

Generally, the concepts underlying these methods can be divided into compensatory and out-ranking methods. Compensatory methods, such as AHP, assumes complete aggregation among criteria. Outranking methods, such as ELECTRE and PROMETHEE, build a preference relation among alternatives evaluated on several criteria.

It should be noted that there are some other methods besides the methods mentioned above. In some developed methods problem of multicriteria decision making is transformed into a scalar variant. Compromise programming is a method of finding a solution that is closest to the ideal solution to an observed problem. This concept is represented in the TOPSIS.

3.1 PROMETHEE Method

The PROMETHEE method was introduced by J. P. Brans (1982) for the first time, by presenting two versions of the method: PROMETHEE I, which includes partial ranking and PROMETHEE II, which includes complete ranking of the alternatives. The continuous variants, PROMETHEE III and PROMETHEE IV, were developed by J. P. Brans and B. Marechal, a few years later. The same authors also developed GAIA, a visual interactive model for graphical presentation of the PROMETHEE methodology. New variants of PROMETHEE V, that include segmented conditions and PROMETHEE VI, were developed in 1992–1994. The PROMETHEE is very useful and adequate in the process of multicriteria decision making in economics and management, which is illustrated by numerous examples of application of this method [3–5].

The PROMETHEE is based on the relationships and function preferences of alternatives that are associated with each individual criterion and the introduction of so-called generalized criteria. Since the types of criteria can be different, expressed in different measurement units or measured on different scales, and since different criteria can have different deviations, generalized criteria need to be introduced into the model. For this purpose, it is necessary to construct the functions of preference associated to each criterion.

Comparing two alternatives $a, b \in A$, with respect to criterion $f_j, j \in \{1, 2, \ldots, m\}$, there are two possible outcomes: the evaluation values of the observed alternatives are the same $f_j(a) = f_j(b)$, or these values are not the same: let's take $f_j(a) > f_j(b)$ without loss of generality. Then alternative denoted by a is preferable.

Let's define $d_j(a, b) = f_j(a) - f_j(b)$, then reference relation P_j with respect to the criterion f_j is defined by $P_j(a, b) = F(d_j(a, b))$, where F is an appropriately selected non-decreasing preference function that satisfies $0 \leq P_j(a, b) \leq 1$.

In theory, six types of function of preference are distinguished [6]: simple criterion, quasi-criterion, criterion with linear preference, criterion with levels of constant preference, criterion with linear preference in the area of indifference, and Gaussian criterion.

Very often, it is convenient to assign modified preference functions by

$$H(d_j(a,b)) = \begin{cases} P_j(a,b), & d_j(a,b) \geq 0 \\ P_j(b,a), & d_j(a,b) \leq 0. \end{cases} \tag{1}$$

After determining the pairs (f_j, P_j) with weighting factors $w_j, j \in \{1, 2, \ldots, m\}$, for each pair of alternatives $a, b \in A$ the preferential index $\pi(a,b)$ associated with that pair can be defined. It represents the degree of preference of alternative a over alternative b:

$$\pi(a,b) = \sum\nolimits_{j=1}^{m} w_j P_j(a,b). \tag{2}$$

Using a preferential index, we introduce the concepts of output and input flow for each alternative $a \in A$, which tell us how much better is that alternative over all others and how much the observed alternative is dominated by the other alternatives, respectively. Let's denote them by $\varphi^+(a) = \frac{1}{n-1} \sum_{b \in A} \pi(a,b)$, $\varphi^-(a) = \frac{1}{n-1} \sum_{b \in A} \pi(b,a)$.

Using specific flows, input and output for all alternatives of the observed problem, it is possible to rank them completely or partially.

These values allow us to introduce the corresponding relations of overall preference (P), indifference (I) and incomparability (R), obtained on the basis of the all criteria. Depending on the way these relationships are introduced, different versions of the PROMETHEE method can be obtained. PROMETHEE I is a partial ranking method. It is based on both introduced flows. As input and output flow do not usually imply the same way of ranking alternatives, in constructing the relations, they are combined in the following way:

$$\begin{aligned} aP^I b &\Leftrightarrow \begin{cases} \varphi^+(a) > \varphi^+(b) \text{ and } \varphi^-(a) < \varphi^-(b) \text{ or} \\ \varphi^+(a) = \varphi^+(b) \text{ and } \varphi^-(a) < \varphi^-(b) \text{ or} \\ \varphi^+(a) > \varphi^+(b) \text{ and } \varphi^-(a) = \varphi^-(b) \end{cases} \\ aI^I b &\Leftrightarrow \varphi^+(a) = \varphi^+(b) \text{ and } \varphi^-(a) = \varphi^-(b) \\ aR^I b &\Leftrightarrow \begin{cases} \varphi^+(a) > \varphi^+(b) \text{ and } \varphi^-(a) > \varphi^-(b) \text{ or} \\ \varphi^+(a) < \varphi^+(b) \text{ and } \varphi^-(a) < \varphi^-(b). \end{cases} \end{aligned} \tag{3}$$

PROMETHEE II provides complete ranking. It is based on a net preference flow that is defined as the difference between output and input flow $\varphi(a) = \varphi^+(a) - \varphi^-(a)$ and in some ways represents the balance of these two flows. For each of the different PROMETHEE methods we can distinguish three basic steps characteristic of all variants: enrichment of preferential structure, enrichment of domination relations, decision making and analysis.

For the purpose of the analysis in this paper, the Visual PROMETHEE software is used.[1] The software was developed by VP Solutions under the supervision one of the

[1] Available in the business, academic, and online versions http://www.promethee-gaia.net/visual-promethee.html, accessed May 18, 2018.

creators of PROMETHEE method, B. Marechal. This software package is currently one of the most complete software packages. It has been used in numerous real-life situations, projects, scientific and professional papers.

4 Selection of Adequate Promotion Service for Sarajevo Canton's Companies by Using PROMETHEE Method

Aiming to explore and provide adequate recommendations on selection of promotion services to develop for clients in Sarajevo Canton (SC), the research was conducted at the Promotion and Marketing Agency of SKYMedia in the period March–April 2018. Respondents were marketing and senior-level managers in the Agency's client companies. The input dataset is obtained by the questionnaire, based on views of experts and clients that consuming promoting services. Based on the first part of the questionnaire, information on companies' characteristics, as well as on currently offered promotion services, were collected. These information allowed us to compare the current presence of different types of promotion services with results provided by the PROMETHEE method.

The second part of the survey provided the crucial information needed to apply the PROMETHEE method. Namely, since the method is based on ranking services on the basis of different criteria, it is necessary to determine the extent to which the criteria are fulfilled for each type of promotion under consideration.

The first step in applying the multicriteria decision making method is the selection of criteria to be used. Usually the criteria are classified into qualitative, quantitative and technical [7]. Quantitative criteria include: availability, frequency, speed of dissemination of information, length of information, geographical flexibility. Qualitative criteria are: the originality of the media, the attention paid to the media, the amount of information that can be conveyed to particular media, how much the media has the ability to build the image of what is being promoted, how emotional the advertisement is, how memorable the message is, how much of added value is provided with a certain message, how active is selected media. The technical criteria include: price, availability of sight advertising for the client, and others [8].

Since there is a lot of possible criteria that influence the choice of the best promotion service, many authors stress out only a few of the most important ones. The priorities are: price, originality and experience with the bidder [9–11].

Keeping in mind the focus of this study, there are outlined some, of the most important criteria that are of particular importance to clients (Table 1).

The next challenge is the choice of services that will be considered. Promotion services are numerous, from those that have been in use for many years, such as posters, radio and television commercials, to the whole range of modern-day types of commercials. Existence of a large number of opportunities complicates the decision making process in media planning and, consequently, the choice of a promotion service. The complexity and challenge of this process is considered in many studies [12, 13].

Table 1. Criteria

Criterium	Criteria expectations
Costs	The lower cost of the advertisement
Continuance	The longer the duration of the advertisement
Availability	The widest range of people to whom the information will be available
Repetition	The greater possibility of repeating information
Flexibility	Simple modifications and corrections in the information transfer process
Originality	As high as possible the originality of the way information is conveyed

Four basic ways of advertising were observed: media, outdoor advertising, digital marketing and guerilla marketing, and for each ways of advertising are considered several types of services (Table 2).

Table 2. Alternatives

A way of advertising	Services
Media	TV commercials
	Radio commercials
	Advertisements in print media
	Advertising in the electronic media
Outdoor advertising	Billboard commercials
	Megaboard commercials
	Wallscape commercials
	City light commercials
	Posters
Digital marketing	Facebook
	Twitter
	Google ads
	Instagram
	Youtube
	A game for computers, smartphones
Guerilla marketing	Kiosk branding
	Elevator branding
	Escalator branding
	Public vehicle branding
	Extensions
	Leaflets
	Events and other promotional activities

Preference tables based on the fulfillment of each of the criteria for each type of advertising were identified. The values in the preference table were calculated as the mean of the ratings given by the respondents. Two types of ranking were applied: first assuming that all criteria were equally important, and second, with modified weighting factors. The weights were defined as the rounded mean value of the grades indicating the importance of the stated criteria by the respondents (Table 3).

Table 3. Criteria weight factors

Cost	Continuance	Availability	Repetition	Flexibility	Originality
4	4	6	4	3	6

In the context of media advertising, the results of current representation show that electronic media are the most represented form of advertising, which is in line with the results of the ranking according to PROMETHEE, and which distinguishes this type of advertising as the best according to the considered criteria.

The best alternative in outdoor advertising, according to the PROMETHEE, is wallscape commercials, which is not in line with the current situation, according to which the billboard commercials are the most commonly used. This result can be explained with the lack of supply for other, better-rated outdoor advertising types and indicates that it is desirable to develop wallscape outdoor advertising services.

Facebook, as the most suitable alternative, is the most represented digital marketing option according to our research. However, promotion through Youtube is the second-ranked option, but according to research it is on the fifth place of the six options offered, which indicates that there is a space to improve the offer of using this option.

The best alternative within guerilla marketing is kiosk branding, while the worst ranked alternatives are leaflets and event activities, which is in contrast to the results of the types of guerilla marketing used so far. Events are the most common form of guerilla marketing, and kiosk branding is the least represented form of guerilla marketing. One reason for this result can lie in the fact that branding services are relatively underrepresented in promotional and marketing agencies, suggesting that the development of such services could be of interest in the future.

It has been evaluated the most common and the highest ranked types of advertising (two for each way of advertising). An analysis of the application of the weighting factors of the criteria was also performed. It has been shown that there is no significant difference in the ranking of alternatives if all used criteria are equally weighted. The results indicate that there is a space for the development of guerilla marketing, which is currently the least represented way of advertising, and in global rankings, the modes of this type of advertising are better ranked than some classic aspects such as megaboard commercials and TV commercials, which are currently more prevalent.

5 Conclusion

The observed case study shows that, even in relatively simple situations of multicriteria decision making, it is not possible to decide by intuitive and experiential reasoning what is the best alternative. Scientific methods, mathematical models and appropriate software packages are of vaste benefit in supporting the choice of the best alternative when we are faced with decisions. Additional analysis and visual presentation of the results significantly contributes to the insight into the interrelationship of alternatives and the understanding of the consequences of choosing an alternative.

By the analysis of the case study results, it is shown that media advertising is the most present and the best ranked advertising option in our research. By the results of PROMETHEE, the best alternative in the outdoor advertising is wallscape commercials, which is currently underused advertising activity. Also, besides the Facebook, as the most adequate and most present type of digital advertising, attention also should be focused on Youtube, in the future. Based on analysis in our research, it is concluded that efficiency of the advertising within guerilla marketing could be improved by the focusing on branding kiosks, escalators and elevators. The ranking of this type of advertising was even better than some of the most commonly used types, such as megaboard or TV commercials. The results obtained by different weightings of showed that there is no significant difference among different ways of criteria weighting.

References

1. Anthony, R.N.: Planning and Control Systems: A Framework for Analysis. Harvard University Graduate School of Business Administration, Boston (1965)
2. Simon, H.A.: The Shape of Automation for Men and Management. Harper and Row, New York (1965)
3. Albadvi, A., Shari, S.A., QahriSaremi, H.: Application of PROMETHEE for market targeting, a case study on the TV market in Iran. ScientiaIranica **14**(3), 221–229 (2007)
4. Karande, P., Chakraborty, S.: Application of PROMETHEE-GAIA method for non-traditional machining processes selection. Manag. Sci. Lett. **2**, 2049–2060 (2012)
5. Mladineo, M., Jajac, N., Rogulj, K.: A simplified approach to the PROMETHEE method for priority setting in management of mine action projects. Croat. Oper. Res. Rev. CRORR **7**, 249–268 (2016)
6. Brans, J.P., Mareschal, B., Vincke, Ph: Operational Research. Elsevier Science Publishers B. V, North-Holland (1984)
7. Rademaker, C.A.: Media Selection for Marketing Communication - An exploratory study among marketing managers, Stockholm School of Economics, Center for Media and Economic Psychology (2011)
8. De Pelsmacker, P., Geuens, M., Van den Bergh, J.: Marketing Communications: A European Perspective. Pearson Education, London (2007)
9. Cagley, J.W., Roberts, C.R.: Criteria for advertising agency selection: an objective appraisal. J. Advert. Res. **24**(2), 27–31 (1984)
10. Henke, L.L.: A longitudinal analysis of the ad agency-client relationship: predictors of an agency switch. J. Advert. Res. **35**(2), 24–30 (1995)
11. Na, W., Marshall, R., Son, Y.: An assessment of advertising agency service quality. J. Advert. Res. **39**(3), 33–41 (1999)
12. Turk, P.B., Katz, H.: Making headlines: An overview of key happenings in media planning, buying, and research from 1985-1991. J. Curr. Issues Res. Advert. **14**(2), 19 (1992)
13. Crosier, K., Grant, I., Gilmore, C.: Account planning in Scottish advertising agencies: a discipline in transition. J. Mark. Commun. **9**(1), 1–15 (2003)

New Technologies in the Financial Industry

Adem Abdić(✉), Adnan Rovčanin(✉), and Ademir Abdić(✉)

School of Economics and Business, University of Sarajevo, Trg oslobođenja 1, 71000 Sarajevo, Bosnia and Herzegovina
{adem.abdic,adnan.rovcanin,ademir.abdic}@efsa.unsa.ba

Abstract. Financial and technological innovations during the last two decades in the financial system were underpinned the emergence of new business models, strategies, processes, products and applications. Further, new entrants into the financial markets and providing financial services changed the dominant operating business models and competitive dynamics of a financial industry. The main purpose of this paper is to explore and analyze evolution of the relationship between finance and technology over the last two centuries. After that, in the paper are identified key drivers of FinTech innovations, are identified key market players and key areas in FinTech. The basic source of data in this paper was the secondary sources provided by other researches those include journal articles, reviews and academic books. The scientific contribution of this paper is reflected in the fact this paper represents one of the pioneering analysis of the FinTech in the B&H. Finally, for further analysis it can be explore the reaction of regulatory bodies on each of the eras of FinTech evolution and on the changes of IT and the nature of financial systems worldwide.

Keywords: Financial market participants · Financial technology · Financial innovations · Financial services · Millennials

1 Introduction

Today's global economy is going through perhaps the most changing period in the modern history of financial markets and financial services. The unstoppable acceleration of technological innovations are enabling a fundamental reinvention of the structure of the financial institutions and their relationships on the financial markets, while simultaneously inviting the entrance of new competitors with the potential to break down the traditional order of the financial industry. All noted could lead to emerge additional risks taking among traditional financial institutions and new entrants (FinTech and/or BigTech companies[1]) in order to maintain their profitability and competitiveness and it could have additional implications for financial system stability of some country.

[1] FinTech Company is used here to describe company whose business models and strategies are based on the technological innovations. BigTech Company is used here to describe large technology companies that expand into the direct provision of financial services or of products very similar to traditional financial products.

I. Karabegović (Ed.): NT 2020, LNNS 128, pp. 981–990, 2020.
https://doi.org/10.1007/978-3-030-46817-0_111

2 Historical Context of Changing Modern Financial Industry

A strong unstoppable wave of financial innovation started from the U.S. in the early 1960s and spread rapidly to other developed countries of the world. Although the dynamics and importance of the innovation process varied from country to country, its key features could be distinguished [1, p. 1]: (a) financial innovations - the development of new financial markets and instruments, (b) securitization - a greater propensity for market setting of interest rates and marketable securities in relation to illiquid bank loans, (c) liberalization - removing restrictions and obstacles in the domestic financial markets, either through deregulation or complete abolition of barriers; (d) globalization - integration of economies around the world through trade and financial flows and (e) growing competition from financial institutions, primarily commercial banks, investment banks and insurance companies. The key result of the process of financial innovations were the emergence of a wide range of new products (primarily financial derivative instruments) that were traded, which led to the expansion and deepening of financial markets. Furthermore, a new wave of international financial integration and deregulation occurred in the early 1970s, characterized by the additional deregulation of capital market surveillance, the deregulation of national financial systems, and strong technological innovations in the field of informatics and information systems [2, p. 290; 3, p. 4; 4, p. 4–5]. According to Saunders and Cornet [5] "The last 30 years have been dramatic for the financial services industry. In the 1990s and 2000s, boundaries between the traditional industry sectors, such as commercial banking and investment banking, broke down, and competition became increasingly global in nature. Many forces contributed to this breakdown in interindustry and intercountry barriers, ***including financial innovation, technology***, taxation, ***and regulation***. Then in 2008-09, the financial services industry experienced the worst financial crisis since the Great Depression". However, it is necessary to emphasize that the interlinkage of the financial systems[2] of countries (finance in general) and technological growth and development (technology in general) has a long history (tradition) and mutually reinforcing from their earliest stages of development. Today, the aforementioned symbiosis between finance and technology is referred by the unique name of financial technology or shortly FinTech. Generally speaking, FinTech refers to the application of information technologies to financial system. In a broad-sense FinTech refers to "the use of technology to deliver financial solutions" [7, p. 2], but in a narrow-sense FinTech refers to the "technology-enabled innovation in financial services that could result in new business models, applications, processes or products with an associated material effect on the provision of financial services" [8, p. 1].

The evolution of FinTech can be presented in three distinct eras such as in the next table.

[2] Financial system consist of three elements: Financial markets, Financial instruments, and Financial participants [6, p. 28].

Table 1. Classification eras of FinTech evolution. (Adapted to [7, p. 3–4])

Era (Range)/Description	Geography	Key players	Shift origin
Era I (1866–1967) The introduction of the telegraph and the laying of the first successful transatlantic telegraph cable in 1866; The development of technology such as the railroads, canals and steamships; The physical trading of securities. (An analogue industry)	Developed countries	Fundamental infrastructure	Globalization
Era II (1967–2008) The introduction of the Fax Machine in 1964, the Automatic Teller Machine (ATM) and the Handheld financial calculator in 1967; The Establishment of SWIFT in 1973; The electronic payment systems; The electronic trading of securities and introduction of "Circuit breakers" on the financial markets; The emergence of the Internet; E-banking; (A digital industry)	Worldwide	Banks and other incumbents on the financial markets	Liberalization, deregulation and technology
Era III (2008–to date) The emergence of the 2008 Global Financial Crises; The rapidity of technological development and the proliferation of new startups and technology companies providing financial services; The general public developed a distrust of the traditional banking system; Thousands or even millions of financial professionals either lost their jobs or were now less well compensated; The emergence of the Millennial Generation; The high level of smartphone penetration and the rapid development of application programming interfaces (APIs).	Developed and developing countries	New startups and established technology and social media companies (FinTech and/or BigTech companies)	2008 Financial crisis and Market reform

3 New Technologies in Financial Sector

3.1 Drivers of the FinTech

The role, position and importance of the traditional financial markets and institutions are significantly changed after the collapse of the Bretton-Woods Agreement which increased the volatility of market risk factors such as exchange rates, interest rates, energy prices and/or prices of key inputs/outputs. Consequently, financial and technological innovations in the financial system/sector were underpinned the emergence of new business models, strategies, processes, products and applications. On the other side, new entrants into the financial markets and providing financial services changed the dominant operating business models and competitive dynamics of a financial industry.

According to [9, p. 26] the current state of FinTech and the evolving landscape has three catalysts/drivers of the financial technological revolution:

1. A changing macroeconomic and regulatory environment
2. The rapid evolution of technology, and
3. Shifting customer expectations.

But, key catalysts/drivers of the financial technology that could influence the market structure of the financial system could be [8, p. 5–10]:

1. On the supply side:
   - The use of application programming interfaces (APIs) over the internet;
   - The mobile banking and smart phones;
   - The cloud computing;
   - The licensing and supervision regulation;
   - The competition aspects in financial regulation;
   - The applications of distributed ledger technology (DLT).
2. On the demand side:
   - The changing customer expectations;
   - The changes in consumer behaviour driven by smart phone technology.

3.2 Market Participants in the FinTech

The key differentiating factors described in the third era of FinTech evolution caused a rapid rate of technology development and changed the landscape of financial systems and identity of the providers of financial services. New startups, established technology and social media companies have challenged the established financial institutions by offering specific, niche services to customers, businesses, and incumbent financial institutions. According to ROFIEG [10, p. 30] contemporary financial participants in the financial markets include: "incumbents (banks, insurers, investment firms, pension funds, financial infrastructures, etc.); big non-financial firms (including the so called BigTechs, such as internet search and advertising companies, device manufacturers, commercial platforms, but also telecoms operators) and by start-ups." They provide technology-enabled financial services [8, p. 1] with the potential to increase market access, the range of product offerings, and convenience while also lowering costs to clients.

3.3 New Digital Technologies and the Financial Technology

The development and applications of the FinTech is dominantly based on the other recent advances in core digital technologies such are [10, p. 27]:

(a) The Artificial Intelligence based on the three related paradigms: (Knowledge representation (KR); Natural language processing (NLP); and Machine learning (ML) and deep learning (DL) which is a form of ML using artificial neural networks (ANN).)
(b) Blockchain/Distributed Ledger Technologies;
(c) Smart contracts;
(d) Internet of Things;
(e) Quantum Computing.

The application of these digital technologies has led to shift paradigm in how traditional financial institutions operate, and has build up new business and revenue models, products and services and/or reduce cost through higher degrees of efficiency and enhancing customer relationships on the financial markets.

While financial markets have developed differently across jurisdictions and they have a different state of development of the financial services industry and the regulatory environment, according to the FSB's analysis [8, p. 1], some key considerations/standpoints of the link between technological innovation and financial innovation on the market structure are:

(a) "To date, the relationship between incumbent financial institutions and FinTech companies appears to be largely complementary and cooperative in nature.
(b) FinTech credit provision and payments is growing rapidly, but is still small as a proportion of overall credit in most jurisdictions.
(c) The competitive impact of BigTech companies may be greater than that of FinTech companies because BigTech companies typically have large, established customer networks and enjoy brand name recognition and trust.
(d) Reliance by financial institutions on third-party data service providers (e.g. data provision, cloud storage and analytics, and physical connectivity) for core operations is currently estimated to be low."

3.4 Key Areas in the FinTech

Based on the presented evolution of FinTech, today, it is possible use the full spectrum of finance and technology-enabled financial services on the financial markets delineated into five areas [7]:

(a) finance and investment,
(b) internal operations and risk management,
(c) payments and infrastructure,
(d) data security and monetization, and
(e) consumer interfaces.

Similarly, it is possible analyse different subsectors of financial services such are [9, p. 30]: ***payments*** (e.g. alternative payment methods, blockchain and cryptocurrencies, QR codes, contactless credit cards, mobile wallets…), ***financing/lending*** (e.g. Donation-based crowdfunding, Reward-based crowdfunding; Crowdinvesting; Crowdlending; Credit and Factoring; Peer-to-peer lending – P2P), ***wealth management*** (e.g. Robo-advise; Social trading; Personal financial management; Investment and banking), ***insurance*** (e.g. Flexible insurance policies to users' needs; "Pay-as-you-go" automobile insurance), and ***digital banking*** (e.g. Prepaid debit cards; Currency exchange facilities; A crypto currency exchange; Peer-to-peer payments, and Direct and virtually unlimited access to bank accounts).

3.5 Risks in the FinTech

Many of the potential risks emerging as a consequence of the use of FinTech are no different from the risks caused by the provision of financial services using more traditional means [10, p. 10]. Nevertheless, regulators and supervisors on the financial markets have identified key risks arising from fintech-related drivers [11, p. 2]:

(a) ***Risks to consumers*** (Lack of consumer understanding; Mis-selling of products and services; Financial exclusion; Data privacy, security and protection).
(b) ***Risks to firms*** (Business model viability; Governance; Technology risk and operational resilience; Data handling; Conduct and AML; Legal).
(c) ***Risks to financial stability*** (Concentration; Alternative channels of financial intermediation; Herd-like behaviour; Use of crypto assets; System-wide vulnerabilities).

Thus, it is important to stressed, although FinTech innovations brings many benefits to consumers, companies and financial institutions the risks that have impact on the financial sector are evolving but remain inherent their business dealing.

3.6 Investments in the FinTech

Figure 1 shows that FinTech sector is took off and attracted a big attention of the masses, both IT engineers and researchers, consumers, regulators and academics alike.

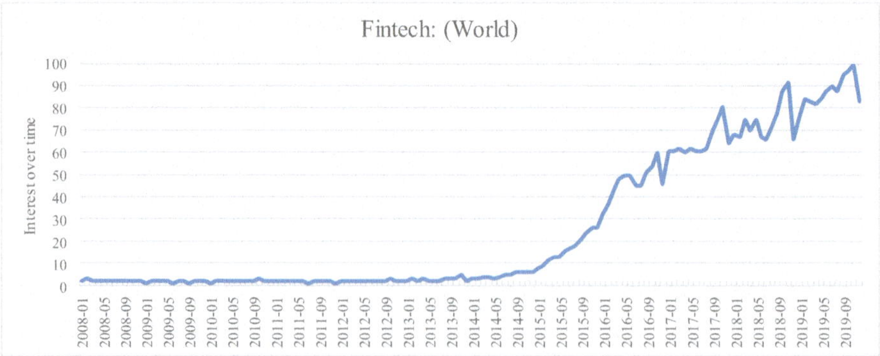

Fig. 1. Google trends for "FinTech" search (2019)

By using innovative and advanced technologies are being introduced new model of business in the financial services industry almost every months. So, in recent years FinTech has become one of the most dynamic and engaging parts of the financial services. Big data and analytics, blockchain, cloud computing, robo-advisors, artificial intelligence (AI), digital payments, digital currencies, crowdfunding, and other forms of peer-to-peer (P2P) financing have become the most interesting areas of FinTech. By using these technologies, FinTech are improving efficiency and quality of financial services and cutting costs [12].

The total global investment in FinTech grew significantly during the period 2010 to 2018 [13]. The largest value was invested in 2018, totalling US$ 55.347 billion, which is an enormous increase of 107.7% compared to the previous year. The total global investment in FinTech in 2018 more than tripled compared to their eight-year average (2010–2017). Figure 2 shows the last ten years by total global investment in FinTech during the period 2010–2018.

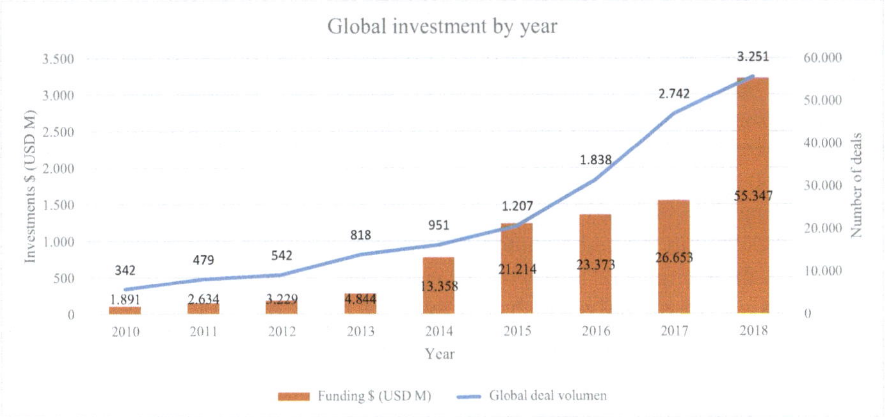

Fig. 2. Global investment by year. (Adapted to Accenture Research analysis on CB [13])

Figure 3 shows the investment in FinTech by segments during the period 2013–2018 [13]. Until 2018 investors invested mostly in Payments (about 30% in average) followed by Lending (about 29% in average). However, in 2018 investors invested mostly in Wealth Management & Asset Management segment (US$ 16.604 billion or 30%), followed by Payments (US$ 12.730 billion or 23%) and Lending (US$ 10.516 billion or 19%).

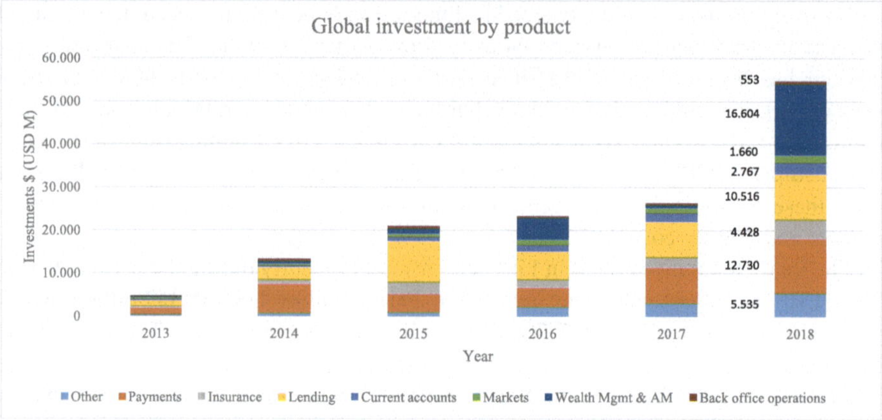

Fig. 3. Global investment by product. (Adapted to Accenture Research analysis on CB [13])

3.6.1 FinTech in Major Regions of the World

According to the data from FinTech100 - Leading Global FinTech Inovators report [14], the FinTech 100 includes FinTech companies from 29 countries: 22 Americas, 11 United Kingdom, 25 Europe, Middle East and Africa, 34 Asia, 7 Australia and 1 New Zeland. However, when comparing the top 10 FinTech companies [Ant Financial (China), Grab (Singapore), JD Digits (China), GoJek (Indonesia), Paytm (India), Du Xiaoman Financial (China), Compass (United States), Ola (India), Opendoor (United States) and OakNorth (United Kingdom)] it is clear that some countries have become more important as origins of FinTech. The companies in the FinTech100 was selected following extensive global research and analysis based on data across a range of dimensions, including the core five factors such as: average annual capital raised, rate of recent capital raising, geographic diversity, sectoral diversity and X-factor [14]. The most companies of the FinTech100 are payments and transactions segment (27), followed by wealth segment (19), insurance segment (17), lending segment (15), neo/challenger banks segment (9), and multiple FinTech sectors (9). Chinese and Asia-Pacific companies dominate the top 10 [14].

3.6.2 FinTech in Bosnia and Herzegovina

FinTech in the Bosnia and Herzegovina makes the first pioneering steps [15]. According to Ahmić [16] in B&H three FinTech companies are registered only. BamCard Company is one of the first FinTech companies in B&H. It's a business card processing company. It's doing business as the only third-party processing and card personalization centre. BamCard provides in the core of the processing and personalization service, as well as the support service Call Center, printing and enveloping. Also, ESP BH is a new FinTech company founded in mid-2019. It is a company whose business is the construction and implementation of a mobile application for payment of goods and services, overhead bills, as well as sending money to other users. OPA Platform (Online Payment Platform) is designed to connect users and other participants in the market chain.

Unfortunately, BH regulation is not familiar with FinTech companies and e-commerce regulations have not yet been implemented. Although the Law for the electronic document and the Law for the electronic signature have been adopted a certification governance body has not yet been established to administer base of the electronic signatures. The existing regulators and supervisors of financial markets are not allowed FinTech companies in the B&H to provide the most used financial services such as crowdfunding and peer-to-peer lending.

FinTech market in B&H mainly relies on the entrepreneurial spirit of the country and receives little government support. However, considering the relatively low cost and high quality of IT labour there is a healthy FinTech product development scene and a strong market in solutions such as chat bots, P2P payments, mobile banking apps and other tools designed to meet the demand for digital transformation [17].

4 Conclusion

Generally FinTech refers to the application of information technologies to provide financial services in the financial sector. When financial institutions (traditional and/or FinTech and BigTech companies) are using established core digital technologies in order to provide new financial products and services, then they are already part of a multi-billion dollar financial industry. New startups, established technology and social media companies have challenged established financial institutions by offering specific, niche services to customers, businesses, and incumbent financial institutions. FinTech includes a wide range of products, technologies, and business models that are changed the traditional financial services industry. It refers to everything from cashless payments, to crowd funding platforms, to wealth management, to virtual currencies and virtually unlimited access to bank accounts.

According to Accenture Research analysis [13], the total global investment in FinTech grew significantly during the period 2010 to 2018. The largest value was invested in 2018, totaling US$ 55.347 billion, which is an enormous increase of 107.7% compared to the previous year. The total global investment in FinTech in 2018 more than tripled compared to their eight-year average (2010–2017).

FinTech is already delivering significant benefits to all participants in the financial system (consumers and investors; traditional financial institutions and start up and established technology companies; and to financial stability and financial inclusion). Unfortunately, like any growing industry, business dealing by using financial technology is not without potential risks. Thus, some regulators and supervisors on the financial markets have struggled to keep up with the fast pace of innovation and identified key risks arising from fintech-related drivers such are: lack of consumer understanding, mis-selling of products and services, data privacy, security and protection, technology risk and operational resilience, and data handling.

References

1. Levich, R.M.: Financial innovations in international financial markets. In: NBER Working Paper Series, 2277, pp. 1–75 (1987)
2. Zbašnik, D.: Financijski rizici globalizacije. Ekonomija/Economics **13**(1), 287–302 (2006)
3. Moles, P.: Financial risk management - sources of financial risk and risk assessment. Edinburgh Business School Heriot-Watt University, Edinburgh, UK (2013)
4. Jorion, P.: Value at Risk: The New Benchmark for Managing Financial Risk, 3rd edn. McGraw-Hill, New York (2007)
5. Saunders, A., Cornett, M.M.: Financial Institutions Management: A Risk Management Approach, 8th edn. McGraw-Hill, New York (2014)
6. Rovčanin, A.: Upravljanje finansijama, 5th edn. Ekonomski fakultet u Sarajevu, Sarajevo (2010)
7. Arner, D.W., Barberis, J., Buckley, R.P.: Fintech and Regtech in a Nutshell, and the Future in a Sandbox, The CFA Institute Research Foundation (2017). https://www.cfainstitute.org/en/research/foundation/2017/fintech-and-regtech-in-a-nutshell-and-the-future-in-a-sandbox. Accessed 14 Jan 2020
8. Financial Stability Board – FSB: FinTech and market structure in financial services: market developments and potential financial stability implications, Switzerland (2017). https://www.fsb.org/2019/02/fintech-and-market-structure-in-financial-services-market-developments-and-potential-financial-stability-implications/. Accessed 12 Jan 2020
9. Arslanian, H., Fisher, F.: The Future of Finance - The Impact of FinTech, AI, and Crypto on Financial Services. Springer International Publishing; Palgrave Macmillan, Cham (2019)
10. European Commission: Expert Group on Regulatory Obstacles to Financial Innovation (ROFIEG): 30 Recommendations on Regulation, Innovation and Finance, Belgium (2019). https://ec.europa.eu/info/files/191113-report-expert-group-regulatory-obstacles-financial-innovation_en. Accessed 13 Jan 2020
11. KPMG International Cooperative: Regulation and supervision of fintech, Swiss (2019). https://assets.kpmg/content/dam/kpmg/xx/pdf/2019/03/regulation-and-supervision-of-fintech.pdf. Accessed 15 Jan 2020
12. Hazik, M., Hassnian, A.: Blockchain, Fintech, and Islamic Finance. Walter de Gruyter Inc., Berlin (2019)
13. Accenture: Accenture Technology Vision 2019 - The Post-Digital Era is Upon Us, Swiss (2019). https://www.accenture.com/_acnmedia/PDF-94/Accenture-TechVision-2019-Tech-Trends-Report.pdf. Accessed 11 Jan 2020
14. KPMG & H2 Ventures: Fintech100 - Leading Global Fintech Innovators report, Swiss (2019). https://h2.vc/wp-content/uploads/2019/12/2019Fintech100.pdf. Accessed 10 Jan 2020
15. Simić, B.: Inovacije u finansijskim uslugama u Bosni i Hercegovini. Banke & Biznis u BiH **206**, 6–7 (2019)
16. Ahmić, A.: Banke i lokalna FinTech scena trebaju biti partneri. Banke & Biznis u BiH **206**, 8–10 (2019)
17. Raiffeisen Bank International AG: CEE Fintech Atlas 2018, Austria (2018). https://www.rbinternational.com/eBusiness/01_template1/829189266947841370-829188968716049154-1397637459049680245-NA-2-EN.html. Accessed 10 Jan 2020

Author Index

I. Karabegović (Ed.): NT 2020, LNNS 128, pp. 991–994, 2020.
https://doi.org/10.1007/978-3-030-46817-0

The manufacturer's authorised representative in the EU is Springer Nature Customer Service Centre GmbH, Europaplatz 3, 69115 Heidelberg, Germany. If you have any concerns regarding our products, please contact ProductSafety@springernature.com

Printed and bound by CPI Group (UK) Ltd, Croydon, CR0 4YY
05/10/2026
02228148-0002